Lecture Notes in Mechanical Engineering

Lecture Notes in Mechanical Engineering (LNME) publishes the latest developments in Mechanical Engineering—quickly, informally and with high quality. Original research reported in proceedings and post-proceedings represents the core of LNME. Volumes published in LNME embrace all aspects, subfields and new challenges of mechanical engineering. Topics in the series include:

- Engineering Design
- Machinery and Machine Elements
- Mechanical Structures and Stress Analysis
- Automotive Engineering
- Engine Technology
- Aerospace Technology and Astronautics
- Nanotechnology and Microengineering
- Control, Robotics, Mechatronics
- MEMS
- Theoretical and Applied Mechanics
- Dynamical Systems, Control
- Fluid Mechanics
- Engineering Thermodynamics, Heat and Mass Transfer
- Manufacturing
- Precision Engineering, Instrumentation, Measurement
- Materials Engineering
- Tribology and Surface Technology

To submit a proposal or request further information, please contact the Springer Editor of your location:

China: Ms. Ella Zhang at ella.zhang@springer.com
India: Priya Vyas at priya.vyas@springer.com
Rest of Asia, Australia, New Zealand: Swati Meherishi at swati.meherishi@springer.com
All other countries: Dr. Leontina Di Cecco at Leontina.dicecco@springer.com

To submit a proposal for a monograph, please check our Springer Tracts in Mechanical Engineering at https://link.springer.com/bookseries/11693 or contact Leontina.dicecco@springer.com

Indexed by SCOPUS. All books published in the series are submitted for consideration in Web of Science.

More information about this series at https://link.springer.com/bookseries/11236

Muhammad Yusri Ismail ·
Mohd Shahrir Mohd Sani · Sudhakar Kumarasamy ·
Mohd Adnin Hamidi · Mohd Shamil Shaari
Editors

Technological Advancement in Mechanical and Automotive Engineering

Proceeding of International Conference in Mechanical Engineering Research 2021

Springer

Editors
Muhammad Yusri Ismail
Faculty of Mechanical and Automotive Engineering Technology (FTKMA)
Universiti Malaysia Pahang
Pekan, Malaysia

Sudhakar Kumarasamy
Faculty of Mechanical and Automotive Engineering Technology (FTKMA)
Universiti Malaysia Pahang
Pekan, Malaysia

Mohd Shamil Shaari
Faculty of Mechanical and Automotive Engineering Technology (FTKMA)
Universiti Malaysia Pahang
Pekan, Malaysia

Mohd Shahrir Mohd Sani
Faculty of Mechanical and Automotive Engineering Technology (FTKMA)
Universiti Malaysia Pahang
Pekan, Malaysia

Mohd Adnin Hamidi
Faculty of Mechanical and Automotive Engineering Technology (FTKMA)
Universiti Malaysia Pahang
Pekan, Malaysia

ISSN 2195-4356 ISSN 2195-4364 (electronic)
Lecture Notes in Mechanical Engineering
ISBN 978-981-19-1456-0 ISBN 978-981-19-1457-7 (eBook)
https://doi.org/10.1007/978-981-19-1457-7

This Springer imprint is published by the registered company Springer Nature Singapore Pte Ltd.
The registered company address is: 152 Beach Road, #21-01/04 Gateway East, Singapore 189721, Singapore

Preface

It is our great pleasure to present the compilation of Proceedings of International Conference in Mechanical Engineering Research (ICMER 2021). The 6th conference in mechanical engineering research is hosted by Faculty of Mechanical and Automotive Engineering Technology (FTKMA) in collaboration with Ningxia University of China, Universiti Teknologi Malaysia (UTM), UCSI University and the University of Mindanao.

ICMER 2021 is a scientific forum of discussion for scientists, researchers and engineers from all over the world to exchange ideas and the research results in the field of automotive technology, advanced fluid, advanced material, energy management and advanced manufacturing.

This conference is the first conference in the history of international conference in mechanical engineering research conducted in virtual through online platform due to the COVID-19 pandemic that hits across the globe. Therefore, we would like to express our sincere appreciation and gratitude to the committee members, co-organizer and reviewers to make this conference successful.

Last but not least, we would like to thank the keynote speakers and participants who contributed their knowledge to this prestigious conference.

Pekan, Malaysia

Muhammad Yusri Ismail
Corresponding Editor

Contents

Energy Management

Automotive Technology

Finite Element Analysis of Automotive Door Hinge

M. I. Hadi, M. R. M. Akramin, and M. S. Shaari

Abstract Door hinges and latches are door retention mechanism elements that play an important role in automobiles by holding the door open in the event of a side impact or rollover collision. Hinges are a group of components that are attached to the vehicle's door and frame, are related to one another, and can rotate along the same axis. Latches are mechanical devices that are used to position the door in a closed position relative to the vehicle body while allowing for controlled release. The standard specific conditions for side door latches and hinges installed on cars to reduce the risk of passengers being thrown out of the vehicle as a result of any impact. The objective of this paper is to identify the weakest point and to perform a structural analysis of automotive door hinge. Computer Aided Design (CAD) software is used to build a CAD model of the hinge and lock. The models of such components is meshed, and boundary conditions is defined, using the commercial meshing program. ANSYS is used to analyses the structural behaviour. Based on the results, the component will be further optimized for the future work.

Keywords Car door hinge · FEA · FEM

1 Introduction

The door hinges allow the car or the doors of the truck to be opened or closed. That is where the door fits on the frame [1]. Each door has two hinges, a upper hinge and a lower hinge. There is a circular pin inside the hinge which enables the hinge to swing around the pin axis [2]. Car hinges and door locks are elements of door retention [3]. Hinge has the capacity to rotate along the same axis and is an assembly made up of components attached to the vehicle's body and the door [4]. For the car locking, nowadays is equipped with two places latch mechanism. First one is 'fully latched position' and the second is 'latched position' [5].

M. I. Hadi · M. R. M. Akramin (✉) · M. S. Shaari
Faculty of Mechanical & Automotive Engineering Technology (FTKMA), Universiti Malaysia Pahang, 26600 Pekan, Pahang, Malaysia
e-mail: akramin@ump.edu.my

M. Y. Ismail et al. (eds.), *Technological Advancement in Mechanical and Automotive Engineering*, Lecture Notes in Mechanical Engineering,
https://doi.org/10.1007/978-981-19-1457-7_1

Table 1 Total ejections: 1995–2003 NASS and FARS report, on an annual basis

	Total occupants	Unejected	All ejection
All crashes	5,023,879	4,969,797	54,082
Rollover	444,267	410,420	33,847
Non-rollover	4,576,612	4,559,377	20,235

The first challenge is to predict the number of deaths in a single year if all vehicles on the road were constructed using prior technologies. On the basis of the 1995–2003 data from the National Automotive Sampling System (NASS) and the FAA (FARS), 5,023,879 vehicle passengers were involved in car accidents on an annual basis. Vehicle expelled occupants carried 54,082 as shown in Table 1. In ejections where the path of ejection was identified, 59% occurred through side glazing and 26% occurred through openings other than side glazing or doors (i.e., convertible tops, sunroofs, windshields, open truck beds). The others, 15%, took place at the door of a truck. The rate of expulsion by doors depends strongly on the occupant's use of the seatbelt. Unbelted occupants account for 94% of door ejections and if it seems to be a small amount (54,082 people about1.08%), i.e. a small number of people being killed as a result of opening a door due to damage [6], it is also very significant. That life is a valuable commodity. We must save lives in every way possible.

The Federal Motor Vehicle Safety Standard 206 "specify standards for side door locks and side door retaining parts, including latches, hinges, and other supporting means, to reduce the risk of passengers being ejected from the vehicle as a result of impact." Already in the early 1960s, passenger ejection was recognized as the leading cause of death in rollovers and a major problem in other crash modes.

Thus, the objective of this paper is to find stresses and deformation of door hinges under different loading conditions in order to enhance the design. The structural analysis is performed using FEA. FEA is used to simulate the over bending test and identify the weakest point area. Last but not least to recommend appropriate design alternatives for increasing strength and reducing its mass.

2 Finite Element Analysis of Door Hinge

The model of an automotive door hinge was created using the 3D modeling of Computer Aided Design software. It is consisting of three parts which are: a door side bracket, a body side bracket, and a circular pin. Figure 1 shows a model of an automotive door hinge for investigation.

The door hinge is made from a variety of materials, including S.G. iron (ductile iron), aluminum, and mild steel. Mechanical, chemical, and physical properties are three essential parameters for material selection. The aluminum and iron are selected for analysis because of its properties and availability. The physical and mechanical properties of an aluminum and cast-iron door hinge are shown in Table 2.

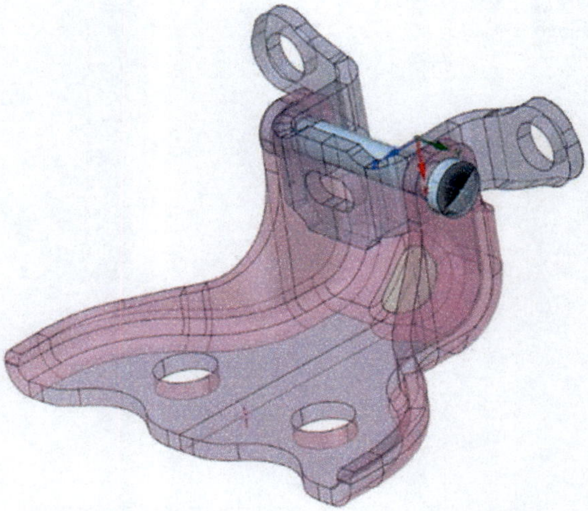

Fig. 1 Modelling of a hinge for an automotive door

The forces are taken from the IS 14225:1995 General specifications for automotive vehicles' locking systems and door retention parts. When a longitudinal load of 11,130 N is added to a door hinge structure, the structure should capable to support the door and undetachable. If the door hinge is subjected to a transverse load of 8930 N, each door hinge mechanism must support the door and undetachable.

The tensile force must be exerted equidistant between the hinge pin's linear core and through the hinge pin's centre line in the longitudinal vehicle direction. According to the original equipment manufacturer, a door hinge assembly must be able to sustain an ultimate normal load of 400 N without breaking. This is based on the door hinge being completely open.

Table 2 Physical and mechanical properties of aluminum door hinge

Properties	Total occupants
Young's Modulus	7.1e + 10 Pa
Density	2770 kg/m^3
Poisson's ratio	0.33
Yield strength	2.8e + 8

Table 3 Physical and mechanical properties of cast iron door hinge

Properties	Total occupants
Young's Modulus	1.1e + 11 Pa
Density	7200 kg/m^3
Poisson's ratio	0.28
Yield strength	8.2e + 8

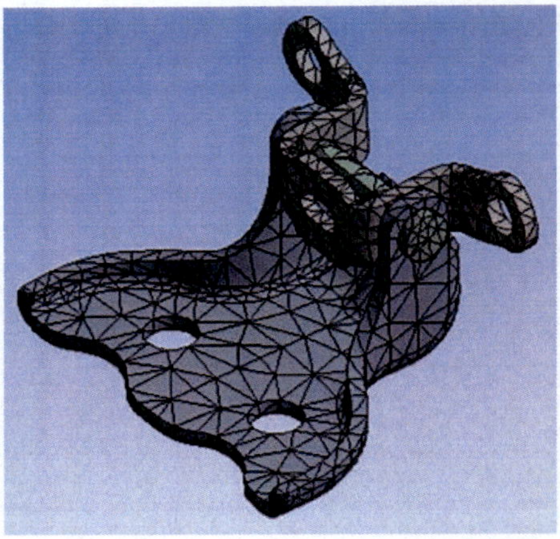

Fig. 2 Mesh of the door hinge

Tables 2 and 3 after the material properties is available, the methodology is shifted to meshing process.

During the meshing process as shown in Fig. 2, the geometry is modelled using .igs file. Later it is imported into FEA software. Once the CAD data for the door hinge structure has been imported, the surfaces of door hinge are developed and meshed. Since all of the dimensions of the door hinge are observable (3D), the part is meshed in a tetra-hedral element.

The forces are taken from the IS 14225:1995 General specifications for automotive vehicles' locking systems and door retention parts. When a longitudinal load of 11,130 N is added to a door hinge structure, the structure should capable to support the door and undetachable. If the door hinge is subjected to a transverse load of 8930 N, each door hinge mechanism must support the door and undetachable.

The tensile force must be exerted equidistant between the hinge pin's linear core and through the hinge pin's centre line in the longitudinal vehicle direction. According to the original equipment manufacturer, a door hinge assembly must be able to sustain an ultimate normal load of 400 N without breaking. This is based on the door hinge being completely open.

3 Results and Discussion

The door hinge is analysed using FEA software. At first, the door hinge is exposed to the maximum normal load on the door panel. The initial design of the door hinge was analysed to determine the maximum stress that would surpass the material's yield strength. Figure 3 shows the total deformation of car door hinge using aluminium. The maximum stress of car door hinge for aluminium is shown in Fig. 4. Figure 5 shows the maximum elastic strain of car door hinge for aluminium. Meanwhile for cast iron, it is shown by Figs. 6, 7 and 8.

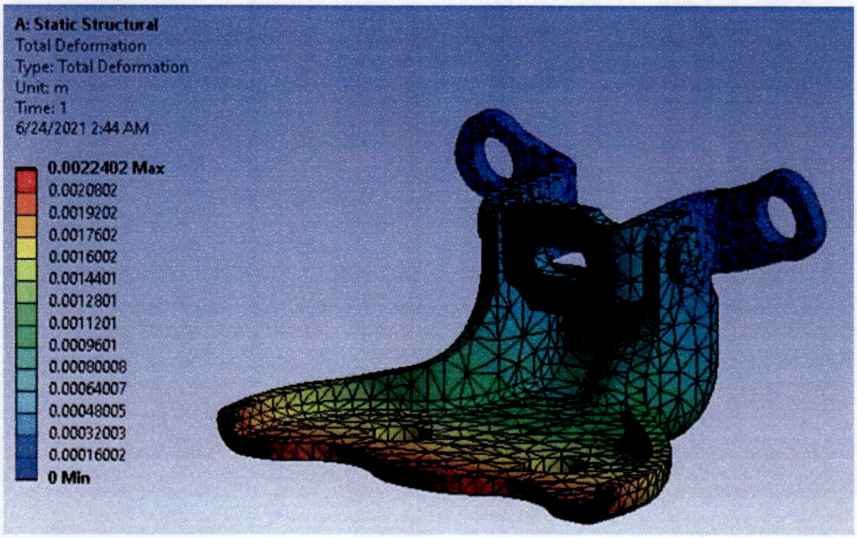

Fig. 3 Total deformation of car door hinge for aluminum

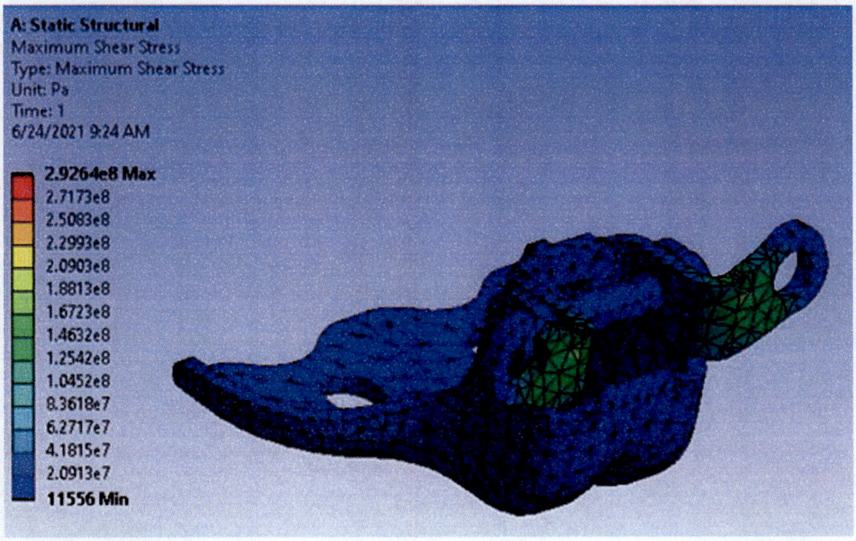

Fig. 4 Maximum stress of car door hinge for aluminum

Maximum deformation of door hinge under extreme condition is 0.22202 mm and maximum stress produced in 2.9264×10^8 Pa for aluminium. For cast iron the maximum deformation of door hinge under extreme condition is 0.14293 mm and maximum stress produced in 3.0105×10^8 Pa.

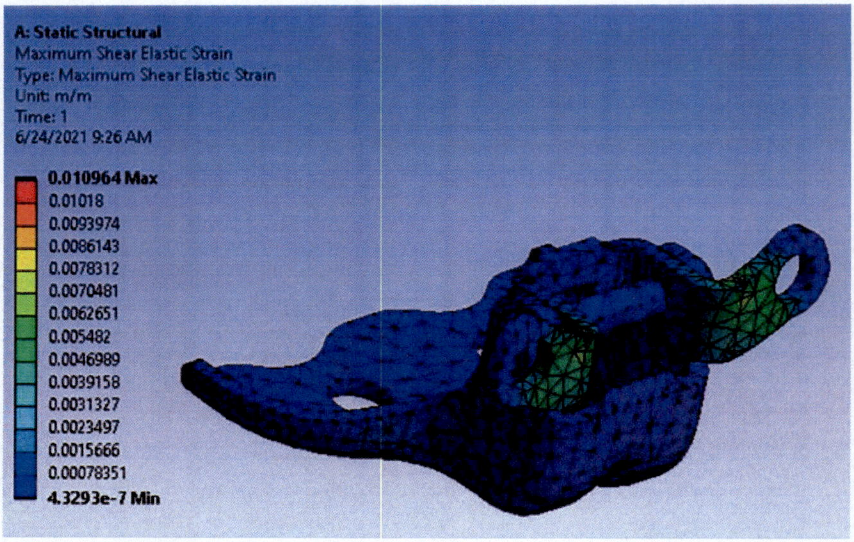

Fig. 5 Maximum elastic strain of car door hinge using aluminum material.

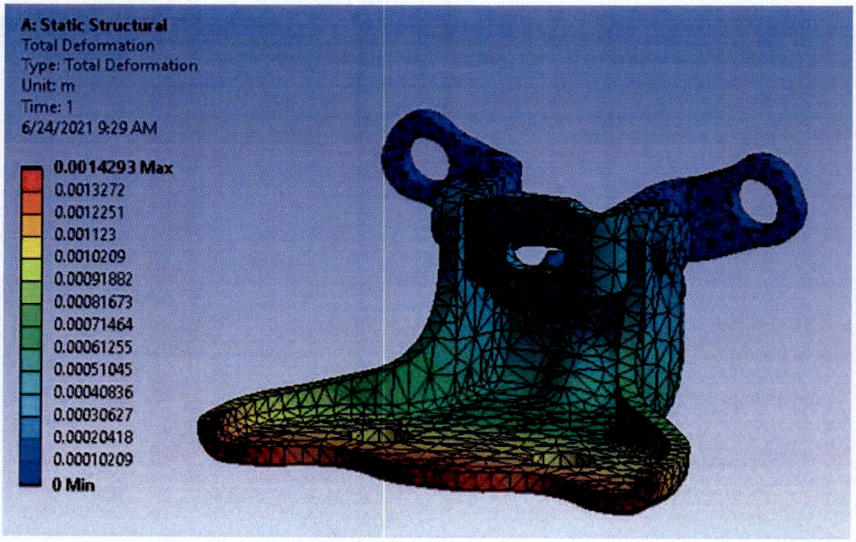

Fig. 6 Total deformation of car door hinge using cast iron material

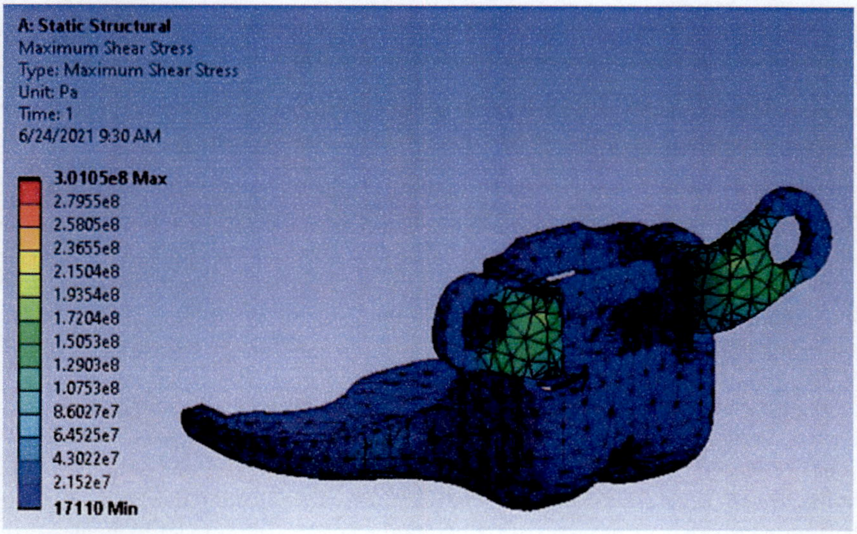

Fig. 7 Maximum elastic strain of car door hinge using aluminum material

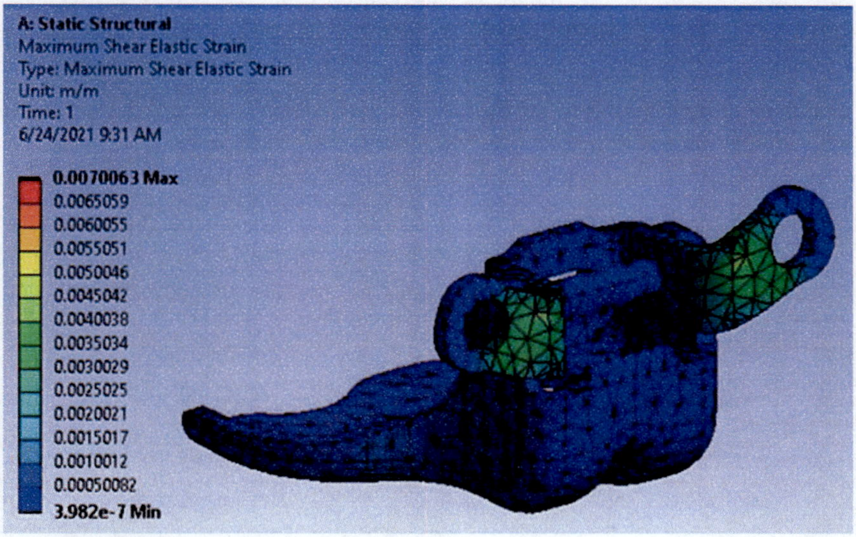

Fig. 8 Total deformation of car door hinge using cast iron material

Table 4 shows the comparison results between two materials. The maximum stress produced using aluminium alloy is exceeding the yield strength value. The yield strength for aluminium alloy is 2.8×10^8 Pa and the maximum stress produced is 2.9264×10^8 Pa. This indicates that the door hinge unable to withstand the maximum

Table 4 Comparison between aluminium and cast iron

Material	Aluminium	Cast Iron
Max. deformation	0.22402 mm	0.14293 mm
Max. stress produced	2.9264×10^8 Pa	3.0105×10^8 Pa

applied load of 400 N. For cast iron, it has zero yield strength in the same sense that there is no yield strength because it does not plastically deform. The maximum stress produced is compared with the ultimate strength of the material. The maximum stress produced for cast iron is not exceeding the ultimate strength. The ultimate strength for cast iron is 8.2×10^8 Pa and the maximum stress produced is 3.0105×10^8 Pa. This indicates that the door hinge could withstand the maximum applied load of 400 N.

Figure 9 shows the safety factor of the door hinge. The safety factor has a straightforward definition. It is defined as the ratio of the material's strength to the highest stress in the component. When the stress at a particular place exceeds the material's strength, the safety factor ratio falls below one, indicating that danger exists. It simply tells us that the stress in a certain location of the model is greater than the material's strength. When the stress in the model is significantly less than the material's strength, the safety factor is more than one, and the model is safe. From the results it shows that the minimum safety factor produced is very crucial since lowest value of safety factor is 0.1631.

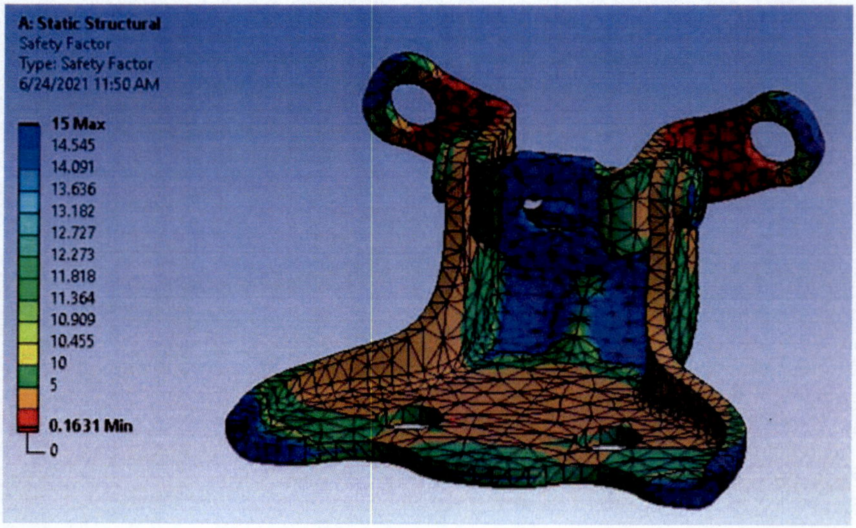

Fig. 9 Safety factor of car door hinge

4 Conclusion

The maximum stress and maximum displacement of the door hinge assembly were determined. The cast iron material shows the less deformation but higher stress compare with aluminium. Based on the yield strength and ultimate tensile strength of the both materials, door hinge made from cast iron is able to withstand the applied load. Thus, it can be concluded that cast iron material is preferable than aluminium.

Acknowledgements Authors would like to acknowledge the Ministry of Higher Education under Fundamental Research Grant Scheme FRGS/1/2019/TK03/UMP/02/21 (university reference RDU1901151) and Universiti Malaysia Pahang for financial supports.

References

1. Liu Z, Gao Y, Yang J (2021) Numerical and experimental-based framework for vibro-acoustic coupling investigation on a vehicle door in the slamming event. Mech Syst Sig Process 158:107759. https://doi.org/10.1016/j.ymssp.2021.107759
2. Liu Z, Gao Y, Yang J, Xu X, Fang J, Xu Y (2021) Effect of discretized transfer paths on abnormal vibration analysis and door structure improvement to reduce its vibration in the door slamming event. Appl Acoust 183:108306. https://doi.org/10.1016/j.apacoust.2021.108306
3. Oliver-Ortega H, Julian F, Espinach FX, Tarrés Q, Ardanuy M, Mutjé P (2019) Research on the use of lignocellulosic fibers reinforced bio-polyamide 11 with composites for automotive parts: car door handle case study. J Clean Prod 226:64–73. https://doi.org/10.1016/j.jclepro.2019.04.047
4. Liu Z et al (2021) Transfer path analysis and its application to diagnosis for low-frequency transient vibration in the automotive door slamming event. Measurement 183:109896. https://doi.org/10.1016/j.measurement.2021.109896
5. Cheon S, Jeong H, Hwang SY, Hong S, Domblesky J, Kim N (2015) Accelerated life testing to predict service life and reliability for an appliance door hinge. Procedia Manuf 1:169–180. https://doi.org/10.1016/j.promfg.2015.09.082
6. Soo VK, Compston P, Doolan M (2016) Is the Australian automotive recycling industry heading towards a global circular economy?—a case study on vehicle doors. Procedia CIRP 48:10–15. https://doi.org/10.1016/j.procir.2016.03.099

Graphene as an Alternative Additive in Automotive Cooling System

Ganesaan Kadirgama, Muhammad Izdihar Bin Razman, Devarajan Ramasamy, Kumaran Kadirgama, and Kaniz Farhana

Abstract The project represents graphene can be used as an alternative additive in the automotive cooling system. Thus, graphene nanofluids have been prepared at 0.1, 0.3 and 0.5% volume concentrations. Afterward, measurement of various thermophysical properties of nanofluid such as thermal conductivity, density, viscosity, and specific has been done. The obtaining data has been analyzed and compared with graphene oxide, titanium oxide, aluminium oxide, silicon carbide, and copper oxide nanofluid to figure out the best nanofluid that can absorb more heat to protect the car engine from overheating. In, summary, the overall best nanofluid among these six would be graphene oxide, with the best thermal conductivity, specific heat capacity, and one of the lowest viscosities. As for comparison among graphene all volume concentrations, the 0.1% graphene nanofluid demonstrated the best with high thermal conductivity and low viscosity.

Keywords Graphene · Nanofluid · Radiator · Automotive cooling system · Comparison

1 Introduction

The engine is the most critical component of the vehicle, and the control of the automobile and the drive of the car air conditioners are both powered by the engine, but the performance of the temperature engine that is too high will also reduce harm and the temperature of the engine that is too low can again increase the consumption

---

G. Kadirgama · M. I. Bin Razman · D. Ramasamy · K. Kadirgama
Faculty of Mechanical and Automotive Engineering Technology, Universiti Malaysia Pahang, 26600 Pekan, Pahang, Malaysia

K. Kadirgama
Center of Automotive Engineering, Universiti Malaysia Pahang, Pekan, Pahang, Malaysia

K. Farhana (✉)
Department of Apparel Engineering, Bangladesh University of Textiles, Dhaka 1208, Bangladesh
e-mail: kaniz.farhana@butex.edu.bd

M. Y. Ismail et al. (eds.), *Technological Advancement in Mechanical and Automotive Engineering*, Lecture Notes in Mechanical Engineering,
https://doi.org/10.1007/978-981-19-1457-7_2

of oil. Therefore, the successful guarantee of ensuring that the engine runs well below normal working temperatures is a cooling device [1].

In addition to the constant expansion of the business engine, the cooling fluid used to match it has also made considerable progress. On the market, current cooling fluid is all containing water coolant (substantially water, ethylene glycol), its boiling point is low, the freezing point is high, fast erosion of cavitation, weak heat conductivity, a significant volume of water vapor is not satisfied with the market requirement. For example, generating water coolant from the steam in the metallic surface of the cylinder periphery, causing timely engine work heating to take place in time, eventually overheating the engine interior and causing its deterioration, and permanent use of water coolant may also trigger corrosion to occur in the cooling system, incrustation scale, the series of problems such as boiling, causing the cooling system to foretell 2 years needs to change once [2, 3].

An atom-thick honeycomb layer of carbon atoms is known as graphene. Because a typical carbon atom has a width of about 0.33 nm, there are about 3 million layers of graphene in 1 mm of graphite, making it the building block for other graphitic materials. Nanographene is a type of graphene that is tailored to certain functions and so has a more complicated manufacturing process than regular graphene. Dehydrogenation is the process of selectively removing hydrogen atoms from organic carbon and hydrogen molecules to produce nanographene. Graphene is tougher than steel but lighter than aluminium, and it is harder than diamond but more elastic than rubber. Graphene is known as one of the strongest substances [4] has been extensively examined. Commercial use of materials and devices based on graphene. Various methods of preparing graphene have been developed to date, which can be divided into chemical vapor deposition of graphene layers, micromechanical exfoliation of graphite using the Scotch-tape peel-off method, epitaxial graphene foils, biomethane synthesis of organic molecular graphene, and GO sheets reduction or deoxygenisation [5]. In the case of using a CNC machine, lubricant mixed with water is used to cool and lubricate the tool and workpiece. It also transports the chip, contaminants, and debris away from the cutting area. Base grease is one of the materials used in lubricants. By adding graphene, it will enhance the tribological properties and thermal conductivity of base grease. In this study, the cooling performance of automotive has been investigated by using graphene nanofluids in various concentrations. Therefore, various thermophysical properties of graphene nanofluids such as thermal conductivity, density, viscosity, and specific heat have been studied comprehensively along with analogy of other's studies.

2 Literature Review

2.1 *Graphene*

In this research, graphene is the most important component as it acts as the additive in the cooling system. The best way to explain graphene is that it is the soft, flaky substance used in pencil lead, a thin and single layer of graphite. Graphite is the rhinestone of carbon element, which means it is constructed with the same atoms, but the crystal arrangement is different, providing different characteristics to the substance. Diamond and graphite, for example, are both sources of carbon, but they have wildly different natures. Although graphite is fragile, diamonds are extremely solid. The atoms of graphene are structured in a hexagonal structure. Interestingly, it takes on certain miraculous properties when graphene is removed from graphite. It is a simple one-atom-thick, the first-ever discovered 2D substance. Regardless of this, kits knew that graphene is also one of the strongest materials. With 130 GPa (gigapascals) of tensile strength, it is more than 100 times stronger than steel. Despite being so thin, graphene's incredible strength is already enough to make it impressive, but its special properties do not stop there. It is also pliable, translucent, superb conductive, and most fluids and gases are seemingly impermeable. It seems almost as if there is no region in which graphene does not excel [6–8].

2.1.1 Graphene Synthesis by Mechanical Exfoliation

The 2D research explosion as a catalyst has been separated from bulk graphite atomically into thin layer materials called graphene in 2004. Sir Konstantin Novoselov and Sir Andre Geim had won the 2010 Nobel Prize in the Physics area for their pioneering task on graphene in this seminal paper and subsequent research. As stated in Sir Andre Geim's Nobel speech, the research outcomes graphene attracted more attention due to its higher electronic efficiency, stability, and tunability at ambient temperatures that are vulnerable to the surroundings. Besides, unlike any other materials, the capability to achieve a broad mean free electron route at submicron distances without dispersion in before isolated graphene Hall bar mechanism that was attached to a microscopically harsh substance encircled by adsorbates and polymer residue was dissimilar with other substrates. Geim addresses surface science in the Nobel Lecture and how to study thin films involves a high vacuum, where properties usually decline as the thickness of material reduces; but this was not the case when Geim and his research group illustrated the effect of the electric field in FLG and began the boom of graphene and two-dimensional material [9, 10].

To describe the emerging optical properties comprehensively, the researchers researched the properties of bulk HOPG samples such as time-resolved reflectivity and time-resolved transmission across fragile samples. They measured the electronic carriers in solid graphite using a 50-fs pump and probe method over a broad range of optical agitation conditions but lower the threshold for the damage of optical conditions. Significant scattering of light was found on the free-standing films while processing the samples, so the samples were taped to glass material, and the tape was then detached, minimizing the light scattering, leaving "mounted films" behind. The mechanical exfoliation strategy takes shape via Scotch tape, but it will not be until 14 years later that the strategy is utilized again to segregate, classify, and describe the effect of the single to FLG ambipolar electric field [9, 11].

The mechanical graphene exfoliation process known as the Scotch tape technique invented in 2004; is simply attaching tape to the surface of some form of bulk graphite, removing the tape away, and then adhering the tape and residues to a material, typically silicon oxide. The tape is pulled off until adhered to the substrate, and a final step of exfoliation leaves behind casually spaced flakes of different thicknesses, specifically monolayer graphene. But, in isolating FLG, the procedure conducted by Geim and Novoselov was quite interesting. To build mesas 5 μm in-depth in square form from 20 μm to 2 mm while the process begins by using an oxygen plasma etch of 1 mm thick HOPG platelets. The etched HOPG was then flipped and dipped on top of the glass into a wet photoresist of 1 μm thick. To solidify the photoresist and catch the mesas, the accumulation is then dried and cured. In phase 4, the dense layer of HOPG staying above the mesas is separated off, remaining the collection of mesas ingrained in the cured photoresist behind. It is in phase 5 where the renowned "Scotch tape process" is introduced. To deduct the number of HOPG layers connected by weak Van der Waals forces step by step, scotch tape is employed frequently to exfoliate mechanically the mesas fixed in the photoresist. The graphite mesas that are mechanically exfoliated are then put in an acetone dip, by dissolving the photoresist in acetone, releasing the flakes. In the water, the glass separates, remaining graphitic flakes in the acetone dip float. In step 7, along with the thin flakes of graphite, a silicon oxide layer material (n^+ doped) of 300 nm is immersed in the acetone dip. With water and propanol, the SiO_2 substrate is stripped and cleaned. Steps 7 and 8 are a single stage where the layer has been immersed, removed as shown in Fig. 1, and then washed with water and propanol in the solution. The water aids catch the narrow flakes on the material in this process. The material is put in a bath of ultrasound of propanol in the final stage, where added ultrasound cleaning eliminates bigger flakes and proceeds to narrow the graphite to possible FLG (~1–3 layers). Flakes below 10 nm thick were found to bind firmly to the SiO_2 surface by VDW and capillary forces, following the material. Lastly, from the ultrasound bath, the SiO_2/graphene composite is extracted, dried, and prepared for fabrication and characterization of the unit [9, 12].

The absorption of water between the occupied FLG graphene films and the SiO_2 substrate is referred to as a "dead sheet," presented in the recording of a standard 1 nm thickness greater than the original graphite interlayer spacing of 0.335 nm. It was successfully isolated and tested the electronic properties of graphene by Geim and his research group who used this mechanical exfoliation technique, even though many scientists who studied the material earlier had made substantial investigations with just a few layers of graphite [9, 13].

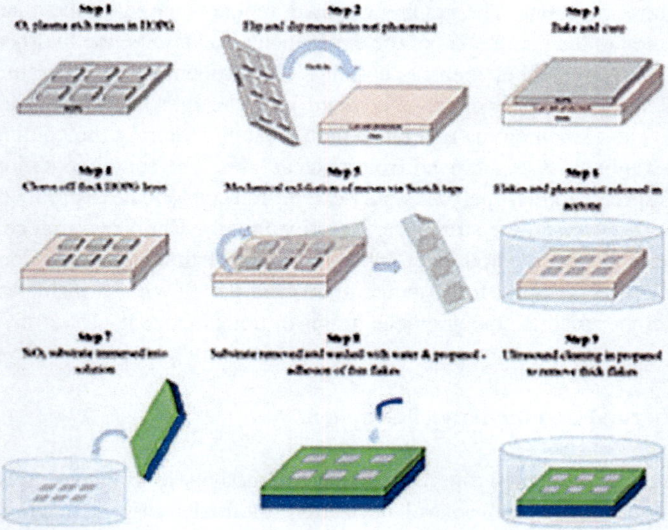

Fig. 1 Mechanical exfoliation of single to few-layer graphene for the measurements of the effect of the electric field in isolation ambipolar by using HOPG and associated microfabrication steps, determined in supporting substance of highly aligned pyrolytic graphite [9]

It was monumental to separate single to few-layer graphene. To determine the superb properties of this new substance at just a few atomic layers, however, electrical measurements were required. To describe the device structure, Hall bar devices with multiterminal were microfabricated using electron-beam lithography. Next, an O_2 plasma etch was used to remove undesirable substrate, connections were established using deposition and lithography of Au (100 nm) on Cr (5 nm) to clean the photoresist polymer using the final lift-off technique. This newly segregated FLG, with 10,000 cm^2/Vs mobilities, $n \approx 5 \times 10^{12}$ cm^{-2} charge density, and mean-free-path of electrons ~0.4 μm, all at room temperature, was involved in exposing the high electronic efficiency and tunability of multiterminal Hall bar devices. The characteristics were exceptional, especially regarding the existence of pollutants and the connected interface with the SiO_2 material. Enhancements in the transport of electrons were illustrated by encapsulated graphene in hexagonal boron nitride (HBN) on two sides, at low and room temperatures, with electrons ballistic transport (no scattering) [9, 14].

2.1.2 Graphene Synthesis by Chemical Vapor Deposition (CVD)

To synthesis graphene by the CVD method, the experimental setup was prepared by using a quartz tube housed within a furnace, precursor (argon and hydrogen), and nucleation (methane) gases, mass flow controllers, and a vacuum pump to control low-pressure conditions. The graphene growth recipe is based on the material used and calibrated to the parameters of the experiment. To dissolve the hydrocarbon gas methane into its integral elements of carbon and hydrogen, high temperature is used, where the surface of the copper is covered with the interactions and depositions of carbon. The carbon atoms reorganize and crystallize during the cooling process to build hexagonal shapes derived from nuclear sites. The actual conditions for the recipe will spread grain fronts to laterally extend and consolidate to produce a continuous sheet (polycrystalline structure). Described by the Ruoff research community, the graphene growth method details the steps of growth of graphene domain on a copper foil. The pressure, time, temperature, and gas flow rates inside the growth tube or chamber mediate the graphene grains or domain size [9, 15, 16].

2.1.3 Thermal Conductivity of Graphene

Thanks to the remarkable properties of heat conductivity of graphene and its potentiality to apply in thermal control utilization, thermal transport in graphene is a flourishing area of study. At room temperature, the general thermal conductivity of graphene within a range of 3000–5000 $Wm^{-1}\ K^{-1}$, an outstanding rate compared with the thermal conductivity of pyrolytic graphite at room temperature of about 2000 $Wm^{-1}\ K^{-1}$. Other studies, however, state that this amount is extravagant and that graphene's thermal conductivity in-plane form for freely suspended samples at room temperature is around 2000–4000 $Wm^{-1}\ K^{-1}$. This figure is still among the largest of any substance known [17, 18].

Graphene is studied as an outstanding conductor of heat, and many investigations have revealed it to have limitless heat conduction capacity based on the sample size, contradicting the micrometer scale of Fourier's law. The researchers found, in both numerical simulations and experiments, that the bigger the graphene fragment could transfer more heat. Graphene could potentially consume an infinite quantity of heat. Logarithmically, increasing thermal conductivity happened due to the stable bonding pattern as like 2D material. Since graphene is substantially stronger than steel to tear and is also fragile and versatile, there may be some attractive real-world applications for its conductivity [17].

Electronics applications based on graphene-enabled thermal management could be vital beneficial from the ability of graphene to expend heat and maintain electronic operation. For smaller and more powerful devices, heat is also a specific factor in micro and nanoelectronics. Thus, for this form of application, graphene and alike materials with excellent thermal conductivity can hold immense promise. The heat conductivity of graphene can be used in various forms, including heat spreaders, thermal interface materials (TIM), thermal greases, nanocomposites based on graphene, and so on [19]. The physical properties of base fluids (EG and water/EG) are tabulated in Table 1 and the properties of nanoparticles are enlisted in Table 2.

Table 1 Different physical properties of base fluid [20]

Property	Ethylene Glycol (EG)	Water/EG (50:50)
Thermal conductivity (Wm^{-1} K^{-1})	0.258	0.380
Dynamic viscosity (mPa.s)	20.9	3.94
Density (kg/m^3)	1113.2	1073.35
Specific heat (j/g.K)	2.347	3.281
Freezing point (°C)	– 12.9	– 37.0
Boiling point (°C)	197.3	107.4

Table 2 Physical characteristics of coolants and tubes [21]

Parameter	Graphene	Carbon nanotubes	Water
Thermal conductivity (Wm^{-1} K^{-1})	500–600	>3000	0.605
Density (kg/m^3)	1500	1000	997.1
Specific heat (j/g.K)	1900	1300	4195

2.1.4 Graphene Nanofluid

Graphene is an auspicious material with excellent electrical, mechanical, and thermal properties in its mixed forms, such as graphene oxide (GO), nanotubes, and nanoparticles, and can significantly improve the properties of host materials and liquids. Graphene has the highest thermal conductivity among the currently known materials, large numbers of scientists study vastly on graphene, and reported the major effect on the improvement of the various automotive cooling systems. The simple technique of synthesis, higher dispersion stability, greater heat conductivity, lower corrosion and erosion, large surface area, and lesser pumping power requirement are all advantages of graphene nanofluid. There have been numerous studies on how to make and stabilize graphene nanofluids, and it is necessary to compile a comprehensive number of methods [18].

3 Methodology

3.1 Preparation of Graphene Nanofluid

3.1.1 Calculation for Graphene Powder Mass

To prepare the nano coolant, graphene powder mass needs to be determined by using this formula (Hussein et al. 2013):

$$\Phi = \left[\frac{\frac{w}{\rho_p}}{\frac{w}{\rho_p} + \frac{w_{bf}}{\rho_{bf}}} \right] \times 100$$

where

w mass of nano particle (Graphene powder).
ρ_p density of nano particle (Graphene powder).
w_{bf} mass of base fluid (Ethylene glycol).
ρ_{pf} density of base fluid (Ethylene glycol).
Φ Concentration of nanofluid.

With the density graphene powder 2.267 g/ml and volume of ethylene glycol used which is 200 ml, the graphene powder mass can be obtained based on the wanted concentration. This is the sample calculation for 0.1% (V/V) concentration:

$$\varnothing = \frac{m_{np}/\rho_{np}}{m_{np}/\rho_{np} + m_{bf}/\rho_{bf}} \times 100\%$$

$$\varnothing = \frac{m_{np}/\rho_{np}}{m_{np}/\rho_{np} + V_{bf}} \times 100\%$$

$$0.1 = \frac{m_{np}/2.267\ \text{g/ml}}{m_{np}/2.267\ \text{g/ml} + 200\ \text{ml}} \times 100\%$$

$$m_{np} = 0.45\ \text{g}$$

3.1.2 Sonication Process

A probe sonicator is utilized in this technique. In a beaker, graphene powder is mixed with ethylene glycol. The probe was then immersed in the nanofluid mixture. To avoid degenerate graphene molecules, the sample should be bathed in cold water at the same time. A sonicator is a powerful piece of laboratory equipment that generates an ultrasonic electric signal to power a transducer as shown in Fig. 2. The electric signal is converted by the transducer employing piezoelectric crystals, which are crystals that respond to electricity by causing a mechanical vibration. The vibration is preserved and amplified by the sonicator until it reaches the probe. The probe goes up and down swiftly in rhythm with the vibration to convey it to the solution.

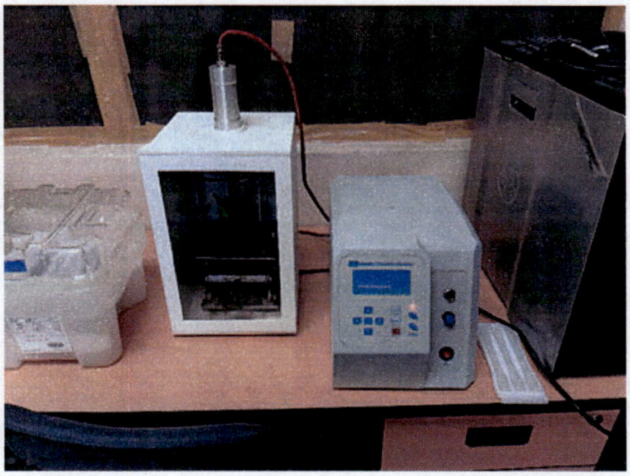

Fig. 2 Probe sonicator setup for ultrasonication

3.2 *Measurement of Thermophysical Properties*

3.2.1 Thermal Conductivity Measurement

For thermal conductivity, C-Therm Technologies TCi-3-A was used, which offers a range of high-pressure cells to safely characterize the thermal conductivity of samples under elevated pressure environments. This machine is conducted for about 8 h to extract the thermal conductivity of one sample concentration as presented in Fig. 3.

The temperature of the sample is raised in a thermal chamber. When electricity is given to the sensor's coil and guard ring at the same time, heat is created. One-dimensional heat conduction from the coil to the heated sample is supported by the guard ring and backing. A voltage drops in the primary sensor coil, which is calibrated to the temperature change, monitor the rate of temperature increase. The rate of

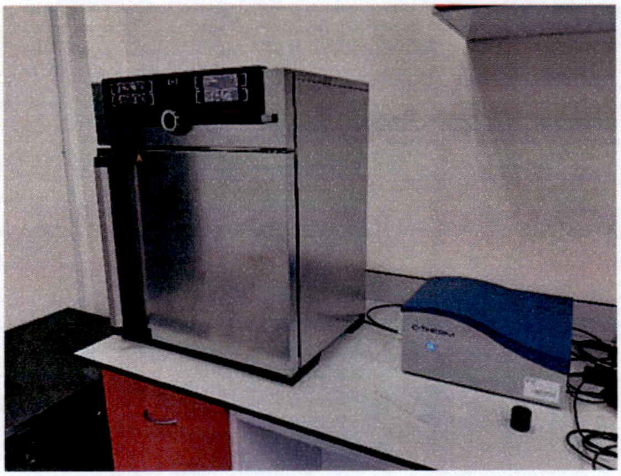

Fig. 3 Setup of C-Therm for measurement of thermal conductivity

Fig. 4 Setup of Anton Paar portable viscometer

growth in the observed temperature is inversely related to the thermal conductivity. The lesser the sample's thermal conductivity, the steeper the temperature rise, and vice versa. The TCi program allows the creation of automated test reports and exports them to Microsoft Excel.

3.2.2 Viscosity and Density Measurement

In this data measurement, Anton Paar Portable Viscometer SVM 30001 is used to measure lubricants, heavy fuels and classify crude oils in terms of kinematic viscosity, dynamic viscosity, and density. This experiment only used around 60 ml of the sample by injecting it using a syringe as displayed in Fig. 4.

3.2.3 Specific Heat Capacity Measurement

For specific heat capacity, IKA Bomb Calorimeter C6000 is used, more specified and easy-to-use equipment as presented in Fig. 5. Around 2.9–3.1 g are used to obtain the specific heat capacity. Before the experiment started, the decomposition vessel is taken out as shown in Fig. 6, to use the crucible (Fig. 7). Then, the measured mass of graphene nanofluid with crucible (Fig. 8) would be inserted back into the decomposition vessel. Lastly, the vessel is inserted inside the bomb calorimeter to obtain the reading as shown in Fig. 9.

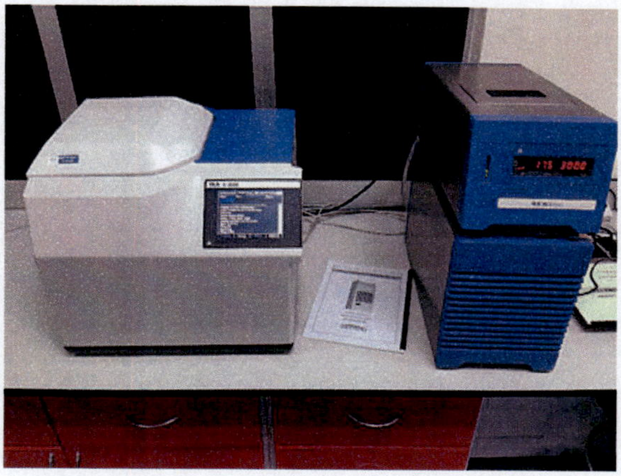

Fig. 5 Setup of bomb calorimeter

Fig. 6 Decomposition vessel

4 Results and Discussion

4.1 Thermal Conductivity and Effusivity

Loading of nanoparticles into the base fluid increases thermal conductivity effectively which is closely related to volume concentration. Herein, graphene also demonstrates this phenomenon as nanofluids [22]. The thermal conductivity of base fluids increases due to adding graphene nanoparticles into it as shown in Fig. 10. And thermal effusivity is also rising as displayed in Fig. 11. However, the increasing volume concentration of nanofluids portrayed lower thermal conductivity and effusivity (Figs. 10 and 11). This could have happened for various reasons; among them, nanoparticle

Fig. 7 Crucible attached to the vessel cap

Fig. 8 Weighing nanofluid with the crucible

size could be responsible for this according to Yang, Cao [23]. Moreover, thermal conductivity is also concerned with temperature. Increasing temperature presents the rising trend of thermal conductivity for all volume concentrations as well as thermal effusivity as shown in Figs. 12 and 13 respectively [24]. A higher thermal conductivity coefficient means better conductivity of nanofluid across the radiator of the car [25]. In the automotive cooling system, the higher the thermal conductivity, the better the nanofluid when absorbing heat and protecting the radiator from overheating.

For the comparison study, the thermal conductivity of aluminium oxide, graphene oxide titanium oxide, silicon carbide-MWCNTs, and copper oxide radiator coolant

Fig. 9 Decomposition vessel before entering the bomb calorimeter

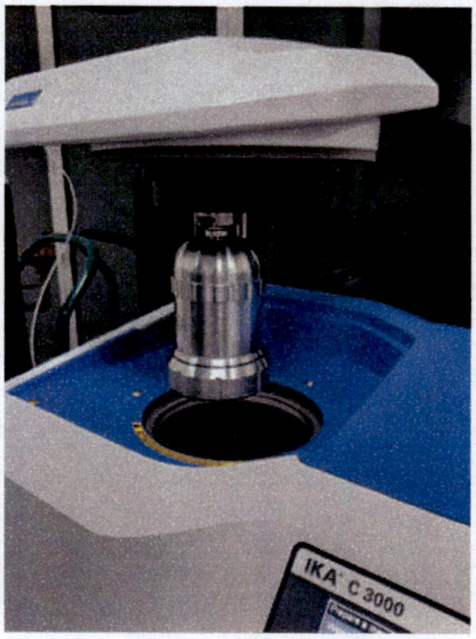

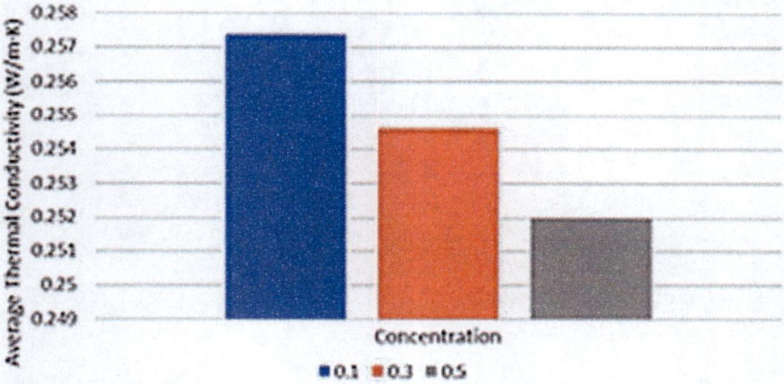

Fig. 10 Thermal conductivity concerning vol. concentrations

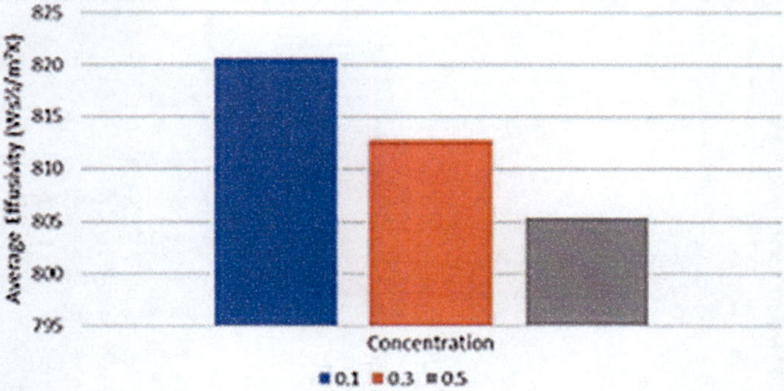

Fig. 11 Effusivity concerning vol. concentration

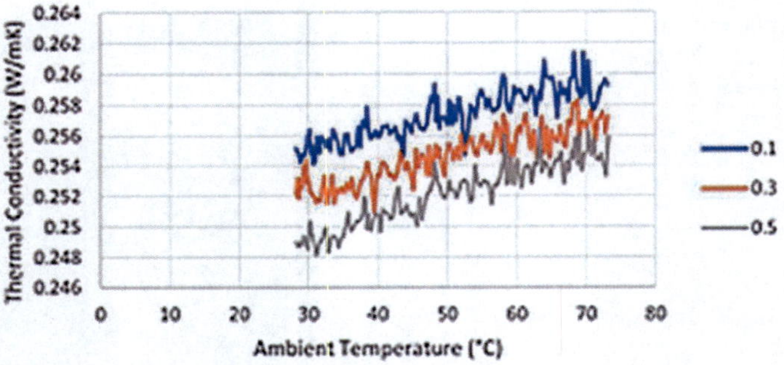

Fig. 12 Thermal conductivity at various temperatures

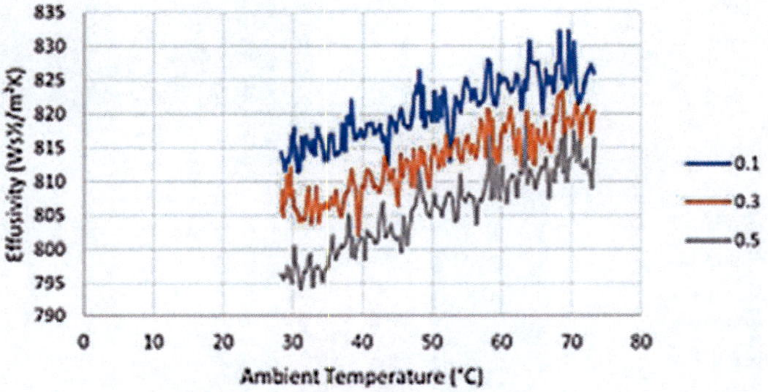

Fig. 13 Effusivity at different temperatures

nanofluids for different volume concentrations is depicted in Fig. 14. The thermal conductivity of the nanofluids increases with increasing volume concentration, as seen in the graph. Thermal conductivity also increases with temperature for both the

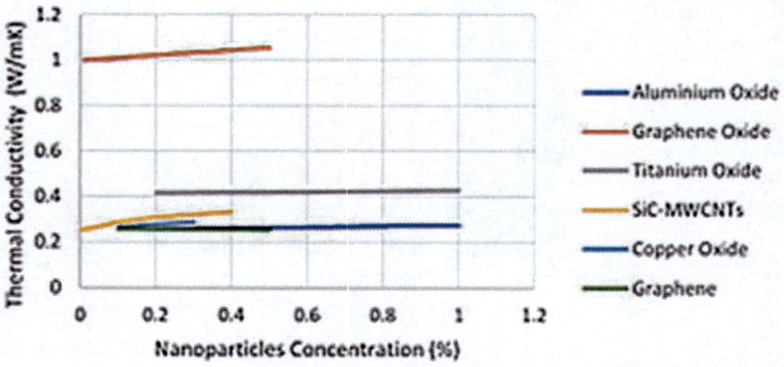

Fig. 14 Thermal conductivity of different nanofluids suspended into ethylene glycol at various vol. concentrations

base fluid and the nanofluid, and the nanofluids have higher thermal conductivity than the base fluid at all volume concentrations. This is because of the particle mobility when the temperature rises. The movement among the particles increases as the fluid temperature rises, increasing the thermal conductivity. The graph lines, however, are not parallel. The explanation for this could be that the thermal conductivity of nanofluids is influenced not only by particle concentration and temperature but also by other factors such as size and form.

In Fig. 14, we can see that the graphene nanofluid has the highest thermal conductivity among other nanofluids. Probably this is because of the nanoparticle size of graphene oxide that is smaller than the other nanoparticles. The value of graphene oxide must be lower than 25 nm, which is the nanoparticle size of graphene powder. As for graphene, the value of thermal conductivity of graphene is almost the same as copper oxide.

Figure 15 shows another graph of 3 nanofluids with water as their base fluids. Different base fluids will also give a different thermal conductivity towards a nanoparticle. A research stated that the base fluid thermal conductivities of ethylene glycol and water at 30 °C can reach 0.254 $Wm^{-1}\ K^{-1}$ and 0.607 $Wm^{-1}\ K^{-1}$ respectively. The thermal conductivity of base fluids would also affect the thermal conductivity of the nanofluids. The thermal conductivity of base fluid rises along with the thermal conductivity of nanofluids. This can be seen in the comparison of Fig. 14 with Fig. 15 where we can see that the value of thermal conductivity of aluminium oxide, copper oxide, and titanium oxide are higher with water as the base fluid than their values of thermal conductivity with ethylene glycol as their base fluid. This proves that not only the thermal conductivity of nanoparticles, but the base fluid of a solution also affects the whole thermal conductivity.

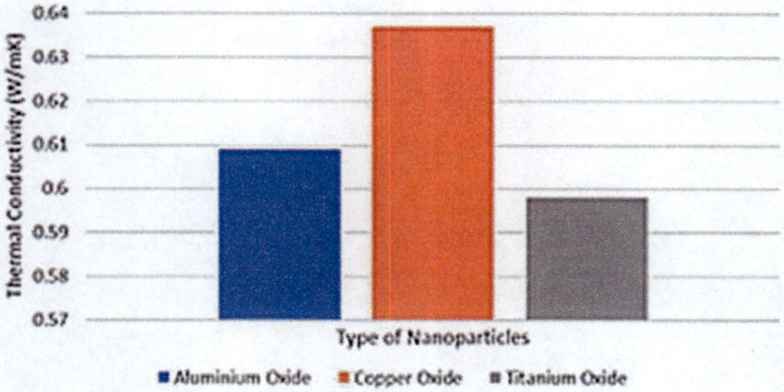

Fig. 15 Thermal conductivity of some nanoparticles with water as base fluid

4.2 Density

The density of nanofluids is equal to volumetric concentrations of nanoparticles and knowledge stated that the higher the concentration of a solution the higher its density [26]. Density may seem like the least important factor among other thermophysical properties, but density also plays a significant role in heat convection. The higher the density of a nanofluid, the slower the heat loss by convection. Herein, the density of graphene nanofluids depicts that the higher volume concentrations are the higher density as shown in Fig. 16. Even though the graphene nanoparticle is extremely light, the nanofluids produced would always give an increase, since there is only a slight difference among the nanofluids. Figure 17 shows a density graph of the nanoparticles themselves without any base fluid. We can see that silicon carbide has the lowest density while copper oxide has the highest density. This density is what we called, bulk density because it came in powdered form.

In comparison with graphene, the graphene bulk density is 2.267 g/cm^3 without any base fluid. But when it is mixed and becomes a nanofluid, the density of

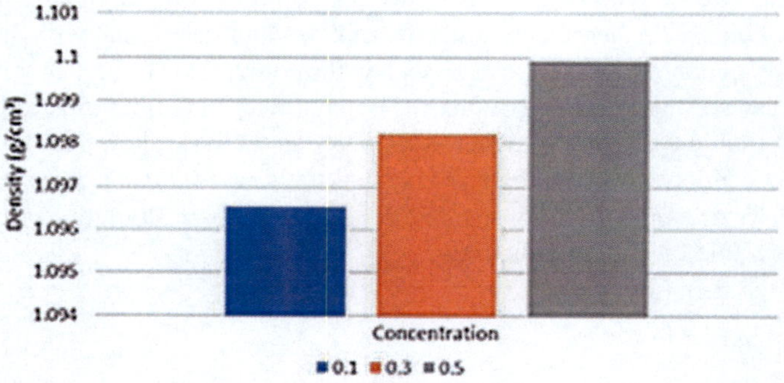

Fig. 16 Density of nanofluids at various vol. concentrations

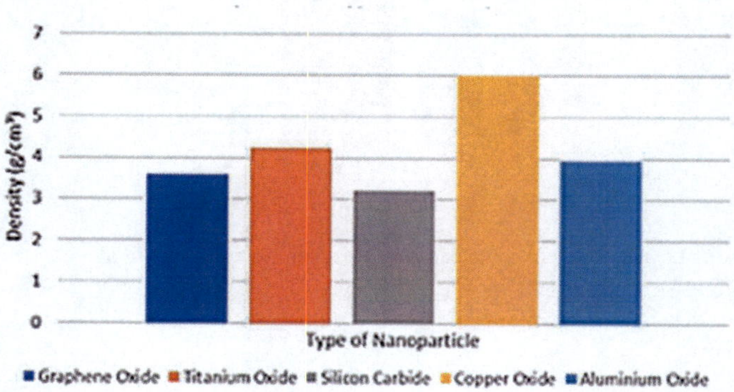

Fig. 17 Density of different nanoparticles

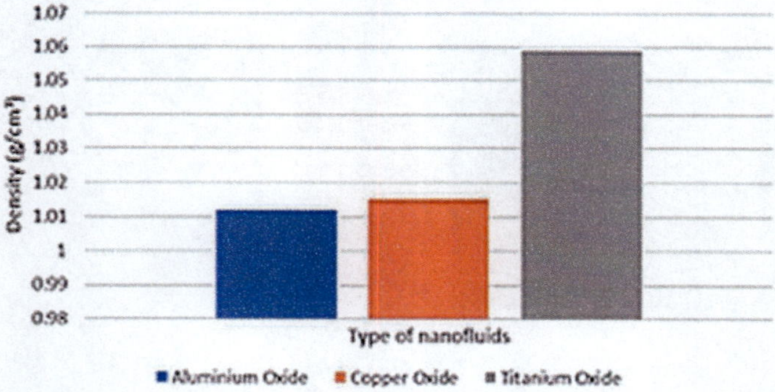

Fig. 18 Density of some nanofluids with water as base fluid

the graphene nanofluid decreases slightly to 1.09992 g/cm^3 for 0.3% nanoparticle concentration. The graphene solution decreases only slightly as it should be, just like other solutions. However, the density difference is not that significant when compared with the effect of temperature changes. When a substance is heated, it expands in volume and loses density. An increase in volume is negligible in solids, resulting in a decrease in density. When the temperature of liquids and gases rises, the volume expands, and the density reduces significantly. This is what affected the density of graphene nanofluid which decreases due to the thermophysical experiment that is conducted until 40 °C. The density of the powdered form does not affect the temperature.

Figure 18 shows some nanofluids with water as base fluid. We can examine that titanium oxide has the highest density and has a huge gap between the other 2 nanofluids. This is due to the temperature of titanium oxide that is at 25 °C while aluminium oxide and copper oxide is at 30 °C. If the titanium oxide is tested at 30 °C as well, the value of the density would be lower and shows only a small difference among the 3 nanofluids. This is because the density of any solution will drop if the temperature increases.

As for the other 2 solutions, the silicon carbide would be predicted to be in the lowest among another nanofluid while graphene oxide would be predicted in between silicon carbide and aluminium oxide, based on Fig. 17.

4.3 *Viscosity*

Viscosity is another important property of nanofluids and increases due to adding nanoparticles into the base fluid. The viscosity of a car coolant is so important due to its flow in radiator tubes. For the nanofluids, the lower the viscosity, the suitable it is to be a nano coolant in a car radiator. Herein, we can see that both dynamic and kinematic viscosity increases along with the concentration as shown in Figs. 19 and 20 respectively. For both graphs, there is a slight difference between the concentration

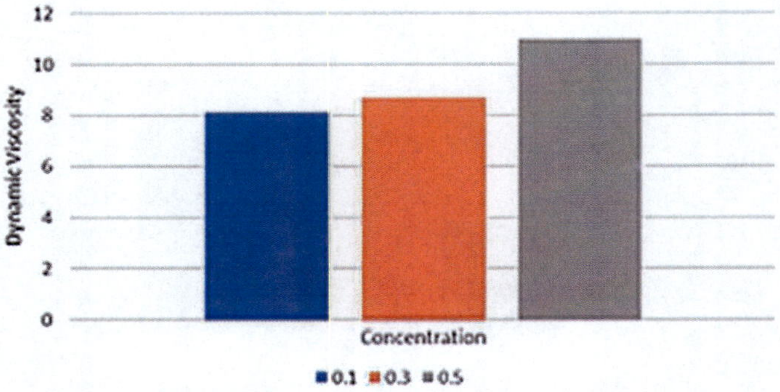

Fig. 19 Dynamic viscosity at different vol. concentrations

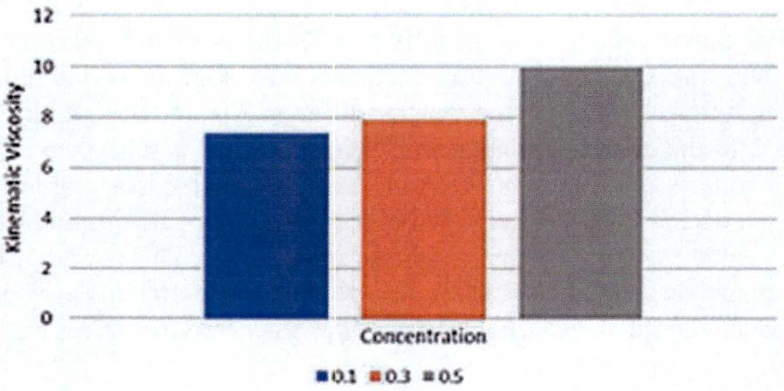

Fig. 20 Kinematic viscosity at varied vol. concentrations

of 0.1 and 0.3%, while for the 0.5% concentration, there is a greater increment when compared to the other concentrations. This proves that 0.1% is the best solution for car radiators due to its lower value in viscosity. The lower the viscosity, the smoother the flow of the nanofluid [27].

Figure 21 shows the viscosity of the solutions with ethylene glycol or ethylene glycol/water as their base fluid in a different range of temperatures. However, there is a range of temperature that all the solutions are stacked on, which is 25–45 °C. From this range of temperature, we can see that copper oxide has the highest viscosity among other solutions. There is only a slight gap in between the solutions but for copper oxide, the gap is too wide, and this is probably because of its nanoparticle concentration or nanoparticle size [27]. The higher the nanoparticle concentration and the smaller the size of nanoparticles, the higher the viscosity of a solution. As for the graph itself, it is shown that the line graph drops. This is a natural phenomenon in which, in most situations, the viscosity of liquids reduces as the temperature rises. When the temperature of a substance rises, so does the mobility of its molecules. The resistance to the flow of a substance or what we called as viscosity, reduces as the mobility of the molecules increases.

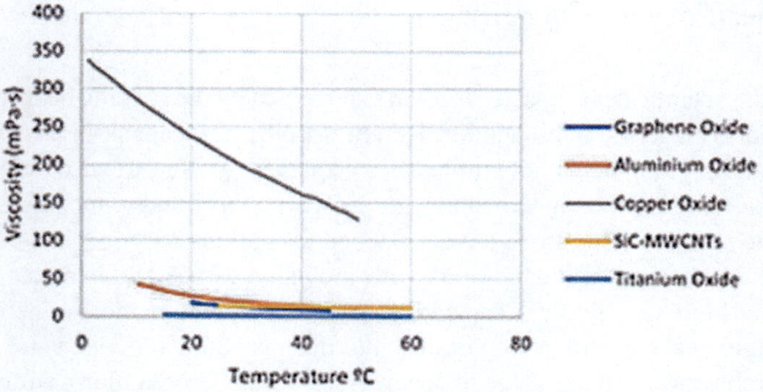

Fig. 21 Viscosity of different nanofluids dispersed into ethylene glycol at different temperatures

In comparison with graphene solution, the viscosity of graphene is almost the same as graphene oxide solution at 40 °C. Other than that, the graphene solution has a higher viscosity compared to titanium oxide but is lower than the other solutions.

Figure 22 shows a graph with water-based nanofluids and their viscosity. There is no specific value of the viscosity of ethylene glycol, but this base fluid is surely more viscous than water due to the increase in the number of hydrogen bonds. From Figs. 21 and 22, we can see a significant gap in between the values of viscosity. The viscosity of the nanofluids with ethylene glycol base is higher compared to the viscosity of the nanofluids with water bases.

The viscosity of any solution is also affected by temperature. Higher temperature leads to a less viscous solution [27]. At lower temperatures, water has a higher viscosity, whereas, at higher temperatures, it has a lower viscosity. For example, water in the freezer loses energy as the temperature drops, the water molecules become more attracted to each other, and the water flows slowly until it freezes to ice. The same goes for any solution including ethylene glycol.

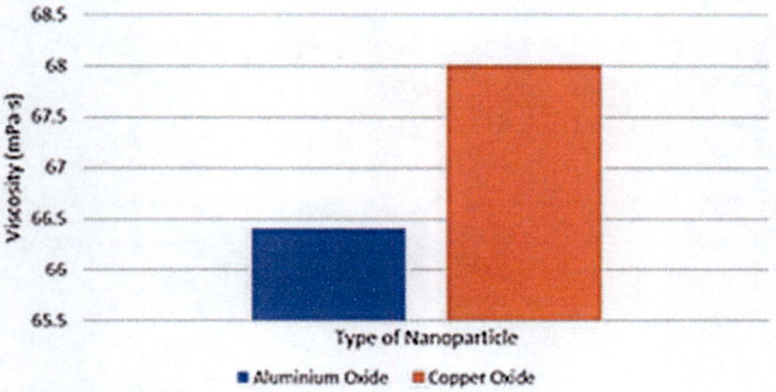

Fig. 22 Viscosity of some nanoparticles with water as base fluid

4.4 Specific Heat Capacity

Nanofluids' specific heat capacity is often determined by the specific heat capacity of nanoparticles dispersed in base fluids. Herein, specific heat capacity is also increased due to loading graphene nanoparticles and displayed a rising trend with increasing volume concentrations as shown in Fig. 23. However, this trend is not similar to other studies such as [28]. This could have happened for various purposes and conditions.

Figure 24 shows a few nanofluids with the value of specific heat capacity. For example, a high specific heat capacitive nanofluid can be made by incorporating high specific heat capacity nanoparticles into the base fluid and vice versa.

Research shows that the value of specific heat capacity goes along with the value of thermal conductivity. By using Fig. 14 as a reference, the predicted value of the specific heat capacity of graphene oxide would be the highest, more than the value of silicon carbide. As for titanium oxide, it would be predicted higher than silicon carbide as well but lower than graphene oxide.

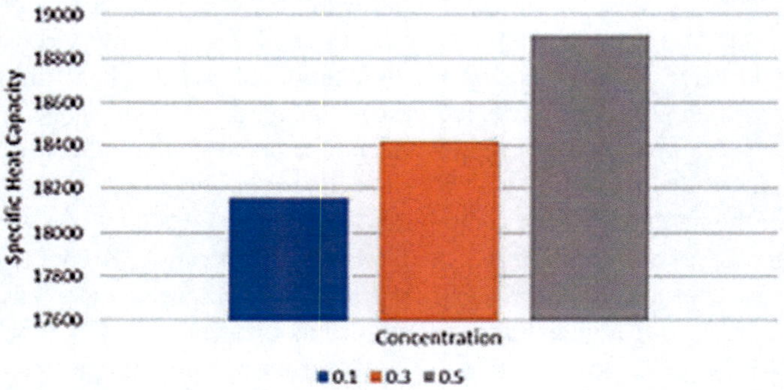

Fig. 23 Specific heat capacity of graphene nanofluids

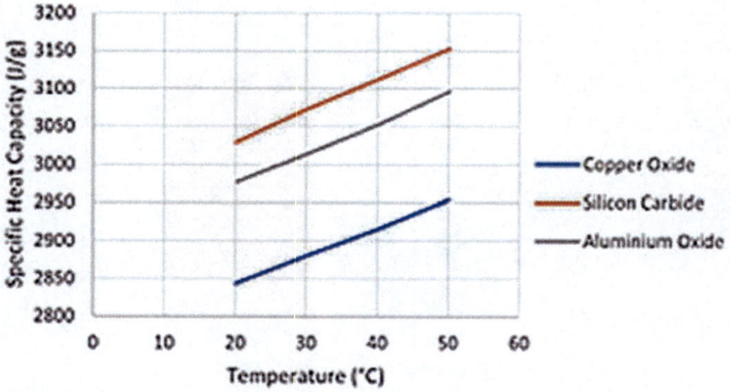

Fig. 24 Specific heat capacity of different nanoparticles with base fluid ethylene glycol

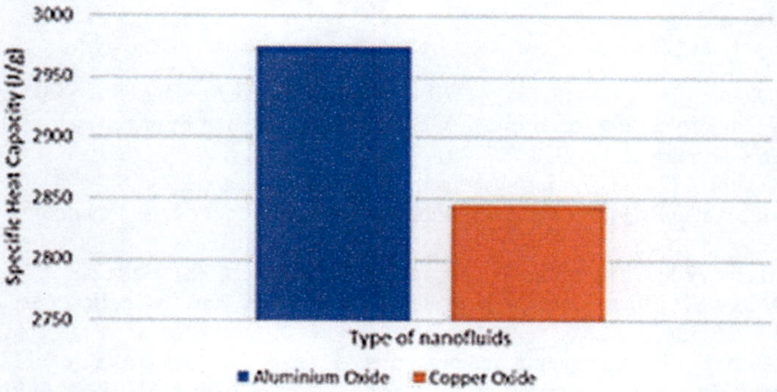

Fig. 25 Specific heat capacity of some nanoparticles with base fluid water-ethylene glycol

Figure 25 shows a graph of some nanofluids with water-ethylene glycol base fluid at 20 °C. It shows that the value of the copper oxide nanofluid increases very slightly from 2843.823 to 2845 J/g when water is added as a base fluid. However, for aluminium oxide, the value of specific heat capacity decreases slightly from 2977.502 to 2975 J/g. The contradiction in the difference of specific heat capacity must be affected by the value of nanoparticle concentration.

5 Conclusion

- The thermal conductivity of graphene is promising to transfer heat in automotive cooling systems and 0.1% vol. con. performs the maximum thermal conductivity. But graphene oxide performed the highest thermal conductivity performance than other oxide nanofluids.
- Graphene nanofluid shows good density behaviour than other nanofluids for the automotive cooling system. However, graphene oxide exhibited better results of density for the cooling systems as well.
- Lower vol. the concentration of graphene shows lower viscosity that is very suitable for automotive cool systems and 0.1% could be better for it. As well as graphene oxide exhibits lower viscosity also preferable for cooling systems.
- Specific heat also increases of graphene nanofluids and graphene oxide shows the highest specific heat capacity, compared to the other nanofluids that are suitable for the automotive cooling system.

Acknowledgements The authors are very thankful to University Malaysia Pahang for providing the financial assistance and laboratory facilities under grant no. RDU190323.

References

1. Orfila O, Saint Pierre G, Messias MJTRPCET (2015) An android based ecodriving assistance system to improve safety and efficiency of internal combustion engine passenger cars. Transp Res Part C: Emerg Technol 58:772–782
2. Prudhvi G et al (2013) Cooling systems in automobiles & cars, 2(4):688–695
3. Raut MS, Walke PJIJoES (2012) Thermoelectric air cooling for cars. Technology 4(5):2381–2394
4. Reich S et al (2002) Tight-binding description of graphene, 66(3):035412
5. Wei D, Liu Y (2010) Controllable synthesis of graphene and its applications. Adv Mater 22(30):3225–3241
6. Geim A (2009) Graphene: status and prospects AK Geim Manchester Centre for Mesoscience and Nanotechnology. University of Manchester. Oxford Road M13 9PL, Manchester, UK, Prospects
7. Geim AK, Novoselov KS (2010) The rise of graphene. Nanoscience and technology: a collection of reviews from nature journals. World Scientific, pp 11–19
8. Mahamude ASF et al (2021) Numerical studies of graphene hybrid nanofluids in flat plate solar collector. In: 2021 International Congress of Advanced Technology and Engineering (ICOTEN). IEEE
9. Hader GGJS (2020) Modelling, C.o.D. Materials, and t. Heterostructures. Synthesis of graphene, p 181
10. Teng C et al (2017) Ultrahigh conductive graphene paper based on ball-milling exfoliated graphene. Adv Func Mater 27(20):1700240
11. Singh K, Ohlan A, Dhawan SJ (2012) Polymer-graphene nanocomposites: preparation, characterization, properties, and applications. Nanocomposites-new trends developments, pp 37–72
12. Bhuyan MSA et al (2016) Synthesis of graphene. Int Nano Lett 6(2):65–83
13. Rao C, Maitra U, Matte HRJGS (2012) Synthesis, characterization, and selected properties of graphene. Graphene: synthesis, properties, and phenomena, pp 1–47
14. Casiraghi C et al (2007) Rayleigh imaging of graphene and graphene layers. Nano Lett 7(9):2711–2717
15. Dzyazko YS, Volfkovich YM, Chaban MO (2021) Composites containing inorganic ion exchangers and graphene oxide: hydrophilic-hydrophobic and sorption properties. Nanomaterials and nanocomposites, nanostructure surfaces, and their applications. Springer, pp 93–110
16. Sharotri N et al (2021) Fundamental of graphene nanocomposites
17. Pinto RV, Fiorelli FASJATE (2016) Review of the mechanisms responsible for heat transfer enhancement using nanofluids. 108, pp 720–739
18. Mahamude ASF et al (2021) Thermal performance of nanomaterial in solar collector: state-of-play for graphene. J Energy Storage 42:103022
19. DiFrancesco ML et al (2020) A hybrid P3HT-graphene interface for efficient photostimulation of neurons. Carbon 162:308–317
20. Contreras EMC et al (2019) Experimental analysis of the thermohydraulic performance of graphene and silver nanofluids in automotive cooling systems, 132:375–387
21. Hosseini SM et al (2016) Performance of CNT-water nanofluid as coolant fluid in shell and tube intercooler of a LPG absorber tower. 102:45–53
22. Balandin AAJ (2011) Thermal properties of graphene and nanostructured carbon materials. Nat Mater 10(8):569–581
23. Yang Y et al (2019) Thermal conductivity of defective graphene oxide: a molecular dynamic study, 24(6):1103
24. Alofi A, Srivastava GJPRB (2013) Thermal conductivity of graphene and graphite, 87(11):115421
25. Leong K et al (2010) Performance investigation of an automotive car radiator operated with nanofluid-based coolants (nanofluid as a coolant in a radiator). 30(17–18):2685–2692

26. Vajjha RS, Das DK (2009) Experimental determination of thermal conductivity of three nanofluids and development of new correlations. Int J Heat Mass Transf 52(21):4675–4682
27. Ahmed SA et al (2018) Improving car radiator performance by using TiO_2-water nanofluid. 21(5):996–1005
28. Selvam C et al (2017) Thermal conductivity and specific heat capacity of water–ethylene glycol mixture-based nanofluids with graphene nanoplatelets 129(2):947–955

A Review on Torque Performance for Different Type of Carrier Fluid in Magnetorheological Brake

Khairul Anwar Abdul Kadir, Nurhazimah Nazmi, Shinichirou Yamamoto, Saiful Amri Mazlan, Nur Azmah Nordin, and Shahir Mohd Yusuf

Abstract Last two decades, the researcher is focused on increasing the braking torque of magnetorheological (MR) brake due to the insufficient braking torque to realize in commercial applications particularly vehicle application. Typical methods for enhancing braking torque include improving the rheological characteristics of the carrier fluid, modifying the structure of the MR brake, and raising the strength of the internal magnetic field. Although MR Fluid (MRF) is frequently used in MR brake as a carrier fluid due to its high responsiveness and ease of fabrication, the sedimentation problem has hindered the application of MRF in MR brakes. As a result, MR Grease (MRG) with high viscosity that reduces sedimentation is predicted to overcome the shortcomings of MRF through its unique self-sealing capabilities that efficiently solve the leakage problem. However, the torque performance of MRG is still low which hinder its performance in MR brake application. This review focuses on the usage of MRG instead MRF as a carrier fluid and the improvement of the MRG's rheological properties which expected to improve the torque performance of MR brake. The factor influencing the rheological properties like the amount, size and shape of magnetic particles are discussed. Furthermore, a few potential additives have been used to improve the rheological properties of MRG and was expected can improved the torque performance of MRG in MR brake are also surveyed and discussed. Therefore, this review will be of significance to MR brake researchers who want to develop a high torque MR brake without occurrence of the sedimentation and leakage.

Keywords Magnetorheological brake · Magnetorheological grease · Rheological properties · Braking torque

---

K. A. A. Kadir · N. Nazmi (✉) · S. A. Mazlan · N. A. Nordin · S. M. Yusuf
Engineering Materials and Structures (eMast) iKohza, Malaysia-Japan International Institute of Technology (MJIIT), Universiti Teknologi Malaysia, Jalan Sultan Yahya Petra, 54100 Kuala Lumpur, Malaysia
e-mail: nurhazimah@utm.my

S. Yamamoto
Department of Bio-Science and Engineering, Shibaura Institute of Technology, Fukasaku 307, Saitama 337-8570, Japan

M. Y. Ismail et al. (eds.), *Technological Advancement in Mechanical and Automotive Engineering*, Lecture Notes in Mechanical Engineering,
https://doi.org/10.1007/978-981-19-1457-7_3

1 Introduction

Magnetorheological (MR) brake is a MR actuator which can effectively shorten the braking distance, reduce the braking lag time, and conveniently realize the integration of various new control technologies. Furthermore, MR brakes have several potential applications in a variety of areas and machines, including the automobile sector, construction, fitness devices, and computer numerical control tools for machines due to outstanding controllability, fast response, low power consumption, and simple structure [1–3]. The MR brake is an electromechanical brake that produces smooth and consistent braking torque by employing MR fluid (MRF) as its working medium. They can offer tunable damping by varying the apparent viscosity and yield stress of the MRF in the presence of an applied magnetic field. However, most MR brakes have low braking torque limitations, making them difficult to apply in commercial applications requiring high torque, such as vehicle braking systems. Increasing braking torque often entails changing the rheological properties of the MR fluid, altering the structure of the MR brake, and increasing the strength of the internal magnetic field [2]. At the meantime, MR brake research has focused on improving the torque performance of MRF brakes, mostly by altering the structure, such as disc-type, drum type, inverted-drum, T-shape rotor, and multiple disk, in order to enhance the effective fluid in MR brakes [4, 5].

Figure 1 illustrates a simple MR brake setup for applications. In this setup, a rotating rotor is contained by a static casing and the space between the rotor and casing is filled by MR fluid. A coil winding is inserted on the casing's border, and when no current is supplied, the MR fluid flows freely. At that same time, a tiny braking torque was produced due to the off-state viscosity of the MR fluid. However, as current is supplied, an electromagnetic field is developed in the working gap, causing magnetic particles to gather and form multiples of chain alignment in the magnetic field direction. Interparticle interaction between magnetic particle chains, on the other hand, causes the formation of a column like or complex structure with controllable yield stress. Shear stress between the rotating disk and the solidified MR

Fig. 1 The MR brake's operational principle

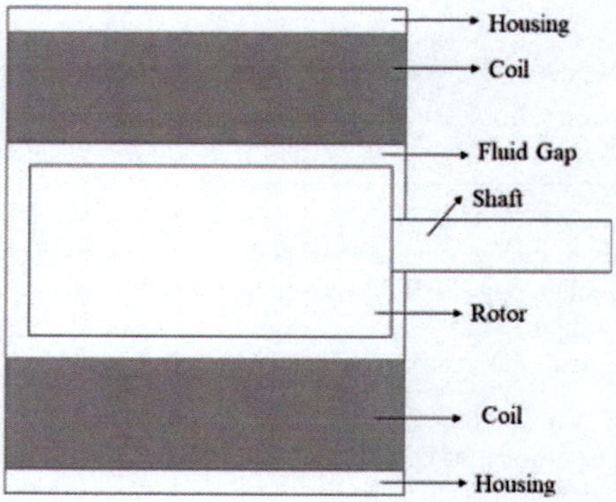

fluid produces the braking torque. An MR rotary brake's braking torque increases rapidly in response to the intensity of an external magnetic field [1].

MR brake actuator was initiated by Carlson et al. [6] from LORD corporation in 1998 by utilizing MRF as a medium and the maximum generated torque up to 4 Nm with applied current, 1 A. Then, Li et al. [7] showed that with increasing of rotary speed, the braking torque of MRF in MR brake was linearly increased. This result has showed that the torque of MRF increase from 1.04 Nm to almost 1.4 Nm when rotary speed of MR brake was increase from 100 to 500 rpm. To responsiveness of the rotational speed during the braking phase, Attia et al. [8] performed an experiment to determine torque performance the brake with application of various voltage inputs. If the rotary speed were fixed at 250 rpm, 2 V were the minimum voltage required of braking torque to stop the rotary shaft. Compared to MRG, the rotary speed does not really affect the performance of MRG in MR brake. As the rotary speed was increased, the braking torque of MRG in MR brake was almost constant [9].

In order to develop compact MR brake with high braking torque, Zhou et al. [10] had designed a new type of MR brake with shearing double disk to reduce the size and weight of MR brake, however capable to achieve 3.5 Nm braking torque when 0.8 A current was supplied. The performance of test the performance of this double disk MR brake for high torque and high power mechanical devices by Wang et al. [11]. The MR brake could supply higher current and generated higher braking torque that was 15 A and 1410 Nm, respectively. Gordaninejad et al. [9] also had test the performance of another MR material that MR grease (MRG) in double plate MR clutches. The result showed that the torque output of MRG could achieve to 0.6 Nm in the absence of magnetic particle and 9.3 Nm with the applied of magnetic field. The study revealed that MRG samples are suitable for torque transmission devices requiring a high performance, stable generated torque with a low torque clutch in the absence of magnetic field.

Other than disk type MR brake, drum type is another major MR brake design. Drum brakes have a cylindrical rotor with a circular magnetic field. According to Kikuchi and Kobayashi [12], drum-type MR brakes have a lower inertia and thus can transduce the MR effect in braking torque easily compared to disc type MR brake. The performance of MRF in ankle foot orthoses equipped with drum-type MR brake tested by Adiputra et al. [13] revealed the high voltage of MR brake contributes to high current which produces a high torque. The usage of MRG in drum-type MR brake has been studied by Dai et al. [14]. The results have shown that the difference in rotational speed reduces with volume fraction at low currents, whereas the difference in rotational speed grows with volume fraction at high applied currents.

To produce high torque MRB with compact size, a T shape MR brake is developed with more complex structures. This MR brake is particularly resistant to particle sedimentation that may occurs at high speed [4]. However, this kind of MR brake would have higher power consumption due to the needs of multi-coil. In an effort to reduce the power consumption MR brake, Nya'Ubit et al. [15] had proposed a compact T-shape MR brakes by using a single-coil and featuring serpentine magnetic flux.

Even though the modification of MR brake can also improve the braking torque of MR brake, however, this method also increases the structure complexity of MR brake which led to increase the manufacturing cost of MR brake. Furthermore, the type of carrier fluid also will affect the application of MR brake particularly in MR brake. Among all the MR material, MRF has gained the most popularity due to its fast response, rapid rheology change, insensitivity to contamination, and simple preparation procedure. Due to this advantage, MRF has been considered in many applications and devices (Table 1).

Table 1 The summary of the MR brake's development

Author	MR material	Result
Carlson et al. [6]	MRF	• 4 Nm when 1 A current was supplied
Li et al. [7]	MRF	• Increase 1.04 Nm to almost 1.4 Nm when rotary speed was increase from 100 to 500 rpm at 0.75 A current supplied
Attia et al. [8]	MRF	• With increased voltage input to the brake, the ability of the MR brake to stop the rotational shaft improves, whereas the brake could not stop the shaft with a 1 V voltage
Zhou et al. [10]	MRF	• Maximum torque is 3.5 Nm when 0.8 A current was supplied
Wang et al. [11]	MRF	• When the current reaches 15 A, the braking torque can achieve 1410 Nm • As axial stress squeezing increase to 1.5 MPa, the braking torque can achieve 2710 Nm • As the temperature increase from 25 to 150 °C, braking torque dropped by 210 Nm, reducing from 1800 Nm to 1590 Nm
Kikuchi and Kobayashi [12]	MRF	• When 1 A current was supplied, the MR brake generated 10 Nm of max braking torque • Suggested a design for the MR brake with a rotor cylinder and multi coils to offer a large range for the efficient MR effect
Nya'Ubit et al. [15]	MRF	• At current 1 A, the predicted braking torque is 1.51 Nm, but the experiment value is 1.91 Nm • The FEMM simulation resulted in a maximum magnetic flux density of 0.45 T in the outer circular part. The magnetic flux may penetrate nearly the whole MRF gap

(continued)

Table 1 (continued)

Author	MR material	Result
Wang et al. [1]	MRF	• The braking torque is almost 980 Nm when the current applied is 2.6 A • Brake torque exhibits considerable decreased with the temperature increasing from 58 to 86 °C • The braking torque decreases from 293.8 to 258.2 Nm with a drop of around 12.1% • Brake torque shows a relatively modest decrease in this period with water cooling approach and the drop rate is just 2.4%
Gordaninejad et al. [9]	MRG and commercial MRF	• MRG clutches can ensure consistent torque output throughout a wide range of operating speeds • The iron particle size of the MRG does not give much effect on torque output • As particle weight percentage increase, the torque output increase from 6.4 to 9.1 Nm when 2 A current was supplied • The torque capacities of the MRG samples are up to 75% more than the commercial MRF
Sukhwani et al. [16]	MRG and MRF	• The off-state braking powers MRG brake is higher of the MRF brake at all the speeds • At current more than 0.5 A the braking power produced by MRG is 1.6–2.45 times lower than the brake powering produced by MRF
Dai et al. [14]	MRG	• The off-state yield stress drops with volume fraction, but the off-state viscosity rises with volume fraction • As current of 1.01 A, rotational speed difference has raised to 6750 r/min • At low applied currents, the rotational speed difference reduces with volume fraction, but at large applied currents, it grows with volume fraction
Singh et al. [17]	MRG	• The torque generated by the designed clutch is 150% greater than the torque produced by the traditional clutch when the same magnetic circuit and material specifications are used • Highest generated torque at 0.52 A for MRG 75 wt% of CIP is 3 Nm while for MRG with 50 wt% of CIP is 2.32 Nm

2 Carrier Fluid

2.1 MR Fluid

MRF suspensions are made up of three different components: carrier liquid such as oils made from petroleum, mineral oils, silicon, polyester, polyether, water, artificial hydrocarbonic oils, magnetic particles, and a small number of additives [18, 19]. In 1948, Rabinow [20] discovered the first MRF. When there is no magnetic field, MRF behaves similarly to Newtonian fluids. As the magnetic field is introduced to MRF, the suspended particles become polarized and align themselves, like chains, with the magnetic field's direction allowing the rheological properties of the MRF, such as apparent viscosity and yield stress, to be controlled by manipulating the current supplied and can form semi solid state from the liquid state within few milliseconds. This is known as the MR effect. As a result, MRF has enormous potential in a wide range of electromechanical applications, including clutches, brakes, valves, dampers, and robotics [2, 18]. Unfortunately, the density mismatch between the magnetic particle and the dispersion liquid frequently leads to significant sedimentation problems, which is an undesired issue for industrial application [21]. As a result, numerous studies focusing on the additive in MRF has been conducted to improve the stability in MRF such as introducing surfactant, introducing the clay additives like claytone APA, baragell and garamite 1958 [22], introducing nano additive like Zinc doped ferrite [23], and used rod-like magnetite/sepiolite nanocomposite particles as magnetic particle [21].

However, the sedimentation rate in MRF can only be decreased, not totally eliminated. As a result, the researchers focused on switching the carrier medium to a solid-like MR material, such as grease, in order to solve the sedimentation issues while having minimal influence on the MR effect. Additionally, MRG has similar behavior to MRF in that it can transfer torque or dissipate energy by supplying a magnetic field, which encouraged the researcher to conduct extensive research on MRG [24].

2.2 MR Grease

MRG is a nonmagnetic grease medium that consisted of magnetic particles with micron size and has a thixotropic crosslink structure. As a result, MRG has a solid form that can effectively avoid the problem of settling and has a simple sealing advantage. In addition, compared to MRF, MRG has superior anti-leakage, oxidation stability, and long-term steady storage characteristics [24]. The study about MRG was found by Rankin et al. [25] in 1999. The results demonstrated that a viscoplastic medium in an MR material will prevent particle sedimentation while not significantly influencing the MR effect of MRG as an MR material.

The impact of temperature on MRG was then investigated by Sahin et al. [26] who discovered that MRG is a temperature-dependent material. As temperature rises, apparent viscosity, shear stress, and yield stress decreases. However, Wang et al. [27] shown that temperature has an effect on the rheological characteristics of MRG in an oscillation mode. The test result showed that the maximum MR effect could reached 90,328% at 70 °C. As a result, it was determined that the MR effect is typically significant when MRG is employed in high-temperature applications. This occurrence happened as a result of the MRG becoming more liquid-like as the temperature is raised, leading particles to move with the applicable of magnetic field.

MRG has been regarded a suitable material to replace MR fluid in controlled waste of energy and torque transmission application such as shock absorbers, clutches, brakes, and engine mounts due to sedimentation in MRF. However, the performance of MRG is highly dependent on the composition of MRG. As a result, the composition of MRG must also be considered prior to application.

3 Improvement of Rheological Properties

3.1 Magnetic Particle

It is worth mentioning that the use of magnetic particles as a filler is a critical component in the production of MRG. Many metals, alloys, and ceramic compounds, such as ferrite-polymer, iron–cobalt alloy, carbonyl iron, nickel–zinc ferrites, iron and metal or metal oxide such as graphite, cobalt, nickel, and ferrite, have been employed as magnetic particles in MR material depending on the research objective. Among these particles, CIPs have been widely used in MR material due to their controllable magnetized and demagnetized properties and high saturation magnetization [28, 29].

In addition, particle size has an impact on MRG performance. The CIP in MRG is often in the 1–10 μm range. Microparticles are favored over nanoparticles because they have a minimal influence when a magnetic field is applied due to Brownian motion. Furthermore, at the microscale, small particles have a higher interfacial friction area between the particle and the medium, resulting in a high damping factor. However, because of the stability issue, CIP sizes larger than 10 μm are not suggested. Even larger sizes can create a greater MR impact [30].

A study was done by Patel et al. [31] about the influence of the difference shape of magnetic particle to the torque performance of MR brake. At 0.8 A current, the braking torque of MRF with flake shape of CIP has generated 17% higher torque compared to MRF with spherical shape of CIP. This event occurred because the flake shape CIP has high surface contact area compared to spherical shape of CIP which assisted the response between CIPs.

Aside from the type and shape of magnetic particle, the contents of magnetic particle is a key factor in improving rheological characteristics and the improved the torque performance of MR brake. This fact has been proved by Acharya et al. [32]

when compared the performance of MRF with different weight percentages of iron powder and commercial MRF in MR brake. They demonstrated that the synthesis MRF and commercial MRF generate almost same torque performance values when the current is 0.8 A. Nonetheless, the ratio of torque performance in the off-state conditions to on-state condition at 0.8 A of current when use commercial MRF is quite high as a result of its low viscous torque. The same phenomenon experience by MRG which has been proved by Gordaninejad et al. [9] demonstrated this result by using 90–99 wt% CIP in MRG as shown in Fig. 2. As the amount of CIP increased, so did the torque performance of MR brakes or clutches. With high content of CIP and higher current supplied that is up to 2 A of current, the torque generate by MRG was higher compared to other 2 studies.

However, research on MRG brakes generally has revealed poor braking torque performance. In general, the MRG has a high viscosity resulting in good distribution stability between the particles and its medium by preventing the particles from settling at the device's base at the off-state condition. Unfortunately, this occasion has also substantially limited magnetic particle mobility inside the matrix phase, decreasing magnetic particle response at the on-state condition. This tendency can result in poor performance, constraining operations with high stress situations, and spending greater energy at the beginning of an application. Therefore, a few types of additives have been suggested in order to improve the torque performance of MRG in MR brake.

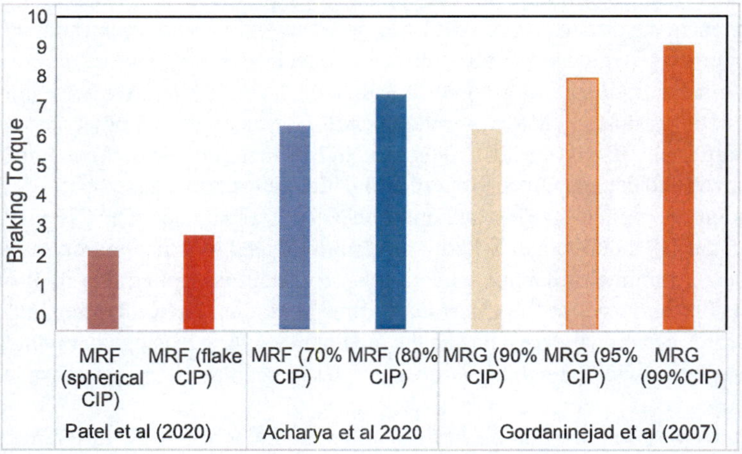

Fig. 2 The studies on the influence of shape and weight percentage of CIP in MRF and MRG to the braking torque

3.2 Additives

There two kind of additive that have been introduced in MRG that are liquid-additive which used oil to dilute the MRG and solid-additive like nanoparticle as additive. Kim et al. [28] has conducted an experiment which using kerosene oil to dilute the MR grease and reducing the initial viscosity of MRG with soft magnetic CIP. The CI particle has also been reported well dispersed in grease. In other work, Mohamad et al. [33] studied the influence of the percentage of silicon and castor oil as additives on the initial viscosity of MRG without sedimentation. This study indicated that a stable MRG may be formed with a 10 wt% oil addition without sedimentation. Furthermore, the viscosity of MRG in the off state has successfully decreased and then risen in the on-state condition. In the same method which done by Mohamad et al. [34] which present the stability and the rheological properties enhancement by using castor oil, hydraulic oil and kerosene oil. This study shows that MRG using kerosene oil has the lowest initial viscosity of any MRG. Furthermore, the addition of oil reduces the apparent viscosity and yield stress of MRG samples as the dilution oil increases.

Other than oil, Mohamad et al. [35] has introduced superparamagnetic nanoparticles γ-Fe_2O_3 in MRG's suspension. The result show enhancement regarding magnetic saturation from 130 to 201 emu/g. Furthermore, the nano additive has also reduced the viscosity of MRG on off state however higher than pure MRG at on state condition. When a magnetic field is provided, the nanoparticles has filled the inter-spaces between micron-sized particles. As a result, the formation of firm and compact chain structures in the MRG is achieved, which indirectly improves MR performance. Recently, Tarmizi et al. [36] has used cobalt ferrite as additive in MRG and revealed that by adding cobalt ferrite into MRG, the magnetization of the MRG increased caused by a strong magnetic properties of cobalt ferrite. Figure 3 showed that the content of CIP and their additive from the work done by Mohamad et al. [35] and Tarmizi et al. [36]. The content of CIP in study done by Tarmizi et al. [36]

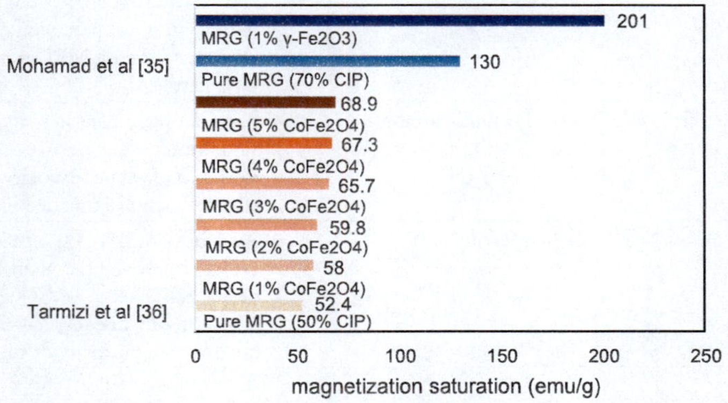

Fig. 3 The magnetization saturation of MRG samples with difference additive

was lower which has reduced the magnetic properties of MRG. However, as the content of cobalt ferrite was increased, the magnetization of MRG samples has also increased. Furthermore, the initial viscosity of MRG was proven to be decreased by 86% compared with pure MRG. The use of this additive has also improved MRG's shear stress and yield stress.

Thus, all the additives above have proved that they can improved the rheological properties of MRG and can be utilized in application. However, the study of the MRG with additive in application particularly in MR brake has not been done yet. Therefore, the braking torque of the MRG brake with additive can been done after this and these are the potential additives can be used in MRG in order to improve the braking torque of MR brake. Table 2 is depicted the summary of the studies which introduce additives in the MRG and improved their rheological properties.

Table 2 Summary of additives in the MRG

Author	Additives	Result
Kim et al. [28]	5 wt% of kerosene oil	• Decrease the viscosity of MRG • Obtaining well dispersed suspension
Mohamad et al. [33]	Castor oil and silicon oil	• MRG containing 15% silicone oil has a greater off-state apparent viscosity than pure MRG • The addition of castor and silicone oils in MRG has reduced their apparent vis-cosity in off-state condition, but considerably enhanced in the on-state condition • A stable MRG was form without manifestation of sedimentation when the addition of oil was less than 10 wt%
Mohamad et al. [34]	Castor oil, kerosene oil and hydraulic oil	• Initial viscosity decreased with the percentage of oil additive • The apparent viscosity, shear stress and yield stress decrease with increment of oil percentage • Oil layer form on MRG with 15 wt% kerosene oil
Mohamad et al. [35]	γ-Fe_2O_3 nanoparticle	• Show an increment in magnetization saturation • Reduced off-state viscosity but slightly increase on-state viscosity
Tarmizi et al. [36]	Cobalt ferrite	• Addition of $CoFe_2O_4$ into MRG has greatly improved its magnetic properties • Addition of $CoFe_2O_4$ has reduced the off-state viscosity of the MRG

4 Conclusion

Many researchers have contributed to the development of the MR braking system. For the optimization of the MR braking system, researchers employed a variety of design of experiment and methodologies. The MR brake is regarded as the future or next generation brake by wire system due to its several features such as rapid reaction time and contactless braking that is suitable for high-speed automobiles. So, research is currently being conducted to make this system practical by making changes to the present design. However, this kind of method has also increased the structure complexity and increased the manufactured cost. Thus, this paper discussed another method to increase the torque performance MR brake that is by increasing the rheological properties of carrier fluid and the potential of MRG as a suitable carrier fluid in MR brake in order to solve the sedimentation and leakage problem in MRF. Until todays, the study of MRG in MR brake was still limited due to the low braking torque generated by MRG which still need to improve. The rheological properties can be increased by increasing the weight percentage of magnetic field, by controlling the initial viscosity of the carrier fluid and introducing the additives. Even though the braking torque can be improved by modified active fluid gap in the structure of MR brake, this way is increased the manufacture cost of MR brake due to its complexity. Therefore, the easier way to improve the torque performance of MRG brake is by enhancing the rheological properties of MRG. Thus, the introducing of the additive in MRG especially solid additives is proved to improve the rheological properties and this occasion was predicted to enhance the torque performance of MRG brake due to the magnetic properties' enhancement in MRG.

Acknowledgements This work was supported by the Ministry of Higher Education under Fundamental Research Grant Scheme (FRGS/1/2020/TK0/UTM/02/57).

References

1. Wang DM, Hou YF, Tian ZZ (2013) A novel high-torque magnetorheological brake with a water cooling method for heat dissipation. Smart Mater Struct 22(2):11–11
2. Hu G, Wu L, Li L (2021) Torque characteristics analysis of a magnetorheological brake with double brake disc. Actuators 10(2):23–23
3. Hu G et al (2020) Performance analysis of rotary magnetorheological brake with multiple fluid flow channels. IEEE Access 8:173323–173335
4. Avraam M et al (2010) Computer controlled rotational MR-brake for wrist rehabilitation device. J Intell Mater Syst Struct 21(15):1543–1557
5. Quoc NV et al (2019) Material characterization of MR fluid on performance of MRF based brake. Front Mater 6(June):1–15
6. Carlson JD et al (1998) Controllable brake. United States Patent
7. Li WH, Du H (2003) Design and experimental evaluation of a magnetorheological brake. Int J Adv Manuf Technol 21(7):508–515
8. Attia EM et al (2017) Theoretical and experimental study of magneto-rheological fluid disc brake. Alex Eng J 56(2):189–200

9. Gordaninejad F, Kavlicoglu BM, Wang X (2007) Study of a magneto-rheological grease (MRG) clutch. Act Passive Smart Struct Integr Syst 2007(6525):65250C-65250C
10. Zhou W, Chew CM, Hong GS (2007) Development of a compact double-disk magneto-rheological fluid brake. Robotica 25(4):493–500
11. Wang N, Liu X, Królczyk G, Li Z, Li W (2019) Effect of temperature on the transmission characteristics of high-torque magnetorheological brakes. Smart Mater Struct 28(5):1–20
12. Kikuchi T, Kobayashi K (2011) Development of cylindrical magnetorheological fluid brake for virtual cycling system. In: 2011 IEEE international conference on robotics and biomimetics, ROBIO 2011, pp 2547–2552
13. Adiputra D et al (2016) Fuzzy logic control for ankle foot orthoses equipped with magnetorheological brake. J Teknologi 78(11):25–32
14. Dai J et al (2019) Investigation of the relationship among the microstructure, rheological properties of MR grease and the speed reduction performance of a rotary micro-brake. Mech Syst Sig Process 116:741–750
15. Nya'Ubit IR et al (2020) Torque characterization of T-shaped magnetorheological brake featuring serpentine magnetic flux. J Adv Res Fluid Mech Ther Sci 78(2):85–97
16. Sukhwani VK, Hirani H (2008) A Comparative study of magnetorheological-fluid-brake and magnetorheological-grease-brake. Tribol Online 3(1):31–35
17. Singh A, Kumar Thakur M, Sarkar C (2020) Design and development of a wedge shaped magnetorheological clutch. In: Proc Inst Mech Eng Part L: J Mater Design Appl 234(9):1252–1266
18. Ahamed R, Choi SB, Ferdaus MM (2018) A state of art on magneto-rheological materials and their potential applications. J Intell Mater Syst Struct 29(10):2051–2095
19. Wu J et al (2019) Simulation and experimental investigation of a multi-pole multi-layer magnetorheological brake with superimposed magnetic fields. Mechatronics 65
20. Rabinow J (1948) The magnetic fluid clutch. Trans Am Inst Electr Eng 67:1308–1315
21. Dong YZ, Han WJ, Choi HJ (2021) Additive effect of rod-like magnetite/sepiolite composite particles on magnetorheology. J Ind Eng Chem 93:210–215
22. Aruna MN et al (2019) Influence of additives on the synthesis of carbonyl iron suspension on rheological and sedimentation properties of magnetorheological (MR) fluids. Mater Res Expr 6(8)
23. Han JK, Lee JY, Choi HJ (2019) Rheological effect of Zn-doped ferrite nanoparticle additive with enhanced magnetism on micro-spherical carbonyl iron based magnetorheological suspension. Colloids Surf A 571(March):168–173
24. Wang K et al (2020) Yield dimensionless magnetic effect and shear thinning for magnetorheological grease. Results Phys 18(May):103328–103328
25. Rankin PJ, Horvath AT, Klingenberg DJ (1999) Magnetorheology in viscoplastic media. Rheol Acta 38(5):471–477
26. Sahin H, Wang X, Gordaninejad F (2009) Temperature dependence of magneto-rheological materials. J Intell Mater Syst Struct 20(18):2215–2222
27. Wang H et al (2019) Effect of temperature on rheological properties of lithium-based magnetorheological grease. Smart Mater Struct 28(3)
28. Kim JE et al (2012) Effect of medium oil on magnetorheology of soft carbonyl iron particles. IEEE Trans Magn 48(11):3442–3445
29. Ahmad Khairi MH et al (2019) Role of additives in enhancing the rheological properties of magnetorheological solids: a review. Adv Eng Mater 21(3):1–13
30. Carlson JD, Jolly MR (2000) MR fluid, foam and elastomer devices. Mechatronics 10(4):555–569
31. Patel SR, Patel DM, Upadhyay RV (2020) Performance enhancement of MR brake using flake-shaped iron-particle-based magnetorheological fluid. J Test Eval 48(3):2393–2411
32. Acharya S et al (2020) Characterization of magnetorheological brake utilizing synthesized and commercial fluids. Mater Today Proc
33. Mohamad N et al (2019) Dilution dependent of different types of redispersing oils on magnetorheological greases. Int J Eng Technol 8(1.7):107–111

34. Mohamad N, Rosli MA, Aziz SAA, Mazlan SA, Ubaidillah, Nordin NA, Yahaya H, Fatah AYA (2019) Intrinsic apparent viscosity and rheological properties of magnetorheological grease with dilution oils. In: 2019: Proceedings of the 6th international conference and exhibition on sustainable energy and advanced materials, pp 16–17
35. Mohamad N et al (2018) Improvement of magnetorheological greases with superparamagnetic nanoparticles. MATEC Web Conf 159:8–12
36. Tarmizi SMA, Nordin NA, Mazlan SA, Mohamad N, Rahman HA, Aziz SAA, Nazmi N, Azmi MA, Tarmizi SMA (2020) Incorporation of cobalt ferrite on the field dependent performances of magnetorheological grease. Sci Total Environ 135577–135577

Study of Engine Performance, Emission and Combustion of Reactivity Controlled Compression Ignition (RCCI) Mode Engine

M. Jamil, M. A. Hamidi, A. F. Yusop, M. F. Zakiyuddin, and M. N. Omar

Abstract Based on research and sufficient evidence, the International Agency for Research on Cancer (IARC), which is part of the World Health Organization (WHO), classified exhaust gas from diesel engines as carcinogenic to humans (Group 1), which has been a factor in the worldwide increase in cancer lung cases. According to the preceding remark, this will become an issue for all diesel transportation, from the smallest, such as a generator used in a night market, to the largest, such as trains. To address this issue, many researchers and scientists study the diesel engine in order to ensure that this internal combustion engine improves in terms of emissions while maintaining performance and fuel efficiency. The diesel engine is known as a combustion that has a thermal efficiency of more than 45%. The most recent technique to reducing gas emissions from diesel engines is to modify the injection system to use dual-fuel Reactivity Control Compression Ignition (RCCI) with main reference fuel (PRF). The study on the RCCI technique shows that it can achieve low NO_x and CO_2 emissions while retaining the high performance of a diesel engine. To minimise HC and CO emissions, the future proposal for this method is to regulate the combustion phasing by regulating the injection at the port injector.

Keywords Reactivity Controlled Compression Ignition (RCCI) · Low Temperature Combustion (LTC) · Primary Reference Fuel (PRF)

1 Introduction

Internal combustion (IC) engines have contributed to modern human progress and industrial growth in terms of mobility. The effects of increased industrialisation and engine manufacturing resulted in pollution in terms of emissions to the entire

M. Jamil · M. A. Hamidi (✉) · A. F. Yusop · M. F. Zakiyuddin · M. N. Omar
Faculty of Mechanical and Automotive Engineering Technology, Automotive Engineering Centre, Universiti Malaysia Pahang, 26600 Pekan, Pahang, Malaysia
e-mail: adnin@ump.edu.my

A. F. Yusop
e-mail: fitriy@ump.edu.my

M. Y. Ismail et al. (eds.), *Technological Advancement in Mechanical and Automotive Engineering*, Lecture Notes in Mechanical Engineering,
https://doi.org/10.1007/978-981-19-1457-7_4

planet. Numerous greenhouse and emissions gases from fuel combustion produced by Internal combustion (IC) engines, such as nitrogen oxides (NO_x), unburned hydrocarbons (HC), soot, carbon monoxide (CO), carbon dioxide (CO_2), and particulate matter (PM), are the primary components of exhaust gas composition. CO and HC will react in the presence of NO_x gas and sunshine, creating ground-level ozone, which is extremely hazardous to the atmosphere [1]. The best way to solve this issue is through public awareness and a multidisciplinary approach by scientific professionals; national and international institutions must address the rise of this threat and provide long-term remedies [2].

In internal combustion engines (ICE), compression ignition (CI) diesel engines have better thermal efficiency due to higher compression ratios (CR) and less throttling losses as compared to spark ignition engines (SI). Because of the significant turbulence and frictional (head) loss when the incoming air must battle its way past the almost closed throttle, SI engines have low efficiency at modest throttle openings (pump loss). The majority of IC engines are inefficient in converting fuel into kinetic energy. The phrase "thermal efficiency" refers to the average thermal efficiency of gasoline engines, which ranges from 20 to 30%. [3].

The mixing activities and spray disintegration are largely reliant on the fuel oxidation and auto-ignition response in typical compression ignition engines. Soot or particulate matter (PM) is generated from (local) rich areas within the combustion chamber (CC) during the mixing-controlled combustion period. Even though it was in the stoichiometric region, the high temperature inside the combustion chamber resulted in NO_x emissions [4]. Many research on diesel engines have been conducted in recent years in order to reduce PM and NO_x emissions [5–7]. Many strategies such as charge boosting device, exhaust gas recirculation (EGR), and ultra-high pressure fuel injection systems have been developed to minimize the engine-out emissions and improve the engine efficiency [8]. The primary factors influencing thermal efficiency, NOx, HC, and PM emissions are fuel oxidation temperature and local equivalency ratio. High temperatures will burn the soot particles, resulting in minimal soot emissions; nevertheless, high temperatures inside the combustion chamber will also produce NO_x [9].

The problem with conventional diesel combustion CDC in stoichiometric condition, the fuel oxidized with air at high temperature resulting in formations of NO_x and it reduces the available oxygen nearer to fuel spray periphery in (CDC) which leads to higher soot emissions [10]. In order to maintain high efficiency while lowering engine emissions in CI engines, a novel technique known as Low Temperature Combustion has been developed. The LTC strategy is a new technique for attaining low levels of soot and NO_x emissions while retaining high thermal efficiency, and it is a beneficial strategy for existing and future IC engines [11]. It also has the potential to reduce particulate matter emissions through improved air-fuel mixing and intake charge dilution, resulting in lower peak combustion temperatures [12].

By providing ample time for fuel-air mixing, the LTC technique may be used to avoid regions with a high fuel-air ratio. Dilution and EGR will be used to lower peak temperatures during combustion. As a consequence, the LTC approach has resulted in excellent thermal efficiency with a lean mixture of operation, optimum combustion phasing at top dead centre (TDC), fast combustion timing, and reduced heat transfer release. However, the LTC approach produces a lot of UHC and CO because the areas are too thin and/or the temperature within the combustion chamber is too low to finish the oxidation process. Furthermore, the premix LTC combustion approach is mostly kinetically regulated, with a limited range of continuous operation. Uniform in-cylinder charge preparation results in volumetric combustion and high pressure-rise rates, as well as engine noise, which is known as engine knocking. Stratifying the and/or T inside the LTC-T window by multiplying the timing and volume of fuel injected, smoke opacity, and NO_x emission generation may be avoided while still allowing for some degree of control over the phasing and rate of combustion.

By limiting fuel reactivity and stratification of fuel mixture reactivity owing to physical and chemical variations in fuel design, a dual-fuel approach can achieve full control of combustion phasing and timing. In general, low reactivity fuel (LRF) will mix extremely well before combustion during the intake-compression stroke, whereas high reactivity fuel (HRF) is premixed to regulate the combustion. Controlling the start of injection (SOI) time helps limit DI of premixing high-reactivity fuel. As a result of the premature application of direct injection spray and the premixing of both fuels prior to autoignition, this will be a dual-fuel combustion mode, which is essentially homogeneous charge compression ignition (HCCI). If the DI is applied too late, particularly around top dead centre, the dual-fuel combustion mode will be at least partially mixing-controlled.

Combustion with late injection has the same requirements as conventional diesel combustion (CDC), and if the amount of premixed combustion of low-reactivity fuel is equivalent to dual-fuel diesel-pilot combustion. In comparison to these two strategies, there is a series of strategies known as reactivity-controlled compression ignition (RCCI), in which the charge is sufficiently premixed to be kinetically controllable and also has sufficient reactivity stratification to allow control over the combustion phasing by controlling the direct injection and amount of the high reactivity fuel (HRF). RCCI provides the capacity to regulate combustion, since direct injection of high reactivity fuel timing may be set to manage combustion phasing [13]. The gasoline surrogate is a main reference fuel (PRF) blend of n-heptane/isooctane/toluene with volume ratios of 19.4/42.2/38.3 respectively [14].

Therefore primary goal of this study is to find out the performance, emission and combustion of reactivity controlled compression ignition (RCCI) mode engine using premixture injection mode of PRF 80 (Primary Reference Fuel). Meanwhile, the sub-goal is to assess the impacts of premixture injection on emissions (CO, NO_x, and HC) and performance (power, torque, in cylinder pressure, and EGT).

Table 1 Chemical and physical properties of PRF80

Primary reference fuel (PRF)	Density (g/cm^3)	Dyn. visc. (mPa·s)	Kin. visc. (mm^2/s)	Weight (g)	Heating value (J/g)
PRF 80	0.67302	0.37650	0.55942	1.25600	47,241

Table 2 Fuel mixture matrices (500 ml each blend)

Primary reference fuel (PRF)	n-heptane	Iso-octane
PRF 80	20%	80%

2 Research Methodology

2.1 Fuel Preparation

The experiment employs two types of fuel: high and low reactivity. As a result, the PRF80 will be utilised for low reactivity, whereas pure diesel B0 will be used for high reactivity. The dual fuel RCCI employed two types of fuel: high and low reactivity fuel. In this project, pure diesel is utilised for high reactivity fuel, whereas PRF80 is used for low reactivity fuel. PRF80's chemical characteristic shown in Table 1. The pure diesel, n-heptane, and iso-octane utilised in this study are accessible at Universiti Malaysia Pahang.

The PRF80's blending procedure involves two types of fuel: iso-octane and n-heptane. Because of its outstanding anti-knock characteristics, pure iso-octane (2, 2, 4-trimethylpentane) has an octane number of 100, whereas n-heptane has an octane number of zero due to its tendency to auto-ignite readily. As a result, in the blending process, an 80:20 combination of isooctane and n-heptane will be utilized as shown in Table 2. The method of fuel blending involves combining the two types of fuel in a beaker and stirring with an IKA RW20 Digital Overhead Stirrer stir machine. The stirring procedure will take around 5 min, with the stirring machine spinning at 150 RPM. After the blending equipment is done, the blended fuel will be stored for at least 2 h before being utilised in the combustion process.

2.2 Engine Specification

The diesel engine used in the experiment and to run the sample fuels was a YANMAR TF120-M. Yanmar Co. Ltd. manufactures the YANMAR TF120-M. The YANMAR TF120-M is a 4-cycle, single-cylinder, water-cooled diesel engine with direct injection (DI) and natural aspiration. The engine has a swept volume of 638 cm^3 and a bore stroke of 92 $\times$ 96 mm. In this research, however, the EGR mode is turned off.

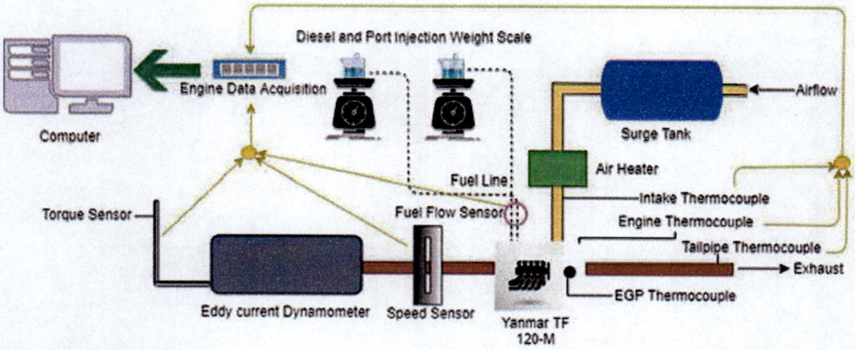

Fig. 1 Engine system flow

The engine was run for the first time using diesel fuel as a baseline reference for other sample fuel baseline characteristics. To guarantee that the engine attained a stable operating state, it was run for 15 min or until the oil temperature reached 60 °C. Engine performance is measured using characteristics such as brake power, torque, fuel consumption, and exhaust gas temperature.

According to Fig. 1, the RCCI engine has two injection systems: port injection and direct injection. The fuel blend will be stored in a port injection fuel tank, while diesel fuel will be stored in a direct injection fuel tank. Outside air will be sucked into the surge tank and routed into the port injection system. The fuel blend is then blended with pre-heated air inside the port injection system before entering the engine.

Inside the engine combustion chamber, the fuel mixture will be mixed with diesel before being injected in a certain direction and the combustion process will take place. As a result of the combustion, waste products including specific gases will be released through the exhaust chamber. The gases will then be analysed using a gas analyzer.

3 Result and Discussion

3.1 Pressure in Cylinder

Figure 2a–d show the result of in cylinder pressure versus cranks position in constant speed at 900 rpm at different load from 0 to 60%. When the load is increased, the pressure rises because the engine first slows down and the torque rises. As a result, the pressure and temperature will rise, making the pressure directly proportional to the load as shown in Fig. 2a–d. The result is that as the power stroke is about to begin, the pressure in the cylinder is exactly proportional to the crank angles at top dead centre (TDC). Also, as the load increases, so does the pressure in the cylinder. The graph shows that the peak pressure for TPS 25 and TPS 50 is somewhat to the

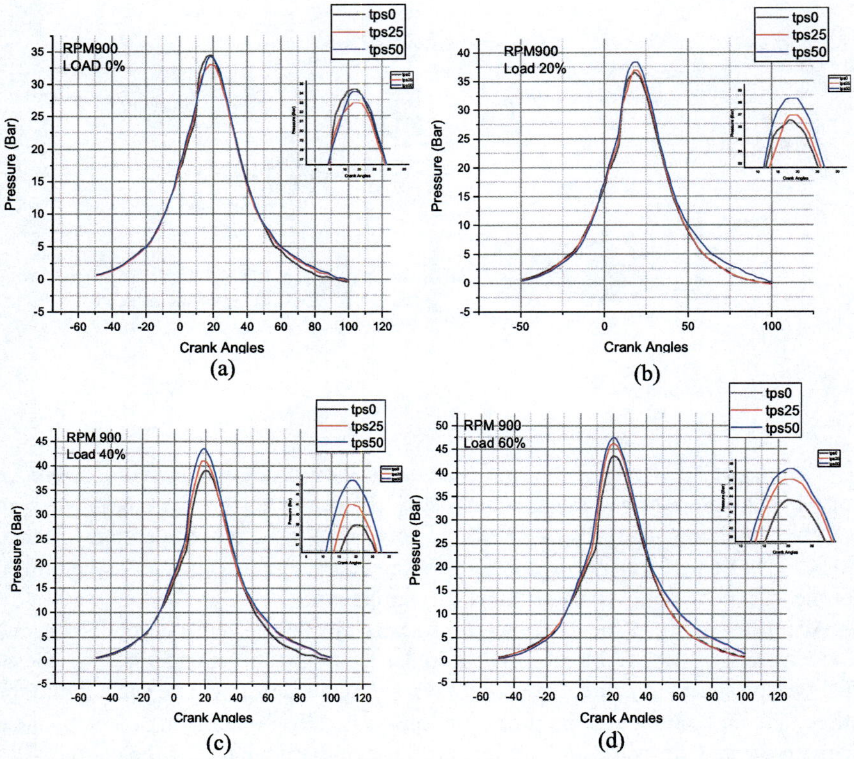

Fig. 2 In cylinder pressure versus Crank angle position for varies TPS condition and load

left of TPS 0 as in Fig. 2c, d. This is owing to the fact that at TPS 25 and TPS 50, the combustion process takes somewhat longer since more fuel is burned, resulting in higher power.

3.2 *Torque*

Figure 3 shows that torque increases as engine speed and load increase. Figure 3a–d shows the output of torque versus rpm for different load from 0% load to 60% load. The torque increases as the load increases because as the load increases, the engine's speed decreases, causing the torque to rise. The torque graph indicates a rising trend as the number of load tests increases. Furthermore, more air and diesel were supplied to allow the engine to remain at each rpm, converting more chemical energy burnt into kinetic energy, implying that the engine's torque output will rise. Furthermore, when the throttle location sensor's % opening increases, the torque decreases significantly as per Fig. 3d. This is because, since the LTC approach was

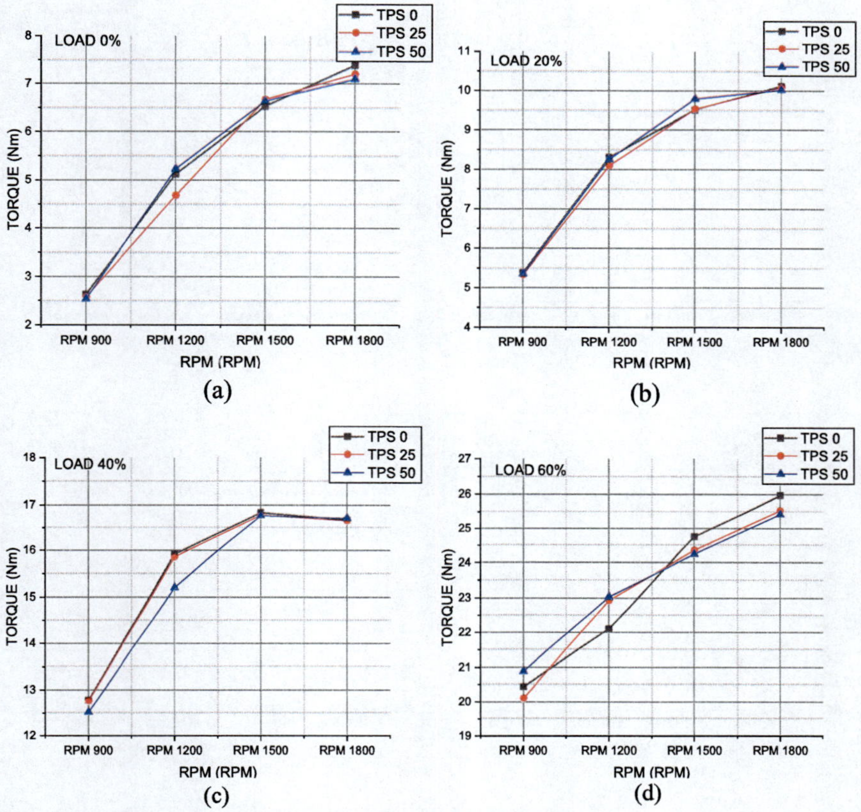

Fig. 3 Torque output versus RPM for varies TPS condition and load

developed to accomplish low emission combustion, the performance of traditional diesel engines has decreased slightly. In the LTC, a rich fuel mixture is utilised to achieve low temperature combustion since not all of the fuel is burnt, resulting in a low temperature. Due to the low temperature combustion and the unburned fuel still remaining in the cylinder, the pressure in the cylinder is lower than the other advanced combustion methods, resulting in reduced torque output of the engine.

3.3 Power

Figure 4 shows the engine speed and load rise, so does the power. The rise in power with increasing load is due to the fact that torque has always increased with increasing load since the beginning. Because power and torque are directly related, increasing engine load increases power as in Fig. 4a–d. As the number of revolutions per minute

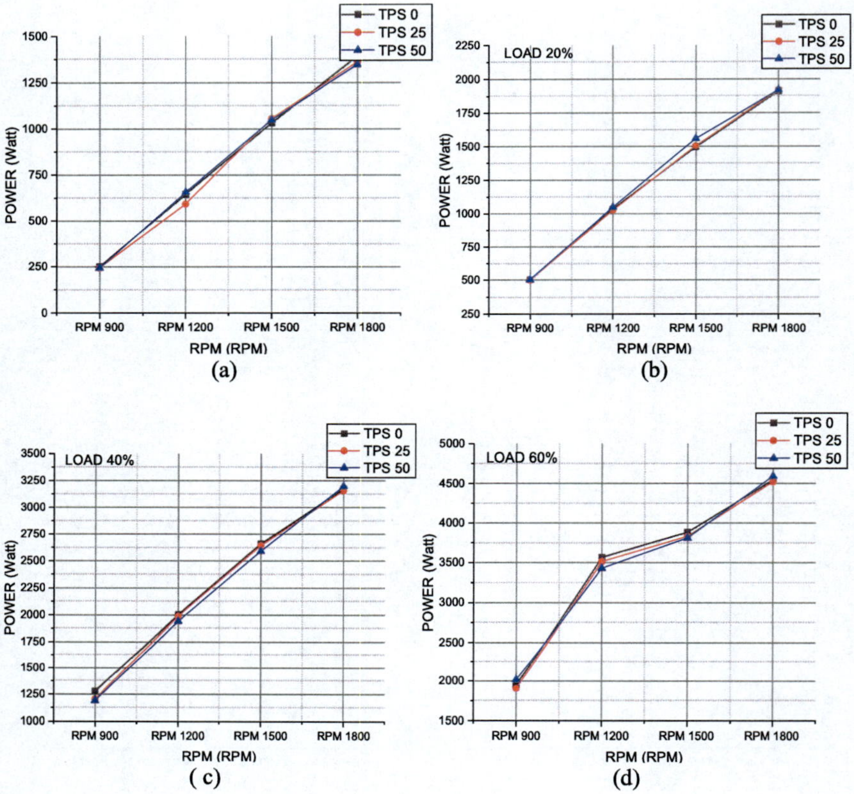

Fig. 4 Power output versus RPM for varies TPS condition and load

grew, so did the power. This is due to the fact that more air and fuel resulted in more energy being released, allowing the engine to deliver more power to the vehicle. Furthermore, when the number of throttle positioning sensor increases, the power value output decreases as in Fig. 4d. This is due to the injection of low reactivity fuel into the combustion chamber. Because the primary goal of the LTC is to reduce emissions, the low temperature of the combustion methods is utilised to minimise the possible value output of the emissions. Because there is still unburned fuel in the cylinder, the lower temperature of the combustion produces low torque, and because power is directly related, power will also drop as the throttle positioning sensor that controls the amount of low reactivity fuel increases.

3.4 *Exhaust Gas Temperature*

EGT increases as engine speed and load increase. When the engine speed increases, more air and fuel are fed into the combustion cylinder, raising the combustion temperature and therefore increasing the EGT as shown in Fig. 5. The relationship between EGT and engine speed is exactly proportional. Furthermore, when the engine load increases, so does EGT. This is due to the fact that torque increases as the load increases. Torque increases when engine speed slows, causing pressure and temperature to rise. As a result, the EGT will rise as the load rises. EGT and engine load are also proportional it can be seen in Figure (a), (b), (c) an (d). Furthermore, as the Throttle Positioning Sensor is increased, EGT will drop (TPS). This is due to the fact that a rich fuel mixture created by opening a TPS will leave unburned fuel in the cylinder. As a result of the unburned fuel, the combustion temperature is low, lowering the EGT value. The relationship between EGT and TPS is inversely proportional.

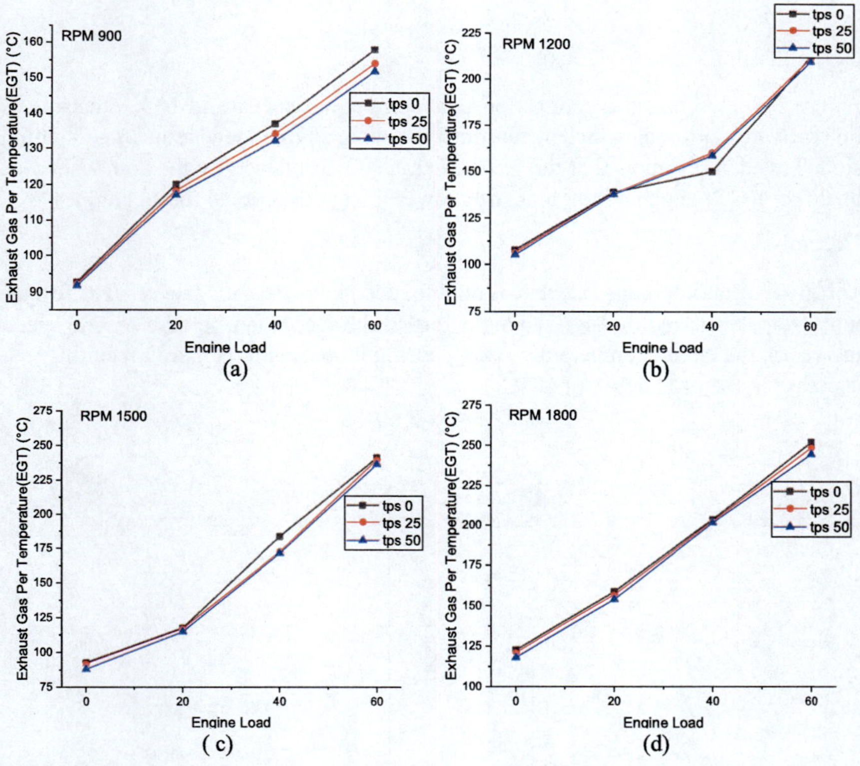

Fig. 5 Exhaust gas temperature versus engine load for varies TPS condition and RPM

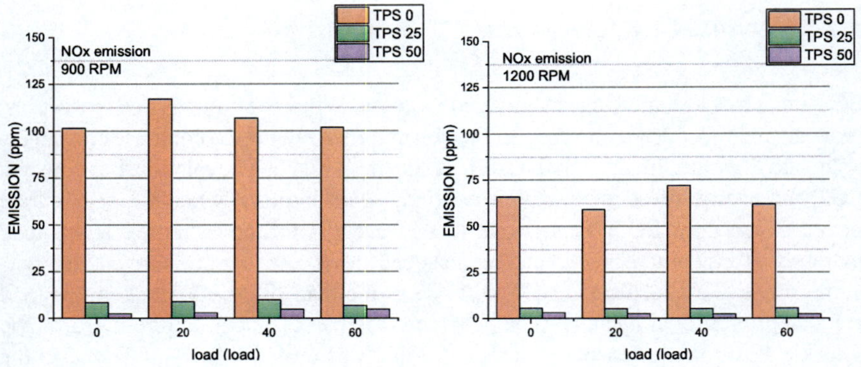

Fig. 6 NO_x emission data based on varies load and TPS condition

3.5 Emission

NO_x emission

Figure 6 shows that the connection between engine speed and NO_x emission is inversely proportional, which means that increasing engine speed reduces NO_x emission. This demonstrates that the gas emission NO_x produced by the combustion of dual fuel RCCI engines is significantly lower than in traditional diesel engines.

HC emission

The graph trends demonstrate that as engine speed increases, the rate of UHC formation decreases based on Fig. 7. This is due to the fact that when the engine speed increases, the engine temperature rises, causing more fuel to be burnt, resulting in a decrease in the production of UHC.

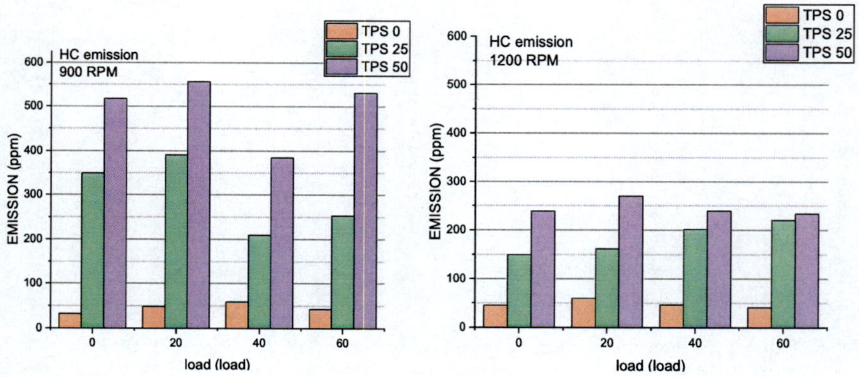

Fig. 7 HC emission data based on varies load and TPS condition

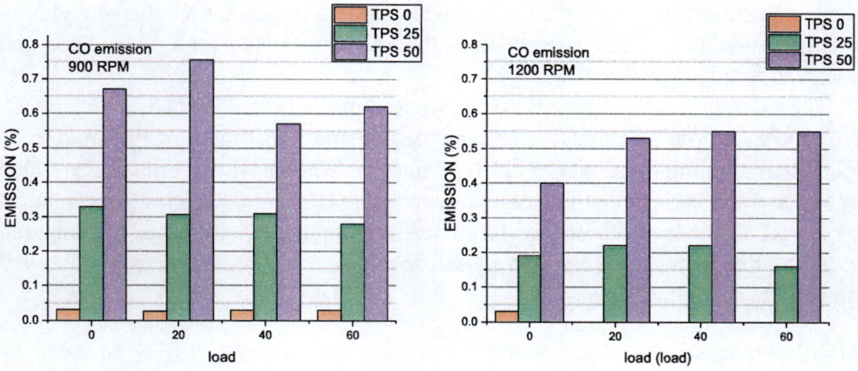

Fig. 8 CO emission data based on varies load and TPS condition

CO emission

Figure 8 shows that the connection between CO emission and engine load is exactly proportional, which means that higher loads result in higher CO emissions. And the link between engine speed and CO emission is inversely proportional, which means that the greater the engine speed, the lower the CO emission. This is due to the fact that at faster speeds, more fuel is used, preventing the production of CO emissions.

4 Conclusion

The usage of the RCCI idea has been investigated experimentally in order to expand the capabilities of a modified diesel engine to function with port injection via the RCCI concept. The usage of the RCCI engine idea has also been investigated experimentally using the PRF80 in order to investigate the capabilities of diesel engines to improve combustion, performance, and pollution. The following conclusion has been reached.

In this experiment, the in-cylinder pressure against crank angle was evaluated for combustion. The engine load and engine speed greatly increase in-cylinder pressure. It also demonstrated an increase in pressure when the RCCI mode was activated as compared to diesel. In cylinder pressure, the engine is more efficient.

The results revealed that the torque value increases with engine load as well as engine speed. When the RCCI mode is on, the graph shows a small drop in torque value. The rich mixture AFR inside the combustion chamber is a disadvantage for the engine since it prevents the engine from completely burning the mixture owing to the low temperature available in the cylinder, resulting in less torque produced. The same holds true for power output. Because of the connection between torque and power. Because of the LTC effect encouraged by RCCI, power is also somewhat lower than when only pure diesel is used. However, the exhaust gas temperature indicates that the combustion within the cylinder is lower than in a typical diesel engine owing to premixed action, and this is regarded as the low temperature combustion approach achieved in this experiment.

The gases emitted by a general combustion engine, whether SI or CI, are NO_x, CO, HC, and CO_2. The RCCI engine research demonstrates that the RCCI approach has the potential to minimise NO_x and CO_2 emissions while having a little higher value of HC and CO emission at low engine speed. The HC and CO emissions produced by this RCCI mode are caused by the engine running in too rich a mode as compared to conventional diesel, resulting in significant HC and CO emissions from unburned fuel inside the cylinder chamber.

Acknowledgements The authors would like to thank the Ministry of Higher Education for providing financial support under Research Acculturation of Early Career Researchers No. RACER/1/2019/TK08/UMP//1 and Universiti Malaysia Pahang for laboratory facilities as well as additional financial support under Internal Research grant RDU192617.

References

1. Li J (2016) Combustion and emissions formation control in reactivity controlled compression ignition (RCCI) engines with various fuels (B. Eng., M.Sc. Harbin Institute of Technology, China) for the Degree of Doctor of Philosophy
2. Manisalidis I, Stavropoulou E, Bezirtzoglou E (2020) Environmental and health impacts of air pollution: a review. Public Health. https://doi.org/10.3389/fpubh.2020.00014
3. An Y, Jaasim M, Raman V, Hernández Pérez FE, Im HG, Johansson B (2018) Homogeneous charge compression ignition (HCCI) and partially premixed combustion (PPC) in compression ignition engine with low octane gasoline. Energy 158:181–191. https://doi.org/10.1016/j.energy.2018.06.057
4. Gan S, Ng HK, Pang KM (2011) Homogeneous charge compression ignition (HCCI) combustion: implementation and effects on pollutants in direct injection diesel engines. Appl Energy 88(3):559–567. https://doi.org/10.1016/j.apenergy.2010.09.005
5. Amaechi JO, Thomas CG, Ochogba CO (2015) Comparative factors in spark ignition (SI) and compression ignition (CI) engines for a sustainable technological economy. Int J Sci Res (IJSR) 4(12):2319–7064. www.ijsr.net. ISSN (Online)
6. Zeng K, Huang Z, Liu B, Liu L, Jiang D, Ren Y, Wang J (2006) Combustion characteristics of a direct-injection natural gas engine under various fuel injection timings. Appl Therm Eng 26(8):806–813
7. Mobasheri R, Peng Z, Mirsalim SM (2011) CFD evaluation of effects of split injection on combustion and emissions in a DI diesel engine. Technical Report, SAE Technical Paper
8. Paykani A, Kakaee AH, Rahnama P, Reitz RD (2016) Progress and recent trends in reactivity-controlled compression ignition engines. Int J Engine Res 17(5):481–524. https://doi.org/10.1177/1468087415593013
9. Jin C, Zheng Z (2015) A review on homogeneous charge compression ignition and low temperature combustion by optical diagnostics. J Chem. https://doi.org/10.1155/2015/910348
10. Goldsborough SS, Hochgreb S, Vanhove G, Wooldridge MS, Curran HJ, Sung CJ (2017) Advances in rapid compression machine studies of low- and intermediate-temperature autoignition phenomena. Prog Energy Combust Sci 63:1–78. https://doi.org/10.1016/j.pecs.2017.05.00

11. Reitz RD, Duraisamy G (2015) Review of high efficiency and clean reactivity controlled compression ignition (RCCI) combustion in internal combustion engines. Prog Energy Combust Sci 46:12–71. https://doi.org/10.1016/j.pecs.2014.05.003
12. Putrasari Y, Lim O (2019) A review of gasoline compression ignition: a promising technology potentially fueled with mixtures of gasoline and biodiesel to meet future engine efficiency and emission targets. Energies 12:238. https://doi.org/10.3390/en12020238
13. Kokjohn SL, Hanson RM, Splitter DA, Reitz RD (2011) Fuel reactivity controlled compression ignition (RCCI): a pathway to controlled high-efficiency clean combustion. Int J Engine Res 12(3):209–226. https://doi.org/10.1177/1468087411401548
14. Cai L, Pitsch H (2015) Optimized chemical mechanism for combustion of gasoline surrogate fuels. Combust Flame 162:1623–1637. https://doi.org/10.1016/j.combustflame.2014.11.018
15. Benajes J, Molina S, García A, Belarte E, Vanvolsem M (2014) An investigation on RCCI combustion in a heavy duty diesel engine using in-cylinder blending of diesel and gasoline fuels. Appl Therm Eng 63(1):66–76. https://doi.org/10.1016/j.applthermaleng.2013.10.052
16. Bhiogade GE, Sunheriya N, Suryawanshi JG (2017) Investigations on premixed charge compression ignition (PCCI) engines: a review. Lecture Notes in Mechanical Engineering, pp 1455–1463. https://doi.org/10.1007/978-81-322-2743-4_139
17. Drews P, Albin T, Heßeler FJ, Peters N, Abel D (2011) Fuel-efficient model-based optimal MIMO control for PCCI engines. IFAC Proc Volumes (IFAC-PapersOnline) 44(1). PART 1. IFAC. https://doi.org/10.3182/20110828-6-IT-1002.01138
18. Egüz U, Leermakers N, Somers B, De Goey P (2014) Modeling of PCCI combustion with FGM tabulated chemistry. Fuel 118(x):91–99. https://doi.org/10.1016/j.fuel.2013.10.073
19. Fukushima N, Katayama M, Naka Y, Oobayashi T, Shimura M, Nada Y, Tanahashi M, Miyauchi T (2015) Combustion regime classification of HCCI/PCCI combustion using Lagrangian fluid particle tracking. Proc Combust Inst 35(3):3009–3017. https://doi.org/10.1016/j.proci.2014.07.059
20. Huang H, Liu Q, Yang R, Zhu T, Zhao R, Wang Y (2015) Investigation on the effects of pilot injection on low temperature combustion in high-speed diesel engine fueled with n-butanol–diesel blends. Energy Convers Manage 106(2015):748–758
21. Sindhu R, Amba Prasad Rao G, Madhu Murthy K (2017) Effective reduction of NOx emissions from diesel engine using split injections. Alexandria Eng J 57(3):1379–1392. ISSN 11100168. https://doi.org/10.1016/j.aej.2017.06.009; https://www.sciencedirect.com/science/article/pii/S1110016817302041

Effect of Primary Reference Fuel on Reactivity-Controlled Compression Ignition Engine Emission Produce

M. F. Zakiyuddin, Muthanna Jamil, M. A. Hamidi, and A. F. Yusop

Abstract Reactivity-Controlled Compression Ignition or also known as RCCI mode engine is a modified of Homogeneous Charge Compression Ignition (HCCI) engine in which have a better control in combustion and wider load range. The strategies of RCCI mode in controlling the combustion is by using dual fuel of different reactivity such as diesel (high reactivity fuel) and gasoline (low reactivity fuel). The objective of this experimental investigation is to evaluate the effect of Primary Reference Fuel towards the emission produce by the RCCI engine mode. Single cylinder CI engine with port injection system is used in this study. Two alkane-based, iso-octane and n-heptane were blend together, PRF80 (80% iso-octane + 20% n-heptane) fuel mixtures were used throughout this study as low reactivity fuel in port injection system and pure diesel as high reactivity fuel in direct injection system. Result found that the performance of RCCI mode engine improve with the use of alkane (iso-octane and n-heptane) in the PRF80 blends especially in comparison to normal mode CI engine (using diesel only). In terms of emission, by using PRF80 in RCCI mode engine, the NOx reduce almost 95% of normal CI engine NOx production. As a result of applying PRF as a low reactivity fuel in an RCCI engine system, knocking resistance may be produced even at high engine compression ratios, resulting in better thermal efficiency and reduced NOx-Soot emissions.

Keywords Reactivity-controlled compression ignition · Primary reference fuel · NOx emissions

M. F. Zakiyuddin · M. Jamil · M. A. Hamidi (✉) · A. F. Yusop
Faculty of Mechanical & Automotive Engineering Technology, Universiti Malaysia Pahang, 26600 Pekan, Pahang, Malaysia
e-mail: adnin@ump.edu.my

M. Y. Ismail et al. (eds.), *Technological Advancement in Mechanical and Automotive Engineering*, Lecture Notes in Mechanical Engineering,
https://doi.org/10.1007/978-981-19-1457-7_5

1 Introduction

1.1 Introduction

The aftereffects of fossil fuel burning in Internal Combustion Engines (ICE) are wreaking havoc on our ecosystem. With the growing demand for automobiles on the road, the transportation industry will be the world's most polluting component by 2030 [1]. At this time, it is believed that roughly 2 billion internal combustion engines are in operation across the world. Most significantly, these engines are utilized to power not only automobiles, but also fully working trucks, trains, ships, and other vehicles.

With a rising awareness of energy pollution and its impact on climate change, as well as adherence to the Kyoto Protocol, which aims to reduce energy pollution from emitting energy sources, the world is receiving a clear message to shift to more sustainable energy production and conservation [2]. The amount of research that has gone into the ICEs has not only made it more efficient, but it has also created ways for reducing hazardous emissions such as oxides of nitrogen NOx, carbon dioxide (CO_2) and particulate matter (PM). While catalytic converters are quite efficient at reducing emissions in gasoline engines, their effectiveness in diesel engines is still limited [3].

Electric vehicles (EV) emit no direct pollutants, many individuals advise driving them as a method to safeguard the environment. However, if we look at the gases released during the manufacture, driving, and recycling of electric cars, we may find that they are not as pure as we assumed [4, 5]. The infrastructure necessary for electric vehicles (EVs) to be as practical and accessible as internal combustion engines (ICEs) still requires a lot of research and scientific discoveries, particularly in the field of battery technology [6].

Reactivity Controlled Compression Ignition (RCCI) engine is the improvement of HCCI mode engine in which have better combustion control and low NOx-PM emission. The strategies of RCCI mode in controlling the combustion is by using dual fuel of different reactivity such as diesel (high reactivity fuel) and gasoline (low reactivity fuel) [7–9]. Cool Ignition that leads to low-temperature combustion (LTC) is known to be highly depending on chemical reaction between the fuel and oxygen [10–13]. Long ignition delay can also help LTC methods minimize engine exhaust emissions by decreasing local equivalency ratios and temperatures. Reduced local equivalency ratios and temperatures diminish particulate (soot) and nitrogen oxides production (NOx) [7].

The significant contribution of this study to the current state of knowledge on ICE and RCCI emission analysis is the demonstration that with engineering solutions and environmental benefits can improve significantly. Moreover, emission produce by PRF80 using single cylinder RCCI mode engine are discussed.

Table 1 The engine specification

	Description	Specification
1	Engine model	YANMAR TF120M
2	Engine type	Horizontal, 4-stroke cycle diesel engine
3	Combustion system	Direct injection + port injection
4	Number of cylinder	1
5	Bore × Stroke (mm)	92 × 96
6	Displacement (L)	0.638
7	Injection timing	17° BTDC
8	Compression ratio	17.7
9	Rate output (kW)	8.94 kW at 2400 rpm
10	Cooling system	Water cooled (radiator type)

2 Methodology

2.1 Engine Laboratory Design

The experiment was conducted using YANMAR TF120M. The specification and design was shown in Table 1 (Figs. 1 and 2).

2.2 Operating Conditions

Direct injection of a high reactive fuel (diesel) and port fuel injection of a low reactive fuel (PRF80l) were used in dual-fuel engine tests. Inagaki et al. have proven the effectiveness of a similar technique [6]. The port injection timing were setup as shown in Table 2. The port injection running at 0% TPS means that the engine only run in direct injection (pure diesel) only. For direct injection will be manually control at the engine body [14].

2.3 Fuel Properties

Alkane group consists of 80% iso-octane and 20% n-heptane mixed together producing PRF80 as low reactivity fuel and pure diesel as high reactivity fuel were used for all tests (Tables 3 and 4).

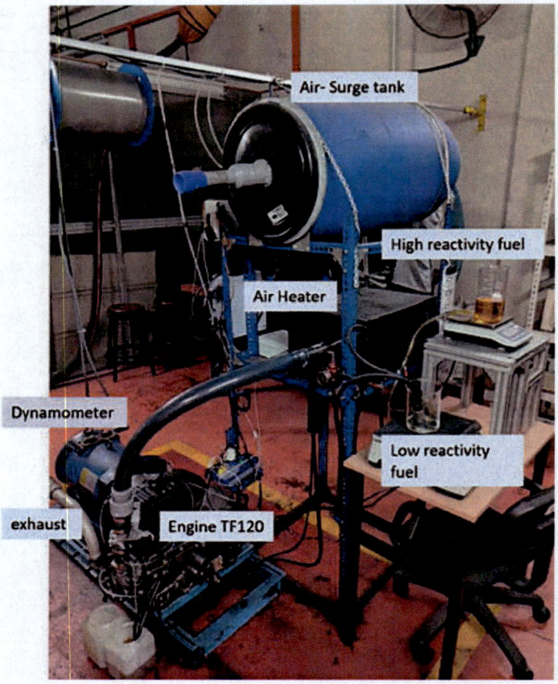

Fig. 1 The engine lab setup

2.4 Reliability Test

The experiment were conducted 3 times per 60 sample collected using gas analyzer to get the reliable average. The reading were taken at optimal temperature and humidity to prevent data collection error.

3 Results and Discussions

The engine experiments were conduct as discussed in operating conditions section. The data were presented as the increasing of load at different port injection (PI) timing opened against various emission produce by the exhaust of the engine. PI @ 0 ms means that the engine run in direct injection (pure diesel) only. The engine speed were keep constant at 1200 RPM.

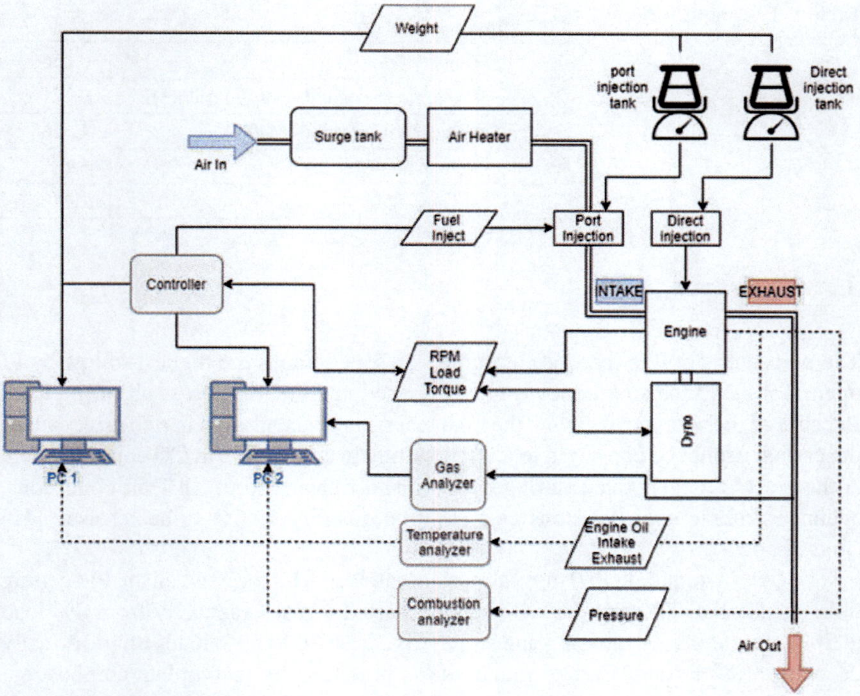

Fig. 2 Diagram of the engine lab setup

Table 2 The engine port injection setup

Engine port injection setup		
1	Port injection open timing @ 0% TPS (Direct injection only) (ms)	0
2	Port injection open timing @ 25% TPS (ms)	3.143
3	Port injection open timing @ 50% TPS (ms)	5.800
4	Engine speed (RPM)	1200
5	Engine load (%)	0, 20, 40, 60, 80

Table 3 PRF80 fuel properties

Properties		
1	Density (g/cm^3)	0.67302
2	Dynamic viscosity (@40 °C) (mPa·s)	0.37650
3	Kinematic viscosity (@40 °C) (mm^2/s)	0.55942
4	Lower heating value (MJ/kg)	47.241
5	RON	80

Table 4 Pure diesel fuel properties

Properties		
1	Density (g/cm^3)	0.856
2	Dynamic viscosity (@40 °C) (mPa·s)	0.271
3	Lower heating value (MJ/kg)	42.526
4	Cetane number	46.1
5	H/C	1.74

3.1 Emission

The percentage of CO_2 produce at normal diesel engine is the highest follow by PI opened at 3.143 ms and the lowest at PI opened at 5.800 ms across all of the load. Because of the absence of air and the high reactants concentration in rich mixes, all of the carbon cannot be converted to CO_2, resulting in an increase in CO concentration. Although CO is generated during operation in rich mixtures, in lean conditions, chemical kinetic considerations cause a little quantity of CO to be released [15] (Figs. 3 and 4).

The CO % produce at PI 0 ms is almost 0 at all load. However, when the PI opening time rises until 5.800 ms, the CO produce started to rise drastically from 0% load to 20% around 0.15% of rising and slowly rise from 20 to 60% load. But drastically decrease at 80% load. Carbon monoxide is produced by incomplete combustion, which occurs when the oxidation process does not complete. This concentration is primarily determined by the air/fuel combination, and it is maximum in rich mixtures when the excess-air factor (λ) is smaller than 1.0 [16]. It can occur, for example, during engine startup and immediate acceleration, when high mixes are necessary.

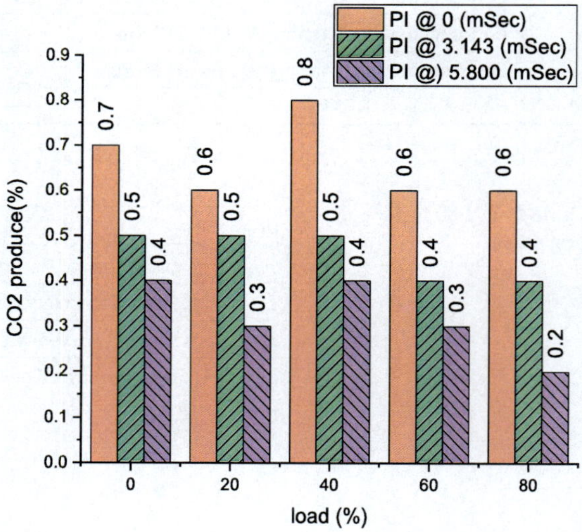

Fig. 3 Percentage of CO_2 produce by different port injection (PI) opening timing at various load and constant engine speed (1200 RPM)

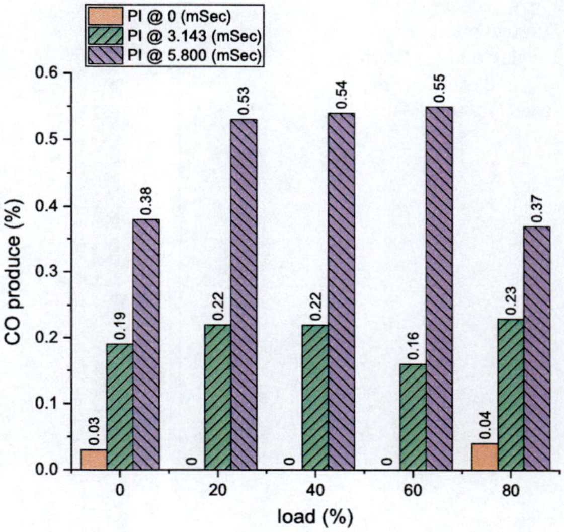

Fig. 4 Percentage of CO produce by different port injection (PI) opening timing at various load and constant engine speed (1200 RPM)

Because of the lack of oxygen and the high concentration of reactants in the rich mixes, all of the carbon cannot be converted to CO_2 and produced CO concentration. Although CO is generated during operation in rich mixes, chemical kinetic factors cause a tiny amount of CO to be released under lean circumstances [17] (Fig. 5).

The NOx produce by normal diesel engine mode is very high across each load. However, as the PI started to increase its injection opening timing at 3.143 and 5.800 ms, the value of NOx dramatically drops less than 10 ppm across each of load. The maximum temperature of the cylinder, oxygen concentrations, and residence

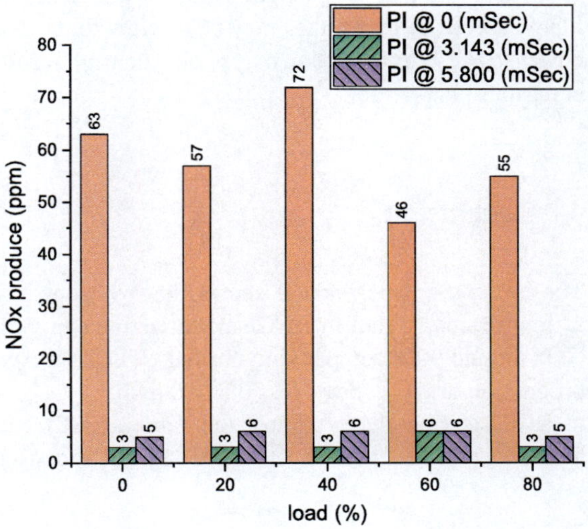

Fig. 5 NOx produce by different port injection (PI) opening timing at various load and constant engine speed (1200 RPM)

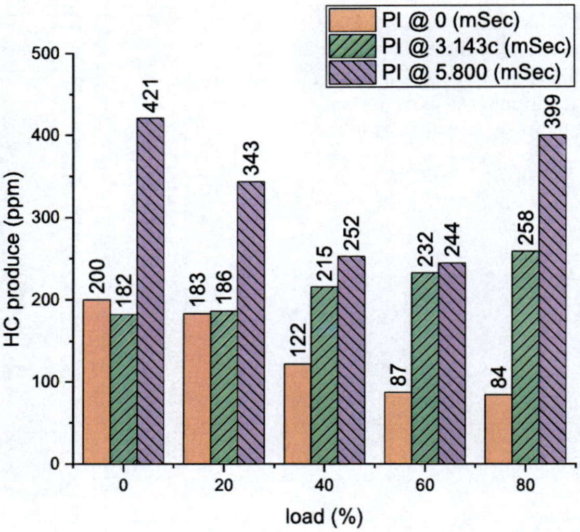

Fig. 6 HC produce by different port injection (PI) opening timing at various load and constant engine speed (1200 RPM)

time all influence the quantity of NOx produced. The majority of the NOx emitted is generated early in the combustion process, when the piston is at the top of its stroke. At this point, the flame temperature is at its highest. The amount of NOx generated triples for every 100 °C increase in combustion temperature [18, 19]. However, in RCCI mode, the NO_x ppm is almost 0 due to LTC (Fig. 6).

From 0% load until 60% load, the HC level decreasing linearly for normal diesel engine and PI at 5.800 ms. However, for PI opened at 3.143 ms, the HC ppm increase linearly from 0% load until 100% load. At 80% load, the HC value for PI opened at 5.800 ms started to rise again. Hydrocarbon emissions are made up of unburned fuels due to inadequate temperature near the cylinder wall. At this point, the temperature of the air–fuel mixture is significantly lower than that of the cylinder center [20]. Flame speeds in lean mixes may be too low for combustion to complete during the power stroke, or combustion may not occur at all, resulting in significant hydrocarbon emissions [17].

4 Conclusion

The presented experimental shows that by using Primary Reference Fuel (PRF) as low reactivity fuel in RCCI mode engine can greatly reduce the production of NOx around 90% compared to normal CI engine. Across several of load range, the production of NOx shows less than 10 ppm.

RCCI engine shows almost the same as HCCI in term of LTC characteristic however with better combustion control and wider load range. This shows that the potential of RCCI engine as the new Internal Combustion Engine (ICE) replacing the normal CI and SI engine capabilities.

However, a lot of research is needed to overcome the higher production of HC and CO produce in RCCI mode compare to normal CI engine.

Acknowledgements The authors would like to thank the Ministry of Higher Education for providing financial support under Research Acculturation of Early Career Researchers No. RACER/1/2019/TK08/UMP//1 and University Malaysia Pahang for laboratory facilities as well as additional financial support under UMP Internal Research grant RDU190358.

References

1. Birol F (2010) World energy prospects and challenges. International Energy Agency
2. Goldemberg J, Johansson T (2004) World energy assessment overview, pp 1–88
3. Brijesh P, Chowdhury A, Sreedhara S (2015) Advanced combustion methods for simultaneous reduction of emissions and fuel consumption of compression ignition engines. Clean Technol Environ Policy 17(3):615–625
4. Loaiza JCV, Sánchez FZ, Braga SL (2016) Combustion study of reactivity-controlled compression ignition (RCCI) for the mixture of diesel fuel and ethanol in a rapid compression machine. J Brazilian Soc Mech Sci Eng 38(4):1073–1085
5. Peters JF, Baumann M, Zimmermann B, Braun J, Weil M (2016) The environmental impact of Li-Ion batteries and the role of key parameters—a review. Elsevier Enhanced Reader *Scopus*
6. Doughty DH, Butler PC, Akhil AA, Clark NH, Boyes JD (2010) Batteries for large-scale stationary electrical energy storage. Electrochem Soc Interface 19(3):49
7. Splitter D, Hanson R, Kokjohn SL, Reitz RD (2010) Improving engine performance by optimizing fuel reactivity with a dual fuel PCCI strategy. In: THIESEL conference on thermo-and fluid-dynamic processes in diesel engines, no. Ci, pp 1–18
8. Reitz RD, Duraisamy G (2015) Review of high efficiency and clean reactivity controlled compression ignition (RCCI) combustion in internal combustion engines. Progr Energy Combust Sci 46:12–71. Elsevier Ltd
9. Jamil M, Hamidi MA (2010) Overview of premixture injection in compression ignition: performance and emissions
10. Agarwal AK, Singh AP, Maurya RK (2017) Evolution, challenges and path forward for low temperature combustion engines. Progr Energy Combust Sci
11. Geng C, Liu H, Ran X, Yao M (2018) The impact of low temperature reforming (LTR) products of fuel-rich n-heptane on compression ignition engine combustion. Fuel 229(May):11–21
12. Singh AP, Agarwal AK (2018) Low-temperature combustion: an advanced technology for internal combustion engines. In: Srivastava DK, Agarwal AK, Datta A, Maurya RK (eds) Advances in internal combustion engine research. Springer Singapore, Singapore, pp 9–41
13. Dempsey A et al (2014) Particulate matter characterization of reactivity controlled compression ignition (RCCI) on a light duty engine. SAE Technical Paper Series, vol 1
14. Inagaki K, Fuyuto T, Nishikawa K, Nakakita K et al (2006) Dual-fuel PCI combustion controlled by in-cylinder stratification of ignitability, SAE Technical Paper 2006-01-0028
15. Faiz A, Weaver CS, Walsh MP, Air pollution from motor vehicles: standards and technologies for …
16. Wu CW, Chen RH, Pu JY, Lin TH (2004) The influence of air-fuel ratio on engine performance and pollutant emission of an SI engine using ethanol-gasoline-blended fuels. Atmos Environ 38(40 SPEC.ISS):7093–7100
17. Zheng M, Mulenga MC, Reader GT, Wang M, Ting DSK, Tjong J (2008) Biodiesel engine performance and emissions in low temperature combustion. Fuel 87(6):714–722

18. Lee CS, Park SW (2002) An experimental and numerical study on fuel atomization characteristics of high-pressure diesel injection sprays. Fuel 81(18):2417–2423
19. Lee DS et al, Aviation and global climate change in the 21st century. Atmos Environ 43:3520–3537
20. Inekwe G, Ajav E (2017) An investigation into the exhaust emissions from tractors operating under different field conditions. NIAE

Emission Characteristics Effect on Rice Bran Oil Enriched with Diesel Fuel on Compression Ignition Engine

M. Norhafana, C. K. Ihsan, M. M. Noor, A. A. Hairuddin, K. Kadirgama, and D. Ramasamy

Abstract Experimental work has been done to investigate emissions characteristics of a single cylinder diesel engine with the rice bran oil (RBO) diesel fuel mixture at various engine speed. The emission parameters evaluated were nitrogen oxide (NO_x, carbon dioxide (CO_2), hydrocarbon (HC) and carbon monoxide (CO). The results with rice bran oil based experiment, (RBO50, RBO75, RBO100) are compared with diesel (RBO00). The results exhibited that CO, CO_2, HC and NO_x emissions are lesser than diesel fuel; Hydrocarbon emissions for both RBO75 and RBO100 were observed at two engine speed (3500 rpm and 2000 rpm). Hydrocarbon emission for RBO75 were highest at 3500 rpm engine speed which is 211 ppm. RBO50 have less and better carbon monoxide (1.2% and 0.32% at 3500 rpm and 2000 rpm respectively) and carbon dioxide emissions (8.3% and 6.9% at 3500 rpm and 2000 rpm respectively) compared with diesel (RBO00) and other fuels mix at both engine speed; 75% load. Higher NO_x emissions in diesel (RBO00) was observed which is 499 ppm and 599 ppm at engine 3500 rpm and 2000 rpm respectively as compared to other fuels; RBO50, RBO75, RBO100. In a nutshell, emission characteristics for rice bran oil were improved compared to diesel and RBO50 can consider as optimum mixture blend in terms of CO_2, CO, NO_X and HC.

M. Norhafana (✉) · M. M. Noor · K. Kadirgama
Faculty of Mechanical and Automotive Engineering Technology, Automotive Engineering Research Group (AERG), Universiti Malaysia Pahang, 26600 Pekan, Pahang, Malaysia
e-mail: norhafana@yahoo.com

M. M. Noor
e-mail: muhamad@ump.edu.my

C. K. Ihsan
Jabatan Kejuruteraan Mekanikal, Politeknik Muadzam Shah, 26700 Muadzam Shah, Pahang, Malaysia

A. A. Hairuddin
Faculty of Mechanical and Manufacturing Engineering, Universiti Putra Malaysia, 43400 Serdang, Selangor, Malaysia

D. Ramasamy
College of Engineering, Universiti Malaysia Pahang, Kuantan, Malaysia

M. Y. Ismail et al. (eds.), *Technological Advancement in Mechanical and Automotive Engineering*, Lecture Notes in Mechanical Engineering,
https://doi.org/10.1007/978-981-19-1457-7_6

Keywords Direct ignition engine · Rice bran oil · Compression ignition · Emission characteristic

1 Introduction

There are many experiments that had been done in the search of an alternative to petroleum fuels. The extinction of fossil fuel drive search for the surrogate for these depleting resource [1]. The main fracture of the prime factor for the depletion is the transportation sector [2]. The petroleum product from fossil fuel contributes the most air pollution and also responsible for world global warming [3]. It is an utmost priority for the world to search for a surrogate to fossil fuel and find an alternative fuel to suit existing engines especially in diesel engines as this engine had been known are widely used to transportation sector. The engine were designed to use mineral derived diesel fuel, so the surrogate alternatives needs to be of similar characteristics [4]. Many alternative fuels had been suggested but most had an issue with current engine which a modification had to be done and it requires a huge cost. Fuels derived from bio origin had been found of its potential as a unique renewable alternative with little or no modification to current engines that may be suitable to be mixed with diesel fuel.

Most biofuels that are available in the world continent may be were derived from domestic product. These fuel had a bonus advantage as it has a huge CO_2 savings with lower fuel emissions [5]. According to the ASTM standards, the fuel properties varies depending on the different source of its origin of feedstock. To directly use these biofuels on diesel engines commonly it will be blended with diesel proportion [6]. The suitable percentage for effective blending ratio depends on the feedstock fuel property [7]. For a suitable blending proportion it is really important to evaluate the blended fuel properties [8, 9].

The main attributes to use biofuels would be to find one with low production cost [10]. Rice bran oil were produced from excess rice bran easily available in most Asian countries with paddy fields. In 2018, global rice production totaled 513 million tons according to Food and Agriculture Organization (FAO) [11]. Between the rice kernels and its outer rice husks, there are a brown layer which is an excess by products during rice milling and polishing process. In Asian countries such as Vietnam, China, Bangladesh and India the excess bran is used as animal feed or is burned as a solid fuel [12]. There are around 16–32 wt% of oil content in rice bran varied by the degree of polishing process and rice type [13]. The free fatty acid (FFA) of RBO is rather high in comparison to other processed cereals and active lipase is present, thus the oil is not directly suitable for food consumption [14]. The oil derived from this rice bran is interesting as biofuels as it is vastly available in rice producing countries.

This research purpose is to examine various rice bran oil and diesel fuel blends on volume basis effect on the emission characteristics of a direct injection diesel engine. Secondly, the most suitable proportion of the blends will be investigated to get the optimal fuel blend proportion. In the discussion, the exhaust emission is thoroughly analyzed.

2 Application Rice Bran Oil (RBO) in Compression Ignition (CI) Engine

Diesel engine use direct injection (DI) compression ignition power cycle. Injected fuels evaporate simultaneously after diesel injection in the engine cylinder. Temperature and pressure, Density, viscosity, surface tension and injection pressure determines the evaporation rate of the fuel [15]. Formation of heterogeneous air and fuel mixture will be resulted when the various conditions are suited [16]. Fuel concentration heterogeneity varies temperature in regions. Air and fuel combustion process and combustion products formation occurs in tandem under these condition. Concentration and fuel types determines the combustion products. Hydrocarbon chains in the fuel decomposed and converted in physicochemical reactions. Hydrocarbon Peroxide (–ROOH) are the combustion base. Fuels disintegration process produces OH^-, O^-, and H^- radicals in physicochemical reactions subsequent chains reacting with fuel cracks such as carbon and sulfur in fuel and nitrogen and oxygen in air. Heat is the consequence of this process and emissions like NO_x, HC, CO and CO_2 are formed [17, 18].

Complete combustion needs satisfactory combustion chamber fuel injection rate of fuel evaporation and satisfactory fuel properties. The evaporation rate of the fuel depends on the air and fuel mixing rate together with their homogeneity. Fuel vapor distribution in the engine cylinder variables (time and dimension) key are the evaporation of the fuel. Along the satisfactory air and fuel equivalence ratios forming directly affect the ignition location, timing and intensity of the air and fuel mixture. High thermal efficiency and power can be produce via appropriate LHV and CN along high degrees of homogeneity in cylinder air and fuel mixture, thus engendering fewer exhaust emissions. The quality of all combustion development are always reflected in the power output and emission data.

In combustion chemistry, the complete carbons in fuel convert to CO_2 during combustion. However, inadequate oxygen concentration and rich fuel mixture will induce more CO partial conversion. They are adversely related where if the CO is high the concentration of CO_2 will be lower. It is much affected by the air and fuel equivalence ratio. Hydrocarbons HC, on the other hand are incomplete combustion chamber fuel combustion consequence. Design and technology of engine and combustion chamber affects the in cylinder operating conditions that have big influence of HC emission together with diesel fuel properties. Most of HC emissions are during the startups of the engines especially on low engine temperature where fuel oxidation process are not optimal and reduced evaporation rate. There are six leading mechanism that drives HC formation which are; exhaust valve leakage, the flame quenching of cylinder wall, air and fuel mixture quality, fuel deceptions caused by carbon deposits, layers of oil and the crevices of combustion chamber [19].

Complex chemical reaction causes NO_x emission especially at high temperature form nitrogen mass in air and oxygen during combustion. Engine temperature for combustion and the engine loads plays the vital influence for NO_x emission concentration. Three reaction mechanism may be used to predict and to explain NO_x formations in internal combustion engines that is (1) The intermediate mechanism, (2) The Fenimore and (3) The Zeldovich. The latter Zeldovich are known as the main important mechanism which relates fuel properties, combustion chamber design, the operating conditions and the characteristics of air and fuel mixture as the main factor affecting NO_x. Formulation.

3 Methodology

Single cylinder compression ignition diesel engine was used in the experimental setup. It was coupled to an eddy current (EC) dynamometers from Dynalec. Engine speed, Torque and load were controlled by Dynalec controllers. Table 1 lists the specification for the engine and Table 2 lists the test conditions. In the experiments, the fuels used are 100% of rice bran oil (RBO100), 15% of diesel and 75% of rice bran oil blends (RBO75), 50% of diesel and 50% of rice bran oil (RBO50) and 100% of diesel (RBO00).

Table 1 Engine specifications

Description	Specification
Type	4-stroke, horizontal cylinder, air-cooled diesel engine
Number of cylinders	1
Combustion system	Direct injection
Bore × stroke	(70 × 55 mm)
Compression ratio	20.1 ± 0.5
Maximum engine power	3.5 kW
Maximum torque	9.28 Nm

Table 2 Test conditions

Parameters	Test condition
Type of fuel	Diesel (RBO00), RBO50, RBO75, RBO100
Engine speed (rpm)	2000, 2500, 3000, 3500
Fuel temperature	27 ± 1 °C
Air temperature	30 ± 1 °C

Table 3 Specifications of the instruments

Particulars	Specifications
Speed measurement	Dynamometer
Temperature measurement	K-type thermocouples range: 100–600 °C Digital temperature indicator
Fuel measurement	Burette and stop watch
Voltmeter	Range: 0–300 V Scale: Nonlinear type
Ammeter	Range: 0–30 A Resolution: 0.2 A
Anemometer	Type: Victor 816B

The performance and operating parameter of the IC engine were measured by different instruments directly or indirectly. All instruments were calibrated prior to the experiments. Table 3 detailed the specifications of said instruments.

Qrotech QRO-401 exhaust gas analyzer was used to measure the test engine emission. O2, NO_x, HC, CO_2 and CO were measured. The range and accuracy of the model are detailed in Tables 4 and 5.

Table 4 SEM emission sensor-electrochemical-multi-range sensors

Sensor		Range	Resolution (ppm)	Accuracy
Carbon monoxide (CO)	Low range	0–2000 ppm	1	2 ppm or 2% of reading
	High range	10,000/20,000 ppm	1	10 ppm or 5% of reading
Nitric oxide (NO)	Low range	0–300 ppm	1	2 ppm or 2% of reading
	High range	2000/4000 ppm	0.1	5 ppm or 5% of reading

Table 5 Infrared (NDIR) sensors

Sensor	Range	Resolution	Accuracy
Hydrocarbon	0–2000 ppm, 2001–15,000 ppm, 15,001–30,000 ppm	1 ppm	4 ppm or 3, 5% of reading, 8% of reading
Carbon monoxide	0–10.00% 10.01–15%	0.01%	0.02% or 3% read. 5% of reading
Carbon dioxide	0.00–16.00% 16.01–20.00%	0.01%	0.3% or 3% read. 5% of reading

3.1 Experimental Setup

Experimental facility arrangement for the study were represented in Fig. 1.

The rice bran oil (RBO) were sourced from local retail shop in Pahang. The oil are imported from Thailand by a company in Singapore and were distributed in Malaysia by GREENLOVE DISTRIBUTION Sdn Bhd. Diesel fuels used are sourced from a local supplier which provides a PETRONAS Euro2 diesel. Test fuel blend were done using electrical magnetic stirrer for the blends concentration of RBO100 fuel (0% vol. mineral diesel + 100% vol. rice bran oil), RBO75 fuel (15% vol. mineral diesel + 75% vol. rice bran oil) and RBO50 fuel (50% vol. mineral diesel + 50% vol. rice bran oil). All fuel mix were continuously stirred for an hour for fine blending. All fuel blends were left static for a minimum of one hour to ensure the different fuel mix reach stability before all testing and experiment [9, 10].Different blending of the different fuel using rice bran oil could induce limitations like higher cetane number, lower volatility, shorter ignition delay, ignitability reduction, and higher lubricity [20]. The fuel blends sample were tested for its density at 15 °C temperature with Specific Gravity/Portable Density Meter (DA-130 N model). Viscosity analysis was done with K23376-KV1000 digital constant temperature kinematic viscosity bath machine. The test was done at 40 °C ± 0.01 steady temperature. Energy content of the fuel affects the power and performance of the engine operated with the diesel-biofuel blends [21, 22]. Limited studies on the analysis of the energy content that have been conducted did not reveal much on the equipment, instrumentation and the detailed procedure used for the analysis [23, 24].

A sample of 10,000 ml of all blends were used without before treatment. The blended fuels were fed to air cooled diesel engine for performance and emission test.

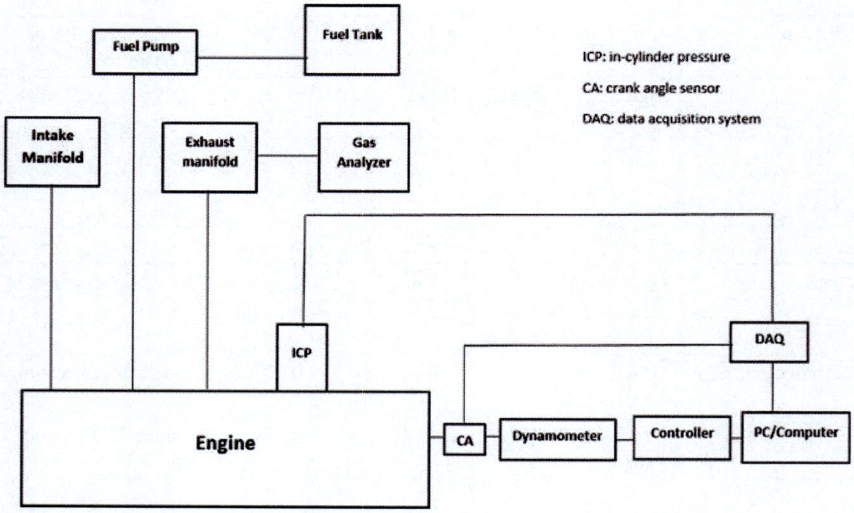

Fig. 1 Experimental setup arrangements

Table 6 Tested fuel properties results

Properties	Testing method	Diesel (RBO00)	RBO50	RBO75	RBO100
Density (kg/m^3)	ASTM D287	839.7	862.62	874.08	897.0
Kinematic viscosity (mm^2/s)	ASTM D445	4.237	3.9782	3.8488	3.59
Energy content (MJ/kg)	ASTM D240	45.714	43.8684	42.9456	41.1
Cetane number	ASTM D4737	46	44.2	43.3	41

The engine injection pressure is varied with engine rotation per minute (RPM) by a governor allowing injection pressure of 20–200 bar depending on the engine speed. For all experiments the engine will be run by diesel first and in each different fuel blend diesel will also be used in intervals to ensure smooth running and fuel line flushing for the different fuel blends. In each testing the engine will be warmed up for 20 min to ensure optimal engine temperature were reached. Measurements of the loads increment range from full load to no loads were tabulated. 30 cc fuel consumption time were recorded for each experiments. Each experiments were repeated three times to make sure the repeatability and validity of the measurements. The exhaust gas temperature and emission data, the ambient temperature and the inlet air flow rate were recorded for each test.

Suitability of every blended fuel usage properties are the most important indicator and needs to be compared with fuel standard [25–27]. Alternative fuel blends need to be studied for its fuel properties and analyzed before each fuel were used on any combustion engines [6, 28]. Table 6 shows the fuel property tests of each fuel blends along the standard test method used. Rice bran oil density is higher compared to diesel fuels in general. For the same engine capacity and power it will increase the specific fuel consumption (SFC) [7]. The denser density of RBO fuel than standard diesel fuel affects fuel sprays penetration and fuel droplet formation [29, 30].

4 Emission Analysis

4.1 Carbon Monoxide (CO)

Carbon monoxide (CO) is the critical exhaust emission that must be minimized. Figure 2a, b shows a variation in CO with brake power at high speed (3500 rpm) and low speed (2000 rpm) respectively. Lowest carbon monoxide was recorded at full load conditions of about 0.7% was observed for RBO100 followed by RBO50 (1.2%), RBO75 (1.57%) and the highest of about 1.7% for RBO00. For RBO00 the high CO may be due to the longer ignition delays (ID) effect form the fuel. Study shown that the in-cylinder air and fuel ratio decreases as a result from increased BSFC and the longer ID thus leaves less air volume for complete combustion [7].

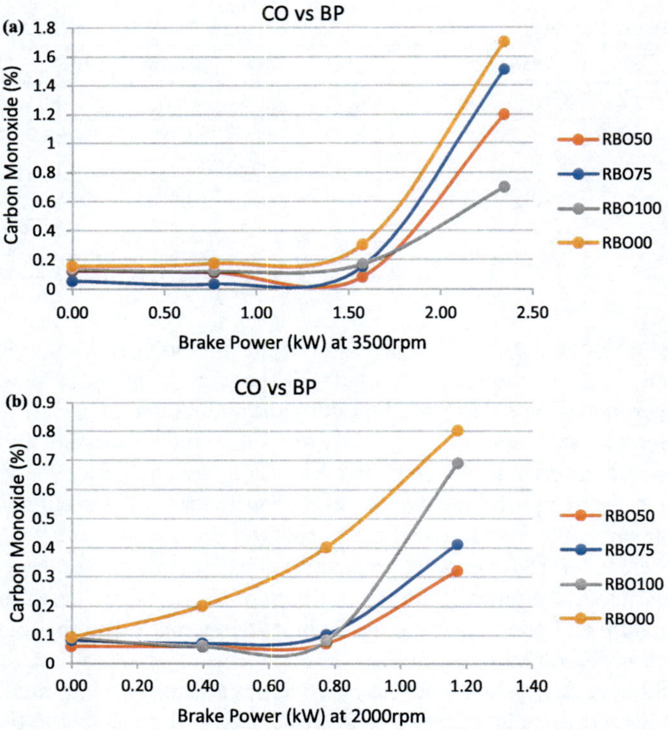

Fig. 2 **a** Variation of carbon monoxide (CO) with brake power (BP) at high speed (3500 rpm). **b** Variation of carbon monoxide (CO) with brake power (BP) at low speed (2000 rpm)

This in returns effect in higher CO emissions. Variation in carbon monoxide (CO) with brake power at low speed which is 2000 rpm be present in Fig. 2b. RBO100 and RBO00 had emissions on the higher side while RBO50 and RBO75 had carbon monoxide emission on the lower side.

4.2 *Carbon Dioxide (CO_2)*

During combustion of carbon chains in fuel, one of the gas emitted are carbon dioxide (CO_2). No direct universal consensus can be derived on the emission of CO_2 in diesel engines surrogated with biofuels [31, 32]. Some researchers have found that when a diesel engines were run with RBO the CO_2 emission were greater than compared to diesel [11, 33, 34]. Figure 3a, b shows a variation in CO_2 with brake power at high speed (3500 rpm) and low speed (2000 rpm) respectively. Figure 2a shows from no load condition to full load conditions graph RBO00 showed a higher value of carbon dioxide emission at all load conditions whereas RBO50 has lower carbon dioxide

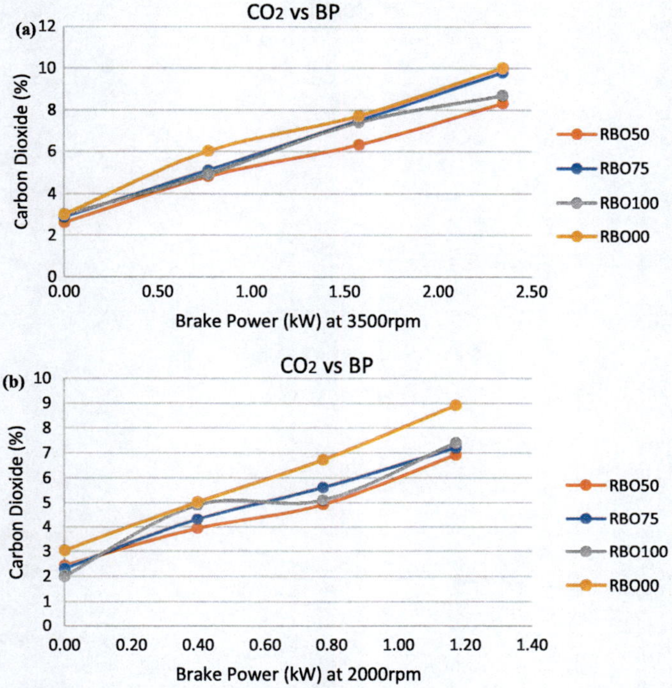

Fig. 3 **a** Variation of carbon dioxide (CO_2) with brake power (BP) at high speed (3500 rpm). **b** Variation of carbon dioxide (CO_2) with brake power (BP) at low speed (2000 rpm)

value. The highest emission of about 10% for both RBO75 and RBO00 at full load conditions while lowest value obtained is about 8.3% for RBO50.

Figure 3b shows a variation in carbon dioxide (CO_2) with brake power at low speed which is 2000 rpm. RBO50 and RBO00 have the same type of variation from no load condition to full load conditions graphs. The only difference is that carbon dioxide emissions for RBO50 were on the lower side whereas RBO00 were on the higher side of the graph. At 2000 rpm, Carbon dioxide emissions for diesel (RBO00) at low speed were found to be higher than RBO100, RBO75 and RBO50.

4.3 Hydrocarbon (HC)

Figure 4a, b shows a variation in hydrocarbon (HC) with brake power at low speed (2000 rpm) and high speed (3500 rpm) respectively. From no load to full load conditions, RBO50 and RBO75 showed the same hydrocarbon emissions. This is partly due to their almost same BSFC. Hydrocarbon emission low of about 28 ppm at part load condition was observed for RBO00; whereas maximum about 211 ppm of

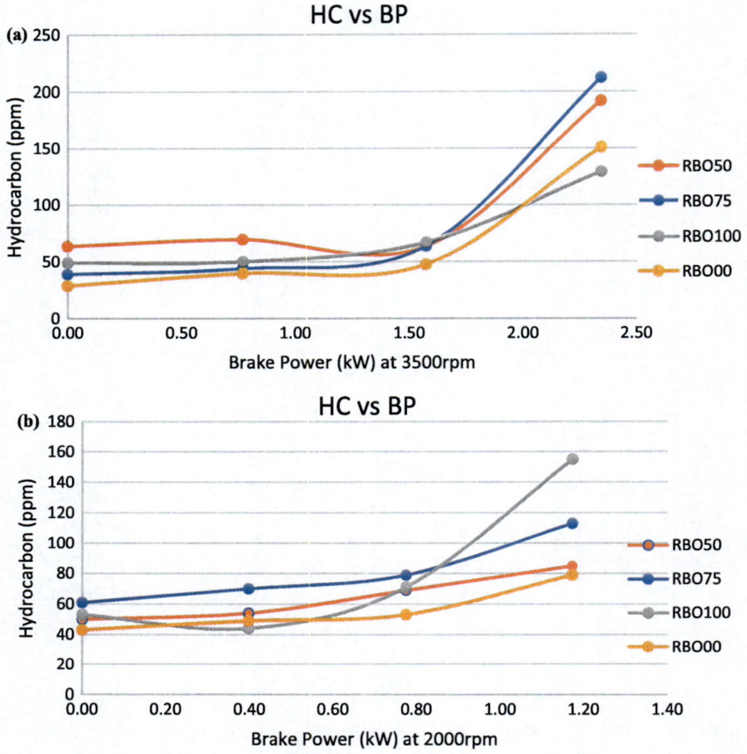

Fig. 4 **a** Variation of hydrocarbon (HC) with brake power (BP) at high speed (3500 rpm). **b** Variation of hydrocarbon (HC) with brake power (BP) at low speed (2000 rpm)

hydrocarbon emission at full load conditions was observed for RBO75. Figure 4b shows that for RBO100 higher hydrocarbon emission is due to its higher BSFC. RBO75, RBO50 and Diesel RBO00 showed same types of hydrocarbon emissions variation from no load to full load conditions.

Burnt gases higher temperature in RBO fuel helps in preventing condensation of higher hydrocarbon, thus decreasing the HC [35, 36]. The higher RBO cetane number results the decrease of the HC emission in arrears to shorter ignition delay [11, 37, 38].

4.4 Nitrogen Oxide (NO_x)

Figure 5a, b displays the disparity in nitrogen oxide (NO_x) with brake power at high speed (3500 rpm) and low speed (2000 rpm) respectively.

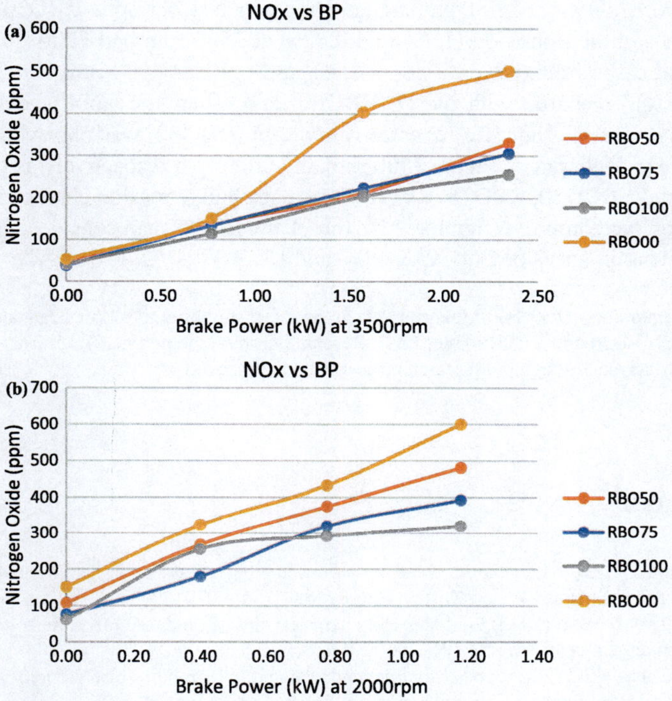

Fig. 5 **a** Variation of nitrogen oxide (NO_x) with brake power (BP) at high speed (3500 rpm). **b** Variation of nitrogen oxide (NO_x) with brake power (BP) at low speed (2000 rpm)

The highest value obtained at full load conditions is about 499 ppm for RBO00 followed by RBO50, RBO75 and RBO100 is the lowest of about 255 ppm. Diesel (RBO00) showed highest nitrogen oxide emissions compared to others blends fuel RBO. Figure 5b displays the disparity in nitrogen oxide as NOx with brake power at low speed which is 2000 rpm. The highest value obtained at full load conditions is about 599 ppm for Diesel (RBO00) trailed by RBO50 RBO75 and RBO100 had the lowest of about 319 ppm.

5 Conclusion

Emissions characteristics of blends of biofuels (RBO); RBO50, RBO75 and RBO100 with diesel fuel in a CI diesel engine with are experimentally investigated. The results with rice bran oil based experiment, (RBO50, RBO75, RBO100) are compared with diesel (RBO00). The results exhibited that CO, CO_2, HC and NO_x emissions are lesser than diesel fuel; Hydrocarbon emissions for both RBO75 and RBO100 were observed at two engine speed (3500 rpm and 2000 rpm). Hydrocarbon emission for

RBO75 were highest at 3500 rpm engine speed which is 211 ppm. RBO50 have less and better carbon monoxide (1.2% and 0.32% at 3500 rpm and 2000 rpm respectively) and carbon dioxide emissions (8.3% and 6.9% at 3500 rpm and 2000 rpm respectively) compared with diesel (RBO00) and other fuels mix at both engine speed; 75% load. Higher NO_x emissions in diesel (RBO00) was observed which is 499 ppm and 599 ppm at engine 3500 rpm and 2000 rpm respectively as compared to other fuels; RBO50, RBO75, RBO100. In a nutshell, emission characteristics for rice bran oil were improved compared to diesel and RBO50 can consider as optimum mixture blend in terms of CO_2, CO, NO_X and HC.

Acknowledgements Universiti Malaysia Pahang financial support under Post Graduate Research Grant PGRS190336 and RDU190386, UMP Research Grant Scheme (RGS), Ministry of Higher Education Malaysia through the research project were appreciated.

References

1. Atabani AE et al (2012) A comprehensive review on biodiesel as an alternative energy resource and its characteristics. Renew Sustain Energy Rev 16(4):2070–2093
2. mat Yasin MH et al (2013) Fuel physical characteristics of biodiesel blend fuels with alcohol as additives. Procedia Eng 53:701–706
3. Bhattacharyya SC (2009) Fossil-fuel dependence and vulnerability of electricity generation: case of selected European countries. Energy Policy 37(6):2411–2420
4. Awad OI et al (2017) Calorific value enhancement of fusel oil by moisture removal and its effect on the performance and combustion of a spark ignition engine. Energy Convers Manage 137:86–96
5. Pandey RK, Rehman A, Sarviya RM (2012) Impact of alternative fuel properties on fuel spray behavior and atomization. Renew Sustain Energy Rev 16(3):1762–1778
6. Dharmaraja J et al (2019) Engine performance, emission and bio characteristics of rice bran oil derived biodiesel blends. Fuel 239:153–161
7. Hoang AT et al (2021) Rice bran oil-based biodiesel as a promising renewable fuel alternative to petrodiesel: a review. Renew Sustain Energy Rev 135:110204
8. Benjumea P, Agudelo J, Agudelo A (2008) Basic properties of palm oil biodiesel—diesel blends. Fuel 87(10):2069–2075
9. Benjumea P, Agudelo J, Agudelo A (2009) Effect of altitude and palm oil biodiesel fuelling on the performance and combustion characteristics of a HSDI diesel engine. Fuel
10. Xu D et al (2021) Physicochemical properties, fatty acid compositions, bioactive compounds, antioxidant activity and thermal behavior of rice bran oil obtained with aqueous enzymatic extraction. LWT 149:111817
11. Ravikumar M, Sivashankar A, Suresh S (2021) Effects of rice bran oil blends and engine load on emission characteristics of a variable compression ratio diesel engine. Mater Today Proc 45:1188–1190
12. Hasan AP, Wakil M, Kafy MA (2014) Prospect of rice bran for biodiesel production in Bangladesh. Procedia Eng 90:746–52
13. Mazaheri H et al (2018) Rice bran oil based biodiesel production using calcium oxide catalyst derived from *Chicoreus brunneus* shell. Energy 144:10–19
14. Zaidel DNA et al (2019) Production of biodiesel from rice bran oil. In: Verma D (eds) Biomass, biopolymer-based materials, and bioenergy, Woodhead Publishing, Sawston, pp 409–447

15. Hoang AT, Le AT (2019) A review on deposit formation in the injector of diesel engines running on biodiesel. Energy Sources Part A Recovery, Utilization Environmental Effects 41(5):584–599
16. Hou D et al (2011) Spray and atomization characterization of a micro-variable circular-orifice (MVCO) fuel injector. SAE Technical Papers
17. Matsubara K et al (2003) Analysis of mixture formation process in a stoichiometric direct injection gasoline engine. SAE Technical Paper 2003-01-0066
18. Effuggi A et al (2008) Mild combustion of methane-derived fuel mixtures: natural gas and biogas. Combust Sci Technol 180(3):481–493
19. Hoang AT, Le AT, Pham VV (2019) A core correlation of spray characteristics, deposit formation, and combustion of a high-speed diesel engine fueled with *Jatropha* oil and diesel fuel. Fuel 244:159–175
20. Khan SA et al (2009) Prospects of biodiesel production from microalgae in India. Renew Sustain Energy Rev 13(9):2361–2372
21. Sumathi S, Chai SP, Mohamed AR (2008) Utilization of oil palm as a source of renewable energy in Malaysia. Renew Sustain Energy Rev 12(9):2404–2421
22. Loh SK et al (2017) First report on Malaysia's experiences and development in biogas capture and utilization from palm oil mill effluent under the economic transformation programme: current and future perspectives. Renew Sustain Energy Rev 74:1257–1274
23. Ali M et al (2017) Prospects of microalgal biodiesel production in Pakistan—a review. Renew Sustain Energy Rev 80:1588–1596
24. Mahmudul HM et al (2017) Production, characterization and performance of biodiesel as an alternative fuel in diesel engines—a review. Renew Sustain Energy Rev 72:497–509
25. Constantino AF et al (2019) Determination of physicochemical properties of biodiesel and blends using low-field NMR and multivariate calibration. Fuel 237:745–752
26. Corral-Gómez L et al (2019) Effect of diesel-biodiesel-ethanol blends on the spray macroscopic parameters in a common-rail diesel injection system. Fuel 241:876–883
27. Fu J, Turn SQ (2019) Characteristics and stability of biofuels used as drop-in replacement for NATO marine diesel. Fuel 236:516–524
28. Zaharin MSM et al (2018) Evaluation on physicochemical properties of iso-butanol additives in ethanol-gasoline blend on performance and emission characteristics of a spark-ignition engine. Appl Therm Eng 144:960–971
29. Tonini S, Gavaises M, Theodorakakos A (2009) The role of droplet fragmentation in high-pressure evaporating diesel sprays. Int J Therm Sci 48(3):554–572
30. Ma L, Roekaerts D (2016) Structure of spray in hot-diluted coflow flames under different coflow conditions: a numerical study. Combust Flame 172:20–37
31. Patel A et al (2016) Sustainable biodiesel production from oleaginous yeasts utilizing hydrolysates of various non-edible lignocellulosic biomasses. Renew Sustain Energy Rev 62:836–855
32. Silitonga AS et al (2016) Pilot-scale production and the physicochemical properties of palm and *Calophyllum inophyllum* biodiesels and their blends. J Clean Prod 126:654–666
33. Xue J, Grift TE, Hansen AC (2011) Effect of biodiesel on engine performances and emissions. Renew Sustain Energy Rev 15(2):1098–1116
34. Hoseini SS et al (2017) Effects of biodiesel fuel obtained from *Salvia macrosiphon* oil (ultrasonic-assisted) on performance and emissions of diesel engine. Energy 131:289–296
35. Venu H, Subramani L, Raju VD (2019) Emission reduction in a DI diesel engine using exhaust gas recirculation (EGR) of palm biodiesel blended with TiO_2 nano additives. Renewable Energy 140:245–263
36. Xiao H et al (2019) Combustion performance and pollutant emissions analysis of a diesel engine fueled with biodiesel and its blend with 2-methylfuran. Fuel 237:1050–1056

37. Liu S et al (2010) Effect of a cetane number (CN) improver on combustion and emission characteristics of a compression-ignition (CI) engine fueled with an ethanol–diesel blend. Energy Fuels 24(4):2449–2454
38. Das P, Subbarao PMV, Subrahmanyam JP (2015) Effect of cetane number and fuel properties on combustion and emission characteristics of an HCCI-DI combustion engine using a novel dual injection strategy. SAE Technical Papers

The Performance of Beta Type Stirling Engine Using Different Fuel

X. H. Ng, R. A. Bakar, K. Kadirgama, Sivaraos, D. Ramasamy, and M. Samykano

Abstract Stirling engine categorized as external combustion engine which defined as a closed-cycle regenerative heat engine to perform the conversion of energy into the mechanical power. The thermal efficiency of the Stirling cycle always is the main criterion, and the literature showed its efficiency of energy conversion is consider relatively as high as the Carnot cycle. Although the Stirling engine consists of great versatility for energy sources, however still inadequate efforts were done for the development of the Stirling engine that is powered by combustion fuel, since generally the engine is fueled by renewable energy which is inapplicable by the public. Therefore, the objectives to fill up the research gaps are to simulate the operation condition of Beta type Stirling engine by manipulated the use of different fuels with the assistant of MATLAB then compared with the outcome of a reference model to validate the outcome and to acquire the optimum performance of the engine, and any index that brings a reputation for the development of the Stirling engine. Compression ratio, and the temperature of the heater that affected by the specifications of Stirling engine design and effective volume of the heater, respectively act as the major element that manipulated the final power output. A higher compression ratio of 18 and power output of 315.88 Watts can be obtained with smaller clearance between the engine primary components, besides the heater temperature that achieves 855.75 K and thermal efficiency of 64.93% is affected by the usage of appropriate combustion fuel as gasoline and bigger effective volume of the heater.

Keywords Engine performance · Stirling engine · Beta type · Combustion fuel · MATLAB · single crank driven mechanisms

---

X. H. Ng · R. A. Bakar (✉) · K. Kadirgama
Faculty of Mechanical and Automotive Engineering Technology, Universiti Malaysia Pahang, 26600 Pekan, Pahang, Malaysia
e-mail: Rosli@ump.edu.my

Sivaraos
Faculty of Manufacturing Engineering, Universiti Teknikal Malaysia Melaka (UTeM), Tunggal, Malaysia

D. Ramasamy · M. Samykano
College of Engineering, Universiti Malaysia Pahang, 26300 Gambang, Pahang, Malaysia

M. Y. Ismail et al. (eds.), *Technological Advancement in Mechanical and Automotive Engineering*, Lecture Notes in Mechanical Engineering,
https://doi.org/10.1007/978-981-19-1457-7_7

1 Introduction

In the current stage of technology all around the world, the issues of emission of harmful gaseous carbon dioxide by machine always link together with the global issues for example climax changes, global warming, etc. As a resident of mother nature, human being required to come out with the changes to resolve this pressing need before condition tend to be even more chronic. Therefore, the efficiency of the machine always is the main concern because we can maximize the outcome with the minimum usage of energy [1–5].

Stirling Engine consists of different types of configurations, and the only difference among those configurations just the ways of setting up the machine with the same theory of operation. We opt for a Beta-type configuration as our targeted research model. Depart from the 'Internal Combustion Machine' that is widely used in the industrial field, the Stirling Engine does not occur any internal combustion process within the combustion chamber, which means the opportunity of the incomplete combustion process that produced harmful gaseous can be cut down. Conversely, the input of the energy only happens on the outer surface of the machine's displacer piston and a different variety of heat sources can be selected makes the Stirling engine becoming more versatile in the selection of burning fuels and flexibility of the machine greatly improved [6–8].

Since the stereotype exists and the public tends to think of the Beta-type Stirling engine (BTSE) as solar energy powered. For the compliance of the commercial market, fuel-powered BTSE becoming requisite due to the low cost of investment and easier access to this technology. Selection of the fuels as the energy source of the Stirling Engine seems to be an interest of conducting this research. Above all, the main interest of this project is to develop a simulation model for the BTSE to gather data of machine peak performance by evaluating different parameters by using different fuels [9–11].

The procedure of developing the script code required to be settled with the usage of MATLAB by the researcher to simulate the real-life operating condition of BTSE, continuous by the comparison with the reference model's research outcome. Besides, investigation on the performance of BTSE that manipulating the combustion fuel and engine specifications, at the end require to obtain the best selection as combustion fuel and suitable parameters for superior performance of BTSE.

2 Methodology

2.1 Geometrical Specifications

See Fig. 1 and Table 1.

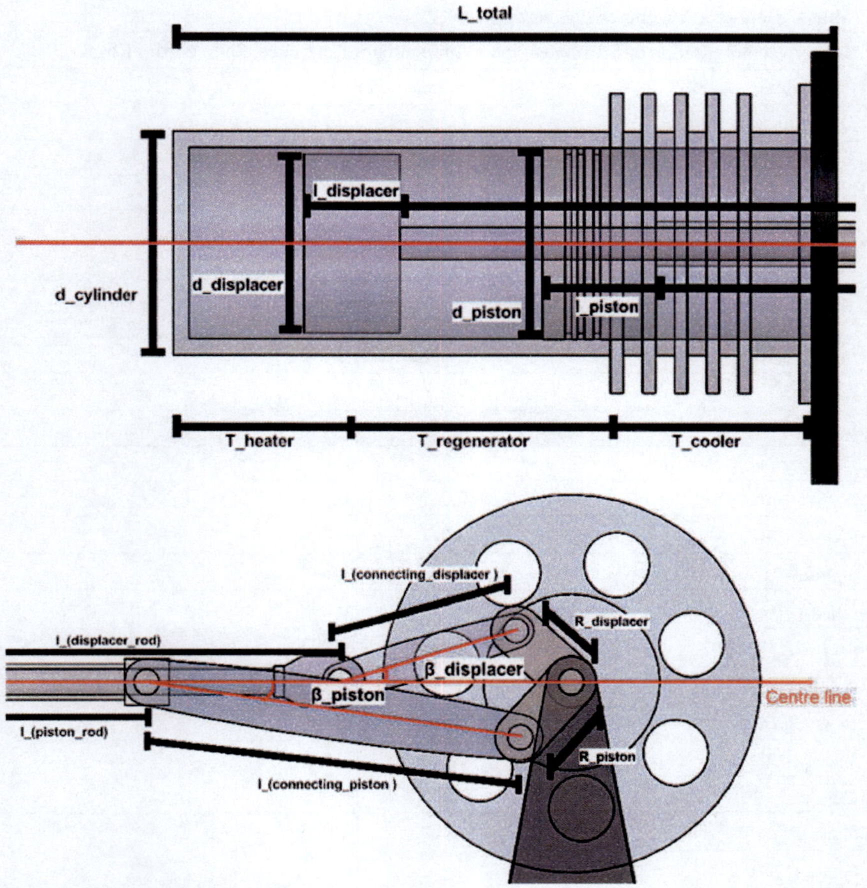

Fig. 1 Section view of single Crank-driven BTSE and its specifications

2.2 Kinematic Evaluation

Reciprocating displacement

$$Y_{displacer} = l_{displacer_{rod}} + l_{displacer} + R_{displacer}\cos(\theta_{displacer}) + l_{connecting_{displacer}}\cos(\beta_{displacer}) \tag{1}$$

$$Y_{piston} = l_{piston_{rod}} + l_{piston} + R_{piston}\cos(\theta_{piston}) + l_{connecting_{piston}}\cos(\beta_{piston}) \tag{2}$$

Table 1 Beta-type Stirling engine parameters

Beta type configuration of Stirling engine	Simulation	Reference: Source from Rosni (2018)
Parameters	Value	
Operational condition		
$T_{cooler}[K]$	300.15	
$L_{cooler}[m]$	0.063	0.09721
$T_{heater}[K]$	Varies	
$L_{heater}[m]$	0.043	0.048
$T_{regenerator}[K]$	$\frac{T_{heater}-T_{cooler}}{\ln\left(\frac{T_{heater}}{T_{cooler}}\right)}$	
$L_{regenerator}[m]$	0.0882	0.152
$\alpha[\text{degree}]$	90°	
$L_{total}[m]$	0.429	0.416
Cylinder characteristic		
$d_{cylinder}[m]$	0.06	0.084
$d^*_{cylinder}[m]$	0.07	0.094
$l_{cylinder}[m]$	0.195	0.344
$l^*_{cylinder}[m]$	0.2	0.349
$c_{displacer_{cylinder}}[m]$	0.01047	0.003
$c_{piston_{cylinder}}[m]$	0.01209	0.002316
Rods Characteristic		
$l_{piston_{rod}}[m]$	0.11312	
$l_{displacer_{rod}}[m]$	0.26931	
$r_{displacer_{yoke}}[m]$		0.025
$r_{piston_{yoke}}[m]$		
$r_{gear}[m]$		0.065
$l_{connecting_{displacer}}[m]$	0.06828	0.0805
$l_{connecting_{piston}}[m]$	0.1439	
$R_{displacer}[m]$	0.0298	0.038 (Contain only 1 crank offset radius)
$R_{piston}[m]$	0.031	
Displacer characteristic		
$d_{displacer}[m]$	0.058	0.081
$l_{displacer}[m]$	0.03	0.182
Power piston characteristic		
d_{piston} [m]	0.06	0.08
l_{piston} [m]	0.0355	0.05

Table 2 Working fluid parameters

Working fluid (helium gas) parameter	
Gas constant, R_{gas}	2076.9 [J/kg K]
Specific heat (constant pressure),$C_{pressure}$	5196.2 [J/kg K]
Specific heat (constant volume),C_{volume}	3115.6 [J/kg K]
Dynamic viscosity, μ	0.00001991 [kg/m s]
Reference temperature,T_0	300 [K]
Prandtl number, Pr	0.6884 [unit]
Molar Mass, $M_{gas,helium}$	0.004 [kg/mol]

where offset angle between crank center to power piston's connecting rod and between displacer's connecting rod, respectively equals to

$$\beta_{\text{piston}} = \tan^{-1}\left[\frac{R_{\text{piston}}}{l_{\text{conneting}_{\text{piston}}}}\right] \tag{3}$$

$$\beta_{\text{displacer}} = \tan^{-1}\left[\frac{R_{\text{displacer}}}{l_{\text{connectingdisplacer}}}\right] \tag{4}$$

Swept volume for expansion and compression

$$V_{\text{swept,expansion}} = \left(\frac{\pi}{4}\right)(d_{\text{displacer}})^2(L_{\text{total}} - c_{\text{displacer-cylinder}} - Y_{\text{displacer}}) \tag{5}$$

$$V_{\text{swept,compression}} = \left(\frac{\pi}{4}\right)\left(d_{\text{displacer}}^2\right)(Y_{\text{displacer}} - l_{\text{displacer}} - Y_{\text{piston}} - c_{\text{piston-cylinder}}) \tag{6}$$

2.3 *Working Fluid*

See Table 2.

2.4 *Heat Source*

See Table 3.

$$\text{Calorific value, C.V.} = \frac{\text{Amount of heat produced}}{\text{mass of the fuel}} \tag{7}$$

Table 3 Properties of the selected combustion fuels

Fuel	Higher Heating Value (HHV) (J/kg) (net calorific value)	Rate of Heat Transfer, $\dot{Q}[W]$	Heater temperature, T_{heater} $[K]$	
			Simulation model	Reference model
Butanol	37,300,000	207.22	746.79	593.44
Diesel	45,600,000	253.33	846.18	658.70
Gasoline	46,400,000	257.78	855.75	664.99
Kerosene	46,200,000	256.67	853.36	663.42

$$\text{Rate of Heat Transfer per second, } \dot{Q}\left[\frac{j}{s} = W\right] \tag{8}$$

$$= \text{Higher Heating Value (Weight of fuel combustion second)}\left[\left(\frac{J}{kg}\right)\left(\frac{kg}{s}\right)\right]$$

Assume 1 kg of commercial fuel required 50 h (180,000 s) to combustion completely,

$$\text{For kerosene, } \left(46,200,000 \text{ J kg}^{-1}\right)\frac{1 \text{ kg}}{180,000 \text{ s}} = \left(256.67 \text{ Js}^{-1}\right),$$

$$\text{where thermal conductivity, } \lambda = 52 \frac{W}{m} K \text{ (Cast iron cylinder wall)} \tag{9}$$

$$R_{total} = \frac{1}{\left(2\pi r_{cylinder,\,outer} L_{heater}\right)h_{air}} + \frac{\ln\left(\frac{r_{cylinder,\,outer}}{r_{cylinder,\,inner}}\right)}{2\pi L_{heater}\lambda} + \frac{1}{\left(2\pi r_{cylinder,\,inner} L_{heater}\right)h_{helium}} \tag{10}$$

$$\dot{Q} = \frac{T_{\infty 1} - T_{\infty 2}}{R_{total}} \tag{11}$$

2.5 Thermodynamics Cycle Evaluation

$$V_{heater} = \left(\frac{\pi}{4}\right)\left(d_{cylinder} - d_{displacer}\right)^2\left(L_{heater}\right) \tag{12}$$

$$V_{regenerator} = \left(\frac{\pi}{4}\right)\left(d_{cylinder} - d_{displacer}\right)^2\left(L_{regenerator}\right) \tag{13}$$

$$V_{cooler} = \left(\frac{\pi}{4}\right)\left(d_{piston} - d_{displacer}\right)^2\left(L_{cooler}\right) \tag{14}$$

2.6 Ideal Adiabatic Analysis

$$pV = mRT, \frac{dP}{p} + \frac{dV}{V} = \frac{dm}{m} + \frac{dT}{T} \tag{15}$$

$$m_{compression} + m_{cooler} + m_{regenerator} + m_{heater} + m_{expansion} = M \tag{16}$$

$$\frac{p\left(\frac{V_{compression}}{T_{compression}} + \frac{V_{cooler}}{T_{cooler}} + \frac{V_{regenerator}}{T_{regenerator}} + \frac{V_{heater}}{T_{heater}} + \frac{V_{expansion}}{T_{expansion}}\right)}{R} = M \tag{17}$$

Mean Effective Temperature was applied in these equations where its equal to the log mean temperature difference between the heater's temperature and cooler's temperature,

$$T_{regenerator} = \frac{T_{heater} - T_{cooler}}{\ln\left(\frac{T_{heater}}{T_{cooler}}\right)} \tag{18}$$

$$p\left(\frac{V_{compression}}{T_{compression}} + \frac{V_{cooler}}{T_{cooler}} + \frac{V_{regenerator}}{\frac{T_{heater}-T_{cooler}}{\ln\left(\frac{T_{heater}}{T_{cooler}}\right)}} + \frac{V_{heater}}{T_{heater}} + \frac{V_{expansion}}{T_{expansion}}\right) = MR \tag{19}$$

$$p\left(\frac{V_{compression}}{T_{compression}} + \frac{V_{cooler}}{T_{cooler}} + \frac{V_{regenerator}\ln\left(\frac{T_{heater}}{T_{cooler}}\right)}{T_{heater} - T_{cooler}} + \frac{V_{heater}}{T_{heater}} + \frac{V_{expansion}}{T_{expansion}}\right) = MR \tag{20}$$

$$\frac{MR}{\left[\left(\frac{V_{compression}}{T_{compression}} + \frac{V_{cooler}}{T_{cooler}} + \frac{V_{regenerator}\ln\left(\frac{T_{heater}}{T_{cooler}}\right)}{T_{heater}-T_{cooler}} + \frac{V_{heater}}{T_{heater}} + \frac{V_{expansion}}{T_{expansion}}\right)\right]} = p \tag{21}$$

2.7 Schmidt Analysis

Expansion and compression volume

$$V_{expansion} = V_{clearance,expansion} + \frac{V_{swept,expansion}}{2}[1 + \cos(\theta + \alpha)] \tag{22}$$

$$V_{compression} = V_{clearance,compression} + \frac{V_{swept,compression}}{2}[1 + \cos(\theta)] \tag{23}$$

Swept volume for expansion and compression process

$$V_{clearance,expansion} = \frac{\pi}{4} d^2_{displacer} c_{displacer_{cylinder}} \tag{24}$$

$$V_{clearance,\ compression} = \frac{\pi}{4} d^2_{piston} c_{piston_{cylinder}} \tag{25}$$

Dead volume and total volume (Fig. 2)

$$V_{dead} = V_{heater} + V_{regenerator} + V_{cooler} \tag{26}$$

$$V_{total} = V_{expansion} + V_{compression} + V_{dead} \tag{27}$$

Substitute $V_{expansion}$ and $V_{compression}$ into p,

$$p = \frac{MR}{\left(\frac{V_{clearance,compression} + \frac{V_{swept,compression}}{2}[1+\cos(\theta)]}{T_{compression}} + \frac{V_{regenerator} \ln\left(\frac{T_{heater}}{T_{cooler}}\right)}{T_{heater} - T_{cooler}} + \frac{V_{heater}}{T_{heater}} + \frac{V_{clearance,expansion} + \frac{V_{swept,expansion}}{2}[1+\cos(\theta+\alpha)]}{T_{expansion}} \right)} \tag{28}$$

$$\text{Let } s = \left[\frac{V_{swept,\ compression}}{2(T_{heater})}\right] + \left[\frac{V_{clearance,\ expansion}}{T_{heater}}\right] + \left[\frac{V_{heater}}{T_{heater}}\right] + \left[\frac{V_{regenerator} \ln\left(\frac{T_{heater}}{T_{cooler}}\right)}{T_{heater} - T_{cooler}}\right] + \left[\frac{V_{cooler}}{T_{cooler}}\right] + \left[\frac{V_{clearance,\ cooler}}{T_{cooler}}\right] + \left[\frac{V_{swept,\ compression}}{2T_{cooler}}\right] \tag{29}$$

$$p = \frac{MR}{\left[s + \left(\frac{V_{swept,expansion} \cos\alpha}{2T_{heater}} + \frac{V_{swept,\ compression}}{2T_{cooler}}\right)\cos(\theta) - \left(\frac{V_{swept,expansion} \sin\alpha}{2T_{heater}}\right)\right]} \tag{30}$$

$$c \sin\beta = \frac{V_{swept,expansion} \sin\alpha}{2T_{heater}} \tag{31}$$

Fig. 2 Right-angled triangle

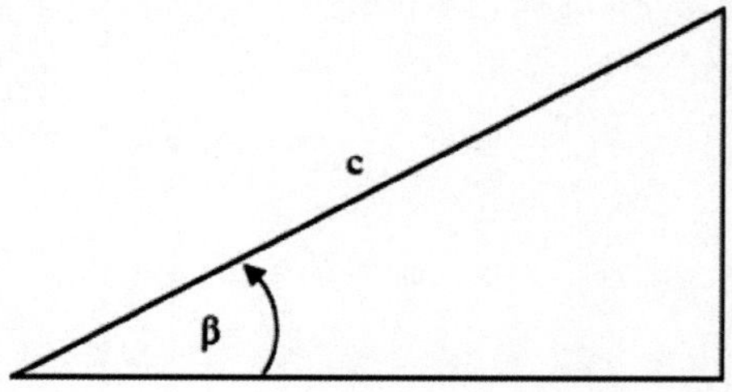

$$c\cos\beta = \frac{V_{swept,expansion}\cos\alpha}{2T_{heater}} + \frac{V_{swept,compression}}{2T_{cooler}} \tag{32}$$

$$\beta = \tan^{-1}\left[\frac{\left(\frac{V_{swept,\ expansion}\sin\alpha}{T_{heater}}\right)}{\left(\frac{V_{swept,\ expansion}\cos\alpha}{T_{heater}} + \frac{V_{swept,\ compression}}{T_{cooler}}\right)}\right] \tag{33}$$

$$c = \frac{1}{2}\sqrt{\begin{array}{c}\left(\frac{V_{swept,\ expansion}}{T_{heater}}\right)^2 + 2\left[\left(\frac{V_{swept,\ expansion}}{T_{heater}}\right)\left(\frac{V_{swept,\ compression}}{T_{cooler}}\right)(\cos\alpha)\right] \\ +\left(\frac{V_{swept,\ compression}}{T_{cooler}}\right)^2\end{array}} \tag{34}$$

Substitute β and c into p,

$$p = \frac{MR}{s(1 + b\cos Ø)}, \quad \text{where } Ø = \theta + \beta \text{ and } b = \frac{c}{s} \tag{35}$$

$$\text{Minimum Pressure, } p_{min} = \frac{MR}{s(1 + b)} \tag{36}$$

$$\text{Maximum Pressure, } p_{max} = \frac{MR}{s(1 - b)} \tag{37}$$

$$\text{Average(Mean) Pressure, } P_{mean}$$

$$= \frac{1}{2\pi}\int_0^{2\pi} PdØ = \frac{MR}{2\pi s}\int_0^{2\pi}\frac{1}{(1 + b\cos Ø)}dØ \tag{38}$$

Based on the integrals' table, $P_{mean} = \frac{MR}{2\pi s}\int_0^{2\pi}\frac{1}{(1+b\cos Ø)}dØ$ can be simply as

$$P_{mean} = \frac{MR}{\left(s\sqrt{1 - b^2}\right)} \tag{39}$$

$$W_{expansion} = \int_0^{2\pi} P\frac{dV_{expansion}}{d\theta}d\theta; \quad W_{compression} = \int_0^{2\pi} P\frac{dV_{compression}}{d\theta}d\theta \tag{40}$$

$$W_{total} = W_{expansion} + W_{compression} \tag{41}$$

Differentiate the equations $V_{expansion}$ and $V_{compression}$

$$\int V_{expansion}d\theta = \int V_{clearance,\ expansion} + \frac{V_{swept,\ expansion}}{2}[1 + \cos(\theta + \alpha)]d\theta \tag{42}$$

$$\frac{dV_{expansion}}{d\theta} = -\frac{1}{2}V_{swept,\ expansion}\sin(\theta + \alpha) \tag{43}$$

$$\int V_{compression}d\theta = \int V_{clearance,\ compression} + \frac{V_{swept,\ compression}}{2}[1 + \cos(\theta)]d\theta \tag{44}$$

$$\frac{dV_{compression}}{d\theta} = -\frac{1}{2}V_{swept,\ compression}\sin(\theta) \tag{45}$$

Substitute $\frac{dV_{expansion}}{d\theta}$ and $\frac{dV_{compression}}{d\theta}$ into $W_{expansion}$ and $W_{compression}$ respectively,

$$W_{expansion} = \int_0^{2\pi} P\frac{dV_{expansion}}{d\theta}d\theta \tag{46}$$

$$W_{compression} = \int_0^{2\pi} P\frac{dV_{compression}}{d\theta}d\theta \tag{47}$$

$$W_{expansion} = \int_0^{2\pi}\left[\frac{MR}{s(1 + b\cos\emptyset)}\right]\left[-\frac{1}{2}V_{swept,\ expansion}\sin(\theta + \alpha)\right]d\theta \tag{48}$$

$$W_{expansion} = -\frac{V_{swept,\ expansion}MR}{2s}\int_0^{2\pi}\frac{\sin(\theta + \alpha)}{1 + b\cos(\beta + \theta)}d\theta \tag{49}$$

$$W_{compression} = \int_0^{2\pi}\left[\frac{MR}{s(1 + b\cos\emptyset)}\right]\left[-\frac{1}{2}V_{swept,compression}\sin(\theta)\right]d\theta \tag{50}$$

$$W_{compression} = -\frac{V_{swept,\ compression}MR}{2s}\int_0^{2\pi}\frac{\sin(\theta)}{1 + b\cos(\beta + \theta)}d\theta \tag{51}$$

According to the appendix stated in the book by Urieli and Berchowitz [1], $W_{expansion}$ and $W_{compression}$ can be reduced to:

$$W_{expansion} = \pi V_{swept,expansion}p_{mean}\sin(\beta - \alpha)\left[\frac{\left(\sqrt{1 - b^2}\right) - 1}{b}\right] \tag{52}$$

$$W_{compression} = \pi V_{swept,\ compression}p_{mean}\sin(\beta)\left[\frac{\left(\sqrt{1 - b^2}\right) - 1}{b}\right] \tag{53}$$

$$\eta_{total} = \frac{W}{W_{expansion}} = \frac{\left(W_{compression} + W_{expansion}\right)}{W_{expansion}} \tag{54}$$

Substitute $W_{expansion}$ and $W_{compression}$ into equation η_{total},

$$\eta_{total} = \frac{\pi V_{swept,compression} p_{mean} \sin(\beta) \left[\frac{\left(\sqrt{1-b^2}\right)-1}{b}\right] + \pi V_{swept,expansion} p_{mean} \sin(\beta - \alpha) \left[\frac{\left(\sqrt{1-b^2}\right)-1}{b}\right]}{\pi V_{swept,expansion} p_{mean} \sin(\beta - \alpha) \left[\frac{\left(\sqrt{1-b^2}\right)-1}{b}\right]} \tag{55}$$

$$\eta_{total} = \frac{V_{swept,\ compression} \sin(\beta) + V_{swept,\ expansion} \sin(\beta - \alpha)}{V_{swept,\ expansion} \sin(\beta - \alpha)} \tag{56}$$

$$= 1 + \frac{V_{swept,\ compression} \sin(\beta)}{V_{swept,\ expansion} \sin(\beta - \alpha)} \tag{57}$$

$$\eta_{total} = 1 - \frac{V_{swept,\ compression}}{V_{swept,\ expansion}} \left(\frac{\tan\beta}{\sin\alpha - \tan\beta\cos\alpha}\right) \tag{58}$$

Substitute β back to the equation η_{total} and simplifying,

$$= 1 - \frac{V_{swept,\ compression}}{V_{swept,\ expansion}} \left\{ \frac{\tan\left(\tan^{-1}\left[\frac{\left(\frac{V_{swept,\ expansion} \sin\alpha}{T_{heater}}\right)}{\left(\frac{V_{swept,\ expansion} \cos\alpha}{T_{heater}} + \frac{V_{swept,\ compression}}{T_{cooler}}\right)}\right]\right)}{\sin\alpha - \tan\left(\tan^{-1}\left[\frac{\left(\frac{V_{swept,\ expansion} \sin\alpha}{T_{heater}}\right)}{\left(\frac{V_{swept,\ expansion} \cos\alpha}{T_{heater}} + \frac{V_{swept,\ compression}}{T_{cooler}}\right)}\right]\right)\cos\alpha} \right\} \tag{59}$$

$$\eta_{total} = 1 - \frac{T_{cooler}}{T_{heater}} \tag{60}$$

Table 4 Volume analysis for a different region of the simulation model and reference model of BTSE

	Simulation output	Reference output
Power piston region		
$V_{clearance,compression}$ $[m^3]$	**3.4148e × 10⁻⁵**	1.1641×10^{-5}
$V_{swept,compression}$ $[m^3]$	1.9603×10^{-4}	**1.9798×10^{-4}**
Displacer region		
$V_{clearance,expansion}$ $[m^3]$	**2.7663×10^{-5}**	1.5459×10^{-5}
$V_{swept,expansion}$ $[m^3]$	2.0302 2.0302×10^{-4}	**2.5495×10^{-4}**
Dead volume		
V_{heater} $[m^3]$	1.3509×10^{-7}	**3.3929×10^{-7}**
$V_{regenerator}$ $[m^3]$	2.7709×10^{-7}	**1.0744×10^{-6}**
V_{cooler} $[m^3]$	**1.9792×10^{-7}**	7.6349×10^{-8}

Bold value indicates the larger value in comparison

3 Result and Discussion

3.1 Data Tabulation of Research Outcome

See Tables 4 and 5.

3.2 Research Outcome with Graphical Illustration

3.2.1 Schmidt Analysis (Pressure–Volume Graph) for Simulation Model and Reference Model

Profile of the graph from Fig. 3 tally with the outline of the Pressure–Volume graph of a general Schmidt analysis for total volume, expansion volume, and compression volume. Therefore, it concluded that the Beta type Stirling engine for simulation model and reference model obeys the principle of the fundamental thermodynamics' mechanism. As discussed before the graph profile was a quasi-elliptical shape (idealized condition) which never occur in the real life, whereas the process was inseparable therefore graph does blend into each other.

A larger area of the total volume graph of the reference model explained the better performance of the rhombic driven configuration and geometrical parameters, at the same time it can generate more power while it is operating [12–14]. Constant temperature heat addition of expansion process of simulation of much shorter than it of a reference model, it can be interpreted as the quantity of heat energy transfer

Table 5 Performance parameters of the simulation model and reference model of BTSE

Schmidt Analysis
p = 450,000 Pa
S = 300 rpm
T_{cooler} = 300.15 K

T_{heater}[K]		Simulation outputs				Reference outputs			
1	Butanol	746.79				593.44			
2	Diesel	846.18				658.70			
3	Gasoline	855.75				664.99			
4	Kerosene	853.36				663.42			
		1	**2**	**3**	**4**	**1**	**2**	**3**	**4**
$r_{compression}$		6.7346				18.0071			
p_{max} [kPa]		866.69	879.75	880.98	880.67	964.25	976.29	977.50	977.20
p_{min} [kPa]		233.65	230.18	229.86	229.94	210.01	207.42	207.16	207.23
W_{total} [J]		50.15	56.18	56.71	56.58	54.27	62.44	63.18	62.99
Q_{in} [J]		83.86	87.07	87.35	87.28	109.82	114.70	115.15	115.04
Q_{out} [J]		-33.70	-30.88	-30.34	-30.70	-55.54	-52.27	-51.98	-52.05
P_{total} [W]		250.76	280.92	283.57	282.91	271.37	312.19	315.88	314.96
$\eta_{thermal}$		59.81	64.53	64.93	64.83	49.42	54.43	54.86	57.76

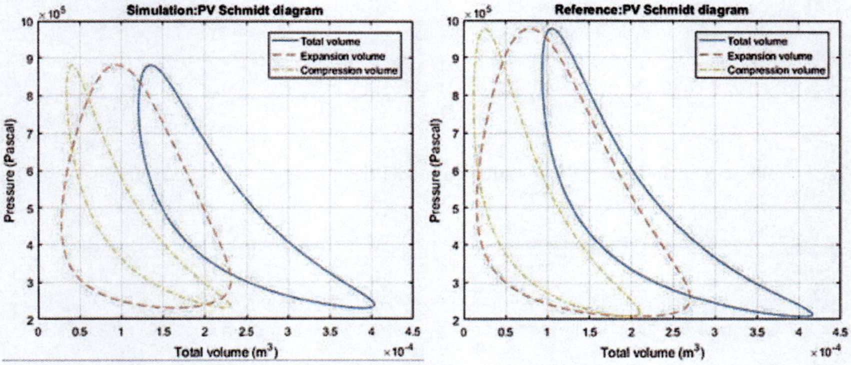

Fig. 3 Pressure against volume for simulation model (left) and reference model (right)

into the simulation model was not much as the amount of energy that energy transfer into the system of a reference model.

3.2.2 Volumetric Displacement Comparison and Cyclic Pressure of Working Fluid with Different Combustion Fuels Comparison

Profile for both graphs of the simulation model and reference model shown in Fig. 4 was the same that means the behaviours of both models operating in the same manner. By theoretical the lowest value of the volume of compression and expansion when piston and displacer at TDC, respectively [15–18]. Conversely, the largest volume for both compression and expansion volume will occur after the crank angle rotated 180°, indicate the position of the displacer and power piston locate at the Bottom Dead Centre (BDC) for every single cycle of reciprocating movement (Fig. 5).

Since the phase angle equals 90°, therefore the difference between crank angle for displacer and power piston compulsory equals 90° as well. The volume of the expansion process lowest at the crank angle of 90°, and the same moment the lowest value of compression process volume during the crank angle of 180°.

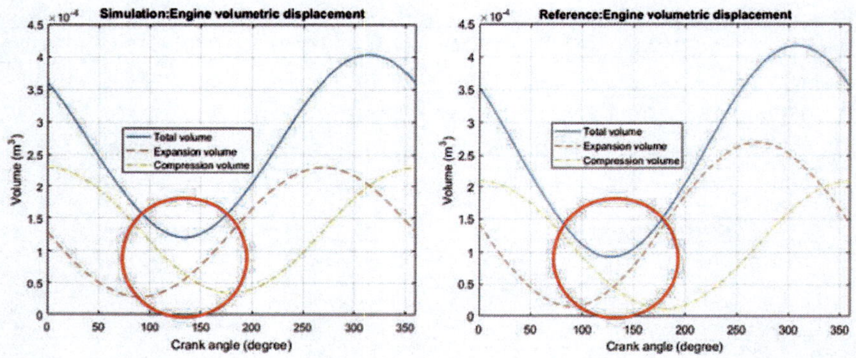

Fig. 4 Volumetric displacement for simulation model (left) and reference model (right)

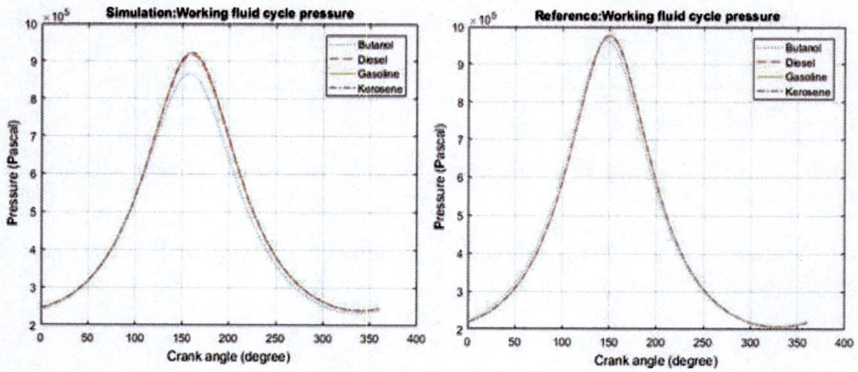

Fig. 5 Cyclic pressure of working fluid with different combustion fuels for simulation model (left) and reference model (right)

Attention focused on the red colour selection in Fig. 4 which the gaps between the total volume graph with the compression volume and expansion volume indicate the clearance volume of power piston and displacer, respectively [19–22]. The graph matches the geometrical parameter in the aspect of displacer-cylinder and piston-displacer top clearance distance. The smaller the clearance volume of expansion and compression process in reference model resulted in causing higher compression ratio of it.

Fundamental of thermodynamics knowledge claimed that the relationship of pressure to volume is inversely proportional, therefore the lowest volume leads to the highest pressure, for simulation model and reference model were approximately at a crank angle of 160° and 150°, respectively which nearby the crank angle of around 135° when the total volume at its smallest value. The largest and smallest total volume always hits on the middle point between their highest and lowest point on the graph, respectively.

3.2.3 Primary Component's Reciprocating Displacement Comparison

In general, the displacement for the movement of displacer top surface and bottom surface travel in the same vector but differ in magnitude for both models, whereas the displacement for power piston top surface travel with the path vertically asymmetric with the profile of displacer top surface movement (Fig. 6).

Displacement of displacer bottom surface and piston top surface almost converge at the crank angle of range from 210° to 245° and 300° to 325° for simulation model and reference model represent the starting to ending process of compression process occurs.

Distance between the top surface of power piston with the bottom surface of displacing equals to the displacer-piston clearance (dead volume), smaller in this value indicates high compression was ongoing and the same moment bigger quantity

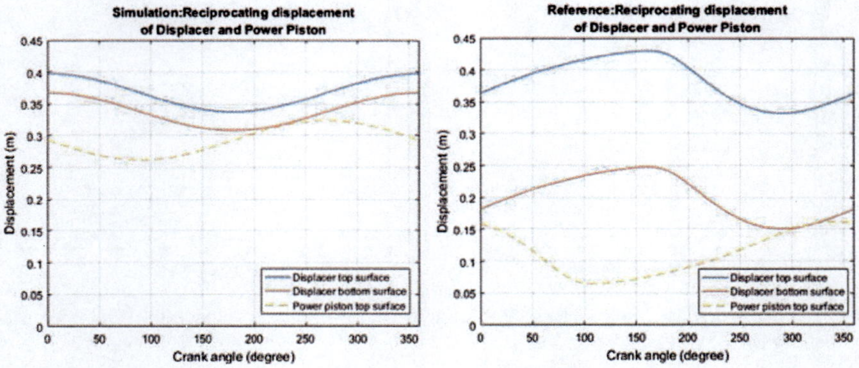

Fig. 6 Reciprocating displacement for simulation model (left) and reference model (right)

of Helium gas as the working fluid transfer into expansion region for completing the reciprocating movement of the engine.

3.2.4 Cyclic Energy Flow and Cyclic Power Flow with Different Combustion Fuels Comparison

90° and 180° were the turning point that the process changed when the crank angle was operated in a certain time. In the case of energy transfer, all the profile of heat energy (work in expansion space, compression space, and its summation) return to the value of 0 at the turning point for simulation model whereas for reference model all graph profile intersects at the turning point. Work done on the system of simulation model divided into three intervals which in the open downward parabola shaped work done process continuous occur with at all crank angle, but the outcome was relatively low [23]. Besides for the evaluation of work done on the reference model, the graph profile fluctuates with the declining trend followed by an uptrend at the crank angle of 180°, even though at the first half of the process was decreasing but the relatively huge magnitude of uprise trend keeps the reference model contains higher performance in comparison with the simulation model (Figs. 7 and 8) [24].

In addition, the power curve of the simulation model showed a similar indication with the total work done graph which discussed in the previous section, three sections of power curve with the peak performance (215 W) at the crank angle around 235°. Nevertheless, for the reference model, a power deficiency gap existed at the crank angle from 90° to 180° means the performance of the Stirling engine equals zero when the crankshaft rotates to the range of crank angle. Although this power curve contains a performance deficiency gap, it able to achieve the highest power output marked at 750 W when crank angle around 240°.

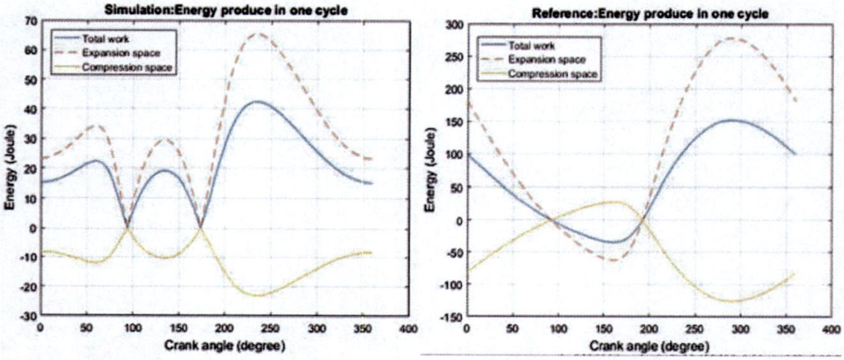

Fig. 7 Energy produced in a single reciprocating cycle for simulation model (left) and reference model (right)

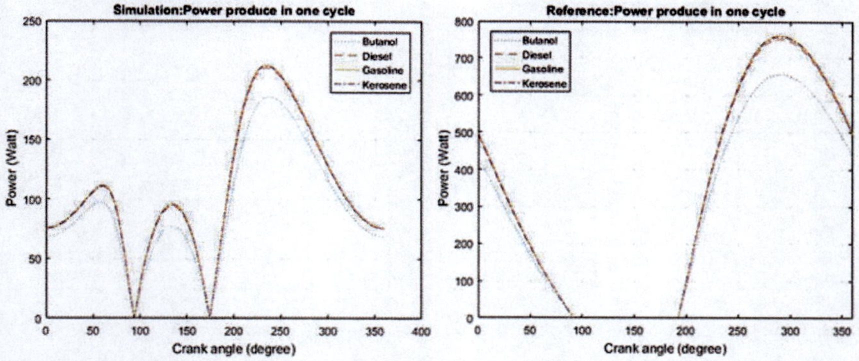

Fig. 8 Cyclic power flow with different usage of combustion fuels for simulation model (left) and reference model (right)

3.2.5 Performance (Power Output) on Different Setting of Phase Angle with Different Combustion Fuels Comparison

Gasoline appears as the most suitable combustion fuel out of four selections of, and no doubt that the Beta type Stirling engine able to achieve peak performance when the phase angle setting was setting up at 90° [25]. The optimal power generated in Watt can be observed from Fig. 9 were tallied with the data outputs in Fig. 10 which for the case of using gasoline as a heat source the maximum power was 283.57 W and 315.88 W, respectively. The phase angle of 90° becomes the demarcation point proved that any phase angle setting brings about an uptrend rising in performance, and vice versa [26].

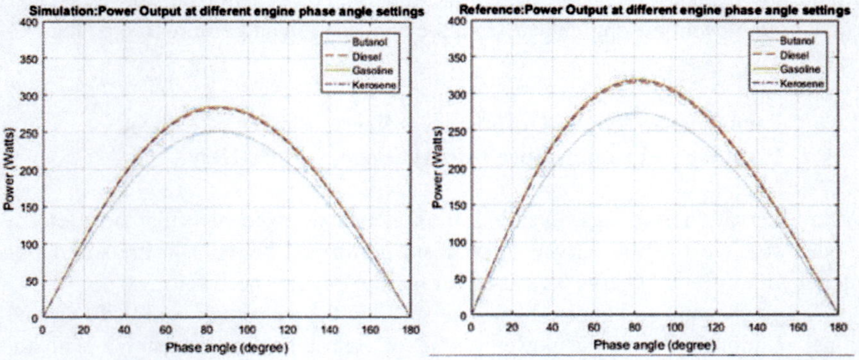

Fig. 9 Power output on different phase angle settings for simulation model (left) and reference model (right)

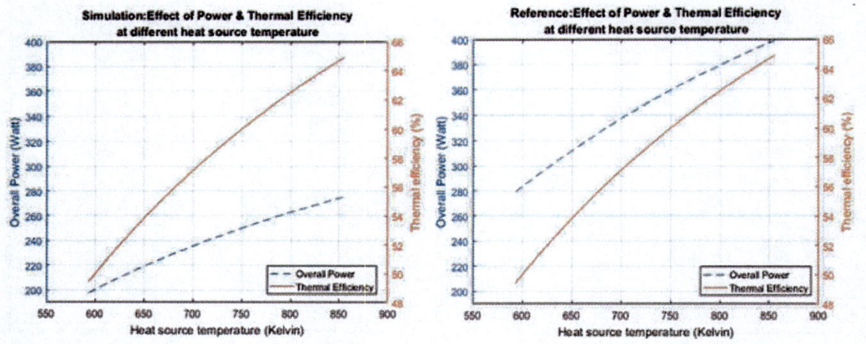

Fig. 10 Power output and thermal efficiency with heat source setting manipulated for simulation model (left) and reference model (right)

3.2.6 Performance (Power Output and Thermal Efficiency) on Different Setting of Heat Sources Comparison

As the graph is shown in Fig. 10, the profile of power curves and thermal efficiency curves for both models behave almost the same which is logarithmically increase along with the uprise of heat source temperature. Thermal efficiency varies from 49 to 64.5% depends on their instantaneous heat source temperature (593.44 to 855.75 K) since the cold terminal temperature was maintained at 300.15 K; conversely for power produced differences were range from 210 to 385 W, and range from 280 to 400 W, respectively for simulation and reference model. Significant disparity determines the performance of which model was more capable to perform greater [27].

Although the fact that higher heat source temperature which transferred into heater region can ensure the rise in overall power produced, the strength of the physical engine component for example the connecting rod, power piston, etc. must take into account to prevent the engine part from the deformation and failure to operate.

3.2.7 Performance (Power Output) on Different Setting of Charge Pressure and Engine Speed Comparison, Respectively

After observation from Figs. 11 and 12, the conclusion can be made that the reference model with the rhombic driven mechanism performed better than the simulation model with different charge pressure settings and engine speed settings.

Charge pressure and engine operating speed in theory contain a directly proportional relationship with the power output of Beta-type Stirling engine with the different driven mechanism. Overall power generated by the usage of gasoline is confirmed to be the most effective combustion fuel among all four of them. Since the charge pressure equals 450,000 Pa and engine speed fixed at 300 rpm for the initial setting, simulations were done by ranging the charge pressure from 250,000 to 750,000 Pa by controlling the pressure by the working fluid was transferred into the

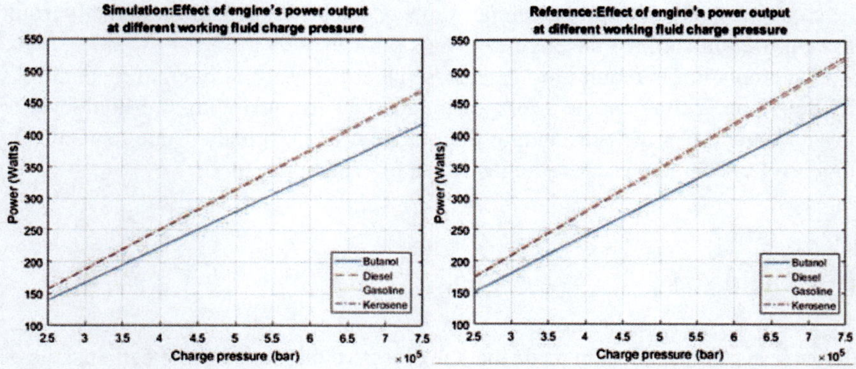

Fig. 11 Power output and thermal efficiency with charge pressure setting manipulated for simulation model (left) and reference model (right)

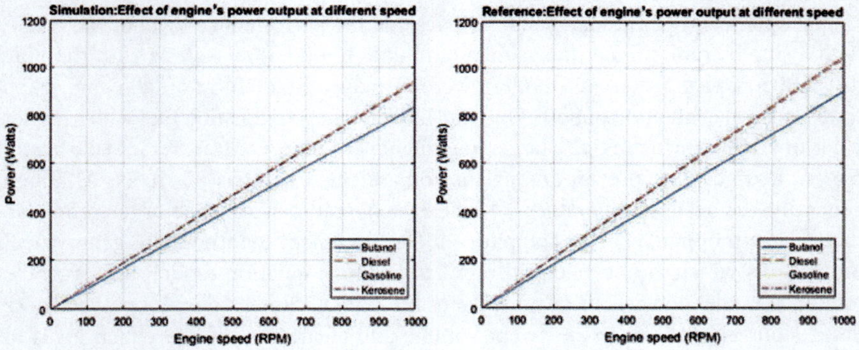

Fig. 12 Power output and thermal efficiency with engine speed setting manipulated for simulation model (left) and reference model (right)

cylinder body, as well as manipulate the engine speed with the usage of lightweight materials that contain higher durability or reduce the dimension size of the component (flywheel).

Charge Pressure's commonly used unit is Pascal (Pa) also can be expressed in $\frac{\text{kg}}{\text{m s}^2}$, higher charge pressure results in a density of the working fluid is bigger since more particles hold in a fixed volume which brings better benefit in providing the bigger ability for heat energy transfer to the system as well as rejection of heat energy out from the system, respectively occurs at expansion region (heater) and compression region (cooler).

Rotational speed measured in the unit of revolution per minute which is defined as the crankshaft completed a cycle of rotational displacement all together finished a series of thermodynamics process from isothermal expansion, heat rejection, isothermal compression, then heat addition. The rise in revolution per minute means the complete cycle of thermodynamics process can be done therefore directly result in the power outcome of the engine.

Contemplation is required to discuss since at the real-life operating conditions the Beta type Stirling engine's component contains its maximum threshold level of physical properties in withstanding the potential of deformation due to relatively high level of charge pressure along with the fast rotational speed.

4 Conclusion

On the basis, the conclusion made that the objective that stated at the earlier stage of research procedures had been achieved which the procedures and results mentioned in Chaps. 3 and 4, respectively provide a basis for how to simulate the operation of BTSE using 'MATLAB' as the computational modeling and simulation software with usage of different parameters.

Nevertheless, operational characteristics and the performance level of the BTSE with a single crank driven mechanism (simulation model) as well as with rhombic driven mechanism (reference model), were displayed in the form of table and graph for the ease of evaluation of both. Our data indicate that combustion fuel with a higher value of HHV able to provide a larger amount of heat energy to transfer into the heater region, and gasoline prevail among the four selected combustion fuels. Although generally, the heater temperature is higher, the overall performance of the reference model beat its opponent in the comparison. The reference model contains geometrical dimensions which is not much different from the simulation model, however, the reference model consists of advantages of the critical dimensions. Moreover, these dimensions able to varies the swept volume and clearance volume which leads to the effect of the compression ratio that played a crucial role in the performance of BTSE. Conversely, the thermal efficiency of simulation remains as it is higher than the reference model in general. Besides, the power generated from the reference model is not sustainable all along the crank angle after the crankshaft rotated for a single revolution even though the peak power generated is relatively higher compared to the outcome of the simulation model. In short, the rhombic driven mechanism is not listed as the primary consideration since the simulation model of BTSE that operates with a single crank driven mechanism does not exist this issue.

Furthermore, the rise in the setting for the effective volume of heater, charge pressure, heater temperature, rotational speed, and compression ratio leads to better performance for BTSE and vice versa with the cylinder-displacer clearance distance and displacer-power piston clearance distance.

Collectively, results provide evidence for a relatively perfect model of BTSE can be produced if combine excellent characteristics from both simulation and reference models. Single crank-driven, geometrical dimensions of reference model for BTSE, helium gas, and gasoline are selected as its elementary design considerations.

Acknowledgements The authors are grateful to Minister of Higher Education for giving grant under No. FRGS/1/2018/TK08/UMP/01/1.

References

1. Urieli I, Berchowitz DM (1984) Stirling cycle machine analysis. Adam Hilger Publishing
2. Syed Shabudeen PS (2010) Fuels and Combustion. ResearchGate. Published 10 Oct 2010. Accessed 28 Dec 2020. https://www.researchgate.net/publication/265602602_Fuels_and_Combustion_CHAPTER_-_4_FUELS_AND_COMBUSTION_41_Introduction_42_Requirements_of_a_Good_Fuel
3. Greenbaum B, Stirling C, Sullivan J. Stirling Engine Fabrication and Design. https://web.wpi.edu/Pubs/E-project/Available/E-project-042816-151447/unrestricted/StirlingEngineFinalReport.JMS1601.pdf
4. The Red King. Uaf.edu. Published 2017. Accessed December 28, 2020. http://ffden-2.phys.uaf.edu/webproj/212_spring_2017/Riley_Bickford/riley_bickford/CompleteCycle.html
5. Mahkamov K (2006) An axisymmetric computational fluid dynamics approach to the analysis of the working process of a solar stirling engine. J Sol Energy Eng 128:45–53
6. Tavakolpour AR, Zomorodian A, Golneshan AA (2008) Simulation, construction and testing of a two-cylinder solar Stirling engine powered by a flat-plate solar collector without regenerator. Renew. Energy 33:77–87
7. Costa SC, Barrutia H, Esnaola JA, Tutar M (2013) Numerical study of the pressure drop phenomena in wound woven wire matrix of a Stirling regenerator. Energy Convers. Manage. 67:57–65
8. Paepe MD, D'Herdt P, Mertens D (2006) Micro-CHP systems for residential applications. Energy Convers Manage 47:3435–3446
9. Thomas B (2008) Benchmark testing of micro-CHP units. Appl Therm Eng 28:2049–2054
10. Li T, Tang D, Li Z, Du J, Zhou T, Jia Y. Development and test of a Stirling engine driven by waste gases for the micro-CHP system. Appl Therm Eng 33–34:119–123
11. Abuelyamen A, Mansour RB, Abualhamayel H et al (2017) Parametric study on beta-type Stirling engine. Energy Convers Manage 145:53–63
12. Salazar JL, Chen W (2014) A computational fluid dynamics study on the heat transfer characteristics of the working cycle of a b-type Stirling engine. Energy Convers Manage 88:177–188
13. Abuelyamen A, Ben-Mansour R, Abualhamayel H, Mokheimer EMA (2017) Parametric study on beta-type Stirling engine. Energy Convers Manage 145:53–63. https://doi.org/10.1016/j.enconman.2017.04.098
14. Eugene AA, Baumeister III T, Sadegh AM (2006). McGraw-Hill 11
15. Zare S, Tavakolpour-saleh AR, Aghahosseini A, Sangdani MH, Mirshekari R (2020). Design and optimization of Stirling engines using soft computing methods: a review. Appl Energy, 116258.https://doi.org/10.1016/j.apenergy.2020.116258
16. Yang H-S, Cheng C-H (2017) Development of a beta-type Stirling engine with rhombicdrive mechanism using a modified non-ideal adiabatic model. Appl Energy 200:62–72
17. Tavakolpour-Saleh A, Zare S, Bahreman H (2017) A novel active free piston Stirling engine: modeling, development, and experiment. Appl Energy 199:400–415
18. Zare S, Tavakolpour-Saleh A (2020) Free piston Stirling engines: a review. Int J EnergyRes 44:5039–5070
19. Uchman W, Remiorz L, Grzywnowicz K, Kotowicz J (2018) Parametric analysis of a beta Stirling engine—a prime mover for distributed generation. Appl Thermal Eng 145:693–704. https://doi.org/10.1016/j.applthermaleng.2018.09.088

20. Remiorz L, Kotowicz J, Uchman W (2018) Comparative assessment of the effectiveness of a free piston Stirling engine-based micro-cogeneration unit and a heat pump. Energy 148:134–147
21. Xiao G, Sultan U, Ni M, Luo Z (2016) Design optimization with computational fluid dynamic analysis of β-type Stirling engine. ResearchGate. Published Oct 2016. Accessed 30 Dec 2020. https://www.researchgate.net/publication/309183405_Design_optimization_with_computational_fluid_dynamic_analysis_of_b-type_Stirling_engine/link/59d4caa9a6fdcc181adc5721/download
22. Karabulut H, Çınar C, Öztürk E, Yücesu HS (2010) Torque and power characteristics of a helium charged Stirling engine with a lever-controlled displacer driving mechanism. Renew Energy 35:138–143
23. Çınar C (2014) Design and manufacturing of a hermetic Stirling engine. Proc IMechE Part E 228:14–20
24. Costea M, Feidt M (1998) The effect of the overall heat transfer coefficient variation on the optimal distribution of the heat transfer surface conductance or area in a Stirling engine. Energy Conv Manage 39:1753–1761
25. Erlich C (2014) Applications of biomass Stirling engines for electrification. A case study of rural areas in Bolivia. Javier Arco Sola
26. Paul CJ, Engeda A (2015) A Stirling engine for use with lower quality fuels. Energy 84:152–160. https://doi.org/10.1016/j.energy.2015.02.109
27. Bataineh KM (2018) Numerical thermodynamic model of alpha-type Stirling engine. Case Study Therm Eng 12, 104e116. https://doi.org/10.1016/j.csite.2018.03.010

Dual Fuel Soy Biodiesel and Natural Gas Swirl Combustion for Toxic Emissions Reduction

Meng-Choung Chiong, Guo Ren Mong, Keng Yinn Wong, Hui Yi Tan, and Nor Afzanizam Samiran

Abstract The dual fuel swirl combustion of soy biodiesel and natural gas was investigated using a generic gas turbine combustor. Liquid fuel spray was generated using air-assisted atomiser with air-to-liquid ratio 2.50. An axial swirler with a vane angle of 45° was utilised to produce the swirling air flow at burner outlet. Liquid fuel spray was then blended with swirling air to form a flammable mixture and setup the flame. The flame colour of soy biodiesel/natural gas was mostly blue, similar to pure biodiesel. When compared to diesel, biodiesel/natural gas combustion reduced nitric oxide emissions by a factor of averaging 3. Empirical model for estimating NO emission was also proposed. Predicted results were closed to experimental data, denoted by R^2 greater than 0.95. This study demonstrates that biodiesel/natural gas combustion with a 20–30% natural gas input power fraction is a promising way to reduce toxic emission, offering a tactical approach to minimise hazardous emissions from neat biodiesel swirl combustion.

Keywords Soy biodiesel · Dual fuel · Natural gas

M.-C. Chiong (✉)
Department of Mechanical Engineering, Faculty of Engineering, Technology, and Built Environment, UCSI University, 56000 Kuala Lumpur, Malaysia
e-mail: chiongmc@ucsiuniversity.edu.my

G. R. Mong
School of Energy and Chemical Engineering, Xiamen University Malaysia, 43900 Sepang, Selangor, Malaysia

K. Y. Wong
School of Mechanical Engineering, Faculty of Engineering, Universiti Teknologi Malaysia, 81310 Skudai, Johor, Malaysia

H. Y. Tan
School of Chemical Engineering, Faculty of Engineering,, Universiti Teknologi Malaysia, 81310 Skudai, Johor, Malaysia

N. A. Samiran
Department of Mechanical Engineering Technology, Faculty of Engineering Technology, Universiti Tun Hussein Onn Malaysia, 84600 Pagoh, Johor, Malaysia

M. Y. Ismail et al. (eds.), *Technological Advancement in Mechanical and Automotive Engineering*, Lecture Notes in Mechanical Engineering,
https://doi.org/10.1007/978-981-19-1457-7_8

1 Introduction

Biodiesel exhibits very similar physical properties with fossil diesel [1, 2]. Furthermore, biodiesel combustion leads to considerably lower soot and particulate matter emissions than that of fossil diesel [1]. Numerous studies on biodiesel compression ignition engine were already performed since the initiation of Kyoto protocol in the late 1990s [3]. Nonetheless, the study of biodiesel combustion under gas turbine condition receives notably lesser attentions. A recent study by Chiong et al. [4] using a swirl burner demonstrated that double bond in biodiesel molecular structure has direct correlation with nitric oxide (NO) emission. Coconut biodiesel with considerably less double bond produced ~50% lower NO than the highly unsaturated soy biodiesel. Spectroscopic analysis of sunflower biodiesel was performed by Chong et al. [5]. It was reported that OH and CH radicals from biodiesel combustion were higher than fossil diesel by factor of 3–4. It was postulated that these radicals contributed to the higher biodiesel NO emission [5].

Higher biodiesel NO emission was also observed by Bolszo and McDonell [6] using a 30 kW Capstone C30 micro gas turbine (MGT) engine. Higher NO emission from soy biodiesel was due to the larger biodiesel fuel droplets that stretched its evaporation time scale. This consequently elevated biodiesel NO emission by ~13 ppm when compared with diesel [7]. By using a Rover IS/60 gas turbine engine, Rehman et al. [8] showed that 25/75 biodiesel/diesel blend produced notably higher NO concentration than 15/85 biodiesel/diesel blend. When engine operated at 21 kW output power, NO concentration produced by 25/75 biodiesel/diesel combustion was nearly 10% higher than that of 15/85 biodiesel/diesel blend. Recently, Kurji et al. [9] showed that swirl combustion of pyrolysis oil/biodiesel blend resulted in constantly higher NO emission when compared with kerosene, regardless of atomising pressure. At stoichiometric condition, NO of saturated biodiesel was 30 ppm higher than NO emission from kerosene combustion. The direct correlation between droplet size and NO emission was ascertained by Hashimoto et al. [10] using a swirl burner. It was unveiled that NO emission reduced by 10 ppm as atomising pressure increased by 0.1 MPa [10].

Owing to the lower biodiesel volatility compared to diesel, it was postulated that carbon monoxide (CO) emission from biodiesel combustion would be higher than fossil diesel [11]. This was evident from the study by Nascimento et al. [12, 13]. At 14 kW MGT output power, they unveiled that the CO emission of castor biodiesel was roughly 50% higher than diesel CO emission [12, 13]. Previous studies constantly show that toxic emissions such as NO and CO from biodiesel gas turbine combustion tend to be higher than fossil diesel. Both NO and CO are toxic emissions that pose a direct threat to human health [14, 15]. Increased NO concentration can harm the human respiratory system and make an individual more susceptible to respiratory infections and asthma. Therefore, these emissions must be reduced to promote the use of biodiesel for future power generation.

Dual fuel combustion was introduced as a potential approach of enhancing the emissions performances of biofuels. Small amount of secondary fuel was introduced to support the combustion of primary fuel [16]. Previous dual fuel combustion studies performed by Queiros et al. [17], Kurji et al. [18], Kun-Balog et al. [19], Jiang and Agrawal [20] resulted in promising emissions reduction. However, studies on dual fuel combustion are still limited to date. Due to the lack of knowledge on biodiesel/natural gas (NG) swirl combustion, this research intends to investigate the emissions of a soybean biodiesel (SME)/NG blend using a generic swirl burner typically used in the gas turbine engine.

2 Experimental Setup

The schematic diagram for experimental setup used in this study is depicted in Fig. 1. The liquid fuel was transferred from fuel to atomiser using Longer BQ50-1J peristaltic pump. An air-assisted atomiser (Delavan: SN type-30610-1) was employed to atomise the liquid fuel. Fuel spray was formed at atomiser outlet when atomising air was also supplied into the air-assisted atomiser. The mass flow rate of atomising air was regulated using Sierra SmartTrak 50 controller. Air-to-liquid (ALR) ratio 2.5 was used in this study.

Before entering the burner plenum, the main combustion air was heated to 250 °C. Mass flow rate of the main combustion air was controlled by another Sierra SmartTrak 50 mass flow controller. Three cable heaters with nominal output power 500 W were used to heat the air. The temperature of main combustion air was regulated

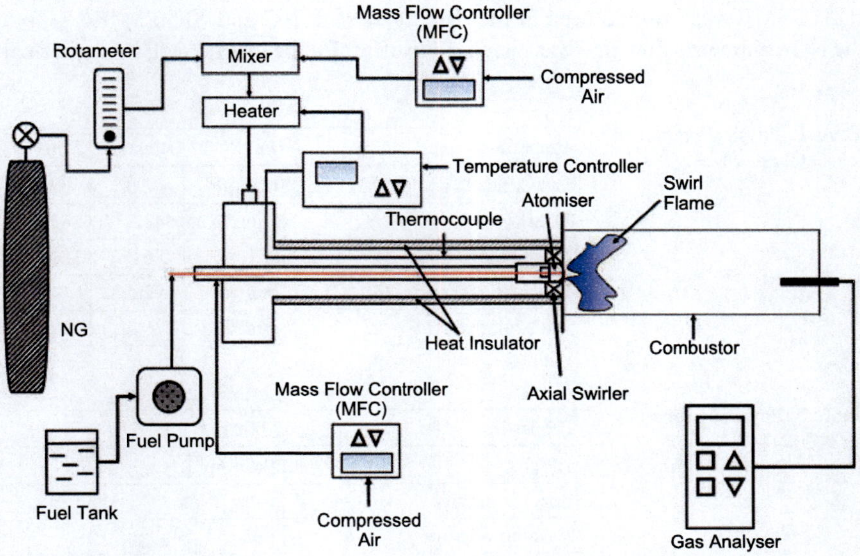

Fig. 1 Experimental setup for SME/NG swirl combustion (adapted from [2])

using a temperature controller. The temperature of the air was measured using a K-type thermocouple, the reading of air temperature was feedback to the temperature controller to accomplish the temperature regulation process.

Swirling flow was generated at burner outlet as main combustion air passed through an axial swirler that was placed concentrically at burner exit. The swirling flow blended with liquid fuel spray to form flammable mixture. Swirler vane angle 45° was used in this study. A flame torch was used to ignite this mixture. A 400 mm long combustor with a diameter of 130 mm made of carbon steel transferred the flue gas from to the exhaust. A gas analyser was positioned at combustor outlet to quantify the emissions. The main combustion air was transported into mixer for mixing with NG in dual fuel operation. NG volume flow rate was regulated using a rotameter.

Table 1 compares physical properties of SME with diesel. SME exhibits higher fuel density than diesel with about 10% lower LHV. Lower LHV for SME is due to the presence of oxygen in its molecular composition. The SME is inherently less volatile when compared with diesel, as denoted by considerably higher SME flash point than fossil diesel. Viscosity for SME is also higher than diesel by about 1.9 mm^2/s. The NG was purchased from Gas Malaysia Berhad through a local industrial gas supplier. Table 2 shows the physical properties of NG.

The baseline fuels were fossil diesel and neat SME. Dual fuel operation SME/NG was studied using three different SME and NG input power fractions, namely 90/10 SME/NG, 80/20 SME/NG, and 70/30 SME/NG. As indicated in Eq. 1, the ratio 90 indicates that SME contributes 90% of total input thermal power, while NG contributes the other 10%. Mass flow rate of SME and volume flow rate of NG are represented by $\dot{m}_{SME}$ and $\dot{V}_{NG}$, respectively. As a result, in the 90/10 SME/NG combustion, the mass flow rate for SME would be lower than neat SME. The flow rate of SME was lowered even further for 80/20 SME/NG and 70/30 SME/NG, but the NG volumetric flow rate was increased to offset for the loss in SME input thermal

Table 1 Physical properties for diesel and SME

Properties	Unit	Diesel	SME
Lower heating value (LHV)	[MJ/kg]	42.6	37.0
Density	[kg/m^3]	843.3	882.0
Flash point	[°C]	76.0	159.0
Kinematic viscosity (40 °C)	[mm^2/s]	2.4	4.3
Molecular weight	[g/mol]	226.0	292.2

Table 2 Physical properties for NG [21]

Properties	Unit	
LHV	[$kCal/m^3$]	8862
Specific gravity	[–]	0.68
Burning velocity	[m/s]	0.28
Autoignition temperature	[°C]	537
AFR_S	[–]	17.3

power. For all fuels examined, constant thermal input power of 9.3 kW was used and global flame equivalence ratio (ϕ) was varied from 0.65 to 0.9.

$$\dot{m}_{SME} LHV_{SME}:\dot{V}_{NG} LHV_{NG} = 90:10 \tag{1}$$

3 Results and Discussions

Figure 2 depicts that SME/NG swirl flames do not fully attach to the burner. However, this does not affect overall stability of the flames, due to the intensified reaction when of NG was added. The yellowish sooty flame brush in diesel swirl flame is due to the presence of aromatic structure in its chemical composition [2, 5]. Biodiesel swirl flame is mainly bluish, owing to the absent of aromatic structure in SME composition [2, 5].

Figure 3a shows that NO emission from 70/30 and 80/20 SME/NG swirl combustion is noticeably lower than those of neat biodiesel and diesel. Lower NO emission from 70/30 and 80/20 SME/NG combustion is mainly due to the higher NG mass fraction that converts a significant part of non-premixed reaction into premixed combustion. Homogeneous NG/air mixture reacts at a considerably quicker reaction rate than SME droplets, since droplet evaporation time scale is absent in the former. This suppresses the creation of reaction zone with high temperature, reducing the thermal NO formation as a consequent [22].

Meanwhile, owing to the reduction in SME thermal power fraction in 70/30 and 80/20 SME/NG, droplet density of SME is decreased in the inner reaction zone. This

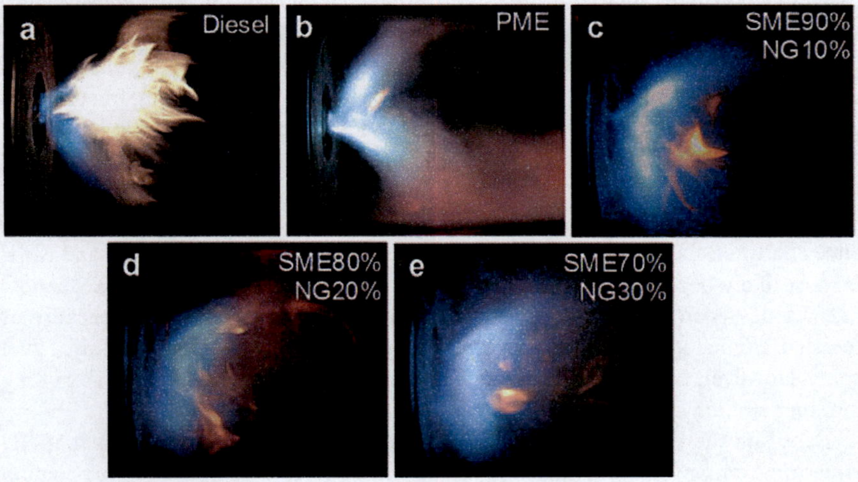

Fig. 2 Swirl flame physical appearances for diesel, SME, and SME/NG ($\phi = 0.65$)

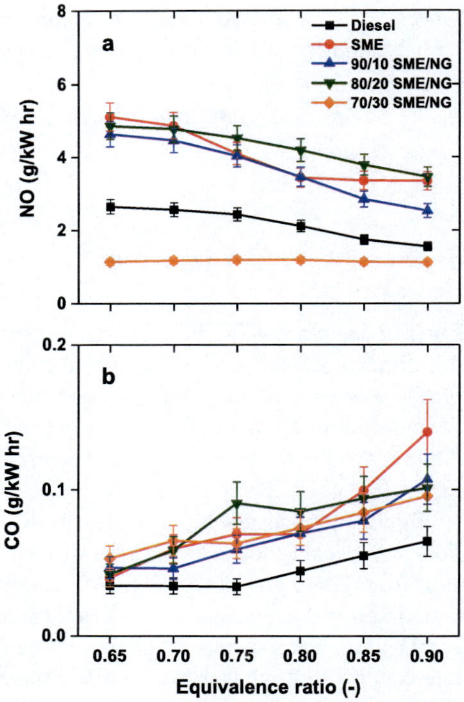

Fig. 3 NO and CO emissions comparison for diesel, SME, and SME/NG

inherently lower the reaction intensity at the central region near the burner's centerline. This comprises another plausible reason for significant reduction in thermal NO production in 70/30 SME/NG. Introduction of gaseous fuel reduces the size of inner reaction zone, denoting the shifting of reaction zone towards the outer edge of swirl flame when gaseous fuel was added [23].

The differences in NO emission between neat SME, 90/10 are only marginal as seen in Fig. 3a. This is primarily due to the lower NG input power fraction and combustion mode is still mainly non-premixed. Present study shows that NG power fraction of 20–30% is necessary for lowering NO emission appreciably in SME/NG dual fuel operation.

CO emissions for all fuels examined are generally low (<0.2 g/kWh). The differences between them are trivial. Similar CO emission between SME/NG and SME is most likely because of reduced atomising air in dual fuel mode when SME flow rate was lessened. Radial dispersion of SME fuel spray was weakened because of reduced atomising air. This causes poorer air penetration in to the central of fuel spray. However, NG addition nullifies the reduction in atomisation quality, resulting in rather similar CO emission with neat SME.

Based on the experimental data in Fig. 3a, the present study proposes that NO emission from SME/NG combustion can be correlated to ϕ using quadratic equation delineated by Eq. 2. The quadratic constants p, q and r can be estimated using empirical models proposed in Fig. 4, where all three coefficients are correlated to NG input power fractions via quadratic functions.

$$\mathrm{NO} = \left(p \cdot \phi^2\right) + (q \cdot \phi) + r \quad (2)$$

Figure 5 shows that Eq. 2 and empirical models in Fig. 4 can estimate the specific NO emission for neat SME and SME/NG swirl flames satisfactorily. R^2 for the estimation curves are very close to 1. This indicates that Eq. 2 and empirical models in Fig. 4 are reliable tool in estimating the specific NO emission for SME/NG combustion. Empirical model was not developed for CO due to differences between neat SME and SME/NG are insignificant (Fig. 3b).

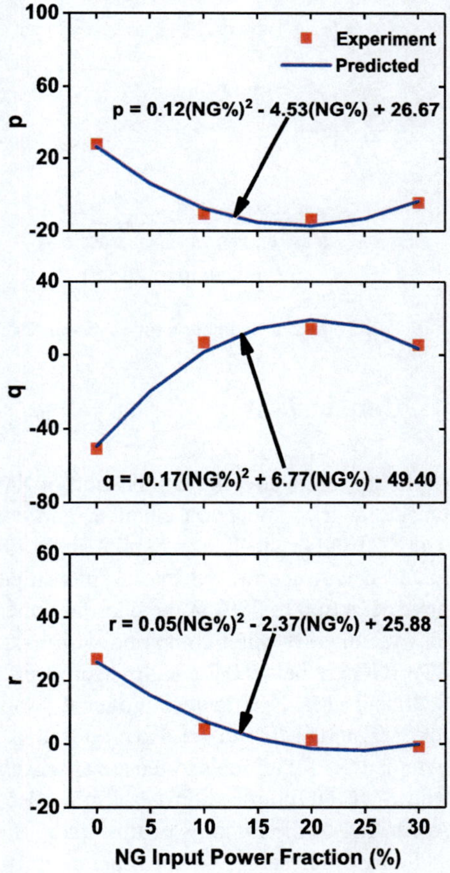

Fig. 4 Prediction of quadratic constants p, q and r for NO empirical model in Eq. 2

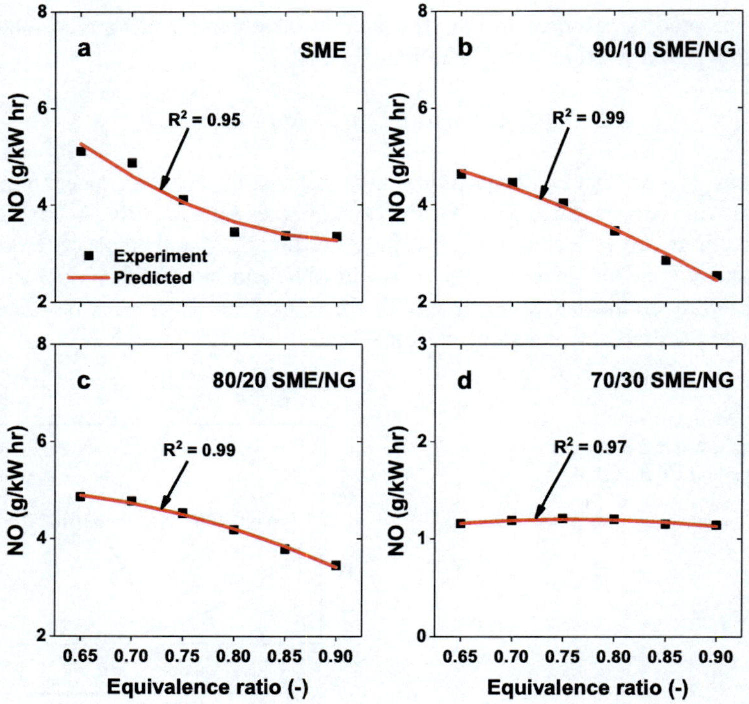

Fig. 5 NO emission estimation for SME and SME/NG

4 Conclusions

In the current study, the combustion of SME/NG are examined using a model gas turbine swirl combustion chamber and compared with pure biodiesel and diesel. The SME/NG swirl flames exhibit slight lift-off from burner base. Flame instability incurred was neutralised by the intensified NG reaction. By increasing NG input power fraction to 30% in the dual fuel operation, the NO emission was reduced up to a factor of 3 when compared with neat SME. CO emission for neat SME and SME/NG are below 0.2 g/kWh. Moreover, disparities in CO emission between neat SME and SME/NG are not significant. Empirical model for estimating NO emission for SME and SME/NG are also proposed in this study, based on the experimental data obtained. R^2 for all fuels tested are approaching 1. This research shows NO emission from SME/NG combustion can be minimised by altering the NG input power fraction. According to the findings of this study, SME/NG is a cleaner alternative to diesel and pure biodiesel for power generation.

Acknowledgements The authors gratefully acknowledge the financial supports from the Centre of Excellence for Research, Value Innovation and Entrepreneurship (CERVIE), UCSI University under the conference funding (CONF-NATN-FETBE-100) and Research Excellence & Innovation Grant (REIG-FETBE-2021/035).

References

1. Chiong MC, Chong CT, Ng JH, Lam SS, Tran MV, Chong WWF et al (2018) Liquid biofuels production and emissions performance in gas turbines: a review. Energy Convers Manag 173:640–658. https://doi.org/10.1016/j.enconman.2018.07.082
2. Chiong MC, Valera-Medina A, Chong WWF, Chong CT, Mong GR, Mohd Jaafar MN (2020) Effects of swirler vane angle on palm biodiesel/natural gas combustion in swirl-stabilised gas turbine combustor. Fuel 277:118213. https://doi.org/10.1016/j.fuel.2020.118213
3. Chiong M-C, Kang H-S, Shaharuddin NMR, Mat S, Quen LK, Ten K-H et al (2021) Challenges and opportunities of marine propulsion with alternative fuels. Renew Sustain Energy Rev 149:111397. https://doi.org/10.1016/j.rser.2021.111397
4. Chiong MC, Chong CT, Ng JH, Tran MV, Lam SS, Valera-Medina A et al (2019) Combustion and emission performances of coconut, palm and soybean methyl esters under reacting spray flame conditions. J Energy Inst 92:1034–1044. https://doi.org/10.1016/j.joei.2018.07.003
5. Chong CT, Chiong MC, Ng JH, Lim M, Tran MV, Valera-Medina A et al (2019) Oxygenated sunflower biodiesel: Spectroscopic and emissions quantification under reacting swirl spray conditions. Energy 178:804–813. https://doi.org/10.1016/j.energy.2019.04.201
6. Bolszo CD, McDonell VG (2009) Emissions optimization of a biodiesel fired gas turbine. Proc Combust Inst 32:2949–2956
7. Bolszo CD, McDonell VG (2009) Evaluation of plain-jet air blast atomization and evaporation of alternative fuels in a small gas turbine engine application. At Sprays 19:771–785
8. Rehman A, Phalke DR, Pandey R (2011) Alternative fuel for gas turbine: esterified jatropha oil-diesel blend. Renew Energy 36:2635–2640
9. Kurji H, Valera-Medina A, Runyon J, Giles A, Pugh D, Marsh R et al (2016) Combustion characteristics of biodiesel saturated with pyrolysis oil for power generation in gas turbines. Renew Energy 99:443–451
10. Hashimoto N, Ozawa Y, Mori N, Yuri I, Hisamatsu T (2008) Fundamental combustion characteristics of palm methyl ester (PME) as alternative fuel for gas turbines. Fuel 87:3373–3378
11. Panchasara HV, Simmons BM, Agrawal AK, Spear SK, Daly DT (2009) Combustion performance of biodiesel and diesel-vegetable oil blends in a simulated gas turbine burner. J Eng Gas Turbines Power 131:031503
12. Nascimento MAR, Lora ES, Corrêa PSP, Andrade RV, Rendon MA, Venturini OJ et al (2008) Biodiesel fuel in diesel micro-turbine engines: modelling and experimental evaluation. Energy 33:233–240
13. Nascimento MAR, Sierra RGA, Silva Lora EE, Rendon MA (2011) Performance and emission experimental evaluation and comparison of a regenerative gas microturbine using biodiesel from various sources as fuel. J Energy Resour Technol 133:022204
14. Turns SR (2012) An introduction to combustion: concepts and applications, vol 499, 3rd edn. Mc-GrawHill
15. Gann R, Friedman R (2015) Principles of fire behavior and combustion. Jones 7 Bartlett Learning
16. Kahila H, Wehrfritz A, Kaario O, Vuorinen V (2019) Large-eddy simulation of dual-fuel ignition: diesel spray injection into a lean methane-air mixture. Combust Flame 199:131–151
17. Queiros P, Costa M, Carvalho RH (2013) Co-combustion of crude glycerin with natural gas and hydrogen. Proc Combust Inst 34:2759–2767

18. Kurji H, Valera-Medina A, Okon A, Chong CT (2017) Combustion and emission performance of CO_2/CH_4/biodiesel and CO_2/CH_4/diesel blends in a Swirl Burner Generator. Energy Procedia 142:154–159
19. Kun-Balog A, Sztankó K, Józsa V (2017) Pollutant emission of gaseous and liquid aqueous bioethanol combustion in swirl burners. Energy Convers Manag 149:896–903
20. Jiang L, Agrawal AK (2014) Combustion of straight glycerol with/without methane using a fuel-flexible, low-emissions burner. Fuel 136:177–184
21. Malaysia Energy Commission (2016) Piped Gas Distribution Industry Statistics 2016
22. Lefebvre AH, Ballal DR (2010) Gas Turbine combustion: alternative fuels and emissions, 3rd edn. CRC Press, Boca Raton
23. Evans MJ, Sidey JAM, Ye J, Medwell PR, Dally BB, Mastorakos E (2018) Temperature and reaction zone imaging in turbulent swirling dual-fuel flames. Proc Combust Inst 000:1–8

Design and Analysis of Composite Materials for Vehicle Engine Mount

A. R. Abd Hamid and T. M. Chin

Abstract This article will aim to replace rubber in engine mounts with alternative materials, preferably composites. The engine mount is the most important component of a vehicle's engine; its job is to isolate and absorb engine vibrations. As a result, the engine mount is subjected to high stresses and adverse conditions. For example, high temperatures, chemical leaks, vibrations, and the engine's mechanical torque. Rubber has a lower life duration when all aspects are considered. As a result, many researchers have tried to extend the mount's lifespan. The purpose of this paper is to demonstrate this. To extend the mount's life and increase performance by decreasing power dissipation without sacrificing comfort, a combination of creative ideas and materials are combined. As a result, expect a stiffer and more durable mount. Three different materials are used to compare three different conceptual designs. Rubber, Polyurethane, and Carbon Fiber-Epoxy Resins are utilized in the Pugh technique to identify the optimal mix of conceptual ideas and materials. To identify the final combination, many analyses are performed, including structural analysis, random vibration analysis, explicit analysis, static deflection, and mount weight (Design 2 with CF-EP). Despite the fact that PU and CF-EP are equivalent, CF-EP warrants further investigation and testing because there is still room for development.

Keywords Design of engine mounts · Analysis of engine mounts · Carbon fiber epoxy resin composite · PUGH method comparison · Alternative to polyurethane mounts

1 Introduction

Engine mount is a component that is attached in between an engine block and the chassis of a car. Bolts and nuts are used to hold them in place. Besides, it dampens the vibrations generated from the engine [1]. It is a crucial component as vibration from

A. R. Abd Hamid (✉) · T. M. Chin
Faculty of Engineering, Technology and Built Environment, Mechanical Engineering Department, UCSI University, 56000 Kuala Lumpur, Malaysia
e-mail: amar@ucsiuniversity.edu.my

M. Y. Ismail et al. (eds.), *Technological Advancement in Mechanical and Automotive Engineering*, Lecture Notes in Mechanical Engineering,
https://doi.org/10.1007/978-981-19-1457-7_9

the mechanical activity within the engine is inevitable no matter how refined and well balanced the engine is. These vibrations are generally caused by the spinning motion of rods, pistons and, gas detonations every second [2]. Besides vibrations, engine mounts are also used to restrict engine movements caused by acceleration and deceleration as well as the torque generated by the engine which produces a rocking motion [3]. Generally, engine mounts are made of rubber because it is affordable and easy to obtain yet has good properties. One of the best materials to be used as a dampener, providing stable support and restrict engine movements [4]. However, rubber has limited life due to it being constantly exposed to harsh environments and chemicals. They tend to break down and lose their function.

The aim of this study is to show that there are alternative materials that can be used in manufacturing an engine mount such as composite-based materials providing a better driving experience in terms of comfort and rigidness of the car and providing ample torque tolerances during power transmission [5]. It will prolong the health of the engine from losing balance and less wear and tear [1]. Hence, this paper will compare CF-EP with Rubber and PU (the current most popular alternative).

2 Methodology and Experimental Setup

2.1 Conceptual Design

Three different bush-type design was modeled with Solidworks. The overall diameter, the thickness (45 mm) as well as design aspect were all adapted by a researcher [6]. Chamfered edges were applied across all 3 models for better dampening [7]. Another design feature that is kept constant is the inner diameter for the M10 bolt [6]. A base structure of an octagonal shape was used in the first model. The ideology for this approach was adapted from those who had researched comparing different types of design [8]. Among the researcher's designs, the hexagon design was the best. Conceptual Design 1 in Fig. 1 with the addition of 2 sides.

Conceptual design 2 adopts the design criteria from conceptual design 1, constructing from an octagonal (8 sides) polygonal base model. However, additional bush support is added to 4 corners of the base model for support and dampening capabilities [8]. Theoretically, this design outperform design 1 but further justification is needed with the aid of simulation analysis. Conceptual Design 2 in Fig. 2.

The third design is a control design for comparison. Similar to existing cylindrical engine mounts but the thickness is increased to the maximum allowable thickness of 45 mm. Conceptual Design 3 in Fig. 3.

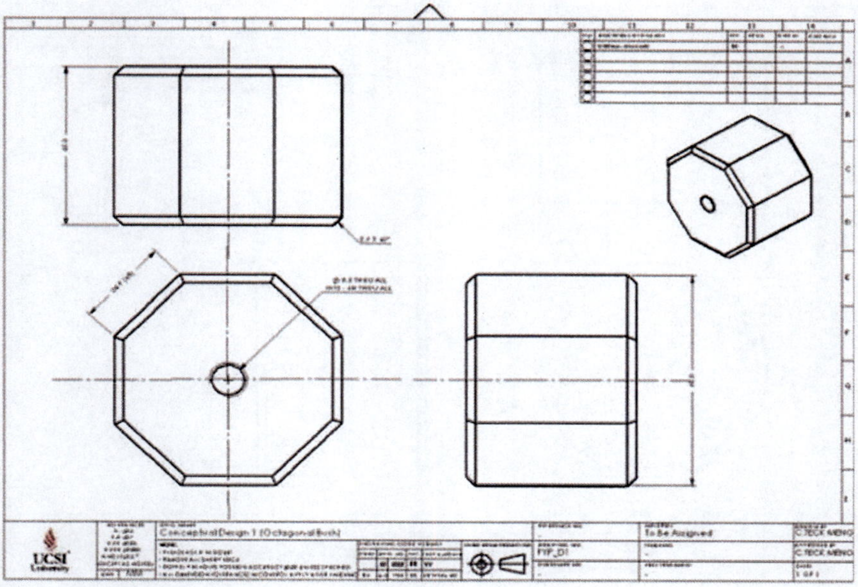

Fig. 1 Conceptual design 1

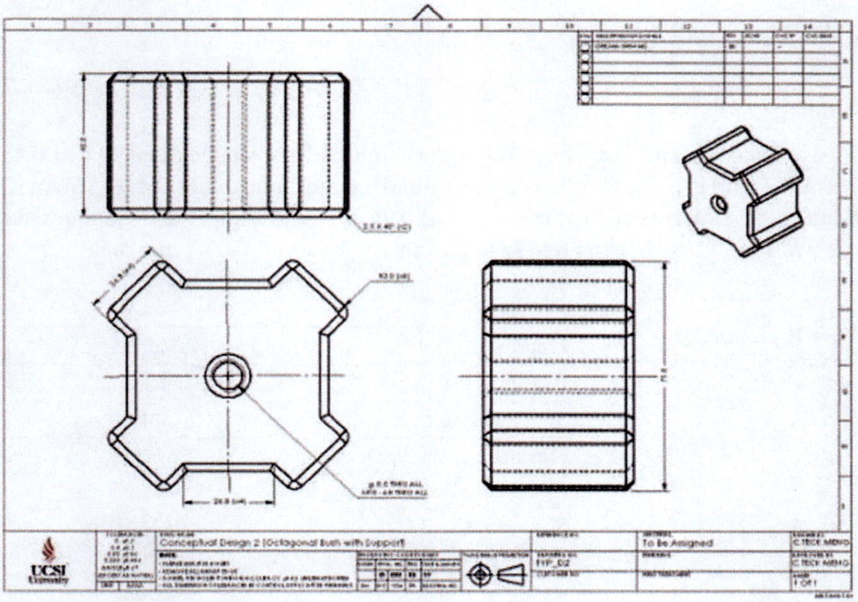

Fig. 2 Conceptual design 2

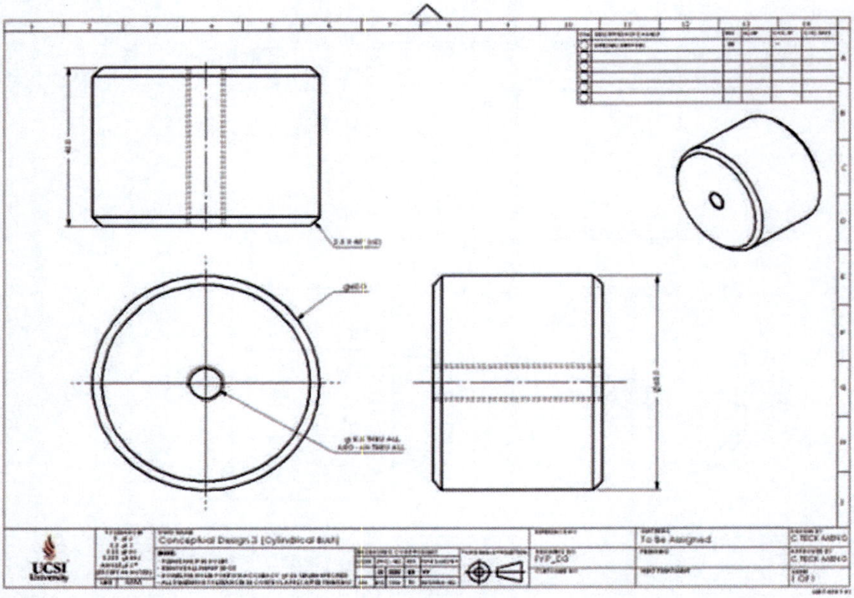

Fig. 3 Conceptual design 3

2.2 *Materials*

Rubber

A base material. The Density of 960 kg/m^3 with 20 MPa tensile strength and 0.45 Poisson's ratio [9]. The Hardness of 30–90 and thermal conductivity of 0.15 W/m K. Young's, and shear modulus, are 0.01 and 0.44 MPa [10]. Shear and Compressive stresses are 10–20 MPa and 6 MPa [11]. Table 1 shows rubber's properties.

Table 1 Mechanical properties of natural rubber (NR)

Property	Value
Density, ρ	960 kg/m^3
Young's modulus, E	0.01 MPa
Maximum compressive stress, σ_{max}	6 MPa
Maximum shear stress, τ_{max}	10–20 MPa
Shear modulus, G	0.44 MPa
Tensile strength	20 MPa
Hardness	30–90
Poisson's ratio	0.45
Thermal conductivity, K	0.15 W/m K

Table 2 Mechanical properties of polyurethane (PU)

Property	Value
Density, ρ	100 kg/m^3
Young's modulus, E	2.1 MPa
Maximum compressive stress, σ_{max}	22–164 MPa
Maximum shear stress, τ_{max}	13.3–99.7 MPa
Shear modulus, G	0.17–4.15 MPa
Tensile strength	50 MPa
Hardness	80–95
Poisson's ratio	0.32
Thermal conductivity, K	0.29 W/m K

Polyurethane

Well-known in automotive industries and does not emit harmful gases during manufacturing. Common for upgrading to replace rubber parts. Which improves the structural rigidity of a car [12]. Its tensile strength and hardness of 50 MPa and 80–95 with thermal conductivity 0.29 W/m.K [13]. The young's and Poisson ratio are 2.5 MPa, and 0.32 [14]. The shear modulus ranges from 0.17 to 4.15 MPa [15]. The maximum compressive and shear stresses is 22–164 MPa and 13.3–99.7 MPa. The density is 100 kg/m^3 [16]. The properties of PU are displayed in Table 2.

Carbon Fiber Epoxy Resin

The popularity of carbon fiber reinforced composite is increasing in the development of automobiles. Especially with the epoxy resin. This combination creates a material that holds an exceptional quality of properties among many materials and contributing to lesser carbon emission [17]. It is lightweight and strong which makes it a good contender to many materials [18]. CF/EP's density is 160 kg/m^3 with a tensile strength measured at 110 MPa [19]. The young's modulus, E was found to be 70,000 N/mm^2 Poisson's ratio rated at 0.1 [20]. The durometer readings for the hardness of CF/EP is measured at a range of 85–90 with a shear modulus of 5600 MPa [21]. Both maximum compressive and shear stress are rated at 34,900 and 35 MPa [22]. Thermal conductivity, 0.79 W/m K [23]. Table 3 below shows its properties.

3 Result and Discussions

Several simulations and comparisons have been taken place to identify the better combination of design and material. Structural Analysis, Random Vibration Analysis, Explicit Analysis, Comparison of weights and Static Structural.

Table 3 Mechanical properties of carbon fiber epoxy resin matrix (CF/EP)

Property	Value
Density, ρ	160 kg/m^3
Young's modulus, E	70 MPa
Maximum compressive stress, σ_{max}	34,900 MPa
Maximum shear stress, τ_{max}	35 MPa
Shear modulus, G	5600 MPa
Tensile strength	110 MPa
Hardness	85–90
Poisson's ratio	0.1
Thermal conductivity, K	W/m K

3.1 *Structural Analysis*

Acknowledging the design nature of 4 bottom mounts with force acting from the top [24]. It secures a 4-cylinder engine of weight between 980 and 1180 N (100–120 kg). Assuming the engine's weight to be 1000 N, an equal distribution of weight to each mount is 250 N. Table 4 summarizes the data for each design.

Table 4 Summary of structural analysis

Conceptual designs			
Rubber			
Max total deformation (mm)	2.6851	1.7229	2.7727
Max directional deformation (mm)	2.2266	1.4704	2.3893
Max equivalent elastic strain (mm/mm)	0.0438	0.0284	0.0508
Max equivalent (Von-Mises) stress (Mpa)	2.5082	1.6837	2.8891
Polyurethane			
Max total deformation (mm)	0.5216	0.3413	0.5446
Max directional deformation (mm)	0.4333	0.2911	0.4699
Max equivalent elastic strain (mm/mm)	0.0084	0.0056	0.0096
Max equivalent (Von-Mises) Stress (Mpa)	2.4848	1.6713	2.6767
Carbon fiber-epoxy resin			
Max total deformation (mm)	0.0179	0.0098	0.0194
Max directional deformation (mm)	0.0152	0.0086	0.0171
Max equivalent elastic strain (mm/mm)	0.0002	0.0001	0.0003
Max equivalent (Von-Mises) stress (Mpa)	4.8289	2.5994	5.3584

Table 5 Summary of random vibration analysis

Conceptual designs			
Rubber			
Min frequency of vibration (Hz)	7.7433	8.1106	7.7925
Max frequency of vibration (Hz)	38.703	32.850	38.801
Response PSD @10 Hz (mm^2/Hz)	0.6066	6.37 × 10–9	0.2842
Max response PSD @ 30 Hz (mm^2/Hz)	1.543 × 10–5	2.125 × 10–11	0.0040
Polyurethane			
Min frequency of vibration (Hz)	0.00035	0.00067	0.00035
Max frequency of vibration (Hz)	0.00176	0.00138	0.0018
Response PSD @10 Hz (mm2/Hz)	0.02975	2.07×10^{-10}	0.0257
Max Response PSD @ 30 Hz (mm^2/Hz)	0.00043	2.94×10^{-12}	0.00036
Carbon fiber-epoxy resin			
Min frequency of vibration (Hz)	0.00215	0.00229	0.00215
Max frequency of vibration (Hz)	0.01116	0.00847	0.01145
Response PSD @10 Hz (mm^2/Hz)	0.02398	2.34×10^{-10}	0.02385
Max response PSD @ 30 Hz (mm^2/Hz)	0.00034	2.33×10^{-12}	0.00034

3.2 Random Vibration Analysis

Considering the psd acceleration value of road [11] as input condition for testing purposes. Table 5 below shows the recorded data.

3.3 Explicit Analysis

Below Table 6 summarizes the collective data for explicit analysis for each individual mounts as well as different applied materials stated previously.

3.4 Static Deflection

Rubber stiffness ranges from 0.6 to 0.8 MPa and up to 4 MPa [26]. PU stiffness averages 3.365 MPa [9]. However, hardness ranges from 55 to 75 which translates to 1290–1944 MPa with conversion chart [27]. CF-EP stiffness range from 6000 MPa [28] and is possible to reach 70 GPa [29]. Obeying Eq. (1), Table 7 shows the results.

in good agreement. Chen et al. [8] used numerical simulation software to calculate the effect of erosion on the three-way pipe, and compared the calculation results with experimental data. The results proved the accuracy of the numerical simulation calculations. In order to ensure the integrity of the pipeline system, Takahashi [9] used the finite element method to study the problem of the thinning of the pipe wall by the high temperature and pressure water and steam flow respectively eroding the pipe.

In the current research, the research parts on erosion and wear are mainly concentrated on 90° elbows and three-way pipes. As a kind of indispensable parts in the coal chemical transportation system, the research on reducing pipes is relatively scarce. In this paper, numerical simulation calculations are used, and the CFD calculation fluid software Fluent is used to select the reducer pipe conveying the outer parts of the coal slurry storage tank as the geometric model, the mathematical model is modeled according to the actual parameters of the factory, and the erosion wear rate and the erosion wear location distribution are simulated and calculated, and the influence of the solid particle size on the erosion wear rate of the reducer is studied. It provides some theoretical basis for pipeline design in actual engineering and research on erosion and wear and maintenance of pipeline system.

2 Numerical Simulation Model

The transportation system of coal slurry is a typical solid–liquid two-phase flow, and the transportation pipeline is generally dominated by turbulent flow. In this paper, the discrete phase model is mainly used to study the changes in the flow field inside the variable diameter pipe and the erosion and abrasion of the pulverized coal slag solid particles on the pipe wall surface. And the turbulence model, multiphase flow model, wall rebound model and erosion model are included in the study.

2.1 Boundary Conditions

In this paper, the reduced diameter pipe conveyed by the coal slurry storage tank to the external components is selected as the geometric model. The model is modeled according to the actual structural parameters of the factory. The inlet straight pipe diameter is D = 250 mm, and the outlet straight pipe diameter d = 150 mm. Considering the prevention of backflow, the geometrical distance between the inlet and outlet of the reducing pipe is suddenly changed to 5 times the pipe diameter, and the taper angle of the reducing pipe is 48°.

Characteristics of coal water slurry

According to the actual production and operation data, the transportation temperature of the system is 48 °C. The liquid phase in coal slurry is mainly water, the density is

Table 6 Summary of explicit analysis

Conceptual designs			
Rubber			
Max deformation over-time (mm)	12.964	5.797	13.024
Max Y-axis deformation (mm)	5.083×10^{-18}	2.361×10^{-18}	5.707×10^{-18}
Polyurethane			
Max Deformation over-time (mm)	1.068×10^{-9}	4.620×10^{-9}	1.399×10^{-9}
Max Y-axis deformation (mm)	1.202×10^{-17}	1.108×10^{-18}	4.556 × 10–18
Carbon fiber-epoxy resin			
Max deformation over-time (mm)	6.677×10^{-10}	2.887×10^{-10}	8.746×10^{-10}
Max Y-axis deformation (mm)	1.855×10^{-18}	1.225×10^{-18}	5.540×10^{-18}

Table 7 Summary of static deflection

Material	Stiffness (MPa)	Static deflection (mm)
Rubber	0.6–0.8/4	62.5
Polyurethane	3.365–1944	0.12
Carbon fiber Epoxy resin	Avg 6000	0.0416

Table 8 Mass of mounts

Conceptual design	Mass (kg)	Mass (g)
Rubber		
1	0.1236	123.6
2	0.1655	165.5
3	0.1188	118.8
Polyurethane		
1	0.0129	12.9
2	0.0172	17.2
3	0.0124	12.4
Carbon Fiber-Epoxy Resin		
1	0.0206	20.6
2	0.0276	27.6
3	0.0198	19.8

3.5 Weight of Mounts

Accumulating volume from SolidWorks and obeying Eq. (2), Table 8 shows the mass of mounts with each design and material.

3.6 Equations

Static deflection is calculated to identify deflection under load. Where, Fg, w, Weight of Engine and K, Stiffness of material. The static deflection, δ can be expressed as;

$$\delta = \left| \frac{F_g}{k} \right| \tag{1}$$

The mass of the mounts with each material can be calculated by using a mass formula. The mass formula can be express as;

$$\text{Mass, } M = \rho \text{ (density)} \cdot v \text{ (volume)} \tag{2}$$

3.7 Discussions

Across the gathered data for static structural and comparing in total deformation alone, both conceptual designs 1 and 2 outperformed design 3. Similar results can be seen across the different implementations of materials. Hence, mounts with CF-EP outperformed PU and followed by Rubber are similar throughout. The random vibration results across 3 different designs are comparable. Relatively, the type of material plays a huge role, design 2 is stiffest where design 1 and 3 shows similar behavior. Considering the materials, PU can dampen and absorb more vibration, followed by CF-EP then Rubber. In explicit analysis, design 2 has the best results. design 1 and design 3 has a comparable outcome, where design 1 deforms lesser as force and velocity is applied. However, the mount with PU deforms lesser due to the core elasticity limits. Hence, PU performed better than CF-EP, and rubber. But CF-EP is strongest compared among them and mounts that were made of rubber have the lowest weight to strength ratio among them. Even though, PU is lighter than CF-EP due to its individual density. They are still comparable as the margin was small.

Table 9 Final comparison of materials through PUGH method

Element	Rubber	Polyurethane	Carbon fiber epoxy resin
Structural analysis	1	2	3
Random vibration	1	3	2
Explicit analysis	1	3	2
Static deflection	1	2	3
Masses	1	3	2
Cost	3	2	1
Time required	1	2	3
Processing temperature	1	2	3
Total	10	19	19

4 Conclusion and Recommendation

The final verdict is to be determined by the following Pugh method in comparing PU and CF-EP head-to-head accompanied by rubber. Table 9 shows a final comparison between materials. The scoring of (1–3) is implemented. The Score of 1 being least desirable and 3 for the most ideal.

In conclusion, Conceptual design 2 produces the best results of the three. In terms of materials, PU and CF-EP outperformed rubber in all tests. The final decision for the best possible match-up would be (Design 2 with CF-EP). In all honesty, PU was comparable to the composite of choice, CF-EP. Aside from the fact that both can be done at home with ready-made products, CF-EP stands out as a composite material and offers more benefits. The reason for this is that the basic construction has the potential to outperform PU in theory, and there are more studies where its properties can be enhanced with additives, ratio configurations, different carbon fibre weaves, and mixture layer thickness. This composite could be an alternative to PU in the automotive industry because it is much lighter, stronger, more resistant to elements, and lasts longer. As a result, CF-EP merits additional research that focuses on motor sports and automobile development.

References

1. Azammi AN, Sapuan SM, Ishak MR, Sultan MT (2018) Conceptual design of automo- bile engine rubber mounting composite using TRIZ-Morphological chart-analytic network process technique. Defence Technol 14(4):268–277

2. Sarkar C, Hirani H, Sasane A (2015) Magnetorheological smart automotive engine mount. Int J Curr Eng Technol 5(1):419–428
3. Adhau A, Kumar PV (2013) Engine mounts, and its design considerations. Int J Eng Res Technol (IJERT) 2(11):763–769
4. Lu M (2010)CAE in automotive engine mount development. In: ASME International Mechanical Engineering Congress and Exposition, vol 44274, pp 95–99
5. Solomon AR (1997) Engine mount system focusing to reduce vehicle shake (No. 973273). SAE Technical Paper
6. Morato Gomes MJ, Pouzada AS (2002) Dynamic behaviour of rubber compounds for engine mounts. In: Key engineering materials, vol 230. Trans Tech Publications Ltd., pp 303–306
7. Battula SK, Raju PRM, Ratnam CH (2018) Engine isolation angular mount with different types elastomers. Mater Today Proc 5(9):19555–19564
8. Kim JJ, Kim HY (1997) Shape design of an engine mount by a method of parameter optimization. Comput Struct 65(5):725–731
9. Schaefer RJ (2010) Mechanical properties of rubber. In: Piersol A, Paez T (eds) Harris' shock and vibration handbook, 6th edn. McGraw-Hill Companies Inc., pp.1–33
10. Kato R, Oka K, Takayama M (2003) The tensile tests of natural rubber bearings focused on the effect of the steel flange plates. In: ASME pressure vessels and piping conference, vol 4157, pp 81–88
11. South JT (2001) Mechanical properties and durability of natural rubber compounds and composites. Doctoral dissertation, Virginia Tech
12. Gabrielle-Desantis (2020) Should you upgrade your motor mounts? [online] MotorBiscuit.https://www.motorbiscuit.com/should-you-upgrade-your-motor-mounts. Last accessed 24 Mar 2021
13. Pleiger Plastics Company.Polyurethane Material Properties|Pleiger Plastics Company. https://pleiger.com/polyurethane-material-properties/. Last accessed 24 Mar 2021
14. Karpiesiuk, J (2020) Young's modulus and poisson's ratio of polyurethane adhesive in lightweight floor system. Mod Approach Mater Sci 2(3)
15. Patel PS, Shepherd DE, Hukins DW (2008) Compressive properties of commercially available polyurethane foams as mechanical models for osteoporotic human cancellous bone. BMC Musculoskeletal Disorders 9(1)
16. Basf.com. https://plasticsrubber.basf.com/global/en/performance_polymers/locations/great-britain/insulation.html#:~:text=Its%20material%20density%20usu-ally%20range,m3%20to%20100%20kg%2Fm3. Last accessed 24 Mar 2021
17. Sprenger S, Kothmann MH, Altstaedt V (2014) Carbon fiber-reinforced composites using an epoxy resin matrix modified with reactive liquid rubber and silica nanoparticles. Composites Sci Technol 105:86–95
18. Yu Z, Li R, Peng Z, Tang Y (2017) Carbon fiber reinforced epoxy resin matrix composites. Mat Sci Adv Compos Mater
19. Carbon/Epoxy Composite Carbon fibre -Epoxy Resin Matrix. https://docs-emea.rs-online.com/webdocs/1153/0900766b81153da2.pdf. Last accessed 24 Mar 2021
20. Performance-composites.com (2021) Mechanical properties of carbon fibre composite materials. http://www.performancecomposites.com/carbonfibre/mechanicalproperties_2.asp. Last accessed 14 June 2021
21. Matweb.com (2021) Overview of materials for epoxy/carbon fiber composite. http://www.matweb.com/search/datasheet.aspx?mat-guid=39e40851fc164b6c9bda29d798bf3726&ckck=1. Last accessed 24 Mar 2021
22. Wang C, Ji X, Roy A, Silberschmidt VV, Chen Z (2015) Shear strength and fracture toughness of carbon fibre/epoxy interface: effect of surface treatment. Mater Des 85:800–807
23. Joven R, Das R, Ahmed A, Roozbehjavan P, Minaie B (2012) Thermal properties of car- bon fiber-epoxy composites with different fabric weaves. In: Conference: SAMPE international symposium proceedings, Charleston, SC
24. Mansour G, Tsongas K, Tzetzis D, Tzikas K (2017) Dynamic mechanical characterization of polyurethane/multiwalled carbon nanotube composite thermoplastic elastomers. Polymer-Plastics Technol Eng 56(14):1505–1515

25. Prażnowski K, Mamala J, Śmieja M, Kupina M (2020) Assessment of the road surface condition with longitudinal acceleration signal of the car body. Sensors 20(21):5987
26. Ooi LE, Ripin ZM (2011) Dynamic stiffness and loss factor measurement of engine rubber mount by impact test. Mater Des 32(4):1880–1887
27. Alonso J (2016) Breaking down bushing stiffness. Edgeautosport.com. https://blog.edgeautosport.com/breaking-down-bushing-stiffness
28. Capela C, Oliveira SE, Ferreira JAM (2018) Fatigue behavior of short carbon fiber reinforced epoxy composites. Composites Part B Eng S1359836818319796
29. Davis DC, Wilkerson JW, Zhu J, Ayewah DOO (2010) Improvements in mechanical properties of a carbon fiber epoxy composite using nanotube science and technology. Composite Struct 92(11):2653–2662

Experimental Investigation of a Diesel Engine Using Waste Plastic Oil Blends

A. M. Norkhizan, A. F. Yusop, and M. A. Hamidi

Abstract Plastic have now become essential materials in the present world and implimentation in the industrial field is persistently rising. The aim of this project is to investigate the effects of waste plastic oil on engine performance and pollutant emissions. The engine test was conducted under constant engine speed of 1800 RPM and varies engine load of 20%, 40% and 60% respectively. The performances of engine were analyzed in term of brake power (BP), brake specific fuel consumption (BSFC) and brake thermal efficiency (BTE). The effect of WPO on brake power was only dominant during low and medium engine load condition. WPO can reduce the BSFC of diesel engine except for high load condition. In addition, the formation of NO_x and CO for WPO fuel blends were greater than D100 under all engine load conditions. Therefore, in conclusion, the implementation of WPO in diesel engine can improve engine performance but limit its effectiveness in terms of pollutant emissions.

Keywords Waste plastic oil · Diesel engine · Performance · Emission

1 Introduction

Increasing world population cause high plastics demand especially for household applications. It is about 299 million tons of plastic produced in year 2013 that is 4% higher than year 2012. In the meantime, it causes the growth of plastic waste. In year 2013, 33 million tons of plastic waste was reported in US. In Europe, 38% of the plastic waste still finally ended up to the landfill which taking a large space, 26% were recycled and 36% were used for energy recovery. Billions of years may require for the plastic to be degrade on its own. Therefore, the disposal of plastic would cause serious environmental issue. The demand for plastic will be quadruple by year 2050 that expected to be around 1400 million tonnes [1]. These waste plastics can be

A. M. Norkhizan (✉) · A. F. Yusop · M. A. Hamidi
Advanced Fluid Focus Group (AFFG), Faculty of Mechanical and Automotive Engineering Technology, Universiti Malaysia Pahang, 26600 Pekan, Pahang, Malaysia
e-mail: muhamadnorkhizan@gmail.com

M. Y. Ismail et al. (eds.), *Technological Advancement in Mechanical and Automotive Engineering*, Lecture Notes in Mechanical Engineering,
https://doi.org/10.1007/978-981-19-1457-7_10

managed by various method such as reducing, reusing, recycling, waste-to-energy and disposal at landfill sites. Recycling is one other the potential solution to solve this plastic waste disposal issue. Other than that, the plastic will be contaminated with soil, dirt, food wastes, aluminium foils and paper labels. In addition, the limitation of recycling plastic is costly and complicated process due to the water pollution and insufficient separation that require massive labor. Although this method can decrease plastic waste volume, more effective method has been introduced. Traditional mechanical recycling methods only limited to certain type of plastics waste. Today, the deterioration of fossil fuel become a serious issue. Therefore, the use of waste plastic as fossil fuel alternative has been introduced. This method also solves excess waste plastics in landfill. The process of converting waste plastic to oil is called pyrolysis process [2–6]. Lots of studies have been done on waste plastic oil which implemented in diesel engines [3, 4, 7–9]. The aim of this study is to explore the effectiveness of waste plastic oil to be used in diesel engine. The performance and emission characteristics of diesel engine were compared between conventional diesel fuel and waste plastic oil blended fuel. Experimental test was carried out on single cylinder diesel engine under constant speed of 1800 RPM and varies engine load conditions (low load, medium load and high load). In addition, three different fuel blends were tested. Table 2 shows the properties of fuel blends that involved in this study.

2 Methodology

Pure diesel was supplied by Rahar Jati Sdn Bhd, Malaysia. WPO was supplied by Mensilin Green Energy Sdn Bhd, Malaysia. The bifuel WPO-diesel were mixed together at several proportion as shown in Table 1 using mechanical stirrer for 10 min with blade speed of 1000 RPM and followed by homogenizer at 2000 RPM.

Table 1 Fuel blend proportion and testing parameters

No.	Blend	Diesel (Vol%)	WPO (Vol%)	Engine speed (RPM)	Load (%)
1	D100	100	0	1800	20 40 60
2	WPO5	95	5	1800	20 40 60
3	WPO10	90	10	1800	20 40 60

Table 2 Properties of diesel, WPO, WPO5 and WPO10

Properties	Diesel	WPO	WPO5	WPO10	Standard
Density, kg/m^3	827.4	788.64	827.95	826.04	ASTM D445
Kinematic viscosity, cSt @ 40 °C	2.87	1.85	2.91	2.83	ASTM D445
Heating value, kJ/kg	42.9	46.49	45.70	45.61	ASTM D240
Cetane number	62	56	–	–	ASTM D613

2.1 Engine Setup

A direct injection diesel engine model Yanmar TF120M is used to perform the experiment. The details specification on the engine was tabulated in Table 3. Eddy current dynamometer which rated with 15 kW was coupled with crank mechanism of the engine. KANE gas analyzer was used to measure the concentration of gas emissions from the exhaust. Table 4 shows the sensitivity and accuracy of KANE gas analyzer. All fuel blends were tested in the engine under various engine load with constant speed as shown in Table 1. Figure 1 shows the schematic diagram of the experimental setup.

Table 3 Engine specifications

Model	Yanmar model TF120M
Type	Horizontal single-cylinder 4-stroke diesel engine
Bore (mm)	92
Stroke (mm)	96
Displace volume (cm^3)	638
Compression ratio	17.7:1
Continuous rating output	10.5 HP @ 2400 rpm
Maximum rating output	12.0 HP @ 2400 rpm
Fuel injection type	Direct Injection
Injection timing	17° BTDC
Max power	7.7 kW @ 2400 rpm
Max torque	161 Nm @ 4500 rpm
Cooling system	Water-radiation
Fuel tank capacity	11L

Table 4 Sensitivity and measurement accuracy of emission sensor

Equipment	Resolution	Accuracy	Range
KANE gas analyser			
Carbon monoxide (CO)	0.01%	±5%	0–10%
Oxygen O_2 (fuel cell)	0.01%	±5%	0–25%
Hydrocarbon (HC)	1 ppm	±5%	0–3000 ppm
Carbon dioxide (CO_2)	1 ppm	±5%	0–16%
Nitric oxide	1 ppm	0–4000 ppm ±4% or 25 ppm	0–5000 ppm

Fig. 1 Engine setup

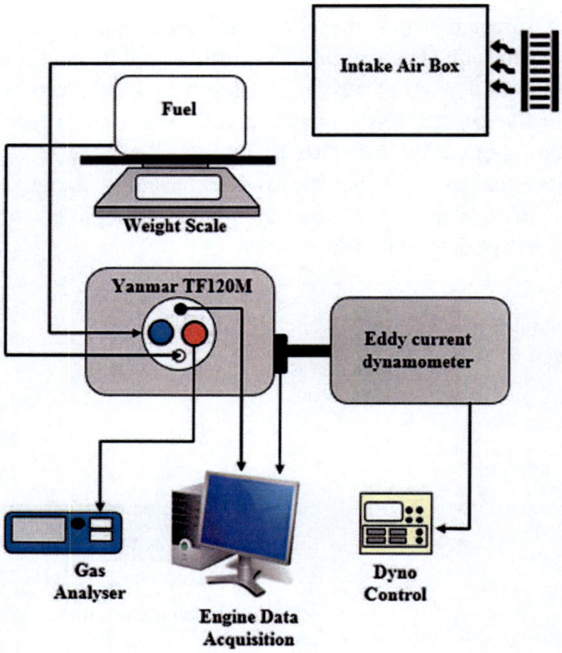

3 Result and Discussion

3.1 Performances of WPO Blends

The performances of diesel engine was examined based on the brake power (BP), brake torque (BT), brake specific fuel consumption (BSFC) and brake thermal efficiency (BTE) on both WPO and conventional diesel fuel (D100). The testing parameters were plotted as a function of engine load in Figs. 2, 3 and 4.

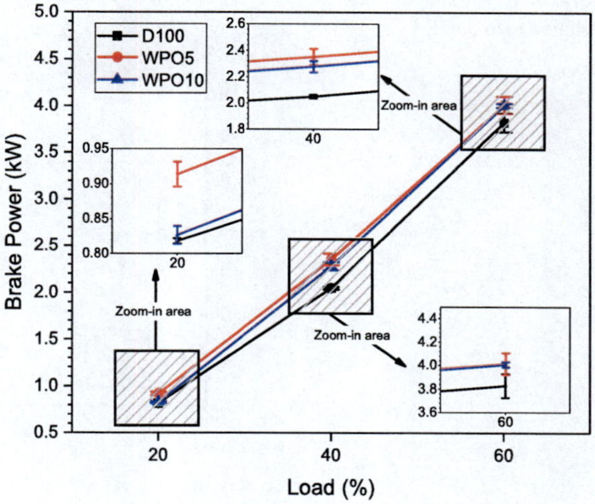

Fig. 2 Brake power under various engine loads condition

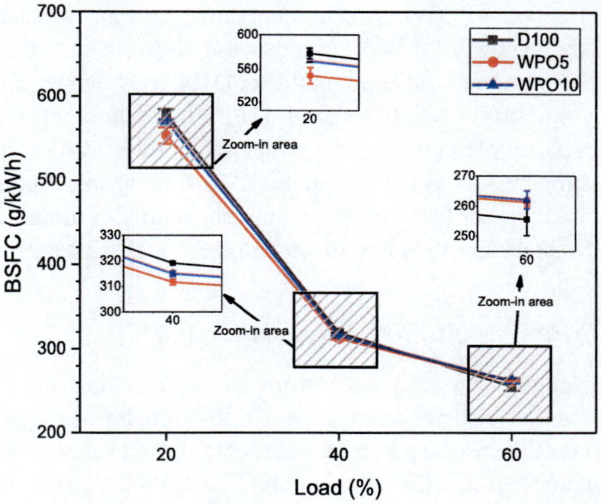

Fig. 3 BSFC under various engine loads condition

Brake Power (BP)

Figure 2 indicates the BP of diesel engine with WPO blends under constant engine speed of 1800 rpm and varies engine load of 7, 14, 21, 28 Nm. It is observed that the BP significantly increase at all engine load conditions. During low load, WPO5 shows the highest BP compared with other blends. The improvement is about 11.51% higher than D100. While WPO10 is closely equivalent with D100. Secondly during medium load, the same outcome is obtained. However, the BP for WPO10 and WPO5 shows no differences to each other. The BP is improved as the WPO blended in the fuels especially WPO5 with 14.35% more than D100. Thirdly during high load condition, WPO5 still gives the highest BP compared to WPO10 and D100. However,

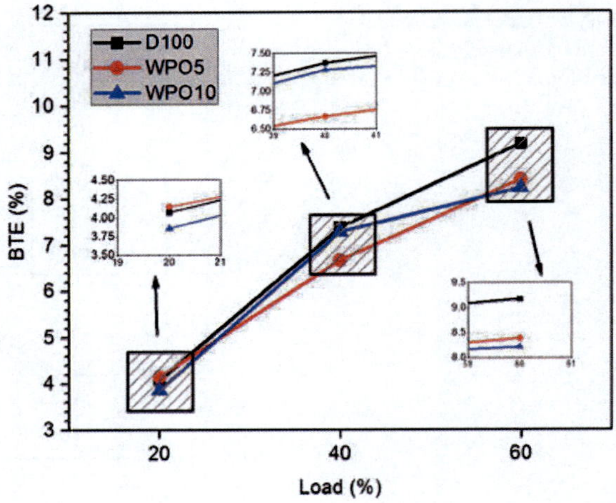

Fig. 4 BTE under various engine load conditions

the differences between WPO5, WPO10 with D100 is below 5%. Therefore, the effect of WPO is only dominant during low and medium load conditions. The higher engine power for WPO blends was believed due to the higher heating value of WPO compared to conventional diesel D100 as shown in Table 2 [10, 11].

Dissimilar with Das et al. [12]. The authors reported that the existence of WPO in diesel blend did decrease the brake power of diesel engine especially for WPO10 compared to WPO20 and WPO30. Increasing amount of WPO did increase the brake power but it is lesser than conventional diesel. Brake power also depends on the thermal efficiency of the engine which closely related to the calorific value of fuel.

Brake Specific Fuel Consumption (BSFC)

In engine tests, the fuel consumption can be measured from the flow rate of fuels per unit time by using a flow meter. In addition, brake specific fuel consumption (BSFC) is one of the useful parameters and often used to analyse the performance of an engine at which lesser BSFC is favourable. Figure 3 shows the brake specific fuel consumption of conventional diesel D100 and waste plastic oil blends, WPO5 and WPO10. In general, the BSFC is reducing as the engine load rise. All engine load conditions give different trend of outcome. During low load, the BSFC for WPO5 was the lowest among other blends. The BSFC for WPO5 was 4.47% lower than conventional diesel D100. Meanwhile, the BSFC for WPO10 and D100 were comparable. In medium load condition, the same outcome was found with percentage reduction of 2.43% lower than conventional diesel D100. However, the BSFC for WPO10 was lower than D100 with percentage reduction of 1.32%. In contrast, different result obtained during high load condition. The BSFC for WPO5 and WPO10 was higher than D100. Which is not preferable. The BSFC was increase up to 2.61% for WPO10 and 2.21% for WPO5. Therefore, the WPO can affect the BSFC of diesel engine except for high load condition. Dissimilar result obtained by Devaraj et al. [13] that waste plastic oil based did increase the BSFC during low load and medium load. This could be happened due to large increase in brake power against load as compared to fuel consumption. The authors conclude that WPO blends provide higher BSFC compared to conventional diesel during all load conditions due to its lower heating value and higher density [13].

Brake Thermal Efficiency (BTE)

Thermal efficiency is often quoted as either brake thermal efficiency (BTE). Thermal efficiency can be defined as the ratio of work produced per cycle to the amount of fuel energy supplied per cycle which is released in the combustion process. BTE is measured by dividing brake power with the amount of fuel energy supplied per cycle. The fuel energy supplied, which can be released by combustion is given by the mass of fuel supplied to the engine per cycle times the heating value of the fuel [14]. Figure 4 shows the brake thermal efficiency of conventional diesel and waste plastic oil blends under various engine load conditions. During low load, the BTE for WPO5 is the highest compared to other blends with 2% greater than conventional diesel D100. Contradict with low load conditions, the BTE for WPO5 is the lowest among other blends. The percentage reduction of BTE for WPO5 was 9.5% lesser than conventional diesel D100. Lastly, the conventional diesel D100 gives the highest BTE when the engine run under high load condition. While WPO10 was the lowest with 10% lesser than conventional diesel D100. The decrement of BTE for WPO blends is due to the lower calorific value of WPO compared to conventional diesel D100 [15].

In contrast with Das et al. [12], the authors reported that presence of WPO in fuel blend did improve the brake thermal efficiency under medium and high loads conditions. It was reported that the presence of WPO does increase the viscosity and lessen the density that occurred to reduce the BTE [16].

3.2 Emission Characteristics of WPO Blends

Nitrogen Oxides (NO_x)

Nitrogen oxide (NO_x) is the most undesired gases from the exhaust where there are regulations that restrict the allowable amount of NO_x. The NO_x formation depends on in-cylinder temperature, in-cylinder pressure, air-fuel ratio, combustion duration and chemical reactions [17]. When combustion temperature reaches more than 1370 °C, the N and O_2 in the air combine to form these oxides of nitrogen. Photochemical smog is the main product of NO_x formation when it is released to the atmosphere. The smog is form from the reaction between exhaust gases (NO_x) with sunlight energy. Higher than normal NO_x emissions may be caused by overheated engine, lean air-fuel mixture, over-advanced spark timing and too high compression [18]. Figure 6 shows the formation of NO_x in diesel engine under different engine load conditions. Under low engine load conditions, the addition of waste plastic oil in diesel blends did increase the NO_x formation. The NO_x formation for WPO10 is 16.67% higher than conventional diesel D100. While the NO_x formation for WPO5 is 8.33% higher than D100. During medium engine load condition, the differences in NO_x formation was clearly seen. As observed in Fig. 5, the NO_x formation for WPO blends was more than convention diesel D100. The NO_x formation for WPO10 and WPO5 was 31.56% and 10.58% higher than conventional diesel D100. In addition, the NO_x formation under high engine load condition higher than medium engine load conditions which is due to higher engine temperatures. Similar to previous engine load conditions, the NO_x formation for WPO10 was the highest among other blends. The NO_x formation for WPO10 and WPO5 were 20.27% and 4.49% higher than conventional diesel D100, respectively.

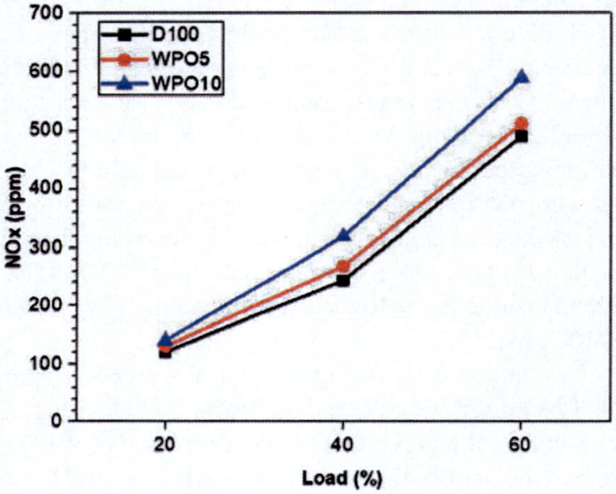

Fig. 5 NO_x formation under various engine load conditions

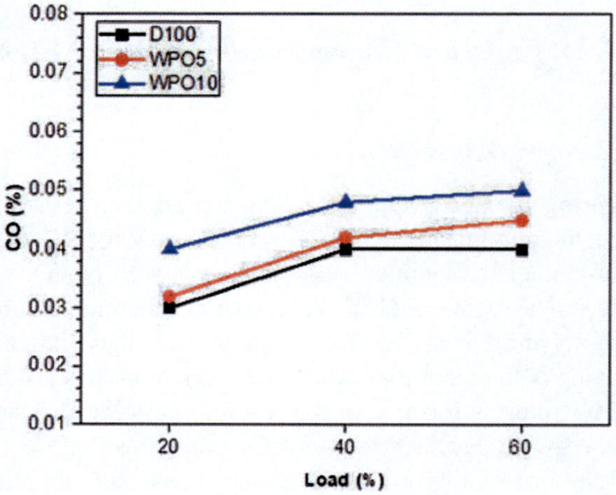

Fig. 6 CO formation under various engine load conditions

Carbon Monoxide (CO) and Carbon Dioxide

The typical content of CO in exhaust gases is about 0. 2% to 50%. The lost chemical energy that was not fully utilized in the engine can also cause CO formation [19]. In conclusion, CO is formed when there are poor fuel-air mixing and incomplete combustion. In summary, excessive CO emissions may be caused by excessively rich air-fuel ratio equipment failure such as faulty injector, dirty air filter and engine oil diluted with the working fuel [18]. As observed in Fig. 6, the CO formation for WPO blended fuels were greater than convention diesel D100. During low engine load condition, the CO formation for WPO10 and WPO5 were 33.33% and 6.67% higher than conventional diesel D100. Meanwhile during medium engine load condition, the percentage of increases were significantly reduced to 20% and 5%, respectively. In addition, under low engine load condition, the percentage increase for WPO5 was more than previous engine load conditions about 12.50% higher than D100.

CO_2 is not considered as a pollutant emission. However, it has close relation with greenhouse effect. Formation of CO_2 is used as the indicator of ideal combustion. Therefore, large formation of CO_2 is desirable. Irregular formation of CO_2 can be occurred by misfires, incomplete combustion, engine mechanical component issues, leaking in exhaust line and improper air-fuel ratio [20]. Figure 7 shows the CO_2 formation for the engine under three different engine load conditions. During low engine load condition, the CO_2 formation for WPO5 and WPO10 were lower than conventional diesel D100. CO_2 formation for WPO10 was the lowest with 16.67% lower than D100. Meanwhile, WPO5 produce lower CO_2 emission with 8.33% lesser than D100. In addition, the same outcome was observed during medium and high engine load conditions.

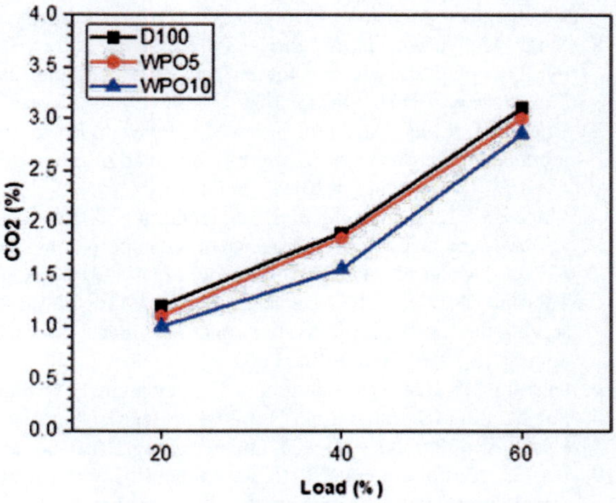

Fig. 7 CO_2 formation under various engine load conditions

4 Conclusion

The performances and emission characteristics of single cylinder diesel engine using diesel blended with waste plastic oil under constant engine speed and varies engine loads were investigated. Addition of WPO in fuel blend did improve the performance of diesel engine especially for brake power and brake specific fuel consumption. Among all three fuel blends, WPO5 shows the highest brake power in all engine load conditions. Brake specific fuel consumption for WPO5 was the lowest compared to other fuel blends especially during low and medium engine load conditions. In addition, the formation of NO_x and CO for WPO fuel blends were higher than D100 under all engine load conditions. Therefore, in conclusion, the implementation of WPO in diesel engine can improve engine performance but limit its effectiveness in terms of pollutant emissions. Waste plastic oil suitable to be implemented in existing diesel engine. However, further tests are required to evaluate the reliability and durability of those engines. WPO has proved to be a possible alternative to fossil fuel in diesel engine.

Acknowledgements The authors would like to thank the Ministry of Higher Education for providing financial support under Research Acculturation of Early Career Researchers No. RACER/1/2019/TK03/UMP/1 and University Malaysia Pahang for laboratory facilities as well as additional financial support under Internal Research grant RDU192605.

References

1. Oasmaa A, Qureshi MS, Pihkola H, et al (2020) Pyrolysis of Plastic Waste: Opportunities and Challenges. J Anal Appl Pyrolysis 104804. https://doi.org/10.1016/J.JAAP.2020.104804
2. Singh TS, Verma TN, Singh HN (2020) A lab scale waste to energy conversion study for pyrolysis of plastic with and without catalyst: Engine emissions testing study. Fuel 277:118176. https://doi.org/10.1016/j.fuel.2020.118176
3. Quesada L, Calero M, Martín-Lara MA, Pérez A, Blázquez G (2019) Characterization of fuel produced by pyrolysis of plastic film obtained of municipal solid waste. Energy 186. https://doi.org/10.1016/j.energy.2019.115874
4. Mangesh VL, Padmanabhan S, Tamizhdurai P, Narayanan S, Ramesh A (2020) Combustion and emission analysis of hydrogenated waste polypropylene pyrolysis oil blended with diesel. J Hazard Mater 386:121453. https://doi.org/10.1016/j.jhazmat.2019.121453
5. Tulashie SK, Boadu EK, Dapaah S (2019) Plastic waste to fuel via pyrolysis: A key way to solving the severe plastic waste problem in Ghana. Therm Sci Eng Prog 11:417– 424. https://doi.org/10.1016/j.tsep.2019.05.002
6. Qureshi MS, Oasmaa A, Pihkola H, Deviatkin I, Tenhunen A, Mannila J, Minkkinen H, Pohjakallio M, Laine-Ylijoki J (2020) Pyrolysis of plastic waste: Opportunities and challenges. J Anal Appl Pyrolysis 104804. https://doi.org/10.1016/j.jaap.2020.104804
7. Ravi S, Karthikeyan A (2020) Assessment of performance and emission characteristics of diesel engine supplied with waste plastic oil propanol and ethylhexyl nitrate blends. Mater Today Proc. https://doi.org/10.1016/j.matpr.2020.10.206
8. Saleem J, Dotto GL, McKay G (2021) Current scenario and challenges in using plastic wastes as oil absorbents. J Environ Chem Eng 9:104822. https://doi.org/10.1016/j.jece.2020.104822

9. Das AK, Hansdah D, Mohapatra AK, Panda AK (2020) Energy, exergy and emission analysis on a DI single cylinder diesel engine using pyrolytic waste plastic oil diesel blend. J Energy Inst 93:1624–1633. https://doi.org/10.1016/j.joei.2020.01.024
10. Hosseini SH, Taghizadeh-Alisaraei A, Ghobadian B, Abbaszadeh-Mayvan A (2017) Performance and emission characteristics of a CI engine fuelled with carbon nanotubes and diesel-biodiesel blends. Renew Energy 111:201–213. https://doi.org/10.1016/j.renene.2017.04.013
11. Mohammadi P, Nikbakht AM, Tabatabaei M, Farhadi K, Mohebbi A, Khatami M (2012) Experimental investigation of performance and emission characteristics of DI diesel engine fueled with polymer waste dissolved in biodiesel-blended diesel fuel. Energy 46:596–605. https://doi.org/10.1016/j.energy.2012.07.049
12. Das AK, Hansdah D, Mohapatra AK, Panda AK (2020) Energy, exergy and emission analysis on a DI single cylinder diesel engine using pyrolytic waste plastic oil diesel blend. J Energy Inst 1–10. https://doi.org/10.1016/j.joei.2020.01.024
13. Devaraj J, Robinson Y, Ganapathi P (2015) Experimental investigation of performance, emission and combustion characteristics of waste plastic pyrolysis oil blended with diethyl ether used as fuel for diesel engine. Energy 85:304–309. https://doi.org/10.1016/j.energy.2015.03.075
14. Atkins RD (2009) An Introduction to Engine Testing and Development. SAE International
15. Kumar S, Prakash R, Murugan S, Singh RK (2013) Performance and emission analysis of blends of waste plastic oil obtained by catalytic pyrolysis of waste HDPE with diesel in a CI engine. Energy Convers Manag 74:323–331. https://doi.org/10.1016/j.enconman.2013.05.028
16. Mani M, Nagarajan G, Sampath S (2011)) Characterisation and effect of using waste plastic oil and diesel fuel blends in compression ignition engine. Energy 36:212–219. https://doi.org/10.1016/j.energy.2010.10.049
17. Heywood JB (1988) Internal Combustion Engine Fundamentals
18. Pickerill K (2009) Today's technician: automotive engine performance classroom manual and shop manual. Cengage Learn.
19. Ozsezen AN, Canakci M (2011) Performance and combustion characteristics of alcohol-gasoline blends at wide-open throttle. Energy 36:2747–2752. https://doi.org/10.1016/j.energy.2011.02.014
20. Awad OI, Ali OM, Mamat R, et al (2017) Using fusel oil as a blend in gasoline to improve SI engine efficiencies: A comprehensive review. Renew Sustain Energy Rev 69:1232–1242. https://doi.org/10.1016/j.rser.2016.11.244

Biocrude Potential Assessment of Macroalgae for Sustainable Biofuel Production

Nida Khan, K. Sudhakar, and R. Mamat

Abstract Biofuel are commonly regarded as sustainable renewable fuels and are offers a feasible solution for social and economic development. Within the biofuel source, macroalgae are quickly becoming a typical contender as a possible third-generation fuel source due to ease of cultivation and fast growth rates. It is the most excellent bioenergy alternative that overcomes the downsides of first- and second-generation biofuels. Macroalgae based biocrude oil through Hydrothermal liquefaction (HTL) is a promising pathway for sustainable biofuel production. This study aims to compare the biochemical composition of 15 different species collected from various literature sources. The biocrude potential from different species are estimated and compared. The results suggested that green seaweed *U. luctuca* (16.81%) contain the highest percentage of biocrude content. This indicates the macroalgae as a promising feedstock for biofuels, bio-chemicals and other bio-products.

Keywords Seaweed · Biocrude yield · Biomass · Biofuels

N. Khan · R. Mamat
Automotive Engineering Centre, Universiti Malaysia Pahang, 26600 Pekan, Pahang, Malaysia

K. Sudhakar (✉)
Faculty of Mechanical and Automotive Engineering Technology, Universiti Malaysia Pahang, 26600 Pahan, Pekang, Malaysia
e-mail: sudhakar@ump.edu.my

Centre of Excellence for Advanced Research in Fluid Flow (CARIFF), Universiti Malaysia Pahang, 26300 Gambang, Pahang, Malaysia

Energy Centre, Maulana Azad National Institute of Technology, Bhopal 462003, India

R. Mamat
School of Mechanical Engineering, Ningxia University, Yinchuan, China

M. Y. Ismail et al. (eds.), *Technological Advancement in Mechanical and Automotive Engineering*, Lecture Notes in Mechanical Engineering,
https://doi.org/10.1007/978-981-19-1457-7_11

1 Introduction

1.1 Background

The use of non-renewable fuels as a primary energy source for several years results in increasing the effect of global warming and the air pollutants that are detrimental to the atmosphere and public health. The combined concerns of climate change and energy scarcity, resulting from modern industry's uncontrolled energy demands, have prompted the development of bioenergy feedstock [1–3]. Biofuel production is widely recognized as a viable means of social and economic growth and is one of the most sustainable renewable fuels. However, not all biomass can be classified as a sustainable resource. The cultivation, harvesting, treatment, processing, and transportation of biomass may have greater environmental consequences than the benefits based on the selected pathways. Biofuels of the first-generation source from food crops or conventional feedstock; the biggest drawback of first-generation biofuel is still a significant concern; it is primarily produced from agricultural resources, which has an adverse effect on economy. Non-edible plant biomass, agro-residues, and lignocellulosic feedstocks are used to make second-generation biofuels. It has lower net energy yield, difficulties with feedstock transport, and high production costs restrict their use. The third generation has gained broad interest as an alternative sustainable source; the most promising bio-feedstock for renewable fuel production are macroalgae and microalgae. Macroalgae is one of the least studied forms of bioenergy sources accessible across the world. Interestingly, when grown in the sea, seaweed has a high average CO_2 sequestration rate (36.7 tonne hectare-1 year-1, seven times that of traditional lignocellulosic biomass), a rapid rate of growth (harvested up to six times per year even without fertiliser), and does not compete with land-based food crops [4, 5].

1.2 Classification and Distribution of Seaweeds

Brown seaweed (Phaeophyceae), red seaweed (Rhodophyceae), and green seaweed (Chlorophyceae) are three distinct types of macroalgae. The majority of the seaweed produced are used for human consumption (approximately 47%) and for the manufacturing of hydrocolloids such as agar, carrageenan, and alginate (over 50%) [6]. Many species of marine macroalgae have been harvested in Asian countries since ancient times for their various health benefits due to their high nutritional and low caloric values. Macroalgae are used as a source of techno-functional polysaccharides (for example, the phycocolloids agar, carrageenans, and alginates) in the food, cosmetics, and pharmaceutical industries in Western countries [7, 8] and as fertilizers and feed ingredients for animal nutrition [9, 10]. Malaysia has a lot of prospects for growing seaweed; therefore, it still has a lot of potentials to become the global leader in seaweed production. Sabah state is Malaysia's primary seaweed cultivating zone.

It provides an ideal climate for cultivating high-value seaweed, including a variety of red and green seaweed species [11].

1.3 Different Conversion Process

Various technological methods to convert biomass to various types of energy depends on the raw material properties and the kind of energy requirement. Direct combustion, pyrolysis, gasification, and current commercial biodiesel production all require a dry feedstock and drying before conversion. Hydrothermal liquefaction (HTL), fermentation, and other wet-conversion techniques, and anaerobic digestion have become famous for biomass with high moisture content to eliminate drying energy requirements. Therefore, the commercial usage of seaweeds, mainly, has grown popularity during the last two decades.

1.4 Significance of Biocrude and Hydrothermal Liquefaction Process

(HTL) process of algae for crude bio-oil (biocrude) manufacture has gained lot of interest because of its safe raw material characteristics, high bio-oil output, and lack of a drying phase. Unlike anaerobic digestion, hydrothermal liquefaction (HTL) is capable of producing high-quality, i.e., high energy densities fuel quickly, generally in a matter of minutes to tens of minutes [12, 13], biocrude yield from HTL is greater than that of other techniques (e.g., pyrolysis) because not only the lipids but also the proteins and carbohydrates in algae mainly transformed into biocrude [14, 15]. In HTL, algae nutrients (N, P, and K) convert into water-soluble salts or acids, reusing algae growth via aqueous phase recovery and utilization [16, 17]. Carbon (71–73%), hydrogen (7–8%), oxygen (10–11%), nitrogen (6–7%), and sulphur content (0–1%) are the primary components of biocrude oil, which results in lower greenhouse gas emissions than traditional biofuel and bioethanol [18]. By employing an appropriate catalyst that can remove oxygen, nitrogen, and double bonds, biocrude can improve to production gasoline, Jet fuel, and other fuels [19]. Two-stage HTL and co-liquefaction have recently investigate as a way to increase yields of macroalgae-derived biocrude [20, 21] (Fig. 1).

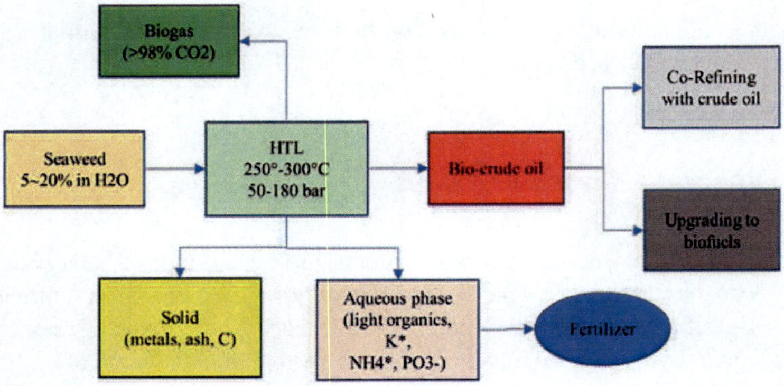

Fig. 1 Process flow of HTL process of seaweed into bio-crude oil

2 Experimental Methodology

2.1 *Study Area and Macroalgae Collection*

The following macroalgae species were chosen for this study: Rhodophyceae (red macroalgae), Heterokontophyceae (brown macroalgae) (brown macroalgae) and Chlorophyceae (green macroalgae) (Table 1). The methodology of the study is presented in Fig 2.

2.2 *Estimation of Biocrude Yield*

The biocrude oil is produced by the conversion of protein, carbohydrates, and lipid using an additive in a complex reaction called hydrothermal liquefaction (HTL). The biocrude oil yield can be calculated by feedstock of biochemical content where:

$$Y_{BIOCRUDE} = (Y_{LIPID} * W_{LIPID}) + (Y_{PROTEIN} * W_{PROTEIN}) + (Y_{CARBOHYDRATE} * W_{CARBOHYDRATE})$$

where $Y_{BIOCRUDE}$, Y_{LIPID}, $Y_{PROTEIN}$, AND $Y_{CARBOHYDRATE}$ are biocrude, lipid, protein, and carbohydrate HTL yields in (wt %) where W_{LIPID}, $W_{PROTEIN}$, and $W_{CARBOHYDRATE}$ are lipid, protein, and carbohydrates contents dry weight (wt %) of macroalgae. The biocrude yields are calculated in the range, which is the upper and lower limits. Biochemical yield conversion factors of 0.80, 0.18, and 0.15 were used as the upper limit for lipids, protein, and carbohydrates, respectively. For the same components, the lower limit conversion factor is 0.55, 0.11, and 0.06. [33]. These

Table 1 Protein, Lipid and Carbohydrate content of the seaweed species

Type of Seaweed	Lipid (%)	Protein (%)	Carbohydrate (%)	Reference
Green Seaweed				
U. lactuca	6.20	20.60	54.30	[22]
Ulva Reticulata	2.30	20.00	58.00	[23]
Codium fragile	1.80	20.90	32.30	[24, 25]
Enteromorpha flexusa	5.60	7.23	39.90	[26]
Caulerpa lentillifera	7.20	9.70	11.80	[26]
Brown Seaweed				
Laminaria japonica	1.50	30.90	59.70	[22]
Hizikia fusiforme	1.76	12.9	61.85	[27, 28]
Saccharina japonica	1.90	8.00	51.90	[23]
Laminaria digitata	1.00	15.00	48.0	[23]
Unduria pinnatifida	1.74	18.16	33.35	[29]
Red Seaweed				
G. chilensis	1.30	13.70	66.10	[24]
Gracilaria Changii	3.30	6.90	41.52	[30]
Gelidium Amansii	0.90	12.20	83.60	[22]
Kappaphycus Alvarezii	1.10	3.40	65.20	[31]
E. spinosum	0.01	6.04	69.66	[32]

conversion factors are based on the yields range of model compounds obtained from HTL performed at 350 °C for 1 h and 10% solids.

3 Results

3.1 Biocrude Content of Red Seaweed

The biocrude content of selected seaweed species analyses and the results are shown below. The level of biocrude was high in red Seaweed *Gelidium Amansii* (15.46%), whereas *E. spinosum* has the lowest biocrude content (4.84%) (Fig. 3).The carbohydrate contents are the highest d 56.5 ± 13.14% in GA, ensuring it be a promising substrate for bioethanol production [34]. Until yet, no study has been done on producing biocrude from Gelidium Amansii using the HTL method (Fig. 3).

3.2 Biocrude Content of Brown Seaweed

Brown algae have a significant carbohydrate component made up of alginic acid, laminarin, mannitol, and fucoidan, with proportions ranging from 10–40 wt%, 2–34 wt%, 5–25 wt%, and 5–20 wt%, depending on the species [35–38]. As shown in Fig. 4, brown seaweed *Laminaria japonica* has the highest biocrude content (15.72%), whereas *Unduria pinnatifida* (4.96%) has the lowest. When the Ni/HZSM-5 catalyst was used, the maximum yield of L. japonica bio-oil (15.32 0.2%) was obtained [39] (Fig. 4).

Fig. 2 Flowchart showing the research methodology

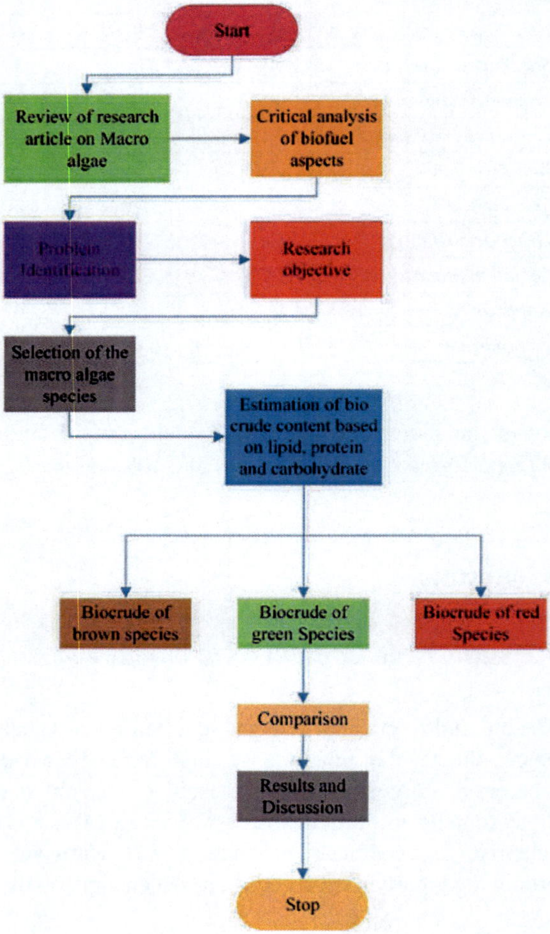

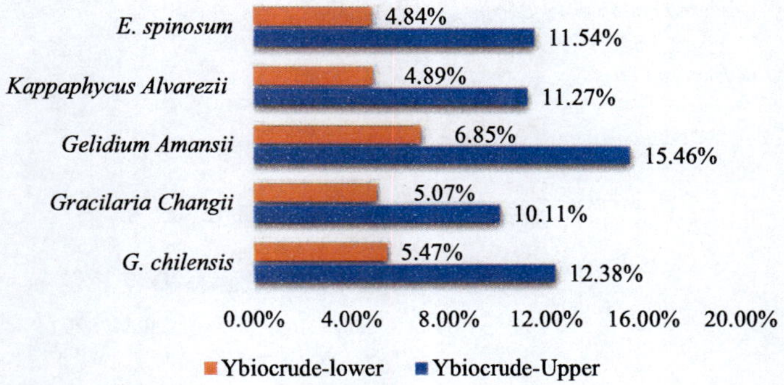

Fig. 3 Percentage of biocrude content of red seaweed

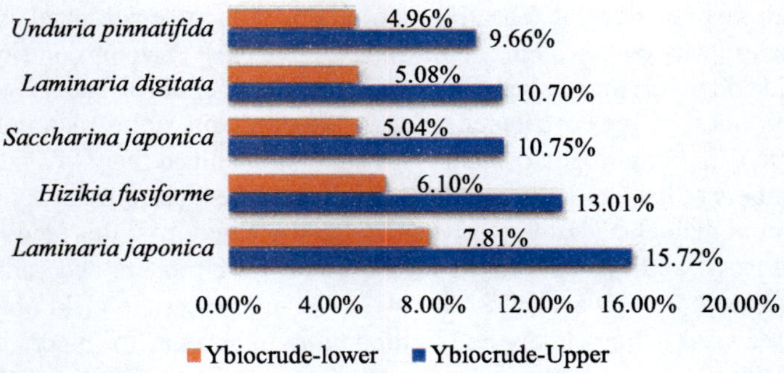

Fig. 4 Percentage of biocrude content of brown seaweed

3.3 *Biocrude Content of Green Seaweed*

The composition of *Ulva lactuca* varies depending on where it grows and when it is harvested. According to [39], overall bio-crude yields were obtained for the genus *Ulva lactuca* (29.9%), respectively. According to our study, *Ulva lactuca* contains the highest percentage of biocrude content (13.09%); (Fig. 5). Regional and seasonal changes in biomass composition are most likely responsible [40–42] in different bio-crude values.

4 Discussion

The above result concludes that the composition of various algae strains has a significant impact on the content of liquefied biocrude and its reaction pathway. [43, 44]. In comparison to microalgae, which have a higher lipid content, macroalgae have a

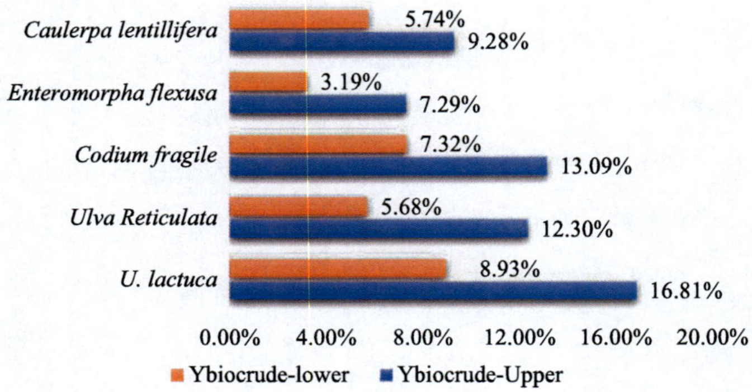

Fig. 5 Percentage of biocrude content of green seaweed

higher carbohydrate content. The ability to produce biocrude is ranked as follows: Lipids take priority over proteins and carbohydrates [45]. Proteins contribute more to bio-crude oil yields than polysaccharides, but at a higher nitrogen cost [46]. It was discovered that proteins contributed more to bio-crude oil yields than polysaccharides, but at a higher nitrogen content. This study determined that *Ulva lactuca* and other members of the Chlorophyceae family have the best performance. In general, the maximum theoretical value of ulva (16.81%) obtained from this study matched Biller and Ross' additive conversion technique of lipid, protein, and carbohydrate components achieved by HTL (18.7%) [47]. The high biocrude yield obtained for different seaweed is most likely due to the failure to account for biocrude created by interactions between biological compounds, in addition to individual additive conversion yields.

5 Conclusion

The use of carbon fuels and hydrocarbon materials in industrial development has a negative effect on the natural. The development of macroalgae for nature-based fuels will solve environmental issues. Although macroalgae biomass is a possible renewable energy source, improvements in biocrude production quality have been studied in response to technical advancements while keeping the cost in mind. In this particular research, biocrude oil production from naturally growing macroalgae species in Malaysia are compared. Overall, observation suggests that green seaweed *Ulva lactuca* might be a viable source for fine chemicals utilized in biorefineries.

5.1 *Significance of the Study*

The study indicates previously undiscovered Malaysian species that could be used in a Macroalgae biorefinery for biofuel production. However, only a few seaweeds are suitable for biorefinery, and the composition will determine the cost of production. Overall, the findings imply that *U. luctuca* seaweeds might develop as a viable source of fine chemicals used in biorefineries. This research findings will assist the growth of macroalgae biofuel industry. The identification of appropriate catalysts, solvents to effectively manage product supply and enhance biocrude characteristics should be the focus of study in the near future.

5.2 *Limitations*

The study is based on limited open-source information on biochemical composition of various marine macroalgae. However, actual data and biochemical analysis of various species is needed to realise the true potential.

5.3 *Future Scope*

One of the most significant outcomes of this research is demonstrating the biocrude oil potential of filamentous macroalgae species. Seaweed's biodiversity is abundant in Malaysia and other tropical nations. Harnessing the macroalgae feedstock potential for export, job creation, and environmental security, will boost the country's self-esteem by demonstrating the country's potential in the International arena. There is a possibility of establishing a distinguished status as a global leader in biocrude oil production on the way to becoming a developed country. In the long term, this would very certainly increase incentive or interest in other nations to follow suit with similar ambitions or goals, assisting in the country's transformation into a green economy.

Acknowledgements The authors are grateful for the financial assistance granted by the Universiti Malaysia Pahang (www.ump.edu.my) under the Doctoral Research Scheme (DRS) PGRS210348.

References

1. Ragauskas AJ, Williams CK, Davison BH, Britovsek G, Cairney J, Eckert CA, et al (2006) The path forward for biofuels and biomaterials. Science (80-)311:484–9. https://doi.org/10.1126/SCIENCE.1114736
2. Khan SA, Rashmi, Hussain MZ, Prasad S, Banerjee UC, Khan SA, et al (2009) Prospects of biodiesel production from microalgae in India. Renew Sustain Energy Rev 13:2361–72

3. Cuellar-Bermudez SP, Garcia-Perez JS, Rittmann BE, Parra-Saldivar R (2015) Photosynthetic bioenergy utilizing CO_2: an approach on flue gases utilization for third generation biofuels. J Clean Prod 98:53–65. https://doi.org/10.1016/J.JCLEPRO.2014.03.034
4. Lee HW, Choi SJ, Park SH, Jeon JK, Jung SC, Joo SH et al (2014) Catalytic conversion of Laminaria japonica over microporous zeolites. Energy 66:2–6. https://doi.org/10.1016/J.ENERGY.2013.05.023
5. Bayramoglu G, Akbulut A, Ozalp VC, Arica MY (2015) Immobilized lipase on micro-porous biosilica for enzymatic transesterification of algal oil. Chem Eng Res Des 95:12–21. https://doi.org/10.1016/J.CHERD.2014.12.011
6. Buschmann AH, Camus C, Infante J, Neori A, Israel Á, Hernández-González MC et al (2017) Seaweed production: overview of the global state of exploitation, farming and emerging research activity. 52:391–406. https://doi.org/10.1080/09670262.2017.1365175
7. Bocanegra A, Bastida S, Benedi J, Rodenas S, Sanchez-Muniz FJ (2009) Characteristics and nutritional and cardiovascular-health properties of seaweeds. J Med Food 12:236–258. https://doi.org/10.1089/JMF.2008.0151
8. Faggio C, Pagano M, Dottore A, Genovese G, Morabito M (2015) Evaluation of anticoagulant activity of two algal polysaccharides. Nat Prod Res 30:1934–1937. https://doi.org/10.1080/14786419.2015.1086347
9. Pereira L (n.d) Seaweed a review of the nutrient composition of selected edible seaweeds
10. Makkar HPS, Tran G, Heuzé V, Giger-Reverdin S, Lessire M, Lebas F et al (2016) Seaweeds for livestock diets: a review. Anim Feed Sci Technol 212:1–17. https://doi.org/10.1016/J.ANIFEEDSCI.2015.09.018
11. Phang SM (2006) Seaweed resources in Malaysia: current status and future prospects. Aquat Ecosyst Heal Manag 9:185–202. https://doi.org/10.1080/14634980600710576
12. Anastasakis K, Ross AB (2011) Hydrothermal liquefaction of the brown macro-alga Laminaria Saccharina: effect of reaction conditions on product distribution and composition. Bioresour Technol 102:4876–4883. https://doi.org/10.1016/J.BIORTECH.2011.01.031
13. Chen H, Zhou D, Luo G, Zhang S, Chen J (2015) Macroalgae for biofuels production: progress and perspectives. Renew Sustain Energy Rev 47:427–437. https://doi.org/10.1016/J.RSER.2015.03.086
14. Minowa T, Yokoyama SY, Kishimoto M, Okakura T (1995) Oil production from algal cells of Dunaliella tertiolecta by direct thermochemical liquefaction. Fuel 74:1735–1738. https://doi.org/10.1016/0016-2361(95)80001-X
15. Guo Y, Yeh T, Song W, Xu D, Wang S (2015) A review of bio-oil production from hydrothermal liquefaction of algae. Renew Sustain Energy Rev 48:776–790. https://doi.org/10.1016/J.RSER.2015.04.049
16. Chen WT, Zhang Y, Zhang J, Yu G, Schideman LC, Zhang P et al (2014) Hydrothermal liquefaction of mixed-culture algal biomass from wastewater treatment system into bio-crude oil. Bioresour Technol 152:130–139. https://doi.org/10.1016/J.BIORTECH.2013.10.111
17. Sawayama S, Minowa T, Yokoyama SY (1999) Possibility of renewable energy production and CO_2 mitigation by thermochemical liquefaction of microalgae. Biomass Bioenerg 17:33–39. https://doi.org/10.1016/S0961-9534(99)00019-7
18. Vardon DR, Sharma BK, Blazina GV, Rajagopalan K, Strathmann TJ (2012) Thermochemical conversion of raw and defatted algal biomass via hydrothermal liquefaction and slow pyrolysis. Bioresour Technol 109:178–187. https://doi.org/10.1016/J.BIORTECH.2012.01.008
19. Kumar K, Ghosh S, Angelidaki I, Holdt SL, Karakashev DB, Morales MA et al (2016) Recent developments on biofuels production from microalgae and macroalgae. Renew Sustain Energy Rev 65:235–249. https://doi.org/10.1016/J.RSER.2016.06.055
20. Wang S, Zhao S, Cheng X, Qian L, Barati B, Gong X et al (2021) Study on two-step hydrothermal liquefaction of macroalgae for improving bio-oil. Bioresour Technol 319. https://doi.org/10.1016/J.BIORTECH.2020.124176.
21. Yuan C, Wang S, Cao B, Hu Y, Abomohra AEF, Wang Q et al (2019) Optimization of hydrothermal co-liquefaction of seaweeds with lignocellulosic biomass: Merging 2nd and 3rd generation feedstocks for enhanced bio-oil production. Energy 173:413–422. https://doi.org/10.1016/J.ENERGY.2019.02.091

22. Jung KA, Lim SR, Kim Y, Park JM (2013) Potentials of macroalgae as feedstocks for biorefinery. Bioresour Technol 135:182–190. https://doi.org/10.1016/J.BIORTECH.2012.10.025
23. Pereira L (2011) A review of the nutrient composition of selected edible seaweeds
24. Ortiz J, Uquiche E, Robert P, Romero N, Quitral V, Llantén C (2009) Functional and nutritional value of the Chilean seaweeds Codium fragile, Gracilaria chilensis and Macrocystis pyrifera. Eur J Lipid Sci Technol 111:320–327. https://doi.org/10.1002/EJLT.200800140
25. Guerra-Rivas G, Gómez-Gutiérrez CM, Alarcón-Arteaga G, Soria-Mercado IE, Ayala-Sánchez NE (n.d) Screening for anticoagulant activity in marine algae from the Northwest Mexican Pacific coast. J Appl Phycol 23:495–503
26. McDermid KJ, Stuercke B (2003) Nutritional composition of edible Hawaiian seaweeds. J Appl Phycol 15:513–524. https://doi.org/10.1023/B:JAPH.0000004345.31686.7F
27. Hwang E, Park S-Y, Sun Z, Shin H-S, Lee D-G, Yi TH (2013) The protective effects of fucosterol against skin damage in UVB-Irradiated human dermal fibroblasts. Mar Biotechnol 163:16:361–70. https://doi.org/10.1007/S10126-013-9554-8
28. Ey C, Hj H, Tj N (2010) Protective effect of a polysaccharide from Hizikia fusiformis against ethanol-induced cytotoxicity in IEC-6 cells. Toxicol In Vitro 24:79–84. https://doi.org/10.1016/J.TIV.2009.08.031
29. Park K-J, Kim BY, Kyoung Park S, Lee J-H, Kim YS, Gil Choi H et al (2012) Morphological and biochemical differences in three Undaria pinnatifida populations in Korea. Algae 189–96. https://doi.org/10.4490/algae.2012.27.3.189.
30. Barrow C, Shahidi F (eds) (2007) Marine Nutraceuticals and functional foods. https://doi.org/10.1201/9781420015812
31. Effects of corn starch and Kappaphycus Alvarezii seaweed blend concentration on the optical, mechanical, and water vapor barrier properties of composite films | Abdul Khalil | BioResources (n.d.) https://ojs.cnr.ncsu.edu/index.php/BioRes/article/view/13217. Accessed 2 Oct 2021
32. Diharmi A, Fardiaz D, Andarwulan N (2019) Chemical and Minerals Composition of Dried Seaweed Eucheuma spinosum collected from Indonesia Coastal Sea Regions. Int J Ocean Oceanogr 13:65–71
33. Biller P, Ross AB (2012) Hydrothermal processing of algal biomass for the production of biofuels and chemicals. Biofuels 3:603–623. https://doi.org/10.4155/BFS.12.42
34. Is T, Kt L (2016) Comparison of different process strategies for bioethanol production from Eucheuma cottonii: an economic study. Bioresour Technol 199:336–346. https://doi.org/10.1016/J.BIORTECH.2015.08.008
35. Davis TA, Volesky B, Mucci A (2003) A review of the biochemistry of heavy metal biosorption by brown algae. Water Res 37:4311–4330. https://doi.org/10.1016/S0043-1354(03)00293-8
36. Anastasakis K, Ross AB, Jones JM (2011) Pyrolysis behaviour of the main carbohydrates of brown macro-algae. Fuel 90:598–607. https://doi.org/10.1016/J.FUEL.2010.09.023
37. Rioux LE, Turgeon SL, Beaulieu M (2007) Characterization of polysaccharides extracted from brown seaweeds. Carbohydr Polym 69:530–537. https://doi.org/10.1016/J.CARBPOL.2007.01.009
38. Percival E (2007) The polysaccharides of green, red and brown seaweeds: their basic structure, biosynthesis and function. 14:103–117. https://doi.org/10.1080/00071617900650121
39. Li J, Wang G, Gao C, Lv X, Wang Z, Liu H (2013) Deoxy-Liquefaction of Laminaria japonica to high-quality liquid oil over metal modified ZSM-5 catalysts. Energy Fuels 27:5207–5214. https://doi.org/10.1021/EF4004208
40. Adams JMM, Ross AB, Anastasakis K, Hodgson EM, Gallagher JA, Jones JM et al (2011) Seasonal variation in the chemical composition of the bioenergy feedstock Laminaria digitata for thermochemical conversion. Bioresour Technol 102:226–234. https://doi.org/10.1016/J.BIORTECH.2010.06.152
41. Khairy HM, El-Shafay SM (2013) Seasonal variations in the biochemical composition of some common seaweed species from the coast of Abu Qir Bay, Alexandria Egypt. Oceanologia 55:435–452. https://doi.org/10.5697/OC.55-2.435
42. Marinho-Soriano E, Fonseca PC, Carneiro MAA, Moreira WSC (2006) Seasonal variation in the chemical composition of two tropical seaweeds. Bioresour Technol 97:2402–2406. https://doi.org/10.1016/J.BIORTECH.2005.10.014

43. López Barreiro D, Prins W, Ronsse F, Brilman W (2013) Hydrothermal liquefaction (HTL) of microalgae for biofuel production: state of the art review and future prospects. Biomass Bioenerg 53:113–127. https://doi.org/10.1016/J.BIOMBIOE.2012.12.029
44. Gai C, Zhang Y, Chen W-T, Zhang P, Dong Y (2015) An investigation of reaction pathways of hydrothermal liquefaction using Chlorella pyrenoidosa and Spirulina platensis. Energy Convers Manag 96:330–339. https://doi.org/10.1016/J.ENCONMAN.2015.02.056
45. Biller P, Ross AB (2011) Potential yields and properties of oil from the hydrothermal liquefaction of microalgae with different biochemical content. Bioresour Technol 102:215–25. https://doi.org/10.1016/J.BIORTECH.2010.06.028
46. Yang W, Li X, Li Z, Tong C, Feng L (2015) Understanding low-lipid algae hydrothermal liquefaction characteristics and pathways through hydrothermal liquefaction of algal major components: Crude polysaccharides, crude proteins and their binary mixtures. Bioresour Technol 196:99–108. https://doi.org/10.1016/J.BIORTECH.2015.07.020
47. Neveux N, Yuen AK, Jazrawi C, Magnusson M, Haynes BS, Masters AF et al (2014) Biocrude yield and productivity from the hydrothermal liquefaction of marine and freshwater green macroalgae. Bioresour Technol 155:334–41. https://doi.org/10.1016/J.BIORTECH.2013.12.083

Advanced Fluid

Thermal–Hydraulic Performance of Water: Ethylene Glycol Mixture Through Guide-Vane Swirl Generator: A Numerical Simulation

M. A. At-Tasneem, W. H. Azmi, and M. A. Ismail

Abstract One of the devices used to improve heat transfer in heat exchangers is the decaying swirl generator. When the inlet flow passes through this turbulator, a swirling flow is created, which thins the boundary layer on the tube wall and improves the flow's force convection. This paper numerically investigated a new design of a decaying swirl generator known as the guide-vane swirl generator (GVSG). The GVSG has a diameter of 15.5 mm and a length of 35 mm, and it is placed at L/D = 10 in a tube with a diameter of 16 mm and a length of L/D = 93.75. It has four vanes, each with a length of L/D = 1 and half of each twisted 180°. With a constant heat flux applied to the tube, the heat transfer enhancement was tested for Reynolds numbers ranging from 4583 to 35,000. Ansys Fluent was used throughout the simulation process. The friction factor obtained a maximum value of 4.06 at the highest Reynolds number tested. The Nusselt number achieved a maximum value of 1.23 at Reynolds number 20,000. As a result, the maximum thermal-hydraulic performance (THP) of GVSG inside the heated plain horizontal tube is 0.81 achieved at Reynolds number 35,000. Since THP was less than unity, it is clear that GVSG causes a greater deficit in pressure drop than they promote heat transfer. As a result, the geometry of the GVSG proposed in this paper, such as its length, diameter, vane length, and twist angle, need to be adjusted to improve the GVSG's THP.

Keywords Heat transfer enhancement · Thermal–hydraulic performance (THP) · Passive technique · Decaying swirl generator

---

M. A. At-Tasneem
Faculty of Mechanical and Automotive Engineering Technology, Universiti Malaysia Pahang, 26600 Pekan, Pahang, Malaysia

W. H. Azmi (✉)
Department of Mechanical Engineering, College of Engineering, Universiti Malaysia Pahang, 26300 Gambang, Pahang, Malaysia
e-mail: wanazmi@ump.edu.my

M. A. Ismail
School of Mechanical Engineering, Universiti Sains Malaysia, Engineering Campus, 14300 Nibong Tebal, Pinang, Malaysia

M. Y. Ismail et al. (eds.), *Technological Advancement in Mechanical and Automotive Engineering*, Lecture Notes in Mechanical Engineering,
https://doi.org/10.1007/978-981-19-1457-7_12

1 Introduction

Heat transfer enhancement techniques are utilized to boost heat transfer while having little effect on the overall system efficiency. It is used in various applications, including air conditioning systems, refrigerators, automotive cooling parts, and others. There are three heat transfer enhancement techniques [1]: active, passive, and compound. The active technique required external power to generate energy for the system [2, 3], whereas the passive technique did not require external energy to generate energy for the system [4, 5]. The compound technique combines two or more active and passive techniques that result in a greater augmentation than using just one passive or active technique individually [6, 7]. In terms of design and implementation, the active technique is more complicated than the passive technique since it requires external power to alter the fluid flow to improve thermal efficiency. External energy is used in the active technique in various ways, including mechanical assistance, electrostatic fields, jet cleaning, suction, and injection. Active techniques are not extensively used in scientific domains since it is difficult to deliver all the external energy in many applications.

In the passive technique, no other external source of energy is required to improve the heat transfer of a system. Instead, the swirl device shape or structural surface of the flow channel is altered to improve a system's thermal efficiency. As a result, swirler inserts or surface modifications are utilized to mix fluid and turbulent flow, resulting in an improved rate of heat transfer. Furthermore, compared to other heat transfer enhancement techniques, this technique has several advantages, including a low cost, quick production, and reduced installation time. It is more convenient since the swirler development is simple and may be employed in the current heat exchange system. Many researchers have carried out the investigation of passive heat transfer enhancement techniques. By inserting the swirler into the tube, swirling flow is created by mixing the flow into the bulk flow and disturbing the layer of boundaries on the surface of the tube due to the swirl devices' changes in shape. Therefore, Table 1 displays a few passive swirler that proved to enhance heat transfer. The friction factor and Nusselt vary depending on the geometry and parameters of the flow and how the swirler blocks and distorts it. The goal of heat transfer enhancement techniques is to provide optimal THP with the lowest friction factor possible. As a result, this study explored the THP of a new design of a decaying swirl generator known as a guide-vane turbulator. A decaying swirl generator, known as the guide-vane swirl generator (GVSG), is designed to be evaluated for its ability to generate swirl flow and improve THP. This design was inspired by Cabral et al. [8], who introduced the static guide-vane type swirler to generate an asymmetric helical flow of air from a laminar flow that Re < 300. As a result, this paper examines the feasibility of utilizing this type of swirl generator for THP with turbulence flow inlet.

Table 1 Passive Swirler

Author	Type of swirler	Name of swirler	Highest Friction Factor Enhancement	Highest nusselt number enhancement	Thermal-hydraulic performance (THP)
Goh et al. [9]	Axial array turbulator	Self-rotating turbulator	1800%	360%	1.12
Liu et al. [10]	Axial array turbulator	Fluid exchangers insert	701%	269%	1.62
Zhang et al. [11]	Surface modification	Rectangular winglet vortex generators	644%	118%	–
Firoozi et al. [12]	Surface modification	Dimple configurations	450%	480%	3.12
Chaurasia et al. [13]	Continuous swirler	Single and double-helical screwtape	1637%	182%	1.58
Pourramezan et al. [14]	Continuous swirler	Twisted conical strip inserts	5122%	1466%	3.94
Saedodin et al. [15]	Decaying swirl generator	Twisted turbulator	13.96%	7.93%	1.04
Jumpholkul et al. [16]	Decaying swirl generator	Free rotating swirl generator	–	54.5%	1.7
Dagdevir et al. [6]	Compound continuous swirler	Perforated, dimpled twisted tape	–	–	1.42
Du et al. [17]	Compound continuous swirler	Transverse rib with regularly spaced helical tapes	1531%	177%	1.26

2 Numerical Modeling

Azmi et al. [18] and Nabil et al. [19] studied the effectiveness of a mixture of water and ethylene glycol (W: EG) at different ratios for mono and hybrid nanofluids in a circular tube for forced convection heat transfer. The results revealed that a W: EG ratio of 60:40% provided the best heat transfer performance compared to 20:80%, 50:50%, 40:60%, and 80:20%. Researchers also employed it to improve the thermophysical properties of several nanofluids [20–23] and the effectiveness of a passive swirler for heat transfer enhancement [24]. As a result, the W: EG mixture at a 60:40% ratio was used as a working fluid. The thermophysical properties of W: EG 60:40% are listed in Table 2.

Table 2 Thermophysical properties

Descriptions	W:EG 60:40% [25]	Copper tube [26]
Thermal conductivity, k (W/mK)	0.4112	387.6
Density, ρ (kg/m^3)	1055.832	8978
Specific heat, C_p (J/kg.K)	0.002322	381.0
Dynamic viscosity, μ (Pa.s)	3498.6	–

The computational domain of a horizontal straight copper tube was divided into three sections: the entrance, the turbulator, and the fully developed section. The thermophysical properties of the copper tube are presented in Table 2. The tube's diameter (D) and length (L) are 0.016 m and 1.5 m (L/D = 93.75), respectively. The lengths of the entrance, turbulator, and fully developed sections are respectively 0.155 m (L/D = 9.7), 0.045 m (L/D = 2.8), and 1.3 m (L/D = 81.25). The heat transfer by forced convection will be measured at the fully developed section where the swirl flow is generated.

Figure 1 depicts the geometry and dimensions of the GVSG. The GVSG has a diameter of 15.5 mm and a length of 35 mm (L/D = 2.26). The diameter of the GVSG is 3% smaller than the diameter of the tube, and the length is L/D = 2.19 of the diameters of the tube. The leading edge of the GVSG's spherical nose is 4 mm in diameter, and it directs the flow to four vanes. The vanes are 1 mm thick and directly correspond to the inlet flow before being twisted 180° and exiting at the trailing edge. The twisted vane length is 8 mm, which is the same as the plain tube's L/D = 0.5.

Figure 2 illustrates mesh generation of the GVSG in the tube. Meshes are divided into three sections. The unstructured tetrahedral mesh was generated in the section where the GVSG was located. The GVSG section's upstream is the inlet section, and the downstream is the fully developed section. A structured hexahedral mesh was used to discretize these sections. The global mesh size is 0.6 mm, with 25 layers

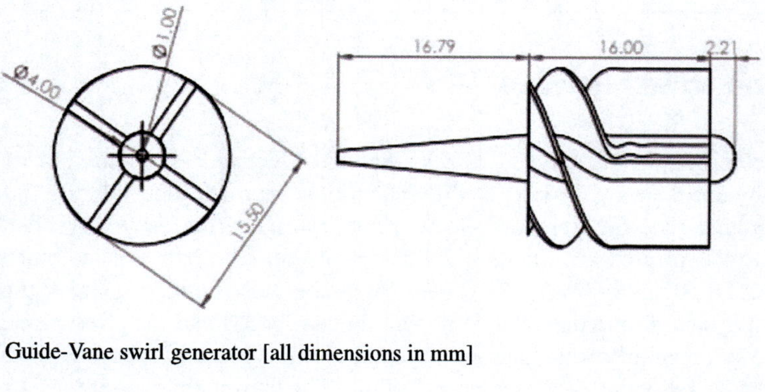

Fig. 1 Guide-Vane swirl generator [all dimensions in mm]

Fig. 2 Mesh generation

of inflation and a smooth transition at a growth rate of 1.15. There are 6,462,128 meshes in total.

2.1 Governing Equations

The transport equations for a three-dimensional, steady, incompressible flow are discretized and solved. The following are the governing equations for continuity, Reynolds-average momentum, and energy:

Continuity equation:

$$\frac{\partial(\rho u_i)}{\partial x_i} = 0, i = 1, 2, 3 \tag{1}$$

Reynolds-average momentum equations:

$$\frac{\partial(\rho u_i u_j)}{\partial x_j} = -\frac{\partial P}{\partial x_i} + \frac{\partial}{\partial x_j}\left[\mu\left(\frac{\partial u_i}{\partial x_j} + \frac{\partial u_j}{\partial x_i}\right)\right], j = 1, 2, 3 \tag{2}$$

Energy equation:

$$\frac{\partial}{\partial x_i}\left(\rho u_i C_p T - k\frac{\partial T}{\partial x_i}\right) = 0 \tag{3}$$

where u_i and x_i is velocity and space component, respectively. Table 3 shows the solver configurations used to discretize and solve the transport equations.

The flow is tested in the turbulence flow regime for Reynolds numbers ranging from 4583 to 35,000 calculated from the uniform input velocity. By assuming that the outlet pressure is equal to zero, the governing equation is solved. The tube is applied to a constant heat flux to measure the flow's force convection heat transfer. Table 4 lists the simulation's detailed boundary conditions.

Table 3 Solver Configuration

Solver	Descriptions
Pressure-velocity coupling	Coupled
spatial discretization	Gradient: least squares cell based Pressure: second order Momentum: second order upwind Specific dissipation rate: second order upwind Reynolds stresses: second order upwind Energy: second order upwind

Table 4 Boundary condition

Viscous model	Reynolds stress-BSL
Inlet condition	Velocity magnitude: 0.63–4.811 m/s [Re = 4583–35,000] Turbulence specification method: intensity and hydraulic diameter Reynolds stress specification method: K or turbulent intensity, I where $I = 0.16Re^{-\frac{1}{8}}$
Outlet	Gauge pressure [Pa]: 0 Turbulence specification method: intensity and hydraulic diameter Reynolds stress specification method: K or turbulent intensity, I where $I = 0.16Re^{-\frac{1}{8}}$
Wall condition	*Entrance* Wall motion: stationary wall Shear condition: no slip Thermal: constant heat flux: 7957 W/m^2 *Turbulator* Wall motion: stationary wall Shear condition: no slip Thermal: constant heat flux: 0 W/m^2 *Fully developed* Wall motion: stationary wall Shear condition: no slip Thermal: constant heat flux: 7957 W/m^2

2.2 Parameter Definitions

The performance of GVSG is measured by the increase in friction factor, the Nusselt number, and, finally, the THP. The friction factor shown in Eq. 4 is calculated by measuring the pressure difference, *dP*, between the tube's inlet and outlet. The flow velocity is taken from the uniform inlet flow, where the Reynolds number is calculated.

$$f = dP\frac{D}{L}\frac{2}{\rho V^2} \tag{4}$$

The Nusselt number is then calculated for both plain tube and plain tube with GVSG inserts using Eq. 5. The heat convection coefficient, h, is calculated by dividing the supplied heat flux by the temperature difference at the wall and centerline velocity, where the velocity is maximum. Both temperatures are taken downstream of the trailing edge in a fully developed flow section.

$$Nu = \frac{hD}{k}, h = \frac{\dot{q}}{(T_s - T_b)} \tag{5}$$

Finally, the THP is calculated using the following equation [27]:

$$THP = \frac{\frac{Nu_{SG}}{Nu_{PT}}}{\left(\frac{f_{SG}}{f_{PT}}\right)^{\frac{1}{3}}} \tag{6}$$

3 Results and Discussion

The decaying swirl generator, known as the guide-vane swirl generator (GVSG), was inserted at L/D = 10, where the entrance length ends, to generate the swirl flow in the fully developed section.

Figures 3 and 4 depicts the friction factor and relative friction factor generated along the flow for Reynolds numbers ranging from 4583 to 35,000, respectively. The friction factor trend for the plain tube and plain tube with GVSG turbulator is similar. When the Reynolds number is the lowest, the friction factor is significantly higher, and it gradually decreases as the Reynolds number increases. Nonetheless, the relative friction factors are more than three times higher than in the plain tube at all Reynolds numbers. At the lowest Reynolds number of 4584, the relative friction factor has the smallest value of 3.26. The relative friction factor increases as the Reynolds

Fig. 3 Friction factor

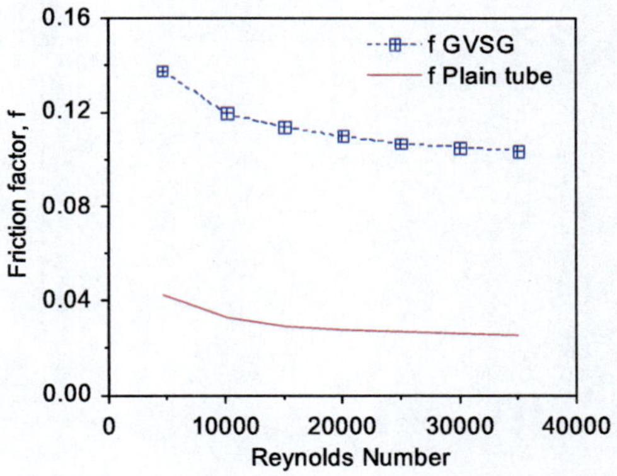

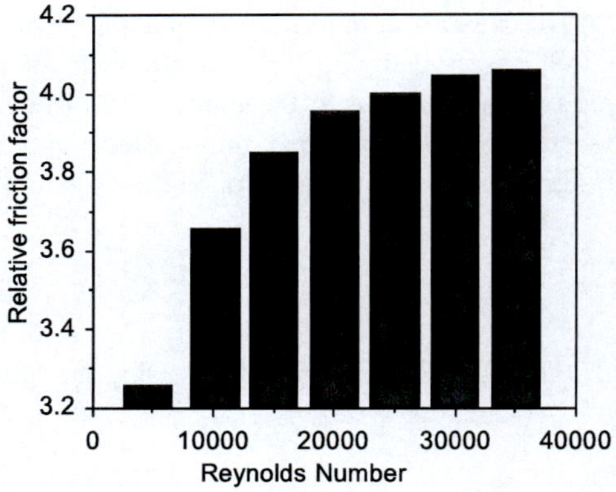

Fig. 4 Relative friction factor

number increases until it reaches its maximum at 35,000. After approaching the twisted vane, the incoming flow of the leading edge is totally obstructed before it is deformed in the direction of the vane. At this point, the pressure drag is extremely high, followed by the skin friction drag, which contributes to the extremely high friction factor. Indurain et al. [28] also numerically computed a twisted turbulator with a vane length similar to the GVSG of L/D = 0.5. Nonetheless, the blade angle is 40°, which is significantly less than the GVSG. The twisted tape was placed into a circular tube with an L/D = 45.56 length. As a result, the highest relative friction factor is 2.9 at the highest Reynolds number of 200,000. This result is lower than the GVSG of 35,000. Nonetheless, the angle is 4.5 times smaller than GVSG, which may have contributed to a lower relative friction factor than GVSG.

Figures 5 and 6 show the Nusselt number and relative Nusselt number of the flow with the GVSG turbulator versus the plain tube in a horizontal orientation. The Nusselt number with GVSG turbulator is higher than the plain tube at all Reynolds numbers investigated. The trend has revealed that the relative Nusselt number has a negative quadratic relationship with the Reynolds number. Nonetheless, the result

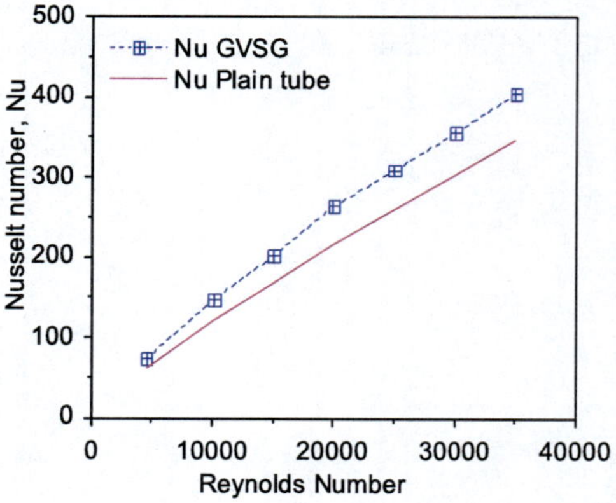

Fig. 5 Nusselt number

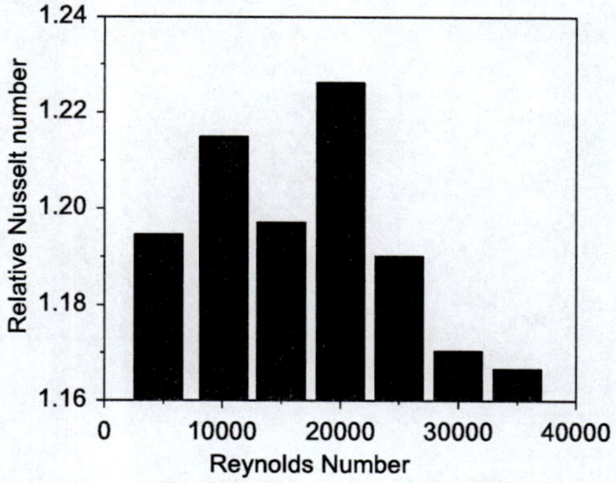

Fig. 6 Relative nusselt number

has shown that the Nusselt number fluctuates with increasing Reynolds number, with the maximum Nusselt number reaching at Reynolds number 20,000 with 1.23. The Reynolds number of 35,000 yielded the smallest Nusselt number of 1.17. This result is relatively similar to the propeller turbulator proposed by Nikoozadeh et al. [29]. The turbulator was numerically tested with the length of the turbulator being L/D = 0.2, six blades, and a blade angle of 45°. The friction factor was increased by 68% to achieve the maximum Nusselt number by 28%.

Figure 7 depicts the THP of the flow with the GVSG turbulator inside the heated tube. The value decreases as the Reynolds number increases, with the maximum value being 0.81 at Reynolds number 4583. The lowest THP was obtained at the highest Reynolds number with 0.73. Every heat transfer enhancement technique necessitates an increase in the Nusselt number. Nonetheless, the significant drawback of the friction factor leads to the relatively low THP of the flow. This value may overutilize pumping power to push the flow, increasing the cost of operation while delivering the same performance as the plain tube. The swirl flow structure inside the tube is visualized to determine the source of this value. Figure 8 illustrates the downstream

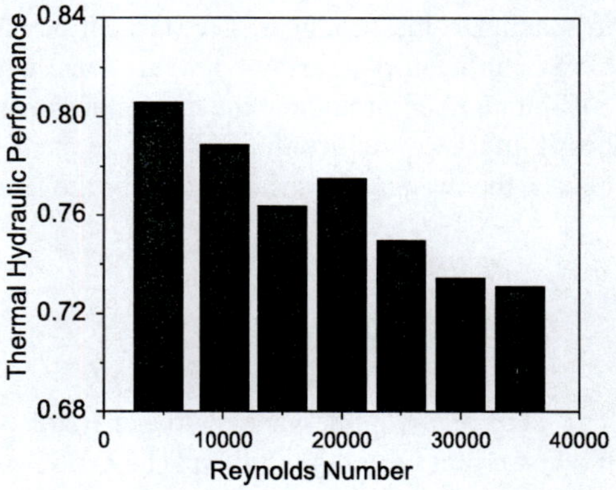

Fig. 7 Thermal-hydraulic Performance (THP)

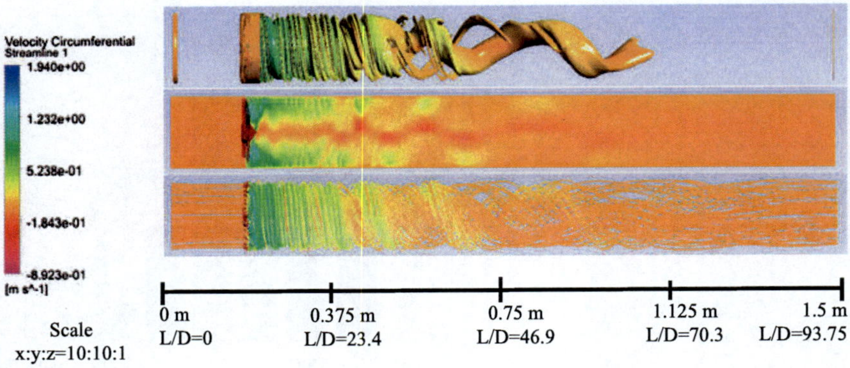

Fig. 8 Vortex core (top), cross-sectional plane (middle) and streamline (bottom) of velocity circumferential

velocity pattern produced after the inlet flow flowed through the GVSG at a Reynolds number of 4583. From top to bottom of Fig. 8, the velocity circumferential is depicted in the form of a vortex core, a cross-sectional plain, and a flow streamline of the fluid. The GVSG is positioned at L/D = 10, as specified in the methodology. As observed in the image, the swirl flow develops after passing through the GVSD. The intensity of the swirl flow formed is greatest near the GVSG's trailing edge and remains high for the first L/D = 13.4 after it passes through the GVSG. Since the GVSG has a longer training edge in the centre of the structure, it developed a lower circumferential velocity point in the middle of the tube and a high value in the outer region. However, as the fluid moves downstream the tube, the circumferential velocity drops continuously until L/D = 46.9. After this point, the circumferential velocity is equivalent to inflow flow, and swirl flow completely vanishes.

The THP is low because it is computed by considering the Nusselt number and pressure drop over the whole 1.5 m tube. Nonetheless, based on the observations, it is clear that the created swirl flow that provided the higher Nusselt number only lasted until L/D = 46.9 and then vanished. Thus, with the same GVSG position, the GVSG geometry may be suited for tube lengths of up to 0.75 m or shorter to maximize the benefit of GVSG in a circular tube for THP. Otherwise, a few GVSG turbulators in array configurations may be employed to increase the local Nusselt number throughout the tube to improve THP for a 1.5 m plain tube. Another possibility is to modify the GVSG's geometry, such as by altering the number of blades, the twist angle, the diameter, or the length.

4 Conclusions

The THP of the guide-vane swirl generator (GVSG) is numerically tested inside the horizontal heated plain tube. The GVSG was installed at the plain tube's inlet to determine the intensity of the swirl flow generated downstream of the tube. The friction factor, Nusselt number, and THP for the presence of GVSG inserts in a tube were measured using pressure drops, wall temperatures, and centerline temperatures. In comparison to the plain tube, the Nusselt number was found to be 22.6% higher. The highest THP, on the other hand, is less than unity because the friction factor is four times that of a plain tube. GVSG's geometry and dimensions allowed for swirl flow downstream of the tube. Nonetheless, the swirl flow's intensity was too low, and it began to deteriorate halfway down the tube. As a result, this GVSG geometry can be used for shorter tubes, or multiple GVSG can be used in an array inside the tube to regenerate the swirl flow at the desired location.

Acknowledgements The authors would like to thank the Center of Excellence for Advanced Research in Fluid Flow (CARIFF) for funding assistance under RDU190385 for inventory and the Advanced Automotive Liquids Laboratory (A_2LL) for providing full access to a high-performance computer.

References

1. Saha SK, Ranjan H, Emani MS, Bharti AK (2020) Active and passive techniques: their applications. Appl Sci Technol 17–72
2. Liu JB, Bayati M, Abbas M, Rahimi A, Naderi M (2019) Mesoscopic approach for simulating nanofluid flow and heat transfer in a finned multi-pipe heat exchanger. Int J Numer Meth Heat Fluid Flow 29(8):2822–2839
3. Bezaatpour M, Goharkhah M (2020) Convective heat transfer enhancement in a double pipe mini heat exchanger by magnetic field induced swirling flow. Appl Therm Eng 167:114801
4. Mousa MH, Miljkovic N, Nawaz K (2021) Review of heat transfer enhancement techniques for single phase flows. Renew Sustain Energy Rev 137(110566):1277–1287
5. Javed S, Ali HM, Babar H, Khan MS, Janjua MM, Bashir MA (2020) Internal convective heat transfer of nanofluids in different flow regimes: a comprehensive review. Phys A Stat Mechs Appl 538:122783
6. Dagdevir T, Ozceyhan V (2021) An experimental study on heat transfer enhancement and flow characteristics of a tube with plain, perforated and dimpled twisted tape inserts. Int J Therm Sci 159:106564
7. Maghrabie HM, Attalla M, Mohsen AAA (2021) Performance assessment of a shell and helically coiled tube heat exchanger with variable orientations utilizing different nanofluids. Appl Therm Eng 182:116013
8. Vidal Cabral R, Damiani Rocha A (2019) Three-Dimensional laminar swirling flow in a pipe. In: 25th ABCM international congress of mechanical engineering, Brazil
9. Goh LHK, Hung YM, Chen GM, Tso CP (2021) Entropy generation analysis of turbulent convection in a heat exchanger with self-rotating turbulator inserts. Int J Thermal Sci 160
10. Liu P, Dong Z, Lv J, Shan F, Liu Z, Liu W (2020) Numerical study on thermal-hydraulic performance and exergy analysis of laminar oil flow in a circular tube with fluid exchanger inserts. Int J Thermal Sci 153
11. Zhang K, Sun Z, Zheng N, Chen Q (2020) Effects of the configuration of winglet vortex generators on turbulent heat transfer enhancement in circular tubes. Int J Heat Mass Transfer 157
12. Firoozi AO, Majidi S, Ameri M (2020) A numerical assessment on heat transfer and flow characteristics of nanofluid in tubes enhanced with a variety of dimple configurations. Thermal Sci Eng Prog 19

13. Chaurasia SR, Sarviya RM (2020) Thermal performance analysis of CuO/water nanofluid flow in a pipe with single and double strip helical screw tape. Appl Therm Eng 166:114631
14. Pourramezan M, Ajam H, Raoufi MA, Abadeh A (2020) Performance evaluation and optimization of design parameters for twisted conical strip inserts in tubular laminar flow Using Taguchi approach. Int J Therm Sci 152:106324
15. Saedodin S, Zaboli M, Rostamian SH (2020) Effect of twisted turbulator and various metal oxide nanofluids on the thermal performance of a straight tube: Numerical study based on experimental data. Chem Eng Process Process Intensification 158:108106
16. Jumpholkul C, Asirvatham LG, Dalkılıç AS, Mahian O, Ahn HS, Jerng DW (2020) Experimental investigation of the heat transfer and pressure drop characteristics of SiO_2/water nanofluids flowing through a circular tube equipped with free rotating swirl generators. Heat Mass Transfer/Waerme- und Stoffuebertragung 56(5):1613–1626
17. Du J, Hong Y (2020) Experimental investigation on thermal-hydraulic characteristics in a traverse rib tube fitted with regularly spaced helical tapes. Int J Heat Mass Transf 154:119726
18. Azmi WH, Usri NA, Mamat R, Sharma KV, Noor MM (2017) Force convection heat transfer of Al_2O_3 nanofluids for different based ratio of water: ethylene glycol mixture. Appl Therm Eng 112:707–719
19. Nabil MF, Azmi WH, Hamid KA, Mamat R (2018) Experimental investigation of heat transfer and friction factor of TiO_2-SiO_2 nanofluids in water: ethylene glycol mixture. Int J Heat Mass Transf 124:1361–1369
20. Sedaghat F, Yousefi F (2019) Synthesizes, characterization, measurements and modeling thermal conductivity and viscosity of graphene quantum dots nanofluids. J Mol Liq 15(278):299–308
21. Awua J, Ibrahim S, Kwaghger A (2017) Investigation of thermal conductivity of palm kernel fibre nanofluid using de-ionized water and ethylene glycol mixed at ration of 50:50 and 60:40. Int J Eng Trends Technol 49:47–53
22. Chiam HW, Azmi WH, Usri NA, Mamat R, Adam NM (2017) Thermal conductivity and viscosity of Al2O3 nanofluids for different based ratio of water and ethylene glycol mixture. Exp Thermal Fluid Sci 81:420–429
23. Samylingam L, Anamalai K, Kadirgama K, Samykano M, Ramasamy D, Noor MM (2018) Thermal analysis of cellulose nanocrystal-ethylene glycol nanofluid coolant. Int J Heat Mass Transf 127:9
24. Jafari M, Farhadi M, Sedighi K (2017) An experimental study on the effects of a new swirl generator on thermal performance of a circular tube. Int Commun Heat Mass Transfer 87:277–287
25. ASHRAE Handbook—fundamental (SI Edition), American Society of Hearing, Refrigerating and Air-Conditioning Engineers Inc. Atlanta, GA
26. Perry RH (1997) Perry's chemical engineers' handbook. 7th edn. McGraw-Hill
27. Promvonge P, Koolnapadol N, Pimsarn M, Thianpong C (2014) Thermal performance enhancement in a heat exchanger tube fitted with inclined vortex rings. Appl Therm Eng 62(1):285–292
28. Indurain B, Beaubert F, Lalot S, Uystepruyst D (2020) Computational fluid dynamics investigation of the thermal performances of a Swirler with profiled blades. Heat Transfer Eng 7632 (2020)
29. Nikoozadeh A, Behzadmehr A, Payan S (2020) Numerical investigation of turbulent heat transfer enhancement using combined propeller-type turbulator and nanofluid in a circular tube. J Therm Anal Calorim 140(3):1029–1044

Photodegradation of Biobased Oils Polymer Blended with High Density Polyethylene upon Ultraviolet Irradiation Exposure

Nurulsaidatulsyida Sulong, Anika Zafiah Mohd Rus, Nurul Syamimi Mohd Salim, Nik Normunira Mat Hassan, and Noraini Marsi

Abstract In the present study, two types of polymers that is biopolymer derived from vegetable oils and High Density Polyethylene (HDPE) was blended together. Biopolymer are covers both biobased and biodegradable polymers while HDPE is thermoplastic polymer that is easily remoulded and recycled. This is an advantage to produce bio-based polymer that can remould and degrade by producing a new polymer blend. The blended polymer of HDPE with 30% of BP via melt mixing technique using Brabender (B) machine and a manually mixed (M) namely as HB and HM respectively was produced. The sample in the form of dumbbell shape from the Injection Moulding machine was then exposed to Ultraviolet (UV) irradiation at a constant temperature of 50 °C for 500, 1000, 2000 and 3000 h irradiation exposure. The FTIR spectroscopy was identified that both HB and HM have the same functional group of C-H group at 2915 cm^{-1} and 2848 cm^{-1} as the main backbone of the polymer blended structure. For HB and HM, peaks at 1697–1714 cm^{-1} were contributed to the C=O group in the amide region. For CH_3 bending the peaks were observed at 1472–1462 cm^{-1}. In addition, the carbonyl Index (CI) was calculated. HB shows the highest CI for 1000H irradiation exposure while 500H for the lowest CI. The HM of 500H exposed upon UV shows the highest while 3000H shows the lowest value of CI. After UV irradiation the CI of HB and HM shows fluctuated trend due to degradation occurring under photo-oxidation circumstances is the chain scission accompanying with the cross-linking and the formation of the carbonyl group.

N. Sulong (✉)
Faculty of Mechanical Engineering, University Technology MARA (UiTM), Johor Branch, Pasir Gudang Campus, Jalan Purnama, 81750 Masai, Johor, Malaysia
e-mail: saidatulsyida@uitm.edu.my

N. Sulong · A. Z. M. Rus · N. S. M. Salim
Sustainable Polymer Engineering, Advanced Manufacturing and Materials Center (SPEN-AMMC), Faculty of Mechanical and Manufacturing, Engineering University Tun Hussein Onn Malaysia, 86400 Parit Raja, Batu Pahat, Johor, Malaysia

N. N. M. Hassan · N. Marsi
Faculty of Engineering Technology, University Tun Hussein Onn Malaysia, Pagoh Higher Education Hub, 84600 Pagoh, Muar, Johor, Malaysia

M. Y. Ismail et al. (eds.), *Technological Advancement in Mechanical and Automotive Engineering*, Lecture Notes in Mechanical Engineering,
https://doi.org/10.1007/978-981-19-1457-7_13

Keywords Biopolymers · High density polyethylene · Thermal degradation · Ultraviolet irradiation · Carbonyl index (CI) · Fourier transformation infrared spectroscopy

1 Introduction

Polyethylene (PE) is a semi crystalline thermoplastic and is the most widely produced polymer in the world, making it an important commercial polymer material nowadays [1, 2]. It also serves as a model thermoplastic system within the polyolefin family because of its simple structure, which is based on ethylene (CH_2–CH_2) repeat units with varying degrees of crosslinking and branching. One of the various types of PE, high-density PE (HDPE) is a linear form with low levels of branching, and consequently a higher density and degree of crystallinity. HDPE is commonly utilized due to its high ductility, toughness, and chemical resistance, for instance in geomembranes, pipelines, and plastic lumber [2]. HDPE also used as plastic bag in supermarket. According to [3], plastic bags are produced from one of three basic polymers: polyethylene, high density polyethylene (HDPE), low density polyethylene (LDPE), or linear low-density polyethylene (LLDPE). Supermarket plastic bags are usually made from HDPE. Plastics are not toxic or harmful, however, plastic bags are manufactured using organic and inorganic additives such as colorants and pigments, plasticizers, antioxidants, stabilizers and metals make them against the degradation process then contributed to a great source of atmospheric pollution and ozone destruction [3].

Renewable materials derived from vegetable oils are now widely developed for obtaining green polymers. Palm oil is one of the vegetable oils which contain unsaturated fatty acids such as oleic acid. The content of oleic acid in palm oil is about 30–40%. Oleochemical industries use oleic acid as the starting material for many important chemical reactions such as olefin metathesis, ozonolysis and epoxidation [4]. When compared to their petroleum-based counterparts, renewable materials can be degradable, sustainable, more environmentally friendly (lower greenhouse gas emission and fossil fuel usage) and/or can be a renewable option to plastic product production. Despite certain limitations, bioplastics remain a front runner as a suitable replacement for petroleum-based plastic [5]. However, bio-based polymers have another problem in term of processes ability and by increasing production capacity, they are still quite expensive compared to commodity polymers and their properties are also often inferior, or at least do not correspond to the expectation of converters or users.

Polymer degradation can be caused by heat (thermal degradation), light (photodegradation), ionizing radiation (radio degradation), mechanical action, or by fungi, bacteria, yeasts, algae, and their enzymes (bio degradation). Photodegradation is degradation of molecule caused by the absorption of photons, particularly those wavelengths found in sunlight, such as infrared radiation, visible light, and ultraviolet (UV) light. However, other forms of electromagnetic radiation can cause photodegradation. Photodegradation includes photodissociation, the breakup of molecules into smaller pieces by photons. A common photodegradation reaction is oxidation. Photodegradation also destroys paintings and other artefacts. Light is an induced polymer degradation, or photodegradation, includes the physical and chemical changes caused by irradiation of polymers with ultraviolet or visible light. In order to be effective, light must be absorbed by the substrate (polymeric system) [6].

In this research, by taking the advantages of HDPE as a versatile materials and bio-based oils polymer itself as a degradable material, there are methods to utilize their potentials by blending the HDPE polymer with the bio-based oils polymer to increase performance and accelerate the degradation process of the blended biopolymer by exposing the blended polymer to under UV weatherometer. In this research, HDPE was blended with 30% of bio-based oils polymer were prepared by two types of blending technique of melt mixing by using Brabender machine named as HB and secondly by manually mixing named as HM. HB and HM were exposed to UV irradiation at constant temperature which is 50 °C at 0H, 500H, 1000H, 2000H and 3000H and tested by using Fourier Transmittance Infrared (FTIR) to observe the stability properties of HB and HB also carbonyl index (CI) was calculated as indicators of biopolymer blended degradation when exposed to UV exposure. Fourier transform infrared (FTIR) spectroscopy has been recently used to study the changes in surface chemistry of polyethylene after weathering. FTIR spectroscopy has been used to monitor carbonyl group formation, vinyl group formation, and changes in crystallinity of weathered polyethylene [7].

2 Methodology

2.1 Materials

HDPE (HI1600) injection moulding grade with a density of 0.94–0.965 g/cm^3 and the HDPE melting point is 120° was supplied by Titan Petchem (M) Sdn. Bhd was used to produce HB. Meanwhile, high density polyethylene (HB0972) with density of 0.957 g/cm^3, melt index of 4 g/10 min and vicat softening point of 124 °C was supplied by Lotte Chemical Titan Sdn. Bhd was used to produce HM.

VeSawit Palm oil was used as the original palm oil to produce the virgin biopolymer (VP) for HB while waste cooking oil was used as the waste palm oil to produce the waste oil polymer (WP) for HM. The conversion of the monomer starts with the catalyst preparation to generate the epoxies from the unsaturated fatty compound, and the second reaction is the opening of the acid catalyst ring of the epoxies to form polyols or bio-monomer [8–17].

2.2 Instruments and Process

Bio-monomer was mixed with cross linking agent that was 4, 4'-Methylene Diphenyl Diisocyanate (MDI) and toluene as solvent agent. The mixture was stirred together for about 30 min at 50 °C and resulting viscous compound was poured into the mould to produced thin films. The thin film (VB) was cut into smaller size and then crushed using Pulverisette14 (Brand Fritch from Germany). The first sieve ring used was 6 mm at 14,000 rpm for 1 min while for the second cycle is using 0.75 mm sieve ring at 10,000 rpm for 2 min to produce biopolymer in particle form. Then, by using BrabenderPlastograph® EC machine (from Germany), HDPE were added with VP at 30% loading (wt /wt of biopolymer) at 130 °C and 25 rpm speed rate for 20 min. The blended biopolymer produced was crushes using Plastic Granulator SLM 50FY to produce smaller particle for HB production by using injection moulding.

Meanwhile, for production of HM, the monomer was heated in an oven at 80 °C until liquidized. Bio-monomer was mixed with cross linking agent that is flexible Polymethane Polyphenyl Isocyanate (Modified polymeric-MDI) with a viscosity at 25 °C of 120–160 cps, specific gravity at 25 °C of 1.18–1.20 g/ml, NCO content, % wt of 26.3–27.3. The liquid bio-monomer and isocyanate was stirred together with proportion ratio of 1:0.5 and called as waste oil biopolymer (WP).After that, HDPE and 30% WP were mixed manually together until become homogenously. HDPE and WP blended is left for 30 min for solidified and ready for injection moulding process.

To produced dogbone tensile sample, HB and HM were then fed to injection moulding machine (Nissei Horizontal Screw Type Injection Moulding NP7 Real Mini from Japan). The mould was designed according to ISO 527 (5A). Barrel temperature set points from nozzle zone which is 170 °C, 170 °C for front zone, 165 °C for middle zone, 160 °C for rear 1 zone, 155 °C for rear 2 zone and 70 °C for feed zone.

FTIR spectroscopy was conducted using Perkin–Elmer (UK) Spectrum One FTIR spectrometer (in the range 4000–400 cm^{-1}, absorbance units) to provide detailed information of the functional groups present at the surface of the samples. Scans were executed at scanning of 32 times at resolution 4 cm^{-1}. The spectra were acquired using the spectrometer of Attenuated Total Reflection (ATR). HB and HM were placed flat on the crystal plate and clamped down for infrared spectroscopy analysis. Materials generated a spectrum or fingerprint that identified the molecular composition of the materials. The Carbonyl Index (CI) was evaluated by following equation to obtain a quantitative comparison of the functional groups formed during the UV irradiation [18, 19].

$$CI = A_{1715}/A_{2850} \tag{1}$$

where A_{1715} is the absorbance of carbonyl stretching vibration at 1715 cm^{-1}, and A_{2850} is the symmetrical stretching vibration of methylene at 2850 cm^{-1}.

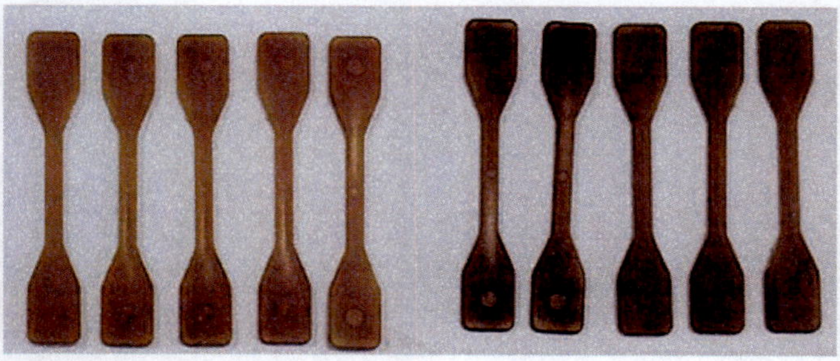

Fig. 1 The sample produced after injection moulding

Fig. 2 QUV Weatherometer machine

HB and HM produced from injection moulding (Fig. 1) were then exposed to UV irradiation in weatherometer UV Lamp Test Chamber Model HD-703 (Haida International Equipment Co., LTD) (Fig. 2) at 50 °C with 500H, 1000H, 2000H and 3000H exposure hours. The UV irradiation of the dumbbell sample was carried out using an array of UV fluorescent lamp emitting light in the region from 280 to 320 nm with a tail extending to 400 nm.

3 Results and Discussion

Figure 3 show the overlay of FTIR spectra for HB and HM at 30% loading of bio-based oils polymer at different UV irradiation exposure hours. The C-H group, C=O group, CH_3 and C-O group (esters group) were assigned in Table 1 and Table 2. From Fig. 3, peaks at 2914 cm^{-1} and 2848 cm^{-1} were refer to C-H group. From previous researcher, the frequency of C-H group is at 3000–2850 cm^{-1} [20]. Peak at 1698–1701 cm^{-1} contribute to C=O group in amide region while 1472–1310 cm^{-1} is assigned to the CH_3 bending peaks and for the strong intensity absorptions near

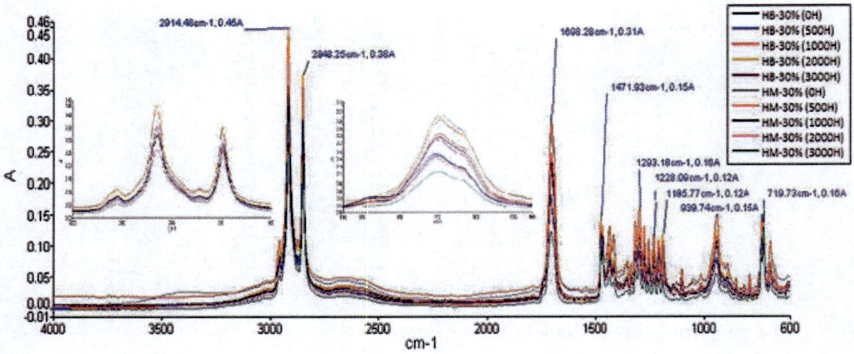

Fig. 3 Overlay of FTIR spectra for HB and HM at 30% loading of BP at different hours

Table 1 Wavenumbers (cm^{-1}) and assignments of IR bands of HB at different UV irradiation exposure hours

Wavenumbers (cm^{-1})/Assignments of IR bands	0 h	500 h	1000 h	2000 h	3000 h
C-H (stretching)	2914/2848	2914/2848	2914/2848	2914/2848	2914/2848
C=O	1698	1701	1698	1698	1699
CH_3 bending	1471	1472	1471	1471	1472
C-O (Esters Group)	1293–1186	1290–1184	1293–1186	1293–1185	1292–1099

Table 2 Wavenumbers (cm^{-1}) and assignments of IR bands of HM at different UV irradiation exposure hours

Wavenumbers (cm^{-1})/Assignments of IR bands	0 h	500 h	1000 h	2000 h	3000 h
C-H (stretching)	2915/2848	2915/2848	2915/2848	2914/2848	2914/2848
C=O	1698	1698	1700	1700	1699
CH_3 bending	1471	1471	1472	1472	1472
C-O (esters group)	1293–1185	1293–1186	1293–1185	1293–1185	1293–1185

1000–1300 cm^{-1}, indicate to esters group (C-O). FTIR spectra verified that nearly all the diisocyanate groups reacted during polymerization and urethane linkages and amide groups were created as the peaks around 2254–2310 cm^{-1} characteristic of isocyanate (-N=C=O) group of the MDI were disappeared [21, 22].

As refer to Fig. 3, the intensity of the absorbance for HB peaks gradually increased with the increasing of UV irradiation exposure hours. The highest intensity for HB is at 2000H exposure while the lowest intensity is at 500H exposure for all peaks. The absorbance intensity of C-H group and C=O group for HB at 2000 h are 0.45A and 0.30A respectively. Meanwhile, at 500H exposure, the absorbance intensity for C-H and C=O is 0.32A and 0.18A respectively.

By increasing the UV irradiation exposure time, the content of the carbonyl and ester groups also increased due to the proportional to the number of chain scissions occurring in HB/HM after weathering. According to Nguyen et al. [23], polyethylene is well known to mainly undergo two chain fracture reactions, Norrish I and Norrish II upon UV irradiation exposure. The short chain undergoes a cross-linking reaction. In the Norrish I reaction, the resulting free radicals can attack the polyethylene, which may lead to chain scission or crosslinking. In the Norrish II reaction, vinyl groups and terminal carbonyl groups are produced, and a chain scission occurs.

For HM peaks, the intensity of the absorbance peaks shows decreasing with increasing in UV irradiation exposure hours. The highest intensity absorbance is at 500H exposure with 0.44A for C-H peak and 0.29A for C=O peak while the lowest intensity is at 3000H exposure with the intensity for C-H is 0.26A and for C=O is 0.12A. This is due to decreases in intensity of C-H bond absorption after UV exposure indicated that a large number of C-H bonds were consumed during the photooxidation reaction. UV radiation either converted them into free radicals by removing the hydrogen atom during the initiation step or they reacted with a peroxy radical to form hydroperoxide and another free radical during the propagation step [24].

Figure 4 illustrates the formation of carbonyl group (CI value) as the function of irradiation duration. Carbonyl Index (CI) is the most commonly used indicator to measure the chemical oxidation of polyolefins such as polyethylene and polypropylene, and carbonyl index is generally considered to reflect the degradation of the mechanical properties of these polymers [25]. In Fig. 4, CI value for HB decrease at 500H exposure then increases at 1000H then gradually decrease at 2000H and 3000H. Meanwhile for HM the CI value decreases at 500H and 1000H then increase back at 2000H while decrease back at 3000H. Increasing in CI value implying easy oxidation reaction occur due to aging time increases. According to [20], this indicates that photo-oxidative aging and cross-linking between molecular chains are a pair of competition reactions in the UV aging process that shows in increasing of CI value. The decreases of CI value due to carbonyl groups formed and were consumed to form macromolecular radicals that further combine to form new double bonds during the

Fig. 4 Carbonyl index (CI) value of HB and HM

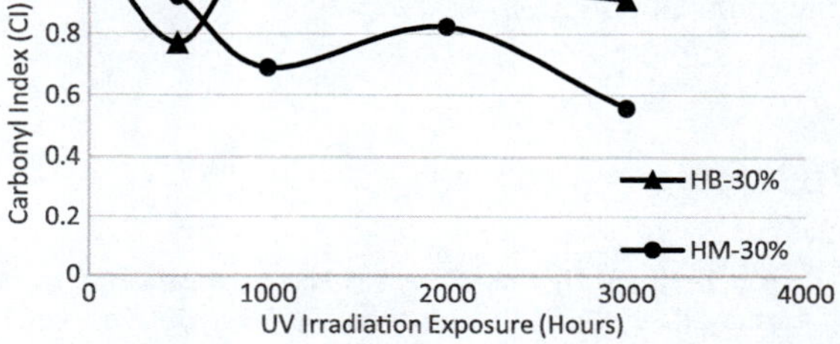

UV irradiation process of polyethylene. During the UV irradiation exposure, the cross-linking structures are produced, which is the reason for the fact that the CI of HB and HM decreases. The carbonyl group is interrupted by high-energy UV light to form free radicals during the degradation process, and the molecular chains with free radicals react with each other to form a new cross-linking structure.

The intensity absorbing and CI value for HB shows parallel to each other, which is at 500H shows lowest intensity with lowest CI value that is 0.7712. Meanwhile, the highest intensity absorption is at 2000H with the CI value is 0.9828. For HM, intensity absorbing, and CI value also shows parallel to each other at the lowest absorbing intensity with lowest CI value that is at 3000H with 0.5566 CI value. For the highest intensity absorbance is at 500H with the highest CI value that is 0.925. This shows, the correlation ship between the intensity absorbance and CI value supports each other during photo-oxidation process. The HB and HM shows different oxidation process occurs effectively at different UV exposure hours due to contradict in materials used which is HB come from VP and HM from WP. Otherwise, the blended technique for HB by using Brabender machine while HM was blended manual with different crosslinking agent, but both shows photodegradation was occur upon UV irradiation exposure.

4 Conclusions

In this study, HB and HM was blended by using different technique with different origin biopolymer which is VP and WP also using different crosslinking agent shows successfully of blending. By blending bio-based oil polymer and HDPE which is produced HB and HM shows increasing properties of bio-based oil polymer itself in terms of processability and end use implementation and improve the drawbacks of HDPE itself which is its decaying at long period of time. HB and HM were exposed to different exposure hour to identify the photodegradation process. HB shows the maximum oxidation process occur at 2000H while HM shows at 500H. UV irradiation exposure may cause photo-oxidative degradation which results in breaking of the polymer chains, produces free radical and reduces the molecular weight, causing deterioration of mechanical properties and leading to useless materials after some period of time exposure make it suitable for end used implementation such as biodegradable packaging, disposal cutlery or crockery, biodegradable plastic bags, mulching film for plantation industries and many more that can reduced environment pollution.

References

1. Zou P, Xiong H, Tang S (2008) Natural weathering of rape straw flour (RSF)/HDPE and nano-SiO_2/RSF/HDPE composites. Carbohyd Polym 73:378–383

2. Fairbrother A, Hsueh HC, Kim JH, Jacobs D, Perry L, Goodwin D, White C, Watson S, Sung LP (2019) Temperature and light intensity effects on photodegradation of high-density polyethylene. Polym Degrad Stab 165:153–160
3. Morales-Mendez JD, Silva-Rodríguez R (2018) Environmental assessment of ozone layer depletion due to the manufacture of plastic bags. Heliyon 4
4. Desnellia B, Mujahidina DY, Permanaa and Radimana, CL (2015) The olefin reaction between crude palm oil fatty acid methyl ester (CPO FAME) and ethylene using grubbs II catalyst
5. Harding KG, Gounden T, Pretorius S (2017) Biodegradable plastics: a myth of marketing. Procedia Manufact 7:106–110
6. Yousif E, Haddad R (2013) Photodegradation and photostabilization of polymers, especially polystyrene. Review
7. Zou P, Xiong H, Tang S (2008) Natural weathering of rape straw flour (RSF)/HDPE and nano-SiO_2/RSF/HDPE composites. Carbohydrate Polym 73:378–383
8. Sulong N, Rus AZM (2013) Influence of TiO_2 on selfclean bio coating. Appl Mech Mater 315:399–403
9. Rus AZM, Abdullah N, Abdullah MFL (2015) Interconnected interface enhanced electrical properties of graphite in bio-based epoxy from insulating to conductor composites. Indian J Sci Technol 8
10. Rahim RA, Rus AZM, Bakhtiar ASA (2013) Preliminary fabrication of polymer membranes from renewable resources. Appl Mech Mater 315:428–432
11. Rus AZM (2010) Polymers from renewable materials. Sci Prog 93:285–300
12. Najibah Ab Latif and Anika Zafiah Mohd Rus (2014) Vibration transmissibility study of high density solid waste biopolymer foam. J Mech Eng Sci 6:772–781
13. Rus AZM, Hassan NNM (2015) Thermal degradation and damping characteristic of UV irradiated biopolymer. Int J Polym Sci 11. Article ID 615284
14. Ghani MKZA, Rus AZM (2013) Influence of hot compression moulding of biopolymer filled waste granulate biopolymer. Appl Mech Mater 315: 448–452
15. Hassan NNM, Rus AZM, Sulong N, Mohid SR (2013) Acoustic study based on sustainable green polymer treated with H_2O. Adv Mater Res 74:281–285
16. Rus AZM, Mohid SR, Sulong N, Marsi N (2013) Biopolymer doped with titanium dioxide superhydrophobic photocatalysis as self-clean coating for lightweight composite. Adv Mater Sci Eng
17. Mohid SR, Rus AZM, Harun NH (2013) Influence of bio polymer composites as heat absorption coating. Appl Mech Mater 315:404–407
18. Wang S, Zhang J, Liu L, Yang F, Zhang Y (2015) Evaluation of cooling property of high-density polyethylene (HDPE)/titanium dioxide (TiO_2) composites after accelerated ultraviolet (UV) irradiation. Solar Energy Mater Solar Cells 143:120–127
19. Barbes L, Radulescu C, Stihi C (2014) ATR-FTIR spectrometry characterisation of polymeric materials. Romanian Rep Phys 66(3):765–777
20. Jiang T, Qi Y, Wu Y, Zhang J (2019) Application of antioxidant and ultraviolet absorber into HDPE: enhanced resistance to UV irradiation. e-Polymers 19:499–510
21. Lu Y, Larock RC (2010) Soybean oil-based, aqueous cationic polyurethane dispersions: synthesis and properties. Prog Organic Coat 69(1): 31–37
22. Macalino AD, Salen VA, Reyes LQ (2017) Castor oil based polyurethanes: synthesis and characterization. In: IOP conference series: materials science and engineering. p 229
23. Nguyen VD, Hao J, Wang W (2018) Ultraviolet weathering performance of high-density polyethylene/wood-flour composites with a basalt-fiber-included shell. Polymers 10:831
24. Cui H, Hanus R, Kessler MR (2013) Degradation of ROMP-based bio-renewable polymers by UV radiation. Polym Degrad Stab 98:2357–2365
25. Rouillon C, Bussiere P-O, Desnoux E, Collin S, Vial C, Therias S, Gardette J-L (2016) Is carbonyl index a quantitative probe to monitor polypropylene photodegradation? Polym Degrad Stab 128:200–208

Effects of Solvents on ZnO Nanoparticles Synthesis via Sol–gel Method

Suraya Sulaiman, Nur Syazwa Zamri, Radhiyah Abd Aziz, Mohamad Farid Mohamad Sharif, Natasha Ahmad Nawawi, and Nur Ayuni Jamal

Abstract Zinc oxide nanoparticles have been prepared by the sol-gel method at different solvent (methanol (MeOH), ethanol (EtOh) and distilled water) and at different calcination temperatures (700 °C, 800 °C and 900 °C). The phase and microstructure of the prepared ZnO powder were investigated. ZnO powder were characterized by using XRD, EDX and FESEM. XRD analysis shows ZnO exhibited a hexagonal (wurtzite) structure with crystallite sizes 34.146 nm, 34.283 nm, and 34.523 nm and FESEM micrographs show that synthesized ZnO has a nanorod-like structure with an average particle size, 113.716 nm, 125.825 nm, and 141.725 nm for solvent methanol, ethanol and distilled water and calcination temperature of 700 °C, 800 °C and 700 °C respectively. The obtained ZnO nanoparticles are homogenous and consistent in size, corresponding to the XRD results that exhibit good crystallinity. EDX analysis shows pure ZnO with different solvents at 700 °C calcination temperatures. The surface of the ZnO also exhibits elements of O and Zn. This result has confirmed that the ZnO nanoparticles has high purity. Based on the analysis from the XRD and FESEM test, the best solvent with the best calcination temperature has been chosen, which is Methanol at 700 °C.

Keywords Zinc oxide nanoparticle · Grain size · Sol-gel method · Solvent

S. Sulaiman (✉) · N. S. Zamri · R. A. Aziz · M. F. M. Sharif
Universiti Malaysia Pahang, 26600 Pekan, Pahang, Malaysia
e-mail: surayas@ump.edu.my

S. Sulaiman
Universiti Teknologi Malaysia, 81310 Johor Bahru, Johor, Malaysia

N. A. Nawawi
Universiti Teknologi Mara, 13500 Permatang Pauh, Pulau, Pinang, Malaysia

N. A. Jamal
Universiti Islam Antarabangsa Malaysia, 50728 Kuala Lumpur, Malaysia

M. Y. Ismail et al. (eds.), *Technological Advancement in Mechanical and Automotive Engineering*, Lecture Notes in Mechanical Engineering,
https://doi.org/10.1007/978-981-19-1457-7_14

1 Introduction

In recent years, there have been many problems related to the global energy crisis and environmental issues [1, 2]. This situation has occurred due to massive consumption of oil, coal and gas, and other sources of non-renewable energy used to burn fossil fuels and generate and develop power [3, 4]. The effect of these activities is the addition of vast quantities of carbon dioxide and other greenhouse gases in the atmosphere. This has a catastrophic environmental impact and has affected our climate. In 2017, approximately 67% energy wasted in the United States [5] and more than 60% of thermal energy is lost in waste heat worldwide [4, 6, 7], indicating 80% of fuel energy is lost in conventional cars as heat. In most cases, waste heat is lost without any economic gain, and thermoelectric technology holds the key to convert this massive amount of waste heat back into useful electricity [1, 8] and provide an encouraging solution to the energy crisis and environmental issues [9].

In thermoelectric technology, thermoelectric devices are widely used to convert a large amount of high-temperature energy or thermal energy, including waste heat, into usable electrical energy. One of the required characteristics is the materials needs to be stable at high temperatures. It is envisaged that an excellent thermoelectric device will convert 15–20% of waste heat into useful electricity [6]. However, the energy conversion efficiency is still low, around 7–8% [1, 8, 9]. The high capital cost of thermoelectric technologies in comparison to conventional counterparts is also a major limitation for the broader applications that must be addressed. The thermoelectric device's efficiency is determined by the materials' electrical resistivity (ρ), thermal conductivity (κ), and Seebeck coefficient (S) [1, 10–12]. A dimensionless quantity known as a figure of merit (ZT) connects these parameters. The ZT value represents the device's ability to transform energy efficiently. The higher the ZT value, the more efficiently the device can convert energy. ZT values for thermoelectric materials have no theoretical limit; however, the ZT value should be minimum of 1 to assure the material's practical appropriateness [13–15].

The selection of thermoelectric materials influences the efficiency of the thermoelectric device. Over the years, conventional thermoelectric materials are bismuth telluride (Bi_2Te_3) for room temperatures up to 200 °C, lead telluride (PbTe) for 400–600 °C, and silicon germanium alloys (SiGe) for 400–1000 °C have been extensively studied, but these materials have low resistance to high air temperatures, toxic, high cost of the included components, and rare elements [15–17]. Either the material or the system begins to breakdown under this heat. Thus, current technologies have been unable to capture energy where most of it is available in the 700–1000 °C. For this reason, zinc oxide (ZnO) is presented due to excellent physical and chemical properties, high electron mobility, inexpensive materials, less toxic, minimal in environmental impact compared to tellurides [15–17] as well as its ability to operate at high temperature ($\geq$700 °C). These advantages make ZnO an ideal potential for thermoelectric applications at high temperatures.

One of the obstacles of its emergence is ZnO itself has lower electrical conductivity [18] and higher thermal conductivity about 49 W/mK at 27 °C and 10 W/mK at

727 °C [19–21], which restricts its use and efficiency in thermoelectric applications. The high thermal conductivity of ZnO is due to its simple crystal structure and light atomic mass. As a results, the thermoelectric performance is poor [15, 22, 23] and ZT value is less than 0.01 [15, 21]. It has been reported that the thermal conductivity of ZnO ceramics can be reduced without compromising electrical conductivity by having particles that are smaller in size than the mean-free path of phonon and larger than the mean free paths of electrons or holes as it correlates directly with the phonon scattering and lattice thermal conductivity of nanostructured materials [24]. However, despite numerous techniques reported on the alternation of grain sizes in ZnO ceramics, there has been limited studies compiling the effect of ZnO ceramic synthesis parameters on its grain sizes, and corresponding thermal conductivity and thermoelectric properties. Therefore, in this paper will investigates the relationship between the effect solvents and calcination temperatures on the size of grain size.

2 Experimental

2.1 Preparation of ZnO

ZnO nanostructure was synthesized using sol-gel method. The schematic diagram of the experiment is shown in Fig. 1 and three different precursor solutions were prepared as summarized in Table 1. The first solution, Sol A was prepared as follows: 10 g of zinc acetate dehydrate ($Zn(CH_3COO)_2.2H_2O$) with purity ≥98% (Sigma-Aldrich) was dissolved in 150 ml Methanol (MeOH) and then stirred using magnetic stirrer at 80 °C until the material dissolved and get a clear solution. After that, 3 g sodium hydroxide (NaOH) with purity ≥97% (Sigma-Aldrich) was added to the solution and continue stir at 80 °C for 2–3 h to achieve a white precipitate gelation solution. The final powder was calcined at different temperatures (700 °C, 800 °C and 900 °C) in air for 1 h to obtain ZnO. Sol B and Sol C was prepared in a similar way but using different solvent which is Ethanol (EtOh) and distilled water.

2.2 Characterization Methods

The phases transition of sample powders was identified and characterized using X-ray diffractometer (XRD) on a Bruker D8 Advance model and microstructures, size and shape of the samples were observed by the FESEM on a JSM-7800F Schottky model.

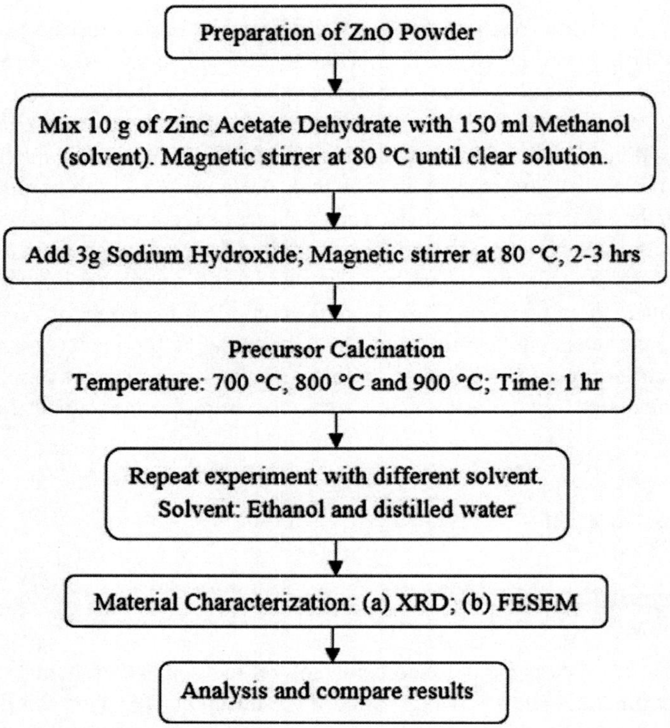

Fig. 1 Schematic diagram of the experiment

Table 1 Parameters of the three precursor solutions

Precursor solutions	Materials	Solvents	Medium
Sol A	Zinc acetate dihydrate	MeOH	NaOH
Sol B	Zinc acetate dihydrate	EtOH	NaOH
Sol C	Zinc acetate dihydrate	Distilled water	NaOH

3 Results and Discussions

3.1 XRD Analysis

Figures 2, 3 and 4 presents the XRD patterns within a scan range of 10–80° for ZnO powder of each solvent at calcination temperatures of 700, 800 and 900 °C for 1 h using furnace. The crystal structure of ZnO is wurtzite hexagonal structure

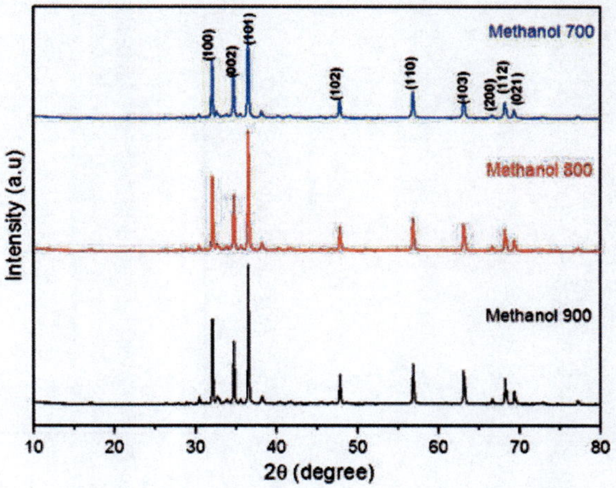

Fig. 2 XRD Analysis at 700, 800 and 900 °C calcination temperature for sol A (methanol)

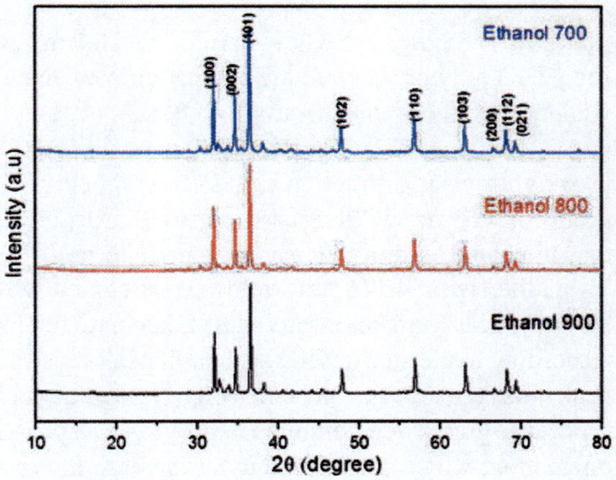

Fig. 3 XRD Analysis at 700, 800 and 900 °C calcination temperature for sol B (ethanol)

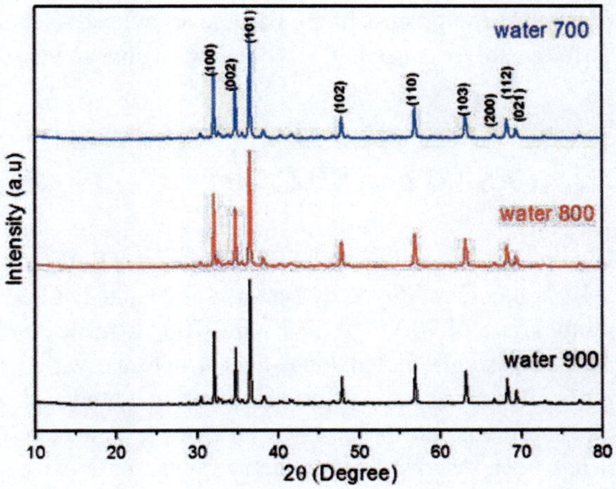

Fig. 4 XRD Analysis at 700, 800 and 900 °C calcination temperature for Sol C (distilled water)

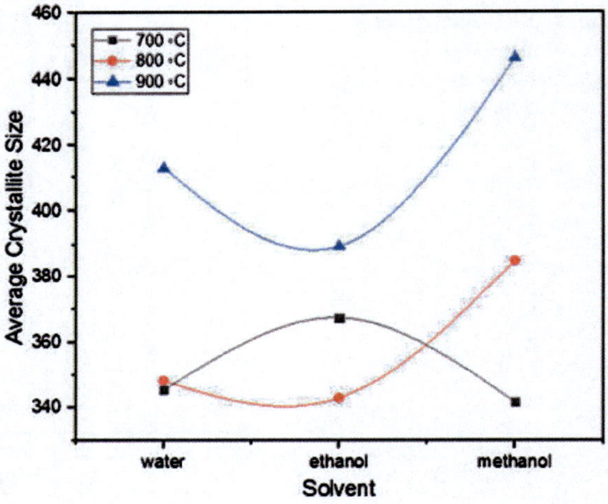

Fig. 5 Average crystallite size graph

[25, 26]. The powder XRD pattern of a standard ZnO gives the characteristic peak for a ZnO poly crystal and the cell parameter can be obtained refinement. The peak is identified to originate from (100), (002), (101), (102), (110), (103), (200), (112), and (201) planes [23]. The synthesized powder, ZnO nanostructure showed a single phase with clear diffraction peaks, correspond to the reported data with the space group of P63mc (JCPDS card no 36-1451) [23, 25, 26]. From results, it shows that methanol with a calcination temperature of 700 °C has the smallest average crystallite size, 34.146 nm, followed by ethanol with a calcination temperature of 800 °C, 34.283 nm, and water with a calcination temperature of 700 °C, 34.523 nm. According to the study, XRD with high peak cause a rise in crystallite size. The ZnO peak looks crisp and clean, with a high crystalline content at temperature 900 °C. This is because greater temperatures improve the evaporation process and the interparticle interaction, which might lead to an increase in crystalline formation. From Fig. 5, it can show that calcination temperature influenced the rate of evaporation and the crystallization process of the particles produced. Hence, it can be concluded that the different solvents used affected the grain size and crystallite size.

3.2 *FESEM and EDX Analysis*

Figures 6, 7 and 8 shows FESEM images and EDX analysis of ZnO nanoparticles for solvent methanol (Sol A), ethanol (Sol B) and distilled water (Sol C) and calcination temperature of 700 °C, 800 °C and 700 °C respectively. The shape is nanorod-like structure, single-dimensional nanostructures, which might make carrier transport more efficient because of decreased grain boundaries, surface defects, disorders and discontinuous interfaces. As a result of zinc acetate being utilized as a precursor, it has been discovered that the synthetic ZnO exist in the form of a nanorod-like

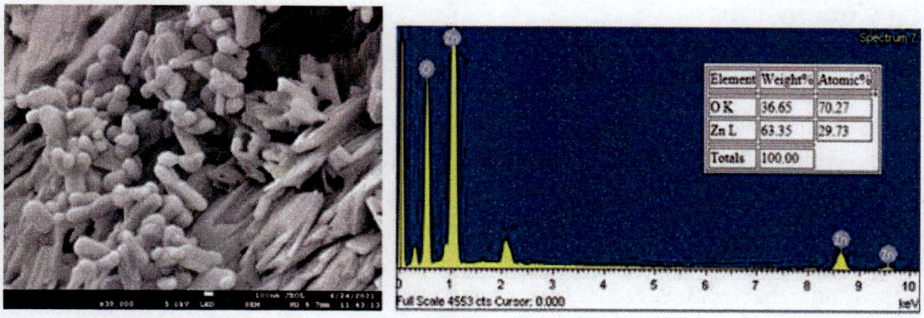

Fig. 6 FESEM and EDX analysis at 700 °C for sol A (methanol)

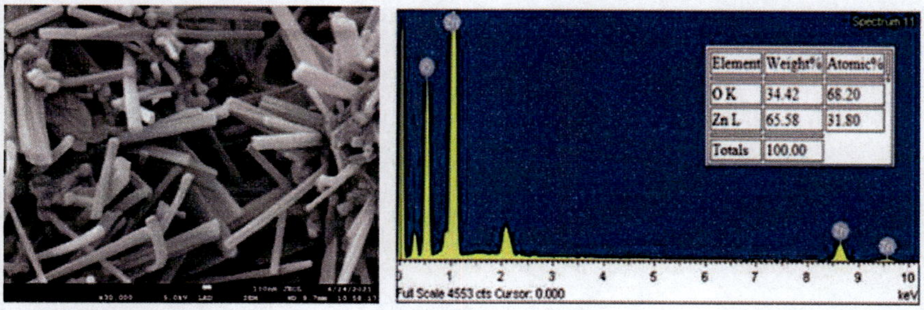

Fig. 7 FESEM and EDX analysis at 700 °C for sol B (ethanol)

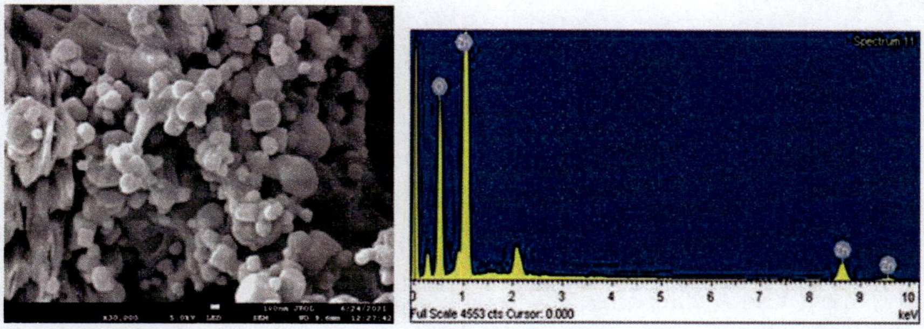

Fig. 8 FESEM and EDX analysis at 700 °C for sol C (distilled water)

structure. ZnO nanoparticles that is formed using zinc acetate is smoother than ZnO nitrate. The results are as shown scanning emission microscopy produced an image with a nanorod-like structure with an average particle size, 113.716 nm, 125.825 nm, and 141.725 nm for methanol at 700 °C, ethanol at 800 °C, and water at 700 °C. An energy dispersive X-ray (EDX) spectroscopy analysis shows pure ZnO with different solvents. The surface of the ZnO also exhibits elements of O and Zn. This result has confirmed that the ZnO nanoparticles has high purity.

4 Conclusions

ZnO nanoparticles were synthesized the sol-gel method at different solvent and at different calcination temperatures. From XRD and FESEM results, it was observed that all the ZnO nanoparticles exhibited the hexagonal, wurtzite structure. The smallest crystallite sizes of ZnO nanoparticles were 34.146 nm followed by 34.283 nm, and 34.523 nm with solvent methanol, ethanol and distilled water and calcination temperature of 700 °C, 800 °C and 700 °C respectively. FESEM micrographs show that synthesized ZnO has a nanorod-like structure with smallest average particle size is 113.716 nm followed by 125.825 nm, and 141.725 nm for methanol at 700 °C, ethanol at 800 °C, and water at 700 °C respectively. The obtained ZnO nanoparticles are homogenous and consistent in size, corresponding to the XRD results that exhibit good crystallinity. Based on the analysis from the XRD and FESEM test, the best solvent with the best calcination temperature has been chosen, which is Methanol at 700 °C. These results indicate that the solution solvent and calcination temperatures have a significant influence on the crystalline structure and morphology on ZnO nanoparticles.

Acknowledgements This paper is supported by Sustainable Research Collaboration Grant Funded by IIUM & UiTM (Grant no. RDU200744) and Faculty of Manufacturing and Mechatronic Engineering Technology, Universiti Malaysia Pahang.

References

1. Nam WH, Lim YS, Choi S-M, Seo W-S, Lee JY (2012) High-temperature charge transport and thermoelectric properties of a degenerately Al-doped ZnO nanocomposite. J Mater Chem 22(29):14633–14638
2. Jia D, Weib Z, Mazzonib S, Mengarellib M, Rajooe S, Zhaoc J, Poua J, Romagnoli A (2018) Thermoelectric generation for waste heat recovery: application of a system level design optimization approach via Taguchi method. Energy Convers Manage 172:507–516
3. Zoui MA, Bentouba S, Stocholm JG, Bourouis M (2020) A review on thermoelectric generators: progress and applications. Energies 13(14):1–32
4. Gayner C, Kar KK (2016) Recent advances in thermoelectric materials. Prog Mater Sci 83:330–382
5. Wolf M, Hinterding R, Feldhoff A (2020) High power factor vs. high zT—a review of thermoelectric materials for high-temperature application. Entropy 21(1058):1–32
6. Funahashi R, Barbier T, Combe E (2015) Thermoelectric materials for middle and high temperature ranges. J Mater Res 30(17):1–14
7. Zhang X, Zhao L-D (2015) Thermoelectric materials: energy conversion between heat and electricity. J Materiomics 1:92–105
8. Jantrasee S, Moontragoon P, Pinitsoontorn S (2016) Thermoelectric properties of Al-doped ZnO: experiment and simulation. J Semiconductors 379:1–8
9. Jouhara H, Khordehgah N, Almahmoud S, Delpech B, Chauhan A, Tassou SA (2018) Waste heat recovery technologies and applications. Therm Sci Eng Prog 6:268–289
10. Koresh I, Amouyal Y (2017) Effects of microstructure evolution on transport properties of thermoelectric nickel-doped zinc oxide. J Eur Ceram Soc 37(11):3541–3550

11. Liu W, Hu J, Zhang S, Deng M, Han C-G, Liu Y (2017) New trends, strategies and opportunities in thermoelectric materials: A perspective. Mater Today Phys 1:50–60
12. Park Y, Cho K, Kim S (2017) Thermoelectric characteristics of glass fibers coated with ZnO and Al-doped ZnO. Mater Res Bull 96:246–249
13. Tritt TM, Subramanian M (2006) Thermoelectric materials, phenomena, and applications : a bird' s eye view. MRS Bull 31(March):188–198
14. Kim HJ, Skuza JR, Park Y, King GC, Choi SH, Nagavalli A (2012) System to measure thermal conductivity and seebeck coefficient for thermoelectrics. NASA/TM–2012-217791 Dec 2012
15. Yin Y, Tudu B, Tiwari A (2017) Recent advances in oxide thermoelectric materials and modules. Vacuum 146:356–374
16. García LM (2016) Preparation and characterisation of nanostructured bulk Bi_2Te_3 thermoelectric materials using ultrasound milling. Cardiff University
17. Ohtaki M (2011) Recent aspects of oxide thermoelectric materials for power generation from mid-to-high temperature heat source. J Ceram Soc Jpn 119(11):770–775
18. Janotti A, Van De Walle CG (2009) Fundamentals of zinc oxide as a semiconductor. Rep Prog Phys 72(126501):1–29
19. Colder H, Guilmeau E, Harnois C, Marinel S, Retoux R, Savary E (2011) Preparation of Ni-doped ZnO ceramics for thermoelectric applications. J Eur Ceram Soc 31(15):2957–2963
20. Tsubota T, Ohtaki M, Eguchi K, Arai H (1997) Thermoelectric properties of Al-doped ZnO as a promising oxide material for high- temperature thermoelectric conversion. J Mater Chem 7:85–90
21. Jood P, Mehta RJ, Zhang Y, Peleckis G, Wang X, Siegel RW, Borca-tasciuc T, Dou SX, Ramanath G (2011) Al-Doped zinc oxide nanocomposites with enhanced thermoelectric properties. Am Chem Soc 11:4337–4342
22. Jood P, Peleckis G, Wang X, Dou SX (2012) Effect of gallium doping and ball milling process on the thermoelectric performance of n-type ZnO. J Mater Res 27(17):2278–2285
23. Han L, Van Nong N, Zhang W, Hung T, Holgate T (2014) Effects of morphology on the thermoelectric properties of Al-doped ZnO. RSC Adv 4:12353–12361
24. Han C, Li Z, Dou S (2014) Recent progress in thermoelectric materials. Sci Bull 59(18):2073–2091
25. Hasnidawani JN, Azlina HN, Norita H, Bonnia NN, Ratim S, Ali ES (2016) Synthesis of ZnO nanostructures using sol-gel method. Procedia Chem 19:211–216
26. Guan W, Zhang L, Wang C, Wang Y (2017) Theoretical and experimental investigations of the thermoelectric properties of Al-, Bi- and Sn-doped ZnO. Mater Sci Semicond Process 66:247–252

A Comparative Study and ANSYS Simulation on Thermal Performance of Shell and Tube Heat Exchanger Operated with Al_2O_3-Water and TiO_2-Water Nanofluids

U. Z. A. Rahman, A. S. A. Abdelhamid, and Mohammed W. Muhieldeen

Abstract Shell and tube heat exchangers operated with nanofluids are extensively studied and applied in the industry due to the capabilities of nanofluids as a more efficient working fluid compared to commonly used fluid such as water and oil. In this study, the shell-and-tube heat exchanger is modelled and analyzed using ANSYS, where temperature difference and pressure drop of alumina in water (Al_2O_3-water) and titanium dioxide in water (TiO_2-water) nanofluids for volumetric concentration from 0 to 2.5% were studied. The results were compared with the findings from existing research and journals. Al_2O_3-water nanofluid showed significantly higher heat transfer enhancement compared to TiO_2-water nanofluid for volume concentration ranging from 1 to 2.5%. Al_2O_3-water nanofluid also showed a temperature difference percentage of 12% for 2.5% volume concentration, while TiO_2-water nanofluid showed 9.15% of temperature difference at 2.5% volume concentration.

Keywords Nanofluids · Thermal enhancement · Shell and tube heat exchanger · ANSYS

1 Introduction

1.1 Research Overview

The application of nanofluids in heat exchangers has been studied numerous times, as researchers are looking into all possibilities of enhancing thermal performance of heat exchangers [1, 2]. One of the main reasons is because heat exchanger is

U. Z. A. Rahman · A. S. A. Abdelhamid · M. W. Muhieldeen
Mechanical Engineering Department, Faculty of Engineering, Technology & Built Environment, UCSI University, 56000 Kuala Lumpur, Malaysia
e-mail: ummi@ucsiuniversity.edu.my

M. W. Muhieldeen (✉)
UCSI-Cheras Low Carbon Innovation Hub Research Consortium, Kuala Lumpur, Malaysia
e-mail: al-gailani@ucsiuniversity.edu.my

M. Y. Ismail et al. (eds.), *Technological Advancement in Mechanical and Automotive Engineering*, Lecture Notes in Mechanical Engineering,
https://doi.org/10.1007/978-981-19-1457-7_15

widely used heat transfer application and is one of main components in engineering devices such as air-conditioning, refrigerators, heat pump, power plants, and other machineries operating at very high temperature. The commonly used working fluid inside shell and tube heat exchanger is distilled water, mainly due to the cost and resource availability. However, the utilization of distilled water does not show a significant heat transfer rate when it comes to heat exchanging applications. As a result, researchers recently began to study and compare alternative fluids such as nanofluids to find the best working fluid for heat exchanging applications. Thermal performance associated properties such as temperature difference, pressure drop, viscosity, volume concentration, volume and mass flow rate, Reynolds and Nusselt numbers are extensively studied ever since the birth of nanofluids in 1990s.

Nanofluids have proven capabilities of providing a higher thermal performance compared to commonly used working fluid such as water and oil, due to the large amount of surface areas offered by nanoparticles, and for metallic nanoparticles, both thermal capacity and conductivity are able to enhance the thermal performance [3, 4]. An experimental study found the thermal conductivity and convective heat transfer of CuO-water and TiO_2-water nanofluids in shell and tube heat exchanger [5]. For volume concentration of 0.02, 0.04 and 0.06%, it has been found that the thermal conductivity and heat transfer overall efficiency were increasing proportionally to the increment of volume concentration [6]. Since several models have been introduced, selection of the model is critical for CFD simulation [7–12]. The optimum volume concentration associated with thermal performance has also been studied comprehensively by numerous researchers [13], for a fixed mass in-flow rate of 2 kg/s, the best thermal performance was found when the volumetric concentration of Al_2O_3-water is at 1% [14], where the thermal conductivity is found to be at 0.62 W/m.K at 1% concentration at room temperature. Similar optimum value was agreed by Nallusamy [15] for Al_2O_3-water with nanoparticles of approximately 50 nm diameter on convective heat transfer.

2 Methodology

The research work starts with modeling the shell and tube heat exchanger via SolidWorks and SpaceClaim. The dimension and parameters of the shell and tube heat exchangers are fixed and obtained from research journals. Next, the model is exported to Ansys where sets of boundary condition were applied to the model prior to the analyses. In Ansys Fluent, simulation on temperature difference and pressure drop versus nanoparticle concentration were carried out. Several comparisons were made between simulation results of Al_2O_3-water and TiO_2-water nanofluids, and between simulation results of Al_2O_3-water and TiO_2-water nanofluids and findings from several research journals.

Table 1 Specifications of heat exchanger

No.	Description	Value
1	Tube OD	30 mm
2	Tube ID	25 mm
3	Shell OD	731 mm
4	Shell ID	635 mm
5	Tube/shell length	7310 mm
6	Number of tubes	21

2.1 *Specification of Heat Exchanger*

Table 1 shows the specifications of the shell and tube heat exchanger. These values are adapted from [16, 17].

2.2 *Meshing*

In the simulation works, the number of elements is 3,426,808. A finer mesh was prepared for a location closer to the wall, to precisely solve for the flow adjacent to the wall. The turbulent flow near a flat wall can be divided into four regions. At the wall, where fluid velocity is zero, and in a thin layer above it, the flow velocity is proportional with the distance from the wall. This region is called the viscous sublayer, or laminar sublayer. Further away from the wall is a region called the buffer layer. In the buffer region, turbulence stresses begin to dominate over viscous stresses, and eventually connects to a region where the flow is fully turbulent, and the average flow velocity is related to the log of the distance to the wall. This is known as the log-law region. Accordingly, the meshing of the of the near wall was made manually by adjusting the quality and the meshing parameters of the near wall. The meshing of shell and tube heat exchanger is shown in Fig. 1.

For the tubes inside the shell and tube heat exchanger, tetrahedral meshing was selected. The tubes are one of the most critical parts, as the nanofluids pass through them. Therefore, tetrahedral meshing has been selected for the tubes to provide more accurate results, as shown in Fig. 2.

2.3 *Boundary Condition*

The enhancement inside a counterflow shell and tube heat exchanger utilizing Al_2O_3-water has been carried out by [18] to study the temperature profile and pressure drop. The boundary conditions set is presented in Table 2.

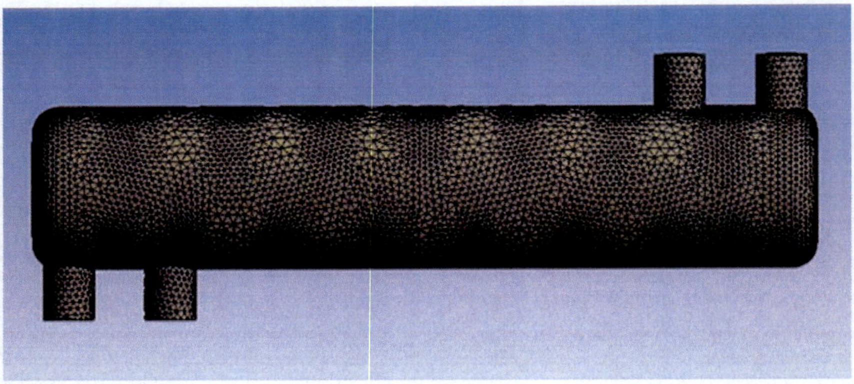

Fig. 1 Meshing of shell and tube heat exchanger

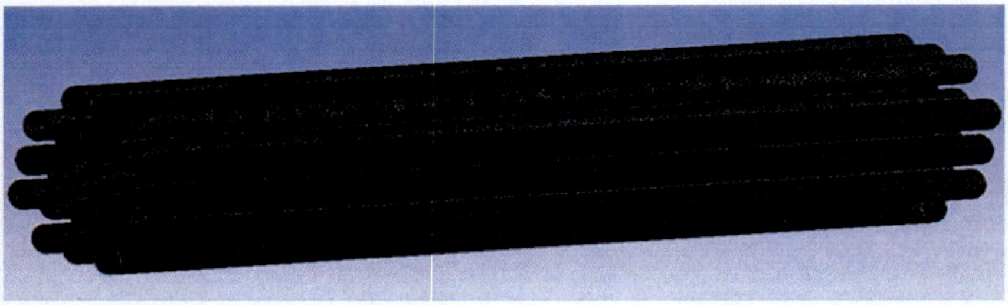

Fig. 2 Meshing of the tubes inside the shell and tube heat exchanger

Table 2 Boundary condition for Al_2O_3 nanofluid [18]

Location	Boundary condition	Shell	Tube
Inlet	Velocity	1.5 m/s	1.2 m/s
Outlet	Pressure	0 Pa	0 Pa
Temperature	Inlet temperature	333 °C	298 °C

CFD simulation of TiO_2-water nanofluid inside shell and tube heat exchanger has been conducted by [19] to study the temperature distribution inside the heat exchanger. The boundary conditions set is presented in Table 3.

Table 3 Boundary condition for TiO_2 nanofluid [19]

Location	Boundary condition	Shell	Tube
Inlet	Velocity	1.0 m/s	2.0 m/s
Outlet	Pressure	0 Pa	0 Pa
Temperature	Inlet temperature	365 °C	298 °C

Table 4 Ansys boundary condition

Location	Boundary condition	Shell	Tube
Inlet	Velocity	1.2 m/s	1.8 m/s
Outlet	Pressure	0 Pa	0 Pa
Wall	No slip	No heat flux	Coupled
Temperature	Inlet temperature	350 °C	298 °C
Inlet	Flow rate	5.5 kg/s	5.5 kg/s

The initial boundary conditions used for the shell and tube heat exchanger to generate temperature profile and pressure drop are listed down in Table 4.

The properties of Al_2O_3-water and TiO_2-water nanofluids used in Ansys were tabulated in Tables 5 and 6, respectively.

Table 5 Properties of Al_2O_3-water nanofluid

Volume fraction (%)	Thermal conductivity (W/m.K)	Density (kg/m^3)	Specific heat (J/kg.K)	Viscosity (Ns/m^2)
0 (pure water)	0.600	998.20	4182.00	0.001003
0.5	0.628	1009.17	4119.51	0.000799
1.0	0.641	1022.69	4060.62	0.000810
1.5	0.653	1036.21	3972.98	0.000819
2.0	0.664	1049.73	3885.36	0.000829
2.5	0.701	1060.21	3799.81	0.000840

Table 6 Properties of TiO_2-water nanofluid

Volume fraction (%)	Thermal conductivity (W/m.K)	Density (kg/m^3)	Specific heat (J/kg.K)	Viscosity (Ns/m^2)
0 (pure water)	0.600	998.20	4182.00	0.001003
0.5	0.615	1012.25	4122.62	0.000910
1.0	0.619	1026.68	4061.32	0.000941
1.5	0.626	1040.50	4000.22	0.000982
2.0	0.642	1056.27	3940.68	0.001010
2.5	0.651	1070.81	3880.77	0.001038

2.4 *Simulation of Temperature Profile and Pressure Drop Using Ansys*

Simulation through Ansys generates a color spectrum streamlines from blue to red. For both temperature profile and pressure drop analysis, the blue color represents lower value, while the red color represents the high value. The inlet and outlet for the flow were set, and the color spectrum is analyzed. The simulations are presented in Figs. 3 and 4.

Fig. 3 Temperature streamlines for temperature profile analysis using Ansys

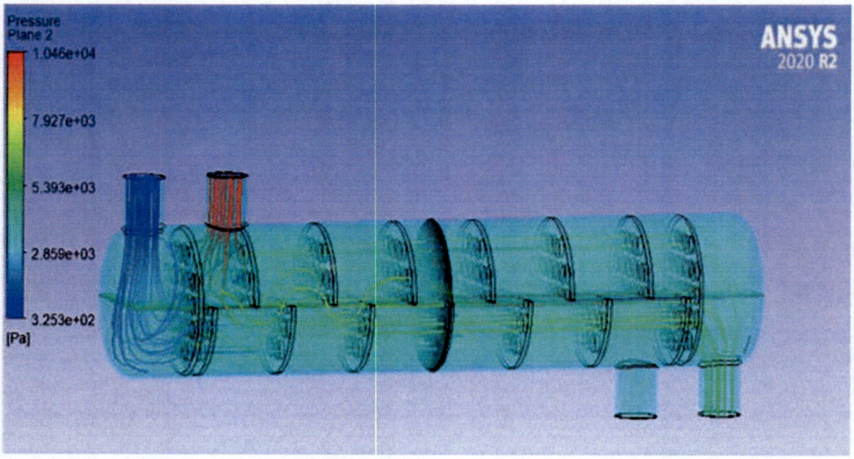

Fig. 4 Pressure streamlines for pressure drop analysis using Ansys

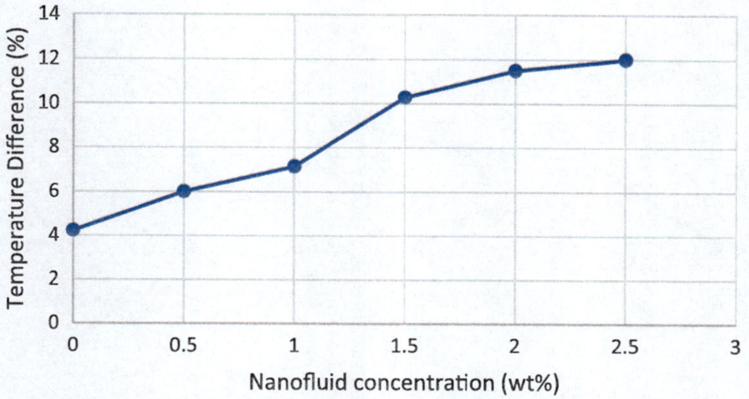

Fig. 5 Temperature difference versus concentration for Al_2O_3-water nanofluid

3 Results and Discussion

3.1 Temperature Difference Versus Nanoparticle Concentration for Al_2O_3-water Nanofluid

Figure 5 shows the temperature difference versus nanoparticle concentrations for Al_2O_3-water nanofluid ranging from 0 to 2.5%. For pure water (at 0% nanoparticle concentration), the temperature difference is 4.22%. The temperature difference keeps on increasing significantly with the nanoparticle concentration increment until it reaches 11.51% temperature difference at volume concentration of 2%. A huge increment is observed for concentration from 1 to 1.5% with around 3.14% temperature difference enhancement. At 2.5% volumetric concentration, the temperature difference starts to show a slower pace from concentration of 2–2.5%.

3.2 Temperature Difference Versus Nanoparticle Concentration for TiO_2-water Nanofluid

Figure 6 shows the temperature difference versus nanoparticle concentrations forTiO_2-water nanofluid ranging from 0 to 2.5%. For the pure water (at 0% nanoparticle concentration), the temperature difference is 4.22%. Starting from the concentration of 0.5–2%, the temperature difference was increasing but with smaller slope. However, when the concentration is between 2 and 2.5%, a major enhancement in the temperature difference was observed; from 7.85% enhancement at 2% concentration to as high as 9.15% enhancement at 2.5% concentration.

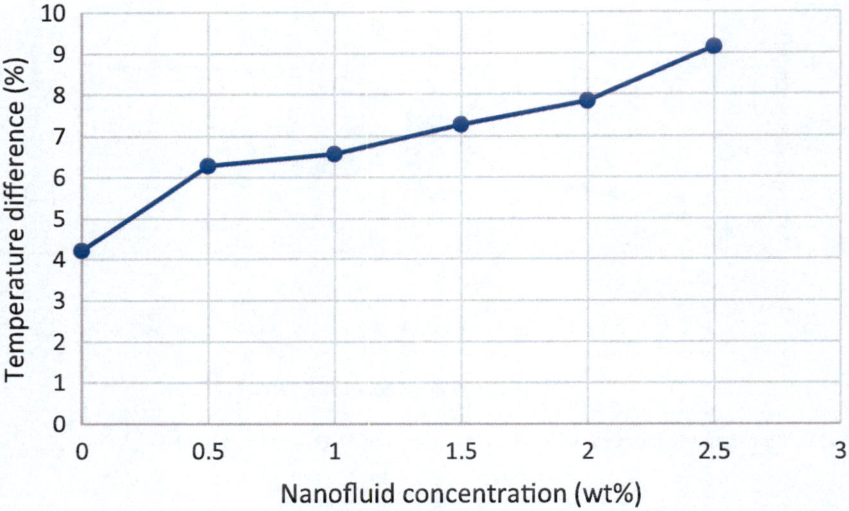

Fig. 6 Temperature difference versus concentration of TiO_2-water nanofluid

3.3 *Comparison of Temperature Difference Versus Nanoparticle Concentration Between Al_2O_3-water and TiO_2-water Nanofluids*

Figure 7 shows the comparison between Al_2O_3-water and TiO_2-water nanofluids for temperature difference at nanoparticle concentrations varying from 0 to 2.5%.

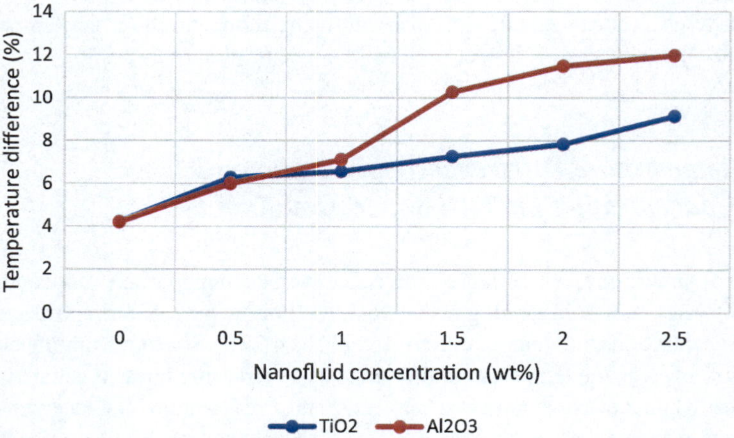

Fig. 7 Comparison of temperature difference versus concentration between Al_2O_3-water and TiO_2-water nanofluids

Generally, the Al_2O_3-water nanofluid showed higher heat transfer enhancement in all concentrations except at 0.5% concentration. At 0.5%, TiO_2-water nanofluid has slightly higher thermal enhancement than Al_2O_3-water nanofluid. Starting from 1% until 2.5% concentration, Al_2O_3-water nanofluid showed obviously higher enhancement than the TiO_2-water nanofluid. Concentration of 2% has the biggest gap in terms of temperature difference; the percentage of enhancement of TiO_2-water nanofluid is recorded at 7.85%, while the Al_2O_3-water nanofluid has 11.51% enhancement, resulting into 3.66% difference between the two nanofluids.

3.4 Comparison of Pressure Drop Versus Nanoparticle Concentration Between Al_2O_3-water and TiO_2-water Nanofluids

Figure 8 shows the pressure drop versus nanoparticle concentrations ranging from 0% to 2.5% for Al_2O_3-water and TiO_2-water nanofluids. It can be observed that the pressure drop is directly proportional to the volume concentrations of both nanofluids. The pressure drop of Al_2O_3-water nanofluid is relatively higher than those of TiO_2-water nanofluid for all values of concentrations.

While analyzing the simulation results in this research work, it can be observed that; even though the pressure drop associated with Al_2O_3-water nanofluid is higher than those associated with TiO_2-water nanofluid, the heat transfer enhancement of Al_2O_3-water nanofluid is greater than TiO_2-water nanofluid. The heat transfer capability managed to override the amount of pressure drop. Therefore, Al_2O_3-water nanofluid is more efficient especially from 1.5 to 2% concentrations.

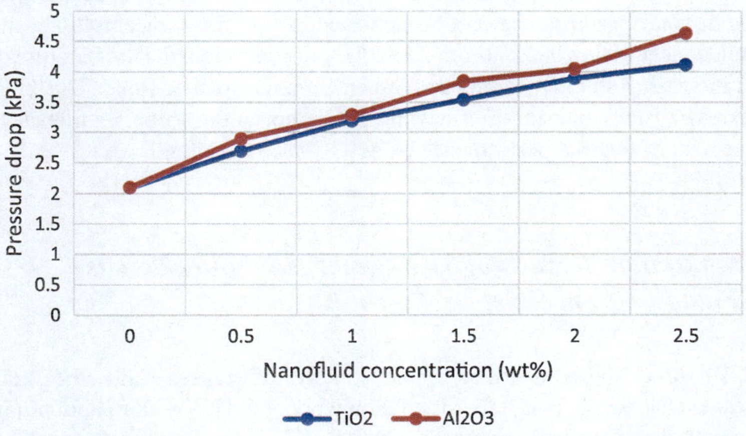

Fig. 8 Comparison of pressure drop versus concentration between Al_2O_3-water and TiO_2-water nanofluids

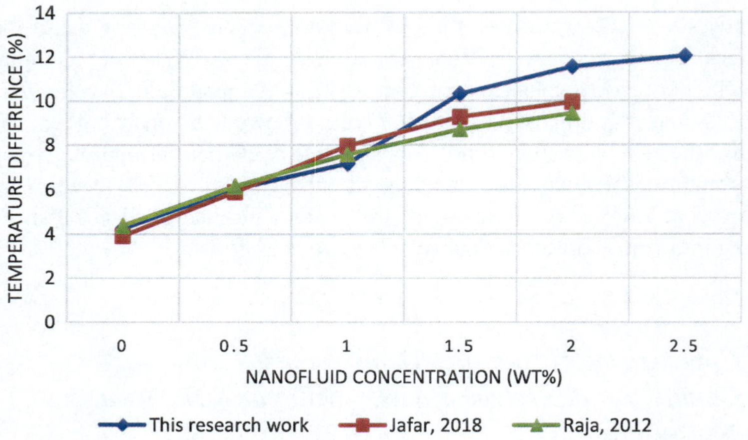

Fig. 9 Comparison of simulation results of Al_2O_3-water nanofluid versus findings from existing journals

3.5 *Simulation Results of Al_2O_3-water Nanofluid Versus Findings from Existing Journals*

Figure 9 shows the differences in temperature difference enhancement inside shell and tube heat exchangers of based on findings by Jaafar [20]. Raja et al. [21] and Ansys simulation results of Al_2O_3-water nanofluid. It is observed that the Ansys simulation is correlated with findings from these journals as the simulation graph is of similar pattern with the existing journals. Jaafar [20] and Raja et al. [21] studied the temperature difference for up to 2% concentration only, while the simulation results projected the thermal enhancement further until 2.5% concentration. The highest thermal performance can be achieved at 1–1.5% concentration, and the optimum concentration is 2%. From 2 to 2.5% concentration, the temperature difference is increasing slightly, and it can be seen from the smaller slope. This indicates that increasing the concentration beyond 2% is not going to be significant to the improvement of thermal performance of Al_2O_3-water nanofluid.

3.6 *Simulation Results of TiO_2-water Nanofluid Versus Findings from Existing Journals*

Figure 10 shows the comparison of temperature differences inside the shell and tube heat exchangers between the simulation works for TiO_2-water nanofluid in this research paper, and the findings from Galvan et al. [22] and Barzegarian et al. [23]. It can be observed that the simulation results are satisfied and aligned with the previous studies, from 0 to 1.5% concentrations. Studies by Galvan et al. [22] and Barzegarian

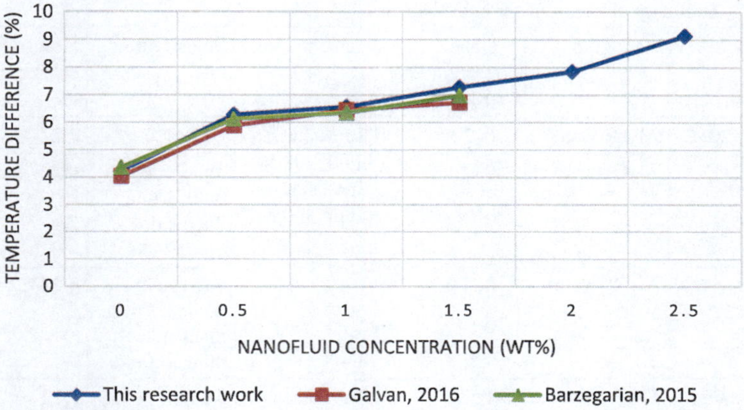

Fig. 10 Comparison of simulation results of TiO_2-water nanofluid versus findings from existing journals

et al. [23] however, were conducted until 1.5% concentration only, while the simulation results projected the thermal enhancement further until 2.5% concentration. The highest thermal performance can be achieved at 2–2.5% concentration, and it is shown from the deepest slope on the graph. The optimum concentration is by far recorded at 2.5%. There are still possibilities for the thermal performance of TiO_2-water nanofluid to increase further for concentration more than 2.5%. This situation could be caused by the stability of TiO_2-water nanofluid at higher concentration.

The summary of heat transfer enhancement from other journals and this simulation work is presented in Table 7.

4 Conclusion

Based on the simulation results of both temperature difference and pressure drop in this study, the findings can be concluded as below:

- Al_2O_3-water nanofluid has the highest heat transfer enhancement for volume concentration between 1 and 1.5%. At this concentration range, the viscosity is almost unchanged, hence allowing higher heat transfer enhancement.
- For Al_2O_3-water nanofluid, 2% is the optimum volume concentration associated with thermal performance. Increasing the volume concentration beyond the optimum value may not be too significant with the thermal performance. This is due to the increase in pressure drop at the respective volume concentration range. The simulation result for pressure drop indicates that for concentration from 2 to 2.5%, the pressure drop is the highest for Al_2O_3-water nanofluid.
- TiO_2-water nanofluid has the highest heat transfer enhancement for volume concentration between 2 and 2.5%.

Table 7 Summary of heat transfer enhancement

Research	Nanofluid	Volume concentration (%)	Temperature	Velocity/flow rate	Heat transfer enhancement (%)
Jaafar [20]	Al_2O_3-water	0–2	Tube side fluid: 298 K Shell side fluid: 320 K	3 LPM	4.22– 9.9
Raja [21]	Al_2O_3-water	0–2		Tube side fluid: 1 m/s Shell side fluid: 2 m/s	4.22– 9.3
This research	Al_2O_3-water	0–2.5		Tube side fluid: 1.2 m/s Shell side fluid: 1.8 m/s	4.22–12
Galvan [22]	TiO_2-water	0–1.5	Tube side fluid: 310 K Shell side fluid: 350 K	3 LPM	4.22– 6.22
Barzegarian [23]	TiO_2-water	0–1.5		2.5–7 LPM	4.22–6.66
This research	TiO_2-water	0–2.5		Tube side fluid: 1.2 m/s Shell side fluid: 1.8 m/s	4.22– 9.15

- For TiO_2-water nanofluid, 2.5% is the optimum volume concentration associated with thermal performance. However, there are still possibilities for the thermal performance to be enhanced for concentration more than 2.5%. This is due to the slight increase in pressure drop at the respective volume concentration range. The simulation result for pressure drop indicates that for concentration from 2 to 2.5%, the pressure drop is low for TiO_2-water nanofluid.

Generally, for each type of nanofluid there will be an optimum concentration at which more heat transfer and less flow resistance occurs in different flow regions can be obtained, depending on the properties of the nanofluid. Pressure drop must be further studied along with the efficiency of the heat transfer of the specific volume concentration associated with the nanofluid. The optimum volumetric concentration of nanofluid can be specified depending on how well this specific volumetric concentration is enhancing the temperature difference efficiency. Both pressure drop and temperature difference must be studied parallelly.

References

1. Huminic G, Huminic A (2012) Application of nanofluids in heat exchangers: a review. Renew Sustain Energy Rev 16(8):5625–5638
2. Okonkwo EC, Wole-Osho I, Almanassra IW, Abdullatif YM, Al-Ansari T (2021) An updated review of nanofluids in various heat transfer devices. J Therm Anal Calorime 145(6)
3. Hasanuzzaman M, Saidur R, Rahim NA (2011) Effectiveness enhancement of heat exchanger

by using nanofluids. In: 2011 IEEE 1st conference on clean energy and technology, CET, pp 98–103
4. Shahrul IM, Mahbubul IM, Saidur R, Khaleduzzaman SS, Sabri MFM, Rahman MM (2014) Effectiveness study of a shell and tube heat exchanger operated with nanofluids at different mass flow rates. Numer Heat Transfer Part A Appl 65(7):699–713
5. Kumar N, Sonawane SS (2016) Experimental study of thermal conductivity and convective heat transfer enhancement using CuO and TiO_2 nanoparticles. Int Commun Heat Mass Transfer 76:98–107
6. Saini H, Shukla VN (2017) Variation in the properties of nanofluids due to change in temperature and concentration of nanoparticles. Int J Eng Res V6(05):763–767
7. Muhieldeen MW, Adam NM, Salman BH (2015) Experimental and numerical studies of reducing cooling load of lecture hall. Energy Buildings 89:163–69
8. Muhieldeen MW, Kuang YC (2019) saving energy costs by combining air-conditioning and air- circulation using CFD to achieve thermal comfort in the building. J Adv Res Fluid Mech Therm Sci 58(1):84–99
9. Salman BH, Hamzah MZ, Purbolaksono J, Inayat-Hussain JI, Mohammed HA, Muhieldeen MW (2011) Determination of correlation functions of the oxide scale growth and the temperature increase. Eng Fail Anal 18:2260–2271
10. Francesca P, Tey WY, Tan LK, Muhieldeen M.W (2020) Investigation on generalised trapezoidal differencing time-marching scheme for modelling of acoustical wave. CFD Lett 12(2):11–21
11. Ng YH, Tey WY, Tan LK, Arada GP, Muhieldeen MW (2020) Numerical examination on two-equations turbulence models for flow across NACA 0012 airfoil with different angle of attack. CFD Lett 12(2):22–45
12. Tey WY, Che Sidik NA, Asako Y, Muhieldeen, MW, Afshar O (2021) Moving least squares method and its Improvement: a concise review. J Appl Comput Mech 7(2):883–889
13. Shahmohammadi P, Beiki H (2016). A numerical investigation of γ -Al_2O_3-water nanofluids heat transfer and pressure drop in a shell and tube heat exchanger. 4(1):29–35
14. Simpson S, Schelfhout A, Golden C, Vafaei S (2018) Nanofluid thermal conductivity and effective parameters. Appl Sci (Switzerland) 9(1)
15. Nallusamy S (2017) Characterization of Al2O3/water nanofluid through shell and tube heat exchangers over parallel and counter flow. J Nano Res 45:155–163
16. Fanaritis JP, Bevevino JW (2008) Designing shell-and-tube heat exchangers. Chem Eng (New York) 83(14):62–71
17. Beldar R, Komble S (2018) Mechanical design of shell and tube type heat exchanger as per ASME section VIII Div.1 and TEMA codes for two tubes. Int J Eng Tech Res 8:1–4
18. Kadhim ZK (2020). Republic of Iraq heat transfer enhancement of a counter flow shell and tube heat exchanger
19. Salman K, Prakash PE, College M (2018) CFD analysis of TiO_2/water nanofluid flow in a double pipe heat exchanger. 5(5)
20. Albadr J (2018) Thermal performance of shell and tube heat exchanger using PG/water and Al_2O_3 Nanofluid
21. Raja M, Arunachalam RM, Suresh S (2012) Experimental studies on heat transfer of alumina/water nanofluid in a shell and tube heat exchanger with wire coil insert. Int J Mech Mater Eng 7(1):16–23
22. Galvan FR, Barranco V, Galvan CC, Batlle S, Fajardo SF, García (2016) A critical review of the application of electrochemical techniques for studying corrosion of mg and mg alloys: opportunities and challenges. Intech i(tourism):13
23. Barzegarian R, Moraveji MK, Aloueyan A (2015) Experimental investigation on heat transfer characteristics and pressure drops of BPHE (brazed plate heat exchanger) using TiO_2-water nanofluid. Exp Therm Fluid Sci

Classification of Lubricants Base Oils for Nanolubricants Applications—A Review

G. Kadirgama, Mohd Kamal Kamarulzaman, D. Ramasamy, K. Kadirgama, and Sakinah Hisham

Abstract Lubricants are derived from biological or non-biological sources. However, the use of stand-alone lubricants lacked desirable tribological properties and had hit their performance limit. A frequent solution to this obstacle is introducing a few but effective additives in the base oil lubricants. These formulations significantly enhance the lubricants, especially in thermal properties, tribological characteristics, and anti-oxidation capability. Advancement in nanotechnology offers the potential to enhance the performance of the lubricant base oil using nanoparticles additives. Introducing lubricant base oils with nanoparticles is critical to improving lubricant characteristics, mainly resistance to wear and friction. Understanding the base oil lubricants commonly used with nanoparticles is vital as the initial guidance in solving these obstacles. Therefore, this review paper aimed to highlight the classification of lubricants base oils for nanolubricants applications. A good understanding of the base oil lubricants leads to the quickly discovering of novel nanolubricants formulations.

Keywords Lubricants · Nanoparticles · Nanolubricant

G. Kadirgama · M. K. Kamarulzaman (✉) · D. Ramasamy (✉)
College of Engineering, Universiti Malaysia Pahang, 26600 Pekan, Pahang, Malaysia
e-mail: kamalkz@hotmail.com

D. Ramasamy
e-mail: deva@ump.edu.my

M. K. Kamarulzaman · D. Ramasamy · K. Kadirgama · S. Hisham
Automotive Engineering Centre, Universiti Malaysia Pahang, 26600 Pekan, Pahang, Malaysia

K. Kadirgama · S. Hisham
Faculty of Mechanical and Automotive Engineering Technology, Universiti Malaysia Pahang, 26600 Pekan, Pahang, Malaysia

M. Y. Ismail et al. (eds.), *Technological Advancement in Mechanical and Automotive Engineering*, Lecture Notes in Mechanical Engineering,
https://doi.org/10.1007/978-981-19-1457-7_16

1 Introduction

Tribology has come from two Greek words, namely "tribos" and "logos". The word "tribos" represents "rubbing", whereas "logos" represents "principle". Therefore, tribology is easily understood as the principle of rubbing. In the standard term, tribology is the science and engineering of wear, friction, and lubrication [1]. It is crucial due to two factors, namely, energy loss and failures of equipment. The high friction condition is the source of energy loss. At the same time, the failure of equipment is critical for the smooth operation of machines such as bearings, gears, engines, and mechanical systems in many fields, including machinery, mining, aerospace, and automotive [2].

The most effective way to solve these limitations is by using lubricant. Efficient lubrication of the moving parts is essential for smoothing the moving surfaces to reduce wear, friction, and energy loss. In addition, lubricants are also significant in preventing surfaces from corrosion, decreasing the wear mechanism, and functioning as a coolant. The majority of the lubricants which had been manufactured are lack desired tribological properties. It was discovered that lubricants are utilized with many additives to enhance their tribological properties [3]. Currently, most of the lubricants have hit their limit of performance. Improvements in lubricants with additives are essential to meet current and future challenges. Therefore, it is crucial to produce lubricants that can achieve energy efficiency in any working conditions and work under extreme situations.

The urgent need for energy efficiency has directed the present research toward utilizing nanoparticles as the additive for lubricants. The addition of nanoparticles in lubricants oil improves the friction properties, extreme pressure properties, and anti-wear properties of the lubricants. Improvements of nanolubricants are reports by many researchers. Lee et al. [4] found 24% reduction of friction coefficient when using commercial mineral oil together with graphite nanoparticles on a disk-on-disk tribotester. Zhang et al. [5] found 14% reduction of wear scar diameter and 17% decreases of friction coefficient when using polyalphaolefin-9 synthetic base oil with graphene sheets nanoparticles. Gupta et al. [6] discovered improvement in anti-wear property by 61.79% when using castor oil with calcium-copper-titanate nanoparticles on four-ball tester. Therefore, an extensive review is essential to discover the suitable lubricants for nanolubricants base oils. This paper reviews the lubricants and their classification that are frequently used with nanoparticle additives to produce novel nanolubricants.

2 Classification of Lubricants

The most basic way to reduce friction and wear between the moving contact bodies is by using liquid lubrication. The lubricant has many essential functions, but primarily, it is the liquid lubricant that separates the contacting mating surfaces of the moving bodies by forming a fluid film between the rubbing surfaces [7]. The fluid film formed reduces the contact between the rubbing surfaces and directly reduces the friction. Besides reducing friction, liquid lubricant also removes wear particles and excessive heat from the system.

The lubricant's performance is influenced by specific thermal-physical properties such as oxidation resistance, thermal stability, flash point, pour point, and viscosity index (VI). Furthermore, the operating characteristics of lubricant also influenced by the lubrication regimes in which is it operates, namely, boundary lubrication, mixed lubrication, and hydrodynamic lubrication [8]. However, the condition is very critical in the boundary lubrication regime. This condition is due to a thin fluid film that can only be preserved, and direct contact between the asperities plays a dominant function in the boundary lubrication regime.

The correct selection of lubricant is critical as it functions to reduce friction and wear of materials. Lubricants have been classified into different types depending on their source and properties of lubricating oils. In general, lubricants are synthesized from two different base oil sources. They are the biological source (vegetable oils, animal fats) and non-biological source (mineral oils, synthetic oils), as shown in Fig. 1. The base oil shows different properties and suitability for different applications [9].

A variation of outcome lubricant properties and performance can be analyzed and predicted by properly selecting base oil for lubricant formulation. An effective lubricant can be synthesized through the base oil characteristics, especially by considering all potential limitations. The American Petroleum Institute (API) classified the quality of the base oil lubricant as described in API 1509 (API 2021) [10]. Important information on a different group of base oils is presented in Table 1.

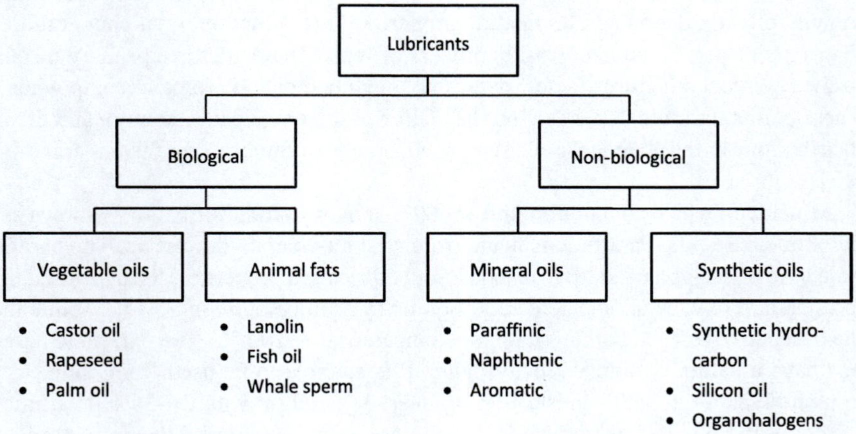

Fig. 1 Classification of lubricants sources

Table 1 American Petroleum Institute (API) classification for quality of base oils

Group	Properties		
	Saturates	Sulphur	VI
Group I	<90%	>0.03%	80 < x < 120
Group II	>90%	<0.03%	80 < x <120
Group III	>90%	<0.03%	>120
Group IV	Polyalphaolefins (PAO)		
Group V	All other base stocks not included in Group I, II, III, or IV		

2.1 *Mineral Oil*

Mineral oil is a petroleum-based fluid. It is produced from crude petroleum oil through fractional distillation and refinement processes. Mineral oil is usually used as a lubricant for the lubrication of bearings, engines, gears, machinery and turbines, which requires its temperature to be moderated. This process is performed to remove the high molecular weight of aromatic (to delay the reduction in viscosity of oil), paraphenes (to avoid deposition of wax during lubrication process), and compounds with nitrogen and sulphur (to avoid corrosion of wearing surfaces) [11].

Generally, mineral oils are different due to the source of the crude petroleum oil, and the refining process had been used. As mineral oils are produced from petroleum crude oil, the main component of the mineral oils is hydrocarbons. The hydrocarbons present in the mineral oils have different viscosity, sulphur content, and chemical structure. Thus, the variations of these factors are the primary root cause that leads to the differences in the mineral oils. Typically, the viscosity of mineral oils at room temperature falls within 5–700 cS. A small quantity of sulphur is required to promote mineral oils' oxidation and lubrication properties [12]. However, if sulphur is more than 1%, it caused a rapid corrosion process of seals. There are three primary forms in the chemical structures factor: paraffinic, naphthenic, and aromatic compounds. These compounds are dependent on the source of the crude petroleum oil, and their distributions in the chemical structure provide unique properties for the mineral oils [13].

Mineral oil which contains around 45–60% of the paraffinic structure, is known as paraffinic base oils. Paraffinic is made from straight-chain hydrocarbon. The paraffinic base oil is good in additive response and lubrication properties. It can be used in the standard oxygen atmosphere condition up to a temperature of 150 °C. While in the absence of oxygen condition up to a temperature of 200 °C. The paraffinic base oils have a naturally high viscosity index. It is suitable to be used from boundary to hydrodynamic (or full) lubrication regimes. Mineral oil with 24–35% paraffinic structure and 65–75% naphthenic structure are identified as naphthenic base oils. The naphthenic oils consist of a cyclic carbon atom structure without any presence of an unsaturated bond. Its structure permits good solubility of additive. In addition, mineral oils with benzene-like structure compounds are known as aromatic base oils. Mineral oils are readily available and low cost. However, some limitation makes the use of mineral oil limited such as loss of viscosity, the potential of an explosion, oxidized at high temperature, non-environmental friendly, and produced from non-renewable resources [14].

2.2 Synthetic Oil

In the last hundred years, the utilization of synthetic lubricant has risen consistently due to the limitation of mineral oil to serve specific applications. Even though mineral oils own many desirable characteristics such as low cost and availability, they also possess specific severe undesirable characteristics: the potential of explosion or combustion with oxidizing elements, oxidation, and loss of viscosity and solidification at low temperatures.

Moreover, synthetic oils are the alternatives for petroleum-based oils. It had been produced using the chemical modification of petroleum-based rather than using entirely from crude oil. The synthetic oils are artificially formulated using uniform hydrocarbon molecules to create lubricant with higher quality than the mineral oils. These enable synthetic oils to operate under extremely high or low temperatures [15]. The chemical structure of synthetic oil is specially formulated for specific applications. Several types of synthetic oil are usually utilized, such as synthetic hydrocarbon lubricants, silicon analogues of hydrocarbons, and organohalogens.

Synthetic Hydrocarbon oils. A lot of hydrocarbon fluids are highly potential to be used as a lubricant. However, due to the economic aspect, the production of lubricant oils had been restricted into a certain relevant range output. As oil refineries produce lubricant oils in mass production, the final cost becomes cheaper. Therefore, applications with high usage of lubricant oils such as engine oils have also become more economical. Synthetic hydrocarbon base oil can be categorized into several types, they are (i) polyalphaolefins [16], (ii) polyglycols [17], (iii) cycloaliphatics [18], and (iv) esters [19]. Synthetic oils synthesized from linear paraffin is known as polyalphaolefins based oils. Polyalphaolefins has a well-defined molecular structure. Therefore, it has a low ability to absorb additives. However, the polyalphaolefins indicates greater performance as compared to the mineral base oils. Normally, the ability of polyalphaolefins to absorb additives can be improved by using naphthenic or ester base oils [20].

Silicon Analogues of Hydrocarbons. The synthetic lubricants consist of several types of silicon analogues of the hydrocarbons group, which are chemically stable. The silicon analogues of hydrocarbons can be grouped into two essential compounds: silicones [21] and silahydrocarbons [22]. Even though the compounds in both groups consist of silicon, they have different molecular structures.

Organohalogens. Organohalogens group with halogenated hydrocarbons are identified as an antioxidant. In the past, specific machinery such as oxygen and air compressors used pure sulphuric acid as a lubricant. As sulphuric acid can be used as a lubricant for steel, there is a significant problem in preventing acids from mixing with contamination and moisture from the compressed gas. One more problem related to the use of sulphuric acid is corrosion. Recently, the corrosion problem can be reduced by using chlorine and fluorine to produce compounds with desirable properties due to their excellent thermal stability and oxidation. Generally, the synthetic lubricants categorized in the organohalogens group including perfluoropolyalkylethers [23], chlorotrifluoroethylene [24], chlorofluorocarbons [25], and perfluoropolyether [26].

2.3 Biological Lubricant

Biological lubricants are commonly used in pharmacological and food production, for example, in kiln or bakery stoves. It is used in the condition where it is critically required for contamination to be minimized. Biological lubricants can be obtained from natural sources such as vegetable oil and animal fats [27]. Samples for vegetable oil sources are castor oil, rapeseed oil, and palm oil, while animal fats are lanolin, fish oil, and whale sperm. As the world is currently facing the depletion of petroleum sources, lubricants have become a viable option for mineral oil lubricants.

Vegetable oils consist of triglycerides. It is renewable and harmless to the environment, biodegradable, high flash point and viscosity index, and good lubrication is most important [28]. Vegetable oils are usually produced from plant sources using the pressing and solvent method [29]. Furthermore, vegetable oils can be grouped into four primary types according to their major fatty acid. They are lauric acid, erucic acid, ricinoleic acid and oleic-linoleic acid [30]. Lauric oil consists of 2% polyunsaturated fatty acid, 7% monounsaturated fatty acid, and 51% lauric acid. One of the lauric oil examples is coconut oil. Erucic oil consists of around 20–50% of erucic acid, a long chain type fatty acid. Example of Erucic oils is mustard oil and rapeseed oil [31]. In addition, canola oil is also an Erucic oil, but it is categorized as low content of erucic oil. Canola oil consists of oleic acid at 58%, polyunsaturated acid at 36%, and less than 2% erucic acid. It also has low stability of oxidation due to the presence of an unsaturated bond [32]. Oil with ricinoleic acid at 89% is identified as ricinoleic oil. An example of ricinoleic oil is castor oil. Castor oil consists of palmitic acid at 1%, stearic acid at 1%, oleic acid at 3%, linoleic acid at 4%, and the significant fraction is ricinoleic acid at 91% [33]. Moreover, some vegetable oils that consist of a percentage of unsaturated fatty acids are categorized under oleic-linoleic oils. They are sunflower oil, cottonseed oil, peanut oil, soyabean oil, olive oil, and palm oil [34].

Work on developing high-performance lubricants for extreme temperature (high and cryogenic) and vacuum or nano and micro-scale electromechanical systems

(NEMS and MEMS) are now underway. Mesogenic lubricants and ionic liquid lubricants are two new types of sophisticated lubricants that have recently become popular [35, 36].

3 Conclusion

The frequently used lubricants base oils for nanolubricants have been reviewed. Common lubricants used in nanolubricants are derived from two sources; biological and non-biological. The biological source consists of vegetable oils and animal fats, while the minerals oil and synthetic oils are grouped into the non-biological source. Mineral oil is produced from petroleum's crude oil. On the other hand, synthetic oil is produced using petroleum-based chemical modification rather than entirely from crude oil. The vegetable oil and animal fats are derived from natural sources and become an alternative option for non-biological lubricants. Mineral oil has several advantages, such as low cost and availability. The ability of synthetic lubricants to work under extremely low or high temperatures makes it a preferable choice. The biological lubricants are commonly used in the condition where critically required for minimum contamination. The right combination of lubricants base oil and nanoparticles additives produced novel lubricants with improved properties and performance. It is essential to reduce friction and wear of materials and influence the lubricant's specific thermal-physical properties such as oxidation resistance, thermal stability, flash point, pour point, and viscosity index.

Acknowledgements The authors sincerely thank the Ministry of Higher Education (Malaysia) for financial support through the University Malaysia Pahang Research Grant RDU 190194, FRGS/1/2018/TK03/UMP/02/26. "Mohd Kamal bin Kamarulzaman" is the recipient of the UMP Post-Doctoral Fellowship in Research.

References

1. Singh A, Chauhan P, Mamatha T (2020) A review on tribological performance of lubricants with nanoparticles additives. Mater Today: Proc 25:586–591
2. Kamarulzaman MK, Abdullah A (2020) Multi-objective optimization of diesel engine performances and exhaust emissions characteristics of Hermetia illucens larvae oil-diesel fuel blends using response surface methodology. Energy Sources, Part A Recovery, Utilization, and Environmental Effects pp 1–14
3. Hisham S, Kadirgama K, Ramasamy D, Samykano M, Harun W, Saidur R (2020) Statistical approach for prediction of thermal properties of CNC and CNC-CuO nanolubricant using Response Surface Methodology (RSM). In: IOP conference series: materials science and engineering, vol 1. IOP Publishing, p 012016
4. Lee C-G, Hwang Y-J, Choi Y-M, Lee J-K, Choi C, Oh J-M (2009) A study on the tribological characteristics of graphite nano lubricants. Int J Precis Eng Manuf 10(1):85–90

5. Zhang W, Zhou M, Zhu H, Tian Y, Wang K, Wei J, Ji F, Li X, Li Z, Zhang P (2011) Tribological properties of oleic acid-modified graphene as lubricant oil additives. J Phys D Appl Phys 44(20):205303
6. Gupta RN, Harsha A (2017) Synthesis, characterization, and tribological studies of calcium–copper–titanate nanoparticles as a biolubricant additive. J Tribol 139(2):021801
7. Sakinah M, Hassan M, Kadirgama K, Kadirgama G, Ramasamy D, Amirruddin A, Rahman M, Noor M (2018) Optimization on wear performance of anti wear additive added biolubricant. In: Improved performance of materials. Springer, pp 1–9
8. Hisham S, Kadirgama K, Mohammed HA, Kumar A, Ramasamy D, Samykano M, Rahman S (2020) Hybrid nanocellulose-copper (II) oxide as engine oil additives for tribological behavior improvement. Molecules 25(13):2975
9. Shafi WK, Raina A, Ul Haq MI (2018) Friction and wear characteristics of vegetable oils using nanoparticles for sustainable lubrication. Tribol Mater Surf Interfaces 12(1):27–43
10. Institute AP (2021) Engine oil licensing and certification system. API 1509 Twentieth Edition, May 2021
11. Awang N, Ramasamy D, Kadirgama K, Samykano M, Najafi G, Sidik NAC (2019) An experimental study on characterization and properties of nano lubricant containing cellulose nanocrystal (CNC). Int J Heat Mass Transf 130:1163–1169
12. Ali MH, Abdullah A, Yasin MM, Kamarulzaman MK (2020) Cyclic pressure variations in a small diesel engine fueled with biodiesel and antioxidant blends. Int J Automot Mech Eng 17(2):7851–7857
13. Ilyas SU, Narahari M, Theng JTY, Pendyala R (2019) Experimental evaluation of dispersion behavior, rheology and thermal analysis of functionalized zinc oxide-paraffin oil nanofluids. J Mol Liq 294:111613
14. Zheng G, Ding T, Huang Y, Zheng L, Ren T (2018) Fatty acid based phosphite ionic liquids as multifunctional lubricant additives in mineral oil and refined vegetable oil. Tribol Int 123:316–324
15. Hackler R, Vyavhare K, Kennedy R, Celik G, Kanbur U, Griffin P, Sadow A, Zang G, Elgowainy A, Sun P (2021) Synthetic lubricants derived from plastic waste and their tribological performance. ChemSusChem
16. González R, Viesca JL, Battez AH, Hadfield M, Fernández-González A, Bartolomé M (2019) Two phosphonium cation-based ionic liquids as lubricant additive to a polyalphaolefin base oil. J Mol Liq 293:111536
17. Takiwatari K, Sato S, Hoshi Y, Nanao H, Mori S (2021) Effect of functional groups of polyglycol oils on their lubrication properties. Tribology Online 16(1):24–30
18. Zareanshahraki F, Asemani H, Skuza J, Mannari V (2020) Synthesis of non-isocyanate polyurethanes and their application in radiation-curable aerospace coatings. Prog Org Coat 138:105394
19. de Sousa Marques MAC, Guimarey MJ, Domínguez-Arca V, Amigo A, Fernández J (2021) Heat capacity, density, surface tension, and contact angle for polyalphaolefins and ester lubricants. Thermochim Acta 703:178994
20. Hong FT, Schneider A, Sarathy SM (2020) Enhanced lubrication by core-shell TiO_2 nanoparticles modified with gallic acid ester. Tribol Int 146:106263
21. Bazan P, Kuciel S, Nykiel M (2019) Characterization of composites based on polyoxymethylene and effect of silicone addition on mechanical and tribological behavior. Polym Eng Sci 59(5):935–940
22. Snyder CE, Pettigrew FA (2020) Silahydrocarbons. In: Synthetics, mineral oils, and bio-based lubricants. CRC Press, pp 227–234
23. Hellman PT, Gschwender L, Snyder CE Jr (2006) A review of the effect of metals on the thermo-oxidative stability of perfluoropolyalkylether lubricants. J Synth Lubr 23(4):197–210
24. Cavdar B, Sharma S, John P (1995) Wettability aspects of friction and wear reduction by a fluorinated sulphonamide additive in a chlorotrifluoroethylene-based fluid. Tribol Int 28(8):501–506

25. Nakagawa M, Kaneko S, Ito S (2016) Demolding in ultraviolet nanoimprinting assisted by a nanoscale lubricating fluid layer of condensed alternative chlorofluorocarbon. Bull Chem Soc Jpn 89(7):786–793
26. Sun J, Li A, Su F (2019) Excellent lubricating ability of functionalization graphene dispersed in perfluoropolyether for titanium alloy. ACS Appl Nano Mater 2(3):1391–1401
27. Kamarulzaman MK, Hafiz M, Abdullah A, Chen AF, Awad OI (2019) Combustion, performances and emissions characteristics of black soldier fly larvae oil and diesel blends in compression ignition engine. Renew Energy 142:569–580. https://doi.org/10.1016/j.renene.2019.04.126
28. Owuna F, Dabai M, Sokoto M, Dangoggo S, Bagudo B, Birnin-Yauri U, Hassan L, Sada I, Abubakar A, Jibrin M (2020) Chemical modification of vegetable oils for the production of biolubricants using trimethylolpropane: a review. Egypt J Pet 29(1):75–82
29. Tulashie SK, Kotoka F (2020) The potential of castor, palm kernel, and coconut oils as biolubricant base oil via chemical modification and formulation. Therm Sci Eng Prog 16:100480
30. Dehghani Soufi M, Ghobadian B, Atashgaran M, Mousavi SM, Najafi G (2019) Biolubricant production from edible and novel indigenous vegetable oils: mainstream methodology, and prospects and challenges in Iran. Biofuels Bioprod Biorefin 13(3):838–849
31. Sajeeb A, Rajendrakumar PK (2019) Comparative evaluation of lubricant properties of biodegradable blend of coconut and mustard oil. J Clean Prod 240:118255
32. Goyal A, Tanwar B, Sihag MK, Kumar V, Sharma V, Soni S (2021) Rapeseed/canola (Brassica napus) seed. In: Oilseeds: health attributes and food applications. Springer, pp 47–71
33. Greco-Duarte J, Cavalcanti-Oliveira E, Da Silva J, Fernandez-Lafuente R, Freire D (2017) Two-step enzymatic production of environmentally friendly biolubricants using castor oil: Enzyme selection and product characterization. Fuel 202:196–205
34. Nogales-Delgado S, Encinar JM, Cortés ÁG (2021) High oleic safflower oil as a feedstock for stable biodiesel and biolubricant production. Ind Crops Prod 170:113701
35. Khemchandani B, Somers A, Howlett P, Jaiswal A, Sayanna E, Forsyth M (2014) A biocompatible ionic liquid as an antiwear additive for biodegradable lubricants. Tribol Int 77:171–177
36. Qu J, Luo H, Chi M, Ma C, Blau PJ, Dai S, Viola MB (2014) Comparison of an oil-miscible ionic liquid and ZDDP as a lubricant anti-wear additive. Tribol Int 71:88–97

An Experimental Evaluation of Specific Heat of Mono and Hybrid Nanofluids

Kaniz Farhana, Kumaran Kadirgama, Danial Mohamed, Abu Shadate Faisal Mahamude, Sivarao Subramonian, Devarajan Ramasamy, and Mahendran Samykano

Abstract The experimental study evaluates the specific heat capacity of diversified mono and hybrid nanofluids. Specific heat is one of the most important attributes of mono and hybrid nanofluids for various heat and thermal applications. Herein, varied mono nanofluids such as CNC, Al_2O_3, and ZnO and only one hybrid nanofluid such as Al_2O_3/CNC have been studied to figure out their specific heat capacity. Standard test method applied to measure the specific heat of mono and hybrid nanofluids by using DSC. Mono nanofluids and hybrid nanofluids present some significant results of specific heat capacity and hybrid nanofluids show a maximum of 126% negativity than mono nanofluids. These experimental values would be a good aspect of the nanofluid applications.

Keywords Specific heat · Mono nanofluids · Hybrid nanofluids · Base fluid

K. Farhana (✉)
Department of Apparel Engineering, Bangladesh University of Textiles, Dhaka 1208, Bangladesh
e-mail: kaniz.farhana@butex.edu.bd

K. Kadirgama · A. S. Faisal Mahamude
Faculty of Mechanical and Automotive Engineering Technology, Universiti Malaysia Pahang, 26600 Pekan, Pahang, Malaysia

K. Kadirgama · D. Mohamed
Centre for Automotive Engineering, Universiti Malaysia Pahang, 26600 Pekan, Malaysia

S. Subramonian
Faculty of Manufacturing Engineering, Universiti Teknikal Malaysia Melaka (UTeM), Durian Tunggal, Malaysia

D. Ramasamy · M. Samykano
College of Engineering, Universiti Malaysia Pahang, 26300 Gambang, Pahang, Malaysia

M. Y. Ismail et al. (eds.), *Technological Advancement in Mechanical and Automotive Engineering*, Lecture Notes in Mechanical Engineering,
https://doi.org/10.1007/978-981-19-1457-7_17

1 Introduction

Nanofluids have acquired enormous examination activities in the different investigations (numerical and experimental) because of their further developed properties, particularly as heat transfer liquids. Various examinations have uncovered the qualities, constraints, and relevance of single-sort nanoparticles and the characteristics of their colloidal solution blends. The disclosure of the synergistic impact of nanoparticles, as identified with their advanced thermophysical properties has been additionally investigated in trial and mathematical investigations of mono and hybrid nanofluids. While the significant spotlight has been on thermal conductivity, furthermore, consistency, the other critical attribute that obtains slight examination consideration is the specific heat, which is a critical heat character in energy frameworks. Appropriate comprehension of the specific heat of mono and hybrid nanofluids is a scientific choice for heat transfer and heat applications on an industrial scale. Specific heat is one of the most important criteria of nanofluids and it depends on the volume fractions of nanofluids. The augmentation of nanoparticles into the base fluid decreases the specific heat capacity of nanofluids [1]. Similarly, Namburu et al. [2] experimentally analyzed the specific heat of SiO_2/water-ethylene glycol (60:40) nanofluids with a volume concentration range of 0–10%. The authors concluded that the specific heat capacity of SiO_2 nanofluids declines as the volume concentrations of nanofluids increase and about 12% decreases than the base fluids at 10% volume concentration of SiO_2 nanofluid. However, the specific heat of nanofluids raises with the improvement of temperature and it does not depend on the particle size of nanofluids as well [3]. Differential scanning calorimetry (DSC) is the most sensible approach that can be applied to evaluate the specific heat capacity (C_p) of both fluids and solid as well [4, 5].

The progression in nanoparticle innovation has expanded the area of exploration of materials explicit heat restraint. Generally, the investigation of the impact of size on the specific heat of nanomaterials has been considered later as well as the investigation of the specific heat of steady colloidal combinations of nanomaterials and customary heat transfer liquids such as water, ethylene glycol. There was an enormous difference between the specific heat capacity of mono and hybrid nanofluids [6]. There are distinctive approaches to order nanofluids; mono and hybrid nanofluids can be ordered by their sort integral material, by base liquid properties, or by nanomaterial properties such as Pak and Cho [7] characterized the nanofluids by the amount of nanomaterial constituents in the liquid. The authors did this predominantly because this way can follow the course of events advancement of nanofluids. Earlier, mono nanofluids are the fundamental materials that scientists started to explore. Notwithstanding, as a long time elapsed, 2-nanoparticles hybrid nanofluids started to ascend in unmistakable quality because of their defined attributes. For all nanofluid arrangements, they examined the examinations that have inspired to investigate the specific heat capacity of nanofluid. Of late, for instance, Balaji et al. [8] studied the specific heat capacity for various volumetric concentrations of hybrid nanofluids at different temperatures using DSC. The author found that specific heat capacity was decreased with the loading of nanomaterials dispersed into the base fluid. The highest decrement in the specific heat capacity was obtained at 31.8% at 0.2 vol% than the base fluid at 20 °C. The maximum enhancement of specific heat capacity was about 6.32% for 0.2 vol% with improvement in the temperature range from 20 to 60 °C. In another study, Muneeshwaran et al. [9] reported the specific heat depends on the material, nanoparticle size and shape, and temperature as well. The specific heat enhances as the diameter of the nanoparticle increases. The literature stated the individualistic evaluation of specific heat of mono and hybrid nanofluids separately and not any comparison between them. Moreover, Bumataria et al. [10] researched the thermal performance of mono and hybrid nanofluids in heat pipes. The study was conducted based on the kinds of heat pipes, transport medium, preparation, and characterization methods of nanofluids. And finally, the authors analysed the influence of size and shape of nanoparticles, mass fractions, filling ratio, heat load and bent angles on the performance of the thermal conductivity, viscosity, heat transfer and stability. The authors reported that mono and hybrid nanofluids have significant effect in therml performance of heat pipes. As consequence, there is some gap to research the definite specific heat capacity of mono and hybrid nanofluids. Therefore, in this study, evaluation of specific heat of mono and hybrid nanofluids have been demonstrated in combined and illustrated the analogy between them. Thus, the experiment follows the materials of Al_2O_3, CNC, ZnO nanoparticles with base fluids of ethylene glycol and water as well as measured the specific heat at different temperatures and analyzed.

2 Methodology of the Experiment

Three types of nanoparticles such as ZnO, Al_2O_3, and CNC (crystal nano-cellulose) have been used to prepare mono and hybrid nanofluids. Al_2O_3 nanoparticles from two different sources have been utilized in the study and defined as Al_2O_3-A and Al_2O_3-B. Besides, only 50% Al_2O_3-B + 50% CNC hybrid nanofluid was prepared due to the availability of Al_2O_3-B nanoparticles. Herein, only one volume concentration of 0.1% is used for mono and hybrid nanofluid. Aqueous ethylene glycol (water: ethylene, 60:40) has been used as a base fluid. The physical characterization of nanoparticles has been done using characterizing techniques such as Transmission electron microscopy (TEM), and Field-emission scanning electron spectroscopy (FESEM) [11–13]. Mono and hybrid nanofluids have been prepared by a two-step method and the calculation of the number of nanoparticles for required volume concentration has been calculated using the following equation [14].

$$\phi = \left[\frac{\frac{w}{\rho_p}}{\frac{w}{\rho_p} + \frac{w_{bf}}{\rho_{bf}}} \right] \times 100 \tag{1}$$

where the volume fraction of nanofluids (%) is ϕ, the mass is w and ρ determines the density. The subscripts bf and p define for nanoparticles and base fluid respectively.

After that, the specific heat capacity (C_p) of both nanofluids and hybrid nanofluids at 0.1% volume concentrations has been calculated applying the Linseis DSC1000 (Differential scanning calorimetry-Germany) following the standard method entitled "ASTM E1269-11(2018) Standard Test Method for Determining Specific Heat Capacity by Differential Scanning Calorimetry". Here the weight of mono and hybrid nanofluids was fixed with 13 mg. Afterward, the specific heat of the samples has been measured using an aluminum crucible in DSC. Figure 1 shows the specific heat measurement tools and the settings of DSC that have been depicted in Table 1. This equipment shows the maximum viable accuracy C_p by utilizing modulated rate of heating temperature profiles. This approach interprets a consistent change in the flow of heat at of the system and the samples can detect the heat gain eminently than a linear system of heating [15]. Moreover, specific heat C_p evaluation could be altered

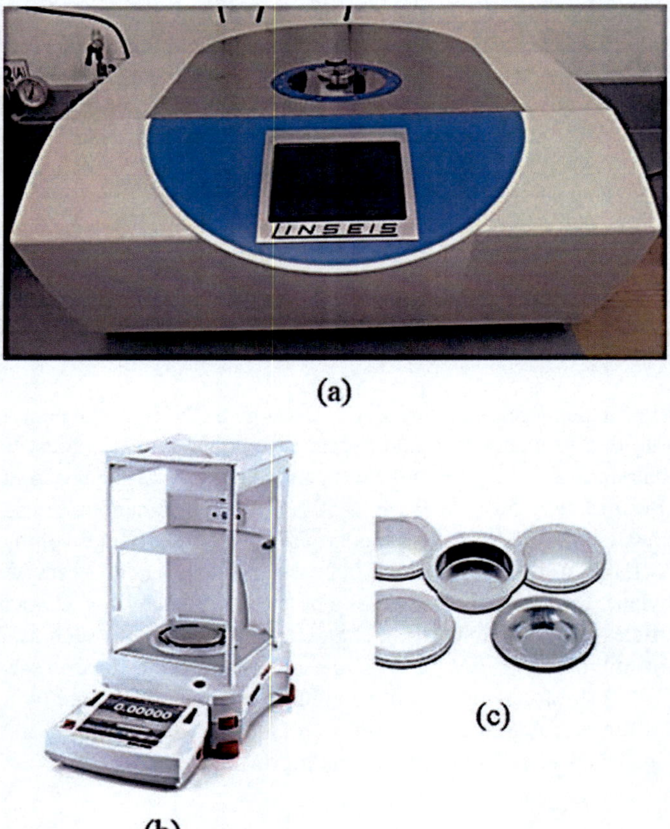

Fig. 1 Equipment of specific heat capacity evaluation **a** DSC Linseis 1000, **b** electrical balance, and **c** crucible made of aliminium [17]

Table 1 Setting of adjusted parameters of DSC

Parameters	Settings
Temperature	25–80 °C
Sampling interval	0.1 s
Preheat temperature	25 °C
High-temperature differential scanning (HDSC)	250 μV
Temperature modulation: period	60 s/amplitude: 3 K
Gas	Nitrogen (20 sccm/min)

by the weight of the substances and the rate of heating owing to the apply of DSC [16].

3 Results and Discussion

The micrographic images of the nanoparticles have been displayed in Fig. 2. CNC nanoparticles were rod-shaped mostly and fewer were round-shaped (Fig. a), Al_2O_3 nanoparticles were round-shaped (Fig. b), ZnO nanoparticles also were round-shaped (Fig. c), and lastly, (Fig. d), portrayed the combined scenario of Al_2O_3-B/CNC. There are few studies (numerical and experimental) had been evaluated for the determination of specific heat capacity of mono and hybrid nanofluids at different temperature and volume concentrations altogether. In this study, Fig. 3 portrays the scenario of specific heat of base fluid, mono, and hybrid nanofluids. Base fluid, mono, and hybrid nanofluids present similar trends of specific heat capacity with the spectrum of temperature from 30 to 80 °C; except Al_2O_3-A mono nanofluid that illustrates the increasing trend of specific heat capacity. However, Al_2O_3-A mono nanofluid exhibited lower specific heat than base fluid at 30 °C temperature and similar at 50 °C temperature [18]. Besides, the lesser specific heat capacity of rigid nanoparticles compared to that of base fluid definitely will reduce the specific heat capacity of the nanofluid blends [19]. On the other hand, Al_2O_3-A mono nanofluid depicts higher specific heat capacity compared to base fluid both at 70 and 80 °C temperatures. However, specific heat capacity at 80 °C temperature was lower than 70 °C temperature. The increment performance of Al_2O_3-A mono nanofluid compared with base fluid has been presented in Fig. 4 in percentage. The specific heat of Al_2O_3-A mono nanofluid was in a surging direction within a range of temperature from 30 to 80 °C. The declining specific and afterword increasing with improving temperature trend of mono and hybrid nanofluids are very similar with a current study of Singh et al. [20]; as the authors studied that the specific heat was minimal with increasing the temperature for mono nanofluids but increase with the augmentation of nanoparticles at room temperature. In another study, O'Hanley et al. [21] investigated the specific heat capacity of mono nanofluids (water-based alumina, silica, copper-oxide) at various volume concentrations with different sizes (diameters) of nanoparticles and

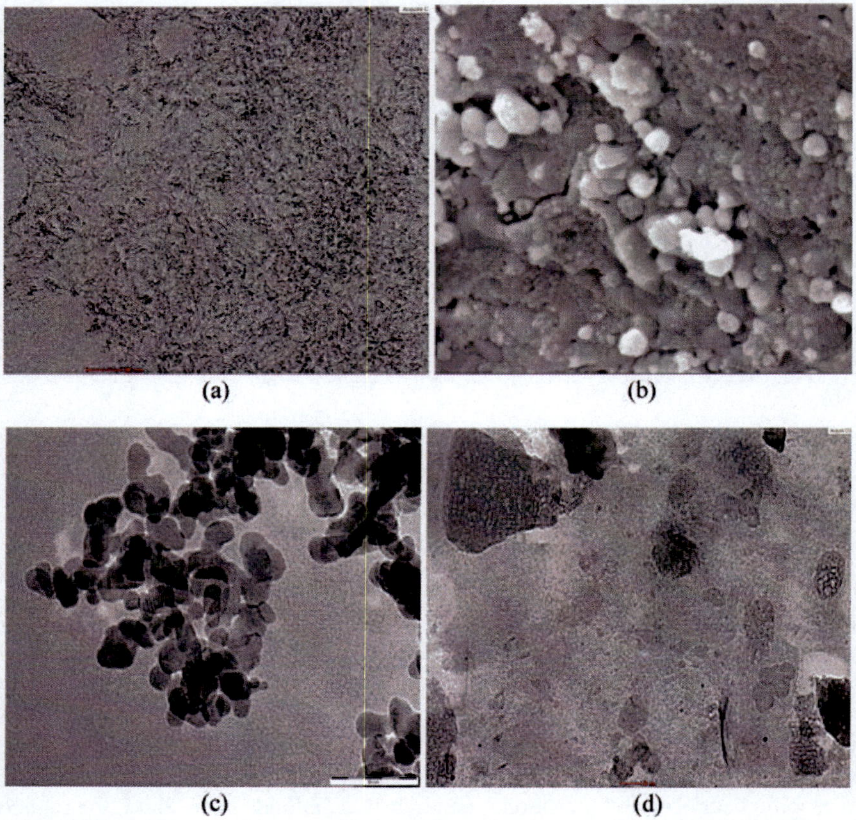

Fig. 2 Characterization schemes; **a** TEM image of CNC (magnification ×50,000), **b** FESEM image of Al_2O_3 (WD 9.9 mm; magnification ×70,000), **c** TEM micrograph of ZnO (magnification × 50,000), and **d** TEM micrograph of Al_2O_3/CNC (magnification ×62,000)

reported the decreasing trend of specific heat capacity with rising volume fractions. In addition, Sekhar and Sharma [18] reported that thermal diffusivity was the criteria to affect the specific heat capacity of mono nanofluids. They also found the raising temperature deducts the effective specific heat capacity of mono nanofluids.

Besides, Fig. 3 presents the positive and negative values of specific heat between the mono and hybrid nanofluids. Base fluid and Al_2O_3-A display the positivity and CNC, ZnO Al_2O_3-B, and hybrid (Al_2O_3-B + 50% CNC) show the negativity attributes of specific heat capacity. This phenomenon of specific heat capacity was also obtained in our previous study such as Farhana et al. [17]. Furthermore, Das et al. [22] studied the negative specific heat in a thermodynamic model and reported that mechanical instability leads to the negative value of specific heat. Herein, mono nanofluids (CNC, ZnO, Al_2O_3-B) smaller resultant data of negativity at 50 °C temperature compared to 30 °C temperature whereas, at 70 and 80 °C temperature, all the mono nanofluids (CNC, ZnO, Al_2O_3-B) represents gradually higher negativity

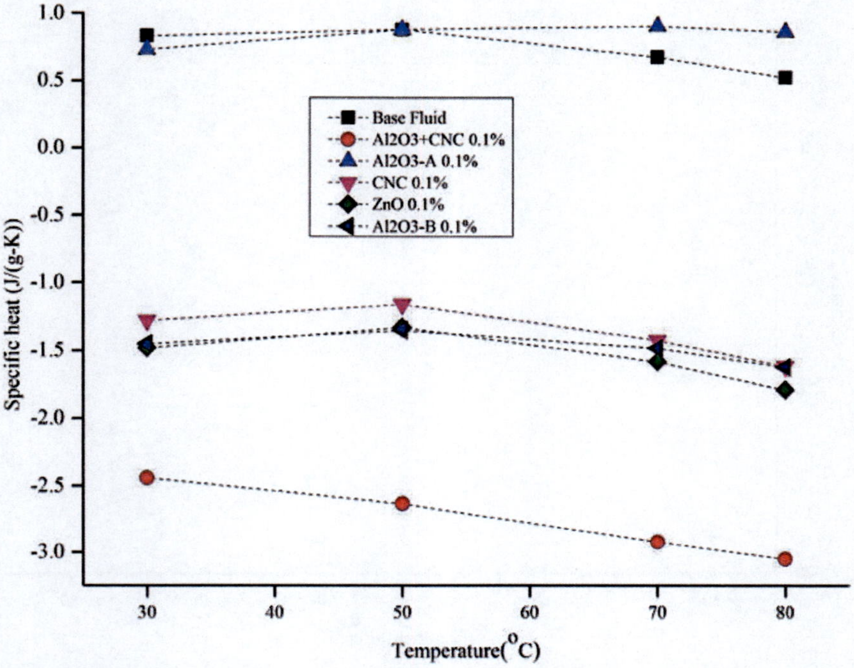

Fig. 3 Evaluation of specific heat capacity of mono and hybrid nanofluid at different temperatures

specific heat capacity [17]. However, hybrid nanofluid determines the progressive improvement of the negative value of specific heat. The enhancement of negativity of specific heat of hybrid nanofluid relatively with mono nanofluids (CNC, ZnO, Al_2O_3-B) has been illustrated in Table 2. The enhancement of specific heat capacity of hybrid nanofluid compared to the CNC, ZnO, and Al_2O_3 mono nanofluids at a varied temperature such as 30, 50, 70, and 80 °C have been tabulated in Table 2. The maximum increase of specific heat is around 126% at 50 °C temperature for CNC mono nanofluid; minimum enhanced of about 70% at 80 °C temperature for ZnO mono nanofluid compared to the base fluid.

4 Conclusion

The study presents the experimental analysis of specific heat of mono and hybrid nanofluids. 0.1% volume concentration of mono nanofluids (CNC, ZnO, Al_2O_3-A, Al_2O_3-B) and hybrid nanofluid (Al_2O_3-B + 50% CNC) was preferred and the DSC tool has been utilized to figure out the specific heat capacity behaviour of mono and hybrid nanofluids. Herein, a dramatic change has been found in the specific heat

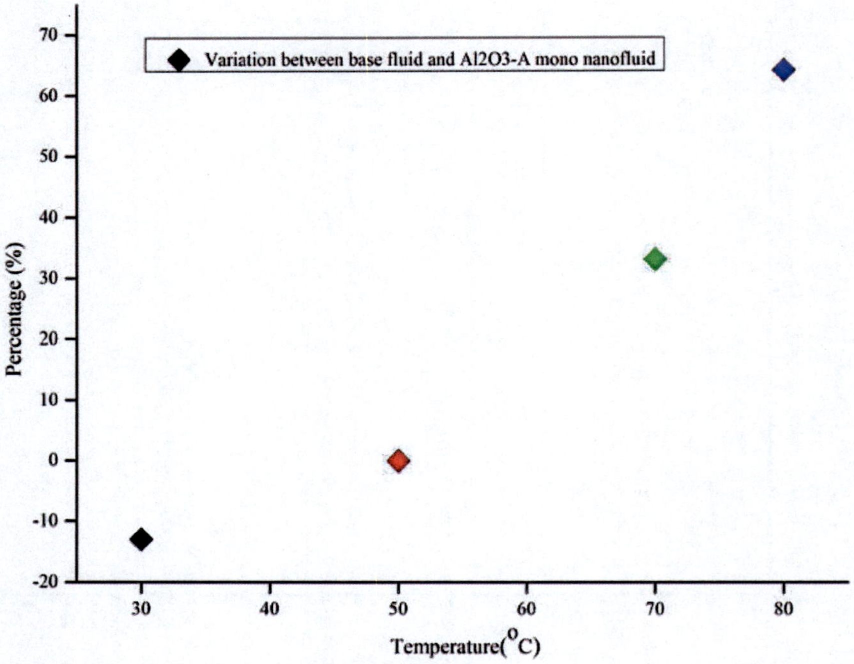

Fig. 4 Analogy of specific heat of base fluid and Al_2O_3-A mono nanofluid

Table 2 Enhancement percentage of hybrid nanofluid compared to various mono nanofluids

Temperature (°C)	CNC (%)	ZnO (%)	Al_2O_3-B (%)
30	90	75	77
50	126	97	94
70	105	84	97
80	89	70	88

capacity of nanofluids (both mono and hybrid) by adding tiny amounts of nanoparticles into the base fluid. Moreover, both positive and negative values of specific heat were obtained after evaluating the nanofluids (mono and hybrid). The negativity of hybrid nanofluid was much higher about 126% than mono nanofluids. This experimental evaluation of specific heat would be useful in various heat transfer application purposes of mono and hybrid nanofluids.

Acknowledgements The authors are grateful to the Minister of Higher Education for giving a grant under No. FRGS/1/2019/TK07/UMP/02/3 and Universiti Malaysia Pahang for assisting laboratory facilities under the grant No. RDU192209.

References

1. Zhou S-Q, Ni R (2008) Measurement of the specific heat capacity of water-based Al_2O_3 nanofluid. Appl Phys Lett 92(9):093123
2. Namburu P et al (2007) Experimental investigation of viscosity and specific heat of silicon dioxide nanofluids. Micro Nano Lett 2(3):67–71
3. Satti JR et al (2016) Specific heat measurements of five different propylene glycol based nanofluids and development of a new correlation. Int J Heat Mass Transf 94:343–353
4. Kodre K et al (2014) Differential scanning calorimetry: a review. Res Rev J Pharm Anal 3(3):11–22
5. O'Neill M (1966) Measurement of specific heat functions by differential scanning calorimetry. Anal Chem 38(10):1331–1336
6. Adun H et al (2021) A critical review of specific heat capacity of hybrid nanofluids for thermal energy applications. J Mol Liq 116890
7. Pak BC, Cho YI (1998) Hydrodynamic and heat transfer study of dispersed fluids with submicron metallic oxide particles. Exp Heat Transfer 11(2):151–170
8. Balaji T et al (2021) Enhanced heat transfer characteristics of water based hybrid nanofluids with graphene nanoplatelets and multi walled carbon nanotubes. Powder Technol 394:1141–1157
9. Muneeshwaran M et al (2021) Role of hybrid-nanofluid in heat transfer enhancement—a review. Int Commun Heat Mass Transfer 125:105341
10. Bumataria RK, Chavda N, Panchal HJ (2019) Current research aspects in mono and hybrid nanofluid based heat pipe technologies. Heliyon 5(5):e01627
11. Baghbanzadeh M et al (2014) Investigating the rheological properties of nanofluids of water/hybrid nanostructure of spherical silica/MWCNT. Thermochim Acta 578:53–58
12. Ghanbarpour M et al (2015) Improvement of heat transfer characteristics of cylindrical heat pipe by using SiC nanofluids. Appl Therm Eng 90:127–135
13. Wei B et al (2017) Thermo-physical property evaluation of diathermic oil based hybrid nanofluids for heat transfer applications. Int J Heat Mass Transf 107:281–287
14. Sundar LS et al (2018) Experimental investigation of Al_2O_3/water nanofluids on the effectiveness of solar flat-plate collectors with and without twisted tape inserts. Renew Energy 119:820–833
15. Demetzos CJ (2008) Differential scanning calorimetry (DSC): a tool to study the thermal behavior of lipid bilayers and liposomal stability. J Liposome Res 18(3):159–173
16. Clas S-D, Dalton CR, Hancock BC (1999) Differential scanning calorimetry: applications in drug development. Pharm Sci Technol Today 2(8):311–320
17. Farhana K et al (2021) Analysis of efficiency enhancement of flat plate solar collector using crystal nano-cellulose (CNC) nanofluids. Sustainable Energy Technol Assess 45:101049
18. Sekhar YR, Sharma K (2015) Study of viscosity and specific heat capacity characteristics of water-based Al_2O_3 nanofluids at low particle concentrations. J Exp Nanosci 10(2):86–102
19. Murshed SS (2012) Simultaneous measurement of thermal conductivity, thermal diffusivity, and specific heat of nanofluids. Heat Transfer Eng 33(8):722–731
20. Singh RK et al (2017) Performance evaluation of alumina-graphene hybrid nano-cutting fluid in hard turning. J Clean Prod 162:830–845
21. O'Hanley H et al (2012) Measurement and model validation of nanofluid specific heat capacity with differential scanning calorimetry. Adv Mech Eng 4:181079
22. Das C, Gupta SD, Mekjian A (2003) Negative specific heat in a thermodynamic model of multifragmentation. Phys Rev C 68(1):014607

Preparation and Characterization of Cross-Linked Chitosan/Cellulose Bionanohybrids

Mostafa Yusefi, Kamyar Shameli, Justin Chan Zhe, and Nor Azwadi Bin Che Sidik

Abstract Bionanohybrids have been extensively studied as the cornerstone of novel materials for manifold applications. This research aims to fabricate layer-by-layer bionanohybrids of chitosan nanoparticles cross-linked natural cellulose using sodium tripolyphosphate cross-linker (Ch-TPP-C) and ionic gelation method. Chitosan coating enhanced the multi-functionality of natural cellulose obtained from rice straw waste. In turn, the functional groups of the cellulose matrix offer exogenous groups for chitosan as a coating agent. The image of scanning electron microscopy (SEM) of Ch-TPP-C showed that the spherical-like chitosan nanoparticles covered the entangled cellulose well. The hydrodynamic size of 128.50 and 205.73 nm was found for chitosan nanoparticles and Ch-TPP-C, respectively. The hydrodynamic size of the samples slightly increased after one month of storage. The results from X-ray diffraction (XRD) and Fourier transforms infrared (FTIR) spectra showed the formation of Ch-TPP-C with multifunctional properties. Therefore, the incorporation of dual polysaccharides of natural cellulose matrix and chitosan coater was successful to synthesize the low-cost bionanohybrids potentially for anticancer drug delivery applications.

Keywords Chitosan · Natural cellulose · Bionanohybrids · Crosslinking agent

1 Introduction

This is very timely that using innovative materials with the desirable characteristics can overcome the global health risk issues. Insights into bionanohybrids have illuminated several significant features for biomedical applications in the past few decades [1, 2]. This has been linked to evidence of bias for developing advanced technologies to promote a healthy society [3, 4]. In this manner, the therapeutic nature of nanomaterials with arsenal of targeted drug delivery systems and advanced medical

M. Yusefi (✉) · K. Shameli · J. Chan Zhe · N. A. Bin Che Sidik
Malaysia-Japan International Institute of Technology, Universiti Teknologi Malaysia, Jalan Sultan Yahya Petra, 54100 Kuala Lumpur, Malaysia
e-mail: myusefi175@gmail.com

M. Y. Ismail et al. (eds.), *Technological Advancement in Mechanical and Automotive Engineering*, Lecture Notes in Mechanical Engineering,
https://doi.org/10.1007/978-981-19-1457-7_18

devices have manifested several advantages over the conventional cancer therapy methods due to their low damage to normal cells, accumulation at the tumor location, and ability to deliver sufficient drug dosage to the tumor [5, 6]. To date, the U.S. Food and Drug Administration (FDA) has approved five nano-carriers include liposomes, albumin nanoparticles (NPs), and polymeric micelles [7]. Natural polymers have taken advantage of pH-switchability as tunable materials in the latter studies in drug delivery applications to replace most of the current nonbiodegradable polymers [8–10].

As the most abundant biopolymer on earth, cellulose indicates appropriate mechanical and physiochemical properties [11]. Despite this, it does possess several issues, including low solubility in solvent fluids [12]. Its physical, thermal, and mechanical characteristics can be enhanced via different surface modifications and coating methods [13, 14]. The most prevalent coating material on natural cellulose matrix is chitosan to fabricate a double polysaccharide complex with high biocompatibility and water-rich structures [15]. Chitosan as a popular polysaccharide is obtained from natural resources such as chitin and the shells of prawns or crabs [16]. Researchers have utilized low molecular weight chitosan and different crosslinkers to synthesize chitosan-based bionanohybrids with desirable properties [17, 18]. In this regard, gelation, reverse micell ar, and emulsification solvent diffusion are frequently used [13]. Ionic gelation is a facile procedure free from using organic solvents that the positive charge of chitosan interacted with the negative charge of sodium tripolyphosphate (TPP) [19]. The morphology of chitosan bionanohybrids is related to the fabrication procedure, mixing ratio of the ingredients, and TPP concentration [20]. Albeit many researchers have synthesized chitosan bionanohybrids by using TPP cross-linker and ionic gelation method, optimization of the fabrication parameters are still ongoing as an important issue in this respect [9, 21]. These parameters not only impact the colloidal stability but also regulate the crosslinking and biocompatibility of bionanohybrids. It was stated that using TPP and chitosan with a ratio around 2:1 indicated the effective production of chitosan bionanohybrids containing appropriate physicochemical properties [22, 23].

This current study aims to fabricate crosslinked chitosan NPs and chitosan crosslinked cellulose bionanohybrids via ionic gelation procedure and TPP crosslinker. The physical, chemical, and morphological properties of the synthesized samples were evaluated by scanning electron microscopic (SEM), dynamic light scattering (DLS), thermogravimetric analysis (TGA), X-ray powder diffraction (XRD), and Fourier-transform infrared spectroscopy (FTIR).

2 Materials and Methods

2.1 Materials

Sodium chloride (NaCl), acetic acid (CH_3COOH) potassium hydroxide (KOH), chitosan (low molecular weight), acetic acid glacial (CH_3COOH) (98%), Tween-80, and TPP were provided from Sigma Aldrich, USA. All the reagents used were of analytical grade and used as received.

2.2 Synthesis of Chitosan Nanoparticles and Chitosan Crosslinked Cellulose Bionanohybrids

Natural cellulose from rice straw waste was obtained as stated earlier [24]. 30 g rice straw waste was mixed with toluene/ethanol 2:1 (v/v) and then placed in a soxhlet instrument for 12 h at 70 °C. This was followed by mixing the dewaxed rice straw with NaCl (1.4%) and CH_3COOH (with pH 4) for 5 h at 70 °C. After washing the sample, potassium hydroxide solution (5%) was used to treat the sample for 12 h. The natural cellulose solution was centrifuged and then freeze-dried.

Ionic gelation method and TPP were used to synthesize chitosan-cellulose (Ch-TPP-C) bionanohybrids and chitosan NPs [9]. Initially, 0.250 g of chitosan powder was mixed with 80 mL of acetic acid (1.0%) and 2% (*v*/*v*) of Tween-80. Then, 0.125 g cellulose was added into the chitosan solution under 9000 rpm stirring of a homogenizer for around 10 min. This was followed by dropwise adding of TPP (0.50 g in 15 mL deionized water) into the solution under vigorous stirring for around 1 h and then the Ch-TPP-C sample was centrifuged and finally freeze-dried.

Chitosan NPs was prepared following the above procedure, although without using the natural cellulose.

2.3 Characterization

Electro-Scan SEM (JSM 7600 F) with an acceleration voltage of 10 kV was used to obtain the SEM images. DLS was carried out using an Anton Paar instrument (Litesizer 500) at room temperature. Thermal analysis was performed by using TGA (STA F3 Jupiter) Q50 V20 at a heating rate of 10 °C/min under the nitrogen atmosphere. XRD (Philips, X'pert, Cu Ka) at room temperature was utilized, which the analysis was accomplished in dispersion 2 angles of 5°–80° with scanning rate of 2 s/step, voltage of 45 kV, a Ni-filtered Cu K radiation (=1.5406 A)°, and a filament current of 40 mA. For FTIR spectroscopy (ThermoNicolet, Waltham), the sample at a ratio of 1:100 w/w was mixed with KBr to prepare a transparent pellet which was analyzed under a transmittance mode in a range between 4000 and 400 cm^{-1} with a 4 cm^{-1} resolution and an accumulation of 128 scans.

3 Results and Discussion

Figure 1a indicates the schematic processes of Ch-TPP-C fabrication. Figure 1b shows the possible interactions between active functional groups presented in TPP crosslinker and low molecular weight chitosan for the synthesis of chitosan NPs and Ch-TPP-C bionanohybrids. The ratio 2:1 (*v*/*v*) between the chitosan and TPP cross-linker was used following our previous study [9]. Also, a different study [16] recommended this ratio to synthesize chitosan NPs with desirable physiochemical features. During the synthesis of Ch-TPP-C, cationic groups presented in chitosan potentially reacted with phosphoric ions in the TPP solution [9].

3.1 Scanning Electron Microscopy (SEM)

Figure 2 indicates the SEM image of Ch-TPP-C. An average diameter of Ch-TPP-C was found to be 51.19 ± 2.8 nm. The ratio between the TPP crosslinker and chitosan powder can control the size of chitosan NPs. In this respect, a separate report [22] stated that a ratio of 1:2 between the chitosan and TPP is suitable for the formation of chitosan NPs with desirable morphology. The layer-by-layer synthesis successfully was used to form chitosan NPs as also coated the natural cellulose to fabricate Ch-TPP-C bionanohybrids with the spherical shape. In addition, the chitosan coating was homogenous with a multi-layered structure and low agglomeration [25, 26]. This can show that the cellulose as solid support is entangled in the chitosan NPs as a coating agent.

3.2 Dynamic Light Scattering (DLS)

The particles size of the sample in a solution can be measured by DLS analysis. As revealed in Fig. 3, the hydrodynamic particle dimension of chitosan NPs and Ch-TPP-C were found to be 128.50 ± 3.16 nm and 205.73 ± 2.95 nm. By using a certain TPP concentration, the sizes of the chitosan NPs could be enhanced because of the particle aggregation [27]. The hydrodynamic size of Ch-TPP-C is bigger than chitosan NPs. This can show that the chitosan effectively coated the entangled cellulose, as also was seen in the SEM image. Table 1 shows the hydrodynamic particle size of the fabricated sample suspensions during 28 days of storage. The hydrodynamic particle size of the chitosan NPs was found to be 151.78 ± 5.20 nm and 166.53 ± 4.18 nm after two and four weeks of storage, respectively. Further results indicated the hydrodynamic size

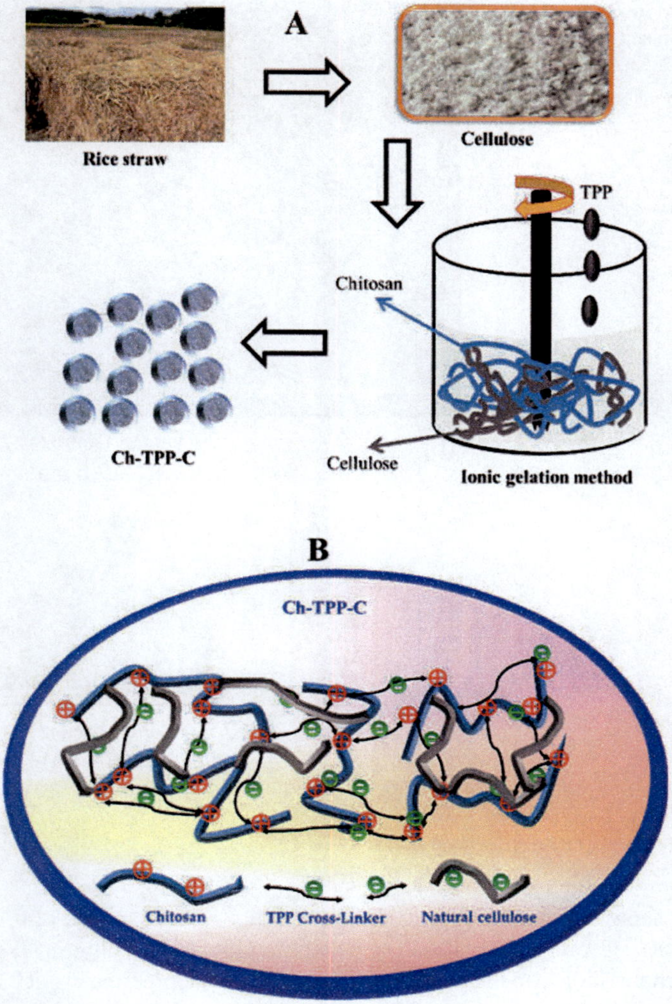

Fig. 1 **a** The schematic processes of Ch-TPP-C fabrication. **b** The interactions between functional groups of low molecular weight chitosan and TPP cross-linker, natural cellulose in the synthesis of Ch-TPP-C bionanohybrids

of Ch-TPP-C was 234 ± 2.81 nm and 245.61 ± 3.32 nm after two and four weeks of storage, respectively. These increases in the hydrodynamic size could be related to the swelling index of the samples over time. The bigger hydrodynamic size of Ch-TPP-C than chitosan NPs was related to the presence of the natural cellulose matrix within the Ch-TPP-C bionanohybrids. The fabricated chitosan NPs and Ch-TPP-C, therefore, with small hydrodynamic sizes can be used in anticancer drug delivery applications.

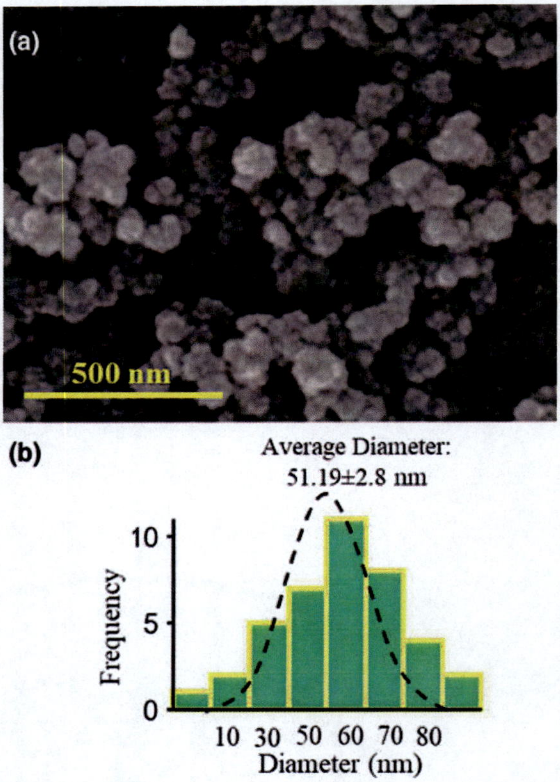

Fig. 2 SEM image and the average diameter of Ch-TPP-C

3.3 Thermogravimetric Analysis (TGA)

TGA can show the thermal properties and stability of the samples [28]. Figure 4 indicates the TGA result of chitosan NPs and Ch-TPP-C. The chitosan NPs and Ch-TPP-C samples indicated the main thermal degradation approximately at 237 °C and 2670 °C with the last residue of 35.72 wt% and 33.86 wt%, respectively, at 800 °C. The main thermal degradation is related to carbonyl as well as carboxyl groups to rapture the bonds of the samples, as stated earlier [29, 30]. It can be understood from the TGA results that the fabricated polysaccharide samples indicated acceptable thermal stability.

3.4 X-Ray Powder Diffraction

Figure 5 shows the XRD result of low molecular weight chitosan, TPP crosslinker, chitosan NPs, and Ch-TPP-C. The low molecular weight chitosan indicated peaks

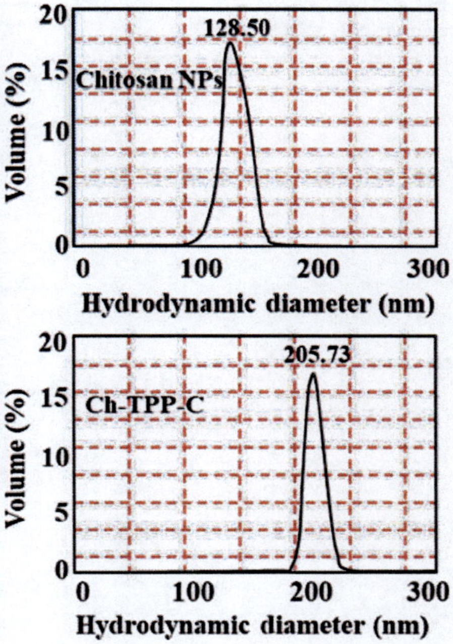

Fig. 3 The hydrodynamic diameter of chitosan NPs and Ch-TPP-C

Table 1 The hydrodynamic particle diameter of the fabricated sample suspensions during four weeks of storage

Sample	Hydrodynamic particle size (nm)		
	As prepared	After 2 weeks	After 4 weeks
Chitosan NPs	128.50 ± 3.16	151.78 ± 5.20	166.53 ± 4.18
Ch-TPP-C	205.73 ± 2.95	234 ± 2.81	245.61 ± 3.32

of crystallinity approximately at $2\theta = 17°$ and $20°$. These peaks are disappeared or have a decreased intensity in the XRD spectra of chitosan NPs because the chitosan was cross-linked with TPP, proving the successful synthesis of the chitosan NPs [31]. Ch-TPP-C showed a pattern attributed to crosslinked chitosan and cellulose with peaks approximately at $2\theta = 22°$ and $15°$, which is in good agreement with the JCPDS card no. 04-0784 [9]. These results, thus, potentially displayed the effective production of chitosan and chitosan-cellulose bionanohybrids crosslinked with TPP.

3.5 Fourier-Transform Infrared Spectroscopy

FTIR analysis is able to identify changes in the chemical structure of different samples via generating an infrared absorption spectrum. Figure 6 displays FTIR spectra of

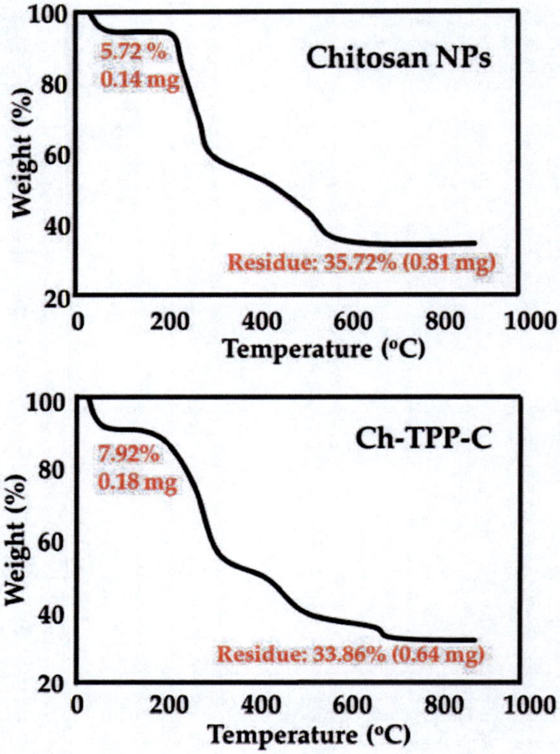

Fig. 4 TGA analysis of chitosan NPs and Ch-TPP-C

molecular weight chitosan, TPP crosslinker, chitosan NPs, and Ch-TPP-C. In the spectra of the TPP crosslinker, an intense peak at 1209 cm^{-1} is related to the stretching vibration of P=O. The PO_2 group with the symmetric and asymmetric stretching vibrations caused showing the peak at 1157 cm^{-1} [32]. The bands at around 1136 and 889 cm^{-1} are related to the symmetric stretching vibration of O–P=O and the asymmetric stretching vibration of the P—O—P group, respectively. Chitosan NPs showed a peak at 3676 cm^{-1} because of the presence of –OH stretching vibration. The band at 3360 cm^{-1} was related to the stretching vibration of N—H as overlapped with O–H due to the presence of the amino groups of low molecular weight chitosan and also the phosphate groups of TPP [33]. This might show the cross-linking between chitosan and TPP. As reported earlier [29], in Ch-TPP-C bionanohybrids, the amine groups of chitosan with carbonyl groups of cellulose can cause the presence of imines with a carbon–nitrogen double bond. The freeze-drying process during the sample preparation possibly moderated vaporization of water on the surface of the chitosan NPs and Ch-TPP-C bionanohybrids, increasing the –OH groups and fabrication of white fluffy polysaccharide NPs [34].

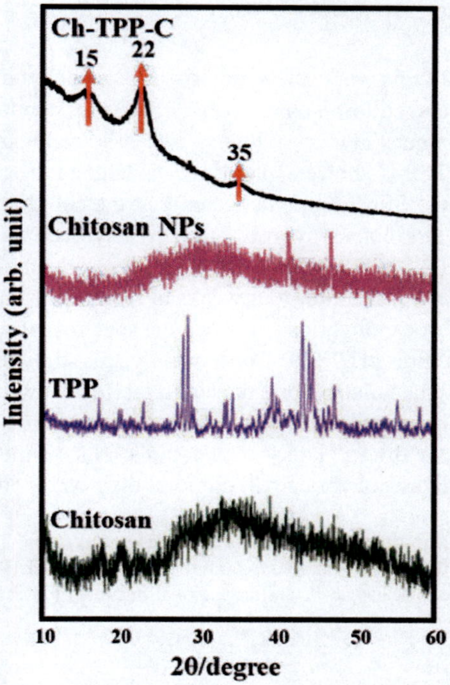

Fig. 5 XRD spectra of low molecular weight chitosan, TPP crosslinker, chitosan NPs, and Ch-TPP-C

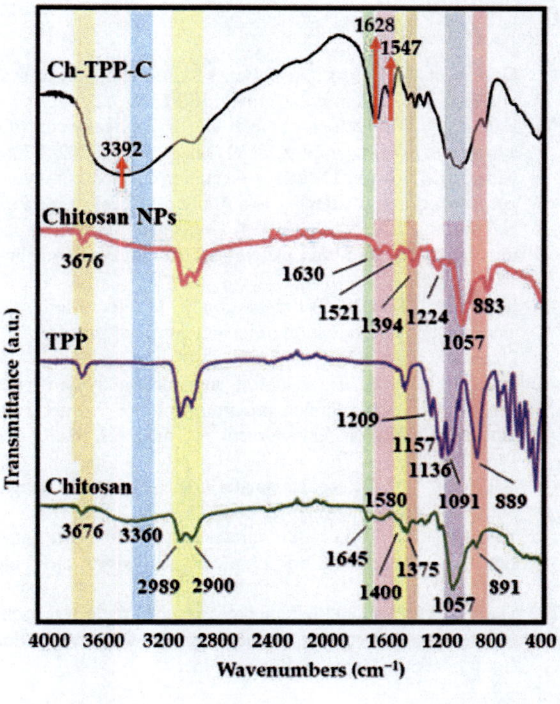

Fig. 6 FTIR results of low molecular weight chitosan, TPP crosslinker, chitosan NPs, and Ch-TPP-C

4 Conclusions

In this work, the synthesis of bionanohybrids containing a natural cellulose matrix and chitosan coater (Ch-TPP-C) was conducted by using facile ionic gelation technique, TPP cross-linker, Tween-80 stabilizer, and acetic acid. SEM analysis of Ch-TPP-C showed spherical NPs with an average size of 51.19 ± 2.8 nm. In addition, the SEM image indicated that the cellulose matrix was coated well with chitosan. The bionanohybrids indicated a hydrodynamic size of 205.73 ± 2.95. Moreover, the DLS results displayed that the hydrodynamic size of the bionanohybrids has slightly increased after four weeks of storage. From the XRD and FTIR data, the Ch-TPP-C bionanohybrids contained the spectroscopy peaks attributed to cellulose crosslinked chitosan by TPP. Consequently, this study revealed that natural cellulose is an appropriate matrix for chitosan to fabricate low-cost polysaccharide bionanohybrids with improved colloidal stability and multifunctional physiochemical properties possibly for drug delivery applications. Not to mention, future studies on the Ch-TPP-C bionanohybrids will pinpoint their use for drug delivery and in-vivo evaluations.

Funding This research was funded by Takasago Thermal Engineering Co. Ltd. grant (#4B422) and (#4B631) from the research management center (RMC) of Universiti Teknologi Malaysia (UTM) and Malaysia-Japan International Institute of Technology (MJIIT).

References

1. Dave V et al (2019) Lipid-polymer hybrid nanoparticles: synthesis strategies and biomedical applications. J Microbiol Methods 160:130–142
2. Bakaraju V et al (2020) An electronic and optically controlled bifunctional transistor based on a bio-nano hybrid complex. ACS Omega 5(17):9702–9706
3. Nasef MM, Güven O (2012) Radiation-grafted copolymers for separation and purification purposes: status, challenges and future directions. Prog Polym Sci 37(12):1597–1656
4. Nakhjavani SA et al (2019) Gold and silver bio/nano-hybrids-based electrochemical immunosensor for ultrasensitive detection of carcinoembryonic antigen. Biosens Bioelectron 141:111439
5. Bae J, Park JW (2015) Preparation of an injectable depot system for long-term delivery of alendronate and evaluation of its anti-osteoporotic effect in an ovariectomized rat model. Int J Pharm 480(1–2):37–47
6. Yusefi M et al (2020) Evaluating anticancer activity of plant-mediated synthesized iron oxide nanoparticles using Punica granatum fruit peel extract. J Mol Struct 1204:127539
7. Shi J et al (2017) Cancer nanomedicine: progress, challenges and opportunities. Nat Rev Cancer 17(1):20–37
8. Rai P et al (2021) Recent advances in the sustainable design and applications of biodegradable polymers. Bioresour Technol 124739
9. Yusefi M et al (2021) 5-fluorouracil encapsulated chitosan-cellulose fiber bionanocomposites: synthesis, characterization and in vitro analysis towards colorectal cancer cells. Nanomaterials 11(7):1691
10. Madhusha C et al (2020) Facile mechanochemical approach to synthesizing edible food preservation coatings based on alginate/ascorbic acid-layered double hydroxide bio-nanohybrids. J Agric Food Chem 68(33):8962–8975

11. Yusefi M et al (2020) Analysis on physiochemical properties of cellulose fiber from rice straw waste. In: IOP conference series: materials science and engineering. IOP Publishing
12. Yusefi M et al (2020) The potential anticancer activity of 5-fluorouracil loaded in cellulose fibers isolated from rice straw. Int J Nanomed 15:5417
13. Ignacz G, Fei F, Szekely G (2018) Ion-stabilized membranes for demanding environments fabricated from polybenzimidazole and its blends with polymers of intrinsic microporosity. ACS Appl Nano Mater 1(11):6349–6356
14. Shekhar S et al (2020) An exploration of electrocatalytic analysis and antibacterial efficacy of electrically conductive poly (D-glucosamine)/graphene oxide bionanohybrid. Carbohyd Polym 240:116242
15. Anirudhan TS, Christa J (2017) Multi-polysaccharide based stimuli responsive polymeric network for the in vitro release of 5-fluorouracil and levamisole hydrochloride. New J Chem 41(20):11979–11990
16. Bozoğlan BK, Duman O, Tunç S (2020) Preparation and characterization of thermosensitive chitosan/carboxymethylcellulose/scleroglucan nanocomposite hydrogels. Int J Biol Macromol 162:781–797
17. Chen Y et al (2018) Bioinspired fabrication of carbonated hydroxyapatite/chitosan nanohybrid scaffolds loaded with TWS119 for bone regeneration. Chem Eng J 341:112–125
18. Wei Y et al (2018) Partially reduced graphene oxide and chitosan nanohybrid membranes for selective retention of divalent cations. RSC Adv 8(25):13656–13663
19. Ruman U et al (2021) Synthesis and characterization of chitosan-based nanodelivery systems to enhance the anticancer effect of sorafenib drug in hepatocellular carcinoma and colorectal adenocarcinoma cells. Nanomaterials 11(2):497
20. Schwaiger D, Lohstroh W, Müller-Buschbaum P (2020) Investigation of molecular dynamics of a PTB7: PCBM polymer blend with quasi-elastic neutron scattering. ACS Appl Polym Mater 2(9):3797–3804
21. Ribeiro EF et al (2020) Chitosan and crosslinked chitosan nanoparticles: synthesis, characterization and their role as pickering emulsifiers. Carbohyd Polym 250:116878
22. Maluin FN et al (2019) Preparation of chitosan–hexaconazole nanoparticles as fungicide nanodelivery system for combating Ganoderma disease in oil palm. Molecules 24(13):2498
23. Yusefi M et al (2021) Facile fabrication of polysaccharide nanocomposites using ionic gelation method. J Res Nanosci Nanotechnol 3(1):37–45
24. Yusefi M et al (2021) 5-fluorouracil loaded magnetic cellulose bionanocomposites for potential colorectal cancer treatment. Carbohydr Polym 118523
25. Abdul Khalil HPS et al (2016) A review on chitosan-cellulose blends and nanocellulose reinforced chitosan biocomposites: properties and their applications. Carbohydr Polym 150:216–226
26. de Mesquita JP et al (2012) Bio-based nanocomposites obtained through covalent linkage between chitosan and cellulose nanocrystals. Carbohyd Polym 90(1):210–217
27. Ding Z et al (2021) Preparation, characterization and biological activity of proanthocyanidin-chitosan nanoparticles. Int J Biolog Macromol
28. Izadiyan Z et al (2021) Anticancer activity of 5-fluorouracil-loaded nanoemulsions containing Fe_3O_4/Au core-shell nanoparticles. J Mol Struct 131075
29. Pech-Cohuo S-C et al (2018) Production and modification of cellulose nanocrystals from Agave tequilana weber waste and its effect on the melt rheology of PLA. Int J Polym Sci
30. Ahmed ME et al (2020) Sustainable antimicrobial modified chitosan and its nanoparticles hydrogels: synthesis and characterization. Int J Biol Macromol 162:1388–1397
31. Vaewbundit S, Siriphannon P (2021) Soft solution in-situ synthesis of chitosan/iron oxide nanocomposites and their magnetic properties. Soft Matter
32. Pan C et al (2020) Study on the relationship between crosslinking degree and properties of TPP crosslinked chitosan nanoparticles. Carbohyd Polym 241:116349

33. Silvestro I et al (2020) Preparation and characterization of TPP-chitosan crosslinked scaffolds for tissue engineering. Materials 13(16):3577
34. Jiang F, Han S, Hsieh Y-L (2013) Controlled defibrillation of rice straw cellulose and self-assembly of cellulose nanofibrils into highly crystalline fibrous materials. RSC Adv 3(30):12366–12375

Effect of Catalyst in the Pyrolysis of Waste Polyethylene Terephthalate (PET) Plastics

Emil Jean D. Loreniana, Justin Dhavee D. Sorongon, and Cresencio P. Genobiagon Jr.

Abstract Pyrolysis is taken into account as the foremost promising thermochemical conversion technology for converting solid wastes into liquid fuels. The present work focused on producing liquid from PET plastic fuel with catalyst employment through the pyrolysis process. Moreover, it aims to differentiate quantitively and qualitatively the liquid yield's improvement concerning the catalyst employed in its benzoic acid content, high heating value, density, and fire and flashpoint. The product yields and compositions are also determined. The experiment takes place in a batch reactor with a heating range of 350–500 °C. The process that used $Ca(OH)_2$ as a catalyst encompasses a maximum liquid yield of 70.6%. In terms of density, the oil produced from three catalysts' employment has comparable value with the commercial light petroleum fuel in a range of 794–884 kg/m^3. The pyrolysis with the Fe_2O_3 catalyst has the lowest flashpoint of 28.2 °C, while the zeolite and $Ca(OH)_2$ catalytic process has the same flash point of 30 °C. Also, the Fe_2O_3 catalytic process has the highest fire point of 40.6 °C among the three catalysts. The pyrolysis with $Ca(OH)_2$ catalyst had a maximum heating value of 9284.05 cal/g when $Ca(OH)_2$ was used as a catalyst. The paper concludes with the addition of catalyst, namely zeolite, $Ca(OH)_2$, and Fe_2O_3, into PET plastic in the pyrolysis process, increased oil yields, and improved its characteristics.

Keywords Pyrolysis · Catalyst · Flash point · Fire point · High heating value · Density

E. J. D. Loreniana · J. D. D. Sorongon
Chemical Engineering Program, College of Engineering Education, University of Mindanao, Matina Campus, New Matina, 8000 Davao City, Philippines

C. P. Genobiagon Jr. (✉)
Mechanical Engineering Program, College of Engineering Education, University of Mindanao, Matina Campus, New Matina, 8000 Davao City, Philippines
e-mail: cresencio_genobiagon@umindanao.edu.ph

M. Y. Ismail et al. (eds.), *Technological Advancement in Mechanical and Automotive Engineering*, Lecture Notes in Mechanical Engineering,
https://doi.org/10.1007/978-981-19-1457-7_19

1 Introduction

Polyethylene terephthalate (PET) is a versatile synthetic polymer that can be thermoformed into various applications, such as containers, disposables, and meditative applications. In the report conducted by Recycling International, PET exploitation will surmount 20 million tons by 2021 and continuously increase per annum [1]. The mass production of PET waste is growing rapidly, with the upsurge of civilization, brisk industrialization, and deviations in the mode of living. Additionally, despite the versatility of PET, it has a fair share of adverse effect in the environment particularly its slow deterioration, causes severe problems in the locale and human health as a whole [2]. Numerous disposal techniques for PET are taken into account but dump or sanitary landfill discarding is the conventional enforcement for the disposal of plastic waste [3]. The dumping of waste plastics in landfills allows habitation of nonentities, stimulating a differing sort of diseases [4].

Moreover, the expense of transit or freight, manual labour, and repairs can escalate reprocessing expenditure [3]. Besides, as a result of fast-paced expansion, areas allotted for dumpsite or landfills, specifically in urban areas, is downsizing. Thus, a better approach and attested process is necessary to discard these PET waste, and the most feasible remedy is pyrolysis.

Pyrolysis refers to a thermochemical conversion of waste plastics without the presence of oxygen with a temperature ranging from 300 to 900 °C to convert plastic waste into liquid oil [5]. The utilization of catalysts helps to reinforce plastic waste's overall pyrolysis process and maximize process performance. Catalysts have a crucial function in fostering process performance, leveling precise output, and lowering operation's temperature and residence time [6]. A broad category of catalysts is utilized in the pyrolysis of PET withal the foremost utilized catalysts are zeolite, ZSM-5 and Y-zeolite, FCC, and MCM-41 [7]. The catalytic reactivity during the process with the employment of homogenous catalysts might incorporate cracking, oligomerization, cyclization, aromatization, isomerization and the like [8].

Several articles documented the employment of micro structured and mesostructured catalysts for the energy recovery of PET plastic waste to convert it into liquid oil and activated char. One experiment pyrolyzed polyethylene (PET) plastics with HZSM-5 catalysts [9]. From that set-up it is observed that HZSM-5 improved oil yield formation with the conformation of aromatics and iso-alkanes compound. Gaca also conducted an experiment in which he pyrolyzed PET waste with altered MCM-41 and HZSM-5 and disclosed that with the application of HZSM-5 as catalyst yields lighter organic compounds (C3–C4) with higher aromatic compound [10]. Another study employed assorted types of catalysts and noted that merging HZSM-5 with SiO_2–Al_2O_3 or MCM-41 resulted to greater liquid oil yield with lesser gaseous products [11]. Lastly, Aguado also conducted an experiment in catalytic pyrolysis of PET with HZSM-5 and observed the build-up of aromatics and other compounds Simultaneously, the application of mesostructured MCM-41 lowered the aromatics attained due to its minimal acid catalytic reaction. Indeed, employment of artificial catalysts improved general process outcome and the quality of products being produced [12]. Nevertheless, besides zeolite catalysts, numerous alternative catalysts are cost and process-efficient.

Catalysts such as zeolite, $Ca(OH)_2$, and Fe_2O_3 have enhanced thermal productivity and modified pyrolytic product conformations [13]. These catalysts have unique catalytic properties that can maximize the generated oil and enrich product quality. However, they have pros and cons to be considered. Some catalyst may not be appropriate at elevated temperatures due to coke development, and some catalyst has high gaseous yield due to the reaction time [14]. Herein, the effect of catalyst on the pyrolysis of PET plastic waste is examined taking into consideration that the type of catalyst employed plays a critical role to establish the viability of the whole process.

In this study, $Ca(OH)_2$, Fe_2O_3, and zeolite have been employed in the catalytic pyrolysis of PET plastics as to differentiate quantitatively and qualitatively the liquid fuel yield's improvement concerning the catalyst used. This paper can be a future reference for PET pyrolysis-related studies. This study aims to determine the best suitable catalyst in PET plastics' pyrolysis based on the overall parameters used in the study. Additionally, the paper focuses mainly on differentiating the employed catalyst's effects in terms of the liquid, gas, and char yield in the process using a batch reactor. The gathered data can be used as a future reference for forthcoming studies relevant to PET waste's catalytic pyrolysis.

The scope of the study is only limited in studying the physical properties of product yield, namely the percent weight of liquid, gas, and char, as well as the liquid's benzoic acid content, high calorific value, density, and the fire point and flashpoint of the liquid.

2 Material and Methods

2.1 Conceptual Framework

The study claims that the application of catalyst regarding its type in the PET plastic pyrolysis has exhibited varying results in terms of pyrolysis product properties and yield. The addition of such material provides changes in the operational process and improves the resulting output. The experiment consists of three types of variables (see Fig. 1). Parameters like temperature and pressure fall in the independent variables since they have a fixed value for the pyrolysis. On the other hand, dependent variables are parameters that change or are highly affected by the moderator's addition and the area of concern for this study. Lastly is the moderator variable, which alters the result and mainly influences the correlation between the two variables.

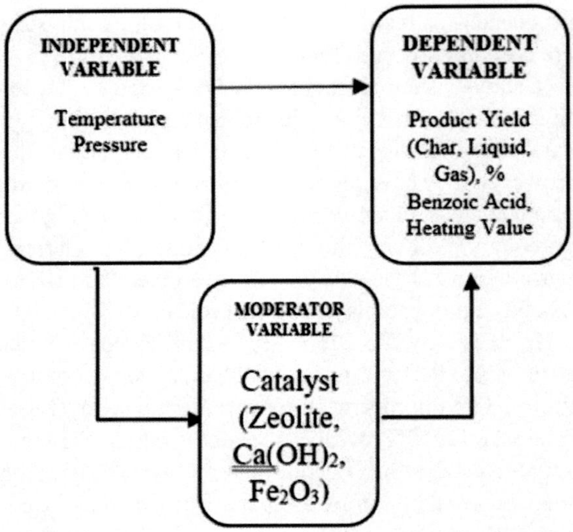

Fig. 1 Conceptual framework

2.2 *Research Design*

The research design employed throughout the study is the experimental method. This method is used to determine the significance and suitability of a particular catalyst in the pyrolysis of PET plastic in terms of the liquid properties (flash point, fire point, density, heating value) and the product yield, and the % of benzoic acid. Moreover, the catalyst's effectiveness to acquire the desired results is based on the pyrolysis product of the plastic alone.

2.3 *Research Material and Equipment*

The materials used in experimenting are shredded waste PET plastics, empty glass containers, face masks, and an LPG tank. The catalyst used was purchase through online stores, and the obtained catalyst is 500 g Fe_2O_3, 500 g $Ca(OH)_2$, and 500 g Zeolite, all manufactured by the Dalkem Corporation. The following are the equipment used in the process: condenser, digital thermometer, and the pyrolysis reactor itself.

Upon completing the experiment, the necessary materials used for the sample's testing are as follows: 0.2 M of NaOH, 0.3 g KHP, and distilled water. In acquiring the analysis of the product, specific equipment is utilized, namely the bomb calorimeter, to determine the high heating value, flash point apparatus for the flash and fire pint, pycnometer for acquiring the density, and analytical balance accurate mass measurement, and the burette.

2.4 Experimental Set-Up

The waste PET plastic bottles are collected on the school premises and in the neighboring environment or establishment. After gathering an adequate amount of PET plastic bottles, it is then washed, dry, and sorted out to separate it from its cup.

Moreover, three catalysts were used for the experiment. These are $Ca(OH)_2$, Fe_2O_3, and zeolite. The catalyst utilized per batch of the experiment consists of five trials and another five trials for the plastic's thermal cracking alone. Both cracking was conducted in a similar operating condition.

2.5 Experimental Procedure

In this study, the collected waste PET plastic bottles are shredded before feeding it in the reactor. A sample of shredded plastic weighing 100 g was placed inside the reactor together with the weighted catalyst. The ratio of the PET to catalyst is 2:1.

The operating parameters are closely monitored to avoid unnecessary accidents or the occurrence of an unfortunate incident. The reactor was checked and ensured that it is tightly sealed before conducting the test and its other parts. Furthermore, as the reaction progress, the following parameters were measured as well: temperature, pressure, and the reaction. Moreover, the process has a residence time of 1 h.

The acquired gaseous product passes through a condenser and cools down by the water after that. The condensed gases were collected in a 500 ml glass container and analyzed using the bomb calorimeter and the flashpoint apparatus. The determination of the benzoic acid content of each sample was also carried out in the experiment.

Sample Preparation for Heating Value determination. The produced liquid product from pyrolysis, which contains some catalyst and the plastic alone, was placed in a tiny vessel and weighed using the analytical balance. The weighted liquid samples have a varying mass but not exceeding 1 g.

The samples gathered were brought to the University Laboratory and placed in the bomb calorimeter with a 2 L of water. A total of 20 testing runs were conducted for the heating value determination. After some time, the values were obtained.

Sample determination for Flash point and Fire point. The samples were tested in the University Laboratory using the flash point apparatus along with digital thermometer for temperature determination, and a burner. An amount of a liquid product is place in a metal cup, and as the temperature rises its temperature was measured as the burner was passed over its surface until some flash is observed and eventually the temperature in which the flame emerges. A total of 20 runs were conducted, five trials for each catalyst employed to increase the accuracy of the result. The flash point and fire point ranges from 28 to 39 °C and 39 to 57 °C respectively.

Sample Preparation for Benzoic Acid Content. The analysis of benzoic acid content of the samples was obtained by using the 0.2 M NaOH solution that was standardized using the 0.3 g of KHP. The standardization and the preparation of the solution were carefully prepared with a supervision of a professor.

After the solution was prepared, the sample was weigh using the analytical balance. The samples were titrated using the NaOH until a slight pinkish color can be observed. For this analysis a total of 20 tests were done. In the case of liquid with catalyst, the benzoic acid content ranges from 0.5 to 11% while the liquid with no catalyst have a content of 11–15%.

Sample Preparation for Density Measurement. The analysis of the density is done using the apparatus called the pycnometer that offers a high accuracy measurement of the said property of the liquid. The empty pycnometer was first place in an analytical balance to determine its weigh which was used as a reference in calculating the density of the sample. Thereafter, it was filled with water, place in an analytical balance and weigh again then emptied. Repeating the procedure for the sample with unknown density but now it was filled with liquid product. Lastly when the value for the weight of Pycnometer with liquid sample was on hand the density is acquired.

2.6 Statistical Treatment of Data

The data acquired was tabulated and analyzed using the calculated mean value of each sample which are presented in a graphical form. These values were used to determine how each catalyst induces changes to the properties and yield of the product. Subsequently, an interpretation is done to elaborate what the values imply.

3 Results and Discussion

3.1 Thermal and Catalytic Pyrolysis Test Result

The test results for pyrolysis of waste PET plastics having a particular type of catalyst and the one with no catalyst is shown in Table 1. It represents the mean value of the product in regards to its % benzoic acid, product yield, high heating value, fire and flash point, and density which are relevant parameters to display the influence of the catalyst.

Percent Benzoic Acid Content and Product Yield. The percentage of the acid contain in the oil with and without catalyst (see Fig. 2). A 12.8% benzoic acid contains in the oil indicate that it is not recommended to process the given plastic mainly because it could clog piping and heat exchanger and if ever chosen it must be well-monitored. Another effect of the high amount of the acid is it troublesome for the transition of one process to another since the clog oil must be firstly removed that might cause problem. To address this, the catalyst is employed to compensate

Table 1 Overall data of the pyrolysis

Catalyst	None	Zeolite	$CaOH_3$	Fe_2O_3
% benzoic acid	12.8%	10.51%	10.5%	1.13%
% wt liquid	39.60%	69.40%	70.60%	62.20%
% wt gas	32.70%	5.00%	11.20%	8.80%
% wt char	27.70%	25.60%	18.20%	29%
Density (kg/m^3)	906.49	887.36	794.80	805.95
High heating value (cal/g)	7987.67	9710.08	9284.05	8983.83
Flash point (°C)	38.4	30	30	28.2
Fire point (°C)	55	40.2	39.8	40.6

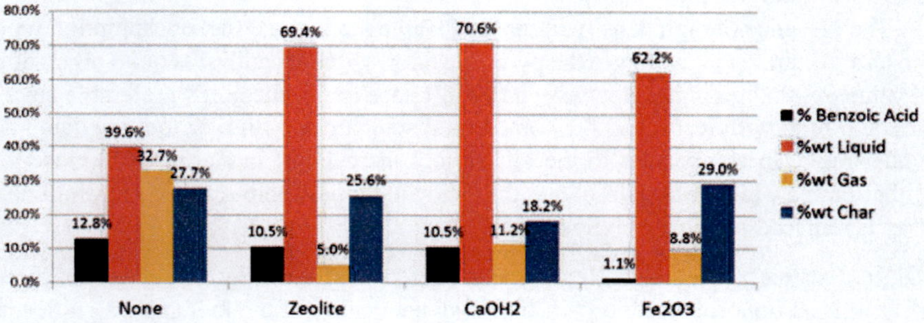

Fig. 2 Weight percentage of the product yield

to the negative effect of the acid. As shown in the figure the catalyst can improve the process by minimizing the amount of the acid and also the damage it can do to the equipment. From the figure it can be clearly seen that the lesser the acid the better.

The presence of benzoic acid in the liquid results in the oil's unfavorable use directly to the internal combustion engine. It is corrosive and may cause deterioration in the quality of fuel, and need serious attention is to be run on an industrial scale [15]. Nonetheless, the addition of catalysts can improve the pyrolysis oil.

Pyrolysis with Zeolite, $Ca(OH)_2$, Fe_2O_3 have 10.51%, 10.45%, 1.13% benzoic acid respectively. It can be shown that catalyst has a visible effect on lowering the acid characteristic of the oil, decreasing the amount of acid by 0.2% for zeolite, 0.21% for $Ca(OH)_2$, and 1.68% for Fe_2O_3.

PET plastic normally has a lower yield of liquid with higher yield of gas compared to other plastic. The experiment yield is only 39.6% which is quite low. However, the addition of catalyst could change the % composition of the product promoting an increase in liquid yield. As shown in the figure, with the presence of zeolite, $Ca(OH)_2$ and Fe_2O_3 the % liquid increase to 69.4%, 70.6%, 62.2% respectively, showing that catalyst not only lower benzoic acid but also enhance the product yield.

PET plastic yield ranges from 23 to 40 wt% and because the experimental value is within range, it clearly shows that PET plastic is not suitable to be used as a material due to low production of the oil, but if gaseous product became a preference this type of plastic is of good choice. In contrast, the experimental yield especially the liquid with the presence of catalyst increases and having a % value more than the expected range. This indicates the high influence of the catalyst to assist the liquid production and improving the performance of the plastic. Additionally, if the PET is used and high liquid yield is desired, it is advisable to have a catalyst considering how distinguishable the value from the original one-plastic alone.

Density. The figure below (Fig. 3) shows the density of pyrolytic oil with and without the presence of catalyst. The density of none, zeolite, $Ca(OH)_2$, Fe_2O_3 were 906.49 kg/m^3, 887.36 kg/m^3, 794.80 kg/m^3 and 805.95 kg/m^3 accordingly. The decrease in the density of oil means that the usage of catalyst is favorable since the value is approaching the commercial standard of gasoline and diesel which is 780 kg/m^3 and 870 kg/m^3 [16].

The presence of high density in fuel is advantageous since fuel consumption will be less. As shown in the figure the pyrolytic of oil without catalyst is quite high with a value of 906.49 kg/m^3. However, it doesn't necessarily mean it's preferable since it is too high with respect to the commercial standard-too high or too low value is undesirable. In comparison to the oil without the catalyst having a value close to either the gasoline or the diesel and thus showing a desirable result if conventional fuel is replaced by the pyrolytic oil.

High Heating Value. High Heating value refers to the energy content of the fuel [17]. It is an important property in assessing the quality of the fuel and a significant criterion that must be considered in utilization of energy. Figure 4 shows the sample's experimental calorific value without a catalyst is 7987.6699 while using zeolite, $Ca(OH)_2$, and Fe_2O_3 the calorific value were 9710.0779 cal/g, 9284.0513 cal/g, and 8983.8340 cal/g respectively. The figure displays the increase in calorific value by applying catalyst, making it almost the same as the commercial-grade, which is 10152.89 cal/g for gasoline and 10,272.33 cal/g for diesel [16].

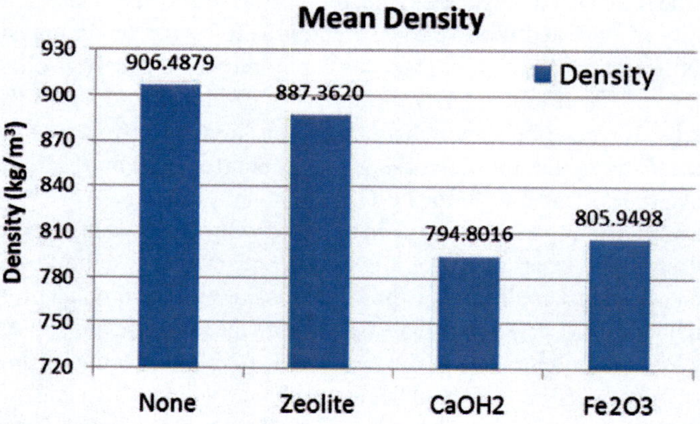

Fig. 3 Density of the liquid yield

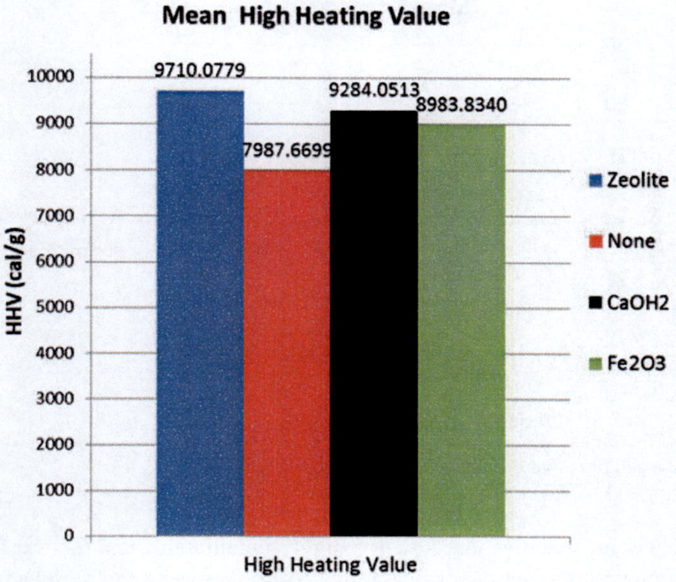

Fig. 4 Experimental high heating value (HHV) of oil yield

The heating value of pyrolytic oil ranges from 30 to 45 MJ/Kg and normally PET plastic have a low calorific value of less than 30 MJ/kg. The resulting value of oil without catalyst is higher than its normal value which is a good one but much lower to the prescribe range, indicating the low efficiency of the pyrolytic oil and limiting its usage as a fuel in engine. On the other hand, the pyrolytic oil with catalyst have a high value that is very close to the gasoline and diesel indicating its potentiality for application.

Flash point and Fire point. A characteristic that refers to the lowest temperature where fuel can disperse to stimulate the formation of an ignitable mixture in the atmosphere, which can ignite, is known as a flashpoint. Correspondingly, it is a vital factor for fuel characterization in assessing fire hazards. On the other hand, the degree to which a liquid fuel will burn for a particular time or sustain a combustion reaction is defined as the fuel's fire point. The prior characteristic is accustomed in evaluating the hazard associated with its usage and storage. Typically, the flashpoint is much lower for any liquid oil than the fire point for about (5–10) °C.

The graph in Fig. 5. shows the important property of the oil acquired at constant temperature. For a diesel fuel its flashpoint value is 55 °C and a fire point 62 °C whereas, the kerosene has a value 38 °C and 46 °C. In terms of the pyrolysis oil its flash and fire point ranges of 30–39 °C and 39–55 °C with respect to the catalyst employed. As shown in the above figure, the oil without the presence of catalyst has higher flash point compared to the ones with catalyst. The flash point of pyrolysis oil with zeolite, $Ca(OH)_2$ and Fe_2O_3 has comparable temperatures ranging 28–30 °C. A low value

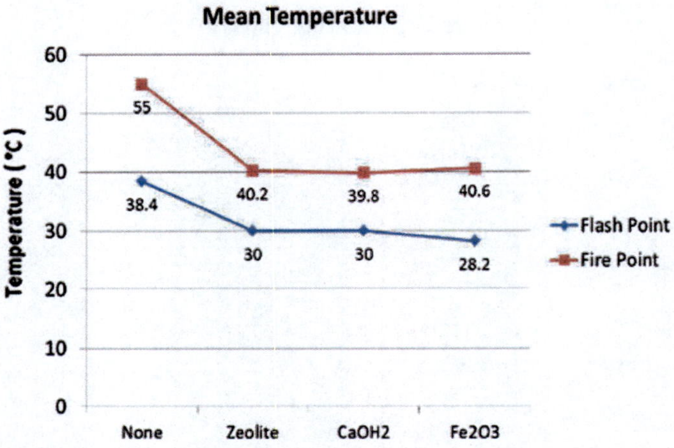

Fig. 5 Flash and fire point of oil yield

of the flash point specifies that a fuel is highly inflammable and the transportation and handling will need extra consideration. Withal, kerosene and diesel a fuel under light petroleum distillate have a higher flash point compared to the pyrolysis oil which specifies that these are easy to handle however additional precautions are still required.

4 Conclusion and Future Works

This study shows that the employment of catalysts influences the pyrolysis of PET plastics. It was observed in the test results of samples, it varies upon the type of catalyst used as well as with samples without catalyst. Yet, despite the variance, the improvement in the pyrolysis oil has been visible. Properties, such as benzoic acid, has been reduced to a significant value—indicating the effectiveness of the catalysts since the presence of acid greatly deteriorates the quality of oil due to its corrosiveness. Furthermore, PET plastics normally have low liquid yield with high gaseous product however, with the used of catalyst in the process, gaseous product decreases and the oil yield increases. In addition, the density decreases and the high heating value (HHV) increases compared to its original value of 7987.67 cal/g. Accordingly, catalyst shows prominent effect on PET plastics by providing huge impact in improving the overall quality of the oil and the process as well.

For further development of this study, the authors recommend to conduct the study in a different heating rate and in a different type of reactor as to investigate the difference of its effect in the process and in the product yield. Additionally, the authors recommend to examine the properties of gaseous product produce during pyrolysis since it could be a good potential for other energy source. Lastly, further analysis in the properties of the liquid yield can be done such as its determining its alkane content and the like.

References

1. Hurley R, Woodward J, Rothwell JJ (2018) Microplastic contamination of river beds significantly reduced by catchment-wide flooding. Nat Geosci 11(4):251–257
2. Chirayil CJ, Mishra RK, Thomas S (2019) Materials recovery, direct reuse and incineration of PET bottles. In: Recycling of polyethylene terephthalate bottles
3. Gandidi IM, Susila MD, Mustofa A, Pambudi NA (2018) Thermal—catalytic cracking of real MSW into bio-crude oil. J Energy Inst 91(2):304–310
4. Alexandra LC (2012) Municipal solid waste: turning a problem into resource waste: the challenges facing developing countries, urban specialist. World Bank, pp 2–4
5. Rehan M, Nizami AS, Shahzad K, Ouda OKM, Ismail IMI, Almeelbi T et al (2016) Pyrolytic liquid fuel: a source of renewable electricity generation in Makkah. Energy Sourc Part A Recov Util Environ Eff 38(17):2598–2603
6. Miandad R, Barakat MA, Aburiazaiza AS, Rehan M, Nizami AS (2016) Catalytic pyrolysis of plastic waste: a review. Process Saf Environ Protect (2016)
7. Ratnasari DK, Nahil MA, Williams PT (2017) Catalytic pyrolysis of waste plastics using staged catalysis for production of gasoline range hydrocarbon oils. J Anal Appl Pyrol 124:631–637
8. Serrano DP, Aguado J, Escola JM (2012) Developing advanced catalysts for the conversion of polyolefinic waste plastics into fuels and chemicals. ACS Catal 2(9):1924–1941
9. Uemichi Y, Hattori M, Itoh T, Nakamura J, Sugioka M. Deactivation Behaviors of Zeolite and Silica-Alumina Catalysts in the Degradation of Polyethylene. Ind Eng Chem Res 37(3):867–872
10. Gaca P, Drzewiecka M, Kaleta W, Kozubek H, Nowińska K (2008) Catalytic degradation of polyethylene over mesoporous molecular sieve MCM-41 modified with heteropoly compounds. Polish J Environ Stud 17(1):25–31
11. Lin YH, Yang MH, Yeh TF, Ger MD (2004) Catalytic degradation of high density polyethylene over mesoporous and microporous catalysts in a fluidised-bed reactor. Polym Degrad Stab 86(1):121–128
12. Aguado J, Sotelo JL, Serrano DP, Calles JA, Escola JM (1997) Catalytic conversion of polyolefins into liquid fuels over MCM-41: comparison with ZSM-5 and amorphous SiO_2–Al_2O_3. Energy Fuels 11(6):1225–1231
13. Lee J, Kwon EE, Park YK (2020) Recent advances in the catalytic pyrolysis of microalgae. Catal Today 355:263–271
14. Kim S, Tsang YF, Kwon EE, Lin KYA, Lee J (2019) Recently developed methods to enhance stability of heterogeneous catalysts for conversion of biomass-derived feedstocks. Korean J Chem Eng 36(1)
15. Vijayakumar A, Sebastian J (2018) Pyrolysis process to produce fuel from different types of plastic—a review. IOP Conf Ser Mater Sci Eng 396:012062
16. Sharuddin SDA et al (2016) Energy conservation and management. University of Malaya, Kuala Lumpur, Malaysia, pp 308–306
17. Ionescu G, Bulmau C (2019) Analytical pyrolysis: estimation of energy potential for solid pyrolysis by-products using analytical methods. Polytechnic University of Bucharest, Romania

Advanced Material

Optimization of WC-Tac-Co for Green Porosity via Metal Injection Moulding

Siti Nur Fatimah Khairudin, Hazriel Faizal Pahroraji, Siti Khadijah Alias, and Mohd Halim Irwan Ibrahim

Abstract MIM is a technique that is often used because it allows the production of complicated components with sophisticated geometry and accurate shape. The primary aim of this paper is to determine the optimal injection moulding parameters for WC–Co cemented carbide with TaC acting as a grain growth inhibitor (GGI) by using the Design of Experiments—Taguchi method. A significant proportion of green components is required to obtain the most efficient green porosity. This study applies the Taguchi technique to optimise the green porosity of injection moulded components depending on injection moulding parameters. These parameters were taken into account: GGI percentage, injection temperature, injection pressure, and injection speed. An orthogonal array of L9 (3^4) experimental base design was chosen as an experimental setup, and the responses were analysed using Minitab version 18. It found that the best parameter combinations for porosity are GGI (0.4 wt%), injection temperature (145 °C), injection pressure (45%), and injection speed (35%). Consequently, by managing the optimum parameter configuration, the intended product's best performance may be attained and maintained.

Keywords Metal injection moulding (MIM) · Taguchi method · WC–Co

S. N. F. Khairudin
School of Mechanical Engineering, College of Engineering, Universiti Teknologi MARA, 40450 Shah Alam, Selangor, Malaysia

H. F. Pahroraji (✉) · S. K. Alias
Centre for Mechanical Engineering Studies, Universiti Teknologi MARA, Johor Branch, Pasir Gudang Campus, 81750 Masai, Johor, Malaysia
e-mail: hazriel@uitm.edu.my

M. H. I. Ibrahim
Faculty of Mechanical and Manufacturing Engineering, Universiti Tun Hussein Onn Malaysia, 86400 Parit Raja, Batu Pahat, Johor, Malaysia

M. Y. Ismail et al. (eds.), *Technological Advancement in Mechanical and Automotive Engineering*, Lecture Notes in Mechanical Engineering,
https://doi.org/10.1007/978-981-19-1457-7_20

1 Introduction

In recent decades, metal injection moulding (MIM) has expanded dramatically to include a broad range of applications that include consumer electronics, automotive, medical, and firearms. The fabrication process of MIM is based on plastic injection moulding technology applied to powder metallurgy. Low-cost forming method for difficult-to-machine and hard-to-cast alloys and metals [1]. MIM (metal injection moulding) is an effective way to produce large quantities of complex geometries [2]. The MIM process is following the near-net-shape process since it allows parts to be directly used in services without secondary processes. Another significant advantage offered by MIM is its ability to reduce losses and recycle expensive material quickly [3, 4]. MIM's method consists of four main steps: mixing the most delicate powder and the binder to create the feedstocks; injecting the components from feedstock; debinding: eliminating the bonding in moulded parts; and sintering: the thermal process to achieve final high-density components. The method modified by powder injection moulding [1, 5, 6].

Cemented carbides have been widely used as rock drill buttons, pressing dies, cutting equipment, and other abrasion-resistant engineering components because they combined toughness, strength, and wear resistance. This alloy's tungsten carbide process provides exceptional hardness and wear resistance, while the binder phase contributes to the composite material's strength and tensile strength [7].

WC–Co hard metal is a sintered material made up of brittle tungsten carbide (WC) crystals tied by a hard-cobalt (Co) binder. Having a balanced property of hardness, strength, toughness, and chemical stability, cemented carbide possesses an incredibly high Young modulus, an outstanding thermal conductivity, and a low thermal expansion coefficient, contributing to its potential application in a broad range of technological fields. Therefore, WC–Co is widely used as mining, machining, drilling, and cutting tools with wear components and less-chip forming tools [8–10].

The grain growth inhibitors (GGI) are widely used to manufacture fine and ultrafine-grained tungsten carbide (WC–Co) materials. WC–Co cemented carbides can be made from MIM and have proven effective at producing small and complex bulk materials. A further benefit of MIM is its ability to produce cemented nanostructured carbides with particle sizes well below 1 m that are tailored to inhibit grain growth. GGI is a standard method for preventing grain growth in the sintering process since it retains a small average size within the sintered product with high sintering activity [11]. The final microstructure plays an essential role in getting high mechanical properties at the end of the process.

The optimisation of process parameters is becoming a primary concern for researchers, like defects, costs, and efficiency in planning or experimentation can be maximized. Optimization strategies used in the MIM industry are most often based on trial and improvement. This approach is extremely time-consuming and expensive due to the many experiments needed [12]. In traditional experiments, the experimental process varies one factor at a time, keeping all other factors constant. A wide range of experiments has revealed that this method does not produce satisfactory results. DOE designs of different robustness were implemented in manufacturing to be able to plan and analyze experimental data with high efficiency. Researchers can identify the quality parameter in an experiment using DOE (Design of Experiment) techniques. The study of DOE in MIM is also carried out by Momei and Alaei, who estimate the factors which affect the density, strength, and hardness of the final MIM by using a full factorial design [13].

The Taguchi method is also essential in creating superior systems, in addition to being one of the keys to quality. Taguchi's robust design is an invention of Dr. Geneichi Taguchi that reduces product variations while optimizing the parameters involved in manufacturing or laboratory environments to achieve desired product quality. Taguchi is mainly used as an approach to design factory experiments with two or more variables having varying levels or values, as well as the possible combinations of those levels across the given parameters [14]. An increasing number of studies have looked at the impact of injection moulding parameters on the mechanical properties of moulded parts and the incidence of moulding defects [15, 16]. Five process parameters (injection speed, melt temperature, mould temperature, metering scale, and hold pressure time) were used in the study to evaluate the micro moulding process and part quality of micro gears. [17]. Amin et al. [18] implementing Taguchi's method by using L27 orthogonal arrays with 7 factors and 3 level design for studying them in optimizing the green density of moulded components.

2 Material and Methodology for MIM Process

2.1 Material Characterisation

The feedstock preparations for this study consisted of WC-6Co and Tantalum Carbide (TaC) as grain growth inhibitors (GGI). Palm stearin (PS) and low-density polyethylene (LDPE) were used for binder formulations. Tables 1, 2, and 3 show the starting materials characteristics.

Table 1 General characteristics of the WC-6Co

Identification	WC-6Co
Tap density (g/cm^3)	14.93
True pycnometer density (g/cm^3)	12.722
	D_{10}: 3.5
Powder size (μm)	D_{50}: 6.8
	D_{90}: 12.7

Table 2 Characteristics of TaC

Identification	TaC
Average particle diameter (μm)	1.09
Tap density (g/cm³)	13.9

Table 3 Properties of the binder components

Types of polymer	Density (g/cm³)	Melting temperature (°C)	Decomposition temperature (°C)
PS	0.89	54.3	398.5–598.8
LDPE	0.95	127	389.6–501.6

2.2 Green Part Preparation

Using a 43% powder loading, the Brabender Plastograph mixer mixes WC-TaC-6Co. Powder loading of 43% was chosen since that is approximately 2–3% below the Critical Powder Volume Concentration (CPVC) of the selected WC-TaC-6CO powder. In this case, the CPVC for the powder was found to be 45% [18]. A homogeneous mixture is obtained after 1 h of mixing at 140° and 40 rpm. The mixing temperature shall be set up within the range of the binders' highest melting point and the lowest degradation point of the binder, fixed at 140 °C at 40 rpm for around 80 min to avoid degrading of the binders. The compound was treated with ambient humidity and temperature before being crushed into smaller pallets for easier injection moulding machine feeding. To obtain the green sample shown in Fig. 1, the feedstock was crushed and fed into a horizontal injection moulding machine modelled Nissei NP7-1F. By implementing Taguchi orthogonal arrays (Statistical Method), density optimization was performed. The injection process was designed with three levels and four parameters. As shown in Table 4, the parameters and the level are given. An analysis of the results was performed by using the Signal to Noise Ratio (S/N ratio).

2.3 Design of Experiment

Process optimization can be enhanced by using Taguchi Method. The Taguchi method is more economical than traditional optimization methods due to its low number of experiments. A systematic approach is used in this method to design and analyse experiments to improve product quality. Taguchi defines quality as the loss of the product. These losses are mostly caused by variation (noise) in the process. As a result, Taguchi parameter design is geared towards reducing variation and meeting the mean target. Two essential tools are used to achieve this, orthogonal arrays (OA) and the signal-to-noise ratio (S/N). Orthogonal arrays are balanced matrices that can investigate a variety of parameters, while S/N ratios are used as quality characteristics indicators. To increase the green intensity and consistency of the sample, Taguchi's

a) Injection machine

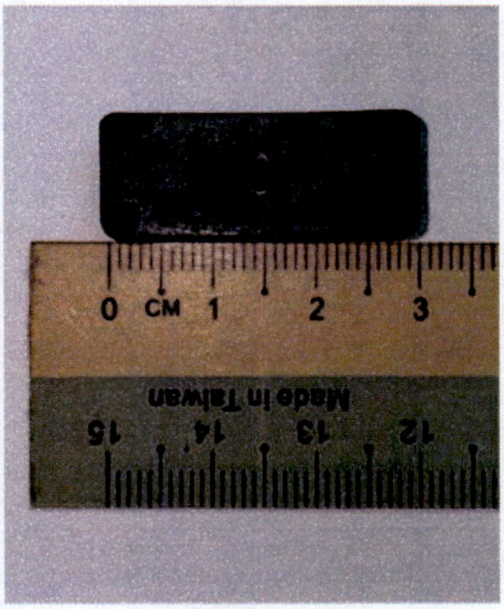

b) Green part

Fig. 1 The green sample

Table 4 Taguchi design injection parameters

Level	GGI (%)	Injection temperature (°C)	Injection pressure (%)	Injection speed (%)
	A	B	C	D
1	0.4	145	45	30
2	0.8	150	50	35
3	1.2	155	55	40

orthogonal array (statistical method) will be used for optimization. A crucial aspect of the procedure is selecting an experimental design. Three-level experimental designs with four parameters considered in the injection moulding are all very significant in screening tests using classical variance analysis (ANOVA). L9 (3^4) orthogonal analytical array used with four significant parameters involved are GGI (A), injection temperature (B), injection pressure (C), and injection speed (D). Tables 4 and 5 display the orthogonal configuration used in this analysis. A total of 27 samples were tested for nine different studies. MIM's optimization of green intensity highlights the effectiveness and superiority of green components. The data used in the noisy signal experiment were collected using Minitab software 18. Optimizing the operation involves developing a signal-to-noise graph based on the product's density to determine the highest possible level. S/N ratios established optimal parameter settings. The ranks of the most influential variables are focused on the software as well. For an engineering study, signal-to-noise characteristics are usually classified into three general groups; the smaller-the better, the nominal-the better, and the larger-the better. In this study, the small- the better is used to find the optimal parameters. Logarithms are designed primarily based on the mean square deviation derived from an equation used to calculate the signal-to-noise ratio for smaller-is-better quality characteristics of composite signals. The S/N ratio equation:

$$\frac{S}{N} = -10\text{Log}\, MSD. \tag{1}$$

An equation for mean square deviation based upon a smaller-the-better characteristic could be written as follows:

$$MSD = \frac{Y_1^2 + Y_2^2 + \cdots + Y_n^2}{n} \tag{2}$$

Table 5 Orthogonal array of L9 structure

Trials	Injection moulding parameter			
	GGI (%)	Injection temperature (°C)	Injection pressure (%)	Injection speed (%)
	A	B	C	D
1	0.4	145	45	30
2	0.4	150	50	35
3	0.4	155	55	40
4	0.8	145	50	40
5	0.8	150	55	30
6	0.8	155	45	35
7	1.2	145	55	35
8	1.2	150	50	40
9	1.2	155	45	30

The S/N equation can be written as:

$$\frac{S}{N} = -10\text{Log}\left(\frac{Y_1^2 + Y_2^2 + \cdots + Y_n^2}{n}\right) \quad (3)$$

2.4 *Porosity*

A substance's porosity is expressed as a portion of the volume occupied by voids, ranging from 0 to 1, or as a percentage that ranges from 0 to 100%. To get the porosity measurement, the green sample is measured using the Mettler Toledo machine. The green part though the density test at a temperature of distilled water at 24 °C. The formula in Eq. (4) was used to calculate the porosity where:

$$\varnothing = \frac{V_V}{V_T} \times 100\% \quad (4)$$

$\varnothing$ Porosity
V_V Volume of voids
V_T Total volume.

3 Result and Discussion

A porosity issue occurs frequently when manufacturing metal components. A phenomenon characterized by the presence of small bubbles or pockets in thicker sections of metal moulded parts. To fabricate components with a porous microstructure, it is necessary to analyse the effect of porosity on the morphology of fracture surfaces. Injection process parameters also affect porosity as well as powder loading. Using the same powder loading, the effects of optimization parameters for injection were investigated.

According to the Taguchi method, the mean signal-to-noise ratio (S/N) was analysed conceptually to graph the effect and visually identify the factors. The capabilities of producing superior green components were explored through the application of green strength optimizations. Metal Powder Industries Federation (MPIF) Standard 42 was used to measure the porosity of the samples.

Data was obtained from S/N ratios using a statistical program for determining optimized factors (Table 6). Signal to noise response curves allows the density data to distinguish between four parameters and three levels. A higher S/N ratio indicates a better setting for selected variables. By incorporating more influential factors, the most important ones are considered. S/N ratios have been determined and summarised in Table 7. As can be seen from Table 6, average S/N ratios are

Table 6 Mean and S/N ratio calculated from density replication of L9 structure experiment

Trials	GGI	Temperature	Pressure	Speed	Means	S/N ratio
1	0.4	145	45	30	0.014	37.08
2	0.4	150	50	35	0.014	37.08
3	0.4	155	55	40	0.670	3.48
4	0.8	145	50	40	0.200	13.98
5	0.8	150	55	30	0.200	13.98
6	0.8	155	45	35	0.130	17.72
7	1.2	145	55	35	0.076	22.38
8	1.2	150	50	40	0.130	17.72
9	1.2	155	45	30	0.190	14.42

Table 7 The response table for the analysis of variation in the S/N ratio

Level	Factor			
	GGI	Temperature	Pressure	Speed
1	25.92	24.42	24.16	21.84
2	15.20	22.97	21.89	25.74
3	18.20	11.93	13.27	11.74
Delta	10.71	12.49	10.88	14.01
Rank	4	2	3	1

shown for each factor level, and the difference between the highest and lowest levels of S/N ratios for the same factor is replaced with the lowest level.

This table also contains the results of a Mean Variability Analysis (Table 8) since this table provides the variability of green part density depending on the level of the factor. The results of Tables 7 and 8 show that the level factors parameter is very similar for analyses of Mean and S/N ratios, which indicates the rank for parameters would deliver the optimum and low fluctuation density of the green compact.

Therefore, the ranking of the most influential factors was successfully established by the response table of the S/N ratio. Porosity was determined by using the porosity

Table 8 Response table for means

Level	GGI	Temperature	Pressure	Speed
1	0.23110	0.09727	0.09130	0.13403
2	0.17717	0.11460	0.13420	0.07350
3	0.13133	0.32773	0.31410	0.33207
Delta	0.09977	0.23047	0.22280	0.25857
Rank	4	2	3	1

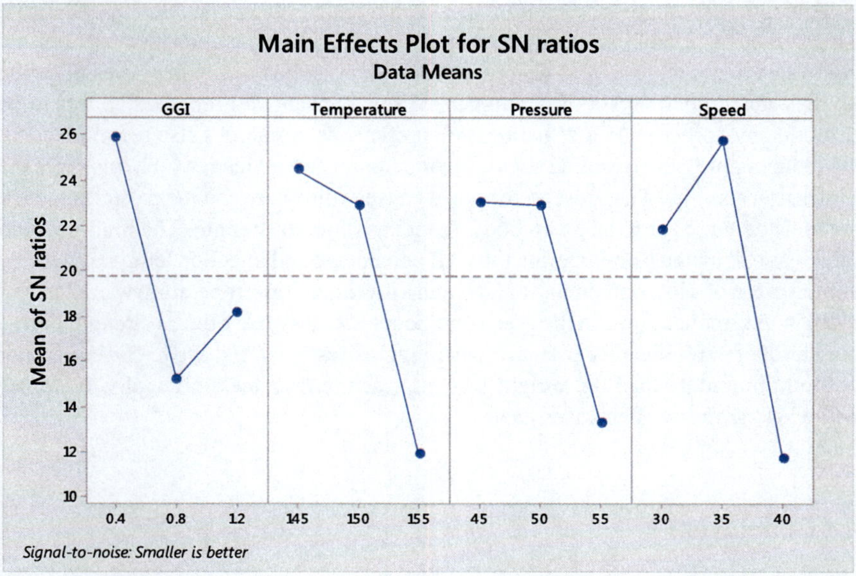

Fig. 2 Porosity variations of green parts at different injection parameters in terms of S/N ratios

equation. As shown in Fig. 2, the parameters with the highest peak values are determined to be the optimal parameters for porosity. In the scenario above, A1 (0.4 wt% of GGI), B2 (145 °C injection temperature), C3 (45% injection pressure), and D2 (35% injection speed) will produce the optimal configuration for obtaining an optimal density for the green part.

As shown in Tables 7 and 8, the two most important variables are injection speed and injection temperature. The injection speed and injection temperature are very important in producing a high-quality part in MIM. Injection speeds of MIM are higher than those for polymers due to the feedstock's higher thermal conductivity [19]. For the filling of a component, the injection speed is typically set at the minimum required to ensure no defects. It is important to choose the injection speed by considering the smallest channel through which the melt must flow. Overly rapid injection speeds may result in frosty or rough surfaces on the part. Surface ripples will appear on a part if the injection speed is too slow. The selection of injection speed should be influenced by all other factors in the process, like melt temperature, shot size, and mould temperature [20].

Choosing a lower injection temperature will result in a defect-free green component because the pores will be blocked and protected. Low moulding temperatures cause internal stresses, which result in a crack. The higher injection temperature will allow more particles to fill the mould cavity because all trapped air pockets will be forced out, resulting in more particles filling the cavity [12]. Metal powders are fused at high temperature, increasing the density and strength of the part [21].

A high injection pressure could cause the binders to separate due to low viscosity. This can be avoided by reducing the injection pressure at the changeover point in a controlled manner [21, 22]. A decrease in porosity resulted as smaller pores filled the voids, causing the pores to grow bigger MIM's mechanical properties appear to be degraded by porosity. As low an injection pressure as possible is also recommended. By reducing pressure, residual stresses in the mould are reduced, which improves the dimension stability. The presence of pores causes voids between the particles, which are considered to form the green body, reducing sintered strength. The highest mean ratio was calculated by comparing the GGI percentage and injection temperature. The combination of a low percentage of GGI and a low injection temperature was expected to fill in the smaller pores in the green component. Transverse rupture strength (TRS) and hardness significantly increase when TaC is less than 0.4 wt%. By increasing TaC addition more than 0.6 weight percent, fracture toughness gradually decreases while WC grain size decreases [23].

4 Conclusion

From Taguchi and analysis of variance results, the following can be concluded:

1. Taguchi produces similar results using analysis of variance. The method has been proven to be effective for optimizing and analysing the process parameters associated with the moulding process. This optimization using the Taguchi method has been demonstrated to be effective.
2. As in perfect scenarios, smaller-is-better was the criterion that was used to predict the green sections with the least amount of porosity. It is determined that the optimal parameters to employ for nominal porosity are GGI (0.4 wt%), injection temperature (145 °C), injection pressure (45%), and injection speed (35%).
3. Continuously achieving the desired product quality can be achieved by controlling the optimal parameter settings. The results of this study will be useful for future research.
4. The interaction between injection speed and injection temperature plays a significant contribution to porosity followed by injection pressure and GGI.

Acknowledgements The author would like to thank the Ministry of Higher Education Malaysia, Universiti Teknologi MARA for awarding research grant 600-IRMI/FRGS 5/3 (166/2019) for the financial support to conduct this research.

References

1. Cao P, Hayat MD (2020) Reactive powder metal injection molding
2. Piotter V, Gietzelt T, Merz L (2003) Micro powder-injection moulding of metals and ceramics. Sadhana Acad Proc Eng Sci 28(1–2):299–306
3. Abolhasani H, Muhamad N (2010) A new starch-based binder for metal injection molding. J Mater Process Technol 210(6–7):961–968
4. Rödiger K, van den Berg H, Dreyer K, Kassel D, Orths S (2000) Near-net-shaping in the hardmetal industry. Int J Refract Met Hard Mater 18(2):111–120
5. German RM (1990) Powder injection molding. Metal Powder Industries Federation, Princeton, NJ
6. Fu G, Loh NH, Tor SB, Murakoshi Y, Maeda R (2004) Replication of metal microstructures by micro powder injection molding. Mater Des 25(8):729–733
7. Sun J et al (2020) A review on binderless tungsten carbide: development and application. Nano-Micro Lett 12(1)
8. Prathabrao M, Amin SYM, Ibrahim MHI (2017) Review on sintering process of WC-Co cemented carbide in metal injection molding technology. IOP Conf Ser Mater Sci Eng 165(1)
9. Jiang A, Wen B, Li Q (2015) Fabrication of WC-TiC-6%Co hard metals by micro-powder injection moulding. IWMECS, pp 543–547
10. Luyckx S (2008) The hardness of Tungsten carbide-cobalt hardmetal. Handb Ceram Hard Mater 946–964
11. Pahroraji HF, Ibrahim MHI, Muniandy P, Amin SYM, Alias SK (2020) Parameter optimization of WC-TaC-6Co green part in injection moulding using taguchi method. IOP Conf Ser Mater Sci Eng 834(1)
12. Tatt TK, Muhamad N, Haron CHC, Sulong AB (2020) Parametric optimization of metal injection moulding process using taguchi method. Int J Integr Eng 12(4):210–219
13. Momeni V, Alaei MH (2019) Optimization of injection parameters in metal injection molding of 4605 low alloy steel. Modares Mech Eng 19(5):1199–1208
14. Foster WT (2000) Basic Taguchi design of experiments. Natl Assoc Ind Technol Conf
15. Shen CYCYK, Liu JJ, Chang CT (2002) Comparison of the results for semisolid and plastic injection molding process. Int Commun Heat Mass Transf 29(1):97–105
16. Mehat NM, Kamaruddin S (2011) Investigating the effects of injection molding parameters on the mechanical properties of recycled plastic parts using the Taguchi method. Mater Manuf Process 26(2):202–209
17. Zhao PSCJ, Mayes RH, Chen G, Xie H (2003) Effects of process parameters on the micro molding process. Polym Eng Sci 43(9):1542–1554
18. Amin AM et al (2017) Green density optimization of stainless steel powder via metal injection molding by Taguchi method. MATEC Web Conf 135
19. Products AP, Industries T (2012) Molding of components in metal injection molding (MIM). Handbook of metal injection molding 1:109–133
20. Ebnesajjad S, Khaladkar P (2018) Manufacturing parts from melt-processible fluoropolymers
21. Engström S (2017) Metal injection molding: a review of the MIM process and its optimization
22. Wahi A, Muhamad N, Jamaludin KR, Rajabi J, Madraky A (2012) Optimization of injection molding parameters by data mining method in PIM process. J Teknol Sci Eng 59:193–196
23. Su W, Sun Y, Yang H, Zhang X, Ruan J (2015) Effects of TaC on microstructure and mechanical properties of coarse grained WC–9Co cemented carbides. Trans Nonferrous Met Soc China 25(4):1194–1199

The Change of Solidification Parameters on Hypoeutectic Aluminum–Silicon Alloy Under Different Cooling Rates

X. H. Ma and N. A. Abd Razak

Abstract Computer-aided cooling curve analysis (CA-CCA) is a cost-saving and easy-to-use thermal analysis tool to identify the solidification characteristics of a material. This paper aims to present the relationship between cooling rate and solidification parameters of aluminum–silicon alloy before semisolid metal processing. An induction furnace was used to melt the as-prepared specimen by heating the respective graphite crucible to 680 °C and left it to be cooled and solidified under three different cooling rate conditions. A double thermocouple method was employed by placing one K-type thermocouple at the crucible center and another one near to the crucible wall. Both thermocouple sensors were connected to a data acquisition device, NI 9219 which was linked to a computer with DASYLab pre-installed inside for data logging, along with OriginPro 2019b for graphical representation. Low, intermediate, and high cooling rate conditions were achieved when the crucible was cooled with its top and bottom surface enclosed with Fiberfrax felt (1.0 °C/s), at ambient temperature (1.3 °C/s), and by a compressed air flow supplied from air compressor (1.9 °C/s), respectively. The increase in cooling rate facilitates the nucleation rate and the growth of crystals within the alloy system and thus leads to a shorter solidification time with all the critical points highlighting the phase transition are shifted accordingly.

Keywords CA-CCA · Cooling rate · Aluminum–silicon alloy · Solidification parameters

1 Introduction

The versatile and useful applications of aluminum have proved to be one of the materials of choice in the diverse manufacturing industries such as packaging, medical, and electrical applications, as well as automotive manufacturing because

X. H. Ma · N. A. Abd Razak (✉)
Department of Mechanical Engineering, College of Engineering, Universiti Malaysia Pahang, Lebuhraya Tun Razak, 26300 Gambang, Pahang, Malaysia
e-mail: azhani@ump.edu.my

M. Y. Ismail et al. (eds.), *Technological Advancement in Mechanical and Automotive Engineering*, Lecture Notes in Mechanical Engineering,
https://doi.org/10.1007/978-981-19-1457-7_21

of its lightweight, formability, corrosion resistance, and high specific strength. It has undeniably contributed to significant economic growth, and new ways of using aluminum in manufacturing processes and manufactured products are currently being investigated worldwide to bring aluminum to its full potential [1–3]. Since the first discovery by David Spencer that the spherical shape of the solid particles imparts special rheological properties to the mixture after a continuous stirring alloy during solidification [4], processing metal alloys in the semisolid state, also well-known as Semisolid Metal (SSM) processing has been the recent popular melt treatment and alloy manipulation technique. It is under intensive research and development due to its significant advantages, for example, low porosity, heat treatability, less energy consumption and prolonged die life [5–8].

The SSM processing is a novel metal forming technology for the production of near net-shaped metal components [3, 5]. This technology deals with semisolid slurries, in which non-dendritic solid particles are dispersed in a liquid matrix. Semisolid die castings typically contain significantly lower levels of entrapped gases and lubricants as compared to conventional High Pressure Die Casting (HPDC) components, due to the laminar flow with high stability (at a solid fraction, $0.5 < f_s < 0.7$) in semisolid metal processing, hence allowing them to be heat-treated to optimize the mechanical properties [6, 8]. SSM processes patented in recent decades are generally classified into two main processing routes, namely, the rheo-route and the thixo-route. In theory, the rheo-route techniques involve a direct cooling from the solidifying melt whereas the thixo-route techniques involve reheating from the solidified billet [3, 5]. More evidence and variation of processing techniques were successively reviewed and detailed in the literature [9–11] to prove the feasibility of the semisolid metal forming through the examination of the microstructure evolution as well as the evaluation of the mechanical properties for alloy under study.

Aluminum alloys with silicon as the major alloying constituent are the most important commercial casting alloys, primarily because of their superior casting characteristics. The typical commercial products of this type of alloy are engine blocks, transmission case, rear axle housing, trailer suspension saddle [2]. The addition of the other elements in an appropriate quantity could enhance the material‘s properties of predominant metal, for instance, copper promotes strengthening and machinability while magnesium contributes to high corrosion resistance and a well-retained low thermal expansion level [2, 12, 13]. Hence, besides strength reinforcement, alloying also imparts significant effects on other characteristics of aluminum alloys.

Thermal analysis can be described as an approach to record and interpret the temperature variation as a function of time during the cooling or heating of the material. Amongst the techniques applied in thermal analysis, computer-aided cooling curve analysis (CA-CCA) is the simplest and convenient to be carried out in the laboratory. It is based on the attribute that the thermal events on a cooling curve are directly related to the phase transformation occurring in a solidifying sample. Depending on the nature of the analysis, different kinds of information such as the latent heat of solidification (liquidus, eutectic and solidus temperatures in particular), solid fraction, dendritic coherency point (DCP), and phase transformation can be determined [14–16]. Solid fraction and dendritic coherency of solidifying alloy is noted worthwhile to have a great impact on the slurry flowability. The laminar flow of molten alloy filling into the die cavity could minimize gas porosity [17–20]. On the other hand, it was also addressed by the authors of the previous studies [9, 21] that when shear stress is applied on a dendritic microstructure, the phenomenon of entrapped liquid between the dendritic arms occurs, hindering the sliding of the globules, hence increasing the viscosity during die filling. Therefore, a proper ratio of the solid particles in a liquid matrix should be prudently regulated in corresponding to the dendritic coherency of the alloy during the solidification process.

Close control of the cast structure in the SSM process is one of the critical requirements for product performance. Therefore, basic information especially the solidification parameters is a must for a better understanding of solidification processes and as a foundation for future development of new materials and improved casting methods. The limited findings in the past open literature about the detailed study on the influence of cooling rate in the context of aluminum–silicon alloy have been giving impetus towards this study. Hence, a thermal analysis experiment is conducted at the present study through the application of the computer-aided cooling curve analysis technique to investigate the effect of cooling rate on solidification parameters of aluminum–silicon alloy prior to semisolid metal processing.

2 Experimental Procedure

The chemical composition of the as-received sample was identified by using optical emission spectroscopy. The spectrometry test was repeated three times and its average value was taken as shown in Table 1.

A mass of 350 g of the specimen was taken from the as-received sample and marked as specimens A, B and C respectively. These specimens were separately put into a graphite crucible with 43 mm in diameter and 110 mm in depth. The crucible was then heated up to a range of 830 °C by an induction heating machine. As the specimens start to melt, the heating process is to be held for five minutes to homogenize the melt. As soon as the crucible was taken out of the copper coil at the induction heating machine by using a pair of tongs, the crucible containing specimen A was then placed on the multiple-layered Fiberfrax felt sheets and the top enclosed with a Fiberfrax lid, but the crucible body is left exposed to ambient air. This was termed as a slow cooling condition. Meanwhile, the crucible with specimen B was left aside exposed to ambient air to be cooled naturally, whereas the crucible with specimen C was cooled by a compressed air flow supplied from the air compressor.

Table 1 Chemical composition of the as-received sample

Elements	Al	Si	Fe	Cu	Mn	Mg	Cr	Ni	Zn	Ti	Sn
Composition (wt%)	81.5	9.76	0.69	1.16	0.31	0.23	0.11	0.54	1.48	0.09	>0.35

Simultaneously, two K-type thermocouples were inserted inside the crucible, one located at the center of the crucible and another one near to the crucible wall. Both of the thermocouples' tips were immersed 90 mm apart from the top surface of the crucible and the gap in between the two thermocouples was about 20 mm. As the cooling process started, the data was recorded through the data acquisition device, NI 9219 and transferred to the user computer. The same experiment was repeated three times to obtain the average value of the thermal parameters for each cooling condition. The experimental setup for thermal analysis on each of the specimens is shown in Fig. 1a–c, respectively.

Data smoothing and plotting of the cooling curve, together with the first derivative curve (also called cooling rate curve) was conducted with the software application of OriginPro 2019b for graphical representation. A zero curve or baseline was constructed on the first derivative curve for the computation of solid fractions. Fraction of solid phase within a solidifying alloy is crucial for understanding problems concerning castability, involving phenomena such as fluidity of solidifying melts, formation of porosities as well as causes of hot tearing [22]. The solid fraction was calculated from the integrated area between the first derivative curve and the baseline curve measured from the liquidus temperature to the solidus temperature.

3 Results and Discussion

The cooling curve for slow cooling employed on specimen A is shown in Fig. 2a. As depicted in the figure, the wall temperature is slightly lower than the temperature at the center of the crucible due to the heat convection from the wall surface to the ambient. The cooling rate of 1.0 °C/s was calculated at the temperature difference between the initial point (680 °C) to the point which is 50 °C prior to the liquidus temperature of specimen A, within the time taken in second, as shown in Eq. (1).

$$Cooling\,rate = \frac{T_o - T_{liq+50\,^\circ\mathrm{C}}}{t_o - t_{liq+50\,^\circ\mathrm{C}}} \tag{1}$$

The cooling curve was then further analyzed by developing its first derivative to more precisely estimate and obtain the essential thermal parameters such as liquidus, eutectic, and solidus temperature. The derivative curve (also called the cooling rate curve) eased the identification of phase transformation that happened within the alloy during the solidification process. The identification method on nucleation, growth of crystal and phase changes of alloy over the solidification period on the thermal profile was well-explained in previous studies [14, 22]. In the case of the present work, the liquidus, eutectic, and solidus temperatures for slow cooling occurred at 578 °C, 496 °C, and 468 °C respectively as shown in Fig. 2b.

A zero curve or baseline was then created under the first derivative of the cooling curve. The baseline curve for slow cooling which is shown in Fig. 2c was used to attain the fraction of the solid phase corresponding to its solidifying temperature. The

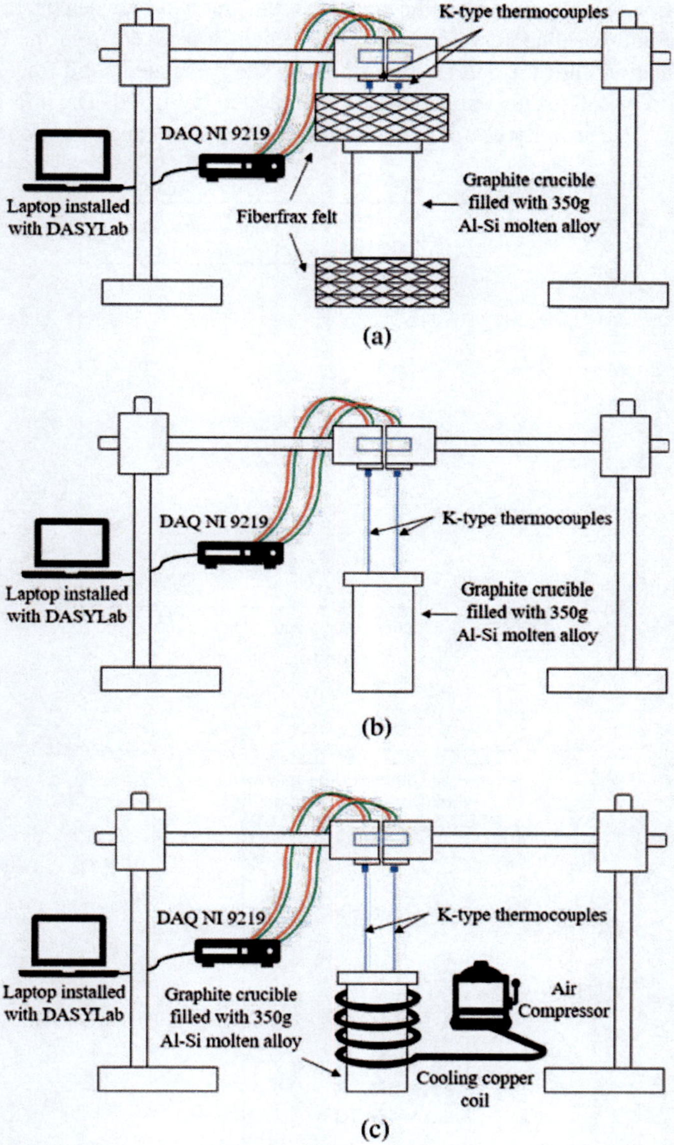

Fig. 1 Thermal analysis experimental setup for **a** slow cooling condition, **b** intermediate cooling condition, and **c** fast cooling condition

solid fraction was obtained from the area under the graph in-between the first derivative curve and baseline curve and its measurement was taken from the beginning of solidification until the end of solidification. The calculated solid fraction corresponding to its solidifying temperature is then plotted in Fig. 2d. The information in this figure is particularly useful to determine the optimal temperature for semisolid

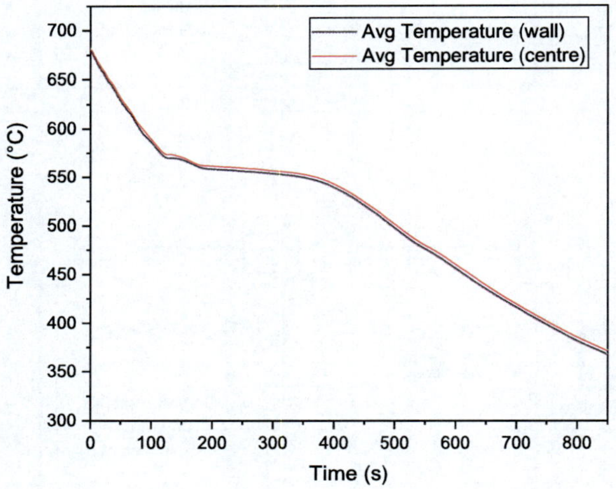

(a)

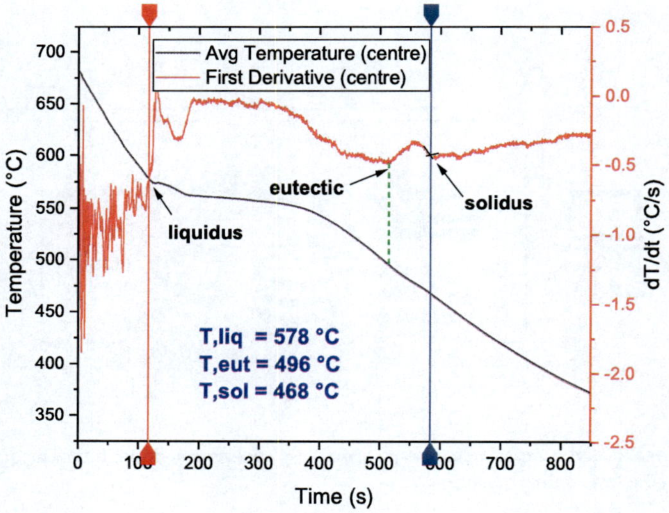

(b)

Fig. 2 For a low cooling rate of 1.0 °C/s **a** cooling curve, **b** first derivative curve and cooling curve as a function of time, **c** baseline of the first derivative, **d** calculated temperature-fraction solid relation, and **e** $T_w - T_c$ curve and DCP

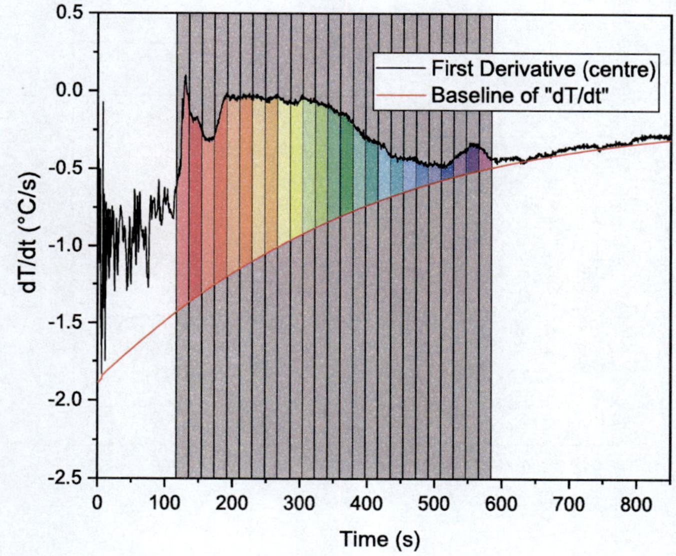

(c)

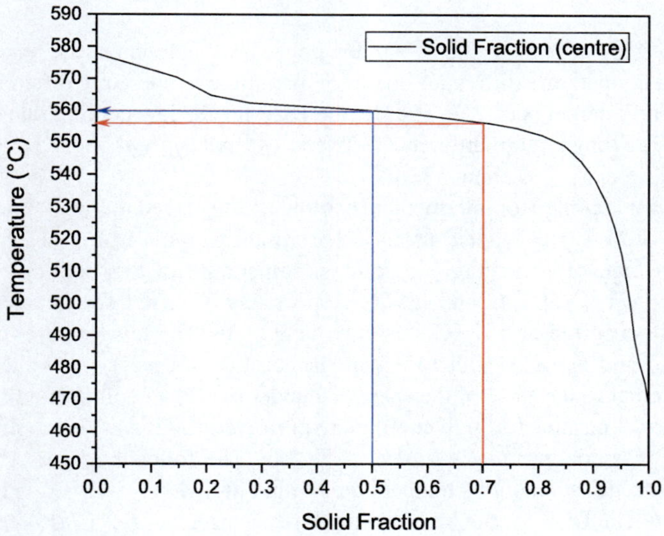

(d)

Fig. 2 (continued)

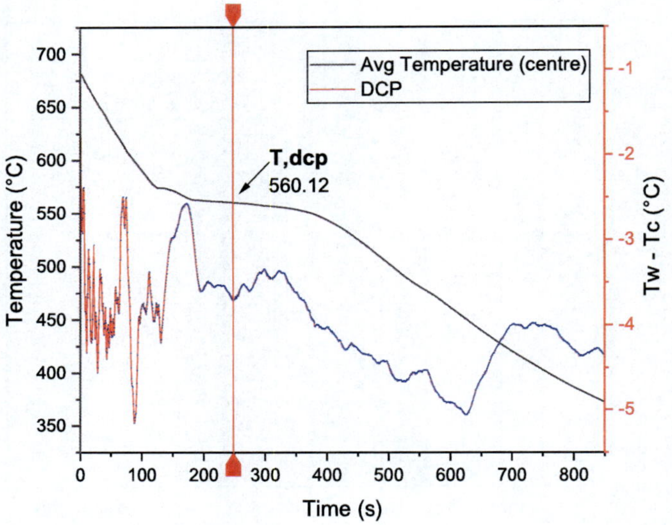

(e)

Fig. 2 (continued)

metal processing. The dendritic coherency point, DCP is found at the first minimal point of the temperature difference curve (or maximum temperature difference) [23] over the solidification period. In Fig. 2e, the DCP for the low cooling rate occurred at 560 °C. The temperature difference between T_{liq} and T_{DCP} is about 18 °C and its duration of time, t_{DCP} is about 132 s.

The thermal profiles for intermediate cooling are displayed in Fig. 3. The cooling rate of 1.3 °C/s was calculated using the aforesaid formula in Eq. (1). Figure 3b outlines the liquidus, eutectic, and solidus temperature of this cooling condition occurring at 571 °C, 507 °C, and 455 °C, respectively. The DCP for the intermediate cooling rate occurred at 557 °C as shown in Fig. 3e. The temperature difference between T_{liq} and T_{DCP} is about 14 °C, and its duration of time, t_{DCP} is about 132 s, which is similar to the t_{DCP} of the specimen under the slow cooling condition.

The thermal profiles for high cooling are portrayed in Fig. 4. The cooling curve with a 1.9 °C/s cooling rate is shown in Fig. 4a. The following figure shows the liquidus, eutectic, and solidus temperatures found at 574 °C, 502 °C, and 456 °C, respectively. The DCP for the high cooling rate occurred at 568 °C as illustrated in Fig. 4e. The temperature difference between T_{liq} and T_{DCP} is about 6 °C, and its duration of time, t_{DCP} is about 21 s.

The data of solidification parameters for all three cooling rates are summarized and depicted in Fig. 5. The determination of solidification parameters especially the liquidus and solidus temperature is a vital key factor as it will become a reference temperature for SSM feedstock billet processing [24]. The cooling rate is known to affect the morphology and distribution of phases in the microstructure of an alloy since the release of latent heat is corresponsive to the metallurgical phases within the alloy [25]. According to Backerud et al. [22] research study, the original particle size depends on the cooling rate. At a high cooling rate, more α-phase forms directly from the liquid. The increase in cooling rate contributed to a shorter solidification time of an alloy. Based on the experimental results, with increasing cooling rate,

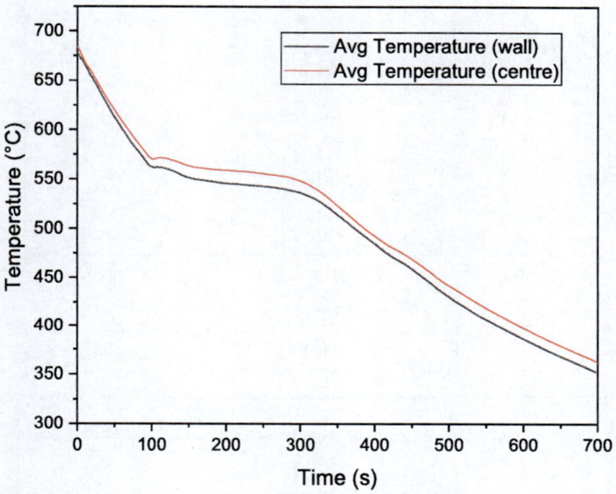

(a)

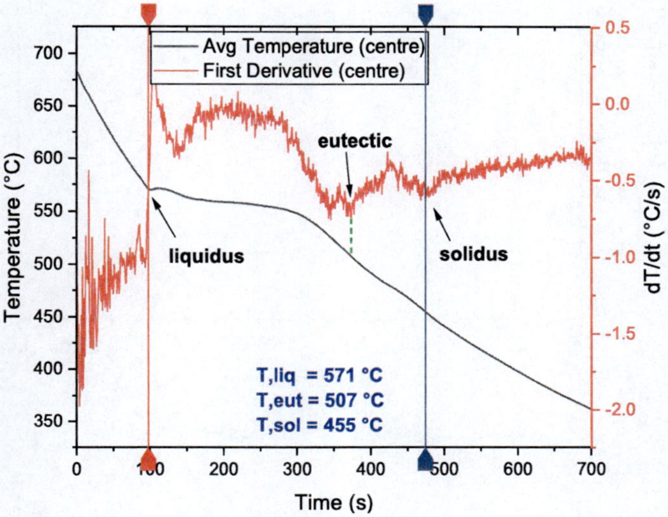

(b)

Fig. 3 For an intermediate cooling rate of 1.3 °C/s **a** cooling curve, **b** first derivative curve and cooling curve as a function of time, **c** baseline of the first derivative, **d** calculated temperature-fraction solid relation, and **e** $T_w - T_c$ curve and DCP

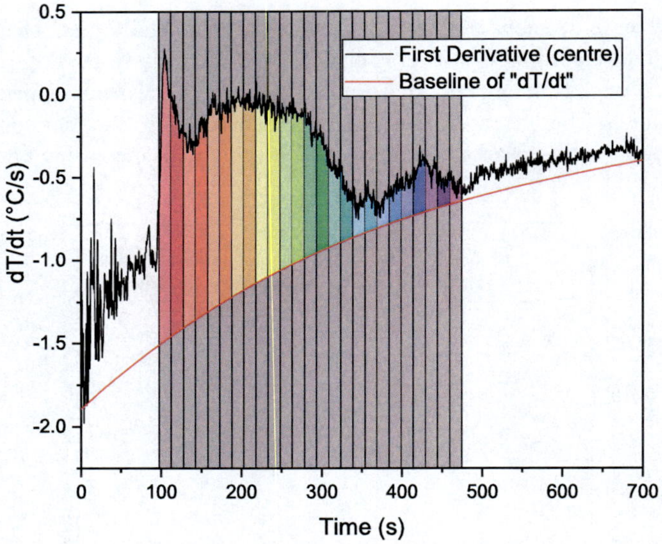

(c)

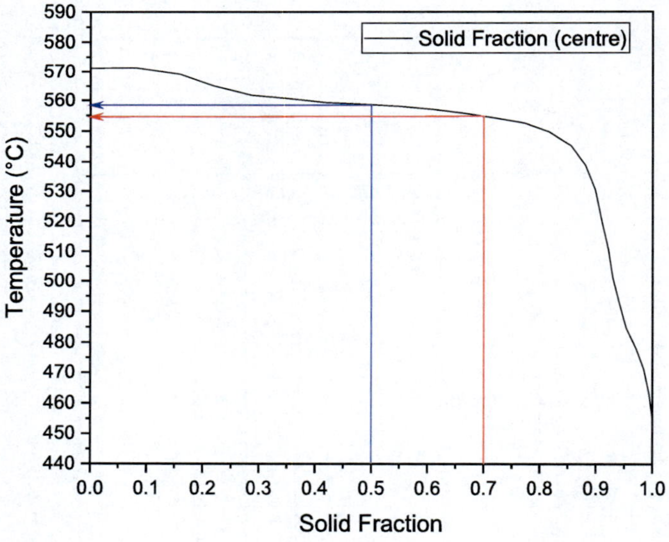

(d)

Fig. 3 (continued)

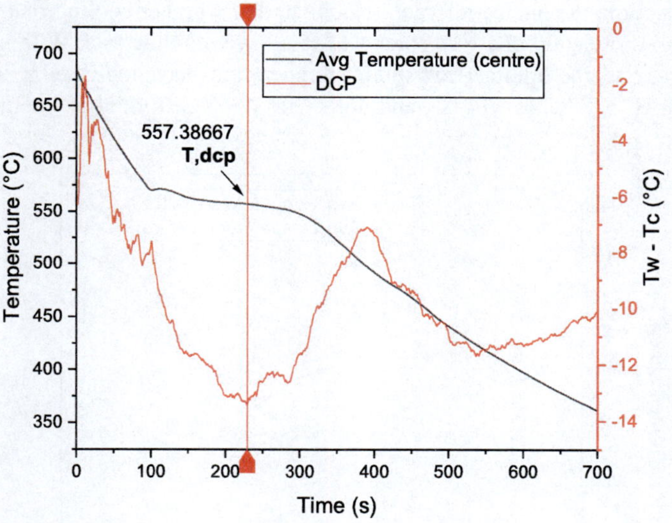

(e)

Fig. 3 (continued)

the solidification of the melt initiates and ceases at a lower temperature, whereas the solidification temperature range slightly increases. Even though the temperature value does not indicate a proportional trend, the liquidus and solidus points of the sample were found at a lower temperature when the cooling rate increase, except for the eutectic point which shows an opposite trend. It cannot be denied that another study [23] addressing several ambiguities still exist upon the effect of cooling rate on the solidification parameters of alloy. A fluctuation also occurs in the dendritic coherency point of the sample with the increase of the cooling rate. T_{DCP} decreases by 3 °C from 1.0 °C/s to 1.3 °C/s and then increase to 11 °C at 1.9 °C/s. In this experiment, the temperature value at 0.5–0.7 solid fraction for all three cooling rate conditions remain in the range of 555–560 °C, as illustrated in Fig. 6. In other word, it could be said that the fraction solid of alloy is not influenced by the cooling rate, which is in agreement with previous finding for Al–Si–Cu alloys [26].

4 Conclusion

According to the experimental results of this study, different cooling rates make noticeable changes in the thermal parameters. At a lower cooling rate condition, the amount of heat released from the specimen is lower and thus resulting in a smaller slope of the cooling curve, and vice versa. The increase of cooling rate results in a shorter solidification time, and consequently, all the critical points highlighting the phase formation are shifted accordingly. At the same time elapsed from the beginning of solidification, the nucleation rate is facilitated at a higher cooling rate, and thus obtaining a more solid phase as compared to a lower cooling rate. With increasing the cooling rate, the liquidus and solidus temperatures have indicated a decreasing trend, that is, 578 °C to 574 °C, and 468 °C to 456 °C from slow cooling to fast

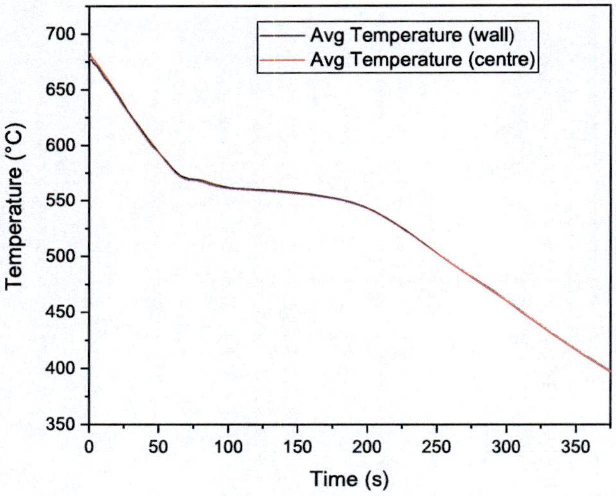

(a)

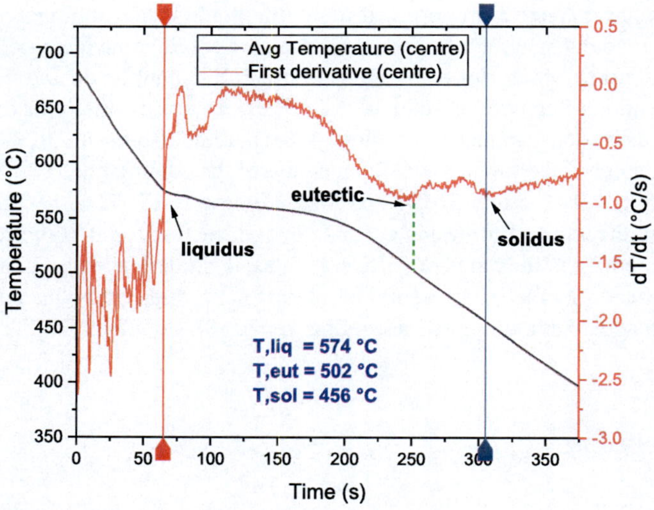

(b)

Fig. 4 For a high cooling rate of 1.9 °C/s **a** cooling curve, **b** first derivative curve and cooling curve as a function of time, **c** baseline of the first derivative, **d** calculated temperature-fraction solid relation, and **e** $T_w - T_c$ curve and DCP

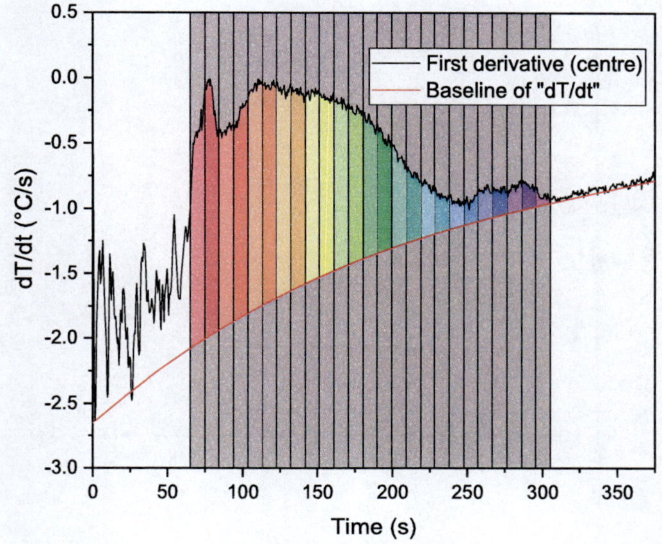

(c)

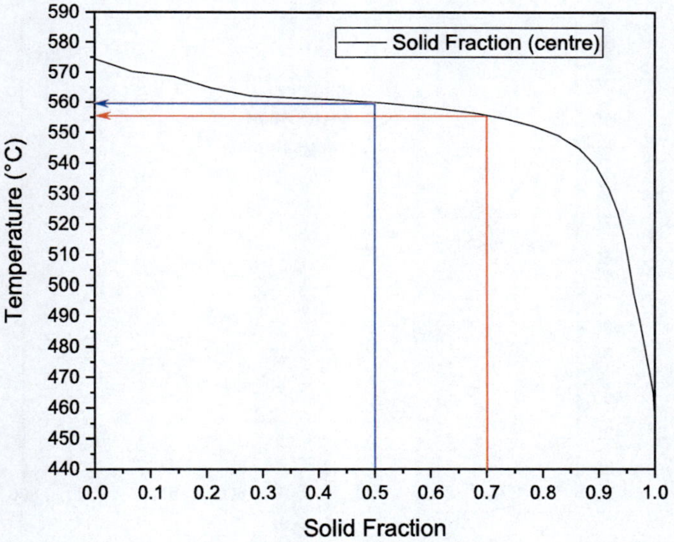

(d)

Fig. 4 (continued)

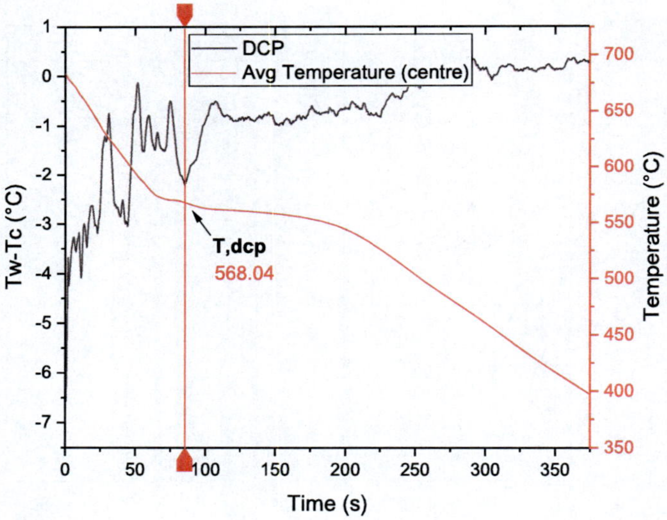

(e)

Fig. 4 (continued)

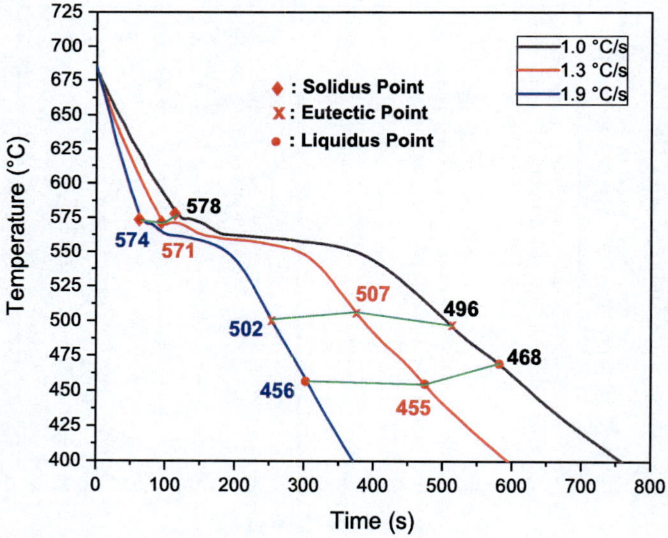

Fig. 5 Solidification behavior at different cooling rates

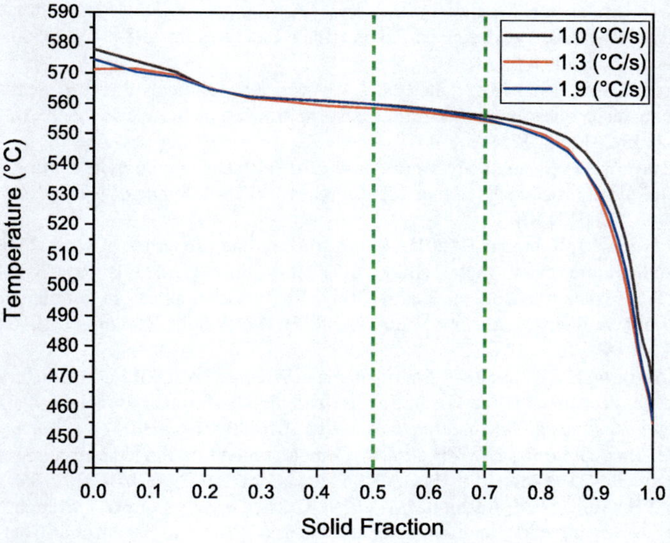

Fig. 6 Solid fraction corresponding to the temperature of the solidifying alloy for three cooling rates

cooling, respectively. This experimental work also concludes that the cooling rate does not affect the fraction solid of the alloy.

Acknowledgements The authors would like to be obliged to the Ministry of Higher Education for providing financial support under Fundamental Research Grant Scheme (FRGS) No. FRGS/1/2019/TK03/UMP/03/4 (University reference RDU1901166) and Universiti Malaysia Pahang for laboratory facilities.

References

1. ASM International Handbook Committee, ASM handbook volume 2: properties and selection—nonferrous alloys and special-purpose materials
2. Kaufman JG, Rooy EL (2004) Aluminum alloy castings: properties, processes, and applications. https://doi.org/10.1017/CBO9781107415324.004
3. Nafisi S, Ghomashchi R (2016) Semi-solid processing of aluminum alloys. Springer International Publishing, Cham. https://doi.org/10.1007/978-3-319-40335-9
4. Spencer DB, Mehrabian R, Flemings MC (1972) Rheological behavior of Sn-15 pct Pb in the crystallization range. Metall Trans 3(7):1925–1932. https://doi.org/10.1007/BF02642580
5. Alexandrou AN, Burgos GR, Entov VM (2000) Semisolid metal processing: a new paradigm in automotive part design. SAE Tech Pap 724. https://doi.org/10.4271/2000-01-0676
6. Hu XG, Zhu Q, Midson SP et al (2017) Blistering in semi-solid die casting of aluminium alloys and its avoidance. Acta Mater 124:446–455. https://doi.org/10.1016/j.actamat.2016.11.032
7. Salleh MS, Omar MZ, Syarif J, Mohammed MN (2013) An overview of semisolid processing of aluminium alloys. ISRN Mater Sci 2013:1–9. https://doi.org/10.1155/2013/679820

8. Omar MZ, Atkinson HV, Palmiere EJ, Howe AA, Kapranos P (2004) Microstructural development of HP9/4/30 steel during partial remelting. Steel Res Int 75(8–9):552–560. https://doi.org/10.1002/srin.200405810
9. Mohammed MN, Omar MZ, Salleh MS, Alhawari KS, Kapranos P (2013) Semisolid metal processing techniques for nondendritic feedstock production. Sci World J 2013:16. https://doi.org/10.1155/2013/752175
10. Ahmad AH, Naher S, Aqida SN, Brabazon D (2014) Routes to spheroidal starting material for semisolid metal processing. Compr Mater Process 5:135–148. https://doi.org/10.1016/B978-0-08-096532-1.00515-X
11. Pola A, Tocci M, Kapranos P (2018) Microstructure and properties of semi-solid aluminum alloys: a literature review. Metals (Basel) 8(3). https://doi.org/10.3390/met8030181
12. Miller WS, Zhuang L, Bottema J et al (2000) Recent development in aluminium alloys for the automotive industry. Mater Sci Eng A 280(1):37–49. https://doi.org/10.1016/S0921-5093(99)00653-X
13. Liu D, Atkinson HV, Kapranos P, Jirattiticharoean W, Jones H (2003) Microstructural evolution and tensile mechanical properties of thixoformed high performance aluminium alloys. Mater Sci Eng A 361(1–2):213–224. https://doi.org/10.1016/S0921-5093(03)00528-8
14. Ihsan-Ul-Haq, Shin JS, Lee ZH (2004) Computer-aided cooling curve analysis of A356 aluminum alloy. Met Mater Int 10(1):89–96. https://doi.org/10.1007/BF03027368
15. Husain NH, Ahmad AH, Rashidi MM (2020) Thermal analysis of 6061 wrought aluminium alloy using cooling curve analysis-computer aided (CCA-CA) method. IOP Conf Ser Mater Sci Eng 788(1). https://doi.org/10.1088/1757-899X/788/1/012018
16. Sudheer R, Prabhu KN (2016) A computer aided cooling curve analysis method to study phase change materials for thermal energy storage applications. Mater Des 95:198–203. https://doi.org/10.1016/j.matdes.2016.01.053
17. Koke J, Modigell M (2003) Flow behaviour of semi-solid metal alloys. J Nonnewton Fluid Mech 112(2–3):141–160. https://doi.org/10.1016/S0377-0257(03)00080-6
18. Fan Z (2002) Semisolid metal processing. Int Mater Rev 47(2):49–86. https://doi.org/10.1179/095066001225001076
19. Modigell M, Koke J (2001) Rheological modelling on semi-solid metal alloys and simulation of thixocasting processes. J Mater Process Technol 111(1–3):53–58. https://doi.org/10.1016/S0924-0136(01)00496-4
20. Flemings MC (1991) Behavior of metal alloys in the semisolid state. Metall Trans A 22(5):957–981. https://doi.org/10.1007/BF02661090
21. Hirt G, Kopp R (2009) Thixoforming: semi-solid metal processing, p 454
22. Backerud L, Chai G, Tamminen J (1990) Solidification characteristics of aluminum alloys. Found Alloy 2:2
23. Farahany S, Ourdjini A (2013) Effect of cooling rate and silicon refiner/modifier on solidification pathways of Al–11.3Si–2Cu–0.4Fe alloy. Mater Manuf Process 28(6):657–663. https://doi.org/10.1080/10426914.2013.763972
24. Rosso M (2012) Thixocasting and rheocasting technologies, improvements going on. J Achiev Mater Manuf Eng 54(1):110–119
25. Marchwica P, Microstructual and thermal analysis of AlSi and MgAl alloys subjected to high cooling rates. Electronic theses and dissertations
26. Veldman NLM, Dahle AK, Stjohn DH, Arnberg L (2001) Dendrite coherency of Al–Si–Cu alloys. Metall Mater Trans A Phys Metall Mater Sci 32(1). https://doi.org/10.1007/s11661-001-0110-1

A Brief Overview on the Utilization of High Strength Steel (HSS) for Automotive Structural Welding Applications

M. N. M. Salleh, M. Ishak, and M. M. Quazi

Abstract High-strength steels are now increasingly used in automotive structural applications owing to their resilience, crashworthiness, and ease of manufacturing. This paper reviews the application of high-strength steels for the automotive structure, describing the condition of the steel after been put in the car structures. Thereafter, the importance of advanced high strength boron steel is highlighted, and its weldability is discussed. Current issues related to welding and changes in microstructure are discussed. It is imperative that Boron steel is gaining widespread attention due to its good mechanical characteristic in car structures and its weldability remains a topic of significant research interest. Lastly, the important considerations are summarized.

Keywords High strength steel · Automotive structure · Boron steel

1 High Strength Steels

Automotive steels have become an important material for body construction of motor vehicles in North America since the 1900s [1]. HSS was developed in 1990s as it possessed higher strength values up to 500–600 MPa for the raw materials, whichever could achieve until 1500 MPa after heat treatment processes [2–5]. HSS have their potential advantages compared to conventional steels, especially in reducing the structural's weights, saving costs in term of transportations, materials, and handling.

In an article reported by Hyunho Yeom, it was important to apply this HSS due to the increasing demands for vehicles' parts and elements with longer service life rather than using conventional steels [6]. According to the steels usage developments, HSS was introduced to be a selective material with thicknesses lower than 2.0 mm

M. N. M. Salleh (✉) · M. Ishak · M. M. Quazi
Joining Welding and Laser Processing Lab, Faculty of Mechanical and Automotive Engineering Technology, Universiti Malaysia Pahang, 26600 Pekan, Pahang, Malaysia
e-mail: mdnaqiuddinsalleh91@gmail.com

M. Ishak
e-mail: mahadzir@ump.edu.my

M. Y. Ismail et al. (eds.), *Technological Advancement in Mechanical and Automotive Engineering*, Lecture Notes in Mechanical Engineering,
https://doi.org/10.1007/978-981-19-1457-7_22

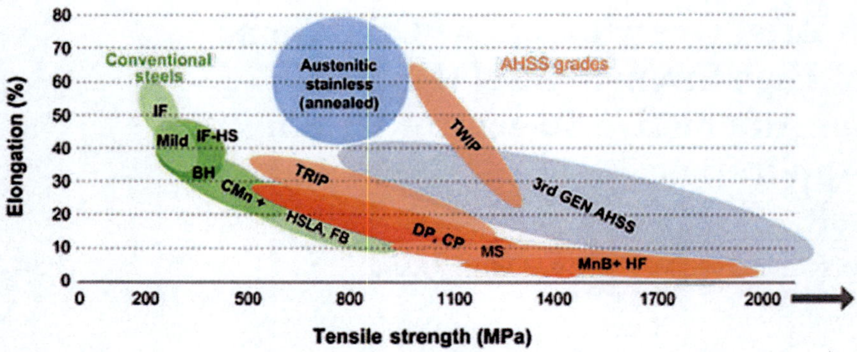

Fig. 1 Illustration of automotive sheet steel grades based on strength and ductility [7]

which could reduce more weight in a car body. For a future steels technology, steel-ductility performance has to be improvised from the first generation of HSS until the third generation [7] as shown in the steel-ductility diagram in Fig. 1.

Over the years, the demand for using HSS in automotive structures are highly increased due to its mechanical properties which are appicable to be applied in automotive structures in terms of passenger safety and stronger structures. In this paper, Boron steel application in automotive welding part will be discussed. This steel has become a selective material that can enhance its mechanical properties after been hot stamped.

However, the manufacturer faces some problems in using this HSS since it has an Al–Si coating which difficuilt to be weld especially by using laser welding. In this paper, the applications of HSS and especially Boron steel for automotive welding application will be discussed. The effect of welding these types of steel and the challenge faced also will be discussed.

2 Application in Car Body Parts

HSS provides the lightweight structures opportunities with improving crash properties in future car bodies, especially the frame such as B-pillars. Figure 2 shows the usages of these materials in body-in-white (BIW) of Chevrolet Colorado which showing the strength results [1]. Parts with purple colors indicate the usage of strength up to 1300 MPa which is 15% for the whole BIW. This steel was the highest strength used especially at B-pillars which need for the dissimilar joints.

From Fig. 2, a variable thickness blank was a single blank with thicknesses varying along the length. This part could be designed and stamped with an appropriate thickness of the material, thus providing additional mass reduction. Later on that, Acura TLX use this product on the door rings as shown in Fig. 3. It uses 52% HSS and AHSS together to help achieve high rigidity to promote a smooth and quiet ride,

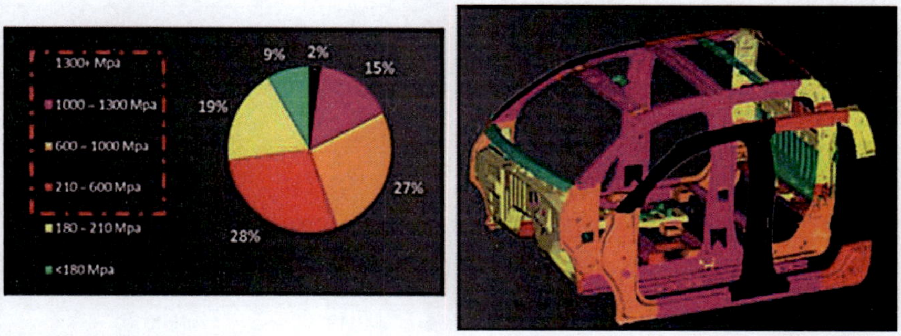

Fig. 2 2015 Chevrolet Colorado BIW cab structures [1]

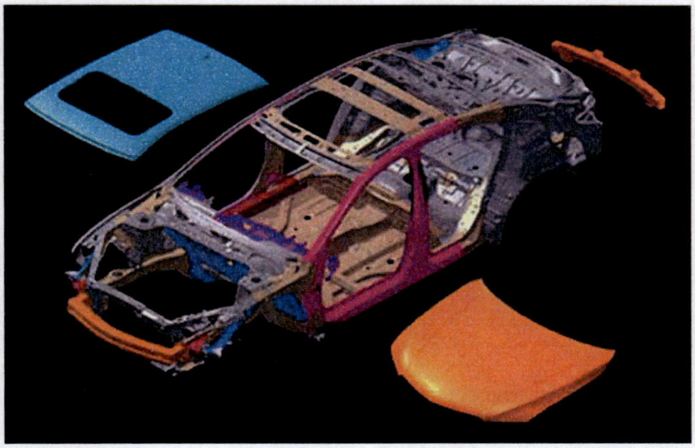

Fig. 3 Example of a one-piece hot stamped door ring (purple) in the 2015 Honda Acura TLX [1]

crisp handling, safety performance, and long-term durability without increasing the cars' weight. This technology was also applied by Honda for their 2016 Honda Pilot to achieve the reduction in mass and increases the rigidity of the car's structure. The B-pillar part was formed by applying new technology known as tailored welded blanks (TWB's).

The application of HSS would contribute to the lighter and thinner material with the unchanged performance such as previous conventional steels in car parts. Implementation of the HSS material in car body part was driven by the forces of weight reduction, crash performances improvement, high strength and stiffness, cost reduction, and sustainability [8, 9]. A cooperative of HSS used by Ford of Europe, Volvo Cars, Jaguar, and Land Rover defined that HSS are steels that yield strength higher than 180 MPa as shown in Table 1.

Common HSS steels used in BIW to replace conventional mild steel are thin sheet dual-phase steel (DP-steel) where its microstructure consists of a mixture of ferrite

Table 1 HSS definition by Ford, Volvo, Jaguar, and Land Rover [10]

Steel class	Abbreviation	Min yield strength (MPa)
Mild steel/forming grades	MS	<180
High strength steel	HSS	=180 < 280
Very high strength steels	VHSS	=280 < 380
Extra high strength steels	EHSS	=380 < 800
Ultra high strength steels	UHSS	=>800

and martensite. Martensitic microtructure is formed within the island within the soft ferritic matrix which gives tensile strength of almost 1000 MPa where it was important for a good elongation and formability [11]. So far, however, recent developments in the field of material usage for BIW have led to a renewed interest in the usage of Boron steel where the characteristic of this material meets the requirements of high strength with light-weight and good formability especially for hot stamping purposes [10, 12–14].

3 Boron Steel in Automotive

Boron steel is one type of advanced high strength steels (AHSS) substituting the usage of HSS such as Twinning-Induced Plasticity (TWIP), Transformation-Induced Plasticity (TRIP) and Dual-Phase (DP) steels [15, 16] in car body structures. According to Shi and Westgate [17] from the welding institute (TWI), materials used in automotive application until April 2008 were 0.8–1.5 mm thick of DP600, DP800, DP1000, TRIP700, uncoated boron steel, Al–Si coated Usibor steel, and Zn coated low carbon steel which is having 600–1550 N/mm^2 tensile strength.

In the automotive industry, this Boron steel sheet was applied to the tailor welded blanks (TWB) technology with having thicknesses ranged about 1.0–1.6 mm, approximately [12, 14, 15, 18]. Boron steel is formerly formed from the low-carbon and low-alloy steels where the hardenability of the carbon steel was increased when the Boron microalloying was done by the content ranged about 0.000005–0.005% in the alloying elements [19].

Low carbon steel with lower than 0.40% of Carbon, the C content is the most effective in producing high strength Boron steel where the Boron element is fully killed and de-oxidised the steel in order to increase the hardenability [20]. The smaller atomic radius possessed by the Boron element and very solubility in iron cause it to concentrate mainly on the boundaries of austenite grains. The atoms in boron are arranged in regions with defects where it decreased the energy of boundaries and

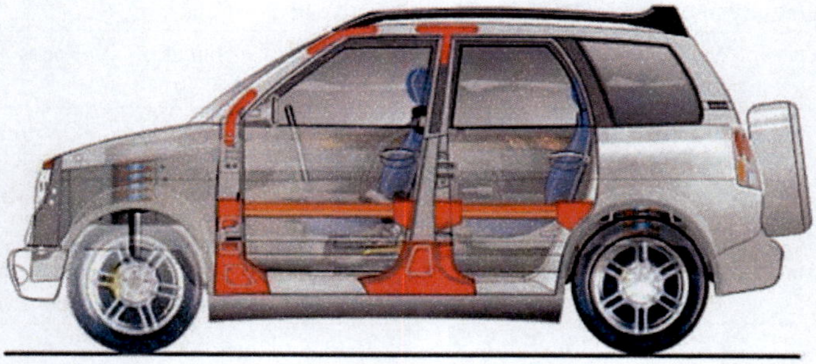

Fig. 4 Boron steel in strengthening bars, gussets, and plates (orange part) used in 2003 Volvo XC90 SUV [23]

the probability of producing crystallization centre which lead to the increment of its hardenability of steels [21]. In terms of its microstructures, it is assumed that boron also promotes primarily the shear transformation of austenite and decelerates the formation of the main phase such as ferrite and granular bainite [21].

Boron alloy steels are used in many applications, as a wear material and as high-strength structural steel. Focusing on automotive application, this steel is used as the structural part, which is the chassis of a car. The types of Boron steel used on vehicles nowadays has extremely high strength, as used by Volvo cars where the strength of the Boron parts has a yield point of about 1350–1400 MPa [22]. The important parts in a BIW are the A- and B-pillars where the position of the car users is around this region. According to Len Watson, 22MnB5 steel in windscreen pillars and gussets in the centre pillars improve the vehicle safety legislation where it improves the crashworthiness with better weight reduction [23]. Figure 4 shows the schematic illustration of Boron steel application (orange part) in the 2003 Volvo XC90 SUV. The aim of the application of Boron is to strengthen the weaker areas of car parts.

The most popular Boron steel type for automotive applications is from Mn-B type. The range of yield strength (YS) and ultimate tensile strength (UTS) of Mn-B steel are 1200 and 1600 MPa, respectively [24]. The usage of this type of Boron steel was originally for the suspension and chassis or called body-in-white (BIW) of a car. TWB technology was a need in producing the parts such as A- and B-pillars where the welding joining is needed before the product is subjected to the hot stamping process for the final product. Table 2 shows the summary of the research work carried out using Boron steel welding for automotive applications in recent years.

From Table 2, it was observed that the most recent high strength steel material used for automotive application was 22MnB5 Boron steel. The application of Boron steel is mostly for the structure of a car which consists of A- and B-pillar, floor sills, cant rails, side-impact door beams bumper beams [10]. Most of the researchers used the laser welding method to weld the steel due to the deep penetration, high speed, small heat-affected zone (HAZ), fine welding seam quality, and superior metallurgical bonding which is a need for automotive industries towards producing lightweight and cost reduction of productions [12–14, 25]. So far, however, Boron steel suffers

Table 2 Summary of boron steel welding for automotive applications

Material	Welding method	Joint type	Thickness (mm)	Application	Focus
22MnB5 (uncoated) [25]	200 W Fibre laser (PW mode)	Similar butt joint	1.6	B-pillar	Optimization parameters for joint strength
N22CB uncoated Boron steel [26]	300 W Nd:YAG laser (PW mode)	Similar butt joint	1.8	A-Pillar, B-pillar	Parameters effect on weld geometry
Al-10%Si coated boron steel [12]	4 kW Nd:YAG laser (CW mode)	Similar butt joint	1.2	Car body	Phase transformation of coated boron steel before and after hot stamping
AlSi10Fe3, ZnNi10 coated 22MnB5 + A6061-T4 [13]	4 kW Nd:YAG laser (CW mode)	Dissimilar lap joint	1.5	Underbody plate	Removing coating and laser welding/brazing ability studies
AlSi10Fe3 22MnB5 coated + AZ31 [27]	4 kW Nd:YAG laser (CW mode)	Dissimilar lap joint	1.2	Gearbox housing	Behaviour of Boron-Mg welding/brazing study
Press hardened 22MnB5 (Al–Si coated) [14]	4 kW disk laser	Similar overlapped joint	0.5, 1.0, 2.0	A-pillar, B-pillar, roof rails, door-beam	Coating problem in laser welded boron
MBW 1500P, 1900P boron steel [10]	4 kW Nd:YAG laser (CW mode)	dissimilar thickness butt joint	1.0, 1.5	A- and B-pillar, floor sills, cant rails	Distortion effect during laser welding
Press hardened 22MnB5 [18]	Hybrid 1 kW fibre laser (CW) + MIG	Similar butt joint	1.5	B-pillar	Gap bridging ability study for hybrid laser-MIG welded boron
22MnB5 (coating removed) [28]	5 kW fibre laser (CW mode)	Similar butt joint	1.6	Car body	Laser ablation, strength post-weld uncoated 22MnB5
Al–Si and Zn coated 22MnB5 [29]	5 kW fibre laser (CW mode)	Similar butt joint	1.6	Car body	Formability of 22MnB5 (Al–Si and Zn coated)
Al–Si and Zn coated 22MnB5 [30]	3.5 kW fibre laser (CW mode)	Similar butt joint	1.6	Car body	Mechanical properties of laser ablated welded boron steel

(continued)

Table 2 (continued)

Material	Welding method	Joint type	Thickness (mm)	Application	Focus
Uncoated boron steel [17]	4 kW Nd: YAG (CW mode)	Similar butt joint	1.5	Stitched BIW	Tensile strength

reduced mechanical properties in the welded region because of the inflow of the Al–Si coating layer during welding [29]. So that, Boron steel can only be welded after the coating layer is removed as done by some researchers with methods such as sandpaper, surface grindings, and laser ablation [13, 15, 25, 27–29].

Boron steel will then be subjected to the hot stamping process after they were joined by laser welding process. A study was done to investigate the effect of the Al–Si content from Boron steel on the phase transformation of this steel after been laser welded and hot-stamped [12]. After laser welded, Al distributed in an irregular circle shape while Si distributed in curved shape along the fusion boundary. After been hot stamped, Al was still remained in an irregular shape, but there is no Si segregation in fusion zone [12]. Al clearly segregated at the boundary between fusion zone and base metal which could affects the mechanical properties and fracture position as shown in Fig. 5.

Boron steel also have been used in dissimilar thickness welding such as was done by Fahlström where 1.0 and 1.5 mm thickness of boron steel was used with butt joint configuration [10]. The intention of this dissimilar welding is to be applied in TWB, where the thicker plate acts as a support frame while the thinner plate could absorb the collision from the crash [29]. Another dissimilar welding between 22MnB5 with lightweight material such as Aluminium and Magnesium also was studied by past researchers [13, 27]. This is the new development for a more weight reduction in a car which could contributes to the increment of fuel economy. It has been estimated that for every 10% of weight removed from a car's weight could increase the fuel

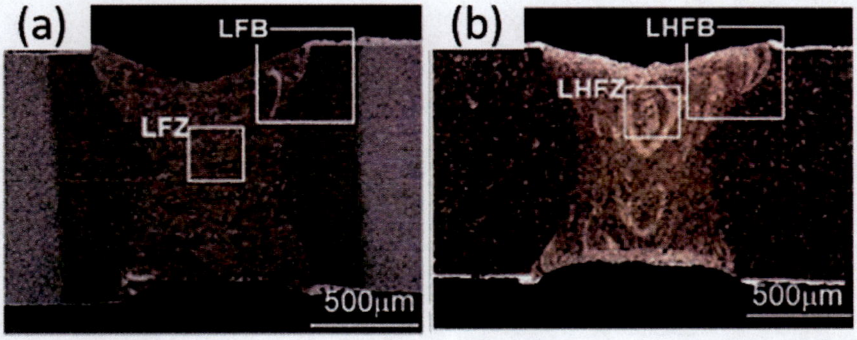

Fig. 5 OM image with a rectangle highlighted for elemental mapping with EPMA for Al. **a** After laser welding. **b** After hot stamping [12]

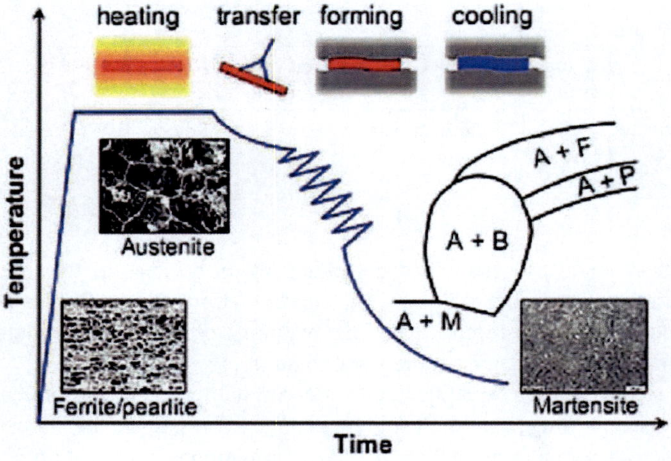

Fig. 6 Schematic illustration of microstructure transformation in hot stamping process [31]

economy by 7% along with the reduction of Green House gas (GHG) emission by 5% [24].

Boron steel also was selected due to its martensitic microstructure behaviour after been hot stamped which could make the parts stronger. Figure 6 shows the illustration of microstructure transformation for a hot stamped AHSS parts from a martensitic steel such as Boron. The microstructures of a AHSS transformed from austenite to martensite completely with the very small spring back problem and it was easy to form due to the increment of ductility and plasticity at high temperature [31]. The yield strength decline, the steel with complex shape can be formed by the dies.

4 Conclusion

The following conclusion can be drawn from this review.

1. High strength steel is important for automotive structures, especially for the structures that need the higher strength of the material.
2. HSS provides the lightweight structures opportunities with improving crash-worthiness.
3. Boron steel is one of the widely HSS material used in car structures.
4. The applications of Boron steels are such as A-Pillar, B-pillar, underbody plate, rails, door beam, and others. The joining combination is in term of single material used and also different materials combinations, joined by welding processes.
5. For the usage of Boron steel and its challenge of weldability to be applied in automotive structures, some recommendations exist to simplify the challenge

in welding this type of steel. To avoid less heat absorption as due to the Al–Si coating, tailor welded blank (TWB) method can be applied where another steel material which not using Al–Si coating could be butt weld joint together with Al–Si coating Boron steel without removing the coating and it will save time and cost in producing a welded product. For example, A-pillar, the B-pillar part can use the TWB method with different materials joining.

Acknowledgements The authors would like to thank the Ministry of Higher Education for providing financial support under Fundamental Research Grant Scheme (FRGS) No. FRGS/1/2018/TK03/UMP/02/9 (University reference RDU 190123) and Universiti Malaysia Pahang for laboratory facilities as well as additional financial support under Postgraduate Research Grant Scheme (PGRS) PGRS1903149 and Internal Research Grant RDU210364.

References

1. Rana R, Singh SB (2016) Automotive steels. Design, metallurgy, processing and applications. Woodhead Publishing
2. Xiao W, Tao W, Zhu B, Zhang Y (2017) Analysis of hot forming of ultra-high strength steel hollow part using variable strength and variable cross technology. In: Advanced high strength steel and press hardening: proceedings of the 3rd international conference on advanced high strength steel and press hardening (ICHSU2016). World Scientific
3. Wang J, Afshan S, Schillo N, Theofanous M, Feldmann M, Gardner L (2017) Material properties and compressive local buckling response of high strength steel square and rectangular hollow sections. Eng Struct 130:297–315
4. Merklein M, Wieland M, Lechner M, Bruschi S, Ghiotti A (2016) Hot stamping of boron steel sheets with tailored properties: a review. J Mater Process Technol 228(Suppl C):11–24
5. Nakagawa T (1993) Recent developments in auto body panel forming technology. CIRP Ann Manuf Technol 42(2):717–722
6. Yeom H, Choi B, Seol T, Lee M, Jeon Y (2017) Very high cycle fatigue of butt-welded high-strength steel plate. Metals 7(3):11
7. Hall J, Fekete J (2017) Steels for auto bodies: a general overview. Automotive steels. Elsevier, pp 19–45
8. Fahlström K, Larsson J (2013) Laser welding of 1900 MPa boron steels. In: The 14th NOLAMP conference. The 14th Nordic laser materials processing conference. Luleå Tekniska Universitet, Gothenburg, 26–28 Aug 2013
9. Geiger M, Vollertsen F, Deinzer G (1993) Flexible straightening of car body shells by laser forming. SAE technical paper
10. Fahlström K (2015) Laser welding of boron steels for light-weight vehicle applications. University West
11. Lindgren LE, Karlsson L (1988) Deformations and stresses in welding of shell structures. Int J Numer Meth Eng 25(2):635–655
12. Yoon TJ, Oh MH, Shin HJ, Kang CY (2017) Comparison of microstructure and phase transformation of laser-welded joints in Al-10 wt%Si-coated boron steel before and after hot stamping. Mater Charact 128:195–202
13. Windmann M, Rottger A, Kugler H, Theisen W (2016) Removal of oxides and brittle coating constituents at the surface of coated hot-forming 22MnB5 steel for a laser welding process with aluminum alloys. Surf Coat Technol 285:153–160

14. Gerhards B, Engels O, Reisgen U, Olschok S (2015) Laser beam welding of press hardened ultra-high strength 22MnB5 steel. Lasers in manufacturing 2015-LIM 2015 und world of photonics congress. International Congress Center Munich, Germany
15. Yaakob KI, Ishak M, Idris SRA (2019) The effect of argon shielding gas flow rate on welded 22MnB5 Boron steel using low power fiber laser. The advances in joining technology. Springer, pp 39–48
16. Spena PR, Cortese L, Nalli F, Májlinger K (2018) Local formability and strength of TWIP-TRIP weldments for stamping tailor welded blanks (TWBs). Int J Adv Manuf Technol, 1–15
17. Shi S, Westgate S (2008) Laser welding of ultra-high strength steels for automotive applications. In: Pacific international conference on applications of lasers and optics. LIA
18. Moller F, Kugler H, Kotschau S, Geier A, Goecke SF (2014) Gap bridging ability in laser GMA hybrid welding of thin 22MnB5 sheets. In: Schmidt M, Vollertsen F, Merklein M (eds) 8th international conference on laser assisted net shape engineering. Elsevier Science B.V., Amsterdam, pp 620–629
19. Azarkevich AA, Kovalenko LV, Krasnopolskii VM (1995) The optimum content of boron in steel. Met Sci Heat Treat 37(1):22–24
20. Bolton G (2014) Ultra high strength steels alloyed with Boron. Cited 20 Dec 2018. Available from: https://www.linkedin.com/pulse/20140811184308-89189426-ultra-high-strength-steels-alloyed-with-boron
21. Kraposhin V, Talis A, Kamenskaya N, Arestov V, Zaitsev A (2018) Arrangement of collective B 12 atoms in the crystal structure of γ-Fe and effect of boron on the hardenability of steel. Met Sci Heat Treat 60(1–2):63–71
22. Materia T (2007) Boron in steel. Cited 25 Dec 2018. Available from: https://www.totalmateria.com/page.aspx?ID=CheckArticle&site=kts&NM=214
23. Watson L (2018) Boron steel in vehicles; implication of HSLA/UHSS and boron steels for rescuers. 2008. Cited 24 Dec 2018
24. Singh M (2016) Application of steel in automotive industry. Int J Emerg Technol Adv Eng 6(7):246–253
25. Yaakob K, Ishak M, Quazi M, Salleh M (2019) Optimizing the pulse wave mode low power fibre laser welding parameters of 22Mnb5 boron steel using response surface methodology. Measurement 135:452–466
26. Salleh MNM, Ishak M, Yamasaki K, Quazi MM, Halil AM (2021) Pulsed Nd: YAG laser parameters effect on welding uncoated advance high strength steel (AHSS) for automotive. J Adv Res Fluid Mech Therm Sci 84(1):91–100
27. Windmann M, Rottger A, Kugler H, Theisen W (2016) Laser beam welding of magnesium to coated high-strength steel 22MnB5. Int J Adv Manuf Technol 87(9–12):3149–3156
28. Lee MS, Moon JH, Kang CG (2014) Effect of the die temperature and blank thickness on the formability of a laser-welded blank of a boron steel sheet with removing Al–Si coating layer. Adv Mech Eng 10
29. Lee MS, Moon JH, Kang CG (2014) Investigation of formability and surface micro-crack in hot deep drawing by using laser-welded blank of Al–Si and Zn-coated boron steel. Proc Inst Mech Eng Part B J Eng Manufact 228(4):540–552
30. Moon JH, Seo PK, Kang CG (2013) A study on mechanical properties of laser-welded blank of a boron sheet steel by laser ablation variable of Al–Si coating layer. Int J Precis Eng Manuf 14(2):283–288
31. Lei C, Cui J, Xing Z, Fu H, Liu H (2012) Research on hot stamping of square cup for advanced high strength steel by numerical simulation. J Comput Theor Nanosci 9(9):1165–1170

Evaluation of Tin Slag Polymer Concrete Column Compressive Behavior Using Finite Element Analysis

M. S. Manda, M. R. M. Rejab, and Shukur Abu Hassan

Abstract In this study, numerical analysis software is used to model the behavior of Tin Slag Polymer Concrete (TSPC) Column under compression. Concrete damage plasticity (CDP) model approach is employed to describe the TSPC property in the finite element (FE) model. FE model is developed based on experimental work data conducted by previous researcher. FE modelling of the TSPC column is performed with purpose to present baseline data for future improvement of the modelling as well as to facilitate future parametric study. The reason is that TSPC is a new material and there was no available previous FE model reported in previous literature as references. The FE model was validated by comparing the simulation results and experimental data for TSPC column under compression. The results indicate that FE model has achieved compressive strength of 37.65 MPa compared with experimental data of 37.62 MPa indicating 0.08% deviation and almost similar location of failure mode. Stress–strain curve indicating that FE model is stiffer than experimental specimen. In conclusion, the stress–strain curve and failure modes for the FE model must be further improved by adjusting CDP parameter in FE model to be able to describe TSPC column specimen accurately. However, the parameters applied can be used as references for future modification on modelling of the TSPC column under compression.

Keywords TSPC column · Compressive behavior · FE · Concrete damage plasticity · Stress–strain curve

M. S. Manda · M. R. M. Rejab (✉)
Faculty of Mechanical and Automotive Engineering Technology (FTKMA), Universiti Malaysia Pahang, 26600 Pekan, Pahang, Malaysia
e-mail: ruzaimi@ump.edu.my

S. Abu Hassan
Faculty of Engineering, School of Mechanical Engineering, Centre for Advanced Composite Materials (CACM), Universiti Teknologi Malaysia, 81310 Skudai, Johor Bahru, Johor, Malaysia

M. Y. Ismail et al. (eds.), *Technological Advancement in Mechanical and Automotive Engineering*, Lecture Notes in Mechanical Engineering,
https://doi.org/10.1007/978-981-19-1457-7_23

1 Introduction

Tin slag is a byproduct of tin production process and according to Omar [8], tin slag waste is a normally occurring radionuclides material (NORM). Tin slag is mostly dumped with tight regulation under Atomic Energy Licensing Berhad (AELB) approval [2]. Its volumes are increasing and Malaysia, as tin producing country is one of the Tin slag wastes contributor in the world to deal with Tin slag waste. Several studies have introduced Tin slag as aggregates in road pavements and cement concrete aggregates. Yusof [14] has reported the potential to use tin slag as aggregates in road pavement while Rustandi et al. [11] and Hashim et al. [6] have confirmed their finding in potential to use Tin slag as aggregates in cement concrete. However, the finding shows that tin slag does not give enough strength enhancements as aggregate in cement concrete because both cement and Tin slag are brittle material. This condition cannot achieve optimum combine properties as promotes by conventional composite material. Therefore, the study in using Tin slag as aggregate in cement concrete material does not attract researchers in the field as there was no continuation reports regarding the study. Despite that, to contribute in solving Tin slag wastes issue, Faidzal et al. [4] has studied the potential of using it as aggregates in polymer concrete. Faidzal has casted the Tin slag particles uniformly as aggregates in polymer concrete by using polyester resin as matrix binder. The finding of the study shows that competitive strength enhancement is achieved. Then, Shakil and Hassan [12] has continued the study by introducing gap graded approach during mixing of tin slag aggregates with polyester to form tin slag polymer concrete specimen. In the study, mixture between coarse and fine Tin slag particles has been casted into TSPC and compressive strength consistently shows a competitive value but with a little bit reduction compared to uniform aggregates. In addition to that, they have also started to apply external strengthening approach on tin slag polymer concrete (TSPC) specimen by wrapping the TSPC specimen with FRP material. The findings show further strength enhancement and has open a way for wide potential in TSPC strengthening study.

Finite element method (FEM) is commonly used to validate experimental results through parametric study [3]. Therefore, to predict TSPC performance, FE model will provide more time and cost saving approach compared to experimental test. Based on previous data by Shakil and Hassan [12], the FE model of TSPC column may be developed to evaluate the performance prior to further experimental process. Other than predicting experimental data, FEM model are also useful in structural design where FE model are often used to help in predicting the structural performance. In this case, FE model of conventional cement concrete and polymer concrete structures are widely available which are usually used in simulation software to facilitate the optimizing process of a concrete structure quickly and at a lower cost compared to experimental process. In the FEM software, there are built in available model to describe material properties such as Mechanical Elastic Model, Drucker-Prager Model, Brittle Crack Model, SOLID65 Model and Concrete Damage Plasticity Model (CDP). Each model is compatible with certain structural model based on its material properties.

Other than that, each model also required their own parameters to allow the model to be functional in FEA software. In case of a concrete material behavior description, the most suitable model is CDP model [13]. In addition to that, Goh et al. [5] also states that CDP model are capable to provide general description to model a concrete material behavior and other type of quasi-brittle material in most types of structures. From observation on the behavior of TSPC cylindrical specimen under compressive load as reported by Shakil and Hassan [12], the TSPC can be categorized as a quasi-brittle material which is different from cement concrete specimen where it shown completely brittle behavior. This situation does not raise an issue because according to Goh et al. [5], CDP model is also capable of describing quasi-brittle material besides completely brittle material. In ABAQUS CDP, the strain elements of a concrete material response towards uniaxial compressive behavior are described as plastic strain, ε_c^{pl}. This value must always positive, ascending and not repeating. Equation (1) described how plastic strain is converted in ABAQUS from experimental stress–strain data.

$$\varepsilon_c^{pl} = \varepsilon_c^{in} - \left[\left(\frac{d_c}{1 - d_c}\right)\left(\frac{\sigma_c}{E_o}\right)\right] \tag{1}$$

Currently, in particular for TSPC, there was no available FE model found in the previous literatures which make the process of TSPC modeling become troublesome as there were no baseline parameters to be referred. The purpose of this study is to evaluate the compressive behavior of TSPC short column specimen using commercial ABAQUS software. FE model of TSPC specimen will be develop base on previous experimental test. The specimen property will be defined using CDP model and the experimental data. The FE model will be validated by comparing the results of numerical simulation and experimental findings focusing on stress–strain curve and failure mode. The aim of this paper is to provide baseline data as future reference for FE model development of TSPC material behavior to be applied in parametric study.

2 Experimental Data

The experimental setup is based on study by Shakil and Hassan [12]. In the study, Tin slags were pre-treated by drying and sieving to produce coarse (4 mm) and fine (2 mm) aggregates according to ASTM C136. The aggregates were mixed together with Polyester resin and Methyl Ethyl Ketone Peroxide (MEKP) hardener with 70:30 aggregates to polyester ratio. The mixture is then casted into 50 mm diameter PVC pipe and cured for 3 days at room temperature. The solidified mixture is then demoulds and cut into 100 mm length to produce final specimen of TSPC short cylindrical column with 50 mm diameter and 100 mm length. The standard used for TSPC specimen preparation is ASTM C192/470 standard specification for molds for forming concrete test cylinders vertically.

Fig. 1 Compression test set-up of TSPC short column specimen [12]

For mechanical testing, Shakil and Hassan [12] performed the test according to ASTM C579-01 standard test methods for compressive strength of polymer concretes. In the compression test, Instron 600 kN Universal Testing Machine is used at a loading rate of 1 mm/min. The stress–strain curve data of the experiment is referred to describe the material property as well as physical properties of the specimen in FE model. Young's Modulus, E_o will be determine from experimental results stress–strain slope on the elastic behavior and the Poisson's ratio, v based on polyester resin as matrix binder in TSPC. Figure 1 shows the compression test set up.

3 Finite Element Modeling

In order to develop the FE model of TSPC specimen, the references has been made on previous similar study on conventional concrete material as there is no available FEM specifically on TSPC behavior under compression. According to Raza et al. [10], ABAQUS offer three material models to describe brittle material such as concrete namely CDP Model, Brittle Crack Model and Smeared Crack Model, however CDP model is preferred because it deals broadly with the three-dimensional nonlinear inelastic behavior of concrete [1]. In addition to that, in a study of plastic damage model, Yu et al. [13] suggested that CDP is the most accurate model to predict the behavior of FRP confined as well as unconfined concrete structure. The CDP model in ABAQUS will theoretically described the TSPC column specimens used in the experiment as well as its behavior under compression.

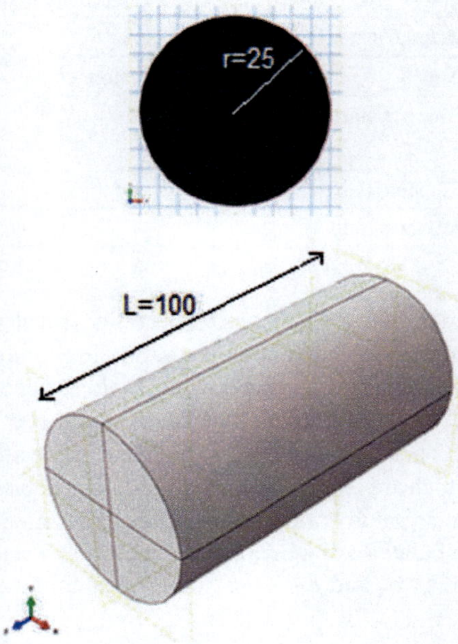

Fig. 2 Specimen circular sketch on part module and extrusion forming TSPC cylindrical specimen

3.1 Part and Model

Part and model are created using 3D modeling space with deformable option type to feature solid shape. In part section, a simple cylindrical specimen is develop by sketching a circular shape and extruded to create a model of solid cylinder to represent short column specimen with 50 mm diameter and 100 mm height. Figure 2 shows the sketch of circular shape before extrusion in part module.

3.2 Material Property

After finish with part module, the cylindrical specimen is generated. Then, in material property module, the material behaviors are described with several parameters. First, density value is filled in general option by calculation based on TSPC specimen mass and volume. Then, in mechanical elasticity parameter, Young's Modulus, E_0 and Poisson's ratio, v of the specimen are added based on slope of experimental stress–strain curve in the elastic region and polyester resin as matrix binder in TSPC specimen. After that, in mechanical plasticity behavior, concrete damage plasticity (CDP) model is used in describing the model. There are three main parameter to be filled in CDP model which are plasticity, compression damage and tension damage. For plasticity parameter, the value is referred to ABAQUS manual [1]. In the manual, f_{bo}/f_{co} is defined as a ratio of initial equibiaxial compressive yield stress to initial uniaxial

Table 1 TSPC model elasticity and plasticity parameter

Material's name	TSPC specimen	Plasticity parameters	
Density	8.91×10^{-10} (tonne/mm^3)	Dilation angle	35
Concrete elasticity		Eccentricity	0.1
		f_{bo}/f_{co}	1.16
E_o (Mpa)	1191.67	K	0.67
Poisson's ratio, v	0.3	Viscosity parameter	1.00×10^{-5}

compressive yield stress. This ratio typically exhibits multi-axial to uniaxial material strengths. K parameter is a ratio of second stress invariant on tensile meridian. The K parameter controls how the yield surface will shape like. Table 1 shows the value used for elasticity and plasticity parameter in CDP.

Then, for compressive behavior, there are another three sub-parameter to be filled which are corresponding yield stress, inelastic strain and compression damage. Yield stress value is referred from the experimental data. After that, inelastic strain ε_c^{in} is calculated using Eq. (2). ε_c is compressive strain based on corresponding compressive stress, σ_c and E_0 is the Young's Modulus of the TSPC specimen.

$$\varepsilon_c^{in} = \varepsilon_c - \frac{\sigma_c}{E_0} \tag{2}$$

Then compression damage parameter, d_c is calculated using Eq. (3). σ_{co} is the yield stress of the experimental stress–strain curve:

$$d_c = 1 - \frac{\sigma_c}{\sigma_{co}} \tag{3}$$

After that, plastic strain ε_c^{pl} is calculated using Eq. (1). The value of plastic strain must be in two conditions which are always positive value and in ascending order. This is the requirement of ABAQUS processor; otherwise error will occur during simulation. Finally, for tension damage parameter, d_t the yield stress value is assumed based on the final crushing stress of the specimen in experimental test (9.95 MPa). The cracking strain is assigned as zero value because there was no splitting tensile data on the experimental setup. Figure 3 shows the tabulation of experimental stress–strain value, inelastic strain value, compression damage parameter and the plastic strain. Plastic strain data are to ensure that ABAQUS will successfully perform the simulation of the FE model described using CDP parameter by keeping the value to be always positive and ascending.

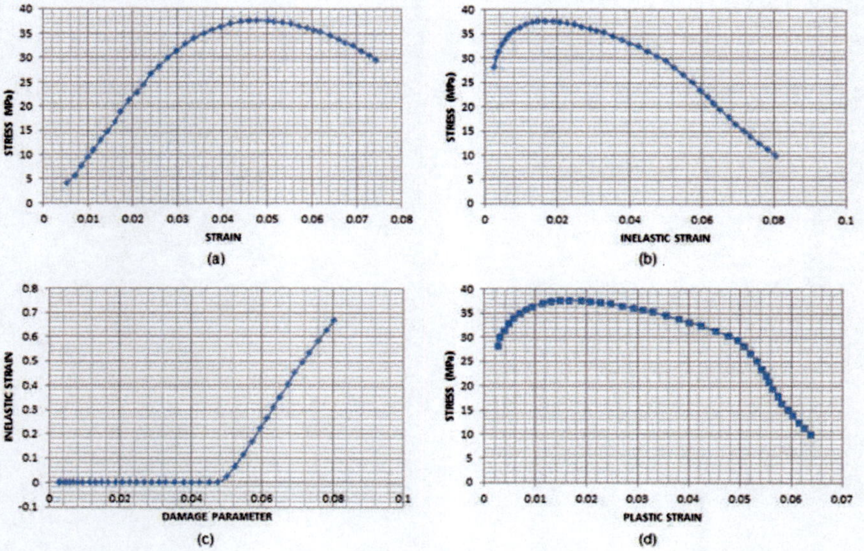

Fig. 3 Parameters of CDP for TSPC model property damage description in ABAQUS. **a** Experimental stress–strain curve. **b** Corresponding experimental stress-inelastic strain curve. **c** Inelastic strain-compression damage parameter. **d** Corresponding experimental stress-plastic strain curve

3.3 Assembly and Step Definition

After describing the property of TSPC model, the part was assembled in assembly module by creating instances. Because there was only one part, there was no translation, assembly and interaction definition required. After that, in step module, ABAQUS has already created initial step. In initial step, the procedure type is assign as static general. Then, step 1 must be created where the procedure type is assigned as dynamic explicit. The reason is that, dynamic explicit will allow the element deletion to occur during simulation to observe the crack pattern that happened.

3.4 Load and Boundary Condition

In load module, the nature and magnitude of load or boundary condition must be defined according to step that has been assigned. Here, the TSPC model is set as fixed at the bottom and uniform pressure is applied on top of the specimen. Figure 4 shows the load and boundary condition that has been set on the TSPC model.

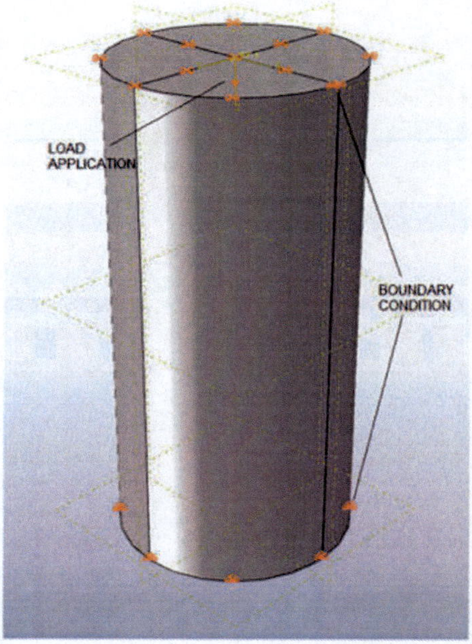

Fig. 4 Load and boundary condition of the TSPC model

3.5 *Mesh and Job Setup*

After that, in the mesh module, first thing to do is setting the model as part because no editing is allow in assembly instances model, otherwise the mesh of the element cannot be assigned. Then, assign element type has been made to modify the element type and its response during simulation. Here, in element control option, the element deletion is allowed by selecting element deletion option to simulate the crack pattern of the TSPC specimen during failure mode. The part is then seed by assigning element size and meshed the part to produce a final specimen model in meshed form. Several elements have been created to compare the effectiveness of TSPC model mesh density towards the experimental results. Then, in job module, a job is created by keep the available option in default state. Figure 5 shows the TSPC FEM model of the specimen.

4 Results and Discussion

According to Micha and Andrzej [7], during modelling of a material using CDP model, mesh size should be taken into consideration as it influence the deviation between experimental and modelling results. Other than that, plasticity parameters may also be varied to observe the modeling response toward experimental results for the sake of achieving precision model. However, the variation in CDP plasticity

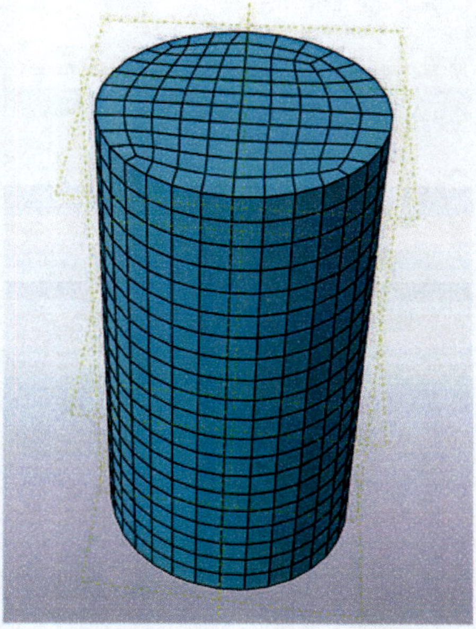

Fig. 5 Element meshing on TSPC FE model of the specimen

parameters such as dilation angle, eccentricity, f_{bo}/f_{co}, K and viscosity parameter must be made within allowable range as stated in ABAQUS user manual [1].

4.1 *Stress–Strain Comparison Between Numerical Model and Experimental*

In this study, several mesh size of the element in TSPC model was simulated to evaluate the response towards uniaxial compression. The result shows that model with smaller element meshing has increase the precision between modelling and experimental results. Experimental results for compressive strength of TSPC specimen from Shakil and Hassan [12] study has achieve 37.62 MPa. In the TSPC model that have been develop, compressive strength from FEM simulation has shown 36.63 MPa for 7 element meshing, 36.78 MPa for 5 element meshing and 37.65 MPa for 3 element meshing. The results of FE modeling and experimental show that smaller the element meshing resulting in more precision data in comparison with experimental data. Table 2 shows the meshing specification of TSPC FE model and percentage of deviation from experimental data in term of compressive strength achievement. Figure 6 shows the stress strain curve of the experimental results and FEM modelling results. From the stress–strain curve, TSPC model depicts more incline slope compare to the actual TSPC specimen on linear elastic region and in hardening behavior indicating that the model has higher stiffness than the actual material. The curve also has shown that smaller meshing element resulting in closer curve shape between experimental and

Table 2 Element meshing specification and percentage of maximum compressive strength deviation from experimental results

Mesh type	Seed size	Total elements	Total nodes	Maximum strength from FEM (MPa)	Maximum strength from experiment (MPa)	% Deviation from experimental (%)
C3D8R	3	8646	9861	37.6530	37.6216	0.08
	5	1920	2373	36.7750		2.25
	7	840	1095	36.6251		2.65

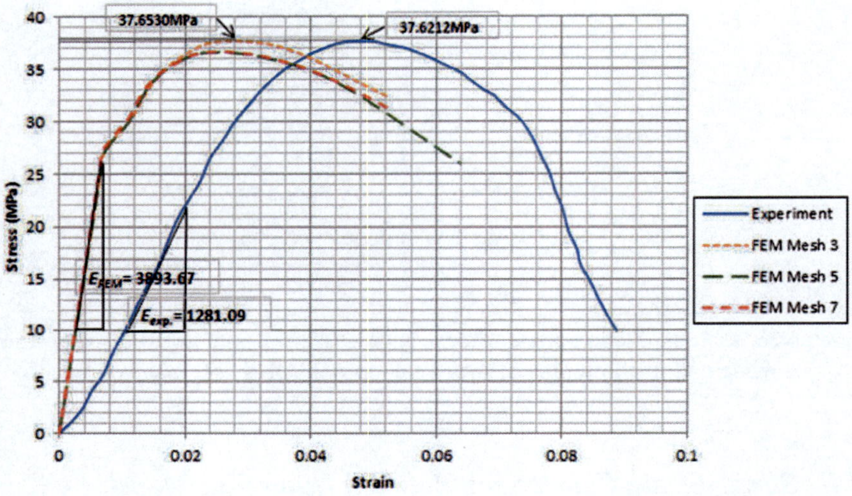

Fig. 6 Stress–strain curve comparison of TSPC FE model and experimental results

FE model stress–strain curve. However the curve shape are not very well match with an average of 19.83% different based on compressive strength increment. Due to the different, Young's Modulus (E_o) of the experimental specimen and FE model has shown quite a large different as the experimental specimen has 1281.09 MPa while the TSPC model has 3893.67 MPa indicating that FE model has higher stiffness compared to experimental specimen. Therefore further modification and addition on tensile damage parameter must be made on the CDP parameter in future study.

4.2 Failure Mode Comparison Between Numerical Model and Experimental

Conventional cement concrete cylindrical specimen failure mode is commonly occurred in diagonal shear. According to Polus and Sumigala [9], both FE model

Fig. 7 Failure mode comparisons between TSPC FE model and experimental

and actual concrete cylindrical specimen has shearing failure mode. However, the results from experimental for TSPC shows that the failure mode of TSPC specimen are likely as sudden crushing and occur at the middle or one third from top of the specimen. Almost similar failure mode are shown in TSPC model when the damage indicator show that it start at a location of one third from the top of the TSPC model. Figure 7 shows the comparison between failure mode of experimental specimen and FE model. These results are due to the similar increment in compressive strength for both experimental and FE model. However, the failure areas of experimental specimen are larger than FE model as shown in Fig. 7. This is the result of higher strain increment on experimental specimen than FE model as shown in stress–strain curve in Fig. 6. This finding requires further improvement on the TSPC model especially in parameters description as well as experimental test to observe how top part of the specimen behaves upon failure.

5 Conclusions

This study is able to produce a FE model for TSPC specimen and then compared with the experimental results. Both stress–strain curve and failure mode from TSPC model has shown some similarity but with lack of precision when compared to the experimental test stress–strain curve and failure mode. Even though the stress–strain curve of TSPC model does not very well match with experimental stress–strain curve, both has achieved almost similar compressive strength. These indicate that the TSPC model may be used for structural design as in a design process; both yield and maximum strength are parameter that will be considered during analysis. However, for research purpose, further modification must be made on the TSPC model to be able to precisely describe the model behavior for parametric study purpose. For failure mode, the location of failure is almost similar with experimental specimen but for TSPC model, the compression damage is much more localized resulting in crack pattern propagation which start at one side of the specimen. This may be further improved by applying tension damage parameter in the TSPC model. However, to obtain tension damage parameter, another test must be performed on the specimen which is splitting tensile test. Finally, as a general conclusion, to enhance the TSPC model in order to closely replicate the behavior of actual TSPC specimen, some of the parameter in property module must be modified. However, the parameter used in this study may be used as references for future modification on modelling of the TSPC column under compression especially in a parametric study.

Acknowledgements The authors are grateful to the Universiti Teknologi Malaysia under the collaborative research grant (CRG UTM-RDU192311) and Universiti Malaysia Pahang (PGRS210339).

References

1. Abaqus 6.14 User Guide (2014) Dassault systèmes. Simulia Corporation, Providence, RI, USA
2. AELB Technical Guidance (2016) Code of practice on radiation protection relating to technically enhanced naturally occurring radioactive material (TENORM) in oil and gas facilities. Lembaga Perlesenan Tenaga Atom, Malaysia. LEM/TEK/58 Sem.1
3. Chellapandian M, Prakash SS, Rajagopal A (2017) Analytical and finite element studies on hybrid FRP strengthened RC column elements under axial and eccentric compression. Compos Struct 184:234–248
4. Faidzal MMY, Hassan SA, Omar B, Zakaria K, Zaharuddin MFA (2018) Particle size effect on optimal mixture ratio of tin slag polymer concrete under compression. J Built Environ Technol Eng 5 (2018). ISSN 0128-1003
5. Goh WI, Mohamad N, Abdullah R, Samad, AAA (2014) Compression test and finite element analysis of foamed concrete cube. J Eng Technol 5(1). ISSN: 2180-3811
6. Hashim MJ, Mansor I, Ismail MP, Sani S, Azmi A, Sayuti S, Ibrahim MZ, Anuar AA, Rahim AAA (2018) Preliminary study of tin slag concrete mixture. Agency nuclear Malaysia: engineering division, industry technology division and waste technology development centre. IOP Conf Ser Mater Sci Eng 298:012014
7. Micha S, Andrze W (2015) Calibration of the CDP model parameters in Abaqus. In: Advances in structural engineering and mechanics (ASEM15), pp 1–11
8. Omar M (2000) NORM waste management in Malaysia. In: Proceeding of international conference on the safety of radioactive waste management. IAEA Spain, pp 89–92
9. Polus P, Szumigała M (2019) Laboratory tests vs. FE analysis of concrete cylinders subjected to compression. In: AIP conference proceedings, vol 2078, 020089
10. Raza A, Rehman AU, Masood AB, Hussain I (2020) Finite element modelling and theoretical predictions of FRP-reinforced concrete columns confined with various FRP-tubes. Structures 26:626–638
11. Rustandi A, Nawawi FW, Pratesa Y, Cahyadi A (2017) Evaluation of the suitability of tin slag in cementitious materials: mechanical properties and leaching behavior. In: International conference on chemistry and material science (IC2MS), vol 299, p 012046

12. Shakil UA, Hassan SA (2020) Behavior and properties of tin slag polyester polymer concrete confined with FRP composites under compression. J Mech Behav Mater 29(1):44–56
13. Yu T, Teng JG, Wong YL, Dong SL (2010) Finite element modeling of confined concrete-II: plastic-damage model. J Eng Struct 32:680–691
14. Yusof MAW (2005) Investigating the potential for incorporating tin slag in road pavements. Ph.D. thesis. University of Nottingham, School of Civil Engineering, USA

Creep Life Prediction of P91 Steel Using Omega Method

S. N. A. Rosli, N. Ab Razak, M. R. Mahazar, and N. A. Alang

Abstract Martensitic P91 steel is desirable for structural components operating at elevated temperatures. It is extensively used in nuclear power plant boilers, pipelines, reactor pressure vessels, and steam generators due to its high creep strength and corrosion resistance. Predicting the P91's creep rupture life is critical for safe operation. Numerous creep laws have been developed throughout the years to anticipate the deformation, propagation of damage, and rupture of materials subjected to the creep phenomena. The Omega method is one of the most widely used in API RP579 on fitness-for-service purposes. In this study, the creep tests have been performed at 600 °C for 160, 180 and 190 MPa. In order to predict the rupture life, the omega method has been employed, which utilised the initial creep strain rate and creep strain. The experimental data has been compared to available literature data for P91 material. The predicted life was always more significant than the experimental result, and it was strongly linked to the omega value. The result shows that the value omega value of the test data are in line with the available data and the initial creep strain rate increased linearly with increased of stress and temperature. The predicted rupture life values are consistent and close to the experimental results.

Keywords P91 · Creep behaviour · Omega method

1 Introduction

Modified 9Cr-1Mo ferritic steel (also known as Grade 91 or P91) is one of the best Cr-Mo alloys for high-temperature applications due to its creep strength and mild oxidation resistance up to 650 °C. It is frequently utilised in construction and piping systems of fossil-fuel power plants [1, 2]. Despite its excellent creep and corrosion resistance at elevated temperatures, P91 steel may experience failure due to

S. N. A. Rosli (✉) · N. Ab Razak · M. R. Mahazar · N. A. Alang
Structural Performance Materials Engineering Focus Group (SUPREME), Faculty of Mechanical & Automotive Engineering Technology (FTKMA), Universiti Malaysia Pahang, 26600 Pekan, Pahang, Malaysia
e-mail: MMD20001@student.ump.edu.my

M. Y. Ismail et al. (eds.), *Technological Advancement in Mechanical and Automotive Engineering*, Lecture Notes in Mechanical Engineering,
https://doi.org/10.1007/978-981-19-1457-7_24

microstructural changes associated with elevated temperatures and prolonged service exposure time or severe local overheating of the component.

Premature failure in power plants always occurs during service, and hence the necessity for life assessment of creep range component is a significant concern. Numerous techniques for life prediction have been developed to ensure safety at elevated temperatures, such as Theta projects [3], the Kachanov approach [4], and the Omega method [5] can analyse these creep curves. Furthermore, the recorded creep strain with time variations of critical components is crucial for predicting the rupture life [6].

The Omega method has now been recognised for inclusion in API RP579 on fitness-for-service [7]. This method has been developed by the MPC Petroleum and Chemical Committee, is widely used in the United States of America to determine the expected lifespan of petroleum and power corporations [6]. The Omega model enables the prediction of a component's remaining life in the creep regime at elevated temperatures and pressures. The Omega parameters, which are made up of a strain rate and a multi-axial damage parameter, are used to forecast the rate of strain accumulation, creep damage accumulation, and remaining time to failure as a function of stress state and temperature [6].

In this work, the objective of this study is to predict the creep life of P91 by using Omega method. The creep tests have been performed at 600 °C for 160, 180 and 190 Mpa. In addition, the creep rupture experimental data has been analysed and compared to available literature data for P91 material utilising the Omega method.

2 Theoretical Framework

Omega method was proposed for predicting rupture life using creep strain versus time curves without the primary and secondary regimes. Omega method was derived assuming the overall creep life was spent during the tertiary stage whereas the secondary is non-existent and primary creep is usually small therefore is negligible [8]. This method was based on the linear relation between the logarithm of creep strain rate ($\dot{\varepsilon}$) and strain (ε). The following equation well expresses their relationship:

$$\dot{\varepsilon} = \dot{\varepsilon}_0 \times e^{\Omega\varepsilon} \quad (1)$$

$$ln\dot{\varepsilon} = ln\dot{\varepsilon}_0 + \Omega\varepsilon \quad (2)$$

where $\dot{\varepsilon}_0$ is the initial creep strain rate, [Inline Image Removed] is the creep strain, and Ω is the parameter that describes the evolution of creep strain. Plotting Eq. (2), the slope of the line is Ω, and the y-axis intercept is $\dot{\varepsilon}_0$. Experimentally, it is found that the $\dot{\varepsilon}_0$ and Ω are both a function of stress and temperature and can be expressed by the power creep law as shown in Eqs. (3) and (4) respectively:

$$\varepsilon^0 = A\varepsilon\sigma^{n\varepsilon} exp\left(-\frac{Q\varepsilon}{RT}\right) \tag{3}$$

$$\Omega = A_\Omega\sigma^{n_\Omega} exp\left(-\frac{Q_\Omega}{RT}\right) \tag{4}$$

A_0, A_Ω are stress coefficients, n_0, n_Ω are stress exponents, and Q_0 is the apparent activation energy. Q_Ω is a value indicating the temperature dependence on Ω. Integrating Eq. (1), the time *t* at a particular creep strain can be calculated.

$$t = \frac{1}{\Omega\dot{\varepsilon}_0}(1 - e^{-\Omega\varepsilon}) \tag{5}$$

The exponential term in Eq. (5) is negligible at rupture. Thus, the creep rupture time, t_r, can be stated mathematically as follows:

$$t_r = \frac{1}{\Omega\dot{\varepsilon}_0} \tag{6}$$

3 Material and Experimental Setup

3.1 Material

P91 steel is used in this work. The material was supplied in the form of a pipe section. The chemical compositions of the material are listed in Table 1. The chemical compositions were examined using Oxford Foundry-Master UV. Metallographic assessment prior to testing has been performed on the P91 steel. The metallographic specimen was sectioned, mounted, ground, and polished. For etching purposes, the surface of the P91 mounting sample was wiped with distilled water and dried by an air pressure gun. Then the sample was swabbed with little drops of Villela Reagent chemical. The etchant was cleaned up after 10 to 15 s by water, and the sample was cleaned up with distilled water. An air pressure gun was used to dry the surface of the P91 mounting samples. Finally, the etched sample was examined under optical microscopic.

Table 1 Chemical composition (wt %) of the P91 pipe used

Element	C	Mn	Si	Ni	Cr	Mo	V	P
Composition (wt%)	0.108	0.380	0.410	0.100	8.790	0.994	0.226	0.003
Element	Al	Ti	Nb	Cu	Co	N	S	
Composition (wt%)	0.005	0.004	0.040	0.100	0.021	0.045	0.005	

3.2 Experimental Procedure

The material was supplied in the form of a pipe section. All specimens were extracted longitudinally from the pipe section, as shown in Fig. 1. Figure 2 shows the creep specimen used with a gauge length of 36 mm. The creep rupture test was conducted in accordance with the ASTM E139-11 recommended procedure [9].

Figure 3 shows the creep machine which contained the main part of the apparatus as well as the data recorder device. The creep specimen was mounted on the creep machine. The temperature and stress of creep were conducted at 600 °C, ranging from 160 to 190 MPa, respectively. The cylindrical-shaped electric furnace with a temperature controller was used to create a high-temperature environment around the specimen. The thermocouples were placed in three different locations on the furnace wall: top, middle, and bottom, allowing for a minor adjustment of the heat in each zone to achieve a uniform temperature throughout the furnace. Temperatures were checked during the test to ensure they stayed within ±2 °C. The temperature of the

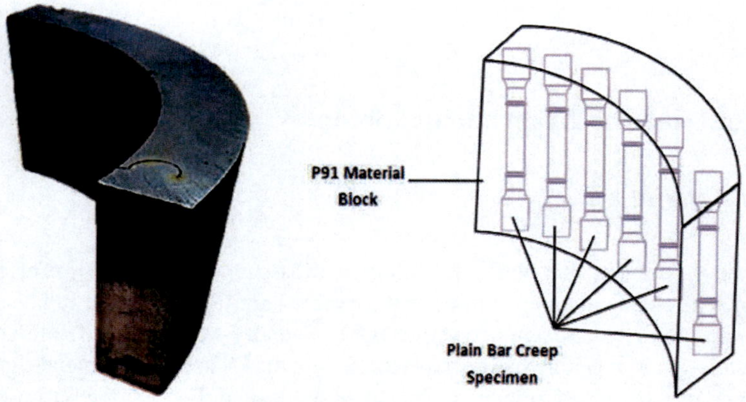

Fig. 1 Schematic diagram of the P91 specimen

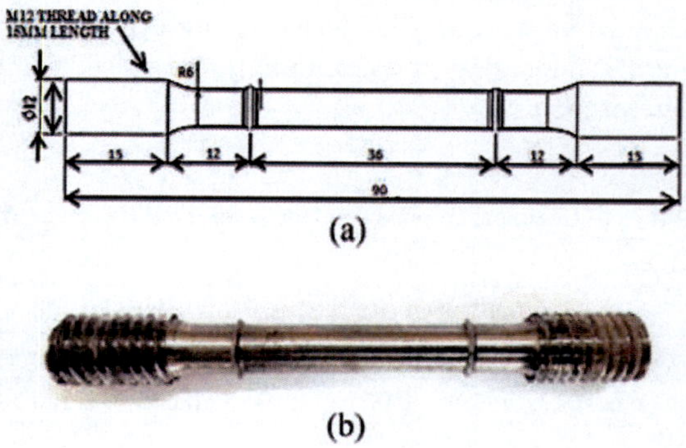

Fig. 2 Uniaxial round bar specimen: **a** dimensions of P91 specimen; and **b** picture of P91 specimen

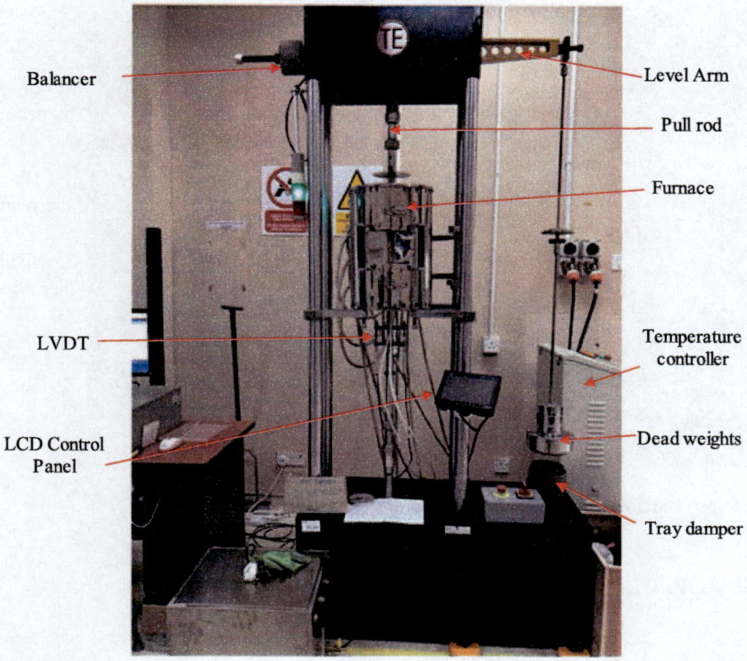

Fig. 3 Constant load creep machine

specimens in the gauge length region is supposed to be represented by the readings from these two thermocouples. The close-up of complete setups of creep specimen with thermocouples is shown in Fig. 4.

The longitudinal elongation of the material was measured using a linear variable differential transformer (LVDT) that was centrally clamped between two aluminium plates. Before testing, the LVDT was calibrated, and the calibration curve was acquired (voltage versus displacement). The data recorder continuously recorded the voltage reading from the LVDT during load up, but the recording duration was increased to 300 s during creep (afterload up). The elongation of the material was determined by converting the LVDT output voltage using the previously acquired calibration curve. The specimen's elongation was monitored during the test, and the test was terminated when the specimen broke.

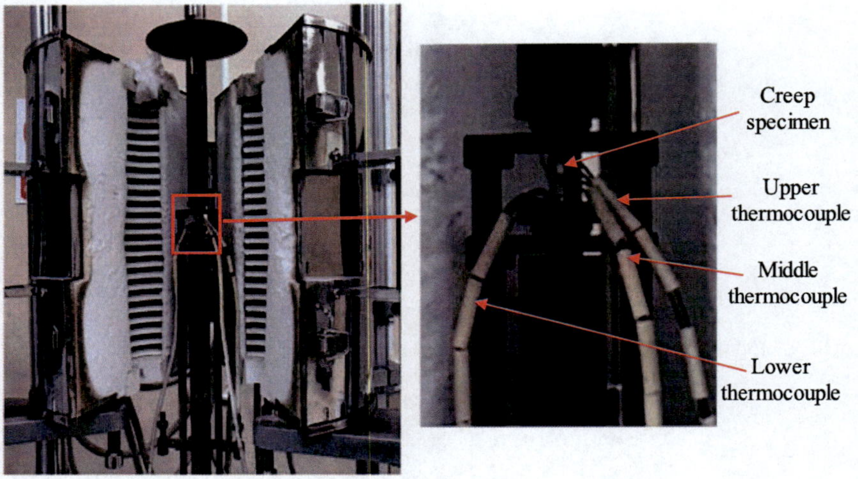

Fig. 4 Creep specimen and thermocouple setup

4 Results and Discussion

4.1 Microstructure of P91

Figure 5 shows the microstructure of P91 steel, which had a structure of tempered martensite with occasional large stringer-like non-metallic inclusions. The martensitic lath structure and prior austenitic grain boundaries (PAGBs) were visibly exposed at the magnification of 500x. The prior austenite grain size was measured to be $\approx 21\ \mu m$.

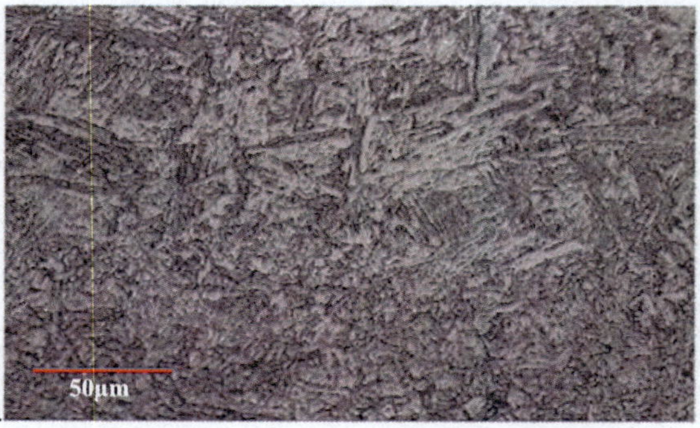

Fig. 5 P91 microstructure

4.2 Result of Creep

Figures 6, 7 and 8 show the creep strain and strain rate for P91 steel testes at 600 °C under the stress of 160 MPa, 180 MPa, and 190 MPa, respectively. According to the classical creep strain versus time curves [10], it can be noted that there are three typical stages, which are referred to as the primary, secondary, and tertiary stages. As shown in Fig. 6 at lower stress (160 MPa), the creep curve is longer and the three creep regions can be seen. This characteristic is similar to those reported for martensitic material [11, 12]. At high stress (190 MPa), creep curves are relatively short, and the primary creep is almost negligible. Figures 6b to 8b shows the progression of creep strain rate over the time. It can be seen that the creep strain rate is almost constant in the secondary region and accelerates at the tertiary phase, and eventually result in a fracture of the specimen.

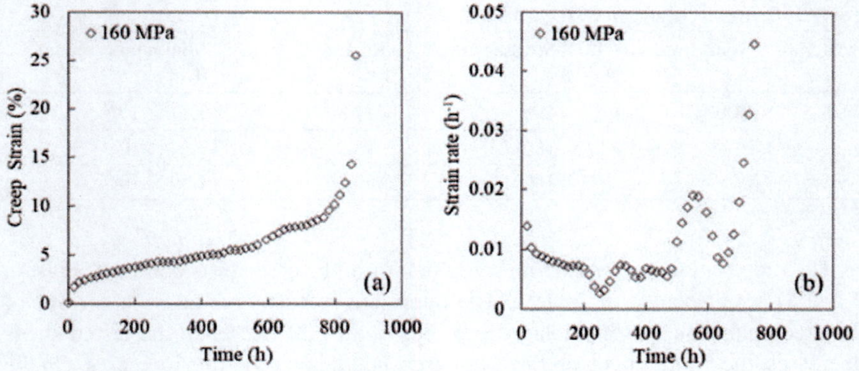

Fig. 6 **a** Creep strain and **b** strain rate data at 160 MPa

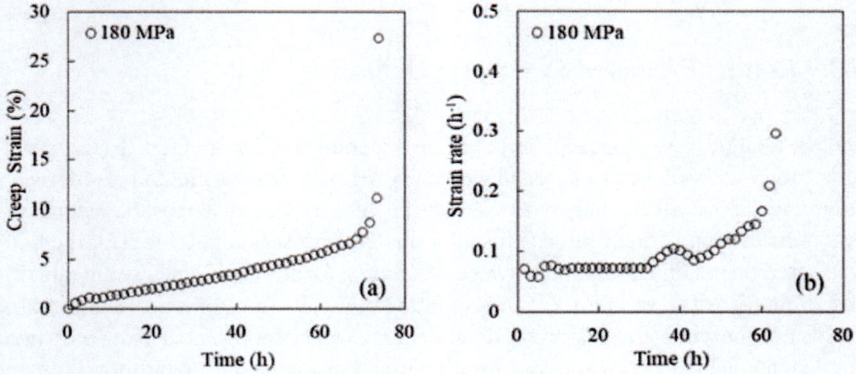

Fig. 7 **a** Creep strain and **b** strain rate data at 180 MPa

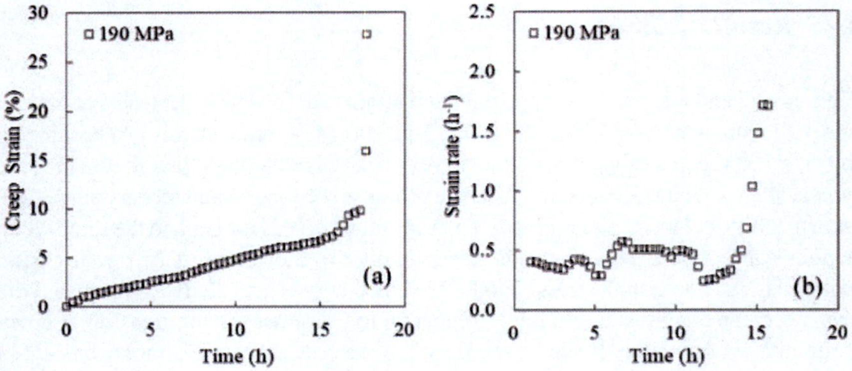

Fig. 8 **a** Creep strain and **b** strain rate data at 190 MPa

Table 2 Creep test results of P91

Material	Temperature T/°C	Stresses σ/MPa	Rupture time t/h	Minimum creep rate ($\dot{\varepsilon}_{min}$)
P91	600 °C	190	17.8	1.77×10^{-3}
		180	73.8	7.24×10^{-4}
		160	860.7	2.59×10^{-5}

The rupture time and minimum creep rate from the creep tests obtained (Figs. 6, 7 and 8) are presented in Table 2. The minimum creep rate is taken as a slope of the creep curve in the secondary creep region. At 600 °C, when the tested stress decreases, the minimum creep rate decreases, but the rupture time increases. On the other hand, as the temperature under the same stress increases, the minimum creep rate increases but the rupture time reduces.

4.3 Results Predicted by Omega Method

Omega method was proposed for predicting rupture life using creep strain versus time curves without the primary and secondary regimes. Omega method was derived assuming the overall creep life was spent during tertiary stage whereas the secondary is non-existent and primary creep is usually small therefore is negligible. This method was based on the linear relation between the natural logarithm of creep strain rate ($\dot{\varepsilon}$) and strain (ε) as shown in Eq. (2) and is plotted in Fig. 9. To summarise, creep strain versus time curves at given stress and temperature can be represented using only two coefficients, which are $\dot{\varepsilon}_0$ and Ω. From the graph of the natural logarithm of creep strain rate versus creep strain curves shown in Fig. 9, Ω is the slope of the line. The value of initial creep strain rate and omega under varying stresses are tabulated in

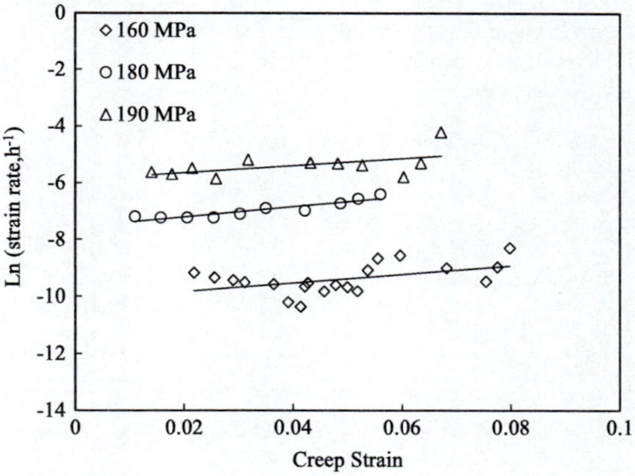

Fig. 9 Natural logarithm of creep strain rate versus creep strain

Table 3 Creep test results of P91

Temperature T/°C	Stresses σ/MPa	Ω	Initial creep strain rate, $\dot{\varepsilon}_0$	Rupture time t/h	Predicted Rupture time t/h
600 °C	190	24.4	1.89×10^{-3}	17.8	19.1
	180	37.1	2.92×10^{-4}	73.8	80.7
	160	26.9	2.39×10^{-5}	860.7	960.1

Table 3. The predicted creep rupture life using Eq. (6) is also presented in Table 3. It is shown that the predicted rupture time using the Omega method gives a higher value than the actual rupture life. Higher value of omega may be regarded to creep softening. For some material, the omega value can be very large up to 200 or more and most of the creep life was spent at very low strains [6].

In this work, the creep test performed was limited due to time and material constraints. However, the creep test data analyses were not limited to the test results from the current work. Thus the available literature data from Ab Razak [13] and Chen [14] were compared and analysed with the present test data. It should be noted that the available data from Ab Razak [13] and Chen [14] are for P91 steel and P91/12Cr1MoV dissimilar joint, respectively.

The stress dependency of Ω and [Inline Image Removed] are plotted in Fig. 10a and b respectively. The available data from Ab Razak [13] and Chen [14] were compared and analysed with the present test data. It is shown that in Fig. 10a, the values Ω of the test data represented by the circle symbol is in line with the available data. Generally, the values of Ω increase with decreasing level of stress and temperature. Meanwhile, in Fig. 10b, the values of $\dot{\varepsilon}_0$ increased linearly with increased stress and temperature. These trends can be directly compared to the temperature and stress dependencies of Ω and $\dot{\varepsilon}_0$ as given by the API-579 Fitness for Service and are also aligned with those observed by Prager [6].

Fig. 10 **a** P91's stress dependency of omega, Ω and **b** P91's stress dependency of initial strain rate, $\dot{\varepsilon}_0$

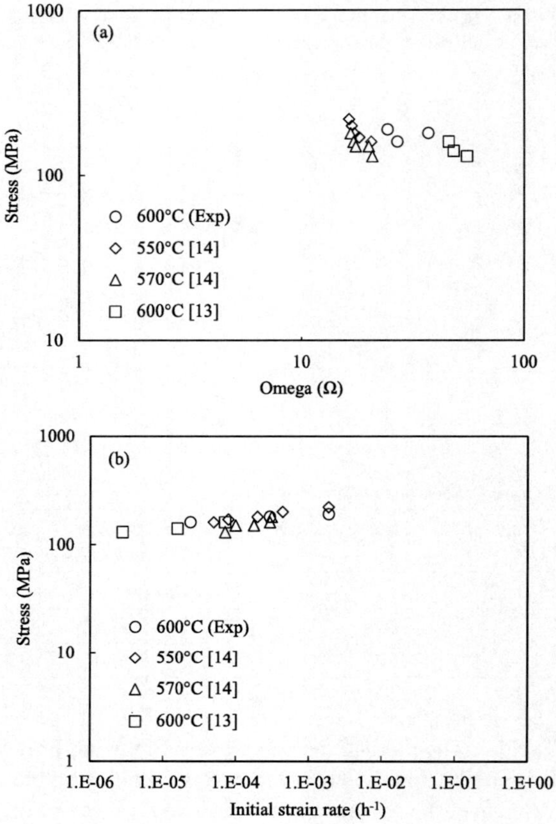

Fig. 11 Plot of predicted versus actual data of creep life P91

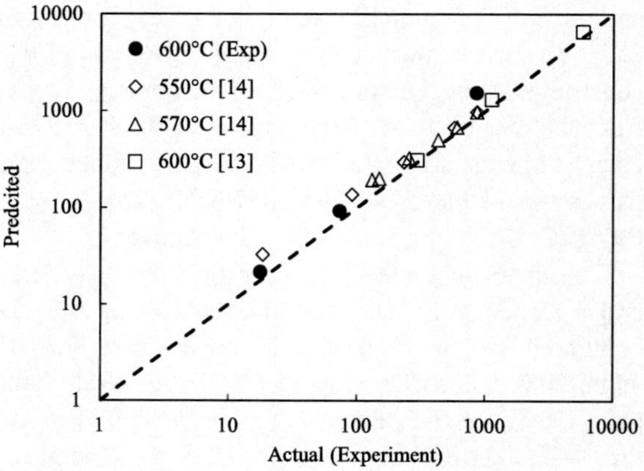

Experimental data and available literature data [13, 14] in Fig. 11 proves that the predicted values are consistent and close to the experimental results. It is implies that the accelerated creep test can be used to predict the long term creep rupture. In addition, the omega method can be used to predict the rupture life of the dissimilar weld joint under varying condition [14]. This also suggests that the simple and reliable Omega method can be used to predict the life time of high temperature components.

5 Conclusions

In this paper, creep tests were conducted on the P91 steel under 600 °C at various stress levels. The actual rupture time magnitudes were compared to the predicted results of creep life using the Omega method. As a result, the following conclusions can be drawn.

- Creep curves can be observed under varying stresses. The primary region is very short with decreasing creep rate; the secondary region demonstrates a constant creep speed and takes the longest time in the creep process; the creep rate accelerates until the ultimate fracture occurs in the tertiary stage.
- The Omega approach may be used to forecast the rupture life of P91 steel in a variety of conditions. The values of Ω for 160 MPa, 180 Mpa, and 190 Mpa of 26.9, 37.1, and 24.4 respectively.
- The predicted life was always more significant than the experimental result, and it was strongly linked to the omega value. The result shows that the omega values of the test data are in line with the available data and the initial creep strain rate increased linearly with an increase in stress and temperature. In addition, the predicted rupture life values are consistent and close to the experimental results.

Acknowledgements Sincerely thanks to Universiti Malaysia Pahang for providing laboratory facilities and financial assistance under grant number FRGS/1/2019/TK03/UMP/03/5 which enabled this research to be carried out. This work was also supported by the Faculty of Mechanical and Automotive Engineering Technology, UMP.

References

1. Farrer J (2004) The alloy tree: a guide to low-alloy steels, stainless steels, and nickel-base alloys. CRC Press
2. Viswanathan R, Bakker W (2001) Materials for ultrasupercritical coal power plants—boiler materials: Part 1. J Mater Eng Perform 10(1):81–95
3. Perez J, Stewart CM (2019) Assessment of the theta projection model for interpolating creep deformation. In: Turbo Expo: Power for Land, Sea, and Air, 2019, vol 58684. American Society of Mechanical Engineers, p V07AT31A002
4. An L (2015) The development of advanced creep constitutive equations for high chromium alloy steel (P91) at transition stress range. University of Huddersfield

5. Haque MS (2019) Modification of the MPC omega model to predict primary and tertiary creep. In: Pressure vessels and piping conference, vol 58974. American Society of Mechanical Engineers, p V06AT06A032
6. Prager M (2000) The omega method–an engineering approach to life assessment. J Pressure Vessel Technol 122(3):273–280
7. Anderson TL, Osage DA (2000) API 579: a comprehensive fitness-for-service guide. Int J Press Vessels Pip 77(14–15):953–963
8. Prager M (1995) Development of the MPC omega method for life assessment in the creep range
9. E139-11 A (2011) Standard test methods for conducting creep, creep-rupture, and stress-rupture tests of metallic materials. In: American society for testing and materials, Philadelphia, Pa, vol 19103
10. Bråthe L, Josefson L (1979) Estimation of Norton-Bailey parameters from creep rupture data. Metal Sci 13(12):660–664
11. Shrestha T, Basirat M, Charit I, Potirniche GP, Rink KK (2013) Creep rupture behavior of Grade 91 steel. Mater Sci Eng A 565:382–391
12. Choudhary B (2013) Tertiary creep behaviour of 9Cr–1Mo ferritic steel. Mater Sci Eng A 585:1–9
13. Ab Razak NA (2018) Creep and creep fatigue interaction in new and service exposed P91 steel
14. Chen H, Zhu G, Gong J (2015) Creep life prediction for P91/12Cr1MoV dissimilar joint based on the omega method. Procedia Eng 130:1143–1147

Effect of Cavity Thickness on Copper Alloy Corrosion Resistance

M. Nasuha, M. M. Rashidi, A. Hadi, Z. Shayfull, and T. M. Sheng

Abstract This investigation inspects on the effect of cavity's thickness during metal casting process on the corrosion resistance of copper alloy product. As the thickness increases, the cooling rate becomes higher due to higher latent heat available in the thicker and larger cavity volume. As such the quantity of Dendritic Arm Spacing, DAS and its Secondary, SDAS per unit area becomes higher. This eventually results in better properties such as the higher hardness and good corrosion resistance because its correlation with DAS and SDAS distribution in the microstructure. The copper alloy used in this project is Nickel Aluminium Bronze (NAB) alloy which consists of elements such as the copper, aluminium, iron, nickel and manganese. Sand casting process has been used and the NAB alloys have been fabricated according to the ASTM B148 UNS 95,800 standards with the usage of 1.1% degassing agent. A range of product cavity's thickness have been fabricated for gating system and proper machining processes have been carried out to prepare the specimens for the immersion test. The specimens were immersed in sea water for a period of 17 weeks and changes in the specimen mass and pH and TDS values of the sea water used was measured. The data analysis revealed that the specimens were not corroded yet for the period of 17 weeks as there are not much changes in the specimen mass. The pH and TDS values are showing changes but these changes are very small comparatively.

Keywords Nickel Aluminum Bronze · Metal casting · Corrosion · Dendritic Arm Spacing

M. Nasuha · M. M. Rashidi (✉) · A. Hadi · T. M. Sheng
Faculty of Mechanical and Automotive Engineering Technology, Universiti Malaysia Pahang, Kampus Pekan, 26600 Pahang, Malaysia
e-mail: mrashidi@ump.edu.my

A. Hadi
e-mail: asnul@ump.edu.my

Z. Shayfull
School of Manufacturing Technology, Universiti Malaysia Perlis, KampusTetap Pauh Utara, 02600 Arau, Perlis, Malaysia
e-mail: shayfull@unimap.edu.my

M. Y. Ismail et al. (eds.), *Technological Advancement in Mechanical and Automotive Engineering*, Lecture Notes in Mechanical Engineering,
https://doi.org/10.1007/978-981-19-1457-7_25

1 Introduction

Among non-ferrous material, copper alloy has become choice of engineering application due to its relative better casting flow ability during process, good strength-cost ratio and corrosion resistance. These advantage enable copper alloy to be employ in oil, gas and selective corrosive environment and application. Nowadays, copper alloy has become relevant in related industries due to their reliability at corrosive environment as competitive as other engineering material.

Generally copper (Cu) alloy with Nickel (Ni) and aluminium (Al) as alloying element has become sole focus of previous studies over the past few years. An alloy which employed at minimum 10% (Ni and Al elements) of total mass is knowingly as Nickel Aluminum Bronze (NAB). So, the as-cast copper alloy with these mixtures will produce different phases in its microstructure after solidification stage. It is due to the influence of both Ni and Al alloying contained in the solidified composition [1].

As such, solidified phase will have dissimilar effect, distribution and uniformity on its microstructure. This effect is also affected largely due to its casting cavity thickness. Previous studies showed there is potential of different alloy's microstructure at different cavity's thickness application on casting cavity due to relative different cooling time [2]. It happens because the thinner the casting cavity part, the shorter the cooling time due to less latent heat availability. Restriction of cooling time will limit the growth of solidified microstructure. Theoretically, decreasing cavity's thickness will produce smaller grain size relative to bigger casting cavity part. So, it is worth to find out whether the effect of large grain size to NAB is equivalent to smaller one (thinner section) especially its corrosion resistance behaviour [3]. Until now there is not much reports published on the effect of corrosion on different NAB's product thickness.

This study aims to examine the effect of grain size on the microstructure, properties and corrosion behaviour of different NAB's product thickness. The product was dimensioned as customized connecting rod. This customized connecting rod is having different thickness at its different section. The connecting rod selection as a product to be analysed is due to its application as part of local engine assembly in nearby location. It is expected that the connecting rod can be used for maritime engine which exposed to a seawater condition. The effect of seawater condition on fabricated part (connecting rod) is one of the important things need to observe [4]. So an immersion corrosion test was conducted to determine material resistance to an aggressive and aqueous environment [5]. Macro and micro structure of the samples were examined by optical and SEM microscope with EDS capability, correspondingly.

2 Methodology

In this study, a connecting rod product of NAB material were produced by commercial foundry facilities (shown in Fig. 1). The experimental mixture elements of Cu were added with composition of Ni, Al and other traces elements of total 10% at minimum [6]. All elements were added in the induction furnace and stirred. After reaching melting state around 1100 °C ± 20, the mixture was ready to be tap. A treatment by degassing agent to minimize gas content in the molten metal was practiced before the mixture was ready to be pour. Then, the mixture was poured into green sand moulds containing connecting rod cavity's dimension. Chemical composition of all kinds of experimental Cu, Ni, Al and traces elements are list in Table 1.

Corrosion samples measuring 15 × 10 × 6 mm were cut from allocated location in Fig. 1 and prepared by fine grinding using silicon carbide papers up to 2400 grit [7]. Immersion testing was conducted in seawater for 17 weeks [8]. Connecting rod was set to be act as an electrode and seawater was chosen as its electrolyte. The

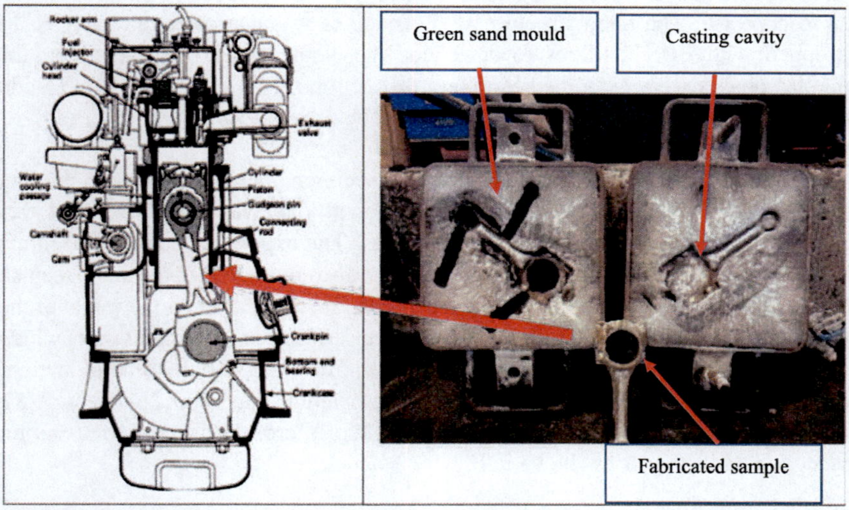

Fig. 1 Schematic representation of marine diesel engine's connecting rod with a metal casting part from fabricated mould

Table 1 Composition of copper alloy mixture

Element	Composition (wt. %) ± 0.1%
Copper	81.3
Aluminium	9.0
Nickel	4.5
Iron	4.0
Manganese	1.2

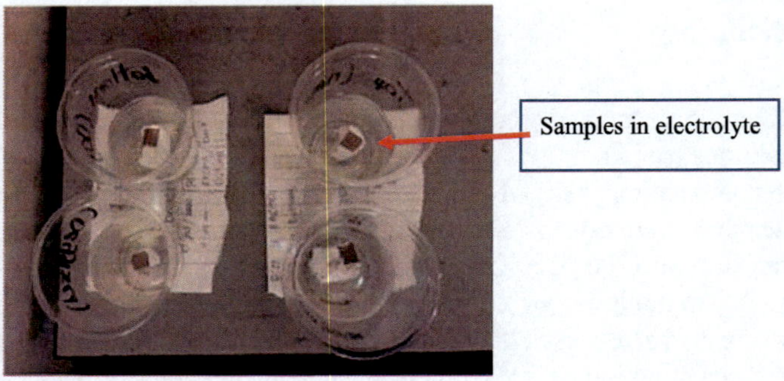

Fig. 2 First day of immersion test

seawater was taken from local beach known as Pantai Sepat, Kuantan which having a clear water in colour. The temperature of the test environment was set at 26–27 °C by using an incubator.

Microstructure of the as-cast samples was etched and observed using optical microscope [9]. The microstructure of the samples was characterized using SEM equipped with EDS facilities to determine the elements distribution. Later, the samples phases were examined by X-ray diffraction (XRD) analysis [10]. To add further, an immersed specimen was also examined by SEM/EDS to determine its microstructure and its morphology.

During the filtration process of immersion specimen, there is a corrosion product accumulated on the filtrating paper. The corrosion product was visible by naked eyes and qualitatively compiled as greenish particles. The exposed specimen (10 mm^2) is then blown by air to and ultra-sonic cleaner to remove the deposited greenish particles. The aim of this process is to gain an accurate result of the final weight of the specimen. The specimen weight might be decrease as compare to initial mass which called as mass reduction comparison mechanism. In order to determine reduction, the final weight of the specimen need to be clean, examined, measured and record [8]. Later, the corrosion product was dried and brought to microstructure and composition inspection and analysis as shown in Figs. 2, 3 and 4.

3 Results and Discussion

After immersion period of time, the seawater electrolyte colour is visually changed from a clear water to a greenish in colour. Figures 2 and 3 indicates the colour changed during immersion test process. The changes in seawater is due to the reaction between seawater (electrolyte) and connecting rod specimen (electrode). The greenish seawater is then filtrated by filtration paper in order to obtain the corrosion product.

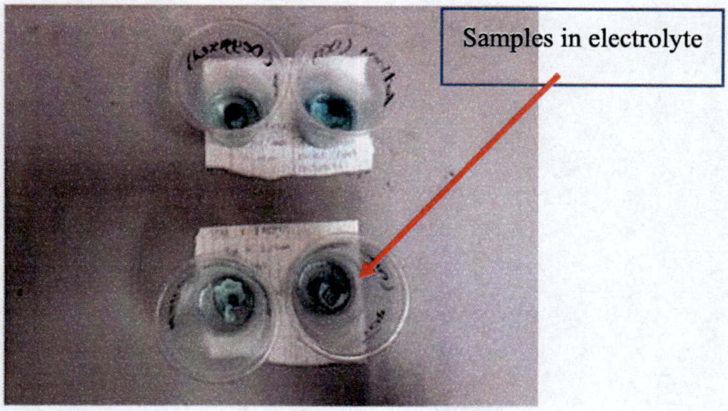

Fig. 3 After 17 weeks of immersion test

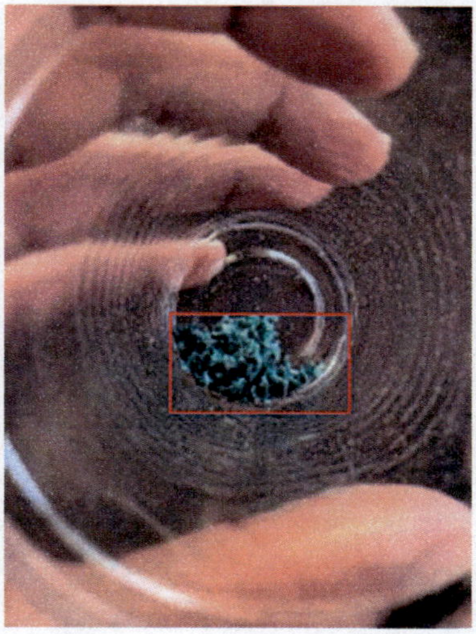

Fig. 4 Precipitated elements covering the immersed specimen

Observation on EDX result in Fig. 5 shows the corrosion product composition that precipitated on the filtrated paper and alloy specimen. There are several elements discovered which are chlorine, oxygen, carbon, magnesium and also copper. Chlorine, oxygen and also magnesium was probably come from the seawater since the electrolyte that was used in immersion is natural seawater. Since there was a reaction between seawater and immersed alloy (electrode), there is an elements of copper alloy presence as a precipitated corrosion product. There was corrosion occurred since there was a difference in mass weight reading during measurement. So, copper alloy and its elements dissolved in the seawater electrolyte. Based on the given Fig. 5, chlorine have a major percentage of weight which is 32.27% since the chlorine is come from natural seawater. In the mean-time, the percentage of copper is only 3.04% which means, a smaller number of connecting rod materials was dissolved in

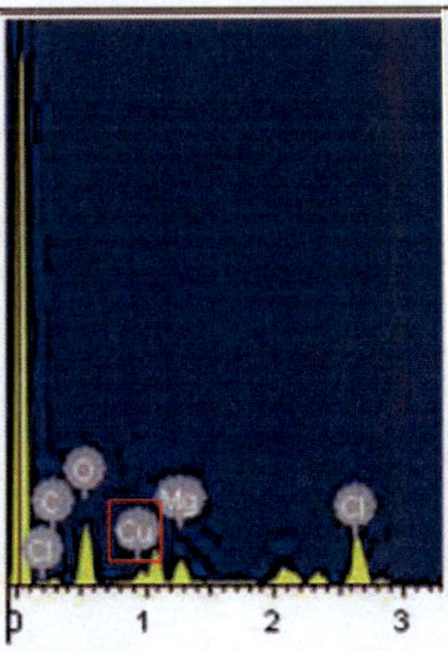

Element	Weight%	Atomic%
C K	14.99	24.25
O K	41.87	50.86
Mg K	7.83	6.26
Cl K	32.27	17.69
Cu L	3.04	0.93
Totals	100.00	

Fig. 5 EDX result of immersion test product composition

seawater electrolyte. The data that was generated during immersion test was translated in terms of mass loss value. The aim of observing mass loss is to determine the corrosion rate each of the part. The details of mass loss result can be found in Table 2.

Figure 6 illustrate the trend fluctuations of the specimen mass. The mass measurement was slightly decreased from its original value in 123 days. Both thicker and thinner part showing mass reduction after exposed to seawater electrolyte. This circumstance is related to the TDS value which will be discussed later.

Generally, the rate of corrosion is a speed measurement at which any metal deteriorates in a specific environment. The rate of corrosion is dependent upon the environmental condition as well as the type and condition of metals (Wharton et al., 2005). The rate of corrosion can be calculated by using Eq. 1 as stated by Baboian [8].

$$CorrosionRate, CR = \frac{k \times W}{A \times T \times D} \tag{1}$$

$$CR = \frac{k \times \left(m_{initial} - m_{final}\right)}{\pi r^2 \times T \times D} + c \tag{2}$$

Table 2 Mass loss of corrosion testing (immersion)

Sample	Initial mass (g)	Final mass (g)	Immersed period (Hour)
Top, 1.5 cm thickness	15.031	14.996	2952
Bottom, 0.75 cm thickness	35.808	35.786	2952

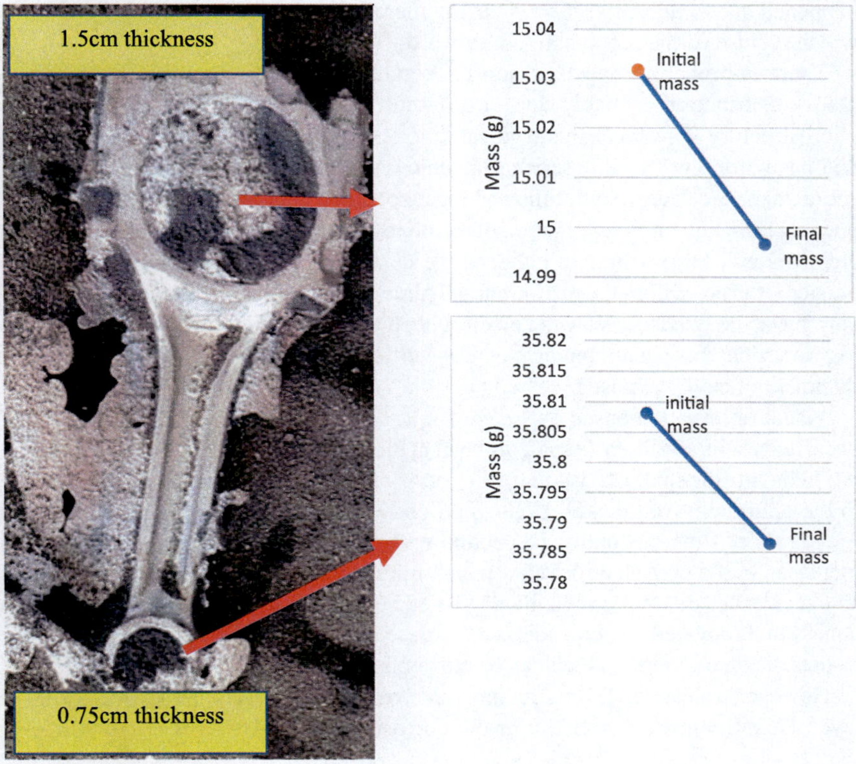

Fig. 6 Mass loss of selected component thickness

Equation 1 can be express as shown in Eq. 2. The parameters involved are as follows; the k is the unit conversion constant with a constant value of 3.45×103, W is the mass loss during immersion test in g, A is the surface area that uncoated and exposed to seawater in cm^2, T is the period of immersion test in hours while D is the density of NAB which is 8 g/cm^3. Mass loss value was taken to calculate the corrosion rate based on immersion test technique.

Table 3 shows the detail of the corrosion rate. Both section part was corroded but at a different level of corrosion rate. Thicker dimension at the top sided of the part is corroded more (0.0054 mm per year, mmpy) compare with a thinner part (0.0032 mmpy). Most probably this occurrence (higher corrosion rate at a thicker part)

Table 3 Corrosion rate of different section of connecting rod

Corrosion rate (mmpy)	
Top, 1.5 cm thickness	0.0054
Bottom, 0.75 cm thickness	0.0032

happened due to thicker section is having more irregular microstructure distribution which require further explanation afterward.

There is probably several reason for condition stated previously. Firstly, irregularity distribution probably due to different level of galvanic series in the alloy microstructure. This mechanism galvanic series happened because different element will have different galvanic series. Galvanic series occurred whenever alloy solidification take place at different time and localize as different phases. So, they created a phenomenon known as segregation of elements during solidification phase. Different elements will have solidified either at the centre of the grain, grain boundary and outside of grain will having different galvanic series. So, when this irregular distribution was exposed to a seawater electrolyte, it will contribute towards different mass loss. As such, this phenomenon encourage different corrosion rate between different thickness of cavity exposed.

Visual observation based on FESEM micrograph in Figs. 7 and 8 shows several occurrences. Firstly, there is spall off layer at the upper side of the corroded specimen. It is believing the spall-off was due to copper-rich alpha phase which is initially anodic to the-aluminum-iron-nickel- rich *kappa* phase. They corroded preferentially for a lower rate of time. Secondly, inspection reveal a possibility of hydrogen level did increases in the corroded location, which influenced the drop of pH level. Thirdly, an occurrences of copper metallic deposited appeared and detected in the corrosion zone which covered the corrode zone. Consequently, it is difficult to examine the re-deposition of copper which can or may exist on the corroded surfaces.

However, a clear corrosion structure (sponge-like) is reveal underneath of spall-off layer. This structure occurred during this corrosion period. After this structure, a layer

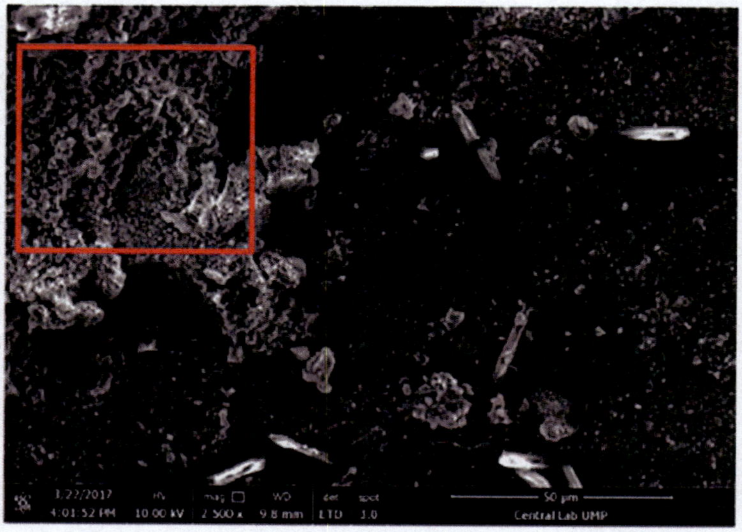

Fig. 7 Micrograph of corroded area. Red rectangle was magnified to 4000x

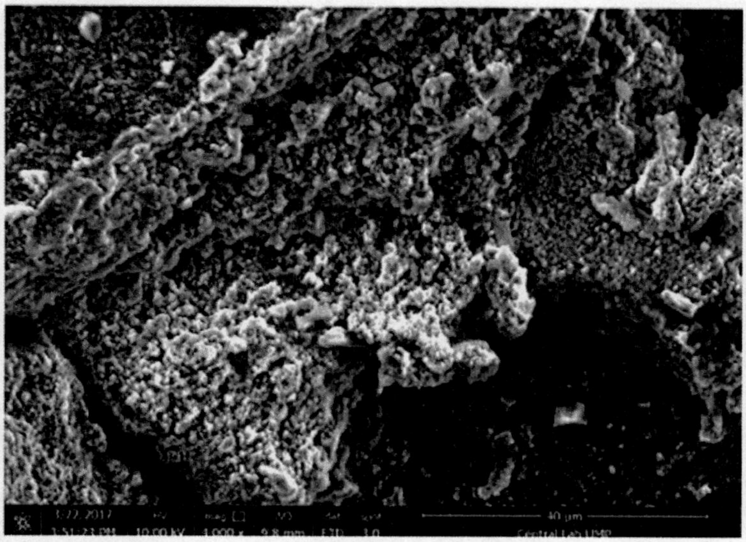

Fig. 8 Micrograph of magnified corrosion film

of alloy was discovered intact without much corroded found. So, there is possibility that corroded film which spall-off improve protection against further corrosion attack. The protection is a prospect attributed to a decrease of anodic dissolution reaction. This phenomenal reduce the ionic transportation across the oxide layer. Thus it will encourage a decrease the cathodic reaction rate around the oxide layer.

Figure 9 shows an analysis of XRD measurement. There is four different type of corroded phase detected namely, Copper Nickel Aluminum Hydroxide Carbonate Hydrate, Oxonium Aluminum Sulfate, Manganese (2+)1H-benzimidazole-5,6-dicarboxcylate and Iron Cloride Hydroxide.

Observation shows the highest peak mostly are the phases that contain an aluminium elements. Electromechanical or galvanic series analysis shows it is in accordance to XRD result. The sequence of corroded elements is started from Al, Fe, Mn, Ni and lastly is Cu element. This is because Al is located at the top of the series and more active compare to the other four materials. The most active material in electromechanical or galvanic series will corrode faster. This indicate that Al, Fe, Mn will corrode faster compare to Ni and Cu and justifies the existence of Al, Fe and Mn elements at corroded phase. These phase appeared before Ni and Cu corroded phase exist.

The value of power hydrogen, pH is one of the parameter affected the seawater electrolyte due to the corrosion mechanism. Table 4 show the result of pH value at different section of the connecting rod. pH value is recorded during initial and final days of testing period. Generally, all of the seawater decreasing in pH value.

The initial value for all seawater sample is nearly to natural value (pH of 7) but still alkaline which is 7.78. The reading was taken during the first day of immersion

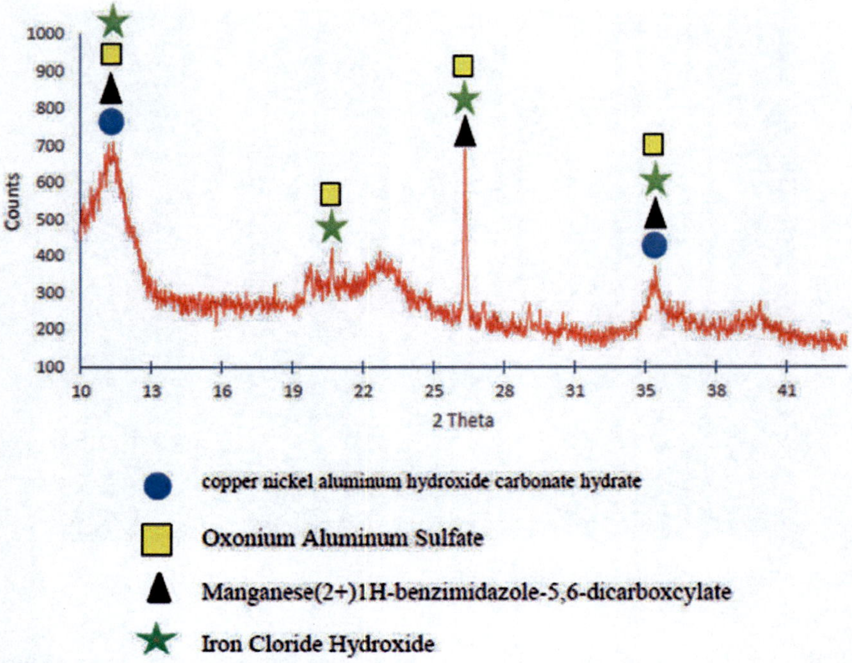

Fig. 9 Corrosion film XRD analysis

Table 4 pH value of seawater electrolyte at different section of connecting rod after 123 days' immersion test

pH value (initial pH 7.78)	
Top, 1.5 cm thickness	6.36
Bottom, 0.75 cm thickness	6.64

test. After that it has been immersed for 17 weeks in control environment (foundry laboratory), it has been found that the pH value is decreasing to more acidic seawater which is 6.34 and 6.64 respectively. This occurrence happened due to the reaction between seawater as electrolyte and the copper alloy as it electrode.

Thicker component will have produced lower pH value (6.36) compare to the thinner section (6.64) at the end of the test session (17 weeks). Thicker part has more reaction to seawater since the pH value (almost 18.2% dropped) is more acidic compare to the thinner section (14.6% dropped). Its mean there is more corrosion occurred which plummeted the pH value at the lowest (Fig. 10).

Another data which was taken concurrent with mass loss and pH value is Total Dissolve Solid, TDS in seawater electrolyte suspension. Table 5 show the result of TDS value at different section of connecting rod. TDS value is recorded during initial and final days of testing period in the electrolyte suspension (seawater).

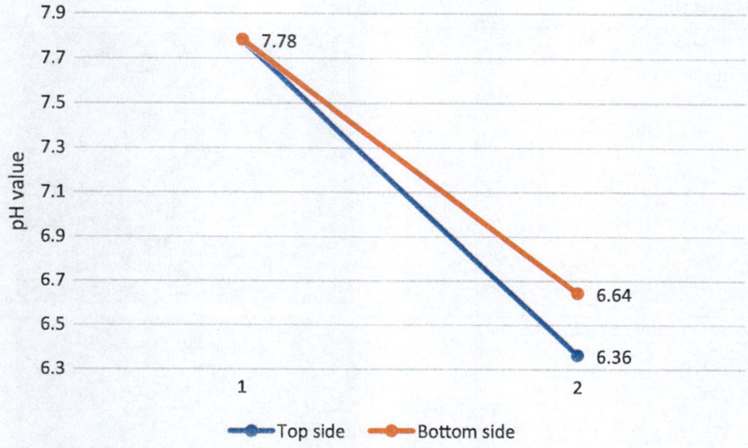

Fig. 10 pH value of initial and final day of immersion test

Table 5 TDS value of seawater electrolyte at different section of connecting rod after 123 days' immersion test

TDS value (initial is 0)	
Top, 1.5 cm thickness	653
Bottom, 0.75 cm thickness	638

The initial TDS value is zero during the first day of immersion test. After 17 weeks, it has been discovered the TDS value is increasing to a suspended solid state which having 653 mg/L (top side) and 638 mg/L (bottom side) respectively. There is a presence of greenish suspended solid in the electrolyte seawater as shown in Fig. 4. This suspended solid existed due to reaction of electrolyte and electrode in the period of time. Relatively there is not much different between top and bottom sided of the component after exposed to corrosive environment. The range is only 15 mg/L which translated to 2% differences.

Further visual inspection reveal the grain and dendritic structure is corroded as reveal in Fig. 11. The dendritic structure was eroded and the visibility of DAS and its secondary was reduced. Before immersion, the dendritic structure geometry is slick and smooth. After the immersion test, the dendritic structure is visibly crystal-look alike. The dendritic shape formation was almost disappeared. A close up micrograph examined at the SDAS distribution show there is a void surrounding the dendritic structure which was exposed by the corrosive electrolyte. So, both thicker and thinner section is affected to seawater in this corrosive environment condition.

It can be said that, thinner section is more resistant to seawater because their DAS and SDAS structure formation is lesser during solidification. As such, there is minimum DAS and SDAS in contact with electrolyte during corrosive environment. So less DAS and SDAS contributed towards element segregation factor. Thus there is slight chances for elements segregation to effect the electrochemical series

Structure	Micrograph
Dendritic arm spacing DAS (after the immersion test)	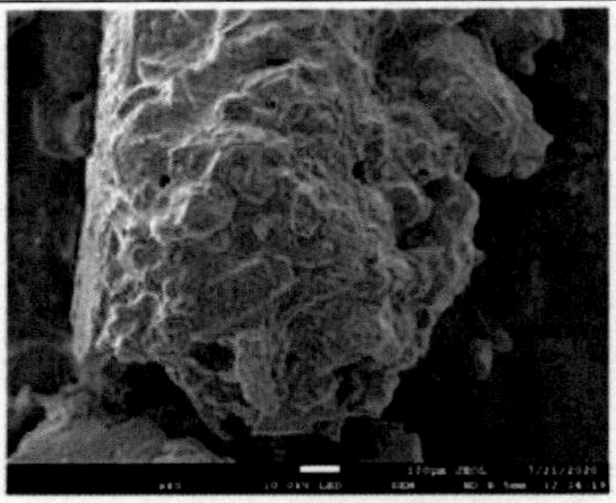
Secondary dendritic SDAS (after the immersion test)	
Secondary dendritic SDAS (after the immersion test)	
Dendritic structure (before the immersion test)	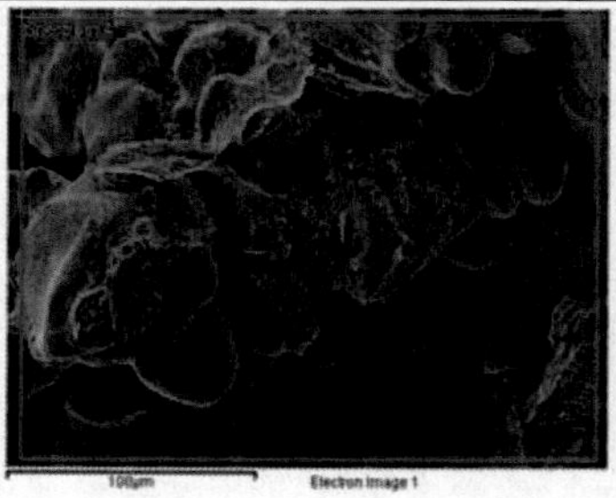

Fig. 4.11 Dendritic structure after and before immersion test

of all phases existed. As a result, less porosity could be trapped inside the alloys microstructure. Finally, less porosity could probably harm the overall corrosion mechanism.

4 Conclusion

In this study, the effect of cavity thickness parameter on corroded NAB alloy was examined and the following result were obtained:

1. Increasing the cavity thickness generates more latent heat inside the cavity. As such there is more time for cooling time available. Then, there was more time for grain to be grew. Bigger grain will have more cross section of its grain boundary. So, there is higher chances of grain boundary to be penetrated by corrosion electrolyte during immersion test. Comparatively, corrosion will be occurred more alongside thicker side of grain boundary relative to a thinner side of cavity.
2. Increasing the cavity thickness permit more corrosion rate compare to a thinner part in this NAB's product cavity.

Acknowledgements The authors would like to express his sincere thanks to Automotive Excellent Centre (AEC), Universiti Malaysia Pahang (UMP) and Ministry of Higher Education (Malaysia) for providing laboratory facilities and financial assistance under project no. RDU190128 @ FRGS/1/2018/TK03/UMP/02/12.

References

1. Nabiałek MG, Dośpiał MJ, Szota M, Pietrusiewicz P, Jdryka J (2011) Investigation of the thermal and magnetic properties of Fe 61Co10Zr2 Fe61Co10Zr 2.5Hf2.5Me2W2B20 (Me = Y, Nb, W, Ti, Mo, Ni) bulk amorphous alloys obtained by an induction suction method. J Alloys Compd 509(7):3382–3386
2. Nabiałek M, Pietrusiewicz P, Dośpiał M, Szota M, Gondro J, Gruszka K, Dobrzańska-Danikiewicz A, Walters S, Bukowska A (2015) Influence of the cooling speed on the soft magnetic and mechanical properties of Fe61Co10Y8W1B20 amorphous alloy. J Alloys Compd 615(S1):S56-S60
3. Barik RC, Wharton JA, Wood RJ, Tan KS, Stokes KR (2005) Erosion and erosion-corrosion performance of cast and thermally sprayed nickel-aluminum bronze. Wear 259(1–6):230–242
4. Gopal G, Suresh Kumar L, Vijaya Bashkar Reddy K, Uma Maheshwara Rao M, Srinivasulu G (2017) Analysis of piston, connecting rod and crank shaft assembly. Mater Today Proc 4(8):7810–7819
5. Jirapure SC, Borade AB (2014) Naval corrosion - causes and prevention. J Eng Sci Res Technol 3(7):1–6
6. Meigh HJ (2000) Cast and wrought aluminium bronze: properties, processes and structure, copper development association/IOM communication, pp 3–4

7. Eslami HS, Hasbullah IM, Hassan J (2011) Effect of forging process on microstructure mechanical and corrosion properties of biodegradable Mg-1Ca alloy. Mater Des 603(32):2596
8. Baboian R (2002) Corrosion test and standards application and interpretation. Second edn. West Conshohocken: ASTM International
9. Nabialek M, Dospial M, Szota M, Olszewski J, Walters S (2011) Manufacturing of the bulk amorphous Fe61Co10Zr 2+xHf3-xW2Y2B20 alloys (where x = 1, 2, 3) their microstructure, magnetic and mechanical properties. J Alloy Compd 509(SUPPL. 1):S155–S160
10. Sandu AV, Baltatu MS, Nabialek M, Savin A, Vizureanu P (2019) Characterization and mechanical proprieties of new TiMo alloys used for medical applications. Materials 12(18), art. no. 2973

pH-Responsive Nanocapsules as Smart Coating for Corrosion Protection: A Review

N. S. Mohamed, J. Alias, N. A. Johari, and A. Zanurin

Abstract A new generation of smart coating contain nano capsule that actively respond to changes in the local environment has triggered great interest among material researchers in the field of anti-corrosion. Many researchers reported that the preparation of pH-responsive nano capsules is usually applying unfriendly chemicals, a complex procedure and time-consuming, which remains as a great challenge for effective corrosion protections. This review presents, the achievement during the last 10 years in the field of pH-responsive nano-capsules, the formulation technique of such nano-capsule, testing and evaluation of the pH-responsive nano-capsule.

Keywords pH response · Nano-capsules · Self-healing

1 Introduction

Nowadays, mankind uses huge number of various types of metal and its alloys in daily life such carbon steel, aluminum alloys, magnesium alloys and many more [1]. These materials are widely used by industrial due to the excellent mechanical properties. However, the problem of corrosion that affect the global economic and generating risks associated with safety and environment [2]. Metal usually reacts with corrosive media in the surrounding environment and resulting in damage and deterioration [3]. In order to control or mitigate the corrosion process, researcher had introduced various method such as organic coatings, corrosion inhibitor and hybrid protective coating [4].

Coatings can provide effective protection against environmental factors such as ultraviolet (UV), heat, oxygen, moisture, and ions in the short term [5]. Mechanical attack during operation can badly damage the barrier effect of the coating. If the damage is not visible and cannot be repaired, the corrosive medium will easily permeate the coating, causing coating failure [6]. At present, damage coating need

N. S. Mohamed (✉) · J. Alias · N. A. Johari · A. Zanurin
Department of Mechanical Engineering, College of Engineering, University Malaysia Pahang, Gambang, 26300 Kuantan, Pahang, Malaysia
e-mail: shuhadamohd11@gmail.com

M. Y. Ismail et al. (eds.), *Technological Advancement in Mechanical and Automotive Engineering*, Lecture Notes in Mechanical Engineering,
https://doi.org/10.1007/978-981-19-1457-7_26

restoration and replacement which are highly costing and troublesome [7]. To solve this problem self-healing coating or smart coating is introduce and emerging which act as inhibiting the corrosion reaction at the damage [8]. Corrosion inhibitors of various sorts have been successfully encapsulated to respond to a trigger mechanism such as pH changes.

There are two categories of self-healing or smart coating anti corrosion based on different working mechanism that is by self-healing polymer materials such as linseed oil [9], bisphenol A [2], hexamethylene dissocynate [10]. When a micro/nano-capsule containing self-healing agents is inserted into a polymer coating, the active agents are released into the damaged region and lead to the interior part of the damage via chemical reaction, filling up the cracks and repairing the damage [11]. The second categories are corrosion inhibitor such as ethyl cellulose [12], mesoporous polydopamine [13], chitosan [14], etc.

In smart coating technologies there are two ways have been developed for introducing corrosion inhibitors into the self-healing coatings, which is direct doping and encapsulation of molecules into the micro/nano-capsules that are distributed evenly in the coatings [11, 14]. Direct doping, on the other hand, has the problem of inhibitor development and unwanted interactions between the matrix and the inhibitors, which degrade the barrier function of the coatings [15].

On the other hand, insertion of corrosion inhibitors into micro/nano-capsules can overcome the weakness of direct doping [16]. Some advance work has been performed in the organic coatings such as benzotriazole (BTA)-loaded nano-capsules [17], crosslinked chitosan based nanofilms [18], thiourea containing [12], polydopamine nano-capsules [19], etc. The smart capsules can quickly respond to the stimuli such as pH response and release corrosion inhibitors immediately, thus forming a protective film on the damage metal or alloys surface (see Fig. 1). In this paper, we evaluate representative studies on the development of smart anti-corrosion coatings technologies based on micro/nano-capsules that respond to pH stimuli over the last ten years.

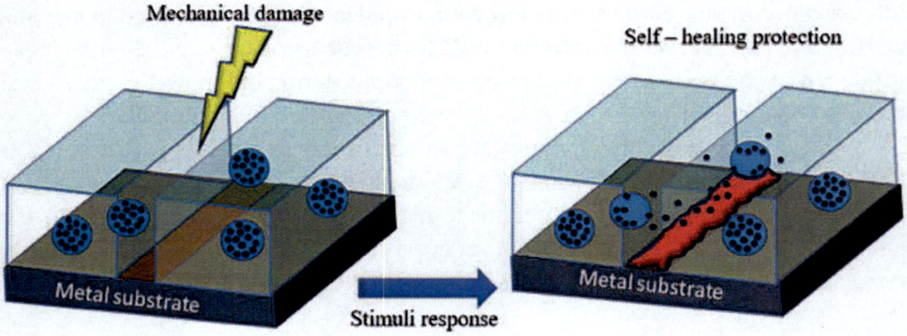

Fig. 1 Smart coating self-healing processes with embedded stimuli-responsive micro/nano-capsules [17]

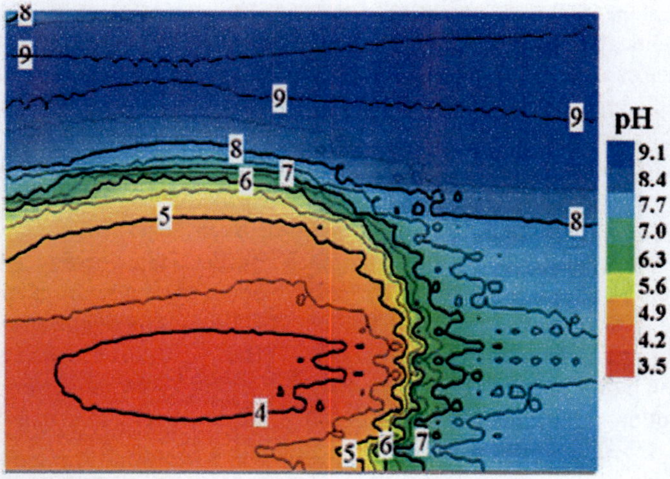

Fig. 2 Corroded galvanized steel pH distribution map [20]

2 Fundamentals of pH Responsive Micro/nano-Capsule

Coating that uses pH gradient as corrosion inhibition triggers and smart self-healing are one of the most effective in use. In an electrochemical system, the pH value might differ dramatically between anodic and cathodic locations. For example, during the corrosion process, the pH gradient of galvanic steel between anodic and cathodic sites can be as large as six units. (see Fig. 2) [20]. The pH map shows acidity on the lower left side and alkalinization on the upper left and the right side of the scanned area. The change in pH is due to the hydrolysis of Zn^{2+} ions resulting from the reduction of pH in the micro anodic area, while an increase in pH in the micro cathodic area due to oxygen reduction. The changes of pH value near to the corrosive zones were considered as stimulus to control the release of corrosion inhibitor [21]. As a result, it's critical to identify and characterize the numerous pH gradients that exist in electrochemical systems so that the most effective pH sensitive trigger for corrosion prevention may be selected [22].

3 Formulation of pH Responsive Micro/nano Capsule

Recently smart coatings for corrosion protection usually consist of organic/plasma electrolyte oxidation (PEO) coating and dispersed micro/nano-capsule for storing inhibitors [23]. There are several types of micro/nano container including mesoporous silica [23], poly-urea formaldehyde [3, 9], halloysite nanotubes [24], zeolites nanoparticles [5] have been developed. Inhibitor carrier or nano-capsule must be able to transport and discharge the inhibitors at the right time and at the right speed to fully exploit the inhibitors' potential so that they can work effectively [25]. Most of the researcher attempting to formulate the pH responsive capsule. Lowering the pH destroys the capsule shell, causing the inhibitors to be released [26].

A sensitive polyelectrolyte molecule can be used to make capsules that can use release inhibitors sensibly in response to pH variations in coating imperfections [11]. Increases and reductions in local pH can deprotonate/protonate polymer molecules, resulting in increased inhibitor release, or accelerate hydrolytic destruction, resulting in increased swelling [5]. The composite nano capsules consisted of nano particles core and polyelectrolyte multilayer film are the typical example of the method used. Other method used for the control release of inhibitors like the formulation of pH responsive layer on the encapsulated inhibitors [27]; for example, the encapsulation of inhibitors incorporated halloysite nanotubes (HNT) by a pH sensitive shell [16] like tetraethyl orthosilicate (TEOS) as shown in (see Fig. 3). When lowering the pH of the solution, it will induce the TEOS film to be destroy, by allowing the stored inhibitors or benzotriazole (BTA) to enter the solution [25].

In a study by [3] employed a layer-by-layer approach to apply multilayers of chitosan (CHI) and alginate (ALG) to a dual functional poly capsule, and benzotriazoles (BTA) were inserted in its layer as shown in (see Fig. 4). The study found that when chitosan protonation occurs in an acidic environment (low pH), its layer disintegrates due to electrostatic repulsion, allowing H^+ diffusion to reach the alginate layer. The high interaction power between alginate and H^+ causes the BTA to be released in this scenario. Another research method discovered by [28] shows the employment of the end stopper at the end of HNT. This approach demonstrated that lowering the pH, increases the dissolution of the end stopper, resulting in the release of inhibitors. Moreover, another study conducted by [29, 30], shows that the incorporation of L-histidine (L-His) in the halloysite nanotubes (HNTs)—reduce graphene

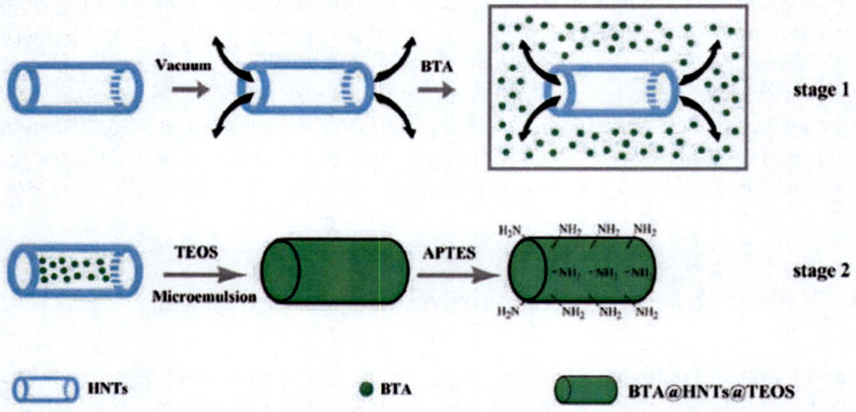

Fig. 3 The functional nanocontainers' assembly technique (BTA@HNTs@TEOS-APTES) [16]

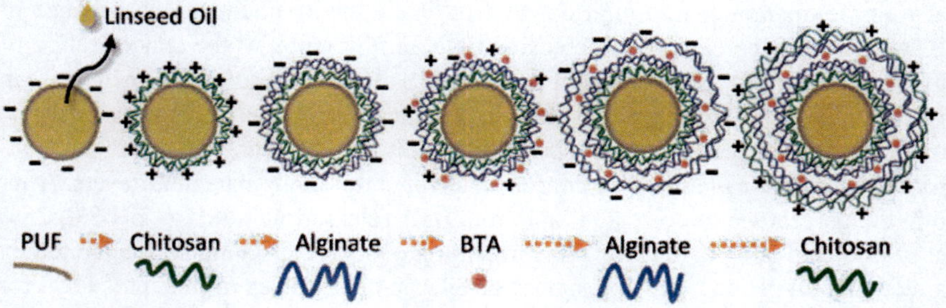

Fig. 4 Schematic diagram shows the assembling of layer- by- layer, CHI/ALG/BTA/ALG/CHI on linseed oil (LO)/poly urea-formaldehyde (PUF) microcapsules [3]

Table 1 The inhibitor-incorporated carrier's pH sensitivity for corrosion protection of metals and alloys was examined in a number of research

Substrate	Carrier	Inhibitor	References
Magnesium alloy	Mesoporous silica	2-mercatobenzothiazole	[23]
Carbon steel	Poly(urea–formaldehyde)	benzothiazole	[3]
Carbon steel	Hollow mesoporous silica	Thiourea	[12]
Cold rolled steel	Melamine formaldehyde	3-mercatobenzothiazole	[32]
X52 steel	Halloysite nanotubes (HNTs)	1-butyl-3-methylimidazolium chloride [Bmim][Cl]	[16]
Carbon steel	Halloysite nanotubes (HNT)	Imidazole and dodecylamine	[33]
Carbon steel	Halloysite nanotubes (HNT)	Dodecylamine	[34]
Mild steel sheet	Mesoporous silica	Molybdate	[35]

oxide (rGO) system indicates the inhibitor was released in a low pH (acidic) environment rather than a basic environment to produce the production of identical or opposite charges because of pH fluctuations.

The electrostatic repulsion between inhibitors and MS was identified as the explanation for excessive inhibitor release in neutral and alkaline environments in the mesoporous (MS) carrier [26]. However, in another study, the acidic environment was found to have a higher rate of inhibitor release in MS [31]. Many studies have looked at the formulation and influence of pH release inhibitors for enhancing the corrosion resistance of metals and alloys, and some of them are included in Table 1.

4 Evaluation of pH Responsive Micro/nano Capsule

UV–vis spectroscopy was used to determine the rate of inhibitor release from the micro/nano-capsule and to determine a suitable pH environment. He [25] perform his experiment in deionize water (pH 1, 3, 5) to monitor the release rate of the inhibitor at room temperature, he found out that in acidic environment the absorbance of BTA gradually increase with increasing time. Furthermore, as the concentration of hydrogen ions rises, the breakdown rate of TEOS film rises, resulting in an increase in BTA release rate. In another study [3], on duo functional poly (urea-formaldehydes) microcapsule, the percentage of corrosion inhibitors BTA higher in the acidic environment due to the pH sensitive conformation of chitosan/alginate multilayers. After 6 h, the percentage of corrosion inhibitors BTA released was 68% at pH 3.0, 25% at pH 5.0, and 0.9% at pH 9.0. The corrosion process is substantially accelerated by a reduction in pH, this active response of BTA in an acidic environment is ideal for use as anticorrosive coatings for metals and alloys.

In a study conducted by [27], in the production of poly(urea–formaldehyde) microcapsules linseed oil packed with BTA using a layer-by-layer process, the amount of BTA released from the multilayer capsule in NaCl solutions (0.05 mol/L) at pH 5.7 is 56%, whereas at pH 9.0 it is about 43%. The release rate of BTA in pH 5.7 was 13.78×10^{-3}%/min while in pH 9.0 was 3.16×10^{-3}%/min. The multilayer microcapsules released about 40–42% of BTA in the first 40 min of immersion in NaCl solution. This high initial release rate is expected because the polyelectrolytes poly(ethylenimine) (PEI) and poly(styrene sulfonate) (PSS) are sensitive to changes in environmental pH and respond quickly, allowing for rapid release of the corrosion inhibitor, which is ideal for active corrosion prevention. The microcapsule works efficiently in the acidic media.

In another study about the pH response of the combination of halloysite nanotubes (HNTs) with graphene oxide (GO), and the BTA as loaded inhibitors were carried out in 3.5 wt.% NaCl solutions at different (pH = 3, 7, 11). The release amount of BTA from (see Fig. 5) display stage 1; an initial fast release and stage 2; a gradual release. The higher release rate of BTA were observed at pH 3 followed by pH 11. The gradual release rate during second stage is good for durable corrosion inhibitors. The results of BTA@HNT-GO nano composites has the pH response ability and can be applied to smart coating [4].

In addition, the self-release of corrosion inhibitors from loaded HNT in 0.1 M NaCl at various pH values was investigated (2, 5, 7 and 11). As a display, the loaded HNT displayed the absorbance in all acidic, neutral, and basic conditions in (see Fig. 6). The highest absorbance intensity was observed in acidic media at pH 2. The absorbance intensity decreasing with increasing pH value. Although self-release inhibitors were detected at all pH values, the efficient release of inhibitors was reported in acidic medium rather than neutral and basic media [33].

5 Conclusion

This review covers recent research on pH responsive micro/nano-capsules for smart self-healing anticorrosion coatings for metals and alloys. The formulation and evaluation of polymeric or inorganic micro/nano-capsules with multifunctional provide insight into the development of a new generation of smart coatings that can self-heal

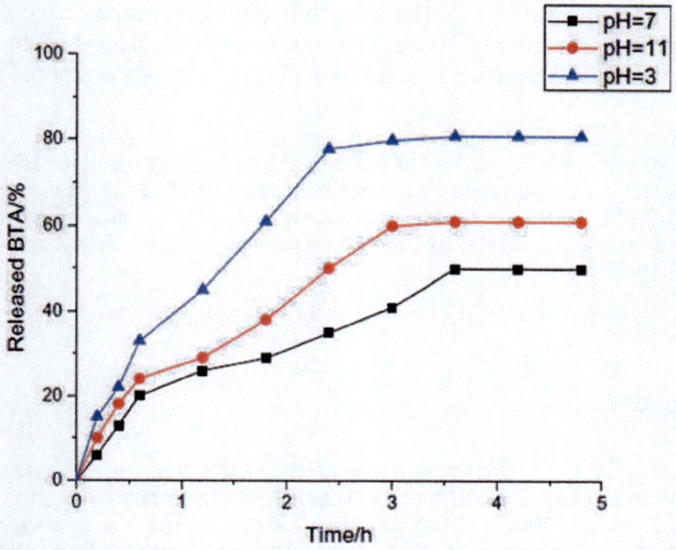

Fig. 5 BTA@HNTs-GO nanocomposite UV–vis spectroscopy [4]

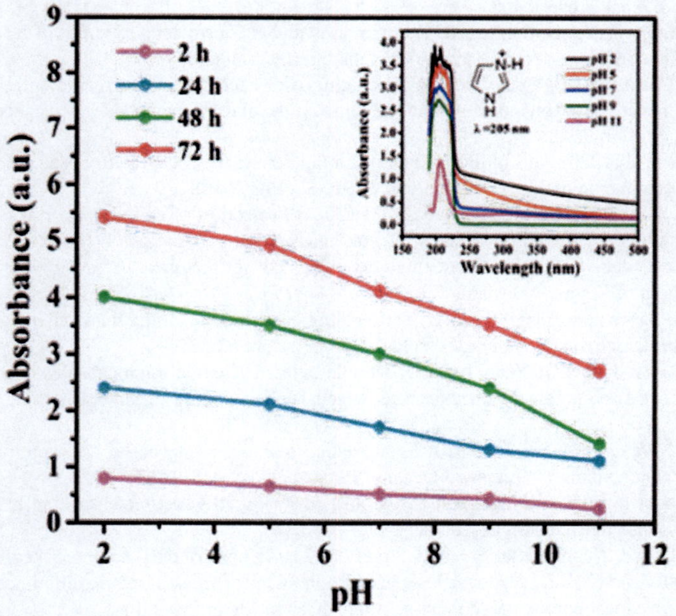

Fig. 6 UV−vis spectroscopy release behaviours of inhibitors from loaded HNTs immersed in different pH solutions for 2, 24, 48, and 72 h [33]

in response to stimuli like pH. The use of pH responsive stimuli nano capsule to respond instantly to external stimuli and in acidic environment is highly suitable for active corrosion protection of metal and alloys in early phases of the corrosion process.

Acknowledgements The authors would like to acknowledge the Ministry of Higher Education for providing financial support under Fundamental Research Grant Scheme (FRGS) No. FRGS/1/2019/TK05/UMP/02/5 (University reference RDU1901128) and University Malaysia Pahang for laboratory facilities as well as additional financial support under Internal Research grant PGRS210373.

References

1. Ou B, Wang Y, Lu Y (2021) A review on fundamentals and strategy of epoxy-resin-based anticorrosive coating materials. Polymer-Plastics Technol Mater 60(6):601–625
2. Zhang C, Wang H, Zhou Q (2018) Preparation and characterization of microcapsules based self-healing coatings containing epoxy ester as healing agent. Prog Org Coat 125:403–410
3. da Cunha ABM, Leal DA, Santos LRL, Riegel-Vidotti IC, Marino CEB (2020) pH-sensitive microcapsules based on biopolymers for active corrosion protection of carbon steel at different pH. Surf Coat Technol 402
4. Chen L, Yu Z, Yin D, Cao K (2021) Preparation and anticorrosion properties of BTA@HNTs-GO nanocomposite smart coatings. Compos Interfaces 28(1):1–16
5. Zhou C et al (2020) Epoxy composite coating with excellent anticorrosion and self-healing performances based on multifunctional zeolitic imidazolate framework derived nanocontainers. Chem Eng J 385
6. Hao X et al (2020) Antifouling and antibacterial behaviors of capsaicin-based pH responsive smart coatings in marine environments. Mater Sci Eng C 108
7. Ress J, Martin U, Bosch J, Bastidas DM (2021) Protection of carbon steel rebars by epoxy coating with smart environmentally friendly microcapsules. Coatings 11(20):1–12
8. Xie ZH, Shan S (2018) Nanocontainers-enhanced self-healing Ni coating for corrosion protection of Mg alloy. J Mater Sci 53(5):3744–3755
9. Siva T, Sathiyanarayanan S (2015) Self healing coatings containing dual active agent loaded urea formaldehyde (UF) microcapsules. Prog Org Coat 82:57–67
10. Sun D, An J, Wu G, Yang J (2015) Double-layered reactive microcapsules with excellent thermal and non-polar solvent resistance for self-healing coatings. J Mater Chem A, 3(8):4435–4444
11. Cai H, Wang P, Zhang D (2020) Smart anticorrosion coating based on stimuli-responsive micro/nanocontainer: a review. J Oceanol Limnol 38(4):1045–1063
12. Wang J et al (2021) A CO2-responsive anti-corrosion ethyl cellulose coating based on the pH-response mechanism. Corros Sci 180
13. Ni X, Gao Y, Zhang X, Lei Y, Sun G, You B (2021) An eco-friendly smart self-healing coating with NIR and pH dual-responsive superhydrophobic properties based on biomimetic stimuli-responsive mesoporous polydopamine microspheres. Chem Eng J 406
14. Wang W et al (2018) pH-responsive Capsaicin@chitosan nanocapsules for antibiofouling in marine applications. Polymer 158:223–230
15. Koochaki MS, Khorasani SN, Neisiany RE, Ashrafi A, Trasatti SP, Magni M (2021) A highly responsive healing agent for the autonomous repair of anti-corrosion coatings on wet surfaces. In operando assessment of the self-healing process. J Mater Sci 56(2):1794–1813

16. Yang YJ, Yaakob SM, Rabat NE, Shamsuddin MR, Man Z (2020) Release kinetics study and anti-corrosion behaviour of a pH-responsive ionic liquid-loaded halloysite nanotube-doped epoxy coating. RSC Adv 10(22):13174–13184
17. Li J, Zeng H, Luo JL (2021) Probing the corrosion resistance of a smart electroless Ni-P composite coating embedded with pH-responsive corrosion inhibitor-loaded nanocapsules. Chem Eng J 421
18. Miras J, Liu C, Blomberg E, Thormann E, Vílchez S, Esquena J (2021) pH-responsive chitosan nanofilms crosslinked with genipin. Colloids Surf A Physicochemical Eng Aspects 616
19. Liang Y et al (2016) Facile synthesis of smart nanocontainers as key components for construction of self-healing coating with superhydrophobic surfaces. Nanoscale Res Lett 11(1)
20. Taryba M et al (2011) The combined use of scanning vibrating electrode technique and micro-potentiometry to assess the self-repair processes in defects on 'smart' coatings applied to galvanized steel. Electrochimica Acta 56(12):4475–4488
21. Fu J et al (2013) Acid and alkaline dual stimuli-responsive mechanized hollow mesoporous silica nanoparticles as smart nanocontainers for intelligent anticorrosion coatings. ACS Nano 7(12):11397–11408
22. Manoj A, Ramachandran R, Menezes PL (2020) Self-healing and superhydrophobic coatings for corrosion inhibition and protection. Int J Adv Manuf Technol 106(5–6):2119–2131
23. Ouyang Y, Li LX, Xie ZH, Tang L, Wang F, Zhong CJ A self-healing coating based on facile pH-responsive nanocontainers for corrosion protection of magnesium alloy. J Magnes Alloys
24. Sun M, Yerokhin A, Bychkova MY, Shtansky DV, Levashov EA, Matthews A (2020) Self-healing plasma electrolytic oxidation coatings doped with benzotriazole loaded halloysite nanotubes on AM50 magnesium alloy. Corros Sci 111:753–769
25. He Y, Xu W, Tang R, Zhang C, Yang Q (2015) PH-Responsive nanovalves based on encapsulated halloysite for the controlled release of a corrosion inhibitor in epoxy coating. RSC Adv 5(110):90609–90620
26. Shahini MH, Taheri N, Mohammadloo HE, Ramezanzadeh B (2021) A comprehensive overview of nano and micro carriers aiming at curtailing corrosion progression. J Taiwan Inst Chem Eng. Taiwan Institute of Chemical Engineers
27. Leal DA, Riegel-Vidotti IC, Ferreira MGS, Marino CEB (2018) Smart coating based on double stimuli-responsive microcapsules containing linseed oil and benzotriazole for active corrosion protection. Corros Sci 130:56–63
28. Xing X, Wang J, Li Q, Hu W, Yuan J (2018) A novel acid-responsive HNTs-based corrosion inhibitor for protection of carbon steel. Colloids Surf A Physicochemical Eng Aspects 553:295–304
29. Xing X et al (2020) A novel method to control the release rate of halloysite encapsulated Na2MoO4 with Ca2+ and corrosion resistance for Q235 steel. Appl Clay Sci 188
30. Jia Y, Qiu T, Guo L, Ye J, He L, Li X (2020) Preparation of pH responsive smart nanocontainer via inclusion of inhibitor in graphene/halloysite nanotubes and its application in intelligent anticorrosion protection. Appl Surf Sci 504
31. Falcón JM, Otubo LM, Aoki IV (2016) Highly ordered mesoporous silica loaded with dodecylamine for smart anticorrosion coatings. Surf Coat Technol 303(Part B):319–329
32. Alizadeh M, Sarabi AA (2019) pH-Responsive MFPTT microcapsules containing dimethyl sulfoxide: preparation, characterization and tuning the release behavior of microcapsule contents. Prog Org Coat 134:78–90
33. Khan A, Hassanein A, Habib S, Nawaz M, Shakoor RA, Kahraman R (2020) Hybrid Halloysite Nanotubes as Smart Carriers for Corrosion Protection. ACS Appl Mater Interfaces 12(33):37571–37584
34. Falcón JM, Sawczen T, Aoki IV (2015) Dodecylamine-loaded halloysite nanocontainers for active anticorrosion coatings. Front Mater 2
35. Saremi M, Yeganeh M (2014) Application of mesoporous silica nanocontainers as smart host of corrosion inhibitor in polypyrrole coatings. Corros Sci 86:159–170

Effect of Artificial Aging on the Microstructure and Mechanical Properties of AJ62 Magnesium Alloys

M. I. M. Ramli, M. A. F. Romzi, J. Alias, and N. A. Abd Razak

Abstract Magnesium (Mg) alloys with addition of strontium, such as AJ62, are die-castable and have good creep resistance at high temperatures. The formation of compounds containing strontium are extremely useful to increase the elevated temperature properties and strontium is an effective grain refiner for magnesium alloys. In this study, the effect of artificial aging of AJ62 magnesium alloys on the microstructure development and mechanical properties was studied. The alloys were solution heat treated, cooled to room temperature, before artificially aged to the room temperature. Different heating times and cooling conditions in the aging parameters were used. Aging time and cooling conditions affect the dendrite and eutectic phases refinement and lead to varying mechanical properties. Refinement of dendrite size enhanced the ductile properties.

Keywords Magnesium alloys · Artificial aging · Microstructure · Ductility

1 Introduction

Magnesium (Mg) is the lightest metal structure, which is desirable in the automotive industry for energy efficiency and system performance [1–4]. Automobile engines that are stronger, lighter, and more efficient are characteristics of a car component that could have a wide range of uses [5–9].

AJ62 magnesium alloys are die-castable alloys with Sr element which offers excellent creep performance. These magnesium alloys were significantly reduced by 24% of weight compared with the equivalent aluminum block [10]. Strontium additions are known to successfully suppress $Mg_{17}Al_{12}$ and could be replaced by a

M. I. M. Ramli (✉)
Faculty of Mechanical and Automotive Engineering Technology, Universiti Malaysia Pahang, 26600 Pekan, Pahang, Malaysia
e-mail: idzwan@ump.edu.my

M. A. F. Romzi · J. Alias · N. A. A. Razak
Department of Mechanical Engineering, College of Engineering, Universiti Malaysia Pahang, 26300 Gambang, Pahang, Malaysia

M. Y. Ismail et al. (eds.), *Technological Advancement in Mechanical and Automotive Engineering*, Lecture Notes in Mechanical Engineering,
https://doi.org/10.1007/978-981-19-1457-7_27

stable Al-Sr compound at the interdendritic regions and allowing the $Mg_{17}Al_{12}$ to precipitate continuously at elevated temperatures. This situation further hinders the dislocation movement and increases the strength of magnesium [11, 12].

Many studies are being undertaken to improve the performance of AJ62 magnesium alloy, particularly the creep performance for automotive applications [7–10]. The heat treatment process is another aspect that may improve the performance of magnesium alloys [13, 14]. Precipitation hardening and strengthening of magnesium alloy by alloying and heating is the heat treatment mechanism to strengthen the strength of Mg alloys and inimize their mechanical asymmetry [15]. A supersaturated Mg alloy can be aged under different conditions, leads to various metastable and stable precipitates formation [16, 17]. Thus, artificial aging can be applied for tailoring the mechanical properties of the alloys. A supersaturated state is generally thermodynamically unstable, thus the aging process may start at room temperature (RT) immediately after solution treatment and quenching. There is, however, a scarcity of information on the effect of different conditions of precipitation strengthening on the AJ62 magnesium alloy.

As a result, the purpose of this research was to determine the effect of artificial aging on the microstructure and mechanical properties of AJ62 magnesium alloys. The strength and ductility of the alloy, as well as the microstructure, including the grain and phase distribution, were evaluated using several ageing parameters.

2 Experimental

The AJ62 material (nominal compositions in wt.%: 6Al–2Sr–balance Mg) used in this study was extracted from a BMW N52 engine block [18]. The chemical composition of the alloy was performed by spectroscopy. The precipitation strengthening process of the sample was conducted according to Table 1, which represents the temperature experienced in the actual condition. Three samples were taken into the heating and cooling process. The aging process differed in heating time and cooling conditions to simulate the temperature experience in application. The first sample, termed S1, was undergoing a solution heat treatment (SHT) process at 150 °C, quenched in

Table 1 Sample arrangement for heat treatment processes including solution heat treatment, quenching and artificial aging

Sample No	Solution heat treatment temperature (°C)	Cooling condition	Aging parameter		
			Temperature (°C)	Heating time (hr)	Cooling condition
S1	150	Quench	120	1	Slow
S2	150	Quench	120	2	Slow
S3	150	Quench	120	1	Rapid

water, and then artificially aged at 120 °C for 1 h, before finally slow cooling to room temperature. The second sample, S2, underwent a SHT condition at 150 °C, quenched in water, and artificially aged at 120 °C for 2 h, then slowly cooled to room temperature. The S3 sample was also given the SHT condition at 150 °C, quenched in water, and artificially aged at 120 °C for 1 h, but then rapidly cooled in water.

Tensile test samples were extracted from the specimen and prepared according to the ASTM E8 subsize sample (Fig. 1). The samples were subjected to tensile tests at room temperature with a crosshead speed of 1 mm min^{-1}. Three samples were made for each composition to minimize errors. The microstructure samples were mounted, polished and chemically etched before going through microstructure observation. The microstructure observation was carried out by an Olympus optical microscope

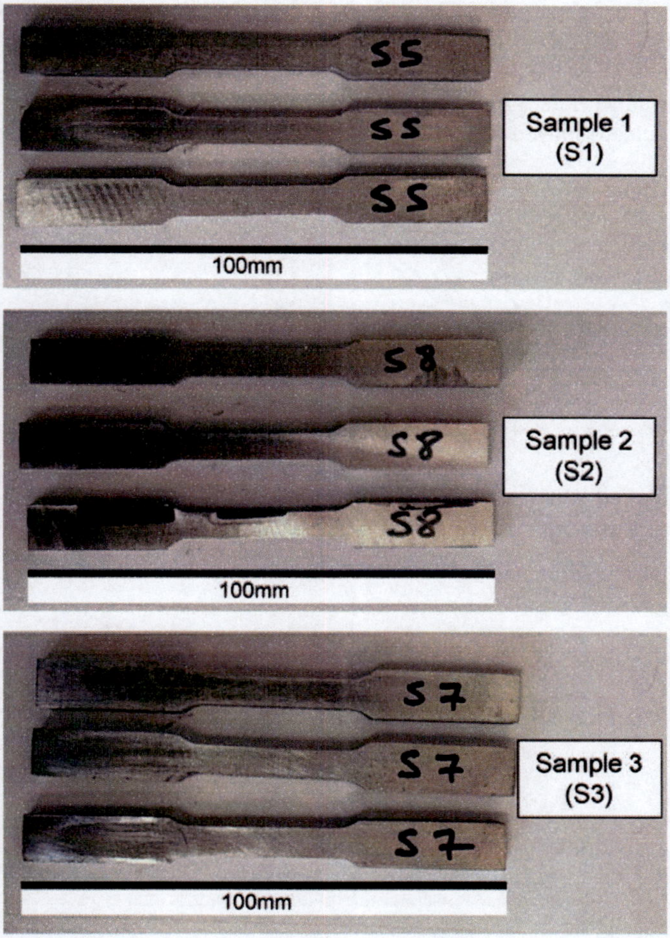

Fig. 1 Subsize tensile samples prepared in accordance with ASTM E8

and a Hitachi TM3030 Plus scanning electron microscope (SEM). Elemental analysis was conducted by energy dispersive x-ray spectroscopy (EDX) to identify the chemical composition of phases of the samples.

3 Result and Discussion

3.1 Microstructure Observation

Figures 2a–c show the optical micrographs of the microstructure of Mg–6Al–2Sr alloys with different heat treatment processes. Figure 2a shows the microstructure of S1 consisted of primary Mg, coarse eutectic $Mg_{17}Al_{12}Sr$ and Al_4Sr lamellar particles formed at the dendritic boundaries and isolated Al-Sr tiny particles, which was confirmed based on the weight percent composition result. The eutectic phases were turned into much finer after a longer artificial aging duration (2 h) as observed in the S2 sample, displayed in Fig. 2b. There is not much difference in the particle size

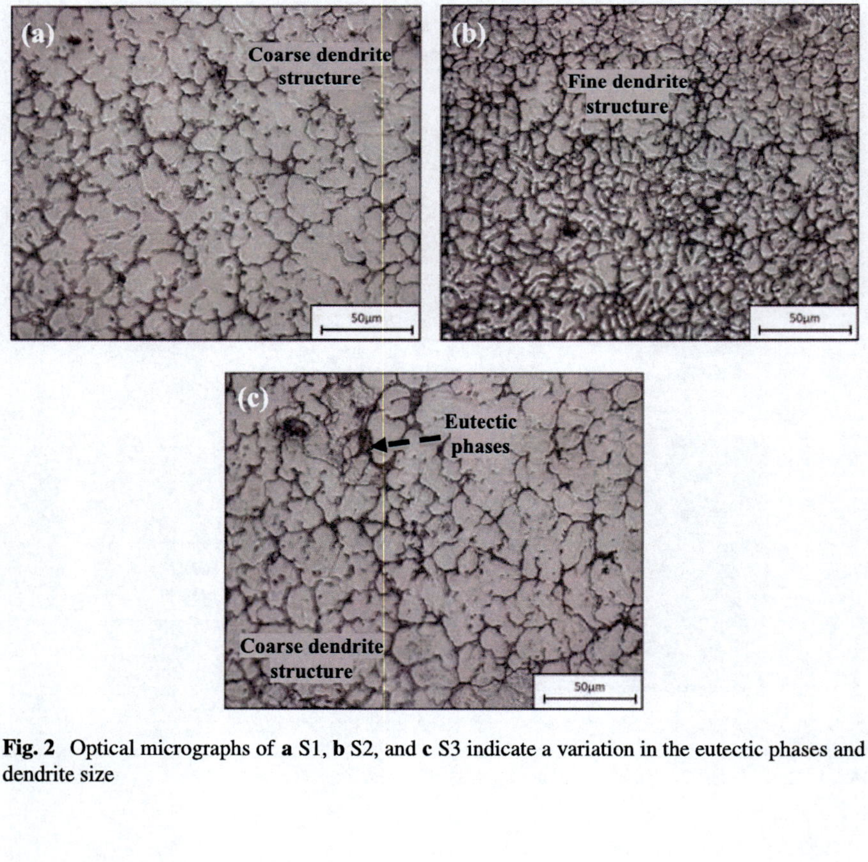

Fig. 2 Optical micrographs of **a** S1, **b** S2, and **c** S3 indicate a variation in the eutectic phases and dendrite size

and distribution after the sample is artificially aged for 1 h and quenched. However, a slightly smaller dendrite size can be observed in the S3 sample as compared to S1 (Fig. 2c).

In this alloy, it was observed that the addition of 2 wt.% of Strontium, Sr, and 6 wt.% of Al, formed a variety of Mg–Al-Sr eutectic and A1-Sr phases. Figure 4a shows the EDX spectrum 1 of a tiny Al-Sr which can be seen in the bright particle (Fig. 3a), with a confirmed composition of 33 wt.% Sr and 3 wt.% Al. This particle was randomly distributed near the eutectic phases and can be observed in the sample, S1. The eutectic phases of the alloy are found in a lamellar structure, which is confirmed to consist of Al_4Sr phases, as well as in a bulk greyish appearance ($Mg_{17}Al_{12}Sr$), and these phases are found in a network distribution along the dendrite boundaries [19]. The coarse size of the eutectic phases is depicted in S1 and S3, while smaller eutectic phases are revealed in S2. The presence of Al_4Sr suggests that there is an insufficient amount of Sr to bind all the aluminium atoms [20, 21], as confirmed by only 1.489 wt.% of Sr in the lamellar phases, based on the wt.% of the phases in spectrum 2 (Fig. 4b). At the same time, however, Sr reacted exclusively with Mg in the bulk greyish particles, as indicated in EDX spectrum 3 (Fig. 4c). Hence, it is very likely that the existence of manganese, Mn led to the presence of Al_8Mn_5 intermetallic particles in the alloy [22–24], however, it has not been easily observed during microstructure observation.

3.2 *Mechanical Properties of AJ62 Alloys*

Figure 5 displays the stress-strain curves of AJ62 magnesium alloys with different heat treatment processes. It was observed that the dendrite refinement was achieved in S2 with a longer artificial aging duration. The ductility of the sample also increased, although the tensile strength of the sample was slightly lower than S1 (Table 2). The addition of strontium is known to improve mainly the elevated temperature properties, especially creep resistance. Although trace strontium addition increased the primary creep strain of the Mg–5Al alloy, the minimum creep rate decreased with the increase in strontium content [25]. Sr also effectively increases the average grain size, which improves their plastic anisotropy [12, 26, 27]. At the same time, the mechanical performance of the Mg-Al-Sr alloy system somehow, may not significantly affected by heat exposure. However, the optimal improvement in mechanical strength can be achieved by modification of the second phase particles [28].

In addition to the well-known dispersion hardening from isolated second-phase particles, intermetallic phases with extensive spatial connections tend to have an extra strengthening impact [21]. Another effect is the lower Al solute concentration in the α-Mg matrix, which results in less solid solution strengthening [14, 29, 30].

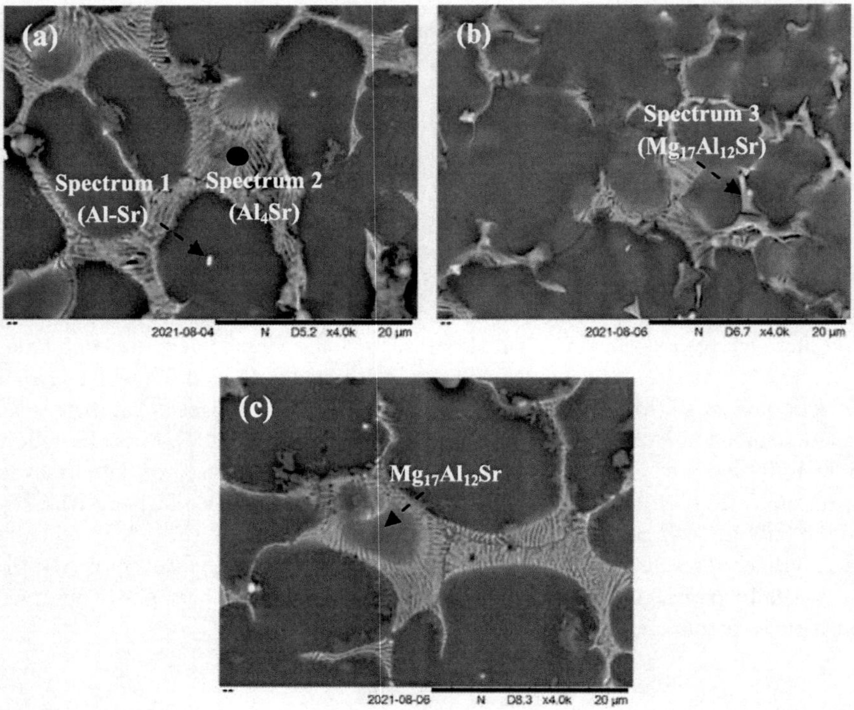

Fig. 3 SEM images of (a) S1, (b) S2, and (c) S3 indicate variation in second phase particle, $Mg_{17}Al_{12}Sr$ size

4 Conclusion

This study generally presented the effect of different artificial aging processes on the microstructure and mechanical properties of the AJ62 magnesium alloys extracted from the BMW N52 engine block. Based on the results, different artificial aging durations resulted in dendrite refinement and eutectic phase size variation. The result indicated that:

- Dendrite refinement and eutectic phase size reduction occurred when artificial aging was conducted for a longer duration.
- The tensile strength of the alloy that has been artificially aged for 1 h and slow cooled (S1) is slightly higher than the samples that have been artificially aged for 2 h and slow cooled (S2), at approximately 2% increment, and artificially aged for 1 h and quenched (S3) (10% increment).
- The ductility of S2 sample has increased due to the refinement of the dendrite size and eutectic phases.

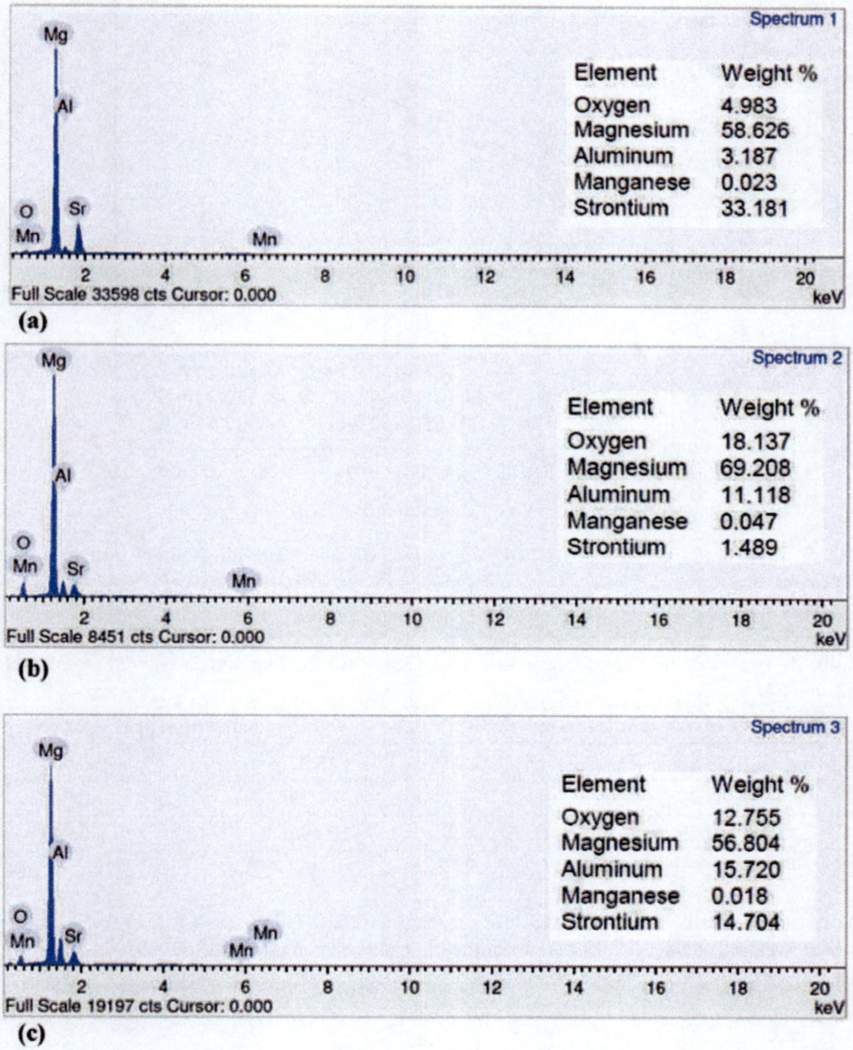

Fig. 4 EDX compositional analysis of second phases and intermetallic phases formed in samples S1, S2 and S3: **a** Spectrum 1, **b** Spectrum 2, and **c** Spectrum 3 of Fig. 3

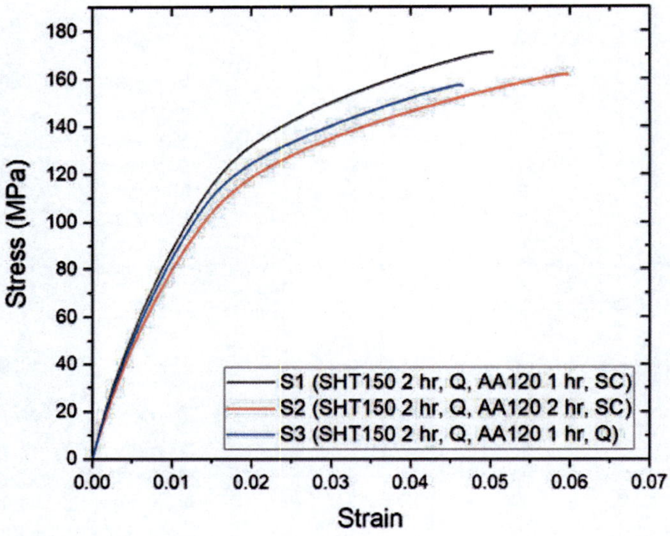

Fig. 5 Stress-strain curves of AJ62 Mg alloys (S1, S2 and S3) with different heat treatment processes

Table 2 The result of ultimate tensile strength (UTS), yield modulus and ductility of samples S1, S2, and S3

Sample	Ultimate tensile strength (UTS) (MPa)	Young's modulus (GPa)	Strain
S1	173.10	115.02	0.051
S2	169.79	101.99	0.060
S3	157.39	107.96	0.046

Acknowledgements The authors would like to thank the Universiti Malaysia Pahang for laboratory facilities as well as financial support under the Internal Research grant PGRS2003191.

References

1. Ramli MIM, Romzi MAF, Alias J (2021) Effect of surface condition on the corrosion behaviour of AZ31 magnesium alloy. Mater Today Proc. https://doi.org/10.1016/j.matpr.2021.02.213
2. Alias J, Harun WSW, Ayu HM (2019) A Review on the preparation of magnesium-based alloys prepared by powder metallurgy and the evolution of microstructure and mechanical properties. Key Eng Mater 796:3–10. https://doi.org/10.4028/www.scientific.net/KEM.796.3
3. Mayyas A, Omar M, Hayajneh M, Mayyas AR (2018) Vehicle's lightweight design versus electrification from life cycle assessment perspective. J Clean Prod 167:687–701. https://doi.org/10.1016/j.jclepro.2017.08.145
4. Powell BR, Krajewski PE, Luo AA (2021) Magnesium alloys for lightweight powertrains and automotive structures. LTD

5. Annamalai S, Periyakgoundar S, Gunasekaran S (2019) Magnesium alloys: a review of applications. Mater Tehnol 53(6):881–890. https://doi.org/10.17222/mit.2019.065
6. Joost WJ, Krajewski PE (2016) Towards magnesium alloys for high-volume automotive applications. SMM. https://doi.org/10.1016/j.scriptamat.2016.07.035
7. Xu T, Yang Y, Peng X, Song J, Pan F (2019) Overview of advancement and development trend on magnesium alloy. J Magnes Alloy 7(3):536–544. https://doi.org/10.1016/j.jma.2019.08.001
8. Viswanadhapalli B, Bupesh Raja VK (2020) Application of magnesium alloys in automotive industry-a review, pp 519–531
9. Song J, She J, Chen D, Pan F (2020) Latest research advances on magnesium and magnesium alloys worldwide. J Magnes Alloy 8(1):1–41. https://doi.org/10.1016/j.jma.2020.02.003
10. Fischersworring-Bunk A, Landerl C, Fent A, Wolf J (2005) The New BMW Inline Six-cylinder Composite Mg/Al Crankcase, 49–58, 2005.
11. Li Y et al (2019) Evolution of β Mg17Al12 in MgAlZnAg alloy over time. Mater Sci Eng A 754:470–478. https://doi.org/10.1016/j.msea.2019.03.094
12. Xu N, Xie W, Dang C, Peng X, Yan Y, Chen Y (2011) Microstructures and mechanical properties of AJ62 magnesium alloy with Y and Nd elements. Mater Sci Forum 686:253–259. https://doi.org/10.4028/www.scientific.net/MSF.686.253
13. Jiang MG, Yan H, Chen RS (2015) Enhanced mechanical properties due to grain refinement and texture modification in an AZ61 Mg alloy processed by small strain impact forging. Mater Sci Eng A 621:204–211. https://doi.org/10.1016/j.msea.2014.10.075
14. Kubásek J, Vojtěch D, Martínek M (2013) Structural characteristics and elevated temperature mechanical properties of AJ62 Mg alloy. Mater Charact 86:270–282. https://doi.org/10.1016/j.matchar.2013.09.018
15. Bhattacharjee T, Suh B-C, Sasaki TT, Ohkubo T, Kim NJ, Hono K (2014) High strength and formable Mg–6.2Zn–0.5Zr–0.2Ca alloy sheet processed by twin roll casting. Mater Sci Eng A 609:154–160. https://doi.org/10.1016/j.msea.2014.04.058
16. Yan Y et al (2017) Effects of Zn concentration and heat treatment on the microstructure, mechanical properties and corrosion behavior of as-extruded Mg-Zn alloys produced by powder metallurgy. J Alloys Compd 693:1277–1289. https://doi.org/10.1016/j.jallcom.2016.10.017
17. Liu Z, Xin R, Wu X, Liu D, Liu Q (2018) Improvement in the strength of friction-stir-welded ZK60 alloys via post-weld compression and aging treatment. Mater Sci Eng A. https://doi.org/10.1016/j.msea.2017.12.008
18. Pekguleryuz M, Celikin M (2010) Creep resistance in magnesium alloys. Int Mater Rev 55(4):197–217. https://doi.org/10.1179/095066010X12646898728327
19. Vojtech D, Kubásek J, Voderová M, Šerák J (2014) Structure and mechanical properties of mg-based alloys for elevated temperature applications. Manuf Technol 14(3):482–487. https://doi.org/10.21062/ujep/x.2014/a/1213-2489/mt/14/3/482
20. Yang Y, Peng X, Wen H, Wei G, Xie W, Lavernia EJ (2014) Microstructure and mechanical behavior of Mg–10Li–3Al–2.5Sr alloy. Mater Sci Eng A 611:1–8. https://doi.org/10.1016/j.msea.2014.05.064
21. L'Espérance G, Plamondon P, Kunst M, Fischersworring-Bunk A (2010) Characterization of intermetallics in Mg-Al-Sr AJ62 alloys. Intermetallics 18(1):1–7. https://doi.org/10.1016/j.intermet.2009.05.017
22. Ren L, Zhang X, Fang L, Hu H (2019) Fine grain microstructure of Mg-Al-Sr alloy solidified under high pressures. Int J Mater Mech Manuf 7(5):214–218. https://doi.org/10.18178/ijmmm.2019.7.5.462
23. Wang Y, Xia M, Fan Z, Zhou X, Thompson GE (2010) The effect of Al8Mn5 intermetallic particles on grain size of as-cast Mg–Al–Zn AZ91D alloy. Intermetallics 18(8):1683–1689. https://doi.org/10.1016/j.intermet.2010.05.004
24. Alias J, Zhou X, Das S, El-Fakir O, Thompson GE (2017) Microstructure evolution and texture development of hot form-quench (HFQ) AZ31 twin roll cast (TRC) magnesium alloy. AIP Conf Proc 1901(1):130005. https://doi.org/10.1063/1.5010565
25. Dobeš F, Dymáček P (2020) Creep behavior of AJ62 Magnesium–Aluminum–Strontium alloy. J Magnes Alloy 8(2):414–420. https://doi.org/10.1016/j.jma.2020.03.001

26. Ning X, Xie WD, Dang CM, Peng XD, Yang Y, Chen YL (2011) Microstructures and mechanical properties of AJ62 magnesium alloy with Y and Nd elements. Mater Sci Forum 686:253–259. https://doi.org/10.4028/www.scientific.net/MSF.686.253
27. Hazeli K, Sadeghi A, Pekguleryuz MO, Kontsos A (2013) The effect of strontium in plasticity of magnesium alloys. Mater Sci Eng A 578:383–393. https://doi.org/10.1016/j.msea.2013.04.101
28. Baril É, Labelle P, Fischersworring-Bunk A (2004) AJ (Mg-Al-Sr) alloy system used for new engine block. SAE Tech. Pap no. 724. https://doi.org/10.4271/2004-01-0659
29. Alias J (2020) Role of aluminium on the microstructure and corrosion behaviour of magnesium prepared by powder metallurgy method. Int J Automot Mech Eng 17(3):8206–8213. https://doi.org/10.15282/ijame.17.3.2020.14.0618
30. Miao J, Sun W, Klarner AD, Luo AA (2018) Interphase boundary segregation of silver and enhanced precipitation of Mg 17 Al 12 Phase in a Mg-Al-Sn-Ag alloy. Scr Mater 154:192–196. https://doi.org/10.1016/j.scriptamat.2018.05.047

Microstructural and Mechanical Characterization of AlSi10Mg Additively Manufactured Material Using Direct Metal Laser Sintering Technique

S. P. Tan, M. A. Ramlan, M. S. Shaari, Akiyuki Takahashi, and M. R. M. Akramin

Abstract AlSi10Mg alloy produced by additive manufacturing (AM) technology using direct metal laser sintering (DMLS) technique has resulted better in handling complex geometry. However, limited studies are performed for this AM method to show the integrity of aluminium alloys produced by DMLS to meet the required industry standard. This study investigates the effect of post-process on microstructure, mechanical properties, and fatigue life behaviour to AlSi10Mg material that DMLS produces. In this study, the specimens were tested with different post-process types: annealing (TS) and heat treatment processes (T5 and T6 conditions). All test results were compared with as-built processed specimens. Scanning electron microscope (SEM) and optical microscope are used to capture the microstructure images. The results showed that the tensile strength of the post-processed was decreased approximately 25% (decreased from 391 to 299 MPa). Still, the ductility was approximately 200% (in-creased from 3.2 to 6.8%) higher than the as-built specimen. This is because spherical silicon particles become coarsened when the specimen ductility is increased after heat treatment. For fatigue behaviour, it shows the as-built and heat-treated specimens are closely similar compared to findings from the literature. Overall, this study showed that the post-process changed the tensile strength and microstructural of AlSi10Mg but only significantly improved fatigue performance.

Keywords AlSi10Mg · Additive manufacturing · DMLS · Fatigue life · Mechanical characterization

---

S. P. Tan · M. A. Ramlan · M. S. Shaari (✉) · M. R. M. Akramin
Faculty of Mechanical and Automotive Engineering Technology, Universiti Malaysia Pahang, 26600 Pekan, Pahang, Malaysia
e-mail: shamil@ump.edu.my

A. Takahashi
Department of Mechanical Engineering, Faculty of Science & Technology, Tokyo University of Science, 2641 Yamazaki, Noda-shi, Chiba-ken 278-8510, Japan

M. Y. Ismail et al. (eds.), *Technological Advancement in Mechanical and Automotive Engineering*, Lecture Notes in Mechanical Engineering,
https://doi.org/10.1007/978-981-19-1457-7_28

1 Introduction

Recently, AM has been further applied into the advanced manufacturing industries by producing end-products (final parts). Several additive manufacturing (AM) techniques and processes are widely used by giant and well-known manufacturing companies such as Boeing, BMW, GE, and many more. Therefore, expanding the AM technology key player and hitting the manufacturing markets worldwide is essential. The AM material also varies from different types of metals such as aluminum alloy or titanium and non-metallic such as nylon or polystyrene. Meanwhile, it is still limited to only a few types of materials. Findings that provide similar outcomes towards AM technique can be found in several studies done by researchers in the field, such as SS316L material [1], TiAl6V4 material [2], and AlSi10Mg material [3–5]. Proportionally, the process and technology driving the AM machinery also evolve hastily. It started from selective laser sintering (SLS) in the mid-1980s, followed by electron beam melting (EBM) in the early 1990s, selective laser melting (SLM), and direct metal laser sintering (DMLS) in the mid-1990s [6].

Despite meeting manufacturing efficiency, the integrity of AM must be further studied to ensure human safety indemnity in AM-produced components, especially for automotive, aerospace, and oil and gas industries. Thus, the static, dynamic and mechanical characteristic of AM has to be well explored and forecasted [7]. Fatigue ehavior of AM material is put into relation confrontation in between AM process [7], built orientation [8, 9], surface roughness and porosity [10, 11], the effect of post-processes [12] and mechanical properties [11]. Furthermore, microstructure analysis studies were carried out to understand the characteristics of AM material on changes after mechanical properties [13, 14]. According to [11], these components influence the mechanical properties particularly heat treatment and surface roughness emphatically affects the fatigue strength. Meanwhile, structural integrity analysis is applied to every AM produced component because cracks emerge from fatigue and fracture behavior when cycling stresses are applied [15–17].

Previous studies have a lack of research on the structural integrity of AM material. Also, most studies on AlSi10Mg material are commonly fabricated by the SLM method [18]. Most research on the DMLS method is still in the initial stage, not as mature as the SLM method. Therefore, the material characteristics and microstructure of AlSi10Mg fabricated by DMLS are investigated in this paper. The aim is to determine the integrity of AlSi10Mg material manufactured by the DMLS technique based on mechanical characterization. In the following, the response of post-treatment on AM material manufactured by DMLS is compared to the as-built model. Several types of experimental work on post-treatment are carried out. The changes in the material properties and fatigue strength affected by the post-process of AlSi10Mg are studied and discussed.

2 Methodology

The effect of post-process on the AM material is conducted in this paper. A flow chart of the methodology in this paper is shown in Fig. 1. First, a DMLS machine fabricated all experimental work specimens. Afterwards, specimens remained as-built (AB) condition and underwent different types of post-process. Several types of post-processes were carried out in this study which are annealing (TS), solution heat treatment (T6), and artificial ageing (T5 & T6) processes.

The preparation and fabrication of the AlSi10Mg samples using the DMLS can be divided into three stages, pre-process, laser sintering, and post-process. Before the product can be printed using the DMLS method, the input files for the components should be produced using the 3D computer-aided design (CAD) software. Next, the model is designed in CAD software before creating an output file known as a stereolithography (STL) file. This file will be the input file for the AM machine for fabrication purposes. The support structures are also included along with the model before the final slicing process is performed. The sliced file then is imported into the EOS M290 via a computer onboard with the machine.

After that, several specimens were further for post-process. Those specimens underwent the annealing process at 300 °C for 2 h before going through T5 and T6

Fig. 1 Experimental workflow chart

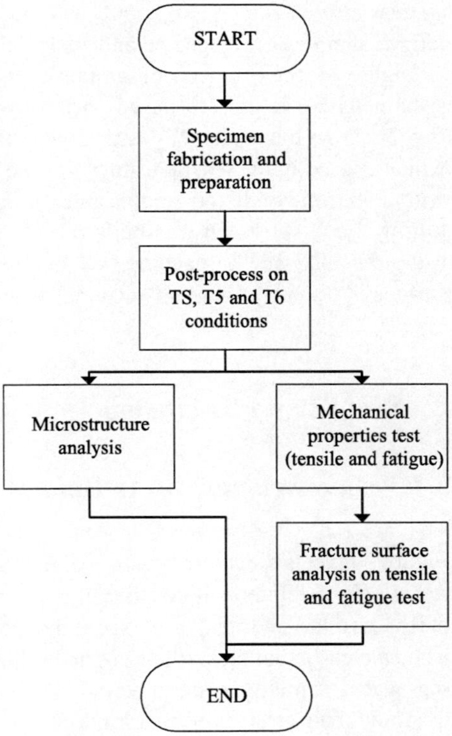

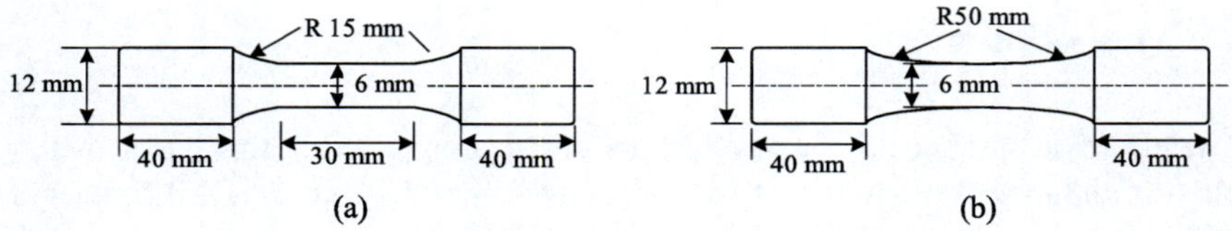

Fig. 2 Geometry design of samples for **a** tensile test **b** fatigue test

processes. For T6, 520 °C for 6 h was applied, then quenched with water at room temperature. Lastly, the temperature for artificial ageing is set at 165 °C, heated for 7 h before being removed, and cooled down at room temperature.

Two types of microstructure analysis were carried out namely scanning electron microscope (SEM) with energy dispersive spectroscopy (EDAX) and optical microscope. First, microstructure characterization was performed to study the microstructure on all specimens. Next, the tensile test is performed to determine the mechanical properties of AM material. This test was conducted by a universal tensile machine (UTM) following the American Society for Testing and Materials (ASTM) E8 standard. Then, an Instron testing machine conducted the fatigue test at room temperature. During the experiment, an ASTM E466 standard stress-controlled testing mode is applied. The test's overhead speed (1.8 mm/min) and strain rate ($1 \times 10^{-3}\ s^{-1}$) are set. For cyclic loading, the test is based on a sinusoidal wave at 15 Hz frequency and a stress ratio of 0.1 ($R = \sigma_{min}/\sigma_{max}$). Then, the fracture surfaces of both tensile and fatigue samples are inspected and analyzed under an optical microscope.

Figure 2 is the geometry of samples used for the tensile test and fatigue test. These specimens are fabricated based on the respective ASTM standards. For example, Fig. 2a shows the geometry design for a tensile sample with a gauge length of 30 mm which is used in the tensile testing. On the other hand, Fig. 2b indicates the sample with a 50 mm continuous radius and a minimum of 6 mm diameter used in fatigue testing. Each specimen was built vertically to build a platform and manufactured using the EOS M290 machine. All samples are inspected for imperfection on the surface like sharp edges and any cracks before going through the test.

3 Results and Discussion

3.1 Microstructure of As-Built vs Heat Treated on AlSi10Mg

Figure 3 reveals the microstructure of the metallographic cross-sectioned of as-built AlSi10Mg specimens. It consists of α-Al columnar fine grains (dendrites) with Si-particles (interdendrites) disclosed at the melt pool and the boundaries. Figure 3a and b show clearly that the particles of as-built AlSi10Mg material have overlapped and segregated at melt pools and boundaries during laser sintering [19, 20]. According to Aboulkhair et al. [19], this kind of fine microstructure formation is because of

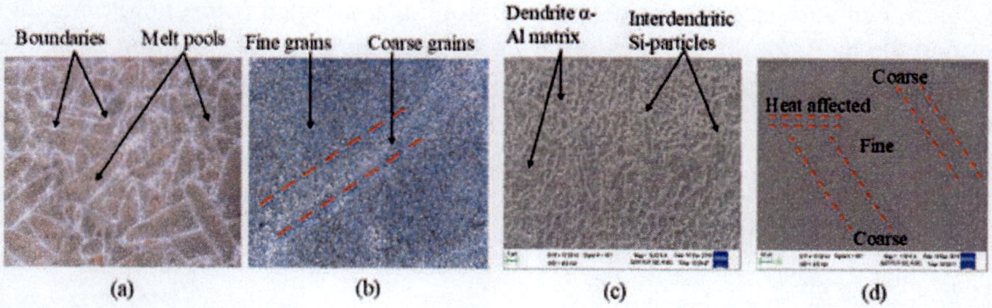

Fig. 3 Microstructure of AlSi10Mg for the as-built specimen. Under the microscope: **a** 500 μm. **b** 50 μm. Under SEM of the as-built specimen. **c** Magnification at 1 μm **d** Magnification at 10 μm

the rapid cooling rate during fabrication processing. The traces of a series of a laser beam that passes on the specimens create heat-affected zones with visible coarser particles, resulting in an inhomogeneous microstructure [20].

Figure 3c and d show the magnified image at the centre of the melt pool. A fine microstructure of cellular dendrites of α-Al and interdendrites of Si-are observed in the melt pool. Observed that, the grain size becomes coarser and segregated rather than spherical near the melt pool boundary. Keeping specimens at high temperatures with long hours caused the coarsening characteristic since overlapping two adjacent melt pools results in a slower solidification rate. The fine grains of the as-built AlSi10Mg specimens are attributed to the even arrangement of the Si particles in the microstructure of the specimen, which is likely due to the precipitation of the silicon phase along the Al-Si cellular boundaries.

Figures 4 are exhibit the microstructure for the heat-treated specimen which was conducted using SEM. The behaviour of microstructure is clearly displayed. Figure 4a and b show a microstructure with a columnar morphology indicating fine cellular-dendritic growth of Al matrix and interdendritic Si-particles during solidification. The grey cellular features are primarily Al particles and are surrounded by white fibrous Si particles. In contrast, Fig. 4c and d show the different areas in the melt pool. Thijs et al. [21] found that the microstructure of AlSi10Mg produced by the AM method is characterized by three regions across the melt pool: a fine and

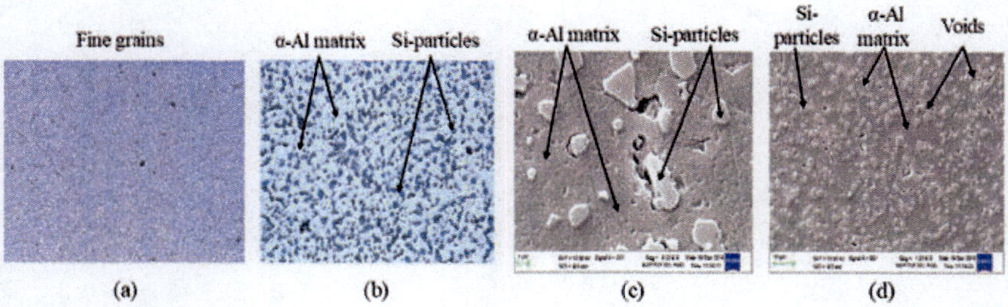

Fig. 4 Microstructure of AlSi10Mg for the heat-treated specimen. Under microscope: **a** 500 μm. **b** 50 μm. Under SEM of heat-treated (T6) SLM AlSi10Mg. **a** 1 μm. **b** 10 μm

a coarse cellular structure inside the melt pool heat-affected (transition) zone and around the melt pool in the previously deposited layers. In the transition zone, the eutectic Al-Si network structure is broken to some extent by coarsening the Si into spheroidized particles, which increases the diffusion rate of the Si [22]. The different thermal histories experienced by three regions have been attributed in contact with readily solidified Al. At the same time, the Al has less heat conductivity due to slower solidification [13, 23].

3.2 *Tensile Test*

Following are the tensile test results of as-built and post-process AlSi10Mg specimens that were carried out at room temperature as shown in Table 1 and Fig. 5. Based on the result, realized that the yield and tensile strength were reduced post-processed. Meanwhile, the ductility of the specimens after post-process was increased.

From the result, the specimens showed brittle behaviour when the strain was just above 3% under quasi-static loading. The AB specimens showed the highest tensile and yield strength of 391.0 MPa and 233.0 MPa respectively yet retained the

Table 1 Experimental tensile test results

Specimens	Yield Strength, 0.2% offset (MPa)	Ultimate Tensile Strength (MPa)	Modulus of Elasticity (GPa)	Strain (%)
AB	233.0	391.0	62.0	3.2
TS	192.0	304.0	79.9	5.6
TS + T5	187.0	299.0	64.3	6.8
TS + T6	227.0	287.0	74.0	6.6

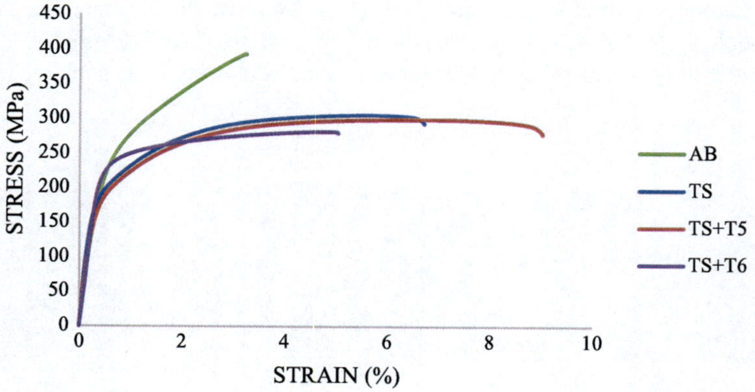

Fig. 5 Comparison of tensile results between as-built and post-processed for AlSi10Mg

lowest ductility of 3.2% in strain. The annealing process (TS) made an incredible impact on the mechanical properties. After the specimen went through the annealing process, great drop in yield strength and tensile strength (192.0 MPa and 304.0 MPa) was observed. However, the ductility is improved, and the strain increases by 5.6%. Meanwhile, the specimens experienced annealing with artificial aging on T5 and T6 condition also found a similar result. Both decreased tensile (at 299.0 MPa and 287.0 MPa) and dramatically increased ductility (at 6.8% and 6.6%), respectively. The annealed specimens have been softening compared to AB. The T6 specimen is the only sample that experienced the peak-hardened solution heat treatment (SHT) process. Even though the specimen has the lowest UTS among all specimens, the yield strength is as high as the AB specimen.

The AB specimen consists of high strength and can be a virtue to grain refinement. This is due to the mechanical properties affected by the grain size [24]. For the specimen after heat-treated, the strength and ductility changes are influenced by several factors such as the change of Si phases in terms of number or size, formation of Mg2Si, the initial hardening rate, recovery rate, etc. [25]. Upon the SHT and artificial ageing, the Si atom is trapped in the Al matrix, and the distance between Si–Si particles increases simultaneously. Meanwhile, the size of Si particles increased, which induced localized stress and strain reduction [26]. Based on the tensile test result, the ductile and brittleness of the material is difficult to categorize by heeding the number of strains only [27]. So, the fracture surface of the specimen investigation is conducted. Alboulkhair et al. [3] claimed that initial failure always starts at a surface or sub-surface imperfection then propagates along the plane opposite to the loading direction until the final crack.

3.3 *Fatigue Test*

Results of data obtained from a series of fatigue experiments are plotted according to the stress-life (*S–N*) method. Table 2 and Table 3 both listed the overall fatigue test results for the as-built and heat-treated specimens respectively. Furthermore, the loading parameters used in fatigue testing are also included in the table. From the fatigue test, the stresses are different for as-built and heat-treated. The loading

Table 2 Overall fatigue test results for as-built AlSi10Mg

Specimens	Cycles to Failure	σ_{max} (MPa)	σ_{min} (MPa)	σ_{mean} (MPa)
90% UTS	1.08×10^3	356.31	35.60	195.97
80% UTS	2.44×10^3	316.72	31.67	174.20
70% UTS	5.99×10^3	277.33	27.71	152.42
60% UTS	1.03×10^4	237.54	23.75	130.65
50% UTS	3.51×10^4	197.95	19.80	108.87

Table 3 Overall fatigue test results for heat-treated AlSi10Mg

Specimens	Cycles to Failure	σ_{max} (MPa)	σ_{min} (MPa)	σ_{mean} (MPa)
90% UTS	3.77×10^4	258.48	25.85	142.16
80% UTS	6.05×10^4	229.26	22.98	126.37
70% UTS	1.99×10^5	201.04	20.10	110.57
60% UTS	5.08×10^5	143.02	17.23	94.78
50% UTS	2.13×10^6	96.24	14.36	78.98

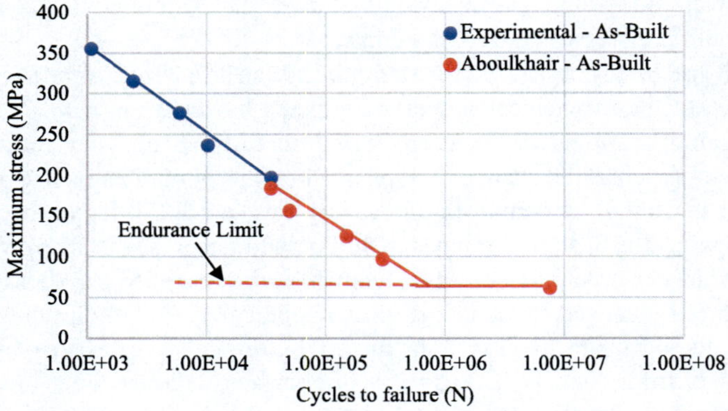

(a)

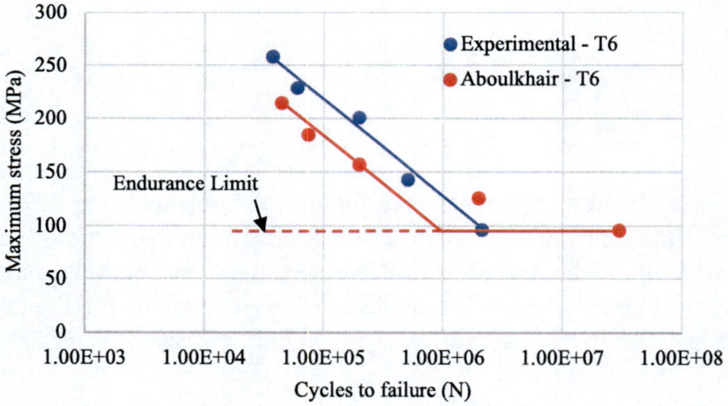

(b)

Fig. 6 Comparison of fatigue behavior between experimental and previous study for **a** as-built [3] and **b** heat-treated specimens

parameters are designed accruing to the UTS. Five specimens underwent fatigue testing for each case for the level of stress from 90 to 50% of the UTS.

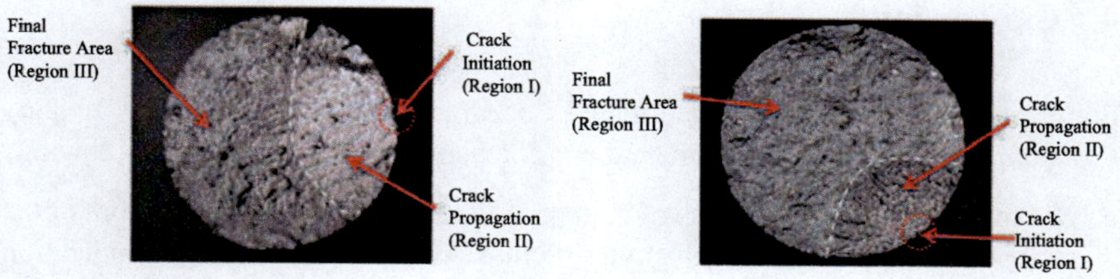

Fig. 7 Fatigue fracture surface of **a** as-built **b** T6

Figure 6 shows the fatigue behaviour of the specimens based on the fatigue data plotted. These data are plotted according to the stress-life (*S–N*) curve and compared to the results of the previous study. The fatigue life behaviour is comparable to the results from Aboulkhair et al. [3]. Since the level of stresses for as-built in the previous study is lower than the experiment, thus the cycles to failure are significantly lower than the experimental results for as-built. However, for the heat-treated, the level of stress is almost similar. Hence, the fatigue life result is better compared to the previous study. One of the reasons is the author performed SHT for T6 in a different way. The specimens only underwent heat treatment for 1 h at 520 °C before water quenched unlike this study; for 6 h. The microstructure is likely not having enough time for the Si to spheroidized. The formation of Si spheroids through spheroidization results in strengthening the material through the heat treatment process. On the other hand, an endurance limit was analyzed based on the *S–N* curve. The performance was increased significantly after heat-treated by T6 condition. In comparison to as-built specimen, heat-treated showed a similar number of cycles in all manner of stress ranges, favourable fatigue behaviour in low-stress range, and with no overrun during the test even performed at 96.24 MPa for maximum stress.

Figure 7 shows the crack propagation and fatigue fracture on the surface of the specimen after undergoing fatigue tests for as-built and heat-treated specimens. It is found that, the crack initiation is the result of defects in the material. The fracture surface is flat around the crack initiation part from a macroscopic appearance viewpoint and this is also found in the study of Mcmillan & Hertzberg [28].

The whole fatigue-fracture surface consisted of four main regions. There are (i) fatigue cracked region, (ii) stretched region, (iii) overload fracture region, and (iv) final fracture region. However, certain researchers considered the second and third regions as the fatigue crack propagation region [27]. In SLM, two types of crack initiation under fatigue loading are categorized [29]. The first type is planar fabrication flaws such as lack of penetration or fusion provide ideal sites for fatigue cracking. The second type is failure due to poor surface roughness during the fabrication of the specimens. The effect of post-process on crack propagation and fatigue fracture inside the test specimen is indicated in Fig. 7b. It is observed that the heat-treated specimen is reasonably flat and uniform on the fracture surface. However, some tiny dimples appear on the fracture surface, and it is restricts the perception of the specimen with any process. It is moreover evident that the heat treatment caused the microstructural coarsening.

4 Conclusions

In this paper, both the microstructure and mechanical properties of AlSi10Mg alloy on as-built and post-process were studied. The main conclusions are the following:

- The microstructure is made up of melt pools with boundaris, while the core of the melt pool is fine and uniform in grain structure. After post-process, spheroidization of Si particles was formed and coarsened.
- For the mechanical properties of this alloy after several types of post-process, it is appealing to increase ductility yet decrease in tensile strength. The as-built consists of the highest tensile strength (391 MPa) but lowest ductility (3.2%) among all specimens. Meanwhile, the T5 thermal processed specimen annealed has the highest ductility (6.8%) but lower tensile strength (299 MPa). This is due to the spheroidization of Si particles observed in the microstructure.
- As observed, when a different SHT approach is applied, heat-treated specimen's cycles to failure have significantly enhanced fatigue life compared to as-built in comparison to the previous study.

Further studies need to be performed to understand fatigue crack behaviour and propagation of the specimens. This is mainly for studies on the different types of specimens with different types of post-process when without annealing. This is important to understand the structural integrity performance of material applied to different applications.

Acknowledgements The author would like to acknowledge the Ministry of Higher Education under Fundamental Research Grant Scheme FRGS/1/2019/TK03/UMP/02/21 (university reference RDU1901151) and Universiti Malaysia Pahang (UMP) for financial supports. Also, the authors would like to thank UMP for allowing the research to be conducted using the High-Performance Computer (HPC).

References

1. Duval-Chaneac MS, Gao N, Khan RHU, Giles M, Georgilas K, Zhao X, Reed PAS (2021) Fatigue crack growth in IN718/316L multi-materials layered structures fabricated by laser powder bed fusion. Int J Fatigue 152:106454
2. Lee S, Ahmadi Z, Pegues JW, Mahjouri-Samani M, Shamsaei N (2021) Laser polishing for improving fatigue performance of additive manufactured Ti-6Al-4V parts. Opt Laser Technol 134:106639
3. Aboulkhair NT, Maskery I, Tuck C, Ashcroft I, Everitt NM (2016) Improving the fatigue behaviour of a selectively laser melted aluminium alloy: influence of heat treatment and surface quality. Mater Des 104:174–182
4. Ch SR, Raja A, Jayaganthan R, Vasa NJ, Raghunandan M (2020) Study on the fatigue behaviour of selective laser melted AlSi10Mg alloy. Mater Sci Eng, A 781:139180
5. Xu Z, Liu A, Wang X (2021) Fatigue performance and crack propagation behavior of selective laser melted AlSi10Mg in 0°, 15°, 45° and 90° building directions. Mater Sci Eng, A 812:141141

6. Venuvinod PK, Ma W (2004) Rapid prototyping: laser-based and other technologies. Springer Science + Business Media, New York
7. Ferro P, Fabrizi A, Berto F, Savio G, Meneghello R, Rosso S (2020) Defects as a root cause of fatigue weakening of additively manufactured AlSi10Mg components. Theoret Appl Fract Mech 108:102611
8. Tridello A, Fiocchi J, Biffi CA, Chiandussi G, Rossetto M, Tuissi A, Paolino DS (2020) Effect of microstructure, residual stresses and building orientation on the fatigue response up to 109 cycles of an SLM AlSi10Mg alloy. Int J Fatigue 137:105659
9. Beretta S, Gargourimotlagh M, Foletti S, du Plessis A, Riccio M (2020) Fatigue strength assessment of "as built" AlSi10Mg manufactured by SLM with different build orientations. Int J Fatigue 139:105737
10. Leon A, Aghion E (2017) Effect of surface roughness on corrosion fatigue performance of AlSi10Mg alloy produced by Selective Laser Melting (SLM). Mater Charact 131:188–194
11. Beevers E, Brandão AD, Gumpinger J, Gschweitl M, Seyfert C, Hofbauer P, Rohr T, Ghidini T (2018) Fatigue properties and material characteristics of additively manufactured AlSi10Mg—effect of the contour parameter on the microstructure, density, residual stress, roughness and mechanical properties. Int J Fatigue 117:148–162
12. Bagherifard S, Beretta N, Monti S, Riccio M, Bandini M, Guagliano M (2018) On the fatigue strength enhancement of additive manufactured AlSi10Mg parts by mechanical and thermal post-processing. Mater Des 145:28–41
13. Fite J, Eswarappa Prameela S, Slotwinski JA, Weihs TP (2020) Evolution of the microstructure and mechanical properties of additively manufactured AlSi10Mg during room temperature holds and low temperature aging. Addit Manuf 36:101429
14. Varmus T, Konecna R, Nicoletto G (2021) Microstructure and fatigue performace of additively manufactured AlSi10Mg. Transp Res Procedia 55:518–525
15. Akramin MRM, Shaari MS, Ariffin AK, Kikuchi M, Abdullah S (2015) Surface crack analysis under cyclic loads using probabilistic S-version finite element model. J Braz Soc Mech Sci Eng 2015 37:6 37(6):1851–1865
16. Shaari MS, Akramin MRM, Ariffin AK, Abdullah S, Kikuchi M (2016) Prediction of fatigue crack growth for semi-elliptical surface cracks using S-version fem under tension loading. J Mech Eng Sci (JMES) ISSN 10(3):2375–2386
17. Zhang W, Hu Y, Ma X, Qian G, Zhang J, Yang Z, Berto F (2021) Very-high-cycle fatigue behavior of AlSi10Mg manufactured by selected laser melting: crystal plasticity modeling. Int J Fatigue 145:106109
18. Jian ZM, Qian GA, Paolino DS, Tridello A, Berto F, Hong YS (2021) Crack initiation behavior and fatigue performance up to very-high-cycle regime of AlSi10Mg fabricated by selective laser melting with two powder sizes. Int J Fatigue 143:106013
19. Aboulkhair NT, Maskery I, Tuck C, Ashcroft I, Everitt NM (2016) The microstructure and mechanical properties of selectively laser melted AlSi10Mg: the effect of a conventional T6-like heat treatment. Mater Sci Eng, A 667:139–146
20. Zhou L, Mehta A, Schulz E, McWilliams B, Cho K, Sohn Y (2018) Microstructure, precipitates and hardness of selectively laser melted AlSi10Mg alloy before and after heat treatment. Mater Charact 143:5–17
21. Thijs L, Kempen K, Kruth JP, Van Humbeeck J (2013) Fine-structured aluminium products with controllable texture by selective laser melting of pre-alloyed AlSi10Mg powder. Acta Mater 61(5):1809–1819
22. Zhuo L, Wang Z, Zhang H, Yin E, Wang Y, Xu T, Li C (2019) Effect of post-process heat treatment on microstructure and properties of selective laser melted AlSi10Mg alloy. Mater Lett 234:196–200
23. Aboulkhair NT, Everitt NM, Ashcroft I, Tuck C (2014) Reducing porosity in AlSi10Mg parts processed by selective laser melting. Addit Manuf 1–4:77–86
24. Li W, Li S, Liu J, Zhang A, Zhou Y, Wei Q, Yan C, Shi Y (2016) Effect of heat treatment on AlSi10Mg alloy fabricated by selective laser melting: microstructure evolution, mechanical properties and fracture mechanism. Mater Sci Eng, A 663:116–125

25. Hadadzadeh A, Amirkhiz BS, Mohammadi M (2019) Contribution of Mg2Si precipitates to the strength of direct metal laser sintered AlSi10Mg. Mater Sci Eng, A 739:295–300
26. Li XP, Wang XJ, Saunders M, Suvorova A, Zhang LC, Liu YJ, Fang MH, Huang ZH, Sercombe TB (2015) A selective laser melting and solution heat treatment refined Al-12Si alloy with a controllable ultrafine eutectic microstructure and 25% tensile ductility. Acta Mater 95:74–82
27. Brooks CR, Choudhury A (2002) Failure analysis of engineering materials. McGraw-Hill Education, New York
28. McMillan JC, Hertzberg RW (1968) Application of electron fractography to fatigue studies. Electron Fractography 89–123
29. Wycisk E, Solbach A, Siddique S, Herzog D, Walther F, Emmelmann C (2014) Effects of defects in laser additive manufactured Ti-6Al-4V on fatigue properties. Phys Procedia 56(C):371–378

Quenching Heat Transfer Characteristics of Copper Rod in Saturated and Various Subcooled Condition

H. Zeol, M. Z. Sulaiman, H. Z. Hui, H. Ismail, and T. Okawa

Abstract This study investigated the quenching performance of a copper rod with 50 mm length and diameter of 20 mm. The specimen was heated to 600 °C as the initial temperature and immersed in a quenching pool of pure water (distilled water) followed with a subsequent quench seven times. Under atmospheric pressure, the experiments are conducted in saturated and various subcooled conditions (90, 80 and 60 °C). The cooling curves (temperature vs time) and the cooling rate curves (°C/s) of the copper cylinder are obtained from the experiment. Results show that the cooling performance for 1st quench and the subsequent quench for saturated and 90 °C subcooled condition shows a different performance related to the formation of the oxide layer at the copper surface that changes the surface characteristic. Vice versa, the cooling performance in 80 °C and 60 °C subcooled conditions has consistent performance for all quench, which is believed to be the domination of the subcooling effect, even though the physical surface appearance shows the same. Overall, the cooling curve of the copper rod was enhanced with the increase of subcooled temperature, especially for 60 °C subcooled conditions. The cooling curves for the subcooled of 90 and 80 °C still maintain the slope with the three-section shape, which is similar for the saturated case, but for the 60 °C subcooled conditions, the cooling curve slope suddenly increased and shifted to the left, showing the drastic decrease of centre temperature and the impact on the highly subcooled condition. The cooling rate curve shows the increasing peak value of cooling rate with increasing the subcooled temperature, which is the highest value during quench in 60 °C conditions. The minimum heat flux (MHF) point temperature rises and occurs faster, and the Critical Heat Flux (CHF) point is achieved early with the increasing subcooled temperature. The highly subcooled condition 60 °C shows no film boiling regime formation and the MHF point location is not visible.

H. Zeol · M. Z. Sulaiman (✉) · H. Z. Hui · H. Ismail
Faculty of Mechanical and Automotive Engineering Technology, Universiti Malaysia, 26600 Pekan, Pahang, Malaysia
e-mail: zuhairi@ump.edu.my

T. Okawa
Department of Mechanical and Intelligent Systems, The University of Electro-Communications, 1-5-1, Chofugaoka, Chofu-shi, Tokyo 182-8585, Japan

M. Y. Ismail et al. (eds.), *Technological Advancement in Mechanical and Automotive Engineering*, Lecture Notes in Mechanical Engineering,
https://doi.org/10.1007/978-981-19-1457-7_29

Keywords Quenching heat transfer · Minimum heat flux · Subcooling

1 Introduction

Quenching heat transfer is a rapid cooling process of a hot object that immerse in a colder fluid bath. Quenching is a widely used process for achieving martensitic and bainitic structures where its related to the heating and subsequent rapid cooling rate procedure to increase the hardness of metal [1]. Quenching is vital as a safety procedure in heavy industry, especially in the nuclear industry, acting as an emergency core cooling system (ECCS) in a nuclear reactor. It plays a significant role as a countermeasure during the loss of coolant accident (LOCA) in nuclear reactors [2]. During quenching, heat transfer rates are restricted when the film boiling occurs, where a stable vapour layer covers the hot object's surface. This occurrence creates a strong resistance to energy transfer between the two surfaces [3]. Cooling curve and cooling rate curve analysis is commonly used to observe the quenching performance of the test object since it is the most comprehensive way to characterise a quench media. The cooling curve is produced when a material that is initially at a temperature above the boiling of the quench media is introduced into the quench media [4]. Leidenfrost point or known as minimum film boiling (MHF) was focused by researchers to determine the performance of the cooling rate during the quenching process, which is with the early time to achieve MHF point, will contribute to the early ends of film boiling regime, then accelerate the cooling rate [5]. Many factors are related with the formation of the stable film boiling regime, and one of the main factor is the liquid temperature, either in saturated or in subcooled condition [6].

Many researchers investigate the effect of the subcooled condition on quenching performance. In 2007, Bolukbasi et al. [7] studied vertical brass cylinder quenching performance with five different dimensions in saturated and various subcooled conditions. The specimen was heated up until 600 °C, then directly quenched in a quenching pool. The result shows that in the saturated condition, three sections shape the graph, representing the film boiling regime, nucleate boiling regime, and natural convection cooling regime. However, the film boiling regime was not occurring at subcooled conditions for all specimens because the film or the vapour blanket was demolished at higher temperatures. They believe the heat transfer occurred at nucleate boiling and the natural convection stage.

Similar to Lotfi et al. [8], they experimented with examining the quenching performance of a silver sphere in distilled water, Ag and TiO_2 water nanofluids with slightly a subcooled condition (90 °C). The result obtained shows that the film boiling started immediately after being immersed in the fluids. They noticed the formation of a thick and smooth vapour blanket covering the sphere during the film boiling regime. Then with decreasing the temperature, the vapour blanket collapse and solid-fluid start to contact each other. This shows the slightly subcooled condition will have a slight impact on the cooling process. In 2017, Young et al. [9] experimented with tabilize the effect of subcooling on quenching performance. They use a zircaloy rod specimen with an initial temperature of 600 °C, then quickly quench into the highly subcooled pure water (25 °C). They found that once the specimen dropped into the quench pool, the temperature centre decreased drastically. They discovered that the liquid subcooling had a significant influence on the quenching curve and phenomena. The liquid subcooling increased led to shifting the quenching curve to the left (decreasing the quenching duration).

Lee et al. [10] investigated the quenching performance of the zircaloy-4 and Cr-alloy-coated cladding tubes in saturated and 50 °C subcooled pure water. Both specimens were heated up to 600 °C and then subsequent quenched into the quench pool three times. They discovered that the cooling curve for both specimens shows the decreasing of surface temperature exponentially in the nucleate boiling regime. Using a high-speed camera, both specimens show vapour film formation in the early stage during quench in saturated fluid, but no vapour film boiling was observed during quench in a subcooled fluid. In 2021, Wang et al. [11] investigate the quenching performance of the FeCrAl rodlets in various subcooled (95, 90, 85 and 80 °C) in distilled water. The specimen was prepared with two types: bare and rough surfaces with 0.2 μm and 1.6 μm, respectively. Both have an initial temperature of 600 °C. They found that the cooling rate performance for both specimens shows increment trending, in which the cooling curve slope becomes steeper and shifts to the left with increasing the liquid subcooling degree. Moreover, the quenching time is shorted, and the MHF increases as the liquid subcooling degree increases.

Kang et al. [12] conducted an experiment to analyse the quenching performance of 2 types of surface of zirconium vertical rod specimen, which are completely wettable surface (CWS) and Bare surface (BZS) with an initial temperature of 800 °C in saturated and various subcooled distilled water. They found that the cooling curve is strongly related to surface hydrophilicity and the subcooling fluids condition. Both specimens show great enhancement of cooling rate with shorted time as the subcooling degrees increase. Direct quenching without stable film boiling regime shows at 60 K subcooled for CWS, and >15 K subcooled for BZS specimen, which indicates the early collapse of vapour film blanket.

Similar to Ebrahim et al. [13], they researched quenching performance in saturated and subcooled liquid using three different cylinder shape specimens: stainless steel (S.S.), Zirconium (Zr), and Inconel-600 rods with an initial temperature of 550 °C. For all types of specimens, they identified that subcooling strongly influences the temperature of the MHF point. As the liquid subcooling increases, the heat transfer increases due to the higher driving temperature difference between the heated surface and the ambient liquid temperature.

In addition, in 2020, Xiong et al. [14] conducted an experiment of quenching performance of two types of the specimen which are FeCrAl alloy and Zircaloy-4 with initial temperature 600 °C in both saturated and various subcooled conditions (80, 85, 90 and 95 °C) of distilled water. They noticed a similar result, which is the cooling curves shifted to the left for both specimens. Using a high-speed camera, they observed that the liquid-vapour interface's evaporation is weak and the vapour film is thin in the large subcooling cases, which confirmed that more portion of heat flux is tabiliz to heat up the subcooled liquid in the film boiling regime.

However, the temperature of the specimen can also significantly influence the formation of a stable film boiling regime, although the surrounding fluids were in highly subcooled condition. In 2015, Hsu et al. [15] experimented with investigating the quenching performance of stainless steel and zircaloy spheres in both de-ionised and natural seawater at 33 °C subcooled conditions. The initial temperature for both specimens was set up for a very high temperature which is 1000 °C. The cooling curves obtained for quenching in de-ionised water show the formation of the stable film boiling regime, which lasts about 16 s for stainless steel and 12 s for zircaloy specimen. Further investigation with a high-speed camera tabilizeson found the small bubble nucleated at the surface. Then the stable vapour blanket is formed, although initially the interface is initially unstable due to the transient contact at the sphere surface, but tabilizes within a short period.

Based on the aforementioned literature, it is found that most of the research in quenching performance in distilled water, seawater and recent nanofluids either in saturated and various subcooled conditions. On top of that, most of the experiments used quenching metal with surface coated treatment or alloy types such as stainless steel, zircaloy, FeCrAl, brass, copper coated with nickel or other metal alloy and coated surface metal as the specimen due to the low thermal conductivity characteristics and to avoid any surface degradation due to the oxidation layer occurred when the metal exposed at the very high temperature during the quenching process [16].

This paper aims to evaluate the quenching heat transfer performance based on a cooling curve using subsequent quenches of untreated surface copper rod metal (without any surface coating) in saturated and subcooled distilled water and qualitative evaluation of the physical surface changes during the subsequent quenched of this copper rod specimen.

2 Methodology

2.1 Experimental Setup

Figure 1 shows the schematic drawing for the copper cylinder rod specimen (50 mm × 20 mm diameter), while Fig. 2 shows the picture for details apparatus setup for the quenching experiment. This experiment involved several main components; 20 L closed stainless steel bath tanks combined with JULABO circulator, S.Y. Electric melting furnace, and a NiDAQ data acquisition system. Ungrounded K-type thermocouples MISUMI with a 1 mm diameter were used and inserted at the centre of the copper rod to record the centre temperature measurement during the experiment. The uncertainty for temperature measurement is ±1.5 °C. National Instruments data acquisition (NiDAQ) was used to measure temperature time during the quenching process. The automatic quenching system was designed by installing the AitTAC pneumatic air cylinder with a compressed air supply to get the same speed of specimen movement during the quenching process for all quenching experiments. 80 mm

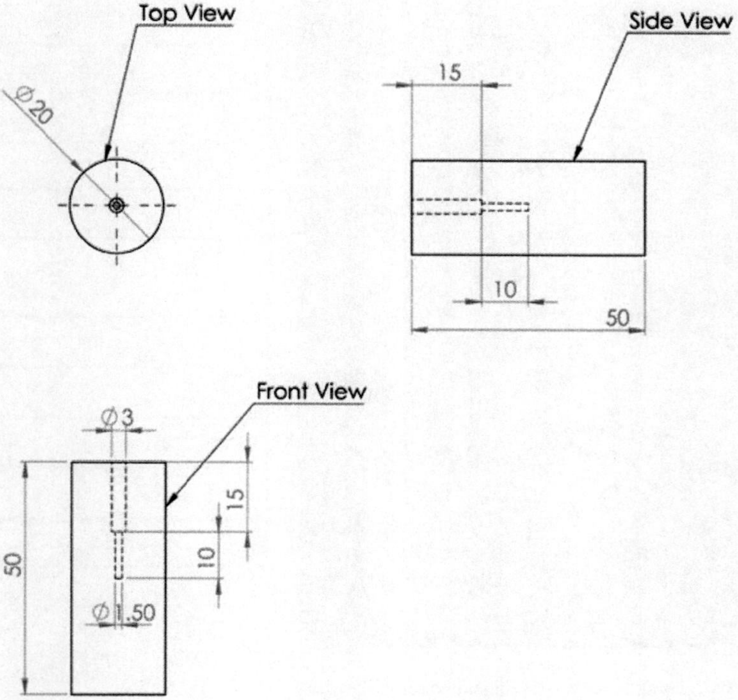

Fig. 1 Details schematic drawing for copper rod specimen

borosilicate was prepared as a quenching pool (180 × 130 mm diameter). The furnace and copper rod were placed directly above the quench pool at the prepared platform on top of the quench pool. The copper rod was placed into the furnace while attached to a pneumatic cylinder. Table 1 shows the details about the apparatus used during the quenching experiment.

2.2 *Experimental Procedure*

The experiment was started with heat up the distilled water in a closed stainless steel tank until it achieved 60 °C, then maintained that temperature. 1500 ml distilled water was prepared in a beaker then placed into a closed stainless steel tank, then waited until it achieved an equilibrium temperature (60 °C) with surrounding distilled water in the tank. Simultaneously, the copper rod was prepared to get a mirror surface. The copper rod's surface was prepared using different sizes of the grit of sandpaper, which are 800, 1000, 2000, 5000 and 7000, then final polishing by using polish metal paste. Lastly, the copper rod was cleaned using acetone and distilled water. After finishing preparing the copper rod surface, the copper rod was fixed at the quench rod and

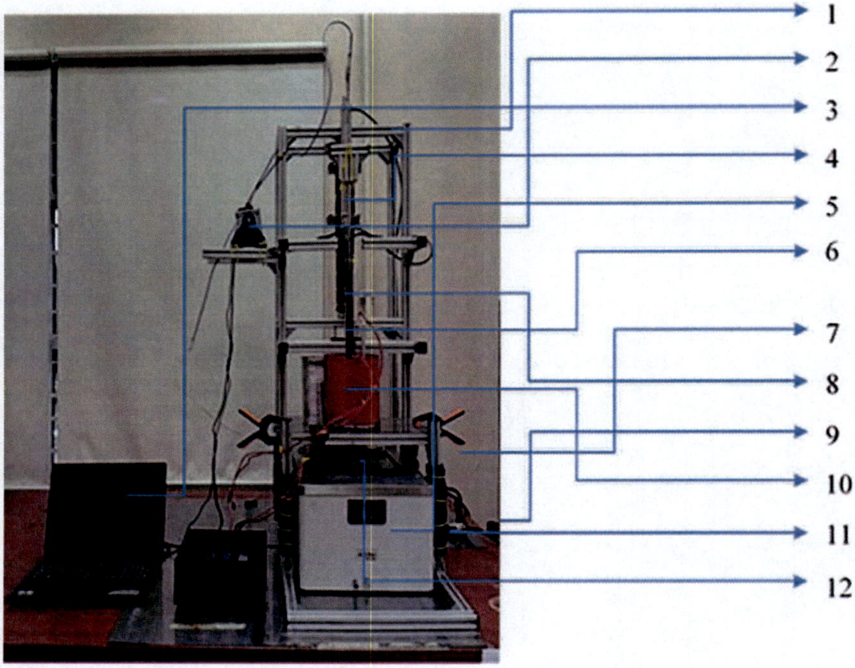

Fig. 2 Quenching heat transfer apparatus setup

inserted into the furnace to heat up until it achieved 600 °C initial temperature. Once the desired temperature was reached, the furnace was turned off and quickly immersed into the quench pool using an air valve controller. The experiment was ended once the thermal equilibrium was achieved between the copper rod and the distilled water in the quench pool. The rod was subsequently quenched repetitively seven times afterwards with the same copper surface without washing the surface during the repetition tests to investigate the surface effect on the rate of heat transfer. The temperature-time data from the quenching process were collected and analysed. After finishing all the procedures, the experiment was repeated by changing the temperature of the quench pool to 80, 90 and 100 °C.

3 Result and Discussion

Figure 3a shows the cooling curves with multiple quenches of copper rod in distilled water at saturated condition, while Fig. 3b shows the cooling rate curves. Both cooling curves and cooling rate curves are divided into three-section shapes, representing the cooling process that involved three main regimes: film boiling regime, nucleate boiling regime, and natural convection cooling regime [17]. These curves also consist

Table 1 Apparatus used during quenching

No	Name	Description
1	Quenching Rig	– Aluminum profile as the main structure for quenching rig
2	NiDAQ Data Logger	– Collecting the temperature data from thermocouple
3	Laptop	– To receive temperature data from the data logger
4	Pneumatic Piston Cylinder	– AirTAC brand – 300 mm stroke with 10 bar maximum pressure
5	JULABO Tank with Circulator	– JULABO brand – Maintain the desired subcooled temperature
6	Quenching Rod	– 8 mm O.D. diameter Stainless Steel 316L pipe – Attached with the specimen and fitted with cylinder piston
7	Fitting coupling	– Joint supply from air compressor supply sources
8	Thermocouple	– MISUMI brand grounded type – Type-K thermocouple with 1 mm probe diameter
9	Air Regulator	– AirTac brand. Control and filter air pressure input – 10 bar of maximum pressure
10	Electrical Furnace	– S.Y. Electric Melting Furnace – 1100 °C maximum running temperature
11	Hand valve pneumatic controller	– AirTac brand – 3 step position control valve, 8 bar maximum pressure
12	Beaker (Pool Quench)	– 2000 ml beaker volume as quench pool

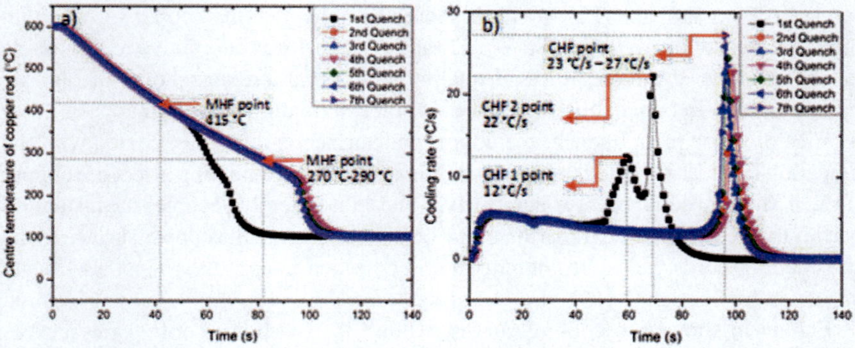

Fig. 3 The quenching performance of copper rod in distilled water at saturated condition; **a** Cooling curve, and **b** Cooling rate curve

of two main reference points known as minimum heat flux (MHF), which represent the transition point from film boiling to nucleate boiling [8], and critical heat flux (CHF) at where the point of the maximum boiling heat transfer [18]. The formation of film boiling regime acting as a resistant layer between the surface and water decreased the heat transfer from the surface to the water, decreasing the cooling rate [7].

It can be seen in Fig. 3a, the 1st quench cooling curve for clean surface shows the fastest cooling time, where the minimum heat flux point (MHF) is located at a temperature which is 415 °C and occur at the quickest time, which is after 43 s of quenching. After that, the subsequent quench using the same unwashed copper rod shows the decreasing of cooling time with a longer time taken for the rod to cool. It can be observed that the cooling curve data scattered almost the same for the 2nd quench until the 7th quench, which is the MHF point temperature is between 270 and 290 °C and the time taken for the MHF to occur is between 80 and 84 s.

As shown in Fig. 3b, it could be clearly seen that double CHF point (double peak of cooling rate) could be observed during the 1st quench. A similar observation was observed in the study of Umehara et al. [19] in quenching of the stainless steel in nanofluids, but the next subsequent quench had only a single CHF point noticed. The CHF 1 point during 1st quench happened at 59 s after the quench and had a peak cooling rate value of about 12 °C/s. The CHF 2 point happened at 69 s, and the peak value of cooling rate recorded about 22 °C/s, which almost had the same reading as the subsequent quench after that (23–27 °C/s). However, the CHF point for the subsequent quench happened at about 98, 38 s later than CHF 1 and 28 s later than CHF 2.

The different performance for the first quench (shown in Figs. 3a and b) and with subsequent quench is suspected to the formation of the oxidation layer at the copper rod surface (as shown in Fig. 4c–h). It should be noted that the copper surface with high thermal conductivity of 386 W/m K at 20 °C [20] was susceptive to thin oxide layer formation at the surface, and qualitative observation was shown in Fig. 4b. Following the first quench with polished surface, oxidation started to appear. After a series of subsequent quench, the copper rod surface is now fully coated with the oxidation layer, as depicted in Fig. 4c–h. It is believed the oxidation affect the copper surface' thermal conductivity characteristic and subsequently becomes the insulation barrier that leads to the degradation of the heat transfer from the copper surface to the surrounding fluids, hence prolonging the time needed to cool the rod as previously discussed by Sher et al. [16].

Figure 5a shows the cooling curves, while Fig. 5b shows cooling rate curves with multiple quenches of copper rod in distilled water at 90 °C, which is in slightly subcooled condition. The cooling curve in Fig. 5a maintained the three-section shape graph. The 1st quench with a clean and polished surface shows the fastest cooling time, where the MHF point was located at a higher temperature, at 463 °C, which is an increment of 10% compared to the MHF of the 1st quench in a saturated condition (see Fig. 3a). Respective to the cooling time enhancement, the time taken to achieve MHF point is about 22 s, a 49% reduction of time taken compared in a saturated condition. As depicted in Fig. 5a, the subsequent quench using the same unwashed

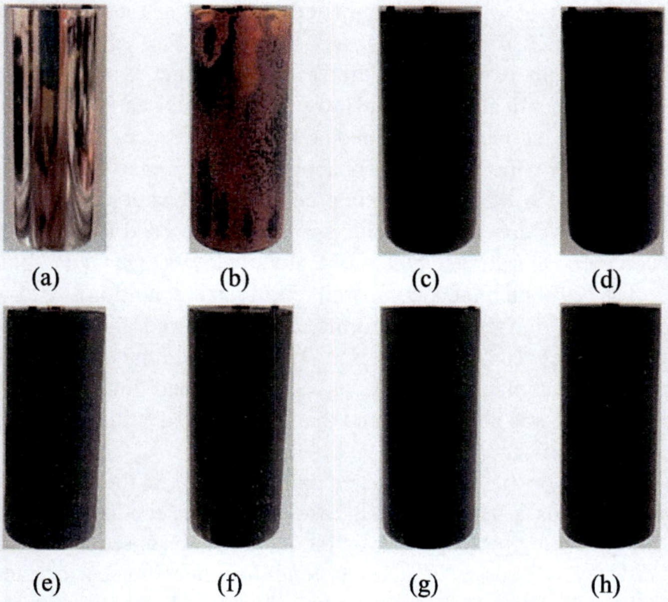

Fig. 4 The surface structure of copper rod during quenching in distilled water at saturated condition **a** after polishing, **b** 1st quench, **c** 2nd quench, **d** 3rd quench, **e** 4th quench, **f** 5th quench, **g** 6th quench and **h** 7th quench

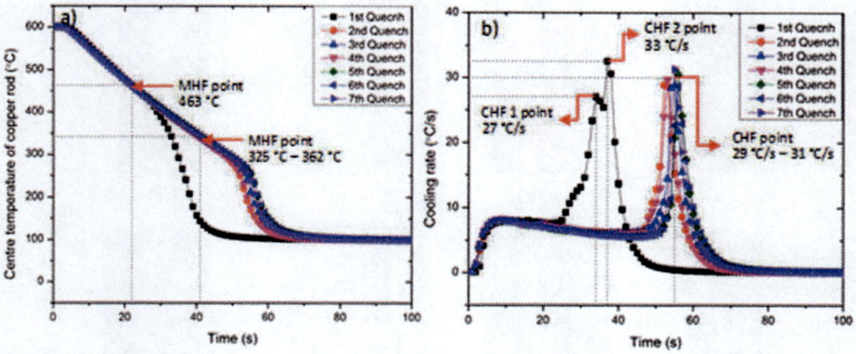

Fig. 5 The quenching performance of copper rod in distilled water at 90 °C subcooled conditions; **a** Cooling curve, and **b** Cooling rate curve

copper rod shows the decreasing of cooling time with a longer time taken to cool the rod. The subsequent quenching from the 2nd quench until the 7th quench also shows minimal changes in the graph pattern. The MHF point temperature for these series of quenching processes is between 325 °C and 362 °C, which between 17 and 20% of increment and the time taken to achieve this MHF point is between 37 and 44 s, reduction of time about 48%–54% compared to the case in a saturated condition, respectively (see Fig. 3a).

As shown in Fig. 5b, two peaks for the CHF point, which is similar to the case in saturated condition could also be seen. However, CHF 1 is nearly unnoticeable. The CHF 1 point during 1st quench happened at 34 s (42% reduction of time compared to saturated condition), just slightly early about 3 s compared to CHF 2 point, which occurred at 37 s (46% reduction of time compared to saturated condition) after the quench. The peak cooling rate value for CHF 1 recorded about 27 °C/s (54% increment compared to saturated condition), slightly lower than CHF 2 point, which is about 33 °C/s (29% increment compared to saturated condition). Afterwards, the subsequent quench from the second to the seventh showed almost the same peak cooling rate value between 29 and 31 °C/s. These denoted the enhancement of 13–21% of cooling rate compared to those in a saturated condition. The time duration to achieve this point was about 55 s, and this is a 44% reduction of time compared to saturated conditions.

As shown in Fig. 6c–h, the fully coated oxidation layer at the copper rod surface could be seen, and this is also obtained in case of saturated condition (see Fig. 4c–h). The different quenching performance for 1st quench with subsequent quench is due to the oxidation layer discussed in saturated condition quenching performance. Based on the cooling curves and cooling rate curve obtained, the cooling process occurs

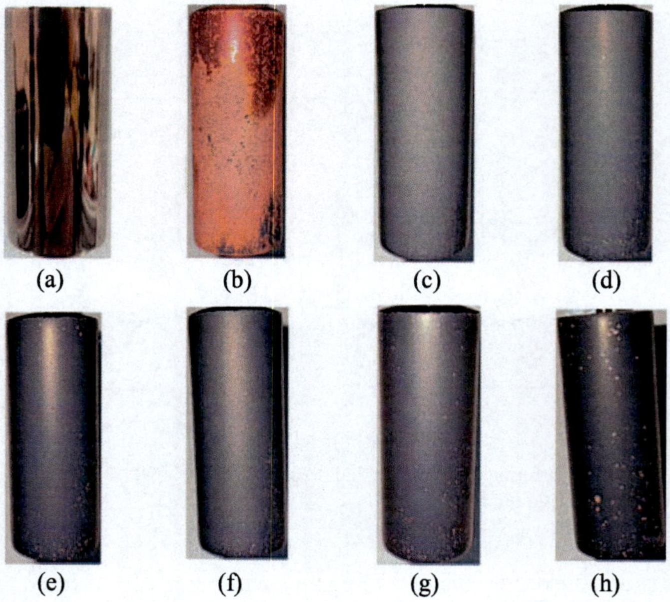

Fig. 6 The surface structure of copper rod during quenching in distilled water at slightly subcooled condition 90 °C **a** after polishing, **b** 1st quench, **c** 2nd quench, **d** 3rd quench, **e** 4th quench, **f** 5th quench, **g** 6th quench and **h** 7th quench

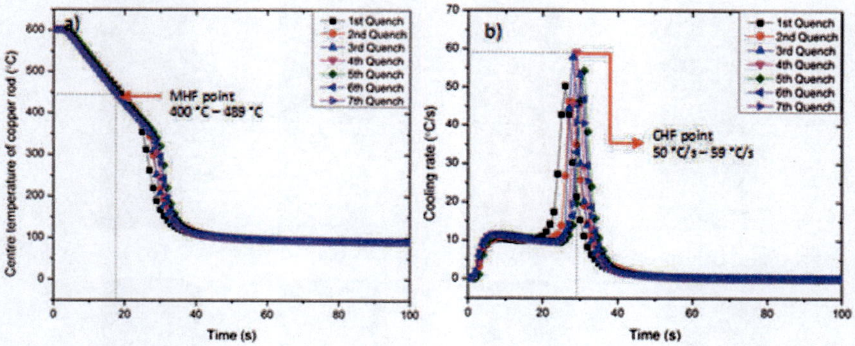

Fig. 7 The quenching performance of copper rod in distilled water at 80 °C subcooled conditions; **a** Cooling curve, and **b** Cooling rate curve

slightly faster than in saturated conditions. It clearly shows the effect of slightly subcooled conditions towards the quenching performance, which is similar to the results obtained by Lotfi et al. [8].

Figure 7a and b shows the cooling curves and cooling rate curves with multiple quenches of copper rod in distilled water at 80 °C subcooled conditions. Figure 7a Cooling curves maintained the three-section shape graph but with a high slope value, representing the high cooling rate. As referred to the curves, there is not much distinct cooling performance between the 1st quench and subsequent quench. In Fig. 7a, the MHF point was recorded at the temperature between 400 and 489 °C, which is an increment of between 5 and 10% compared to quench in 90 °C subcooled conditions (see Fig. 5a). The time taken to achieve MHF point is about 15–18 s, which is a 32–59% reduction of time compared to 90 °C subcooled conditions. The 1st quench performance has not much different from the subsequent quench that is in contrast with the result in saturated (see Fig. 3a) and 90 °C subcooled conditions (see Fig. 5a).

As shown in Fig. 7b, there is no formation of the double peak cooling rate value (double CHF point) during the 1st quench, which in contrast to the result obtained in saturated (see Fig. 3b) and 90 °C subcooled conditions (see Fig. 5b). The peak value of cooling rate for all quenching nearly similar performance ranging between 50 and 59 °C/s. It shows a significant increment of about 44–46% compared to 90 °C subcooled conditions. The quenching time taken to achieve the CHF point ranges between 26 and 31 s after the quench, which is a 24–44% reduction of time compared to 90 °C subcooled conditions (see Fig. 5b).

As shown in Fig. 8, the oxidation layer also occurred with the same pattern as shown in Fig. 8c–h. However, the cooling performance shows a different pattern with saturated (see Fig. 3a and b) and 90 °C subcooled conditions (see Fig. 5a and b). These phenomena are due to the domination of the subcooled condition factor towards the quenching performance. In 80 °C subcooled conditions with higher subcooling conditions, more heat is needed for sensible heat of the fluids surrounding, thinning the vapour film of vapour blanket during the film boiling regime [21]. Since in the

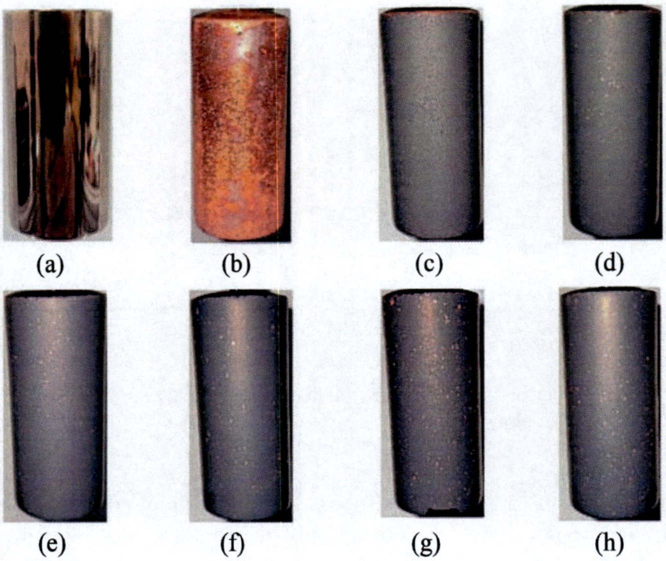

Fig. 8 The surface structure of copper rod during quenching in distilled water at subcooled condition 80 °C **a** after polishing, **b** 1st quench, **c** 2nd quench, **d** 3rd quench, **e** 4th quench, **f** 5th quench, **g** 6th quench and **h** 7th quench

present case, the 600 °C initial temperature of the specimen was insufficient to supply heat to sustain the formation of stable film boiling for a longer time. It is proven in previous research conducted by Nishio et al. [18] that the MFH temperature is proportional to the liquid subcooling, which is the higher liquid subcooling requires a higher temperature of the test specimen. Under a higher liquid subcooling condition, more heat is needed to sufficiently heat the surrounding fluid to form a stable vapour film.

Figure 9a and b shows the cooling curves and cooling rate curves with multiple quenches of copper rod in distilled water at 60 °C, which is in highly subcooled condition. The cooling curve in Fig. 9a depicted a steeper negative slope value representing the high cooling rate. From the cooling curves recorded, there is no clear visible three-section shape as obtained in the saturated (see Fig. 3a), 90 °C subcooled conditions (see Fig. 5a), and 80 °C subcooled conditions (see Fig. 7a). Once the specimen dropped into the highly subcooled water pool, the centre temperature decreased suddenly. Then, the boiling regime was changed quickly to the single-phase natural convection regime. In other words, the stable film boiling regime did not seem to be observed clearly for the present highly subcooled water condition, which is no stable vapour blanket happened. The vapour blanket is suspected of collapsing directly once the copper rod specimen quench in the distilled water, leading to the liquid-solid directly contacting each other [14]. The time taken for the copper rod specimen to cool down from the initial temperature of 600 °C to the saturation point of water of 100 °C takes about 22 s. With the large value of liquid subcooling, more portions

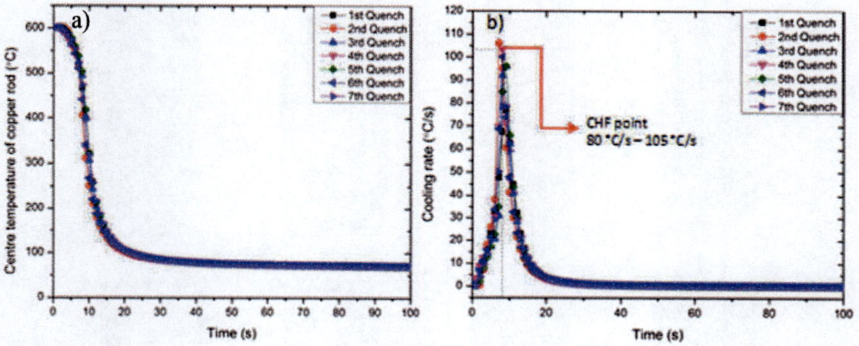

Fig. 9 The quenching performance of copper rod in distilled water at 60 °C subcooled conditions; **a** Cooling curve, and **b** Cooling rate curve

of the heat flux are used to heat the liquid surrounding, then the evaporation on the liquid-vapour interface is weak, and the vapour film is thin [14].

As depicted in Fig. 9b, the peak cooling rate was recorded at a very high value between 80 and 105 °C/s and its occurred less than 10 s, which is an extremely high cooling rate. As refer to Figs. 3b, 5b, 7b and 9b, the peak value of cooling rate trending shows increasing and the time to achieve CHF point shortened with increase the subcooled temperature. The increment was expected due to the rise in the subcooled temperature. The vapour film collapses early at high temperature, then re-establish the liquid-solid contact with rapid boiling with a sharp rise peak of cooling rate value at the copper rod surface [22].

Although the oxidation layer appearance as shown in Fig. 10 of the copper surface still have the same with others condition, the factor of the highly subcooled condition becomes dominant factors, then contribute to the curves obtained. The same issues were discussed for the quenching performance in 80 °C subcooled conditions, hence confirmed again with the result obtained by Nishio et al. [18].

Figure 11a shows the compilation of 1st cooling curves performance, while Fig. 11b shows the compilation of 1st cooling rate curves of copper rod in distilled water for all conditions, which is in saturated, 90 °C, 80 °C, and 60 °C subcooled conditions. As could be referred to in the cooling curves in Fig. 11a, the quenching performance sequential shift to the left as the fluids temperature decreases. The cooling curves for the subcooled of 90 and 80 °C still maintain the three-section shape similar to the saturated case. Quenching performance in saturated conditions shows the longer time taken to cool the rod. The MHF point temperature was recorded at 415 °C and occurred at 43 s after quenching. When the liquid is slightly subcooling 10 °C (90 °C subcooled conditions), the quenching time is markedly shortened to about 22 s at the temperature of 463 °C, equivalent to a relative acceleration by a factor of 49% to the saturated case. At the highest subcooling degree tested (60 °C subcooled conditions), the quenching time is below 20 s, corresponding to a relative cooling time enhancement by over 75% compared to saturated conditions.

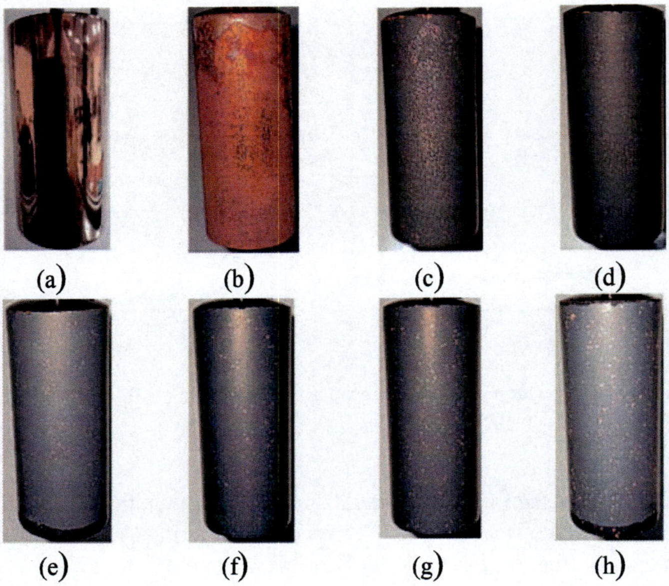

Fig. 10 The surface structure of copper rod during quenching in distilled water at highly subcooled condition 60 °C **a** after polishing, **b** 1st quench, **c** 2nd quench, **d** 3rd quench, **e** 4th quench, **f** 5th quench, **g** 6th quench and **h** 7th quench

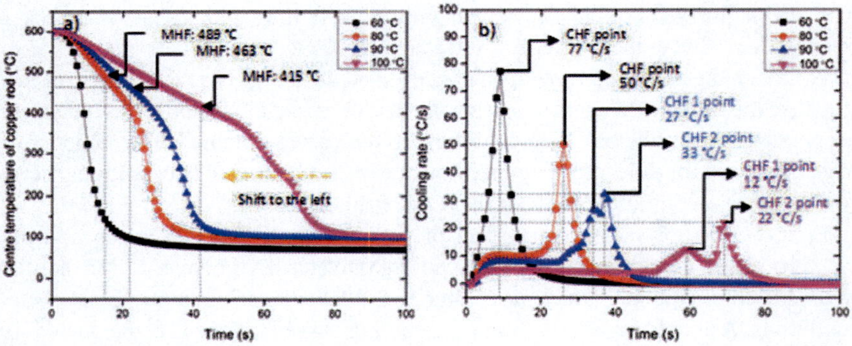

Fig. 11 The compilation of the 1st quench performance of copper rod in distilled water at saturated, 90, 80, and 60 °C subcooled conditions; **a** Cooling curve, and **b** Cooling rate curve

The cooling rate curve in Fig. 11b shows the same trend, which is with increasing the subcooled temperature, the value of the peak cooling rate becomes higher and the time taken to achieve the CHF point shortened. The highest value of the peak cooling rate is recorded about 77 °C/s during 1st quench in 60 °C subcooled condition, which is the increment of about 71% compared to CHF 2 in saturated condition and the time taken to achieve the CHF point during 1st quench in 60 °C subcooled conditions is about 9 s, which is 85% reduction of time compared to CHF 2 in a saturated condition. The double CHF (double peak cooling rate) was only visible during the 1st quench in saturated and 90 °C subcooled conditions, while for the subcooled 80 and 60 °C conditions, the double CHF was invisible. The quenching heat transfer enhancement upon increasing the subcooling degree is significantly attributed to the increasing temperature difference between the copper rod and liquid pool. Similar result with Fan et al. [23], they discovered that the cooling curves obtained were shifted to the left with sequentially increasing the slope value. They agreed that this enhancement is due to the temperature difference between the specimen and surrounding fluid.

4 Conclusion

In this study, the quenching heat transfer characteristics in different quenched conditions have been explored by performing the quenching experiments of a heated copper rod in distilled water at saturated and various subcooled conditions (60, 80, and 90 °C). A rod with 20 mm in diameter and 50 mm in length at the initial temperature of 600 °C was tested. The following results were obtained:

(a) The quenching performance in saturated shows the longer time taken to cool the rod. The three-section shape is clearly shown with the formation stable film boiling regime. There are two CHF points noticed in the cooling rate curve during 1st quench in saturated.
(b) In the slight subcooled condition of fluids (90 °C), the performance of the cooling rate has shown an enhancement with an increment of the peak cooling rate value and shortened the time to achieve CHF point, and double CHF point observed in the first quench.
(c) Both cooling performances of saturated and 90 °C subcooled conditions show a gap of cooling rate between the 1st quench and the subsequent quench, which is believed due to the formation of the oxidation layer at the copper rod surface.
(d) Cooling curve slopes in 80 °C subcooled conditions increase significantly, incrementing the peak cooling rate value and reducing the time to achieve CHF compared to saturated and 90 °C subcooled cases. Cooling curves data in 80 °C subcooled shows almost the same for all quench, although the same surface structure is obtained, which is due to the domination of the subcooled factors towards the quenching performance, with the low initial temperature of the specimen (600 °C).
(e) The cooling curve for the highly subcooled condition (60 °C) did not maintain the three-section shape, which the MHF point did not seem to be observed. The centre temperature decreases suddenly during quenching. The peak cooling rate and the time taken to achieve CHF point recorded the highest and faster value compared to others subcooled and saturated.

Acknowledgements Financial support by the Malaysian Ministry of Height Education under FRGS (FRGS/1/2019/TK07/UMP/02/2) and Universiti Malaysia Pahang (www.ump.edu.my) under PGRS2003148 are gratefully acknowledged. We wish to express our gratitude to all the staff in Universiti Malaysia Pahang that involved in the experimental work. Last but not least, thank you to the first author's wife, Nur Fatehah Binti Md Shakur for the continuous moral support and motivation during his study journey.

References

1. Babu K, Prasanna Kumar TS (2011) Effect of CNT concentration and agitation on surface heat flux during quenching in CNT nanofluids. Int J Heat Mass Transf 54(1–3):106–117. https://doi.org/10.1016/j.ijheatmasstransfer.2010.10.003
2. Fu BR, Ho YH, Ho MX, Pan C (2016) Quenching characteristics of a continuously-heated rod in natural seawater. Int J Heat Mass Transf 95:206–213. https://doi.org/10.1016/j.ijheatmasstransfer.2015.11.093
3. Kim H, DeWitt G, McKrell T, Buongiorno J, wen Hu L (2009) On the quenching of steel and zircaloy spheres in water-based nanofluids with alumina, silica and diamond nanoparticles. Int J Multiph Flow 35(5): 427–438. https://doi.org/10.1016/j.ijmultiphaseflow.2009.02.004
4. Boyer HE, Archambault P, Moreaux F, Kobasko NI (1992) Techniques of quenching. In: Liscic B, Tensi HM, Luty W (eds) Theory and technology of quenching. Springer, Berlin, Heidelberg. https://doi.org/10.1007/978-3-662-01596-4_10
5. Lee GC, Kim SH, young Kang J, Kim MH, Jo HJ (2019) Leidenfrost temperature on porous wick surfaces: decoupling the effects of the capillary wicking and thermal properties. Int J Heat Mass Transf 145:118809. https://doi.org/10.1016/j.ijheatmasstransfer.2019.118809
6. Gylys J, Skvorcinskiene R, Paukstaitis L, Gylys M, Adomavicius A (2016) International journal of heat and mass transfer film boiling influence on the spherical body's cooling in sub-cooled water. Int J Heat Mass Transf 95:709–719. https://doi.org/10.1016/j.ijheatmasstransfer.2015.12.051
7. Bolukbasi A, Ciloglu D (2007) Investigation of heat transfer by means of pool film boiling on vertical cylinders in gravity. Heat Mass Transf und Stoffuebertragung 44(2):141–148. https://doi.org/10.1007/s00231-007-0238-7
8. Lotfi H, Shafii MB (2009) Boiling heat transfer on a high temperature silver sphere in nanofluid. Int J Therm Sci 48(12):2215–2220. https://doi.org/10.1016/j.ijthermalsci.2009.04.009
9. Young C, Kim S (2017) Parametric investigation on transient boiling heat transfer of metal rod cooled rapidly in water pool. Nucl Eng Des 313:118–128. https://doi.org/10.1016/j.nucengdes.2016.12.005
10. Lee KG, In WK, Kim HG (2019) Quenching experiment on Cr-alloy-coated cladding for accident-tolerant fuel in water pool under low and high subcooling conditions. Nucl Eng Des 347(November 2018):10–19. https://doi.org/10.1016/j.nucengdes.2019.03.017
11. Wang Z, Zhong M, Deng J, Liu Y, Huang H, Zhang Y (2021) Annals of nuclear energy experimental investigation on the transient film boiling heat transfer during quenching of FeCrAl. Ann Nucl Energy 150:107842. https://doi.org/10.1016/j.anucene.2020.107842
12. young Kang J, Lee GC, Kim MH, Moriyama K, Park HS (2018) Subcooled water quenching on a super-hydrophilic surface under atmospheric pressure. Int J Heat Mass Transf 117:538–547. https://doi.org/10.1016/j.ijheatmasstransfer.2017.09.006
13. Ebrahim SA, Chang S, Cheung FB, Bajorek SM (2018) Parametric investigation of film boiling heat transfer on the quenching of vertical rods in water pool. Appl Therm Eng 140(March):139–146. https://doi.org/10.1016/j.applthermaleng.2018.05.021
14. Xiong J,Wang Z,Xiong P, Lu T,Yang Y (2020) Experimental investigation on transient boiling heat transfer during quenching of fuel cladding surfaces. Int J Heat Mass Transf 148(xxxx):119131. https://doi.org/10.1016/j.ijheatmasstransfer.2019.119131

15. Hsu SH, Ho YH, Ho MX, Wang JC, Pan C (2015) On the formation of vapor film during quenching in de-ionised water and elimination of film boiling during quenching in natural sea water. Int J Heat Mass Transf 86:65–71. https://doi.org/10.1016/j.ijheatmasstransfer.2015.02.049
16. Sher I, Harari R, Reshef R, Sher E (2012) Film boiling collapse in solid spheres immersed in a sub-cooled liquid. Appl Therm Eng 36:219–226. https://doi.org/10.1016/j.applthermaleng.2011.11.018
17. Jan J, Company FM, Mackenzie DS (2020) On the development of parametrical water quenching heat transfer model using cooling curves by ASTM D6200 quenchometer on the characterisation of heat transfer rate in various boiling regimes using quenchometers and its application for quenching process. no. July 2020. https://doi.org/10.1007/s11665-020-04803-z
18. Nishio S, Uemura M, Sakaguchi K (1987) Film boiling heat transfer and Minimum-Heat-Flux (MHF)-point condition in subcooled pool boiling. JSME Int J 30(266):1274–1281. https://doi.org/10.1299/jsme1987.30.1274
19. Umehara Y, Okawa T, Enoki K (2019) Evaluation of the performance of nanofluid as quenching coolant. Tetsu-To-Hagane/J Iron Steel Inst Japan 105(11):1050–1058. https://doi.org/10.2355/tetsutohagane.TETSU-2019-035
20. Carvill J (1993) 3 - Thermodynamics and heat transfer. In: Carvill J (ed) Mechanical engineer's data handbook. Butterworth-Heinemann, pp 102–145. https://doi.org/10.1016/B978-0-08-051135-1.50008-X
21. Li JQ et al (2018) An experimental study of the accelerated quenching rate and enhanced pool boiling heat transfer on rodlets with a superhydrophilic surface in subcooled water. Exp Therm Fluid Sci 92(November 2017):103–112. https://doi.org/10.1016/j.expthermflusci.2017.11.023
22. Li JQ, Mou LW, Zhang YH, Yu JQ, Fan LW, Yu ZT (2018) Pool boiling heat transfer and quench front velocity during quenching of a rodlet in subcooled water: Effects of the degree of subcooling. Exp Heat Transf 31(2):148–160. https://doi.org/10.1080/08916152.2017.1397819
23. Fan L, Li J, Zhang L, Yu Z, Cen K (2016) Pool boiling heat transfer on a nanoscale roughness-enhanced superhydrophilic surface for accelerated quenching in water. Appl Therm Eng 109:630–639. https://doi.org/10.1016/j.applthermaleng.2016.08.131

The Effect of Trawl Activities to Subsea Pipelines of East Coast Peninsular Malaysia: A Risk Analysis

Ahmad Faizal Ahmad Fuad, Mohd Hafizi Said, Khalid Samo, Mohd Hairil Mohd, Fatin Alias, and Mohd Asamudin A. Rahman

Abstract Subsea pipeline is prone to the damage due to fishing activities. Bottom trawling is one of the threats to the pipeline integrity. Fishermen used bottom trawling method to catch large volume of fish and benthic along the trawl route on the seabed. This causes the trawl net to be stuck on the subsea pipeline if their activity is close to the oil and gas platform. This paper evaluates the risk of subsea pipeline fish trawling operations on the east coast of offshore Peninsular Malaysia. On-site surveys were used to establish the characteristics of trawling devices used by local trawlers in the region. Based on site survey that was conducted, the frequency of fish trawlers crossing the pipelines is estimated. The pull-over load estimate for the otter board was determined using the Det Norske Veritas Germanischer Lloyd (DNVGL) algorithm. The severity and frequency index of the risk matrix was developed based on literature review. The findings showed that the pull-over load of otter board would not damage the pipelines. The risk presented to the pipelines by the operations of fish is considered low and moderate.

Keywords Risk Assessment · Trawl · Subsea pipeline

1 Introduction

The east coast of Peninsular Malaysia located opposite the South China Sea and adjacent to the states (Kelantan, Terengganu and East Johor). The waters of east coast Peninsular Malaysia are an important fishing area with an exclusive economic zone. The total fisheries catch in east coast Peninsular Malaysia waters in 2018 is

A. F. Ahmad Fuad · M. H. Said
Nautical Science Programme, Faculty of Maritime Studies, Universiti Malaysia Terengganu, Kuala Nerus, 21030 Terengganu, Terengganu, Malaysia

K. Samo · M. H. Mohd · F. Alias · M. A. A. Rahman (✉)
Maritime Technology and Naval Architecture Programme, Faculty of Ocean Engineering and Technology & Informatics, Universiti Malaysia Terengganu, Kuala Nerus, 21030 Terengganu, Terengganu, Malaysia
e-mail: mohdasamudin@umt.edu.my

M. Y. Ismail et al. (eds.), *Technological Advancement in Mechanical and Automotive Engineering*, Lecture Notes in Mechanical Engineering,
https://doi.org/10.1007/978-981-19-1457-7_30

182,341 metric tons [1]. This catch is 41.3% of the total catch of waters in Peninsular Malaysia and the main method is the trawling where fish nets are pulled by using boats. There are two methods of trawling, namely bottom trawling [2] and mid-water trawling [3]. The method used in east coast Peninsular Malaysia waters is the bottom-trawling, similar with reported data from Sabah and Labuan waters [4]. In addition, previous study from Althaus et al. [5] describes the impact of bottom trawling on habitats and ecological communities. Hobday et al. [6] studied the effect of ecological risk assessment for fishing over 2000 species and more than 200 habitat types. Currently, the number of fish trawl in east coast Peninsular Malaysia in 2017 is 1004 and 3736 in 2018. The number in 2018 had increased 3.72 times than in 2017.

In addition to fisheries industries, oil and gas industries is the major industries in east coast of Peninsular Malaysia with oil platforms located from sea border with Thailand (Cakerawala block) to the sea border with Indonesia (Natuna block).

Subsea pipelines are installed on the seabed of the east coast of Peninsular Malaysia to transport oil and gas at Kerteh Terengganu. However, the same sea areas that contain these subsea pipelines are also a fishing ground. The most commonly used fishing method in the area, namely bottom-trawling method would definitely contact the trawl gear and the subsea pipelines. Subsequently, it may affect the subsea pipelines, which in the worst scenario it would cause the pipe to leak and resulted a marine pollution. The contact is categorized as the third-party resulting from human activities at sea such as trawling, anchoring, and dropped objects [7].

To determine the severity of the impact, DNV GL had developed a guideline to compute the pull-over load and hooking of trawl gear on the subsea pipelines [8]. DNV GL also had developed a risk assessment method to assess pipeline protection against accidental external loads [9]. Several studies have been carried out on the effect of trawling equipment on subsea pipelines [10–14]. Meanwhile, Kristoffersen et al. [15], Hinton and Procter [16], Gong and Xu [17], Wan et al. [18] studied performance of X65 pipeline in the time of impact.

Recently, Herlianto et al. [19] demonstrated the influence of subsea pipeline consequently of the load trawl gear pull-over. This external interference may lead to a failure on the pipeline. For instance, it made unpleasant lateral buckling and create unnecessary bending moments and strains. In another study, Yu et al. [20] carried out an experiment on a pipeline deformation due to an anchor collision and measured the results of the collision path numerically.

Similarly, Longva et al. [21] performed a fishing gear load test on subsea pipeline The friction of the board-pipe, the tension of the wire between the board and the trawl net, the towing line drag characteristics, and the direction of over-trawling were all elements that influenced load, according to the results. Currently, a new approach was proposed by Wu et al. [22] to measure the possibility of hooking trawl boards via simulation tools along with statistical data. In 2017, DNV revised its proposed intrusion procedure between trawl gear and pipelines [23].

To date, no research performed on the operations of fish trawlers on subsea pipelines offshore east coast Peninsular Malaysia. So, this study is to elucidate the effect of trawler pull-over load on subsea pipelines at the east coast of Peninsular Malaysia to Kerteh shore refinery terminal.

2 Materials and Methods

The measurement of morphological parameters included the scour depth that was conducted by various researchers. Validation among the current model and the earlier simulation results of the sediment scour model will be presented to determine the local scour under complex form. Equations (1)–(6) used to compute of the force of the pull-over load [8].

The equation for Pull-Over Load of an otter board or trawl door [24] as below:

$$F_P = C_F.V(m_t k_w)^{1/2} \tag{1}$$

where:

F_P is Pull-Over Load of an Otter board

k_w is warp line stiffness

V is trawling velocity

m_t is steel mass of board/beam with shoes

C_F is an empirical coefficient.

C_F is determined by:

$$C_F = 8.0 \cdot \left(1 - e^{-0.8\overline{H}}\right) \tag{2}$$

where $\overline{H}$ is a dimensionless height:

$$\overline{H} = \frac{H_{sp} + OD/2 + 0.2}{B} \tag{3}$$

where:

H_{sp} is the span height (negatively for the partly buried trenched pipeline)

OD is the pipeline outer diameter including coating

B is half of the trawl board height.

The warp stiffness, k_w is assumed as:

$$k_w = \frac{3.5.10^7}{L_W} \tag{4}$$

where L_W is the length of warp line in meter.

Pull-Over Force of Clump Weight:

$$F_P = 3.9 \cdot m_t \cdot g \cdot \left(1 - e^{-1.8 \cdot h'}\right) \cdot \left(\frac{OD}{L_{clump}}\right)^{-0.65} \tag{5}$$

Table 1 Frequency index of fishing vessel crossing

Index	Frequency index	Data range
0	Extremely low	0–900
1	Very low	901–1800
2	Low	1801–2700
3	Slight low	2701–3600
4	Moderate	3601–4500
5	Slight high	4501–5400
6	High	5401–6300
7	Very high	6301–7200
8	Extremely high	7201–8100

where:

F_P maximum horizontal pipeline pull-over force from clump weights.

OD is the outer diameter of pipe including coating.

L_{clump} is the distance from reaction point to centre of gravity of the clump weight.

m_t is the steel mass and g is the gravitational acceleration.

$$L_{clump} = 0.55 \cdot \left(0.5 \cdot D_{drum} + A_{clump}/0.5 \cdot 0.53 + 0.28\right) \tag{6}$$

Table 1 shows the frequency index developed in this study and used a 9-point scale with an increment of 900 across groups varying from 0 to 8100 frequencies. Based on a DNV-RP-F111 [8], the table is developed by changing Table Criteria for score assessment, shipping loss threats, emergency anchoring and dragged anchors from anchored ships. For the number of crossings less than 90,000, from 90,000 to 180,000 and over 180,000 respectively, the table has 3 ratings as follows: 0, 1, and 2. The adaptation was taken from the DNV GL table by taking 1% of the scores of each category. The pipeline impact of the trawl gear at each crossing will increase the chance of impact to 100% as a result of this adjustment.

Table 2 shows the Risk Matrix table for pipeline fishing operations to determine the risk index's results. The severity and frequency indices were combined to calculate the total of the frequency and severity for a specific pipeline. Meanwhile, Table 3 demonstrates the product of the summation and the associated risk rating. There are five stages of the danger matrix, from extremely low to maximum high danger.

3 Results and Discussion

There are 27 pipelines of various lengths and grade in the waters of the east coast of Peninsular Malaysia. However only 7 pipelines were selected in this study based

Table 2 Risk Matrix Table for Fishing Activities on Pipeline

	Severity	0	1	2	3	4
Frequency		No Damage	Slight Damage	Minor Damage	Local Damage	Major Damage
0	Extremely Low	0	1	2	3	4
1	Very Low	1	2	3	4	5
2	Low	2	3	4	5	6
3	Slight Low	3	4	5	6	7
4	Moderate	4	5	6	7	8
5	Slight High	5	6	7	8	9
6	High	6	7	8	9	10
7	Very high	7	8	9	10	11
8	Extremely High	8	9	10	11	12

Table 3 Risk matrix table of fishing activities impact on pipeline

Meaning	Risk Matrix	Colour
Very Low Risk (VL)	0-3	
Low Risk (L)	4-5	
Moderate (M)	6-7	
High Risk (H)	8-10	
Very High Risk (VH)	11-12	

on length of 100 km and above. This length was chosen because it would have a significant number of crossings compared to less than 100 km. Figure 1 shows that location on chart of Pipeline ID 01, pipeline PL02, Pipeline PL03, Pipeline PL04, Pipeline PL05, Pipeline PL06 and Pipeline PL07 respectively. The details of 7 pipelines are shown in Table 4. Note that numbering of these pipeline was not an actual name to protect its confidentiality.

3.1 Trawl Gear Specification

The type of otter trawl gear used on Peninsular Malaysia's east coast is the conventional otter trawl gear with a polyvalent or rectangular board. This type of trawl gear is made up of a pair of otter boards, a warp line, and a net. Trawlers employ two types of otter boards on the eastern coast of Peninsular Malaysia, namely steel and

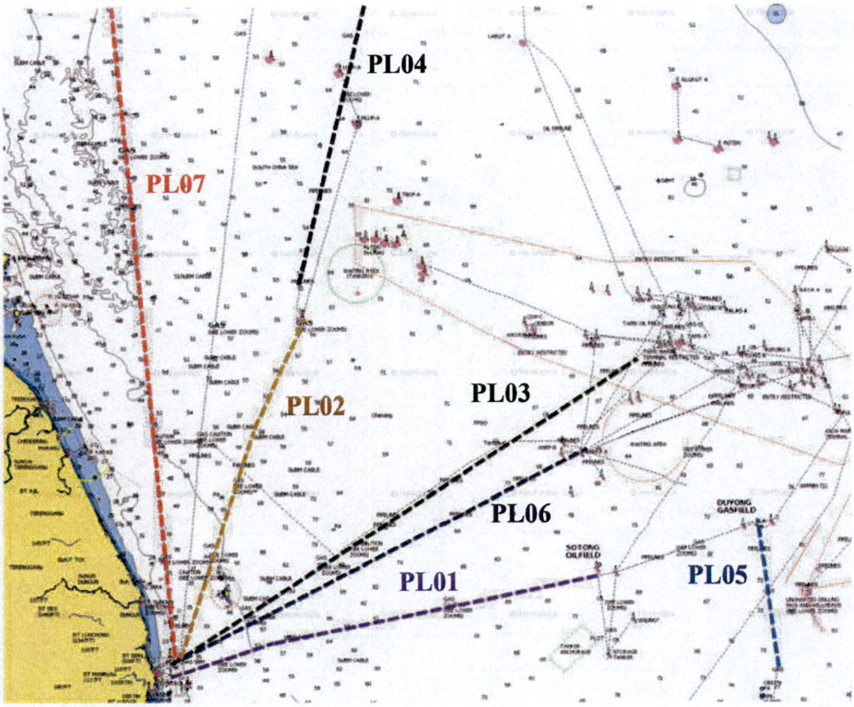

Fig. 1 Pipeline layout for the present study in east coast peninsula Malaysia. Pipeline PL01 (Kerteh Terminal to Sotong SCPA), PLO2 (RDPA Resak to Kerteh Terminal), PLO3 (ANDRA Angsi to Kerteh Terminal), PLO4 (Bunga Raya to Resak), PLO5 (West Natuna to Duyong CPP), PLO6 (ANDRA Angsi to Kerteh Terminal), and PLO7 (PLTS – TGAS)

Table 4 Pipelines of East Coast Offshore Peninsular Malaysia

No	Pipeline ID	Overall Diameter OD (mm)	Wall thickness (mm)	Length (km)	Material Grade	Yield strength (kN/m^2)
1	PL01	762	20.62	155.7	5LX-60	4.2×10^5
2	PL02	711.2	25.1	135	5LX-65	4.5×10^5
3	PL03	830.8	28.5	164	5L-X65	4.5×10^5
4	PL04	610.0	21.7	164	5L-X65	4.5×10^5
5	PL05	457.2	17.5	100	5L-X60	4.2×10^5
6	PL06	610.0	23.8	165.5	5L-X65	4.5×10^5
7	PL07	711	18.3	290	5L-X65	4.5×10^5

steel-reinforced timber, as depicted in Fig. 2. These otter-boards require a strong fish trawler capable of towing 3 knots of heavy fishing gear. As a result, the turbine's horsepower ranges from 350 and above. The majority of trawler engines have horsepower ranging from 350 to 500 horsepower. Then, the thickness of the wrap line is 2.5 cm. The foreign fish trawl vessels that illegally fishing in east coast Peninsular Malaysia were using a pair trawl technique. In this method, no otter trawl board was used, instead two vessels are used to open and tow the trawl net. However, rolling bobbins (Fig. 3) is used as a weight and roller at bottom of the trawl net.

Fig. 2 Steel vee door and wooden door reinforced by steel frame used by fishermen in Terengganu

Fig. 3 The ground gear of pair trawl nets fitted with rolling bobbin used by a foreign fish trawl

3.2 Calculation of Pull-Over Load and Pull-Over Force

Equations (1)–(4) is used to calculate Pull-Over Load. Pull over calculation is specifically for the single otter board application, which being used by the local fishing trawler. Trawl depth, trawling speed of 3 knots (1.54 m/s), 300 kg steel vee door standard (Table 5) and span height on the seabed are the inputs from equations. The pipelines have all been buried. However, the span height is zero because the pipes are buried beneath the seabed [25]. Water depth and diameter of the pipe that determine the pull-over load for a particular pipe as shown in Table 6. Pipeline PL03 receives the highest pull-over load because it has the largest pipe diameter and located in deeper water compared to other pipelines. Pipeline PL05 receives the lowest pull-over load because it has the smallest diameter although has the same depth as PL03. The result of pull-over load for each pipeline is compare with the level force of damage in Table 7. None of the pull-over load force for each pipeline achieves its slight damage force. Therefore, the otter trawl board's pull-over load not affect the pipelines. Nevertheless, the yield power of the pipelines may be lesser than the current condition because of age factor and corrosion [26].

The Eqs. (5), (6) were used to calculate pull-over force. The calculation is to determine the force of rolling bobbins used by the foreign fish trawl, illustrated in

Table 5 Type of Fish Trawl Otter-board used in Peninsula Malaysia

Trawler categories	Types of otter-boards	Dimensions in cm (L × W × T)	Material used	Weight (kg) of one otter board
350 HP and above	Steel Vee Door	190 × 106 × 6.5	Steel	300 kg
350 HP and above	Common Flat Wooden Door	230 × 115 × 5	Steel Frame and Wood	250 kg
350 HP and above	Steel Vee Door	165 × 110 × 5	Steel	210 kg

Table 6 Results of Pull-over Load for Otter Board and Pull-over Force for Rolling Bobbin for each Pipeline

Pipeline ID	Max. water depth (m)	Fishing Boat Crossing Frequency per year	Pull-over load otter board (N)	Pull-over force clump weight (N)
PL01	69	447	44,408	5.30
PL02	68	1275	41,925	5.55
PL03	73	4129	44,796	5.01
PL04	68	202	40,723	6.13
PL05	73	17	34,978	7.39
PL06	73	4129	39,304	6.13
PL07	50	253	50,653	5.55

Table 7 Results on level of damage according to the given force

Pipeline ID	Thickness	Level of damage	Force (kN)	Yield Strength (kN/m^2)	Pull-over load otter board F_P (N)	Pull-over load rolling bobbin F_P (N)
01	20.62	Global	372.178	4.20E + 05	44,408	5.30
		Local	274.933			
		Minor	157.832			
		Slight	101.827			
02	25.1	Global	453.040	4.50E + 05	41,925	5.55
		Local	334.667			
		Minor	192.123			
		Slight	123.951			
03	28.5	Global	514.407	4.50E + 05	44,796	5.01
		Local	380.000			
		Minor	218.148			
		Slight	140.741			
04	21.7	Global	391.672	4.50E + 05	40,723	6.13
		Local	289.333			
		Minor	166.099			
		Slight	107.160			
05	17.5	Global	315.864	4.20E + 05	34,978	7.39
		Local	233.333			
		Minor	133.951			
		Slight	86.420			
06	23.8	Global	429.575	4.50E + 05	39,304	6.13
		Local	317.333			
		Minor	182.173			
		Slight	117.531			
07	18.3	Global	330.304	4.50E + 05	50,653	5.55
		Local	244.000			
		Minor	140.074			
		Slight	90.370			

Table 8 Results of Risk Matrix for Each Pipeline

Pipeline ID	Yield Strength (kN/m^2)	Freq. of Crossing	Pull-over load Otter board (kN)	Freq. Index	% YS pull-over load	Severity Index	Risk matrix
PL 01	4.2×10^5	1978	44.408	2	10.57	0	2(L)
PL 02	4.5×10^5	1225	41.925	1	9.32	0	1(VL)
PL 03	4.5×10^5	4129	44.796	4	9.95	0	4(L)
PL 04	4.5×10^5	202	40.723	0	9.05	0	0(VL)
PL 05	4.2×10^5	17	34.978	0	8.33	0	0(VL)
PL 06	4.5×10^5	4129	39.304	4	8.73	0	4(L)
PL 07	4.5×10^5	253	50.563	0	11.26	0	0(VL)

Fig. 3. The findings presented that the pull-over force of the rolling bobbins far less than pull-over the force of the otter trawl board. Furthermore, the weight of rolling bobbins is only 2.5 kg each compared to the weight of the otter board of 300 kg. Therefore, none of the pull-over force of the rolling bobbin may damage the pipelines. However, snagging destroy fishing tools, interrupt fishing activities and can also trigger crew injuries [10]. On the other hand, no fish trawlers in the sample region had snagged the pipelines with their fishing gear. This may be attributed to the muddy seabed being buried under the pipelines. Consequently, Pipelines were not seen as a snagging hazard by the fishermen.

3.3 Frequency Index, Severity Index and Risk Matrix for Fishing Activities

The result of the severity index and frequency and correspond risk matrix are shown in Table 8. The score of risk matrix for pipelines ID 04, 05, and 07 score are 0 (Very Low). The score for pipelines ID PL01, PL02, and PL03 are 2 (Very Low), 1 (Very Low), and 4 (Low) respectively. The difference of the risk matrix score is due to the difference in frequency of crossing by vessels, while the severity remains the same for all pipelines. The ranking of risk are as follows, PL03 (4), PL06 (4), PL07 (2), PL02 (1), PL04 (0), PL05 (0), and PL07 (0). The frequency of crossing for PL03 and PL06 are high because the pipelines are located along the coastal shipping route and both pipelines are located side by side. While pipelines PL04, PL05, and PL07 are located outside the shipping route.

4 Conclusions

This research was conducted to investigate the influence of trawler pull-over load on subsea pipelines at the east coast of Peninsular Malaysia to Kerteh shore refinery terminal. Therefore, seven pipelines were assessed in this study. The threat to the pipelines comes mainly from local and foreign trawlers. The local trawlers are using the otter trawl method, while the foreign trawlers are using pair trawl method. Generally, the impact force from the local trawler is much higher than a foreign trawler because local trawlers are using the heavy otter board (250 kg), while the pair trawl is using rolling bobbin (2.32 kg) as part of the ground gear. The impact force from both the otter board and rolling bobbin would produce no damage to the pipeline based on the criteria depicted in the risk matrix. The ranking of risk are as follows, PL01 (4), PL02 (4), PL03 (2), PL04 (1), PL05 (0), PL06 (0), and PL07 (0). The east coast Peninsular Malaysia waters is frequently trawled and the vast areas available as the fishing grounds by trawlers both local and illegal foreign fishers.

Acknowledgements The research was funded by the Ministry of Higher Education Malaysia FRGS/1/2018/TK01/UMT/02/1 (VOT-59503).

References

1. Department of Fisheries Malaysia, Landings of marine fish: tonnage class and fishing gear group (2018)
2. Goode SL, Rowden AA, Bowden DA, Clark MR (2020) Resilience of seamount benthic communities to trawling disturbance. Mar Environ Res 161:105086. https://doi.org/10.1016/j.marenvres.2020.105086
3. Seafish (2018) Basic fishing methods. Hull, UK
4. Ahmad Fuad AF, Said MH, Samo K, Rahman MAA, Mohd MH, Zainol I (2020) Risk assessment of fishing trawl activities to subsea pipelines of Sabah and Labuan waters. Sci World J. https://doi.org/10.1155/2020/6957171
5. Althaus F, Williams A, Schlacher TA, Kloser RJ, Green MA, Barker BA, Bax NJ, Brodie P, Schlacher-Hoenlinger MA (2009) Impacts of bottom trawling on deep-coral ecosystems of seamounts are long-lasting. Mar Ecol Prog Ser 397:279–294. https://doi.org/10.3354/meps08248
6. Hobday AJ, Smith ADM, Stobutzki IC, Bulman C, Daley R, Dambacher JM, Deng RA, Dowdney J, Fuller M, Furlani D, Griffiths SP, Johnson D, Kenyon R, Knuckey IA, Ling SD, Pitcher R, Sainsbury KJ, Sporcic M, Smith T, Turnbull C, Walker TI, Wayte SE, Webb H, Williams A, Wise BS, Zhou S (2011) Ecological risk assessment for the effects of fishing. Fish Res 108:372–384. https://doi.org/10.1016/j.fishres.2011.01.013
7. The Hydrographer RMN: Kertih Kawasan Galian Minyak Tapis MAL655. Pulau Indah (2007)
8. DNVGL-RP-F111: interference between trawl gear and pipelines (2017a) Health and safety executive, London, UK
9. DNVGL-RP-F107: risk assessment of pipeline protection (2017b) DNV, Høvik, Norway
10. Rouse S, Kafas A, Catarino R, Peter H (2018) Commercial fisheries interactions with oil and gas pipelines in the North Sea: considerations for decommissioning. ICES J Mar Sci 75:279–286. https://doi.org/10.1093/icesjms/fsx121

11. Shuai Y, Shuai J, Zhang X (2018) Experimental and numerical investigation of the strain response of a dented API 5L X52 pipeline subjected to continuously increasing internal pressure. J Nat Gas Sci Eng 56:81–92. https://doi.org/10.1016/j.jngse.2018.05.037
12. Underwood MJ, García-Seoane E, Klevjer TA, Macaulay GJ, Melle W (2020) An acoustic method to observe the distribution and behaviour of mesopelagic organisms in front of a trawl. Deep Res Part II Top Stud Oceanogr. https://doi.org/10.1016/j.dsr2.2020.104873
13. Zheng J, Palmer A, Brunning P, Gan CT (2014) Indentation and external pressure on subsea single wall pipe and pipe-in-pipe. Ocean Eng 83:125–132. https://doi.org/10.1016/j.oceaneng.2014.03.028
14. Vestrum O, Kristoffersen M, Polanco-Loria MA, Ilstad H, Langseth M, Børvik T (2018) Quasi-static and dynamic indentation of offshore pipelines with and without multi-layer polymeric coating. Mar Struct 62:60–76. https://doi.org/10.1016/j.marstruc.2018.08.001
15. Kristoffersen M, Børvik T, Westermann I, Langseth M, Hopperstad OS (2013) Impact against X65 steel pipes—an experimental investigation. Int J Solids Struct 50:3430–3445. https://doi.org/10.1016/j.ijsolstr.2013.06.013
16. Hinton BRW, Procter RPM (1983) The effect of strain-rate and cathodic potential on the tensile ductility of X-65 pipeline steel. Corros Sci 23:101–123. https://doi.org/10.1016/0010-938X(83)90110-5
17. Gong S, Xu P (2016) The influence of sea state on dynamic behaviour of offshore pipelines for deepwater S-lay. Ocean Eng 111:398–413. https://doi.org/10.1016/j.oceaneng.2015.11.013
18. Wan H, Du C, Liu Z, Song D, Li X (2016) The effect of hydrogen on stress corrosion behavior of X65 steel welded joint in simulated deep sea environment. Ocean Eng 114:216–223. https://doi.org/10.1016/j.oceaneng.2016.01.020
19. High T, Subsea P, Karunakaran D (2017) Lateral buckling induced by trawl gears pull-over loads on high temperature/high pressure subsea pipeline. In: Proceedings of the ASME 2012 31st international conference on ocean, offshore and arctic engineering. pp 1–6
20. Yu JX, Zhao YY, Li TY, Yu Y (2016) A three-dimensional numerical method to study pipeline deformations due to transverse impacts from dropped anchors. Thin-Walled Struct. 103:22–32. https://doi.org/10.1016/j.tws.2016.02.006
21. Longva V, Sævik S, Levold E, Ilstad H (2013) Dynamic simulation of subsea pipeline and trawl board pull-over interaction. Mar Struct 34:156–184. https://doi.org/10.1016/j.marstruc.2013.09.004
22. Wu X, Longva V, Sævik S, Moan T (2015) A simplified approach to estimate the probability of otter board hooking at pipelines. J Offshore Mech Arct Eng 137:061702. https://doi.org/10.1115/1.4031670
23. DNVGL-RP-F107: recommended Failure Rates for Pipelines (2017c). DNV, Høvik, Norway
24. Jun CW, Sohn JH, Park SH (2016) Effect analysis of design parameter of trawl door by considering interaction with seabed. Int J Precis Eng Manuf 17:1515–1522. https://doi.org/10.1007/s12541-016-0177-7
25. Taghizadeh Edmollaii S, Edalat P (2017) Accidental limit state of submarine pipeline: trawl gears pull-over loads and effect of free span. Int J Marit Technol 8:47–58. https://doi.org/10.29252/ijmt.8.47
26. Rouse S, Kafas A, Hayes P, Wilding TA (2017) Development of data layers to show the fishing intensity associated with individual pipeline sections as an aid for decommissioning decision-making. Underw Technol 34:171–178. https://doi.org/10.3723/ut.34.171

Effect of Pullulan Amount on ZnO NPs Via Sol–Gel Technique

Eleen Dayana Mohamed Isa, Kamyar Shameli, Nurfatehah Wahyuny Che Jusoh, Roshasnorlyza Hazan, and Nor Azwadi Che Sidik

Abstract Zinc oxide nanoparticles (ZnO NPs) is a very attractive materials due to their favourable properties that can be applied in various applications. These desirable characteristics of ZnO NPs can be tailored based on the synthesis process. With the increasing focus towards environmentally friendly process, green synthesis has been gaining popularity as the preferable approach. This study focused on the ZnO NPs' synthesis through sol–gel process in the presence biopolymer, pullulan. The impact of pullulan's amount on the ZnO NPs' characteristics were studied. Based on the results obtained, the general trend observed was that the particles size decreased with increasing pullulan amount. The determined band gap of ZnO NPs was found to be approximately between 3.26 and 3.28 eV. Overall, these results indicate that the properties of ZnO NPs is dependent on pullulan amount. These green synthesized ZnO NPs can be applied across various fields such as pharmaceuticals, cosmeceuticals, environmental and others.

Keywords Biopolymer · Pullulan · Zinc oxide nanoparticles

1 Introduction

Zinc oxide nanoparticles (ZnO NPs) very attractive metal oxide material due to its capability to be applied in multiple applications across various fields. The main properties that contribute to ZnO NPs application in various fields are its non-toxicity,

E. D. M. Isa · K. Shameli (✉) · N. W. C. Jusoh · N. A. C. Sidik
Malaysia-Japan International Institute of Technology, Universiti Teknologi Malaysia, Jalan Sultan Yahya Petra, 54100 Kuala Lumpur, Malaysia
e-mail: kamyar@utm.my

N. W. C. Jusoh
Advanced Materials Research Group, Center of Hydrogen Energy, Universiti Teknologi Malaysia, 54100 Kuala Lumpur, Malaysia

R. Hazan
Malaysian Nuclear Agency, 43000 Bangi, Kajang, Selangor, Malaysia

M. Y. Ismail et al. (eds.), *Technological Advancement in Mechanical and Automotive Engineering*, Lecture Notes in Mechanical Engineering,
https://doi.org/10.1007/978-981-19-1457-7_32

high stability, low cost, high melting point and others [1]. It is a n-type semiconductor with a broad direct band gap width (~3.37 eV), large excitation binding energy (60 meV) and activated by UV [2]. It exists in three crystal structures such as hexagonal wurtzite, cubic zinc blende and cubic rocksalt where in ambient condition, wurtzite structure is the most common as it is thermodynamically stable [3, 4]. The most common ways in synthesizing ZnO NPs are through physical, chemical and green synthesis, commonly known as biosynthesis, method. Among these methods, biosynthesis emerged as the preferable method and has been gaining popularity in these recent years due to its environmentally friendly approach, simple and fast.

Synthesis technique that utilized natural products is the common definition of green synthesis [5]. Previous report showed that ZnO NPs produced via green synthesis exhibited an improvement in terms of size and morphology over physical and chemical synthesis [6]. The some of the most attractive features of green synthesis are the feasible and easy fabrication process, low to almost non-toxic to the environment and cheap [7, 8]. These features are aligned with green chemistry principle thus make them a preferable method in synthesizing ZnO NPs [9, 10]. Overall, there are three main biomaterials that have been used in ZnO NPs production which are plant extracts, microorganisms and biopolymers [5]. The most common biomaterials being used is plant extract. This extract can be obtained through any parts of the plants. Plant extracts have various phytochemical compounds which can act as, either or all, reducing, stabilizing and capping agents [11]. A study reported on the fabrication of ZnO NPs via precipitation method using the *Butea monosperma* seed's extract. By mixing the zinc salt precursor and extract, precipitation occurred and this precipitate was reported to be ZnO NPs. ZnO NPs with good dispersion was obtained with higher concentration of the extract [12]. Besides plant extract, microorganism was also used as biomaterials but to a lesser extent than plant extract. It is not being used as frequently due to its high cost and time. Saravanakumar and colleagues described the generation of ZnO NPs using fungi. The synthesized ZnO NPs exhibited spherical shape with mean particle size of approximately 30 nm [13]. Another biomaterial that is relatively low its usage in ZnO NPs production is biopolymer. There various classes of biopolymers. A work that utilized biopolymer, specifically mixture of xanthan gum and konjak gun, in ZnO NPs production. Uniform ZnO NPs was obtained with 0.8 wt% of polysaccharide [14].

In this article, the ZnO NPs were produced via sol–gel technique in the presence of pullulan where it serves the capping agent. To the best of our knowledge, our group are the first to use biopolymer pullulan in the synthesis of ZnO NPs. The effect pullulan's amount on s of ZnO NPs were investigated.

2 Materials and Methodology

2.1 Materials

Pullulan powder and analytical grade chemical, zinc nitrate hexahydrate ($Zn(NO_3)_2 \cdot 6H_2O$), were obtained from China and local company, R & M Chemicals respectively.

2.2 Synthesis of Zinc Oxide Nanoparticles

The synthesis technique was adopted from our previous work [15]. In a beaker labelled A, 50 mL of water was measured and specific amount of pullulan was added and dissolved in the water. In separate beaker labelled B, 10 mL of water was measured and 4.5 g of $Zn(NO_3)_2 \cdot 6H_2O$ was dissolved in the water. Both solutions were combined, stirred and heated until the formation of gel. The gel was transferred to ceramic crucible and calcined at 600 °C for 1 h. The amount of pullulan was then varied to 1, 2.5, 5 and 7.5 g and the samples obtained were named 2ZNP, 5ZNP, 10ZNP and 15ZNP respectively. Figure 1 shows the flowchart of ZnO NPs synthesis.

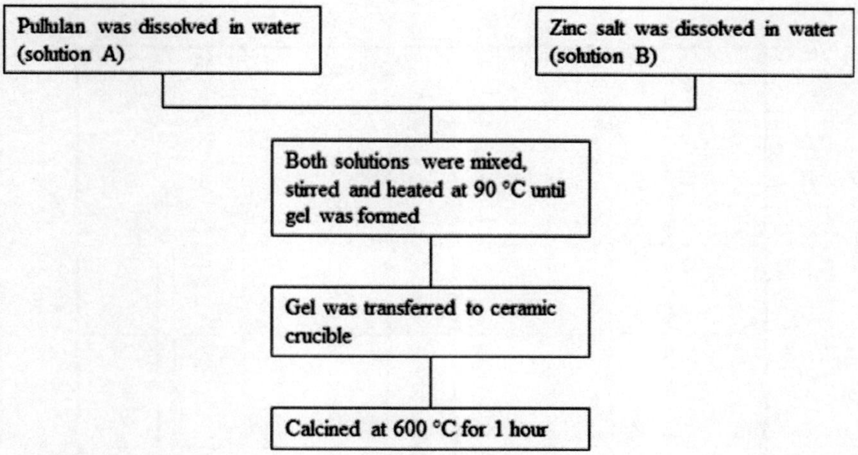

Fig. 1 Flowchart of ZnO NPs synthesis

2.3 Characterizations

PANalytical X'pert Pro MPD diffractometer with Cu Ka radiation (k = 1.5406) in the range from 20° to 80° was used to analyze the X-ray diffraction (XRD) pattern of synthesized ZnO NPs. Fourier Transform Infrared (FTIR) spectra were recorded using Shimadzu IRTracer-100 via potassium bromide (KBr) pellet method over the range of 400–4000 cm^{-1}. The ratio between sample and KBr salt is 1:10. Shimadzu UV-2600 spectrophotometer with an integrating sphere accessory was used to collect the data. For absorbance measurement, 0.125 mg/mL of ZnO NPs solution was prepared and used for measurement. For diffuse reflectance measurement, dried samples of ZnO NPs were used with barium sulphate as the reference.

3 Results and Discussion

A series of green synthesized ZnO NPs was produced in the presence of pullulan and the impact of pullulan amount on ZnO NPs characteristics were investigated. The synthesized samples were analyzed through XRD and Fig. 2 shows the XRD pattern. All the samples exhibited characteristic ZnO diffraction peaks at $2\theta = 31.83°$, 34.48°, 36.31°, 47.58°, 56.63°, 62.89°, 66.40°, 67.97° and 68.88° and all these peaks agree to (100), (002), (101), (102), (110), (103), (200), (112) and (201) crystal plane of

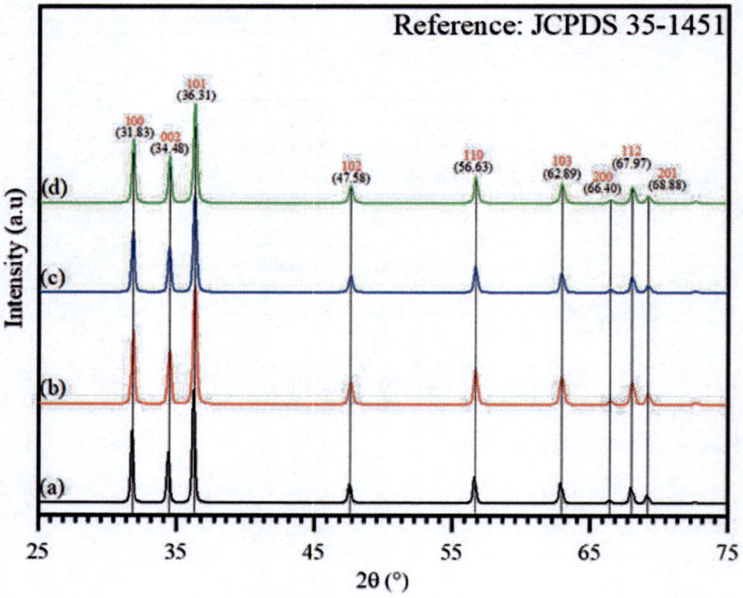

Fig. 2 The XRD diffractogram of ZnO NPs. **a** 2ZNP, **b** 5ZNP, **c** 10ZNP and **d** 15ZNP

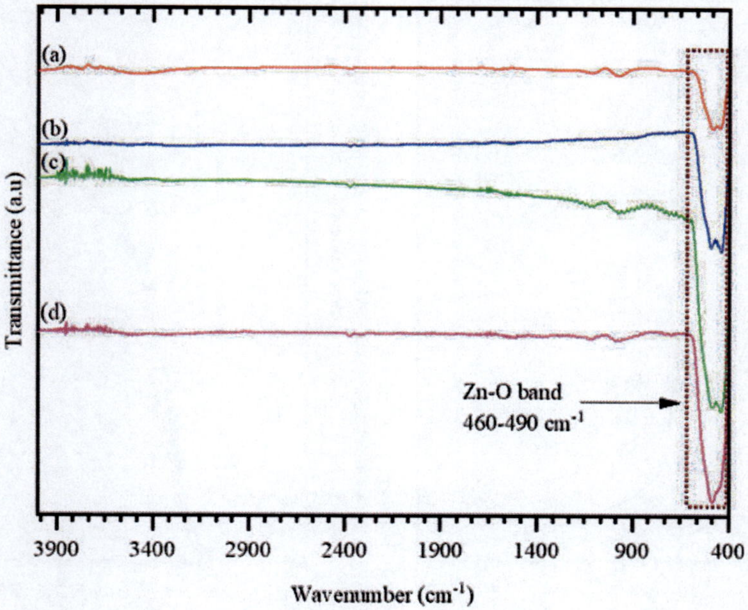

Fig. 3 FTIR spectra of **a** 2ZNP, **b** 5ZNP, **c** 10ZNP and **d** 15ZNP

hexagonal wurtzite of ZnO [16]. Furthermore, the synthesized ZnO NPs are pure as no other peak was observed in the diffractogram. By using Scherrer formula, the ZnO NPs' crystallite size were calculated [17]. The average crystallite size calculated were 37.93, 28.46, 31.04 and 34.15 nm for 2ZNP, 5ZNP, 10ZNP and 15ZNP respectively. With the increment of pullulan mass from 1 to 2.5 g, the crystallite size decreased while further increased in pullulan amount resulted in an increased of crystallite size.

FTIR analysis was conducted to confirm and analyzed the produced ZnO NPs. Figure 3 represents the FTIR spectra of all synthesized samples. There is only one prominent peak present around the region of 460–490 cm^{-1}. This peak represents the Zn–O bond of ZnO NPs. There was no peak representing pullulan present and this shows that the pullulan is absent after calcination process. In terms of peak intensity, a general trend of increasing intensity with pullulan amount was observed.

UV–Visible spectrophotometer was used to analyzed the optical properties of ZnO NPs and the analysis was conducted between the 250–750 nm. The absorbance spectra of ZnO NPs were depicted as Fig. 4A. Sharp absorbance peak appears around 369–371 nm range. Similar as the XRD result, there were no additional peak observed indicating the samples is pure [18]. With the increased of pullulan amount, there were slight blue shift and this may indicate a particle size's reduction which in accordance with the XRD results. The single prominent absorbance peak at around 369 nm is because of electronic transition occurred between valance band and conduction band of ZnO NPs [19]. The ZnO NPs' band gap was determined through Tauc's plot (Fig. 4B). The band gap was determined through extrapolation of linear portion.

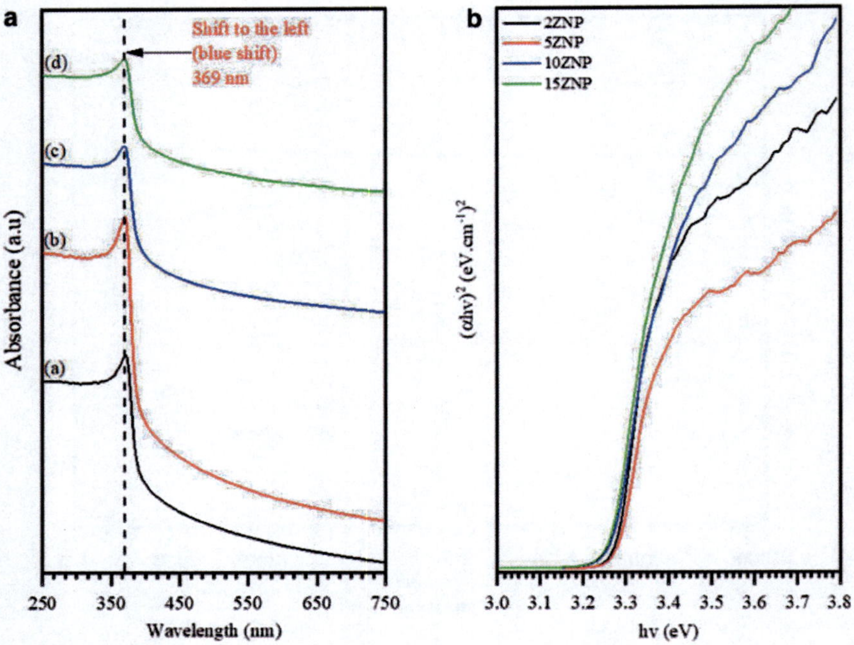

Fig. 4 **A** UV–Vis absorbance spectra of (a) 2ZNP, (b) 5ZNP, (c) 10ZNP and (d) 15ZNP and **B** Tauc's plot of ZnO NPs

The band gap values obtained were 3.274, 3.281, 3.266 and 3.261 eV for 2ZNP, 5ZNP, 10ZNP and 15ZNP respectively. Similar as XRD results, the band gap value increased as the amount of pullulan increased from 1 to 2.5 g while further increment of pullulan amount resulted in the decreased of band gap value.

4 Conclusion

Green synthesized ZnO NPs were successfully obtained via sol–gel technique and the pullulan amount's impact on the characteristics of ZnO NPs were studied and compared. The produced ZnO NPs samples exhibited hexagonal wurtzite structure with no impurities. Based on UV–Vis and XRD analysis, the general conclusion can be drawn is that as the amount of pullulan increased the crystallite size decreased. However, substantial decreased in crystallite size was observed as the pullulan amount increased from 1 to 2.5 g. Further increased resulted in slight increment of crystallite size. However, further analysis is need to prove this. The band gap determined for the synthesized sample ranging from 3.274, 3.281, 3.266 and 3.261 eV for 2ZNP, 5ZNP, 10ZNP and 15ZNP respectively. This synthesized ZnO

NPs can be used for applications such as catalyst, photocatalyst, antibacterial agents and others.

Acknowledgements Takasago Thermal Engineering Co. Ltd. Grant (R.K.130000.7343.4B422, R.K.1300007343.4B472) and Fundamental Research Grant Scheme (R.K130000.7843.5F031) funded this research. Deepest gratitude to UTM and Malaysia-Japan International Institute of Technology (MJIIT) for the research facilities.

References

1. Rahman A, Harunsani MH, Tan AL, Khan MM (2021) Zinc oxide and zinc oxide-based nanostructures: biogenic and phytogenic synthesis, properties and applications. Bioprocess Biosyst Eng 44(7):1333–1372
2. Qi K, Cheng B, Yu J, Ho W (2017) Review on the improvement of the photocatalytic and antibacterial activities of ZnO. J Alloy Compd 727:792–820
3. Wojnarowicz J, Chudoba T, Lojkowski W (2020) A review of microwave synthesis of zinc oxide nanomaterials: reactants, process parameters and morphologies. Nanomaterials 10(6)
4. Ong CB, Ng LY, Mohammad AW (2018) A review of ZnO nanoparticles as solar photocatalysts: synthesis, mechanisms and applications. Renew Sustain Energy Rev 81:536–551
5. Mohamed Isa ED, Shameli K, Che Jusoh NW, Mohamad Sukri SNA, Ismail NA (2021) Variation of green synthesis techniques in fabrication of zinc oxide nanoparticles—a mini review. IOP Conf Ser Mater Sci Eng 1051(1)
6. Khandel P, Yadaw RK, Soni DK, Kanwar L, Shahi SK (2018) Biogenesis of metal nanoparticles and their pharmacological applications: present status and application prospects. J Nanostruct Chem 8(3):217–254
7. Shanan Abed M, Abed AS, Mohammad Othman F (2020) Green synthesis of silver nanoparticles from natural compounds: glucose, eugenol and thymol. J Adv Res Fluid Mech Therm Sci 60(1):95–111
8. Mudhafar M, Zainol I, Jaafar CNA, Alsailawi HA, Majhool AA (2020) Microwave-assisted green synthesis of Ag nanoparticles using leaves of Melia Dubia (Neem) and its antibacterial activities. J Adv Res Fluid Mech Therm Sci 65(1):121–129
9. Bandeira M, Giovanela M, Roesch-Ely M, Devine DM, da Silva Crespo J (2020) Green synthesis of zinc oxide nanoparticles: a review of the synthesis methodology and mechanism of formation. Sustain Chem Pharm 15
10. Kumar H, Bhardwaj K, Kuca K, Kalia A, Nepovimova E, Verma R, Kumar D (2020) Flower-based green synthesis of metallic nanoparticles: applications beyond fragrance. Nanomaterials 10(4)
11. Mohd Yusof H, Mohamad R, Zaidan UH, Abdul Rahman NA (2019) Microbial synthesis of zinc oxide nanoparticles and their potential application as an antimicrobial agent and a feed supplement in animal industry: a review. J Anim Sci Biotechnol 10:57
12. Ali SG, Ansari MA, Alzohairy MA, Alomary MN, Jalal M, AlYahya S, Asiri SMM, Khan HM (2020) Effect of biosynthesized ZnO nanoparticles on multi-drug resistant pseudomonas aeruginosa. Antibiotics 9(5):14
13. Saravanakumar K, Jeevithan E, Hu X, Chelliah R, Oh D-H, Wang M-H (2020) Enhanced anti-lung carcinoma and anti-biofilm activity of fungal molecules mediated biogenic zinc oxide nanoparticles conjugated with β-D-glucan from barley. J Photochem Photobiol, B 203:111728
14. Liu J, Wu X, Wang M (2019) Biosynthesis of zinc oxide nanoparticles using biological polysaccharides for application in ceramics. J Electron Mater 48(12):8024–8030
15. Isa EDM, Shameli K, Jusoh NWC, Hazan R (2020) Rapid photodecolorization of methyl orange and rhodamine B using zinc oxide nanoparticles mediated by pullulan at different calcination conditions. J Nanostruct Chem 11(1):187–202

16. Mohamed Isa ED, Che Jusoh NW, Hazan R, Shameli K (2021) Photocatalytic degradation of methyl orange using pullulan-mediated porous zinc oxide microflowers. Environ Sci Pollut Res 28(5):5774–5785
17. Taghavi Fardood S, Ramazani A, Moradi S, Azimzadeh Asiabi P (2017) Green synthesis of zinc oxide nanoparticles using Arabic gum and photocatalytic degradation of direct blue 129 dye under visible light. J Mater Sci Mater Electron 28(18):13596–13601
18. Mahamuni PP, Patil PM, Dhanavade MJ, Badiger MV, Shadija PG, Lokhande AC, Bohara RA (2019) Synthesis and characterization of zinc oxide nanoparticles by using polyol chemistry for their antimicrobial and antibiofilm activity. Biochem Biophys Rep 17:71–80
19. Sukri S, Shameli K, Wong MMT, Teow SY, Chew J, Ismail NA (2019) Cytotoxicity and antibacterial activities of plant-mediated synthesized zinc oxide (ZnO) nanoparticles using Punica granatum (pomegranate) fruit peels extract. J Mol Struct 1189:57–65

Effect of Porosity and Permeability Characteristics on the Silver Catalyst of the Hydrogen Peroxide Monopropellant Thruster Performances

Muhammad Shahrul Nizam Shahrin, Norazila Othman, Nik Ahmad Ridhwan Nik Mohd, and Mastura A. B. Wahid

Abstract Thruster performances can be optimized by adjusting the geometrical configurations and its operating conditions. Attributes such as pressure loss across catalyst bed, mass flowrate, and velocity are essential parameter to its performance. The presence of silver catalyst as porous medium affecting the flow and reaction inside the chamber due to its properties such as porosity, permeability, and inertial resistance. Hence, this work is motivated by the scarcity of the available literature on this topic specifically in regards with the hydrogen peroxide monopropellant thruster. The effect and relationship of porosity and permeability characteristics on the silver catalyst to the thruster performances is yet unknown. To investigate the effect and also to find the relationship between thruster performances and porosity—permeability properties of a silver catalyst, a study is done using Computational Fluid Dynamics (CFD). In this study, species transport model with EDM turbulent-chemical interaction has been used with the realizable k-ε turbulent to simulate the reaction inside the catalyst bed chamber. Extensive analysis provided in the current work which considers the performances in term of mass flowrate, velocity, pressure loss, and thrust. From this study, it shows that the porosity of the bed is least important compared to the permeability and inertial resistance of the silver catalyst and these two factors are independent from the effect of porosity but linked between one another.

Keywords Hydrogen peroxide monopropellant thruster · Pressure drop across catalyst bed · Silver screen catalyst · Viscous resistance · Inertial resistance

1 Introduction

Utilization of silver catalyst in the hydrogen peroxide monopropellant thruster is not uncommon. That was contributed by the fact that silver catalyst is widely available on the market, traditionally reliable and has a very good reputation, easy to handle

M. S. N. Shahrin · N. Othman (✉) · N. A. R. N. Mohd · M. A. B. Wahid
School of Mechanical Engineering, Universiti Teknologi Malaysia, 81310 Skudai, Johor, Malaysia
e-mail: norazila@mail.fkm.utm.my

M. Y. Ismail et al. (eds.), *Technological Advancement in Mechanical and Automotive Engineering*, Lecture Notes in Mechanical Engineering,
https://doi.org/10.1007/978-981-19-1457-7_33

and to work with in terms of safety and storage, and it comes with reasonable price range. By definition, catalyst is a material that enhance the rate of reaction without being used in the process or reaction; where the activation energy usually lowered with the presence of any catalytic material. Normally, silver that is used as a catalyst is in metallic or an alloy form while ionic form usually not use since it will produce precipitation.

Solid state catalyst normally paired with the hydrogen peroxide monopropellant thruster known as heterogenous type catalyst. These catalysts can be in various forms such as woven mesh or screen, powder, grain particle, or flakes. Besides, silver or other metal type of catalyst sometime used as a coating to another cheaper metal to reduce the total cost since the reaction rate will not be influenced by the core structure rather than the outer skin of the catalyst material as the reaction take place at the interphase layer only.

Reaction zone for hydrogen peroxide decomposition process with the silver catalyst can be divided into 4 main zones which are; low temperature zone, nuclear boiling zone, film boiling zone, and high temperature reaction zone [1]. Silver catalyst have its own advantages and disadvantages despite being popular choice to be used as a hydrogen peroxide catalyst. Among the advantages are; high decomposition efficiency, compactness, less complicated fabrication process, and available in many types of form [1]. While there are also disadvantages of this material such as; low melting point, and easily to deform [2]. Kang et al. [1], stated that there are two main factors that responsible to the deformation of the silver catalyst such as; compression load from the propellant injection which will turn it into lump and washed to the end of the catalyst bed, and undesired distribution of catalyst pack as the catalyst loss its mechanical strength which contribute to other problems such as catalyst attrition and channeling.

Despite any shape and type, silver catalyst is porous to enable the propellant flows through it and undergo reaction while flowing. Porosity can be defined as ratio coefficient between a void of a material to the total volume of that material, while permeability in simple term mean the ability of a fluid to get through porous solid. So, by definition, porosity and permeability are two different parameters as both are used to measure distinct behavior. However, permeability still is a characteristic of a porous medium which tell us the capacity and ability to transmit fluid, while porosity is a ratio of geometrical shape.

Besides, permeability of each catalyst is dependent on distinct factors including the arrangement of the catalyst, geometric irregularity, and size distribution [3]. So, at same porosity value, different catalyst will have different permeability value [3]. Hence direct correlation between porosity and permeability is not unambiguous [3].

In this study, permeability parameter is viewed in term of inverse permeability which also known as viscous resistance, so this term will be used onwards. When a fluid flowing across the porous medium, the fluid will be experiencing a friction with the surface of the solid porous material. When this happened, the energy of the flow loses as to overcome the friction.

Another characteristic of the porous material is the inertial resistance. Inertial resistance is some form of energy loss because of the fluid needed to exert some of its kinetic energy to leave off the obstacles where it usually lost via turbulence in the body of fluid. Inertial resistance typically proportional to the squared of velocity, while viscous resistance is proportional to the velocity. So, inertial resistance tends to dominate the high flow rates. According to Soares et al. [4], dependency and effects of viscous and inertial resistances are intricate but both resistances are greatly associated to the geometrical structures of the porous material.

There are two principal operational indicators for the thruster combustion which are, pressure drop across catalyst, and pressure drop across the propellant inlet [5]. Both have large effect to the hydrogen peroxide decomposition process stability and structure weight. For this reason, pressure drop across catalyst bed has become important criteria in influencing the performances of the thruster in term of velocity, flowrates, and also thrust. While both effecting the performances, in this study, only the pressure across catalyst bed is observed as the inertial and viscous resistance is a properties of porous silver catalyst which expecting to have less relevance with the pressure drop across the inlet. Hence, it is important to acknowledge the effect of the porosity characteristics in relationship with the thruster performances since in thruster rocket application, exceptional precision is required. Wholly, it will affect the overall mission and the economy of the spacecraft system either directly or indirectly.

In addition, increase in porosity also means increase in total surface area of the catalyst [6]. This mean that high porosity tends to have better performances since the decomposition of the hydrogen peroxide is greater due to larger surface area. However, higher surface area also means that the catalyst become more packed and tighter, hence, will contribute to higher pressure loss [7].

According to Soares et al. [4], porosity has a large influence on the inertial resistance. They suggested that as the fluid temperature rises, pressure drop decreases. This means that at lower porosity, temperature seems to affect the pore arrangement but for higher porosity the temperature seems does not affecting the pressure drop [4]. Furthermore, at a given bulk volume-average velocity, the pressure drops decreased with increased porosity [4]. In non-organized systems, in which the flow field is notably chaotic, viscous and inertial resistances are strongly related to the internal structure of the material [4]. As a result of the drag forces (and other forces) experienced by these materials, some turbulence and recirculation may be experienced in these regions, substantially impacting the inertial resistance.

According to Soares et al. [4], a first-order dependence of the pressure drop on the air velocity was recorded for most cases, and a higher order relation between the pressure drop and the velocity of the fluid was also observed, especially in low porosity value. According to Zhong et al. [8], the pressure difference rises with the flow rate and it appears quite nonlinear as the flow rate becomes large enough. The viscous effect typically influences small flow velocity but small flow velocity can be used as the boundary for the transition from viscous flow to inertial flow is still unknown [8].

The use of computational method especially the computational fluid dynamics (CFD) to simulate the flow inside the porous medium has become so prominence as it could help observing and understanding these phenomena which usually failed to be captured via experimentation. Hence, the work on this paper is motivated by the scarcity of the available literature on this topic specifically in regards with the hydrogen peroxide monopropellant thruster. Thus, to investigate the relationship and correlation between thruster performances attributes such as flowrate, velocity, pressure drop, and thrust with the porosity, permeability, and inertial resistance of the porous silver catalyst this paper is presented.

2 Methodology

The numerical simulation of the monopropellant thruster is an intricate work involving lot of physical concept and mathematical equations. The purpose of numerical simulation is to supply the research with elementary insight without have to deal with expensive experiment in the first place. In this study, the movement of the research is guided by the designed flowchart as shown as in Fig. 1. The thruster model is first created in a 2-Dimensional (2D) shape based on typical hydrogen peroxide monopropellant thruster design. This design was adopted from the Othman's [9] work using 90 wt% hydrogen peroxide concentration with the silver screen catalyst as depicted in Fig. 2.

Once the 2D model is completed with several adjustments and simplifications, then a suitable mesh grid is applied. The simplifications were made to reduce the complexity of the simulation especially in the part where it will not directly affect

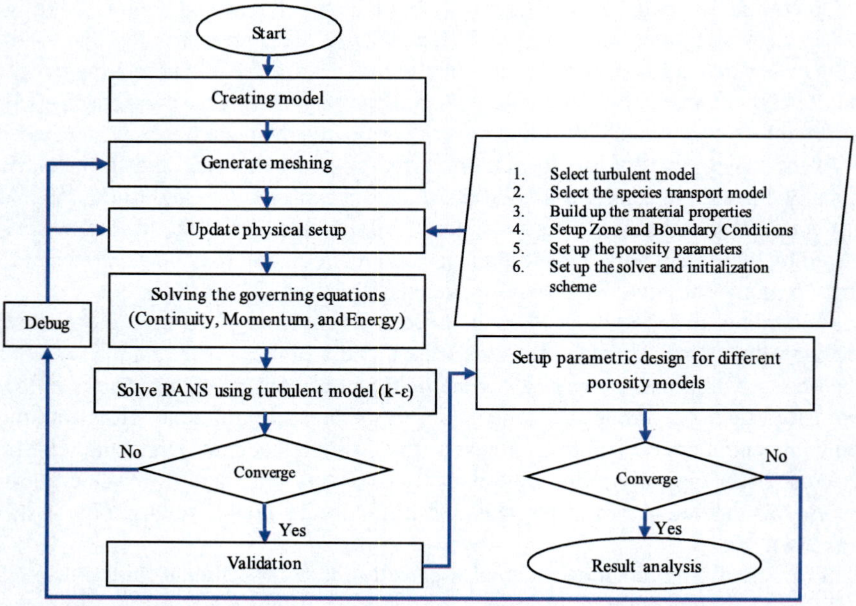

Fig. 1 Research flowchart

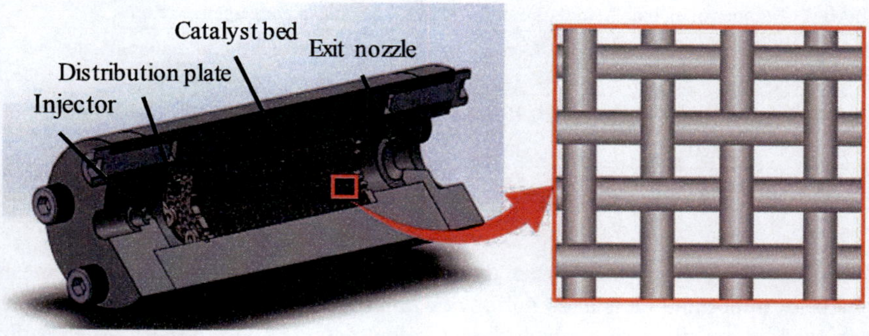

Fig. 2 Basic internal components of the thruster with silver screen mesh catalyst (left) [9], illustration of a ply of silver screen mesh before being stacked together into catalyst bed (right)

the integration of the results. As an example, in the 2D model, the distribution plate is not modelled because it is made of perforated plate and it is thin enough to be considered as a part of catalyst since both are porous in nature, instead, a boundary is applied as a porous medium zone starting from the plate. The mesh generated is shown in Fig. 3, showing part of the catalyst bed and nozzle sections having a uniform quadrilateral grid.

The grid independent test is done to find the suitable mesh size in order to have minimum iterations at the fastest speed as possible without sacrificing the coherence of the results. Initially, the mesh grid size is set at a range of 0.20–1.50 mm per element producing 153,116 numbers of elements, then the size is reduced to 0.10–1.00 mm, and further down to 0.10–0.50 mm for the third run creating total number of elements of 368,870 and 1,209,482 respectively. The first mesh can produce a result at speed of 0.963 s of the wall-clock time per iteration, while the second and third run at speed of 2.285 and 6.753 s. After comparing all three results from this test, second mesh is chosen as a basis of this study. Considering that the speed to achieve

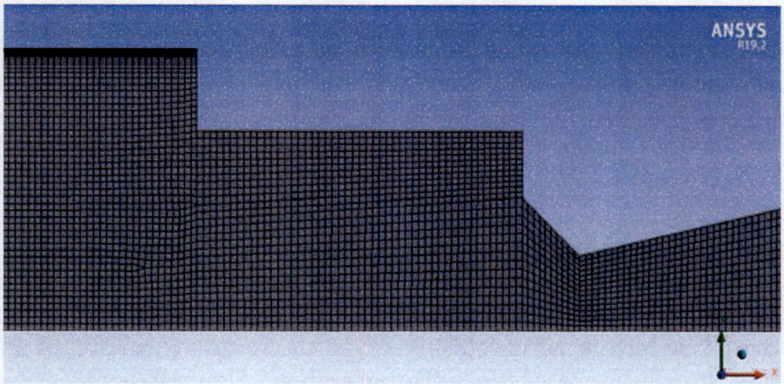

Fig. 3 Close-up view of the mesh created showing part of catalyst bed and nozzle area

Table 1 Number of element and duration for meshes to achieve convergence criteria

Mesh	Number of elements	Average wall-clock time per iteration (s)	Maximum temperature at catalyst bed (K)
Mesh 1	153,116	0.963	1120
Mesh 2	368,870	2.285	1164
Mesh 3	1,209,482	6.753	1164

stable result is slightly longer than the first mesh but not as slow as the third mesh and most importantly it is in compliment with the consistent maximum temperature achieved by the simulation which shown in the second and third mesh. The results of the test represented in Table 1.

The boundary conditions for this model including propellant inlet, air inlet, outlet and wall. The model is cut into half since it is geometrically symmetric to reduce the computing cost. The 2D porous zone is used instead of 1D solution, so no interfaces for porous jump boundary condition is activated, instead catalyst bed region is set as a porous medium zone where the properties in this region including the porosity, viscous and inertial resistances are varied accordingly and will be explained in next sub-section of this article. While, the inlet pressure and temperature are fixed at 1.50 MPa and 300 K. The mass fraction of the propellant inlet is set to be 0.9 of hydrogen peroxide and 0.1 mass fraction of water. Hydrogen peroxide is considered to be at 90% concentration and additive materials (which usually available in laboratory because of impurity and stabilizers used for the solution) is considered not presence. In addition, few assumptions were made such as the structure of the catalyst in the porous zone is considered to be isotropic while velocity formulation using the superficial velocity, and the axial heat conduction and radiation heat transfer are disregard. Once the boundary conditions are set, the governing equations are then selected.

Several preliminary setups were conducted with some adjustment and corrections done accordingly, and once the successful solution is found, full parametric test is then repeated using the said solution. Here, steady state, pressure-based solver is set up. The application of the turbulent model for this simulation are based on those preliminary runs and also insight from the literatures. According to Vestnes [10], for similar simulation, k-ω model was used in a multiphase simulation. However, Wang et al. [11] implement standard k-ε model with introduction of Eddy-Dissipation Method (EDM) in a single-phase simulation. After comparing both models, we found that k-ε model is much more suited our case as to increase the accuracy and convergence of the simulations. COUPLED solver is used in this study as it could expedite and stabilize the convergence of the results with all the discretization scheme is at second order magnitude. For the turbulent-chemistry interaction, EDM is selected.

The catalytic decomposition take place on the surface for reaction in the catalyst. So, species transport model is suited to model the surface reaction. The surface reaction supposedly take place in one mixture of species; hence it should be a single-phase homogeneous reaction. According to Vestnes [10], heterogeneous reaction should be modelled through volumetric reaction through the interphase interaction models. The fluid flow inhibits the formation of a velocity and diffusion boundary layer and the decomposition is thereby reaction limited [10]. On the other hand, the porous media model in the software used, does not include the volume of the catalyst material.

2.1 Governing Equations

Hydrogen peroxide, H_2O_2, undergo decomposition reaction inside the thruster chamber specifically in the catalyst bed section. The decomposition reaction with the presence of silver catalyst water steam, H_2O, and oxygen gas, O_2, in the exothermic reaction. Equation (1) shows the reaction equation where the energy release in form of heat from the reaction denoted as Q. Whilst, the physics of the flow is governed by the continuity, momentum, and energy equations which shown in Eqs. (2)–(6). The continuity equation described in Eq. (2) shows the mass conservation law with presence of isotropic porosity term denoted as γ; while ρ and $\vec{v}$ represent the density and velocity vector.

In momentum equation, as described in Eq. (3), p and $\overline{\overline{\tau}}$ are representing the pressure and turbulent stress tensor. Since the effect of gravity is ignored the gravity term is removed from the equation. Additional momentum source, S_i is added and expended further in Eq. (4). In this equation, the first term of the right-hand side represents the permeability term, α, or viscous resistance, $\frac{1}{\alpha}$, and viscosity of the fluid, μ, coincided in the same term. The second term represents the inertial resistance factor, C_2. Equation (4) is the source term that contributes to the pressure gradient in the porous region, hence, leading to pressure loss by means of viscous and inertial resistances. v in this equation represent the velocity magnitude of the flow inside the porous medium.

The energy equation shows in Eq. (5) where; E, k_{eff}, T, h_j, $\vec{J_j}$, and $\overline{\overline{\tau}}_{eff}$, are representing total energy, effective conductivity, temperature, enthalpy, diffusion flux of the species, and effective tensor stress for turbulent respectively. This equation also equipped with additional energy source due to the chemical interaction, S_h. While Eq. (6) shows the species transport equation where Y_j, $\vec{J_j}$, R_j, and S_j each represent the local mass fraction of the species, species diffusion flux, species production rate through chemical reaction, and rate of creation of the species.

$$2H_2O_2 \xrightarrow{\text{Catalyst}} 2H_2O + O_2 + Q \tag{1}$$

$$\frac{\partial(\gamma\rho)}{\partial t} + \nabla\cdot(\gamma\rho\vec{v}) = 0 \tag{2}$$

$$\frac{\partial}{\partial t}(\gamma\rho\vec{v}) + \nabla\cdot(\gamma\rho\vec{v}\vec{v}) = -\gamma\nabla p + \nabla\cdot\left(\gamma\overline{\overline{\tau}}\right) + S_i \tag{3}$$

$$S_i = -\left(\frac{\mu}{\alpha}v_i + C_2\frac{1}{2}\rho|v|v_i\right) \tag{4}$$

$$\frac{\partial}{\partial t}(\rho E) + \nabla\cdot(\vec{v}(\rho E + p)) = \nabla\cdot\left(k_{eff}\nabla T - \sum_j h_j\vec{J_j} + (\overline{\overline{\tau}}_{eff}\cdot\vec{v})\right) + S_h \tag{5}$$

$$\frac{\partial}{\partial t}(\rho Y_j) + \nabla\cdot(\rho\vec{v}Y_j) = -\nabla\cdot\vec{J_j} + R_j + S_j \tag{6}$$

In this study, porosity, viscous resistance, and inertial resistance of the porous medium catalyst are the varied parameters. It is run using 2 levels full factorial DOE technique using 3 factors and the value for each variable represented in Table 2. Here, the porosity is limited to the range of 0.2–0.8, coupled with viscous resistance, and inertial resistance from minimum of $5.67 \times 10^6\text{m}^{-2}$ to maximum $5.67 \times 10^{10}\text{m}^{-2}$, and $6.09 \times 10^3\text{m}^{-1}$ to $6.09 \times 10^5\text{m}^{-1}$ respectively. Total number of 8 cases are set up with other input parameters are let at constant value.

Finally, after parametric designs were run. The results are then analyzed using post-processing and data visualization technique. Results are presented in term of temperature, pressure, velocity, and density contours. Next, recorded results then analyzed in the factorial contour plots in order to show the effect of the porosity, viscous resistance, and inertial resistance to the thruster performances (mass flowrate, exit velocity, pressure drop across catalyst bed, and thrust). From this results and analysis, the correlation and relationship between porosity properties and thruster performance can be established.

Table 2 Variables for each case at maximum and minimum levels

Designation	Porosity	Viscous resistance (m^{-2})	Inertial resistance (m^{-1})
Case 1	0.2	5.67E + 06	6.09E + 05
Case 2	0.8	5.67E + 10	6.09E + 03
Case 3	0.2	5.67E + 10	6.09E + 05
Case 4	0.8	5.67E + 10	6.09E + 05
Case 5	0.2	5.67E + 10	6.09E + 03
Case 6	0.2	5.67E + 06	6.09E + 03
Case 7	0.8	5.67E + 06	6.09E + 05
Case 8	0.8	5.67E + 06	6.09E + 03

3 Results and Discussion

3.1 Simulation Results

The simulation of the hydrogen peroxide thruster is done with the initial setup based on input for Case 1. Once the result of that simulation is found reliable, the remaining cases are then run. The results of the Case 1 are shown in Figs. 4, 5, 6 and 7, showing the temperature, pressure, velocity, and density contours. From Fig. 4, temperature can be seen reaching 1150 K at the front part of the catalyst bed, where it is the place reaction first took place. Downstream the bed, temperature reduced to about 900 degree and one the flow exits from the exit nozzle, temperature cooled down by the surrounding. From this, it is safe to conclude that the temperature inside the catalyst bed ranging from 900 to >1000 K which is compliment with the experimental findings such as by work of Jung et al. [12] that acquired temperature of 1020 K for the 90 wt% hydrogen peroxide and work of Vestnes [10] that achieved >950 K for 87.5 wt% concentration.

Meanwhile, pressure contour as in Fig. 5, shows that the pressure variation along the thruster during the reaction take place. The highest pressure shown there is at the propellant inlet zone where the pressure accumulated and build up before entering the porous medium catalyst bed. The high pressure in that region due catalyst bed that act as an obstacle to the flow. Once the flow entering the porous zone, it has to overcome layers of silver catalyst obstacles over and over again until it reaches the end of the bed. Hence, the pressure gradually dropped over the entire length of the catalyst bed.

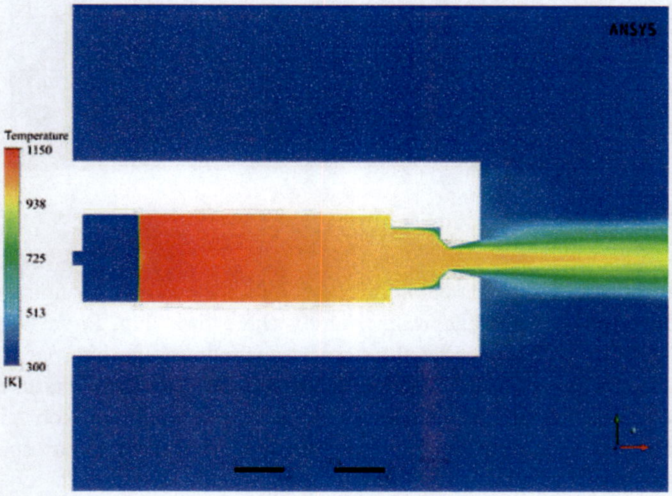

Fig. 4 Temperature contour

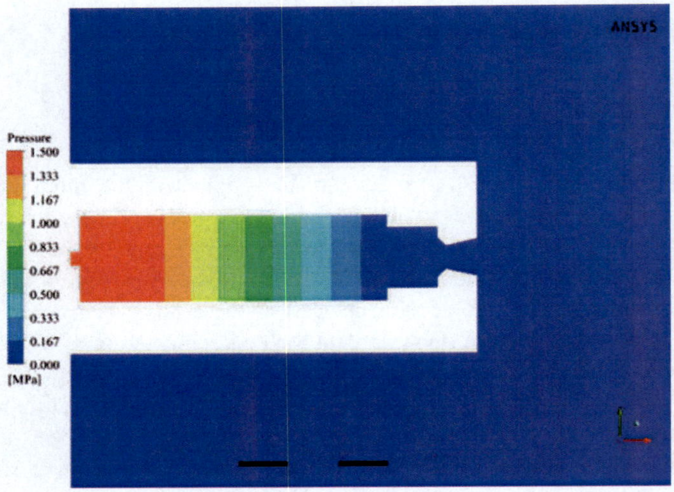

Fig. 5 Pressure contour

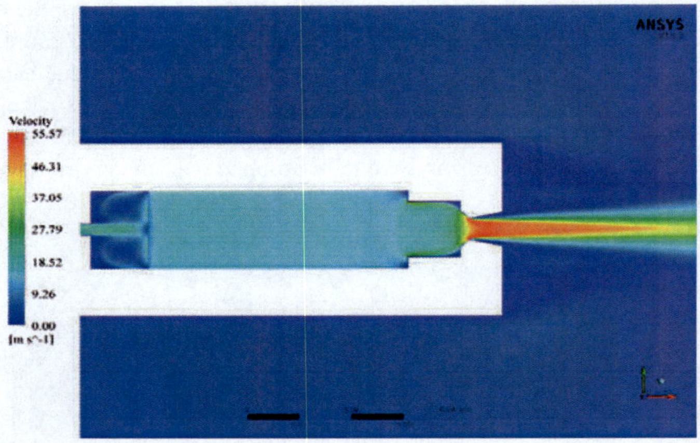

Fig. 6 Velocity contour

In the meantime, hydrogen peroxide reacts with the catalyst, changing it phases from liquid to oxygen gas and water steam, where from Fig. 6, we can see the velocity intensity decreases a bit through downstream as the hydrogen peroxide releases some energy from the reaction. However, the velocity magnitude appears to be quite uniform as this simulation is done with assumption of isotropic porous medium. Hence, velocity moves quite dominantly in axial direction especially in porous catalyst bed region. As the flow moves to the downstream region, the flow velocity reached its maximum value due to area contraction in the nozzle area especially at the throat. Some circulation happened before the catalyst bed as the flow

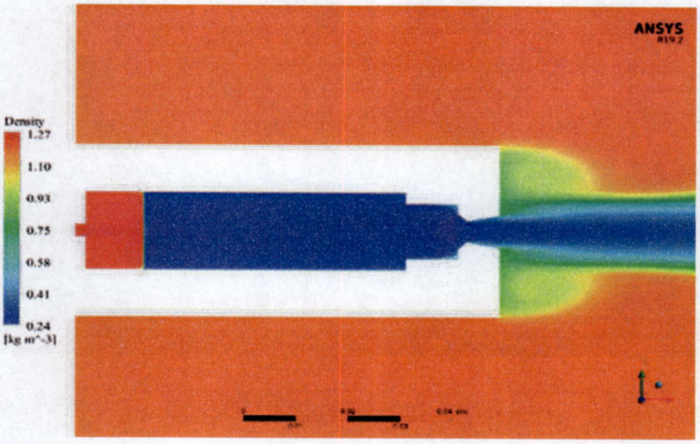

Fig. 7 Density contour

velocity tried to enter the bed area but deviated back due to porous wall which only certain amount of volume pass through at a time.

In Fig. 7, the density contour clearly shows the different of the flow densities, depicting the phase of the fluid inside the thruster changes as the decomposition of hydrogen peroxide liquid turn into gas. As we can see, pressure steadily reduced across the catalyst bed as its loss its momentum as more energy needed to overcome the porosity of the bed. After exiting the bed, it regains some of its momentum by the decomposition reaction that occurred but it is not enough to overcome the overall loss. This too can be observed in the velocity contour where the flow starts to pick up again since the phase of the products are not liquid anymore; which mean they are more flexible and can be squeezed through the nozzle increasing the velocity magnitude even more. The pressure drop across the catalyst bed are recorded by measuring the pressure different at the beginning and end of the catalyst bed. This behavior and effects of the porous silver catalyst parameters influencing the performances such as pressure drop, mass flowrate, exit velocity, and hence the thrust produced.

So, in this study, the relationship of the porous medium properties (viscous resistance, inertial resistance, and porosity) is investigated; the interaction has been calculated in order to analyze the behaviors. From Figs. 8, 9, 10 and 11, we can see that these three parameters and their relationship correlates one another at certain degree are described in contour plots.

In Fig. 8, the contour plot of the mass flowrate of the flow showed against all the three properties. The mass flowrate is measured after the catalyst bed at the throat nozzle in order to see effects of the properties to the mass flowrate. From this figure, the porosity seemed does not so influential to the mass flowrate compared to viscous and inertial resistance. With increase porosity, the mass flowrate moves linearly in Fig. 8a, b shows that it does not affect the mass flowrate of the flow. In Fig. 8c, the inertial resistance seemed having more influence since it moves from wide range of value [100,000–600,000 (m^{-1})] while viscous resistance effective changes are

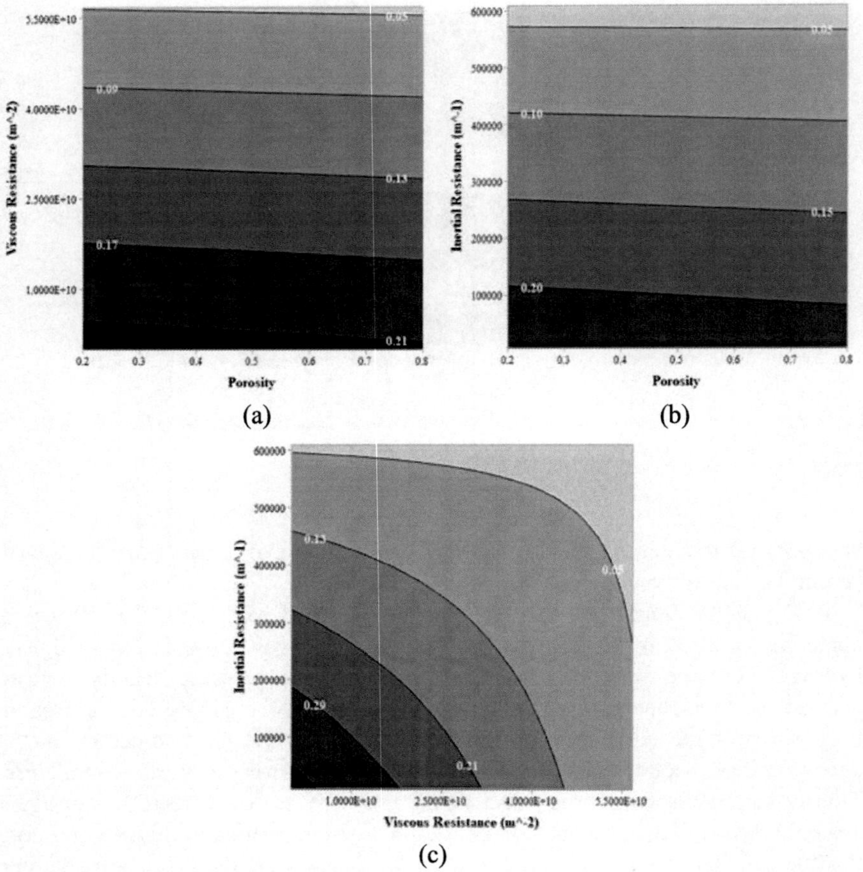

Fig. 8 Contour plot of mass flowrate (kgs^{-1}) versus viscous resistance, inertial resistance, and porosity

concentrated at the narrower range of the upper bound values [concentrated at 1–5.5 ($\times 10^{10} m^{-2}$)].

As the inertial resistance increases, the mass flowrate decreases. Similarly, increments of the viscous resistance shows a decrement in the mass flowrate. This behavior very much expected as both of the resistances intensify, the energy needed to overcome the shear with the catalyst walls increases, hence, reducing the amount of fluid able to flow at the moment; unless if greater (inlet) pressure is exerted to increase the momentum. From this comparison, small change in viscous resistance is more impactful compared to the inertial resistance which need greater value to have the same amount of changes of the mass flowrate, where this relationship is clearly depicted in the Fig. 8c.

The effect on the velocity is kind of resemble the effect on the mass flowrate. Similarly, porosity gave very little impact to the velocity of the flow while viscous

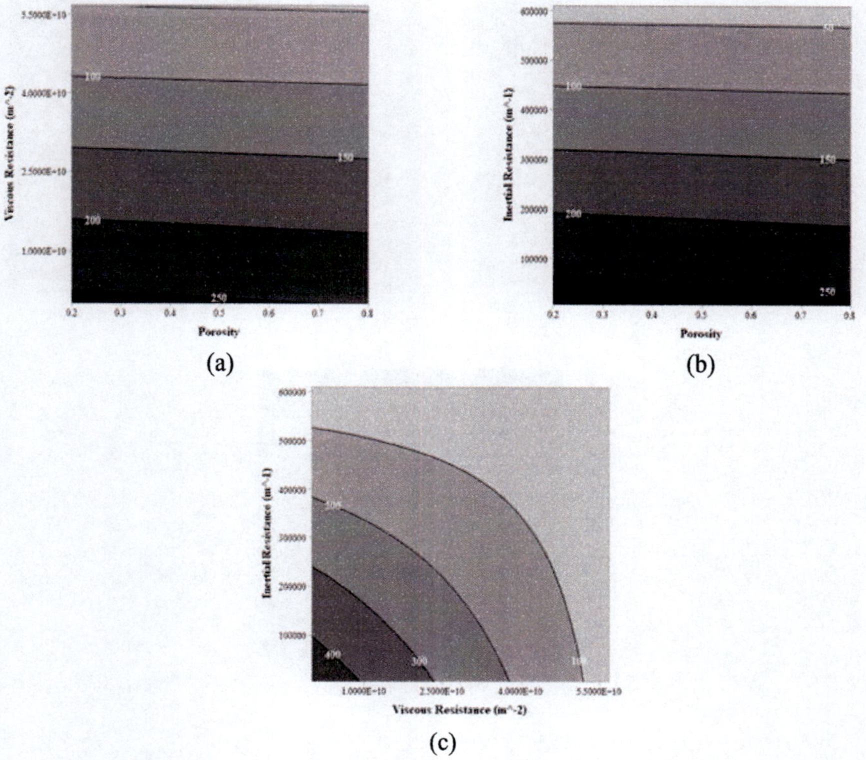

Fig. 9 Contour plot of velocity (ms^{-1}) versus viscous resistance, inertial resistance, and porosity

and inertial resistances dominate the influence on the flow velocity. As the viscous and inertial resistances increases, velocity reduced. In this study, the exit velocity was found ranging from 50 to 250 m/s by tweaking the both of the properties. This relationship is shown in Fig. 9a–c.

The porosity effect on the pressure drop across catalyst bed a bit skewed compared to the previous performances as it shows a better impact of the porosity to the pressure drop as shown in Fig. 10a, b. However, based on these figures, the significant of the porosity is still much less compared to viscous and inertial resistances. The increase in both viscous and inertial resistances result in increase of pressure drop. This is inversely proportional to the effect on mass flowrate and velocity as with higher pressure loss also means that the flow lost its momentum, hence, become slower and less mass flowing through. The relationship of viscous and inertial resistances to the pressure drop shown in Fig. 10c.

The effect of viscous resistance to the thrust performance is more significant compared to the effect of porosity to the production of thrust. From Fig. 11a, viscous resistance increases resulting in decreasing thrust, while with increase of porosity, thrust behave much subtler with little decrement.

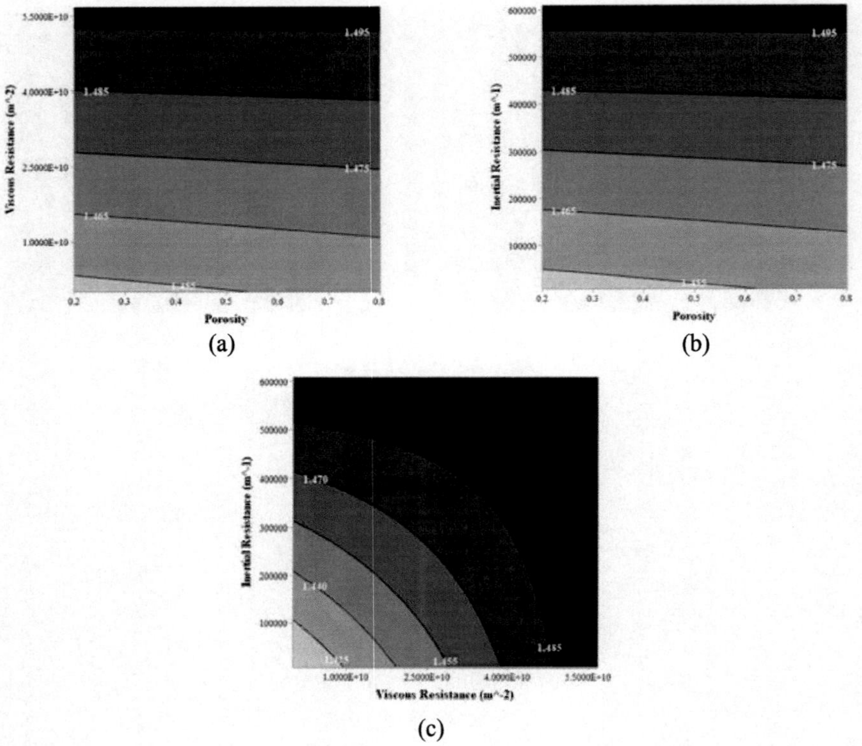

Fig. 10 Contour plot of pressure drop (MPa) versus viscous resistance, inertial resistance, and porosity

The effect of inertial resistance compared to the effect of porosity to the thrust performance is somewhat similar with the effect of viscous resistance against the porosity, as shown in Fig. 11b. This also mean that both viscous and inertial resistances give higher impact to the thruster performance, hence considered more significant to the porosity itself even though both parameters only presence with the use of material that are porous and have porosity index.

The relationship between inertial and viscous resistances are expected. As both is a source of sink to the momentum equation, the lower the value result in lower pressure loss, which turn into higher thrust produces. Both parameters behave similarly in such manner at various range of porosity.

However, both parameters are dependable since the change in geometry not only will change only one parameter but all of three. Since the values of inertial and viscous resistance is hardly measured due the geometrical irregularities, an experiment should be sufficed to fine the values for each catalyst used. As discussed before, other factor such as porous medium type, viscosity of the fluid, and compaction pressure will also influence the value of the parameters. An experiment shall give us a right value of the specific catalyst for a particular thruster. So, this research is conducted in an

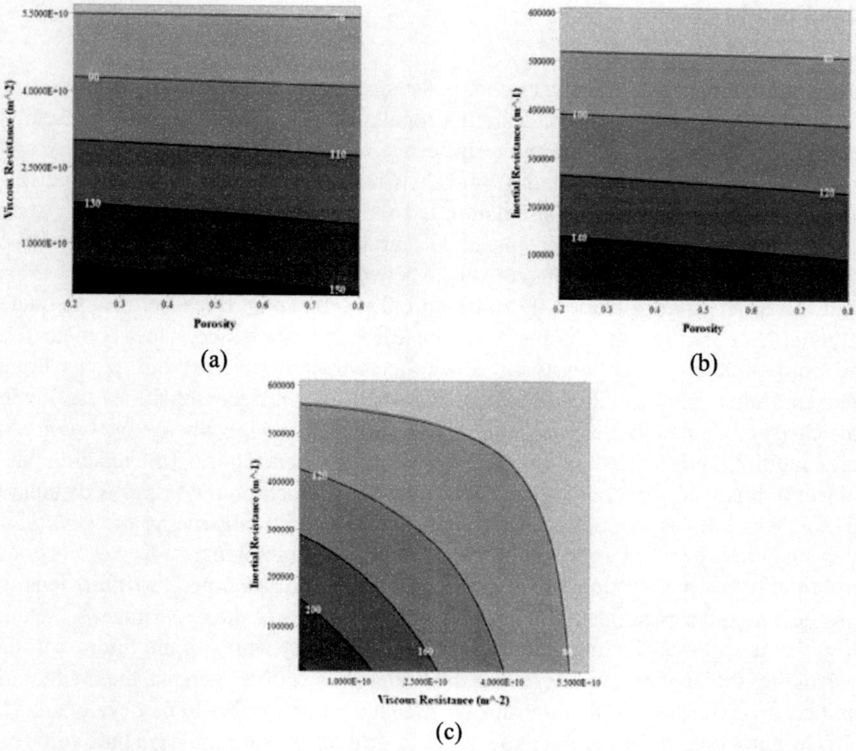

Fig. 11 Contour plot of thrust (N) versus viscous resistance, inertial resistance, and porosity

understanding that permuting these values is quite an extreme design feature as it is physically hard to change the values of viscous and inertial resistances. However, with the results produced for the 100 N class thruster, certain values of porosity, inertial and viscous resistances can be estimated early on. So, it is an important aspect to look at as we deal with the porous catalyst material, and as to foresee the effect of the material porosity in our design.

With continuous study of porous medium in hydrogen peroxide monopropellant thruster, it is expected that an objective function between all the parameters that contribute to the performances can be clearly formulated.

Within this study, there also a limitation that take place including, (1) the test samples are mainly tight porous media, which have a viscous resistance on the order of $10^6 - 10^{10} \mathrm{m}^{-2}$, (2) the structure of the actual silver catalyst is not uniform (anisotropic) which does not reflect true properties of the material, (3) only porosity, viscous and inertial resistances are considered in this test while influences of other parameters like diameter and length of the catalyst are not considered, (4) single-phase instead of multiphase simulation, and (5) effect of temperature variations is not investigated.

4 Conclusions

The simulation on the porosity, viscous resistance, and inertial resistance of the silver catalyst for the hydrogen peroxide monopropellant thruster application are presented within this study. To investigate the influence of the silver catalyst's porosity and permeability characteristics and to find their relationship with the thruster performances, a single-phase CFD simulation is conducted. By utilizing the steady state solver, and realizable k-ε turbulent model, the simulation is done by changing the porosity from 0.2 to 0.8, viscous resistance between 5.67×10^6 and 5.67×10^{10} m^{-2}, and inertial resistance from 6.09×10^3 to 6.09×10^5 m^{-1}. Because of inadequate amount of literatures, the finding of this research is limited since clear comparison cannot be made. In addition, to cut down the computational cost and to accelerate the research, many simplifications were adapted. Plus, the decomposition reaction in the catalyst bed match the interpretation of multiphase model, where flow is considered multiphase if more than one state of material presence during the reaction take place. In this study, the liquid–gas flow in the decomposition process was deducted to single-phase gas reaction. This we believed somehow influencing the results of this study in aspect of its accuracy, however still, the general bias of the results does not contrary with our estimation as other indicators such as temperature plot, density profile, and other parameter still aligned with the results of the experiment adopted. Yet, this is still a preliminary study and hopefully many works in the future will be discussing this matter and improving the finding. To further increase the credential and accuracy of this result, a multiphase model simulation should be developed.

The numerical and experimental investigation of the silver catalyst in the hydrogen peroxide monopropellant thruster in term of porosity and permeability characteristics are generally found in extremely limited in the open literature. Most of the works available are experimental work or development of the thruster unit itself making its hard to make a one to one comparison in term of relationship with the thruster performances. The temperature recorded in the simulation has become a basis in reflecting its accuracy. The decomposition of hydrogen peroxide at 90 wt% concentration is recorded at 1050 K which can be considered in accordance to the literatures explained previously. The mass flowrate is recorded at range of 0.04–0.4 kgs^{-1} depending on the design variables as shown before in Table 2. Pressure drop recorded at >1.4 MPa for all cases, while velocity recorded at range of 43 ms^{-1} to about 450 ms^{-1}, and thrust produces >66 N up until 359 N depending on variables.

Hence, we found that the effect of porosity of the silver catalyst is insignificant to the mass flowrate, velocity, pressure drop across catalyst bed, and thrust; While, the viscous resistance and inertial resistance shows a noticeable relationship in influencing the performance of the thruster. This phenomenon coincides with the Darcy's Equation as described in Eq. (4) where both of the terms are the main source of the momentum sink while at variation of porosities they still predominate the effect of the thruster performances.

Even though porosity have slimmer effect onto the performances of the thruster, but it is very clear that both viscous and inertial resistances are derivative of the porous medium properties itself. In other word, the presence of the viscous and inertial resistances is very much dependent on the presence of porous medium even though changes in porosity does not necessarily change the value of both resistances. So, as for now, it is safe to conclude that porosity does not directly correlates to the viscous and inertial resistance in terms their relationship to the thruster performances.

Acknowledgements This study is a component of the research work on the hydrogen peroxide monopropellant thruster optimization at the Liquid Propulsion Laboratory, School of Mechanical Engineering, Universiti Teknologi Malaysia (UTM). This research is supported by the Ministry of Higher Education (FRGS award No. 5F099). The authors of this article would like to acknowledge the university for providing laboratory facilities and apparatus needed in delivering this study.

References

1. Kang H et al (2016) Effect of H_2O_2 injection patterns on catalyst bed characteristics. Acta Astronaut 2017(130):75–83
2. Pędziwiatr P et al (2018) Decomposition of hydrogen peroxide—kinetics and review of chosen catalysts. Acta Innovations 26:45–52
3. Jildeh Z et al (2018) Thermocatalytic behavior of manganese (IV) oxide as nanoporous material on the dissociation of a gas mixture containing hydrogen peroxide. Nanomaterials 8(4)
4. Soares C et al (2015) Evaluation of resistances to fluid flow in fibrous ceramic medium. Appl Math Model 39(23–24):7197–7210
5. Kuan C-K, Chen G-B, Chao Y-C (2007) Development and ground tests of a 100-Millinewton hydrogen peroxide monopropellant microthruster. J Propul Power 23(6):1313–1320
6. Khaji Z et al (2020) Catalytic effect of platinum and silver in a hydrogen peroxide monopropellant ceramic microthruster. Propul Power Res 9(3):216–224
7. Essa K et al (2017) Development and testing of an additively manufactured monolithic catalyst bed for HTP thruster applications. Appl Catal A 542(February):125–135
8. Zhong W et al (2018) Determination of permeability and inertial coefficients of sintered metal porous media using an isothermal chamber. Appl Sci 8(9)
9. Othman N et al (2011) Design and testing of a 50 N hydrogen peroxide monopropellant rocket thruster. J Mekanikal 33:70–81
10. Vestnes F (2016) A CFD-model of the fluid flow in a hydrogen peroxide monopropellant rocket engine in ANSYS fluent 16.2
11. Wang W, Xia BZ, Sun HH (2013) Simulation study of HTP decomposition within silver screen catalyst bed of the monopropellant gas generator. Appl Mech Mater 404:436–441
12. Jung S, Choi S, Kwon S (2017) Design optimization of green monopropellant thruster catalyst beds using catalytic decomposition modeling. In: 53rd AIAA/SAE/ASEE Joint Propulsion Conference. American Institute of Aeronautics and Astronautics, Atlanta, GA

Numerical Simulation of the Effect of Surface Roughness on the Throttling Characteristics for Multi-stage Pressure Reducing Valves

Guan Wang, Jianfei Deng, Linyuan Kou, and Xuejun Zhu

Abstract The multi-stage pressure reducing valves are increasingly used in technical engineering fields such as residual oil hydrogenation. The overall performance of the valve can be affected by a numbers of factors, and the effect of roughness is still not fully understood. Using computational fluid dynamics technology, the influence of roughness on the throttle characteristics of the spool of a series multi-stage pressure-reducing regulating valve is studied. Under the same inlet velocity condition, numerical simulation of different roughness is carried out, and its influence on the internal flow of the series multi-stage pressure reducing valve is analyzed. The findings depict that the maximum pressure difference in the valve and the friction pressure difference increase with the increase of the roughness and after 2 mm, the increase decreases with the increase of the relative roughness, and finally gradually stabilizes; the average velocity at the valve outlet and the average internal velocity decrease progressively with the increase of roughness, and the relationship is approximately linear; at the roughness of 1 mm, the maximum wall shear stress is 5.6 times that when the roughness is 0 mm. In addition, the flow resistance coefficient increases linearly with the increase of roughness. The research results can provide theoretical support for the structural design of the series multi-stage pressure reducing valve.

Keywords Multi-stage pressure reducing valves · Roughness · Throttling characteristics · Flow resistance coefficient · Computational fluid dynamics

---

G. Wang (✉) · J. Deng · L. Kou · X. Zhu
School of Mechanical Engineering, Ningxia University, Yinchuan 750021, Ningxia, China
e-mail: belonging@163.com

G. Wang · L. Kou · X. Zhu
The Key Laboratory of Ningxia Intelligent Equipment CAE, Yinchuan 750021, Ningxia, China

M. Y. Ismail et al. (eds.), *Technological Advancement in Mechanical and Automotive Engineering*, Lecture Notes in Mechanical Engineering,
https://doi.org/10.1007/978-981-19-1457-7_34

1 Introduction

With the development of energy conservation and emission reduction projects, high-pressure differential pressure reducing valves have been widely used in engineering fields. Among them, multi-stage pressure reducing regulating valves are widely used in various industrial systems with the advantages of stable flow control and high pressure resistance [1]. The multi-stage reduced pressure throttling element is the key pressure control structure of the multi-stage reduced pressure regulating valve, which with narrow flow path and high pressure. In the actual industry, the regulating valve will cause damage to the surface due to long time use, forming a surface with unevenness as a result that affecting the internal flow field of the valve, and the roughness is usually used to characterize the internal uneven surface, therefore, it is important to study the effect of roughness on the internal flow field of the valve to adapt to the harsh working conditions and extend the service life in the engineering field.

At present, research on the pressure-reducing performance of valves has always been a focus of attention in the study of valves. Hou et al. [2] conducted a parametric analysis of a high-pressure pressure reducing valve and analyzed the effects of relative angle, orifice plate thickness, orifice plate number and plate hole diameter on the control performance of a multi-stage pressure-reducing regulating valve, and the results showed that when the relative angle is set to 180°, the steam flows through the porous crowned valve core, the maximum decompression pressure can be obtained, and the turbulence is the lowest. Gao et al. [3] designed a new axial flow regulating valve for flow regulation in small-diameter pipelines at low pressure of gas, and also used computational fluid dynamics (CFD) method to simulate the internal flow and flow regulation characteristics of the designed regulating valve, through the simulation analysis, the flow characteristic curve is obtained, and the results showed that the flow regulation characteristics of the designed regulating valve matched the design curve. Based on the standard K-ε turbulence model, Huang et al. [4] studied the flow resistance performance of four spool structure (flat bottom, small arc, large arc, and wavy) regulating valves and found that the flow resistance performance of the large arc and wavy spool structure regulating valves was 9.71% less than that of the flat bottom spool regulating valve, while the flat-bottom spool flow resistance performance taken certain advantages when the spool was at full opening state. Yao et al. [5] numerically simulate the flow characteristics of the valve orifice at different openings and verified the flow characteristics of the valve orifice by experiments.

In the history of valve development, most researchers have studied the internal flow field of valves; however, neither experimental nor numerical simulation techniques are currently available to study the internal flow field with reference to the real physical conditions of valves, which include the effects caused by friction. Many studies have been conducted experimentally for frictional losses inside pipes or fittings [6–12], in addition, Song [13] investigated the effect of periodic structural surface roughness on the flow field and pressure drop inside a low Reynolds number circular pipe using numerical simulations, where the considered periodic roughness exhibits sinusoidal, triangular and rectangular forms, and a theoretical model based on a modified perturbation theory was developed to quantify the effect of roughness on the fully developed Stokes flow in the pipe. Saidi [14] studied the effect of roughness on asymmetric vortex flow around a cone under subsonic conditions using numerical simulations to investigate the interrelationship between boundary layer separation, vortex structure and surface roughness. Kang et al. [15] investigated the along-range drag coefficient of non-Newtonian power-law fluid turbulent flow in four shape cross-section pipes (circular cross-section, triangular cross-section, rectangular cross-section and trapezoidal cross-section), and the results showed that the along-range drag coefficient in different shape cross-section pipes increased with the increase of relative roughness, and the increase reduced with the increase of relative roughness, and finally gradually stabilized.

The analysis of the above literature showed that surface roughness has a great influence on the internal flow of the valve, however, the effects of surface roughness on the internal flow field of high pressure differential pressure reducing valves have not been studied systematically enough. This paper presents CFD technology to study the effect of surface roughness on the internal flow field of the high-pressure differential pressure reducing valve, and analyze the internal pressure and velocity characteristics of the valve. This study can provide a theoretical basis for improving the throttling effect of the throttling element, and also has some reference significance for the design of other series multi-stage pressure reducing regulating valves.

2 Numerical Simulations

2.1 Control Equations and Turbulence Models

Fluid control equations and turbulence models have an important influence on the numerical simulation analysis of the internal flow field. When analyzing the internal flow of multi-stage pressure reducing valve, it is required to solve the mass conservation equation and momentum conservation equation first。

Continuity equation:

$$\frac{\partial \rho}{\partial t} + \frac{\partial(\rho u)}{\partial x} + \frac{\partial(\rho v)}{\partial y} + \frac{\partial(\rho w)}{\partial z} = 0 \tag{1}$$

Momentum equation:

$$\frac{\partial u}{\partial t} + u\frac{\partial u}{\partial x} + v\frac{\partial u}{\partial y} = -\frac{1}{\rho}\frac{\partial p}{\partial x} + \nu\frac{\partial^2 u}{\partial x^2} + \nu\frac{\partial^2 u}{\partial y^2} \tag{2}$$

The flow state of the medium inside the valve through the multi-stage pressure reducing valves is turbulent flow because the Reynolds number value is greater than 10^6. The RNG K-ε turbulence model is similar to the standard K-ε model, which takes into account the influence of vortices on turbulence and improves the accuracy of vortex calculation, while the RNG K-ε turbulence model is proven to be more accurate in describing the complex flow inside the high-pressure pressure-relief valve [16], so this numerical simulation the RNG K-ε turbulence model is used, and its corresponding equations for turbulent kinetic energy K and turbulent dissipation rate ε are as follows:

$$\frac{\partial(\rho k)}{\partial t} + \frac{\partial(\rho k u_i)}{\partial x_i} = \frac{\partial\left[\alpha_k \mu_{eff} \frac{\partial k}{\partial x_j}\right]}{\partial x_i} + G_k - \rho\varepsilon \tag{3}$$

$$\frac{\partial(\rho\varepsilon)}{\partial t} + \frac{\partial(\rho\varepsilon u_i)}{\partial x_i} = \frac{\partial}{\partial x_j}\left[\alpha_\varepsilon \mu_{eff} \frac{\partial\varepsilon}{\partial x_j}\right] + \frac{C_{1\varepsilon}\varepsilon}{k} G_k - C_{2\varepsilon}\rho\frac{\varepsilon^2}{k} - R_\varepsilon \tag{4}$$

where R_ε is:

$$R_\varepsilon = \frac{C_\mu \rho \eta^3 (1 - \eta/\eta_0)}{1 + \beta\eta^3} \frac{\varepsilon^2}{k} \tag{5}$$

The effective turbulent viscosity μ_{eff} is:

$$\mu_{eff} = \rho C_\mu \frac{k^2}{\varepsilon} \tag{6}$$

In the formula, G_k is the turbulent kinetic energy generation term caused by the mean velocity gradient, α_k, α_ε and $C_{2\varepsilon}$ is an empirical constant, $\alpha_k = \alpha_\varepsilon = 1.393$, $C_{2\varepsilon} = 1.68$ $C_{1\varepsilon}$ is the value of the empirical constant correction, $C_{1\varepsilon} = 1.42$; $C_\mu = 0.0845$; $\eta = SK/\varepsilon$, $\eta_0 = 4.38$, $\beta = 0.012$.

2.2 Surface Roughness Modeling

For turbulent wall boundary flow, the effect of wall roughness can be modeled using the modified "law of the wall" for roughness.

In ANSYS Fluent, the wall law is defined with the mean velocity modified by roughness as:

$$\frac{u_p u^*}{\tau_w/\rho} = \frac{1}{K} \ln\left(E \frac{\rho u^* y_p}{\mu}\right) - \Delta B \tag{7}$$

where u_p is the average velocity at the near-wall surface, y_p is the distance from the center of mass of the near-wall unit to the wall, τ_w is the wall shear stress, ρ is the fluid density, K is the von Karman constant of 0.4187, E is the empirical constant of 9.793, and μ is the dynamic viscosity of the fluid.

The frictional velocity u^* can be written as:

$$u^* = C_\mu^{1/4} k^{1/2} \tag{8}$$

where C_μ is a constant, k is the turbulent kinetic energy.

In addition, ΔB is related to the dimensionless roughness height K_s^+, where $K_s^+ = \rho K_s u^*/\mu$, where K_s is the physical roughness height.

For the smooth regime ($K_s^+ \leq 2.25$):

$$\Delta B = 0 \tag{9}$$

For the transitional regime ($2.25 < K_s^+ \leq 90$):

$$\Delta B = \frac{1}{K}\left[\frac{K_s^+ - 2.25}{87.75} + C_s K_s^+\right] \times \sin\{0.4258(\ln K_s^+ - 0.811)\} \tag{10}$$

In the fully rough regime ($K_s^+ > 90$):

$$\Delta B = \frac{1}{K}\ln\left[1 + C_s K_s^+\right] \tag{11}$$

where C_s is the roughness constant, which is taken as 0.5 in this numerical simulation.

2.3 Geometric Structures and Computational Models

The geometric model used for this numerical simulation analysis is a series multi-stage pressure reducing regulator valve with a nominal throughput of DN100 and a spool diameter of 73 mm. Figure 1a shows the structure of the series multi-stage pressure-relief regulating valve, the main components include the valve body, spool, spool sleeve, stem and so on. The fluid flow direction is left in and right out, the fluid medium flows into the valve pressure reducing structure from the notch at the bottom of the spool, and flows out from the top four round holes on the spool sleeve to complete the pressure reduction of the fluid, the main pressure reducing area is the pressure reducing structure composed of the spool and the spool sleeve, as shown in Fig. 1b, the pressure reducing structure consists of three notches as well as the outlet, each notch consists of four stages of pressure reducing structure, as shown in Fig. 1c.

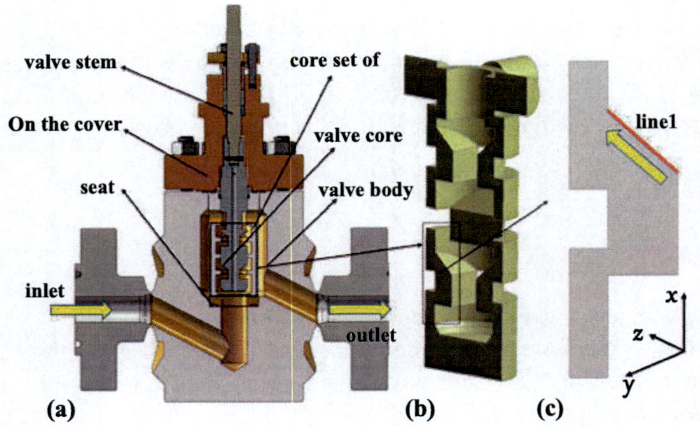

Fig. 1 Geometric model and internal flow channel model of tandem multi-stage pressure reducing control valve

2.4 Boundary Conditions and Grid-Independence

This analysis takes into account the influence of wall friction factor and analyzes the influence of roughness on the internal flow field of the pressure reducing valve. With the inlet velocity of 10 m/s, the default roughness thickness value k_s is equal to the equivalent roughness e, and the variation range is 0–3 mm, and the roughness constant C_s is kept at the default value of 0.5, which is considered as a uniform distribution of wall roughness. Due to the symmetry of the structure, half of the actual flow field is used for calculation, the symmetry center plane is used as the symmetry surface, and all other surfaces are set as smooth and slip-free wall surfaces, See Table 1.

When meshing, both mesh quality and mesh size are important conditions that affect the accuracy of the calculation. ANSYS mesh is used to divide the mesh, and because of the complexity of the buck structure part, automatic meshing is used for meshing. Before the numerical simulation, the mesh needs to be checked for independence. By changing the size of the cell size, the number of meshes in the computational region is adjusted to obtain a suitable mesh division. The original structure was selected for the grid independence check, as shown in Fig. 2, and the relative error of the mass flow rate at the outlet was 0.02% at the grid sizes of 1.25 and 1 mm, so the numerical simulation was considered possible. Since the complexity of

Table 1 Boundary conditions for computational domain

Boundary	Type	Conditions
Inlet	Velocity-inlet	$V_0 = 10\mathrm{m/s}$
Outlet	Pressure outlet	$P = 0$
Wall	Smooth non-slip wall surface	$u = v = w = 0$
Symmetry	Symmetry plane	

Fig. 2 Grid independence verification

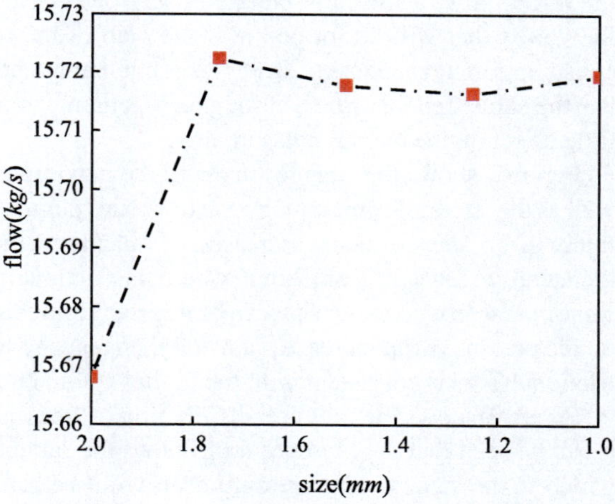

the buck structure part is considered, the grid size of 1 mm is used for this numerical simulation, the total number of grids is 1,469,466.

3 Results and Discussion

Figure 3 shows the pressure distribution in the plane of symmetry for different flow channel inclinations. When the fluid enters the throttling structure, its flow cross section contracts rapidly. At this time, the fluid velocity increases and the pressure decreases. According to the law of conservation of energy, when the fluid flows in

Fig. 3 X-direction pressure drop distribution diagram

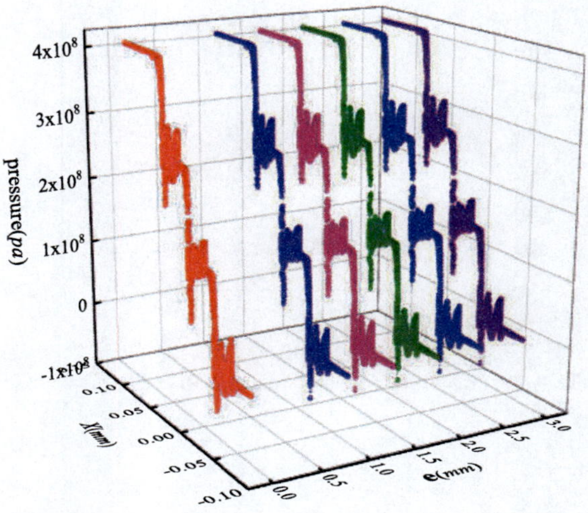

the throttling structure, the pressure cannot return to its original value because of the vortex that will be formed near the wall of the bend, thus causing the energy consumption. It can be seen from the graph that the pressure drop in the X direction has the same trend at different roughness, and it can be seen that the roughness has little effect on the overall pressure drop.

Figure 4 shows the change curves of the maximum pressure drop in the valve as well as the frictional pressure drop at different roughness, and it can be seen from the figure that as the roughness increases, the maximum pressure drop increases first and then tends to level off from 2 mm, where the frictional pressure drop also realizes the same characteristics, and what can be guessed from this is that the effect of roughness on the pressure drop increases first and then tends to level off, which means that the frictional This is consistent with the findings of the literature [15].

Figure 5 shows the pressure distribution of line1 in the X-direction, from which it can be seen that the pressure drop shows the same trend at the near-wall surface, which is the same as the pressure drop with the same trend in the X-direction at different rough-nesses as described earlier, where, at the end of line1 (the direction indicated by the yellow arrow in Fig. 1), the pressure at the end decreases as the roughness increases, indicating that at the same local pressure The friction along the line leads to more pressure loss, resulting in different end pressures.

Figure 6 shows the average velocity at the outlet of the valve and the internal average velocity, from the figure can be seen, with the gradual increase of roughness, the average velocity at the outlet and the average velocity inside the valve in gradually decreasing, and the approximate linear relationship, after linear fitting to obtain:

$$V_{out} = -0.02325e + 17.23521 \quad (12)$$

$$V_{valve} = -0.35643e + 35.33786 \quad (13)$$

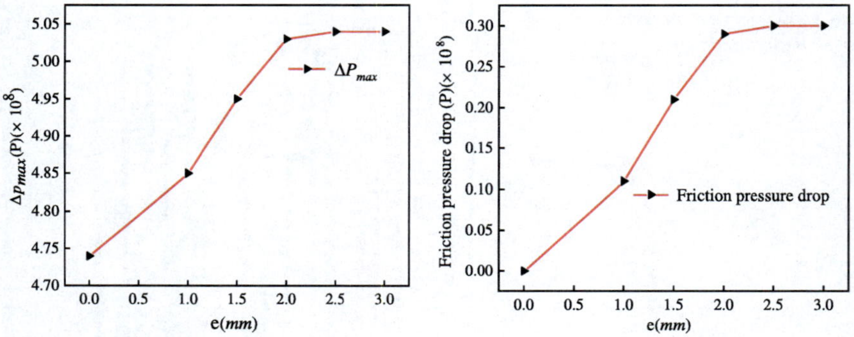

Fig. 4 Maximum pressure drop in the valve and friction pressure drop under different roughness

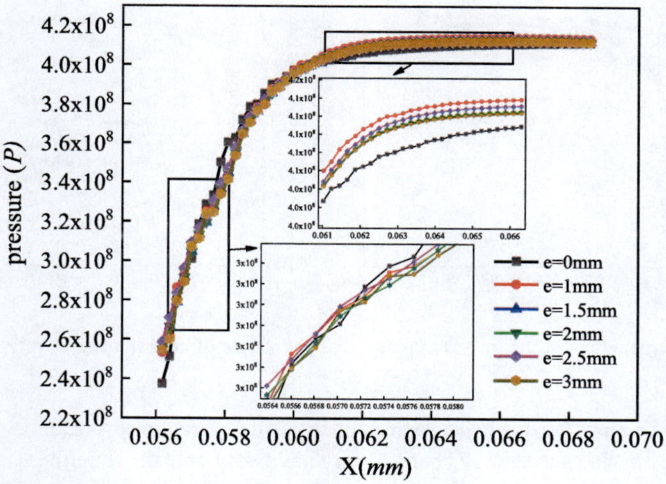

Fig. 5 Different roughness line1X direction pressure distribution chart

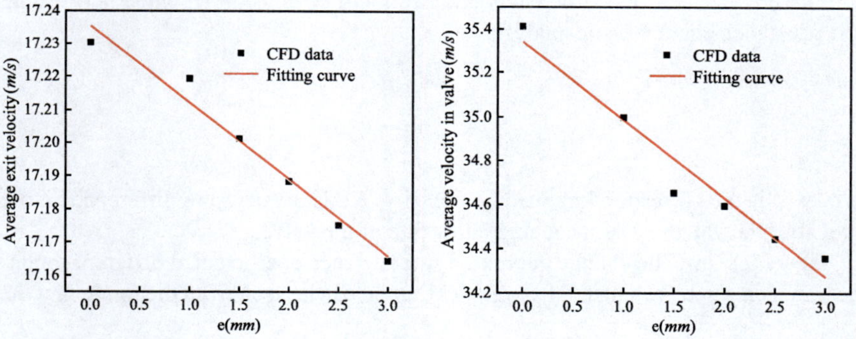

Fig. 6 Average velocity at the valve outlet and internal average velocity

where V_{out} and V_{valve} represent the average velocity at the outlet and the average velocity inside the valve, respectively, and e represents the roughness.

Figure 7 shows the distribution of turbulent kinetic energy and turbulent dissipation rate of line1 in the x-direction. What can be seen from the figure is that the roughness has a great influence on the turbulent intensity of the flow field at the near-wall surface, and the turbulent kinetic energy and turbulent dissipation rate increase substantially at the roughness height of 1 mm, however, they start to decrease again at 2 mm, which shows that the influence of roughness on the flow field at the near-wall surface is affected by the roughness height At the same time, it further shows that the frictional pressure drop caused by the previously described roughness will become limited when it increases to a certain height.

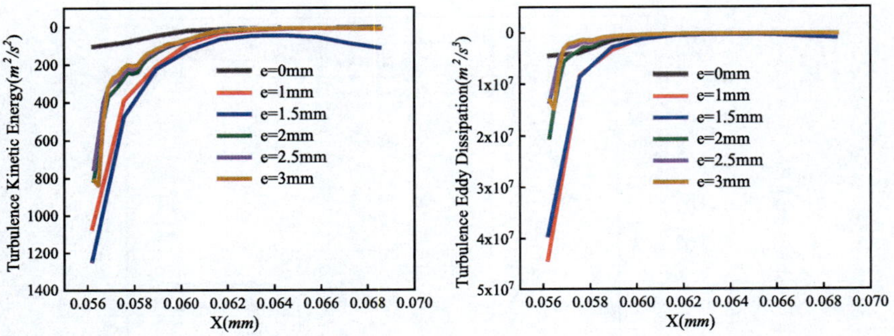

Fig. 7 Turbulent kinetic energy and turbulent dissipation rate of line1 in the x-direction

Figure 8 shows the wall shear stress of line1 in X direction. As can be seen from the figure, the wall shear force changes very obviously with the roughness, especially at 1 mm, the wall shear force increases suddenly, and its maximum value is 5.6 times when the roughness is 0 mm, and keeps the same trend at other roughness heights, and does not change too much.

The flow resistance coefficient equation should be considered when performing the resistance characteristic analysis:

$$\xi = \frac{2\Delta P}{\rho v_a^2} \tag{14}$$

where: ρ is the corresponding medium density, ΔP is the pressure difference before and after the valve, v_a is the average flow rate of the valve.

Figure 9 shows the change curve of flow resistance coefficient at different roughness, which can be calculated from Eq. (3), and it can be seen from the figure that the

Fig. 8 Line1 different roughness X direction wall shear stress

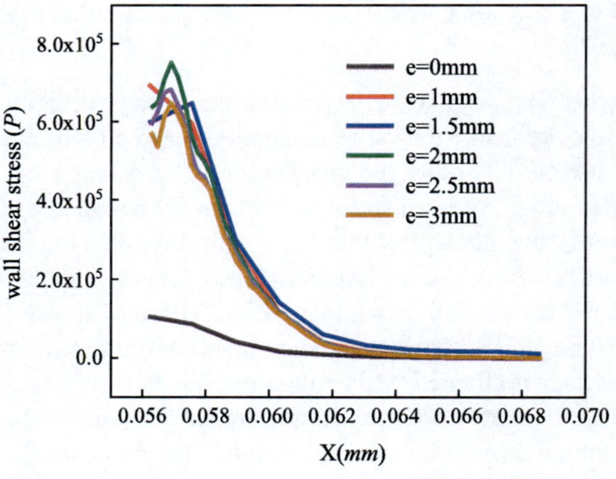

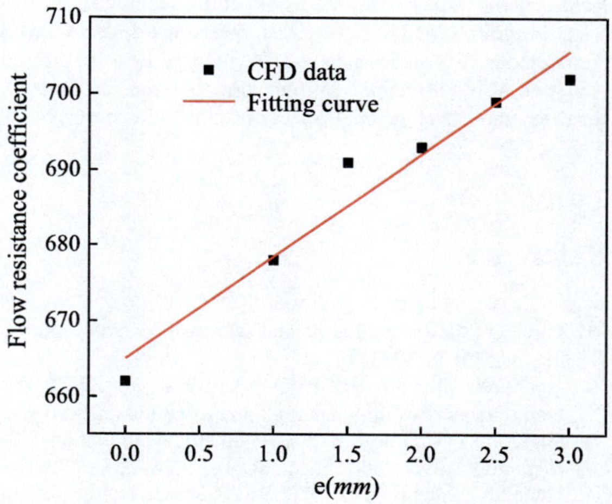

Fig. 9 Variation curve of flow resistance coefficient at different roughness

flow resistance coefficient shows a linear increase with the increase of roughness, and by linear fitting we get:

$$\xi = 13.5429e + 664.9286 \tag{15}$$

where ξ represents the flow resistance coefficient and e represents the roughness.

4 Conclusion

Numerical simulation was used to study the effect of internal wall roughness on the flow characteristics of tandem multi-stage pressure relief valves. The throttling characteristics were investigated under steady-state operating conditions, with the following conclusions:

1. The maximum pressure difference and the friction pressure drop inside the valve increase with the increase of roughness height, and it was found that after the roughness height is 2 mm, the increase diminutions with the increase of relative roughness, and finally gradually stabilizes.
2. With the progressive increase of roughness, the average velocity at the valve outlet and the internal average velocity are progressively decreasing, and the relationship is approximately linear, $V_{out} = -0.02325e + 17.23521$, $V_{valve} = -0.35643e + 35.33786$ are obtained by linear fitting.
3. The roughness has a great influence on the wall shear stress, especially at 1 mm location, the maximum value of wall shear stress is 5.6 times of that at 0 mm. In addition, the flow resistance coefficient increases linearly with the increase of roughness, and $\xi = 13.5429e + 664.9286$ is obtained by linear fitting.

Acknowledgements The authors gratefully acknowledge research support from the National Key R&D Program (2018YFB2004000), the Ningxia Youth Top Talent Project (2020), the Ningxia Autonomous Region Science and Technology Research (Support) Project (Key Technologies of High-end Valve Structural Optimization and Erosion Performance Research), the Ningxia Key Research and Development Project of China (Western Light, 2017).

References

1. Qiu YQ (2010) Calculation and selection design of multi-stage pressure regulating valve. Sci Technol Inf 19:30–31
2. Hou CW, Qian JY, Chen FQ, Jiang WK, Jin ZJ (2018) Parametric analysis on throttling components of multi-stage high pressure reducing valve. Appl Therm Eng 128
3. Gao JN, Li GZ, Luo XY (2020) Numerical simulation of working characteristics of axial flow regulating valve. Fluid Mach 48(10):12–16+58
4. Huang K, Wang HA, Li ZQ (2018) Research on flow resistance characteristics of throttle valve with different spool structure. Mod Mach 01:35–40
5. Yao SW, Liu Y, N SL, Yin FL (2015) Design of multi-stage cage sleeve pressure reducing valve and study of flow characteristics of valve port. Hydraul Pneumatics 08:34–37. (in chinese)
6. McKeon BJ, Swanson CJ, Zagarola MV et al (2004) Friction factors for smooth pipe flow. J Fluid Mech 511:41–44
7. Ito H (1960) Pressure losses in smooth pipe bends
8. Das SK (1996) Non-Newtonian liquid flow through globe and gate valves. Advances in engineering fluid mechanics: multiphase reactor and polymerization system hydrodynamics. Gulf Professional Publishing, pp 487–505
9. Telis RJ, Polizelli MA, Gabas AL et al (2005) Friction losses in valves and fittings for viscoplastic fluids. Can J Chem Eng 83(2):186–193
10. Fester VG, Kazadi DM, Mbiya BM et al (2007) Loss coefficients for flow of Newtonian and non-Newtonian fluids through diaphragm valves. Chem Eng Res Des 85(9):1314–1324
11. Mbiya BM, Fester VG, Slatter PT (2009) Evaluating resistance coefficients of straight-through diaphragm control valves. Can J Chem Eng 87(5):704–714
12. Cabral RAF, Telis VRN, Park KJ et al (2011) Friction losses in valves and fittings for liquid food products. Food Bio-Prod Process 89(4):375–382
13. Song S, Yang X, Xin F et al (2018) Modeling of surface roughness effects on Stokes flow in circular pipes. Phys Fluids 30(2):023604
14. Saidi N, Cerdoun M, Khalfallah S et al (2020) Numerical investigation of the surface roughness effects on the subsonic flow around a circular cone-cylinder. Aerosp Sci Technol 107:106271
15. Kang YP, Zhang H, Xu Y (2018) Influence of roughness on the coefficient of drag along the turbulent flow of power-law fluids. Gas Therm Power 38(08):7–12
16. Han X, Zheng MG, Yu YQ (2011) Hydrodynamic characterization and optimization of contra-push check valve by numerical simulation. Annal Nucl Energy 38(6)
17. Chen FQ, Qian JY, Chen MR, Zhang M (2018) Turbulent compressible flow analysis on multi-stage high pressure reducing valve. Flow Measur Instrum 61
18. Yu D, Revell A, Sinha J, Hahn W (2019) A computational fluid dynamics (CFD) analysis of fluid excitations on the spindle in a high-pressure valve. Int J Press Vessels Piping 175
19. Srikanth C, Bhasker C (2008) Flow analysis in valve with moving grids through CFD techniques. Adv Eng Softw 40(3)
20. Jin ZJ, Chen FQ, Qian JY, Zhang M (2016) Numerical analysis of flow and temperature characteristics in a high multi-stage pressure reducing valve for hydrogen refueling station. Int J Hydrogen Energy 41(12)

Influence of Temperature and pH Value in 3.5% NaCl Solution on Electrochemical Performance of 316L Stainless Steel

G. Wang, Z. K. Zou, P. Zhang, Y. Wu, L. Y. Kou, and Y. Q. Xu

Abstract Using potentiodynamic polarization, AC impedance and Mott-Schottky curve measurement methods, the electrochemical behavior and passivation characteristics of 316L stainless steel in 3.5% NaCl solution were studied. The experimental research results show that: The passivation film on the sample surface is relatively stable in the low-temperature alkaline solution. As the temperature of the solution increases and the pH decreases, the pitting potential of 316L stainless steel decreases accordingly, the corrosion current density increases, and the passivation interval decreases; The donor concentration and acceptor concentration near the surface oxide film increase; The change trend of electrochemical impedance spectroscopy is obviously different, and the pitting corrosion resistance of stainless steel becomes worse.

Keywords 316L stainless steel · Pitting potential · Electrochemical performance · Passivation film · pH value

1 Introduction

The high-salt wastewater produced by coal chemical companies is characterized by high salinity, large emissions, and high chloride ion content [1]. 316L stainless steel has the advantages of high strength, resistance to intergranular corrosion, stress corrosion, pitting corrosion, welding, and excellent processing performance. It is widely used in engineering construction, and the research on it is also continuously deepening. The main reason for the corrosion resistance of 316L stainless steel is that the passivation film formed on its surface isolates the further contact between the substrate and the corrosive medium, so that the substrate is protected [2, 3].

G. Wang (✉) · Z. K. Zou · P. Zhang · Y. Wu · L. Y. Kou · Y. Q. Xu
School of Mechanical Engineering, Ningxia University, Yinchuan 750021, Ningxia, China
e-mail: belonging@163.com

G. Wang · L. Y. Kou · Y. Q. Xu
The Key Laboratory of Ningxia Intelligent Equipment CAE, Yinchuan 750021, Ningxia, China

M. Y. Ismail et al. (eds.), *Technological Advancement in Mechanical and Automotive Engineering*, Lecture Notes in Mechanical Engineering,
https://doi.org/10.1007/978-981-19-1457-7_35

Guo et al. [4] found that the outer layer of the passivation film is mainly composed of iron oxides FeO, Fe_2O_3 and a small amount of Cr_2O_3, and the inner layer is mainly composed of chromium oxides Cr_2O_3, $Cr(OH)_3$ and metals Fe and Cr. Existing research shows that the corrosion resistance of stainless steel is related to its passivation ability. Strehblow [5] and Hoppe [6] studied the influence of pH on Fe–Cr alloy passivation film. They found that the film formed more stably in alkaline solution and the amount of Fe^{3+} increased significantly. In acid solutions, iron oxides usually have poor stability, because the anions in the corrosive solution have a stronger ability to penetrate the passivation film of FeCr alloy. On the contrary, the chromium oxide layer has poor stability in alkaline solution [7]. Arya's [8] research found that in a 3.5% NaCl solution, as the pH value of the solution increases, the nitrogen stainless steel exhibits a lower donor density and a stable passivation film. And consistent with low passivation current density (i_{pass}), higher pitting potential (E_{pit}) and polarization resistance (R_p). Ferreira [10] studied the relationship between the temperature and the polarization curve of AISI 304 and 316 stainless steel in NaCl solution and found that the pitting potential decreases with increasing temperature.

In summary, a large number of scholars have conducted research on the corrosion behavior of steel under different temperature and different pH systems, and significant results have been obtained. However, there are few reports on the corrosion behavior under the combined action of the two conditions, and further research and discussion are needed.

The use environment of stainless steel is more complicated, and the difference in acidity and alkalinity is large. The thickness of stainless-steel passivation film is generally in the range of a few to tens of nanometers, so electrochemical methods are usually used to study the properties of the passivation film. In order to reduce the harm caused by corrosion to equipment in the coal industry, extend the service life of materials and equipment, and improve economic benefits. This selection of 316L stainless steel specimen, using an electrochemical potentiodynamic polarization curve, electrochemical impedance spectroscopy and Mott-Schottky curve. The effects of temperature and pH on the pitting resistance and passivation performance of 316L stainless steel in NaCl solution were studied.

2 Experimental Part

2.1 *Materials and Samples*

The experimental material is 316L stainless steel, and its chemical composition (mass fraction, %) is: C 0.014, Si 0.44, Mn 1.15, P 0.027, S 0.004, Cr 16.15, Ni 10.01, Mo 2.02, N 0.55, Fe balance. Solder the back of the sample with the copper wire by soldering, and seal it in epoxy resin, exposing a working surface with an area of 1 cm^2. Then use water sandpaper to polish to 1000# in turn, rinse with absolute ethanol and blow dry, then place it in a desiccator for later use. The pitting potential is carried out according to GB/T 17899-1999 "*Stainless steel pitting potential measurement method*".

2.2 *Electrochemical Measurements*

The test solution used 3.5% (mass fraction, the same below) NaCl solution. Adjust the pH value of the prepared NaCl solution to 4, 7 and 10 with 1 mol/L NaOH solution and 1 mol/L HCl solution. During the test, a constant temperature water bath was used to keep the temperature of the solution in the electrochemical test at 30 °C, 40 °C, and 50 °C, respectively.

The electrochemical measuring instrument adopts PARSTAT4000 electrochemical workstation, and the software system adopts VersaStudio. The three-electrode system is uniformly used in the process of measuring the impedance and polarization curve (Tafel curve) and Mott-Schottky curve of the sample. The reference electrode is a saturated calomel electrode (SEC), the auxiliary electrode is a Pt electrode, and the working electrode is a sample.

All electrochemical tests are performed after the working electrode is polarized in the test solution for 10 min. The electrochemical impedance test has a measured potential of 0 V, a sine wave signal amplitude of 10 mV, and a measurement frequency of 100 kHz–10 MHz. After that, the working electrode is scanned with a dynamic potential, the scanning range is −0.7~2 V (relative to the open circuit potential), and the scanning rate is 1 $mV \cdot s^{-1}$. The measured polarization curve parameters were extracted, and the pitting corrosion potential and corrosion current density were fitted. The test frequency of the Mott-Schottky curve is 1000 Hz, the amplitude is 10 mV, and the potential is swept from −0.6 to 2 V. Finally, the corrosion morphology was observed through a Zeiss metallurgical microscope.

3 Results and Discussion

3.1 *Potential Polarization Curve*

Put the 316L stainless steel into the NaCl solution of different pH values with a mass fraction of 3.5% at 30 °C for the polarization curve test. The results are shown in Fig. 1a. As the acidity in the corrosive medium decreases, the corrosive solution rises from a weakly acidic pH = 4 to a weakly alkaline pH = 10. As the H^+ concentration in the solution began to decrease, the concentration of Fe^{2+} and Fe^{3+} in the solution in the pits decreased, and the concentration of Cl^- decreased, and the resistance of the metal film in the pits increased, which slowed down the corrosion growth of the pits. Secondly, the increase of OH^- reduces the self-acidification of the sample, reduces

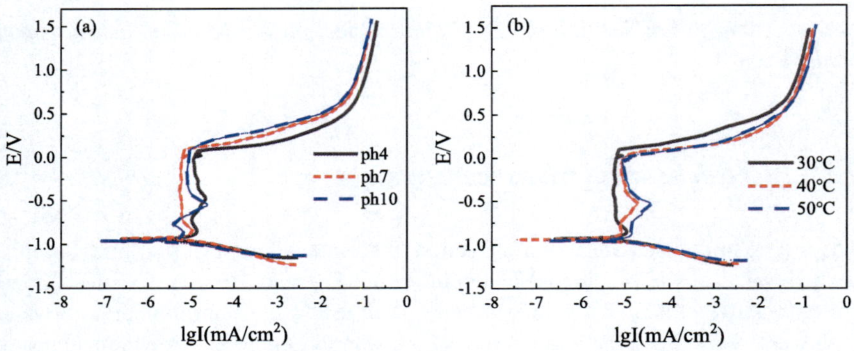

Fig. 1 Polarization curves of 316L stainless steel under different conditions. **a** 30 °C, **b** pH = 7

the local corrosion sensitivity and the degree of hydrogen evolution corrosion, and slows the dissolution of the metal in the corrosion hole.

On the other hand, according to the point defect theory, the doped electrons in the film are mainly oxygen vacancies. The oxygen vacancy can react with the OH^- ion solution to promote the growth of the passivation film. Therefore, the higher the pH of the solution, the stronger the stability of the passivation film [10]. The pitting potential under alkaline conditions is higher than that under acidic conditions, indicating that the formed passivation film has strong pitting resistance under alkaline conditions.

The electrochemical parameters of the polarization curve were fitted with Tafel, and the results are shown in Table 1. As the pH value increases, the pitting corrosion potential (E_{corr}) of the sample increases from −933.56 to −884.15 mV, and the self-corrosion current density (I_{corr}) gradually decreases from 6.59 to 1.4 mA·cm^{-1}.

The self-corrosion current density reflects the size of the metal corrosion rate [11]. The reduction of self-corrosion current density and the increase of pitting corrosion potential indicate that the increase of pH indicates that the Cl^- corrosion resistance of 316L stainless steel is increased, and the passivation potential range is increased, which improves the corrosion resistance of stainless steel.

Figure 1b shows the potentiodynamic polarization curves of 316L stainless steel measured in 3.5% NaCl solutions at different temperatures at pH = 7.The self-corrosion current density (I_{corr}) increases, and the pitting corrosion potential decreases. The Tafel fitting result of the polarization curve is consistent with this trend, as shown in Table 2. This indicates that the increase in temperature accelerates the reaction rate on the electrode surface, which increases the corrosion rate of 316L

Table 1 Tafel fitting results of different pH values at a temperature of 30 °C

pH	E_{corr}/mV	I_{corr}/mA · cm^{-1}
4	−933.56	6.59
7	−924.99	5.89
10	−884.15	1.4

Table 2 Tafel fitting results in different temperature solutions at pH = 7

Temperature/°C	E_{corr}/mV	I_{corr}/mA · cm^{-1}
30	−924.99	5.89
40	−933.56	6.5
50	−956.61	7.16

stainless steel. Although the increase in temperature accelerates the corrosion rate of 316L stainless steel, the shape of the polarization curve is basically the same, indicating that the corrosion mechanism has not changed at different temperatures.

Studies have shown that the main components of the passive film of 316L stainless steel are Cr_3O_2, FeO and NiO. The ions have a variety of behaviors in the corrosion process: combine with cations, reduce ion mobility, or inhibit ion adsorption. With the increase of temperature, the lower valence metal oxides or hydroxides on the surface of the passivation film, such as $Fe(OH)_2$, Cr_2O_3, etc., are oxidized to Fe_2O_3, CrO_3, which leads to an increase in the valence state, which in turn induces The increase of the anion concentration on the surface of the passivation film accelerates the intrusion of anions into the passivation film, and the pitting potential decreases, making corrosion more likely to occur [12].

At the same time, the temperature increase increases the activity of Cl^-, and a large amount of Cl^- is adsorbed on the surface of the sample. It is often believed that Cl^- enters the passivation film by adsorption or under the action of an electric field prior to the defective position of the passivation film, causing the destruction of the passivation film. As a result, small corrosion holes are formed at some points of the newly exposed base metal, which leads to the decrease of the resistance of the passive film on the surface of the stainless steel and the deterioration of the protection. The decrease in the stability of the passivation film increases the sensitivity of pitting corrosion, which increases the ion concentration difference between the inside and outside of the 316L stainless steel pits, which intensifies corrosion. The passivation film is gradually destroyed, causing the metal matrix to directly react with the corrosion solution, which further intensifies the corrosion dissolution process [13].

3.2 *AC Impedance Curve*

Figure 2a shows the effect of pH on the Nyquist diagram of 316L stainless steel in a 3.5% NaCl solution at 30 °C. In the figure, the abscissa represents the real part of impedance Z, and the ordinate represents the imaginary part of impedance Z. The capacitive reactance arc does not show a strict semicircular shape but has a slight deviation. The main reason is the dispersion effect between the interface of the material and the NaCl solution [14]. The Nyquist diagrams in different pH values all show a single capacitive reactance arc, and with the increase of pH, the radius of the capacitive reactance arc in the high frequency range increases, and the corrosion reaction resistance of the sample increases. The reason may be that as the

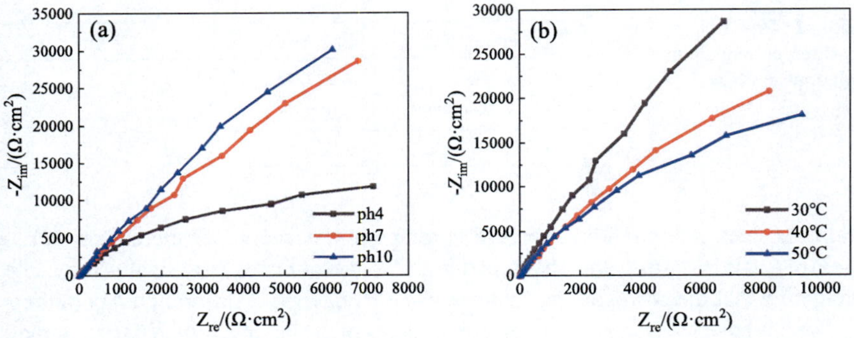

Fig. 2 Nyquist spectrum of 316L stainless steel under different conditions. **a** 30 °C and **b** pH = 7

pH value increases, the concentration of OH^- in the solution increases, and there is a competitive adsorption of OH^- and Cl^- in the solution, which reduces the dissolution rate of Cl^- on the passivation film. The sample in the lower pH solution maintains the integrity of the passivation film and increases the film resistance. It is reflected in the Nyquist diagram that the capacitive arc radius increases with the increase of pH, and the corrosion resistance of metals increases.

From Fig. 2b, it can be seen from the Nyquist spectrum that the curves show similar capacitive reactance arcs at different temperatures. And the impedance spectrum arc is larger at 20 °C, and the radius of the impedance arc gradually decreases with the increase of temperature, indicating that the stability of the passivation film on the surface of the sample deteriorates. And in the figure, it can be seen that there is no shrinkage of the impedance spectrum arc in the real part, and it can be inferred that no pitting corrosion occurs. As the temperature increases, the resistance value decreases, indicating that the temperature increase increases the corrosion rate of 316L stainless steel.

The influence of pH on the Bode diagram of 316L stainless steel when the temperature is 30 °C and the influence of temperature on the Bode diagram of 316L stainless steel when pH = 7 is shown in Fig. 3. Figure 3a, b show the relationship between phase angle and frequency; Fig. 3c, d show the relationship between impedance and frequency.

It can be seen from Fig. 3a that in the low frequency region, as the pH increases, the phase angle gradually increases, and the wave crest moves to the high frequency region. The impedance modulus reflects the polarization impedance of the electrode [15]. It can be seen in Fig. 3c that as the temperature decreases, the impedance modulus of stainless-steel decreases, indicating that the stability of the sample's passive film becomes worse.

The results show that the passivation film formed under low-temperature alkaline conditions is smoother and more uniform. It is worth noting that in the low frequency region, the lower value of the phase angle is significantly reduced, indicating that the impedance is the pure resistance of the passivation film. However, in a wide

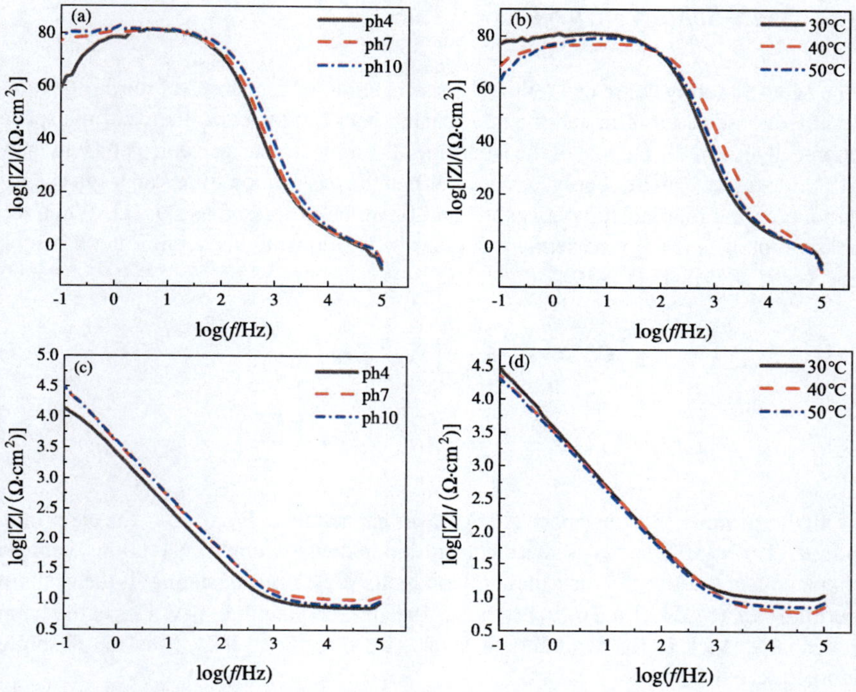

Fig. 3 The Bobe pattern of 316L stainless steel under different conditions. **a** 30 °C, **b** pH = 7

frequency range, the phase is still close to 90°, which indicates that the passivation oxide film exists on the surface in the solution under all conditions.

As the temperature increases, it can be found that in the low-frequency region in Fig. 3b, the low-frequency phase angle decreases, and its peak frequency region moves toward the high-frequency region. Figure 3d shows that the impedance modulus decreases with the increase of temperature, and the polarization resistance of the sample's passivation film decreases, and the greater the frequency, the more obvious the difference. Therefore, it can be inferred that both the increase in temperature and the decrease in pH make the stability of the passivation film worse.

Based on the analysis results of Nyquist and Bode diagrams, it can be seen that the electrochemical reaction process is fast in the high frequency region of the impedance spectrum, while the diffusion process is slow. When the temperature rises, the diffusion speed of the particles in the solution increases, and the metal cations produced after the metal matrix is dissolved continue to diffuse into the solution. In order to maintain the anions generated by the hydrolysis in the neutral solution in the etch pit, the anions generated by the hydrolysis also continuously diffuse to the pit, and this diffusion process affects the speed of the electrode reaction process. During the diffusion process, the ions constantly hovering around the passivation film will destroy the stability of the passivation film and increase the pitting sensitivity.

3.3 Mott-Schottky Curve

The Mott-Schottky curve can reflect the semiconductor properties of the 316L stainless steel passivation film and the relationship between the capacitance of the space charge layer and the change of the electrode. Thereby, the surface corrosion behavior of stainless steel can be deeply explored. When the passivation film is an n-type semiconductor, the relationship between C and E can be expressed as Eq. (1). When the passivation film is a p-type semiconductor, the relationship between C and E can be expressed as Eq. (2) [16–19].

$$C^{-2} = \frac{2}{N_D q \varepsilon \varepsilon_0}\left(E - E_{FB} - \frac{kT}{q}\right) \tag{1}$$

$$C^{-2} = \frac{2}{-N_A q \varepsilon \varepsilon_0}\left(E - E_{FB} - \frac{kT}{q}\right) \tag{2}$$

In the formula, C is the space charge layer capacitance, F/cm^2; q is the electronic power (1.6×10^{-19}), N_D is the electron donor density, cm^{-3}; N_A is the electron donor concentration, cm^{-3};εis the semiconductor dielectric constant;ε_0 is the vacuum permittivity, (8.85419×10^{-12} F/m); E is the applied potential, mV; E_{FB} is flat band potential, mV; k is Boltzmann's constant, (1.38×10^{-23} J/K); T is the absolute temperature, K.

Figure 4 shows the Mott-Schottky curve of 316L stainless steel measured under different conditions. Figure 4a is a graph showing the influence of different pH values in a NaCl solution with a molecular mass of 3.5% at 30 °C. Figure 4b is a graph showing the influence of different temperatures in a NaCl solution with pH = 7 and a molecular mass of 3.5%. It can be seen that in the −0.5~0.2 V potential curve, the slope of the curve is positive, and as the temperature increases, the peak point of the curve moves to the left. In this potential range, the passivation film exhibits n-type semiconductor characteristics. In the 0.5–2 V potential curve, the slope of the

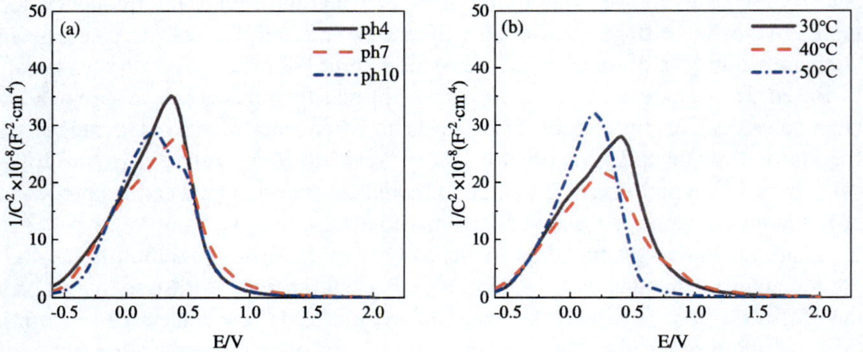

Fig. 4 Mott-Schottky curve of 316L stainless steel under different conditions. **a** 30 °C, **b** pH = 7

curve is all negative, and the passivation film in this potential interval exhibits p-type semiconductor characteristics. The n-p type semiconductor conversions all occur in the 0.2~0.5 V range.

The electron energy band theory shows that when the number of electrons in the conduction band of an oxide exceeds the number of holes in the valence band, the oxide is defined as an n-type semiconductor; On the contrary, when the number of holes in the oxide valence band exceeds the number of electrons in the conduction band, the oxide exhibits p-type semiconductor characteristics. The passivation film exhibits different semiconductor characteristics, depending on the dominant defects in the passivation film. If metal ions are missing or cation vacancies are present in the passivation film, such a passivation film generally exhibits p-type semiconductor characteristics, but on the contrary, it exhibits n-type characteristics.

According to the point defect model (PDM) [20], when the passivation film is in a Cl^- containing medium, oxygen holes react with Cl^- to form a new oxygen hole/metal ion hole pair. The generated oxygen holes continue to react with other Cl^- to generate a large number of metal ion holes. As the pH decreases, the collision probability between Cl^- and the passivation film increases, thereby promoting the reaction and accelerating the generation of metal ion holes. The metal ion holes will quickly accumulate in the area between the substrate and the passivation film, hindering the growth of the passivation film. Therefore, as the pH decreases, the 316L stainless steel passivation film can be dissolved at a lower potential, resulting in the rupture of the passivation film, pitting corrosion, and a decrease in corrosion resistance.

As the temperature increases, the electron acceptor density of the passivation film increases. This is because as the temperature increases, the low-valence metal oxide on the surface of the passivation film formed by 316L stainless steel is oxidized, which promotes the increase of the valence state. The increase in acceptor density can cause the pitting corrosion potential to decrease, which makes corrosion more likely to occur. The donor current density reflects the number of carriers in the passivation film, that is, the point defects in the space charge layer. Therefore, the higher the carrier content, the more point defects in the passivation film. These defects will be the location where pitting of the passivation film occurs. It can also be determined that the greater the donor density, the higher the probability of corrosion and cracking of the passivation film, and also hinders the growth of the passivation film. As a result, 316L stainless steel has a higher sensitivity, which also explains why the anode polarization curve will shift positively as the temperature rises.

3.4 Corrosion Morphology Observation

The corrosion morphology of 316L stainless steel at different pH values and different temperatures in 3.5% NaCl is shown in Fig. 5. It can be seen that at pH = 7 and the temperature is 30 °C, slight corrosion appears on the surface of the sample, and the pitting pits on the surface are smaller in diameter and smaller in number;

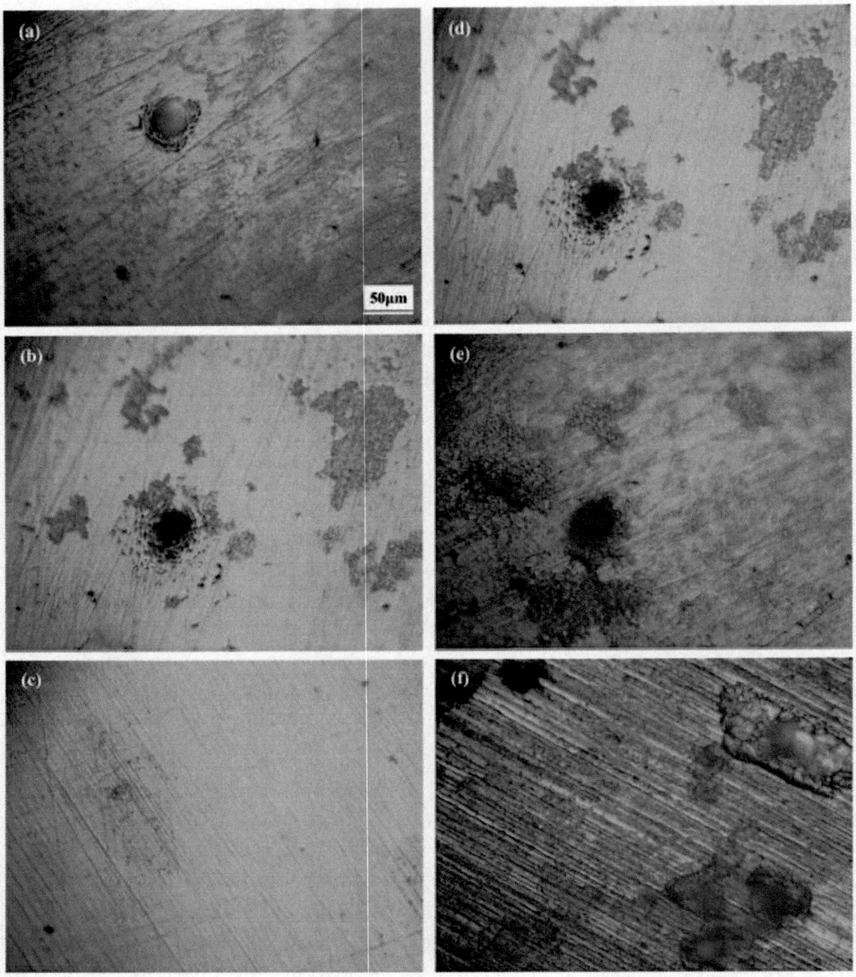

Fig. 5 Corrosion morphology of 316L in 3.5% NaCl solution under different conditions **a** pH = 4, 30 °C, **b** pH = 7 and 30 °C, and **c** pH = 10, 30 °C **d** pH = 7 and 30 °C, **e** pH = 7 and 40 °C, and **f** pH = 7 and 50 °C

As the temperature of the solution increases, the corrosion intensifies, the number of corrosion pits on the metal surface increases, the diameter increases, and the corrosion is more obvious; When the temperature is 30 °C, the pH value increases and the corrosion is aggravated. When pH = 4, the corrosion is the most serious.

This is because the increase in temperature accelerates the reaction rate of Cl^- in the solution, and the decrease in pH makes the concentration of H^+ in the solution start to increase. As a result, the concentration of Fe^{2+} and Fe^{3+} in the solution in the corrosion hole increases, the concentration of Cl^- increases, and the resistance of the metal film in the corrosion hole decreases, which intensifies the corrosion growth of the corrosion hole. This phenomenon is consistent with the results of electrochemical tests.

4 Conclusions

1. According to the polarization curve results, as the temperature of the solution increases and the pH decreases, the self-corrosion potential of 316L stainless steel shifts negatively, the pitting corrosion potential decreases, and the corrosion current density increases; Electrochemical impedance spectroscopy test results show that 316L stainless steel can form a stable passivation film in a 3.5% NaCl solution, and the lower the temperature, the higher the pH, the more stable the passivation film formed.
2. The Mott-Schottky curve results show that in NaCl solutions under different pH and temperature conditions, when the potential is lower than 0.2 V, the passivation film exhibits n-type semiconductor properties. When the potential is 0.5–2 V, the passivation film exhibits p-type semiconductor properties. Temperature and pH have a certain influence on the semiconductor properties of the passivation film.
3. Combined with the observation of the corrosion morphology, the passivation performance of the passivation film on the surface of 316L stainless steel in the low-temperature alkaline solution is better than that in the high-temperature acid solution.

Acknowledgements The authors gratefully acknowledge research support from the National Natural Science Foundation of China (No. 52165020), the Ningxia Youth Top Talent Project (2020), the Ningxia Autonomous Region Science and Technology Research (Support) Project (Key Technologies of High-end Valve Structural Optimization and Erosion Performance Research), the Ningxia Key Research and Development Project of China (Western Light, 2017).

References

1. Li T, Sen G, Xueshuang Z (2010) Conditionality of waste water zero discharge in coal chemical industry. Environ Prot Chem Indus 30(05):371–375
2. He W, Knudsen OØ, Diplas S (2009) Corrosion of stainless steel 316L in simulated formation water environment with CO_2–H_2S–Cl. Corros Sci
3. Li M, Zeng C, Lin H et al (2002) Corrosion behavior of 316 stainless steel in F^-/Cl^- acid solution. J Chin Soc Corros Protect (03):35–39
4. Guo Q, Liu J, Yu M et al (2015) Effect of passive film on mechanical properties of martensitic stainless steel 15–5 pH in a neutral NaCl solution. Appl Surf Sci 327:313320
5. Haupt S, Strehblow H-H (1995) A combined surface analytical and electrochemical study of the formation of passive layers on Fe/Cr alloys in 0.5 M H_2SO_4. Corros Sci 37(1)

6. Hoppe H-W, Haupt S, Strehblow H-H (1994) Combined surface analytical and electrochemical study of the formation of passive layers on Fe/Cr alloys in 1 M NaOH. Surf Interface Anal 21(8):514–525
7. Ding J, Zhang L, Lu M, Wang J, Wen Z, Hao W (2014) The electrochemical behavior of 316L austenitic stainless steel in Cl^- containing environment under different H_2S partial pressures. Appl Surf Sci 289
8. Arya SB, Raja VS, Tiwari AN (2013) A comparative study of semiconducting behavior of passive film of high nitrogen and Ni and Mn free stainless steels in 3.5 wt. % NaCl. Adv Mater Res 2655:626–631
9. Ferreira MGS, Moura e Silva T, Catarino A, Pankuch M, Melendres CA (2019) Electrochemical and laser Raman spectroscopy studies of stainless steel in 0.15 M NaCl solution. J Electrochem Soc 139(11)
10. Cheng YF, Luo JL, Wilmott M (2000) Spectral analysis of electrochemical noise with different transient shapes. Electrochim Acta 45(11):1763–1771
11. Fengping W, Wanli K, Hemin J (2008) Principles, methods and applications of corrosion electrochemistry. Chemical Industry Press, Beijing
12. Hakiki NE, Da Cunha Belo M (2019) Electronic structure of passive films formed on molybdenum-containing ferritic stainless steels. J Electrochem Soc 143(10)
13. Xia W, Chenhui G, Yuxiang C et al (2016) Effect of heat treatment on corrosion behavior of X80 steel in Yingtan soil simulated solution. Heat Treat Met 41(08):148–153
14. Yu C, Xu C et al (2014) Effect of temperature on electrochemical corrosion behavior of 316L stainless steel in borate buffer solution. Corros Protect 35(04):344–347–351
15. Xin W, Junhua D, Jian T et al (2012) Influence of temperature on piting corrosion resistance of Cr26Mol ultra pure high chromium ferrite stainless steel in 3.5% NaCl solution. Acta Metall Sin 48(04):502–507
16. Alves VA, Brett CMA (2002) Characterisation of passive films formed on mild steels in bicarbonate solution by EIS. Electrochimica Acta 47(13)
17. Sikora J, Sikora E, Macdonald DD (2000) The electronic structure of the passive film on tungsten. Electrochimica Acta 45(12)
18. Jian L, Yi W, Jibo J et al (2012) Electrochemistry behavior of Rebars with different grain size and Mott-Schottky research of passive films. Acta Chim Sin 70(10):1213–1220
19. Yunlian Z, Meilun S, Zhiyuan C (2006) Mott-Schottky investigation of passivation film of rebar in simulated concrete pore solution. Mater Mech Eng 07:7–10
20. Moayed MH, Newman RC (2005) Evolution of current transients and morphology of metastable and stable pitting on stainless steel near the critical pitting temperature. Corros Sci 48(4)

The Influence of Coal Water Slurry Particle Size on the Erosion of Reducing Pipe

G. Wang, Q. F. Gao, J. F. Deng, W. H. Wang, Y. X. Zhang, X. J. Zhu, and Y. Q. Xu

Abstract Coal chemical industry transportation of coal slurry is a typical solid–liquid two-phase flow, this process is accompanied by the erosion and abrasion of the pulverized coal slag on the inner wall of the pipeline. This phenomenon will have a significant impact on the service life of the transportation components and become a hidden engineering hazard. Aiming at the erosion problem of the coal chemical transportation system, the variable diameter pipe in the commonly used pipeline in this working condition is taken as the research object, based on the actual working condition, using the discrete phase model, track and solve the particle motion trajectory information under the Lagrange framework to obtain particle motion information. And combining this information with the erosion model, the effect of the particle size of the cinder and coal powder on the erosion of the inner wall of the pipeline is analyzed. The results show that as the particle size increases, the maximum erosion wear rate decreases. When the particle size is less than 500 μm, the maximum erosion wear rate decreases faster, when the particle size is greater than 500 μm, the maximum erosion wear volume decreases slowly. This study can provide a certain theoretical basis for the design of coal chemical transportation pipelines and the research on erosion, wear and maintenance.

Keywords Erosion · DPM model · Particle size · Solid–liquid two-phase flow

1 Introduction

Sufficient energy has provided the country with a safe, stable and sustainable economic development. Our country is rich in coal resources, and the coal chemical industry can achieve clean use of coal resources, which meets the current economic

G. Wang (✉) · Q. F. Gao · J. F. Deng · W. H. Wang · Y. X. Zhang · X. J. Zhu · Y. Q. Xu
School of Mechanical Engineering, Ningxia University, 750021, Yinchuan, Ningxia, China
e-mail: belonging@163.com

G. Wang · X. J. Zhu
The Key Laboratory of Ningxia Intelligent Equipment CAE, Yinchuan 750021, Ningxia, China

M. Y. Ismail et al. (eds.), *Technological Advancement in Mechanical and Automotive Engineering*, Lecture Notes in Mechanical Engineering,
https://doi.org/10.1007/978-981-19-1457-7_36

development and environmental protection requirements. However, during the transportation of coal slurry in the coal chemical industry, the erosion of the transportation pipeline is a serious problem in the coal chemical transportation industry. In this transportation system, the solid particles of coal slag and coal powder with smaller particle size will flow with the carrying fluid and impact the inner walls of transportation pipelines, valves and other components. Since this is a long-term and continuous process, many components in the transportation system are facing severe erosion damage, resulting in an extremely short service life. Especially in the pipeline with sudden geometric shape, the force is more complicated and the stress level is high, the erosion damage is serious and uneven, and the phenomenon of perforation is very easy to occur. Small or subtle changes in operating conditions may cause significant damage caused by erosion.

At present, the research on the erosion of solid–liquid two-phase flow by domestic and foreign researchers mainly adopts experimental research and numerical analysis methods.

In terms of experimental research, Ye [1] designed a substantial experiment for the pipeline material of the coal liquefaction system. Through the processing of the experimental data, they proposed an erosion mathematical model associated with the experimental data. The model is mainly used in the erosion prediction of the coal liquefaction pipeline; Li [2] used HSV technology to capture the movement of a single solid particle in the fluid, and obtained the movement trajectory of a single particle in the flow channel; Gandhi et al. [3] designed a solid–liquid two-phase flow erosion test and found that the factor that has the greatest influence on the size of the wear is the fluid velocity.

In terms of numerical simulation, Huang and Shi [4] numerically simulated the erosion phenomenon of the mixture of C4, methyl tert-butyl ether and methanol in the process of flowing through the three-way pipeline with the method of computational fluid dynamics. The results show that the maximum wear occurs from the double outlet branch pipe to the center of the pipe. Zhang and Dong [5] simulated the wear in the reducing pipe in the coal liquefaction system with the CFD method, and found that wear rate decreases with the increase of the ratio of the length to the inner diameter of the conical pipe, decreases with the increase of the bending diameter ratio, and first decreases and then increases with the increase of the pipe diameter. Fu and Gao [6] used ANSYS-FLUENT to numerically simulate coal liquefaction elbows, and analyzed the sum relationship between the diameter of the elbow pipe and the bend diameter ratio and erosion. The analysis results show that as the bending diameter ratio and pipe diameter increase, the secondary flow generated by the fluid gradually decreases, and the erosion wear rate decreases. When the bending diameter is relatively small, it is found that the maximum wear rate occurs on the outer arch wall and side wall of the elbow, and when the bending diameter ratio is large, erosion occurs on the outer arch wall surface. Shao [7] compiled different erosion mathematical models into FLUENT in the form of custom functions, and obtained the maximum erosion rate of the pipeline and the distribution of erosion wear positions through sufficient iterative calculations. Comparing the obtained data with the experimental data, it is found that the numerical calculation and the experimental data are

998 kg/m^3, and the viscosity is 0.001003 Pa s; the density of the coal slurry mixed solution is 1220 kg/m^3, and the viscosity is 0.0012 Pa s; the density of pulverized coal particles is 1728 kg/m^3, the particle size distribution is between 0.05 and 2 mm, the particle size is set to be 0.05 mm, 0.1 mm, 0.35 mm, 0.5 mm, 1 mm and 2 mm respectively, the shape factor of the coal powder is 0.5, and the particle mass The flow rate is 0.01 kg/s; when the coal water slurry is at 48 °C, it can be approximated as a Newtonian fluid, so it can be calculated as an incompressible fluid. The coal-water slurry satisfies the continuity equation and the momentum equation during the flow of the tube.

Wall material characteristics

This article selects Inconel 718 alloy as the pipe material for erosion simulation. The alloy is a nickel–chromium–iron alloy with a density of 8240 kg/m^3, a hardness of 363 N/mm^2, and a melting point of 1260–1320 °C. At 1000 °C, it has strong oxidation resistance. The material also has the characteristics of corrosion resistance, easy processing and good weldability under high and low temperature environments.

Simulation condition setting

Since coal slurry can be approximately regarded as an incompressible fluid at 48 °C, the standard "k-ε" model is used for the turbulence model; the DPM model and the Euler multiphase flow model are selected; The inflow is in the way of velocity inlet and out of the pressure outlet, the inlet velocity is set to V = 4 m/s and adjusted to the escape condition. A particle rebound model is set on the wall, and considering the influence of the wall surface roughness, the roughness is 10 μm, and the standard wall function is used to process the vicinity of the wall. The random walk model is used to consider the interaction between solid particles and fluid vortices, and the coupling of pressure and velocity uses the PC-SIMPLE algorithm. The discretization of momentum, turbulent kinetic energy and turbulent dissipation rate all adopt the first-order upwind style, and the second-order difference scheme for the pressure discrete equation.

2.2 Turbulence Model

The two-equation turbulence model proposed by Launder and Spalding [10] determines the turbulence length and time scale by solving two independent transport equations. The robustness, economy and reasonableness of simulation accuracy of this model have made it the mainstream model for current engineering flow calculations. Its robustness, economy and reasonableness of simulation accuracy have made it the mainstream model of current actual engineering flow calculations. This paper uses this model to simulate and solve the turbulent flow. The equation is as follows:

$$\frac{\partial(\rho k)}{\partial t} + \frac{\partial(\rho k u_i)}{\partial x_i} = \frac{\partial}{\partial x_j}\left[\left(\mu + \frac{\mu_t}{\sigma_k}\right)\frac{\partial k}{\partial x_j}\right] + G_k + G_b - \rho\varepsilon + S_k \quad (1)$$

$$\frac{\partial(\rho\varepsilon)}{\partial t}+\frac{\partial(\rho\varepsilon u_i)}{\partial x_i}=\frac{\partial}{\partial x_j}\left[\left(\mu+\frac{\mu_t}{\sigma_\varepsilon}\right)\frac{\partial\varepsilon}{\partial x_j}\right]+C_{1\varepsilon}\frac{\varepsilon}{k}(G_k+C_{3\varepsilon}G_b)-C_{2\varepsilon}\rho\frac{\varepsilon^2}{k}+S_\varepsilon \quad (2)$$

where k is the turbulent kinetic energy; ε is the turbulent energy dissipation rate; μ_t is the turbulent viscosity, $\mu_t=\rho C_\mu\frac{k^2}{\varepsilon}$; G_k is the turbulent kinetic energy generated by the average velocity gradient; G_b is the turbulent kinetic energy generated by buoyancy; C_μ, $C_{1\varepsilon}$, $C_{2\varepsilon}$, $C_{3\varepsilon}$ are constants, and their values are $C_\mu=0.09$, $C_{1\varepsilon}=1.44$, $C_{2\varepsilon}=1.92$; $\sigma_k,\sigma_\varepsilon$ are the turbulent Prandtl numbers of turbulent kinetic energy and turbulent energy dissipation rate, and their values are respectively $\sigma_k=1.0$, $\sigma_\varepsilon=1.3°$.

2.3 Discrete Phase Model

In this paper, the simulation calculation of the trajectory of the solid particles in the two-phase flow of coal slurry uses the integration of the movement of the solid particles in the Lagrange coordinate system. When building a particle tracking model and calculating the erosion wear rate, several assumptions need to be made:

(1) The solid particles inside the fluid are assumed to be independent of each other, ignoring the interaction between the particles;
(2) The crushing of solid particles will not be considered;
(3) The geometrical changes caused by particles hitting the reducing pipe are ignored.

The characteristics of the discrete phases are dynamically analyzed, and the interaction between the solid and liquid phases is fully coupled to each other. When the particles are in suspension, according to Newton's second law, the forces received by the solid particles in the fluid are balanced with each other, thereby establishing the governing equation of the particle motion of the solid particles:

$$m_i\frac{du_i}{dt}=\left(\sum F_i\right) \quad (3)$$

where F_i is represents the force on solid particles, mainly including drag force, additional mass force, inertial force, gravity, pressure gradient force, buoyancy, Saffman lift, Magnus force, etc. This article mainly considers the force that has a greater effect on the movement of particles:

Drag force

$$F_d=\frac{3\mu}{4\rho_p d_p^2}C_d\mathrm{Re}_s\left(\vec{u}-\vec{u_p}\right) \quad (4)$$

where C_d [11] is the drag coefficient

$$C_d = \frac{24}{\mathrm{Re}_s}\left(1 + b_1 \mathrm{Re}_s^{b_2}\right) + \frac{b_3 \mathrm{Re}_s}{b_4 + \mathrm{Re}_s} \tag{5}$$

$$b_1 = \exp\left(2.3288 - 6.4581\varphi + 2.4486\varphi^2\right) \tag{6}$$

$$b_2 = 0.0964 + 0.5565\varphi \tag{7}$$

$$b_3 = \exp\left(4.905 - 13.8944\varphi + 18.4222\varphi^2 - 10.599\varphi^3\right) \tag{8}$$

$$b_3 = \exp\left(1.4681 + 12.2584\varphi - 20.7322\varphi^2 + 15.8855\varphi^3\right) \tag{9}$$

φ is the shape factor, and its value is determined by the ratio of the actual particle surface area to the spherical surface area of the same volume.

Pressure gradient force

$$F_p = \left(\frac{\rho}{\rho_d}\right)\nabla P \tag{10}$$

Buoyancy

$$F_b = \left(\frac{\rho_p - \rho}{\rho_p}\right)g \tag{11}$$

Saffman lift [12]

$$F_S = m_p \frac{2K v^{\frac{1}{2}} \rho d_{ij}}{\rho_p d_p (d_{lk} d_{kl})^{\frac{1}{4}}} \left(\vec{u} - \vec{u_p}\right) \tag{12}$$

Wall collision rebound recovery equation
When solid particles collide with the pipe wall, energy transfer and capacity loss will occur. During the collision, heat and noise will be generated and the wall material may be deformed, so the rebound speed must be less than the impact incident speed. Most studies mainly use the ratio of particle velocity components before and after the collision to measure the process, and define the ratio of the velocity before and

after the collision as the coefficient of restitution. Its value depends on the impact angle, particle properties, wall material properties and fluid properties. This article uses the model proposed by Grant and Tabakoff [13]:

$$e_t = 0.988 - 1.66\alpha + 2.11\alpha^2 - 0.67\alpha^3 \tag{13}$$

$$e_n = 0.993 - 1.76\alpha + 1.56\alpha^2 - 0.49\alpha^3 \tag{14}$$

2.4 *Erosion Model*

The erosion of materials is usually expressed by erosion rate (or wear ratio), that is, the mass or volume of material erosion caused by particles per unit weight. For now, there are many versions of the mathematical model of erosion, and the influencing factors considered are also focused. Different models may even differ by several orders of magnitude. However, it is generally believed that the particle mass flow, particle size, impact angle and particle impact velocity have a greater impact on the erosion wear rate.

In this paper, the particle abrasion and deposition (PEA) model [14] is used to characterize the erosion effect of particles and the like on the pipe wall. The fluid velocity, particle impact angle and particle shape are considered in the model. The erosion model equation is:

$$R = \sum_{p=1}^{N_p} \frac{m_p C\left(d_p\right) f(\gamma) v^{b(v)}}{A} \tag{15}$$

where R is the erosion rate, the unit is kg/(m^2 s); m_p is the particle mass flow rate; $C\left(d_p\right)$ is the particle size function; $f(\gamma)$ is the impact angle function; $b(v)$ is the particle Speed exponential function; A is the area of the impact unit.

3 Results and Discussion

3.1 *Flow Field Velocity and Pressure Analysis*

The flow field analysis of the coal slurry flow in the reducing pipe model is carried out, and after sufficient iteration, the velocity and pressure distribution cloud diagram of the reducing pipe is obtained.

Figures 1 and 2 show the internal pressure cloud diagram of the reducing pipe, the internal pressure of the pipeline first increases and then decreases along the pipeline axis. In the inlet section of the reducing pipe, the fluid pressure does not change much. When the coal water slurry passes through the reducing pipe due to the sudden change of the pipe diameter, its pressure decreases rapidly. The pressure drop at this point is as high as 36,690 Pa. In addition, a negative pressure appeared at the end of the reduced diameter, and then the pressure gradually decreased along the straight pipe section of the outlet, but the pressure drop was relatively small at 12,225 Pa.

After analysis, it can be seen that the sudden drop in pressure at the reducing pipe is due to the reducing pipe section acting as a throttle at this time. The pressure of the coal water slurry is reduced and the flow velocity increases. The negative pressure at the end of the variable diameter is because the inclination angle is large, so the secondary flow phenomenon of the coal water slurry occurs due to its own inertia. The subsequent slight drop in pressure is due to the fact that the coal water slurry

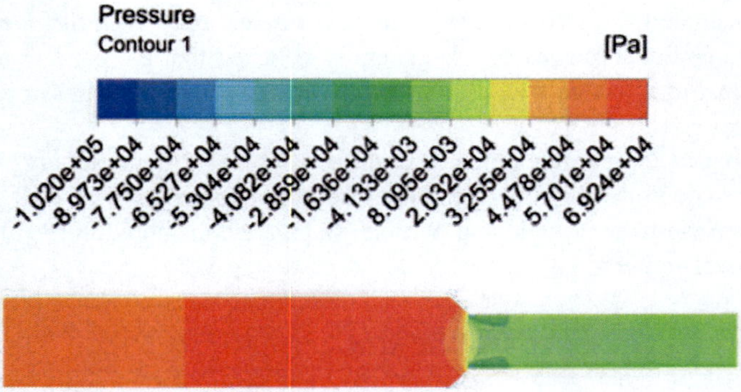

Fig. 1 Internal pressure diagram of reducer

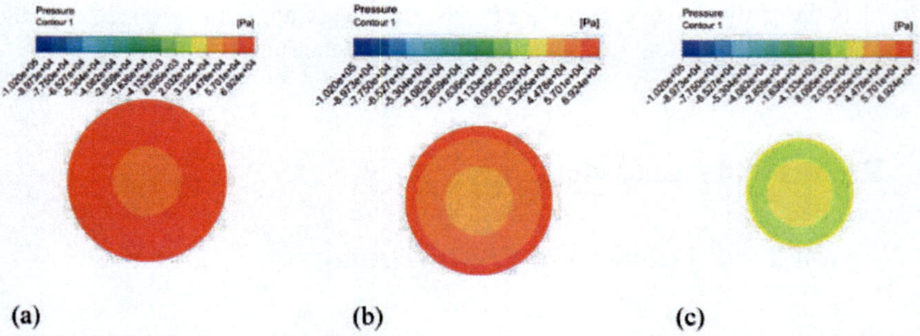

Fig. 2 Pressure diagrams at different sections of the reducing pipe (**a** is the entrance section of the reducing section, **b** is the middle section of the reducing section, **c** is the exit section of the reducing section)

needs to overcome the frictional resistance of the pipe wall during the flow of the pipeline and convert kinetic energy into internal energy, so the pressure gradually decreases.

As shown in Figs. 3 and 4, the internal velocity cloud diagram of the diameter reducing pipe, the flow rate of coal slurry in the axial direction increases as the length of the pipe becomes longer. Before the diameter of the pipe section, the pipe diameter remains constant at a certain speed. Due to the sudden change of the geometric flow path at the reducer, the flow velocity increases rapidly. This is due to the throttling effect of the reducer. After reaching the straight pipe section at the outlet, the flow velocity in the pipe reaches the maximum, but the maximum flow velocity at the outlet pipe section tends to gradually narrow. The analysis shows that part of the kinetic energy of the fluid is dissipated due to the viscous force of the fluid. It can be seen that, due to the existence of the boundary layer of the pipeline, the coal slurry flows slowly in a laminar flow near the wall of the pipeline, and the flow rate is low.

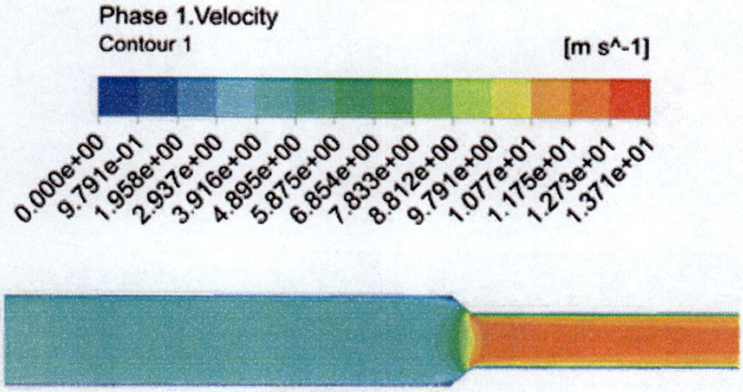

Fig. 3 Internal speed diagram of reducing pipe

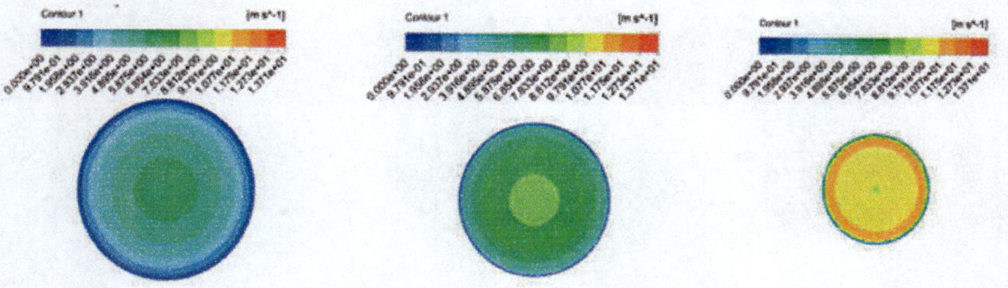

(a) (b) (c)

Fig. 4 Velocity diagrams at different sections of the reducing pipe (**a** is the entrance section of the reducing section, **b** is the middle section of the reducing section, and **c** is the exit section of the reducing section)

3.2 Turbulence Intensity Analysis of Flow Field

Figures 5 and 6 are the turbulence intensity cloud diagram of the reducing pipe, it can be seen from the figure that the turbulence intensity gradually increases along the axis of the reducing pipe. In the inlet straight pipe section, due to the viscous force of the coal water slurry, the turbulence intensity gradually decreases. When passing through the reducing pipe, as the cross-sectional area of the flow decreases, the intensity of turbulence increases, the flow conditions change, and the flow velocity increases. In addition, a secondary flow vortex occurs at the junction of the variable diameter section and the outlet pipe, and the turbulent kinetic energy increases rapidly. The turbulent kinetic energy at this place has a greater impact on the trajectory of the particles. Therefore, the frequency of particles hitting the wall surface will increase, and the erosion rate will also increase.

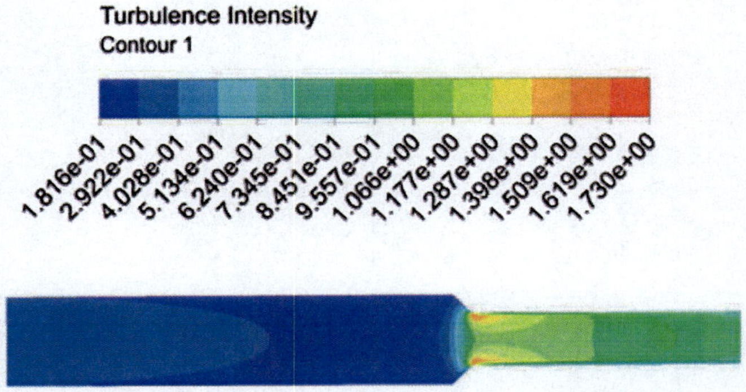

Fig. 5 Turbulence intensity diagram inside the reducing pipe

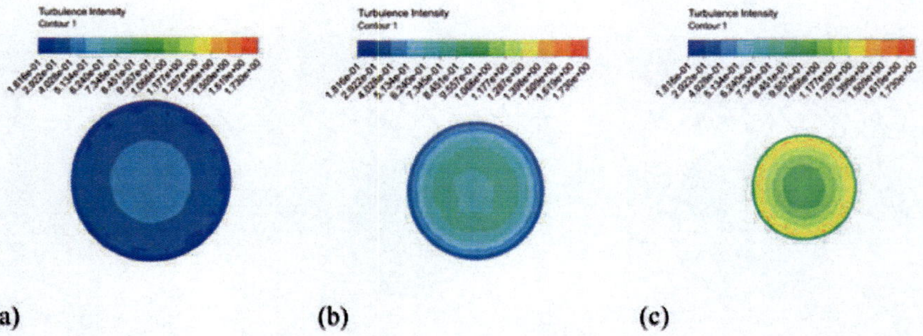

Fig. 6 Turbulence intensity at different sections of the reducing pipe (**a** is the entrance section of the reducing section, **b** is the middle section of the reducing section, and **c** is the exit section of the reducing section)

3.3 Analysis of Particle Mass Concentration and Movement Trajectory

The particle size of pulverized coal particles in coal water slurry is small, and the movement of the particles is mainly affected by the drag force of the fluid. From Fig. 7, it can be seen that the pulverized coal particles of coal slag are scattered throughout the pipeline. It can be seen from the figure that in the straight inlet pipe section, the pulverized coal particles flow evenly along the axis of the pipeline under the action of the liquid. At the near wall surface of the pipeline, some particles will cause erosion and abrasion to the pipeline. From the above-mentioned changes in the flow rate of the coal water slurry, it can be seen that when the coal water slurry enters the reducing pipe, the flow rate of the coal water slurry increases rapidly due to the change of the good cross-sectional area of the flow channel. Therefore, it can be seen that the pulverized coal particles will have a strong impact on the wall of the reducing pipe. With the flow of coal water slurry in the pipeline, the turbulent pulse action gradually weakened, and the concentration distribution of pulverized coal particles in the straight pipe section gradually returned to a uniform state.

Some particle trajectories obtained by Lagrange tracking method are shown in Fig. 8. It can be seen from the particle trajectory diagram that the pulverized coal slag particles follow the fluid well, and their trajectories are roughly parallel to the pipeline axis. At this time, the concentration, impact angle, and impact velocity of pulverized coal particles near the wall of the pipeline are relatively small. Therefore,

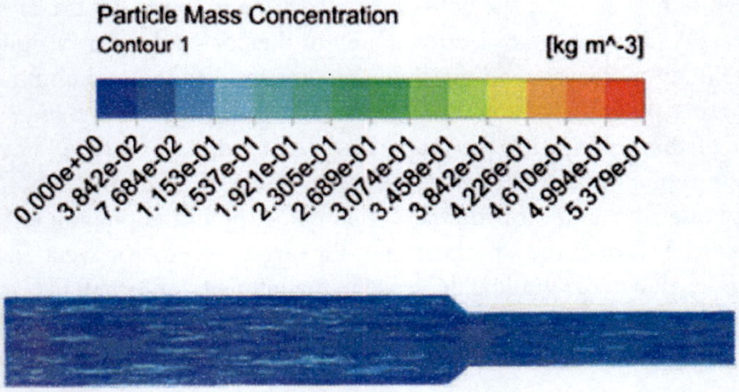

Fig. 7 The particle mass concentration diagram inside the reducing pipe

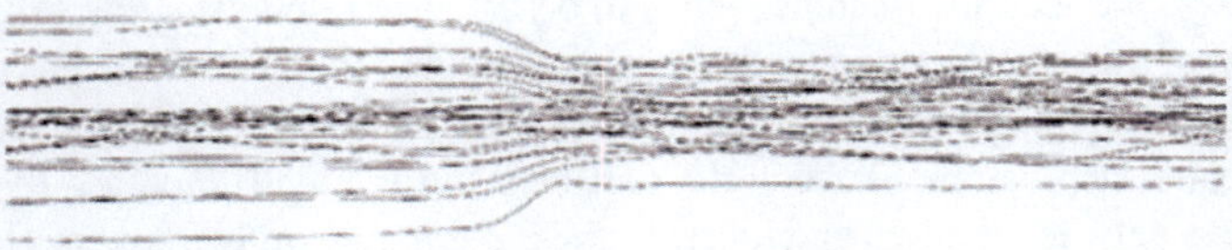

Fig. 8 Particle trajectory diagram

there are few particles in the straight pipe section at the inlet that directly hit the pipe wall along the direction of flow, so the erosion and wear of the coal particles in the straight pipe section at the inlet of the pipe is very small. When the coal water slurry enters the tapered pipe section, the trajectory of the pulverized coal particles will fluctuate greatly with the movement of the fluid. At the wall of the pipeline, the pulverized coal slag particles will cause a certain erosion and wear on the pipeline. When the particles leave the reducing pipeline and flow into the straight pipe section of the outlet, Because of the turbulent pulsation in the fluid flow process, there will still be large fluctuations in the trajectory of the pulverized coal particles. Therefore, the pulverized coal particles in the straight pipe section of the outlet still produce erosion and wear on the wall surface.

3.4 Analysis of the Influence of Different Particle Sizes on Erosion Wear Rate

Figure 9, the areas with greater erosion and wear are concentrated at the junction of the reducing pipe and the outlet straight pipe section, and with the increase in particle size Increase, the wear area points of the straight outlet pipe section increase, but they are all scattered at the straight outlet pipe section.

From the velocity cloud diagram Fig. 1 and particle trajectory diagram Fig. 8 of the coal slurry, it can be seen that the reason for the lower erosion and wear rate at the inlet straight pipe section is: Although the frequency of particles hitting the wall surface is higher here, the collision speed here is lower, so the erosion wear rate is lower; And as the cross-sectional area of the coal water slurry flow changes, the flow velocity increases, and the flow velocity reaches the maximum at the end of the reducing pipe, and the particle movement trajectory here changes with the movement of the fluid. After the pulverized coal and slag particles flow into the straight pipe section at the outlet, the flow rate of the solid particles is still slowly increasing under the action of inertia, which can explain the phenomenon that the larger the particle size of the solid particles, the larger the erosion wear and the wear area. However, due to the gradual decrease in the intensity of turbulence, the erosion wear on the straight pipe section is gradually reduced, and it is significantly smaller than the erosion wear on the connection between the reducing pipe section and the outlet straight pipe section.

Figure 10 is the relationship between the particle size and the maximum erosion wear rate. As the particle diameter increases, the erosion rate of the reducer decreases significantly, and then the decrease tends to be flat. After analysis, it is believed that the reason for this phenomenon is: When the pulverized coal particle size is small, the coupling between the pulverized coal particles and the fluid medium is better, which causes frequent collisions between the particles and the wall. Increase the number of collisions between the two and increase the erosion rate of the reducer [15]; when the diameter of the particles increases, the particles are more affected by gravity and

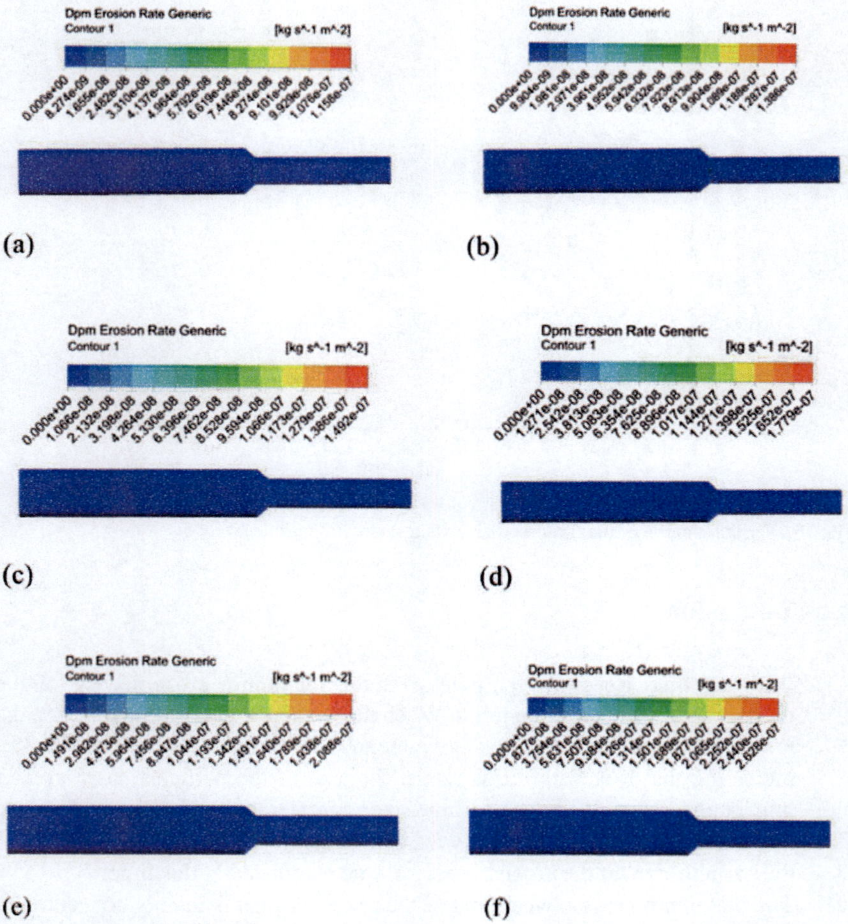

Fig. 9 Erosion and wear diagram of the wall surface of the reducing pipe (the particle size of diagram **a** is 2 mm, the particle size of diagram **b** is 1 mm, the particle size of diagram **c** is 0.5 mm, the particle size of diagram **d** is 0.35 mm, the particle size of diagram **e** is 0.1 mm, and the particle size of diagram **f** is 0.05 mm)

are difficult to be carried by the fluid. The number of collisions between particles increases, and the kinetic energy loss of the particles themselves becomes larger. Eventually, the number and strength of the collision between the solid particles and the wall surface are reduced, and the erosion rate of the inner wall of the reducer is reduced.

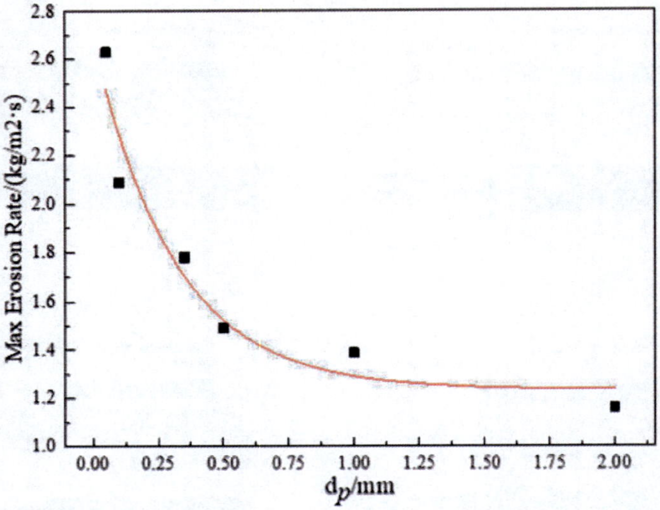

Fig. 10 Variation curve of maximum erosive wear rate with particle size

4 Conclusion

(1) With the increase of the particle size, the maximum erosion wear of the reducer shows a decreasing trend. When the particle size is within 0.5 mm, the maximum wear decrease rate is greater, and the decrease rate reaches 77.3%, indicating that it is within this range, the size of the particles has a greater impact on erosive. When the particle size is greater than 0.5 mm, the decrease rate of the wear volume gradually tends to be flat, indicating that the impact of the particle size on the erosive wear volume is reduced at this time;

(2) The maximum erosion wear area of the reducing pipe is mainly concentrated at the connection between the reducing pipe section and the straight outlet pipe section, where the velocity of the solid particles is relatively high, and the impact frequency of the particles on the wall is relatively high.

(3) As the particle size increases, the erosion wear rate of the straight pipe section at the outlet also increases, and the scatter points in the erosion wear area also increase.

Acknowledgements The authors gratefully acknowledge research support from the National Natural Science Foundation of China (52165020), the National Key R&D Program (2018YFB2004000), the Ningxia Youth Top Talent Project (2020), the Ningxia Autonomous Region Science and Technology Research (Support) Project (Key Technologies of High-end Valve Structural Optimization and Erosion Performance Research), the Ningxia Key Research and Development Project of China (Western Light, 2017).

References

1. Ye J (2013) Test and numerical study on erosion and wear of coal liquefaction pipeline materials. Master, Zhejiang Sci-Tech University
2. Li YL (2015) Solid–liquid two-phase CFD-DEM method in variable curvature flow channel and its application in large desulfurization pumps. Jiangsu University
3. Gandhi BK, Singh SN, Seshadri V (1999) Study of the parametric dependence of erosion wear for the parallel flow of solid–liquid mixtures. Tribol Int 32(5):275–282
4. Huang Y, Shi ZX (2005) Application of CFD in the study of tee erosion wear. Chem Equip Technol (01):65–67
5. Zhang HJ, Dong JH (2011) A variable-diameter elbow in the high-pressure oil-coal slurry transportation pipeline and its wear prediction. J Hebei Univ Technol 40(06):54–58
6. Fu L, Gao BJ (2009) Numerical simulation and abrasion prediction of the flow field in the elbow of the oil-coal slurry pipeline. Chem Mach 36(05):463–466
7. Shao D (2016) Numerical simulation research on erosion and wear of coal chemical pipeline. East China University of Science and Technology
8. Chen X, Mclaury BS, Shirazi SA (2006) Numerical and experimental investigation of the relative erosion severity between plugged tees and elbows in dilute gas/solid two-phase flow. Wear 261(7–8):715–729
9. Takahashi K (2010) Experimental study of low-cycle fatigue of pipe elbows with local wall thinning and life estimation using finite element analysis. Int J Press Vessels Pip 87(5):211–219
10. Launder BE, Spalding DB (1972) Lectures in mathematical model of turbulence. Academic
11. Haider A, Levenspiel O (1989) Drag coefficient and terminal velocity of spherical and nonspherical particles. Powder Technol 58(1):63–70
12. Saffman PG (1965) The lift on a small sphere in a slow shear. J Fluid Mech 22(2):385–400
13. Grant G, Tabakoff W (1975) Erosion prediction in turbomachinery resulting from environmental solid particles. J Aircraft
14. Tsuji Y, Morikawa Y (1982) LDV measurements of an air-solid two-phase flow in a horizontal pipe. J Fluid Mech 120(1):385–409
15. Teng XS, Meng JZ (2021) FLUENT-based simulation analysis of oil pipeline elbow erosion and research on prevention and control measures. Petrochem Corros Prot 38(03):19–22

Study on Nozzle Baffle in Shield Machine Remote Pressure Maintaining System

G. Wang, Y. X. Zhang, Z. C. Wu, L. Y. Kou, X. Shang, Q. F. Gao, W. H. Wang, and Y. Q. Xu

Abstract The research on the mechanism of nozzle baffle at home and abroad mainly focuses on the theoretical research of nozzle baffle gap and back pressure, and lacks engineering application. This paper takes the remote pressure maintaining system of shield machine as the research object. Aiming at the problem of baffle inclination caused by impurities in the nozzle baffle mechanism during shield machine construction and maintenance, a simulation experiment was carried out on the nozzle baffle mechanism. Draw the characteristic curve of nozzle baffle clearance and back pressure; The relationship between baffle angle and back pressure under different nozzle baffle clearance in engineering practice is analyzed; The linear regression model is established to prove that the problem of shield remote pressure maintaining system is not caused by the inclination of baffle, which provides a theoretical basis for the maintenance of shield remote pressure maintaining system.

Keywords Nozzle baffle · Remote pressure maintaining system · Baffle inclination

1 Introduction

From the origin of pneumatic technology in 1776 to today's pneumatic technology breaking through the traditional dead zone, pneumatic technology at home and abroad has realized low consumption, high speed, rapid response, miniaturization and oil-free. Due to its unique characteristics, pneumatic technology has become indispensable in all walks of life. As the core component, the nozzle baffle mechanism has been deeply studied based on the nozzle baffle structure at home and abroad [1–4].

G. Wang · Y. X. Zhang · L. Y. Kou · X. Shang · Q. F. Gao · W. H. Wang · Y. Q. Xu (✉)
School of Mechanical Engineering, Ningxia University, Yinchuan 750021, Ningxia, China
e-mail: 469821959@qq.com

Y. X. Zhang · Z. C. Wu
Yinchuan InauTech Controls Co., LTD, Yinchuan 750000, Ningxia, China

G. Wang · L. Y. Kou · X. Shang · Y. Q. Xu
The Key Laboratory of Ningxia Intelligent Equipment CAE, Yinchuan 750021, Ningxia, China

M. Y. Ismail et al. (eds.), *Technological Advancement in Mechanical and Automotive Engineering*, Lecture Notes in Mechanical Engineering,
https://doi.org/10.1007/978-981-19-1457-7_37

Lu Xiang-hui et al. simulated the flow field of a double nozzle flapper valve nozzle applied to the electro-hydraulic servo front stage when the structural parameters changed [5]. Wu Yu-kui et al. proposed a nozzle baffle servo valve driven by piezoelectric [6]; Wu Zheng-jiang made an experimental study on the flow characteristics of the flapper valve port of hydraulic nozzle.

At present, the research on nozzle baffle mechanism at home and abroad focuses on optimizing structural parameters, and remarkable results have been achieved. However, there is a lack of reliable mathematical model establishment method for nozzle baffle gap and back pressure. In engineering application, the gap distance with the best linear correlation between nozzle baffle structure back pressure and nozzle baffle gap is selected in engineering application. However, in practical application, it is difficult to keep the baffle level with the nozzle, thus affecting the pressure change in the backpressure chamber, Taking the pressure maintaining system of shield machine as the research object, this paper studies the influence of baffle inclination on nozzle baffle structure, and studies the influence of baffle inclination on back pressure under different working conditions. In engineering practice, the linear relationship between baffle inclination and back pressure chamber pressure under different baffle inclination and different working conditions is studied to ensure that the nozzle baffle structure is not affected when the baffle inclination is not affected. This paper takes the pressure maintaining system of shield machine as the research object.

The pressure maintaining system of shield machine can keep the pressure constant during construction. The remote pressure maintaining system of shield machine can realize remote and high-precision control. As its core, the nozzle baffle mechanism has a vital impact on the whole system [7].

2 Composition and Working Principle of Nozzle Baffle Mechanism

Nozzle baffle mechanism is the most common length flow or length pressure converter in pneumatic control. Its function is to convert small displacement into pressure change. The working principle of the nozzle baffle mechanism is shown in Fig. 1. The compressed air of the air source enters the backpressure chamber through the constant orifice and flows into the atmosphere from the gap between the nozzle and the baffle; The air flow channel formed by the nozzle and baffle can be regarded as a variable orifice. The change of the gap s between the nozzle and baffle will cause the change of the air flow through the gap. Under working conditions, the air supply pressure PS is fixed. Within a certain gap range, when the baffle is close to the nozzle, the gas flow through the nozzle baffle gap will decrease and the pressure Po of the backpressure chamber will increase; Conversely, when the baffle leaves the nozzle, the pressure Po will decrease. In this way, the backpressure Po has a one-to-one correspondence with the baffle clearance s, that is, the conversion from air pressure

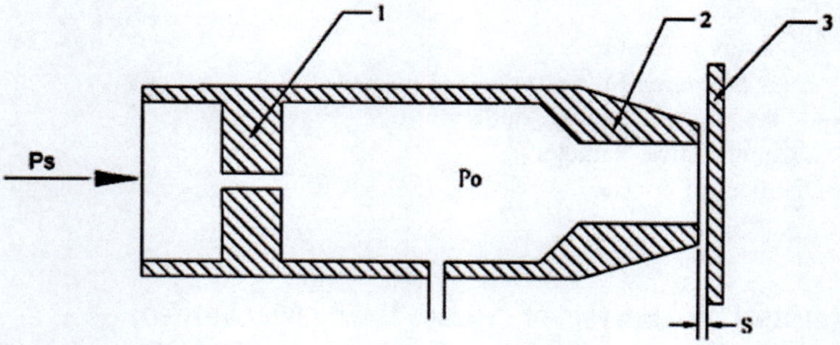

Fig. 1 Schematic diagram of nozzle baffle. 1—constant orifice; 2—nozzle; 3—baffle

Ps to Po can be realized by accurately controlling the nozzle baffle spacing [8–10]. However, in engineering practice, due to the manufacture, installation and different working conditions of the baffle, there will be an inclined angle between the baffle and the nozzle, as shown in Fig. 2.

The air passing through the variable orifice formed by the nozzle baffle and the matched constant orifice shall be turbulent, and the mass flow passing through shall be

$$G = \alpha_1 \frac{\pi}{4} d^2 \sqrt{p_B (p_s - p_o)} = \alpha_2 \pi D S \sqrt{p_{DP} (p_o - p_{DP})} \tag{1}$$

The relationship between the distance between the nozzle and the baffle and the nozzle back pressure is obtained.

$$S = \frac{\alpha_1 d^2}{4\alpha_2 D} \sqrt{\frac{p_o (p_s - p_o)}{p_{DP} (p_o - p_{DP})}} \tag{2}$$

Fig. 2 Structural diagram of nozzle baffle

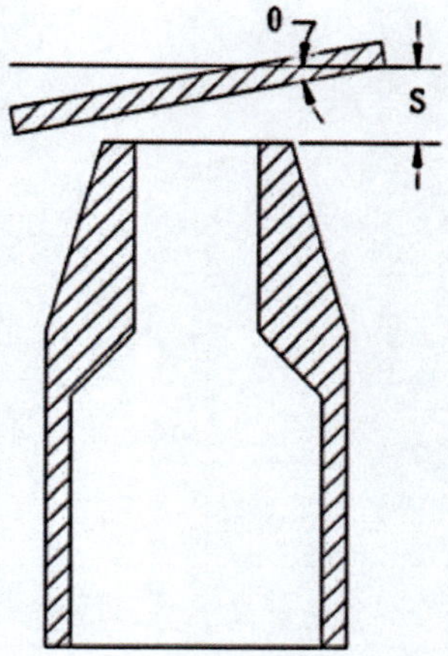

α—Flow coefficient
p_s—Air source pressure
p_o—Back pressure of the nozzle
p_{DP}—Pressure at the nozzle outlet
d—Constant orifice diameter
D—Diameter of nozzle.

3 Simulation Analysis of Nozzle Baffle Mechanism

3.1 Nozzle Model

According to the size of the nozzle baffle mechanism in the shield machine remote pressure maintaining system, the nozzle diameter is 1 mm, the constant orifice diameter is 0.1 mm and the length is 0.75 mm, the simulation experimental model is established. The nozzle baffle structure in the shield machine pressure maintaining system is shown in Fig. 3. Because the nozzle baffle structure is a symmetrical structure, in order to make the next calculation fast, the Design modeler is used to simplify the model and establish the model of the fluid area. To establish a two-dimensional

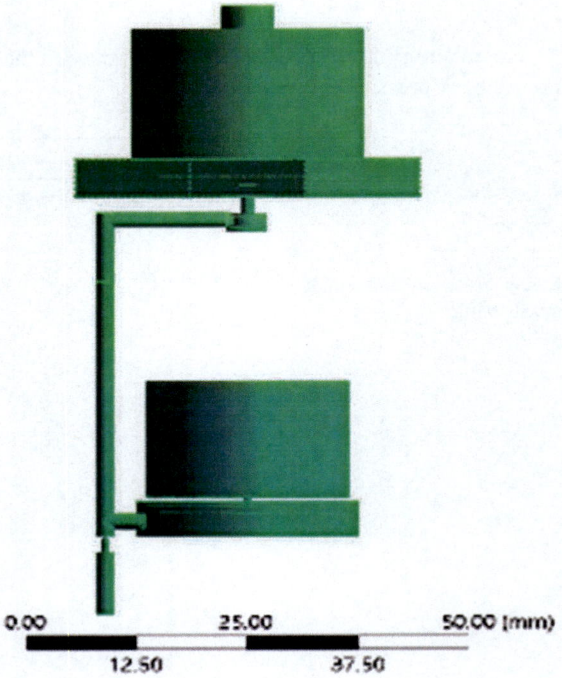

Fig. 3 Three-dimensional nozzle baffle model

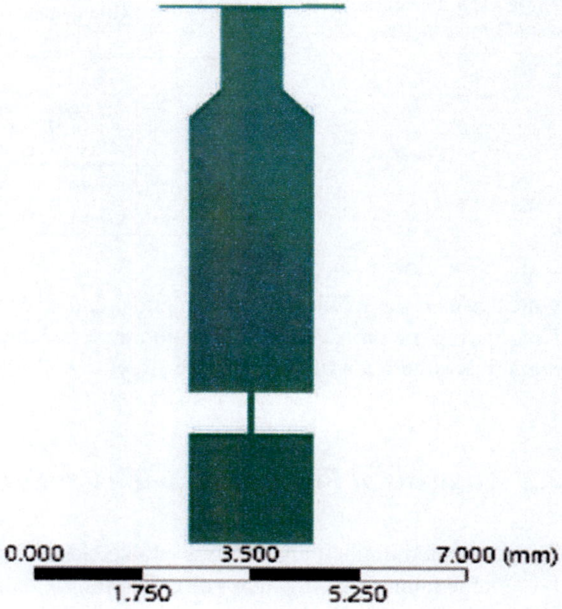

Fig. 4 Two-dimensional nozzle baffle model

model to ensure the complete definition of the geometric size of the calculation area, as shown in Fig. 4.

3.2 Grid Independence Test

In the simulation of fluid field, the quantity and quality of mesh generation have a great impact on the final results of numerical simulation. In order to improve the accuracy of subsequent numerical simulation, the error caused by grid sensitivity must be excluded in grid modeling. At the same time, the number of grids must be reduced as much as possible on the premise of excluding grid sensitivity, so as to improve the calculation efficiency under the condition of ensuring accuracy.

The preliminary grid model after pretreatment is used to calculate the number of different grids. The fluid medium is air, the uncoupled implicit algorithm is adopted, and the turbulence model is standard k-ε. The boundary conditions of the model are shown in Table 1.

The grid independence test results are shown in Table 2. When the number of grids reaches 140,488, the change of back pressure chamber pressure is very small

Table 1 Boundary conditions of the model

Turbulence model	Import (Pa)	Export (Pa)	Operating (Pa)
Standard k-ε	140,000	0	101,325

Table 2 Grid independence test of nozzle baffle

Nozzle parts (mm)	Grid number	Back pressure (Pa)	Relative error (%)
0.01	18,302	65,432.1	10
0.008	44,009	66,318	8
0.006	56,407	70,079	3
0.005	140,488	71,817	1
0.0025	178,818	70,374	2

with the increase of the number of grids, and the relative error is less than 5%. Considering the problems of calculation time and amount, the number of grids for numerical simulation of nozzle baffle structure is 140,488.

3.3 Analysis of Flow Field Characteristics of Nozzle Baffle

The uncoupled implicit algorithm is adopted, and the turbulence model is standard k-ε. As the numerical simulation model of nozzle baffle, the calculation results are shown in Fig. 5. Under the same working conditions, the relationship between back pressure cavity pressure and gap of different baffle gaps.

The air supply pressure is P_S, enters the air chamber (back pressure chamber) through a constant orifice, and then discharges through the nozzle baffle gap. Change

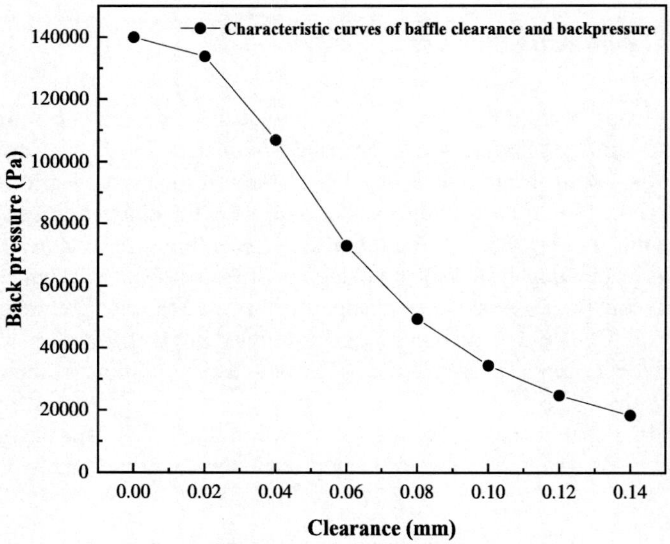

Fig. 5 Characteristic curve of baffle clearance and backpressure change

the gap s between the baffle and the nozzle. Even if the ratio of gas resistance value between variable orifice and constant orifice changes, the pressure in the gas chamber can be adjusted.

When the baffle covers the nozzle (S = 0), the gas resistance of the variable throttle orifice tends to infinity, and the air chamber pressure is equal to the air source pressure.

When the gap between the plate and the nozzle is $S \geq 1/4D$, the nozzle orifice is fully opened. Due to D > d, the gas resistance of the variable throttle orifice tends to zero, and the origin pressure is equal to the pressure at the variable throttle orifice, which is similar to the law of proportional ventilation chamber [11].

Changing the nozzle baffle clearance leads to the pressure change in the back pressure chamber. Through the fluent simulation experiment results and consulting the pressure change data of the back pressure chamber under different clearances in relevant literature, the calculation results of the simulation model are reliable [8].

4 Influence of Nozzle Baffle Inclination on Back Pressure in Engineering Practice

In the remote pressure maintaining system of shield machine, the use range of nozzle baffle is the section with the best linearity. Therefore, three values of the initial distance between baffle and nozzle are selected in this experiment: 0.04, 0.06 and 0.08 mm. Change the angle between baffle and nozzle, and get the relationship between baffle inclination, back pressure and baffle Inclination under different initial nozzle baffle distances, as shown in Table 3.

Table 3 Relationship between baffle inclination and back pressure at different initial distances

Angle	Clearance 0.04 (mm)	Clearance 0.06 (mm)	Clearance 0.08 (mm)
0	106,928	72,803	49,346
0.2	116,479	80,896	54,724
0.4	123,955	89,983	60,869
0.6	132,413	99,703	67,875
0.8	133,086	109,542	75,791
1.0		117,677	84,576
1.2		120,791	93,243
1.4			102,500
1.6			108,049

5 Analysis of Experimental Results and Establishment of Linear Regression Model

5.1 *Relationship Between Baffle Angle and Pressure in Back Pressure Chamber*

Changing the inclination of the nozzle baffle will reduce the distance between the baffle and the nozzle. When the baffle is close to the nozzle, the gas flow through the gap between the nozzle baffle will decrease and the pressure Po of the back pressure chamber will increase; Until one side of the baffle comes into contact with the nozzle, the flow area changes and the pressure in the back pressure chamber also changes. The simulation results show that there is a linear relationship between the inclination of the baffle and the pressure in the back pressure chamber before the baffle contacts the nozzle.

Through the analysis of FLUENT simulation results, there is only one independent variable in the change trend between baffle angle and back pressure, and the univariate linear model is selected for regression analysis. Mathematical modeling is carried out by using Origin software, and the linear fitting results are shown in Table 4.

When the gap is 0.04 mm, the equation of nozzle baffle Angle and back pressure is $P = 41965.5\theta + 10735.1$, and Pearson correlation coefficient $r = 0.99891$ in regression analysis, indicating that there is an extreme correlation between the series and the coefficients. Determination coefficient $R^2 = 0.99783$, indicating that the fitting effect is very ideal, so we can get a reliable relationship model between nozzle baffle Angle and back pressure when the clearance is 0.04 mm, the linear fitting effect is shown in Fig. 6.

The results show that in the nozzle baffle mechanism, the different inclination angles of the baffle have a great impact on the back pressure. The main reason for the change of the back pressure is that the gap between the baffle and the nozzle decreases, the inclination of the baffle changes, the pressure in the back pressure chamber increases, and the inclination of the baffle has a linear relationship with the back pressure.

Similarly, linear fitting is carried out for the gaps of 0.06 mm and 0.08 mm respectively to obtain the relationship between baffle inclination and pressure in back pressure chamber under three different nozzle baffle gaps, as shown in Tables 5 and 6.

When the gap is 0.06 mm, the equation between nozzle baffle angle and back pressure is $P = 45718.2\theta + 72241.7$, Pearson correlation coefficient $r = 0.9995$ in regression analysis, indicating that there is an extreme correlation between series

Table 4 Regression equation of gap 0.04 mm

	Numerical	Standard error
Intercept	107,354.1	517.8
Slope	41,965.5	1383.8

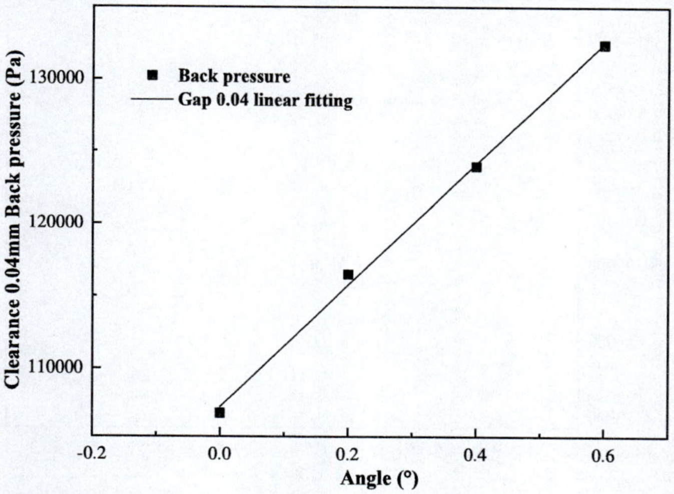

Fig. 6 Gap 0.04 mm linear fitting

Table 5 Regression equation of gap 0.06 mm

	Numerical	Standard error
Intercept	72,241.7	437.2
Slope	45,718.2	722.0

Table 6 Regression equation of gap 0.08 mm

	Numerical	Standard error
Intercept	46,794.6	1157.6
Slope	38,315.9	1383.6

and coefficient; The judgment coefficient $R^2 = 0.999$ shows that the fitting effect is very ideal, so we can get a sufficiently reliable relationship model between nozzle baffle angle and back pressure when the gap is 0.06 mm, the linear fitting effect is shown in Fig. 7.

When the gap is 0.08 mm, the equation between nozzle baffle angle and back pressure is $P = 35315.9\theta + 46794.6$, Pearson correlation coefficient $r = 0.99611$ in regression analysis, indicating that there is an extreme correlation between series and coefficient; The judgment coefficient $R^2 = 0.99094$ shows that the fitting effect is very ideal, so we can get a sufficiently reliable relationship model between nozzle baffle angle and back pressure when the gap is 0.08 mm, the linear fitting effect is shown in Fig. 8.

The experimental results show that the inclination Angle θ of the nozzle baffle has a linear relationship with the pressure in the back pressure chamber under different working conditions, which is caused by the decrease of the baffle clearance.

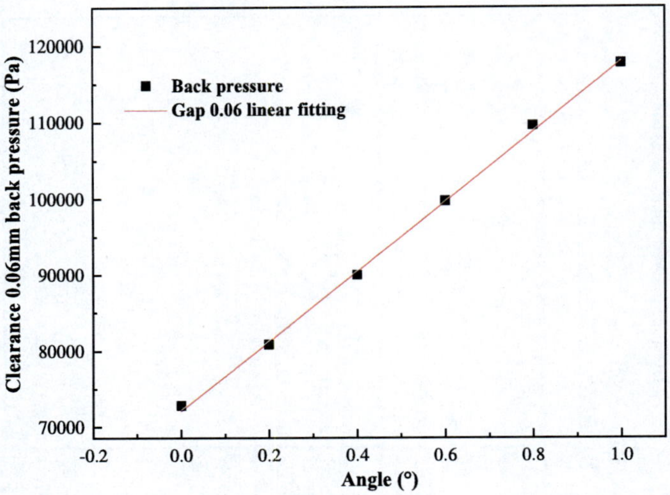

Fig. 7 Gap 0.06 mm linear fitting

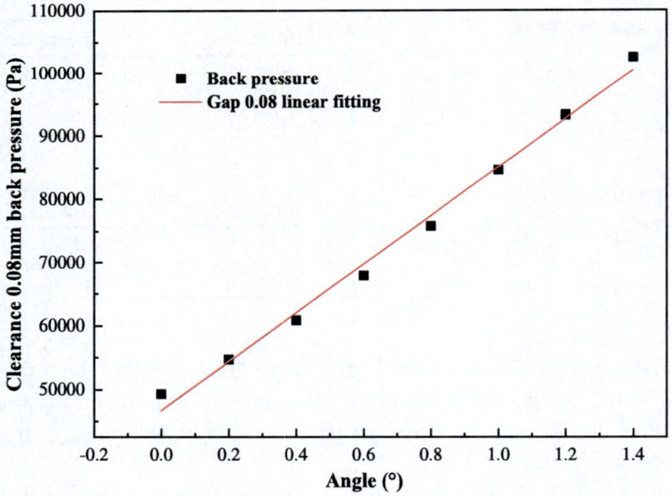

Fig. 8 Gap 0.08 mm linear fitting

5.2 *Relationship Between Nozzle Baffle Clearance, Baffle Inclination and Pressure in the Back Pressure Chamber*

According to the simulation result analysis, the variation trend among the baffle clearance, baffle Angle and back pressure, and the two independent variables and

Table 7 Fitting equation of back pressure without baffle clearance and angle

	Numerical	Standard error
Intercept	169,123.7	1990.9
Clearance	−155,420.0	3243.3
Angle	40,607.2	1243.3

one dependent variable among the three, the multiple linear model is selected for regression analysis (Table 7).

The regression equation of gap, Angle and back pressure is $\mathrm{P} = 169{,}123.7 - 1554200\theta + 40{,}607.2093\mathrm{h}$. The determination coefficient $R^2 = 0.99334$ indicates that the fitting effect is very ideal, so we can get a reliable mathematical model of the relationship among clearance, Angle and back pressure.

In engineering practice, different nozzle baffle structures form different inclined plane angles, but the relationship among the three of the same nozzle baffle structure is shown in Fig. 9.

The results show that in practical engineering application, different baffle angles can meet the linear requirements under certain working conditions. Changing the baffle angle will lead to the decrease of nozzle baffle gap and the change of pressure in the back pressure chamber. Under certain working conditions, the relationship between them is linear. Similarly, there is a linear relationship between the angle of the baffle and the back pressure under different working conditions.

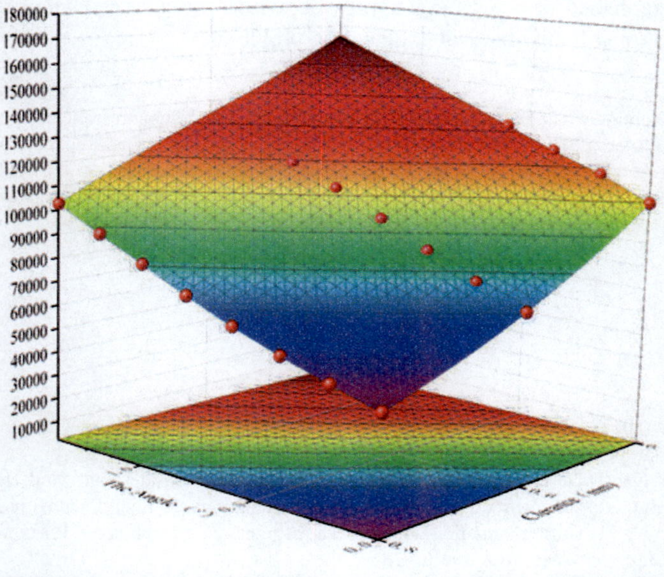

Fig. 9 Correlation between nozzle baffle clearance, baffle inclination and backpressure chamber pressure

6 Conclusions

(1) The shield machine fills the air cushion chamber with compressed air to balance the water and soil pressure of the slurry chamber and tunnel face. The working environment of the shield machine is complex, and the impurities of dust, slurry and oil sometimes enter the pressure retaining system of the shield machine, resulting in the inclination of the baffle. The simulation shows that when the baffle is inclined, the inclination of the baffle is still linearly related to the pressure in the back-pressure chamber, to meet the working conditions and improve the working reliability of air cushion silo has an important impact on the safety, quality, construction period and cost of construction, and has a broad application prospect.

(2) In the manufacturing process of nozzle baffle, the manufacturing accuracy of nozzle baffle is required to be high. Through fluent simulation, it is concluded that within the working pressure range of nozzle baffle and under different working conditions, the inclination angle of nozzle baffle is always linear with the pressure in the back-pressure chamber. Therefore, the baffle accuracy can be reduced, the manufacturing cost can be reduced and the production efficiency can be improved.

(3) In the maintenance of shield machine, the maintenance workers will focus on the maintenance of nozzle baffle, and the baffle will be replaced every time. The simulation experiment shows that the inclination of baffle will not affect the working performance of nozzle baffle mechanism. The linear regression model is established by using Origin software, which provides a reliable theoretical basis for solving practical engineering problems.

Acknowledgements The authors gratefully acknowledge research support from the Ningxia Youth Top Talent Project (2020), the Ningxia Autonomous Region Science and Technology Research (Support) Project (Key Technologies of High-end Valve Structural Optimization and Erosion Performance Research), the Ningxia Key Research and Development Project of China (Western Light, 2017).

References

1. Ning C (2017) Introduction and improvement of pneumatic technology. Chemical Industry Press, Beijing, pp 60–86
2. Wang X (2007) Development trend of pneumatic technology. Mod Manuf Mod Trend 5:1
3. Lu Y (1991) development direction of pneumatic technology. Hydraul Pneumatic 2–3
4. Chen Q (1991) Current situation and prospect of pneumatic technology. Pneumatic Branch China Liq Seven Seal Ind Assoc 12–18
5. Lu X, Gao D (2013) calculation and analysis of flow field in two-stage electro-hydraulic servo valve and double nozzle baffle valve. China Mech Eng 23(16):1951–1956
6. Wu Y (2004) Experimental research on nozzle baffle piezoelectric servo valve. Jilin University

7. Yu W. An automatic air pressure maintaining test bench for shield machine. China 4:26. 201821438151.52019
8. Zhao S (2006) Simulation experiment of nozzle baffle mechanism based on fluent. Mach Tool Hydraul 189–191
9. Liang J, Yuan J et al (2008) Characteristic analysis of pneumatic single nozzle flapper valve. Hydraul Pneumatic Seal 13–17
10. Zhang Y (2006) model construction and Simulation of jet pipe servo valve. Northwest University of Technology, Xi'an, pp 21–43
11. Wang H (2014) Fluid mechanics as I understand it. National Defense Industry Press, Beijing

Geometry and Kinematics Analysis of Seven-Bar Three-Axis Fixed Compound Mechanism

Xing Zhenwei and Wang Yutan

Abstract A two-degree-of-freedom seven-bar three-axis fixed composite mechanism was introduced. The mechanism was divided into two closed loops to provide the position and direction of movement for the end effector components. Two four-bar analytic equations were constructed through complex number operations, the closed kinematics equations of the mechanism were established, and then converted into a matrix format. The Jacobian matrix was further deduced, and the geometrical analysis and solution of the mechanism model were carried out, and the mathematical model was analyzed. The Jacobian matrix was used to establish a mathematical model, derive the working space range of the mechanism, establish a kinematics model, and obtain the end-effector position, speed and acceleration curve graph. The purpose was to accurately calculate the energy parameters of the mechanism at work. The simulation results showed that when the end-effector moved to the lowest position, the speed was zero, the acceleration change was small, and the impact was minimal. The corrugated cardboard box flap can be folded to the horizontal direction without causing damage to the surface of the box. The correctness and feasibility of the mathematical model was verified by energy parameters, which provided the basic principle analysis for the organization and provided a theoretical reference for practical applications.

Keywords Three-axis fixed · Kinematics · Geometric analysis · Working space

1 Preface

With the rapid development of robots, people are interested in using a closed-loop mechanism as a robot manipulator. The use of a closed-loop manipulator can eliminate the gear drive train and promote the advantages of other transmissions usually required by an open-loop manipulator. In recent years, many two-degree-of-freedom planar linkage mechanisms have been proposed, and their geometry and kinematics

X. Zhenwei · W. Yutan (✉)
School of Mechanical Engineering, Ningxia University Yinchuan, Yinchuan 750021, China
e-mail: wang_yt1108@163.com

M. Y. Ismail et al. (eds.), *Technological Advancement in Mechanical and Automotive Engineering*, Lecture Notes in Mechanical Engineering,
https://doi.org/10.1007/978-981-19-1457-7_38

have been studied. Especially the application of the two-degree-of-freedom closed-loop planar linkage mechanism began in the 1980s. The two-degree-of-freedom closed-loop seven-bar planar mechanism has also begun to attract attention in the field of robotics. H Xie [1] proposed that the double-crank seven-bar mechanism is a two-degree-of-freedom mechanism that converts two input rotary motions into output linear motions for the design of hybrid power transmission devices. Keming Wang [2] proposed a novel type. The two-degree-of-freedom non-coupling rotating parallel mechanism is analyzed and all the odd heterogeneous types are analyzed. MG Zalyubovskii [3] proposed a design method for a space seven-link mechanism without redundant constraints, and derived. To determine the expression of the reasonable ratio of the rod length, Shrinivas [4] proposed a comprehensive analytical method for the plane variable topology seven-bar mechanism. This method uses variable topology to analyze the motion between two dead points. This method is effective and effective for solving the motion space. It is non-iterative. GM Gadad [5] proposed an analysis method of a two-position variable topology seven-bar mechanism for function generation.

This paper proposes a seven-bar three-axis fixed composite mechanism. The mechanism consists of two loops [7]. One loop uses the contraction characteristic of the connecting rod to maintain the freedom of the mechanism. Through the contraction of the connecting rod, the two fixed axes are located at the same point, forming a composite hinge. One loop of the mechanism is a four-bar mechanism composed of a crank, a connecting rod and a frame, which is used to change the position of the end effector of the mechanism, and the other loop presents a five-bar mechanism to form a seven-bar mechanism[8]. Therefore, the two inputs of the seven-bar mechanism are decoupled, one provides position and the other provides direction. The mechanism provides a non-circular motion track, which can be used for box folding or a closed-loop manipulator.

2 Equivalent Model of Seven-Bar Compound Mechanism

Please Compound mechanism with seven-bar three-fixed shafts with two degrees of freedom are shown (see Fig. 1). One axis of the mechanism is fixed at node *A*, and the other two fixed axes are at node *D* to form a compound hinge. Link 1 to link 7 is represented by l_1, l_2, l_3, l_4, l_5, l_6, l_7, the length of *DG* is l_{33}, and the extension of *FG* is l_{77}. The seven-bar mechanism can be divided into two closed-loops [9], the first closed-loop *ABCDA* is the drive loop and the second closed-loop *DEFGD* is the directional loop to control the direction of the end-effector. The second loop is a five-bar mechanism. One link is reduced at the compound hinge *D*. Link 3 is shared by the two loops. If link 3 is fixed in the first loop, the second loop becomes a four-bar mechanism. In this mechanism, the working space is determined by the end-effector. The position of the end-effector is controlled by the first loop. At the same time, the crank 1 and connecting link 2 in the first closed-loop are used to form

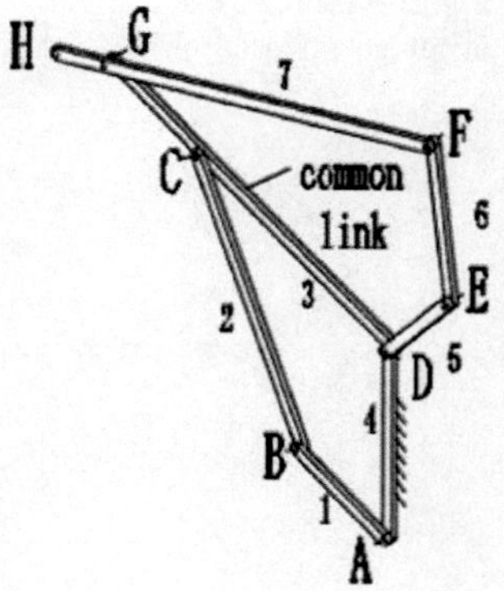

Fig. 1 Seven-bar three-fixed axis compound mechanism

the four-bar mechanism and the second loop. The five-bar mechanism constitutes a seven-bar and three-fixed-axis compound mechanism [10].

The seven-bar three-axis fixed composite mechanism is mainly composed of two closed-loop composite mechanisms. According to the different driving methods and convenience, the mechanism can be divided into three driving modes, namely crank-crank, Crank-rocker, rocker-crank. There are crank components in the three driving modes, which use the crank revolving motion to drive the entire mechanism to make complex motions to meet different design requirements. In addition, there is a rocker-rocker drive mode, which requires additional power for the mechanism, which increases the complexity of the mechanism. This article mainly discusses the crank-crank driving mode, that is, component 1 and component 5 are used as the power parts of the mechanism.

3 The Geometric Analysis of the Mechanism and the Jacobin Matrix

The origin of the coordinate system is at A,(see Fig. 2). the angular displacements of crank 1, crank 5, and end-effector are represented by θ_1, θ_5, and θ_7 respectively[11]. In the first closed-loop, crank 1 and link 3 is denoted as B and C respectively, and the coordinates are denoted as

$$B = \begin{bmatrix} l_1 \cos\theta_1 \\ l_1 \sin\theta_1 \end{bmatrix}, C = \begin{bmatrix} l_3 \cos\theta_3 \\ l_4 - l_3 \sin\theta_3 \end{bmatrix} \tag{1}$$

The angular displacement θ_1 of crank 1 is the input angle of the first closed-loop, and the angular displacement θ_3 of the output connecting link 3 is the output angle of the first closed-loop [12]. Assuming that all connecting links are rigid bodies, nodes B and C always maintain a fixed distance from the entire connecting link movement. Then there is as follows (Table 1)

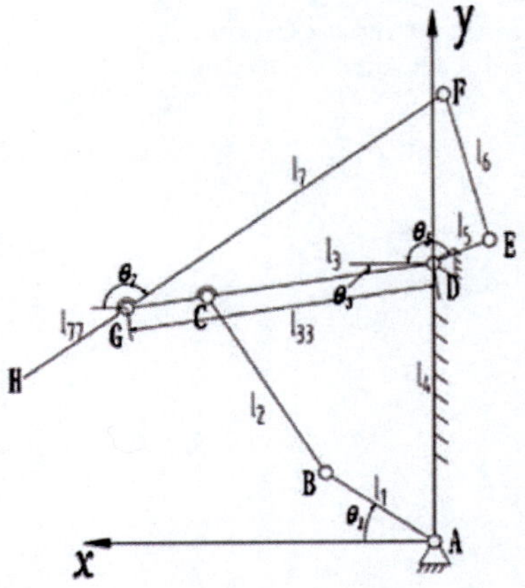

Fig. 2 Kinematic diagram of compound mechanism

Table 1 Table related parameters of kinematic analysis

Symbol	Explanation
l_1 through to l_7	Length of the respective links of the seven-bar mechanism
l_{33}	Length of DG link
l_{77}	Length of the extension bar of link 7
A, *B*, ..., *H*	Coordinate vector of the nodes in the seven-bar mechanism
$\theta_1, \theta_3, \theta_5, \theta_7$	Angular displacement of corresponding links in the seven-bar mechanism
$\dot{\theta}_1, \dot{\theta}_3, \dot{\theta}_5, \dot{\theta}_7$	Angular velocities of the corresponding links in the seven-bar mechanism
$x_{\mathrm{H}}, y_{\mathrm{H}}$	Cartesian coordinates of the H point of the end-effector
$\dot{x}_H, \dot{y}_H$	Speed of the end-effector H point of the seven-bar mechanism
J	Jacobian matrix of seven-bar mechanism

Substituting Eq. (1) into Eq. (2), can obtain the angular displacement θ_3 as

$$\theta_3 = 2\arctan\left(\frac{A \pm \sqrt{A_2 + B_2 - C_2}}{B - C}\right) \tag{2}$$

The corresponding expressions for the corresponding element in Eq. was shown as

$$A = 2l_3l_4 - 2l_1l_3\sin\theta_1;\ B = 2l_1l_3\cos\theta_1;\ C = l_2^2 - l_4^2 - l_1^2 - l_3^2 + 2l_1l_3\sin\theta_1$$

In Eq. (2), once the input angular displacement θ_1 is determined, the output angular displacement θ_3 can be determined. Once the position of the output link 3 is determined, the second closed-loop will become another four-bar mechanism [13, 14]. Its derivation is as follows. Similar to the analysis of the first closed-loop four-bar mechanism, the coordinates of nodes *E* and *F* can be expressed as

$$E=\begin{bmatrix} l_5\cos\theta_5 \\ l_4+l_5\sin\theta_5 \end{bmatrix} \tag{3}$$

$$F=\begin{bmatrix} l_{33}\cos\theta_3 - l_7\cos\theta_7 \\ l_4 - l_{33}\sin\theta_3 + l_7\sin\theta_7 \end{bmatrix} \tag{4}$$

The angular displacement θ_5 of the connecting crank 5 is the input angle of the second closed-loop, and the angular displacement θ_7 is the output angle of the second closed-loop. Because the distance between nodes E and F is constant, there is

$$(F-E)^T(F+E)-l_6^2=0 \tag{5}$$

Substituting Eq. (4) into Eq. (5), the angular displacement θ_7 can be obtained as

$$\theta_7 = 2\arctan\left(\frac{D\pm\sqrt{D^2+E^2-F^2}}{E-F}\right) \tag{6}$$

The corresponding expressions for the corresponding element in Eq. was shown as

$$D=2l_{33}l_7\sin\theta_3+l_5l_7\sin\theta_5-2l_4l_7,\ E=2l_{33}l_7\cos\theta_3-2l_5l_7\cos\theta_5$$
$$F=l_6^2+2l_{33}l_4\sin\theta_3+2l_5l_7\sin\theta_5-l_4^2-l_7^2-l_5^2-l_{33}^2$$

The angular displacement θ_3 is obtained from the first closed-loop. In this mechanism, point H is the end-effector point, and the angular displacement θ_7 can be obtained from Eq. (6), so the coordinate of node G is expressed as follows

$$G=\begin{bmatrix} l_{33}\cos\theta_3 \\ l_4-l_{33}\sin\theta_3 \end{bmatrix} \tag{7}$$

$$H=\begin{bmatrix} xH \\ yH \end{bmatrix} = G+\overrightarrow{GH} = \begin{bmatrix} l_{33}\cos\theta_3 - l_{77}\cos\theta_7 \\ l_4-l_{33}\sin\theta_3 - l_{77}\sin\theta_7 \end{bmatrix} \tag{8}$$

$$\mathrm{v}_H = \begin{bmatrix} \dot{x}_H \\ \dot{y}_H \end{bmatrix} = \begin{bmatrix} -l_{33}\sin\theta_3 & l_{77}\sin\theta_7 \\ -l_{33}\cos\theta_3 & -l_{77}\cos\theta_7 \end{bmatrix} \tag{9}$$

From Eq. (9), $\mathrm{v_H}=\begin{bmatrix}\dot{x}_H & \dot{y}_H\end{bmatrix}^T$ is the speed of the end-effector H point, and $\dot{\theta}_3$,$\dot{\theta}_7$ is the angular acceleration of connecting link 3 and connecting link 7, respectively. The speed of the end-effector is related to $\begin{bmatrix}\dot{\theta}_3 & \dot{\theta}_7\end{bmatrix}^T$ and has nothing to do with the driving space $\begin{bmatrix}\dot{\theta}_1 & \dot{\theta}_7\end{bmatrix}^T$. In order to express the speed of the end-effector H point with the driving space $\begin{bmatrix}\dot{\theta}_3 & \dot{\theta}_7\end{bmatrix}^T$, the following Eq. (10) is derived

$$\begin{bmatrix} \dot{\theta}_3 \\ \dot{\theta}_7 \end{bmatrix} = \begin{bmatrix} M & 0 \\ Q & P \end{bmatrix} \begin{bmatrix} \dot{\theta}_1 \\ \dot{\theta}_5 \end{bmatrix} \tag{10}$$

The corresponding expressions for the corresponding element in Eq. was shown as

$$M = \frac{\frac{4l_1l_3\cos\theta_1\sigma_{17}+8l_1l_3\cos\theta_1\sigma_{16}}{2\sigma_{14}} - \sigma_{19}}{\sigma_{13}} - \frac{(\sigma_{14}-\sigma_{15})(\sigma_{15}+2l_1l_4\cos\theta_1)}{\sigma_{13}^2}$$

$$Q = \frac{(2l_1l_{33}\sigma_{12}\sigma_1 - 2l_4l_{33}\sigma_{12}\sigma_1)}{\sigma_2^2} - \frac{\frac{4l_7l_{33}\sigma_{10}\sigma_7\sigma_1 - 4l_4l_{33}\sigma_{12}\sigma_1\sigma_6}{2\sigma_4} - 2l_7l_{33}\sigma_{10}\sigma_1}{\sigma_2}$$

$$P = \frac{(2l_4l_5\cos\theta_5 - \sigma_{11})\sigma_3}{\sigma_2^2} - \frac{\frac{4l_5l_7\sin\theta_5\sigma_7 + 4l_5l_7\cos\theta_5\sigma_6 - 4l_4l_5\cos\theta_5\sigma_5}{2\sigma_4}}{\sigma_2} - \sigma_9$$

$$\begin{aligned}
&\sigma_2 = l_4^2 + l_5^2 - l_6^2 + l_7^2 + l_{33}^2 - 2l_4l_7 + 2l_4l_{33}\sigma_{10} - 2l_7l_{33}\sigma_{10} - \sigma_8 + \sigma_9; \\
&\sigma_3 = \sigma_4 - \sigma_{11} + 2l_7l_{33}\sigma_{12}\sigma_4 = \sqrt{\sigma_7^2 - \sigma_5^2 + \sigma_6^2}; \\
&\sigma_5 = l_4^2 + l_5^2 - l_6^2 + l_7^2 + l_{33}^2 + 2l_4l_{33}\sigma_{10} - \sigma_8; \\
&\sigma_6 = 2l_4l_7 + 2l_7l_{33}\sigma_{10} - \sigma_9\sigma_7 = \sigma_{11} - 2l_7l_{33}\sigma_{12}; \\
&\sigma_8 = 2l_4l_5\sin\theta_5; \sigma_9 = 2l_5l_7\sin\theta_9\sigma_{10} = \sin((\sigma_{14} - \sigma_{15})/\sigma_{13}) \\
&\sigma_{11} = 2l_5l_7\cos\theta_5; \sigma_{12} = \cos((\sigma_{14} - \sigma_{15})/\sigma_{13}); \\
&\sigma_{13} = -l_2^2 + l_3^2 + 2l_3l_4 - 2l_1l_3\sin\theta_1 + l_4^2 - \sigma_{18} + l_1; \\
&\sigma_{14} = \sqrt{\sigma_{17}^2 - \sigma_{16}^2 + 4l_1^2l_3^2\cos^2(\theta_1)}; \sigma_{15} = 2l_1l_3\cos\theta_1; \\
&\sigma_{16} = l_3^2 - l_2^2 + l_4^2 - \sigma_{18} + l_1; \sigma_{17} = 2l_3l_4 - \sigma_{19}; \\
&\sigma_{18} = 2l_1l_4\sin\theta_1; \sigma_{19} = 2l_1l_3\sin\theta_1
\end{aligned}$$

According to Eqs. (9) and (10), we can know

$$v_H = J\dot{\theta} \tag{11}$$

$J = J_1J_2$ is the Jacobin matrix of the institution

$$J_1 = \begin{bmatrix} -l_{33}\sin\theta_3 & l_{77}\sin\theta_7 \\ -l_{33}\cos\theta_3 & -l_{77}\cos\theta_7 \end{bmatrix} J_2 = \begin{bmatrix} M & 0 \\ Q & P \end{bmatrix} \tag{12}$$

4 Movement Space of the Seven-Bar Mechanism

The first closed-loop of the seven-bar mechanism controls the working space, and the second closed-loop controls the direction of the mechanism [6]. In the first closed-loop circuit is the crank connecting link mechanism, the crank 1 rotates once, and the rotation range from the output connecting link 3 is:

$$\theta_{3\,\min} = \arcsin \frac{l_3^2 + l_4^2 - (l_1 + l_2)^2}{2l_3l_4} \tag{13}$$

$$\theta_{3\,\max} = \arcsin \frac{l_3^2 + l_4^2 - (l_1 - l_2)^2}{2l_3l_4} \tag{14}$$

In the seven-bar mechanism, only crank 1 and crank 5 can achieve full rotation. The second closed-loop controls the direction of the end-effector. The rotation ranges of the output connecting link 7 is as shown in Fig. 3.

$$\theta_{7\,\min} = \arccos \frac{l_3^2 + l_7^2 - (l_6 + l_5)^2}{2l_3l_7} \tag{15}$$

$$\theta_{7\,\max} = \arccos \frac{l_3^2 + l_7^2 - (l_6 - l_5)^2}{2l_3l_7} \tag{16}$$

The motion trajectory of the end effector of the seven-bar three-axis fixed composite mechanism is shown in Fig. 4. The shape is crescent, and its trajectory has no sudden change from top to bottom, indicating that its motion state is continuous, which is conducive to the folding of the carton flap.

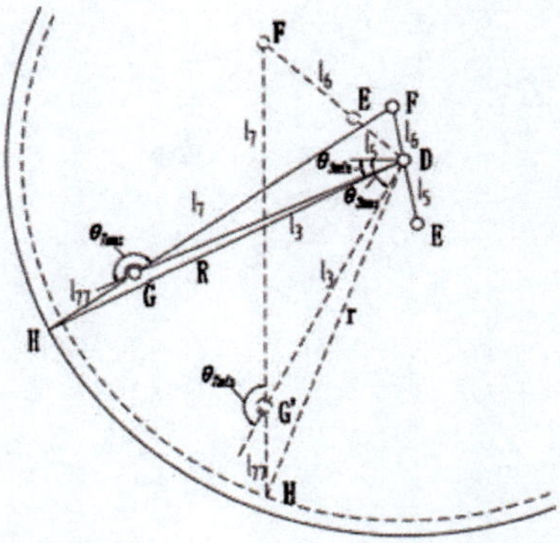

Fig. 3 Radius and limit angle of two concentric circles

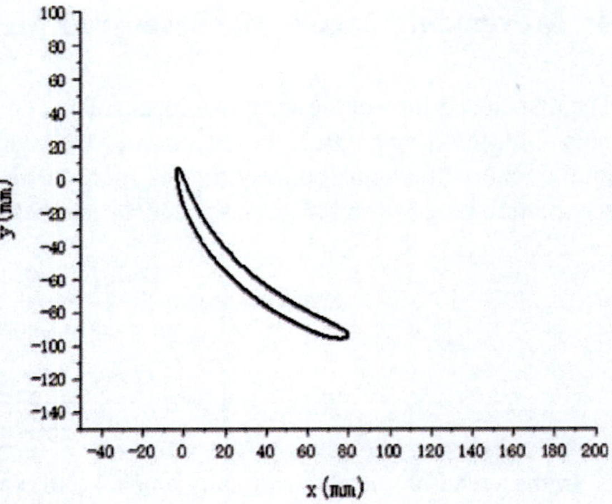

Fig. 4 Workspace of the end-effector

5 Kinetic Analysis

5.1 A Establishment of Mathematical Model

The force of each member is shown (see Fig. 5). In the complex coordinate system, [6] The modulus r_i of the complex vector of each member is a constant, the argument θ_i is a variable, the distance from the center of mass to the rotation pair is r_{ci}, and the mass is m_i, The moment of inertia around the center of mass is J_i, the external

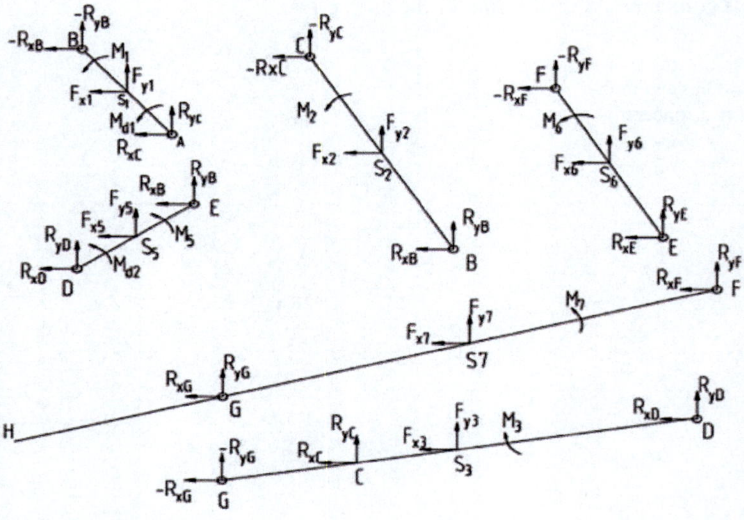

Fig. 5 Force analysis diagram of rods

forces acting in the center of mass are F_{xi} and F_{yi}, the external moment is M_i, the restraining reaction forces of the rotating pair are R_x and R_y, the driving moment is M_{di}, and the moment of inertia is M_i, available from theoretical mechanics [15]:

The force analysis of crank AB can be obtained:

$$\begin{cases} R_{xA} - R_{xB} + F_{x1} = M_1 \mathrm{Re}\ddot{S}_1 \\ R_{yA} - R_{yB} + F_{y1} - m_1 g = M_1 \mathrm{Im}\ddot{S}_1 \\ M_{d1} + M_1 + R_{xA} r_{c1} \sin\theta_1 - R_{yA} r_{c1} \cos\theta_1 \\ +R_{xB}(r_1 - r_{c1}) \sin\theta_1 - R_{yB}(r_1 - r_{c1}) = J_1 \ddot{\theta}_1 \end{cases} \tag{17}$$

The force analysis of the component BC can be obtained:

$$\begin{cases} R_{xB} - R_{xC} + F_{x2} = M_2 \mathrm{Re}\ddot{S}_2 \\ R_{yB} - R_{yC} + F_{y2} - m_2 g = M_2 \mathrm{Im}\ddot{S}_2 \\ M_2 + R_{xB} r_{c2} \sin\theta_2 - R_{yB} r_{c2} \cos\theta_2 \\ +R_{xC}(r_2 - r_{c2}) \sin\theta_2 - R_{xC}(r_2 - r_{c2}) = J_2 \ddot{\theta}_2 \end{cases} \tag{18}$$

The force analysis of the component DG can be obtained:

$$\begin{cases} R_{xD} - R_{xG} + F_{x3} + R_{xC} = M_{33} \mathrm{Re}\ddot{S}_3 \\ R_{yD} - R_{yG} + F_{y3} + R_{yC} - m_{33} g = M_{33} \mathrm{Im}\ddot{S}_3 \\ M_3 + R_{xD} r_{c3} \sin\theta_3 - R_{yD} r_{c3} \cos\theta_3 + R_{xG}(r_{33} - r_{c33}) \sin\theta_3 \\ -R_{yG}(r_{33} - r_{c33}) \cos\theta_3 - R_{xC}(r_{33} - r_{c33}) \sin\theta_3 \\ +R_{yC}(r_{33} - r_{c33}) \cos\theta_3 = J_3 \ddot{\theta}_3 \end{cases} \tag{19}$$

The force analysis of crank DE can be obtained:

$$\begin{cases} R_{xD} - R_{xE} + F_{x5} = M_5 \mathrm{Re}\ddot{S}_5 \\ R_{yD} - R_{yE} + F_{y5} - m_5 g = M_5 \mathrm{Im}\ddot{S}_5 \\ M_5 + R_{xD} r_{c5} \sin\theta_5 - R_{yD} r_{c5} \cos\theta_5 + \\ R_{xE}(r_5 - r_{c5}) \sin\theta_5 - R_{yE}(r_5 - r_{c5}) \cos\theta_5 = J_5 \ddot{\theta}_5 \end{cases} \tag{20}$$

The force analysis of the component EF can be obtained:

$$\begin{cases} R_{xE} - R_{xF} + F_{x6} = M_6 \mathrm{Re}\ddot{S}_6 \\ R_{yE} - R_{yF} + F_{y6} - m_6 g = M_6 \mathrm{Im}\ddot{S}_6 \\ M_6 + R_{xE} r_{c6} \sin\theta_6 - R_{yE} r_{c6} \cos\theta_6 + \\ R_{xF}(r_6 - r_{c6}) \sin\theta_6 - R_{rF}(r_6 - r_{c6}) \cos\theta_6 = J_6 \ddot{\theta}_6 \end{cases} \tag{21}$$

The force analysis of the component FG can be obtained:

$$
\begin{cases}
R_{xF} + R_{xG} + F_{x7} = M_7 \mathrm{Re}\ddot{S}_7 \\
R_{yF} + R_{yG} + F_{y7} - m_7 g = M_7 \mathrm{Im}\ddot{S}_7 \\
M_7 + R_{xF} r_{c7} \sin\theta_7 - R_{yF} r_{c7} \cos\theta_7 - \\
R_{xG}(r_7 - r_{c7}) \sin\theta_7 + R_{yG}(r_7 - r_{c7}) \cos\theta_7 = J_7 \ddot{\theta}_7
\end{cases}
\tag{22}
$$

Substituting Eqs. (17)–(22) to merge into a matrix:

$$
\left[\begin{array}{cccccccccccccc}
1 & 0 & -1 & 0 & 0 & 0 & 0 & 0 & 0 & 0 & 0 & 0 & 0 & 0 \\
0 & 1 & 0 & -1 & 0 & 0 & 0 & 0 & 0 & 0 & 0 & 0 & 0 & 0 \\
r_{c1}\sin\theta_1 & -r_{c1}\cos\theta_1 & (r_1 - r_{c1})\sin\theta_1 & -(r_1 - r_{c1})\cos\theta_1 & 0 & 0 & 0 & 0 & 0 & 0 & 0 & 0 & 0 & 0 \\
0 & 0 & 1 & 0 & -1 & 0 & 0 & 0 & 0 & 0 & 0 & 0 & 0 & 0 \\
0 & 0 & 0 & 1 & 0 & -1 & 0 & 0 & 0 & 0 & 0 & 0 & 0 & 0 \\
0 & 0 & r_{c2}\sin\theta_2 & r_{c2}\cos\theta_2 & (r_2 - r_{c2})\sin\theta_2 & -(r_2 - r_{c2})\cos\theta_2 & 0 & 0 & 0 & 0 & 0 & 0 & 0 & 0 \\
0 & 0 & 0 & 0 & 1 & 0 & 1 & 0 & 0 & 0 & 0 & 0 & -1 & 0 \\
0 & 0 & 0 & 0 & 0 & 1 & 0 & 1 & 0 & 0 & 0 & 0 & 0 & -1 \\
0 & 0 & 0 & 0 & -(r_3 - r_{c3})\sin\theta_3 & (r_3 - r_{c3})\cos\theta_3 & r_{c3}\sin\theta_3 & r_{c3}\cos\theta_3 & 0 & 0 & 0 & 0 & (r_{33} - r_{c33})\sin\theta 3 & -(r_{33} - r_{c33})\cos\theta_3 \\
0 & 0 & 0 & 0 & 0 & 0 & 1 & 0 & -1 & 0 & 0 & 0 & 0 & 0 \\
0 & 0 & 0 & 0 & 0 & 0 & 0 & 1 & 0 & -1 & 0 & 0 & 0 & 0 \\
0 & 0 & 0 & 0 & 0 & 0 & r_{c5}\sin\theta_5 & -r_{c5}\cos\theta_5 & (r_5 - r_{c5})\sin\theta_5 & -(r_5 - r_{c5})\cos\theta_5 & 0 & 0 & 0 & 0 \\
0 & 0 & 0 & 0 & 0 & 0 & 0 & 0 & 1 & 0 & -1 & 0 & 0 & 0 \\
0 & 0 & 0 & 0 & 0 & 0 & 0 & 0 & 0 & 1 & 0 & -1 & 0 & 0 \\
0 & 0 & 0 & 0 & 0 & 0 & 0 & 0 & r_{c6}\sin\theta_6 & -r_{c6}\cos\theta_6 & (r_6 - r_{c6})\sin\theta_6 & -(r_6 - r_{c6})\cos\theta_6 & 0 & 0 \\
0 & 0 & 0 & 0 & 0 & 0 & 0 & 0 & 0 & 0 & 1 & 0 & 1 & 0 \\
0 & 0 & 0 & 0 & 0 & 0 & 0 & 0 & 0 & 0 & 0 & 1 & 0 & 1 \\
0 & 0 & 0 & 0 & 0 & 0 & 0 & 0 & 0 & 0 & r_{c7}\sin\theta_7 & -r_{c7}\cos\theta_7 & -(r_7 - r_{c7})\sin\theta_7 & (r_7 - r_{c7})\cos\theta_7
\end{array}\right]
\begin{bmatrix}
R_{xA} \\ R_{yA} \\ R_{xB} \\ R_{yB} \\ R_{xC} \\ R_{yC} \\ R_{xD} \\ R_{yD} \\ R_{xE} \\ R_{yE} \\ R_{xF} \\ R_{yF} \\ R_{xG} \\ R_{yG}
\end{bmatrix}
=
\begin{bmatrix}
m_1 \mathrm{Re}\ddot{A} + m_1 r_{c1}\ddot{\theta}_1 \cos(\theta_1 + \pi/2) + m_1 r_{c1}\dot{\theta}_1^2 \cos(\theta_1 + \pi) \\
m_1 \mathrm{Im}\ddot{A} + m_1 r_{c1}\ddot{\theta}_1 \sin(\theta_1 + \pi/2) + m_1 r_{c1}\dot{\theta}_1^2 \sin(\theta_1 + \pi) \\
J_1\ddot{\theta}_1 - M_{d1} - M_1 \\
m_2 \mathrm{Re}\ddot{B} + m_2 r_{c2}\ddot{\theta}_2 \cos(\theta_2 + \pi/2) + m_2 r_{c2}\dot{\theta}_2^2 \cos(\theta_2 + \pi) \\
m_2 \mathrm{Im}\ddot{B} + m_2 r_{c2}\ddot{\theta}_2 \sin(\theta_2 + \pi/2) + m_2 r_{c2}\dot{\theta}_2^2 \sin(\theta_2 + \pi) \\
J_2\ddot{\theta}_2 - M_2 \\
m_3 \mathrm{Re}\ddot{C} + m_3 r_{c3}\ddot{\theta}_3 \cos(\theta_3 + \pi/2) + m_3 r_{c3}\dot{\theta}_3^2 \cos(\theta_3 + \pi) \\
m_3 \mathrm{Im}\ddot{C} + m_3 r_{c3}\ddot{\theta}_3 \sin(\theta_3 + \pi/2) + m_3 r_{c3}\dot{\theta}_3^2 \sin(\theta_3 + \pi) \\
J_3\ddot{\theta}_3 - M_3 \\
m_5 \mathrm{Re}\ddot{D} + m_5 r_{c5}\ddot{\theta}_5 \cos(\theta_5 + \pi/2) + m_5 r_{c5}\dot{\theta}_5^2 \cos(\theta_5 + \pi) \\
m_5 \mathrm{Im}\ddot{D} + m_5 r_{c5}\ddot{\theta}_5 \sin(\theta_5 + \pi/2) + m_5 r_{c5}\dot{\theta}_5^2 \sin(\theta_5 + \pi) \\
J_5\ddot{\theta}_5 - M_{d2} - M_5 \\
m_6 \mathrm{Re}\ddot{E} + m_6 r_{c6}\ddot{\theta}_6 \cos(\theta_6 + \pi/2) + m_6 r_{c6}\dot{\theta}_6^2 \cos(\theta_6 + \pi) \\
m_6 \mathrm{Im}\ddot{E} + m_6 r_{c6}\ddot{\theta}_6 \sin(\theta_6 + \pi/2) + m_6 r_{c6}\dot{\theta}_6^2 \sin(\theta_6 + \pi) \\
J_6\ddot{\theta}_6 - M_6 \\
m_7 \mathrm{Re}\ddot{F} + m_7 r_{c7}\ddot{\theta}_7 \cos(\theta_7 + \pi/2) + m_7 r_{c7}\dot{\theta}_7^2 \cos(\theta_7 + \pi) \\
m_7 \mathrm{Im}\ddot{F} + m_7 r_{c7}\ddot{\theta}_7 \sin(\theta_7 + \pi/2) + m_7 r_{c7}\dot{\theta}_7^2 \sin(\theta_7 + \pi) \\
J_7\ddot{\theta}_7 - M_7
\end{bmatrix}
$$

5.2 *Simulation Result Analysis*

In order to verify the feasibility and reliability of the mathematical model of the overall equation of the mechanism, assuming that the working resistance of the end effector Fr = 420 N, the length, mass and moment of inertia of each member of the seven-bar and three-axis fixed composite mechanism are shown in Table 2 [16, 17]. The simulation simulation test obtains the test result of the folded corrugated cardboard box.

Table 2 Motion parameters of each member

Link	Link length(mm)	Quality(Kg)	Moment of inertia (Kg·m2)
l_1	160	1.19	0.31
l_2	350	2.37	0.68
l_3	346	2.35	0.67
l_{33}	506.8	3.35	0.98
l_5	150	1.13	0.32
l_6	360	2.44	0.70
l_7	684.5	4.46	1.33
l_{77}	180	1.31	0.35

Figure 6 shows the test results of a seven-bar three-axis fixed composite mechanism. It can be seen that the curve is called periodic change. Link 1 is the first to move on one side of the end effector. The link force of rod 1 is zero at the beginning of the test [18]. As the end effector contacts the corrugated cardboard, the resistance continues to increase, until the corrugated cardboard is folded 90 degrees from the vertical state and then the horizontal state, the connecting rod force reaches the maximum, and the connecting rod force of the rod 1 is only 0.2 s from the minimum to the maximum. Second. At this time, the connecting rod 2 is used as the driving force of the directional loop. when the end effector moves to the lowest position, the connecting rod force changes, first increases and then decreases. After that, the connecting rod force of rod 1 decreases, and the increase in connecting rod force of rod 2 drives the end effector to move upward into the next motion cycle.

Figure 7 is the motion curve of the seven-bar three-axis fixed composite mechanism. The position, speed and acceleration curves of the end effector are all periodic curves [19]. When the end effector moves from the top to the bottom and then back to the top, the speed decreases first and then increases. Although the acceleration keeps increasing, it decreases sharply at the position turning back. When the end effector runs to the lowest position, the speed at this time is zero, and the acceleration fluctuation is very small. Using this feature, the carton can be folded. In the process of sealing and folding the corrugated cardboard, if the speed is too high, the

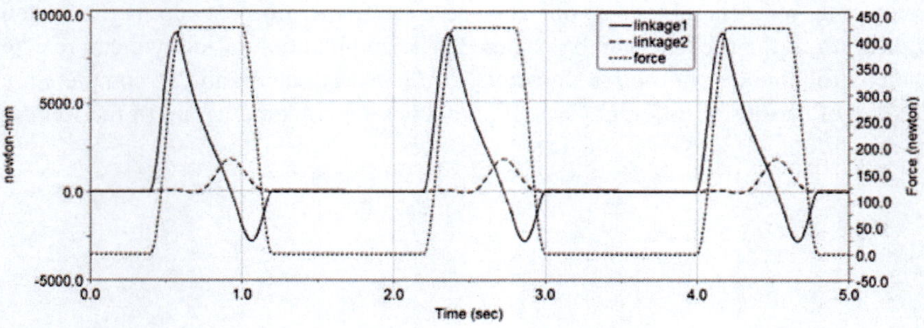

Fig. 6 Test results of carton folding

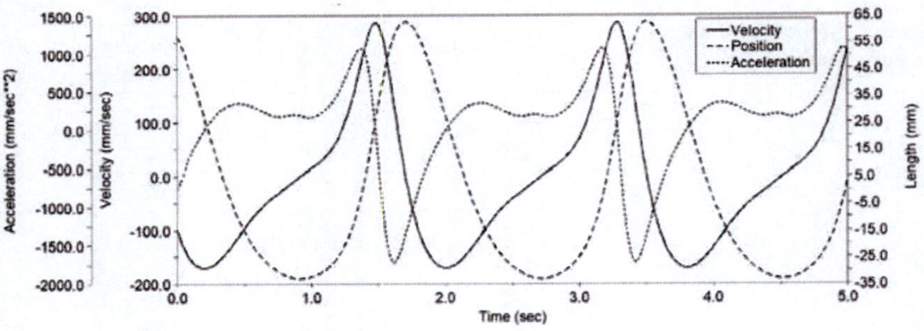

Fig. 7 Motion curve of control point H

carton lid will be folded into the carton. If the acceleration fluctuation is too large, the end effector will damage the surface of the carton, which does not meet the design requirements of the product. Therefore, the seven-bar three-axis fixed composite mechanism has the characteristics of zero speed at the lowest position and small acceleration fluctuations, which can realize the folding of the corrugated cardboard box, which can realize the folding of the flap of the carton to the horizontal direction and the end execution The device will not damage the surface of the carton to meet the design requirements.

6 Application Scenarios of Institutions

The seven-bar three-axis fixed composite mechanism can be applied to a closed-loop manipulator, and provides the required mechanism position and direction, which can be used for carton folding [20]. The mechanism device is installed on the column of the corrugated cardboard box alignment adjustment device to complete the folding and clamping of the cardboard box. The seven-bar three-axis fixed composite mechanism is a two-degree-of-freedom seven-bar composite mechanism with excellent structural rigidity, tough anti-interference ability, stability and good control. The device is composed of a stepping motor bracket, a stepping motor, a coupling, a connecting rod, a crank and so on. The speed parameter of the stepping motor and the length of the rocker 3 can be changed. The combination of the two can realize the controllable output motion trajectory of the end effector, and can complete the folding of cartons of different sizes. Figure 8 is a schematic diagram of the folding of the carton.

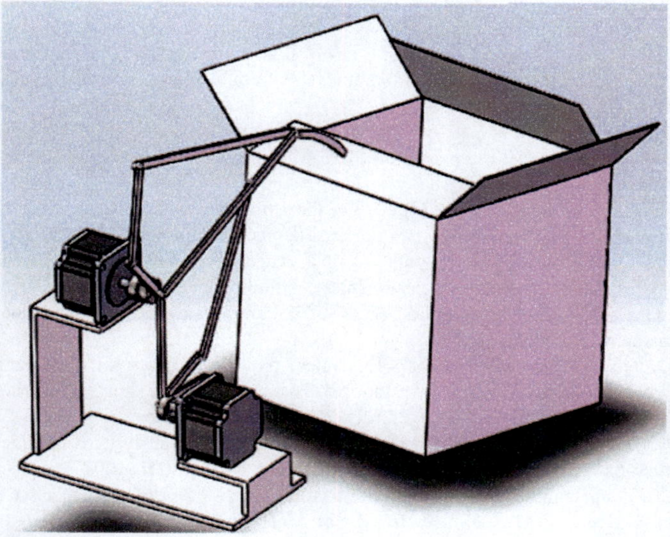

Fig. 8 Carton folding diagram

7 Conclusion

Based on the mechanism principle of the two-degree-of-freedom seven-bar three-axis fixed composite mechanism and the analysis of the composition characteristics of the plane linkage mechanism, the mechanism is a hybrid system composed of a rigid system and a plane mass point system. The vector method is used to analyze the dynamic parameters of the seven-bar three-axis fixed composite mechanism, and the mathematical model is established. In the traditional mechanical calculation process, the components and torque are ignored. The main reason is that the inertia and large scale of the seven-bar three-axis fixed composite mechanism bring errors in the calculation parameters. The design of the mechanism is affected, but the establishment of mathematical models solves the problem that the existing equations cannot obtain accurate calculation results during the movement process, and provides strong support for the research of the mechanism. In addition, through the mechanical equations used in this article, the characteristics of the components can be reflected in the mathematical model, so as to more accurately analyze the forces of each component. It can be seen that the seven-bar three-axis fixed composite mechanism is feasible and easy to operate by establishing a simulation model.

Acknowledgements This work was finally supported by the National Natural Science Foundation of China(Grant No.31660239).

References

1. Xie H (2012) Kinematic simulation and analysis of double-crank seven-bar mechanism. Sci Mosaic
2. Wang K, Zhang Y, Jing X (2019) Kinematics and singularity analysis of a novel uncoupled 2-DOF rotational parallel mechanism. In: IFToMM international conference on mechanisms, transmissions and applications
3. Zalyubovskii MG, Panasyuk IV (2020) On the study of the basic design parameters of a seven-link spatial mechanism of a part processing machine. Int Appl Mech 56(1)
4. Shrinivas S, Balli et al (2003) Synthesis of a planar seven-link mechanism with variable topology for motion between two dead-center positions. Mech Mach Theory
5. Gadad GM, Balli SS, Daivagna UM (2005) Triad and dyad synthesis of planar seven-link mechanisms with variable topology
6. Wang J, Nie L, Zhao D et al (2017) Equivalent five-bar linkages for the singularity analysis of two-dof seven-bar linkages. In: ASME international design engineering technical conferences & computers & information in engineering conference
7. Zhao X, Chen J, Wu J et al (2013) Kinematics modeling and analysis of modified elliptical gear-seven-bar seedling mechanism. China Mech Eng 24(08):1001–1007
8. Li N (2017) Kinematics analysis and optimization design of a typical planar six-bar linkage mechanism based on ADAMS. Dev Innovation Mech Electr Prod 30(01):62–63+110
9. Nie L (2020) Singularity and branch identification of a 2 degree-of-freedom (DOF) seven-bar spherical parallel manipulator. Mech Sci 11(2)
10. Zuo F, Wu B, Lu M et al (2019) Kinematics simulation and optimization design of six-bar pusher mechanism. Appl Autom (12):39–41
11. Lavi R (2012) Mechanism design. Encycl Comple Syst Sci 14(3–4):5510–5523
12. Chen X, Tang Y (2020) Dynamic modeling, response, and chaos analysis of 2-dof hybrid mechanism with revolute clearances. Shock Vibr 2020:1–20
13. Cheng S, Chen H (2019) Dynamic analysis of reciprocating compressor with two revolute joint clearances. In: IOP conference series: materials science and engineering, vol 677, no 3, p 032079 (5pp)
14. Chu ZB, Li-Dan, Huang, QX (2011) Dynamic study of 8-R seven bar III degree of freedom linkage mechanism. Switzerland (480–481):1480–1484
15. Wu X, Wang K, Wang Y et al (2021) Kinematic design and analysis of a 6-DOF spatial five-Bar linkage[J]. Mech Mach Theory 158(1):104227
16. Lu K, Yuan Y (2012) Kinematics and dynamics simulation analysis of six-bar mechanism based on ADAMS. J Henan Univ Technol (Nat Sci Ed) 31(05):555–560
17. Zhao X, Guo J, Li K et al (2020) Optimal design and experiment of 2-DoF five-bar mechanism for flower seedling transplanting. Comput Electron Agric 178(11):105746
18. Wang Y, Wen X, Engineering CO et al (2016) Kinematic analysis and optimization method of six bar linkage in shaping machine. Mech Eng
19. Nukulwuthiopas W, Laowattana S, Maneewarn T (2003) Dynamic modeling of a one-wheel robot by using Kane's method. In: IEEE international conference on industrial technology. IEEE
20. Yao W, Dai JS (2008) Dexterous manipulation of origami cartons with robotic fingers based on the interactive configuration space. Trans ASME J Mech Des 130(2):022303–1–022303–8

Energy Management

Economic Analysis Comparison Between Payback Period and Net Present Value for Office Building Energy Consumption

Z. Noranai, N. M. Sobri, and M. Z. M. Bosro

Abstract Energy saving is one of the challenges of today. In recent years, growing concerns about the environmental impacts of energy consumption and global warming has doubled the importance of this issue. This study has outlined five energy saving actions as proposed to decrease the office building energy consumption. The proposed energy saving actions was identified as reducing lamps number, changing the lamps specification, installing the lamps sensor switch, controlling the setting of air conditioner temperature and applying the sustainable energy management system. The KWSP building in Kedah, Malaysia, was the target office building and the energy audit was conducted in a specific area of the building. The proposed energy saving actions was analyzed by comparing the result of the payback period (PP) and the net present value (NPV) method where the main goal is to select which energy saving actions resulted more economically helpful. The economic analysis was conducted for each of the proposed actions and the result shows that the most profitable investment in terms of the PP is Action 4 or Action 5, Action 1, Action 2 and Action 3. For the NPV analysis, the sequence has been slightly changing with Action 4, Action 5, Action 1, Action 3 and Action 2. In overall, all the energy saving action is suitable to be implemented in an office building as the initial investment can be recovered in less than 2 years.

Keywords Building energy consumption · Payback period · Net present value

Z. Noranai (✉) · N. M. Sobri
Faculty of Mechanical and Manufacturing Engineering, Universiti Tun Hussein Onn Malaysia, 86400 Batu Pahat, Johor, Malaysia
e-mail: zamrin@uthm.edu.my

M. Z. M. Bosro
Faculty of Civil Engineering and Build Environment, Universiti Tun Hussein Onn Malaysia, 86400 Batu Pahat, Johor, Malaysia

M. Y. Ismail et al. (eds.), *Technological Advancement in Mechanical and Automotive Engineering*, Lecture Notes in Mechanical Engineering,
https://doi.org/10.1007/978-981-19-1457-7_39

1 Introduction

Office buildings deal with many people every day and their working schedule is not only limited to specific hours during a day. The large number of people and heat generated from their activities, office equipment and lights increase cooling demand and energy consumption of the office buildings [1]. According to power technology newsletter [11], Malaysia will need RM 33 billion worth of investments to achieve 2025 target of renewable energy sources. The expected investments would not only come from the government but from public-private partnerships and private financing. Since 2000, Malaysia's energy intensity (energy/GDP) has been rising. This implies that over time Malaysia uses more energy to produce a unit of GDP and this provides a compelling reason for Malaysia to improve its efficiency of energy use. Energy efficiency offers an effective and efficient energy policy instrument to address the energy supply security issue as well as energy-related environmental issue in the country. At the same time, energy efficiency is also one of the ways that will lead the country to a sustainable energy path [12].

Moreover, energy uses from the electrical appliances is the lifeblood of our style of living nowadays and it is difficult to survive without the electric energy [2]. Energy-efficient buildings play an essential role in energy resources preservation and climate change mitigation [3]. According to [4, 5] buildings consume an average of 30% of energy consumption and is expected to grow rapidly about one-third of the world's energy and the worldwide energy consumption for buildings. Consequently, it is essential to find appropriate solutions to improve energy performance of office building. As such, this paper has conducted the economic analysis comparison between the five energy saving actions to improving the energy performance of an office building. Several of the proposed action is taken from previous study conducted by [6–8] as shown in next subtopic.

2 Literature Review

Energy saving actions is the proposed method in this research to decrease the energy consumption in KWSP building. There are many good actions that can be implement to reduce the electricity usage. Therefore, the energy saving actions is planned in this study to be applied in the KWSP building.

2.1 Reducing Lamps Number

In areas with excessive amounts of light, the number of bulbs is lowered. It is recommended to remove any superfluous lamps from areas with low occupants or that are rarely utilized.

2.2 Changing the Lamps Specification

The most significant end-use for reducing energy usage in a facility is lighting. After air conditioning, lighting is the most energy-intensive electrical equipment. The majority of old offices use inefficient electrical equipment, resulting in higher electricity expenses. As a result, it is recommended that the type of lamp be changed to more energy-efficient lighting in order to reduce energy usage. The formula below can be used to assess the potential energy and cost savings from replacing an old lamp with a new one.

$$\text{Energy consumption saving(kWh)} = \text{N} \times (\text{OP} - \text{NP}) \times \text{t} \quad (1)$$

$$\text{Energy cost saving (RM)} = \text{N} \times (\text{OP} - \text{NP}) \times \text{t} \times \text{TRF} \quad (2)$$

where N = quantity of light fixture, OP = Old lamp power rating (Watts), NP = New lamp power rating (Watts), t = operating time (hours) and TRF = Tariff charges, assuming RM0.509/kWh.

2.3 Installing the Lamps Sensor Switch

Occupant sensors can detect indoor activities within a particular location. They provide simplicity by turning on lights when someone walks into the room and save electricity by turning off lights when the last person leaves. Occupancy sensors must be set in areas where inhabitants or occupant behaviour can be detected. Ultrasonic and infrared occupancy sensors are both available. Ultrasonic sensors sense sound, but infrared sensors detect heat and movement. Task lights are turned on by a person's movement in a room and shut off automatically a few minutes later when the person departs the area in this kind of application.

2.4 Controlling the Setting of Air Conditioner Temperature

In most buildings, the air conditioning system consumes the most energy. The same can be said for the KWSP building, which accounts for 63.2% of overall energy use. All of the building's indoor sections are air conditioned for at least 8 h every day. One of the elements that contributes to total cooling load and electric consumption is the temperature setting point of the air conditioner. Lower temperature settings result in increased cooling demand, while higher temperature settings result in decreased cooling load. As a result, it was recommended that the air conditioner be set to a higher temperature within the recommended range. There is an interior design condition for comfort cooling in the office building, as described below, according to the MS 1525:2014 regulations [13].

i. The temperature range for a conditioned space area that delivers the acceptable degree of comfort is 24–26 °C.
ii. The recommended minimum dry bulb temperature is 23 °C.
iii. A conditioned room area with a relative humidity (RH) range of 50% to 70% meets the recommended degree of comfort.

The Malaysian government has encouraged a minimum temperature of 24 °C in its offices, whereas the World Health Organization advises a maximum air temperature of 24 °C for worker comfort.

2.5 Applying the Sustainable Energy Management System

The focus of the energy management system is on an efficient and long-term energy management model. The objective is for one organization to be able to efficiently control and manage its energy use. It can also assist in lowering energy costs and efficiently optimizing energy consumption. Basic energy knowledge has been distributed to the personnel, but no systematic energy management system has been built in the KWSP building. The current trend in office energy consumption is considered wasteful for several reasons:

(a) There was no comprehensive energy monitoring system or a systematic energy management system that established the energy strategy, objectives, and action plan.
(b) Insufficient knowledge on how to operate and maintain inefficient electricity components.
(c) No previous energy audit programme has been carried out; as a result, no one is concerned about the importance of energy-saving actions and solutions.

The energy management matrix in Fig. 1 depicted the current state of the KWSP building's energy management system. Obviously, there is no formal energy policy in this building, but there are certain unspoken norms. Aside from that, there is no official company commitment to energy management. Informal contacts are employed in the marketing segment to promote energy efficiency. At this time, only low-cost actions have been taken. As a result, a sustainable energy management system (SEMS) is developed in this research to improve the current condition and achieve a given target in the energy matrix.

The economic analysis was conducted by using two methods known as the payback period (PP) and net present value (NPV). The payback period is calculated from ratio of invested cost to the annual return [9]. In all other words, the shorter the payout, the more appealing the investment. Likewise, the longer the payout, the less appealing it is. Following is the net present value (NPV) method. It is defined as the sum of the present values of the annual cash flows minus the initial investment. The

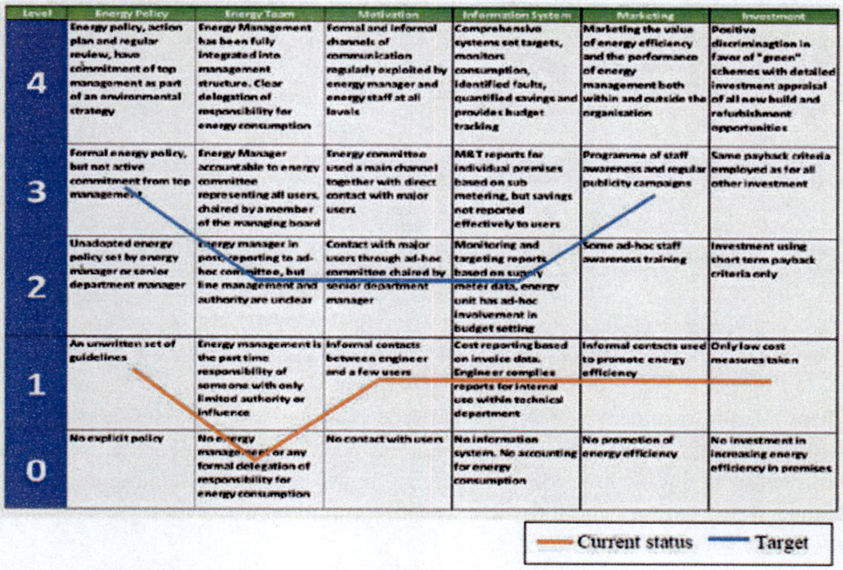

Fig. 1 Energy management matrix for KWSP building, Kedah

cash flows discounted or adjusted by incorporating the uncertainty and time value of money [10]. The investment to be consider economical when the net present value has to be positive or at worst zero.

3 Methodology

This study aims to compare the economic analysis towards the proposed energy saving actions in the office buildings. The targeted office building was KWSP building in Kedah, Malaysia. By gathering data from their electricity bills, the actual energy cost of a building was analyzed. The energy cost for the chosen building has been determined, and the building has undergone a pre-audit. The energy audit was conducted in a specific area of the building, such as the staff office, service counter, and other high-traffic areas. This area was chosen because, in conjunction with lighting, air conditioning, and other electrical equipment, it requires the use of computers. The building envelope characteristics, such as windows, doors, and insulation, have been identified based on the recognition of the office building. The proposed energy saving actions which have been identified is (1) reducing lamps number, (2) changing the lamps specification, (3) installing the lamps sensor switch, (4) controlling the setting of air conditioner temperature and (5) applying the sustainable energy management system. The economic analysis was conducted by using the payback period (PP) and the net present value (NPV) method for each of the proposed actions. The equation for calculating the PP and the NPV was following study by [1] and described as Eqs. 3 and 4 as follows.

$$\text{Payback period (PP)}=\frac{\text{Initial investment}}{\text{Cash inflows}} \tag{3}$$

and

$$\text{Net present value(NPV)} = \text{Initial investment} + \sum_{n=1}^{n=\text{end of project}} \frac{\text{Cash flow at n year}}{(1+r)^n} \tag{4}$$

where r is the assumed interest rate with the constant value of 10% and n is the number of years. An effective comparison has been made to ranked and select the best energy saving actions for an office building based on the economic analysis method stated before, where the main goal is to find which is more economically advantageous.

4 Results and Discussion

4.1 Payback Period (PP)

The payback period (PP) for each proposed energy saving actions in this study was calculated by using Eq. 1. The result of the PP method is important as it helps in revealing the time to recover the initial investment in the number of years it takes. The result was tabulated as in Table 1.

Referring to the table, Action 4 and Action 5 resulted shorter payback period of the initial investment followed by Action 1, Action 2 and Action 3. It is because, both of 4th and 5th energy saving actions are the zero cost actions and does not need any initial investment to applied in the office building. By controlling the setting of the air conditioner temperature (Action 4), the cooling load from the air conditioner compressor will using less electricity energy as the lower temperature settings result in increased cooling demand, while higher temperature settings

Table 1 Payback period for proposed energy saving actions

Item	Action 1	Action 2	Action 3	Action 4	Action 5
Initial investment (RM)	2000	6000	1500	0	0
Cash inflows (RM)	4855	4222	886	20,855	4130
Payback period (years)	0.4	1.42	1.69	0	0

result in decreased cooling load. Besides that, the Malaysian government promotes a minimum temperature of 24 °C its offices, while the World Health Organization recommends a maximum air temperature of 24 °C for worker comfort. As for Action 5, adopting the sustainable energy management system will provide benefit to the office building as the company can manage and control the energy usage in effective methods. It can also assist in lowering energy costs and efficiently optimizing energy consumption. Furthermore, by minimizing the number of lamps in the office building (Action 1), the payback period also seems practical and economically viable since the investment on Action 1 is returned less than a year. More than that, the office building also able to generate positive cash inflows by implementing all the proposed energy saving actions. In overall, all the proposed energy saving actions could recovered the initial investments in less than 2 years.

4.2 Net Present Value (NPV)

Table 2 shows the result of the net present value (NPV) conducted on the five proposed energy saving actions. NPV method used to assess the present value of an investment's future cash flows. Because it is the money spent at the start of the year, the first investment is considered negative cash flow. Cash inflow, on the other hand, is considered positive cash flow because it is the money received at the end of the year. Result of the net present value might be positive or negative signifying it is a worthwhile investment or not.

From the table above, Action 4, Action 5 and Action 1 shows the most profitable investment compared to Action 3 and Action 2. It is because, their NPV value for the 1st year is positive indicating that the investments made today will give positive return within the 1st year. Furthermore, the reason of the positive NPV value is because, the 4th and 5th energy saving actions are the zero cost actions which does not affected by the 10% interest rate. For Action 1, although the initial investment is slightly higher than Action 3, their cash inflows generated is higher between both Action 2 and Action 3. As such, the amount of interest rate that applied on the initial investments has been recovered through their yearly cash inflows thus resulting the positive NPV value.

Table 2 Net present value for proposed energy saving actions

Item	Action 1	Action 2	Action 3	Action 4	Action 5
Initial investment (RM)	−2000	−6000	−1500	0	0
Cash inflows (RM)	4855	4222	886	20,855	4130
Interest rate (r)	10%				
NPV 1st year (RM)	2414	−2162	−695	18,959	3755
NPV 2nd year (RM)	6426	1327	38	36,195	7168

5 Conclusions

In conclusion, all the proposed energy saving actions can be applied in an office building as the initial investment can be recovered in less than 2 years. For the economic analysis using the PP method, the sequence of the most profitable investment to the lesser is Action 4 or Action 5, Action 1, Action 2 and Action 3. However, when referring to the NPV analysis, the sequence has been slightly changing with Action 4, Action 5, Action 1, Action 3 and Action 2. The difference outcomes between both of the economic analysis method are mostly due to the charged of the 10% interest rate against the initial investments which shows that Action 4 has resulted the most economically advantageous.

Acknowledgements The author acknowledges the funding of this research by the Ministry of Higher Education Malaysia (MOHE) through the Fundamental Research Grant Scheme (FRGS/1/2018/TK07/UTHM/02/6) and research grant No. K091, Centre for Energy & Industrial Environment Studies (CEIES), Faculty of Mechanical & Manufacturing Engineering (FKMP) and Universiti Tun Hussein Onn Malaysia (UTHM).

References

1. Fathi S, Kavoosi A (2021) Effect of electrochromic windows on energy consumption of high-rise office buildings in different climate regions of Iran. Sol Energy 223:132–149
2. Noranai Z, Yusof MZM (2011) Study of energy efficiency opportunities in UTHM. World Acad Sci Eng Technol 77:745–751
3. Luo XJ, Oyedele LO (2021) A data-driven life-cycle optimisation approach for building retrofitting: a comprehensive assessment on economy, energy and environment. J Build Eng 43 (2021)
4. Omar NAA, Joharudin NFM, Ahmad AZS, Noranai Z, Batcha MFM, Taweekun J (2020) Energy consumption and potential saving in MSI complex. J Adv Res Fluid Mech Therm Sci 68:145–151
5. Dong Z, Liu J, Liu B, Li K, Li X (2021) Hourly energy consumption prediction of an office building based on ensemble learning and energy consumption pattern classification. Energy Build 241
6. Noranai Z, Mohamad MHH, Salleh H, Yusof MZM (2014) Energy saving actions for university public library: a case study of UTHM library. Appl Mech Mater 660
7. Noranai Z, Azman ADF (2017) Potential reduction of energy consumption in public university library. IOP Conf Ser Mater Sci Eng 243
8. Abdul Omar NA, Apandi N, Abd Samad AF, Rahim N, Qader Mustafa AA, Arif S, Noranai Z (2020) Potential energy saving by ACMV temperature setting in university building. IOP Conf Ser Mater Sci Eng 824
9. Chandra S, Yadav A (2020) Site selection based on thermo mechanical decay and payback period of solar PV system: need of present scenario. Mater Today Proc 43:287–292
10. Knoke T, Gosling E, Paul C (2020) Use and misuse of the net present value in environmental studies. Ecol Econ 174
11. Power Technology Newsletter. https://www.power-technology.com/comment/malaysia-needs-us8-billion-investment-to-achieve-20-renewable-energy-target-by-2025/. Last accessed 10 Sept 2021

12. Ministry of Energy. https://www.mestecc.gov.my/web/wp-content/uploads/2019/04/13.-National-Energy-Efficiency-Action-Plan-english-only.pdf. Last accessed 09 Sept 2021
13. Department of Standards Malaysia (2014) MS 1525:2014, energy efficiency and use of renewable energy for non-residential buildings—code of practice (second revision)

A Kinetic Mechanism Based on Lens Law Concept of Hybrid Generator

Saiful Bahari Shaari, Zulkifli Mohamed, and Hanif Ramli

Abstract The hybrid generator is one of the renewable energy projects that is supposed to alleviate the issue of electricity availability and cost. It can be regarded as a green technology because the solenoid's magnetic induction is the system's primary source of electricity generation. Unlike a diesel generator, it generates power by burning diesel fuel and emitting CO2 into the atmosphere. Other forms of generators, such as solar and wind turbines, have efficiencies that are affected by weather and season. A hybrid generator system produces electricity by turning magnetic induction force into mechanical energy, which is subsequently converted back into electrical energy by the generator. Simulating a combined magnet solenoid to show how the system reacts when magnetic flux from the solenoid exists is part of developing the hybrid generator prototype. To generate power, this prototype uses translation motion that is subsequently transferred to rotational motion. The prototype's design will serve as a guide for developing the system during the fabrication process. Furthermore, while replacing their energy source with a magnet and solenoid, the existing generators concept is referenced to. The solenoid will serve as an energy source, generating mechanical energy and transferring it to the generator via a crankshaft for conversion into electrical energy. The current prototype of a hybrid generator can be improved to boost its capabilities and efficiency in order to generate more electricity. The relevance of this research is to present new alternate energy that is superior and can solve the problem of generating power using the existing generator.

Keywords Hybrid generator · Green technologies · Prototype

S. Bahari Shaari (✉) · Z. Mohamed · H. Ramli
School of Mechanical Engineering, College of Engineering, Universiti Teknologi MARA (UiTM), 40450 Shah Alam, Selangor, Malaysia
e-mail: saifulbahari@uitm.edu.my

M. Y. Ismail et al. (eds.), *Technological Advancement in Mechanical and Automotive Engineering*, Lecture Notes in Mechanical Engineering,
https://doi.org/10.1007/978-981-19-1457-7_40

1 Introduction

Nonrenewable energy is derived from sources that will deplete or deplete in our lifetimes. Coal, petroleum, and natural gas make up most renewable energy sources [1]. Carbon is the most common component of fossil fuels. Year after year, the world's demand for fossil fuels has increased, resulting in an unfavorable influence on global warming [2]. The atmosphere will be affected if the entire globe burns fossil fuels, because carbon dioxide will be released into the atmosphere. Imagine if carbon dioxide emissions continue unchecked for years, causing the ozone layer to deplete and the world to warm, causing glaciers to melt, rising sea levels, and possibly catastrophic disasters. As a result, steps must be done to limit global warming and its environmental impact while keeping the cost of power low and using renewable energy to generate it [3]. This project will define how to generate electrical power from green or renewable energy using magnetic mechanisms as element sources from a low voltage input to produce a high output voltage [4]. The idea is similar to that of a generator that generates power using oil and diesel. Air oxidizes in the combustion chamber of a combustion generator that uses a heat engine and fossil fuel as a fuel source [5]. Internal combustion generates force and moves mechanical energy because of the high pressure and high temperature it produces. The brushless motor will be rotated, and electricity generated using mechanical energy. However, in terms of the electric generator concept, this device will move mechanical energy from an internal combustion system to an electromagnet system. The generator's electromagnet system combines concepts from the faraday law and solenoid systems [6]. Furthermore, at both ends of the shaft, one at the north end and the other at the south end, use the Permanent Magnet, Neodymium magnet N52. The electric current in a circular loop (winding wire) created by adding a low voltage to the system creates a magnetic field, which is magnetic energy that attracts and repels the shaft with the magnet. This system will generate the force required to rotate the brushless motor by moving mechanical energy. The revolutions per minute (RPM) of the moving piston and shaft, rotation speed of the piston, and beginning torque of the piston are all parameters that must be considered in order to generate energy.

2 Methodology of Project

The generator is powered by a hybrid system, which uses a low input current to produce a higher output. The existing generator generates electricity through a combustion method. Meanwhile, the hybrid generator will provide the electromagnet force that will transport the mechanical energy utilizing the lowest voltage input. The horizontal mechanical energy will be converted to rotational energy to rotate the brushless motor, which can generate electricity. All the generator's notions and hypotheses are combined with combine. The electricity will be generated by the generator. This hybrid generator will use the AC power input for the initial current to build the electromagnet in the first stage. The brushless motor divides the current into two portions when it generates it. The first section is the storage section, in which the electricity is stored in the battery, which may then be used to continuously flow power to the generator, resulting in the creation of an electromagnet. The output of the electricity supply is the subject of the second part. For the flow of the mechanism process, the output of the generator is directly proportional to the current input and the size of the solenoid and crankshaft, as illustrated in Fig. 1.

The motor-generator in the hybrid generator system transfers mechanical energy from the solenoid into electrical energy. It converts mechanical energy from the shaft's translation motion as a result of the force created by a solenoid and transfers it to the motor-generator through a rotating shaft. The system's primary sources are a magnet and a solenoid. Figure 2 shows a solenoid, which is a wire coil used as an electromagnet. It's a machine that transforms electrical energy into mechanical energy. From electric current, a solenoid generates a magnetic field, which is then used to generate linear motion. According to Lenz Law, an induced emf's direction will be such that if it causes a current to flow in a conductor in an external circuit, the current will generate a field that opposes the change that was created, as seen in Fig. 3. The Lenz law, which allows the magnet to travel forward and backward by adjusting the flux of the solenoid, will be used to determine the location of the magnet cylinder shaft shown on Fig. 4. To begin, there is the gravitational effect. The horizontal location of the cylinder magnet and solenoid reduces gravitational forces while rotating the cylinder. If the cylinder is vertical, for example, the movement

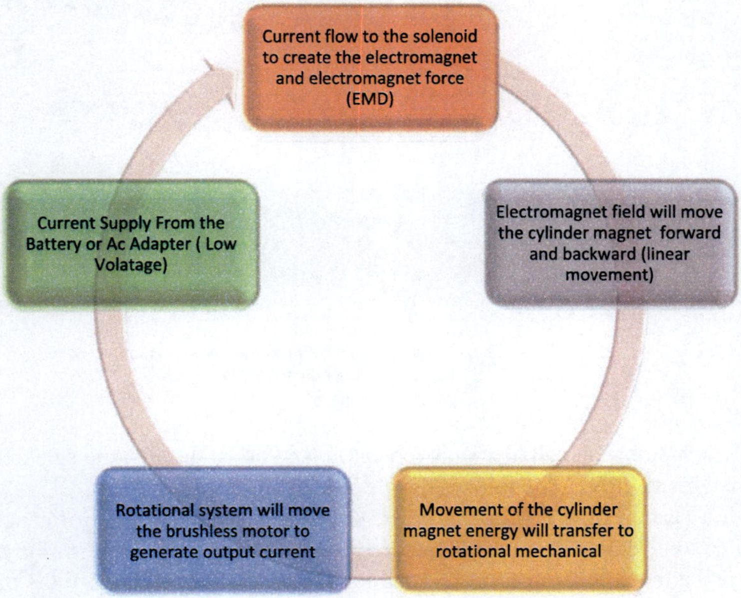

Fig. 1 Mechanism to generate electricity

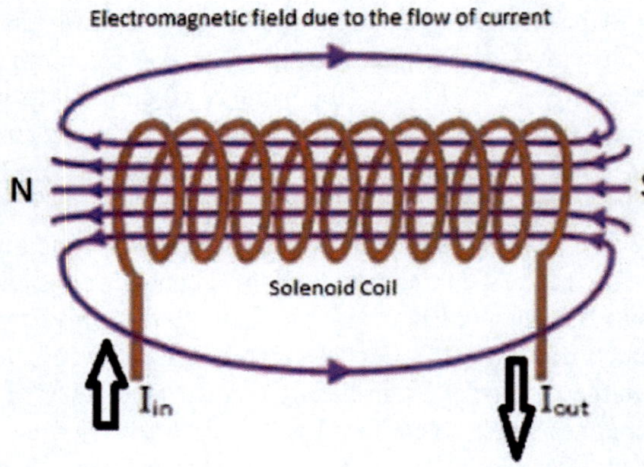

Fig. 2 Direction on the electromagnetic field on the solenoid coil

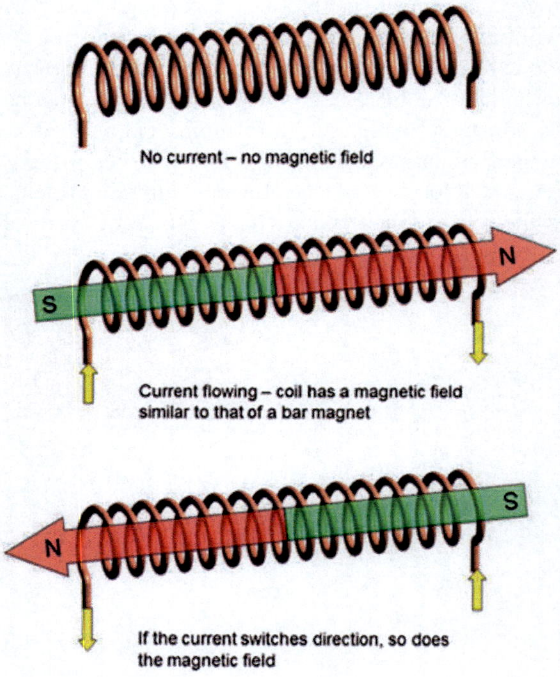

Fig. 3 Direction of current flow on the solenoid

is inefficient since the cylinder's movement up and down through the solenoid will alter the gravity effect. The full system are shown on Fig. 5.

With a current of 0.3 A as the input, the hybrid generator's performance was simulated to calculate its output. The force produced by the solenoid was first estimated using the number of turns (N) and the length of the gap between the magnet and the metal (g) as variables. The number of turns was 100, 200, 300, and 400, and the gap length was 0.8 mm, 0.9 mm, and 1 mm, respectively. Where current

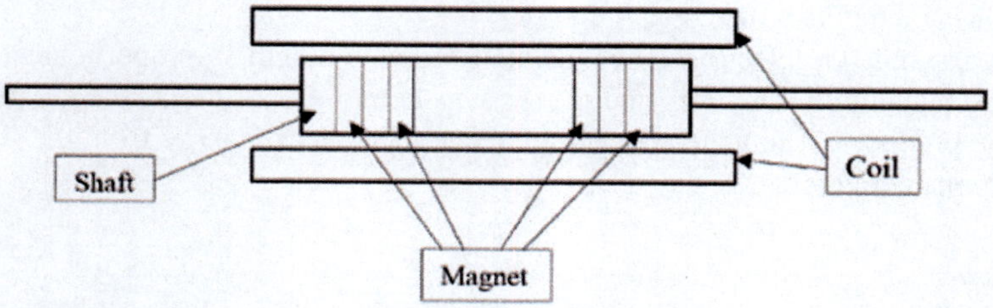

Fig. 4 Arrangements of solenoid magnetic system

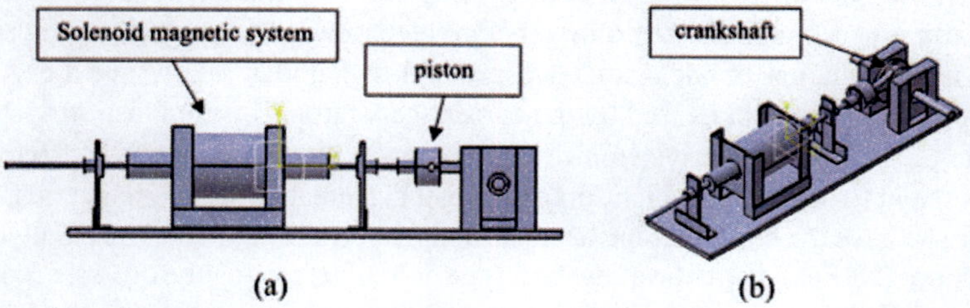

Fig. 5 Full system view **a** front view **b** isometric view

(I), magnetic constant (0), and the solenoid's cross-sectional area are all constants (A). The solenoid will produce a magnetic field that exerts a force on the shaft when current is provided to it. Lorentz force refers to the interaction of magnetic and electric forces on a shaft. After that, the force was estimated by inserting the magnet's size and other attributes into the equation of:

$$\Phi_B = \int\limits_{\Sigma}\int\limits_{t} \mathbf{B}(r, t) \cdot d\mathbf{A}, \tag{1}$$

where

dA = Element of the surface $\sum$ enclosed by the wire loop

B = The magnetic field.

Wire Loop acquires an electromotive force (EMF) according to Faraday's Law. The most widespread version of this law states that the induced electromotive force in any closed circuit is equal to the rate of change of the magnetic flux.

$$\varepsilon = -N\frac{\Phi_B}{dt} \tag{2}$$

where

ε = Electromagnet Force (EMF)

Φ_B = Magnetic Flux 10^{-7}.

As a result, the electromagnet magnetic force strength is proportional to the rate of magnetic flux motion, and the greater the number of loops, the greater the voltage produced. The generator will generate mechanical energy from the (EMF) Electromagnet Force.

3 Results and Discussion

In brief, the goal of this research is to develop an electromagnet energy producing system that uses magnetic force to generate electricity. The generator is designed using a combination of ideas and two principles: Faraday's Law and Lenz Law. The combustion system is used by the current generator. The combustion system is invented for the magnetic system to generate power in this project. The experiment study is held to determine which sort of magnet can create a strong magnetic field in order to achieve the best outcome in electromagnet power delivery. The Neodymium Iron Boron (NdFeb) magnet was the best type of magnet among the others, according to the experiment study part. Furthermore, the electromagnet type was used on this generator, which creates an electromagnet field at the winding wire that may move the permanent magnet cylinder when current is supplied.

3.1 Speed Test

Because the system would create the rotational speed, the tachometer was employed in a hybrid system in the speed test study. The result will be displayed in revolutions per minute on the tachometer (RPM). The speed data will be collected in two ways: first, unloaded or without connecting the generator to the magnetic engine, and second, loaded or with the generator connected to the magnetic engine. Each hour, the speed test will be performed with three different current inputs.

3.1.1 Speed Test for Magnetic Engine Without Generator (Unloaded)

The system can function at a constant pace, according to the results in Table 1. The system can run at a modest speed for 0.1 A of current input, but it can create continuous speed, and the speed will grow as the current input increases. The system's speed is determined by the current input. The revolutions per minute (RPM) of the system will rise as the current input increases. When employing 0.1 amperes, the average speed was 241.25 rpm. When using 0.2 amperes, the average speed was 425.125 rpm, but when using 0.3 amperes, the average speed was 543.125 rpm. The rpm obtained is proportional to the current inputs. When the engine is loaded, the speeds are projected to drop.

Table 1 Speed test without generator

Current Input (amp)	Speed in RPM								
	Time in hour								
	1	2	3	4	5	6	7	8	Average
0.1	242	239	243	240	241	239	244	242	241.25
0.2	423	417	433	429	425	430	424	420	425.13
0.3	541	545	543	540	546	545	542	543	543.13

Table 2 Speed test with generator

Current Input (amp)	Speed in RPM								
	Time in hour								
	1	2	3	4	5	6	7	8	Average
0.1	128	125	126	130	124	128	128	125	126.75
0.2	234	232	230	235	238	234	236	234	234.125
0.3	282	288	300	294	290	292	288	285	289.88

3.1.2 Speed Test for Magnetic Engine with Generator (Loaded)

The speed of the system with the generator has decreased dramatically as shown in Table 2. The average speed obtained with a current input of 0.1 amperes was 126.75 rpm, whereas the average speed obtained with a current input of 0.2 amperes was 234.125 rpm. Finally, when the input current was 0.3 A, 289.88 rpm were produced. Because the engine was connected to the generator, the speed was reduced. Because the solenoid provides force and turns the crankshaft, the force from the solenoid plays a major part in producing power output. The brushless motor inside the generator will be rotated using the power generated by the crankshaft revolution. Because the generator will generate electrical energy from mechanical energy, it is the most important component in this system. Because it takes a certain amount of energy to start the rotation of a crankshaft, the speed drops rapidly when it is connected to the generator. As a result, the system must be adjusted to provide more power, which necessitates the use of two or more solenoid systems to generate additional force and torque. It can also boost the system's performance.

3.1.3 Power Input Output Test

The goal of this testing is to determine the maximum amount of power that the engine and generator can produce given the current input. The preliminary data indicates that the output power is more than the input power, proving the stored power hypothesis in a magnetic system (Table 3).

Table 3 The current output produces by the generator

Current Input (amp)	Power output (amp)
	Magnetic system
0.1	0.0–0.3
0.2	0.9–1.4
0.3	1.4–1.8

Fig. 6 Running system of hybrid generator

Figure 6 depicts the overall operation of the system, as well as an example of electrical appliances used to demonstrate the stability of power output in a hybrid system. To keep the system running under stable conditions, the generator produced current and supplied it to the UPS battery storage. The UPS, which stands for uninterruptible power supply, is a device that provides electrical energy to the system. The system's generator will provide 600 watts of power and will be able to charge the UPS battery storage.

4 Conclusion

The electromagnetic force (EMD) may be created by using the two main Lenz laws and the faradays law in this study to move the mechanical system. Furthermore, the low current input will generate electromagnet energy via the solenoid system, which will move the kinematic system horizontally. The mechanical part required to transfer horizontal movement is a single crankshaft, which will convert horizontal movement to rotational action in order to rotate the brushless motor and provide current output. The goal was achieved because the generator can provide power at only 0.3A, allowing the system to produce more power output. Not only that, but the system has also been demonstrated by the fact that the electric appliances have been turned on. With 0.3A, this system can turn on a variety of electric equipment such as lamps, computers, and phones, as well as charge them. As a result, this system is backed by a UPS battery that is linked to a generator, allowing the system to run smoothly because the UPS battery can store a huge amount of energy and give it to the electric appliances. The magnetic engine can maintain a constant speed and rotate the generator to generate enough power to charge the UPS battery. The magnetic effect is determined by the diameter of the winding wire in addition to the amount of winding wire. The length of the winding in this project is 0.143 m, while the diameter of the winding is 0.06 m. The magnetic field intensity increases as the diameter of the winding increases. Meanwhile, the polarity at the winding wire is used to attract and repel the cylinder magnet shaft. The permanent cylinder in this cylinder was a Neodymium magnet.

Acknowledgements The research team wishes to thank the Universiti Teknologi MARA under the grant numbers 600-RMC/GPK 5/3 (206/2020) for the financial support and the Research Management Institute (RMI) of UiTM for the management assistant

References

1. Elliot A (2018) Solving energy challenges in remote communities. Third W (February):1–7
2. Shahzad U (2015) The need for renewable energy sources. ITEE J ISSN Int J Inf Technol Electr Eng 16–18 (Online). Available: http://blog.wwf.ca/blog/2015/12/10/the-need-for-renewable-energy-in-the-arctic/
3. Newell RG, Raimi D, Aldana G (2019) Global energy outlook 2019: the next generation of energy. Resour Futur July:46
4. Mohtasham J (2015) Review article-renewable energies. Energy Procedia 74(August):1289–1297. https://doi.org/10.1016/j.egypro.2015.07.774
5. Vourvoulias A (2019) 5 Advantages and 5 disadvantages of solar energy. Greenmatch. https://www.greenmatch.co.uk/blog/2014/08/5-advantages-and-5-disadvantages-of-solar-energy
6. Mishra P (2017) Main components, working, advantages and disadvantages. Sol Power Plant (Online). Available: https://www.mechanicalbooster.com/2017/12/solar-power-plant.html#comments

Performance Testing of Pico Hydropower Turbine Prototype

Hema Vharman Ganasan, Mohd Zarhamdy Md Zain, Mastura Ab Wahid, Mohamed Hussein, and Azman Jamaludin

Abstract Present day, insufficiency of energy supply is a major concern around the world. Many renewable energy sources are been in used to overcome this problem. Hydropower is one of the most reliable and sustainable energy sources that can generate electricity. Nowadays, pico hydropower are getting more attention than conventional large hydropower plants due to the high expenditure and environmental concerns. Pico hydropower is an alternative solution that offer low cost and high efficient. Due to its cost-effective factor, pico hydropower turbines can be best introduced to remote areas with insufficient electricity power supply like rural areas in Sabah and Sarawak. The purpose of this study is to design and testing a pico hydro turbine. A prototype model for the design concept of the pico hydro turbine with simpler mechanisms has been developed and tested. The prototype is able to generate power approximately to 11.3 V from the performance testing. The overall highest efficiency of the pico hydro turbine based on the test results is 90.45% at 9.84L/min water flow rate. Thus, this prototype can provide enough power to power LED light bulbs and table fan. The tests results find that the design concept of the pico hydro turbine can become a mini electricity generator.

Keywords Pico-hydro turbine · Rural electrification · Power output · Cost effective

---

H. Vharman Ganasan · M. Z. Md Zain · M. Ab Wahid (✉) · M. Hussein
School of Mechanical Engineering, Faculty of Engineering, Universiti Teknologi Malaysia, UTM Johor Bahru, 81310 Johor, Malaysia
e-mail: masturawahid@utm.my

A. Jamaludin
Engineering Department (Rotating), Integrated Regional Satelite Office (West Coast), Group Technical Solution, Project Delivery & Technology (PD&T) Division Petrolium Nasional Berhad (PETRONAS), Kuala Lumpur, Malaysia

M. Y. Ismail et al. (eds.), *Technological Advancement in Mechanical and Automotive Engineering*, Lecture Notes in Mechanical Engineering,
https://doi.org/10.1007/978-981-19-1457-7_41

1 Introduction

Around 25% of the world population lacked access to electricity, with the figure rising to around 36% in developing nations and Asia in 2016 [1–3]. As a member of the Association of Southeast Asian Nations (ASEAN), Malaysia is concentrating on sustainable and dependable energy sources for rural electrification. Hydropower is a type of energy that converts water into useable electricity and is one of the oldest renewable energy sources. In 1882, a water-wheel on the Fox River in Wisconsin was the first to use flowing water to generate electricity [4]. Hydropower is the most commonly utilized renewable energy source in the world, accounting for 19% of all electricity generated by power plants [5]. Hydropower was first used in Malaysia in July 1900, when the Raub-Australian gold mine business erected a hydroelectric station on the bank of the Sempam River near Raub, Pahang [6, 7]. However, hydropower for home use in Malaysia was only commercially accessible in the late 1970s [6]. Even though Malaysia has profited from major hydropower plants, no attempts have been made to apply the technology in pico hydropower [8].

Pico is a scientific word for a tiny feature, especially in the context of size processes. Pico hydro is a hydropower production system that may produce energy with a maximum output of up to 5 kilowatts, as shown in Table 1. Due to its small size, it is an excellent device for rural electrification and a cost-effective solution to generate power. A Pico hydro can power some household items like radios, televisions, refrigerators, and food processors. The pico hydro turbine is also known as environmentally friendly because it reduces carbon dioxide emissions, reduces reliance on fossil fuels, and prevents deforestation.

Many underdeveloped nations rely heavily on pico hydropower. Pico hydropower research and innovation are extremely encouraging in the development of green technologies [17, 18] .

Highly inaccessible rural areas in some states of Malaysia, make it a hurdle to widespread distribution infrastructure. Other alternative solutions or off-grid power technologies should be explored to overcome this burden to ensure that sufficient supply is stable. Since most of the rural villages are located close to the water supply, the introduction of pico hydro turbines is one alternative. Besides, the pico hydro turbine has no dams and reservoirs, reducing the installation and cost of installation. It is also gentle on the environment and reduces maintenance and replacement costs.

Table 1 Classification of hydropower generation [9]

Hydro generator	Capacity	Feeding
Large	More than 100 MW	National power grid
Small	Up to 25 MW	National power grid
Mini	Below 1 MW	Micro power grid
Micro	Between 6 and 100 kW	Small community or remote industrial areas
Pico	Up to 5 kW	Domestic and small commercial loads

This research aims to analyze and test the performance of a pico hydropower turbine prototype as well as develop a low cost Pico hydro turbine prototype. The final design is tested to obtain the maximum output voltage generated by the Pico hydropower turbine.

2 General Pico Hydropower Principles

The section entails doing a utility analysis of the Pico hydropower system's fundamentals, comprising the application and generation system. The highest electricity production of pico hydropower is five kilowatts. Because of its compact size, this form of hydropower offers cost and installation advantages. Pico- hydro technological innovation can make power available even in the most remote locations.

The production of electrical energy from the potential energy of flowing water across a height difference is known as hydro energy. The water's gravitational potential energy is transformed to mechanical energy, which is subsequently converted back to electrical energy via a generator.

$$P = \eta Q H \rho g \tag{1}$$

where η is the efficiency of the energy conversion system, Q is total volumetric flow, H is head (actual height difference between the free surfaces of the reservoirs or channels upstream and downstream of turbine), ρ is water density and g is the gravitational constant (9.81 m/s). The parts of a pico hydropower turbine consist of a penstock, turbine and generator[16].

The water from the reservoir is piped to the water turbine via the penstock. the penstock accelerates and strengthens the flow of water to operate the impeller of the water turbine. Different penstock shape and size helps to increase the flow rate entering the turbine, such as by having a spiral flow [10] increase the turbine rotation from 50 to 90 rpm and different opening of water baffle [11] increases the efficiency of turbine to 5.93%. In terms of economy, the penstock diameter design with a head loss of less than 10% gives optimum cost per kW [15]. When designing a turbine, two elements must be considered: head and river. The head is the vertical distance difference between water levels while the flow stream is the amount of water flowing for a certain period of time. Reactions and impulse turbines are the two available types of water turbines. Thakes declared in 2009 that the turbine reacted into the axis by the pressure of the reduction water on the turbine, while the impulse turbines moved in the moment of high water jet to the turbine [9]. Both varieties are routinely used for small and large-scale hydropower. Low, medium, and high turbine heads are defined as less than 10 m, 10–50 m, and greater than 50 m, respectively [12] as shown in Table 2.

Table 2 Impulse and reaction turbine

Turbine type	Head classification		
	High (>50 m)	Medium (10–50 m)	Low (<10 m)
Impulse	Pelton turgo Multi-jet pelton	Crossflow turgo pelton	Crossflow
Reaction		Francis pump-as-turbine	Propeller kaplan

The generator is an essential part of the Pico-Hydropower, the torque and rotation of the runner and shaft directly or through a mechanical transmission rotate a generator, and the mechanical power is transferred to electrical power.

3 Pico Hydropower Design

When designing this device, several stages of concept development and design process should be followed. The stage of concept development and design process will ensure the overall progress is organized and structured.

3.1 Pico Hydropower Concept Design

The project design specification should be constructed to make the component selection process more manageable. Then, concept generation will provide a unique concept to the specification. Then, proceed with the functional decomposition blocks shown in Fig. 1 and then the morphological chart is built. Three designs were constructed based on the morphological chart. Evaluation matrix is introduced to choose the final design. After finishing the design selection, design development is carried out. In the design development process, it involves a combination of hardware and software of the design. For the hardware part of this device, it consists of the design for the pico hydropower turbine and electronic component selection.

3.2 Design Concept

The design concepts start with the morphology chart consisting of several designs based on the function decomposition block shown in Fig. 1. As shown in Fig. 1, under water flow function, several options of water source such as water from nozzle, rectangular duct and water channel is chosen. Under turbine function, several types of turbine is chosen such as Pelton and Cross-flow from impulse turbine and Francis turbine under the reaction turbine. To convert this rotational energy to electrical

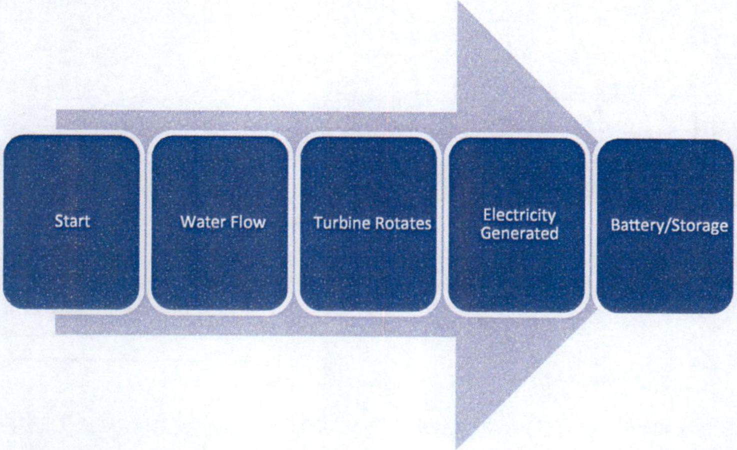

Fig. 1 Function decomposition block

energy function, DC dynamotor, brushless generator and permanent magnet DC motor are needed as a generator and the power generated is stored in a rechargeable battery.

The constructed morphology chart is used as a reference to develop different design concepts that use different options from each function. For the first concept design, a circular casing with a water inlet and outlet. Underneath this casing, the Pelton turbine is placed and connected with a shaft. Pelton turbine is selected because impulse turbines perform better in low water flow rate This small shaft is directly connected to the DC Dynamotor where the rotation of the shaft will generate electricity in the dynamotor. The generated electricity is transferred through connecting wires to the rechargeable battery as storage. It is suitable for the Pelton turbine where it operates at high impact water flow. In the second design concept, there is also an impulse turbine, a cross-flow turbine. A rectangular nozzle is used as a water inlet to rotate this turbine, and a brushless generator is used to convert mechanical energy to electrical energy. A rechargeable Li-ion battery is used to store the electricity. For the final concept design, the spiral casing is used together with Francis turbine. Francis turbine is a reaction turbine, and reaction turbine is effective in free flow and low head. The Francis is connected with a shaft where the rotation of the shaft generates electricity in the DC Permanent Motor. The polymer Li-ion battery is used for electricity storage.

After the conceptual design, the evaluation matrix is used to select the best design for the ten criteria listed, such as user-friendly, easy to manufacture, ease for maintenance, portability, power generation, safe to use, durability, manufacturing cost, simplicity, and affordability. Design 1 scored the highest compared to Design 2 and 3 and met most of the criteria listed. After finishing the design selection, design development is carried out. The design development process involves a combination of hardware and software of the design. The hardware part of this device consists of

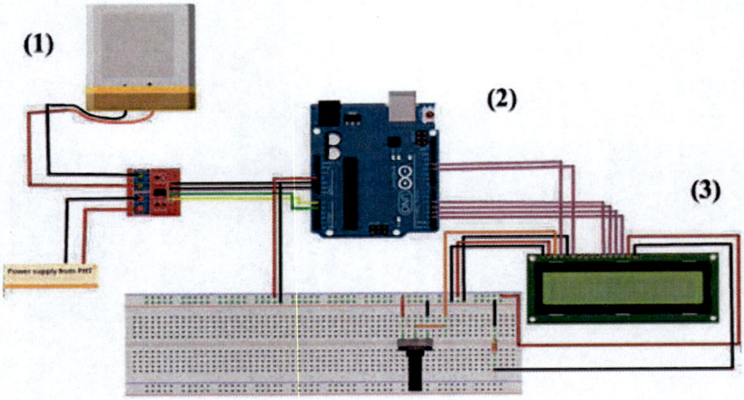

Fig. 2 Schematic drawing; (1) rechargeable Li-po 8000mAh battery, (2) Arduino Uno, (3) LCD 16 × 2 display

the design for the pico hydropower turbine and electronic component selection. The electronic component selections involved are Arduino Uno, LCD 16 × 2 display, MAX 471, lithium polymer rechargeable battery and power source. Figure 2 shows the schematic diagram of the hardware to measure and record the output from the pico hydropower turbine.

3.3 Pico Hydropower Turbine Prototype

After some modifications and several improvements based on the evaluation are made to the final design of the Pico hydropower turbine, as shown in Fig. 3. The final design uses a DC dynamometer to generate DC voltage output for this system. The DC dynamometer is driven by a Pelton turbine with a diameter of 30 mm and connected to a circular penstock.

The prototype is connected to the Arduino UNO to display the turbine's output power and the battery percentage. This Arduino UNO will display voltage, current and power whenever the turbine rotates and produced output power at a specific water flow rate.

3.4 Assembled Pico Hydropower Turbine

The Pico hydropower turbine is then assembled with electrical parts such as Ardunio UNO, breadboard and Max 471. MAX 471 is a high-side power-line monitoring is especially useful in battery-powered systems, since it does not interfere with the

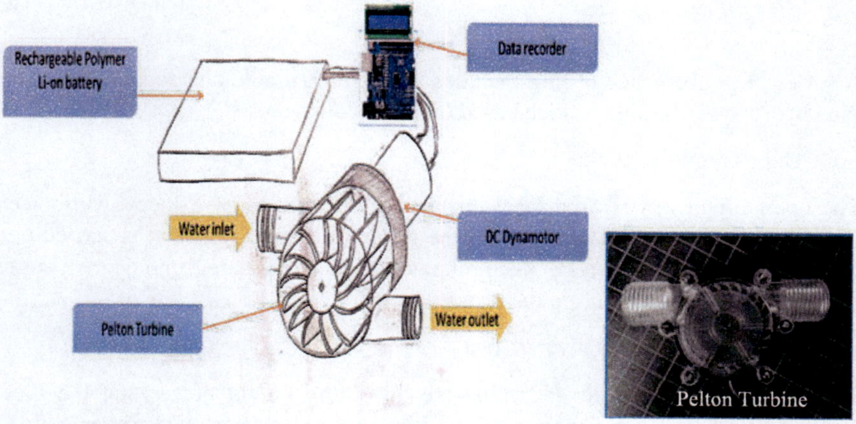

Fig. 3 The prototype of pico hydropower turbine

ground paths of the battery chargers. All the mechanism is tested for design acceptance whether it is fully functional and the voltage and current is in range with the acceptance of the parts. The Arduino UNO is programmed to utilize the software on the computer to display the Pico hydropower turbine prototype's output and the battery charging percentage. The following are the procedures of Pico hydropower turbine prototype mechanisms:

(a) When the power generated at the generator of Pico hydropower turbine, the power is set to on the programming code of the Arduino UNO to be displayed on the LCD display.
(b) The LCD displays voltage, current and power.
(c) The power generated is stored in the rechargeable polymer lithium-ion battery. The charging percentage of the battery is displayed too. The battery needs to be charged for 8 h approximately to be full (Fig. 4).

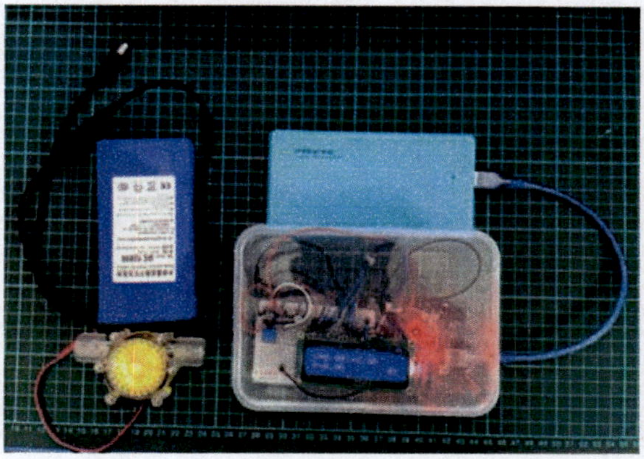

Fig. 4 Pico hydropower turbine prototype assembly

3.5 Performance Test

A series of performance testing includes all the measurable characteristics on the Pico hydropower turbine is done based on the following tests [13, 14].

(a) Open Circuit Test

The open-circuit test is used to evaluate the performance of a Pico hydropower turbine while no load is connected to the generator output. The test is carried out using regular distributed water. The output voltage is measured immediately after the water tap is opened during the test.

(b) Maximum Power Delivered Test

The goal of the test is to determine the maximum output power that the Pico hydropower turbine can generate when utilizing regular distributed water. A fluctuation in water flow rate is employed as the technique. The maximum output power is measured at the highest water flow rate.

(c) Battery Charging Ability Test.

This test is conducted to assess the Pico hydropower turbine's capacity to store energy. A 12.6 V, 8000mAh polymer lithium-ion rechargeable battery was thus put to the test. The time it takes for the battery to reach its full capacity is recorded.

4 Result and Discussion

The prototype of Pico hydropower turbine can generate a 11.52 V DC as a maximum output voltage during open circuit test as shown in Fig. 5. From the plotted graph, the output voltage for the open circuit test range is between 11.30 V DC and 11.52 V DC when no load is connected. The overall voltage range is around 11.39 V DC. Some power is loss in the mechanical system where the shaft rotates to produce electromagnetisms to generate electricity in the generator.

The test were conducted with different flow rates by employing various angle of normal water distribution tap. A series of test were performed three times a day and Fig. 6 shows the results obtained from the maximum power delivered test and it can be seen clearly that at flow rate of 9.84 L/min, the system produce maximum voltage output. The system also can generate power even at the low water flow 3.57 V DC at 4.15 L/min. From Fig. 6, we can conclude that the water flow rate can affect the power delivery of the prototype.

The power generated from the turbine is stored in the rechargeable 12.6 V, 8000mAh polymer lithium-ion battery. The LCD displays the battery percentage during charging the battery. It takes approximately 8 h for the battery to charge from 0 to 100% completely. The current generated by the turbine also must be taken into considerations. Current is vital in determining the ability of the turbine to charge the

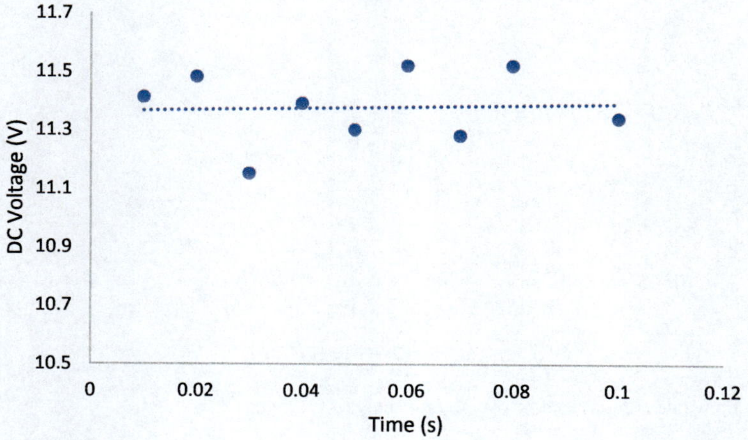

Fig. 5 The DC voltage and time

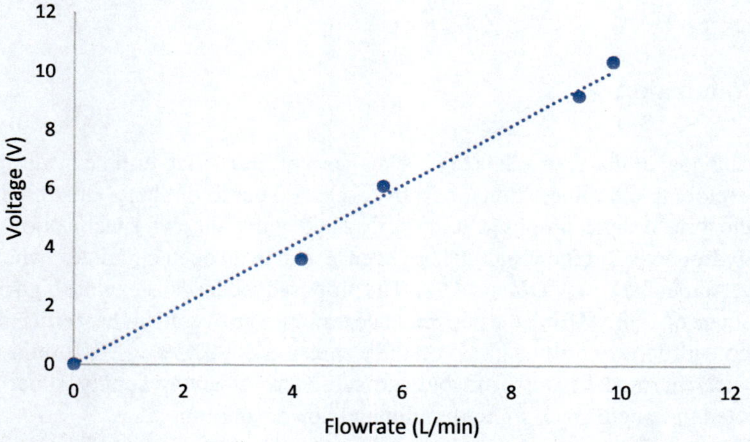

Fig. 6 The changed of voltage output over water flow rate

battery. When there is no load during open circuit test the current reading is 4.2 A. When the load is connected, the reading drops to 2.22 A as some of the current is used to charge the battery. The Pico hydropower turbine's efficiency can be determined by calculating the input and output power, as shown in Fig. 7. The overall highest efficiency of the pico hydropower turbine based on the test results is 90.45%. This efficiency is valid for 9.84 L/min water flow rate.

The Pico hydropower turbine's efficiency was also tested for different flow rates. As can be seen from Fig. 7, the efficiency is linearly related to the flowrate up to 9.84 L/min where it achieve the 90.45% and the lowest efficiency is about 31.37%

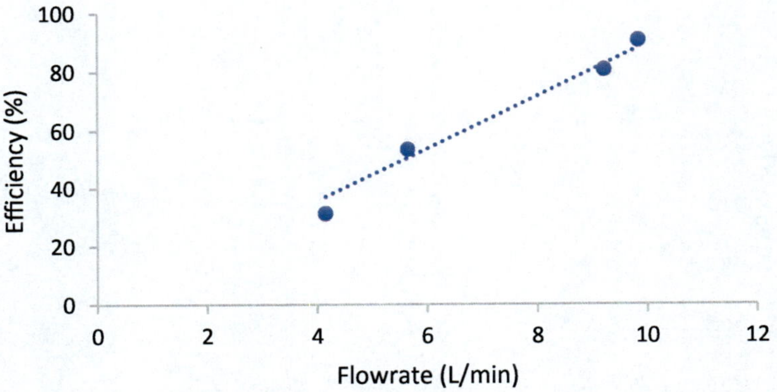

Fig. 7 Flow rate (L/min) against overall efficiency (η %)

for 4.15 L/min. The minimum flowrate is 4.15 L/min as any flowrate below this value cannot generate any electricity.

5 Conclusion

The main goal of this research is to develop a prototype of Pico turbine by designing and developing a turbine. A final design is selected based on the evaluation matrix from the three designs proposed at the early stage with their own specifications. The Pico Hydropower turbine prototype has been tested on an open circuit test where the average output is 11.37 V DC and 4.2A. The proposed model can generate high power and voltage of 11.37 V from the performance testing. The overall highest efficiency of the pico hydropower turbine based on the test results is 90.45% at 9.84L/min water flow rate. The model is also suitable and sufficient for giving supply to networks connected in remote areas without traditional power supply.

Acknowledgements The author would like to acknowledge Universiti Teknologi Malaysia for providing continuous support and funding under the UTM prototype research vote number Q.J130000.2851.00L24.

References

1. Khan MJ, Iqbal MT, Quanicoe JE (2008) River current energy conversion systems: progress, prospects and challenges. Renew Sustain Energy Rev (12):2177–2193
2. Gaiser K, Erickson P, Stroeve P, Delplanque JP (2016) An experimental investigation of design parameters for pico-hydro Turgo turbines using a response surface methodlogy. Renew Energy 85:406–418

3. Wiliamson SJ, Strark BH, Booker JD (2013) Performance of a low-head pico-hydro Turgo turbine. Appl Energy (102):1114–1126
4. Mohibullah M, Amran MR, Mohd Iqbal AH (2004) Basic design aspects of micro hydro power plant and its potential development in Malaysia. In: Proceedings of the IEEE conference on national power & energy, pp 220–223
5. Kaldellis JK (2007) The contribution of small hydro power stations to the electricity generation in Greece: technical and economic considerations. Energy Policy 35(4):2187–2196
6. Oh TH, Pang SY, Chua SC (2010) Energy policy and alternative energy in Malaysia; issues and challenges for sustainable growth. J Renew Sustain Energy Rev 14:1241–1252
7. Martin A, Brian K, Sam A (2010) Remote community electrification in Sarawak. Malays J Renew Energy (35)L1609–1613
8. Ahmada S, Kadir MZAA, Shafiea S (2011) Current perspective of the renewable energy development in Malaysia. Renew Sustain Energy Rev 15:897–904
9. Paish O (2002) Micro-hydropower: Status and prospects. Proc Inst Mech Eng Part A J PO (216):31–40
10. Hidayat MN et al (2020) Design and analysis of a portable spiral vortex hydro turbine for a Pico Hydro Power Plant. In: IOP conference series: materials science and engineering (732)
11. Hu Z et al (2020) Performance of vertical axis water turbine with eye-shaped baffle for pico hydropower. Front Energy 1–14
12. Fraenkel P, Paish O, Bokalders V, Harvey A, Brown A, Edwards R (1991) Micro-hydro power: a guide for development workers. In: In: Fraenkel P (ed) Stockhholm Environment Institute. Immediate Technology Publications in association with the Stockholm Environment Institute, London
13. Haidar AMA, Senan MFM, Norman A, Radman T (2012) Utilization of pico hydro generation in domestic and commercial loads. Renew Sustain Energy Rev (16):518–24
14. Harvey A, Brown A, Hettiarachi P, Inversin A (1993) Micro hydro design manual; a guide to small-scale water power schemes. Intermediate Technology Publications
15. Williams AA, Simpson R (2009) Pico-hydro-reducing technical risks for rural electrification. Renew Energy 34(8):1986–1991
16. Zainuddin H, Yahaya MS, Lazi JM, Basar MFM, Ibrahim Z (2009) Design and development of pico-hydro generation system for energy storage using consuming water distributed to houses. Proc World Acd Sci Eng Technol (59):154–9
17. Loots MV, Dijk B, Barta SV, Vuuren J (2015) Bhagwan, A review of low head hydropower technologies and applications in a South African context. Renew Sustain Energy Rev 50:1254–1268
18. Karuppusamy P (2020) Performance analysis of multiple pico hydro power generator. J Electr Eng Autom 02:92–101

Unsteady Free Convection with Volumetric-Radiation Using LBM

Raoudha Chaabane, Abdelmajid Jemni, Nor Azwadi Che Sidik, and Hong Wei Xian

Abstract In the last decade, coupled convection and radiative engineering problems has emerged as a major area of study in the engineering applications and sciences. The present paper deals with a transient (steady and unsteady) combination between free convection and volumetric radiation in a closed enclosure in thermal incompressible fluid flows in a bi-dimensional, differentially heated enclosure with emitting, scattering and absorbing volume. In the equation of energy, the radiative heat flux was simulated using the CVFEM, (Control- Volume-Finite-Element-Method) in the RTE (Radiative Transfer Equation). The Navier–Stokes equations which describe natural convection were solved with a D2Q9 temperature and velocity flow distributions using Lattice Boltzmann BGK Method. Separate particle distribution functions in the LBM were used to calculate the density, velocity and thermal fields. The obtained findings in the current work were compared with literature and good validation are demonstrated. The streamlines, the horizontal and vertical velocities, midline temperature, and the isotherms were highlighted and discussed under various parameters such as Rayleigh number, convection radiation number, in both steady and also unsteady cases. The LBM-CVFEM was ascertained to be a reliable numerical technique for future engineering applications dealing with transient coupling heat transfer problems.

Keywords Lattice-Boltzmann · Incompressible navier–stokes equations · Transient convection-radiation · CVFEM

R. Chaabane (✉) · A. Jemni
Laboratory of Thermal and Energetic Systems Studies (LESTE), National School of Engineering of Monastir, University of Monastir, Monastir, Tunisia
e-mail: raoudhach@gmail.com

N. A. C. Sidik · H. W. Xian
Malaysia—Japan International Institute of Technology (MJIIT), Universiti Teknologi Malaysia, Jalan Sultan Yahya Petra, 54100 Kuala Lumpur, Malaysia
e-mail: azwadi@utm.my

M. Y. Ismail et al. (eds.), *Technological Advancement in Mechanical and Automotive Engineering*, Lecture Notes in Mechanical Engineering,
https://doi.org/10.1007/978-981-19-1457-7_42

Abbreviations

List of Symbols

c	Speed of lattice
e	Internal-energy
F	Term of external force
G	Acceleration of gravity
H	Height/width of the enclosure
Nu	Nusselt-number
Pr	Prandtl-number
R	Constant of ideal gas
Ra	Rayleigh-number
RC	Convection Radiation coefficient, $\sigma H T_{ref}^4 / kt_c$
$\vec{u}$	Vector of macroscopic velocity
$\vec{q}_R$	Radiative—heat flux (Wm^{-2})
c_p	Specific—heat capacity ($m^2s^{-1}\ K^{-1}$)
k_T	Thermal—conductivity ($Wm^{-1}\ K^{-1}$)
I	Radiative—intensity ($Wm^{-2}sr^{-1}$)
S	Position (m)
k_a	Absorption parameter (m^{-1})
k_d	Scattering parameter (m^{-1})
$I_b(s)$	Black-body radiative intensity ($Wm^{-2}sr^{-1}$)
$\vec{n}_w$	Wall unit normal vector
N_φ	Polar directions control angles number
N_{θ_p}	Azimuthal direction control angles number
$\Delta\Omega^{mn}$	Control-solid angle
ΔV_{ij}	Control-volume
δV_{ij}	Sub-volume
N	Node
n	Refractive index of the medium (n = 1)
G_l	Centroids
x, y	Direction
$\Delta x, \Delta y$	Regular steps
L_x	The x-dimensions of the calculation -domain
L_y	The y-dimensions of the calculation -domain
A^N	Surface of the control -volume ΔV_{ij}
$\Phi^{mnm'n'}$	Averaged scattering phase function
$\nabla.\vec{q}_R$	Radiative information
G	Incident radiation (Wm^{-2})
f_k	Velocity Particle Distribution Function PDF (ms^{-1})
f_k^{eq}	Velocity Equilibrium Particle Distribution Function (ms^{-1})
g_k	Thermal Particle Distribution Function (K)
g_k^{eq}	Thermal Equilibrium Particle Distribution Function (K)

$\vec{r}$	Lattice node
t	Time parameter (s)
Δt	Time step (s)
k	Direction
c_k	Discrete velocity vector
$\Delta\vec{r}$	Lattice link
$T(\vec{r}, t)$	Temperature (K)
T_m	Mean temperature
T_c	Cold temperature
T_h	Hot temperature

Greek Symbols

Ω_k	Collision operator
ξ	Dimensionless time
θ	Non-dimensional temperature
α	Thermal -diffusivity
β	Extinction -coefficient (m^{-1})
β_T	Thermal–expansion coefficient
ν	Kinematics -viscosity
τ_T	Thermal relaxation-time
τ_v	Velocity relaxation-time
ω	Albedo
σ	Constant of Stefan-Boltzmann (Wm^{-2} K^{-4})
τ	Relaxation -time (s)
w_k	D2Q9 lattice weights
ρ	Density (kgm^{-3})
$\Phi(\overrightarrow{\Omega}' - \overrightarrow{\Omega})$	Scattering phase-function
$\overrightarrow{\Omega}$	Outgoing-direction of propagation
$\overrightarrow{\Omega}'$	Incoming-direction
ε_w	Wall-emissivity
θ_p	Polar-angle
φ	Azimuthal-angle

Subscript

b	Black-body
ref	Reference-value
eq	Equilibrium

1 Introduction

Combined natural convection and radiation heat transfer have numerous applications in the area of high temperature heat exchangers, industrial furnaces, fibrous and foam insulations, combustors, space-crafts, glass processing, electronics cooling, optical textile fiber processing, building comfort, thermal insulation and fire protection [1–28]. Several researches which related to convection were reported either without [1–14] or with [15–30] radiation heat transfer. Dixit et al. [1] study the effect of high Rayleigh number on free convection using the mesoscopic approach, namely the Lattice Boltzmann Method (LBM). A non-uniform mesh for the simulation of natural convection in a square cavity was introduced in [2]. Seta et al. [3] highlight an LBM calculation of free convection inside porous-media. Different geometries subjected to convection phenomena were explored in [4–14]. In other side, radiation effect was analysed in [15–30] where authors study combined surface or volumetric radiation with convection on different geometries either inside two dimensional or three-dimensional enclosures.

Nowadays, research of the analysis of numerical resolution of solar collectors, printed circuit boards, heat exchange and fluid flow phenomena in the presence of radiation, becomes a crucial tool for the numerical simulations proving valuable complement to experimental investigation in many problems in industry.

In recent past, the (LBM) Lattice Boltzmann Method has evolved as a good alternative numerical model for the resolution of many engineering problems [31–40]. The proposed approach has accomplished exceptional achievement in a wide engineering area, for instance, complex fluid phenomena and multiphase flow. Therefore, the used approach that has the prospective to become a useful Computational Fluid Dynamics (CED) platform that is higher over the current continuum based CFD tools was applied for calculating the convective part of radiative associated to the convective heat transfer. In fact, LBM receives countless of the benefits of kinetic theories and molecular-dynamics, due to its microscopic derivation. The advantages of LBM take account of efficient implementation for parallel computation, simple calculation procedure and stability, easy handling of boundary conditions inside complex geometries [31–38]. It is known that, LBM is considered as second order accurate in space and also in time. Besides, boundaries conditions can be appropriately and suitably treated which makes this approach very competitive for complex geometries.

This numerical technique has been applied successfully to simulate fluid flow different engineering problems. He et al. [31], Succi [32], Benzi et al. [33], Chen et al. [34], Het et al. [36], Qian et al. [36] and Mohamad [37] introduce the new model based on temperature for this approach in their important works. In their paper [35], Chaabane et al. present a numerical study of boundary treatments in the case of heat transfer conduction via LBM.

Knowing that, those mesoscopic methods are in high speed development and present a novel influential method for fluid flow simulation and especially an efficient tool to transport problems multiphase and single flows [31, 32]. Moreover, this proposed tool has been used in a varied range of engineering applications of free convection without dealing subjected to volumetric thermal radiation [36–38].

Combined steady-state and transient free convection associated with surface radiation in an open enclosure was analyzed by Hinojosa et al. [15]. Conductive convective radiative problem inside a slot was examined by Balaji et al. [16]. Ha et al. [17] studied the interaction between free convection and conduction in a differentially heated cubic cavity with the presence of heat-generating from a cubic conducting bulk. Radiation effect on free convection inside a square geometrical cavity was explored by Yucel et al. [41]. Sediki et al. [19] analyzed numerically the coupling between gas-radiation and a laminar mixed case of convection inside a vertical-circular tubes. The effects of surface radiation in side-vented open cavities on natural convection were investigated by Singh et al. [20]. Ridouane et al. [42] evaluated the effect of radiation on free convection in an enclosure having a square shape and heated from below. Sharma [22] considered the interaction of surface-radiation conjugated to turbulent free convection in an open cavity filled with air.

Radiative convective study with a more complicated external excitation, namely, a periodic temperature were investigated by Singh and Venkateshan [23]. Free-convection stability state in the presence of surface radiation in a cavity with emitting diffusely boundary wall, was investigated by Janallah et al. [24].

Bianco et al. [28] highlight the effect of radiative excitation on an engineering problem dealing with a vertical-convergent and symmetric heated channels. Dixit and Babu [1] extended the application of the thermal Lattice Boltzmann Method based on the BGK approach to simulate a problem that resolve a free convection analysis in a square cavity with a high Rayleigh number. A non-uniform mesh was applied in a specific LBM engineering study by Kuznik et al. [2] in order to simulate a free convection case inside a square enclosure aiming to reduce the computational cost. Jami et al. [7] used the mesoscopic approach to highlight the natural convection inside a cavity containing a heat generating cylindrical obstacle. Also, radiative convective test containing a bulk obstacle was piloted by Mezrhab et al. [43]. In their paper, they applied the FVM (Finite-Volume-Method) in order to solve the related governing differential-equations.

Mondal et al. [27, 28] studied the effect of volumetric-radiation on the natural convection in a square shaped enclosure using a mesoscopic approach model based on lattice Boltzmann method with a non-uniform lattices exhausting an hybrid solver. In their study, LBM was used to calculate the velocity-field and also to solve the density-fields in a participating media and the FVM is applied to solve radiative information. However, combined convection volumetric radiation in a 2D enclosure has been studied explicitly in Banda et al. [44]. Where, they used a non-oscillatory relaxation scheme in the LBM with the DOM (Discrete-Ordinate-Method) to calculate the radiative-information.

Tan and Howell [45] investigated volumetric radiation effect on natural convection in a two dimensional cavity containing an emitting, absorbing and scattering medium. In their paper, radiant transport is analyzed based the exact integral formulation. While energy is solved using a finite difference method. The product-integral method was used to solve the momentum equations. Then, a non-linear SOR tool is used to resolve the obtained algebraic equations. In their work, a high moderate Rayleigh number was applied ranging from 103 to 105. A specific radiation-conduction number is applied varying from zero to infinity.

Besides, CVFEM showed a successful trend in solving radiative information inside multi-dimensional cavities, in addition to axisymmetric and non-axisymmetric radiative engineering problems in more real geometry. This numerical approach has the prospective to become a useful talented tool for engineering problems including radiative transfer effects in multi-dimensional complex engineering geometries with in structured and unstructured grids and meshes. Those medium can be an anisotropic or an anisotropic one. As detailed in [37, 38, 46–48], this approach highlight many advantages.

From literature review, the researchers consider surface radiation and basic steady cases of the domain and thermal behavior inside the cavity. However, in many engineering applications, the energy provided to the system varies with time and gives rise to unsteady free convection flow associated with radiative effect. In addition, thermal and dynamical behaviors of a fluid subjected to a given complex excitation in unsteady cases are not easy to predict on the basis of the results obtained for a steady sate. The aim of this work is the study of coupling volumetric radiation and natural convection in cavities in unsteady (transient) cases.

In this paper, we focus on the multidimensional transient natural convective and radiative transfer, which is an active research subject because of their practical and interesting engineering applications. This problem involves the resolution of the Navier–Stokes equations and the radiative transfer equation (RTE).

The constitutive medium in the hole of these schemes dynamically contributes in the radiative transfer owing to the absorption, scattering and emission of radiation.

It is proved that the directional nature of the (IRF) Intensity Radiation Field increase the computational cost of the RTE determination. Knowing that, this difficulty can borders detail in the numerical process, we aim to improve an innovative hybrid numerical method with a fundamental analysis of these complex phenomena in an unsteady mode.

In order to improve the computational efficiency and to highlight the step change in temperature and velocity profiles, the LBM has been proposed in order to analyze the transient free convection coupled to volumetric radiation in a two-dimensional cavity having a scattering, absorbing and emitting medium.

The radiative information is solved via CVFEM. Many results were discussed presenting the mid-line temperature, the streamlines, the vertical and the horizontal velocities and the isotherms and this for various parameters such as conduction–radiation parameter and Rayleigh number. Two issues were critically examined. We discussed the combined steady case and the transient case of the free convection coupled volumetric radiation. Besides the obtained finding are compared with literature aiming to highlight the accuracy and the efficiency of the proposed LBM-CVFEM numerical tool.

2 Formulation

Let us consider a two dimensional isotopically scattering, emitting and absorbing medium confined by two horizontal non-conducting walls, and two vertical isothermal boundaries at different temperature, T_h, and T_c, ($T_h > T_c$) respectively (Figs. 1 and 2). We consider and assume constant thermo-physical properties, except

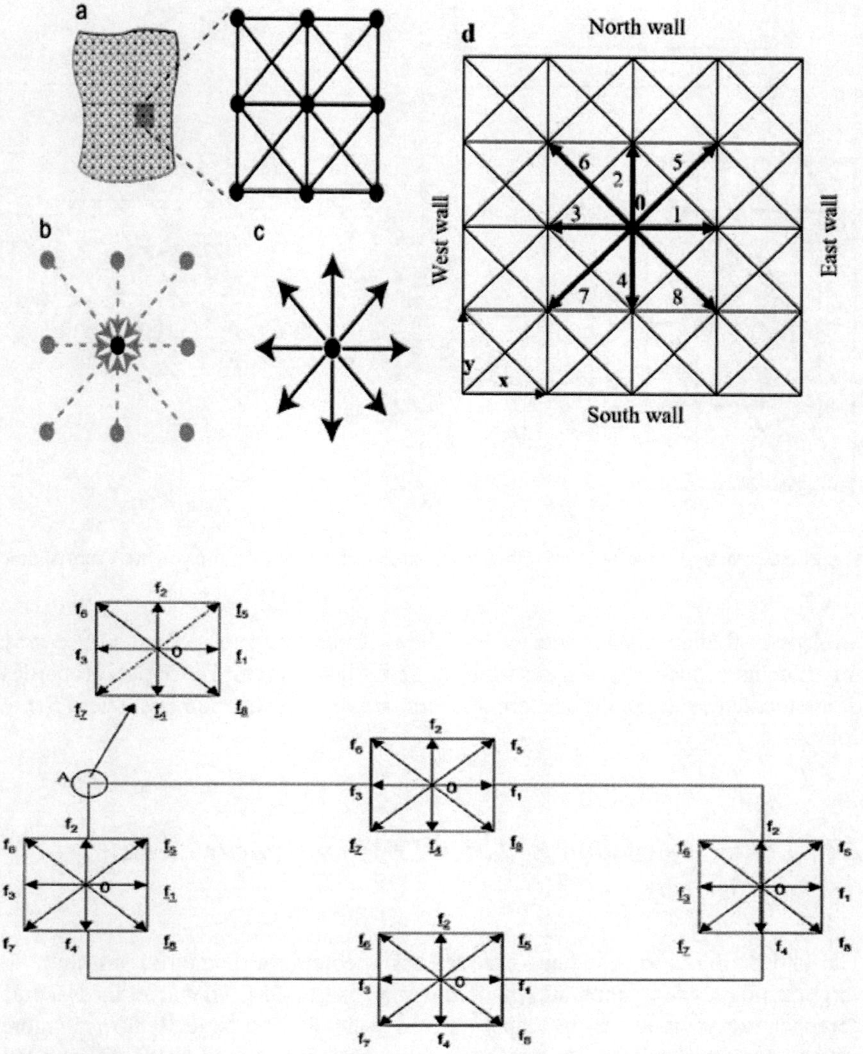

Fig. 1 **a** Part of (D2Q9) lattice, **b** streaming steps, **c** propagation step, **d** D2Q9 lattice diagram, **e** boundary conditions with unknown and known populations

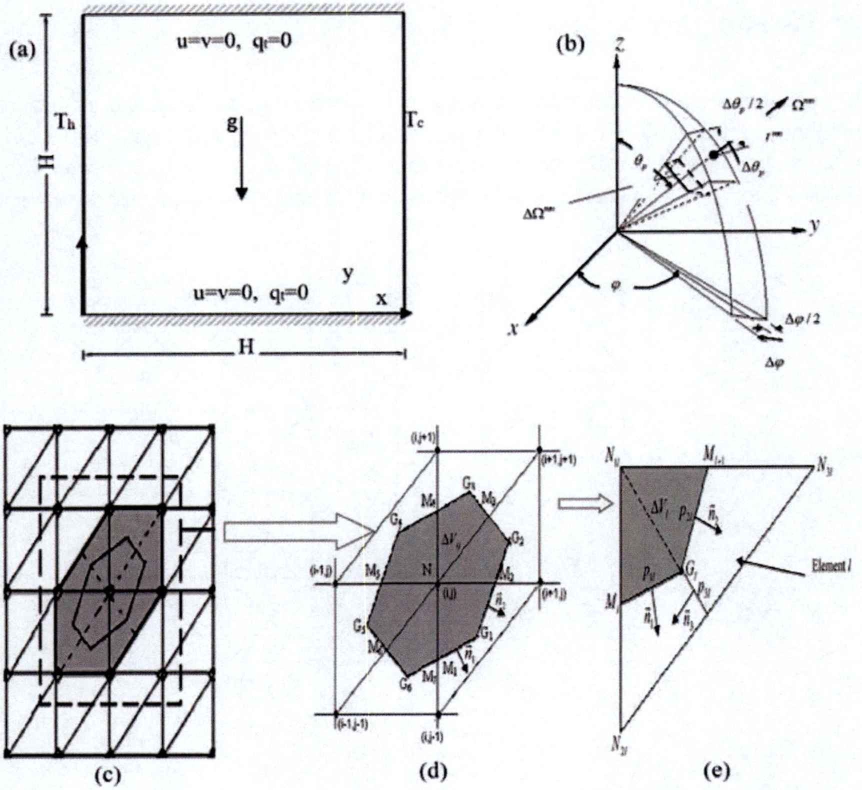

Fig. 2 **a** Geometry, **b** angular discretization, **c** spatial discretization, **d** control volume, **e** sub-volume

for density. Radiatively, the side walls of the enclosure were considered diffuse-gray for a homogeneous scattering, absorbing and emitting medium. The optical properties of the medium, such as the scattering albedo ω and the extinction coefficient β, are constant.

2.1 *Lattice Boltzmann Equation LBM for Velocity Fields and Density*

The lattice Boltzmann method (L B M) has becoming a thoughtful substitute to CFD techniques for computational fluid dynamics [31–38]. The novel mesoscopic approach is especially very well appropriated to simulate numerically flows in more realistic and complex geometries. Moreover, it is forthrightly implemented on parallel computing machines [32]. It is known that, LBM tools are developed from the theory of the lattice gases t (Fig. 1), even though it can also be derived straight from the simplified approach of the Boltzmann B G K equation [33].

In order to compute dynamic and thermal fields with LBM, we use a two different PDFs.

The governing lattice Boltzmann equation used in order to compute the density and velocity fields are based on the PDF f_k written as [1, 31, 32],

$$\text{For } k = 0, .., 8 : f_k(\vec{r} + \vec{c_k}\,\Delta t, t + \Delta t) = f_k(\vec{r}, t) + \Omega_k \tag{1}$$

On the right-hand side of Eq. (1), the collision term Ω_k highlights the single time relaxation approach Bhatangar Gross Krook (BGK) calculation [33] given by the following formulation:

$$\Omega_k = \Delta t\, F - \frac{\Delta t}{\tau_v}[f_k(\vec{r}, t) - f_k^{eq}(\vec{r}, t)] \tag{2}$$

where τ_v is the dynamic relaxation time and f_k^{eq} is the (LEDF) Local Equilibrium Distribution Function and F is the external force which is given by,

$$F = \left[\frac{\vec{G}.(\vec{c_k} - \vec{u})}{RT}\right] f_k^{eq} \tag{3}$$

In Eq. (3), $\vec{G}$ is the external force that acts per unit mass in the direction opposite to gravity and is written as [31]:

$$\vec{G} = \rho g\, \beta_T\, (T - T_m)\, \vec{j} \tag{4}$$

The boussinesq density ρ is explained as:

$$\rho = \overline{\rho}[1 - \beta_T(T - T_m)] \tag{5}$$

where T_m is the mean temperature, $\overline{\rho}$ is the density of the fluid, g represents the acceleration of the gravity while and β_T is the volumetric thermal-expansion coefficient.

For the specific D2Q9 lattice (Fig. 1b) highlighted in this paper, we compute the relaxation time τ_v as [31–39]

$$\tau_v = \frac{3\nu}{c^2 dt} + \frac{1}{2} \tag{6}$$

where R is the ideal gas constant and the lattice speed is defined as $c = \sqrt{3RT}$.

We can deduce the kinetic viscosity, ν, after solving the following equations

$$Ra = \frac{g\beta_T \Delta T H^3}{\alpha\nu} \tag{7}$$

$$g\beta_T = \frac{u^2}{\Delta T H} \tag{8}$$

$$\Pr = \frac{\nu}{\alpha} \tag{9}$$

where Pr is Prandtl coefficient and Ra is the Rayleigh coefficient. T_c is the cold wall temperature and T_h is the hot wall temperature. $\Delta T = (T_h - T_c)$ and H is height and the enclosure. g is the acceleration generated by gravity and $u = Ma\ c/\sqrt{3}$ [31].

The velocities in the specific D2Q9 (Fig. 1b) lattice are written as:

$$c_0 = (0, 0) \tag{10}$$

$$\text{For } k = 1, 4, \quad \vec{c}_k = (\cos((k-1)\pi/2), \sin((k-1)\pi/2).c \tag{11}$$

$$\text{For } k = 5, 9, \quad \vec{c}_k = \sqrt{2}(\cos((2k-1)\pi/4), \sin((2k-1)\pi/4).c, \tag{12}$$

$$c = \Delta x/\Delta t = \Delta y/\Delta t \tag{13}$$

The equilibrium function for the density distribution function is given as

$$f_k^{eq} = f_k^{eq}(\rho, u) = w_k\rho(1 + \frac{c_k u}{c_s^2} + \frac{(c_k u)^2}{2c_s^4} - \frac{uu}{2c_s^2}) \tag{14}$$

The weights are written as

$$w_0 = 4/9,\ w_{1,2,3,4} = 1/9,\ w_{5,6,7,8} = 1/36 \tag{15}$$

2.2 *LBM for Thermal Field*

Associating the volumetric radiation, the governing thermal lattice Boltzmann equation based on the thermal PDF g_k, is written as [38, 46],

$$\begin{aligned} g_k(x + \Delta x, y + \Delta y, t + \Delta t) = (1 - (\frac{\Delta t}{\tau_T}))g_k(x, y, t) + (\frac{\Delta t}{\tau_T})g_k^{eq}(x, y, t) \\ - (\frac{\Delta t}{\rho c_p})w_k \nabla.\vec{q_R} \end{aligned} \tag{16}$$

where $\overrightarrow{q_R}$ is the radiative heat flux. τ_T define the relaxation time and g_k^{eq} is the thermal EDF (Equilibrium Distribution Function) writtena as

$$g_k^{eq} = g_k^{eq}(T) = w_k T(1 + \frac{c_k u}{c_s^2}) \tag{17}$$

where c_s is the lattice speed of sound in media, and it equals to $c/\sqrt{3}$.

The internal energy, ε, the macroscopic density ρ and the velocity $\vec{u}$ are given as:

$$\rho(\vec{r}, t) = \sum_k f_k(\vec{r}, t) \tag{18}$$

$$u(\vec{r}, t) = \sum_k c_k \; f_k(\vec{r}, t)/\rho(\vec{r}, t) \tag{19}$$

$$\rho(\vec{r}, t) \; \varepsilon = \sum_k g_k(\vec{r}, t) \tag{20}$$

Finally the temperature can be calculated from state equation written as, $\varepsilon = RT$.

2.3 *Boundary Conditions*

In Lattice Boltzmann Method, a bounce-back boundary conditions [49] were applied on all solid boundaries for the density distribution functions. However, for the thermal boundary conditions, the two horizontal walls are adiabatic and the two vertical walls are maintained at constant temperatures (T_c and T_h).

2.4 *Computation of Radiative Information*

For the Navier–Stokes equations, the transient energy conservation equation consisting of conduction and radiation can be expressed as

$$\rho c_p \frac{\partial T}{\partial t} = k_T \nabla^2 T - \nabla . \vec{q}_R \tag{21}$$

It is assumed that the thermal conductivity k_T of the emitting, absorbing, scattering medium is independent of temperature.

The radiative heat flux $\vec{q}_R$ is given by:

$$\vec{q}_R = \int_{4\pi} I \vec{\Omega}\, d\Omega \tag{22}$$

where I is the radiative intensity.

The divergence of radiative heat flux is given by

$$\vec{\nabla}.\vec{q}_R = k_a(4\pi I_b - G) \tag{23}$$

where $I_b = \sigma T^4/\pi$ define the intensity of the black-body, G define the incident radiation and k_a is the absorption coefficient.

To summarize, the RTE can be rewritten as

$$\vec{\nabla}.(I(s, \vec{\Omega}).\vec{\Omega}) = k_a I_b(s) - (k_a + k_d) I(s, \vec{\Omega}) + k_a I_b(s) + \frac{k_d}{4\pi} \int_{\Omega'=4\pi} I(s, \vec{\Omega'}) \Phi(\vec{\Omega'} \rightarrow \vec{\Omega}) d\Omega' \tag{24}$$

$I(s, \vec{\Omega})$ the radiative intensity function of direction $\vec{\Omega}$ and position s. k_d is the scattering coefficient and $\Phi(\vec{\Omega'} - \vec{\Omega})$ is the scattering phase function from the incoming $\vec{\Omega'}$ direction to the outgoing direction $\vec{\Omega}$.

When the wall bounding the physical domain is assumed grey and emits and reflects diffusely, the radiative boundary condition for Eq. (24) can be expressed as

$$I_w(\vec{\Omega}) = \frac{\varepsilon_w \sigma T_w^4}{\pi} + \frac{1 - \varepsilon_w}{\pi} \int_{\vec{\Omega'}.\vec{n}_w < 0} I_w(\vec{\Omega'}) \left| \vec{\Omega'}.\vec{n}_w \right| d\Omega' \text{if } \vec{\Omega'}.\vec{n}_w > 0. \tag{25}$$

$\vec{n}_w$ is the wall unit normal vector and ε_w represents the wall emissivity.

3 The CVFEM for RTE

The CVFEM (Control Volume Finite Element Method) has been proved to be very efficacious in the solution of transient conductive radiative transfer in multidimensional cavities [46, 47], as well as for the axisymmetric radiative problems [50], and also for the solution of radiative transfer problems in complex geometries [48]. Chaabane et al. [50, 51] have analysed different benchmark problems of transient combined conductive radiative heat transfer and they demonstrating the efficiency of the numerical method. Therefore, this approach will be applied to solve the RTE in the present proposed coupling problem.

In the CVFEM, the spatial domain is divided into a finite number of control volumes inside the 2D rectangular cavity filled with gray semitransparent medium (Fig. 2a).

In angular domain, $\Delta\Omega^{mn}$ define the control solid-angle and itis given by (Fig. 2):

$$\Delta\Omega^{mn} = \int\limits_{\Delta\theta}\int\limits_{\Delta\varphi} \sin\theta \, d\theta \, d\varphi \tag{26}$$

However, spatial domain is sub-divided into three-node triangular elements (Fig. 2b).

The Radiative Transfer Equation (RTE) integrated over both solid angle and control volume gives:

$$\begin{aligned}\int\limits_{\Delta v_{ij}}\int\limits_{\Delta\Omega^{mn}} \vec{\nabla}.(I(s,\vec{\Omega}).\vec{\Omega})\, d\Omega\, dv &= -\int\limits_{\Delta\Omega^{mn}}\int\limits_{\Delta v_{ij}} (k_a + k_d) I(s,\vec{\Omega}) dv d\Omega \\ &+ \int\limits_{\Delta\Omega^{mn}}\int\limits_{\Delta v_{ij}} k_a I_b(s) d\Omega dV \\ &+ \int\limits_{\Delta\Omega^{mn}}\int\limits_{\Delta v_{ij}} \frac{k_d}{4\pi} \int\limits_{\Omega'=4\pi} I(s,\vec{\Omega'}) P(\vec{\Omega'} \to \vec{\Omega}) d\Omega' d\Omega dV\end{aligned} \tag{27}$$

The divergence theorem, the skew positive coefficient up wind (SPCU) and step schemes were used to calculate terms on the left-hand side in Eq. 27.

The final algebraic equation of the RTE is given as [37, 38]:

$$\begin{aligned}\gamma_{1ij}^{mn} I_{ij-1}^{mn} + \gamma_{2ij}^{mn} I_{i+1j}^{mn} + \gamma_{3ij}^{mn} I_{i+1j+1}^{mn} + \sum_{(m',n')=(1,1)}^{(N_\theta,N_\varphi)} \alpha_{ij}^{mnm'n'} I_{ij}^{m'n'} \\ + \gamma_{4ij}^{mn} I_{ij+1}^{mn} + \gamma_{5ij}^{mn} I_{i-1j}^{mn} + \gamma_{6ij}^{mn} I_{i-1j-1}^{mn} = \beta_{ij}^{mn}\end{aligned} \tag{28}$$

Finally, the algebraic Eq. (27) is transformed to the following matrix form [46, 47].

$$AI = b \tag{29}$$

The (CCGS) conditioned conjugate gradient squared method is used to resolve the obtained matrix system [48, 51].

The obtained coupled equations are solved iteratively in order to yield the temperature and the radiation fields.

4 Results—Discussions

4.1 Transient Free Convection

We deal with a free convection in H wall high cavity subject to a horizontal temperature difference walls as shown in the Fig. 2a. The south and north horizontal boundaries are insulated. The right and left vertical walls are subjected to cold temperature and hot temperature, respectively.

Obtaining finding for the free convection without volumetric radiation are validated against literature results [1, 52–55].

To quantify the results, data of LBGK solution are listed in Table1. The comparison includes the average Nusselt number Nu_{av} along the hot wall, the maximum

Table 1 Simulation results of free convection flow

	$u_{max}/V(y/H)$ for $x/H = 0.5$	y_{max}	$v_{max}/V(x/H)$ for $y/H = 0.5$	x_{max}	Nu_{av}
$Ra = 10^3$					
Present work	3.7084	0.8125	3.7165	0.1796	1.1231
De Vahl Davis	3.649	0.813	3.697	0.178	1.118
Hortmann et al	(–)	(–)	(–)	(–)	(–)
Barakos et al	4.0768	0.806	4.1301	0.181	1.114
A D'Orazio et al	3.653	0.812	3.700	0.179	1.11
$Ra = 10^4$					
Present work	17.1192	0.8203	19.7022	0.1171	2.2417
De Vahl Davis	16.178	0.823	19.617	0.119	2.243
Hortmann et al.	161,802	0.8265	19.6295	0.1193	2.2442
Barakos et al.	16.262,470	0.818	19.71719	0.119	2.245
A D'Orazio et al.	16.237	0.820	19.680	0.117	2.23
$Ra = 10^5$					
Present work	35.7812	0.8515	69.30,993	0.0703	4.5696
De Vahl Davis	34.73	0.855	68.59	0.066	4.519
Hortmann et al.	34.7399	(0.8558	68.6396	0.0657	4.5216
Barakos et al.	35.1725	0.859	68.7462	0.066	4.510
A D'Orazio et al.	34.822	0.852	68.712	0.063	4.504
$Ra = 10^6$					
Present work	64.74174	0.8,515,625	221.5886	0.0390	8.94,467
De Vahl Davis	64.63	0.850	219.36	0.0379	8.800
Hortmann et al.	64.8367	0.8505	220.461	0.0390	8.8251
Barakos et al.	64.8813	0.859	220.7651	0.039	8.806
A D'Orazio et al.	64.867	0.852	221.186	0.039	8.76

horizontal velocity u_{max}/V in the vertical mid-plane $x/H = 0.5$, maximum vertical velocity $v_{max}(x/H)$ in the horizontal mid-plane $y/H = 0.5$ for different moderate and high Rayleigh numbers ($Ra = 10^3 - Ra = 10^6$). A very good concordance is achieved.

In Figs. 3 and 4, the steady state dimensional mid-plane temperature are highlighted for different Rayleigh number. The horizontal velocity and the vertical velocity are plotted in Figs. 5 and 6 respectively.

Local Nusselt-number, Nu (Fig. 7), and its averaged value, Nu_{av}, are used to evaluate the heat transfer rate along the hot wall, and they are written as

$$Nu(y) = \frac{H}{T_h - T_c}\left(uT - \frac{\partial T}{\partial x}\right)_{x=0} \tag{30}$$

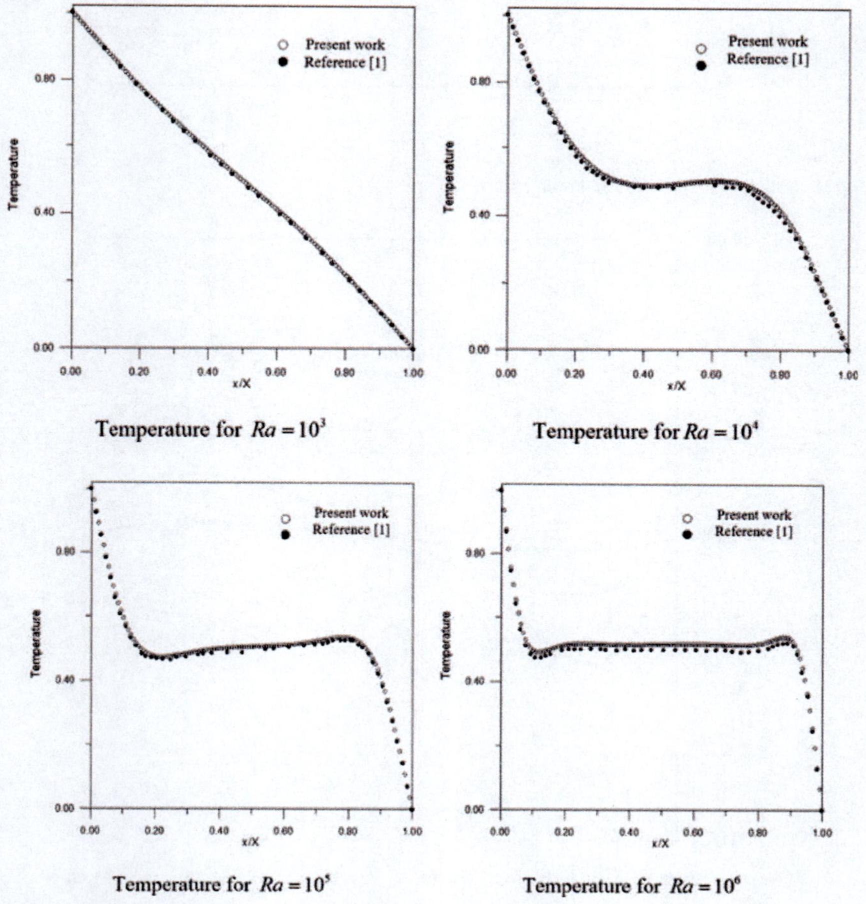

Fig. 3 Steady state dimensional midplane temperature for different Rayleigh number

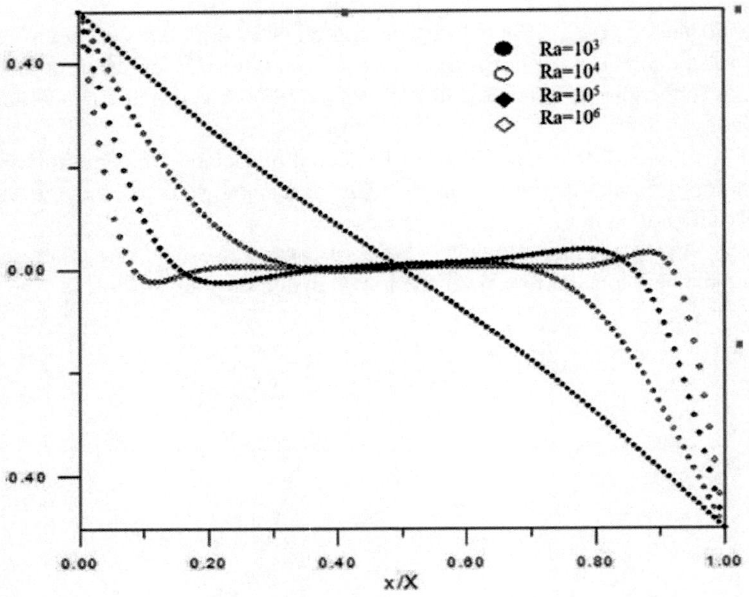

Fig. 4 Steady state dimensional temperature

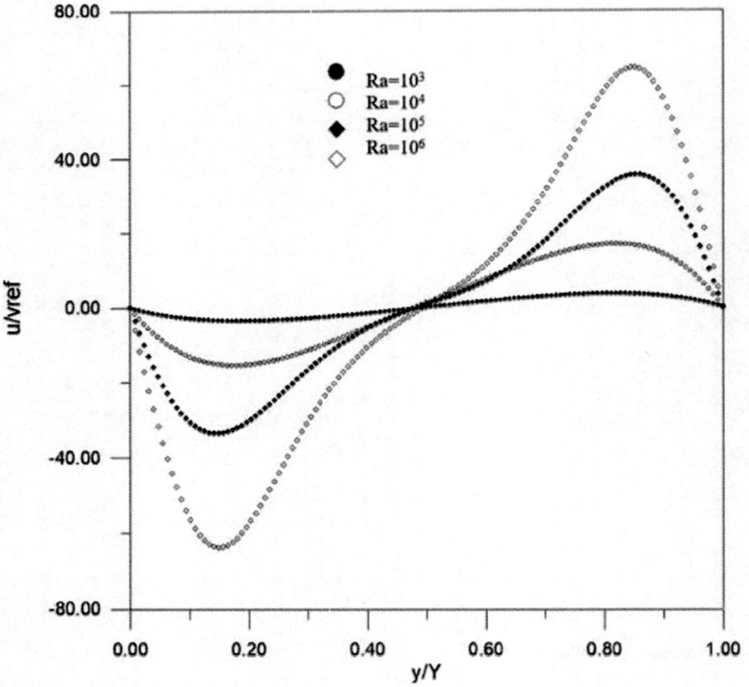

Fig. 5 Horizontal velocity plots u in the vertical mid plane for $\Delta T = 1$

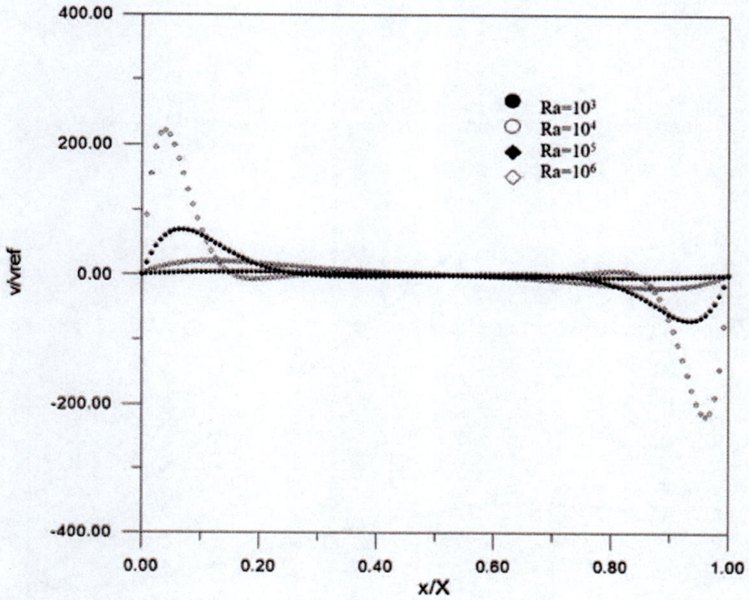

Fig. 6 Vertical Y-velocity plots for $\Delta T = 1$

$$Nu_{av} = \frac{1}{H}\int_{y_1}^{y_2} Nu(y)\,dy \tag{31}$$

The steady state streamlines lines and the isotherms patterns are plotted in Figs. 8 and 9 respectively, in the absence of volumetric radiation for $Ra = 10^3$,$Ra = 10^4$,$Ra = 10^5$ and $Ra = 10^6$.

For $Ra = 10^3$, we notice that the steady state streamlines are uniformly spaced throughout the whole cavity and the fluid flows from the hot side to the cold wall side.

The temperature is slightly higher at the top north side of the hot wall and decreases gradually towards the cold south wall. The non-dimensional temperature at the center of the cavity is approximately 0.5. For a larger Rayleigh number of $Ra = 10^4$.

The steady state streamlines are uniformly spaced within the enclosure and the steady state isotherms are horizontal in the middle of the medium.

The steady state isotherms are very dense at the bottom of the hot wall and at the top of the cold wall. Similar results are reported in literature [1, 52–55] as well. The non-dimensional temperature is approximately 0.5 at the center of the cavity.

For $Ra = 10^5$ and without radiation, the streamlines and the isotherms are observed similar along the vertical midlines. Similar profiles have been reported in literature [1, 52–55]. The temperature field is stratified with horizontal isotherms in the core region of the cavity. But the isotherms are very dense near the bottom

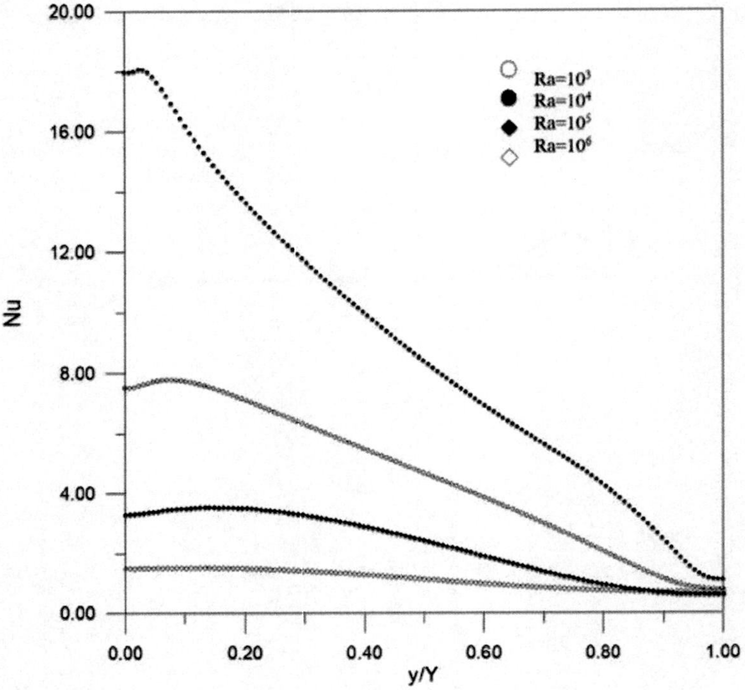

Fig. 7 Local Nusselt numbers at the hot wall for different Rayleigh number

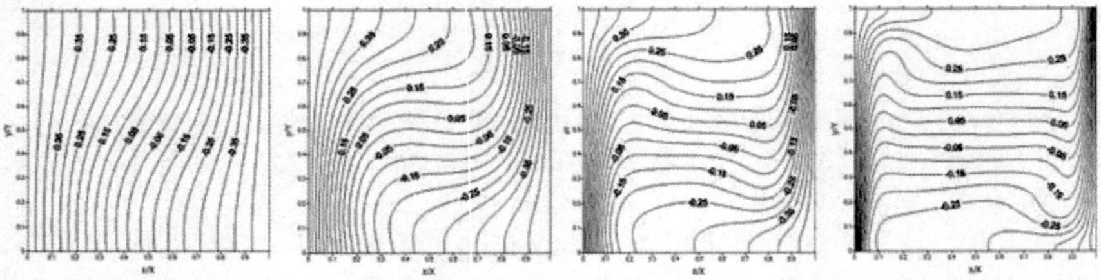

$Ra = 10^3$ $Ra = 10^4$ $Ra = 10^5$ $Ra = 10^6$

Fig. 8 Steady state isotherms for various Rayleigh number

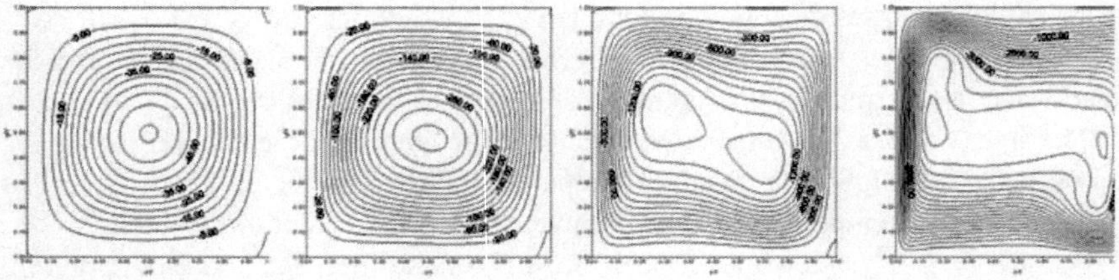

$Ra = 10^3$ $Ra = 10^4$ $Ra = 10^5$ $Ra = 10^6$

Fig. 9 Steady state streamlines for various Rayleigh number

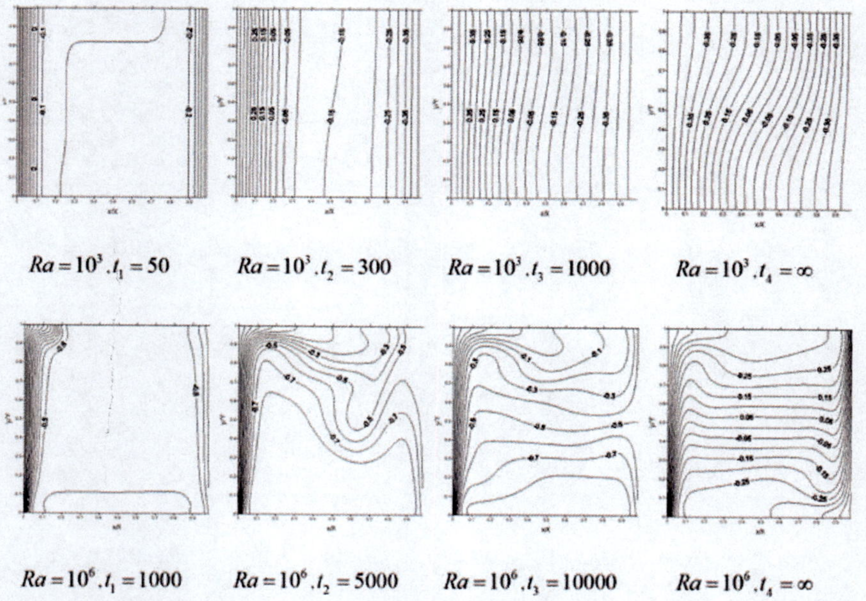

Fig. 10 Transient progression of isotherms for $Ra = 10^3$ and for $Ra = 10^6$

of the hot wall and the top of the cold wall. In the case of $Ra = 10^6$ and without radiation, the streamlines and the isotherms are observed similar along the vertical midlines. Similar profiles have been reported in literature [1, 52–55].

The temperature field is stratified with horizontal isotherms in the core region of the cavity. But the isotherms are very dense near the bottom of the hot wall and the top of the cold wall. The transient evolution of isotherms for $Ra = 10^3$ and $Ra = 10^6$ are depicted in Fig. 10. The temporal evolution of the coupled transient convection volumetric radiation is shown in Fig. 11 where we present the transient evolution of streamlines for $Ra = 10^3$, $Ra = 10^4$, $Ra = 10^5$ and $Ra = 10^6$.

4.2 Transient Natural Convection-Radiation

In this part of our work, we consider radiation effect in a non-scattering participating fluid where total reflection is imposed at the top and bottom cavity walls. The left and right cavity walls are assumed to be diffusive to radiation. We consider the following parameters Prandtl number $Pr = 0.7$ and absorption coefficient $\omega = 0$ with Rayleigh number $Ra = 10^4$.

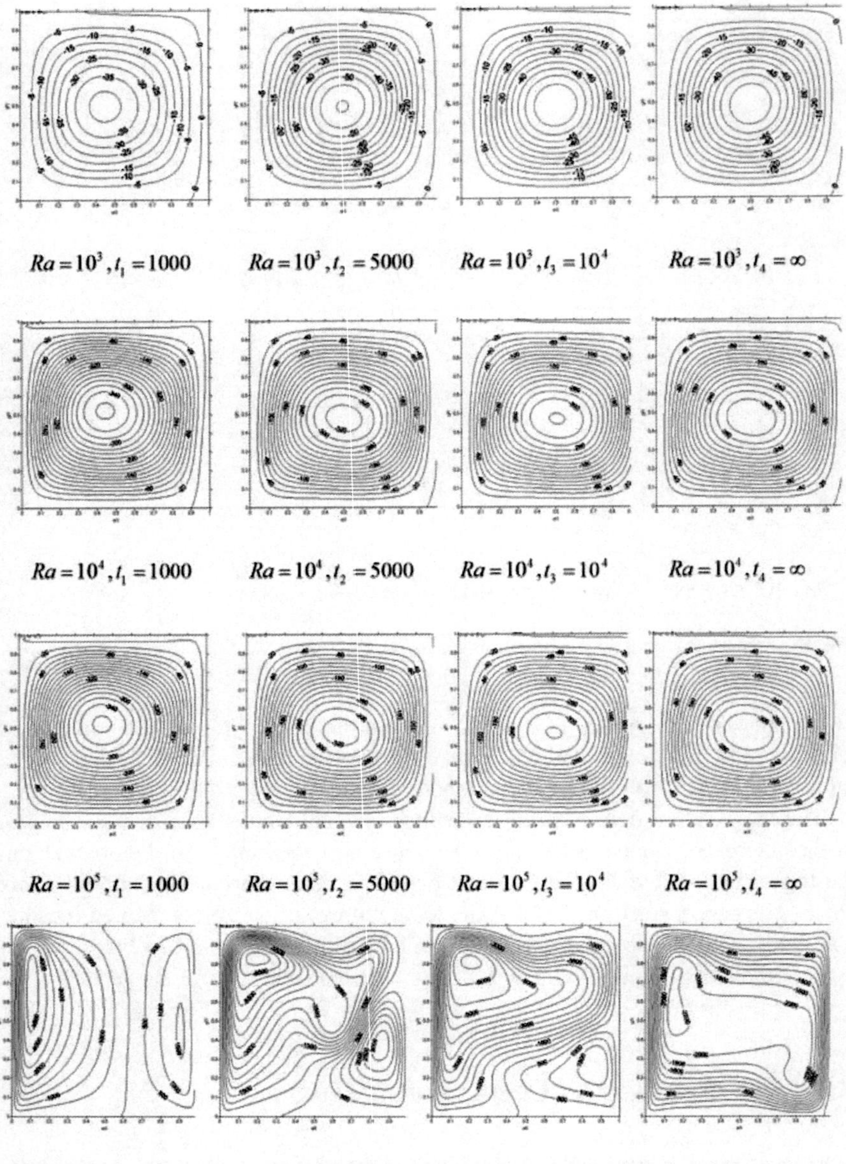

Fig. 11 Transient evolution of streamlines for $Ra = 10^3$, $Ra = 10^4$, $Ra = 10^5$ and $Ra = 10^6$

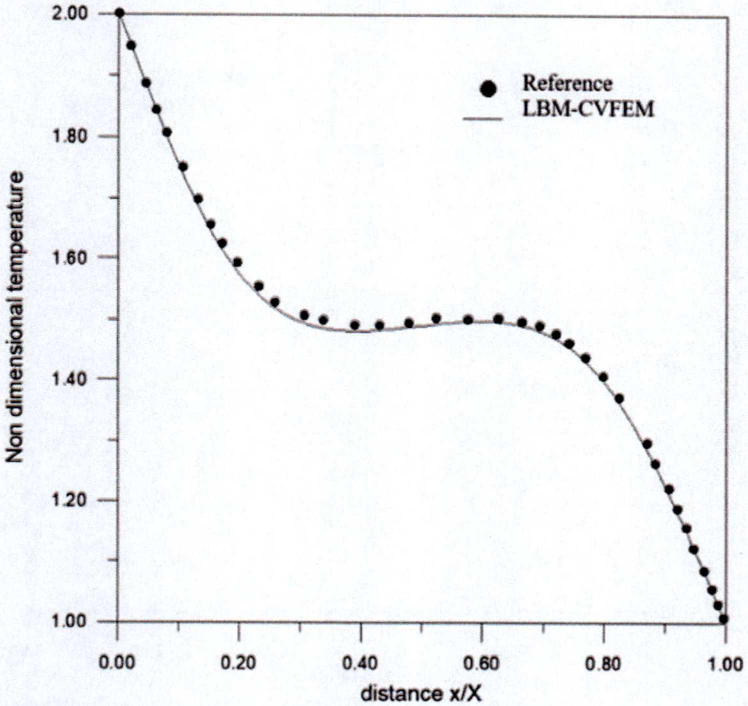

Fig. 12 Steady state non-dimensional temperature for $Ra = 10^4$, $RC = 0.0$,$\omega = 0$, $\beta = 1$ and $\varepsilon = 1$

Based on the hybrid numerical numerical proposed tool LBM-CVFEM, in Fig. 12 we show the steady state non-dimensional cross sections of the temperature at mid-height cavity (y/Y = 0.5) for $RC = 0$, $Ra = 10^4$, $\omega = 0$, $\beta = 1$ and $\varepsilon = 1$, differences in the obtained results are very small compared to references ones [45].

At this stage, we examines the steady state non-dimensional cross sections of the temperature at mid-height cavity (y/Y = 0.5) for $RC = \sigma H T_{ref}^4 / k T_c = 0.1$, $Ra = 10^4$, $\omega = 0$, $\beta = 1$ and $\varepsilon = 1$ (Fig. 13).

In this case, the hot and cold black wall temperatures are maintained at 2 and 1, respectively.

A (4 × 8) directions are considered in the CVFEM to calculate the radiative heat flux. A good very agreement with literature is noticed [45].

The accuracy and the efficiency of the LBM-CVFEM results are very promising. Then, the effect of radiation on the steady state non-dimensional mid-plane temperature for $Ra = 10^4$, $\omega = 0$, $\beta = 1$ and $\varepsilon = 1$, is highlighted in Fig. 14.

Steady state isotherms obtained using the LBM-CVFEM for convection–radiation parameter $RC = 0.0$ and $RC = 0.1$, a Rayleigh number $Ra = 10^4$,a scattering albedo $\omega = 0.0$, an extinction coefficient $\beta = 1.0$ and emissivity of the walls $\varepsilon = 1.0$ are depicted in Fig. 15.

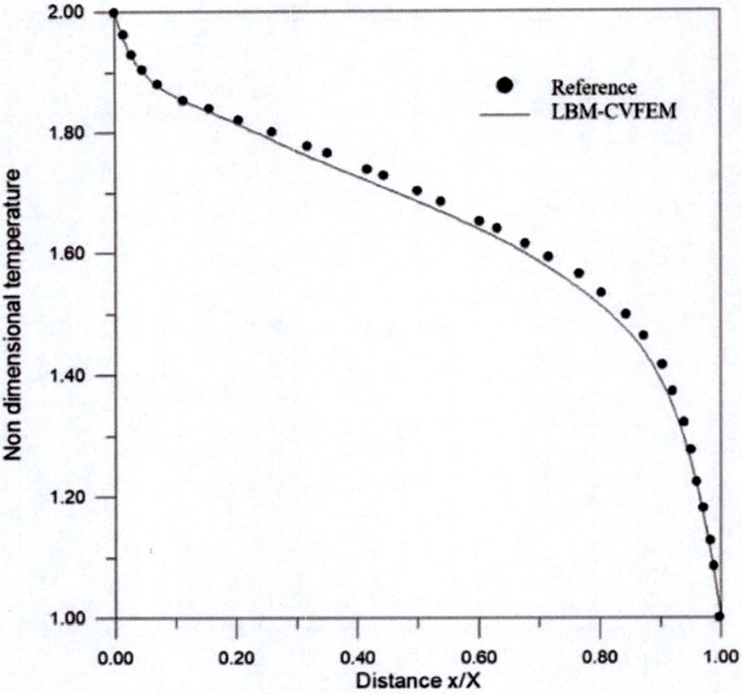

Fig. 13 Steady state non-dimensional temperature for $Ra = 10^4$, $RC = 0.0$, $\omega = 0$, $\beta = 1$ and $\varepsilon = 1$

The isotherms present become almost parallel to the cold and hot walls. Dense isotherms occur in a region close to the cold wall, indicating a high temperature gradient there.

For better insight, we show in Fig. 16 that the transient cross-sections of the temperature at mid-height cavity (y/Y = 0.5) for a wall emissivity $\varepsilon = 1$, $Ra = 10^4$, $\omega = 0$ and $\beta = 1$.

5 Conclusions

In this work, a LBM CVFEM for coupled transient free convection coupled with volumetric radiation systems in a rectangular cavity, was studied numerically.

The energy, mass and momentum equations are resolved with a D2Q9 lattices for velocity distributions of flow and temperature variables based on LBM. The radiation coupling is accounted in the energy equation by considering the numerical technique based on the control volume finite element method (CVFEM) in the participating media.

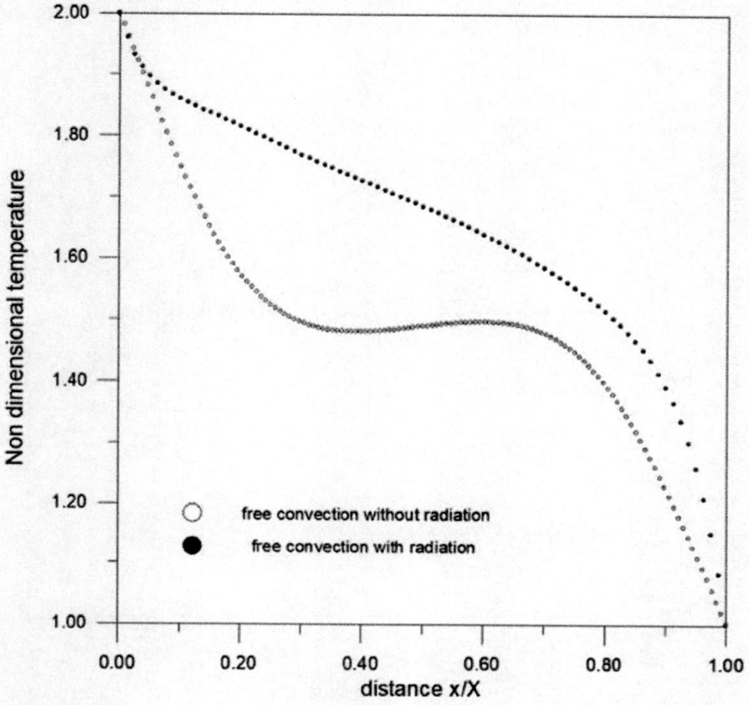

Fig. 14 Radiative effect on the steady state non-dimensional mid-plane temperature for $Ra = 10^4$, $\omega = 0$, $\beta = 1$ and $\varepsilon = 1$

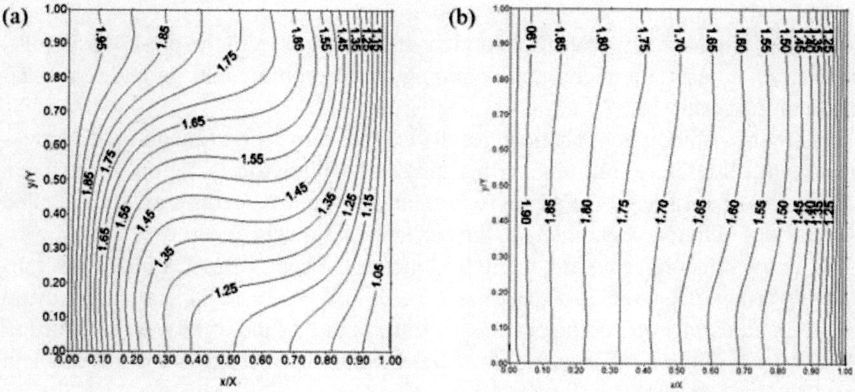

Fig. 15 Steady state isotherms for convection–radiation parameter **a** $RC = 0.0$, **b** $RC = 0.1$, $Ra = 10^4$, a scattering albedo $\omega = 0.0$, an extinction coefficient $\beta = 1.0$, emissivity of the walls $\varepsilon = 1.0$

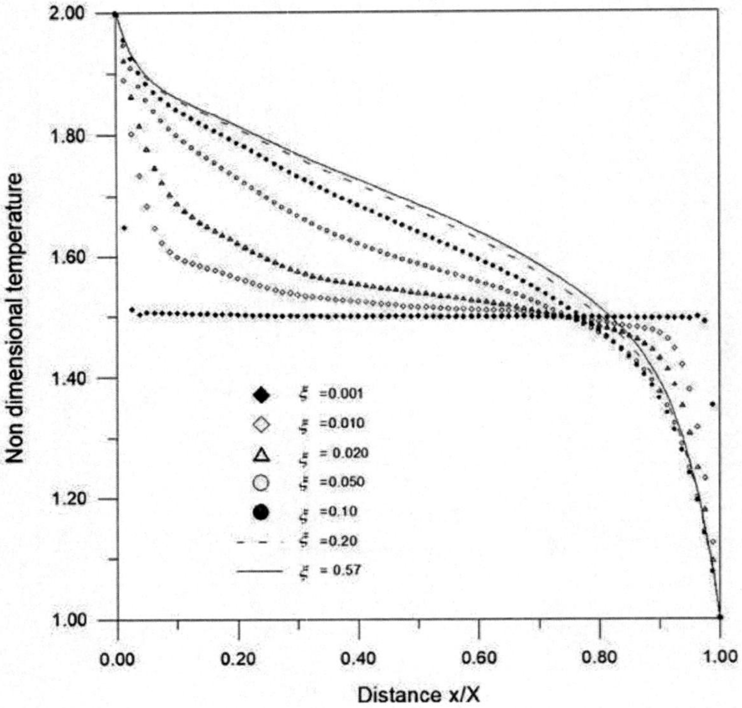

Fig. 16 Transient dimensionless temperature for y/Y = 0.5 for a wall emissivity $\varepsilon = 1$, $Ra = 10^4$, $\omega = 0$ and $\beta = 1$

Steady and unsteady (transient) laminar free convection in the presence of volumetric radiation in the enclosure containing an absorbing, emitting and scattering medium is considered.

In absence of radiation, obtained results for free convection found in good agreements with litterature. Introducing the effect of volumetric radiation, it has been shown that the LBM-CVFEM reproduce all the known features of the coupled convection-radiation phenomena in the participating media.

With the effect of radiation, the hybrid numerical code LBM-CVFEM was validated for convection–radiation parameter $RC = 0.1$, $Ra = 10^4$, a scattering albedo $\omega = 0.0$, an extinction coefficient $\beta = 1.0$, emissivity of the walls $\varepsilon = 1.0$. Further investigation of the models using three-dimensional thermal flows is underway and results will be presented in future.

In the overall evaluation, the LBM-CVFEM proves to be a promising analysis tool for practical engineering computations with a chronometer dynamic and thermal evolution.

Acknowledgements Authors wish to thank Universiti Teknologi Malaysia for supporting the manuscript using Takasago research grant (R.K130000.7343.4B472).

References

1. Dixit H, Babu V (2006) Simulation of high Rayleigh number natural convection in a square cavity using the lattice Boltzmann method. Int J Heat Mass Transf 49(3–4):727–739
2. Kuznik F et al (2007) A double-population lattice Boltzmann method with non-uniform mesh for the simulation of natural convection in a square cavity. Int J Heat Fluid Flow 28(5):862–870
3. Seta T, Takegoshi E, Okui K (2006) Lattice Boltzmann simulation of natural convection in porous media. Math Comput Simul 72(2–6):195–200
4. Kuznik F, Rusaouen G (2007) Numerical prediction of natural convection occurring in building components: a double-population lattice Boltzmann method. Numer Heat Transfer, Part A: Appl 52(4):315–335
5. Mahmood MA et al (2020) Natural convection heat transfer in a concentric annulus vertical cylinders embedded with porous media. J Adv Res Fluid Mech Therm Sci 66(2):65–83
6. Ji Y (2014) CFD modelling of natural convection in air cavities. CFD Lett 6(1):15–31
7. Jami M et al (2007) Lattice Boltzmann method applied to the laminar natural convection in an enclosure with a heat-generating cylinder conducting body. Int J Therm Sci 46(1):38–47
8. Bajorek S, Lloyd J (1982) Experimental investigation of natural convection in partitioned enclosures
9. Bilski S, Lloyd JR, Yang K (1986) An experimental investigation of the laminar natural convection velocity field in square and partitioned enclosures. In: International heat transfer conference digital library. Begel House Inc. Year
10. Fu W-S, Perng J-C, Shieh W-J (1989) Transient laminar natural convection in an enclosure partitioned by an adiabatic baffle. Nume Heat Transfer 16(3):325–350
11. Bouafia I et al (2021) Conjugate natural convection in a square porous cavity filled with a nanofluid in the presence of two isothermal cylindrical sources. J Adv Res Fluid Mech Therm Sci 80(1):147–164
12. Shu C, Peng Y, Chew Y (2002) Simulation of natural convection in a square cavity by Taylor series expansion-and least squares-based lattice Boltzmann method. Int J Mod Phys C 13(10):1399–1414
13. Ibrahim Z et al (2020) Numerical modelling for natural oil convection of NonGuided winding cooling arrangement. CFD Lett 12(4):54–67
14. Mebarek-Oudina F et al (2021) Natural convection enhancement in the annuli between two homocentric cylinders by using ethylene glycol/water based titania nanofluid. J Adv Res Fluid Mech Therm Sci 80(2):56–73
15. Hinojosa J et al (2005) Numerical study of transient and steady-state natural convection and surface thermal radiation in a horizontal square open cavity. Numer Heat Transfer, Part A 48(2):179–196
16. Balaji C, Venkateshan S (1995) Combined conduction, convection and radiation in a slot. Int J Heat Fluid Flow 16(2):139–144
17. Ha MY, Jung MJ (2000) A numerical study on three-dimensional conjugate heat transfer of natural convection and conduction in a differentially heated cubic enclosure with a heat-generating cubic conducting body. Int J Heat Mass Transf 43(23):4229–4248
18. El-Maghlany WM et al (2013) Numerical study of mixed convective cooling in a ventilated cavity utilizing a guide baffle. CFD Lett 5(4):197–207
19. Sediki E, Soufiani A, Sifaoui MS (2003) Combined gas radiation and laminar mixed convection in vertical circular tubes. Int J Heat Fluid Flow 24(5):736–746
20. El Ayachi R et al (2008) Combined effect of radiation and natural convection in a square cavity differentially heated with a periodic temperature. Numer Heat Transfer, Part A: Appl 53(12):1339–1356
21. Ferhi M, Djebali R, Abboudi S (2019) Conjugate natural convection in a partitioned square cavity filled with Al_2O_3-Water nanofluid based on experimental correlations: a lattice Boltzmann investigation. CFD Lett 11(3):1–27
22. Sharma AK et al (2007) Conjugate turbulent natural convection with surface radiation in air filled rectangular enclosures. Int J Heat Mass Transf 50(3–4):625–639

23. Singh S, Venkateshan S (2004) Numerical study of natural convection with surface radiation in side-vented open cavities. Int J Therm Sci 43(9):865–876
24. Jaballah S, Sammouda H, Belghith A (2007) Effect of surface radiation on the natural-convection stability in a two-dimensional enclosure with diffusely emitting boundary walls. Numer Heat Transfer, Part A: Appl 51(5):495–516
25. Bahlaoui A et al (2007) Coupled natural convection and radiation in a horizontal rectangular enclosure discretely heated from below. Numer Heat Transfer, Part A: Appl 52(11):1027–1042
26. Laouer A, Djeghiour R (2019) Lattice Boltzmann simulation of magneto hydrodynamic free convection in a square enclosure with non-uniform heating of the bottom wall. J Adv Res Fluid Mech Therm Sci 59(1):13–28
27. Mondal B, Mishra SC (2008) Simulation of natural convection in the presence of volumetric radiation using the lattice Boltzmann method. Numer Heat Transfer, Part A Appl 55(1):18–41
28. Mondal B, Li X (2010) Effect of volumetric radiation on natural convection in a square cavity using lattice Boltzmann method with non-uniform lattices. Int J Heat Mass Transf 53(21–22):4935–4948
29. Ali FM, Khamat ANA, Junoh MM (2021) Dual solutions in mixed convection stagnation-point flow over a vertical stretching sheet with external magnetic field and radiation effect. J Adv Res Fluid Mech Therm Sci 80(2):22–32
30. Lavanya B (2020) Radiation and chemical reaction effects on MHD convective flow over a porous plate through a porous medium with heat generation. J Adv Res Fluid Mech Therm Sci 68(1):11–21
31. He X, Chen S, Doolen GD (1998) A novel thermal model for the lattice Boltzmann method in incompressible limit. J Comput Phys 146(1):282–300
32. Succi S (2001) The lattice Boltzmann equation: for fluid dynamics and beyond. Oxford university press
33. Benzi R, Succi S, Vergassola M (1992) The lattice Boltzmann equation: theory and applications. Phys Rep 222(3):145–197
34. Chen S, Doolen GD (1998) Lattice Boltzmann method for fluid flows. Annu Rev Fluid Mech 30(1):329–364
35. Chaabane R, Askri F, Nasrallah SB (2010) Numerical modelling of boundary conditions for two dimensional conduction heat transfer equation using lattice Boltzmann method. Int J Heat Technol 28(2)
36. He X, Luo L-S (1997) Lattice Boltzmann model for the incompressible Navier-Stokes equation. J Stat Phys 88(3):927–944
37. Qian Y-H, d'Humières D, Lallemand P (1992) Lattice BGK models for Navier-Stokes equation. EPL Europhys Lett 17(6):479
38. Mohamad A (2007) Applied lattice Boltzmann method for transport phenomena, momentum, heat and mass transfer. Can J Chem Eng 85(6):946–946
39. Perumal DA, Kumar GV, Dass AK (2021) Application of lattice boltzmann method to fluid flows in microgeometries. CFD Lett
40. Sidik NAC, Jahanshaloo L (2014) The use of lattice boltzmann numerical scheme for contaminant removal from a heated cavity in horizontal channel. CFD Lett 6(3):94–100
41. Yücel A, Acharya S, Williams M (1989) Natural convection and radiation in a square enclosure. Numer Heat Transfer 15(2):261–278
42. Ridouane E et al (2004) Interaction between natural convection and radiation in a square cavity heated from below. Numer Heat Transfer Part A Appl 45(3):289–311
43. Mezrhab A et al (2006) Computation of combined natural-convection and radiation heat-transfer in a cavity having a square body at its center. Appl Energy 83(9):1004–1023
44. Banda MK, Klar A, Seaïd M (2007) A lattice-Boltzmann relaxation scheme for coupled convection–radiation systems. J Comput Phys 226(2):1408–1431
45. Tan Z, Howell JR (1991) Combined radiation and natural convection in a two-dimensional participating square medium. Int J Heat Mass Transf 34(3):785–793
46. Chaabane R, Askri F, Nasrallah SB (2011) Parametric study of simultaneous transient conduction and radiation in a two-dimensional participating medium. Commun Nonlinear Sci Numer Simul 16(10):4006–4020

47. Chaabane R, Askri F, Nasrallah BS (2011) A new hybrid algorithm for solving transient combined conduction radiation heat transfer problems. Therm Sci 15(3):649–662
48. Grissa H et al (2010) Prediction of radiative heat transfer in 3D complex geometries using the unstructured control volume finite element method. J Quant Spectrosc Radiat Transfer 111(1):144–154
49. Zou Q, He X (1997) On pressure and velocity boundary conditions for the lattice Boltzmann BGK model. Phys Fluids 9(6):1591–1598
50. Chaabane R, Askri F, Nasrallah SB (2011) Application of the lattice Boltzmann method to transient conduction and radiation heat transfer in cylindrical media. J Quant Spectrosc Radiat Transfer 112(12):2013–2027
51. Chaabane R, Askri F, Nasrallah SB (2011) Analysis of two-dimensional transient conduction–radiation problems in an anisotropically scattering participating enclosure using the lattice Boltzmann method and the control volume finite element method. Comput Phys Commun 182(7):1402–1413
52. Hortmann M, Perić M, Scheuerer G (1990) Finite volume multigrid prediction of laminar natural convection: bench-mark solutions. Int J Numer Meth Fluids 11(2):189–207
53. de Vahl Davis G (1983) Natural convection of air in a square cavity: a bench mark numerical solution. Int J Numer Meth Fluids 3(3):249–264
54. Barakos G, Mitsoulis E, Assimacopoulos D (1994) Natural convection flow in a square cavity revisited: laminar and turbulent models with wall functions. Int J Numer Meth Fluids 18(7):695–719
55. D'Orazio A, Corcione M, Celata GP (2004) Application to natural convection enclosed flows of a lattice Boltzmann BGK model coupled with a general purpose thermal boundary condition. Int J Therm Sci 43(6):575–586

Numerical Study of Magneto-hydrodynamic Free Convection Heat Transfer and Fluid Flow

Raoudha Chaabane, Abdelmajid Jemni, Nor Azwadi Che Sidik, and Hong Wei Xian

Abstract Magneto-Hydro-Dynamic associate to convection heat transfer mode in a two-dimensional cavity filled with conducting fluid having a partially open-ended wall, is discussed in the present work. This engineering application is crucial to understand dynamic and thermal behavior of flow in frequent conclusive practicable and industrial applications in renewable and sustainable energy fields, specifically in collectors based on solar energy. However, scare open literature is find dealing with this important subject. Simulation of this coupled complex engineering physical problems in Magneto-Hydrodynamic (MHD) configurations is a great challenge for traditional CFD numerical approaches. In this paper, we apply the lattice Boltzmann method (LBM) in order to resolve a given engineering numerical simulation struggle. Knowing that, LBM is becoming in the last decades, an essential attractive tool for existing CFD approaches aiming to resolve complex numerous fluid flow problems with complex numerous difficulties. We intend in this work to afford a short review of researches dealing with application involving MHD convective simulations based on LBM aiming to predict the dynamic and thermal behavior recognizing more openings horizon for coming research. In addition, novel findings were highlighted within this work. First, the proposed Lattice Boltzmann model is validated and good agreement is found. Using an in-house LBM code, we introduce the effect of the Rayleigh number, Prandtl number, Hartmann number and geometrically, the position of the partially open-ended wall side in the MHD enclosure. Parametric findings show that all the coefficients influence significantly flow field and also temperature field patterns. This paper provides very useful beginnings for the applicable future MHD convective engineering studies.

R. Chaabane (✉) · A. Jemni
Lab. d'Etudes des Systèmes Thermiques et Energétiques (LESTE), Département de Génie Energétique, Ecole Nationale d'Ingénieurs de Monastir, Université of Monastir, Monastir, Tunisie
e-mail: raoudhach@gmail.com

N. A. C. Sidik (✉) · H. W. Xian
Malaysia–Japan International Institute of Technology (MJIIT), Universiti Teknologi Malaysia, Jalan Sultan Yahya Petra, 54100 Kuala Lumpur, Malaysia
e-mail: azwadi@utm.my

M. Y. Ismail et al. (eds.), *Technological Advancement in Mechanical and Automotive Engineering*, Lecture Notes in Mechanical Engineering,
https://doi.org/10.1007/978-981-19-1457-7_43

Keywords Heat transfer · Convection · LBM · MHD · Partially · Open side

Abbreviations

List of Symbols

Ha	Hartmann-number
B	Magnetic-field intensity, T
ρ	Density, kg m^{-3}
u, v	Velocities, m s^{-1}
g	Acceleration of the gravity, m s^{-2}
k	Thermal-conductivity, W m^{-1} K^{-1}
Ra	Rayleigh-number
Pr	Prandtl-number
H	The characteristic of length scale, m
e_i	Discrete particle speeds, m s^{-1}
ΔT	Temperature difference, K
f	Density distribution functions
g	Internal energy distribution functions
T	Temperature, K
γ	Angle, rad
Nu	Local Nusselt-number
x, y	Cartesian coordinates, m
T_m	Average temperature, K

Greek Symbols

α	Thermal diffusivity, m^2 s^{-1}
β_T	Volumetric thermal expansion coefficient, K^{-1}
τ	Relaxation-time, s
σ	Electrical-conductivity, S m^{-1}
μ	Dynamic-viscosity, kg m^{-1} s^{-1}
ν	Kinematic-viscosity, m^2 s^{-1}

Subscript

avg	Average
H	Hot

C Cold

Highlights

1. The mesoscopic approach LBM issued for highlighting a parametric study of Magnetohydrodynamic convective heat transfer in a two-dimensional medium.
2. The effect of variable mixed boundary conditions is investigated on a MHD cavity.
3. A non-coupled numerical approach (LBM) for a heat transfer hybrid mode bringing additional advantages on parallel computing platforms with more future complex geometry in applications associated to Renewable and Sustainable Energy.

1 Introduction

Convection [1–15] inside a close or in an open cavity is an important research prototype in many practical thermal science and physical engineering applications such as cooling systems of solar collectors, solar thermal receivers, electronic components, thermal insulation systems, building and thermal insulation systems, nuclear reactor systems, food storage industry and geophysical fluid mechanics, building ventilation and energy saving [16–23].

In literature, experimental investigations dealing with this engineering case are scarce [24–26]. For example, numerical efforts were likewise talented in **porous** media [27–34].

Natural convection inside enclosed geometries has been extensively documented in the last decades; however, free convection flows induced by open enclosures have received little attention. It is known that this particular geometry is of great experimental and practical importance because it can model and help to understand thermal performance of thermo-syphons and solar-tower power plants.

Furthermore, specifying appropriate boundary conditions at the aperture plane is a seeked task in numerical engineering simulation. Convection inside a cavity with an open ended wall side was numerically investigated in [16, 19, 21, 24–26, 34–41]. In his work [16], the author study natural convection from open cavities or heated plates attached with parallel vertical strips. He demonstrates that the average Nusselt number is not very sensitive to the inclination angle. Besides he shows that the flow becomes unstable at high Rayleigh numbers and at low inclination angles.

Free convection in cavity with open wall is simulated by Lattice Boltzmann Method (L-B-M) in A. A. Mohamad work's [19]. They demonstrate that when the aspect ratio increases, the rate of heat transfer deceases asymptotically and also they prove that heat transfer can reach conduction limit for a large aspect ratio.

In Ref. [22], the author provide a review dealing with convection phenomena inside cavities. This work is a compilation that helps to highlight the wide range of applications where the knowledge of convection is advantageously used.

Many varieties of geometries with various imposed external conditions (initial conditions, boundary conditions, inclinations, fluid nature, radiative properties, heat source distributions) are examined.

To the best knowledge of the authors of this work, the present problem dealing with convection combined to magneto-hydrodynamic effect in a given open-ended cavity associated with mixed boundary conditions using the mesoscopic approach TLBM (Thermal Lattice Boltzmann Method) has not yet explored.

Magneto-Hydro-dynamic-convection was developed in [9, 42–59] where the goal of these studies was to demonstrate magnetic field effects on thermal patterns inside different geometries with closed or open side cavities.

Magnetic field effect combined to free convection expected a significant attention in various scientific and engineering applications such as metal casting, fluids crystal growth, geothermal energy extractions, fusion reactors and convection associated to magnetic effect [9, 42–62].

Afifah et al. [9] analyzed magneto-viscous effects and thermomagnetic convection of magnetic fluids by both numerical and experimental approaches for deeper understanding of the magnetic fluid. Many papers in literature introduce the effect of magnetic field on free convection flow in a conducting fluid with different directions of the magnetic field [42–47, 50–53, 55–62].

Double-diffusive natural convection with Hydro-magnetic effect inside a two dimensional cavity in the presence of internal heated obstacle is introduced in Ref. [48].

Three-dimensional natural convection in a cubical enclosure with MHD effect in studied in [42, 49, 54].

Moreover, new attention has been intensively concentrated on the applications involved mixed boundary-conditions [56–59, 63–68] on the side walls of cavities with total or partially open side.

Authors in Refs. [56–59, 63–68] introduce the effect of different spatial or side-wall temperature variation on different locations of boundaries of the MHD cavity.

In the air of this short, and after many lectures, we notice that there is no article source or recent publication dealing with MHD convection (Magneto-Hydro-Dynamic Convection) heat transfer coupling problem inside a cavity with a side a total or an open side wall.

Besides, a given application is omnipresent in various energy and civil engineering. For existing numerical tools, in order to appropriately solve and treat such energy engineering, existing numerical tools find great difficulties and the problem present a prodigious challenge to resolve, particularly inside non-basic geometries.

In industrial energy applications, the proposed coupling situation inside closed, totally or partially opened wall side complex geometries play a crucial starring role; unluckily, we find a worry scare in numerous simulations dealing and resolving adequately related test cases in energy engineering systems.

Therefore, the aim of this study is to link such great scientific gap in those specific applications. In the present work, the generic engineering problem will be solved by Lattice Boltzmann (LB) method [32] which will be adopted as a numerical tool to deal with this coupling problem associating mixed boundary conditions.

A short review on the mesoscopic approach based LB models for heat transfer with convection is highlighted in this work and more details can be explored in literature given below.

The proposed mesoscopic approach can resolve all combined equations considering isothermal or non-isothermal mixed variety of boundary conditions by the two dimensional nine velocities D2Q9 model for both thermal and dynamic flow.

After demonstrating good validation with existing literature and aiming to cure the limitation associated with this numerical energy engineering area we aim to elaborate new findings that can characterize heat transfer mode associating magneto-convection inside cavity with a central partially open side wall which is a novel engineering problematic.

2 LBM

Over the two last decades, the mesoscopic tool based on Lattice Boltzmann Method (LBM) prove a successful trend to simulate and resolve numerically many complex industrial and engineering applications which are dubious for existing conventional numerical tools [11, 69–90].

Literate show that algorithms used in LBM tools are much cooler and easier to be implemented particularly when we deal with non-basic geometries.

Nowadays, this mesoscopic approach is faithfully applied to resolve coupled momentum and energy governing equations for different flows [71–76, 78, 80, 81, 84]. Besides, it is proved that is the best efficient computational approach [69–84].

3 Mathematical Formulation

Figure 1 highlights the considered geometry in this paper. Many cases will be studied based a rectangular closed, open-ended side and central partially open east wall side H height cavity.

Various excitations can be imposed along the left west vertical wall such as sinusoidal, linear or constant isothermal temperature.

It is to be noted that then left opening side-wall can be correlated with the a constant temperature of the conducting fluid, Tc. Moreover, the top north and bottom south boundaries are insulated.

Further, a horizontal uniform strength magnetic field is applied to the Newtonian, incompressible conducting fluid.

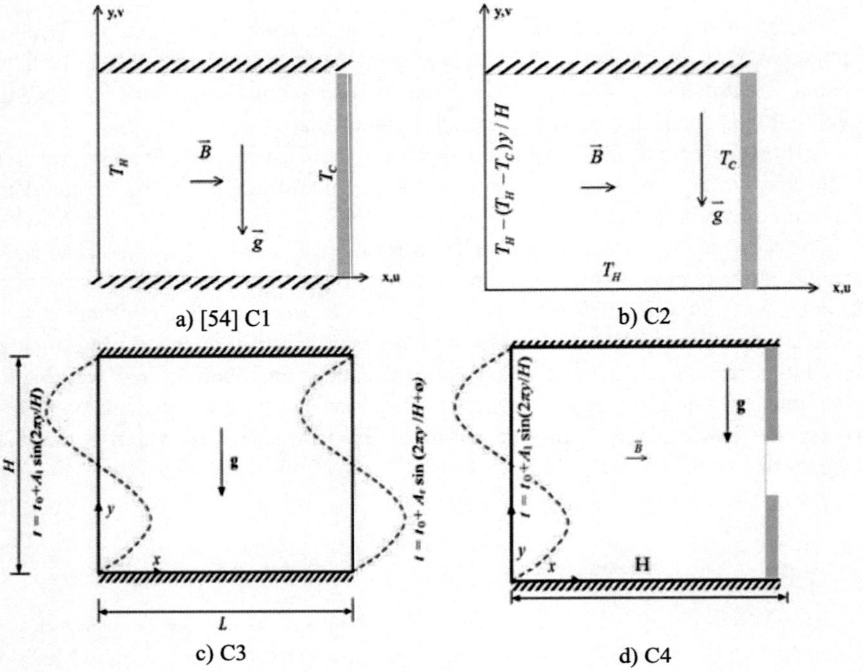

Fig. 1 Geometries and boundary conditions

In the present work, we consider a two dimensional, steady, laminar flow (liquid gallium), steady, laminar where we neglect radiation effects, Joule heating and the viscous dissipation.

Besides, we deal with conducting fluid having constant thermo-physical properties except the Boussinesq sense approximation density.

The considered heat transfer non-dimensional equations in the central partially open-ended side-wall MHD cavity are:

$$\frac{\partial u}{\partial x}+\frac{\partial v}{\partial y}=0 \tag{1}$$

$$\rho\left(\frac{\partial u}{\partial t}+u\frac{\partial u}{\partial x}+v\frac{\partial u}{\partial y}\right)=-\frac{\partial p}{\partial x}+\mu\left(\frac{\partial^2 u}{\partial x^2}+\frac{\partial^2 u}{\partial y^2}\right)+\frac{\mathrm{Ha}^2\mu}{H^2}(v\sin\gamma\cos\gamma-u\sin^2\gamma) \tag{2}$$

$$\rho\left(\frac{\partial v}{\partial t}+u\frac{\partial v}{\partial x}+v\frac{\partial v}{\partial y}\right)=-\frac{\partial p}{\partial y}+\mu\left(\frac{\partial^2 v}{\partial x^2}+\frac{\partial^2 v}{\partial y^2}\right)+\frac{\mathrm{Ha}^2\mu}{H^2}(u\sin\gamma\cos\gamma-v\cos^2\gamma)+\rho g\beta_T(T-T_m) \tag{3}$$

$$\frac{\partial T}{\partial t} + u\frac{\partial T}{\partial x} + v\frac{\partial T}{\partial y} = \alpha\left(\frac{\partial^2 T}{\partial x^2} + \frac{\partial^2 T}{\partial y^2}\right) \tag{4}$$

where *u*, *v*, *p*, *T*, $\text{Ha} = HB\sqrt{\sigma/\mu}$ are fluid velocities, pressure, temperature and Hartmann number, respectively. Their definitions can be found in [64].

The applied external horizontal magnetic field is uniform and with B0 constant magnitude. It is to be noted that Boussinesq model is applied to the variation of density of the conducting fluid.

Moreover, we neglect the induced magnetic field which is generated by the electrically conducting fluid motion, viscous dissipation and Joule heating.

Consequently, the LB-PDF (Particle Distribution Function) for dynamic and thermal fields are given as [69–84]:

$$\begin{aligned} f_k(\mathbf{r} + \mathbf{c_k}\Delta t, t + \Delta t) &= f_k(\mathbf{r}, t) \\ &+ \frac{\Delta t}{\tau_v}[f_k^{eq}(\mathbf{r}, t) - f_k(\mathbf{r}, t) +] + \Delta t\, F \end{aligned} \tag{5}$$

$$g_k(\mathbf{r} + \mathbf{c_k}\Delta t, t + \Delta) = \left(1 - \left(\frac{\Delta t}{\tau_T}\right)\right) g_k(\mathbf{r}, t) + \left(\frac{\Delta t}{\tau_T}\right) \; g_k^{eq}(\mathbf{r}, t) \tag{6}$$

where f_k the PDF and f_k^{eq} is the LEDF (Local Equilibrium Distribution Function) in Eq. (6), the external force term *F* is given by [58, 59].

$$\mathbf{F} = \left[\frac{(\rho\beta_T g(T - T_m)\mathbf{j}) \cdot (\mathbf{c}_k - \mathbf{u})}{RT}\right] f_k^{eq} \tag{7}$$

where the unit vector *j* is in a direction opposite to gravity, *g* is the acceleration due to gravity, ρ is fluid density at T_m mean temperature, *R* is the constant of the ideal gas and β_r is the parameter involving the volumetric thermal expansion.

g_k is the thermal PDF related to the internal energy evolution ***e***.

In the D2Q9 lattice model (Fig. 2), the dimensionless dynamic and thermal relaxation time are defined respectively, as [58, 59]:

$$\tau_v = 1/2 + (3\nu/c^2\Delta t) \tag{8}$$

$$\tau_T = 1/2 + (3\alpha/c^2\Delta t) \tag{9}$$

where ν is the dynamic kinematic viscosity and α is the thermal diffusion parameter, diffusion coefficient) and Δt is the time step of the lattice.

It is to be noticed that the buoyancy force-term is must be added to Eq. (7), as following,

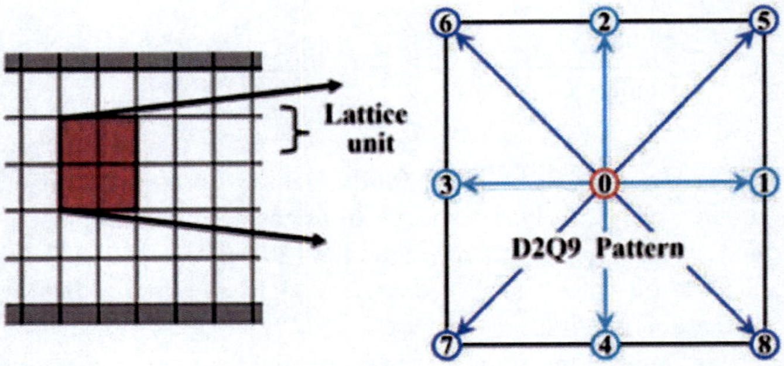

Fig. 2 LBM D2Q9 lattices

$$F_\alpha = 3\, w_k\, g\, \beta_T\, (T_{hot} - T_{cold}) \tag{10}$$

In the considered mesoscopic approach, the total force F is written as:

$$F = F_x + F_y \tag{11}$$

$$F_x = 3 w_k \rho \left[R(v \sin\gamma \cos\gamma) - u \sin^2\gamma) \right] \tag{12}$$

$$F_y = 3 w_k \rho \left[g\beta_T (T - T_m) + R(u \sin\gamma \cos\gamma) - v \cos^2\gamma \right] \tag{13}$$

where $R = \mu \mathrm{Ha}^2$ and γ is magnetic field direction.

The lattice speed is defined as $c = \sqrt{3RT}$ and c_s is the speed of the sound $c_s = \sqrt{RT}$.

The dynamic and thermal Equilibrium Function distribution are given respectively as

$$f_k^{eq} = w_k \rho \left(1 + \frac{\mathbf{c}_k \cdot \mathbf{u}}{c_s^2} + \frac{(\mathbf{c}_k \cdot \mathbf{u})^2}{2c_s^4} - \frac{\mathbf{c}_k \cdot \mathbf{u}}{2c_s^2} \right) \tag{14}$$

$$g_k^{eq} = w_k \rho e \left(1 + \frac{\mathbf{c}_k \cdot \mathbf{u}}{c_s^2} \right) \tag{15}$$

w_k is the weight related to every discrete direction.

Achieving collision and streaming processes, density, velocity and temperature are computed respectively as:

$$\rho(\mathbf{r}, t) = \sum_k f_k(\mathbf{r}, t) \tag{16}$$

$$\mathbf{u}(\mathbf{r}, t) = \sum_k \mathbf{c}_k f_k(\mathbf{r}, t)/\rho(\mathbf{r}, t) \tag{17}$$

$$T(\mathbf{r}, t) = \sum_k g_k(\mathbf{r}, t) \tag{18}$$

In order to describe convective heat transport phenomena, we study the evolution of the he local Nusselt-number and its average value at the bottom south side-wall defined as;

$$NU_y = -\frac{H}{\Delta T}\frac{\partial T}{\partial x} \tag{19}$$

$$NU_{avg} = \frac{1}{H}\int_0^H NU_y dy \tag{20}$$

4 Results and Discussion

We have validated the D2Q9 LBM code for convection inside open cavity with adiabatic horizontal boundaries with non-conducting fluid with isothermal uniform west vertical temperature [56]. Figure 3 applied for configuration C1 (Fig. 1) show

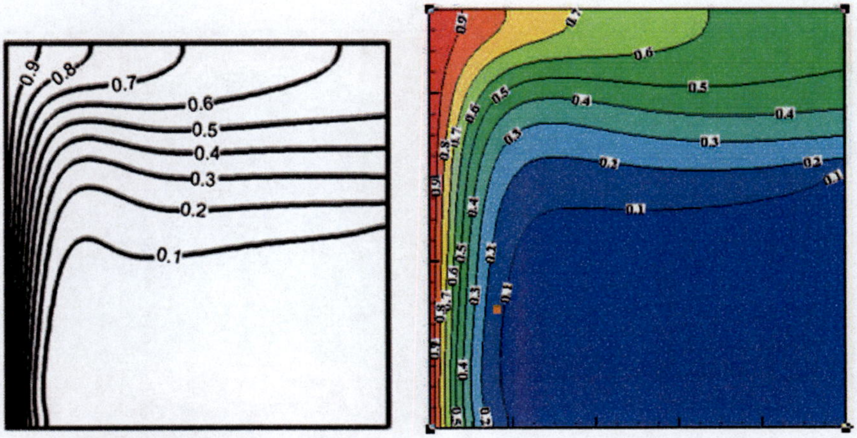

a) Reference [56] b) Present work

Fig. 3 Validation: isotherms at Pr = 0.71 for Ha = 0 and Ra = 105 between Ref. [56] and present work good agreement are obtained for isotherms and streamlines with Pr = 0.71, Ha = 0 and Ra = 105 (Fig. 4).

Besides, a steady state isotherms of linearly heated sidewalls MHD cavity with a moderate Hartmann number of 50, Pr = 0.025 (liquid gallium) and Ra = 10^5 is validated. After grid assessment, good agreement is reached comparing literature finding [43] in Fig. 5. In this case, magnetic field effect is studied in an open liquid gallium cavity with linearly heated west wall for Ha = 50 and Ra = 10^5.

Moreover, the behavior of the Nusselt south wall N_{ub} in the case of two linear vertical walls is highlighted in following test case. Figure 6 demonstrates that the local Nusselt number at the south wall N_{ub} shows an oscillatory behavior with the horizontal distance x/X which symmetric about the centerline of the south side-wall.

For different Hartman numbers, local Nusselt number variation with distance at south sidewall in the case of linearly heated boundaries (Pr = 0.025 and Ra = 105) is highlighted in Fig. 6.

In Fig. 7, we consider a MHD convective case inside an open liquid gallium cavity with linearly heated left vertical wall (C2) for Ha = 150 for different Rayleigh numbers.

As yet shown in Fig. 4, when the Rayleigh number increases, the thermal boundary become narrower.

Figure 8 present isotherms and the streamlines plots for a high moderate Rayleigh number Ra = 10^6 and high Hartmann number Ha = 150 when the cavity is filled with liquid gallium (Configuration C2). Open right-side wall effect shown in Fig. 5a is noticeable in streamlines behavior.

In this stage, we aim to introduce addressed to the mesoscopic approach studies inside MHD open cavities subjected to open cavity with a partially sinusoid ally

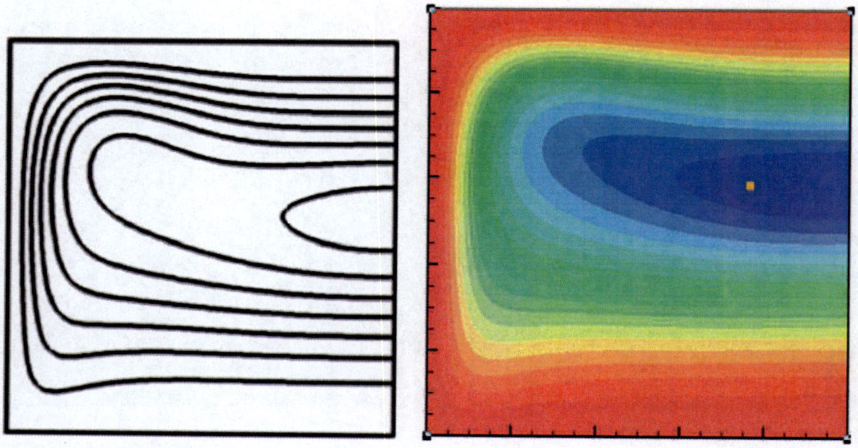

a) Reference [33] b) Present work

Fig. 4 Validation: streamlines at Pr = 0.71 for Ha = 0 and Ra = 105 between Ref. [56] and present work

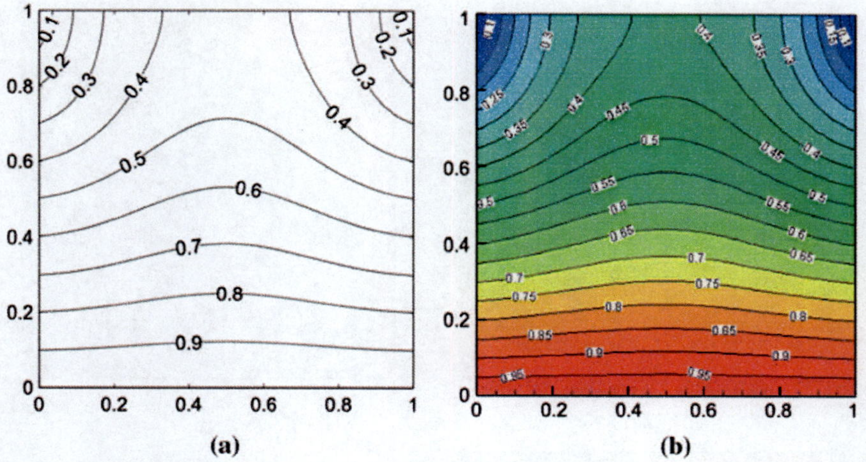

Fig. 5 Validation: isotherms of linearly heated side walls MHD cavity for Ha = 50, Pr = 0.025 and Ra = 105, **a** Ref. [43] and **b** present work

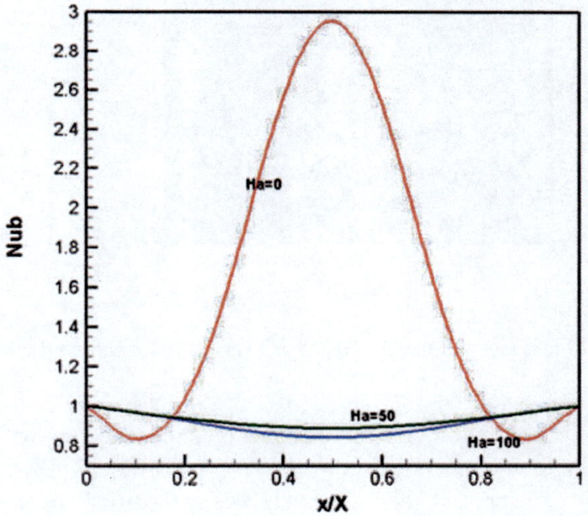

Fig. 6 Local Nusselt number variation along bottom wall, test case of linearly heated side walls (Pr = 25×10^{-3} and Ra = 10^5)

heated left wall (Fig. 1). The objective of the present work is consequently the prediction of the thermal and dynamic behavior of heat transfer in a such decisive energy engineering system.

Besides, we target to examine LBM ability to treat a given non-basic configuration (C4) considering a partially open MHD liquid gallium enclosure subjected to a sinusoid-ally excitation on the west vertical wall with two adiabatic horizontal walls for a Rayleigh number of Ra = 105 and Ha = 50.

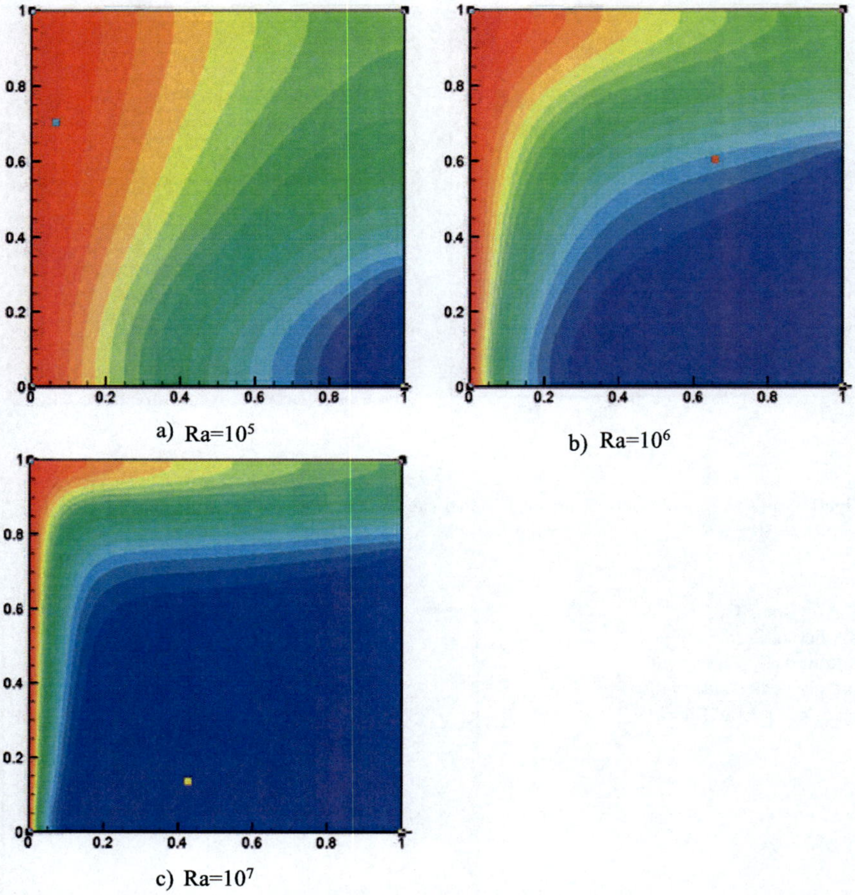

Fig. 7 Isotherms at Pr = 0.025 and Ha = 150 for various Rayleigh numbers

Seeking brevity, more details on papers dealing with sinusoidal heating wall can be found in simulation work [57–59, 63–68].

The present geometry is a H/4 central partially open-ended sidewall liquid gallium enclosure shown in Fig. 1 (C4). The two horizontal sides are insulated. Details of the Lattice Boltzmann right open sidewall are handled in [21].

First, we consider liquid gallium configuration (C3) of Fig. 1. The vertical sidewalls of the enclosure have spatially vertical varying sinusoidal temperature distributions while the horizontal boundaries are insulated, test case is applied for Ha = 150 and Ra = 106. Figure 9 depicts heat transfer rate within isotherms and streamlines evolution inside the MHD cavity. Dynamic and thermal behavior of configuration C4 is illustrated in Fig. 10.

We notice that the flow within the cavity takes place owing to the thermal buoyancy effects caused by the sinusoidally heated right vertical boundary.

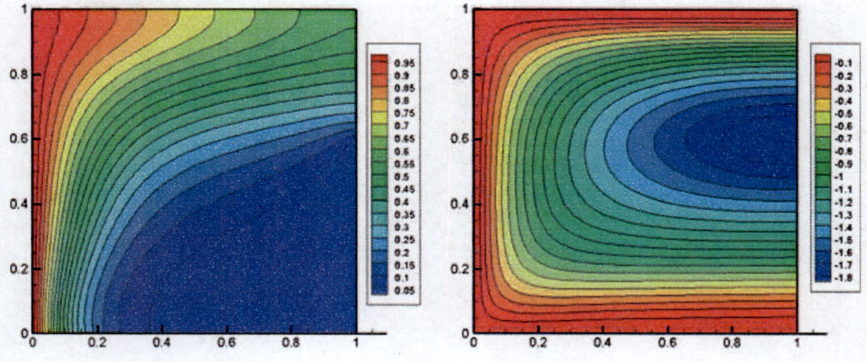

(a) (b)

Fig. 8 Isotherms (**a**) and isotherms (**b**) at Pr = 0.025, Ra = 106 and Ha = 150

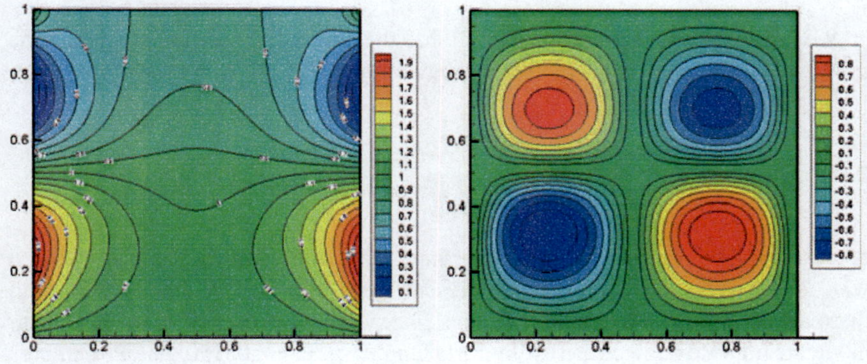

(a) (b)

Fig. 9 Isotherms (**a**) and streamlines (**b**) of sinusoidally heated side walls in a partially open MHD cavity for Ha = 150, Pr = 0.025 and Ra = 10^5

In addition, in configuration C3, the flow is characterized by a symmetric multi-cellular behavior in which the re-circulating eddies or cells of relatively high velocity are formed within the cavity and this in the presence of a high Hartmann Ha = 150 and a high Rayleigh number Ra = 105.

The presence of the central partially open sidewall in configuration C4 change the flow and heat transfer structure. One dominant cell is noticed in the core region of the enclosure and a little cell that occur at upper left vertical side.

As predicted, because of the partially open side effects, the temperature field was sketched by a noticeable drop in its behavior near the open side of the cavity and these both in flow and heat behaviors.

Besides, we highlight that the temperature contour maps lose the sinusoidal behavior as they move to the partially open side.

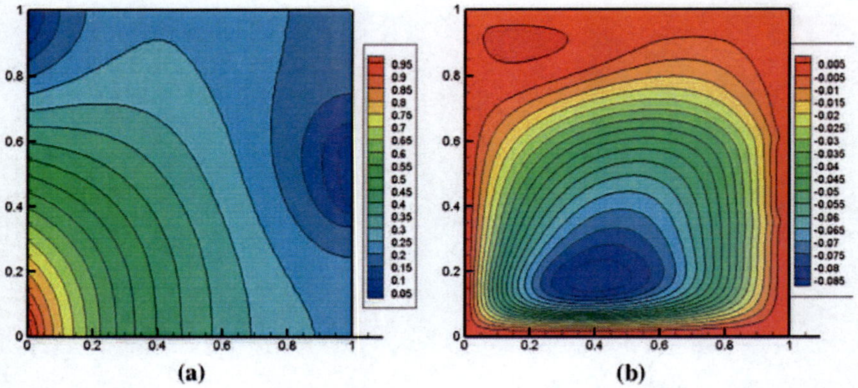

Fig. 10 Isotherms (**a**) and streamlines (**b**) of sinusoidally heated right side wall partially open MHD cavity for Ha = 150, Pr = 0.025 and Ra = 10^6

We infer that a partially open sidewall has the tendency to control efficiently the movement of the fluid in such given configuration (C4).

5 Conclusions

A mesoscopic approach in an MHD partially open cavity with sinusoidally heated wall is not yet study in literature. Present work aims to predict dynamic and thermal behavior in a given energy engineering system. The enclosure is subjected to moderate high Ra number of 105, high Ha numbers (Ha = 50, 150) and Pr numbers of (0.7 and 0.025). A second crucial test case was discussed involving MHD convection inside open cavity with sinusoidal heated west wall which is filled with liquid gallium with Ha = 150 for Ra = 105. Vertical left wall is subjected to spatially varying sinusoidal temperature distributions while horizontal walls are adiabatic and the right wall is a central partially open one. Both for flow and heat transfer simulation, Lattice Boltzmann Method was considered for numerical simulation. After good validation, the present cod is extended to deal with the present complex geometry in order to highlight the ability and the workability of LBM to deal with a given mixed boundary condition configuration. The present work revealed capability of LBM to simulate various boundary conditions inside partially open-ended cavity with sinusoidal heating vertical boundary. An analysis of the opening mass flow was highlighted in the dynamic and thermal behavior of the streamlines and the isotherms.

Acknowledgements Authors wish to thank Universiti Teknologi Malaysia for supporting the manuscript using Takasago research grant (R.K130000.7343.4B472).

References

1. Amber I, O'Donovan TS (2018) Natural convection induced by the absorption of solar radiation: a review. Renew Sustain Energy Rev 82:3526–3545
2. Miroshnichenko IV, Sheremet MA (2018) Turbulent natural convection heat transfer in rectangular enclosures using experimental and numerical approaches: a review. Renew Sustain Energy Rev 82:40–59
3. Biswal P, Basak T (2017) Entropy generation vs energy efficiency for natural convection based energy flow in enclosures and various applications: a review. Renew Sustain Energy Rev 80:1412–1457
4. Haddad Z et al (2012) A review on natural convective heat transfer of nanofluids. Renew Sustain Energy Rev 16(7):5363–5378
5. Mohammed KA et al (2017) Review of forced convection nanofluids through corrugated facing step. Renew Sustain Energy Rev 75:234–241
6. Lomascolo M et al (2015) Review of heat transfer in nanofluids: conductive, convective and radiative experimental results. Renew Sustain Energy Rev 43:1182–1198
7. Murshed SMS et al (2011) A review of boiling and convective heat transfer with nanofluids. Renew Sustain Energy Rev 15(5):2342–2354
8. Dhaidan NS, Khodadadi JM (2015) Melting and convection of phase change materials in different shape containers: a review. Renew Sustain Energy Rev 43:449–477
9. Afifah AN, Syahrullail S, Sidik NAC (2016) Magnetoviscous effect and thermomagnetic convection of magnetic fluid: a review. Renew Sustain Energy Rev 55:1030–1040
10. Che Sidik NA, Aisyah Razali S (2014) Lattice Boltzmann method for convective heat transfer of nanofluids—a review. Renew Sustain Energy Rev 38:864–875
11. Laouer A, Djeghiour R (2019) Lattice Boltzmann simulation of magnetohydrodynamic free convection in a square enclosure with non-uniform heating of the bottom wall. J Adv Res Fluid Mech Therm Sci 59(1):13–28
12. Ji Y (2014) CFD modelling of natural convection in air cavities. CFD Lett 6(1):15–31
13. Ghouizi J et al (2020) Numerical simulation of natural convection in a cavity filled with a nanofluid who's wall containing the heat source is inclined. J Adv Res Fluid Mech Therm Sci 76(1):1–16
14. Ferhi M, Djebali R, Abboudi S (2019) Conjugate natural convection in a partitioned square cavity filled with Al_2O_3-water nanofluid based on experimental correlations: a lattice Boltzmann investigation. CFD Lett 11(3):1–27
15. Selvan MM (2010) Mixed convection in a lid-driven cavity utilizing nanofluids. CFD Lett 2(4):163–175
16. Mohamad AA (1995) Natural convection in open cavities and slots. Numer Heat Transf Part A: Appl 27(6):705–716
17. Singh DK, Singh SN (2016) Combined free convection and surface radiation in tilted open cavity. Int J Therm Sci 107:111–120
18. Wang Z et al (2011) Combined heat transfer by natural convection–conduction and surface radiation in an open cavity under constant heat flux heating. Numer Heat Transf Part A: Appl 60(4):289–304
19. Mohamad AA, El-Ganaoui M, Bennacer R (2009) Lattice Boltzmann simulation of natural convection in an open ended cavity. Int J Therm Sci 48(10):1870–1875
20. Bruce JM (1978) Natural convection through openings and its application to cattle building ventilation. J Agric Eng Res 23(2):151–167
21. Mohamad AA, Bennacer R, El-Ganaoui M (2010) Double dispersion, natural convection in an open end cavity simulation via lattice Boltzmann method. Int J Therm Sci 49(10):1944–1953
22. Baïri A, Zarco-Pernia E, García de María JM (2014) A review on natural convection in enclosures for engineering applications. The particular case of the parallelogrammic diode cavity. Appl Therm Eng 63(1):304–322
23. Polat O, Bilgen E (2002) Laminar natural convection in inclined open shallow cavities. Int J Therm Sci 41(4):360–368

24. Chehouani H, Ait Haj Said A, El Fagrich M (2015) Heat transfer study of free convection through a horizontal open ended axisymmetric cavity using holographic interferometry. Exp Therm Fluid Sci 60:308–316
25. Montiel-González M et al (2015) Theoretical and experimental study of natural convection with surface thermal radiation in a side open cavity. Appl Therm Eng 75:1176–1186
26. Hosseini R, Heyrani-Nobari MR, Hatam M (2005) An experimental study of heat transfer in an open-ended vertical eccentric annulus with insulated and constant heat flux boundaries. Appl Therm Eng 25(8):1247–1257
27. Arpino F, Massarotti N, Mauro A (2011) Efficient three-dimensional FEM based algorithm for the solution of convection in partly porous domains. Int J Heat Mass Transfer 54(21):4495–4506
28. Beckermann C, Viskanta R, Ramadhyani S (1988) Natural convection in vertical enclosures containing simultaneously fluid and porous layers. J Fluid Mech 186:257–284
29. Song M, Viskanta R (1994) Natural convection flow and heat transfer within a rectangular enclosure containing a vertical porous layer. Int J Heat Mass Transfer 37(16):2425–2438
30. Bourouis A, Omara A, Abboudi S (2016) Upward and downward conjugate mixed convection heat transfer in a partially porous cavity. Int J Numer Methods Heat Fluid Flow
31. Wang W et al (2015) Conjugate heat transfer analysis of an impinging receiver design for a dish-Brayton system. Sol Energy 119:298–309
32. Beckermann C, Ramadhyani S, Viskanta R (1987) Natural convection flow and heat transfer between a fluid layer and a porous layer inside a rectangular enclosure
33. Chamkha AJ, Ismael MA (2013) Conjugate heat transfer in a porous cavity filled with nanofluids and heated by a triangular thick wall. Int J Therm Sci 67:135–151
34. Haghshenas A, Nasr MR, Rahimian MH (2010) Numerical simulation of natural convection in an open-ended square cavity filled with porous medium by lattice Boltzmann method. Int Commun Heat Mass Transfer 37(10):1513–1519
35. Saleem M et al (2011) Entropy generation in Marangoni convection flow of heated fluid in an open ended cavity. Int J Heat Mass Transfer 54(21):4473–4484
36. Boetcher SKS, Sparrow EM (2009) Buoyancy-induced flow in an open-ended cavity: assessment of a similarity solution and of numerical simulation models. Int J Heat Mass Transfer 52(15):3850–3856
37. Yücel C et al (1993) Mixed convection heat transfer in open ended inclined channels with discrete isothermal heating. Numer Heat Transfer Part A: Appl 24(1):109–126
38. Bakar NA, Roslan R, Akhir MKM (2020) The effects of internal heat generation or absorption on mixed convection in a lid-driven rectangular cavity using finite volume method. CFD Lett 12(12):38–54
39. El-Maghlany WM et al (2013) Numerical study of mixed convective cooling in a ventilated cavity utilizing a guide baffle. CFD Lett 5(4):197–207
40. Bakar NA, Roslan R, Karimipour A (2020) Magnetic field effect on mixed convection heat transfer in a lid-driven rectangular cavity. CFD Lett 12(1):13–21
41. Douha M et al (2020) Study of laminar naturel convection in partially porous cavity in the presence of nanofluids. J Adv Res Fluid Mech Therm Sci 79(1):91–110
42. Fidaros D, Grecos A, Vlachos N (2017) Development of numerical tool for 3D MHD natural convection. Annex xx:73–74
43. Sathiyamoorthy M, Chamkha A (2010) Effect of magnetic field on natural convection flow in a liquid gallium filled square cavity for linearly heated side wall(s). Int J Therm Sci 49(9):1856–1865
44. Alchaar S, Vasseur P, Bilgen E (1995) Natural convection heat transfer in a rectangular enclosure with a transverse magnetic field
45. Garandet JP, Alboussiere T, Moreau R (1992) Buoyancy driven convection in a rectangular enclosure with a transverse magnetic field. Int J Heat Mass Transfer 35(4):741–748
46. Rudraiah N et al (1995) Effect of a magnetic field on free convection in a rectangular enclosure. Int J Eng Sci 33(8):1075–1084
47. Cowley MJM (1996) Natural convection in rectangular enclosures of arbitrary orientation with vertical magnetic field. Magnetohydrodynamics 32(4):390–424

48. Teamah M (2006) Hydro-magnetic double-diffusive natural convection in a rectangular enclosure with imposing an inner heat source or sink. Alexandria Eng J 45(4):401–415
49. Ozoe H, Okada K (1989) The effect of the direction of the external magnetic field on the three-dimensional natural convection in a cubical enclosure. Int J Heat Mass Transfer 32(10):1939–1954
50. Ece MC, Büyük E (2006) Natural-convection flow under a magnetic field in an inclined rectangular enclosure heated and cooled on adjacent walls. Fluid Dyn Res 38(8):564–590
51. Al-Najem N, Khanafer K, El-Refaee M (1998) Numerical study of laminar natural convection in tilted enclosure with transverse magnetic field. Int J Numer Methods Heat Fluid Flow
52. Jalil J, Al-Tae'y K (2007) MHD turbulent natural convection in a liquid metal filed square enclosure. Emirates J Eng Res 12(2):31–40
53. Gelfgat AY, Bar-Yoseph P (2001) The effect of an external magnetic field on oscillatory instability of convective flows in a rectangular cavity. Phys Fluids 13(8):2269–2278
54. Aleksandrova S, Molokov S (2004) Three-dimensional buoyant convection in a rectangular cavity with differentially heated walls in a strong magnetic field. Fluid Dyn Res 35(1):37–66
55. Kahveci K, Öztuna S (2009) MHD natural convection flow and heat transfer in a laterally heated partitioned enclosure. Eur J Mech B Fluids 28(6):744–752
56. Kefayati G et al (2013) Investigation of Prandtl number effect on natural convection MHD in an open cavity by lattice Boltzmann method. Eng Comput
57. Kalabin E, Kanashina M, Zubkov P (2005) Natural-convective heat transfer in a square cavity with time-varying side-wall temperature. Numer Heat Transfer Part A: Appl 47(6):621–631
58. Sheremet MA, Pop I (2014) Natural convection in a square porous cavity with sinusoidal temperature distributions on both side walls filled with a nanofluid: Buongiorno's mathematical model. Transp Porous Media 105(2):411–429
59. Kazmierczak M, Chinoda Z (1992) Buoyancy-driven flow in an enclosure with time periodic boundary conditions. Int J Heat Mass Transfer 35(6):1507–1518
60. Parvin S et al (2020) Soret and Dufour effects on magneto-hydrodynamics Newtonian fluid flow beyond a stretching/shrinking sheet. CFD Lett 12(8):85–97
61. Behbahani F, Khairuddin U, Yusof R (2020) Application of magneto rheological damper system in warren truss bridge structure. J Adv Res Fluid Mech Therm Sci 76(1):17–29
62. Djebali R, Abboudi S, Abbassi MA (2019) Transition thresholds and routes to unsteadiness of magneto-convective flows in tall cavities at low-Prandtl number fluids. CFD Lett 11(3):55–71
63. Deng Q-H, Chang J-J (2008) Natural convection in a rectangular enclosure with sinusoidal temperature distributions on both side walls. Numer Heat Transfer Part A: Appl 54(5):507–524
64. Martínez DO, Chen S, Matthaeus WH (1994) Lattice Boltzmann magnetohydrodynamics. Phys Plasmas 1(6):1850–1867
65. Bilgen E, Yedder RB (2007) Natural convection in enclosure with heating and cooling by sinusoidal temperature profiles on one side. Int J Heat Mass Transfer 50(1):139–150
66. Sarris I, Lekakis I, Vlachos N (2002) Natural convection in a 2D enclosure with sinusoidal upper wall temperature. Numer Heat Transfer: Part A: Appl 42(5):513–530
67. Varol Y, Oztop HF, Pop I (2008) Numerical analysis of natural convection for a porous rectangular enclosure with sinusoidally varying temperature profile on the bottom wall. Int Commun Heat Mass Transfer 35(1):56–64
68. Saeid NH, Yaacob Y (2006) Natural convection in a square cavity with spatial side-wall temperature variation. Numer Heat Transfer Part A: Appl 49(7):683–697
69. Boon JP (2003) The lattice Boltzmann equation for fluid dynamics and beyond: by Sauro Succi (Clarendon Press, Oxford, 2001) ISBN 0 19 850398 9. Eur J Mech B Fluids 22(1):101
70. Mohamad A (2007) Applied lattice Boltzmann method for transport phenomena, momentum, heat and mass transfer. Can J Chem Eng 85(6):946–946
71. Kao PH, Yang RJ (2007) Simulating oscillatory flows in Rayleigh-Bénard convection using the lattice Boltzmann method. Int J Heat Mass Transfer 50(17):3315–3328
72. Guiet J, Reggio M, Vasseur P (2011) Natural convection of nanofluids in heated enclosures using the lattice Boltzmann method. Comput Therm Sci: Int J 3(4)

73. Chaabane R, Askri F, Nasrallah SB (2011) Parametric study of simultaneous transient conduction and radiation in a two-dimensional participating medium. Commun Nonlinear Sci Numer Simul 16(10):4006–4020
74. Chaabane R, Askri F, Nasrallah SB (2011) Analysis of two-dimensional transient conduction–radiation problems in an anisotropically scattering participating enclosure using the lattice Boltzmann method and the control volume finite element method. Comput Phys Commun 182(7):1402–1413
75. Karani H, Huber C (2015) Lattice Boltzmann formulation for conjugate heat transfer in heterogeneous media. Phys Rev E 91(2):023304
76. Chaabane R, Askri F, Ben Nasrallah S (2011) Application of the lattice Boltzmann method to transient conduction and radiation heat transfer in cylindrical media. J Quant Spectrosc Radiat Transfer 112(12):2013–2027
77. Yue L et al (2021) A lattice Boltzmann model for the conjugate heat transfer. Int J Heat Mass Transfer 165:120682
78. Chaabane R et al (2017) Numerical study of transient convection with volumetric radiation using an hybrid lattice Boltzmann BGK-control volume finite element method. J Heat Transfer 139(9):092701
79. Chaabane R et al (2017) Analysis of Rayleigh-Bénard convection with thermal volumetric radiation using lattice Boltzmann formulation. J Therm Sci Technol 12(2):JTST0020
80. Chen S, Yan YY, Gong W (2017) A simple lattice Boltzmann model for conjugate heat transfer research. Int J Heat Mass Transfer 107:862–870
81. Pareschi G et al (2016) Conjugate heat transfer with the entropic lattice Boltzmann method. Phys Rev E 94(1):013305
82. Lu JH, Lei HY, Dai CS (2017) A simple difference method for lattice Boltzmann algorithm to simulate conjugate heat transfer. Int J Heat Mass Transfer 114:268–276
83. Gao D et al (2017) A modified lattice Boltzmann model for conjugate heat transfer in porous media. Int J Heat Mass Transfer 105:673–683
84. Hu Y et al (2015) Full Eulerian lattice Boltzmann model for conjugate heat transfer. Phys Rev E 92(6):063305
85. Sidik NAC, Jahanshaloo L (2014) The use of lattice Boltzmann numerical scheme for contaminant removal from a heated cavity in horizontal channel. CFD Lett 6(3):94–100
86. Perumal DA, Kumar GV, Dass AK (2012) Lattice Boltzmann simulation of viscous flow past elliptical cylinder. CFD Lett 4(3):127–139
87. Sidik NAC et al (2010) Simulation of rarefied gas flow in slip and transitional regimes by the lattice Boltzmann method. CFD Lett 2(2):66–74
88. Perumal DA, Kumar GV, Dass AK (2021) Application of lattice Boltzmann method to fluid flows in microgeometries. CFD Lett
89. Perumal DA, Dass AK (2010) Simulation of incompressible flows in two-sided lid-driven square cavities: Part II—LBM. CFD Lett 2(1):25–38
90. Sidik NAC, Jahanshaloo L (2012) Lattice Boltzmann numerical scheme for transient hydrodynamics of solid particles in an enclosure. CFD Lett 4(3):102–111

Microwave Hybrid Heating as an Alternative Method for Soldering—A Brief Review

N. M. Maliessa and S. R. A. Idris

Abstract The paper reviews the Microwave Hybrid Heating (MHH) method as well as the effect of MHH towards the interfacial reaction and the shear strength at the solder/Cu joint. Previously, reflow soldering process was performed to solder electronic component. Due to its high defect rate, processing time and energy consumption, MHH method are getting more attention among electronics manufacturers to perform industrial process as it is beneficial in modern microtechnology. MHH method has faster heating rate, improve heating uniformity, reduces the chance of thermal runaway, reduce processing temperature, and reduce hazards to human and environment. This approach has proven to yield scallop-like and angular trapezoid structure of Cu_6Sn_5 and Cu_3Sn in the intermetallic compound (IMC). The IMC thickness shows a competitive result (5.337 and 5.717 μm) compared to reflow soldering. However, not many studies were done on the shear strength of the solder joint.

Keywords Microwave hybrid heating · Sn–Ag–Cu · Susceptor

1 Introduction

Microelectronics assembly especially in a surface mount technology uses reflow soldering as the joining method. Reflow soldering can be defined as 'the joining of mating surfaces that have been tinned and/or have solder between them, placing them together, heating them until the solder fuses, and allowing them to cool in the joined position' [1]. It is used as a technique to provide electrical, thermal, and mechanical connectivity between two metallic surfaces. However, it is known that reflow soldering requires more processing time, consumes higher energy and produced higher solder joint defect rate [2] compared to other conservative methods.

Nowadays, microwave energy has been applied as an alternative to the process that consumed high energy that is being used in the industries [3]. Formerly, microwaves

N. M. Maliessa · S. R. A. Idris (✉)
Faculty of Mechanical and Automotive Engineering Technology (FTKMA), Universiti Malaysia Pahang, 26600 Pekan, Pahang, Malaysia
e-mail: rabiatull@ump.edu.my

M. Y. Ismail et al. (eds.), *Technological Advancement in Mechanical and Automotive Engineering*, Lecture Notes in Mechanical Engineering,
https://doi.org/10.1007/978-981-19-1457-7_44

were used for communications [4]. However, an experiment by Percy Spencer in 1950 about the heating of materials by microwave energy got recognized [5]. Since then, new processes for industrial manufacturing were recognized such as tempering, curing of wood, processing of medical waste and others [4]. There are also emergent interest of the use of microwaves for sintering [6], brazing [7], joining [8] and melting [9].

Researchers stated that microwave materials processing provides a faster, neater, low cost and adaptable process [3, 10–12]. Nevertheless, this method can only be applied on microwave absorbing materials (hydrocarbons), ceramics and ceramic composites [10] which limits the application of microwave for soldering where the materials are metals. In addition, several processing challenges were developed from the unlike mechanism of energy transfer in microwave heating [12]. In microwave heating, electromagnetic energy is converted to thermal energy. The effectiveness of the conversion depends on the dielectric nature of the materials. Conducting materials (metals) do not competently convert microwave energy to heat since they are microwave reflector [13, 14]. Thus, inhomogeneity in the electromagnetic field will result in non-uniform heating.

Nonetheless, an interest of high temperature microwave for processing of polymers and ceramics has been growing since the application of microwave rapid heating managed to reduce processing time and reduces problems such as irregular and localized microstructural coarsening with its uniform heating [15, 16]. It is known that the heating mechanism of microwave energy is undeniably different from conventional furnace heating as microwave heating involved volumetric heating while conventional furnace uses resistant heating [3, 17]. There are only few studies and research regarding the application of microwave on metallic materials due to the known fact that metals reflect microwave. Due to that factor, a new heating method namely Microwave Hybrid Heating (MHH) was introduced to process metallic materials by microwave.

Understanding the intermetallic compound (IMC) formation is one of the important factors in a flip-chip interconnection as it is the result of joining process between the solder and the substrate. IMC consists of a homogeneous phase of two or more materials that forms just before soldering, after soldering or during service. In electronic manufacturing, IMCs formed when in contact with all common base materials, coatings and metallization. Commonly, Cu_6Sn_5, Cu_3Sn and Ag_3Sn are the IMCs that could be found at the interface.

With this new MHH method, it is desired to know the effect of MHH towards the solder reliability. Other than comprehending the IMC formation, shear strength is also one of the important determinant of the solder joint reliability. In this paper, a review of existing literature on the usage of microwave hybrid heating (MHH) as a substitute for conventional furnace for reflow soldering is carried out. The effect of MHH on the interfacial reaction and shear strength between SAC solder/Cu joint is also presented.

2 Microwave Hybrid Heating as Joining Method

There are two classes of microwave hybrid heating (MHH) which are; combination of conventional heating and microwave heating or by using microwave susceptors [11]. Figure 1 presents the classes of MHH. The susceptor acts as an absorber of microwave energy which then converts it into heat. It supply heat by means of either conduction, convection or radiation and is useful for initial heating [18]. According to Haque [19], the most commonly used materials as susceptor are silicon carbide (SiC), graphite (C) and magnetite (Fe_3O_4) and they were chosen due to their high dielectric loss factor and excellent refractory properties [20].

This hybrid-heating concept was applied to achieve two-directional heating mechanism, which is radiant heating and volumetric heating. Radiant heat was obtained from the susceptors while volumetric heat was obtained by the heat transformation from the absorbed microwave energy within the sample's particles [21]. To simplify, the heat is transferred from surface into the material (volumetric heating) as well as the sample itself absorbs microwave energy and is heated from within with the help of susceptor (radiant heating). The experimental setup for MHH were illustrated in Fig. 2 [13]. There are a few parameters that can be manipulated in order to achieve desirable solder joint such as exposure time, amount of susceptor, and power. However, exposure time and amount of susceptor are the parameters that were frequently used.

Few studies have been made on the joining of metallic materials interfacial reaction by using microwave hybrid heating as the heat source. Mild steel to-mild steel was joined by MHH method as reported by Bansal et al. [22] to study the metallurgical bonding of the samples. A joint without voids or cracks was observed between the samples with a dense and homogeneous structure [22]. On another note, Oghbaei and Mirzaee [23], made a report on the effect of MHH over conventional sintering on the particle size of 92.5W-6.4Ni-1.1Fe alloy. They mentioned in their study that

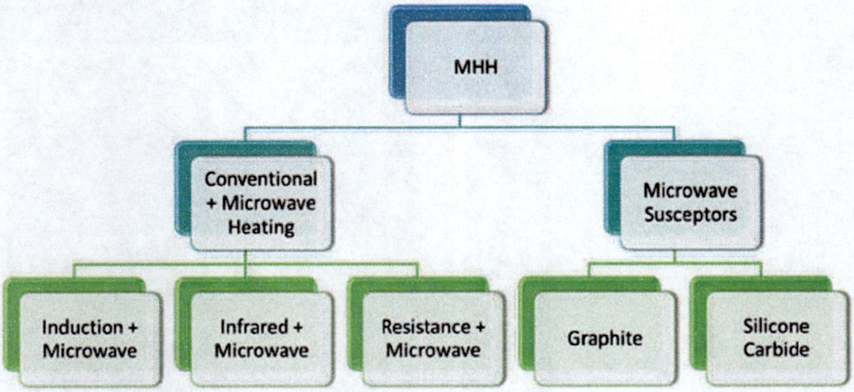

Fig. 1 Classes of microwave hybrid heating method [11]

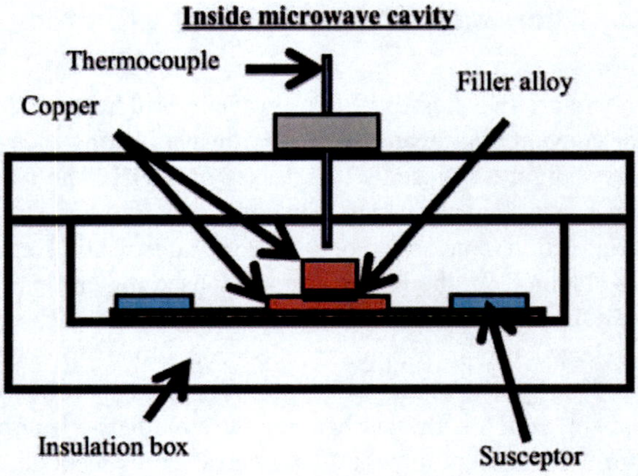

Fig. 2 The experimental setup for microwave hybrid heating (MHH) [13]

MHH method has shown to produce samples with more uniform grain size distribution and higher density (2010) [23]. The SEM micrograph of conventional and microwave sintered 92.5W-6.4Ni-1.1Fe alloy were displayed in Fig. 3.

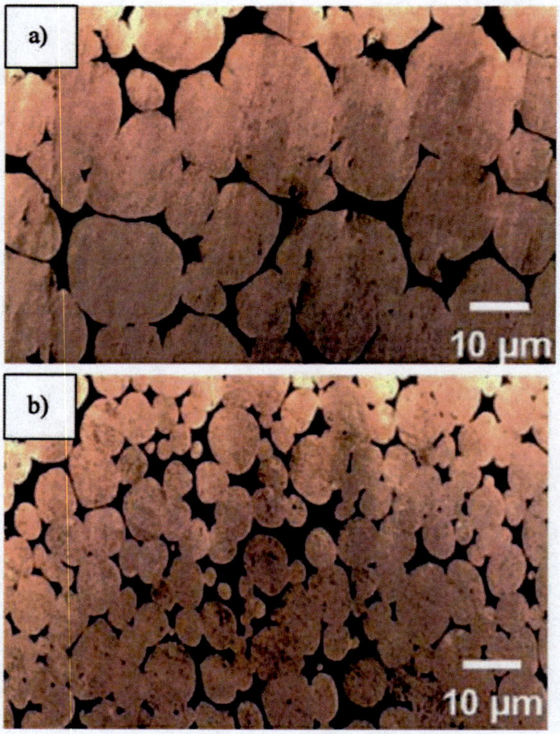

Fig. 3 SEM micrograph of **a** conventional and **b** microwave sintering of 92.5W-6.4Ni-1.1Fe alloy [23]

A two-directional sintering process of aluminum, magnesium and Sn-3.5Ag solder were carried out with the help of SiC susceptor in 2004 by [24]. It has been proven that this setup has resulted a higher density with minimum grain size and distribution observed on Sn-3.5Ag solder. Other than that, superior tensile properties were observed as a result of this method as it helps to reduce the extent of high temperature thermal exposure.

Several advantages of MHH over conventional heating or direct microwave heating is that it has faster heating rate that leads to energy saving and improve heating uniformity, reduces the chance of thermal runaway, reduce processing temperature, and reduce hazards to human and environment [11].

3 Interfacial Reactions at Sn–Ag–Cu/Cu Joint

The presence of intermetallic compound (IMC) indicates that there is chemical reactions occur between the solder and substrate [25]. A thin and uniform layer of IMC is desired to achieve good metallurgical bonding [26]. Without IMCs, no bonding between the solder and the substrate as there is no metallurgical interaction occurs.

The IMC formation is controlled by a diffusion process, either in a solid state or liquid state condition. At first, the Sn from solder started to diffuse towards metal substrate thus, forming IMC. After the IMC are formed, their growth is then governed by a thermally activated diffusion process at solid state. According to Smith and Hashemi [27], there are several factors that influence the diffusion rate such as environment temperature, diffusion distance and concentration difference between compounds.

3.1 Types of Intermetallic Compound

There are many factors affecting the microstructure behavior thus, full control of the microstructure cannot be expected. Factors such as the diffusion rate of the substrate, difference in processing method and temperature may affect microstructure formation and growth. Hence, by understanding the behavior of IMC, topmost microstructure control can be taken. Common IMC between SAC/Cu was discussed in the following section.

Cu_6Sn_5 and Cu_3Sn IMC

During the soldering of SAC solder with Cu substrate, after the flux is oxidized, the Cu (from substrate) starts to liquefy to the molten solder. At this point, the initial dissolution rate is very high. After a short time, the molten solder layer next to the contacted Cu becomes highly saturated with the dissolved Cu throughout the interface. Then, the solid IMC starts to form at local equilibrium solubility (in

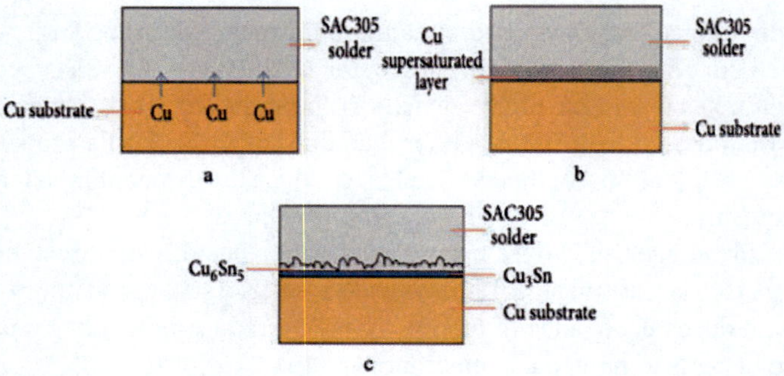

Fig. 4 The interfacial reaction of SAC/Cu during soldering; **(a)** dissolution of Cu substrate, **(b)** supersaturation of molten solder with Cu, and **(c)** formation of IMC [28]

between layer of solder and the contacted metal). Figure 4 illustrates the schematic diagram of the interfacial reaction between SAC/Cu during soldering [28].

The common IMC that can be found in Sn–Ag–Cu/Cu solder joints are Cu_6Sn_5 and Cu_3Sn. Reaction between Cu and Sn formed Cu_6Sn_5 while reaction between Sn and Ag yields Ag_3Sn as well as Cu_3Sn. Scallop-shape Cu_6Sn_5 first formed at Sn/Cu interface and it was formed by the dissolution of Cu, followed by chemical reaction. Cu_3Sn will only be found at the eutectic point when Cu content is high enough for the formation of Cu_3Sn [29]. It was formed by diffusion and by reaction type growth [30].

Similar findings were discovered from previous studies [31–33]. Figure 5 displays the micrographs of cross section for SAC305/Cu joints with microwave hybrid heating with graphite as susceptor (4 and 6 g) for 120 s. IMC layer consisting of scallop-shape Cu_6Sn_5 and Cu_3Sn were observed for both samples soldered with 4 and 6 g or graphite powder. The formation of scallop Cu_6Sn_5 has been observed usually in wetting reaction between Sn-based solder and Cu [34–36]

Experiment on soldering of Sn–Ag–Cu solder onto copper (Cu) plate by MHH with the use of graphite (4 and 6 g) as susceptor were performed in 2014 by Lutfi et al. [3]. In the study, the effect of MHH towards the intermetallic compound formation were investigated. It was shown that scallop-like and angular trapezoid structure of Cu_6Sn_5 as well as Cu_3Sn were found in the IMC. The hardness value of Cu_6Sn_5 after heated in 4 g of susceptor showed a higher and consistent results compared to 6 g of susceptor.

Ag_3Sn

Another intermetallic compound that can be found in Sn-based solder is Ag_3Sn. Thus, when using SnAg type solder, there is a chance for Ag_3Sn to form regardless the type of substrate used. As the temperature decrease during soldering, Cu_6Sn_5 IMCs begins to precipitate at the interface. This process consumes both Cu and Sn and thus, increases the Ag content near the solder/substrate interface [37]. When the Ag composition are concentrated at the solder/substrate interface, Ag starts to precipitate in the form of Ag_3Sn IMCs.

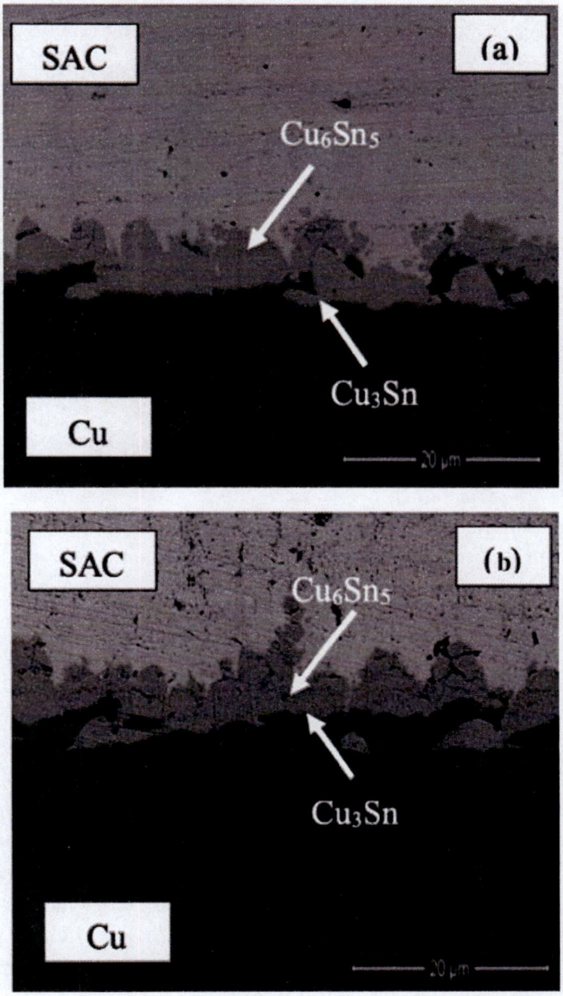

Fig. 5 Microstructure of Sn-3.0Ag-0.5Cu/Cu joint interface heating in 4 and 6 g graphite powder for 120 s [3]

In a study by Choi et al. [38], an Ag_3Sn particles with a long rod shape were observed near the IMC layer between Cu/electroless plated Ni/evaporated Au and Sn-3.5Ag-1.0Cu solder. Same result was also obtained by several other researcher such as Frear. In this study, conventional reflow soldering method was used which is bump metallurgy method. The sample were heated at 250 °C for 60 s using rosin mildly activated flux followed by air cooling on the aluminum block.

According to Lee et al. [39], there are three types of Ag_3Sn that could form which are particle-like, needle-like, and plate-like. Large Ag_3Sn platelets were reported to generate brittle fracture at the interface and provide crack initiation sites. It could also influence the fracture pattern in tensile and shear tests of the solder joints [34]. Since then, more in-depth study regarding Ag_3Sn were conducted. A series of experiments

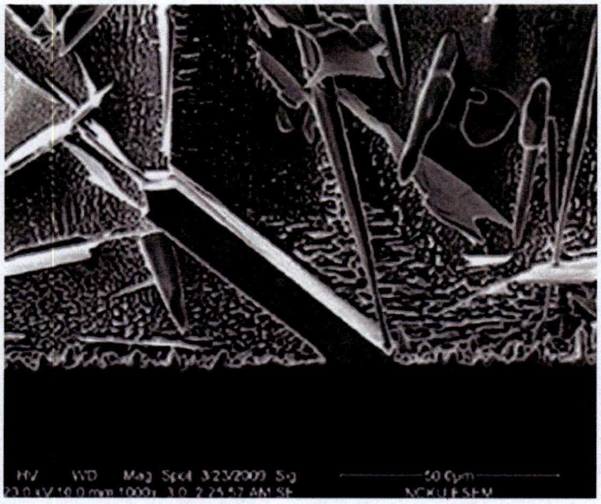

Fig. 6 SEM micrograph of large plate-like and needle-like Ag_3Sn [39]

conducted by Gong et al. [37] to study the entire formation process of Ag_3Sn plates through reflow soldering. It was found that Ag_3Sn plates were formed starting at the middle of reflow cooling stage when the temperature starts to decrease. Shen et al. [40] also conducted a study on the growth mechanism of Ag_3Sn during solidification. Using thermal analysis and phase equilibria calculations, it was concluded that bulk Ag_3Sn were discovered with a slow cooling rate. The bulk Ag_3Sn was controlled by the formation of Ag_3Sn crystal nuclei at the beginning of the eutectic reaction due to the movement of Ag atoms during slow solidification of the alloy (Fig. 6).

Since the type of Ag_3Sn formed was influenced by the cooling rate during solidification process, the effect of cooling rate on the type of Ag_3Sn formed was also investigated. According to Lee et al. [39], rapid cooling will generate small particle-like Ag_3Sn in large amount while slow cooling will generate large plate-like or pillar-like Ag_3Sn. To conclude, the formation and growth of Ag_3Sn can either improve or worsen the solder joint performance. Small particles Ag_3Sn can improve solder joint performance, while huge Ag_3Sn may results in solder joint failure.

IMC Thickness at SnAgCu/Cu Joint

From metallurgical viewpoint, it is important to understand the knowledge of solder/substrate interaction to optimize the reliability of the solder interconnections [41]. One way to improve the reliability of solder joint is by identifying the average thickness of the IMC layer. Appropriately thin IMC layer is desirable because it can acts as a barrier from unfavorable chemical reaction between solder and substrate [42]. However, IMC will eventually grow under certain circumstances such as during soldering or isothermal aging. A thick IMC may degrade the reliability of the solder joints due to its inherent brittle nature and their tendency to generate structural defects.

In the study by Lutfi et al. [15], different type of susceptor (graphite and silicon carbide) were used to join Cu-Sn-Ni–P/Cu in order to study the intermetallic compound formation (IMC) layer and microstructure by microwave hybrid heating

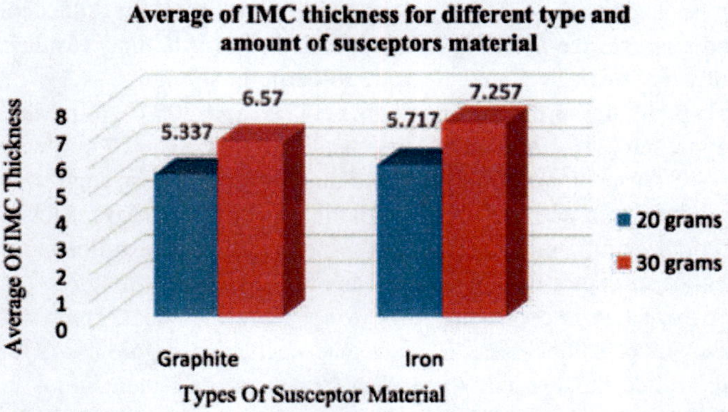

Fig. 7 The average IMC thickness for different type and amount of susceptor material [26]

(MHH). It was found that no voids or cracks were observed between the samples for both susceptors. Cu-7.0Ni-9.3Sn-6.3P paste type solder alloy was used in this experiment. Pure Cu (99.99%) plate with the dimension of 15 mm × 15 mm × 3 mm were used as a substrate with a pure Cu rod with 8 mm diameter and 5 mm height. As a result, Cu_6Sn_5 IMC thickness of 16.5 μm and 50.5 μm was obtained with the use of SiC and graphite as the susceptor respectively.

In addition, a successful joining of Sn-4.0Ag-0.5Cu (SAC405) solder with Cu substrate was obtained from the study by Faisal and Saliza in 2017 [26]. In this study, SAC405 solder ball with diameter of 900 μm was sandwiched between 45 × 50 × 1 mm Cu substrate and was soldered for 80 s. Domestic microwave oven was used with operating frequency of 2.4 GHz and 800 W. 2 types of susceptor was used in their experiment which were graphite powder and iron powder of 20 and 30 g. After soldering process with MHH method, a uniform scallop type of IMC layer was observed in both susceptor. However, the higher the amount of susceptor, the thicker the Cu_6Sn_5 IMC layer obtained. The average thickness was measured using image analyzer and the thickness for 20 g and 30 g of susceptor were 5.337 μm and 6.570 μm for graphite powder meanwhile 5.717 μm and 7.257 μm for iron powders, respectively. Figure 7 shows the average thickness of intermetallic compound for different type and amount of susceptor.

4 Solder Joint Shear Strength

Solder joint reliability is one of the most critical issues in the development of technologies for electronic devices. One of the important parameters of a reliable solder joint is the shear strength between the solder and the substrate. Solder joint strength determines its endurance towards extreme temperature, the frequency of on/off power cycling, the mechanical stresses, and other factors throughout its service life.

According to Nasir et al. [43], the strength and reliability of the solder joint can be defined and characterize by the IMC thickness. A thin IMC may result in a strong solder joint while thick IMC may degrade solder joint strength.

In a study by Maliessa, the shear strength between SAC305/Cu joint were obtained by MHH approach [44]. SAC305/CU were soldered for 5, 6 and 7 min with different amount of SiC (susceptor). It was observed that highest shear strength was obtained by soldering for 7 min with 4 g of SiC with the value of 26.71 MPa. However, while MHH method is gaining more attention in soldering, there is insufficient information on the effect of susceptor or other parameters towards shear strength.

The lap shear test is commonly used to evaluate the shear, creep and thermal fatigue behavior of solder joints in electronic packaging regardless of the type of solder and surface finish used. According to few studies, there is no standardize standard for lap shear test for solder joint, however several standards such as ASTM D1002, D3163, D3164 and D5656 has been used as guide and had been adjusted to suit the requirement of the studies [31, 45].

There were several other mechanical tests that were used to evaluate the solder joint reliability. Drop test often conducted to portable electronic products that are liable to accidental drops which may cause internal circuit board impairment. Bending test were carried out to analyze the fatigue failure.

5 Discussion

Previously, microwave technology was applied on process involving metals such as sintering. In 2005, hybrid microwave sintering were introduced [24]. Since then, more process has started implementing microwave hybrid heating technology. Soldering with microwave technology has shown significant improvement on the intermetallic compound formation and thickness. Unfortunately, there are inadequate information on the mechanical and thermal properties of the solder joint.

While MHH has been getting more attention, there are still more challenges and difficulties to apply this approach in microelectronics assembly. More studies should be done to further investigate the mechanical properties and thermal properties such as shear strength, microhardness and thermal fatigue in order to determine the microwave reliability in soldering. As microwave energy is not easy to control, a lot of other parameters could affect the microstructural properties, mechanical properties and thermal properties of the solder joint.

6 Conclusions

Application of microwave for soldering has emerged as a potential process. Microwave hybrid heating (MHH) is an effective method to process materials with low dielectric loss materials (ceramics) to metals that are known to reflect microwave.

This method offers several advantages over direct microwave heating and conventional heating process such as faster heating rate, energy efficient and improve heating uniformity. In MHH, different susceptor materials may provide different outcome on intermetallic compound. Few common susceptors that were used in MHH are graphite and silicon carbide. A desirably thin IMC layer (5.337 μm) were observed in MHH compared to conventional soldering process. The shear strength of the solder joint with MHH are observed to have competitive results (26.71 MPa). Yet, only a few studies are available on shear strength of the solder joint by microwave hybrid heating and mostly concentrating on tensile strength and hardness.

Acknowledgements The authors would like to thanks the Ministry of Higher Education for providing financial support under Fundamental Research Grant Scheme (FRGS) No. FRGS/1/2018/TK03/UMP/02/21 (University reference RDU190152) and Universiti Malaysia Pahang for laboratory facilities.

References

1. Minges ML, Dostal C (1989) Electronic materials handbook, vol 1. Packaging. ASM International, Materials Park, OH, p 55
2. Lau CS, Abdullah M, Khor C (2013) Optimization of the reflow soldering process with multiple quality characteristics in ball grid array packaging by using the grey-based Taguchi method. Microelectron Int
3. Lutfi M et al (2014) Effect of microwave hybrid heating on the formation of intermetallic compound of Sn-Ag-Cu solder joints. In: 36th international electronics manufacturing technology conference. IEEE
4. Mishra RR, Sharma AK (2016) Microwave–material interaction phenomena: heating mechanisms, challenges and opportunities in material processing. Compos A Appl Sci Manuf 81:78–97
5. Shukla M, Ghosh S (2016) Microwave joining of ceramics: an overview
6. Karayannis VG (2016) Microwave sintering of ceramic materials. IOP Conf Ser: Mater Sci Eng
7. Tamang S, Aravindan S (2019) Brazing of cBN to WC-Co by Ag-Cu-In-Ti alloy through microwave hybrid heating for cutting tool application. Mater Lett 254:145–148
8. Samyal R, Bagha AK, Bedi R (2020) Microwave joining of similar/dissimilar metals and its characterizations: a review. Mater Today: Proc 26:423–433
9. Xu L et al (2017) Study on characteristics of microwave melting of copper powder. J Alloy Compd 701:236–243
10. Srinath M, Sharma AK, Kumar P (2011) A new approach to joining of bulk copper using microwave energy. Mater Des 32(5):2685–2694
11. Yahaya B et al (2014) Microwave hybrid heating of materials using susceptors-a brief review. Adv Mater Res
12. Thostenson E, Chou T-W (1999) Microwave processing: fundamentals and applications. Compos A Appl Sci Manuf 30(9):1055–1071
13. Clark DE, Folz DC, West JK (2000) Processing materials with microwave energy. Mater Sci Eng A 287(2):153–158
14. Metaxas AA, Meredith RJ (1983) Industrial microwave heating. IET
15. Lutfi M et al (2016) Interfacial reaction analysis of Cu-Sn-Ni-P/Cu joint using microwave hybrid heating. Key Eng Mater

16. Sethi G, Upadhyaya A, Agrawal D (2003) Microwave and conventional sintering of premixed and prealloyed Cu-12Sn bronze. Sci Sinter 35(2):49–65
17. Babaghorbani P, Gupta M (2008) Enhancing the mechanical response of a lead-free solder using an energy-efficient microwave sintering route. J Electron Mater 37(6):860–866
18. Chandrasekaran S, Basak T, Srinivasan R (2013) Microwave heating characteristics of graphite based powder mixtures. Int Commun Heat Mass Transfer 48:22–27
19. Haque KE (1999) Microwave energy for mineral treatment processes—a brief review. Int J Miner Process 57(1):1–24
20. Zhao C et al (2000) Hybrid sintering with a tubular susceptor in a cylindrical single-mode microwave furnace. Acta Mater 48(14):3795–3801
21. Somidin F, Saud N, Mohd Salleh MAA (2015) The microstructure evolutions of Sn-0.7 Cu solder using microwave-assisted sintering method at various exposure times. Adv Mater Res
22. Bansal A, Sharma AK, Das S (2013) Metallurgical and mechanical characterization of mild steel-mild steel joint formed by microwave hybrid heating process. Sadhana 38(4):679–686
23. Oghbaei M, Mirzaee O (2010) Microwave versus conventional sintering: a review of fundamentals, advantages and applications. J Alloy Compd 494(1–2):175–189
24. Gupta M, Wong W (2005) Enhancing overall mechanical performance of metallic materials using two-directional microwave assisted rapid sintering. Scripta Mater 52(6):479–483
25. Laurila T, Vuorinen V, Kivilahti J (2005) Interfacial reactions between lead-free solders and common base materials. Mater Sci Eng R Rep 49(1–2):1–60
26. Mohamad Faizal S, Saliza Azlina O (2016) Intermetallic compound formation on lead-free solders by using microwave energy. AIP Conf Proc
27. Smith WF, Hashemi J, Presuel-Moreno F (2006) Foundations of materials science and engineering. McGraw-Hill Publishing
28. Lee LM, Mohamad AA (2013) Interfacial reaction of Sn-Ag-Cu lead-free solder alloy on Cu: a review. Adv Mater Sci Eng 2013
29. Sona M, Prabhu K (2013) Review on microstructure evolution in Sn–Ag–Cu solders and its effect on mechanical integrity of solder joints. J Mater Sci: Mater Electron 24(9):3149–3169
30. Kivilahti J (2002) The chemical modeling of electronic materials and interconnections. JOM 54(12):52–57
31. Hu X et al (2014) Microstructure and shear strength of Sn37Pb/Cu solder joints subjected to isothermal aging. Microelectron Reliab 54(8):1575–1582
32. Wang X, Wang Z, Shang JK (2008) Effect of displacement rate on lap shear test of SAC solder ball joints. In: 2008 international conference on electronic packaging technology & high density packaging. IEEE
33. Yoon J-W, Jung S-B (2006) Effect of isothermal aging on the interfacial reactions between Sn–0.4Cu solder and Cu substrate with or without ENIG plating layer. Surf Coat Technol 200(14–15):4440–4447
34. Kim K, Huh S, Suganuma K (2003) Effects of intermetallic compounds on properties of Sn–Ag–Cu lead-free soldered joints. J Alloy Compd 352(1–2):226–236
35. Sun P et al (2006) Intermetallic compound formation in Sn-Co-Cu, Sn-Ag-Cu and eutectic Sn-Cu solder joints on electroless Ni (P) immersion Au surface finish after reflow soldering. In: Conference on high density microsystem design and packaging and component failure analysis, 2006. HDP'06. IEEE
36. Mookam N, Kanlayasiri K (2012) Evolution of intermetallic compounds between Sn-0.3Ag-0.7Cu low-silver lead-free solder and Cu substrate during thermal aging. J Mater Sci Technol 28(1):53–59
37. Gong J et al (2010) Formation of Ag_3Sn plates in SnAgCu solder bumps. Mater Sci Eng A 527(10–11):2588–2591
38. Choi WK et al (2002) Interfacial microstructure and joint strength of Sn–3.5 Ag–X (X = Cu, In, Ni) solder joint. J Mater Res 17(1):43–51
39. Lee H-T et al (2009) Effect of cooling rate on Ag_3Sn formation in Sn-Ag based lead-free solder. In: 2009 11th electronics packaging technology conference. IEEE

40. Shen J, Chan YC, Liu S (2008) Growth mechanism of bulk Ag_3Sn intermetallic compounds in Sn–Ag solder during solidification. Intermetallics 16(9):1142–1148
41. Liu J et al (2008) Recent development of nano-solder paste for electronics interconnect applications. In: 2008 10th electronics packaging technology conference. IEEE
42. Chen W-H et al (2013) IMC growth reaction and its effects on solder joint thermal cycling reliability of 3D chip stacking packaging. Microelectron Reliab 53(1):30–40
43. Nasir NM et al (2016) Improvement of Sn-0.7 Cu lead free solder joints on shear strength with addition of titanium oxide (TiO_2) particles. Mater Sci Forum
44. Maliessa N, Idris SRA (2019) Effect of different amount of silicon carbide on SAC solder-Cu joint performance by using microwave hybrid heating method. IOP Conf Ser: Mater Sci Eng
45. An T, Qin F (2014) Effects of the intermetallic compound microstructure on the tensile behavior of $Sn_{3.0}Ag_{0.5}Cu$/Cu solder joint under various strain rates. Microelectron Reliab 54(5):932–938

Effect of Opening Ratios with and Without Louvers in Cross Ventilation Using CFD

Lip Kean Moey, Saleh Mohammed Saleh Alyazidi, Vin Cent Tai, Joseph Wu Kai-Seun, Prasath Reuben Mathew, and Ahmed Nurye Oumer

Abstract As the world marches forward implementing concepts of sustainable buildings, higher reliance on natural ventilation can be obtained through louvers. In this research of cross ventilation in a generic isolated building, the leeward opening sizes were manipulated to 1:1, 1:0.25 and 1:0.5, with louver angles of 0°, 15°, 30°, 45° and no louvers. An Atmospheric Boundary Layer (ABL) condition was applied at the inlet of the flow domain and a 3D-steady Reynolds-Averaged Navier–Stokes (RANS) equation was solved with the Shear Stress Transport (SST) k-ω turbulence model. Mesh sensitivity analysis and model validation were performed as per best practices. The results show that as the size of the leeward opening decreases, the acceleration through the louver blades increases. In the absence and presence of louvers, as the windward-louver (W-L) ratio increased from 1:0.25 to 1:1, its dimensionless flow rate (DFR) increases. Highest DFR was obtained when the W-L ratio was 1:1 and the louver angle was 0°, second to louver angle of 15°, followed by the configuration without louvers present. Their respective DFR values were 0.588, 0.544 and 0.522. As the louver angle increased from 0° to 45°, the DFR reduced for all opening W-L ratios.

Keywords Louver · Cross ventilation · Opening ratio · CFD

L. K. Moey (✉) · V. C. Tai
Faculty of Engineering, Built Environment & Information Technology, Centre for Modelling and Simulation, SEGi University, Selangor, Malaysia
e-mail: moeylipkean@segi.edu.my

S. M. S. Alyazidi · J. W. Kai-Seun · P. R. Mathew
Faculty of Engineering, Built Environment & Information Technology, SEGi University, Selangor, Malaysia

A. N. Oumer
Department of Mechanical Engineering, College of Engineering, University Malaysia Pahang, Pahang, Malaysia

M. Y. Ismail et al. (eds.), *Technological Advancement in Mechanical and Automotive Engineering*, Lecture Notes in Mechanical Engineering,
https://doi.org/10.1007/978-981-19-1457-7_45

1 Introduction

The advancement of technology and rapid economic growth can lead to a plethora of benefits and drawbacks to the planet. According to Moey et al. [1], Southeast Asian countries such as Malaysia and Indonesia still depend on fossil fuels as their main source of energy and account for 75% of their primary energy mix. Similarly, they also suggest that 46% of the electricity consumed in Malaysia is by the industrial sector. This fact is alarming as heightened dependence on non-renewable energy can increase greenhouse gas emissions [2]. Hence, it is crucial to seek alternative ways to reduce energy consumption and enforce the usage of renewable energy across the world. Natural ventilation serves as a passive method to cool buildings as it does not require continuous cost to harness it as compared to mechanical ventilation.

Natural ventilation is vital in removing stale polluted indoor air and exchanging it with fresh outside air. In addition, this process can occur by single-sided ventilation, cross ventilation and stack ventilation. Cross ventilation is effective in providing fresh air into a building enclosure since fresh air enters through the windward façade (facing the wind) while the air exits through the leeward façade or the downwind opening. Besides, the driving force for natural ventilation can be either wind induced or due to buoyant forces as a result of temperature difference across the ventilated space [3]. In that note, wind induced ventilation occurs due to a pressure difference across the openings in a building and its efficiency increases when the openings in a building are located at a higher position or closer to the roof due to increased pressure with elevation [4–6].

The performance of natural ventilation systems in a building is important not only for indoor air quality, but also for building energy prediction [7]. It can be affected by external factors such as wind velocity, wind direction and turbulence intensity while internal factors could range from building façade configuration such as windowsill, opening position and size configuration. Sacht and Lukiantchuki [8] studied the effect of opening size on natural ventilation by using Computational Fluid Dynamics (CFD) for two cases which have a window opening area of 10 and 25% of the total floor plan area with varying wind incidence angle. The results showed that the air change rate per hour (ACH) increases with opening size along with the change of pressure coefficient. Similarly, another case study for varying opening size and configuration was done by Yusoff [9] at a mosque in Melaka, Malaysia. The Venturi effect was studied by carrying out three different types of cases using field test measurements and CFD simulations. Based on both studies, it can be said that opening size is crucial for the airflow distribution and indoor air velocity as a larger inlet opening size would lead to a better ventilation rate. Other than that, a small inlet to outlet opening area ratio may help increase air velocity. Thus, it is essential to study the influence of opening size on the ventilation rate, the difference in pressure coefficient and flow distribution in a building.

Isolated buildings equipped with louvers have also become of interest in recent years to understand its potential in enhancing natural ventilation capabilities. Kosutova et al. [10] carried out a research based on a cubic building of 0.15 m in dimension with four types of louver opening configuration and were termed as "Centre", "Top", "Down" and "Up-Down" which have three slats that are at an angle of 15°. A wind tunnel test along with Particle Image Velocimetry (PIV) represents the experimental method used for model verification with CFD simulation. The Reynolds Averaged Navier Stokes (RANS) turbulence model used comprised of Reynolds Stress Model (RSM), Shear Stress Transport (SST) k-ω, and Renormalisation Group (RNG) k-ε. Overall, the RSM turbulence model showed the best agreement with experimental data. Moreover, the centre-centre and top–bottom configuration had the best air exchange efficiency due to better flow mixture throughout the building. As for the variable dimensionless flow rate, "Top" opening configuration was the highest.

Another study involving louvers by Zheng et al. [11] using CFD was carried out to determine the effect of shaded louvers on the performance of natural ventilation. The simulations were carried out for a simplified building model and a coupled indoor-outdoor airflow model. A sensitivity analysis was done by testing the effects of different turbulence model along with near wall function. The CFD studies were conducted with RANS models of SKE, Realisable k-ε (RKE), RNG, SST and RSM. The authors deduced that RKE and SST had the best performance when compared with the experimental results. The shaded louvers and the effect of its rotation angle was also determined to see which configuration lead to the best ventilation rate. The rotation angle was manipulated from 0° to 75° with an increment of 15°. The 0° shaded louver had a flow rate of 284 m^3/h while the non-shaded configuration had a lower flow rate of 276 m^3/h. Similarly, the ventilation rate decreases with an increase of rotation angle which is concurrent to the reduction of discharge coefficient. However, in this study only a single opening position was considered and investigated.

Based on the previous studies regarding natural ventilation, the opening size plays a pivotal role in the distribution of air in a building and has an influence on several performance parameters such as ventilation rate and pressure coefficient. In conjunction with that, prior research conducted on buildings equipped with louvers at the time of writing did not consider the impact of opening sizes paired with varying louver angles. Thus, the purpose of this paper is threefold: (i) to carry out model validation with Ramponi and Blocken [12], (ii) investigate the impact of varying leeward opening size on the dimensionless flow rate (DFR), and (iii) investigate the impact of varying louver angles on DFR. It should be noted that the flow considered in this study is steady, incompressible and isothermal. This paper is organized as follows: Sect. 2 explains the methodology of this paper. Section 3 provides the numerical results as well as the discussion of the results obtained, whilst Sect. 4 concludes the paper.

2 Methodology

2.1 Model Geometry and Simulation Cases

The building geometry used in this study is based on the reference case from Karava et al. [5] and used in the CFD study by Ramponi and Blocken [12]. The building model has a dimension of 100 mm × 100 mm × 80 mm (length × width × height), which corresponds to a reduced scale of 1:200 and a full-scale size of 20 m × 20 m × 16 m. Similarly, the building has a wall thickness of 2 mm whereby the two openings on the windward and leeward face are opposite and symmetrical to each other on the symmetry plane shown in the figure below. In addition, the simulation cases are superimposed with the reference case geometry. The building model shown in Fig. 1 is one in which the size of the inlet opening stays constant while the size of the outlet opening varies from a big to a small opening ratio. The isometric view for each case is shown in Fig. 2.

The first five cases have a windward-leeward (W-L) opening ratio of 1:1. Cases 6–10 have a W-L ratio of 1:0.25, meanwhile cases 11–15 have a W-L ratio of 1:0.5. The leeward opening for the first five cases have a dimension of 46 mm × 18 mm; cases 6–10 have leeward opening dimension of 23 mm × 9 mm; while cases 11–15 have a leeward opening dimension of 32.52 mm × 12.73 mm. The dimensions are summarized in Table 1.

The thickness of the louver is 0.5 mm. The distance between louver blades are set to $0.2H_0$, where H_0 represents the height of the leeward or windward opening. This is shown in Fig. 3. By setting the louver blade distance to $0.2H_0$, this will standardize the louver blade positioning as the leeward opening changes in dimension.

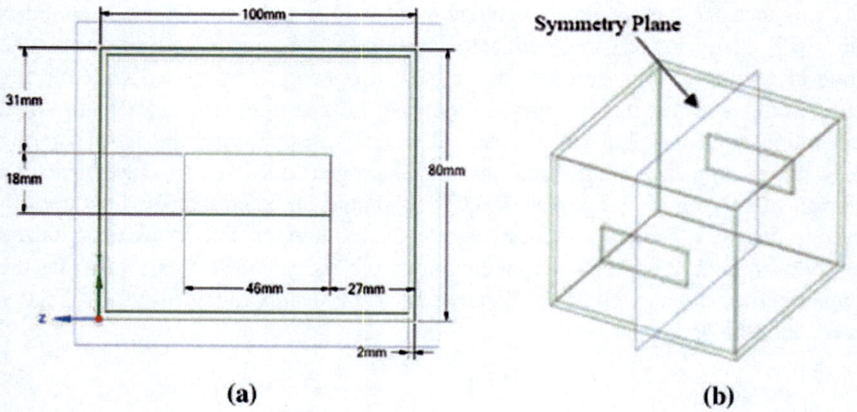

Fig. 1 **a** Model dimensions. **b** Symmetry plane in the building model

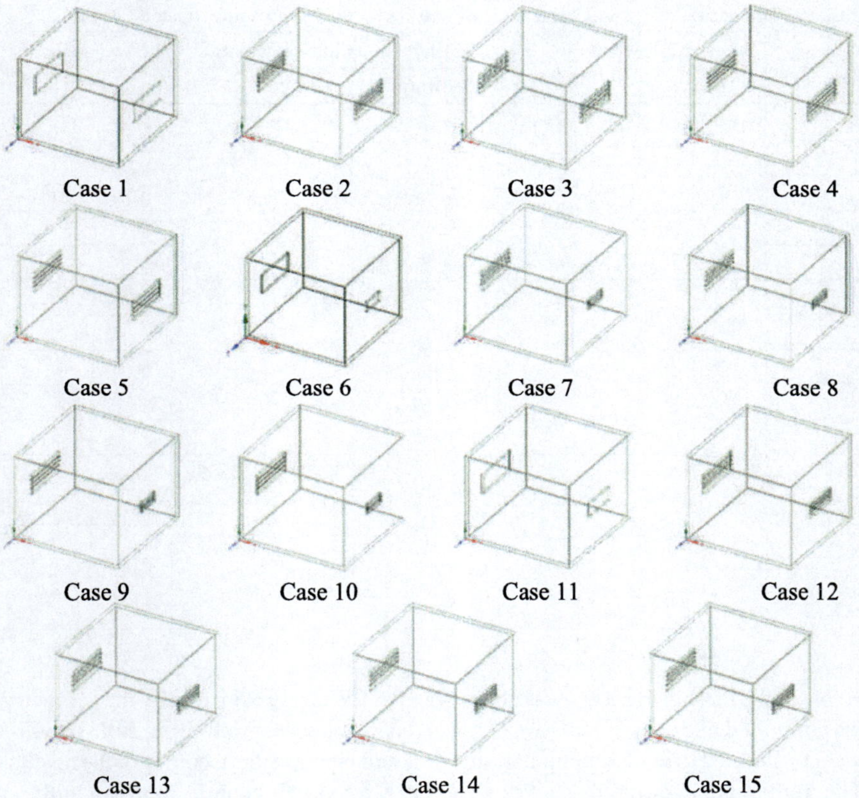

Fig. 2 Isometric view of varying louver angle configurations paired with different opening sizes

2.2 *Computational Domain*

The dimension of the flow domain or computational domain was in accordance with the reference case. Firstly, a Body of Influence (BoI) structure was constructed around the house with the dimension of 0.625H, where H = 0.08 m is the building height of the model. Then, the flow domain dimension was constructed whereby the upstream length from the inlet plane to the model windward wall was set to be 3H (0.24 m), to ensure that the inlet, incident, and approaching flow profiles are horizontally homogeneous [13]. Then, the downstream length from the model leeward wall to the outlet plane was set to be 15H (1.2 m) to ensure fully developed flow while 5H (0.4 m) was set at the lateral side and top of the domain. Besides, the blockage ratio of the building model to the flow domain area was less than 3% which is in accordance with the CFD guideline by Frank et al. [14]. The dimension of the flow domain is 1.54 m × 0.9 m × 0.48 m (L × W × H) as seen in Fig. 4.

Table 1 Dimension of openings and angle of louvers for all 15 cases investigated

Case	Windward-leeward ratio	Opening, mm (width × height)		Louver angle
		Windward	Leeward	
1	1:1	46 × 18	46 × 18	None
2				0°
3				15°
4				30°
5				45°
6	1:0.25		23 × 9	None
7				0°
8				15°
9				30°
10				45°
11	1:0.5		32.52 × 12.73	None
12				0°
13				15°
14				30°
15				45°

Next, the meshing process was done using ANSYS 2019 R3 whereby the geometry was imported and scope sizing was imposed on the house, ground and BoI. The BoI was constructed to reduce computational time and increase the accuracy of the results. The normal angle of the mesh was set to be 12° and then a uniform prism with 10 layers was imposed on the house region which encompasses the louvers as well (Fig. 5). Based on the building height of 0.08 m and a reference velocity of 6.97 m/s, the Reynolds number can be calculated. It was found to be in the range of 38,000 which indicates a turbulent flow. The y+ value used was 200 which resulted in a first cell height of 0.01 m thus placing it within the log-law region. Lastly, the cells of the mesh were converted from tetrahedral to poly-hexcore as it can reduce computational time by up to 10–50% with lower cell count [15].

2.3 Atmospheric Boundary Layer (ABL) Condition

Based on the wind tunnel measurements of the incident air profile at the building model for mean wind speed and turbulent intensity, a set of equations are used to define the boundary conditions at the inlet plane. It is known as the Atmospheric Boundary Layer (ABL) condition. The set of equations below are inserted as a user defined function (UDF) into FLUENT before the simulation to define the velocity profile, turbulent kinetic energy, and specific dissipation rate. The equation of ABL friction velocity, U_{ABL} is shown in Eq. 1. The reference velocity, U_{Ref} is 6.97 m/s,

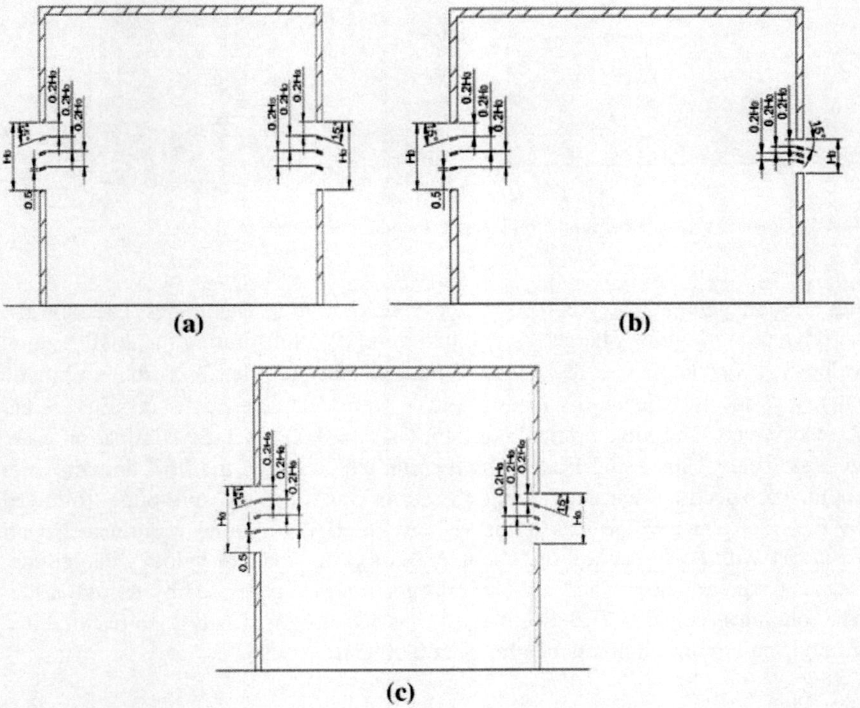

Fig. 3 Section view of the 15° louver with **a** W-L ratio of 1:1; **b** W-L ratio of 1:0.25; **c** W-L ratio of 1:0.5

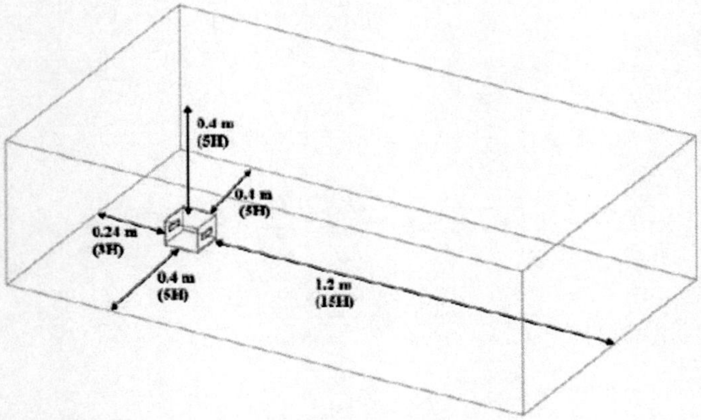

Fig. 4 Dimension of computational domain

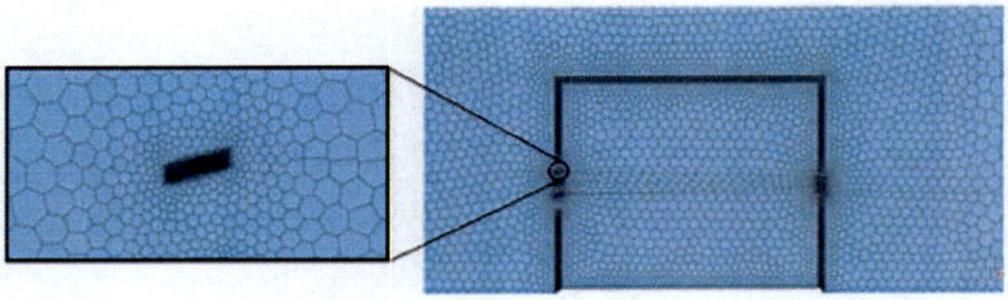

Fig. 5 Detailed view of prism inflation layer surrounding the house and louver

the reference height, Z_{Ref} is 0.08 m, the Von Karman constant, k is 0.42, and the aerodynamic roughness height *Zo* is 0.025 mm [12]. Substituting the ABL friction velocity, U_{ABL} into Eq. 2 to obtain the inlet velocity profile, U. Turbulent kinetic energy, k and turbulence dissipation rate, ε were calculated by using Eqs. 3 and 4, respectively, and the empirical constant $C_{MU} = 0.09$. Specific Dissipation Rate, ω can be calculated using Eq. 5. The top and side walls of the flow domain were set to zero specific shear stress while the outlet was set as pressure-outlet followed by the inlet plane which was set as velocity-inlet. The ground was set as a wall boundary with no-slip condition. The inlet settings included the velocity magnitude, turbulent kinetic energy, and turbulent dissipation rate generated by the ABL file. The roughness constant, C_s was set at 0.5 thus substituting it into Eq. 6 resulted in a ground sand grain roughness height, k_s of 0.0006 m.

$$U^*_{ABL} = \frac{U_{Ref} \times \kappa}{\ln\left(\frac{Z_{Ref}}{Z_o} + 1\right)} \tag{1}$$

$$U = \frac{U^*_{ABL}}{\kappa} \ln\left(\frac{Z_{Ref}}{Z_o} + 1\right) \tag{2}$$

$$k = \frac{\left(U^*_{ABL}\right)^2}{\sqrt{C_{MU}}} \tag{3}$$

$$\varepsilon = \frac{\left(U^*_{ABL}\right)^3}{k\left(Z_{Ref} + Z_o\right)} \tag{4}$$

$$\omega(z) = \frac{\varepsilon(z)}{C_{MU} k(z)} \tag{5}$$

$$k_s = \frac{9.793 Z_o}{C_s} \tag{6}$$

2.4 Solver Settings

In this study, CFD simulation was done using ANSYS 2019 R3. The FLUENT solver of this CFD simulation was set to be pressure-based with the steady time function. The two-equation eddy-viscosity model used was the Shear Stress Transport (SST) k-ω turbulence which is a part of the Reynolds-averaged Navier–Stokes (RANS) equation [12]. The material used was air (fluid) and with a density of 1.225 kg/m^3 and dynamic viscosity of 1.7894×10^{-5} kg/m s. The SIMPLE scheme was chosen as the pressure–velocity coupling and the Green-Gauss Node Based gradient was selected for the spatial discretization. Four discrete schemes of pressure, momentum, turbulent kinetic energy, and turbulence dissipation were selected as second-order discretization scheme as it is more accurate than first-order discretization [12]. Hybrid initialization was used before running the simulation. The convergence criteria set in this study were achieved when all scaled residuals stabilize and had reached a minimum value of 10^{-5} for k, 10^{-6} for x, y and z velocities as well as 10^{-4} for continuity and ω.

2.5 Mesh Independence Analysis and Model Validation

As part of any CFD research, it is essential to carry out a mesh independence study to ensure that the results are independent of the grid resolution while balancing computational time and accuracy. Three different grid sizes were generated to represent coarse, medium, and fine meshing, respectively. The cell count generated were 459,503 (coarse), 815,086 (medium), and 1,413,817 (fine). Once the simulation was complete, post-processing was done to plot the mean streamwise wind speed ratio (U/U_{ref}) as the basis of comparison between the grids. A horizontal line between the inlet and outlet opening was drawn on the mid-plane of the house, extending slightly upstream and downstream of the building's inlet and outlet. The reference wind speed, (U_{ref}) of 6.97 m/s was used. The PIV wind tunnel data from Karava et al. [5] along with the CFD data using the same turbulence model from Ramponi and Blocken [12] were simultaneously compared as seen in Fig. 6. For grid sensitivity analysis, the coarse, medium and fine grids show good agreement with an average percentage error of 2.84%. As for model validation, it was observed that the medium mesh with 815,086 cell count was closest to the CFD turbulence model from Ramponi and Blocken as well as the PIV from Karava. Therefore the medium mesh was used from henceforth.

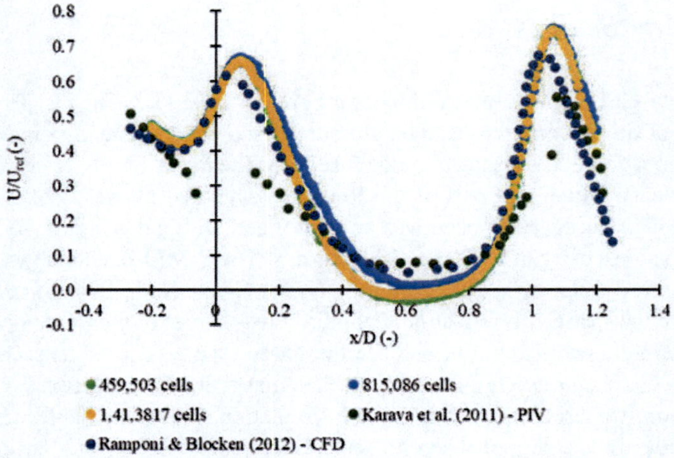

Fig. 6 Simulation results of three different cell counts, PIV experiment result by Karava et al. [5] and Ramponi and Blocken [12]

3 Results and Discussion

3.1 *Dimensionless Velocity Ratio, U/U_{ref}*

Referring to the U/U_{ref} contours present in Table 2, the flow that enters into the building through the windward opening without louvers will flow downward toward the ground of the house. In the presence of a 0° louver, the downward flow is reduced. When a 15° or 30° louver is used, the flow of air has an upward trajectory. The 45° louver angle also produces an upward jet deflection at the windward opening, but its magnitude is reduced due to the reduction of effective opening area.

Moving on, there is a flow acceleration observed between the louver blades at the windward and leeward region. This is attributed to the decrease in effective area due to the presence of louvers [10]. The flow acceleration is most pronounced at the leeward for the building with a W-L ratio of 1:0.25. Therefore, as the area of the leeward opening decreases, the velocity of flow through the louver blades increases. High velocity observed when the leeward opening area is reduced is in harmony with results obtained by Nalamwar et al. [16].

3.2 *Pressure Coefficient (C_p)*

Pressure coefficient, C_p is the dimensionless ratio of the static pressure of a particular region over the free stream static pressure [17]. C_p may be calculated using Eq. (7), where the density of air, ρ is taken to be 1.225 kg/m^3; the reference velocity, U_{ref} is

Table 2 Dimensionless U/U_{ref} contours for different windward-leeward (W-L) ratios and opening configurations of no louver, 0° louver, 15° louver, 30° louver and 45° louver

W-L ratio	Legend	No louver	0°	15°	30°	45°
1:1	U/U_{ref} (-) 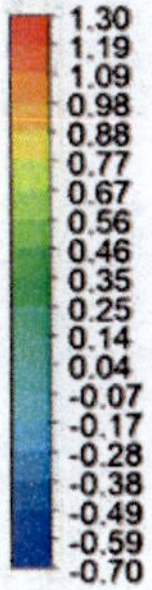	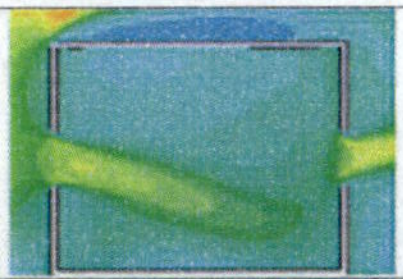	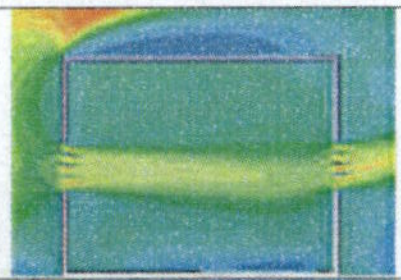	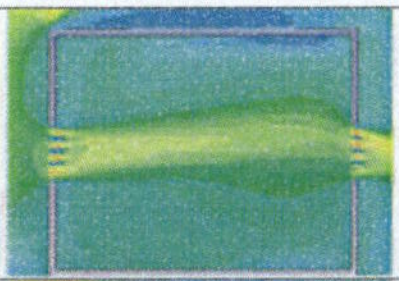	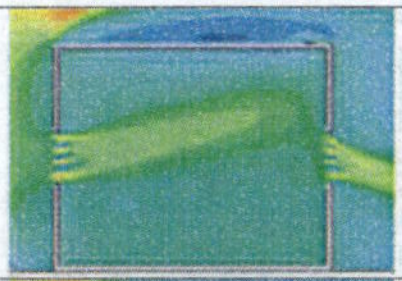	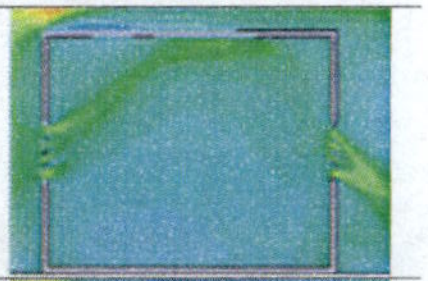
1:0.25			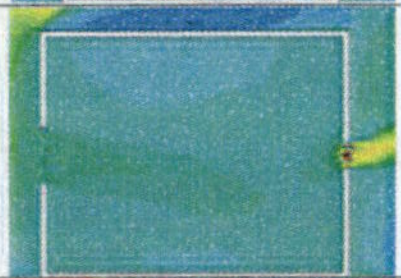	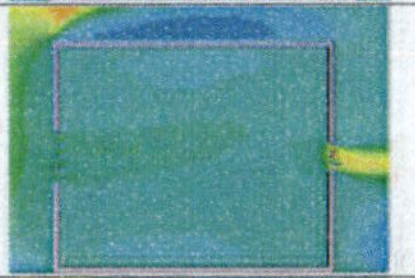		
1:0.5			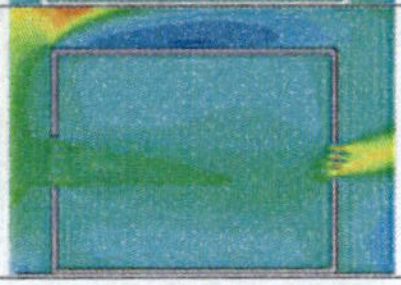	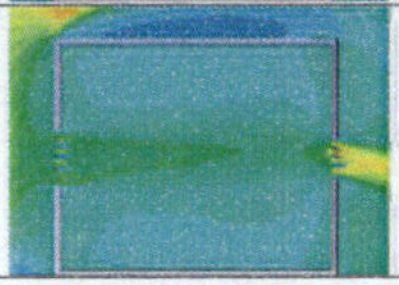	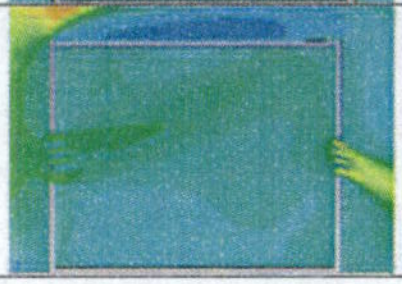	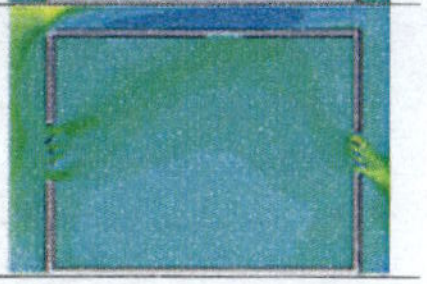

6.97 m/s; the local static pressure, P and reference static pressure, P_0 are taken from the Fluent console.

$$Pressure\ Coefficient,\ C_p = \frac{P - P_0}{\frac{1}{2}\rho U_{ref}^2} \tag{7}$$

On top of that, as the W-L ratio decreased from 1:1 to 1:0.25, the internal pressure coefficient within the building decreased [18]. This was observed in the C_p contours in Table 3. Cases with W-L ratios of 1:0.25 had higher internal pressure coefficients, followed by cases with W-L ratios of 1:0.5 and 1:1. By taking the C_p at the windward opening and subtracting with the C_p at the leeward opening, the ΔC_p can be obtained. It was observed that, as the louver angle increased from 0° to 45°, the ΔC_p increased as well.

It is worthwhile to note that in cross ventilation, as ΔC_p increases, the volume flow rate will increase [19]. However, the same cannot be applied for cross ventilation in the presence of louvers. Configuration 1:0.25 with a louver angle of 45° had the highest ΔC_p as shown in Fig. 7 but the lowest dimensionless flow rate (DFR) as observed in Fig. 8. This is because as the louver angle increases, the effective opening region decreases thus offsetting the higher ΔC_p.

3.3 Dimensionless Flow Rate (DFR)

The Fluent console was used to obtain the volume flow rate in the building at the windward opening. DFR can be obtained by dividing the volume flow rate through the building, $\dot{V}$ by the reference velocity, U_{ref} (6.97 m/s) and the windward opening area, A_o (0.000828 m^2)

$$\text{DFR} = \frac{\dot{V}}{U_{ref} A_o} \tag{8}$$

As observed in Fig. 8, in the absence and presence of louvers, the highest DFR obtained was for W-L ratio 1:1, followed by 1:0.5 and finally 1:0.25. This is because a reduction in the effective opening area causes a reduction in the DFR. Such results are similar to that of Moey et al. [20] when the author manipulated opening ratios. Similarly, as the louver angle increased from 0° to 45°, the DFR also reduced attributed to the aforementioned effective opening area reduction reasoning.

Subsequently, for opening configurations 1:0.25 and 1:0.5, the highest DFR was observed in the absence of louvers. Moving on, as the louver angle increased from 0° to 45°, the DFR reduced as well [10, 11, 21]. This phenomenon, however, was not observed for opening configuration 1:1 whereby the descending order of DFR was observed for the louver at 0°, followed by the 15° louver, no louver, the 30° louver and finally the 45° louver.

Table 3 Pressure coefficient, C_p contours for different windward-leeward (W-L) ratios and opening configurations of no louver, 0° louver, 15° louver, 30° louver and 45° louver

W-L ratio	Legend	No louver	0°	15°	30°	45°
1:1	C_p (-) 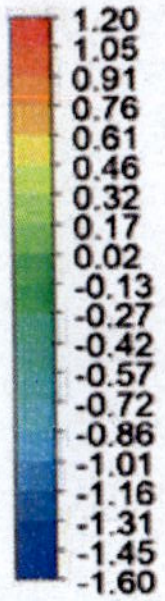					
1:0.25		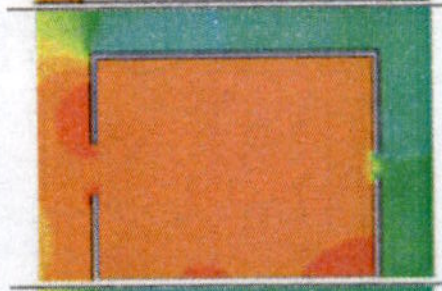		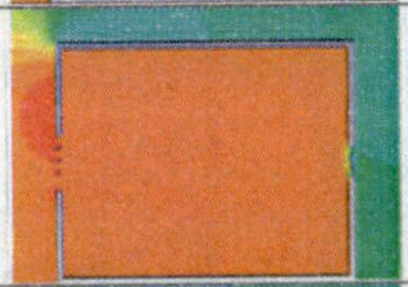	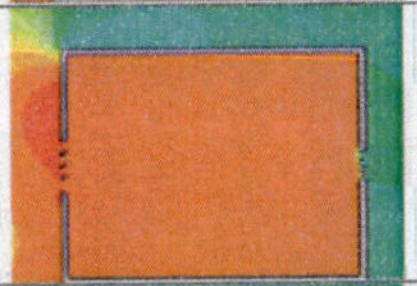	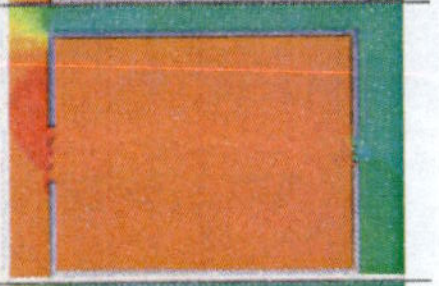
1:0.5						

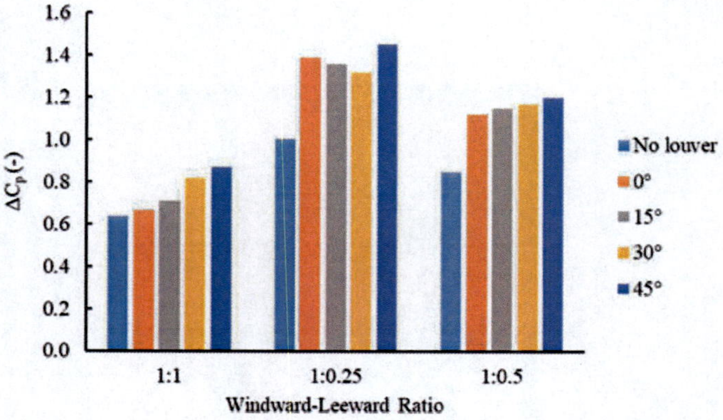

Fig. 7 Graph of ΔC_p for different windward-leeward opening ratios of 1:1, 1:0.25 and 1:0.5 for configurations with no louver, 0° louver, 15° louver, 30° louver and 45° louver

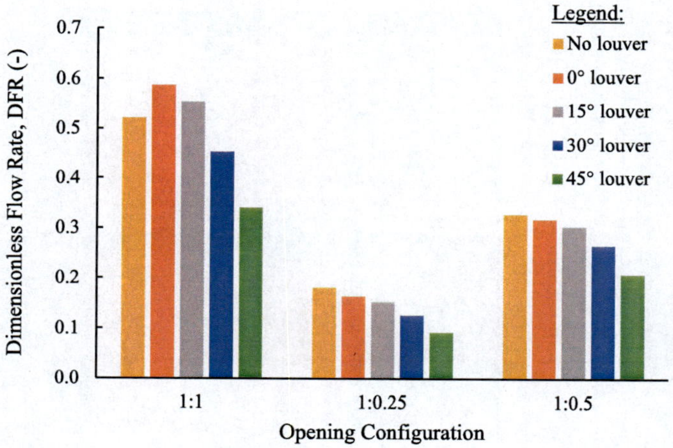

Fig. 8 Graph of dimensionless flow rate, DFR for different windward-leeward opening ratios of 1:1, 1:0.25 and 1:0.5 for configurations with no louver, 0° louver, 15° louver, 30° louver and 45° louver

This anomaly of higher DFR in the presence of 0° and 15° louver was only observed when the opening configurations had a windward-leeward opening ratio of 1:1. The 0° louver had a 12.6% higher DFR than the configuration without louvers, and the 15° louver had a 6.1% higher DFR than the configuration without louvers. For a W-L ratio 1:1, the DFR values for 0°, 15° and no louver respectively were 0.588, 0.544 and 0.522.

4 Conclusion

The investigation of different leeward opening sizes and varying louver angles on a generic isolated building equipped with louvers has been performed. The windward-leeward (W-L) opening ratios of 1:1, 1:0.25 and 1:0.5 were studied as well as louver angles of 0°, 15°, 30° and 45°, alongside the configuration without louvers. Numerical simulations were performed by applying an Atmospheric Boundary Layer (ABL). The reference grid of 815,086 poly-hexcore cells was selected after performing grid sensitivity analysis and model validation. The first objective of model validation was performed by comparing the reference grid with the Particle Image Velocimetry (PIV) from Karava et al. [5] in conjunction with the CFD results by Ramponi and Blocken [12]. The 3D Reynolds-Averaged Navier–Stokes (RANS) equation was solved using the Shear Stress Transport (SST) k-ω turbulence model.

The second objective of investigating the effects of the W-L ratio on the DFR has been performed and has been thoroughly discussed. The results show that, manipulating the W-L opening ratio affects the velocity ratio, U/U_{ref}, internal pressure coefficient and the DFR. The flow acceleration was most pronounced at the leeward for the building with a W-L ratio of 1:0.25. Therefore, as the size of the leeward opening decreases, the acceleration through the louver blades increases. In the absence and presence of louvers, as the W-L ratio increases from 1:0.25 to 1:1, its DFR increases.

Subsequently, the third objective of investigating the louver angles on the DFR has been performed. Highest DFR was obtained when the W-L opening ratio was 1:1 and the louver angle was 0°, second to louver angle of 15°, followed by the configuration without louvers present. Their respective DFR values were 0.588, 0.544 and 0.522. As the louver angle increased from 0° to 45°, the DFR reduced for all opening W-L ratios.

Future studies should incorporate a variety of opening designs, wind incidence angles, the distance between louver blades, a variety of wind speeds, air change effectiveness (ACE) and nexus between energy usage towards the louver installation in order to improve the understanding of the impact of openings and louvers on cross ventilation.

References

1. Moey LK, Goh KS, Tong DL, Chong PL, Adam NM, Ahmad KA (2020) A review on current energy usage and potential of sustainable energy in Southeast Asia countries. J Sustain Sci Manag 15(2):89–107
2. Gielen D, Boshell F, Saygin D, Bazilian MD, Wagner N, Gorini R (2019) The role of renewable energy in the global energy transformation. Energy Strateg Rev 24:38–50. https://doi.org/10.1016/j.esr.2019.01.006
3. Bangalee MZI, Lin SY, Miau JJ (2012) Wind driven natural ventilation through multiple windows of a building: a computational approach. Energy Build 45:317–325. https://doi.org/10.1016/j.enbuild.2011.11.025

4. Kasim NFM, Zaki SA, Ali MSM, Ikegaya N, Razak AA (2016) Computational study on the influence of different opening position on wind-induced natural ventilation in urban building of cubical array. Procedia Eng 169:256–263. https://doi.org/10.1016/j.proeng.2016.10.031
5. Karava P, Stathopoulos T, Athienitis AK (2011) Airflow assessment in cross-ventilated buildings with operable façade elements. Build Environ 46(1):266–279. https://doi.org/10.1016/j.buildenv.2010.07.022
6. Moey LK, Chan KL, Tai VC, Go TF, Chong PL (2021) Investigation on the effect of opening position across an isolated building for wind-driven cross ventilation. J Mech Eng Sci 15(2):8141–8152
7. Yang C, Shi H, Yang X, Zhao B (2010) Research on flow resistance characteristics with different window/door opening angles. HVAC R Res 16(6):813–824. https://doi.org/10.1080/10789669.2010.10390936
8. Sacht H, Lukiantchuki MA (2017) Windows size and the performance of natural ventilation. Procedia Eng 196:972–979. https://doi.org/10.1016/j.proeng.2017.08.038
9. Yusoff WFM (2020) The effects of various opening sizes and configurations to air flow dispersion and velocity in cross-ventilated building. Jurnal Teknologi 82(4):17–28. https://doi.org/10.11113/jt.v82.14537
10. Kosutova K, van Hooff T, Vanderwel C, Blocken B, Hensen J (2019) Cross-ventilation in a generic isolated building equipped with louvers: wind-tunnel experiments and CFD simulations. Build Environ 154:263–280. https://doi.org/10.1016/j.buildenv.2019.03.019
11. Zheng J, Tao Q, Li L (2020) Numerical study of wind environment of a low-rise building with shading louvers: sensitive analysis and evaluation of cross ventilation. J Asian Archit Build Eng 19(6):541–558. https://doi.org/10.1080/13467581.2020.1758113
12. Ramponi R, Blocken B (2012) CFD simulation of cross-ventilation for a generic isolated building: impact of computational parameters. Build Environ 53:34–48. https://doi.org/10.1016/j.buildenv.2012.01.004
13. Blocken B, Stathopoulos T, Carmeliet J (2007) CFD simulation of the atmospheric boundary layer: wall function problems. Atmos Environ 41(2):238–252. https://doi.org/10.1016/j.atmosenv.2006.08.019
14. Franke J, Hirsch C, Jensen AG, Krus HW, Schatzmann M, Westbury PS, Miles SD, Wisse JA, Wright NG (2004) Recommendations on the use of CFD in predicting pedestrian wind environment. In: Cost action C, vol 14
15. Zore K, Parkhi G, Sasanapuri B, Varghese A (2019) ANSYS mosaic poly-hexcore mesh for high-lift aircraft configuration. In: 21st annual CFD symposium, pp 1–11
16. Nalamwar M, Parbat D, Singh DP (2017) Study of effect of windows location on ventilation by CFD simulation. Int J Civ Eng Technol 8(7):521–531
17. Swami M, Chandra S (1987) Procedures for calculating natural ventilation airflow rates in buildings
18. Tecle A, Bitsuamlak GT, Jiru TE (2013) Wind-driven natural ventilation in a low-rise building: a boundary layer wind tunnel study. Build Environ 59:275–289. https://doi.org/10.1016/j.buildenv.2012.08.026
19. Meroney RN (2009) CFD prediction of airflow in buildings for natural ventilation. In: Proceedings of the eleventh Americas conference on wind engineering, Puerto Rico
20. Moey LK, Sing YH, Tai VC, Go TF, Sia YY (2021) Effect of opening size on wind-driven cross ventilation. Int J Integr Eng 13(6):99–108
21. Chandrashekaran D (2010) Air flow through louvered openings: effect of louver slats on air movement inside a space. University of Southern California

Applications of Graphene Nanomaterials in Energy Storage—A State-of-Art Short Review

Kaniz Farhana, Kumaran Kadirgama, Sivarao Subramonian, Devarajan Ramasamy, Mahendran Samykano, and Abu Shadate Faisal Mahamude

Abstract The study presents the usage behavior of graphene in the energy field. Graphene has been comprehensively studied in the energy-related application due to higher conductivity and mechanical flexibility. The architecture of graphene permits it to strengthen and facilitate its application in the energy arena. Herein, the application of graphene in various energy storages such as fuel cells, dye-sensitized solar cells, batteries, nuclear power plants, and thermoelectric has been studied neatly. Graphene reacts towards these substances chemically, mechanically, and electrically to a great extend and appears with the excellent output of these objects. In the future graphene could be applied to the others field of energy and science successfully.

Keywords Graphene · Fuel cell · Solar cell · Battery · Thermoelectric

1 Introduction

Graphene has a large theoretical specific surface area of about 2600 $m^2\ g^{-1}$ with superior electrical and thermal properties. Thermal conductivity of graphene of about ~5000 $W\ m^{-1}\ K^{-1}$ [1] and electrical conductivity is around ~1738 S/m that make an impressive effect in the energy field [2]; as for heat transfer application, thermal conductivity is the main influential criteria while electrical conductivity plays the

K. Farhana (✉)
Department of Apparel Engineering, Bangladesh University of Textiles, Dhaka 1208, Bangladesh
e-mail: kaniz.farhana@butex.edu.bd

K. Kadirgama
Faculty of Mechanical and Automotive Engineering Technology, Universiti Malaysia Pahang, 26600 Pekan, Pahang, Malaysia

S. Subramonian
Faculty of Manufacturing Engineering, Universiti Teknikal Malaysia Melaka (UTeM), Durian Tunggal, Malaysia

D. Ramasamy · M. Samykano · A. S. F. Mahamude
College of Engineering, Universiti Malaysia Pahang, 26300 Gambang, Pahang, Malaysia

M. Y. Ismail et al. (eds.), *Technological Advancement in Mechanical and Automotive Engineering*, Lecture Notes in Mechanical Engineering,
https://doi.org/10.1007/978-981-19-1457-7_46

prime role in electronics and electrical application as well. Graphene is a type of two-dimensional carbon nanomaterial [3]. Graphene became a famous star for scientists in several fields since their experimental implementation in 2004; therefore graphene-based materials and equipment were extensively investigated. The combination of graphene or graphene oxide (GO) with various types of organic or inorganic species was used to make graphene-based hybrid materials [4, 5]. In various energy storage and/or conversion fields, inorganic or organic hybrid graph-based materials have been extensively examined. Commercial use of materials and devices based on graphene. Various methods of preparing graphene have been developed to date, which can be divided into chemical vapor deposition of graphene layers, micromechanical exfoliation of graphite using the Scotch-tape peel-off method, epitaxial graphene foils, biomethane synthesis of organic molecular graphene, and GO sheets reduction or deoxygenation [6].

The application of graphene in energy management and conversion has some advantages in point of energy storage development and transfer and the achievement it brings with it multifaceted innovative strategies. A great deal of research is needed to demonstrate that all graphene products such as foam, paper, and fiber, have a major role to play in enhancing performance, both current and earlier investigations on energy storage, energy transfer, and energy conversion improvement. Even recently, graphene is shown to have greater performance than the conventional generator in practice, which exhibits maximum stress of 200%, without evident electrical properties degradation [7]. About the energy transfer, the battery application with graphene fiber significantly increases the rate of charge and discharge with an improved storage capacity of 763 F g^{-1} [7]. One of the most important causes for the wide use of graphene in the field of energy engineering is the flexibility and the application to various uses and conditions of installation [8]. More and more flexible graphic devices for converting and storing energy are formed and implemented in various fields of industrial and economic improvement [9]. Many more flexible graphic devices for converting and storing energy are formed and implemented in various fields of economic and industrial development [10]. The use of ionic and non-ionic liquids, combined with graphene-based films, networks, 3D structures, and solar cells, has already achieved competing power outputs. We have studied all the literature statements that emphasize the great scientific attainments in this field and investigate higher characteristics related to harvester pattern, power generation, the type of fluid, the methodology of tests, and the mechanism of transduction. The blue energy concept is also considered [11].

This review is a unique study that reflects the major advances that have been achieved in the field of energy harvesting of graphene-based materials in contact with ionic properties (including graphic films, graphic grids, graphic membranes, 3D graphics composites, and tribology structures). Several mechanisms have been thoroughly analyzed for the transduction of power. This review systematically highlights energy production and different types of applications in the energy-related field. Finally, future directions for research and innovative applications are proposed for these harvesters. Graphene in our everyday life, covering solar energy, bioenergy,

energy storage, and other aspects such as household commodities (batteries), electronic products [12], etc. are unprecedented in the field of energy. It can be concluded that graphene in tomorrow's energy applications will become more portable and flexible.

2 Application of Graphene Nanoparticles

2.1 *Graphene in High-Performance Fuel Cells*

Graphene metal hybrid nanomaterials play an important role in the world market as alternative electrical items that can serve the fuel cells [13]. The objective of fuel-cell research is the electrical approach to obtain green energy from chemicals. Methanol/ethanol/formic acid oxidation and oxygen reduction are commonly used reactions. The reactions are made to generate current on the electrode surface. Different electrocatalysts include metal nanoparticles (e.g., Pt, Pd, Au), bimetallic nanoparticles (e.g., PtAu, PtAg, PtCu, and PdCu), and metal oxides because these reactions are less efficient. In many of these catalytic fuel cell reactions, intermediate poisoning devices such as carbon monoxide is developed which interfere with catalyst long-term stability. Other problems include low current intensity and high costs for the catalyst [14]. Graphene as support is shown to decrease the toxic effect significantly and offer catalysts a longer life. The primary needs for the graphene nanomaterial catalyst are the uniform loading of nanoparticles to ensure that the entire area is high, the components are intact bindable and stabilizers are used minimally. Catalyst poisoning, high catalytic current, stable catalytic current with recurring cycles are major advantages of the graphene nanomaterial composite catalysts [15]. Electrochemical (E-RGO) carbon fiber paper or natural extract of Eucalyptus leaves (EL), i.e., El-RGO, coated with reduced graphene oxide (RGO). In both cases, the graphene-coated fiber performs better than the bare CFP. Both power output by time. The reduced GOs are significantly higher both in Fig. 1c and the calculated power of the linear sweeping voltammetry (Fig. 1d). The improved efficiency is related to increased electron transfer between the anode surface and the microorganisms, as confirmed from measurements of the electrochemical impedance in Fig. 1a and cyclic voltammetry in Fig. 1b. The graphene or composite with other materials like polyaniline [16], manganese oxide [14], or zeolite [17]. Higher performance than uncoated anodes has also been demonstrated. Graphene is additionally used in MFC as a compound with other polymers [18]. It shows that the higher performance and costs than of membranes fiber which has been used to reduce the commercial cost.

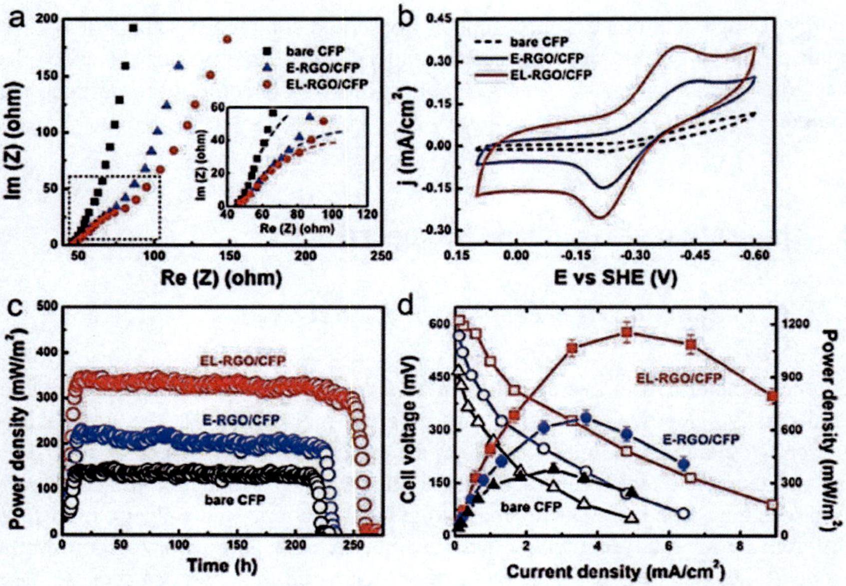

Fig. 1 **a** Measurements of the impedance of the coated (E-RGO/CFP) and uncoated anode bare carbon fiber paper (CFP), **b** cyclic voltammograms of the bare and coated CFP at 50 mV s^{-1}, **c** discharging power of the MFCs using the different anodes, and **d** scheme of i-p curves of the different cells [16]

2.2 *Graphene in Dye-Sensitized Solar Cells*

Renewable energy such as solar energy is one of the most available and finest sources of energy and has shown the potential of substituting fossil fuels, thereby protecting the environment against serious fossil fuel impacts [19]. Solar cells with dye sensitization are promising direct solar-power harvesting technologies. They consist of three compositions: TiO_2 photoanode adsorbed in teal, electrode counter, and electrolyte iodide as shown in Fig. 2. The iodides may be oxidized into a triiodide on the TiO_2 photo-anode under a light beam. This causes an electron to be released into TiO_2. The electron moves to the counter electrode utilizing the external circuit, where the triiodide is met, reducing the triiodide to iodide. A platinum catalyst is needed to accelerate the process. Graphene is considered to replace such a costly platinum catalyst with accepted performance and stability [20]. The photoelectrode is the most important part of the dye-sensitized solar cell (DSSC) since the photons received from the sole are converted to electricity for the absorption of light. An efficient photoanode is necessary to ensure, fill factor, IPCE improvement, and efficient efficiency of DSSCs. The photoanode should give a better use of light and superb collection and injections of electrons [21].

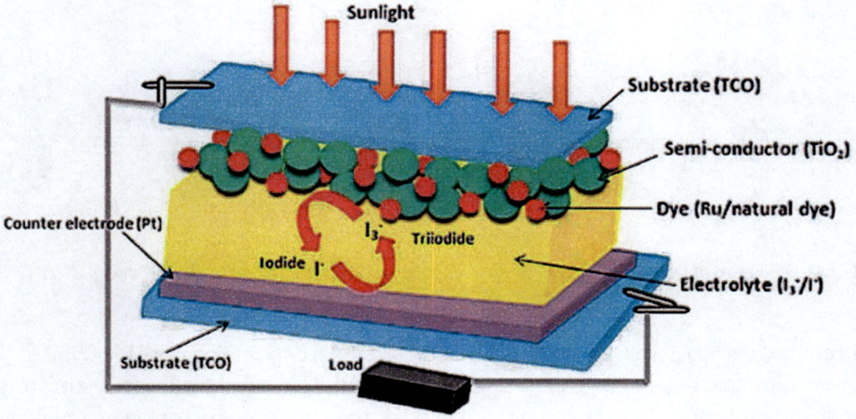

Fig. 2 Schematic diagram of dye-sensitized solar cells [20]

2.3 Graphene in Batteries

The entire world's global oil demand is expected to reach 1500 million tons by 2030. This is a sharp inconsistency between the demand on the market and energy constraints [22]. Vehicles for renewable energy are strategic products for solving the problem of emissions; where 30% of all vehicles converted into renewable energy, 22% of its total oil demand would be saved around the world [23]. The battery is the focus of renewable energy vehicles (accounting for approximately half of the total cost) [24] which must have high energies and high power densities at low cost and long life, with fast charging capacity. Energy density allows for long distances while power density affects the vehicle's acceleration and climb. Therefore, structural models and energy system design are key factors for renewable batteries [25]. Li-ion batteries are the main elements for mobile communication and transport technology development. However, Li-ion batteries have a substantial self-healing system, which controls their lives and forms environmental problems and reliability. The integration of graphene with the material for change in a hydrocarbon phase allows its thermal conductivity to be increased by more than two magnitude orders while maintaining its latent heat-keeping capacity. The sensitive, latent storage combined with the improved heat conduction outside the battery pack significantly reduces the temperature increase in a typical Li-ion battery pack. A significant decrease is achieved. The combined heat storage/heat drive strategy described here can transform Li-and ion's other battery types into a thermal management approach [26]. Graphene-enhanced Li-ion batteries show incredible features such as long life, higher capacity and quicker charging, flexibility, and lightness for usage in wearable electronics [27].

A mesoporous $Cu_2V_2O_7$/rGO composite microsphere structure has been synthesized in combination with heat treatment as LIB anodes. The $Cu_2V_2O_7$/rGO composite has a better reversible capacity, rating capacity, and superb cycling stability

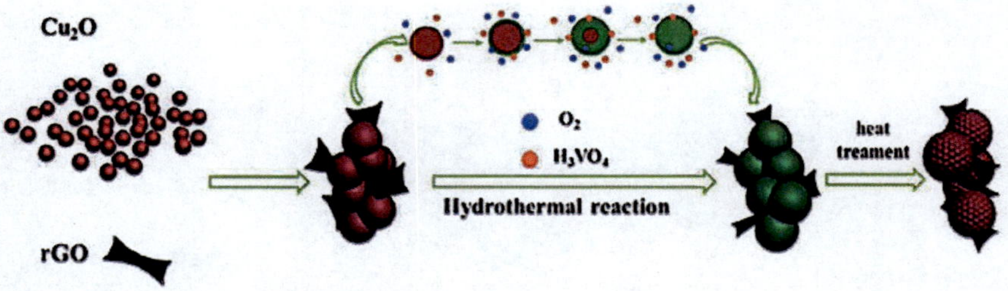

Fig. 3 Diagram of the making of $Cu_2V_2O_7$/rGO mesoporous microspheres with Cu_2O pattern [28]

in comparison to pure $Cu_2V_2O_7$. After 200 cycles at a current density of 100 mA g^{-1} a specific capacity of 883 mAh g^{-1}, was maintained. The excellent electrochemically efficient influence of rGO nanosheets, which is the primary cause for the magnificent performance of CVOG, can be attributed to the synergistic storage energy effect, as shown in Fig. 3 [28].

The goal is achieved in two effective ways. First, thanks to its higher thermal conductivity, wide surface area, and chemical stability, graphene has been demonstrated to be an effective support for LIB transitional metal vanadate [29]. In addition, during repeated lithium-ion insertion and extraction the flexibility of graphene could lull volume change. On the other hand, electrode material morphology controls are also for improving electrochemical properties. Their bigger surface areas, large pores, tunable door sizes, and shape offer a great benefit for the energy storage and conversion of mesoporous materials [30]. SnS_2 nanomaterials consisting of carbon nanofibers, including carbon nanotubes (CNFs), graphene (G), decreased graphene oxide (rGO), are commonly employed as an instrument for improving the conductivity of the SnS_2@carbon nanocomposites in electronic form, resulting in increased electrical activity [16]. Different types of nanocomposite SNS_2@carbon, for example, SnS_2NF on G-C_3N_4 NSs were synthesized [31], SnS_2/MoS_2/CFC [32], SNS_2/CNF [33] SnS_2, NSsG nanoscroll/NS aerogels [34], carbon@SnS_2 core–shell microspheres [35], layer-by-layer SnS_2/G hybrid NS [36], MoO_3/SnS_2 core–shell nanowires [37], SnS_2NF/CNTs network [38], SnS_2/CNTS@PANI [39], SnS_2/G [40].

In particular, rGO is superior for its 2-D conductivity, useful to expedite Li^+ diffusion. The strategy of improved cyclic electrical and rate performance by improving Li^+ diffusion properties, however, is reported rarely through superior electronic conductivity for 2-D structural rGO additives. One study mentioned a synthesis with hydrothermal methods of nanosheet (NS), nanoflower SnS_2 (NF), and nano-flower SnS_2NF@rGO, and the electrochemical cyclic and rate performance as anode material for LIBs have been investigated. More significantly, comparing electrochemical cyclic and rate performance, electrochemical independence (EIS), internal resistance, electronic conductivity, and LY^+ kinetic diffusion was examined in the mechanism of enhanced Li store performance by increasing the Li^+ diffusión property through the addition of high electronic rGO conductivity [41].

2.4 Graphene in Nuclear Power Plants

Remediation of aqueous wastewater and polluted water with man-generated radionuclides, the most toxic of which are the transuranic elements important task of cleaning up legacy nuclear facilities. The Recent accident involving the release of radionuclide Fukushima Daiichi Atomic Energy Environment the water used to cool down in Japan and contamination. The need for effective treatment is emphasized through its reactor cores. Water contaminated with radionuclide methods [42]. During the last decade, interest in the use of new carbon materials for environmental applications has increased significantly. The environmental characteristics and possible applications of carbon nanomaterials include carbon nanotubes (CNT), nanodiamonds, completeness, graphs, and other carbon phases. These nanomaterials' form, size, and free area are among the major advantages [43]. The use of CNTs in the management of nuclear waste has been described. Graphene oxide (GO), which has been confirmed to be non-toxic and biodegradable, is well studied for all carbon nanomaterials [44]. It can be formed in bulk in an environmentally secure way that makes it an environmentally sound material [16].

The viability of using graphene-oxide nanofluid in different chemical coolant environments for improving CHF during ERVC was investigated. The thin wire heater has been tested by the pool boiling system CHF from horizontal to vertical or at $0 \ll$ bis $< 90°$. The heater orientation has been controlled. Graphene-oxide nanofluid dispersion stability was observed in surface charge or the zeta potentials before CHF experimentation, under the chemical conditions of flooding water including boric acid, lithium hydroxide, and trisodium phosphate. Finally, integral effects of nanosheets and chemicals of graphene-oxide on CHF limits have been performed. The results showed that nanofluids with graphene oxide were extremely stable under ERVC coolant chemicals and increased CHF limits at a minimum of about 40% (90° angle) and at the maximum of around 200% (0° angle) compared to pure water [45]. The representative pool performs water boiling phenomena and nanofluid graphene oxides as high as 80% CHF in heat flow conditions as shown in Fig. 4. There are obvious differences between the two fluids in boiling phenomena. The findings show that the efforts of the present work are aimed at understanding physically why CHF is diverse and affected.

For their ability application in nuclear fuels and the removal of nuclear pollution, to recover the uranium from the sweater and nuclear waste, it's essential to develop some appropriate materials. This study shows promising uranium adsorption results for different aqueous solutions including seawater simulated by graphene oxide-amidoxime hydrogel (AGH). We have a high absorption capacity (Q_m = 3984 mg g^{-1}) in aquatic solutions for Uranium species and a high percentage of removals in ppm or ppb level. AGH has enhanced uranium selectivity in the occupancy of high consolidation of competing ions such as Mg^{2+}, Ca^{2+}, Ba^{2+}, and Sr^{2+}. AGH efficiently and selectively binds uranium at low levels of uranium in simulated seawater. The results presented here show that the AGH is a potential adsorber

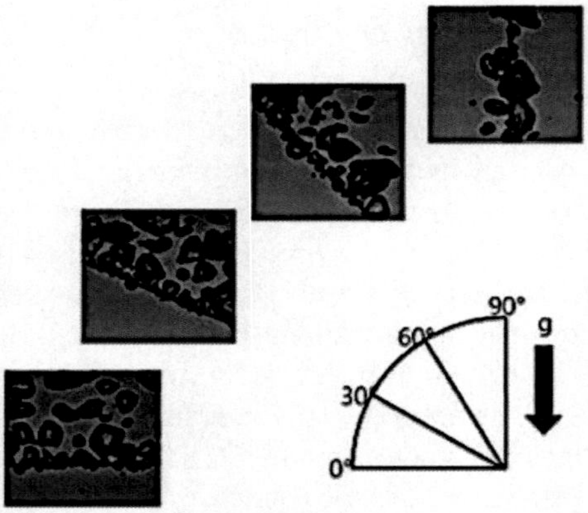

(a) Boiling phenomena of water at 80% CHF for water

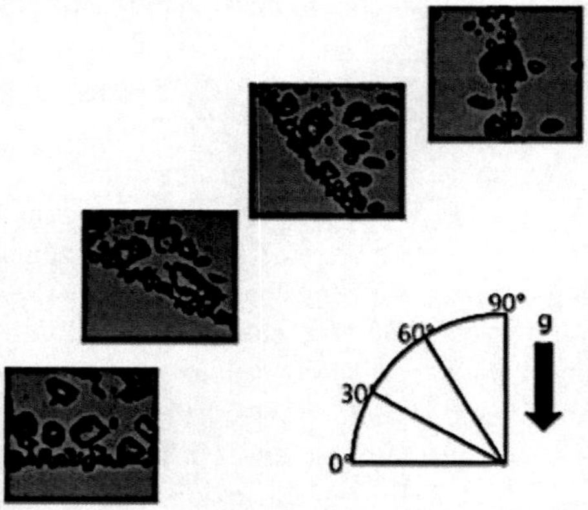

(b) Boiling phenomena of graphene-oxide nanofluid at 80% CHF for water

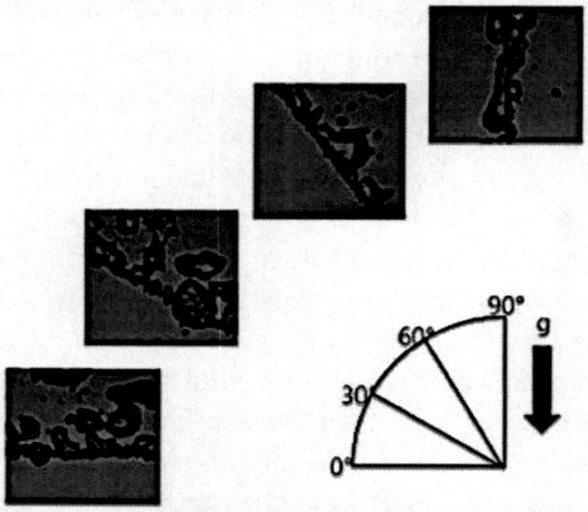

(c) Boiling phenomena of graphene-oxide nanofluid at 80% CHF for graphene-oxide nanofluid

Fig. 4 Influence of heater direction on pool boiling development. **a** Base fluid: distilled water, **b** base fluid: graphene oxide nanofluid [45] for nuclear industrial wastewater remediation and the adsorption of marine uranium [30].

The removal in nitric acid solutions of actinides and rare earth elements (REEs), produced during nuclear processes, is a difficult task. It is of defined attention to isolate actinides and REEs from thick solutions of acids to obtain appropriate sorption materials by modifying oxidized forms of graphenes with organic compounds. This is due to the large surface area of oxidized graphs, chemically stability in highly acidic media, and the possibility of utilizing various organic hydrophobic reagents for actinides and REEs for modifying nitric acid solutions. However, when oxidized graphene is applied as a transport, owing to the formation of stable dispersions or suspensions into fluidic solutions there is a problem with the separation of the solid sorption substances from the suspension during the sorting process [46].

The different methods are applied to coagulate oxidized graphene and to separate ionic strength solutions by loading varied rigid particles to solutions, to obtain graphene-based composites and magnetic particles [48]. Efficient phase segregation during the sorption mechanism can also be ensured by selecting the interaction situations between the oxidized graphene suspension and the hydrophobic ligands in the form of a compact sorption material, which is coagulated in the ligand mix. This makes it particularly important to choose the method of extraction, including the actions for the component mixing during the sorption process. In the case of suspension, the best process is by adding oxidized graphene and ligand directly to the radionuclides solution to achieve an additional sorbent and extract element in one stage. This enables the extraction time to be shortened and sorption conditions to vary easily to ensure that the element isolation is complete, as shown in Fig. 5 [47].

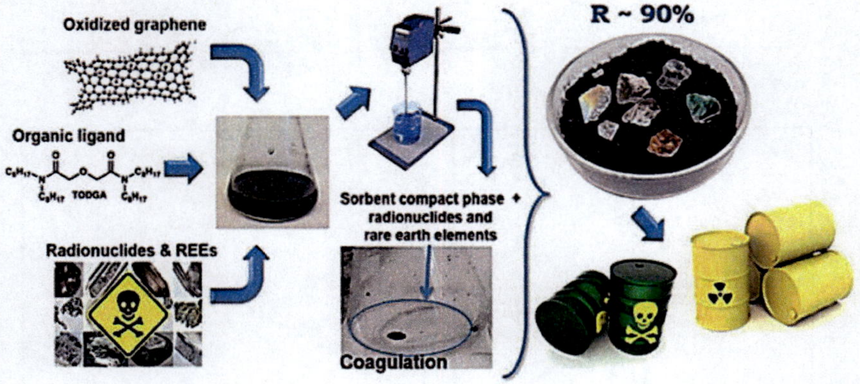

Fig. 5 Formation and characterization of oxidized graphene for rare earth elements and actinides remediation in nitric acid solutions form the waste of nuclear [47]

2.5 Graphene in Thermoelectric

Recently, the physical and engineering communities attracted significant attention to the thermoelectric properties of graphene and graphene nanostructures. The analysis of Seebeck and Nernst's effects in fundamental physics is more important for clarifying some details of the electronic belt structure of graphene, the conductance measures cannot be tested alone because of the ambitious nature of this gapless material. Graphene has the two major disadvantages of a thermoelectric conversion without the gap. This leads to an excellent thermal conductor because of the opposite contributions of electrons, holes, and a small Seebeck coefficient. The thermoelectrical ZT is therefore very small for a two-dimensional graph sheet. However, numerous studies have recently shown that adequate nanostructuring and graphene bandgap engineering can simultaneously reduce structural thermal conductance and improve the Seebeck coefficient without significantly damaging electronic conductance. Therefore, ZT was predicted to be high enough in various graphene nanostructures to make it attractive for the conversion of energy [49]. Figure 6a investigated the decomposition comportations of PANi and fabric-coated graphite nanocomposites.

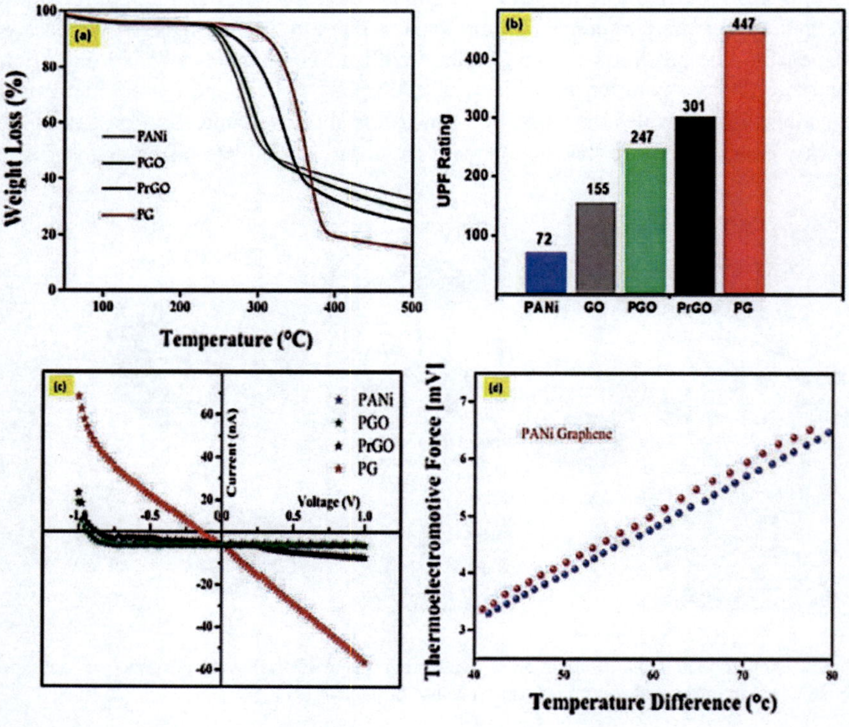

Fig. 6 **a** Scheme of TGA, **b** rating of UPF, **c** plotting of I-V, **d** thermopower of PANi and graphite derivatives coated cotton fabric [50]

The temperature is higher than the room temperature, and the stability is slightly lower. Increasing temperature makes the thermal stability of PANi. Due to PANi decomposition, weight loss of approximately 60–65 in percentage and 210 °C is found. Thermostat degradation takes place at approximately 220 °C in the PGO and PrGO, respectively a weight loss of 65–70% and 73–75%. Due to the presence of Graphene PG, the thermal stability is considerably good, despite the temperature of around 280 °C. Figure 6b Lapsphere UV1000F shows the UV protection capabilities in the visible range. The UPF 447 was excellent in UV protection with the PG-coated fabric. The PANi-cotton-made UPF-values 72, GO-155, PGO-247, and PrGO-301 are outstanding UV-shield [50].

3 Discussion

To increase the current intensity, longevity and proportionally decrease the cost of the fuel cell, graphene nanomaterials show tremendous performance in the fuel cell to produce green energy by making high-performance catalysts [15]. According to Zhou et al. [16], the maximum power density was obtained by 200–400 mW/m^2, and cell voltage was 300–600 mN by using graphene in the fuel cell. Besides, graphene exhibits superb performance as the catalyst to accelerate the photoelectrode process to produce electrons by harvesting solar spectrum in dye-sensitized solar cells [21]. In the case of batteries, graphene signifies the extended life of batteries as well as enhances the charging capacity and flexibility. Graphene makes anodes combinedly in batteries and takes a part in electrochemical reactions to improve the efficiency of the batteries by increasing conductivity [28]. Moreover, graphene also participates in the diffusion mechanism to enhance the storage performance of batteries. On the other hand, to mitigate the wastewater from nuclear power plants, graphene performs tremendous non-toxic and biodegradable performance to save the environments from transuranic elements of contaminated water [42]. Also, graphene exhibits excellent stability in pool boiling systems with coolant chemicals and enhances the performance up to 80% so far [45]. Due to high surface area, conductivity, and chemical stability, graphene shows this notable phenomenon to nuclear power plants also. Recently, graphene depicts good performance in the thermoelectric field owing to the electric belt structure of graphene. Moreover, the nanostructure and bandgap engineering may decrease structural thermal conductance and improved the Seebeck coefficient for the betterment of electronic conductance. The overall scenario of this study of the application of graphene in various areas of energy storage has been portrayed in Fig. 7. Figure 7 also depicts the possible mechanisms that have been occurred during the application of graphene.

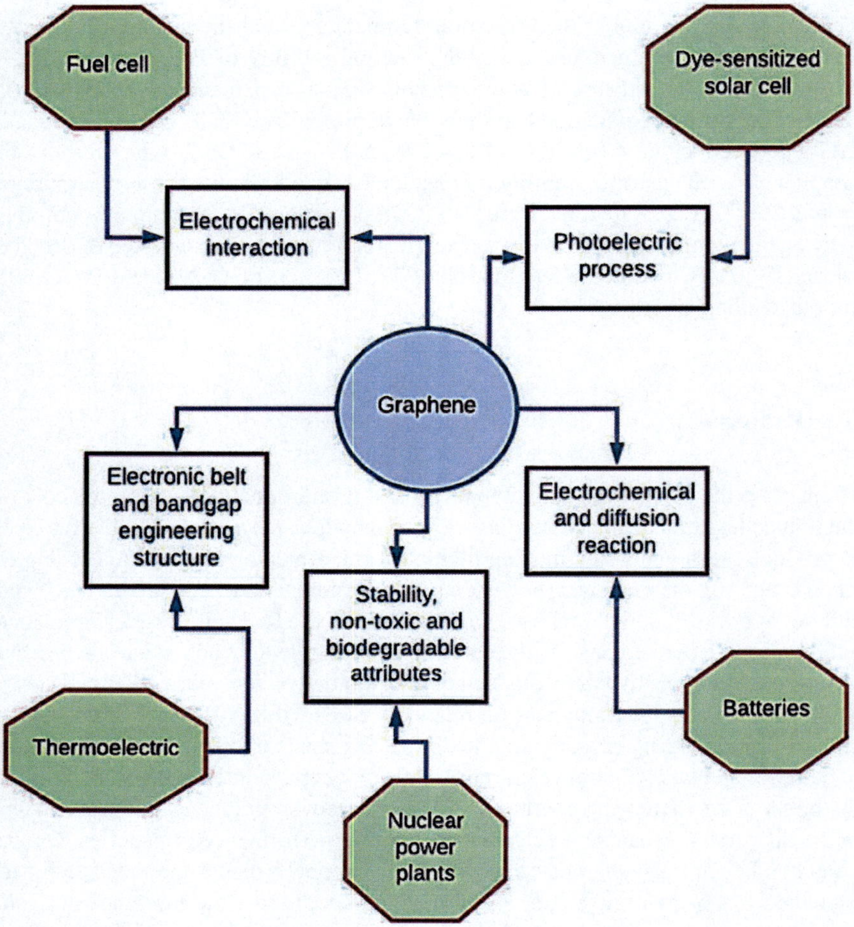

Fig. 7 Schematic representation of graphene application with a scientific orientation [15, 21, 28, 42, 45].

4 Conclusion

The application behavior of graphene in energy storage has been reviewed widely. Various energy storages such as high-performance fuel cells, dye-sensitized solar cells, batteries, nuclear power plants, and thermoelectric have preferred and discussed their performance by applying graphene nanomaterials; and these energy storages showed outstanding results according to their characteristics. Day-by-day, graphene is increasing the wings of utilization purposes continuously, and shortly, the application of graphene could be found in many areas of science and technology.

Acknowledgements The authors are grateful to the Minister of Higher Education for giving a grant under No. FRGS/1/2018/TK03/UMP/02/26 and Universiti Malaysia Pahang for the financial support provided under the grants RDU190323.

References

1. Mahamude ASF et al (2021) Thermal performance of nanomaterial in solar collector: state-of-play for graphene. J Energy Storage 42:103022
2. Lee J-H, Park S-J, Choi J-WJN (2019) Electrical property of graphene and its application to electrochemical biosensing. Nanomaterials 9(2):297
3. Hader GGJS (2020) Synthesis of graphene, p 181
4. Zhu Y et al (2010) Graphene and graphene oxide: synthesis, properties, and applications. Adv Mater Res 22(35):3906–3924
5. Mahamude ASF et al (2021) Numerical studies of graphene hybrid nanofluids in flat plate solar collector. In: 2021 international congress of advanced technology and engineering (ICOTEN). IEEE
6. Wei D, Liu Y (2010) Controllable synthesis of graphene and its applications. Adv Mater 22(30):3225–3241
7. Zhao X et al (2019) A review of studies using graphenes in energy conversion, energy storage and heat transfer development. Energy Convers Manage 184:581–599
8. Renteria JD, Nika DL, Balandin AAJ (2014) Graphene thermal properties: applications in thermal management and energy storage. Appl Sci 4(4):525–547
9. Cui Y, Han BJ (2012) Graphene-based hybrid materials and their applications in energy storage and conversion. Chin Sci Bull 57(23):2983–2994
10. Kiessling D et al (2013) Novel nanographene/porphyrin hybrids–preparation, characterization, and application in solar energy conversion schemes. Chem Sci 4(8):3085–3098
11. Tarelho JP et al (2018) Graphene-based materials and structures for energy harvesting with fluids—a review. Mater Today 21(10):1019–1041
12. Stoppa M, Chiolerio AJS (2014) Wearable electronics and smart textiles: a critical review. Sensors 14(7):11957–11992
13. Hou J et al (2011) Graphene-based electrochemical energy conversion and storage: fuel cells, supercapacitors and lithium ion batteries. Phys Chem Chem Phys 13(34):15384–15402
14. Zhang C et al (2016) Binder-free graphene and manganese oxide coated carbon felt anode for high-performance microbial fuel cell. Biosens Bioelectron 81:32–38
15. Kumar VS et al (2020) One-pot synthesis of Pd20-xAux nanoparticles embedded in nitrogen doped graphene as high-performance electrocatalyst toward methanol oxidation. Int J Hydrogen Energy 45(1):1018–1029
16. Zhou S et al (2019) Biosynthetic graphene enhanced extracellular electron transfer for high performance anode in microbial fuel cell. Chemsphere 232:396–402
17. Paul D et al (2018) Modification of carbon felt anode with graphene oxide-zeolite composite for enhancing the performance of microbial fuel cell. Sustain Energy Technol Assessments 26:77–82
18. Holder SL, Lee C-H, Popuri SRJ (2017) Simultaneous wastewater treatment and bioelectricity production in microbial fuel cells using cross-linked chitosan-graphene oxide mixed-matrix membranes. Environ Sci Pollut Res 24(15):13782–13796
19. Elsaid K et al (2020) Environmental impact of desalination processes: mitigation and control strategies. Sci Total Environ 740:140125
20. Roslan N et al (2018) Dye sensitized solar cell (DSSC) greenhouse shading: new insights for solar radiation manipulation. Renew Sustain Energy Rev 92:171–186

21. Hore S et al (2006) Influence of scattering layers on efficiency of dye-sensitized solar cells. Solar Energy Mater Solar Cells 90(9):1176–1188
22. Bruce PG et al (2012) Li–O_2 and Li–S batteries with high energy storage. Nat Mater 11(1):19–29
23. Johansson BJ (2013) A broadened typology on energy and security. Energy 53:199–205
24. Armand M, Tarascon J-MJ (2008) Building better batteries. Nature 451(7179):652–657
25. Bradley TH, Frank AAJ (2009) Design, demonstrations and sustainability impact assessments for plug-in hybrid electric vehicles. Renew Sustain Energy Rev 13(1):115–128
26. Goli P et al (2014) Graphene-enhanced hybrid phase change materials for thermal management of Li-ion batteries. J Power Sources 248:37–43
27. Wang D et al (2009) Self-assembled TiO_2–graphene hybrid nanostructures for enhanced Li-ion insertion. ACS Nano 3(4):907–914
28. Wang Y et al (2017) Enhanced cyclic performance of $Cu_2V_2O_7$/reduced graphene oxide mesoporous microspheres assembled by nanoparticles as anode for Li-ion battery. J Alloys Compd 724:421–426
29. Mahmood N et al (2014) Graphene-based nanocomposites for energy storage and conversion in lithium batteries, supercapacitors and fuel cells. J Mater Chem A 2(1):15–32
30. Wang F et al (2016) A graphene oxide/amidoxime hydrogel for enhanced uranium capture. Sci Rep 6(1):1–8
31. Huang Y et al (2019) Binder-free flower-like SnS_2 nanoplates decorated on the graphene as a flexible anode for high-performance lithium-ion batteries. J Alloys Compd 774:601–609
32. Lin X et al (2019) In-situ growth of 1T/2H-MoS2 on carbon fiber cloth and the modification of SnS_2 nanoparticles: a three-dimensional heterostructure for high-performance flexible lithium-ion batteries. Chem Eng J 356:483–491
33. Chen G et al (2019) Flexible free-standing SnS_2/carbon nanofibers anode for high performance sodium-ion batteries. Mater Lett 234:121–124
34. Zhang Y et al (2018) Graphene nanoscroll/nanosheet aerogels with confined SnS_2 nanosheets: simultaneous wrapping and bridging for high-performance lithium-ion battery anodes. Electrochim Acta 278:156–164
35. Yu G et al (2018) Carbon@ SnS_2 core-shell microspheres for lithium-ion battery anode materials. Ionics 24(10):2915–2923
36. Xia J et al (2018) Layer-by-layered SnS_2/graphene hybrid nanosheets via ball-milling as promising anode materials for lithium ion batteries. Electrochim Acta 269:452–461
37. Hu C et al (2018) Hierarchical MoO_3/SnS_2 core–shell nanowires with enhanced electrochemical performance for lithium-ion batteries. Phys Chem Chem Phys 20(25):17171–17179
38. Ren Y et al (2017) Three-dimensional SnS_2 flowers/carbon nanotubes network: extraordinary rate capacity for sodium-ion battery. Mater Lett 186:57–61
39. Wang G et al (2015) Two-dimensional SnS_2@ PANI nanoplates with high capacity and excellent stability for lithium-ion batteries. J Mater Chem A 3(7):3659–3666
40. Qu B et al (2015) Origin of the increased Li+-storage capacity of stacked SnS_2/graphene nanocomposite. ChemElectroChem 2(8):1138–1143
41. Liu J et al (2019) Li+ diffusion kinetics of SnS_2 nanoflowers enhanced by reduced graphene oxides with excellent electrochemical performance as anode material for lithium-ion batteries. J Alloys Compd 794:285–293
42. Romanchuk AY et al (2013) Graphene oxide for effective radionuclide removal. Phys Chem Chem Phys 15(7):2321–2327
43. Mauter MS, Elimelech M (2008) Environmental applications of carbon-based nanomaterials. Environ Sci Technol 42(16):5843–5859
44. Zhang X et al (2011) Distribution and biocompatibility studies of graphene oxide in mice after intravenous administration. Carbon 49(3):986–995
45. Park SD et al (2012) Pool boiling CHF enhancement by graphene-oxide nanofluid under nuclear coolant chemical environments. Nucl Eng Des 252:184–191
46. Liu Q et al (2014) Direct synthesis of nitrogen-doped carbon nanosheets with high surface area and excellent oxygen reduction performance. Langmuir 30(27):8238–8245

47. Ali I et al (2021) Preparation and characterization of oxidized graphene for actinides and rare earth elements removal in nitric acid solutions from nuclear wastes. J Mol Liq 335:116260
48. Ren X et al (2014) Impact of Al_2O_3 on the aggregation and deposition of graphene oxide. Environ Sci Technol 48(10):5493–5500
49. Dollfus P, Nguyen VH, Saint-Martin JJ (2015) Thermoelectric effects in graphene nanostructures. J Phys Condens Matter 27(13):133204
50. Panbude A et al (2021) Incorporation of polyaniline on graphene-related materials for wearable thermoelectric applications. Mater Lett 130576

Enhanced Smoke Wire Technique with Control Dripping Valve in a Small Scaled Quasi-atmospheric Boundary Layer Wind Tunnel

Nurizzatul Atikha Rahmat, Mohammad Rozaki Ramli, Mujahid Husaimi Che Hassan, Kamil Khalili Haji Abdullah, and Khairun Adhani Khairunizam

Abstract In the previous work, fabricated smoke wire technique in an atmospheric boundary layer wind tunnel displayed several flaws during experiment such as manual-dripped solution, leaking problem, utilisation of single heated wire, and an ineffective wire heating system in which the electrical circuit did not operated with desired optimum output to heat the wire efficiently. Therefore, present study fabricates an improved smoke wire technique with a control dripping valve to control the dripped-liquid solution quantity and frequency and aims to perform a qualitative investigation to visualize the flow pattern around a simple two-dimensional rigid body namely rectangular and cylinder. The experiment was conducted in a shorter test section of the quasi-atmospheric boundary layer wind tunnel. The wind tunnel has a working section of 0.3 m height and 0.3 m width with a streamwise length of 1 m. The enhanced fabrication successfully produced a continuous and high-quality smoke lines which utilised 10 lines of nichrome wires in a series circuit compared to a single wire in the previous work. The smoke visualization for the combination of 0.4 mm nichrome wire (type C) with 0.6 mm nozzle size at 25.98 V (3.5 A) was found to be the best condition for a continuous smoke streamline. As a result, from the two-dimensional flow experiment past rigid body, a pair of tip vortex structures, horseshoe vortex, and the downwash flow can be evidently seen.

Keywords Smoke wire technique · Boundary layer wind tunnel · Flow pattern · Qualitative study

1 Introduction

Over a century, wind tunnels have been actively used to investigate aerodynamic responses of aircraft and vehicle [1] which eventually extended to other fields, such

N. A. Rahmat (✉) · M. R. Ramli · M. H. C. Hassan · K. K. H. Abdullah · K. A. Khairunizam
Faculty of Mechanical and Automotive Engineering Technology, Universiti Malaysia Pahang, 26600 Pekan, Pahang, Malaysia
e-mail: izzatulatikha@ump.edu.my

M. Y. Ismail et al. (eds.), *Technological Advancement in Mechanical and Automotive Engineering*, Lecture Notes in Mechanical Engineering,
https://doi.org/10.1007/978-981-19-1457-7_47

as the investigation of wind turbine efficiency in wind energy extraction [2, 3], wind-based research in agriculture and forestry [4], and even sports [5]. One of its popular functions and has been widely used among researchers, wind engineers and climatologists is the observation of flow pattern around various shapes of buildings [6] and rigid structures [7] in order to evaluate wind pressures on building walls and study the flow separation and vortex shedding [8]. These factors are critical and have always been a top priority in the fields of wind engineering and climatology to ensure a robust structural building design for the safety of pedestrians.

The study of flow pattern consists of studies that investigate the internal and external fluid flow on a solid object. In Fluid Mechanics, 4th Edition book, external flows involve both low and high-Reynolds number flows [6]. In 1999, Sumner et al. investigates the fluid behavior of side-by-side circular cylinders in steady crossflow by using Reynold's number from 500 to 3000 [9]. V. Talele et al. in their paper used two different methods utilizing computational fluid dynamic (CFD) and artificial neural network (ANN) to predict the flow pattern around square and rectangular bluff body for high Reynolds number [10]. In 2004, D. Sumner, et al. used a low-speed, closed-return wind tunnel to analyze the wake structure of a finite circular cylinder of small aspect ratio [11]. Chashechkin, et al., on the other hand, uses a flow tank to investigate the flow pattern around a thin strip horizontally towed at constant velocity in a continuously stratified liquid [12]. These two approaches are very well-known and widely used for flow pattern visualization.

Investigation on flow pattern is usually done to examine an array of shapes either a simple shape [13], complex shape such as large-span shallow shell roofs [14], triangular-shaped spur dikes [15], and asymmetric-lattice collar shapes [16], or a compound shape. Compounded shapes are done when two or more different objects are placed at certain distances between each other and the difference in their flow patterns are compared. We can see these studies on the flow around two-dimensional hills [17] and in the numerical study of flow patterns in buildings with different heights [18].

These flow visualization experiment utilizing wind tunnel to understand the flow pattern around structures and its aerodynamic interactions with wall can be in quantitative or qualitative measurements, which widely used in engineering, physics, medical science, meteorology, oceanography, and even sports aerodynamics [6]. Quantitative measurements such as hot-wire anemometer, are thought to be limited in measuring flow conditions at discrete times within the flow area. On the other hand, qualitative measurements provide a macroscopic view of the total flow region. Several well-known methods for qualitative measurement are Particle Image Velocimetry (PIV), Laser-Doppler Anemometer (LDA), as well as Smoke Wire Technique (SWT). However, PIV and LDA comes with a high cost and difficult to be conducted as they are very intricate and sophisticated compared to the SWT that is more cost-effective and flexible. This technique is done by introducing smoke into the airflow that fits air currents, allowing the flow pattern to be visualized. The smoke is generated by the electrical current evaporating a wire-heated glycol-based solution. The streams of smoke will be emitted in the air stream, creating a flow pattern, similar to how the smoke liquid is heated. This method allows very fine smoke to be injected into the flow area.

In 1986, the SWT has been mentioned as one of the classical tracer techniques [19]. It is defined where fine aerosol filaments are produced by heating a wire that is coated with oil. This shows that the technique has been used since the early 1980s as a method of flow visualization. They are mostly used in low-speed wind tunnels, but it has also been used in transonic speeds. One main limitation of this technique has been identified by Wu et al. where they found that at a higher speed, the smoke wire will cause the airflow to become fixed vortex pair [20]. In their setup, it is observed that the wire will act like a circular cylinder and changes the unseparated flow to fixed vortex pair when the wind speed is set to 14 ft/s (4.27 m/s). Thus, for their experiments, they limit the wind speed to 13 ft/s (3.96 m/s). The smoke wires are also known to burn as it is heated by flowing electricity through them similarly to how a filament in a light bulb works. Thus, in 2018, N. Gao and X. H. Liu further improved the technique by using a capacitor as the power source [21]. With the capacitor, they used a microcontroller to time their camera to capture the image at the right time. The wire will not be heated any longer than necessary to prevent them from burning out which allows a higher amount of power to be used on the wire which will produce a larger amount of smoke. This, however, would not work as well for video footage.

Several other improvements have been done to the technique. In a 1988, N. B. Mathur et al. have experimented with producing colored smoke [22]. They would use liquid paraffin mixed with wax-based dyes. The rest of the setup has been kept the same by using stainless steel wires heated by passing electricity through them. The technique has been improved from time to time. In 2006, Sharul et al. experimented with several different smoke fluids instead of paraffin oil and different wire setup using nichrome wire instead of stainless-steel wire [23]. This significantly improved the quality of smoke produced by the SWT. The quality of smoke can also be improved by introducing a solution or smoke fluid delivery system to ensure a continuous smoke line. However, until recently this part of the SWT is still manually or semi automatically conducted [23, 24].

The literature around the SWT mainly pointed out the importance of the photography technique. Wu and Tsung-Ju also pointed out in their 1992 paper that a huge amount of light was needed to get a better photo of the smoke flow [20]. This was done by using 2 1000-W lights. For photography, a high shutter speed is needed to capture the smoke flow properly. N. Gao and X. H. Liu used a shutter speed of 1/200 s [21] while Wu and Tsung-Ju in 1992 used a shutter speed of 1/1000 s [20]. For video recording, a high-speed recording was used. Yun Liu et al., in their 2013 paper recorded the smoke flow at 1000 fps (frames per second) [25].

Other useful techniques in addition to the SWT also have been used. J. Soria, et al., in 1990 used the silhouette cinematographic technique [26]. This technique used 50/50 semi-silvered mirrors at 45° to the camera to reflect light which will cast a shadow on the background. This will allow the camera to capture the silhouette of the smoke rather than the smoke itself. On the other hand, J. J. Serrano-Aguilera et al., in their 2016 journal used a laser sheet to highlight a layer or the smoke flow to make it more visible [27].

The SWT also has been used outside of wind tunnels. In 1999, Cornaro et al. put the smoke wire behind a jet to visualize the flow structure onto different surfaces [28]. On the other hand, in their 2013 paper, Yun Liu, et al. put a smoke wire underneath a flapping mechanical insect wing to see the airflow pattern produced by the flapping wings. This is done in an isolating box to eliminate external airflow.

With this background, present study aims to build a quasi-atmospheric boundary layer wind tunnel (BLWT) and enhance the previously fabricated SWT. In the previous work, the fabricated SWT in an atmospheric BLWT displayed several flaws during experiment such as manual-dripped solution, leaking, utilisation of single heated wire, and an ineffective wire heating system in which the electrical circuit did not operated with desired optimum output to heat the wire efficiently. Therefore, present study enhanced the existed SWT with an automatic control dripping valve (CDV) to produce continuous as well as high quality smoke lines. Present work utilized 10 lines of nichrome wires in a series circuit compared to previous arrangement i.e., manual dripping method with single heated wire. In addition, the best wire types, sizes, designs, and voltage levels to use are also analysed. Fabricated enhanced SWT is then used to observe a flow pattern around two-dimensional rigid bodies, i.e., rectangular and cylinder.

2 Experimental Set-up

2.1 Quasi-atmospheric Boundary Layer Wind Tunnel

Present study was conducted in the laboratory of the Faculty of Mechanical and Automotive Engineering Technology, University Malaysia Pahang, Malaysia. Figure 1 depicts a three-dimensional rendering of the wind tunnel. Figure 2a, b show the side and top views of the complete dimension of the wind tunnel, respectively. The body of the wind tunnel is 2.2 m long, 0.53 m tall, and 0.47 m wide. The wind tunnel has a working section of 0.3 m height, 0.3 m width and a streamwise length of 1 m at a velocity of 1.3 m/s. The nozzle section and diffuser section are made by wood, honeycomb from straws, test section by acrylic, and a CDV are five main parts of this wind tunnel.

2.2 Enhanced Smoke Wire Technique

Figure 3 depicts the three-dimensional drawing of the entire CDV which consists of four main components, namely, nichrome cable, tank/reservoir, pump, and piping

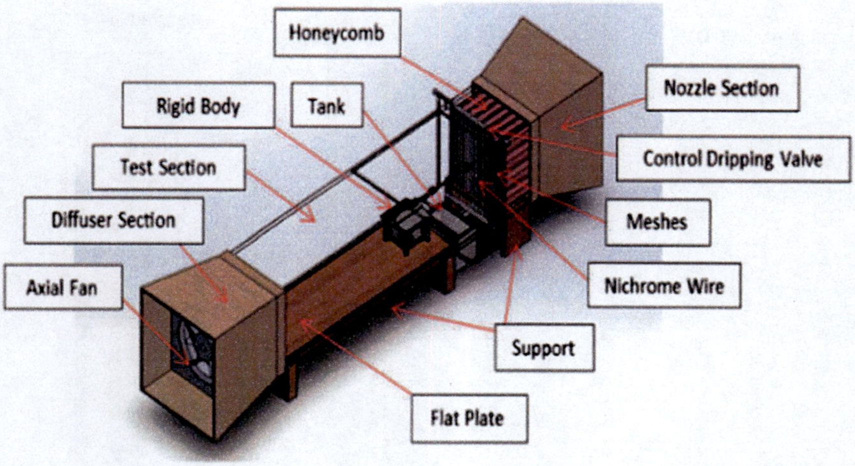

Fig. 1 Three-dimensional drawing of the wind tunnel

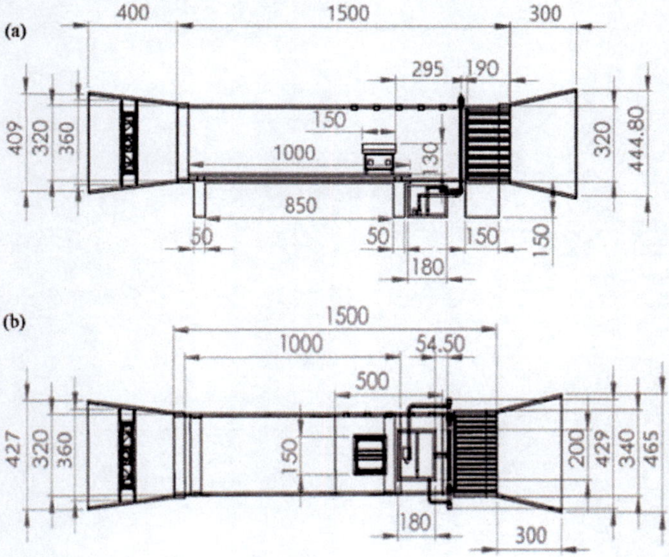

Fig. 2 **a** Side and **b** top view of the wind tunnel (all units in mm)

device. Figure 4 shows the schematic drawing of a dropping nozzle, excess fluid catcher, isometric view of CDV tank, and top view of the CDV.

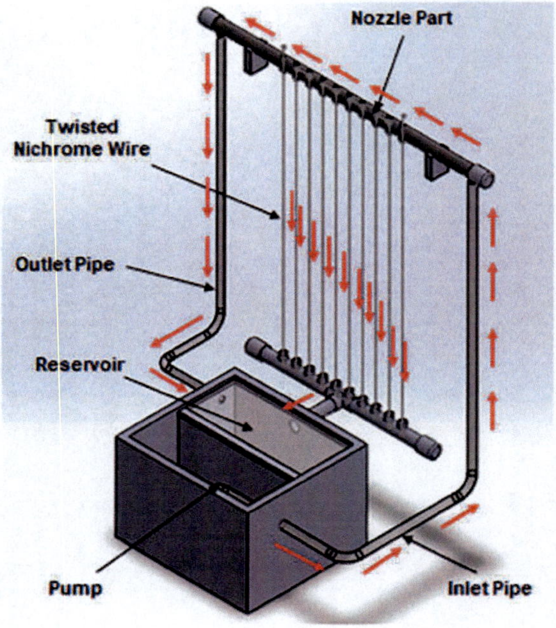

Fig. 3 Three-dimensional drawing of the CDV (all units in mm)

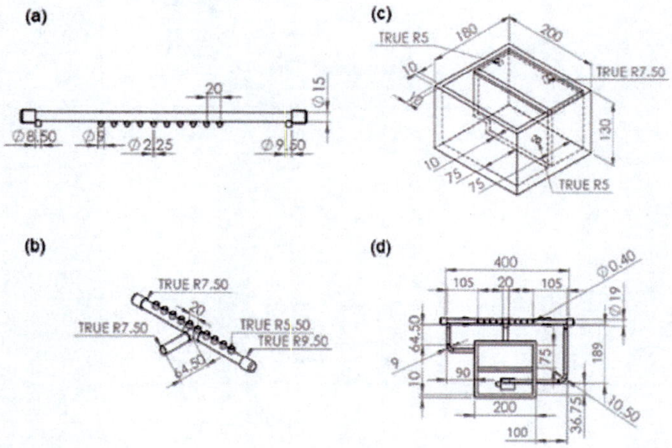

Fig. 4 Schematic drawing of a **a** dropping nozzle, **b** excess fluid catcher, **c** isometric view of CDV tank, and **d** top view of the CDV (all units in mm)

3 Results and Discussions

Present study conducted three different types of experiment referred as Experiment 1, Experiment 2, and Experiment 3 which summarized in Table 1. Experiment 1, 2, and

Table 1 Types of experiments that were conducted in the present work

Type	Title	Detail
Experiment 1	Enhancement of SWT	Refer Table 2
Experiment 2	Selection of nichrome wire and nozzle diameter	Refer Table 4
Experiment 3	Flow visualization	Cylinder and rectangular

Table 2 Comparison of previous and enhanced design of the SWT

Case	Previous design	Enhancement
A	Manual dripping system	• Fully automated dripping system by redesigning the CDV to increase efficiency • Design improved by adding an open area in the CDV to make the nichrome wire installation easier
B	Single smoke wire	• 10 lines of smoke wire (nichrome) to obtain better smoke intensity • Power supply was increased from 300 to 1000 W to heat all 10 smoke wires
C	Tank location	Tank relocated to the side of the wind tunnel for an easy access and installation
D	No tensioner system	A tensioner system was added to increase downward tension of nichrome wires

3 discuss in detail on the enhancement of the SWT, the selection on the combination of nichrome wire and nozzle diameter, and the flow pattern behind two-dimensional rigid body, respectively.

3.1 Experiment 1: Enhancement of SWT

In the previous design of the SWT, several flaws are observed which contribute to the inefficacy of the experiments. Table 2 summarizes all enhancement works, referred to Case A–D, that have been done from the previous design.

Experiment 1. Case A: Fully Automated CDV System

In present design enhancement, the SWT is improved by adding the fully automated solution delivery system, CDV to ensure the SWT runs continuously and automatically. In the previous design, the CDV operated with a manual/semi-automated system because the fog liquid did not circulate continuously in the system in which the location of the pump inlet was set at the bottom of the PVC pipe for the upper part of the CDV as well as the overflow pipe as shown in Fig. 5. When the pump inlet and the overflow pipe are at the same level, the liquid fog did not have enough time to fill in the PVC pipe because it will flow through the overflow pipe first. An

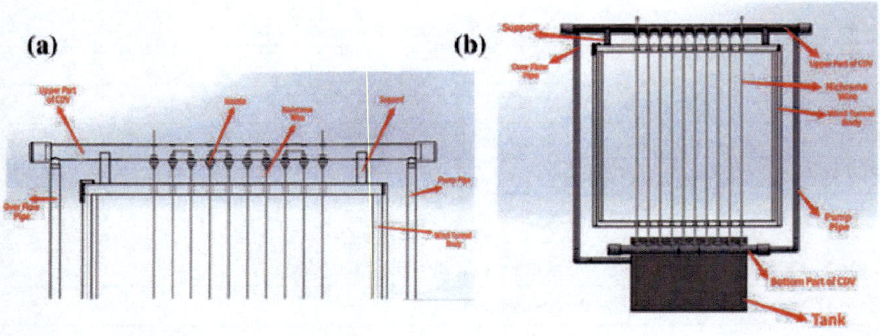

Fig. 5 **a** Components for upper part and **b** front view of the previous CDV design

enhancement was made to the previous design by changing the pump inlet location and the overflow pipe location.

Furthermore, the nichrome wire was hard to install in the previous design because there was no open area at the top part of the PVC pipe in the CDV system. An enhancement was made by adding an open area in the PVC pipe at the upper part of the CDV to make the nichrome wires installation easier. Figure 6 show the technical drawing of the improved CDV system and the open part in the PVC pipe on the upper part of the CDV in the enhanced design. Figure 7 shows the real picture of the fabricated enhanced CDV.

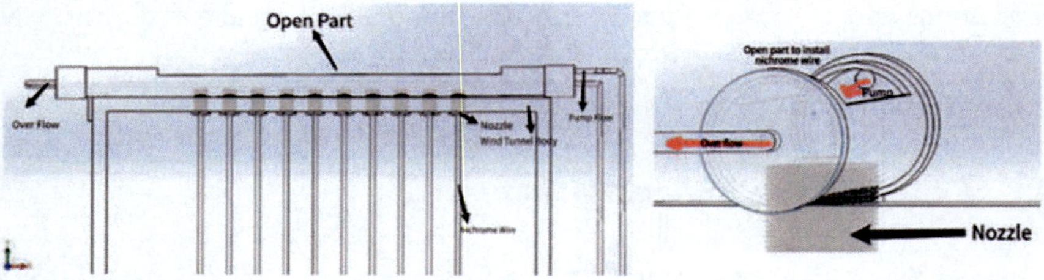

Fig. 6 The open area in the PVC pipe on upper part of the enhanced CDV design

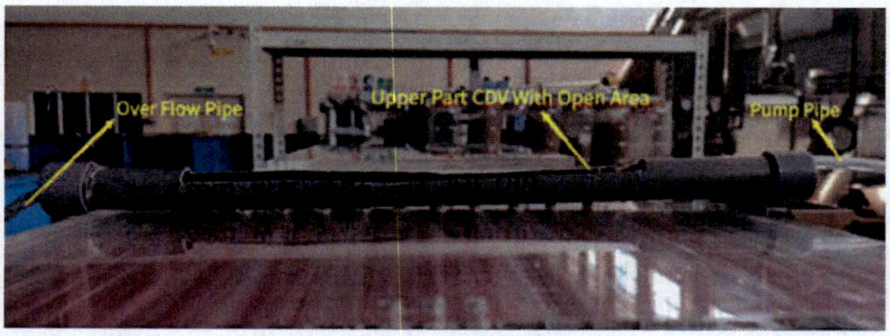

Fig. 7 Real picture of the fabricated upper part of the enhanced CDV

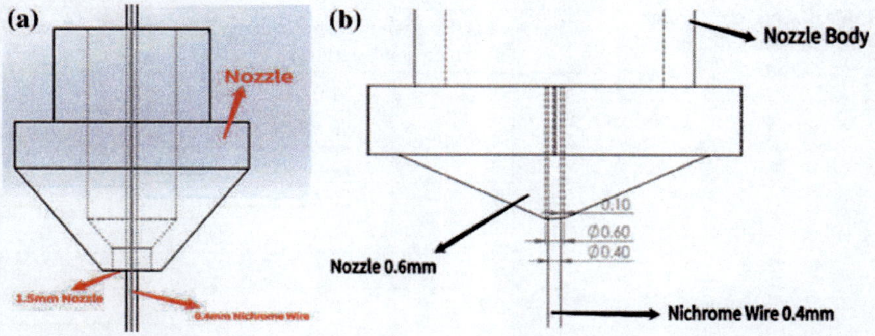

Fig. 8 Schematic drawing of the 0.4 mm diameter nichrome wire with **a** 1.5 mm diameter nozzle in previous design, and **b** 0.6 mm nozzle diameter in the enhanced design

To obtain the best dripping frequency, the nozzle gap between nichrome wire and nozzle wall is an important part to focus on in which a perfect combination of the nichrome wire and nozzle diameter is crucial. Figure 8a shows the nozzle and the nichrome wire gap of the previous design which resulted in excessive dripping of the CDV. Figure 8b shows the nozzle and the nichrome wire of enhanced design which provide optimum drip frequency.

Experiment 1. Case B: Number of Smoke Wire

In the previous study, a 300 W power supply was utilized and only able to heat a single smoke wire. Since the smoke intensity was insufficient for better flow visualization during the flow pattern experiment, present work increased the number of smoke wires from single to 10 lines of smoke wires that connected in series. Therefore, a 1000 W power supply is used which should be able to supply the power needed to heat all 10 smoke wires. Further calculation 10 lines of smoke wire is shown Table 3.

$$R = r * l$$

$$P = \frac{V^2}{R}$$

Table 3 Power calculation

Parameter	Value	Unit
Resistivity of nichrome wire, r	2.67	Ohms/ft
Total resistance, R	45.38	Ohms
Voltage, V	152.02	Volt
Power, P	510.79	Watts

Table 4 Sets of nichrome wire and nozzle diameter combination

Case	Nichrome wire diameter (mm)	Nozzle diameter (mm)	Tolerance (mm)
A	0.5	1.0	0.25
		0.8	0.15
		0.6	0.05
B	0.4	0.8	0.20
		0.6	0.10
		0.5	0.05
C	0.2	0.5	0.15
		0.4	0.10

Experiment 1. Case C: Tank Location

In the previous CDV design, the tank was located under the wind tunnel which resulted in difficulty to operate the wind tunnel. The liquid fog pump was hard to be installed as well as the liquid fog refill process. A new location of the tank was set which is to the side of the wind tunnel to ensure the operation of the wind tunnel easier. Figure 9 depicts the previous and new location of the tank.

Experiment 1. Case D: Tensioner System

When the wind tunnel is running, the nichrome wire will be heated continuously that can cause an expansion and tension loss. Tension loss corrupts the flow visualization as the smoke wire is not straight. Therefore, a tensioner system where a weight is attached at the bottom part of the CDV that will pull the nichrome wire downward is added to the SWT, as shown in Fig. 10. The attached weight will constantly put the nichrome wire in tension regardless of the expansion of the wire when heated.

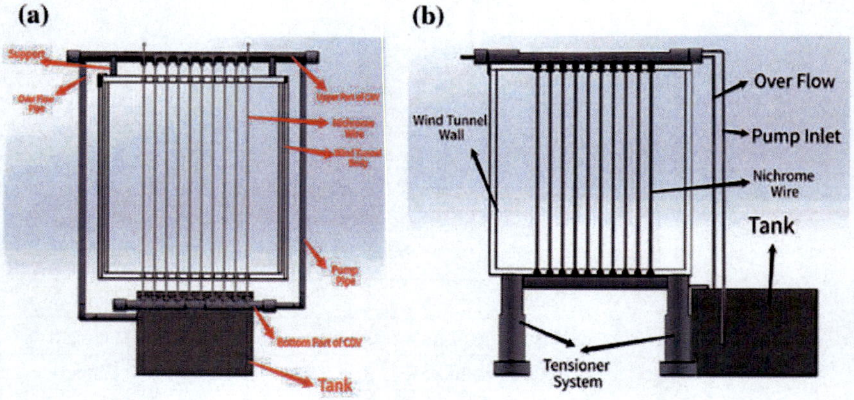

Fig. 9 The location of the tank in the **a** previous, and **b** present enhanced CDV design

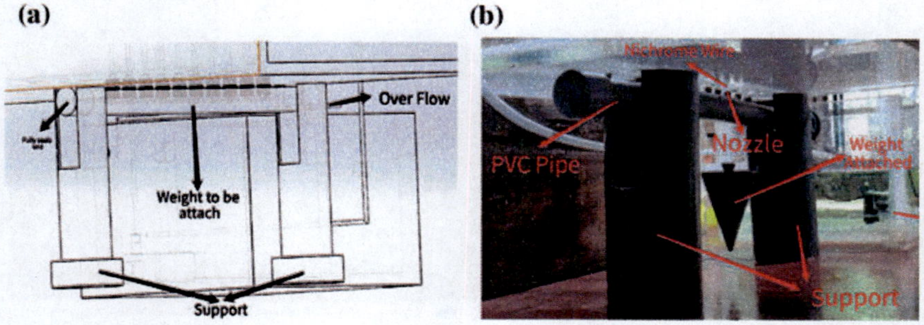

Fig. 10 **a** Schematic drawing, and **b** real picture of the tensioner system

Experiment 1. Summary

Experiment 1 discussed several improvements that have been made to the original wind tunnel assembly. Although the wind tunnel itself did not get any significant modification, the SWT system, on the other hand went through a significant overhaul to increase the quality and intensity of the generated smoke. The original and improved designs of the BLWT and SWT system are shown in Figs. 11 and 12, respectively.

From these improvements, the present wind tunnel operated fully automated with the new enhanced CDV. By using 1000 W power supply, all 10 lines of nichrome wires can be heated continuously without failure. The intensity of smoke from 10 heated nichrome wires also drastically increases and more concentrated. Furthermore, with the addition of a tensioner system, the nichrome wires can stay in tension when heated. Figure 12 shows the instantaneous picture of smoke produced in the BLWT by the improved SWT with enhanced CDV system during operation.

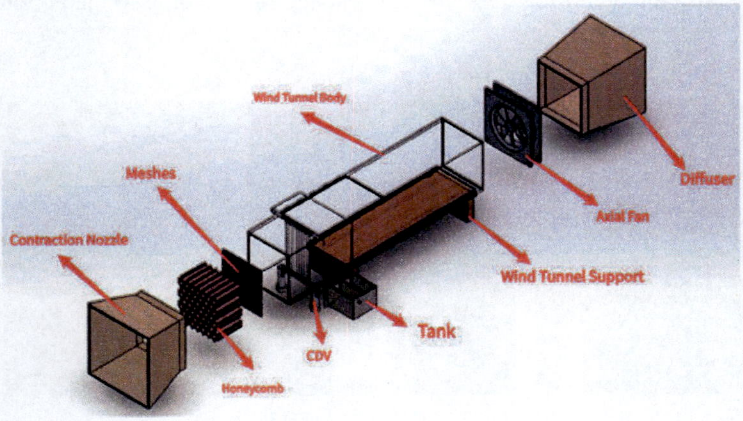

Fig. 11 Exploded view of the BLWT with enhanced SWT

Fig. 12 The instantaneous picture of **a** side and **b** front view of smoke generated by enhanced SWT

3.2 Experiment 2: Selection of Nichrome Wire and Nozzle Diameter

In order to obtain the best diameter combination of nichrome wire and nozzle, present study tested all possible combinations of the nichrome wire and nozzle size which summarized in Table 4. Tolerance in Table 4 is defined as the gap between nichrome wire and nozzle in mm. In this experiment, only a single nichrome is utilized to observe the smoke intensity.

Experiment 2. Case A

For Case A, 3 sets of experiments were conducted by combining the 0.5 mm nichrome wire diameter with 1.0 mm, 0.8 mm, and 0.6 mm nozzle diameter, as shown in Fig. 13a–c, respectively. As seen in Fig. 13a, due to rapid dripping, it is observed that the smoke was not produced when the wire is heated. Meanwhile, the combination with 0.8 mm nozzle diameter shows only partial smokes was produced at the upper part because the liquid fog does not have enough time to drop to the bottom part

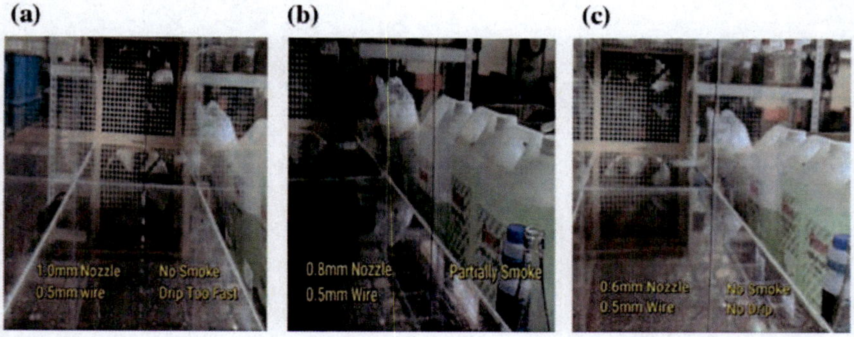

Fig. 13 Smoke conditions with the combinations of 0.5 mm wire diameter with **a** 1.0 mm, **b** 0.8 mm, and **c** 0.6 mm nozzle diameter

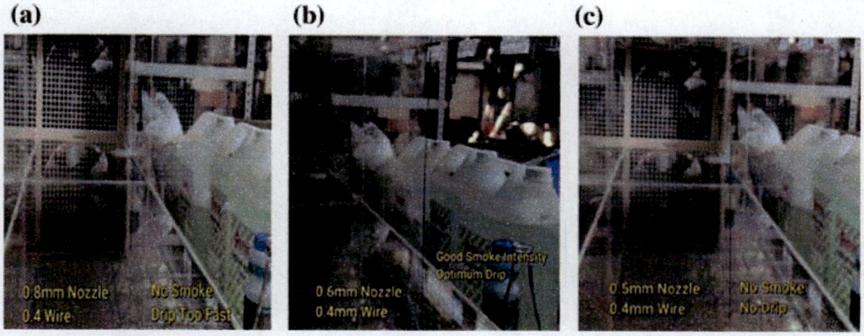

Fig. 14 Smoke conditions with the combinations of 0.4 mm wire diameter with **a** 0.8 mm, **b** 0.6 mm, and **c** 0.5 mm nozzle diameter

of the wire. On the other hand, no smoke is observed with 0.6 mm nozzle diameter combination due to no liquid fog drip, as shown in Fig. 13c.

Experiment 2. Case B

Similarly, 3 combination sets were conducted in Case B between 0.4 mm nichrome wire diameter with 0.8 mm, 0.6 mm and 0.5 mm nozzle diameters as shown in Fig. 14a–c, respectively. As seen in Fig. 14a, smoke is not produced due to rapid dripping in which the nichrome wire did not have enough time to heat the liquid fog. In addition, no smoke is observed in the third combination as well. However, a combination between 0.4 mm nichrome wire diameter with 0.6 mm displays a good smoke intensity from top to the bottom part of the wire, as shown in Fig. 14b.

Experiment 2. Case C

For Case C, 2 sets of combination between 0.2 mm nichrome wire with 0.5 mm and 0.4 mm nozzle diameter were conducted. Figure 15a, b show the result of these combinations. As shown in the figures, no smoke is observed for both combinations since the drip is too slow in which almost no drip at all.

Experiment 2. Summary

From experiment 2, different combination of nichrome wire diameter with different nozzle size displays various results on the smoke intensity. The best combination of the smoke wire and nozzle size to produce good smoke intensity with continuous dripping frequency was 0.4 mm nichrome and 0.6 mm nozzle diameter, respectively.

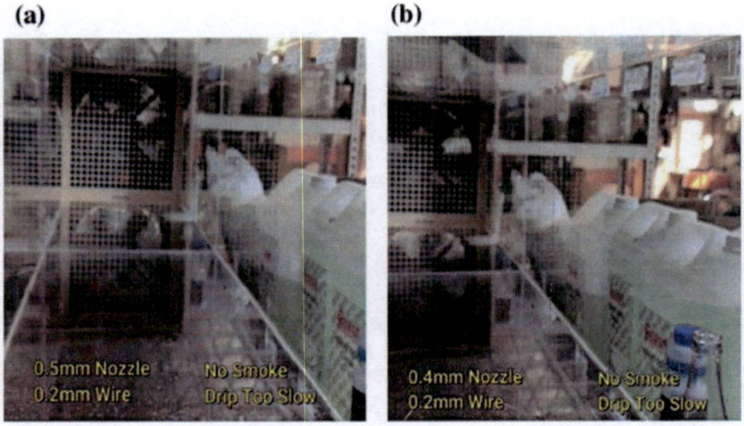

Fig. 15 Smoke conditions with the combinations of 0.2 mm wire diameter with **a** 0.5 mm, and **b** 0.4 mm nozzle diameter

3.3 *Experiment 3: Flow Visualization Past Rigid Body*

To validate the effectiveness of the improved SWT with enhanced CDV system, a flow visualization experiment past rigid body was conducted. In this experiment, a two-dimensional cylinder and rectangular were installed in the upwind of the wind tunnel test section. Figures 16 and 17 show the instantaneous pictures of flow pattern around cylinder and rectangular bluff body, respectively. In the enhanced SWT system, the nichrome wire was heated at 3.5 A shows crisp, high intensity and continuous flow streamlines. From the two-dimensional flow experiment past rigid body, one can clearly observed the pair of tip vortex structures that interact with Karman vortex formation which responsible for the formation of the downwash flow near the top free end of the body in both figures. In real situation, this formation of the downwash flow is a serious problem for stacks operating at low exhaust-to-wind velocity ratios

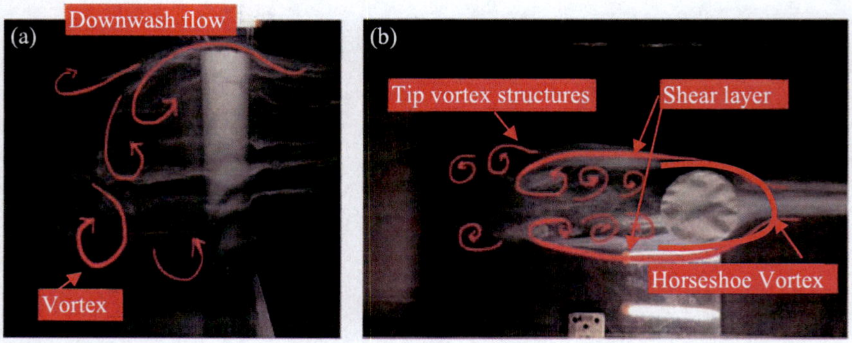

Fig. 16 Flow visualization around a cylinder from **a** side and **b** top views

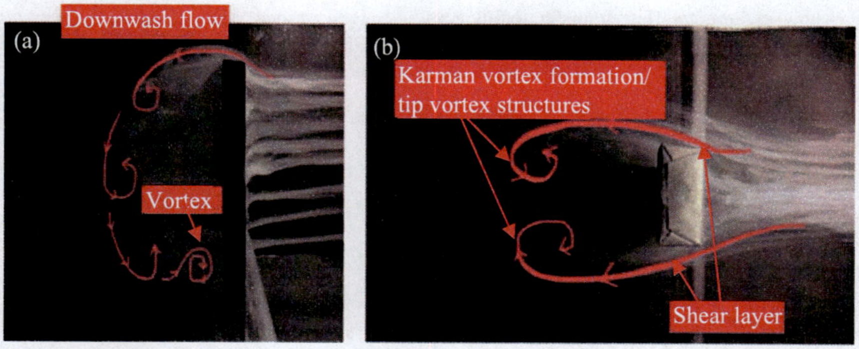

Fig. 17 Flow visualization around a rectangular from **a** side and **b** top views

since pollutants can be brought towards the ground. In addition, the shear layer at the sides of rigid body caused vortex pair to form and remain further downwind can be observed in both figures. In particular for cylinder shape, the horseshoe vortex can be evidently seen at the bottom near the surface.

4 Results and Discussions

A quasi-atmospheric BLWT with an improved SWT was designed and fabricated. In addition, a new CDV with the addition of a tensioner system that operates fully automated was enhanced with 10 lines of nichrome wires using a 1000 W power supply. This fully automated smoke fluid delivery system CDV was introduced to help to enhance the intensity, produce high quality and continuous smoke flow in the wind tunnel, compared to the previous design. Present work also identifies the best wire combinations, diameters, designs, and voltage levels to use. The best combination of nichrome wire and nozzle diameter was found to be 0.4 mm nichrome wire diameter and 0.6 mm nozzle size. This combination manages to provide continuous dripping with continuous smoke from all 10 heated wires. In the flow visualization experiments, shear layer at the sides of cylinder and rectangular bluff body can be observed clearly which caused a vortex pair to form and remain further downwind. The horseshoe vortex can also be evidently observed at the bottom near the surface.

In future work, flow visualization experiment utilizing a high-speed camera is a must to ensure the best quality of the instantaneous pictures during the wind tunnel experiments. The lighting when interpreting the results also plays a big role to ensure the best quality of the flow visualization. Therefore, flow visualization experiment of present enhanced SWT with CDV should be done in a dark room utilizing high speed camera to capture its flow pattern to minimize light pollution when interpreting the results.

Acknowledgements The authors gratefully acknowledge the research grant and financial support provided by the Ministry of Higher Education, MOHE (under the FRGS grant number: FRGS/1/2019/TK07/UMP/02/7 (RDU1901208) and Universiti Malaysia Pahang, UMP (under UMP grant number: RDU190375).

References

1. Witkowski D, Lee A, Sullivan J (1989) Aerodynamic interaction between propellers and wings. J Aircr 26(9):829–836
2. Bayati I, Belloli M, Bernini L, Zasso A (2017) Aerodynamic design methodology for wind tunnel tests of wind turbine rotors. J Wind Eng Ind Aerodyn 167:217–227
3. Roy S, Saha UK (2015) Wind tunnel experiments of a newly developed two-bladed Savonius-style wind turbine. J Appl Energy 137:117–125
4. Van de Ven TAM, Fryrear DW, Spaan WP (1989) Vegetation characteristics and soil loss by wind. J Soil Water Conserv 44:347–349
5. Blocken B, Toplar Y (2015) A following car influences cyclist drag: CFD simulations and wind tunnel measurements. J Wind Eng Ind Aerodyn 145:178–186
6. Cengel Y, Cimbala J (2004) Fluid mechanics: fundamentals and applications, 4th edn. McGraw-Hill, New York
7. Rahmat NA, Hagashima A, Ikegaya N, Tanimoto J (2016) An experimental study on aerodynamic interaction between a boundary layer generated by a smooth and rough wall and a wake behind a spire. Engineering sciences reports, vol 37, no 2. Kyushu University, pp 19–26
8. Rahmat NA, Hagashima A, Ikegaya N, Tanimoto J (2018) Experimental study on effect of spires on the lateral nonuniformity of mean flow in a wind tunnel. Evergreen 5(1):1–15
9. Sumner D, Wong SST, Price SJ, Paidoussis MP (1999) Fluid behaviour of side-by-side circular cylinders in steady cross-flow. J Fluids Struct 13:309–338
10. Talele V, Mathew VK, Sonawane N, Sanap S, Chandak A, Nema A (2021) CFD and ANN approach to predict the flow pattern around the square and rectangular bluff body for high Reynolds number. Mater Today Proc 47:3177–3185
11. Sumner D, Heseltine JL, Danserou OJP (2004) Wake structure of a finite circular cylinder of small aspect ratio. Exp Fluids 37:70–730
12. Chashechkin YD, Mitkin VV (2004) A visual study on flow pattern around the strip moving uniformly in a continuously stratified fluid. J Vis 7(2):127–134
13. Stefano M, Gianluca B (2007) Influence of the free surface on the flow pattern around a rectangular cylinder. In: The 9th international symposium on fluid control, measurement and visualisation, FLUCOME
14. Hongying J, Huixue D, Qianying M, Jun-Hai Z (2019) Airflow patterns around obstacles with large-span shallow shell roof: wind tunnel measurements and direct simulation. Math Probl Eng 1–11
15. Bahrami-Yarahmadi M, Pagliara S, Yabarehpour E, Najafi N (2020) Study of scour and flow patterns around triangular-shaped spur dikes. KSCE J Civ Eng 24:3279–3288
16. Raeisi N, Ghomeshi M (2021) A laboratory study of the effect of asymmetric-lattice collar shape and placement on scour depth and flow pattern around the bridge pier. Water Sci Technol Water Supply 1–16
17. Ferreira AD, Silva MCG, Viegas DX, Lopes AG (1991) Wind tunnel simulation of the flow around two-dimensional hills. J Wind Eng Ind Aerodyn 38:109–122
18. Ali HA (2016) Numerical study of flow pattern in buildings with different heights. Civil Eng J 2(3):95–101
19. Settles G (1986) Modern developments in flow visualization. AIAA J 24(8):1313–1323

20. Wu T-J (1992) Visualisation of stall characteristics of airfoils using the smoke-wire technique. The University of Texas at Arlington, ProQuest Dissertations Publishing
21. Gao N, Liu XH (2018) An improved smoke-wire flow visualization technique using capacitor as power source. Theor Appl Mech Lett 8(6):378–383
22. Mathur NB, Ramesh G, Yajnik KS (1988) Continuous coloured smoke-wire technique for flow visualisation. J Aeronaut Soc India 40:131–137
23. Sharul SD, Mohd Arief MN, Muhammad KK (2006) An improved smoke-wire flow visualization technique. In: Proceedings of the 4th WSEAS international conference on fluid mechanics and aerodynamics, Greece, pp 231–236
24. Ismail AT, Kamaruddin NM (2020) Development of a flow visualisation technique in wind tunnel for hydrokinetic turbine application. Mater Sci Eng 920:1–6
25. Liu Y, Cheng B, Deng X (2013) An application of smoke-wire visualization on a hovering insect wing. J Vis 16(3):185–187
26. Soria J, Chiu WK, Norton MP (1990) A study of unsteady laminar boundary layer flow on a flat plate using a smoke-wire/silhouette flow visualization technique. Exp Therm Fluid Sci 3:291–304
27. Serrano-Aguilera JJ, Hermenegildo García-Ortiz J, Gallardo-Claros A, Parras L, del Pino C (2016) Experimental characterization of wingtip vortices in the near field using smoke flow visualizations. Exp Fluids 57(8):11
28. Cornaro C, Fleischer AS, Goldstein RJ (1999) Flow visualization of a round jet impinging on cylindrical surfaces. Exp Fluids 20(2):66–78

A Numerical Simulation of Heat Transfer Characteristic of Twisted Tube in an Annular Heat Exchanger

Abdallah Talal Banat, Teng Kah Hou, Tey Wah Yen, I. A. Idowu, and Mohammed W. Muhieldeen

Abstract Heat transfer performance of heat exchanger system play significant role toward energy conservation. In this research, numerical simulation using ANSYS was conducted to evaluate the performance of square twisted tube and oval twisted tube for heat transfer enhancement. A comparative models were developed to evaluate the geometry alternation on heat transfer coefficient and pressure drop. The analysis revealed that the square twisted tube heat exchanger has better performance and smaller friction factor compared to the oval twisted tube heat exchanger. Moreover, a computational fluid dynamics (CFD) simulation with a realizable k-ε was used to investigate the influence of twist pitch length on the heat transfer performance and pressure drop. The geometrical parameters include two different cross-section shapes and two different twist pitch lengths of the square twisted tube heat exchanger. The twist pitch lengths used in the investigation are 160 mm and 200 mm. The numerical results shown that the tube with the smallest pitch length offers a higher friction factor and Nusselt number. In addition, heat transfer mechanism of twisted tubes, velocity, streamlines, Nusselt number distribution, and temperature distribution are presented. Heat transfer enhancement of the square twisted tube heat exchanger is associated with the secondary flow production by the curved tube wall. In the oval and square twisted tubes, a secondary flow is formed which offers a longer flow path, hence an enhanced heat exchanger performance. It concluded that alternation of geometry on tube surface is potential to improve the overall performance with acceptable pressure drop.

Keywords Heat transfer · Twisted tube · CFD · Twist pitch · Nusselt number

A. T. Banat · T. K. Hou (✉) · T. W. Yen · M. W. Muhieldeen
Mechanical Engineering Department, Faculty of Engineering, Technology and Built Environment, UCSI University, 56000 Kuala Lumpur, Malaysia
e-mail: TengKH@ucsiuniversity.edu.my

T. K. Hou · M. W. Muhieldeen
UCSI-Cheras Low Carbon Innovation Hub Research Consortium, Kuala Lumpur, Malaysia

I. A. Idowu
Low Carbon Eco Innovatory (LCEI), School of Civil Engineering and Built Environment, Liverpool John Moores University, Liverpool, UK

M. Y. Ismail et al. (eds.), *Technological Advancement in Mechanical and Automotive Engineering*, Lecture Notes in Mechanical Engineering,
https://doi.org/10.1007/978-981-19-1457-7_48

1 Introduction

Heat exchanger represents a system that mainly operates to transmit thermal energy from hot to cold fluid between two flows. Heat exchangers have numerous uses in manufacturing, domestic, and commercial industries [1]. The heat exchange between the two fluids occurs by convection heat transfer mechanism, in which the fluid flows over the tube [2]. The heat exchanger is categorized according to the heat transfer methodology, design characteristics, and flow arrangement. The annular heat exchanger, or double pipe, is among the most common forms of the heat exchanger under the classification of design and construction [3]. Thermal energy in a heat exchanger is produced by moving fluid flows with different temperatures parallel to each other and separated by a physical boundary in the shape of a tube. It tends to cause a forced convection and heat transfer [4]. Annular heat exchangers are used for the cooling fluid's process and sensible heating at a heat transfer maximum area of 50 m^2. This type is suitable for small duties requirements, typically, when one or both fluids are at high temperatures and pressures [5]. The heat exchanger can also be categorized according to the fluid's motion. This category includes parallel flow, counter-flow, and cross-flow. Thermodynamically, the counter flow is considered as the most superior flow comparing to other types [6]. It is the most efficient arrangement, in which it produces the highest temperature gradients in each fluid. In addition, it produces minimal thermal stress in the wall comparing to other flow types [7].

The limited resources and the purpose of conserving energy, cost, and material have moved researchers to search for new methods to improve the efficiency and performance of the exchangers [8]. Researchers around the world are searching for approaches to reduce the pressure drop and boost the heat transfer rate to save energy and limit natural resources to obtain high-performance heat exchangers [9]. From the past researcher's enhancement attempts, studies have found that there are two main heat transfer enhancement techniques that operate to boost the heat transfer rate, which is the active and passive technique, the active method need external power, while the passive method needs no external force [10]. Due to its simplicity, low cost, and reasonable augmentation with appropriate pressure drop, the passive enhancement technique is considered to be a fundamental and effective heat transfer enhancement. For financial, operational, and environmental considerations, it is a vital field. The passive enhancement of heat transfer is focused on the adoption of instruments to cause a swirl flow and velocity in the secondary flow field. Many forms of the passive methods include inserting twisted tapes, wire coils, fins, and using a twisted or corrugated tube [11].

The twisted tube, which is classified as a swirl flow device, is considered as a passive heat transfer augmentation device. According to its geometrical features, a secondary flow is produced by the twisted tube that can maximize the mean velocity, temperature difference of the working fluid, and Reynold's number [12]. The secondary or swirling flow is typically associated with high turbulence, which enables the mixing of various fluids, which also enhances the heat transfer rate. The produced swirl flow will provide a higher heat transfer rate but will include a pressure drop penalty. In the twisted tube, the primary benefit is that it does not need any special attention during assembling, cleaning, and maintenance [13].

Promthaisong et al. [14] studied numerically the effect of using a helical oval tube as a passive enhancement approach to boost the performance of a heat exchanger. Reynolds number is ranged from 5000 to 20,000. Promthaisong et al. used four turbulent models for the validation test, which is RNG k-ε, standard k-ε, SST k-ω, and realizable k-ε turbulence models. Among all models, the validation test illustrates that the realizable k-ε turbulent model values are the nearest to the values obtained from the standard correlation. The realizable k-ε model is proven the most accurate model in this investigation. While several studies discussed about the flow energy in different applications using RNG k-ε, standard k-ε, SST k-ω in CFD analysis [15–19].

The study of the effect of the square twisted tube on the overall performance of the heat exchanger was not considered in previous studies. In the present study, the square cross-sectional of twisted tubes is investigated on the heat transfer enhancement. There is limited literature review reported on the heat transfer and fluid flow characteristics inside square and oval twisted tube which has create a research gap where attract engineer and researchers' interest. In current research, we developed a twisted tube heat exchanger model using CFD simulation to be compared with a smooth circular tube heat exchanger on the performance investigation. The study focusses on turbulent flow model between Reynolds number range of 10,000–16,000. Heat transfer and pressure drop characteristics of a twisted tube heat exchanger were investigated throughout the research.

2 Methodology and Experimental Setup

2.1 Physical Models

The cross-section shapes of the heat transfer enhanced tubes named square twisted tube and oval twisted tube. These different cross-section geometries of the twisted tubes are used for the aim of comparing the heat exchanger performance with the smooth circular tube. The geometrical parameters of the oval twisted tube are the effective heat transfer length, long axis OD, short axis OD, and twist pitch length *P*. The twist pitch, *P* is the distance between the two positions along the tube's length where the tube's section orientations correspond to each other. Along with one pitch distance, the cross-section rotates by 360. The main twist pitch length to be used in the study is 200 mm for the oval twisted tube and 160 mm and 200 mm for the square twisted tube. The total length and effective heat transfer length of the twisted tubes are 1300 mm and 800 mm, respectively for all models. The diameter and length of the outer tube or shell are 40 mm and 800 mm, respectively. All models include cold water as the working fluid flowing through the tube. The hot water flows inside the shell and

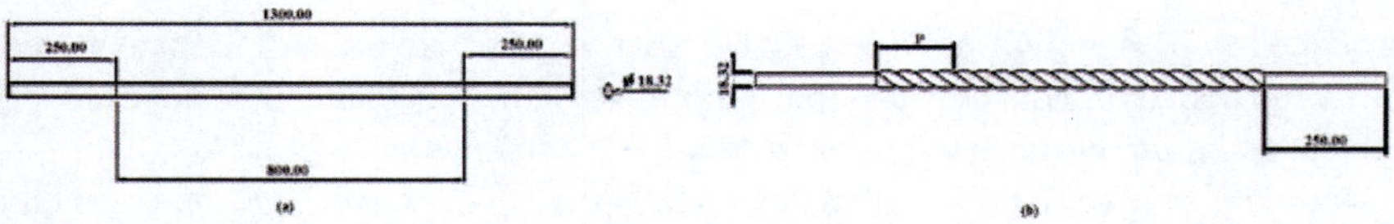

Fig. 1 Smooth circular tube heat exchanger (**a**) and square twisted tube exchanger with P = 160 mm and P = 200 mm (**b**)

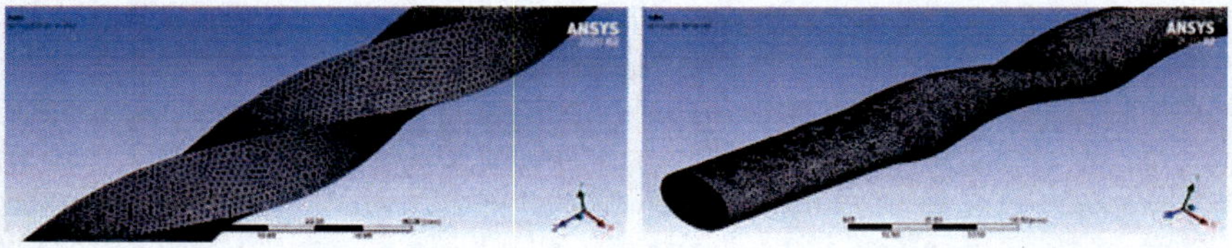

Fig. 2 Zoomed-in mesh of square twisted tube (left) and oval twisted tube (right)

around the tube. The performance of the twisted tubes cases will be compared with the smooth tube. The smooth tube heat exchanger, shown in Fig. 1, has a diameter and effective heat transfer length of 18.32 mm and 800 mm, respectively. Figure 2 shown the twisted tube whereas details of the twisted tube geometry presented in Table 1 in Appendix.

2.2 *Mesh Generation*

In this study, the tetrahedral mesh method has been used for all cases domain. Tetrahedral elements can fit better complex and arbitrary shaped geometries with less computational effort than the hexahedral method. Also, an inflation meshing has been added. The geometry selected is the tube body. The boundary selected is all faces with neglecting the inlets and outlets of the tube faces. For the sizing function, body and face sizing function are the only functions used in this investigation to handle the most critical parts that have a major impact on the results. Body size has been applied to all bodies to control the number of elements. Face sizing is also applied to all parts that are associated with the fluid flow, which include the inlet and outlet of the tube, the intermediate wall of the tube, and the inlet and outlet of the shell. The element sizing for the tube's body, shell body, tube's face (inlet and outlet), intermediate face (wall), and shell face (inlet and outlet) are 1.8 mm, 2.5 mm, 1.6 mm, 1.4 mm, and 2 mm, respectively. The resulting meshing to produce high-quality results are shown in Fig. 2.

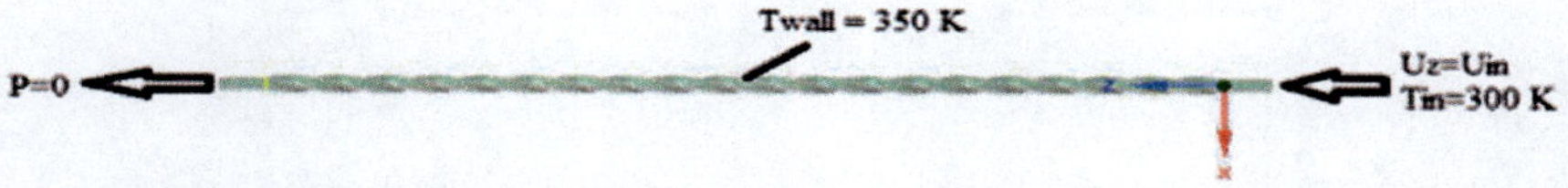

Fig. 3 Boundary conditions assigned for the tubes

2.3 *Setup and Boundary Condition*

The following assumptions are taken into consideration during this investigation.

1. The flow is a single-phase and steady-state.
2. The working fluid of water is Newtonian and incompressible fluid with constant thermophysical properties due to the small temperature gradients around the tube.
3. Thermal radiation, natural convection, and viscous dissipation are neglected.
4. The flow inside all tubes is in turbulent regions.

The realizable k-ε model is employed throughout the research due to being beneficial by performing superior performance for turbulence flow in terms of rotational flow, separation, recirculation, and boundary layer under a high-pressure gradient. Figure 3 shows the boundary conditions assigned for the tube for all cases. The shell's wall thermal conditions are set with zero heat flux due to the assuming of zero heat loss to the surroundings.

2.4 *Numerical Validation*

The numerical results of fluid flow and heat transfer of an oval twisted tube are numerically compared with the available experimental data to ensure the accuracy and reliability of the current numerical results. Figure 4 shown that the current model results is well validated with the reported literature review in Tang et al. [20] on the experimental results which has RMSE of >95%.

3 Results and Discussion

3.1 *Velocity*

In the tube's inlet, water flows through the pipe at a cold temperature value of 300 K, and constant 10,000 Reynold flow. The velocity causes a turbulent flow to be generated, which assists to enhance the heat transfer performance across the heat exchanger. The hydraulic diameter of all geometries is the same, in which to

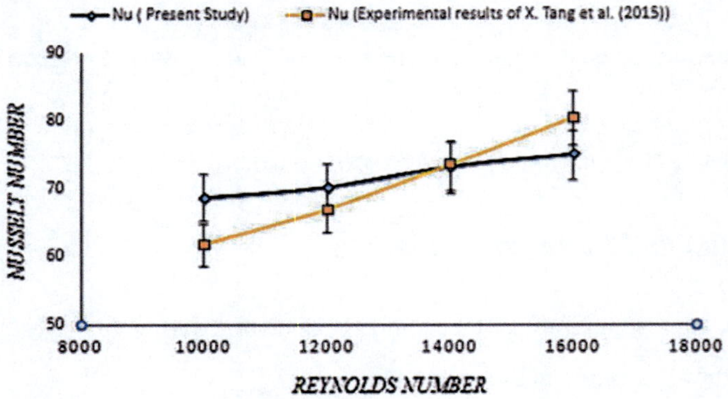

Fig. 4 Validation of numerical and experimental data

ensure that the Reynolds number will be the same for all cases, in a turbulent flow condition. Even though the hydraulic diameter and Reynolds number are the same, but the inlet dimensions of the square twisted tube and oval twisted tube are different. The square and oval tubes have dimensions of (18.32 mm × 18.32 mm) and (15 mm × 24.4 mm), respectively. The inlet area of the square twisted tube and oval twisted tubes are 335.62 mm^2 and 287.46 mm^2, respectively. The mass flow rate value of the square twisted tube with pitch length P = 160 mm and P = 200 mm, and the oval twisted tube with pitch length P = 200 mm are 0.18360 kg/s, 0.18360 kg/s, and 0.15651 kg/s, respectively. From Fig. 5, it can be found that the velocity is enhanced due to the twisting geometry of the three cases, in which the working fluid enters the tube, and its velocity is significantly enhanced. From Fig. 5c, the velocity of the fluid in the oval twisted tube is the most enhanced at the outlet extended region comparing with the other cases.

To ensure the homogeneous flow and consistency of the simulation, Fig. 6 prove the fully developed flow region for the accuracy of the calculation. It is also can be seen that the maximum velocity is increased in square, rectangular and oval twisted tube respectively, in which it occurs due to the geometrical shape of the twisting. The

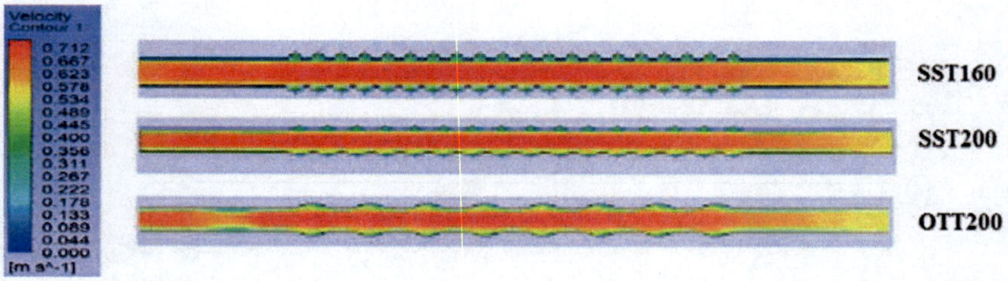

Fig. 5 Velocity contours of water flow of the different twisted tube cases. **a** Square (SST160); **b** rectangular (SST200); **c** oval (OTT200)

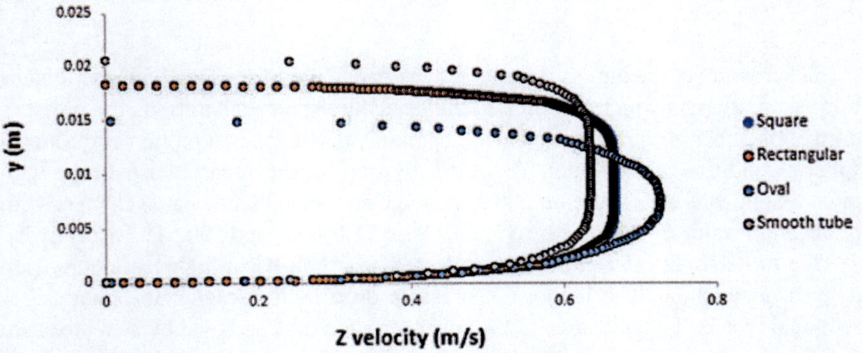

Fig. 6 Velocity profile for 4 different geometry of heat exchanger tube

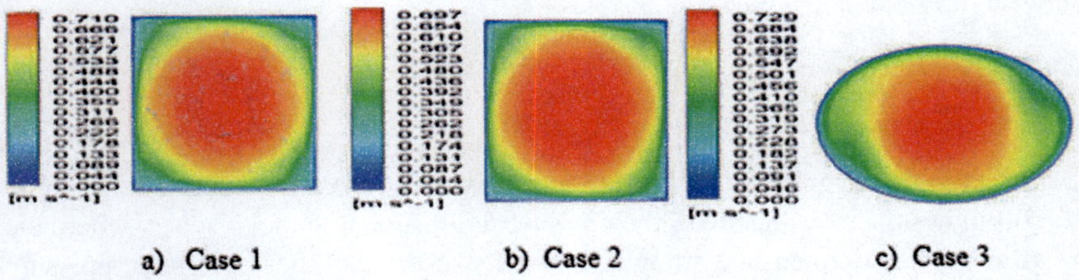

Fig. 7 Axial velocity contours at z = 0.7 m, Re = 10,000 of the different twisted tube. **a** Square; **b** rectangle; **c** oval (OTT200)

oval twisted tube has the highest maximum fluid velocity and the smooth circular tube have the lowest maximum velocity comparing with all cases.

Axial velocity contours of 3 different cases on the cross-section of z = 0.7 m of the intermediate wall is shown in Fig. 7 at Re = 10,000. Due to the cross-sectional shape, it has been found that case 3 has the highest axial velocity. Comparing with the square twisted tube cases, the oval twisted tube has a higher core flow zone, the red region. In other words, this explains that there are more fluids flow in a swirl motion and the mixture between fluid at the wall region and the core fluid are enhanced.

3.2 *Pressure Drop*

Pressure drop is analysed to indicate the hydraulic performance of the cases. Two square twisted tube heat exchanger with P = 160 mm and P = 200 mm, oval twisted tube with P = 200 mm is analysed in the pressure drop term to be compared with the smooth circular tube, in which the flow of all cases is through the same distance. The pressure drop ΔP is obtained from Eq. 1, in which it is the difference between the inlet and outlet pressure.

$$\Delta P = P_{in} - P_{out} \tag{1}$$

The pressure of the outlets is set to be zero-gauge pressure as the boundary condition assumption, in which the fluid discharge at atmospheric condition. The pressure drop of the tubes is equivalent to the inlet pressure due to the assumption of the atmospheric condition at the outlet. The pressure drop for the square twisted tube with pitch length values of 160 mm, 200 mm, and oval twisted tube heat exchanger with pitch length value of 200 mm are 571.119 Pa, 559.148 Pa, and 590.519, respectively.

Due to its small inlet cross-sectional area, the shell has a higher pressure drop than the inner tube. The higher the pressure drop is, the smaller the inlet cross-sectional area is. The inlet area of the square twisted tube and oval twisted tubes are 335.62 mm^2 and 287.46 mm^2, respectively. The pressure drop in the oval twisted tube heat exchanger is the highest comparing with other cases due to its inlet area conditions, in which it has the smallest inlet area. Equation 2 explains the relationship between pressure and area.

$$P = \frac{F}{A} \tag{2}$$

From Fig. 8, the graph of 'pressure versus distance across the inner tube' at Re = 10,000 shows the relationship of the pressure proportional to the distance, where the pressure drop is defined in a straight line. This shows that there is a huge pressure drop for all twisted tube cases, as the pressure is decreasing constantly as it moves from the inlet to the outlet of the heat exchanger. The graph shows that the oval twisted tube heat exchanger P = 200 mm (case 3) has the highest pressure drop among all cases. The smooth circular tube has the lowest pressure drop compared with the twisted tube cases due to being a normal heat exchanger with no passive enhancement technique used. The square twisted tube with P = 160 mm (case 1) has a higher pressure drop than the rectangular twisted tube with P = 200 mm (case 2), which clearly illustrate that the twisting pitch is having an inversed proportional relationship with the pressure drop, in which the higher the twisting pitch is, the lower the pressure drop is. This also can explain why the friction factor of case 1 is

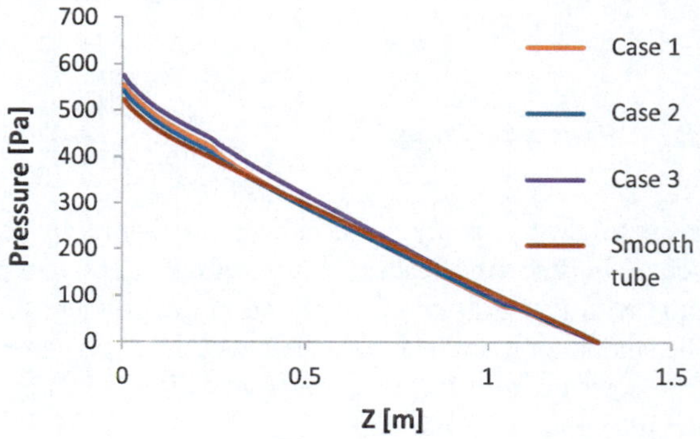

Fig. 8 Graph of pressure versus distance across the inner tube at Reynolds number 10,000

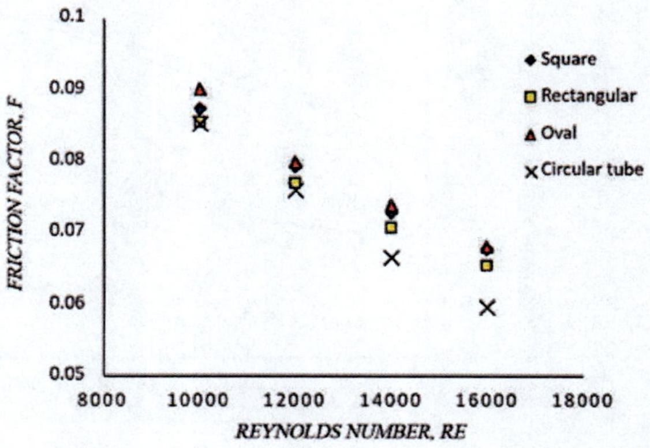

Fig. 9 Friction factor versus Reynolds number graph

higher than case 2, as shown in Fig. 9. Finally, all twisted tube cases have almost the same pressure drop due to the usage of the passive enhancement technique of twisting the tube. Interesting to point out that the pressure drop is not significant compared to different twisted geometry.

3.3 Heat Transfer Performance Analysis

Three cases are used in this study to compare their thermal performance. Several characteristics were necessary to analyse the performance through the comparison of the heat transfer rate, Q. The total heat transfer rate of the twisted tubes can be obtained from Eq. (3). The m_c, $c_{p,c}$, $T_{c,in}$, and $T_{c,out}$ are the tube's inlet mass flow rate, the specific heat capacity of the cold fluid, temperature of the cold fluid inlet, and temperature of the cold fluid outlet.

$$Q = m_c c_{p,c}\left(T_{c,in} - T_{c,out}\right) \tag{3}$$

In this study, the cold fluid stream in the inner tube is investigated to compare how effective are the twisted tubes going to heat the cooling fluid comparing with the normal smooth tube. The shell's heat transfer rate is neglected in this study since the inlet and outlet temperature are constant due to the adiabatic wall boundary condition assigned.

The maximum heat transfer rate should be calculated to determine the heat exchanger effectiveness for all cases. The maximum heat transfer rate is the maximum amount of heat that can be transmitted in the heat exchanger. The heat exchanger effectiveness is the measurement of how effective heat is transmitted from the outer tube fluid, shell, to the inner tube fluid. The maximum heat transfer rate and effectiveness can be obtained from Eqs. (4) and (5).

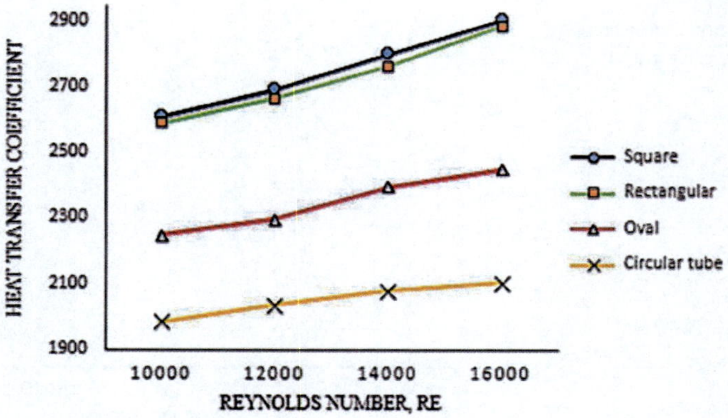

Fig. 10 Total heat transfer rate versus Reynolds number for all twisted tube cases compared with the smooth circular tube

$$Q_{max} = C_{min}\left(T_{h,in} - T_{c,in}\right) \tag{4}$$

$$\epsilon = \frac{Q}{Q_{max}} \tag{5}$$

Figure 10 shown the heat transfer rate versus Reynolds number under effect of different geometry twisted tube. It can be clearly seen that the total heat transfer rate has a directly proportional relationship with the Reynolds, in which the increase of the velocity will aid to increase the total heat transfer rate. On top of that, due to the secondary flow formed by the twisted tube cases, the total heat transfer rate is enhanced by Square twisted tube > rectangular twisted tube > Oval twisted tube > smooth circular tube under same cross-sectional condition. It is strongly believed due to the recirculation motion near the twisted boundary condition, the Brownian motion of the fluid recirculate in high turbulence condition in where eventually increase the heat transfer performance near the boundary condition. On top of that, the recirculation momentum energy drive from one twisted zone to another twisted zone whereas overall performance was expected to increase. These findings could be one of the novel contributions for future engineer to design new heat exchanger geometry for ultimate heat transfer performance.

4 Conclusion

This research is concluded that by varying the geometry of heat exchanger tubing in annular heat exchanger is improves the overall heat transfer performance. These findings could implement in various engineering application such as pharmaceuticals, food storage, material processing, and air-conditioning systems. The research findings concluded that:

1. Twisted tube is enhancing the overall heat transfer performance of heat exchanger
2. The pitch geometry after twisted tube increase the overall heat transfer performance due to the recirculation process near the boundary wall of the heat exchanger tubing surface.
3. The square twisted tube heat exchanger outperforms the oval twisted tube heat exchanger in terms of thermal and hydraulic performance.
4. Velocity of the flow has positive impact toward heat transfer enhancement for both twisted tube and smooth tube cases.

The results of these investigations could be design aid for economic and enhancement of heat exchanger performance. However, more experimental work is required to strengthen the findings.

Acknowledgements The authors gratefully acknowledge UCSI University REIG-FETBE-2021/013 grant support and Faculty of Engineering and Technology, Liverpool John Moores University for support to conduct this research work.

Conflict of Interest The authors declare that there is no conflict of interest regarding the publication of this paper.

Appendix

See Table 1.

Table 1 Geometrical parameters of the different twisted tube cases

Cases	Cross section	Shape	a – b (mm)	A (mm^2)	C (mm)	D_h (mm)	*P* (mm)	S (mm)
1	Square		18.32	335.62	73.28	18.32	160	800
2	Rectangular		18.32	335.62	73.28	18.32	200	800
3	Oval		15–24.4	287.46	62.77	18.32	200	800

References

1. Fadhil NK (2018) Numerical investigation of shell and tube heat exchanger with twist annular. In: IOP conference series: materials science and engineering. IOP Publishing, UK
2. Jiji LM (2009) Heat convection. Springer Science & Business Media, Berlin
3. Teng KH et al (2016) Mitigation of heat exchanger fouling in industry using catalytic materials. Desalin Water Treat 57(1):22–27
4. Vignesh S, Moorthy VS, Nallakumarasamy G (2017) Experimental and CFD analysis of concentric dimple tube heat exchanger. Int J Emerg Technol Eng Res (IJETER) 5(7):18–26
5. Thulukkanam K (2013) Heat exchanger design handbook. CRC Press, Boca Raton
6. Hou TK et al (2017) Industrial heat exchanger: operation and maintenance to minimize fouling and corrosion. In: Heat exchangers—advanced features and applications
7. Shah RK, Sekulic DP (2003) Fundamentals of heat exchanger design. Wiley, New York
8. Hou TK (2018) Fouling and its mitigation on heat exchanger surfaces by additives and catalytic materials. Liverpool John Moores University, United Kingdom
9. Teng KH et al (2017) Retardation of heat exchanger surfaces mineral fouling by water-based diethylenetriamine pentaacetate-treated CNT nanofluids. Appl Therm Eng 110:495–503
10. Jayan N et al (2020) Heat transfer augmentation approach in double pipe heat exchanger. Eng Phy 7(7):791–794
11. Kareem ZS et al (2015) Passive heat transfer enhancement review in corrugation. Exp Therm Fluid Sci 68:22–38
12. Farnam M et al (2018) Heat transfer intensification of agitated U-tube heat exchanger using twisted-tube and twisted-tape as passive techniques. Chem Eng Process Process Intensification 133:137–147
13. Eswar Raja Babu P (2015) Heat transfer analysis on twisted tube heat exchanger technology, vol 5, no 4
14. Promthaisong P, Jedsadaratanachai W, Eiamsa-ard S (2018) Numerical simulation and optimization of enhanced heat transfer in helical oval tubes: effect of helical oval tube modification, pitch ratio, and depth ratio. Heat Transf Eng 39(19):1665–1685
15. Salman BH, Hamzah MZ, Purbolaksono J, Inayat-Hussain JI, Mohammed HA, Muhieldeen MW (2011) Determination of correlation functions of the oxide scale growth and the temperature increase. Eng Fail Anal 18:2260–2271
16. Francesca P, Tey WY, Tan LK, Muhieldeen MW (2020) Investigation on generalised trapezoidal differencing time-marching scheme for modelling of acoustical wave. CFD Lett 12(2):11–21
17. Ng YH, Tey WY, Tan LK, Arada GP, Muhieldeen MW (2020) Numerical examination on two-equations turbulence models for flow across NACA 0012 airfoil with different angle of attack. CFD Lett 12(2):22–45
18. Tey WY, Sidik NAC, Asako Y, Muhieldeen MW, Afshar O (2021) Moving least squares method and its improvement: a concise review. J Appl Comput Mech 7(2):883–889
19. Muhieldeen MW, Kuang YC (2019) Saving energy costs by combining air-conditioning and air-circulation using CFD to achieve thermal comfort in the building. J Adv Res Fluid Mech Therm Sci 58(1):84–99
20. Tang X et al (2015) Experimental and numerical investigation of convective heat transfer and fluid flow in twisted spiral tube. Int J Heat Mass Transf 90:523–541

Study on the Effects of Tube Arrangements to the Heat Transfer Performance of Evaporator Chiller System Based on Industrial Standards

Hamad Ali Hamad Bin Hatrash, Ir. Noor Idayu Binti Mohd Tahir, and Mohammed W. Muhieldeen

Abstract Chiller system is one of the main components of the HVAC system and can be cooled by water or air. The core components of vapor compression chiller systems are compressors, condensers, expansion systems and evaporators. The type of evaporator usually used in the HVAC chiller system is a shell and tube evaporator. Hence, this project aimed to study the effects of industrial standard tube arrangements which are triangular (30°), rotated triangular (60°), square (90°) and rotated square (45°) on the heat transfer performance of evaporator chiller system and selecting optimized tube arrangement design. The methods were used are designing of the four geometries of the shell and tube evaporator and the ANSYS Fluent for the simulation of the designs. The boundary condition of this study was, the inlet mass flow rate of the cold fluid (Refrigerant 134a) was 2.5 kg/s, whereas the mass flow rate of hot fluid (water) was 3.3 kg/s. Also, the hot fluid temperature at the inlet was 12.2 °C and the cold fluid inlet temperature was kept at −15 °C. Based on the study findings, the shell and tube evaporator with tube arrangement at 45° transfer more heat transfer than the other three designs in the CFD. The shell and tube evaporator at tube arrangement 30° was chosen to be the optimized design based on the overall performance. Thus, this study will benefit the HVAC chiller system evaporators. Thus, this study will benefit the HVAC chiller system evaporators.

Keywords Tube arrangements · Heat transfer · Ansys fluent · Chiller · CFD

H. A. H. B. Hatrash · Ir. N. I. B. M. Tahir · M. W. Muhieldeen (✉)
Mechanical Engineering Department, Faculty of Engineering, Technology and Built Environment, UCSI University, 56000 Kuala Lumpur, Malaysia
e-mail: al-gailani@ucsiuniversity.edu.my

Ir. N. I. B. M. Tahir
e-mail: nooridayu@ucsiuniversity.edu.my

M. W. Muhieldeen
UCSI-Cheras Low Carbon Innovation Hub Research Consortium, Kuala Lumpur, Malaysia

M. Y. Ismail et al. (eds.), *Technological Advancement in Mechanical and Automotive Engineering*, Lecture Notes in Mechanical Engineering,
https://doi.org/10.1007/978-981-19-1457-7_49

1 Introduction

1.1 Background

The HVAC system is designed for the comfort and convenience of occupants to meet environmental requirements. The system includes heating, air conditioning, and ventilation. Also, for different forms of construction, such as manufacturing, commercial, residential, and institutional buildings, HVAC systems are more common [1]. HVAC systems can be identified by appropriate processes and distribution facilities [2]. The chiller system is one of the main components of the HVAC system. The chiller system can be cooled by water or air. Air-cooled chillers are mounted outside the unit and condenser coils are cooled by air fans. Water-cooled chillers are typically mounted internally and recycled into an external cooling system.

The heat reject cooling tower or cooling system can be used as the external cooling mechanism more effectively [3]. The core components of vapor compression chiller systems are compressors, condensers, expansion systems and evaporators [4]. The chiller system's main components are interconnected to create a traditional closed-loop cooling system.

The compressor releases the compressed gas refrigerant through the discharge network into the condenser that cools and condenses the refrigerant during normal operation of the vapor compression cooling system. The concentrated refrigerant is transferred through the condenser to the expansion valve so that the refrigerant is cooled by expansion valves operation before reaching the evaporator inlet as a two-phase mixture of liquid and steam refrigerant. The two-phase refrigerant mixture is distributed around the evaporator cover tube package. The refrigerant flows between the tube and cools the heat by absorbing fluid that passes through the interior of the tube bundle [5].

The evaporator and other main parts such as the compressor, condenser and expansion unit of the cooling system are critical components [6]. Several studies discussed about the modeling using ANSYS simulation as a platform to design the model required [7–12]. When the array is large enough and the tubes are arranged in a repeated pattern, the flow pattern is produced [13]. The temperature differential between the fluid evaporation and the steam condensation rate, the evaporation temperature, the irrigation pressure, the liquid viscosity, the head tube length, and the evaporator tube arrangement will be parameter. There are two general types of tube arrangement, the rectangular array (in-line arrangement) and the triangular array (staggered arrangement). The triangular arrangement enables the installation of more tubes in a certain volume and is thus widely used in the designs due to this benefit. The in-line configuration can be used in a variety of cases, such as airside issues [14]. For the dirty shell side services, a square layout is typically used. Since the square arrangement is an in-line pattern, it produces lower turbulence [15].

This study aimed to study on the effects of tube arrangements to the heat transfer performance of evaporator chiller system based on industrial standard. Figure 1 shows the industrial standard tube arrangements.

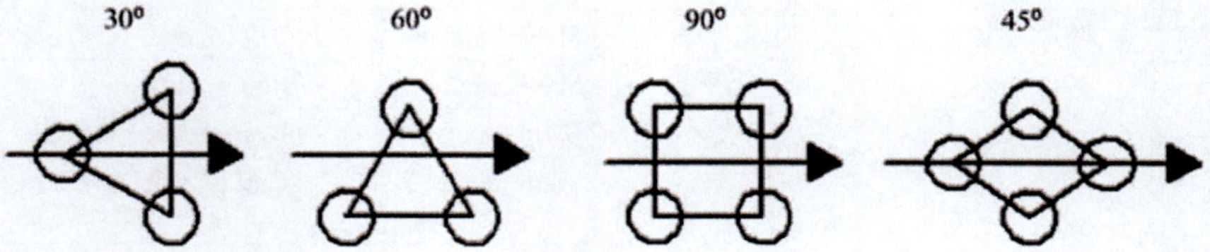

Fig. 1 The industrial standard tube arrangements

2 Methodology

The objectives of carrying out this research is to study the tube arrangements and their effect on heat transfer performance. Therefore, there are four major steps in the methodology of this study, which including "Initial study, Pre-processing, Solution, and Post-processing". In the initial study, it is very important to obtain data input for this study. Also, data input was obtained from an industrial standard TEMA [16], ASME [17], industrial catalogue [18], case studies.

In the pre-processing step, the geometry of the shell and tube has been designed using SOLIDWORK software. The geometry has been created and imported to the ANSYS software in the geometry section, then it saved automatically, and it is ready to be exported to the mesh section. Further, after the meshed, the meshed file is exported back to fluent mesh mode, the domain has been proceeded to setup stage where the parameters and condition setting are to be applied here. Thus, the domain is continued with the processing stage after all the inputs have been applied and numerical and graphical analysis is carried out.

Additionally, the post-processing step shows the result of the analysis. Thus, if the results are not reached or out of the target range, the process algorithms must go back to the pre-processing stage to define the problem and adjust inputs or limits before the result convergence is achieved. Lastly, the data from the results have been analysed by comparing the results of different geometric tube arrangements and the results of this analysis have been drawn.

2.1 *Initial Study*

Data that has been collected was related to the shell specifications and tube specifications from the industry as Table 1.

Table 1 Input data

Shell-side parameter	Shell diameter	217.27 mm
	Shell length	850.9 mm
	Fluid type	Refrigerant 134a
	Material	Carbon steel SA516 GR70
	Thickens of the shell	7.04 mm
	Nozzle diameter	64.87 mm
Tube parameter	Tube diameter	19.05 mm
	Tube length	850.9 mm
	Thickness of the tube	0.71 mm
	Fluid type	Water
	Material	Cooper
	Number of tubes	37
	Tube pitch	1.25 times the OD
	Tube arrangements	Square (90°), rotated square (45°), triangular (30°), rotated triangular (60°)

2.2 Mathematical Approach

During this chapter, the mathematical equation that has been used is the Reynold number equation to determine if the flow is turbulent or laminar as shown in Eq. (1).

$$Re = \frac{P v D}{u} \tag{1}$$

Whereby

Re Reynolds number
P Fluid density in kg/m^3
v Fluid velocity in m/s
D Tube diameter in m
u Fluid dynamic viscosity in kg/(m s).

2.3 Pre-processing

Geometry Design
The geometry method is simply the development of a three-dimensional (3D) environment for a shell and tube evaporator using computer-aided design (CAD) SOLIDWORKS software.

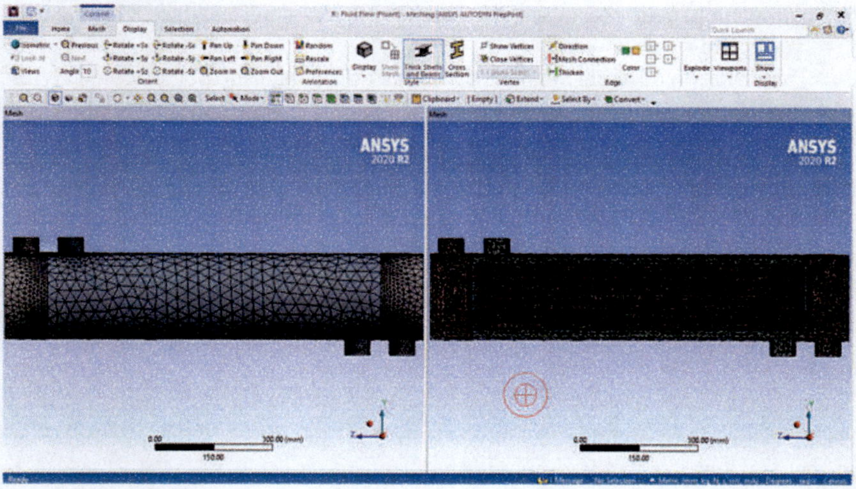

Fig. 2 Generated mesh

Meshing

The method was used to set up the entire geometry is automatic (Tetrahedrons) mesh. Also, based on the grid refinements study the maximum and minimum element size was specified. The physical preference, and the solver preference have been set as a default, which is CFD and FLUENT respectively. Figure 2 shows the generated mesh.

Setup (ANSYS Fluent)

The Fluent solver Pressure based type was selected, absolute velocity formation, gravity on the y-axis and steady time was selected for the simulation.

1. The energy calculation was on and the viscous was set as realizable k-e, enhance wall treatment (k-epsilon 2 equation).
2. Water-liquid, and refrigerant 134a was selected for the fluid domains. Carbon steel SA516 GR70 and copper was selected as material for the solid domains during this simulation.
3. The boundary condition. The inlets were defined as mass flow rate inlets and outlets were defined as pressure outlets (zero-gauge pressure). The inlet mass flow rate of the cold fluid (Refrigerant 134a) was 2.5 kg/s, whereas the mass flow rate of hot fluid (water) was 3.3 kg/s. The hot fluid temperature at the inlet was 12.2 °C and the cold fluid inlet temperature was kept at −15 °C.
4. In Fluent Setup, the final step is the mesh interface. Both sides of the system to use the same interface.

Solution

In the solution stage, the mathematical method used to find the result, each solution method has a different result.

1. Turbulent Kinetic energy was the second-order upwind, and the energy was also the second-order upwind.
2. There are two different methods of initialization which are Standard and Hybrid initialization, in this research the method used is Hybrid initialization.

Post-processing
Analyze results using different parameter such as weight, distance, vector plots, streamlines, temperature contours, etc.

3 Results and Discussion

The purpose of this simulation study was to study the effect of tube arrangements on the heat transfer in the evaporator chiller system using ANSYS FLUENT to examine the pressure difference, change in velocity and temperature difference in the evaporator chiller system.

3.1 Result

Pressure
The water flows into the tube's head inlet and hits the bottom of the inner wall, there is a sudden increase in pressure due to the high-pressure water entering through the inlet. Also, a pressure drop occurs in the tubes due to turbulence. Thus, there is a further pressure drop per length as the water flows through the tubes because the tubes are narrow. Although, the pressure further decreases as the water leaves through the tube's head outlet. As shown in Figs. 3, 4, 5, and 6, the pressure contours for all the arrangements of the shell and tube evaporator.

Based on Figs. 3 and 4, the outlet pressure is almost similar which are 910 Pa and 905.9 Pa respectively, and the outlet pressure difference is 4.1 Pa. Based on the observation, Fig. 3 has the lowest pressure drop, which is 2.96% among all the other tube arrangement designs. Further, based on Fig. 5, the pressure drops were the highest among all the shell and tube evaporators that were observed at 10%. Additionally, Fig. 6 has a pressure drop that was observed as 5.98%. Hence, in all the tube arrangements, the least pressure drop occurs in the shell and tube evaporator with the arrangement of 30° as shown in Fig. 3. In Table 2 shows the summary of the pressure drop of water for each of the arrangements.

Temperature
The water enters through the inlet at a temperature of 12.2 °C for all the shell and tube evaporator with different tube arrangements. Also, the water enters the tube through the head and the tubes have a turbulator in them due to the turbulent flow that occurs.

Fig. 3 Pressure contour of evaporator with tube arrangement at 30°

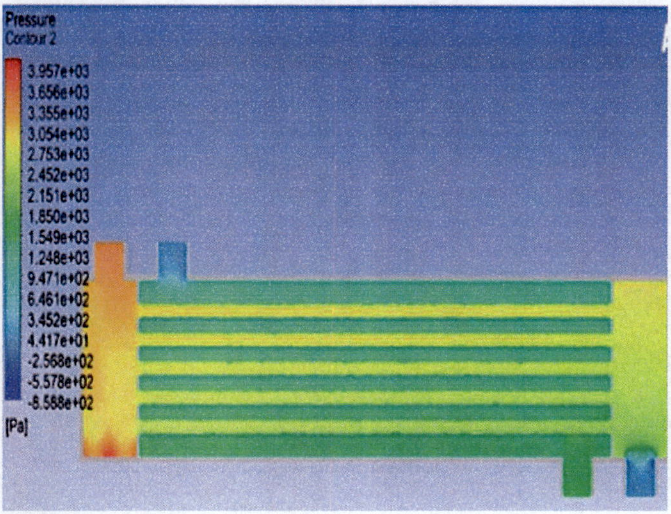

Fig. 4 Pressure contour of evaporator with tube arrangement at 45°

Thus, the turbulent flow increases the amount of heat transfer occurring between the refrigerant R134a; the water inside the shell and the tube evaporator. Further, there is also turbulence in the shell which helps the refrigerant entering the shell nozzle to flow over the tubes containing the water and efficiently cool the water by releasing heat energy into the surroundings. Though, the cooled water leaves the tube and exits through the outlet, as the heat transfer occurs from the higher temperature zone to the

Fig. 5 Pressure contour of evaporator with tube arrangement at 60°

Fig. 6 Pressure contour of evaporator with tube arrangement at 90°

Table 2 Summary of the tubes side pressure drop for all arrangements

Tube arrangements	Tube inlet pressure	Tube outlet pressure	Pressure drop (%)
30°	3.450e+03 Pa	9.100e+02 Pa	2.96
45°	3.450e+03 Pa	9.059e+02 Pa	3.40
60°	3.450e+03 Pa	8.436e+02 Pa	10
90°	3.450e+03 Pa	8.817e+02 Pa	5.98

low-temperature zone. As shown in Figs. 7, 8, 9, and 10, the temperature contours for all the arrangements of the shell and tube evaporator.

Based on Fig. 7, an inlet temperature at the tube side of 12.2 °C and an outlet temperature of 10.41 °C, as the decrement in the temperature of water was 14.7%. Also, in Figs. 8, 9 and 10, the outlet temperature was 10.60 °C, 10.53 °C and 10.50 °C, respectively. Hence, the temperature difference between tube arrangements of 30°, 45°, 60° and 90° were 1.79 °C, 1.6 °C, 1.67 °C and 1.7 °C, respectively. However, based on all the tube arrangements, the most reduction in temperature occurs in the tube arrangement at 30°. Hence, it is the most efficient one, as it has reduced the water temperature to its lowest among all the other heat exchanger designs of

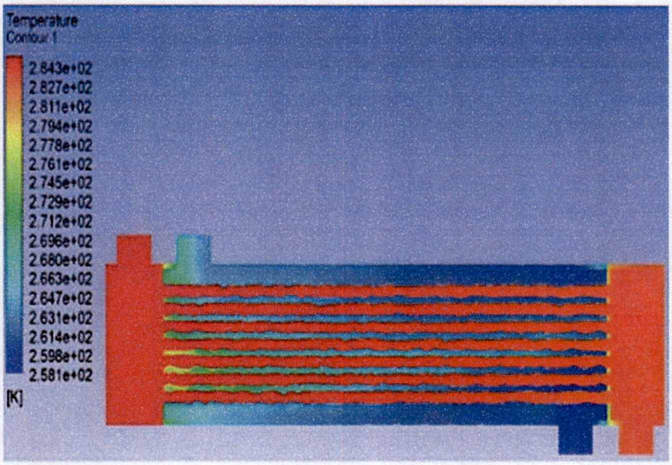

Fig. 7 Temperature contour of evaporator with tube arrangement at 30°

Fig. 8 Temperature contour of evaporator with tube arrangement at 45°

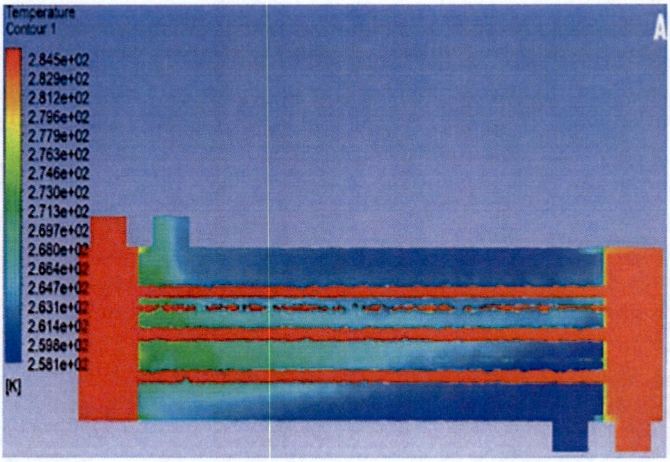

Fig. 9 Temperature contour of evaporator with tube arrangement at 60°

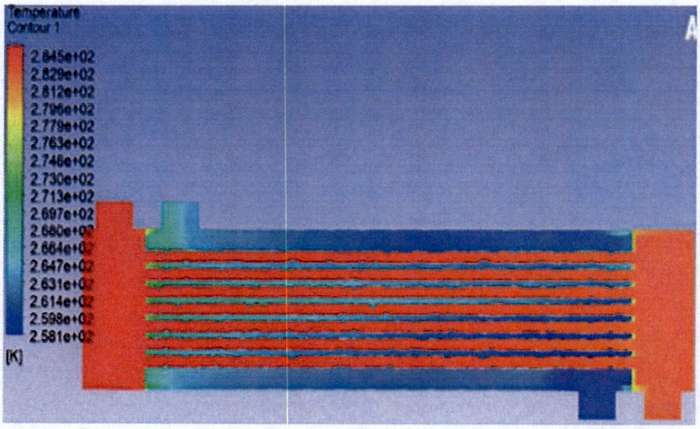

Fig. 10 Temperature contour of evaporator with tube arrangement at 90°

tube arrangement. Table 3 shows the change in temperature of water for all the tube arrangements.

Comparison of CFD Result with Analytic Result for the Tube Side

The counter fluid flow was made of shell and tube exchanger for the maximum transfer of heat energy. Hence, by using Eq. (2), the LMTD value was calculated for the shell and tube evaporators with different tube arrangements (counter flow).

$$LMTD = \frac{(Thi - Tco) - (Tho - Tci)}{\ln \frac{(Thi - Tco)}{(Tho - Tci)}} \tag{2}$$

Table 3 Summary of the tube side temperature difference for all arrangements

Arrangements	Tube inlet temperature	Tube outlet temperature	Decrease in temperature (%)
30°	12.2 °C	10.41 °C	14.7
45°	12.2 °C	10.60 °C	13.11
60°	12.2 °C	10.53 °C	13.69
90°	54 F (12.2 °C)	50.9 F (10.50 °C)	13.93

whereby

Thi Temperature hot inlet;
Tco Temperature cold outlet;
Tho Temperature hot outlet;
Tci Temperature cold inlet.

The heat transfer in the shell and tube evaporator with different tube arrangements exchanger (analytical) was calculated for the tube side using Eq. (3) and was compared with the CFD simulation values for the tube side as the objective two was mentioned comparing the difference between the analytical result and the CFD result achieved. Table 4 shows the comparison of the CFD results and the analytic results of the heat transfer.

$$Q = UA\Delta T_{LM} \tag{3}$$

whereby

Q Heat flow rate [W];
U overall heat transfer coefficient [W m^{-2} K];
A the eat flow area [m^2];
ΔT_{LM} the logarithmic mean temp difference [K].

Based on Table 4, the least difference between analytic and CFD heat transfer occurs in tube arrangement 90° and the maximum was 9.37% that occurred at 60° of the tube arrangements. Although, among all the designs, the maximum heat transfer

Table 4 Validation comparison between CFD and analytic results of heat transfer

Arrangements	LMTD	Analytic heat transfer	CFD heat transfer Values	Validation: difference between analytical and CFD values (%)
30°	52.64297	50,601.2245	52,235.8235	3.23
45°	52.75796	50,711.7595	55,125.25945	8.7
60°	52.4218	50,388.6308	55,111.19961	9.37
90°	52.10266	50,081.8662	51,000.6745	1.83

that was observed in tube arrangement at 45° is 55,125.25945 W in CFD and as in analytic was 50,711.7595 W heat transfer. ANSYS CFD and analytic results were validated by comparison and found in good arrangement with all the designed tube arrangements were observed as below 10% difference between CFD and analytic. Also, after completing CFD analysis results, the observation of the study shows that CFD Analysis was a good tool to avoid costly and time-consuming.

Objective of the study, the heat exchanger design must be chosen by comparing the overall performance of the heat exchanger, such as the pressure drops, velocity and temperature reduction. According to [19], low pressure drop is good for the heat exchangers and it turns to reducing the operation cost, in this study the pressure drops least occurred on the tube-side was in the tube arrangement at 30°, as the heat exchanger was acceptable that had the least pressure drop. Besides, the velocity increase percentage in the tube-side also was low, which was 19.9% among the other three designs of the tube arrangement in the heat exchanger, as it helped for the transfer of heat energy inside the evaporator. Also, the decrement in the temperature of the hot stream satisfies the maximum temperature difference among all the heat exchangers with the tube arrangement at 30°. Moreover, it transfers a good heat transfer rate after the tube arrangement at 45°. Besides, it is the most used in the industrial standard operations.

However, based on all the aspects of the heat exchanger results and discussions, the study came to the better outcome result that had the best-optimized design of tube arrangement which was at 30°.

4 Conclusion

The first objective was to analyse the heat transfer performance of selected tube arrangement designs using ANSYS. Thus, this objective was achieved by using two types of software which were SOLIDWORKS, for the purpose of designing the geometry for all the four designs for the shell and tube evaporators for the different arrangements which were 30°, 60°, 45° and 90°. Also, the ANSYS FLUENT software for CFD simulation. Hence, the simulation of the all designs of the shell and tube evaporator by using ANSYS Fluent solver. More, the refrigerant 134a and the Carbon steel SA516 GR70 were created in ANSYS Fluent through user define with their properties. Though, the model used in Fluent was the viscous model (K-epsilon) second equation due to the turbulence flow; as the results in terms of pressure contour, velocity contour and temperature contour were collected from Postprocessing of Fluent for the analysis.

Moreover, the second objective was to compare the CFD results with the analytical results in terms of the heat transfer rate. Also, the comparison has been done on the tube side, based on the result that was found, among all the designs the maximum heat transfer was in the tube arrangement at 45°, which was 55,125.25945 W in the CFD and as well as in the analytic which was 50,711.7595 W heat transfer.

Furthermore, the third objective was to optimize the tube arrangements designed for evaporator operation. Also, in order to achieve this objective, the heat exchanger design must be chosen according to the overall performance of the heat exchanger, such as the pressure drops, velocity and temperature reduction. Thus, the shell and tube evaporator with a 30° tube arrangement has the lowest pressure drops, which is good in the process of the heat exchanger, and the highest temperature reduction, which means it reduces the water temperature to the minimum, so it was the optimized design.

However, the shell and tube evaporator with tube arrangement at 45° transfer more heat transfer than the other three designs in the CFD and the analytical results. Also, it can be concluded that the shell and tube evaporator with tube arrangement at 30° was the optimized design among the other three tube arrangements, based on the overall performance.

Acknowledgements The authors would like to acknowledge CERVIE of UCSI University for conference funding.

References

1. Seyam S (2018) Types of HVAC systems. In: HVAC system. IntechOpen, pp 49–66
2. Lu L, Cai W, Xie L, Li S, Soh YC (2005) HVAC system optimization—in-building section. Energy Build 37(1):11–22
3. Daikin (2014) Chiller application guide, p 93
4. Hanson S, Schwedler M, Bakkum B (2011) Applications engineering manual. Hesong Z (2007) Process design of shell and tube heat exchanger. Chem Eng Des 5
5. Kulankara S, Buckley ML, Yanik MK, Johnson Controls Technology Co (2015) Compact evaporator for chillers. U.S. Patent 8,944,152
6. Ruíz AAB (2016) Fundamental of heat and mass transfer. Available at: http://repositorio.unan.edu.ni/2986/1/5624.pdf
7. Muhieldeen MW, Adam NM, Salman BH (2015) Experimental and numerical studies of reducing cooling load of lecture hall. Energy Build 89:163–169
8. Muhieldeen MW, Kuang YC (2019) Saving energy costs by combining air-conditioning and air- circulation using CFD to achieve thermal comfort in the building. J Adv Res Fluid Mech Therm Sci 58(1):84–99
9. Salman BH, Hamzah MZ, Purbolaksono J, Inayat-Hussain JI, Mohammed HA, Muhieldeen MW (2011) Determination of correlation functions of the oxide scale growth and the temperature increase. Eng Fail Anal 18:2260–2271
10. Francesca P, Tey WY, Tan LK, Muhieldeen MW (2020) Investigation on generalised trapezoidal differencing time-marching scheme for modelling of acoustical wave. CFD Lett 12(2):11–21
11. Ng YH, Tey WY, Tan LK, Arada GP, Muhieldeen MW (2020) Numerical examination on two-equations turbulence models for flow across NACA 0012 airfoil with different angle of attack. CFD Lett 12(2):22–45
12. Tey WY, Sidik NAC, Asako Y, Muhieldeen MW, Afshar O (2021) Moving least squares method and its improvement: a concise review. J Appl Comput Mech 7(2):883–889
13. Schottelkotte B (2015) Evaporator system basics. The Dupps Company, p 65. Available at: http://www.dupps.com/Evaporator%20Basics.pdf
14. Thulukkanam K (2013) Heat exchanger design handbook. CRC Press, Boca Raton

15. Lewis MJ (2016) Physical properties of foods and food processing systems. Elsevier, Amsterdam
16. TEMA book of standards, 10th edn.
17. ASME homepage, https://www.asme.org/codes-standards, last accessed 12/5/2021
18. Industrial tube size, http://www.cuilong.com/pro-e.asp?type=Cuilong+Evaporator+Tube, last accessed 12/1/2021
19. Towler G, Sinnott R (2012) Chemical engineering design: principles, practice and economics of plant and process design. Elsevier, Amsterdam

Effect of Air Filter Pressure on Fuel Consumption and Cost of Gas Turbine in Southern Power Generation, Malaysia

A. H. Fauzi and M. Z. Sulaiman

Abstract The effect of air filter pressure and fuel consumption for gas turbine generating Block 1 in the Southern Power Generation (SPG) power plant is presented. The prime mover for the generating block is the GE 9HA.02 gas turbine, and the power plant is the latest combined cycle gas turbine (CCGT) commissioned in January 2021 and the world's first commercial operation of the GE 9HA.02 fleet globally. Fuel consumption of the gas turbine is the primary concern as it which significantly affected by the gas turbine performance, which later translates to the power plant revenue to operate at optimum cost. Note that the fuel consumption of the CCGT is closely related to the Air Filter House (AFH) condition located at the most upstream component to protect the gas turbine from erosion, corrosion and fouling; as well as to achieve the required performance, efficiency, and life expectations. The present work aims to evaluate the value of pressure drop in the AFH and the fuel consumption. These two related parameters are significant for mitigation measures to achieve a cost-effective power plant operation. The operation data for both parameters based on the actual CCGT plant operation has been analysed from March to June 2021. Consecutively over the four months of operation, the AFH pressure drop had increased from 666.80 to 741.12 Pa (Pascal), translating to a total increment of 74.32 Pa or an average of 18.58 Pa every month. Separately, fuel consumption increased from 120,460.61 to 123,614.13 m^3/h, a total increment of 3153.52 m^3/h or an average of 788.38 m^3/h for every month, which later translated to an average increment of fuel cost amounting to RM 767.86/h. The present results reveal that the AFH pressure drop has directly impacted the fuel consumption over the analysis period. On average, an increment of 1 Pa of the AFH pressure drop will increase fuel cost amounting to RM 41.33/h. It is expected that current air filtration elements can last within 41 months to achieve their

A. H. Fauzi · M. Z. Sulaiman (✉)
Faculty of Mechanical and Automotive Engineering Technology, Universiti Malaysia Pahang, 26600 Pekan, Pahang, Malaysia
e-mail: zuhairi@ump.edu.my

A. H. Fauzi
e-mail: amirhasdi@tnb.com.my

A. H. Fauzi
Southern Power Generation Sdn. Bhd., 81700 Pasir Gudang, Johor, Malaysia

M. Y. Ismail et al. (eds.), *Technological Advancement in Mechanical and Automotive Engineering*, Lecture Notes in Mechanical Engineering,
https://doi.org/10.1007/978-981-19-1457-7_50

allowable pressure limit but are subjected to environmental and operating philosophy changes. The analysis results could be the basis for early filter replacement, proper selection of filter elements and frequent conduct of online and offline compressor washing as recommended by the manufacturer.

Keywords Combine cycle gas turbine (CCGT) · Air filter pressure · GE9HA.02

1 Introduction

Power plants are built to meet the demand for electricity in relation to the economic and population growth of a country [1–4]. Improving efficiency and reducing pollutants are the critical concerns for any design of power generation plants [5]. The Government of Malaysia has set several policies and planning criteria in managing the energy trilemma, including energy security, sustainability and affordability [3].

Operating a power plant is challenging for the operator to ensure an uninterrupted power supply to the National Grid [6]. The operator has to comply with the grid requirement, which is governed under the Power Purchase Agreement (PPA) and in accordance with the instruction of the Grid System Operator (GSO) [7]. In fulfilling the obligations, the operator has to ensure the power plant achieves optimum performance during the service period at cost effective.

In the CCGT, the gas turbine consumes a large amount of fuel, whether gas, distillate or any available fuel source, as it is the prime mover to the generator in generating the power [8]. Commercially, fuel consumption is a significant concern whereby less consumption of energy will ensure better operating revenue.

The Air Filter House (AFH) is a significant component assembled at the most upstream of the gas turbine system. A Series of filter elements are installed in the AFH for effective filtration before entering the gas turbine [9]. Over the period, the filtration efficiency will be compromised and resulting in the increment of AFH pressure drop, which subsequently affects the performance of the gas turbine [10–16]

Furthermore, if AFH pressure drop increases and results in degrading the performance of the gas turbine, it would mean that the gas turbine fuel consumption will steadily increase with time. Therefore, a study to evaluate the effect of AFH pressure drop on fuel consumption is essential for the power plant operator to establish mitigation measures for operating cost efficiency.

In the AFH, the resistance of filter media against airflow results in a difference in static pressure between the input and output faces of the filter, which is called a pressure drop. Pressure drop is a crucial parameter in the choice of filters and can cause irreversible mechanical resistance problems beyond a specific value. This parameter strongly impacts the performance of a filter medium, and a rapid increase in the pressure drop is associated with the partial or total clogging of the filter medium. As a result, the clogging of a Gas Turbine (GT) filter house will obstruct the system's operation due to the reduction infiltration area caused by the accumulation of dust particles on the surface of the filter. The clogging will also cause a decrease in the air mass flow downstream the filter house, which will ultimately diminish the produced load of the GT system [16]. Meanwhile, the effect of ambient temperature and humidity on gas turbine performances have been discussed in previous studies [17–21].

In Malaysia, the Southern Power Generation (SPG) Power Plant is a subsidiary of Tenaga Nasional Berhad (TNB) is located in Pasir Gudang, Johor, Malaysia as shown in Fig. 1. In this power plant, a total generating capacity of 1440 MW (2 × 720 MW) with the single shaft combined cycle gas turbine (CCGT) power plant in two generating blocks, Block 1 and Block 2. Each comprises one GE 9HA.02 gas turbine, one GE once through (OT) heat recovery steam generator (HRSG), one GE STF-D650 steam turbine and a single GE W88 generator, as shown in Fig. 2.

Fig. 1 Courtesy of SPG. View of SPG power plant

Fig. 2 Courtesy of general electric. View of single shaft CCGT configuration

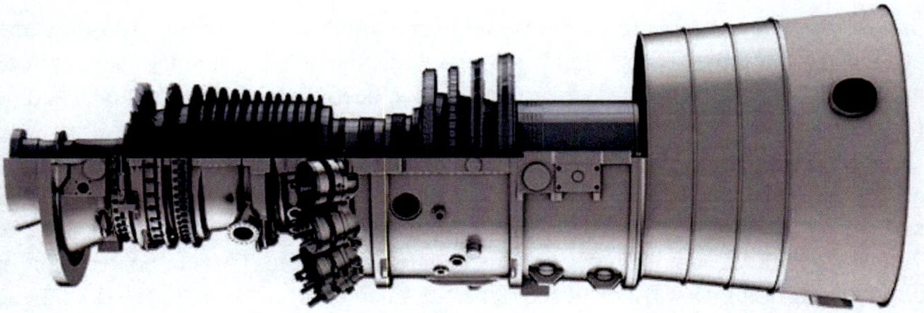

Fig. 3 Courtesy of general electric. Side cross section view of GE 9HA gas turbine [24]

The newly commissioned power plant has started its commercial operation for Block 1 and Block 2 on Jan. 1 2021 and Feb. 19 2021, respectively. It was a remarkable milestone to the industry, where GE had officially announced the first commercial operation of GE 9HA.02 globally at the SPG power plant [22].

The GE 9HA.02 gas turbine is the latest model and the most advanced technology of the GE H-class gas turbine fleet, the world's largest electric power [13, 23]. It has a design efficiency rating of up to 64% and a maximum output of 826 MW in a single shaft combined cycle configuration. Other main design features consist of fourteen stages of compressor blades, sixteen combustors (a can annular type), four stages turbine blades. It is an air cooled type where the cooling air is then extracted from compressor discharge at stages 8th, 10th and 11th for rotor and hot gas path component cooling [24, 25] (Fig. 3).

In this power plant, two sources of fuel are available, natural gas as a primary fuel and distillate fuel for emergency operation. The power plant consumes approximately 100,000 GJ (Gigajoules) of gas per day for each generating block to generate 738 MW gross power and 720 MW net power. The gross power includes auxiliary plant consumption (18 MW) and net export (720 MW).

During the in-service operation, a large volume of ambient air is consumed by the gas turbines. The Air Filter House (AFH) was designed for 815 m^3/s intake airflow and filtered through filter elements to remove the airborne particles and unnecessary debris that potentially compromise the performance of the gas turbines [14–16, 26, 27].

A pressure drop reduction of 50 Pa will result in approximately 0.1% improvement in machine power output [10, 15]. The gas turbine must consume more fuel, reducing power output for the compressor to overcome the inlet system losses. Normally, the pressure losses throughout the inlet filtration system range from 2 to 6 in H_2O (498–1495 Pa) [27]. The operating alarm setting for the AFH pressure drop is 1500 Pa, and the high alarm is 2000 Pa.

The AFH was manufactured by G+H Schallschutz GmbH [28] and filtration elements are manufactured by Parker Hannifin [29]. Three stages of filter elements are used, namely coalescer filter, pre-filter and fine filter. The coalescer filter has no

Fig. 4 View of air filter house (AFH) taken in SPG power plant

filtration efficiency, whereas the pre-filter is specification is F7 and the fine filter is E10 (Fig. 4).

Respective to weather conditions, the maximum temperature and humidity in Pasir Gudang for March, April, May and June is 31 °C, 31 °C, 31 °C and 30 °C; 78%, 77%, 76% and 76% respectively [30].

In the present study, these two elements of AFH pressure drop and fuel consumption for the gas turbine in Block 1 has been evaluated for the first time. For this purpose, the operating database for the AFH pressure drop and fuel consumption have been taken and analysed from March 2021 until June 2021. Since the study is the first academic research conducted on SPG and to the first commercial GE 9HA.02 gas turbine (Block 1), the result will be significant to SPG for technical and commercial mitigation measures.

2 Methodology

2.1 Data Collection

Block 1 operation data for four months, from Mar. 1 2021, until Jun. 30 2021 were acquired from the power plant PI system [31]. The data consist of five parameters; (i) ambient temperature, (ii) relative humidity, (iii) Air Filter House (AFH) pressure drop, (iv) fuel flow rate and (v) gross power.

Daily operational data will be recorded on an hourly basis. In order to have consistent data, the average of 6 h of data (6 measurement points) is proposed and

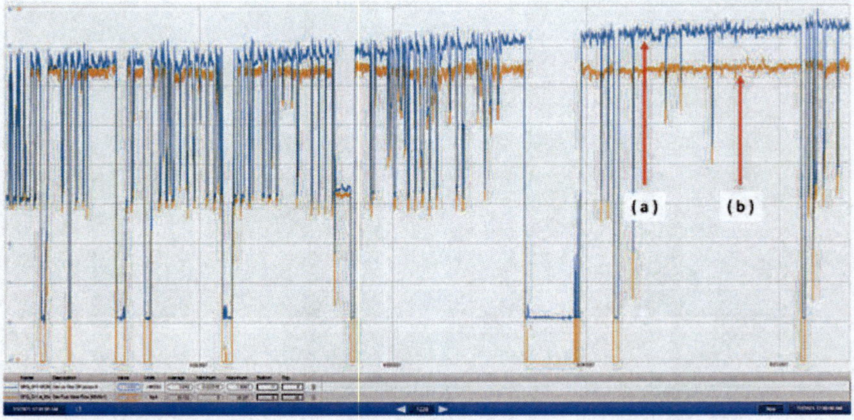

Fig. 5 Trending from Mar. 1 to Jun. 30, 2021, for Block 1. **a** AFH pressure drop (represented by blue line) and; **b** fuel consumption flow rate (represented by the brown line)

was measured on an hourly basis at a time from 12:00 pm to 5:00 pm for the four months duration.

Figure 5 shows Block 1 AFH pressure drop trends and fuel consumption flow rate from Mar. 1 2021, until Jun. 30 2021. The AFH pressure drop has increased more significant compared to the fuel consumption flow rate over the period.

The AFH pressure drop and generation output are dependent variables according to the actual plant baseload operation, generating 738 MW gross power at the sampling date. The gross generated power varies according to the national grid frequency response of the net power export [32].

From the four months database, the selection of data sample will be chosen in the mid of the month but subjected to the consistency of the independent variables (temperature and humidity) reference. Adjustment to the sampling date is required to meet the set criteria, as shown in Fig. 5. In the present analysis, the reference ambient temperature and humidity values for data analysis are 31 °C ± 1.0 °C and 65% ± 5% were used.

Data analysis was conducted using Microsoft Excel to identify the consistent sampling value that meets the criteria based mentioned above. The filtered sampling data for Block 1 during the period are presented in Table 1.

2.2 *Fuel Consumption and Cost Analysis*

Fuel supply (fuel mass flow rate) to the gas turbine is measured by a flow meter recorded in kilogram per second (kg/s). Assuming the gas supply to the power plant at standard conditions (at 1 atm pressure at 60 °F/15.56 °C) [33], the volume (volumetric flow rate) of fuel consumption was calculated using the following formula.

Table 1 Selected data for Block 1

Date	Ambient temperature (°C)	Relative humidity (%)	AFH pressure drop (Pascal)	Gas flow rate (kg/s)	Gross power (MW)
30 March 2021	31.81	64.35	666.80	24.18	737.46
28 April 2021	31.95	63.61	694.68	24.26	738.84
29 May 2021	31.46	66.03	732.52	24.47	736.78
22 June 2021	31.83	68.24	741.12	24.82	740.17

$$V = \frac{\dot{m}_g}{\rho_g} \times 3600 \quad (1)$$

$$SG = \frac{\rho_g}{\rho_a} \quad (2)$$

where:

V Volumetric flow rate, m^3/h.
$\dot{m}_g$ Fuel mass flow rate, kg/s.
ρ_g Molecular Weight of Natural Gas, kg/m^3.
ρ_a Air density, kg/m^3.
SG Gas specific gravity.

The properties of those gas were based on provided data from the Energy Commission in the Piped Gas Distribution Industry Statistics 2016 [34]. The gas specific gravity is 0.59 for supply from Kerteh, Terengganu. Meanwhile the standard air density is 1.225 kg/m^3 (at sea level, 0 m.a.s.l., 15 °C) [33].

2.3 *Fuel Cost Analysis*

Further analysis was conducted to identify the fuel consumption cost increment from March to June 2021. Fuel cost is calculated from the following formula.

$$\text{Fuel Cost} = \frac{V \times \text{GHV} \times \text{Gas Price}}{1000}$$

where:

V Volumetric flow rate, m^3/h.
GHV Gross Heating Value, MJ/m^3.

According to Energy Commission base fuel tariff prices for TNB [4], the most recent available gas pricing from January 2019 to December 2020 is RM 27.20/mmBtu, equivalent to RM 25.78/GJ. Based on Ref. [34], the Gross Heating Value (GHV) applied is 9024 kcal/Sm^3 or equivalent to 37.78 MJ/m^3.

2.4 *Description of Equipment*

GE 9HA.02 gas turbine consists of five main components: air intake, compressor, combustor, turbine and exhaust, where the hot gas discharge into the heat recovery steam generator (HRSG). Detailed specifications of the gas turbine components are described in Table 2.

Figure 6 shows the arrangement of Air Filter House (AFH) and gas turbine including the main components and direction of air flow into gas turbine.

3 Result and Discussion

Based on the analysis conducted to the acquired database for Block 1, the results are translated into comparison charts; (i) Relation of Air Filter House (AFH) pressure drop and Time, (ii) Relation of fuel consumption and time and; (iii) Relation of fuel consumption and AFH pressure drop.

3.1 *Relation Between the Air Filter House (AFH) Pressure Drop Value and Time*

Figure 7 describes the relation of AFH pressure drop over the analysis period. The pressure drop had increased from 666.80 to 741.12 Pa with a total increment of 74.32 Pa or 11.15%. Meanwhile, the average increment is 18.58 Pa per month. Based on the average increment numbers, the expected remaining lifetime for the filter elements to reach the alarm limit at 1500 Pa [27] is 41 months. It is expected to be replaced in November 2024 and subjected to changing environmental conditions and operation and maintenance philosophy.

3.2 *Relation of the Fuel Consumption and Time*

Block 1 fuel consumption shows an increment amount, as depicted in Fig. 8. The total volumetric flow rate had increased by 2.62% or 3153.52 m^3/h for that particular

Table 2 Equipment descriptions

Machine/equipment	Description/specification
Gas turbine	
1. Manufacturer	General Electric (GE)
2. Model	GE 9HA.02
3. Output (Design)	557 MW (Simple cycle) 826 MW (Combined cycle)
4. Compression ratio (X:1)	23.8
5. Number of combustor cans	16
6. Number of compressor stages	14
7. Number of turbine stages	4
8. Exhaust temperature	645 °C
9. Exhaust energy	2559 MM kJ/h
Air Filter House (AFH)	
1. Manufacturer	G+H Schallschutz GmbH
2. Humidity class (Design)	Class G (Dry)
3. Humidity annual mean (Design)	65%
4. Air intake volume flow (Design)	815 m^3/s
5. Air intake volume flow (ISO)	615 m^3/s
6. Wind speed (Design)	32.5 m/s
7. Main components	Weatherhoods Inlet Bleed Heating (IBH) elements Coalescer Pre filter Fine filter Transition duct Roller shutter
8. Filter elements—Coalescer	792 pcs
9. Filter elements—Pre-filter	704 pcs
10. Filter elements—Fine filter	704 pcs
Filter elements	
1. Manufacturer	Parker Hannifin
2. Coalescer	No filtration efficiency
3. Pre-filter	F7 (MERV 13)—85% $\leq$ filtration efficiency for 1μ–3μ particles size [35]
4. Fine filter	E10 (MERV 16)—95% $\leq$ filtration efficiency for 1μ–3μ particles size [35]

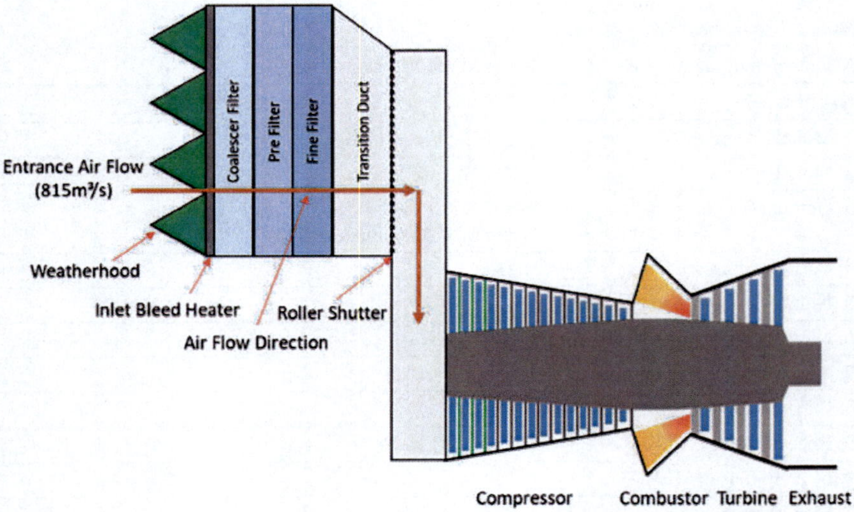

Fig. 6 Arrangement of Air Filter House (AFH) and gas turbine assembly, main components and airflow direction

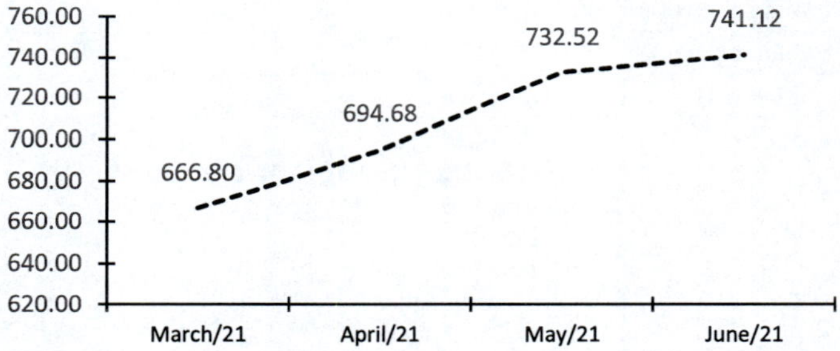

Fig. 7 Relation between the AFH pressure drop value (Pascal) and month for Block 1

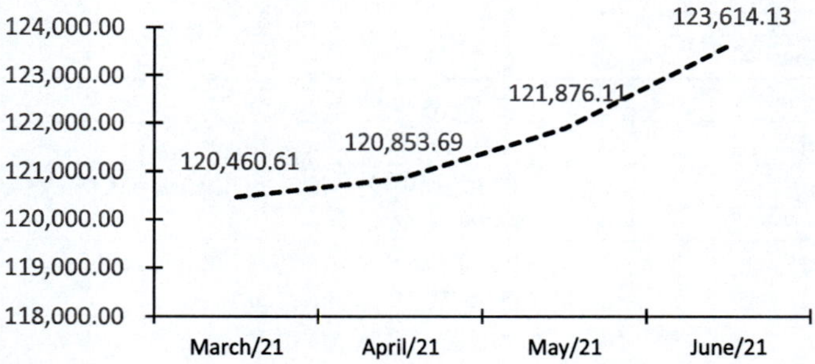

Fig. 8 Relation between the fuel consumption (m^3/h) and Month for Block 1

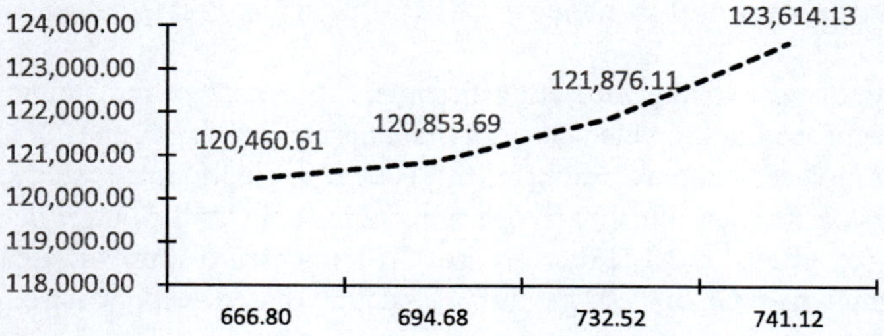

Fig. 9 Relation between fuel consumption (m^3/h) and AFH pressure drop value (Pascal) for Block 1

duration. From a commercial perspective, the total increment is RM 3071.43 or an average of RM 787.86 every month over the period.

3.3 Relation of Fuel Consumption and Air Filter House (AFH) Pressure Drop Value

Figure 9 indicates that fuel consumption increases with the AFH pressure drop increment, proving the statements described in [10–16]. Over the analysis period, a total increment of 74.32 Pa AFH pressure drop has been attributed to the increment of fuel cost amounting to RM 3071.43. An average increment of 1 Pa will increase fuel cost amounting to RM 41.33 per hour operation of the gas turbine.

4 Conclusion

The effect of air filter pressure on fuel consumption of GE 9HA.02 gas turbine in Block 1 has been evaluated. The study was structured to generate technical information on the effect of air filter pressure drop on fuel consumption over the operating period from March to June 2021. The findings are summarised as follow:

- Based on the ambient temperature ranging between $30 \leq T_{ambient} \leq 32$ °C and relative humidity ranging between $60 \leq H \leq 70\%$, the air filter house (AFH) pressure drop and fuel consumption increase in the four months duration.
- Air filter pressure drop was increased 74.32 Pa or 11.15% from March 2021 at an average of 18.58 Pa. Meanwhile, fuel consumption was increased to 3153.52 m^3/h or 2.62% at an average of 788.38 m^3/h.
- From a commercial perspective, at the estimated value of the gas price at RM 25.78/GJ and Gross Heating Value (GHV) 37.78 MJ/m^3, the consumption of the fuel was increased to RM 3071.43/h or an average of RM 767.86/h. In relation to

pressure drop, every 1 Pa increment will result in fuel cost increment amounting to RM 41.33/h.

- In view of the air filter operating lifetime, the estimated remaining operating lifetime of the filtration elements is 41 months starting from June 2021, which is due for replacement in November 2024. However, the lifetime is subjected to the changes of the environment and operating philosophy of the power plant.
- Thus, the effect of Air Filter House (AFH) pressure drop towards the fuel consumption for GE 9HA.02 gas turbine is defined based on the acquired operating data. Both parameters are proportional to each other.
- SPG and the power plant operator is recommended to acknowledge the impact of AFH pressure drop on fuel consumption during the operating period until the subsequent replacement of the air filtration elements. Mitigation measures such as early replacement of filter elements and frequently online and offline compressor washing as recommended by the manufacturer are suggested to minimise the commercial impact.

Acknowledgements The authors would like to thank Southern Power Generation Sdn. Bhd. (SPG), TNB Power Generation Sdn. Bhd., SIPP Energy Sdn. Bhd. and Tenaga Nasional Berhad for providing the facility and operating database needed to carry out the study.

References

1. International Energy Agency (IEA) (2020) Electricity market report—December 2020, vol 1, pp 1–103
2. Department of Statistics Malaysia (2018) Overview electricity sector in Malaysia. Department of Statistics Malaysia, pp 1–2
3. Suruhanjaya Tenaga (2021) Report on Peninsular generation development plan 2020, vol 2020. Suruhanjaya Tenaga, p 18
4. Energy Commission (2019) Energy Commission annual report 2019, pp 1–234
5. Ibrahim TK et al (2016) The optimum performance of the combined cycle power plant: a comprehensive review. Renew Sustain Energy Rev 79:459–474. http://doi.org/10.1016/j.rser.2017.05.060
6. Suruhanjaya Tenaga (2015) The national grid. Geogr J 6(6):52. http://doi.org/10.2307/1781136
7. Suruhanjaya Tenaga (2020) Electricity supply act 1990, 2020th edn.
8. Sanaye S, Meybodi MA, Shokrollahi S (2008) Selecting the prime movers and nominal powers in combined heat and power systems. Appl Therm Eng 28(10):1177–1188. https://doi.org/10.1016/j.applthermaleng.2007.08.003
9. Hutten IM (2016) Air filter applications. In: Handbook of nonwoven filter media. https://doi.org/10.1016/b978-0-08-098301-1.00008-3
10. Sennett P (2007) Air filtration: perfect air filtering for gas turbines. Filtr Sep 44(10):20–22. https://doi.org/10.1016/S0015-1882(07)70320-8
11. Kurz R, Brun K (2009) Degradation of gas turbine performance in natural gas service. J Nat Gas Sci Eng 1(3):95–102. https://doi.org/10.1016/j.jngse.2009.03.007
12. Giampaolo A (2006) Gas turbine handbook—principles and practice, 3rd edn. The Fairmont Press, Inc., Lilburn

13. Kotowicz J, Brzęczek M, Job M (2018) The thermodynamic and economic characteristics of the modern combined cycle power plant with gas turbine steam cooling. Energy 164:359–376. https://doi.org/10.1016/j.energy.2018.08.076
14. Effiom SO, Abam FI, Ohunakin OS (2015) Performance modeling of industrial gas turbines with inlet air filtration system. Case Stud Therm Eng 5:160–167. https://doi.org/10.1016/j.csite.2015.03.008
15. Winsper T (2012) Gas turbines: changing the stages of air filtration. Filtr Sep 49(6):34–36. https://doi.org/10.1016/S0015-1882(12)70288-4
16. Abdul-Wahab SA, Omer ASM, Yetilmezsoy K, Bahramian M (2020) Modelling the clogging of gas turbine filter houses in heavy-duty power generation systems. Math Comput Model Dyn Syst 26(2):119–143. https://doi.org/10.1080/13873954.2020.1713821
17. Erdem HH, Sevilgen SH (2006) Case study: effect of ambient temperature on the electricity production and fuel consumption of a simple cycle gas turbine in Turkey. Appl Therm Eng 26(2–3):320–326. https://doi.org/10.1016/j.applthermaleng.2005.08.002
18. El-Shazly AA, Elhelw M, Sorour MM, El-Maghlany WM (2016) Gas turbine performance enhancement via utilising different integrated turbine inlet cooling techniques. Alexandria Eng J 55(3):1903–1914. https://doi.org/10.1016/j.aej.2016.07.036
19. Omar Kamal SN, Salim DA, Mohd Fouzi MS, Hong Khai DT, Yusri Yusof MK (2017) Feasibility study of turbine inlet air cooling using mechanical chillers in Malaysia climate. Energy Procedia 138:558–563. http://doi.org/10.1016/j.egypro.2017.10.159
20. Pinilla Fernandez DA, Foliaco B, Padilla RV, Bula A, Gonzalez-Quiroga A (2021) High ambient temperature effects on the performance of a gas turbine-based cogeneration system with supplementary fire in a tropical climate. Case Stud Therm Eng 26:101206. http://doi.org/10.1016/j.csite.2021.101206
21. Hashmi MB, Majid MAA, Lemma TA (2020) Combined effect of inlet air cooling and fouling on performance of variable geometry industrial gas turbines. Alexandria Eng J 59(3):1811–1821. https://doi.org/10.1016/j.aej.2020.04.050
22. G.E. (GE) (2021) Press release GE announces first commercial operation of GE's 9HA.02 technology globally at southern power generation's track 4A power plant in Malaysia
23. Kotowicz J, Brzęczek M (2018) Analysis of increasing efficiency of modern combined cycle power plant: a case study. Energy 153:90–99. https://doi.org/10.1016/j.energy.2018.04.030
24. GE Gas Power (n. d.) 9HA gas turbine: 9HA.01 and 9HA.02. https://www.ge.com/gas-power/products/gas-turbines/9ha
25. GE Gas Power (n. d.) How a gas turbine works. GE gas power. https://www.ge.com/gas-power/resources/education/what-is-a-gas-turbine
26. Salilew WM, Abdul Karim ZA, Baheta AT (2021) Review on gas turbine condition based diagnosis method. Mater Today Proc. http://doi.org/10.1016/j.matpr.2020.12.1049
27. Melissa Wilcox KB, Baldwin R, Garcia-Hernandez A (2010) Guideline for gas turbine inlet air filtration systems, p 122
28. Air filter systems for gas turbines. G+H GROUP
29. Gas turbine filtration division. Parker
30. Pasir Gudang, Johor, Malaysia weather averages. Monthly average high and low temperature. Average precipitation and rainfall days. World Weather Online
31. PI system—connecting data, operations & people. OSIsoft
32. Mo H, Sansavini G, Xie M (2018) Performance-based maintenance of gas turbines for reliable control of degraded power systems. Mech Syst Signal Process 103:398–412. https://doi.org/10.1016/j.ymssp.2017.10.021
33. Ulazia A, Sáenz J, Ibarra-Berastegi G, González-Rojí SJ, Carreno-Madinabeitia S (2019) Global estimations of wind energy potential considering seasonal air density changes. Energy 187. http://doi.org/10.1016/j.energy.2019.115938
34. Suruhanjaya Tenaga (2017) Piped gas distribution industry outlook 2016, pp 1–44
35. ANSI/ASHRAE standard 52.2 (2017) Method of testing general ventilation air-cleaning devices for removal efficiency by particle size, pp 1–64

Simulated Performance of an Improved District Cooling System (DCS) in Tronoh, Perak, Malaysia

Jue Hao Teo, J. C. E. Yong, Mohammed W. Muhieldeen, J. Y. Chan, A. G. Olasunkanmi, and C. L. Siow

Abstract District cooling system (DCS) is a popular cooling solution for many institutional applications in Malaysia due to the energy-saving advantages as compared to traditional individual on-site cooling production. However, the problem with DCS is that it's designed on fixed parameters automation prior to its commissioning and system performance is inefficient by the inability to adapt to uncertainties during operation. Objective of this DCS study is to propose a feedback control algorithm to be implemented in a DCS in Malaysia, as well as assess the proposed algorithm for further improvement. A case study of an existing DCS in Tronoh, Perak is performed in MatLab Simulink. The existing algorithm and the proposed algorithm of improved scheduling are implemented, and the simulation results have compared. The motivation of the new algorithm developed is to see further energy-reduction optimization, especially at the TES charging hours at low demand. The simulation results show improvements in the system efficiency of increased 22.6% (weekdays) and 48.2% (weekends) with reduced overall cooling output, and system energy savings of between 18.7% (~17 MWh at weekdays) and 32.1% (~21 MWh at weekends). Simulated model compared with historical data shows that the simulation model cannot replicate the exact conditions and output values of the actual DCS, but the trend of

J. H. Teo · J. C. E. Yong (✉) · M. W. Muhieldeen · J. Y. Chan
Mechanical Engineering Department, Faculty of Engineering, Technology and Built Environment, UCSI University, 56000 Kuala Lumpur, Malaysia
e-mail: jonathanyong@ucsiuniversity.edu.my

M. W. Muhieldeen
e-mail: al-gailani@ucsiuniversity.edu.my

M. W. Muhieldeen
UCSI-Cheras Low Carbon Innovation Hub Research Consortium, Kuala Lumpur, Malaysia

A. G. Olasunkanmi
Civil Engineering Department, Faculty of Engineering, Technology and Built Environment, UCSI University, 56000 Kuala Lumpur, Malaysia

C. L. Siow
Faculty of Mechanical Engineering, Universiti Teknologi Malaysia, Skudai, Johor, Malaysia

M. Y. Ismail et al. (eds.), *Technological Advancement in Mechanical and Automotive Engineering*, Lecture Notes in Mechanical Engineering,
https://doi.org/10.1007/978-981-19-1457-7_51

the output data is sufficiently accurate to model the improvements of implementing feedback control.

Keywords District cooling system · Gas district cooling plant · Feedback control system · Simulink

Nomenclature

Symbol	Description	Dimension/Unit
CL_{total}	*Total cooling capacity supplied by the plant*	kW
$CL_{district}$	*Cooling capacity that meets the district cooling demand*	kW
$CL_{EC,total}$	*Total cooling capacity supplied by the EC*	kW
CL_{SAC}	*Total cooling capacity supplied by the SAC*	kW
$CL_{TES,d}$	*Total cooling capacity supplied by the TES during discharging*	kW
E_{total}	*Total energy usage of the system*	kWh
E_{EC}	*Total energy consumed by the EC*	kWh
E_{SAC}	*Total energy consumed by the SAC*	kWh
$E_{savings}$	*Total energy savings*	kWh
$E_{total,e}$	*Total energy consumed by the system using existing control strategy*	kWh
$E_{total,p}$	*Total energy consumed by the system using proposed control strategy*	kWh
$\eta_{overall}$	*Overall efficiency of the system*	%

1 Introduction

District cooling system (DCS) is a centralized cooling system that distributes thermal energy from a central source, such as a chilled water plant, to residential, commercial and/or industrial customers via a distribution medium instead of generating thermal energy at the individual buildings or facilities [1]. The history of DCS in Malaysia dates back as far as 20 years. In Cyberjaya, the first DCS plant was built back in 1999 as part of Malaysia's Cyberview green initiatives. Since then, several DCS plants have been built across Malaysia. DCSs have many advantages over conventional on-site cooling at individual buildings and facilities. These advantages include lower energy consumption and higher energy efficiency due to scale of economy, electric peak reduction and better energy management through better thermal energy storage opportunities, lower operating and maintenance costs, reduced environmental impact through reduced pollution and easier to implement renewable energy sources or waste energy harvesting, improved reliability of cooling delivery, as well as freeing up building space at end users' sites for usage other than cooling [2, 3].

The gap to bridge improvement of DCS lies in the initial designs of DCS which have fixed parameters (pre-scheduled) without consideration of the uncertainties that exists in the system, there can be deviations in the actual system performance from the theoretical performance [4]. This presents an issue where DCS systems with rigid parameters lack the adaptability to cater to the variations caused by uncertainties. Furthermore, based on the previous studies, there is a need for automation of DCS through development of intelligent control strategies in favour of conventional controllers which require manual adjustments based on experience [5]. Uncertainties discussed in previous studies are found to be tied to develop the control system for DCS, while uncertainties in operation and control stage have not been widely discussed or researched. Some of the research studies shows that the change in condenser water temperature, chiller loading prediction and change in chiller scheduling are to allow the system to adapt the variation in cooling demand from the district and the changes in weather conditions [6, 7]. Other proposed control strategies include model predictive control for predictions of disturbances weather forecasts and district demands, and chiller loading estimation and shifting the load to efficient chillers [8, 9].

Novel improvements of DCS have also been highlighted in previous research. Damien et al. [6] proposed a DCS located in Eastern Paris uses sustainable condenser cooling strategy, Eveloy and Ayou [3] summarized the improvements to DCS in terms of sustainability, and Gang et al. [4] discussed DCS waste heat recovery measures to power processes requiring heat energy. Existing DCS in Malaysia have also made use of thermal energy storage (TES) which uses a chilled water or ice storage tank that is charged throughout the night when the electricity tariff is cheaper, and discharges to meet the load during the day [10]. DCS such as the University Teknologi MARA (UiTM) DCS and the Universiti Malaysia Sarawak (UNIMAS) DCS uses a TES system [11, 12]. Based on the studies that have been done related to DCS, it shows a noticeable trend in DCS design and control optimization to achieve higher COP of the system [6], reduce electricity usage, increase cost savings of system implementation, reduce emissions, and implement green and sustainable technologies to the system [13–17]. The prediction of uncertainties in operation control of the DCS are rarely studied. The uncertainties discussed common throughout majority of the studies reviewed are identified to be the variation in cooling demand and weather conditions. Dusuki [18] found on predictive control for plant operation have focused on the control of TES charging. Hence, there is a need to investigate the implementation of a feedback control system which can adapt to uncertainties in the DCS plant operation under Malaysian context.

Many researchers have done several optimization studies on DCS in Malaysia. However, the design or control strategies of DCS in Malaysia under the context of uncertainties or fluctuating cooling demand are not discussed previously. Hence, this study proposes to assess an improved DCS with feedback control system to improve the system response to uncertainties during operation. The proposed system, with a more comprehensive scheduling with feedback mechanism, aims to make more efficient the DCS in terms of energy usage and, therefore, reducing energy wastage and cost. Simulink tool and validation on the system performance has been carried out and compared with an existing system. The system design is based on a case study of an existing DCS in Tronoh, Perak, Malaysia.

2 Simulation Method

2.1 *Overview of the DCS Case Study*

The proposed improvements of DCS have been developed by DCS operation control. The cogeneration gas district cooling plant (GDCP) at Universiti Teknologi Petronas (UTP) is selected for this case study. The proposed control algorithm will implement additional operating modes, while the mode selection will be based on the varying cooling demand. The plant cooling capacity output, the efficiency of the system and the energy savings are the variables used to gauge the margin of improvements. The simulation for both existing and the improved model of the DCS has used the cooling demand profile based on the historical data adapted from past studies [19].

The UTP GDCP is a plant that produces electrical power for the district and the plant through gas turbine driven generators, which is concurrently producing chilled water for district cooling purpose. Steam has been generated using the excess heat from the gas turbines via a heat recovery steam generator (HRSG). The steam is used in steam absorption chillers (SAC) for chilled water production, coupled with electric chillers, to be supplied to the district for cooling. This GDCP also uses a TES system [20]. The plant also takes advantage of the electricity tariff with a TES system. The system consists of a chilled water tank which is charged by the electric chillers during off-peak hours, and the stored chilled water is supplied to the district during peak hours. The plant equipment is summarized in Table 1.

Table 1 Equipment and specifications of UTP GDCP [21]

Plant equipment	Make and model	Units	Specifications
Gas Turbine (GT)	Solar Taurus 60S	2	4.2 MW each
Heat Recovery Steam Generator (HRSG)	Vickers Hoskin	2	12 ton/h each
Steam Absorption Chiller (SAC)	Hitachi	2	1250 RT each
Electric Chillers (EC)	Dunham Bush	4	325 RT each
Auxiliary Gas Boiler (GB)	Vickers Hoskin	1	6 ton/h
Thermal Energy Storage (TES) tank	–	1	10,000 RTh

2.2 Existing and Proposed Feedback-Control Strategies

The existing control strategy for the chilled water production of the UTP GDCP involves scheduling the operation into two modes, which are tabulated in Table 2, where the operation of different modes and behaviours of the plant equipment are scheduled accordingly.

The proposed strategy uses the existing TES charging schedule to take advantage of the electricity tariff. Beyond that, the algorithm uses measured cooling load data from the district and the measured available TES tank capacity to determine the state of the SACs and ECs. The control algorithm can be separated into two main modes, which are TES charging mode and TES discharging mode. During TES charging, the ECs run at full capacity to charge the tank, while the operation of the SACs are dependent on the district load. In case of the cooling demand exceeding the SACs capacity, the ECs will divert some of the cooling capacity to meet the district load. During TES discharging, the TES tank discharges to meet the district load. Once the demand exceeds the TES capacity, the SACs switch on to meet the load. The ECs will only operate to meet the load when the sum of the maximum cooling capacity of the SAC and the available cooling capacity of the TES tank is not sufficient to satisfy the demand. The SACs and ECs operate with a variable capacity set point based on the demand to produce the exact cooling power needed and minimize wastage of cooing energy. The system is a closed loop system, utilizing measured data at each timestep to predict the behaviour of the system at the next timestep, making the system self-adjusting and more flexible to uncertainties such as cooling demand changes due to weather changes, occupancy schedule changes, etc. These two control strategies will be the basis of constructing the simulation model.

Table 2 Existing operating schedule of UTP GDCP [21]

Operating mode	Operating period	Description
1	Weekdays, 7 am to 5 pm	EC not used (unless required), SAC fully operational, TES tank discharge, GT fully operational
2	Weekdays, 5 pm to 9 pm	EC fully operational, SAC fully operational, TES tank charging, GT fully operational
3	Weekdays, 9 pm to 7 am	EC fully operational (draw power from GT), SAC not used, TES tank charging, only one GT operational
4	Weekends, 7 am to 5 pm	EC not used (unless required), SAC fully operational, TES tank discharge, only one GT operational
5	Weekends, 5 pm to 7 am	EC fully operational (draw power from GT), SAC not used, TES tank charging, only one GT operational

2.3 Parameter Calculations

For the existing system, the input parameter which governs the mode selection of the control system is the time. For the proposed system, the input will be the measured cooling demand from the district. The measured outputs for both systems are the total cooling capacity supplied to the district, CL_{total} [kW], the energy usage of the system, E_{total} [kWh], and the overall efficiency of the system, $\eta_{overall}$ [%], which are defined by Eqs. (1), (2), and (3) respectively. The total energy usage and the energy savings can be calculated using Eq. (4). These parameters will be logged in the

$$CL_{total} = CL_{EC,total} + CL_{SAC} + CL_{TES,d} \quad (1)$$

$$E_{total} = E_{EC} + E_{SAC} \quad (2)$$

$$\eta_{overall} = \frac{CL_{district}}{CL_{total}} \times 100\% \quad (3)$$

$$E_{savings} = \frac{E_{total,e} - E_{total,p}}{E_{total,e}} \times 100\% \quad (4)$$

2.4 DCS Modelling in Simulink

The GDCP equipment are modelled in the MATLAB program using Simulink graphic modelling environment. A model of the DCS which operates with the existing control strategy is built and simulated using collected input data. The chilled water mass flow rate at the HEX is obtained by the summation of the supply chilled water mass flow rates from the TES during discharging, SAC, and EC, and is used to calculate the total output cooling capacity of the system at each time step. The TES subsystem receives charging chilled water mass flow rate value from the four EC and calculates the added capacity to the tank and the available capacity of the tank at each time step. Once the tank is at full capacity, there will be feedback to the EC to stop charging the tank. During discharging, the tank outputs supply chilled water mass flow rate value to be summed at the HEX. Once the TES has fully discharged its available capacity, it will stop discharging and the mass flow rate will be zero.

The EC and SAC subsystems model the inputs and outputs of one unit of each equipment. The controller for both existing and proposed algorithms are designed to treat the two EC and four SAC subsystems as a single EC and a single SAC. This is to simplify the modelling and simulation process, which differs from the individual control of each unit of EC and SAC in practice [22, 23]. Two flow switches assist the controller to divert flow between the TES and the HEX. In the improved system model as in Fig. 2, the TES subsystem remains the same, with only an addition of a discharge capacity data which is fed to the SAC controller. The cooling supply and demand as well as the oversupply values from the HEX, the available tank capacity of the TES, the cooling supply and electricity usage of the EC and SAC, and the heat energy input to the SAC are logged from the simulations of both models at each timestep. The logged data is extracted and tabulated for comparison and analysis of the system performance of the two models. The justification of the improvement in the scheduling logic is tabulated in Table 3 and the overall block diagram of the Simulink modelling sequence is as in Fig. 1.

3 Result and Discussion

3.1 Improved Cooling Efficiency

The existing algorithm causes the cooling supply of the system model to surge up and down at certain time periods. This is due to the feedback to the EC to divert full cooling capacity to the HEX when the district is undercooled, causing the cooling supply to the HEX to surge up and overcooling occurs. Figure 3 shows the cooling output graphs of the Simulink system models simulated under weekday and weekend conditions with 15-min time intervals. From the hourly instantaneous cooling power output from the two models, it is observed that the proposed algorithm enables the plant model to supply cooling power closer to the value of the cooling demand compared to the existing algorithm. This can be achieved due to the variability of the chilled water supply from the SAC and the EC which follows the required capacity calculated from the feedback of cooling supply and demand gap from the last timestep instead of running at maximum capacity and following the predefined schedule.

In short, the improved model is able to mirror the demand of cooling load as feedbacked with only slight delay and occasional oversupply, caused by the drastic increased in demand during peak hours and adjustment at the start-charging of TES. As shown in Table 4, the total cooling output of the system when operating under the proposed algorithm is up to 21.3% less than when operating under the existing system in the weekday scenario, and 41.4% less in the weekend scenario. In contrast, the cooling demand met by the cooling power supplied when operating under the proposed algorithm as compared to the existing algorithm increased by up to 2.5% and 16.7% for weekday and weekend scenarios, respectively.

Overall, the proposed algorithm allows the improved model to achieve high cooling delivery efficiencies compared to the existing algorithm. Furthermore, the proposed algorithm produced a more consistent system efficiency for both scenarios, indicating that the system can adapt to the change in the district cooling demand and reduce oversupply of cooling power. It is observed that the system efficiency produced by the existing algorithm is quite low compared to the historical data range from 69 to 88.4% [22, 24].

Table 3 Summary of the improvements of the proposed algorithm

Improvements	Justification	Logic
1. Measures cooling load demand of the district and feedback to the controller	Control the operation of SAC and EC based on demand instead of fixed schedule to cut unnecessary cooling	Using the measured cooling demand from the previous time step minus the TES output to calculate the output chilled water flow rates required by each of the equipment
2. Measures TES available capacity and feedback to controller	Measure the cooling capacity available during charging and discharging to predict the required capacity from the SAC and EC and where the chilled water is sent to	When charging, the measured available capacity determines the on or off state of the EC. When discharging, the discharged capacity is compared with the cooling demand to determine the on or off state of the SAC
3. Logic for switching off SAC when TES discharging capacity meet cooling demand	Switching off SAC during low load demand to save energy and prevent over cooling, allowing the TES discharging to meet the demand	Measure the difference between the TES output and the cooling demand from the previous time step, if the cooling demand is lower or equal to the TES output, the SAC output chilled water flow rate is set to zero
4. Logic for switching off the EC when TES is fully charged	Switching off EC to prevent unnecessary chilled water production during TES charging to reduce electricity usage	Once the measured TES capacity reaches maximum storage capacity, the output cooling capacity setpoint for the EC are set to zero
5. Logic for diverting EC from charging TES to meeting district load	Allows the EC to switch to meet district load in case of high demand during TES charging period	The SAC controller takes the cooling demand required and compares it with the maximum combined capacity of the SAC (2500 RT). If the required capacity is more than that, feedback signal is sent the EC controller to trigger the EC to switch to district cooling mode
6. Variable EC and SAC cooling output based on required cooling capacity	Variable output allows to controller to control the cooling output of the system more accurately based on the cooling demand, to prevent over cooling and save energy	The measured output capacity of the TES is constantly monitored and compared with the measured district cooling demand. When the district demand is more than the TES output, the capacity difference is used to calculate the chilled water flow rate setpoint of the SAC and EC for the next time step

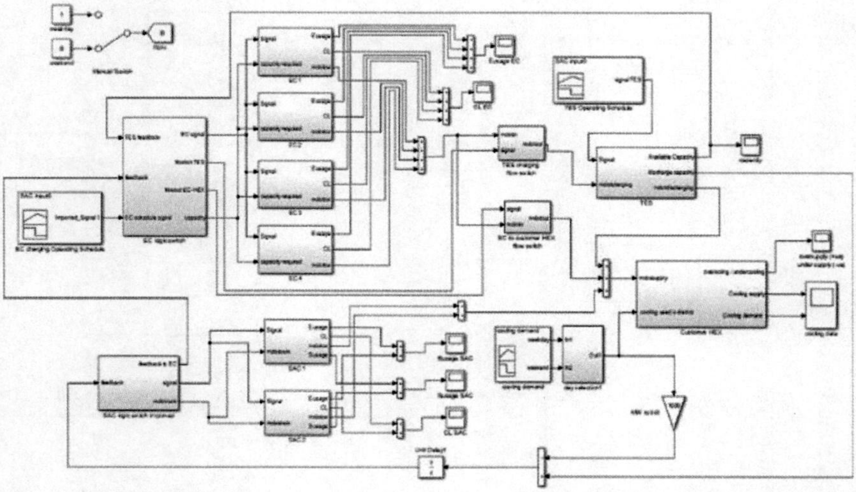

Fig. 1 Block diagram of UTP GDCP simulink model using proposed control

3.2 Reduction in Over and Undersupply

From Fig. 4, it is observed that the system operating under the existing algorithm is over supplying the most during the daytime, specifically from around 7 am to 5 pm in both scenarios which is caused by the constant TES discharging and SAC supply capacities which are higher than the cooling demand. There are also multiple spikes in the graph which is correlates to the fluctuating of the on and off states of the EC to meet the load when the system is undersupplying. In the weekday scenario, there is a period of overcooling from 5 to 9 pm due to the decline in cooling demand from the district, while the cooling supply mainly from the SAC remains constant. Consideration of individual EC outputs, when the EC is supplying to the district, reduces the overcooling marginally as shown by the smaller upward spikes in the graphs compared to full capacity EC output. In the charts of the proposed algorithm, the large area of oversupply during the TES discharging period and the spikes are eliminated, with only a spike in over and under cooling caused by the beginning and end of the TES discharge phase. The over and under cooling under the proposed algorithm is due to the lagging of the system when responding to change in cooling demand.

In short, the little oversupply in cooling load (kW) implies less wastage of energy; while conventional scheduling of DCS has an oversupply of loading, especially during the peak time where huge redundancy is created. These supplies are not used to charge the TES and are wasted for sake of service security. Nevertheless, close mirroring of demand may be risk in developers and consumers point-of-view. Lack of confidence and potential lagging in system response to be overcome together.

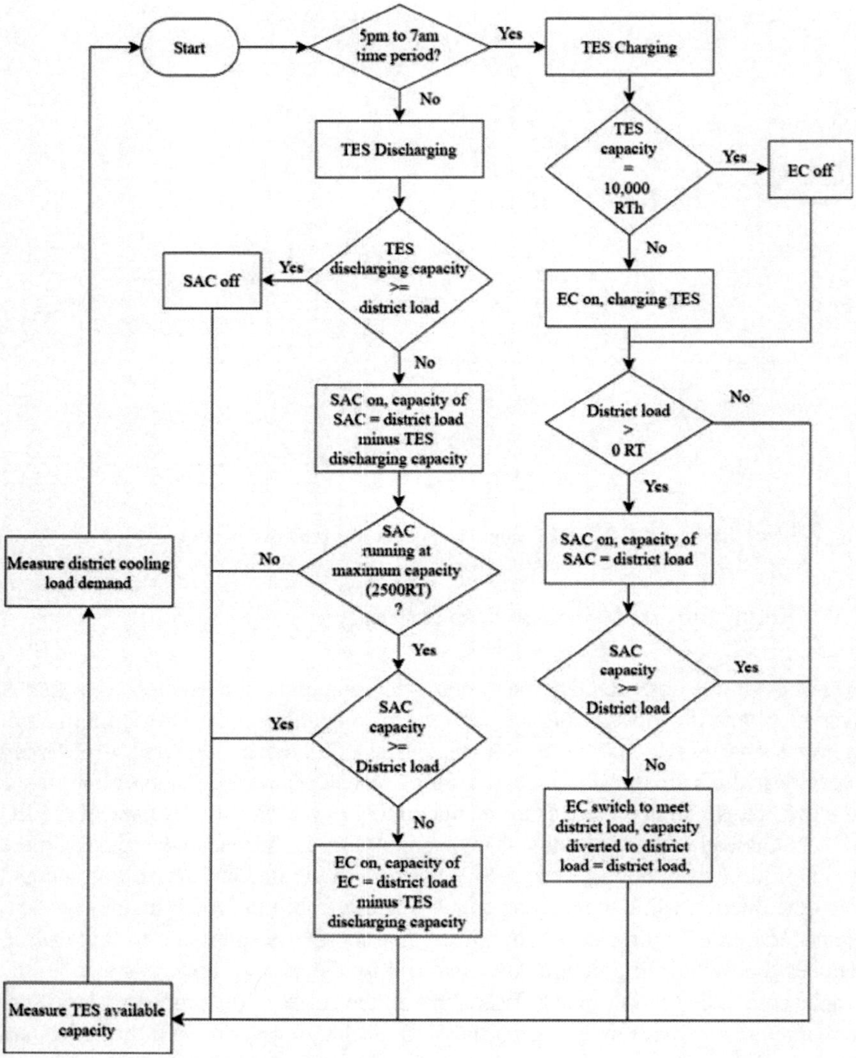

Fig. 2 Proposed DCS control algorithm flowchart

3.3 *Energy Savings*

The energy usage outputs tabulated in Table 5 from the system models show that the system uses up to 18.7% less energy in the weekday scenario and up to 32.1% less energy in the weekend scenario when the system operates on the proposed algorithm as compared to the existing algorithm.

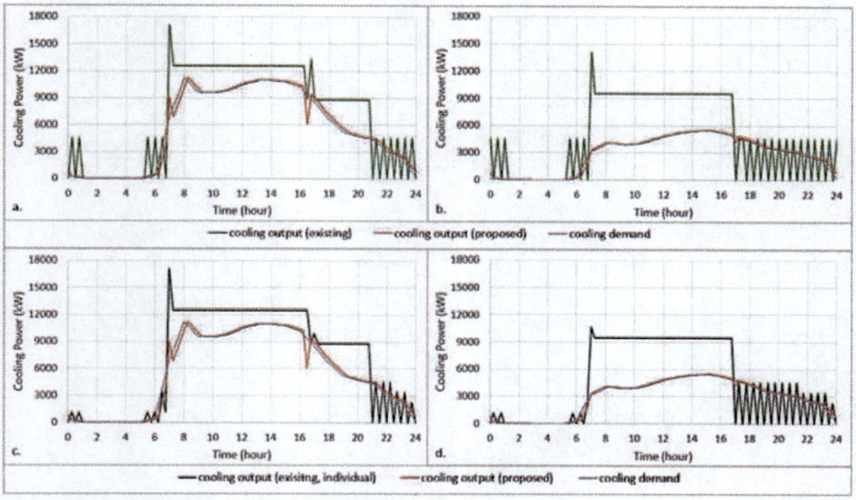

Fig. 3 Cooling output of simulink system model with existing and proposed algorithms and the district cooling demand under **a** weekday scenario and **b** weekend scenario without individual EC output control; **c** weekday scenario and **d** weekend scenario using individual EC output control

Table 4 Total cooling energy supplied, total cooling demand met by system, and overall system efficiency operating under existing and proposed control algorithms

Scenario	Control algorithm	Total cooling energy supplied (kWh)	Total cooling demand met (kWh)	System efficiency (%)
Weekday	Existing	173,826.4	129,848.9	74.7
	Existing (Individual EC output)	169,342.0	130,176.9	76.9
	Proposed	136,758.8	133,030.9	97.3
Weekend	Existing	120,488.9	58,540.5	48.6
	Existing (Individual EC output)	111,630.8	58,243.2	52.2
	Proposed	70,599.4	68,302.2	96.7

The system operating on the existing algorithm uses marginally less energy when the cooling supply of the EC to the district is individually modulated, but the energy expenditure is still 15% higher in the weekday scenario and 28% higher in the weekend scenario compared to operating under the proposed algorithm. The main energy savings come from the reduced steam usage in the SAC due to the SAC operating at less than full capacity throughout the day based on the cooling demand as opposed to full capacity following a fixed schedule.

Since the electricity consumption of the SAC is independent of the output capacity and is assumed to be fixed in the model, the longer period of the SAC being switched

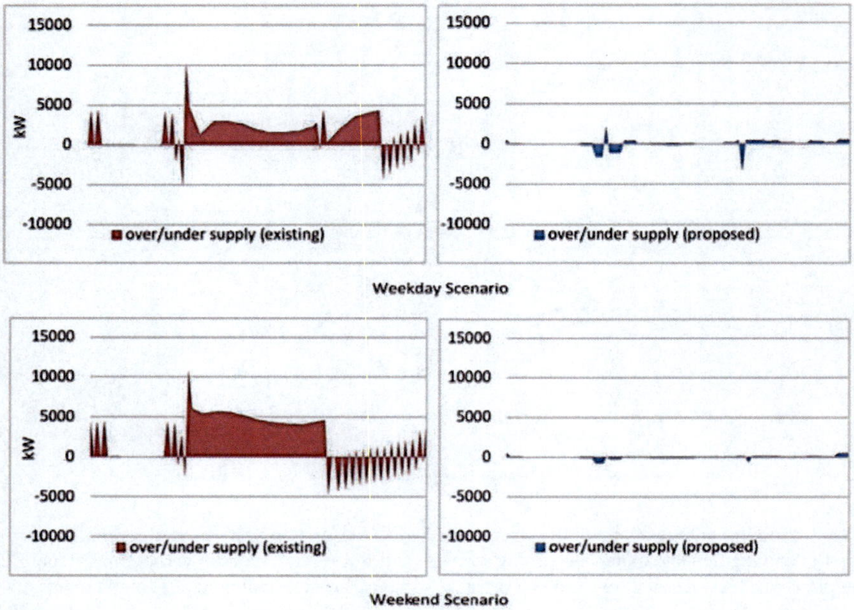

Fig. 4 Over/under supply of cooling power to the district under existing and proposed algorithms for weekday and weekend scenarios over 24 h

Table 5 Total cooling energy comparison of existing control algorithm and new proposed algorithm

Day	Equipment	Energy usage under existing control algorithm (kWh)	Energy usage under existing control algorithm with individual EC output (kWh)	Energy usage under proposed control algorithm (kWh)
Weekday	SAC (Electricity)	576.80	576.80	999.10
	SAC (Steam)	82,480.44	82,480.44	67,342.02
	EC	9509.02	5475.00	6942.07
	Total	92,566.27	88,532.24	75,283.19
Weekend	SAC (Electricity)	412.00	412.00	906.40
	SAC (Steam)	58,914.60	58,914.60	42,210.92
	EC	6214.53	2540.94	1402.57
	Total	65,541.13	61,867.54	44,519.88

on is what contributes to the increase in electricity usage. The EC uses more electricity for both weekday and weekend scenarios when the system runs on the existing algorithm except for the weekday scenario where the system individually moderates the EC cooling power delivery to the district. The application of the proposed algorithm reduces daily EC electricity consumption by a maximum of 2.57 MWh and 4.81 MWh for weekday and weekend scenarios, respectively. The savings achieved by the EC is much higher than the additional electricity used to operate the SAC for longer periods, and hence the trade-off is beneficial.

3.4 DCS Model Discussion

3.4.1 High Efficiency of the Improved Model

From the analysis of the data, it is shown that the proposed algorithm allows the improved model to achieve high cooling delivery efficiencies compared to the existing algorithm. The high efficiency is achieved by allowing the system components to adapt to the supply requirements based on the closed loop feedback from the output of the system. The system can deliver cooling power to the district at the required time to match the cooling demand profile of the district, reducing the overcooling and undercooling effects. The effectiveness of the system to match the cooling supply with the demand allows the system to achieve very high efficiencies. The system is also able to operate at high efficiency consistently for the two different historical cooling demand profiles of the district. The improved efficiency and performance consistency leads to reduction in wasted cooling power and improves overall energy savings of the system.

3.4.2 Ideal DCS Cooling Supply Trend

The DCS model is an idealized model of the UTP GDCP, where the chillers and the TES are modelled based on the theoretical principles of operation. Based on model comparison with historical measured data of the GDCP. In practice, the EC does not switch on and off as frequently as in the system model and instead have a steady output to meet the cooling demand when the system is under supplying. There is also a noticeable change in the EC output patterns between 2005 and 2011 [22, 23], suggesting that the EC schedule was adjusted by the operators based on the cooling requirements in both cases.

Since there is no recorded feedback system implemented by both studies at the times of the data collected in both years, any adjustment to the schedule will be based on the historical data of the district cooling profiles of UTP and the daily requirements [22]. Furthermore, Ramli [24] stated that the UTP management team liaise with the chargeman responsible for the chilled water supply prior to any activities (or lack thereof) in the campus which would affect the chilled water supply prior to the day of the activities. The plant operators then adjust the equipment scheduling accordingly to optimize the cooling supply to the district. The scheduling used for simulating the system model based on operating information collected from multiple sources is not optimized for the simulated cooling demand. This can explain the low system efficiency produced by the existing algorithm used during the weekend scenario. Hence, using the schedule adjusted for the cooling demand in the simulation algorithm will produce more accurate system outputs.

The model comparison with historical daily TES discharge patterns also show that the actual TES discharge cooling capacity varies throughout the whole discharge period instead of a steady discharge rate as used in the model. The historical data of the tank capacity however shows a linear reduction over time during the discharge period. The varying discharge capacity suggests a varying factor of the TES discharge chilled water temperature or mass flow rate. The temperature change is due to the changes in the ambient temperature as well as the heat transfer of the chilled water within the tank. The change in mass flow rate suggests that there is a varying factor in the chilled water pumps as opposed to the fixed average mass flow rates assumed in the model. In-depth modelling with actual data at each time step will allow for a more accurate modelling of the cooling supply trend.

3.4.3 Strength and Weakness of the Model

The model is simple, and the simulation time is short due to the simplicity. The model takes the combined chiller control approach to simulate the four EC as one and the two SACs as one, further simplifying the simulation process. The model is also capable of taking feedback from real time data instead of the collected data generated in the signal generator blocks in the software. In practice, the EC can be switched on and off and operated individually as observed in the model comparison, this individual control is not considered in the model. Furthermore, certain data used to construct the system model components such as the TES discharge rate are based on the historical average or the rated values, which may not accurately model the current condition of the equipment, allowing the model to output only a theoretical trend of the data. However, the performance of the model is sufficient to meet the objectives of this study.

4 Conclusion

This study shows the potential of applying feedback control to DCS by simulation of the UTP GDCP with the existing control strategy and the proposed feedback control strategy. The existing algorithm has been developed from the collected plant equipment operational data and the feedback control algorithm has developed from that. The models are simulated using historical cooling demand profile. The following conclusions have drawn from the present work:

1. The proposed algorithm is able to match the cooling demand profile closely and minimize oversupply and undersupply of cooling power to the district at each time step.
2. The proposed algorithm improves the efficiency of cooling delivery of the DCS by 22.6% on weekdays and 48.2% on weekends. Energy savings achieved by the improved model are 18.7% (37 MWh less) and 32.1% (49.9 MWh) for weekday and weekend scenarios, respectively.
3. The cooling supply from the plant equipment of the system model follows the trend of historical data and it is sufficiently accurate to model the improvements of implementing feedback algorithm to a DCS system.

Technical implementation and electronic configurations will be the next stage of research. Constraints of incompatible instalment, varying loading strain on system and risk on a major facility as DCS will require careful deployment of any improved DCS rescheduling.

Acknowledgements The authors would like to acknowledge CERVIE of UCSI University for conference funding.

References

1. ASHRAE Handbook (2020) HVAC systems and equipment. ASHRAE, United States of America, p 178
2. Dincer I, Zamfirescu C (2011) Sustainable energy systems and applications. Springer US, United States of America, pp 389–394
3. Eveloy V, Ayou DS (2019) Sustainable district cooling systems: status, challenges, and future opportunities, with emphasis on cooling-dominated regions. Energies 12:235
4. Gang W, Wang S, Xiao F, Gao D (2016) District cooling systems: technology integration, system optimization, challenges and opportunities for applications. Renew Sustain Energy Rev 53:253–264
5. Moustakidis S, Meintanis I, Halikias GD, Karcanias N (2019) An innovative control framework for district heating systems: conceptualisation and preliminary results. Resources 8:27–41
6. Damien C, Cynthia N, Guillaume B, Pascal S, Dominique M (2015) Dynamic modelling of a district cooling network with MODELICA. In: 14th conference of International Building Performance Simulation Association, Hyderabad, India
7. Bhaskoro PT, Gilani SIUH, Aris MS (2011) Simulation of intermittent transient cooling load characteristic in an academic building with centralized HVAC system. IPCBEE 8:291–296
8. Coffey B, Haves P, Ma Y, Borrelli H (2010) Development and testing of model predictive control for a campus chilled water plant with thermal storage. Paper presented at the ACEEE summer study on energy efficiency in buildings
9. Powell KM, Cole WJ, Ekarika UF, Edgar TF (2013) Optimal chiller loading in a district cooling system with thermal energy storage. Energy 50:445–453
10. Zainal Abidin WF (2010) Performance evaluation of thermal energy storage for district cooling plant
11. Aziz MBA, Zain ZM, Baki SRMS, Hadi RA (2012) Air-conditioning energy consumption of an education building and its building energy index: a case study in engineering complex. UiTM Shah Alam, Selangor. Paper presented at the 2012 IEEE control and system graduate research colloquium

12. Abdullah MO, Yii LP, Junaidi E, Tambi G, Mustapha MA (2013) Electricity cost saving comparison due to tariff change and ice thermal storage (ITS) usage based on a hybrid centrifugal-ITS system for buildings: a university district cooling perspective. Energy Build 67:70–78
13. Muhieldeen MW, Kuang YC (2019) Saving energy costs by combining air-conditioning and air- circulation using CFD to achieve thermal comfort in the building. J Adv Res Fluid Mech Therm Sci 58(1):84–99
14. Muhieldeen MW, Adam NM, Salman BH (2015) Experimental and numerical studies of reducing cooling load of lecture hall. Energy Build 89:163–169
15. Muhieldeen MW, Wong QY, Rahman UZA, Tey WY (2020) Energy saving by applying different wall thermal insulations on a room at Malaysian institution. J Adv Res Fluid Mech Therm Sci 65(1):130–139
16. Muhieldeen MW, Lim YR, Govinda S, Tey WY (2020) Investigation of the effect of awning using sunlight sensor to reduce cooling load in the room. J Adv Res Fluid Mech Therm Sci 67(1):136–145
17. Muhieldeen MW, Yang LZ, Lye LC, Adam NM (2020) Analysis of optimum thickness of glass wool roof thermal insulation performance. J Adv Res Fluid Mech Therm Sci 76(3):1–11
18. Dusuki NS (2010) Design of two stage charging system for thermal energy storage (TES) tank for GDC UTP
19. Okitsu J, Naono K, Khamis MFI, Haruna AA, Zakaria N (2015) Towards an architecture for integrated gas district cooling with data center control to reduce CO_2 emission. Sustain Comput Inf Syst 6:39–47
20. Majid MAA, Rangkuti SC, Waluyo J (2007) Operating modes of a thermal energy storage system of a gas district cooling plant. In: Applications and design in mechanical engineering, Kangar, Perlis, Malaysia
21. Majid MAA, Sulaiman SA (2011) Gas district cooling in Malaysia, leveraging abundant cleaner-burning natural gas. Lambert Academic Publishing
22. Majid MAA, Gilani SIUH, Rangkuti C, Hassan S (2006) A case study on electricity and chilled water production of a gas district cooling plant. Jurutera, pp 20–27
23. Amear S, Ariffin S, Nordin A, Buyamin N, Amin M, Majid A (2013) Performance analysis of absorption and electric chillers at a gas district cooling plant. Asian J Sci Res 6:299–306
24. Ramli N (2011) Process reliability analysis of gas district cooling by using production chilled water data

Polyethylene Bubble Aluminium SB250-FR+ for Reduced Energy Consumption Building: An Experimental Study

Mateus De Sousa, Mohammed W. Muhieldeen, Jayden Lau, Wah Yen Tey, Teng Kah Hou, and U. Z. A. Rahman

Abstract The excessive energy consumption through the last decades especially from the air conditioning due to the negative awareness behaviour of consumers has reflected not only on the economy but also on the weather and environment. The aim of this study is to save the energy inside the buildings by applying potential thermal insulation to reduce the usage of the air conditioning and save the cost of the electricity bill. Thermal insulation has been used in this study is Polyethylene Bubble Aluminium (SB250-FR+ with 4 mm) on the roof of the guardhouse of UCSI University to reduce the cooling load inside the buildings. The devices that have been used in this study are TSI VELOCICALC to measure the air temperature and air velocity, with Infrared Thermometer to measure profile temperature of the walls, windows, and roof. The data have been collected for 30 days from 9 am–5 pm with and without insulation. The data present shows, the insulation has helped to reduce the load inside the guardhouse. The results show that the heat inside the room has been reduced by is 26.6% and this could help to reduce the usage of the air-condition unit inside the room and save monthly around RM 80. The potential insulation has approved to save the energy inside the buildings that will help to reduce the heat tension of the occupants.

Keywords Thermal insulation · Energy saving · Cooling load · Thermal comfort · Tropical climate

M. De Sousa · M. W. Muhieldeen (✉) · W. Y. Tey · T. K. Hou · U. Z. A. Rahman
Mechanical Engineering Department, Faculty of Engineering, Technology and Built Environment, UCSI University, 56000 Kuala Lumpur, Malaysia
e-mail: al-gailani@ucsiuniversity.edu.my

M. De Sousa
Mechanical Engineering Department, Faculty of Engineering, Dili Institute of Technology, Rua: Aimeti-laran, Dili, Timor-Leste

M. W. Muhieldeen · T. K. Hou
UCSI-Cheras Low Carbon Innovation Hub Research Consortium, Kuala Lumpur, Malaysia

J. Lau
YH Laminated Products Sdn. Bhd., Krubong, Malaysia

M. Y. Ismail et al. (eds.), *Technological Advancement in Mechanical and Automotive Engineering*, Lecture Notes in Mechanical Engineering,
https://doi.org/10.1007/978-981-19-1457-7_52

1 Introduction

Urbanization and increasing living standards have resulted in increased energy consumption in Malaysia, this lead to increase the usage of the air condition to reach to comfort condition [1, 2]. However, China continued to experience an increase in terms of energy use from 10% in the 1970s and increased by 38.6% in 2015. Heat energy consumption accounted for 60% and heat areas accounted for around 25% of total energy consumption. According to national statistics in China in 2015 China's energy consumption reached 4.98 tons of standard coal. This contributes to energy by 37.6% of total energy consumption. Therefore, improving energy-saving technology is very important to implement due to reduce the use of consumption of buildings [3]. According to the Malaysian weather, the outdoor temperature could be in range between 29 and 34 °C under shade and it could be reach to 41 °C under the direct sun, this could affect the indoor environment to make the occupants feel uncomfortable while the temperature inside without using the air condition system to 29 °C [1–12]. However, the humidity could raise the daily temperature comparing to the seasonal temperature, where the humidity of Malaysia could reach between 70 and 90% through the entire of the year, where this may affect the indoor air quality (IAQ) [13]. Venkiteswaran et al. [14] conducted study on a modern office in privet institution in Malaysia, where it was observed that A/C provides comfort in the room such as cooling the room because cold temperatures can be evenly distributed, but this could increase the cost of the electricity that is consumed by the institution. Air conditioning (AC) systems overcool indoor thermal conditions in different offices in Asia's hot and humid climate regions. Overcooling has a negative impact on the health of citizens and due to enormous energy usage [15]. On the other hand, Halim et al. [16] found, buildings accounted for 40% of global primary energy consumption. In hot climate countries, the cooling system used more than half of the building's electricity. According to the Energy Information Association in 2018, global energy demand is expected to rise by 64% before 2040 due to a significant increase in residential, manufacturing, commercial, and urban construction due to industrial development and population growth [17]. To choose the right materials, engineers not only have to understand well about the location of the houses and surrounding temperatures, but they also have to understand and know the radioactive and convective heat transfer coefficients when each material is chosen as the ceiling or roof materials. In Malaysia, the most common roofing materials include Asphalt Shingles, Metal Shingles, Wood Shingles, Slate Shingles, and Clay Shingles. These five materials are always being considered as roofing materials in Malaysia. These five materials are commonly used due to their superior product for successful and durable roofing projects. They also have the price affordable by the people. They are also able to maintain a cooler building under the hot surrounding temperature. Especially for clay roof, they will allow a varying degree of airflow and increase the insulating ability of the entire roof [18].

There is no fixed temperature and humidity for absolute thermal comfort as it depends on various factors such as the clothing of the occupant and the intensity of the activity carried out in other words the metabolic of the occupants. it is also recorded that the indoor temperature range of the building studied with solar passive design is within the thermal comfort standard set by ASHRAE [19].

Thermal insulation is the barrier to transfer the heat into/from the buildings, where it could help to increase the comfort level inside the buildings by less usage of air condition units. While this could happen when used in a proper way which it will lead to save the energy for the heating or cooling system that have been installed inside the buildings [20]. Thermal insulation improves the quality and life span of structures by decreasing energy utilization, and as a related outcome, improved energy use [21]. The selection of insulation material is represented by significant boundaries, including the normal outside air temperature, the thermal conductivity of the structures and the expense of the insulation material. Increments in the thickness of the insulation material will steadily diminish the energy utilization for heating; in any case, the insulation thickness has an ideal worth that limits the absolute investment cost, and assurance of this ideal worth is basic for financial investigation [10].

This is study is focused on how to reduce the tension of heat load inside the buildings by applying thermal insulation Polyethylene Aluminium Bubble with 4 mm thickness, which have never been tested at any study before, on the roof of guardhouse of UCSI University, Kuala Lumpur campus to reduce the cooling load and increase in energy consumption which is increasing every year.

2 Experimental Work

2.1 *Thermal Insulation*

The material was chosen for this study is (Polyethylene Aluminium Bubble SB250-FR+), as shown in Fig. 1, which was being used to determine the reduction of the heat in the guardhouse. The insulation has placed directly on the outdoor surface of the roof of the guardhouse. Is use an insulation against radian heat transfer and can be used to its greatest advantage in instances where radiation is the predominant means of heat transfer, SB250-FR+ is an excellent reflector of all long wave radiant heat that strikes it double sided reflective of up to 99% of all radiant heat. Fire attributes SB250-FR+ is classified as class "O" in accordance with fire test on building materials, SB250-FR+ is also flame retardant, waterproof, nontoxic, bacteria free, fungus and corrosion resistant. Thickness of the polyethylene aluminium SB250-FR+ is 4.00 mm $\pm$ 0.5 mm.

The guardhouse room was exposed to the sun for the majority of the day, with dimensions of (2.4 m length, 1.5 m width and 2.9 m height) and the roof with SB250-FR+ insulation as shown in Fig. 2.

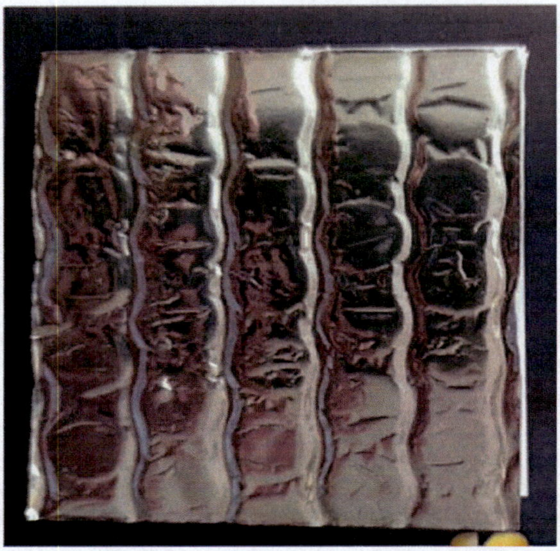

Fig. 1 Polyethylene aluminium bubble SB250-FR+

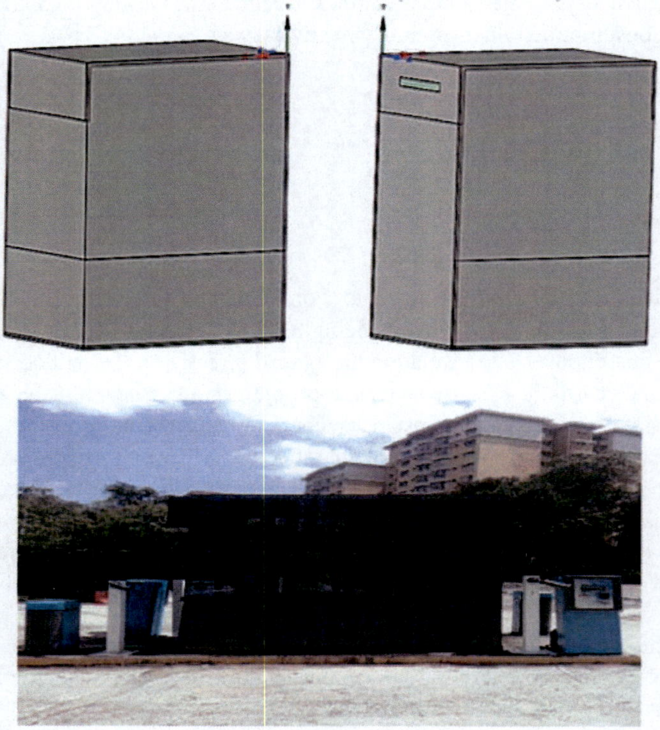

Fig. 2 Geometry and actual model of the guardhouse

2.2 Measuring Equipment

There are two devices have been used in this study, which are VILOCICALC TSI 9515 and Infrared Thermometer. VILOCICALC is used to measure air temperature and air velocity inside the guardhouse as shown in Fig. 3. While the Infrared thermometer is used to measure the walls, roof, and window profile temperature, as shown in Fig. 4.

The VILOCICALC TSI 9515 features are high accuracy over a wide velocity range [0–6000 ft/min (0–30 m/s)], simultaneously measures temperature and velocity, data logging and LogDat2™ downloading software included and large digital display. While the infrared thermometer features are high accuracy within 25 mm around the spot point, laser to point the location of the measure point and thermocouple could be attach to the device.

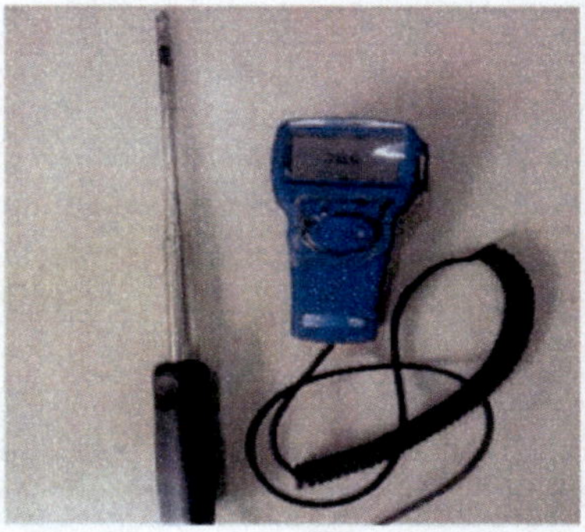

Fig. 3 VILOCICALC TSI 9515

Fig. 4 Infrared thermometer

3 Result

3.1 Temperature Distribution in the Guardhouse

The data has been collected for 30 days with and without insulation. It shows the temperature has been increased from 10 am to 1 pm. The outside temperature increased from 27 to 35.7 °C, while it shows some reduction around 3 pm due some cloudy days. Figure 5 illustrated the difference in the heat that is transferring from the roof to the room comparing with other walls. It has been noticed that the roof keeps increasing during the day while the right and left walls have been affected by the sun direction. Due to the sun direction on the morning time, the right wall temperature that is facing east, the temperature shows increasing at 12 pm and 1 pm, while the left wall that is facing west, the temperature shows increasing started at 4 pm and 5 pm. Therefore, this could affect the room temperature. However, the roof temperature kept increasing from morning till evening time and this shows how the roof is an important part in the room that is required to be isolated by proper thermal insulation.

Figure 6 shows the different between outside, inside temperature with and without insulation SB250-FR+ indicated that gap between roof without insulation and roof with insulation are average 8 °C. The data shows that, the temperature does not increase significant from 10 am to 1 pm. From 1 to 4 pm the temperature steady is because the heat flux from the widow directly to inside the room.

The difference temperature between outdoor and indoor without insulation shows that the from 10 am to 1 pm the different reach to 4.3 °C, this is because the heat flux through the wall, from at 2 pm the roof was heated during the heat peak at the same

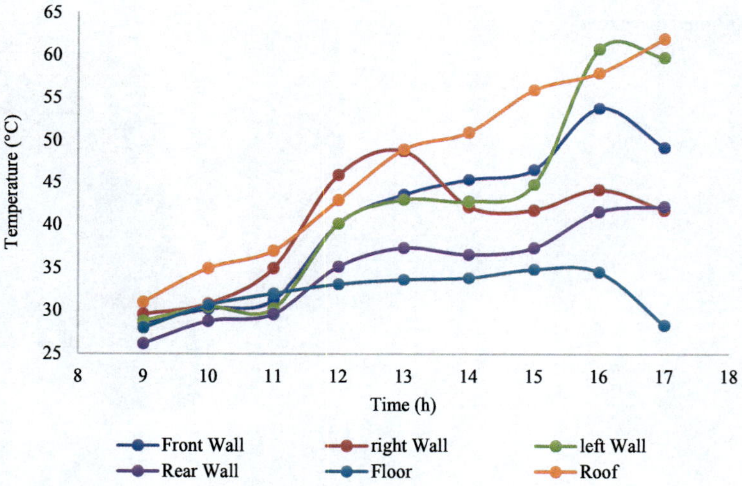

Fig. 5 Surface temperature of the walls and roof of the guardhouse

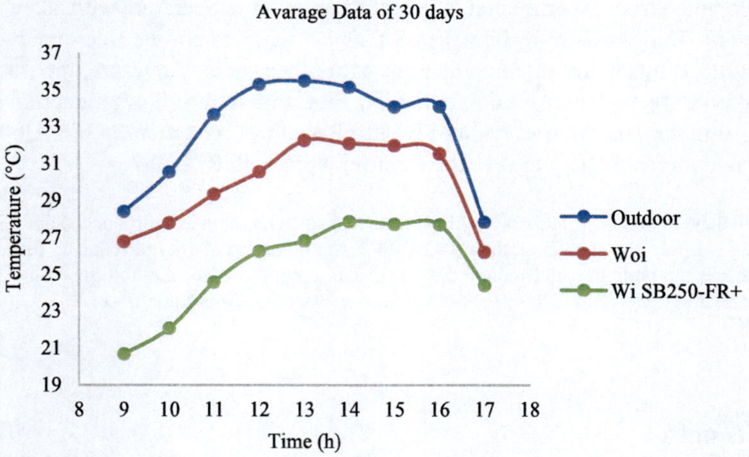

Fig. 6 Guardhouse using polyethylene aluminium SB250-FR+

time sunlight directly enter to inside the guardhouse from west side and the front of guardhouse that make the heat increase at that time. At the 4 pm the different of temperature was 2.5 °C.

3.2 *Cost Analysis*

The power consumption was calculated by considering the air condition unit is operating for 10 h a day and there is 30 days in a month. The results show that the application of Polyethylene Aluminium SB250-FR+ insulation shows significant reduction as justified by the temperature reduction. The electricity cost was calculated based on the electricity tariff by Tenaga Nasional Berhad. The first 200 kWh was 21.80 cent/kWh per month, for the next 100 kWh was 33.4 cent/kWh per month. The results show that by applying Polyethylene Aluminium SB250-FR+ insulation could save the electricity cost by RM 80 per month. While this reduction also, depend on temperature reduction that clearly presented after applying the thermal insulation, that lead to reduce the usage of the air conditioning unit and help to reduce the power consumption.

4 Conclusion

The objective of this study has been achieved by applying Polyethylene Aluminium SB250-FR+ insulation material on the roof of the guardhouse of UCSI University, the data have been collected for 30 days from 9 am to 5 pm. The data measurements have been obtained from experimental in guardhouse are air temperature with and without insulation. The results show how the roof is playing an important role to reduce the thermal load inside the room comparing with other walls. Therefore, the potential insulation Polyethylene Aluminium SB250-FR+ with 4 mm shows reduction in the heat inside the guardhouse, which SB250-FR+ shows reduction 26.6%. However, this reduction could help to save the electricity bill with RM 79.7.

Acknowledgements The authors would like to thank YH Laminated Products Sdn. Bhd. by donating the Polyethylene Aluminium SB250-FR+, to find the possibility of reducing the temperature inside the buildings and reduce the power consumption. Also, the authors would like to appreciate the support from CERVIEW UCSI University to fund this conference.

References

1. Muhieldeen MW, Kuang YC (2019) Saving energy costs by combining air-conditioning and air-circulation using CFD to achieve thermal comfort in the building. J Adv Res Fluid Mech Therm Sci 58(1):84–99
2. Muhieldeen MW, Adam NM, Salman BH (2015) Experimental and numerical studies of reducing cooling load of lecture hall. Energy Build 89:163–169
3. Zhu Y et al (2018) Analysis of heat transfer and thermal environment in a rural residential building for addressing energy poverty. Appl Sci (Switzerland) 8(11)
4. Muhieldeen MW, Wong QY, Rahman UZA, Tey WY (2020) Energy saving by applying different wall thermal insulations on a room at Malaysian institution. J Adv Res Fluid Mech Therm Sci 65(1):130–139
5. Muhieldeen MW, Lim YR, Govinda S, Tey WY (2020) Investigation of the effect of awning using sunlight sensor to reduce cooling load in the room. J Adv Res Fluid Mech Therm Sci 67(1):136–145
6. Ran J, Tang M, Jiang L, Zheng X (2017) Effect of building roof insulation measures on indoor cooling and energy saving in rural areas in Chongqing. Procedia Eng 180:669–675
7. Wahhad AMA (2016) Effects of selected shading devices on an office room temperature distribution mechanical engineering. Selangor, UPM
8. Muhieldeen MW, Yang LZ, Lye LC, Adam NM (2020) Analysis of optimum thickness of glass wool roof thermal insulation performance. J Adv Res Fluid Mech Therm Sci 76(3):1–11
9. Nyers J, Komuves P (2015) Optimum of external wall thermal insulation thickness using total cost method. Obuda University, Budapest
10. Muhieldeen MW, Lye LC, Kassim MSS, Tey TWY, Teng KH (2021) Effect of rockwool insulation on room temperature distribution. J Adv Res Exp Fluid Mech Heat Transf 3(1):9–15
11. Muhieldeen MW, Adam NM, Salleh DE, Tang SH, Kwong QJ (2008) Student behavior that leads to energy abuse at a teaching institution in Malaysia. International seminar in sustainable environment and architecture (9th SENVAR 2nd ISESEE 2008: humanity and technology), Uitm Shah Alam
12. Muhieldeen MW, Adam NM, Salleh E, Tang SH, Ghezavati H (2009) CFD simulation on use of polyethylene single bubble to reduce radiant heat on lecture hall. International Advanced of Technology Congress (ATCi), PWTC, Malaysia
13. Daghigh R, Adam N, Sopian K, Sahari B (2009) Thermal comfort of an air-conditioned office through different windows-door opening arrangements. Build Serv Eng Res Technol 30:49–63
14. Venkiteswaran VK, Liman J, Alkaff SA (2017) Comparative study of passive methods for reducing cooling load. Energy Build 142:2689–2697

15. Cheong KWD, Djunaedy E, Chua YL, Tham KW, Sekhar SC, Wong NH, Ullah MB (2003) Thermal comfort study of an air-conditioned lecture theatre in the tropics. Build Environ 38:63–73
16. Halim NHA, Ahmed AZ, Zakaria NZ (2011) Thermal and energy analysis of ceiling and pitch insulation for buildings in Malaysia. In: 3rd international symposium and exhibition in sustainable energy and environment, 1–3 June 2011, Melaka, Malaysia
17. Nandapala K, Chandra MS, Halwatura RU (2020) A study on the feasibility of a new roof slab insulation system in tropical climatic conditions. Energy Build 208:109653
18. Pásztory Z (2021) An overview of factors influencing thermal conductivity of building insulation materials. J Build Eng 102604
19. Nazi WIWM, Wang Y, Chen H, Zhang X, Roskilly AP (2017) Passive cooling using phase change material and insulation for high-rise office building in tropical climate. Energy Procedia 142:2295–2302
20. Willoughby L, Eur Ing C, Mimeche, Honfsoe, Honfiplante, Hon FDAS (2003) Pilkington insulation. Plant engineer's reference book, 2nd edn. Oxford, Butterworth-Heinemann
21. Zach J, Slávik R, Novák V (2016) Investigation of the process of heat transfer in the structure of thermal insulation materials based on natural fibres. Procedia Eng 151:352–359

Improvements of the Cyclone Separator Performance for Wood Waste Combustion by an Aggregation Chamber

Charlito L. Cañesares

Abstract Reducing fine particles emission from wood waste fuel-driven boilers is one of the most challenging problems associated with exhaust pollution control. Cyclone separator (CS) is the most popular and cheaper air pollution control equipment. But it has limitations for fine particles ($PM_{2.5}$). This research aims to improve the design of CS and evaluate its performance for the effective control of fine particles. An aggregation chamber (AC) installed at the entrance of CS can enhance the system's performance. Mixing mists and flue gases in the AC resulted in the build-up and growth of clusters with sizes that can be easily captured in CS by inertial methods. Samples were collected over three different flue gas velocities from three sampling port locations in the AC. The experimental research was performed to explore the performance of CS modified with an AC. A total of 28 experiments were conducted. Results were all investigated using Scanning Electron Microscopy and Energy Dispersive X-ray spectroscopy (SEM–EDX). The study proved the effectiveness of AC in enhancing fine particles to 242 μm at 24.4 m/s flue gas velocity for concurrent flow. Fine particles that could have been emitted to the atmosphere were reduced by 82.3%. Thus, collection efficiency was significantly improved with the installation of AC at the entrance of CS.

Keywords Aggregation chamber · Cyclone separator · Fine particles · Flue gas · Mist

1 Introduction

Wood waste is a renewable fuel source belonging to the family of biomass fuels. Some popular biomass fuels are wood waste, rice husk, and sawdust. They are reactive, and as a result, combustion is not a significant issue. The combustion of these biomass fuels, the source of air pollution, is the primary concern, particularly on effectively mitigating this pollution.

C. L. Cañesares (✉)
Department of Mechanical Engineering, University of Mindanao, Matina, Davao City, Philippines
e-mail: clcanesares@umindanao.edu.ph

M. Y. Ismail et al. (eds.), *Technological Advancement in Mechanical and Automotive Engineering*, Lecture Notes in Mechanical Engineering,
https://doi.org/10.1007/978-981-19-1457-7_53

Small-scale industries, particularly wood processing industries, use wood waste as fuel in the operation of their boilers. These wood waste fuel-driven boilers emit particulate matter (PM) of various particle sizes, resulting in air contamination if no mitigation measures are taken. Specifically, air contamination is caused by fuel-burning equipment used in power plants and industrial plants that emit various PMs, including dust, smoke, and solid and liquid droplets.

Particulate matter can be defined as solid and liquid droplets suspended in gas that the wind can carry. It can cause damage to the food source, water source, and the air that people breathe. It can reduce visibility and destroy the aesthetic appearance due to the continuous deposition of particulates [1–3]. They are classified according to their size, ranging from coarse particles to fine particles. Fine particles are detrimental to human health, especially ultrafine particles, which can cause health risks such as respiratory problems [4–7]. Given the risk as indicated, governments institute measures to mitigate the impact of the problem. Thus, the government requires all wood waste fuel-driven-boilers that emit flue gases to the atmosphere to be installed with air pollution control devices (APCDs).

Various types of APCDs can be chosen upon from cheaper to expensive ones. The more expensive they are, the more efficient they become. In small-scale wood-processing industries, cyclone separators (CS) are recommended. They are economical, inexpensive to construct, and easy to maintain because they do not have any moving parts though they have limitations for fine particles [8, 9]. Innovations have been made to enhance CS performance in reducing fine particles emission. Particle size enlargement can be achieved using capillary and viscous forces by adding a binder liquid [10, 11]. Droplets of liquid water, as the binder, can be produced by injecting water at high pressure through specially designed nozzles [12]. When droplets contact solid particles, a liquid layer forms at the particle surface [10, 13]. The spraying of mist liquid (water) to the dust-laden flue gases upstream of the CS can improve its performance in collecting the growth particles and reducing fine particulates emission to the atmosphere.

This research aims to develop a highly efficient and affordable APCD that could reduce the PM emissions from wood waste fuel-driven boilers. Since the centrifugal force plays a significant role in cyclone performance and since spray of water (mists) enhances the size of the particle, the collection of the growth particles of this inertial device is higher than that of the particle without the liquid spray. An aggregation chamber is installed at the entrance of CS, thus enhanced the system's performance. Mixing mists and dust-laden flue gases in the AC in parallel flow resulted in the build-up and growth of clusters with sizes that can be easily captured in CS by inertial methods. The mixing of mists and flue gases is also investigated to determine if gas-to-particle adhesion and coagulation lead to the build-up of clusters inside the AC. Subsequently, the elemental composition and particle size distribution of the particles extracted from the three sampling ports of the AC can be validated using SEM–EDX.

The scope of the work presented in this paper is limited to studying a 30-cm cyclone with geometry modeled by Karagoz [14]. While pressure drop is calculated and reported, it is not within the scope of this paper to perform detailed studies of

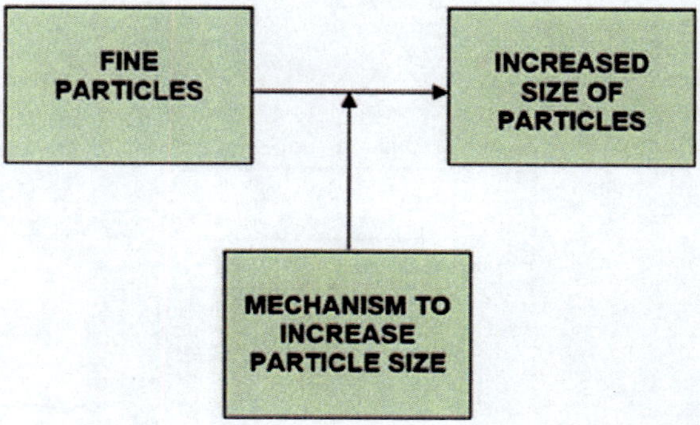

Fig. 1 Conceptual framework

pressure drop in CS. The temperature and velocity of mists are maintained during the conduct of the experiments. The process of droplet collision and aggregation is independent of the flue gas temperature. The CS is only characterized as a system that would reduce particulate emissions.

2 Materials and Methods

2.1 Conceptual Framework

Cyclones are efficient for particles larger than 10 μm. Conversely, they are not very efficient for particles 1 μm or less in diameter. However, fine particles can aggregate to form larger, heavier groups, which together have enough mass to be captured in the cyclone. A heavier mass can improve the performance of CS. This perception was achieved by injecting a spray of water (mists) placed in the ducting upstream of the CS. The concept of this study for enhancing the size of particles is shown in Fig. 1.

2.2 Flue Gas Supply System

Flue gas was supplied from the actual combustion of wood waste fuel burned in the furnace. The composition of wood waste fuels, moisture-free, before combustion is as follows: Carbon 49%, Hydrogen 5.9%, Nitrogen 2.5%, Oxygen 40.7%, and Ash 1.9%. The Gross Heating Value of wood waste fuel is 20,140 kJ/kg [15–18]. Wood waste fuel was manually fed at the rate of 4 kg in every 30-min-duration of the experiment per sample. A total of twenty-eight (28) samples were extracted throughout the conduct of this study. The furnace temperature was maintained between 200 and 230 °C.

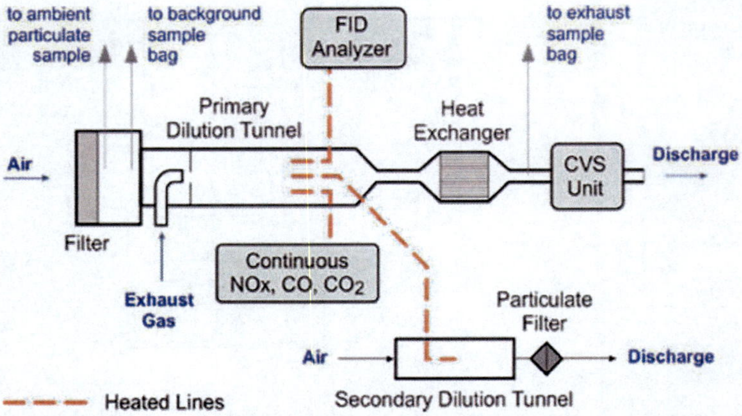

Fig. 2 Sampling train for exhaust gases

2.3 *Nozzle Design*

The type of nozzle used in this study was a fixed geometry hydraulic nozzle. This nozzle covers a wide range of the spray of 60° cone angle, spray velocity, and droplets size distribution. The fixed geometry nozzle atomizes the water by forcing high-pressure water through a small orifice. The nozzle size is 1 mm in diameter. An automatic pump having a capacity of 0.34 L per minute and pressure of 3 kg/cm^2 was used to produce a velocity of 4 m/s with droplets Sauter mean diameter of nearly 39.4 μm.

2.4 *Sampling Train*

As shown in Fig. 2, a sampling train was used to extract samples through the sampling ports [19].

2.5 *Experimental Method*

A furnace, with a dimension of 60 cm diameter and 60 cm high, was used as the miniaturized source of air pollution for wood waste combustion. Fuel was fed manually in the furnace at 4 kg in every 30-min per experiment per sample. A total of twenty-eight (28) samples were extracted throughout the conduct of this study. An exhaust blower was used to suck the flue gases from the furnace.

The aggregation chamber was a round pipe of 10-cm diameter and 150-cm long. Provision was made to accommodate two additional sampling ports. Samples were extracted to three sampling ports of varying lengths. The flue gas stream going to AC would be mixed with water from the peristaltic pump. The experimental setup included a water supply circuit. The water discharged from the supply tank goes to the nozzle at the test section inlet. Sample particles could be extracted in a 4-cm diameter sampling port mounted by the side of the round pipe of varying lengths. An external vacuum pump sucked the filtered gas stream and collected it in the bottom of the filter box of the sample train. The velocity of the flue gas could be adjusted by varying the damper openings [20]. The residence time of coalescing process would depend on the distance traveled by the coalescing particulate and mists. For all experiments, water was used as a mist liquid. A water injection temperature of 22 °C had been maintained for all experiments. The configuration consists of flue gases flowing in the same direction, with the mists issuing from a fixed geometry hydraulic nozzle of 1-mm diameter.

2.6 Sampling Method

Samples were collected over three different flue gas velocities from three sampling port locations in the AC. A total of 28 experiments have been conducted. The sampling period lasted for 30 min per sample. Sampling could not be performed simultaneously at three locations since the probe of the sampling train had to be moved from one sampling port to another. Sample particles were withdrawn from the centerline of the aggregation chamber through three sampling ports by a sampling probe.

2.7 Characterization of Samples

Characterization of the resulting water and flue gas mixture was done using SEM–EDX [21]. The SEM–EDX technique helped in the identification of critical characteristics of particles. It offered the ability to gather information about finer particles and could readily distinguish between clusters and aggregation of particles in addition to the chemical analysis available by EDX. The strength of this analysis technique was its ability to gather statistically significant data on the size, morphology, and composition of the particles.

The surface morphology and elemental composition of each sample were obtained using SEM–EDX detector. Images of the surfaces of sample material are captivated, thus produced images of its shape, size, and topography. The EDX technique allowed one to establish the elemental composition of sample material, thus determining the purity of the sample. The sample was coated with silver to avoid the charge effect and to reduce surface changing. Later, the sample was placed in a sample holder, and the images were captured under various magnifications.

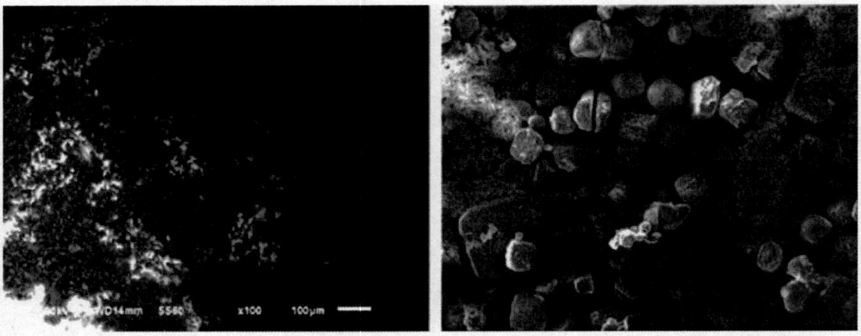

Fig. 3 SEM images without mists at ×100 and ×800 magnification

3 Results and Discussions

The experimental data obtained during the investigation are presented and discussed. As benchmark data, it is necessary to investigate the interactions between the flue gas and mist (water) occurring inside the aggregation chamber (AC). There was no established experimental data set for AC found in the literature review. Few papers found related mostly experimental studies wherein cooler ambient air carries the particles and not a hotter flue gas.

3.1 Analysis of Samples

The samples obtained on Petri discs were analyzed by SEM–EDX technique. Figure 3 shows images of samples acquired using SEM–EDX at a magnification of ×100 and ×800. The figure shows sample #1, in which particles have not been treated with mists. It is necessary to enhance the image to ×800 to obtain the equivalent diameter of the PM quickly by the use of AutoCAD. The result estimates an average diameter of 8.34 μm. It indicates that without mists sprayed on the flue gases, the particles obtained are in the range of 1.5 μm to 16.99 μm, and the particles are approximately evenly distributed.

3.2 Analysis Without Mists

The surface morphology and microstructure of the samples were obtained using SEM–EDX. The different elements associated with the specimen were identified. The atomic percentage of each element present was recorded to determine the major element that composes the samples. Figure 4 is a standard EDX spectrum recorded

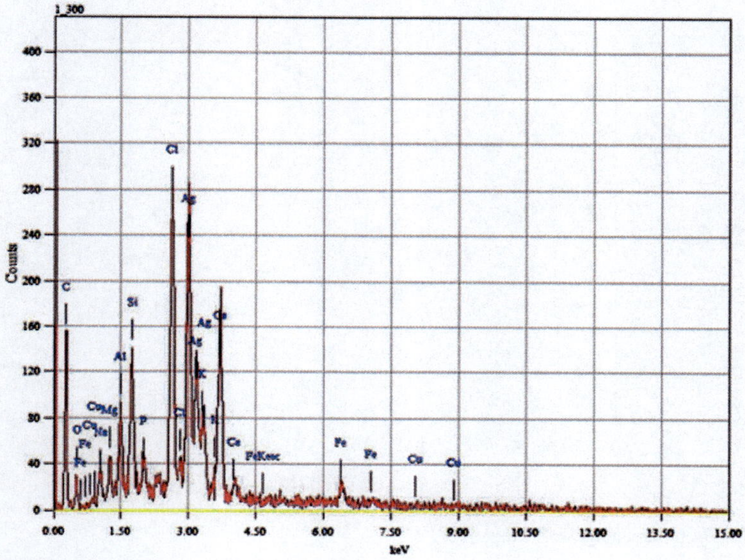

Fig. 4 Result of SEM–EDX analysis without mist

on the examined sample shown in Fig. 3, in which particles have not been treated with mists.

In the left portion of the presented spectrum, one can see four peaks located between 0 and 4.5 kV. The highest element, located on the left part of the spectrum at 0.2 kV, comes from carbon. The hardly visible maximum element, located at 0.5 keV, is connected with the oxygen characteristic line.

The results indicate that the major elements are mainly composed of carbon, oxygen, silver, chlorine, calcium, silicon, and aluminum and small quantities of sodium, magnesium, potassium, phosphorus, iron, and copper. Carbon, whose percentage is almost 75% of the total, inherited its elemental share from chemical compounds CO_2 and CO of the product of combustion. Silver is assumed to have been generated from the silver conductive tabs wherein the sample was attached. The silver tab was mounted on an aluminum stud. The source of aluminum is thought to be the experimental stage used for the SEM–EDX analysis. Calcium, magnesium, and potassium are the primary elemental constituents of ash. The detection of potassium is an indication of the presence of alkali metal in the flue gas. Potassium components play a significant role in deposit formation. It acts as a glue, bonding the individual fly ash particles together [21].

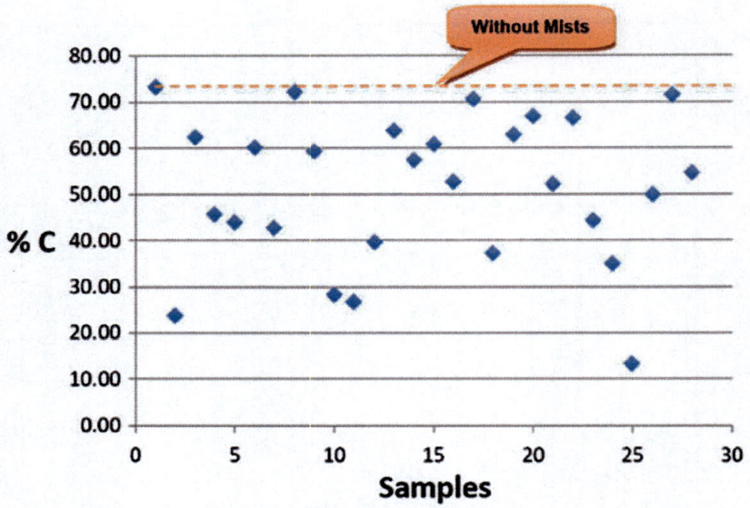

Fig. 5 Carbon percentages of samples 2–28

3.3 *Analysis with Mists*

Figure 5 shows the graph representing the percentage carbon present in sample #1 to sample #28.

The dashed line represents the percentage carbon of sample 1 where the mists was not applied. It is interesting to note that carbon was reduced in almost all samples when the mists was applied in the AC. The biggest reduction is in sample 25, where the sampling location was 180 cm from the entrance of AC.

3.4 *Effect of Concurrent Flow Orientation*

The sampling of PM was performed at three different locations identified as XP1, XP2, and XP3, corresponding to the distance measured from the entrance of AC which is 120 cm, 150 cm, and 180 cm, respectively. A graph in Fig. 6 will show the particle diameters for the concurrent flow of varying flue gas velocities at three different AC locations.

It can be seen that as the AC length increase particle size also increase. There is a little significant increase of particle size when damper opening is between one-third and two-thirds at locations XP1 and XP2. The graph indicates an abrupt particle size increase when damper opening is between two-thirds and fully open at three sampling locations. It means that Vel_3 should be the minimum flue gas velocity to produce a large particle size considering pump operating conditions and nozzle design as constants. However, at XP3, a fully-open damper can produce the highest

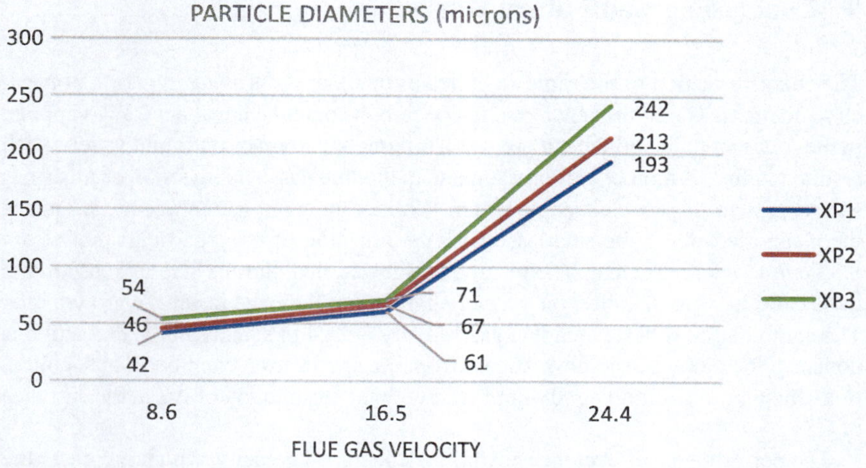

Fig. 6 Particle diameters of concurrent flow in three locations

particle diameter of 241.89 μm for concurrent flow. For concurrent flow, the sampling locations are insignificant to the growth of particles corresponding to Vel_1 and Vel_2. However, at Vel_3, it is evident that there is a significant difference in the growth of particles based on the sampling locations.

3.5 *Cyclone Separator Efficiency*

Measurement of particle concentrations was done using the sampling train. Samples were extracted both upstream and downstream of the cyclone separator. The mass of particles collected upstream of CS without AC was 201.56 mg. The mass of particles collected downstream of CS without AC was 156.48 mg. The efficiency was computed and found to be 22.4%. It means that 45.08 mg of PM were collected in the cyclone separator.

Another test was made when AC was installed at the entrance of CS. Mass of PM collected without mists was 508 mg, and mass of PM collected with mists was 90 mg. The efficiency was computed and found to be 82.3%. The system successfully contained 418 mg of a PM that could have been emitted to the atmosphere. There was a significant improvement in its performance when the mists were applied to the cyclonic separation system.

4 Conclusions and Future Works

The characterization of the samples indicates that there was an aggregation of particles and mists. The significant reduction of carbon content when the mist was applied in the AC is an indication that there was a mixing of carbon dioxide and water which resulted in the dilution of carbon elements in the flue gas. The injection of mists has shown growth of particle size diameter. Without the aggregation agent, i.e., mists, the mean diameter is the smallest. After the injection of mist, particles grew from 8.34 μm to a larger extent. The result also showed that particle size increased with increasing flue gas velocity; the greater aggregation flow, the faster the growth rate. The combination is best when flue gas velocity is 24.4 m/s, and sample extraction is done at a location 180-cm downstream from the aggregation chamber. Thus, collection efficiency was significantly improved with the installation of AC at the entrance of CS.

The performance of cyclone separators depends on velocity, which increases the collection efficiency, as confirmed by many studies. Therefore, the present study can be extended to a comprehensive analysis under high inlet velocities with different nozzle orientations. A cross-flow and counter-flow can also be recommended for future development.

Acknowledgements This research is funded by the Commission of Higher Education (CHED) and the University of Mindanao (UM). My grateful appreciation also goes to DENR-Environmental Management Bureau XI, Air Pollution Division, for the utilization of the Sampling Train equipment.

References

1. The EPA website. [Online]. Available: https://www.epa.gov/pm-pollution, last accessed 2019/11/21
2. The EPA website. [Online]. Available: https://www.epa.gov/environmental-topics/health-topics, last accessed 2020/5/16
3. US Environmental Protection Agency (EPA) (2020) Policy assessment for the review of the particulate matter national ambient air quality standards
4. Kreyling WG, Semmler-Behnke M, Moller W (2006) Ultrafine particle-lung interactions: does size matter? J Aerosol Med 19(1):74–83
5. Seaton A, MacNee W, Donaldson K, Godden D (1995) Particulate air pollution and acute health effects. Lancet (British Edition) 345(8943):176–178
6. Qihong D, Linjing D, Yuguo L (2019) Particle deposition in the human lung: health implications of particulate matter from different sources. J Environ Res 169:237–245
7. Knaapen AM, Borm PJ, Schins R (2004) Inhaled particles and lung cancer. Part A: mechanisms. Int J Cancer 109(6):799–809
8. Kim JC, Lee KW (1990) Experimental study of particle collection by small cyclones. Aerosol Sci Technol 12:1003–1015
9. Fankun W, Chao H, Chengliang J (2011) Experimental study on cyclone separator. In: 2nd international conference on mechanic automation and control engineering, MACE proceedings
10. Cameron IT et al (2005) Process systems modeling and applications in granulation: a review. Chem Eng Sci 3723–3750

11. Fengxian F, Sihong Z, Mingxu S (2019) Numerical investigation of PM2.5 size enlargement by heterogeneous condensation for particulate abatement. J Process Saf Environ Prot 125:197–206
12. Hapgood KP, Litster J, Smith R (2003) Nucleation regime map for liquid bound granules. AIChE J 49(2):350–361
13. Iveson SM et al (2001) Nucleation, growth and breakage phenomena in agitated wet granulation processes: a review. Powder Technol 117:3–39
14. Karagoz I, Avci A, Sendogan O (2013) Design and performance evaluation of a new cyclone separator. J Aerosol Sci 59:57–64
15. Basu P (2006) Combustion and gasification in fluidized beds. CRC Press, Boca Raton, FL
16. Patel B (2012) Biomass characterization and its use as solid fuel for combustion. Iranica J Energy Environ 3(2):123–128
17. Demirbas A (2004) Combustion characteristics of different biomass fuels. Prog Energy Combust Sci 30(2):219–230
18. Al-Shemmeri T, Yedla R, Wardle D (2015) Thermal characteristics of various biomass fuels in a small-scale biomass combustor. Appl Therm Eng 85:243–251
19. Cui C, Cai W, Chen H (2018) Airflow measurements using averaging pitot tube under restricted. Build Environ 139:17–26
20. Fan H, Wu D (2021) Numerical simulation of the flow characteristics of a flue gas damper. J Xi'an Jiaotong Univ
21. Egerton RF (2005) Physical principles of electron microscopy: an introduction to TEM, SEM, and AEM. Springerlink, 202

Soil Characteristic Study to Improve Heat Conductivity Capability in Ground Heat Exchanger

A. M. Aizzuddin, A. A. Asrudin, T. M. Yusof, and W. H. Azmi

Abstract The use of vapor compression air conditioning has contributed to the global warming effect by increasing greenhouse gas emissions. Renewable energy from geothermal sources, specifically ground heat exchangers (GHE), has great potential in building applications. The underlying concept of pipe called GHE utilises the ground as an unlimited thermal reservoir for cooling and heating purposes. Because of the temperature differences between underground and surrounding air, the air in the underground cools in the summer and warms in the winter. Thermal conductivity of the ground or soil is among a parameter that contribute to the GHE's performance. Therefore, the purpose of this research is to investigate the effect of hybrid soils without moisture on the performance of the GHE system. The hybrid soils consist of two elements, which are native soil with three grain sizes and bentonite. The native soil grain sizes are 0.154–0355 mm, 0.355–0.6 mm, and 0.6–1 mm. Bentonite has been introduced into all native soil grain sizes, which ranges from 0 to 100%. The native soil and bentonite were mixed consistently, and the thermal conductivity was measured by using a thermal property analyzer device. The study shows that the grain size 0.6–1 mm of native soil has the highest thermal conductivity at 20% bentonite, which is 0.269 W/m K compared to other grain sizes. The performance of the GHE system was evaluated based on simulation of mathematical model which shows that pipe length of 16 m gives significant effect of temperature reduction. In short, the performance of GHE has increased once the thermal conductivity of hybrid soil increased.

Keywords GHE performance · Hybrid soil · Thermal conductivity · Bentonite

A. M. Aizzuddin · A. A. Asrudin · T. M. Yusof (✉) · W. H. Azmi
College of Engineering, Universiti Malaysia Pahang (UMP), Lebuhraya Tun Razak, 26300 Gambang, Kuantan, Pahang, Malaysia
e-mail: myusof@ump.edu.my

M. Y. Ismail et al. (eds.), *Technological Advancement in Mechanical and Automotive Engineering*, Lecture Notes in Mechanical Engineering,
https://doi.org/10.1007/978-981-19-1457-7_54

1 Introduction

The use of energy is increasing day by day around the world, and the amount of fossil fuel is running out eventually and will not be ample to fulfill the demands of global energy in the future. Out of all the energy demand, one-third of it comes from the cooling and heating space [1]. Some of their time indoors, some of them may work from home, and some of them may shop via window shopping. Because of these factors, the need for building and residential thermal comfort is determined by their current requirements. Air conditioning, ventilation, and heating systems account for up to 65% of total energy consumption in residual buildings around the world [2]. In the service, industrial, residential and engineering sectors, heating and cooling have high demand and are responsible for a total of 50% energy demand of the European Union (EU28) in 2015 [3]. For the residential sector, the demands for heating and cooling systems can reach up to 75%, while space heating demands over 53% [4]. For the past years, the development and technology of the GHE have been used in modern countries, especially in Europe, due to its capabilities to provide cooling and heating, especially during the winter and summer seasons. GHE is essential as the medium for cooling and heating of buildings. The performance of the GHE depends on a few parameters, especially the soil or the ground. The characteristics of the soil also depend on the surroundings since it is a spongy medium type [5]. Most of the pipes that are being installed in GHE are positioned horizontally or vertically. Horizontal pipes have advantages of being less expensive to install and more efficient at absorbing ambient air to avoid ground heat deficits [6]. There are few types of GHE and the general one is the open system, as shown in Fig. 1, closed system and miscellaneous system [7]. In open system GHE, ambient air flows through the pipe buried in the ground for pre-cooling or pre-heating. Therefore, the air will be cooled or heated by using natural process of air conditioning before entering the residence house or building.

Basically, the concept of cooling or heating in GHE is taken by extracting the ambient air as the working fluid, whereby heat transferring process occurred during

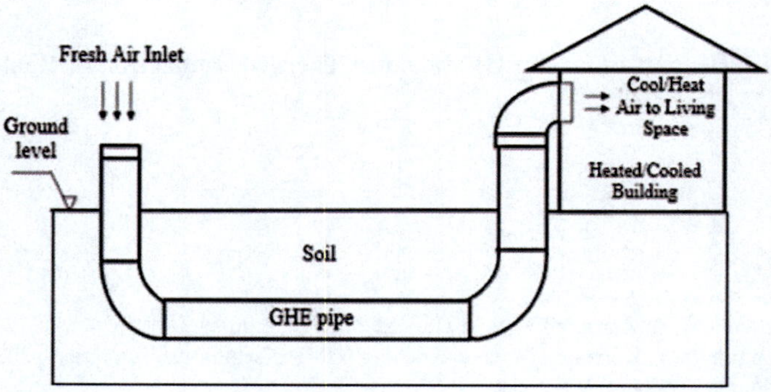

Fig. 1 Basic principle for open system GHE [8]

the air flows through the exchanger and supply the cool or hot air to the buildings. There are numerous report regarding on concept, investigation and real practice of the GHE have been reported with reliable data. The capability of the GHE mostly depends on ground temperature at the proposed geographical location. Previous study conducted by Florides and Kalogirou [7] showed that the temperature of the ground below at a certain depth remains almost constant throughout the year. This study was obtained that the constant ground temperature as low as 21.3 °C. This is because the surface of the ground has variations in temperature that decline as the ground depth increases due to the high thermal inertia of the soil. The actual temperature of the ground was measured by using a drilled borehole in Nicosia, Cyprus. It showed that the temperature is almost constant throughout the year at a depth of 5 m and below [7]. In comparison to the ambient temperature, this constant temperature of underground soil is significantly higher in the winter and lower in the summer [9, 10]. As a result, soil can act as a heat sink in the summer and a heat source in the winter [11].

Generally, pipe length corresponds to the area of the GHE between the ground layer and the working fluid. Different shapes of underground heat exchangers have different pipe lengths. In addition, the length of GHE pipe is one of the main factors that can affect the initial cost of the GHE system [12]. The length of the pipe can increase the area of contact between the soil and the pipe, which increasing in the rate of heat exchanger. This concept was validated by Noorollahi et al. [13]. Whereby, the authors reported that they were increased the length of the pipe and they found that the number of heat pumps used has reduced the usage of electrical energy. However, long trenches must be excavated to install the pipes, which tends to increase the initial investment. Numerous researchers found that the performance of GHE systems is highly depends on the rate of heat transfer between the air and the soil. The rate of heat transfer between air and soil is determined primarily by the thermal properties of the soil. Thus, the length of pipe required for the GHE system can be significantly reduced by burying the pipes in soil with a high thermal conductivity close to them [14]. The thermal conductivity of soil can also be improved by adding certain additives such as bentonite, quartz, and metal particles [15]. A research done by Zhang et al. [16] showed that by introducing bentonite into main soil, it can increase the thermal conductivity of the sand-bentonite mixtures. Wang et al. [17] discovered that a sand-bentonite mixture used as a backfilling material has increased heat transfer rate by up to 31% when compared to standard sand-clay material. Cuny et al. [18] tested three different types of coating soils for different sections of the buried GHE pipe, including in-situ earth, sand, and a mixture of sand-bentonite, and found that the performance of the GHE system with the sand-bentonite coating was the best.

From the best knowledge of authors, the previous studies for thermal conductivity of soil for GHE application were not extensively covered in terms of performance increment and related to the soil thermal conductivity. Basically, the GHE performance has been related to the thermal conductivity of soil and pipe materials. However, pipe materials have no significant effect on the performance of the GHE system [19, 20]. Therefore, hybrid soil between native soil and bentonite has been investigated in this study. The hybrid soil is tested without moisture condition at room temperature. In this experiment, there are three different range of grain sizes of native soil, which are 0.154–0.355 mm, 0.355–0.6 mm and 0.6–1 mm. As a result, the highest thermal conductivity of hybrid soil and bentonite has been obtained and the GHE performance has been investigated based on simulation of mathematical model.

2 Methodology

The purpose of this study was to investigate the effect of hybrid soil on GHE performance. The performance and efficiency of GHE are mainly depends on soil thermal conductivity in the region of the GHE pipe. Thus, the hybrid soil was investigated in order to improve the thermal conductivity. The hybrid soil in this study is composed of two types of soil: native soil as the primary soil and bentonite as the backfilling material. Native soil was used as the primary soil in this experiment, with grain sizes ranging from 0.154–0.355 mm, 0.355–0.6 mm and 0.6–1 mm. The native soil contains grain with varying sizes and manual sieve method was used to obtain correct categories of grain sizes as shown in Fig. 2. In order to categories the grain size of native soil, four sieves with different sizes were used: 0.154 mm, 0.355 mm, 0.6 mm and 1 mm. The grain size range from 0.154 to 0.355 mm was obtained using a sieve with a size of 0.154 mm and 0.355 mm, while the grain size range from 0.355 to 0.6 mm was obtained using a sieve with a size of 0.355 and 0.6 mm, and the grain size range from 0.6 to 1 mm was obtained using a sieve with a size of 0.6 mm and 1 mm. The three grain sizes of native soil are represented in Fig. 3. After classifying the grain size into three ranges, each range was mixed with bentonite at a composition ratio of 0–100% by mass with 10% increment.

Meanwhile, bentonite was used as backfilling material in this experiment due to its characteristics. Although bentonite has low thermal conductivity, it has an excellent backfilling material capability for improving thermal conductivity [21]. The characteristics of bentonite, which can fill up the pores surrounded by native soil and influencing the heat conduction capability of the entire mixture, and increasing

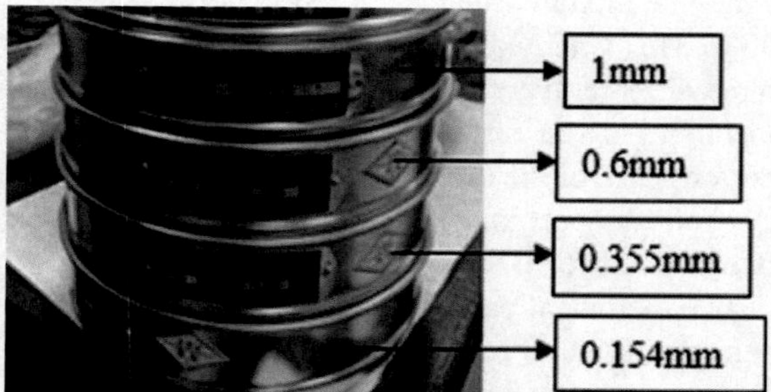

Fig. 2 Sieve in four different range

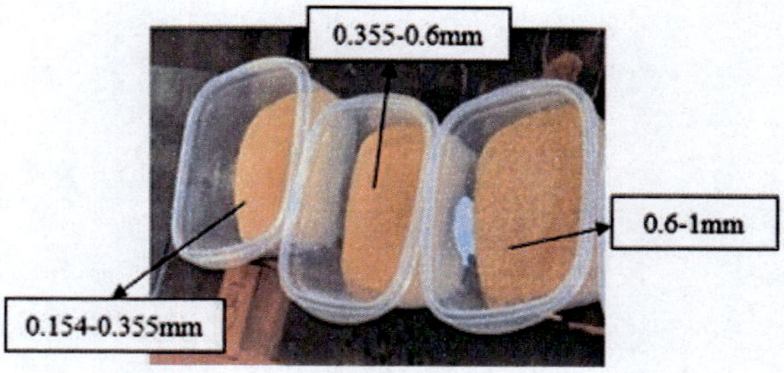

Fig. 3 Native soils in three different range

thermal conductivity. However, in this study, both native soil and bentonite were tested without moisture conditions at room temperature.

For the preparation of hybrid soil which consists of native soil for grain size of 0.154–0.355 mm and bentonite mixtures, both native soil and bentonite were put into a tray as shown in Fig. 4a and placed into an oven at 105 °C of temperature. After 12 h of drying, the samples were cooled to room temperature [17]. Then, native soil and bentonite were mixed properly at various percentages of bentonite by mass, including 0%, 10%, 20%, 30%, 40%, 50%, 60%, 70%, 80%, 90%, and 100% as shown in Table 1. Each sample was placed in a bottle with a total mass of 60 g. Figure 4b shows a sample of hybrid soil composed of native soil and bentonite.

In this investigation, the thermal conductivity of each sample was measured using a KD2 Pro Thermal Properties Analyzer, as shown in Fig. 5. The needle of the KD2 Pro Thermal Properties Analyzer was dipped into the hybrid soils and left for about two minutes in vertical position as shown in Fig. 5 to obtain the thermal conductivity value, which was recorded. This procedure will be repeated three times for each sample to obtain the average of thermal conductivity for each sample. Precautions are necessary during this process, as the needles are extremely sensitive and delicate. The needles should be dipped slowly into the hybrid soils until completely covered, as some of the samples have a large grain size, which may damage the needles. Then, the same procedure was repeated for grain sizes 0.355–0.6 mm and 0.6–1 mm.

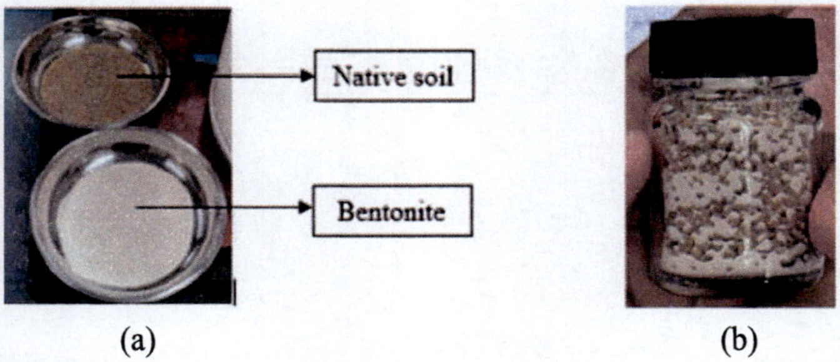

Fig. 4 Native soil and bentonite. **a** Preparation to place in oven and **b** samples of hybrid soil between native soil and bentonite

Table 1 List of samples of hybrid soil between native soil and bentonite

Samples	Grain size 0.154–0.355 mm (%)	Bentonite (%)	Samples	Grain size 0.355–0.6 mm (%)	Bentonite (%)	Samples	Grain size 0.6–1 mm (%)	Bentonite (%)
1	100	0	12	100	0	23	100	0
2	90	10	13	90	10	24	90	10
3	80	20	14	80	20	25	80	20
4	70	30	15	70	30	26	70	30
5	60	40	16	60	40	27	60	40
6	50	50	17	50	50	28	50	50
7	40	60	18	40	60	29	40	60
8	30	70	19	30	70	30	30	70
9	20	80	20	20	80	31	20	80
10	10	90	21	10	90	32	10	90
11	0	100	22	0	100	33	0	100

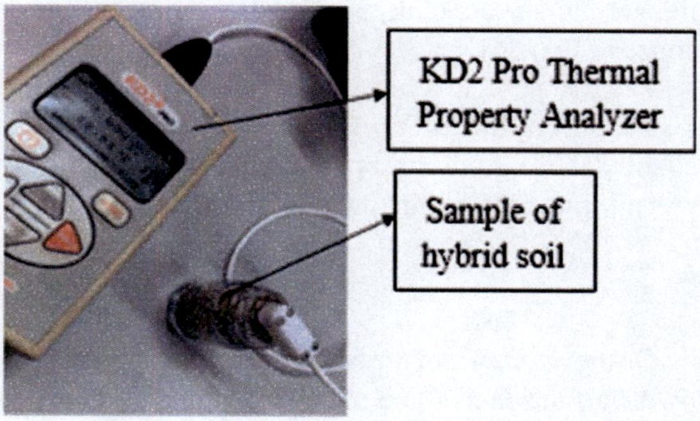

Fig. 5 Measurement of thermal conductivity by using KD2 pro thermal property analyzer

The investigation of the effect of hybrid soil thermal conductivity on GHE performance is conducted using simulation of mathematical model. Simulation of the GHE system is an interesting method for predicting and analysing the GHE performance based on available data and input parameters. It takes less time to analyse performance of the system, and it can be repeated with different input settings. The mathematical model of heat transfer from the air inside the GHE pipe to the surrounding ground can be analyzed using schematic diagram as shown in Fig. 6 [22]. The figure illustrates the temperature of the air (T_{air}) entering the pipe and heat will be transferred to the surrounding ground as the air flows inside the pipe at a constant ground temperature ($T_{z,t}$). Several assumptions have been made in this case: (i) the ground surrounding the pipe has an infinite heat capacity rate or is treated as a thermal reservoir; and (ii) the presence of a pipe in the area around the pipe has no effect on the temperature profile. Thus, the temperature at the surface is constant in the y direction, (iii) the pipe has a uniform cross sectional area, (iv) the air convection flow develops hydrodynamically and thermally within the pipe, and (v) the thermal properties of

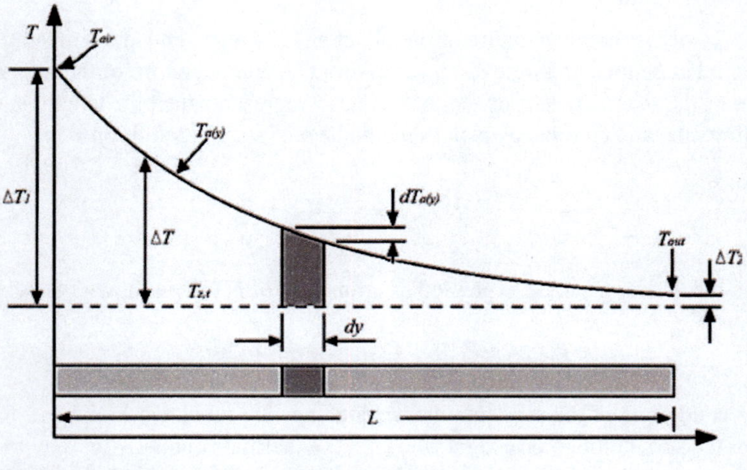

Fig. 6 Schematic of heat transfer process in GHE pipe [22]

the ground surrounding the pipe are homogeneous and its thermal conductivity is constant [23, 24].

The hot ambient air enters the pipe at T_{air}, which decreases at $T_{a(y)}$ as the pipe's distance from the inlet (y) increases. The hot air leaves the pipe at T_{out}, which is slightly above ground temperature, $T_{z,t}$. Thus, Eq. 1 shall be used to express the heat transfer rate received by the ground for any length in the differential section of d_y:

$$dQ = U\pi D\left[T_{air(y)} - T_{z,t}\right]dy \tag{1}$$

On the other side of the heat transfer equation, Eq. 2 shows the heat rate transmitted by the air inside the pipe in the differential section of the d_y [9]:

$$dQ = -\dot{m}_{air}c_p\left[dT_{air(y)}\right] \tag{2}$$

The temperature change of air is a negative quantity, implying that as the pipe length increases, the temperature decreases. A negative sign is therefore added to Eq. 2, resulting in a positive heat transfer rate. The heat transfer between the earth and the air within the pipe is theoretically equal to the heat loss as the air flows through it. As a result, Eqs. 1 and 2 are equated to produce Eq. 3 as shown below:

$$U\pi D\left[T_{air(y)} - T_{z,t}\right]dy = -\dot{m}_{air}c_p\left[dT_{air(y)}\right] \tag{3}$$

By integrating both sides and rearranging Eq. 3 to solve $T_{a(y)}$ yields:

$$T_{air(y),L} = T_{z,t} + \left(T_{air(y),0} - T_{z,t}\right)\left(e^{-UA_{sur}/\dot{m}_{air}c_p}\right) \tag{4}$$

As a result, Eq. 4 can be written as Eq. 5 for $y = L$.

$$T_{air.out} = T_{z,t} + \left(T_{air.in} - T_{z,t}\right)\left(e^{-UA_{sur}/\dot{m}_{air}c_p}\right) \tag{5}$$

where T_{air} out is the temperature of the outlet air, $T_{z,t}$ represents the temperature of the ground at depth z and time t, $T_{air,in}$ represents the temperature of the inlet air, U represents the overall heat transfer coefficient, y represents the pipe length, $\dot{m}$ is the mass flowrate, and cp represents the specific heat of air. By definition:

$$NTU = \frac{UA_{sur}}{\dot{m}_{air}c_p} \tag{6}$$

As a result, Eq. 5 can be expressed as a function of NTU, as shown below:

$$T_{air.out} = T_{z,t} + \left(T_{air.in} - T_{z,t}\right)e^{-NTU} \tag{7}$$

The conditions of the pipe and its surroundings are tabulated in Table 2 for the input of the simulation to investigate the effect of thermal conductivity of hybrid soil between native soil and bentonite on GHE performance. In the simulation, polyvinyl

Table 2 Conditions of pipe and its surrounding for the input of the simulation [20]

Input name	Value	Units
Thermal conductivity of PVC pipe, k_p	0.18	W/m K
Outer diameter of pipe, OD	114.3	mm
Inner diameter, ID	101.8	mm
Pipe length, L	25	m/s
Velocity of air, V	10.73	m/s
Flowrate, $\dot{m}$	0.1	kg/s
Thermal diffusivity, α	0.046	m^2/day
Ground temperature, $T_{z,t}$	24	°C
Air inlet temperature, $T_{air.in}$	35	°C

chloride (PVC) pipe with a thermal conductivity of 0.18 W/m K was applied. The ground temperature applied is based on a depth of 2 m below the ground surface [25–27]. Furthermore, in this simulation, the outer and inner diameters of the pipe, pipe length, air velocity, flowrate, thermal diffusivity, ground temperature, and air inlet temperature are all set to be constant.

3 Results and Discussion

The native soil has been mixed with bentonite in three grain sizes: 0.154–0.355 mm, 0.355–0.6 mm and 0.6–1 mm. Then each grain size of native soil and bentonite was fully mixed with different percentages of bentonite by mass, ranging from 0 to 100% with an increment of 10%. The KD2 Pro Thermal Analyzer was used to measure the thermal conductivity of each sample. Figure 7 shows the thermal conductivity of hybrid soil for 0.154–0.355 mm, 0.355–0.6 mm and 0.6–1 mm grain size of native soil and different percentage of bentonite. For grain size of 0.154–0.355 mm, the value of thermal conductivity at 0% bentonite is 0.189 W/m K. The thermal conductivity increased to 0.204 W/m K when 10% bentonite was added. The increment in thermal conductivity of bentonite from 0 to 10% is approximately 7.35%. Meanwhile, the thermal conductivity was decreased from 10 to 100% of bentonite, which is from 0.204 to 0.119 W/m K. Based to this graph, the presence of 10% bentonite as a backfilling material for grain size 0.154–0.355 mm of native soil can improve the thermal conductivity.

For the grain size ranging from 0.355–0.6 mm and 0% bentonite, the thermal conductivity of hybrid soil is 0.193 W/m K as shown in Fig. 7, which is relatively close to the thermal conductivity at 0% native soil with a grain size of 0.154–0.355 mm. The thermal conductivity of bentonite increases steadily from 0 to 20%, with a thermal conductivity of 0.249 W/m K, which is greater than the thermal conductivity of 0.154–0.355 mm grain size. The increment of thermal conductivity with presence of bentonite from 0 to 20% is about 22.49%. From the graph, the thermal conductivity

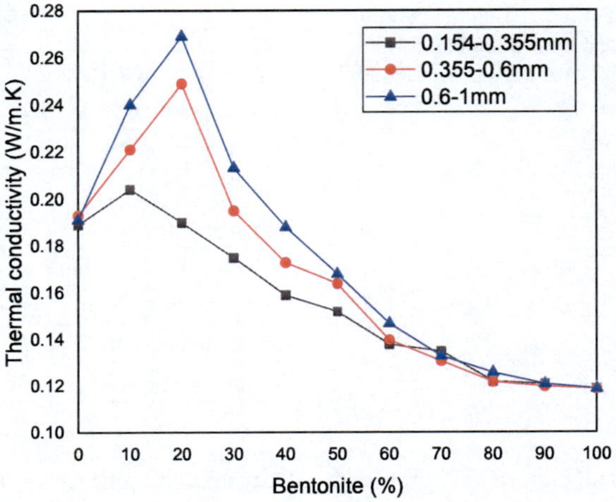

Fig. 7 Thermal conductivity of hybrid soil for 0.154–0.355 mm, 0.355–0.6 mm and 0.6–1 mm grain size of native soil

start to decrease from 20 to 100% of bentonite, which is from 0.249 to 0.119 W/m K. Thus, the presence of 20% bentonite as a backfilling material for grain size 0.355–0.6 mm of native soil can improve the thermal conductivity.

The graph also shows that the thermal conductivity of hybrid soil for grain size ranging from 0.6 to 1 mm with bentonite at 0% is 0.191 W/m K, which is almost the same as the thermal conductivity of native soil at 0% bentonite for grain sizes 0.154–0.355 mm and 0.355–0.6 mm. The thermal conductivity then increases steadily about 29% from 0 to 20% of bentonite with a thermal conductivity of 0.269 W/m K, which is greater than the thermal conductivity of native soil with a grain size of 0.154–0.355 mm and 0.355–0.6 mm. After that, the thermal conductivity start to decreased from 20 to 100% of bentonite, which is from 0.269 to 0.119 W/m K. Clearly, the presence of 20% bentonite as a backfilling material for grain size 0.6–0.1 mm of native soil can increase the thermal conductivity.

In comparison of thermal conductivity for all grain sizes of native soil, there is a significant difference in thermal conductivity between three grain sizes of native soil with bentonite percentages ranging from 10 to 40%. Furthermore, the thermal conductivity of the three grain sizes of native soil follows the same trend with the presence of bentonite from 10 to 100%. This trend is occurred because of when the proportion of bentonite increases, the amount of native soil decreases, and most native soil grains are surrounded by bentonite, which has a low thermal conductivity and thus weakens heat conduction between solid grains. Based on the graph, the highest thermal conductivity for 0.355–0.6 mm and 0.6–1 mm grain sizes is at 20% bentonite, while the highest thermal conductivity for 0.154–0.355 mm is at 10%. In addition, the grain size of 0.6–1 mm of native soil has a highest thermal conductivity of 0.269 W/m K when compared to the grain sizes 0.154–0.355 mm and 0.355–0.6 mm of native soil. The highest thermal conductivity for each grain size was used to investigate the effect of hybrid soil thermal conductivity on GHE performance, which were 0.204 W/m K, 0.249 W/m K, and 0.269 W/m K.

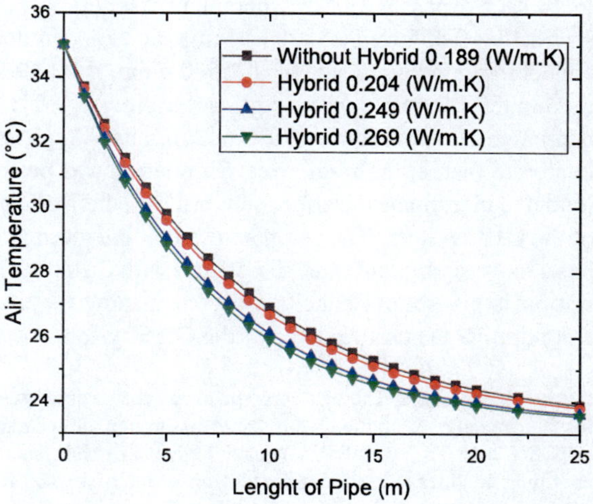

Fig. 8 Length of pipe against air temperature for three thermal conductivity

Simulation to analyse the performance of the GHE were conducted by analysing the air temperature inside the pipe by implementing value of thermal conductivity without hybrid which is 0.189 W/m K and three different thermal conductivity values for hybrid soil: 0.204 W/m K, 0.249 W/m K, and 0.269 W/m K. The air temperature data was simulated from 0 to 25 m of pipe length and plotted in a graph as shown in Fig. 8. The graph indicates that the thermal conductivity without hybrid and three thermal conductivity values of hybrid soil follow the same trend. The graph demonstrates that the air temperature gradually decreases from air inlet temperature of 35 °C to 25.06 °C, 24.88 °C, 24.48 °C and 24.34 °C of 0.189 W/m K, 0.204 W/m K, 0.249 W/m K and 0.269 W/m K respectively for 0–16 m of pipe length. Then, it remains nearly constant between 16 and 25 m of pipe length for all thermal conductivity. The thermal conductivity of 0.269 W/m K has better exit temperature which is at 24.34 °C compared to exit air temperature of 25.06 °C, 24.88 °C and 24.48 °C for thermal conductivity of 0.189 W/m K, 0.204 W/m K and 0.249 W/m K respectively. Therefore, thermal conductivity of 0.269 W/m K has a better effect on the GHE performance compared to thermal conductivity of 0.189 W/m K, 0.249 W/m K and 0.204 W/m K. Thus, it can be concluded that the highest the thermal conductivity of hybrid soil, the greater the impact on air temperature.

4 Conclusion

The hybrid soil was used in this study to investigate the performance of the GHE system using simulation. The hybrid soil is made up of two components: native soil as the main soil and bentonite as the backfilling material. There are three grain sizes of native soil, which are 0.154–0.355 mm, 0.355–0.6 mm and 0.6–1 mm. All three grain sizes of native soil were fully mixed with a percentage of bentonite ranging from 0 to 100% by mass with 10% increment. The highest thermal conductivity for grain size 0.154–0.355 mm is 0.204 W/m K at 10% bentonite, while the highest thermal conductivity for grain size of 0.355–0.6 mm is 0.249 W/m K at 20% bentonite, and the highest thermal conductivity for grain size 0.6–1 mm is 0.269 W/m K at 20% bentonite. However, grain size 0.6–1 mm has a highest thermal conductivity at 20% bentonite than other grain sizes. Simulation was performed by using mathematical model to investigate the effect of hybrid soil thermal conductivity on the performance of the GHE system. The result shows that the thermal conductivity of 0.269 W/m K has a more significant effect compared with 0.204 W/m K and 0.249 W/m K. Therefore, it is possible to conclude that increasing the thermal conductivity of the soil can improve the performance of the GHE system.

Acknowledgements The authors thanked the Ministry of Higher Education (MOHE) for financially supported this study through Fundamental Research Grant Scheme (FRGS/1/2018/TK03/UMP/03/4) and Universiti Malaysia Pahang (www.ump.edu.my) for providing facilities to conduct this study under research grant RDU190143 and PGRS2003208.

References

1. Kumar Agrawal K et al (2019) Effect of soil moisture contents on thermal performance of earth-air-pipe heat exchanger for winter heating in arid climate: in situ measurement. Geothermics 77:12–23
2. Sedaghat A, Habibi M, Hakkaki-Fard A (2020) A novel ground thermal recovery system for horizontal ground heat exchangers in a hot climate. Energy Convers Manage 224:113350
3. Fleiter T et al (2017) Profile of heating and cooling demand in 2015. Heat Roadmap Europe Deliverable 3(1)
4. Lee M et al (2021) Comparative heating performance evaluation of hybrid ground-source heat pumps using serial and parallel configurations with the application of ground heat exchanger. Energy Convers Manage 229:113743
5. Meng X et al (2021) Studies on the performance of ground source heat pump affected by soil freezing under groundwater seepage. J Build Eng 33:101632
6. Neupauer K, Pater S, Kupiec K (2018) Study of ground heat exchangers in the form of parallel horizontal pipes embedded in the ground. Energies 11(3):491
7. Florides G, Kalogirou S (2007) Ground heat exchangers—a review of systems, models and applications. Renew Energy 32(15):2461–2478
8. Ozgener L (2011) A review on the experimental and analytical analysis of earth to air heat exchanger (EAHE) systems in Turkey. Renew Sustain Energy Rev 15(9):4483–4490
9. Naili N et al (2015) Energy and exergy analysis of horizontal ground heat exchanger for hot climatic condition of northern Tunisia. Geothermics 53:270–280
10. De Paepe M, Janssens A (2003) Thermo-hydraulic design of earth-air heat exchangers. Energy Build 35(4):389–397
11. Yang D et al (2019) A demand-oriented approach for integrating earth-to-air heat exchangers into buildings for achieving year-round indoor thermal comfort. Energy Convers Manage 182:95–107
12. Noorollahi Y, Arjenaki HG, Ghasempour R (2017) Thermo-economic modeling and GIS-based spatial data analysis of ground source heat pump systems for regional shallow geothermal mapping. Renew Sustain Energy Rev 72:648–660

13. Noorollahi Y et al (2016) Numerical modeling and economic analysis of a ground source heat pump for supplying energy for a greenhouse in Alborz province, Iran. J Clean Prod 131:145–154
14. Agrawal KK et al (2018) A review on effect of geometrical, flow and soil properties on the performance of earth air tunnel heat exchanger. Energy Build 176:120–138
15. Agrawal KK, Misra R, Agrawal GD (2020) Improving the thermal performance of ground air heat exchanger system using sand-bentonite (in dry and wet condition) as backfilling material. Renew Energy 146:2008–2023
16. Zhang W et al (2015) Investigation on influential factors of engineering design of geothermal heat exchangers. Appl Therm Eng 84:310–319
17. Wang H, Cui Y, Qi C (2013) Effects of sand–bentonite backfill materials on the thermal performance of borehole heat exchangers. Heat Transfer Eng 34(1):37–44
18. Cuny M et al (2018) Influence of coating soil types on the energy of earth-air heat exchanger. Energy Build 158:1000–1012
19. Bansal V et al (2010) Performance analysis of earth–pipe–air heat exchanger for summer cooling. Energy Build 42(5):645–648
20. Aizzuddin A, Yusof T, Azmi W (2021) Investigation of pipe materials and thermal conductivity of soil on the performance of ground heat exchanger operating under Malaysia climate. In: IOP conference series: materials science and engineering. IOP Publishing
21. Omer AM (2016) Experimental investigation of the performance of a ground source heat pump system for buildings heating and cooling. Int J Innov Math Stat Energy Policies 4(1):10–44
22. Derbel HBJ, Kanoun O (2010) Investigation of the ground thermal potential in Tunisia focused towards heating and cooling applications. Appl Therm Eng 30(10):1091–1100
23. Mongkon S et al (2014) Cooling performance assessment of horizontal earth tube system and effect on planting in tropical greenhouse. Energy Convers Manage 78:225–236
24. Benhammou M, Draoui B (2015) Parametric study on thermal performance of earth-to-air heat exchanger used for cooling of buildings. Renew Sustain Energy Rev 44:348–355
25. Yusof TM, Anuar S, Ibrahim H (2014) Study of effect of thermal diffusivity on ground temperature for Malaysian climate. In: Applied mechanics and materials. Trans Tech Publications
26. Yusof T, Anuar S, Ibrahim H (2014) Numerical investigation of ground cooling potential for Malaysian climate. Int J Autom Mech Eng 10
27. Yusof T, Anuar S, Ibrahim H (2015) A review of ground heat exchangers for cooling application in the Malaysian climate. J Mech Eng Sci 1426–1439

Boiler Efficiency Analysis Using Direct and Indirect Method

Wan Mohd Fakhri Wan Zainus and Natrah Kamaruzaman

Abstract Boiler efficiency could be analyzed using two different methods, direct and indirect. The direct method provides the overall efficiency based on input and output while indirect method shows the losses in each process which is more valuable in identify the inefficient process. Therefore, in this study, the effect of different boiler load on the efficiency of the boiler is analyzed using both methods. In addition, the analysis using indirect method also shows the effect of boiler load on the losses in dry flue gas and moisture in air. Data of the temperature, pressure and flow rate of the steam, flue gas, air and fuel are measured onsite. The flue gas analysis was performed to obtain the percentage of flue gas components. Direct efficiency and indirect efficiency based on heat losses at different points were calculated. From the study, it is found that, the efficiency of the boiler increased with the increment of boiler load and reach the maximum efficiency of 83.3% for direct method and 89.8% for indirect method at 55% load. Operating above 55% load will results in the decrement of boiler efficiency for both direct and indirect method. As the load increase, the amount of excess air is increasing which increase the combustion efficiency. However, as the excess air quantity increased further, the heat loss due to the excess air is higher than the heat input by the combustion therefore decreased the efficiency of the boiler. This result could give a clue to the boiler operator to tune the excess air rate when operating the boiler more than 55% load.

Keywords Boiler · Efficiency · Direct and Indirect Method

1 Introduction

Boilers energy losses is one of the problems faced by boiler operators in many industries as this process involved a phase changed process. In order to reduce the losses, many researchers tried to find the suitable solution in between the process to increase the efficiency of the boilers [1–4]. Increasing boiler efficiency will reduce

W. Mohd Fakhri Wan Zainus · N. Kamaruzaman (✉)
Universiti Teknologi Malaysia, 81310 Skudai, Johor, Malaysia
e-mail: natrahk@utm.my

M. Y. Ismail et al. (eds.), *Technological Advancement in Mechanical and Automotive Engineering*, Lecture Notes in Mechanical Engineering,
https://doi.org/10.1007/978-981-19-1457-7_55

the fuel consumption and lower the flue gas emission. Boiler's efficiency could be determine using two methods. The simplest method is called direct method or also known as the 'input-output' method [5]. It relates the steam output and the heat input for determining the efficiency [6]. Another method is called indirect method which is also known as the heat loss method. Within this method, the measurement of efficiency is done by subtracting the fraction of heat loss coming from 100. ASME PTC 4-2013 is the standard for the indirect method of measuring boiler production. In the efficiency determination process, the criteria for indirect method would not include a blowdown loss. Therefore, the efficiency determine by the indirect method is normally higher than the direct method. However, indirect method is one of the best ways to determine the inefficient of the process within the boiler operations [7]. It shows which loss is high and could possibly influence the efficiency of the boilers.

Study done by Retirado-Mediaceja et al. [8] shows than the difference of the performance between these two methods is around 2.7% and the main heat loss is coming from the flue gas. Based on the study of exergy analysis in boiler by Behbahaninia et al. [9], the second largest heat loss was the loss due to the dry flue gas leaving the boiler. The main reason for these losses is the excess air supplied to achieve complete combustion process. Saidur et al. [10] found that by controlling the excess air, the rate of heat transfer could be increasing, and the combustion efficiency could be increased. However, study by Erbas [7] for steam drum boiler found that the main losses is cause by the conversion of hydrogen gas in the fuel to water after combustion. In the study by Luo Chao et al. [11] they discovered that the boiler efficiency could be increased with the increment of exhaust gas temperature.

Based on the previous research, boiler efficiency is also changing with the change of boiler load [12–14]. Therefore, the design of the boilers is normally specific for a certain load operation. Operating boiler at other load beside the optimum one would resulting an inefficient process and increase the operating cost. However, adjusting the certain parameters during non-optimum load operation could help reduce the losses in the process [15].

The purpose of this study is to determine the efficiency of the boiler in Petronas Chemical MTBE based on their operating load conditions. In addition, this study also intended to find the main losses in the process for the plant and optimum load operation with highest boiler's efficiency.

2 Methodology

The study is performed at Petronas Chemical MTBE Gebeng Kuantan, Malaysia. The following Fig. 1. shows the PC MTBE watertube boiler used in the study.

The efficiency calculation is divided into two different methods: direct and indirect efficiency as per following:

Fig. 1 PC MTBE Watertube boiler

2.1 Direct Method

The direct method determines the boiler efficiency from the fuel flow and the steam flow generated after the steam drum. The advantage of direct method is it could obtain the boiler efficiency in a short time by plant works however it does not provide clues to the operator as to why the efficiency of the system is lower. The following equations is used to calculate the efficiency.

$$\eta = \frac{Heat\,output}{Heat\,input} \times 100\% \tag{1}$$

or

$$\eta = \frac{Q \times (h_2 - h_1)}{(q \times GCV)} \times 100\% \tag{2}$$

where

Q = *Quantity of steam generated per hour* (kg/h)

q = *Quantity of fuel used per hour* (kg/h)

h_2 = *Enthalpy of steam at output of boiler* (kCal/kg)

h_1 = *Enthalpy of feed water* (kCal/kg)

GCV = *Gross Calorific Value* (kCal/kg)

2.2 Indirect Method

Indirect efficiency is also known as the measurement process for heat loss efficiency. Indirect efficiency of the boiler could be calculated from the following Eq. (3) based on ASME PTC 4–2013.

$$\text{Boiler Indirect Efficiency} = 100-(\text{i}+\text{ii}+\text{iii}+\text{iv}+\text{v}+\text{vi}+\text{vii}) \tag{3}$$

where,

(i) The percentage of heat loss due to dry flue gas.

$$\frac{m \times C_p \times (T_f - T_a)}{GCV} \times 100 \tag{4}$$

(ii) The percentage heat loss due to evaporation of water formed due to hydrogen in fuel.

$$\frac{9 \times H_2 \times [584 + C_p \times (T_f - T_a)]}{GCV} \times 100 \tag{5}$$

(iii) The percentage heat loss due to evaporation of moisture present in the fuel.

$$\frac{M \times [584 + C_p \times (T_f - T_a)]}{GCV} \times 100 \tag{6}$$

(iv) The percentage of heat loss due to moisture present in the air.

$$\frac{AAS \times humidity\,factor \times C_p \times (T_f - T_a)}{GCV} \times 100 \tag{7}$$

(v) The percentage of heat loss due to unburnt in fly ash.
These percentage will not be calculated because it is referred to as fuel gas boiler

(vi) The percentage of heat loss due to radiation and other unaccounted loss.
These percentage will not be calculated because it is referred to as fuel gas boiler.

(vii) Percentage heat loss due to radiation and other unaccounted loss.
The actual radiation and convection losses are difficult to assess because of the emissivity of various surfaces and the shape of the boiler, airflow pattern, etc. In general, it can be assumed between 1 and 2%.
where:
m = mass of dry flue gas in kg/kg of fuel, or
mass of flue gas = mass of actual air supplied + mass of fuel supplied
$C_{p,\,flue}$ = *Specific heat of flue gas*(0.962964 kJ/kg °C)

$C_{p,sh\ steam}$ – *Specific heat of superheated steam*(1.88406 kJ/kg °C)
T_f = *Flue gas temperature,*(°C)
T_a = *Ambient air temperature*(°C)
GCV = *Gross Calorific Value of the fuel (kJ/kg)*
H_2—*kg of* H_2 *in 1 kg of fuel*
M—kg of moisture of fuel
AAS = *Actual air supplied*

2.3 Data Collection

The data collection is performed onsite for difference boiler load setup. Fuel consumptions rate, feed water flow rate, steam flow rate, feed water temperature and pressure, steam and pressure temperature outlet, fuel specification and properties, excess air, and ambient temperature were the parameters measured.

The boiler should be operated at stable conditions and drains, vents, and blowdown valve must be closed before the measurement is performed. All the parameters were taken at the boiler and at the Distributed Control System (DCS) room. A steady-state operation shall be maintained for at least 30 min. Then, an average value during the operation will be defined. The emissions reading will be carried at 110% Maximum Continuous Rate (MCR) loads until low loads.

2.4 Boiler Setup

The setup for the boiler is shown in the following schematic diagram in Fig. 2.

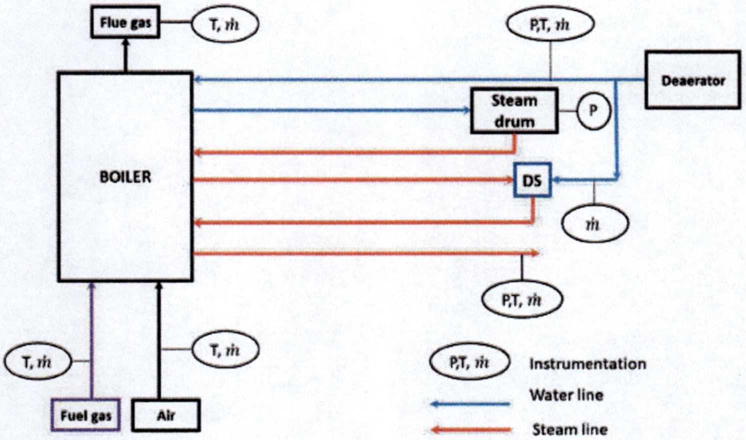

Fig. 2 Schematic diagram of the boiler

Table 1 General setup of boiler

Boiler type	Gas fired water tubed boiler
Fuel type	Methane
Steam Pressure	4450 kPag
Feedwater temperature entering the economizer	121 °C
Ambient temperature	32 °C
Relative humidity	88%
Boiler load	10%, 40%, 55%, 100%, 110%

The following Table 1 shows the working specification of the boiler:

The calculation for boiler efficiency using the direct and indirect method will be carried out using the MATLAB program.

3 Results and Discussion

3.1 *The Effect of Boiler Load on Boiler Efficiency Using Direct Methods*

Figure 3 shows the boiler efficiency for different boiler load using direct method calculation. The efficiency is increasing as the load increased. The efficiency reached maximum point with value of 83.33% when the load is 55%. However, the boiler efficiency started to drop after 55% load. This is because the fuel-fired value has impacted the efficiency value, according to the direct method formula. This efficiency decreases as the enthalpy of the feed water increase.

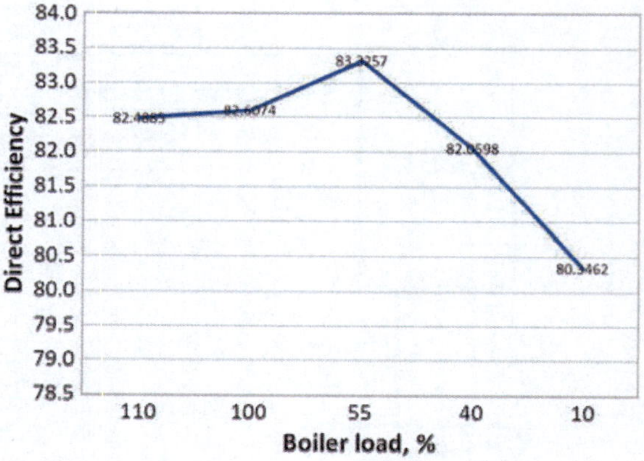

Fig. 3 Boiler efficiency using direct method for different boiler load

3.2 *The Effect of Boiler Load on Boiler Parameters for Indirect Methods*

The following figures show the relation between boiler load with dry flue gas losses Fig. 4a and moisture in air losses respectively Fig. 4b. Based on Fig. 4a percentage of dry flue gas loss is 23.7% at 40% boiler load. When the boiler load was increased to 55%, the percentage loss decreased to a minimum value of 7.8%. After 55% load, the loss due to the dry flue gas increase till the maximum of 9.2% at 100% boiler load. Figure 4b illustrates the percentage of heat loss due to moisture in the air. The lowest heat loss is 0.34% at the boiler loads setting of 55%.

Figure 5 shows the effect of dry flue gas loss on the boiler efficiency The indirect efficiencies are decreasing as the percentage of losses increased. The results obtained in this study agreed with the result from [16]. The increment of flue gas losses is due to the temperature and flow rate of the flue gas itself. By reducing the temperature of the flue gas, it could reduce the losses to minimum [11, 17].

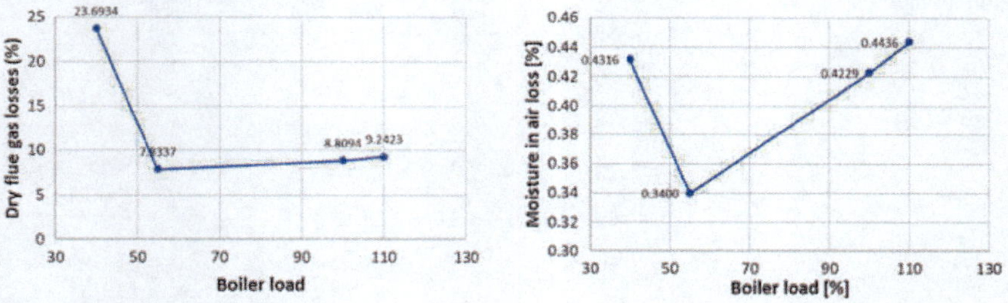

Fig. 4 The Graph of Boiler Load versus **a** Dry flue gas loss **b** Moisture in air loss

Fig. 5 Graph of Dry Flue Gas Loss

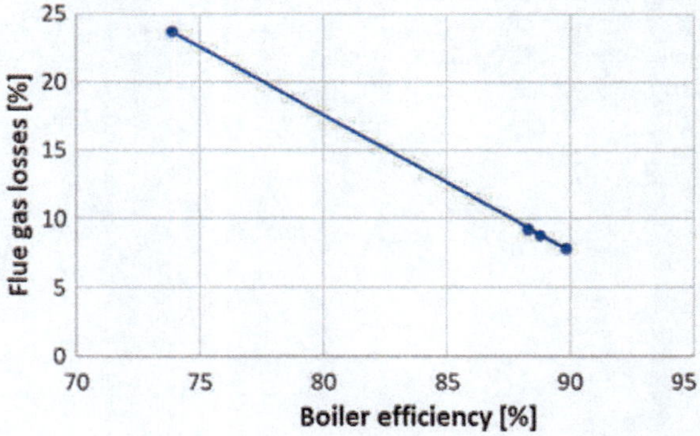

3.3 The Effect of Boiler Load on Boiler Efficiency for Indirect Methods

Figure 6 shows that the boiler's efficiency was 88.9% at a load of 40% MCR. Boiler efficiency increased to the greatest level of 90.7% at 55% MCR load. Furthermore, boiler efficiency decreased slightly until 110% load.

The high excess air quantity beyond what is theoretically required will lowers the flame temperature and result in a lower heat transfer. Vapor in the humidity form in the upcoming air, is superheated as it passes over the boiler. Meanwhile, this heat passes up the chimney, it must be involved as a boiler loss. Loss due to moisture in air – air enter the boiler furnace is not dry and still containing water, resulting incomplete combustion. The following Fig. 7. shows the relation between boiler efficiency with the amount of excess air. Based on the figure, combustion efficiencies increase with

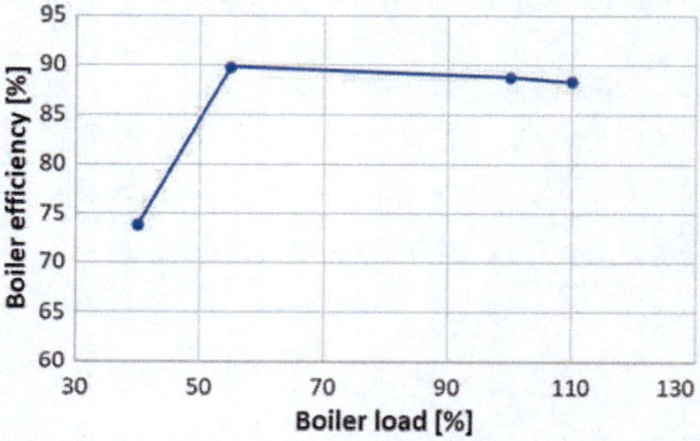

Fig. 6 Boiler efficiency for different boiler load using indirect method

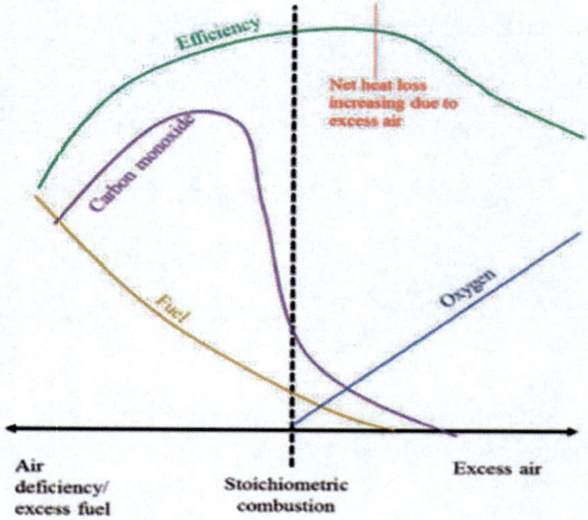

Fig. 7 Air–fuel ratio's effect on combustion efficiency and generation of CO (Miller 2011)

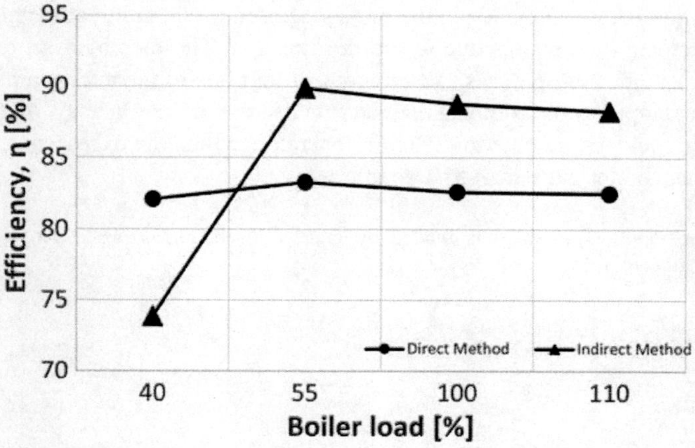

Fig. 8 Efficiency difference between direct and indirect method

the excess air until heat loss in the amount of excess air reaches the heat loss to a point where the net efficiency begins to drop. Therefore, it is important to control the amount of excess air supplied to the boiler based on the operating load.

3.4 Comparison for Direct and Indirect Methods

Figure 8 shows a comparison of two boiler efficiency calculation methods based on boiler load. The direct method is represented by the orange line, whereas the indirect method is represented by the blue line. In general, both direct and indirect method has a similar trend of boiler efficiency against the boiler load. The maximum efficiency was determined at boiler load 55% which is the same as per suggested operating load. However, the efficiency of direct method is lower compared to the efficiency of indirect method because of direct method did consider the effect of external factors and conditions such as blowdown losses and startup and shutdown losses (Wienese 2001).

4 Conclusion

In this study, analysis of boiler performance for different operating load is explored. The calculation of the boiler efficiency is carried out using the direct and indirect method. In general, the efficiency obtained using indirect method is slightly higher compared to the direct method. However, both methods show a similar optimum boiler load condition which is at 55% load. Operating the boiler at this load will resulting the highest efficiency in the process. Based on the indirect method calculation, the main heat loss is due to the dry flue gas. The increment of excess air supply for combustion process caused the heat loss due to the excess air is higher than the heat input by the combustion therefore decreased the efficiency of the boiler. This result could give a clue to the boiler operator to tune the excess air rate when operating the boiler more than 55% load.

Acknowledgements This study is funded by grant R.K130000.7343.4B472 from Takasago Thermal Engineering.

References

1. Zhang Q, Yi H, Yu Z, Gao J, Wang X, Lin H, Shen B (2018) Energy-exergy analysis and energy efficiency improvement of coal-fired industrial boilers based on thermal test data. Appl Therm Eng 144:614–627
2. Fan Y, Si P (2021) The Study of Numerical Simulation of Oxygen Enriched Burner System. CFD Lett 2(4):197–207
3. Showers G (2002) Boiler operation efficiency: insights and tips. Heat/Pip/air Cond Eng 74(11):53–56
4. Al-Attab KA, Badshah N, Zainal Alauddin ZA (2021) Design and performance analysis of a biomass fueled monotube boiler for humid air turbine cycle. J Adv Res Fluid Mech Therm Sci 72(1):111–123
5. Basu P, Kefa C, Jestin L (2012), Boilers and burners: design and theory. Springer Science & Business Media
6. Shah S, Adhyaru DM (2011) Boiler efficiency analysis using direct method. In: 2011 Nirma university international conference on engineering: current trends in technology, NUiCONE, 2011—Conference Proceedings (2)
7. Erbas O (2021) Investigation of factors affecting thermal performance in a coal—fired boiler and determination of thermal losses by energy balance method. Case Stud Therm Eng 26:101047
8. Retirado-Mediaceja Y, Camaraza-Medina Y, Sánchez-Escalona AA, Laurencio-Alfonso HL, Salazar-Corrales MF, Zalazar-Oliva C (2020) Thermo-exergetic assessment of the steam boilers used in a Cuban thermoelectric facility. Int J DesNat Ecodyn 15(3):291–298
9. Behbahaninia A, Ramezani S, Lotfi Hejrandoost M (2017) A loss method for exergy auditing of steam boilers. Energy 140(1):253–260
10. Saidur R, Ahamed JU, Masjuki HH (2010) Energy, exergy and economic analysis of industrial boilers. Energy Policy 38(5):2188–2197
11. Chao L, Ke L, Yongzhen W, Zhitong M, Yulie G (2017) The effect analysis of thermal efficiency and optimal design for boiler system. Energy Procedia 105:3045–3050
12. Kitto XJ, Stultz S (Eds) (2005) Steam—its generation and use. The Babcock and Wilcox, Company, Ohio
13. Singer YJ (ed) (1991) Combustion—fossil power. Combustion Engineering Inc., Windsor
14. Rayaprolu ZK (2013) Boilers. A practical reference, CRC Press, Taylor & Francis Group, USA
15. Barma MC, Saidur R, Rahman SMA, Allouhi A, Akash BA, Sait SM (2017) A review on boilers energy use, energy savings, and emissions reductions. Renew Sustain Energy Rev 79(May):970–983
16. Patro B (2016) Efficiency studies of combination tube boilers. Alex Eng J 55(1):193–202
17. Qi G, Zhang S, Liu X, Guan J, Chang Y, Wang Z (2017) Combustion adjustment test of circulating fluidized bed boiler. Appl Therm Eng 124:1505–1511

A Review of Active Day Lighting System in Commercial Buildings with the Application of Optical Fiber

Lokesh Udhwani and Archana Soni

Abstract The sunlight has traditionally been the primary source of daylight. However, the advent of modernization has made people highly dependent on artificial lights. The technological development in the field of electric fixtures, dimming technologies, electronic ballasts, and light sources has nearly replaced the requirement of natural light in a room. This is not only harmful to the environment because of huge energy consumption but also harms human health. The social infrastructures such as hospitals, banks, institutions, multi-level buildings, malls, offices, etc. must make the maximum use of natural light. This paper reviews the active daylighting systems with the application of fiber optics to make them more efficient and cost-effective. A design of sun-pipes with quartz optical fiber and a control switch for attaining all the targets of high efficiency in the climate of India is proposed through this study.

Keywords Optical fiber · Daylight · Office building · Natural light

1 Introduction

Solar energy has been the topic of various researches and its sustainability has now created consciousness for using renewable energy sources. Many available systems use solar light as an energy source. The comprehensive studies of solar energy sources present all their facets and their application. These studies also comprehend the course of actions to generate power through sunlight, daylight systems, mental as well as physical illness in humans with its effectiveness in treating this illness, non-imaging field, etc. Since the beginning of the construction of the buildings (i.e. before the sixteenth century), these daylight systems were brought into use for illuminating the indoor part of the buildings by constructing roof skylights and windows. During the twentieth century, electrical lighting came into existence for large buildings and especially for spaces constructed below ground level [60]. Since then, the use of electrical lighting has put back the natural daylighting system. In the past few years,

L. Udhwani (✉) · A. Soni
Energy Centre, Maulana Azad National Institute of Technology, Bhopal, India
e-mail: udhwanilokeshmanit@gmail.com

M. Y. Ismail et al. (eds.), *Technological Advancement in Mechanical and Automotive Engineering*, Lecture Notes in Mechanical Engineering,
https://doi.org/10.1007/978-981-19-1457-7_56

it has been accounted that these electrical lightings consume approximately 40% of the total energy consumption in the buildings. In this 40% energy consumption, 19% of the energy is consumed by the commercial buildings only. This statistic shows the wastage of energy which can be reduced if electrical lighting is replaced by daylight systems which will also retrench environmental hazards and improve human health. A task 21 project was executed by the "International Energy Agency" (IEA) keeping various researches over daylight activities as the source. In the year 1995, this project was conducted in Australia, Europe, and North America. The project comprised of methodology to evaluate the daylight systems, an innovation of daylight system and also the ways to control the lighting in the building. The research of IEA task 21 focused majorly on the shaded skylights, roller blinds, diffusing reflecting louvers, light shelves, etc. These techniques work as passive devices to improve daylight availability. These objects are modulated by a system used for guiding the light through non-energy consuming optical parts and optical materials which guide the daylight [38].

The quality of the solar light, as well as the effectiveness of the guiding system, maybe distinguishable because of the climate, weather conditions, latitude, and geographical location. In general, only a few hours of light can be obtained by the static skylight, as sunlight has a dynamic distribution and also different azimuths angles and elevations in respect to a swap in the seasons. The change in the azimuth angles will also change the distribution of the light incident over the device and decrease the efficiency of the light-guiding system. For attaining efficacy from the dynamic daylight system, studies over active devices and optical components are efficacious. Some of these basic optical components comprise prism structure, hot mirror, IR filer, cold mirror, dish concentrator, compound parabolic concentrator, parabolic concentrator, light guide, optical fiber made up of different materials, Fresnel and linear Fresnel lens, and high reflectivity light pipes. For reorientation of daylight, active devices with heliostats are used which comprise a linear solar tracker and tracking system. The daylight index is used for determining the daylight system's performance. The ratio stating the irradiance flux from indoor to outdoor on the same horizontal plane is termed as the daylight factor. According to the "International Commission of Illumination" (ICI), 2% of the daylight factor (DF) is required by a daylight system after which it can be considered up to the par for acclimatizing the standard overcast. As the daylight index will hike, additional electrical lighting sources can be switched by the natural daylight system. Other methods to assess the daylight are "useful daylight illuminance" (UDI) and "daylight autonomy" (DA). As these methods are dynamic daylight metrics, therefore they are brought in the application for measuring public office space and windows opening. Static daylight matrices are based on the static sky condition whereas dynamic daylight metrics are based upon real weather conditions, time series, and weather events. For evaluating the occupant's solar comfort and availability of work light over the work plane, dynamic daylight matrices are brought in an application [46].

A daylighting system comprises scads of instruments and devices such as solar panels, Fresnel lens, parabolic dish, mirrors, lenses, optical fibers, natural luminaries, etc. As per the requirement, these are used in different shapes and sizes in different daylight systems to provide natural light.

In the present paper, different types of optical fibers are studied, observed, and reviewed which were used in office buildings in India. The study is in accordance with the composite climate of India. This study is predominant the energy conservation as the office buildings in India use the maximum energy produced in the country which is approximately estimated to 40%. The paper sets forth the types of developed daylight systems with optical fibers. The progression of the daylight system is also studied with studies from several researchers. The implementation of these systems in office buildings of India and the out-turn of the daylight system on human health is detailed further in this paper. Few systems which are so far in the application are also detailed with their efficiency and effectiveness. This study also adds to the knowledge of the effects of daylight in saving energy resources.

2 Daylight

Daylight is present as a single source of light for centuries. For providing and distributing daylight to the interiors of the buildings, architecture was made through walls and through opening large spaces. Electrical lighting has provided architects to be more creative and in designing these spaces. The newly developed daylight systems have contributed and summed up to a great extent in designing an energy-efficient and user-friendly environment for the building. The buildings must be integrated with these systems at the time of construction for attaining a full length of efficiency from these systems.

Design strategies cannot be separated from daylight strategies. With the help of daylight systems, artificial lighting can be replaced which will further reduce the consumption of energy resources. But cooling and heating loads are highly influential, therefore integrating daylight systems in the buildings requires the proper details of the amount of light that will be needed with the size of the area, and also professionals and many specialties will be required. The first step in designing a daylight system begins with selecting the site for constructing the building and it goes on till the completion of the project occupied [1]. There are various objectives for planning the daylighting which takes place at each stage of the construction of the building:

- **Conceptual Design**: The fundamental design of a building such as an aperture, proportion, or shape gets influenced or influences the design of the daylighting [28].
- **Design Phase**: With the evolvement of the design of the building, the strategies of the building should also advance with it for all parts. The design related to the interior finishing and facades with the integration of the services as well as the systems are all parts of the integration of the daylight system in the building [18].
- **Final/Construction Planning**: The strategies made for integrating daylight system in the buildings highly affects the selection of products and material that

are used for the construction. When the plans related to the construction of the building are made, the plan related to the daylight must be made at that time only [45].

- **Commissioning and Post-Occupancy**: The calibration of the lighting systems must be done after the completion of the construction of the building. After that, the maintenance of the system starts [22].

2.1 Availability Daylight

The strategies related to the daylight system utilize the sunlight, skylight, ground, and building. The presence of natural light is a factor upon which the strategies of these systems depend and these strategies are determined by the condition and latitude of the site of the building, even a slight obstruction can hamper the effect of the system. Another important factor that highly affects the daylight strategies is the climate which therefore makes determining the conditions of the climate and seasons especially the probability of availability of sunlight and ambient temperature a fundamental step. All of this together makes the understanding of the availability of daylight as well as climate conditions of the site a significant factor to study and observe [23]. A daylighting system needs to address all of these conditions. A comparison of the characteristics of daylight and electric lighting is presented in Table 1.

Apart from this, daylight is cost-free during the time it is available whereas electric lighting consumes a lot of energy and is exorbitant. However, the components used for the utilization of daylight need to be purchased. On the other hand, electric lighting can be designed in accordance with the need of the consumer which nowadays seems very essential for office buildings.

2.2 Effect of Day Light Over Human Health

There are innumerable ill-effects on human health caused by the absence of daylight. Direct sunlight, is cardinal for human health. Sunlight consists of Vitamin D which is required for bone strength and development especially for a newborn and old people. People who spend all day in their offices often don't come under direct sunlight for days and this has a profound effect on their health. The need of supplements has reached its peak and one of the most common reasons is not getting sufficient sunlight [40]. In addition, there are many skin problems that have developed in such people because of not getting adequate sunlight. These people have spent so much time away from sunlight that when they come under its direct contact, they often find it uncomfortable. Weak Eyesight is one of the major problems which occurs in people who spend their time in artificial lights. Short-sightedness and myopia are the names of few common eye diseases from which people not getting exposed to

Table 1 Characteristic of daylight and electric lighting

Characteristic	Daylight	Electric lighting
Description	It provides a spectral power distribution in continuity comprising of a component for short wavelength during the availability of natural light	It is applied for obtaining a typical lighting with mainly discontinuous but also with some continuous spectral power distribution
Temporal and absolute photometric and colorimetric characteristics. Illuminance and correlated colour temperature (CCT)	Correlated color temperature dawn and dusk, spectral power distribution and intensity is available with temporary variations	It is available with static as well dynamic intensities which are pre-programmed and are available during all the hours of the day
Spatial light distribution indoors	Through skylight or windows: indoor can be illuminated. When the sky is clear, sun-patches, parallel beams and distinct shadow appears When the condition of sky is over cast, transition is smooth from light to dark	Light is places at somewhere top which lights the surfaces horizontally Parallel beams and patches are not present
Flicker and spectral fluctuation	The stability of these light systems lasts only for a short period of time. Spectral or flicker fluctuations are not present	Spectral fluctuations or/and display flicker can be displayed by the source
Polarization	The daylight attained from a specific region of the sky may be partially polarized but the direct sunlight is not polarized	In the configuration of the lamp, partial polarization is integrated for providing direct transmission or specular reflection

daylight suffer. As per recent studies, insufficient daylight has a profound impact on the eyesight of youngsters nowadays [25]. Further, sufficient daylight can increase health in several ways like improving circadian function which enhances the sleep quality, mood, and cognitive abilities of a person. It also affects the mental health [56] of a person. Overall, spending time in daylight will improve health whereas insufficient daylight will become the root cause of many diseases such as cancer, cardiac problems, etc. [26, 33].

3 Daylight System

The requirement of a daylight system is very essential in terms of energy consumption, human health (both physical and mental), cost efficiency, etc. Daylighting is a process in which sunlight is used as a source to decrease the use of electrical lighting. The buildings with high performance must integrate these daylighting systems to decrease the use of energy resources and also to decrease the side effects that occur to humans because of the absence of daylight. Many programs are available to attain these objectives and these programs can also help the experts in finding the best performing building components like shades, blinds, or glazing.

If daylight enters the building through the windows throughout the day then the effectiveness of the daylight is said to be at the maximum. Glazing helps in achieving a high effective daylighting. Interior design, ground reflectance, sky views, door height, and orientation can be in combination for feasibility at each portion of the room. There are many factors over which the appropriate choices for choosing the glazing in a façade are dependent. These factors constitute the concept of the architecture of the building, user comfort needs, the requirement of the building, the transmittance of the window, climate condition, orientation, and the location [31].

As the common depth of the daylighted zone is evaluated as 1.5–2 times the window's head height, therefore the zone of the daylighting will get deeper when the window height is increased. For mitigating the solar heat gain's effect, there is a need for more control in larger windows. Likewise, the visible window transmittance decreases if the glass area of the window is more. If the windows are south-facing, then the good provision of illumination will be present. The windows which are facing north will provide reduce the heat gain and even the quality of the daylight will be consistent. But the overall fact remains the same that there will be a continuous variation in the quantity of daylight throughout the day [9]. For mitigating the excessive amount of daylight in the workplaces, glare control or personal shading systems should be integrated. The potential to meet the need of the visual connection from the outdoor is increased by the designs of the daylight.

3.1 Parameters of Daylight

The ratio stating the irradiance flux from indoor to outdoor on the same horizontal plane is termed as the daylight factor. The DF is signified in form of a percentage. Consider the illuminance at any point in a room at 180 lux whereas the 20,000 lux of luminance at outside the room, then the available DF at that specific point is 0.9% [17].

3.2 Daylight Factors of an Office Block

In the year 1895, Trotter introduced the DF's concept and stated it as an indicator through which the performance of the daylight system can be assessed. The evaluation of the DF is made based on the daylight illumination ratio of a given plane at a specific point to the illumination that incident over a horizontal plane because of a hemisphere that is unobstructed from the sky.

Both from the exterior and interior values of the illumination, the direct sunlight is exempted in the daylight factor. There are three available components under which the DF is divided. These components are internal reflected components, external reflected components, and sky components. The abstract of all provides the overall DF.

3.3 *Elements of Daylight System*

The most significant components of the daylight systems are detailed below:

3.3.1 Sunlight Source

The daylight system uses sunlight as the primary source. For the evaluation of the efficacy of the available luminous of the direct radiation, the equation which is used is exhibited below. This equation is the ratio of the direct luminance to the direct normal irradiance [19, 31]:

$$K_b = \frac{E_{bn}}{DNI} = \frac{K_m \cdot \int_{380\text{nm}}^{780\text{nm}} G_\lambda \cdot V_\lambda \cdot d\lambda}{\int G_\lambda \cdot d\lambda} \tag{1}$$

Here, the direct luminance is signified by E_{bn}, direct normal irradiance is signified by DNI, for photopic vision: luminous efficacy constant is signified by Km, spectral irradiance is signified by G_λ, eye's spectral sensitivity is signified by V_λ and λ signifies the wavelength. V_λ is represented numerically.

3.3.2 SSFLR

The "sunlight-source lightfastness rating" (SSLFR) is configured of central LFR which is conventional. This comprises the elements such as electric motor, transmission means, mirror row, primary reflector system, secondary reflector system, mobile structural system, and fixed structural system [6]. On a mobile frame, arrangements of mirrors in rows are made which together configures the primary reflector system. The location of the secondary reflector system is made at a specific height in reference to the primary reflector system. It also comprises a reflector cavity within which the OFBs place. There are 3 possible movements for this SSFLR [8].

3.3.3 Optical Fiber Bundles

Tens of thousands of 'optical fibers' comprise together to compose an OFB. These optical fibers are enclosed with the use of epoxy glue which is used for the filling

[48]. Thus, which every optical fiber, optical losses are present which combine to create optical losses for the bundle [27]. Looking through the theoretical point of view, for every single optical fiber, attenuation depends upon the length of the fiber and wavelength of the light. For every individual optical fiber, the transmissivity is expressed as:

$$T(\lambda) = \frac{G_{out,of}}{G_{in,of}} = 10^{\frac{-L.dBloss}{10}} \quad (2)$$

Here, transmissivity for a particular wavelength is signified by T(λ), the outcome from the solar irradiance is signified by the Gout,of, input solar irradiance is signified by Gin,of, fiber length is denoted by L, dBloss denotes each wavelength's attenuation (or loss/km).

3.3.4 Solar Luminaries

The use of natural light with maximum effectiveness in the building is dependent on solar luminaries. An optical element that is made for diffusing light consisting of luminance at a uniform level over a working plane is called a solar luminaire. The work of a solar luminaire is to reduce the transmission glare and losses to the minimum and also utilize the available light to the maximum. The hybrid solar luminaries distribute direct sunlight and also use electric lamps [52]. Polymethyl methacrylate (PMMA) or silica glasses are the materials that are used for making solar luminaires. Approximate efficiency of 80% can be attained by a solar luminaire [50].

3.3.5 Light Control System

A photocell and dimmable electric lamp is an important constituent of the Hybrid solar luminaires. These constituents are substitutes of the natural light for compensating the fluctuations which are available in the collected sunlight's intensity. The reason for the fluctuations occurs because of the movement of the solar collector or because of the change in the coverage of the cloud [12].

4 Optical Fiber in Daylight System

Based on the materials used in making the optical fibers, they are chosen in the applications of daylight systems. The choice for the optical fibers is also made on the basis of the easiness they absorb the needed frequencies, the easiness with which they are drawn into continuous strands, etc. But the attenuation of a signal when it is transmitted through a long-distance is considered as the most important factor for using it in a daylight system.

4.1 Types of Optical Fiber in Daylight System

In general, 3 main types of optical fibers are brought in the application for daylight system.

The glass material made with silica is used in the optical fibers. For 1550 nm of wavelength, 0.2 dB/km tune is attenuated for this material. The material is less feasible for use in daylight systems because of its fragile nature, high bending radii, and high cost for manufacturing [5].

Highly purified liquid is used in this 2nd type of optical fiber. On a usual basis, this highly purified liquid comprises methanol/ethanol and water. The attenuation constants for this material are approximately 2 dB/km. the issues with these materials are that their versatility of application decreases very highly and it is highly difficult to join the cables [59].

"Plastic optical fibers" (POF) are the 3rd type of materials and it uses "polymethyl methacrylate" (PMMA). They provide very high flexibility and the bending radii are also low which is very much suited for the wiring purpose in the building [34, 49].

One more option available for utilizing in making the optical fiber for the daylight system is "silica optical fiber" (SOF) but the burden of the high cost of the silica makes it hard to use it in the system [4].

After all the above-mentioned 3 options are brought in comparison, the most suitable option that can be used in the system of daylight is POF. However, due to the available heating effect, the break-down of fiber occurs at a certain point because of the infrared radiations of the sun. For reducing the effects of ultraviolet and infrared radiations over the fiber, proper filtering should be integrated. Table 2 presents a change in technology adopted over different periods of time.

4.2 Progress of Optical Fiber in Daylight System

Since the beginning of 1950, the supply of electrical lighting started. This brought a huge revolution that changed and affected the use of natural indoor lighting. The houses or buildings slowly got adapted to electrical lighting and natural lighting was left out without considering the side effects of its absence [30]. Especially in India which is such a highly populated country, buildings are constructed by occupying the maximum space and electrical lighting is used as an alternative. This has given a boost to many health issues. The use of daylight will reduce these health issues as well as reduce the energy consumption in office buildings of India. The daylight system started decades back to avoid these types of problems. At the beginning of the 1970s, optical fibers were first brought in the application for channeling the light from

Table 2 Technology adopted over different periods of time

Technology with the year of invention	Details	Cost efficiency	References	Technology with the year of invention
Solatube was 1st installed in the year 1987	It used sun pipe with dome. It provides healthy daylight and is one of the best to use in residential building	The purchasing as well as installation is cost efficient	[16]	Solatube was 1st installed in the year 1987
Monodraught was 1st installed in the year 1974	It used sun pipe with sun catcher. It can light a part of the floor for certain amount of time. Artificial light is needed in addition	Similar to suntube, the purchasing as well as installation cost is very efficient	[51]	Monodraught was 1st installed in the year 1974
Heliobus was 1st installed in the year 1995	It uses heliostat with sun pipe system. Light duct is used as light guidance media. It is device to be installed on roof and its limit is up to 5 stories	The purchase as well as installation cost is very high	[35]	Heliobus was 1st installed in the year 1995
Himawari was 1st installed in the year 1970	It uses Fresnel lens and quartz optical fiber with fiber optics channel. The sunlight is first collected and then concentrated with the help of many Fresnel lenses	The purchase as well as installation cost is high. Though the use of Fresnel lens have reduced the cost slightly	[59]	Himawari was 1st installed in the year 1970

(continued)

Table 2 (continued)

Technology with the year of invention	Details	Cost efficiency	References	Technology with the year of invention
Parans was 1st installed in the year 2004	It uses skylight along with mirror and light ducts. This system comprises of many Fresnel lenses which together combines to form an array. The Fresnel lenses are integrated over the common system available for tracking the sun and it concentrate the light from the sun	The purchase as well as installation cost is very high and in addition high maintenance is required	[58]	Parans was 1st installed in the year 2004

the sun which is present at the collector to integrated light fixtures which are present in the building. Harnessing the power of the sun through collecting and guiding the sunlight with the help of optical fiber shows the effectiveness of this method. This shows the ability to substitute the electric lighting and sorts many of the problems such as consistency, controllability, heat gain, and glare. Very large buildings can be facilitated with natural lighting by transmitting the sunlight with the help of optical fibers without bringing any change in the floor plans or even without any need for weatherproofing which are generally needed for clear story windows or skylights [61]. In this paper, the concept of a fiber optic daylight system is explained with its applications. The advantages and disadvantages are also explained in detail with the progress made by various researchers in the field. The work of few authors over the years in the field of daylight systems are exhibited in Table 3:

4.3 Some Noteworthy Contributions

Some noteworthy contributions for the advancement of the daylight systems are detailed below:

The author classified the day-lighting system as integrated, hybrid, light diffracting, light transport, and light guiding systems and reviewed them. After that, these classified systems were compared with respect to their ease of transportation, light output, efficiency, availability, maintenance, durability, and integration. The outcome was revealed by the authors as the system is efficiently redirecting the light for reaching up to 10 m distance. This is because of the present optical losses occurring after reflection. The system exhibited poor performance in west and east orientation and also not suitable to mount at places with high latitudes. High ceilings are required for the system [41].

The research succeeded in analyzing and presented the main existing system of holographic solar energy. The research also emphasized the optical element's characteristics. It was revealed by the author that the most extensively used types are phase and volume when operated with either reflection or transmission [14].

The authors of the research stated that it is possible to diffract different light manipulations forms when compared to that of the lens, prism, mirror, or other optical elements. In architecture, a large variety of uses are allowed for laminated glass capable of direct lights. The author added that the technical feasibility and performance of the proposed study can be proved by 2 buildings needed for demonstration and field test of samples in the laboratory [39].

The advancement in the hybrid light guidance system was studied by the authors. The authors consecutively delivered electrical light and daylight in the building. In the buildings, the lights were first combined and then the distribution of them was made through luminaries [36].

An author described a computer program through which the daylight system can be analyzed. The author confirmed few findings with the help of this computer program [32].

Table 3 Work over the past years in the field of daylight system

Year	Author	Work	References
2011	Venu Gopal and Madhav Annamdas	Authors provided their research with the full overview of the fiber sensing system with their applications. They also detailed few of the commonly available fiber optics	[3]
2016	António Barrias, Joan R. Casas and Sergi Villalba	Authors provided their research with the latest advancement of the products used in the daylight systems with the help of extensive range of lab experiments. They also provided with reviews referring to various applications in civil engineering structures	[7]
2013	Irfan Ullah and Seoyong Shin	The authors proposed in their study with daylight systems which works on the basis of optical fiber with high efficiency. Their system comprises of distribution, transmission, direction and focusing. With the help of Fresnel lens and parabolic trough, the light was first focused and then guided via trough CPC and collimating device	[57]
2008	Harry Braun	Worked over the biological effect of the daylight over the human body's health and infection control	[10]
2007	Danny H. W. Li and Ernest K. W. Tsang	Authors studied the implications of energy and performance of the daylight system in the office buildings. The authors selected 35 commercial building over which they conducted a survey. They also presented the parameters which affects the designs of the system	[32]

(continued)

Table 3 (continued)

Year	Author	Work	References
2019	R. Urbano Gutiérreza, J. Dua b, N. Ferreiraa, A. Ferreroc, S. Sharplesa	The researchers studied and examined the new louvre screen's performance and the design of the daylight system in the office buildings. They measure the screen with 3 types of different material finishes as used in louvres: very generally utilized specular aluminium; and there are 2 ceramic finished materials which are used for reducing the environmental affects over the system. The outcome supported their research's aim as the system provided visual comfort and satisfactory level of daylighting	[59]
2019	Anja Jamrozika, Nicholas Clementsa, Syed Shabih Hasana, Jie Zhaoa, Rongpeng Zhanga, Carolina Campanellaa, Vivian Loftnessd, Paige Portera, Shaun Lya, Selena Wanga, Brent Bauer	In order for providing view and daylight with minimizing the glare, the author tested the effect of 2 new designed shading systems: automatically tinting windows and windows integrated with manual control motorized mesh shades	[26]
2006	Anca D. Galasiu and Jennifer A. Veitch	The authors provided an investigation with peer review for particular problems related to the utilization of daylight in commercial buildings. The study specifically preferred luminous and physical condition of daylight in the offices. The authors suggested that their project is very efficient in terms of energy saving and is also appropriate for using it in office buildings	[23]
2017	Alex Vlachokostas and Nicholas Madamopoulos	The author measured a "dynamic liquid filled prismatic louver" both as an energy harvesting as well daylight redirecting system. When the proposed system is fixed, it is a categorized as a passive design and therefore it does not need any power to control the façade. This simplifies the approach	[62]

(continued)

Table 3 (continued)

Year	Author	Work	References
2006	Veronica Garcia Hansen	The author aimed to enhance the existing daylight systems and make them efficient in the maximum possible way by experimental measurement and theoretical modeling. The author also reported that the passive mirror light coupled with the LCPs have great potential as a daylight system for deep plan buildings	[21]
2007	S. Sharples and D. Lash	Paper of the authors was in relation to the researches which were conducted in last 15 years over day-lighting within the atrium buildings. The author reported that the modeling of computers will continue and increase in dominating the daylight in such buildings	[50]
2015	Irfan Ullah and Allen Jong-Woei Whang	The authors proposed a new technique for attaining collimated lights. This light will illuminate the FB in an uniform pattern. The authors reported that their design lowers the heat issue in the optical fiber and thus the system is enhanced	[58]
2016	Ngoc-Hai Vu, Thanhtuan Pham and Seoyong Shin	The authors proposed a novel design of a "modified optical fiber day-lighting system" with its experimental analysis and optical simulation. The design was made to integrate the indoor lighting system. It constituted of 3 sub -systems which were distribution, collimation and concentration. The authors revealed through their report that their design is considerably better for utilizing in solar energy application because of its low price and high efficiency	[63]
2019	Biljana Obradovic, Barbara Szybinska Matusiak	The author targeted to select the best suited day-lighting system when needed at high latitudes through literature review	[45]

(continued)

Table 3 (continued)

Year	Author	Work	References
2019	Yun Kyu Yi	The authors proposed a technique for integrating 2 distinct quantity, quality and performances within a single evaluable target. The author tried to understand the reason for the design of the façade of the building by using it as a case study. The author presented a design which is capable of satisfying the performance of the daylight system	
2019	Xiaodong Chen, Xin Zhang, Jiangtao Du	The authors experimented upon humans to study different types of glazing with their considerable effects like transmittance or colour. The experiment was conducted to analyze its effect upon the working performance, mood and self-reported satisfaction. The area of study was chosen as Beijing. From 17-11-17 to 15-01-18, seven systems of glazing were examined. The results were negative in terms of any effect over the participants from glazing. But when they were exposed to glazing for longer period, the impact can be seen over them. but mood was not seen even after a longer period	[11]
1998	Ross McCluney	The author studied the advanced fenestration and day-lighting system. The author also exhibited its high performance glazing and reflecting system. The results show high performance and low operating cost	[38]
2011	Carlos E. Ochoa, Myriam B. C. Aries & Jan L. M. Hensen	The authors examined the present state for art related to the lighting for buildings. The authors also discussed the past advancements with the approaches made for the main modeling and also described the simulation of lightning in the proposed design. Many models were investigated for the paper. All the models were integrated with the algorithm of present geometrical optics	[46]

(continued)

Table 3 (continued)

Year	Author	Work	References
2010	David Daum and Nicolas Morel	The author examined the influence and the interplay of the blind as well as the light control designs. The investigation was based on factors like light energy, heating and cooling loads of a room in offices. The authors reported 50% of energy consumption reduction from their model through annual simulation. The advanced controller exhibited a further 60% of reduction in energy power consumption	[15]
2018	Jiajie Zhu, Wolfram Jahn & Guillermo Rein	The authors examined the concentration hypothesis with the help of simulating lightning with material properties, a geometry model was made. Various weather situations were also modeled. The results exhibited that the peak heat flux will be more than 4000 Wm^{-2}	
2019	Jabar H. Yousifa, Hussein A. Kazemb, Nebras N. Alattara, Imadeldin I. Elhassan	The aim of the research of the authors was to compare all the energy data prediction systems of PV/T with utilizing various ANNs methods. Starting from 1970s, there are various studies which focused upon photovoltaic thermal collectors. These studies targeted to increase its efficiency and also for producing a new hybrid system for heat production and electricity production	

The authors compared the day-lighting performance at a point placed at 50 cm from a south-facing opening in 1:10 models painted for realistic reflectance values, of two light shelves and an overhang, using measurements performed during 1 year. The results were obtained at a point placed at 50 cm from a south-facing vertical opening in 1:10 models. Thelight shelves and overhang produced the same shading effect, as all were designed for a vertical shade angle of 50° [13].

4.4 Advantages and Disadvantages of Optical Fiber in Daylight System

Tables 4 and 5, presents the advantages and disadvantages of fiber based daylight system respectively

Table 4 Advantages of optical fiber based daylight system [2]

Factors	Benefits
Energy consumption	No transportation of energy is made to the fiber which makes since the source of light is isolated from the output of light. Even heat, current, IR or UV radiations are not led through the fiber [20]
Safety	Presence of electricity brings many safety hazards with it such shock, fire or even explosions, etc. In such surroundings where these risks are to be eliminated, optical fiber based daylight systems are very effective
Maintenance	Typically, there is no need to replace the light fittings of the optical fiber based daylight system. This is very beneficial at those places where fitting the lamps are not easily accessible
Cold light	Since these types of light do not consist of any type of IR or UV, therefore it is suitable for those places which requires cold light
4th-level heading	*Lowest Level Heading*. Text follows

Table 5 Disadvantages of optical fiber based daylight system [44]

Factors	Disadvantages
Control	Unlike traditional lights, these systems do not come with switches so it can't be switched of when required. Most of them takes several minutes in getting warm and similarly it requires several minutes at time of cooling down
Limitation	The basic and most important limitation of the optical fiber based daylight systems is that it requires daylight which may not be always available or sufficiently available to power the system especially in the India
Cost efficiency	Since India is a developing country, cost is big factor for utilizing these types of systems. Due to high cost, still office buildings in India avoid installing daylight systems

5 Conclusion

In this paper, various aspects of daylight in respect to optical fiber are observed and reviewed. The progress of the daylight system in India is still very weak. Due to high cost and insufficient efficiency, the office buildings in India avoid using these systems. The health impact on the employees is visible now. The study concludes that a more efficient and cost-effective optical fiber daylight system should be developed for applications in office buildings in India. In addition, the owners of the office buildings should be educated about the benefits of the daylight system and the harmful effects that are caused due to the excess use of artificial light over the physical and mental health of humans.

A design of a sun-pipe with quartz optical fiber and a control switch for attaining all the targets of high efficiency in the climate of India is proposed through this study. This design will also reduce the problem of the amount of light needed by any individual or in accordance with the requirement of space with the help of the added control switch.

References

1. Anas Bin Othman M, Azfahani Ahmad N, Md Ajis A (2017) Daylight strategies for architectural studio facilities: the literature review. IOP Confe Ser Earth Environ Sci 67(1)
2. André E, Schade J (2002) Master's thesis daylighting by optical fiber, p 103
3. Annamdas VGM (2012) Review on developments in fiber optical sensors and applications. Int J Mater Eng 1(1):1–16
4. Asiabanpour B, Estrada A, Ramirez R, Downey MS (2018) Optimizing natural light distribution for indoor plant growth using PMMA optical fiber: simulation and empirical study. J Renew Energy 2018:1–10
5. Aslian A, Shakibaei Asli BH, Tan CJ, Mahamd Adikan FR, Toloei A (2016) Design and analysis of an optical coupler for concentrated solar light using optical fibers in residential buildings. Int J Photoenergy
6. Barbón A, Sánchez-Rodríguez JA, Bayón L, Barbón N (2018) Development of a fiber daylighting system based on a small scale linear Fresnel reflector: theoretical elements. Appl Energy 212(Oct 2017):733–745
7. Barrias A, Casas JR, Villalba S (2016) A review of distributed optical fiber sensors for civil engineering applications. Sensors (Switzerland) 16(5)
8. Bellos E (2019) Progress in the design and the applications of linear Fresnel reflectors—a critical review. Thermal Sci Eng Prog 10(February):112–137
9. Boyce P, Hunter C, Howlett O (2003) The benefits of daylight through windows sponsored by: capturing the daylight dividend program the benefits of daylight through windows (Jan 2003):1–88
10. Braun H (2008) Photobiology: the biological impact of sunlight on health and infection control. Phoenix Project Foundation 1–18
11. Chen X, Zhang X, Du J (2019) Glazing type (Colour and transmittance), daylighting, and human performances at a workspace: a full-scale experiment in Beijing. Build Environ 153(Dec 2018):168–185
12. Christoffersen J, Petersen E, Johnsen K (1997) An experimental evaluation of daylight systems and lighting control. Right Light 4 2(Nov 1997):245–254

13. Claros ST, Soler A (2001) Indoor daylight climate-comparison between light shelves and overhang performances in Madrid for hours with unit sunshine fraction and realistic values of model reflectance. Sol Energy 71(4):233–239
14. Collados MV, Chemisana D, Atencia J (2016) Holographic solar energy systems: the role of optical elements. Renew Sustain Energy Rev 59:130–140
15. Daum D, Morel N (2010) Assessing the total energy impact of manual and optimized blind control in combination with different lighting schedules in a building simulation environment. J Build Perform Simul 3(1):1–16
16. David Babarinde T, Alibaba HZ (2018) Achieving visual comfort through solatube daylighting devices in residential buildings in Nigeria. Int J Sci Eng Res 9(1)
17. Debnath R, Bardhan R (2016) Daylight performance of a naturally ventilated building as parameter for energy management. Energy Procedia 90(Dec 2015):382–394
18. Doulos L, Tsangrassoulis A, Topalis FV (2005) A critical review of simulation techniques for daylight responsive systems. In: Proceedings of the European conference on dynamic analysis, simulation and testing applied to the energy and environmental performance of buildings (DYNASTEE), Athens, Greece, pp 125–139
19. Edwards L, Torcellini P (2002) A literature review of the effects of natural light on building occupants a literature review of the effects of natural light on building occupants. Contract (July):55
20. Estrada A (2017) Utilization of natural light in indoor applications By (Aug)
21. Galasiu AD, Veitch JA (2006) Occupant preferences and satisfaction with the luminous environment and control systems in Daylit offices: a literature review. Energy Build 38(7):728–742
22. Garcia-Hansen VR (2006) Innovative daylighting systems for deep-plan commercial buildings. Faculty of Built Environment and Engineering (Queensland University)
23. IEA SHC (2000) Daylight in buildings—a source book on daylighting systems and components. IEA SHC Task 21—ECBCS Anexo 29 262
24. Jamrozik A, Clements N, Hasan SS, Zhao J, Zhang R, Campanella C, Loftness V, Porter P, Ly S, Wang S, Bauer B (2019) Access to daylight and view in an office improves cognitive performance and satisfaction and reduces eyestrain: a controlled crossover study. Build Environ 165(August):106379
25. Kawasaki A, Wisniewski S, Healey B, Pattyn N, Kunz D, Basner M, Münch M (2018) Impact of long-term daylight deprivation on retinal light sensitivity, circadian rhythms and sleep during the Antarctic winter. Sci Rep 8(1):1–12
26. Knoop M, Stefani O, Bueno B, Matusiak B, Hobday R, Wirz-Justice A, Martiny K, Kantermann T, Aarts MPJ, Zemmouri N, Appelt S, Norton B (2020) Daylight: what makes the difference? Light Res Technol 52(3):423–442
27. Lawless S, Gorthala R (2018) Design and development of a fiber-optic hybrid day-lighting system. J Solar Energy Eng Trans ASME 140(2)
28. Leslie RP, Radetsky LC, Smith AM (2012) Conceptual design metrics for daylighting. Lighting Res Technol 44(3):277–290
29. Li DHW, Tsang EKW (2008) An analysis of daylighting performance for office buildings in Hong Kong. Build Environ 43(9):1446–1458
30. Lingfors D, Volotinen T (2013) Illumination performance and energy saving of a solar fiber optic lighting system. Opt Express 21(S4):A642
31. Littlefair PJ (1990) Review paper: innovative daylighting: review of systems and evaluation methods. Light Res Technol 22(1):1–17
32. Littlefair PJ (1995) Light shelves: computer assessment of daylighting performance. Light Res Technol 27(2):79–91
33. Manfredini R, Fabbian F, Cappadona R, Modesti PA (2018) Daylight saving time, circadian rhythms, and cardiovascular health. Intern Emerg Med 13(5):641–646
34. Maskarenj MS, Avasare M, Ghosh PC (2014) Analysis of plastic optical fiber based daylight system suitable for building applications. Appl Mech Mater 492(January):101–105
35. Mayhoub MS (2014) Innovative daylighting systems' challenges: a critical study. Energy Build 80(September):394–405

36. Mayhoub MS, Carter DJ (2010) Towards hybrid lighting systems: a review. Light Res Technol 42(1):51–71
37. Mccluney R (1998) Advanced fenestration and daylighting systems. In: Daylighting '98, an international conference on daylighting technologies for energy efficiency in buildings
38. Munaaim MAC, Al-Obaidi KM, Ismail MR, Rahman AMA (2014) Empirical evaluation of the effect of heat gain from fiber optic daylighting system on tropical building interiors. Sustainability (Switzerland) 6(12):9231–9243
39. Müller HFO (1994) Application of holographic optical elements in buildings for various purposes like daylighting, solar shading and photovoltaic power generation. Renew Energy 5(5–8):935–941
40. Münch M, Wirz-Justice A, Brown SA, Kantermann T, Martiny K, Stefani O, Vetter C, Wright KP, Wulff K, Skene DJ (2020) The role of daylight for humans: gaps in current knowledge. Clocks Sleep 2(1):61–85
41. Nair MG, Ramamurthy K, Ganesan AR (2014) Classification of indoor daylight Enhancement Systems." Lighting Research and Technology 46(3):245–67
42. Obradovic B, Matusiak BS (2019) Daylight transport systems for buildings at high latitudes. J Daylighting 6(2):60–79
43. Ochoa CE, Aries MBC, Hensen JLM (2012) State of the art in lighting simulation for building science: a literature review. J Build Perform Simul 5(4):209–233
44. Qin X, Zhang X, Qi S, Han H (2015) Design of solar optical fiber lighting system for enhanced lighting in highway tunnel threshold zone: a case study of Huashuyan tunnel in China. Int J Photoenergy
45. Ruck (2000) Daylight in building design, planning for daylight at the conceptual design phase. Build Des 1–20
46. Sapia C (2013) Daylighting in buildings: developments of sunlight addressing by optical fiber. Sol Energy 89:113–121
47. Sharples S, Lash D (2007) Daylight in atrium buildings: a critical review. Archit Sci Rev 50(4):301–312
48. Song J, Jin Z, Zhu Y, Zhou Z, Yang Y (2015) Development of a fiber daylighting system based on the parallel mechanism and direct focus detection. Sol Energy 115:484–493
49. Song J, Zhu Y, Tong K, Yang Y, Reyes-Belmonte MA (2016) A note on the optic characteristics of daylighting system via PMMA fibers. Sol Energy 136(October):32–34
50. Soydan Y, Engin T (1979) Lighting building interiors requiring lighting at day times by conveying daylight (Apr)
51. Su Y, Yu X, Zhang L, Karagianni M, Khan N (2012) Energy saving potential of monodraught TM sunpipes installed in a supermarket. Energy Procedia 14(December):578–583
52. Ullah I, Lv H, Whang AJW, Su Y (2017) Analysis of a novel design of uniformly illumination for Fresnel lens-based optical fiber daylighting system. Energy Build 154:19–29
53. Ullah I, Shin S (2014) Highly concentrated optical fiber-based daylighting systems for multi-floor office buildings. Energy Build 72:246–261
54. Ullah I, Whang AJW (2015) Development of optical fiber-based daylighting system and its comparison. Energies 8(7):7185–7201
55. Urbano Gutiérrez R, Du J, Ferreira N, Ferrero A, Sharples S (2019) Daylight control and performance in office buildings using a novel ceramic Louvre system. Build Environ 151(Oct 2018):54–74
56. Veitch J, McColl S (2001) Full-spectrum fluorescent lighting: a review of its effects on people: A Criticak Review (659):53–111
57. Vlachokostas A, Madamopoulos N (2017) Daylight and thermal harvesting performance evaluation of a liquid filled prismatic façade using the radiance five-phase method and energyplus. Build Environ 126(October):396–409
58. Vu N-H, Pham T-T, Shin S (2016) Modified optical fiber daylighting system with sunlight transportation in free space. Opt Express 24(26):A1528
59. Vu NH, Shin S (2016) A large scale daylighting system based on a stepped thickness waveguide. Energies 9(2):1–15

60. Weisbuch C (2018) Historical perspective on the physics of artificial lighting. C R Phys 19(3):89–112
61. Werring CG (2009) Design and application of fiber optic daylighting systems. http://hdl.handle.net/2097/1473
62. Yi YK (2019) Building facade multi-objective optimization for daylight and aesthetical perception. Build Environ 156(Oct 2018):178–190
63. Yousif JH, Kazem HA, Alattar NN, Elhassan II (2019) A comparison study based on artificial neural network for assessing PV/T solar energy production. Case Stud Therm Eng 13(Dec 2018):100407
64. Zhu J, Jahn W, Rein G (2019) Computer simulation of sunlight concentration due to façade shape: application to the 2013 death ray at Fenchurch Street, London. J Build Perform Simul 12(4):378–387

Numerical Simulation and Flow Characteristic Analysis of Labyrinth Control Valve

G. Wang, W. H. Wang, J. F. Deng, Q. F. Gao, Y. X. Zhang, S. Y. Bao, X. J. Zhu, and N. N. Gou

Abstract For the labyrinth trap regulating valve, combined with the principle of multistage step-down principle, the boundary conditions are set according to the actual working conditions. The $k - \varepsilon$ turbulent model is used to simulate the internal flow field of the labyrinth valve, and the influence of valve opening on the flow field is studied. According to the relevant design theory of labyrinth disk, the flow resistance coefficient and flow coefficient under different valve openings are calculated, and the flow characteristic curve is fitted and compared with the ideal curve. The results show that the flow resistance coefficient decreases linearly with the increase of opening. And the change of opening has little effect on the average velocity of the flow field. The research results provide an important reference for the design of labyrinth valve.

Keywords Labyrinth regulating valve · Multistage step-down · Numerical simulation · Flow characteristic

1 Introduction

As the control element in the fluid pipeline, the regulating valve's main function is to adjust and control the changes in the pressure, temperature, flow and other process parameters of the liquid medium by changing the valve stem stroke. The labyrinth pressure regulating valve, with its strong pressure reduction ability, lower noise, reasonable safety and life, has an irreplaceable role in the application of electric power, petroleum and chemical industries. The study of the pressure reduction mechanism and flow characteristics of the labyrinth control valve has important guiding significance for its design.

G. Wang (✉) · W. H. Wang · J. F. Deng · Q. F. Gao · Y. X. Zhang · S. Y. Bao · X. J. Zhu · N. N. Gou
School of Mechanical Engineering, Ningxia University, Yinchuan, Ningxia 750021, China
e-mail: belonging@163.com

G. Wang · X. J. Zhu
The Key Laboratory of Ningxia Intelligent Equipment CAE, Yinchuan, Ningxia 750021, China

M. Y. Ismail et al. (eds.), *Technological Advancement in Mechanical and Automotive Engineering*, Lecture Notes in Mechanical Engineering,
https://doi.org/10.1007/978-981-19-1457-7_57

Many scholars have made research in this area. Khatri [1] studied the hydraulic performance of drip irrigation in the laboratory and found that the local resistance is very small when the water flow is laminar. Gilaad et al. [2] studies pointed out that the local head loss caused by the change of the cross-sectional shape and flow direction of the flow channel is the main form of energy dissipation in the labyrinth channel. Liu [3] took a variety of typical shapes of runners as the research object, and obtained the local resistance coefficient of the irrigator runner and its variation rules with runner structure, working water pressure, and flow field characteristics by using hydraulic tests. Studies have shown that the local resistance growth coefficient of the vortex flow channel is the highest, and the energy dissipation efficiency is high. The rectangular bypass flow channel is the opposite; when the working water pressure changes greatly, the local resistance characteristics of the emitter do not change too much. This can ensure the irrigation uniformity of the sprinkler to a certain extent; by comparing the calculated results according to experience with the actual calculation results, it is found that if it is simply divided into several conventional runners, only the algorithm of "multiplying the local resistance coefficient" is used. Calculated by considering the adjacent influence between the runners, the head loss results obtained generally have deviations. Wei et al. [4] measured the flow characteristics of labyrinth channels of different shapes and sizes through hydraulic tests, which provided a certain reference value for the design of labyrinth drip irrigation. Zhang et al. [5] used Fluent software to simulate the flow in the rectangular labyrinth channel of the drip irrigation emitter, and improved the channel with a circular arc structure based on the obtained streamline. The research results show that after optimization, the vortex zone and low-speed zone of the runner are basically eliminated, and the mainstream velocity distribution is relatively uniform, which improves the anti-match performance of the drip irrigator. Yu [6] used Fluent software to study the influence of five key parameters in typical labyrinth flow channels such as tooth shape, body shape, and rectangle on the flow. The results show that the upper base width and rotation angle have a significant effect on the flow regime index. Yao et al. [7] studied the flow and safety of the labyrinth valve in supercritical thermal power, and studied the structural strength and safety of the labyrinth valve under the working condition of high temperature and high pressure steam, which provided a reference for the selection and design of the labyrinth valve in medium high temperature and high pressure. As for the study of fluid–structure coupling of labyrinth valves, Dr. Bathe from MIT made an outstanding contribution. Under his leadership, an academic exchange meeting on fluid–structure coupling was held every year in the United States, and it realized the calculation of fluid–structure coupling problem in commercial software. Li [8] described the turbulence model used in the simulation of fluid–solid thinning in the study of wind-induced vibration, and verified that the flutter derivative could be easily obtained in the study of Feng [9] established a nonlinear aerodynamic elastic calculation model based on the Euler equation of compressible fluid and the elastodynamic equation of linear structure. The solution of The Euler equation on the dynamic grid was described by ALE dynamic description. Davis et al. [10] studied the influence of spool shape on the inherent flow characteristics of the valve through simulation and experimental methods. However, in practical application, the working flow characteristics of the valve are affected by the pressure drop at both ends of the valve, and often deviate from the inherent flow characteristics.

Scholars at home and abroad have made relevant researches on the labyrinth regulating valve by means of experiment and simulation analysis. Great progress has been made in the theoretical and numerical simulation research on the fluid–structure coupling problem, which lays a foundation for the application of the labyrinth regulating valve in practice. However, for the high pressure differential maze valve, more studies focus on the principle introduction and qualitative verification analysis, the research on the strength and safety of the valve structure only for specific models and working conditions, the lack of more systematic research, especially the safety of labyrinth trap control valve. Therefore, this paper aims at the high pressure difference labyrinth trap control valve under specific working conditions, combined with the principle of multistage throttling step-down, the introduction of $k - \varepsilon$ turbulence models, valve opening is studied on the influence of flow velocity and pressure in the valve, the calculated model of labyrinth regulator, flow resistance coefficient of resistance curve along with the change of the opening of the confluence, in order to better provide guidance for the valve design.

2 Mathematical Model of Flow Calculation

2.1 Physical Model

The labyrinth control valve is mainly composed of cover, valve body, seat, throttling components, valve core components, as shown in Fig. 1. Guide ring and other parts. Power is provided through the actuator to overcome the medium pressure and provide the pressure for valve closing.

The core component of the labyrinth regulating valve is the labyrinth core package, as shown in Fig. 2. Labyrinth type valve core is made of many pieces of labyrinth type platter bonded with binder under specific conditions, or with no electrolytic nickel brazing material, after pressure, into the vacuum furnace heating sintering. The disc maze runner is made of electric corrosion machining. The maze piece can be designed to be thin, can accurately control the flow, good regulation performance.

When the fluid through the right angle bend, labyrinth regulating valve flow direction of the sudden change on the flow state has a great impact, making the turbulence changes more violent, turbulence will consume energy, thus increasing the resistance of the valve, reduce the flow velocity of the fluid through the valve, thus reducing the vibration of the valve. As shown in Fig. 3, the high-pressure fluid flows into the spool through the curved flow channel through the surrounding labyrinth core package. When flowing through the curved flow channel, the kinetic energy and pressure energy are converted to each other, making the high-pressure fluid depressurize to a certain extent. At the same time, each section of the flow path bears a close

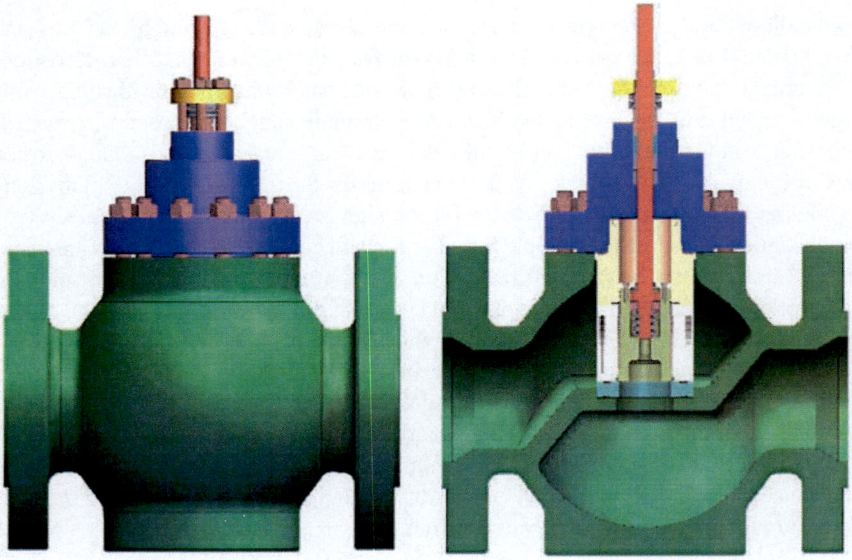

Fig. 1 The structure diagram of the labyrinth control valve

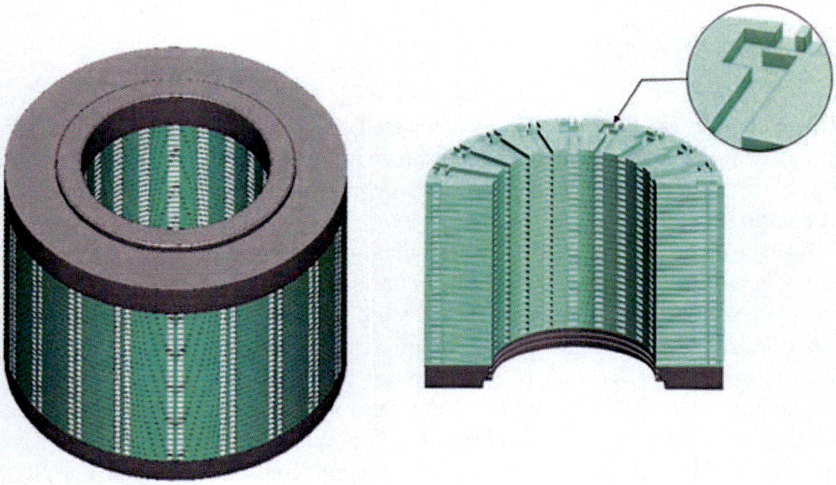

Fig. 2 Labyrinth control valve labyrinth core package

but small differential pressure. Avoid local pressure sharp drop caused by cavitation, reduce the noise caused by cavitation, and improve the service life of the valve.

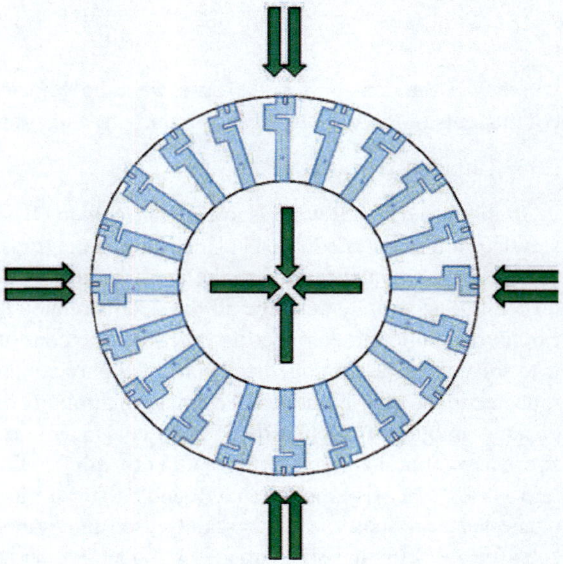

Fig. 3 Schematic diagram of labyrinth flow channel and circulation

2.2 *Control Equations and Turbulence Models*

(1) Turbulent transport equation

In terms of physical structure, turbulence is a flow formed by the superposition of various eddy currents of different scales. The size of eddy currents and the direction distribution of rotation axis are uncertain. Small-scale vortexes are mainly determined by viscous forces and large-scale vortices are mainly determined by boundary conditions. The large size vortex obtains energy from the main vortex and transmits the energy to the small size vortex through the interaction between vortices, and finally by the action of fluid viscosity. The small-scale vortices are disappearing, and the mechanical energy is converted into the heat of the fluid. Considering the incompressible flow in Cartesian coordinate system, the governing equation of instantaneous turbulence can be expressed as [10]:

$$div\,\boldsymbol{u} = 0 \tag{1}$$

$$\frac{\partial u}{\partial t} + div(u\boldsymbol{u}) = -\frac{1}{\rho}\frac{\partial p}{\partial x} + vdiv(gradu) \tag{2}$$

$$\frac{\partial v}{\partial t} + div(v\boldsymbol{u}) = -\frac{1}{\rho}\frac{\partial p}{\partial y} + vdiv(gradv) \tag{3}$$

$$\frac{\partial w}{\partial t} + div(w\boldsymbol{u}) = -\frac{1}{\rho}\frac{\partial p}{\partial z} + vdiv(gradw) \tag{4}$$

where ρ is the density, t is the time, $\boldsymbol{u}$ is the velocity vector, u, v, and w are the components of the velocity vector in the x, y, and z directions.

(2) Turbulence model

In this paper, the Reynolds averaging method (RANS) in the indirect numerical simulation method is adopted [11]. Although the transient Navier–Stokes equations can describe turbulence, due to the nonlinearity of the analytical method, it is still difficult to accurately describe all the details related to three-dimensional time and cannot be applied to engineering practice. The core of Reynolds averaging method is to solve the time-homogenized Navier–Stokes equation. The transient pulsation is reflected in the time-homogenized equation through a certain model, and the problem of large amount of calculation of DNS is also avoided. According to the Reynolds stress assumption and different treatment methods, the turbulence model is divided into vortex viscosity model and Reynolds stress model. The two equation model of the vortex-viscose model is widely used in engineering, and its most basic two equation model is the standard $k - \varepsilon$ model, which is divided into the equation of turbulent kinetic energy k and dissipation rate ε. There are also improved models RNG $k - \varepsilon$ and Realizable $k - \varepsilon$ models.

The $k - \varepsilon$ two-equation model is an equation that introduces the dissipation rate ε of turbulent kinetic energy k on the basis of turbulent kinetic energy, which was proposed by Launder and Spalding [12] in 1974, where the dissipation rate of turbulent kinetic energy is defined as:

$$\varepsilon = \frac{\mu}{\rho}\overline{\left(\frac{\partial u_i'}{\partial X_k}\right)\left(\frac{\partial u_i'}{\partial X_k}\right)} \tag{5}$$

The dynamic viscosity can be expressed as a function of k and ε,

$$\mu_t = \rho C_\mu \frac{k^2}{\varepsilon} \tag{6}$$

C_μ is an empirical constant.

In the standard $k - \varepsilon$ model, the transport equations corresponding to the two unknowns are:

$$\frac{\partial(\rho k)}{\partial t} + \frac{\partial(\rho k u_i)}{\partial x_i} = \frac{\partial}{\partial x_j}\left[\left(\mu + \frac{\mu_t}{\sigma_k}\right)\frac{\partial k}{\partial x_j}\right] + G_{k+}G_b - \rho\varepsilon - Y_M + S_k \tag{7}$$

$$\frac{\partial(\rho\varepsilon)}{\partial t} + \frac{\partial(\rho\varepsilon u_i)}{\partial x_i} = \frac{\partial}{\partial x_j}\left[\left(\mu + \frac{\mu_t}{\sigma_\varepsilon}\right)\frac{\partial\varepsilon}{\partial x_j}\right] + C_{1\varepsilon}\frac{\varepsilon}{k}(G_k + C_{3\varepsilon}G_b) - C_{2\varepsilon}\rho\frac{\varepsilon^2}{k} + S_\varepsilon \tag{8}$$

$$G_k = \mu_t \left(\frac{\partial u_i}{\partial x_j} + \frac{\partial u_j}{\partial x_i} \right) \frac{\partial u_i}{\partial x_j} \; G_b = \beta g_i \frac{\mu_t}{\Pr_t} \frac{\partial T}{\partial x_i} \tag{9}$$

$$C_{1\varepsilon} = 1.44,\; C_{2\varepsilon} = 1.92,\; C_{3\varepsilon} = 0.09,\; \sigma_k = 1.0,\; \sigma_\varepsilon = 1.3$$

Among them, u_i, u_j are transient velocity, G_k is the generation term of turbulent kinetic energy k caused by the average velocity gradient, and G_b is the generation term of turbulent kinetic energy caused by buoyancy [13], $C_{1\varepsilon}$, $C_{2\varepsilon}$, $C_{3\varepsilon}$ is an empirical constant, Y_M represents the contribution of compressible turbulent pulsation expansion, S_k and S_c is user-defined source item.

2.3 Calculation Model and Boundary Conditions

The model adopts a hybrid structure grid division method, as shown in Fig. 4. In order to improve the grid accuracy, the minimum grid size is set to 1 mm and the maximum is 10 mm, and Smooth Elements Globally performs grid smoothing processing, so that the grid quality is at a higher level. After division, the number of grids is about 5 million. Import the mesh into the CFD software, and the boundary conditions are related to the boundary. Among them, the pressure inlet is set as 7.2 MPa, the pressure outlet is 0.6 MPa, and the fluid medium is 25 °C normal temperature water.

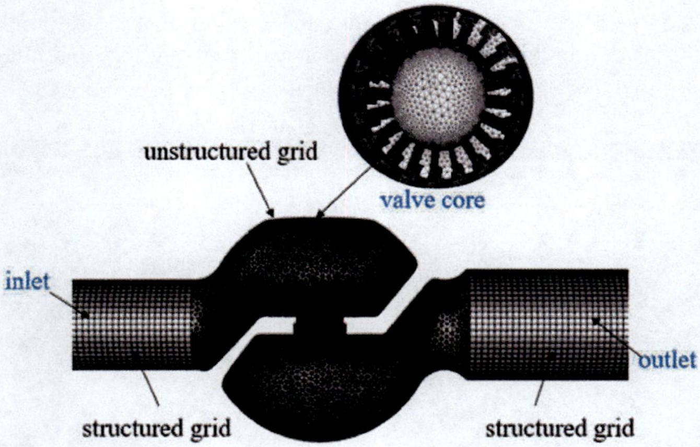

Fig. 4 Calculation model of labyrinth control valve

3 Results and Discussion

3.1 Velocity Field Analysis

As can be seen from the velocity contour, the flow velocity of the fluid is uniform in the flow passage of the valve body, but at the right angle turn, the flow state changes suddenly due to the obstruction, the kinetic energy is dissipated, and vortex is generated at the turn. When the fluid enters the split flow path from the total flow path, with the gradual increase of the area of the flow path (the section before and after each turn is a sudden expansion effect), the flow velocity at the outlet of the throttle groove is gradually reduced. Obviously, Fig. 5 illustrates under the same inlet pressure condition, the velocity of the fluid medium flowing through the maze channel at small opening is obviously greater than that at large opening under the condition of 30 and 70% valve opening. After passing through the labyrinth valve passage, the fluid medium velocity decreases with different opening degrees due to fluid hedging and kinetic energy consumption, but the opening degree has no significant influence on the relative size of the fluid velocity reduction. In other words, different valve openings have little effect on the change in velocity under the same inlet pressure.

However, the number of maze slots on the maze disc is generally designed according to the even number. The medium in the throtting groove flows into the throtting component in the central position of the hedge, consumption of kinetic energy, effectively avoid the direct erosion of high-speed medium flow on the inner wall of the parts. According to the empirical data, when the liquid medium flow rate in the valve is less than or close to 30 m/s, the erosion of the throttle element is minimum. Considering that the valve has near-wall pressure loss and other factors

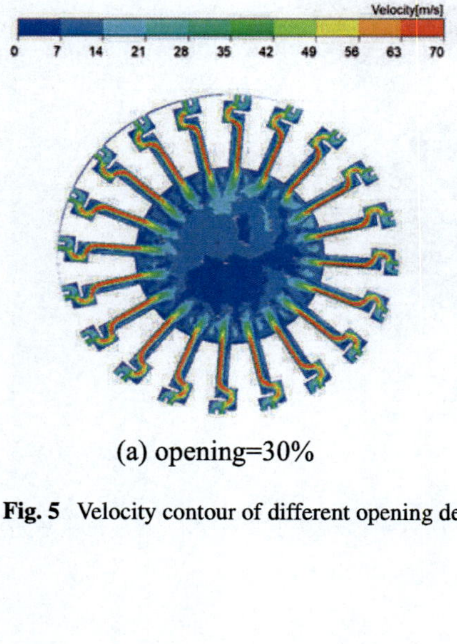

(a) opening=30%

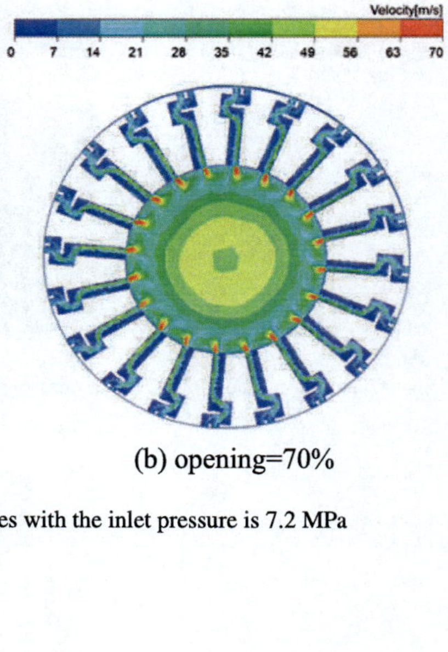

(b) opening=70%

Fig. 5 Velocity contour of different opening degrees with the inlet pressure is 7.2 MPa

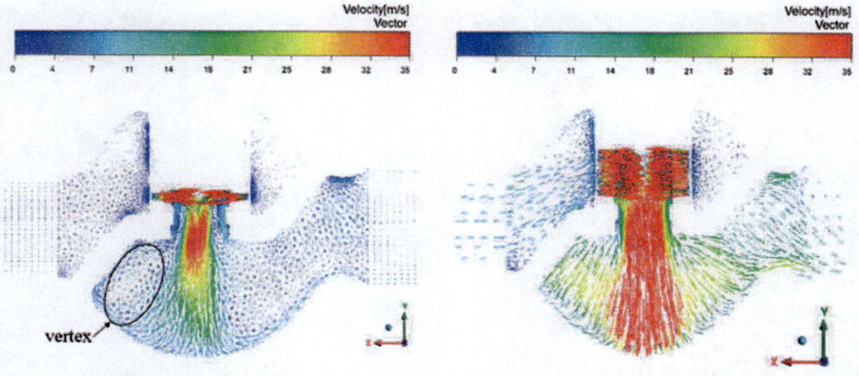

(a) opening=10% (b) opening=80%

Fig. 6 Velocity vortex contour with different opening degrees when the inlet pressure is 7.2 MPa

under actual conditions, and the current velocity is the local maximum flow velocity in the throttling groove, it can also be met.

It can be found through the calculation results: Fig. 6 shows, when the valve opening is small (10%), the speed at the edge of the spool opening is larger, but because the velocity distribution area is smaller, the impact on the valve body is smaller. At this point, you can see that the backflow area of the spool is formed. The center pressure in the reflux area is low and the vortex with small intensity is formed. At this time, because the valve opening is small, the high pressure in front of the valve flows through the reducing valve into low pressure flow, and the valve depressurization effect is obvious.

As valve opening increases, the backflow area begins to shrink. The mechanical energy of the fluid is mainly consumed by these vortices, and the vortices gradually weaken in the process of downstream development, and the flow gradually tends to be uniform, and the reflux area gradually disappears.

When the valve opening is 80% or more than 80%, it can be seen from the velocity cloud diagram that the fluid velocity distribution through the spool is more uniform, and there is no vortex on the back of the spool.

3.2 Pressure Field Analysis

For the fluid in the valve, when the medium pressure is lower than its saturated steam pressure, not only blocking flow, but also lead to the occurrence of flash phenomenon. According to the related theory of fluid mechanics, when the high pressure medium flows through a certain resistance element, the static pressure energy and dynamic pressure energy are converted to each other, and the increase of the flow rate will lead to the decrease of pressure. Therefore, increasing the resistance coefficient of

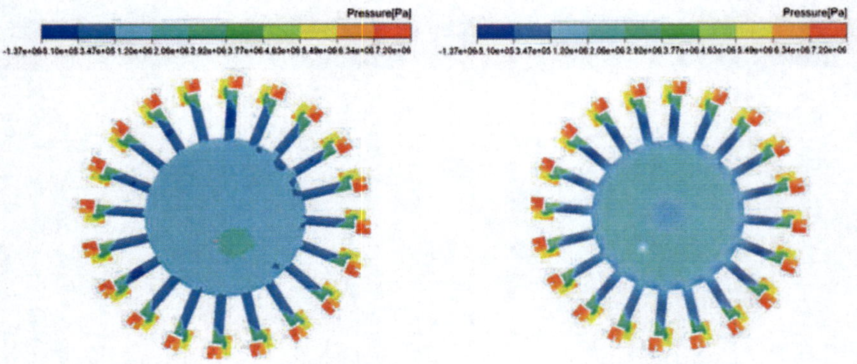

(a) opening=30% (b) opening=70%

Fig. 7 Pressure contour of different opening degrees when the inlet pressure is 7.2 MPa

the medium can achieve the purpose of high pressure drop. However, in the limited body space to create enough flow resistance for the medium, can not use a single step-down structure. Therefore, the labyrinth type regulating valve adopsts the principle of multistage step-down, the total pressure difference of the valve with the step-down method, so that each level of pressure difference $\Delta P_n < \Delta P_c$. (ΔP_n, the pressure drop of grade n, ΔP_c, the critical pressure drop corresponding to the blocking flow), can prevent the cavitation of the medium.

Labyrinth type core package is a labyrinth type high pressure difference control valve internal core parts, is made of multi-layer metal plate by vacuum brazing. All disc surface by spark erosion into many throttling groove, high pressure medium flows through a series of right-angle bend, due to its powerful role of flow resistance, have complete control of velocity pressure Shared by throttling groove step by step, so as to realize multi-stage decompression, greatly ease the elevation difference in pressure valve vibration, noise and cavitation phenomenon.

For a continuous fluid, the pressure drop is proportional to the square of the increase in velocity when the flow rate is too high. As can be seen from Fig. 7, the pressure drop in the series part of the medium is large. After flowing into the parallel shunt, the area of the flow channel gradually increases, the flow rate decreases, and the pressure drop at each stage gradually decreases. Therefore, in order to avoid cavitation, it is necessary to ensure that the minimum pressure of the medium in the series flow path is not lower than the saturated steam pressure of the medium under the working condition when the maze flow channel is designed in this case.

3.3 *Research on Flow Characteristics*

After unit conversion, the flow coefficient calculation formula is:

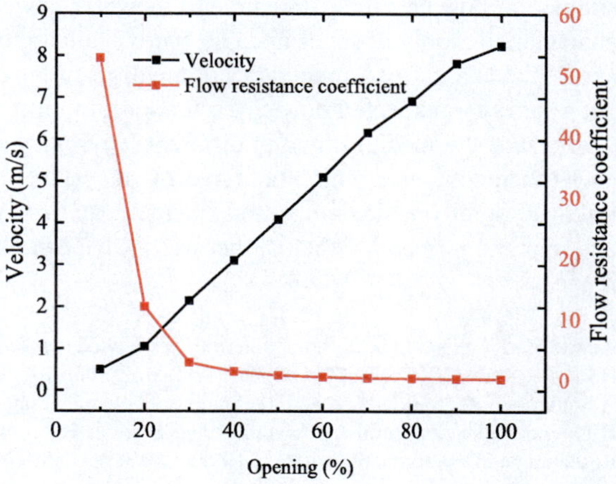

Fig. 8 Velocity and flow resistance coefficient change with opening

$$Cv = 1.156 \times 3.6 \times Q\sqrt{\frac{G}{\Delta P}} \quad (10)$$

In the formula, Q is the mass flow of the medium, kg/s; G is the specific gravity of the medium (at room temperature, the specific gravity of water is 1); ΔP is the pressure difference KPa before and after the valve.

Establish three-dimensional models at 5%, 10%...100% openings respectively, and set the boundary conditions as pressure inlet and pressure outlet. After numerical calculation, the flow coefficients at different openings are obtained in Fig. 8.

It can be seen that the simulation flow characteristic curve of the labyrinth valve has linear flow characteristic. But at low opening, the flow resistance coefficient is larger. Analysis of the reasons, mainly due to the low valve opening model and the actual model error. But on the whole, the simulated flow characteristics can reflect the actual ideal flow state more truly.

At present, the requirements of pressure difference and flow rate under different working conditions can be met by changing the size of the throttle groove and the series of turning. High pressure difference labyrinth regulating valve can meet the low opening when small flow high pressure drop, high opening when large flow low pressure drop flow characteristics, to meet the practical engineering application.

4 Conclusion

The flow velocity in the labyrinth valve core will not increase rapidly with the decrease of the valve opening. At the same inlet pressure and with little opening, the fluid velocity through the labyrinth valve is almost not affected by the valve opening.

The design of the labyrinth regulating valve (theory) flow characteristic curve is a parabolic flow characteristic, by comparison can be known: At large opening, the simulation flow characteristic curve of the valve is well in line with the design flow characteristic, only at small opening there is a large deviation, the surface through a reasonable and certain regularity of the distribution of labyrinth flow channel, can get a variety of expected flow characteristics, to meet the actual demand.

Because the medium flow in different labyrinth passages has little influence on each other, compared with other types of valves, the maze valve can only work in a relatively small opening range, and the maze valve can work in the minimum opening and maximum opening, that is, the maze valve can work in the maximum opening range.

Acknowledgements The authors gratefully acknowledge research support from the National Key R&D Program (2018YFB2004000), the Ningxia Youth Top Talent Project (2020), the Ningxia Autonomous Region Science and Technology Research (Support) Project (Key Technologies of High-end Valve Structural Optimization and Erosion Performance Research), the Ningxia Key Research and Development Project of China (Western Light,2017).

References

1. Khatri K, Phillips A, Gitlin H, Wu I (1979) Hydraulics of microtube emitters. J Irrig Drainge 105(2):163–173
2. Gilaad Y, Krystal L, Zanker K (1974) Hydraulic and mechanical properties of drippers. San Diego, California, USA
3. Li S, Wang T, Xu X (2018) A study on the vibration characteristics of high pressure drop sleeve trap. J Vibr Shock 37(4):147–152
4. Wei Z, Tang Y, Zhao W, Lu B (2005) Experimental study on the structure and hydraulic performance of the labyrinth channel of drip irrigation emitters. Trans Chinese Soc Agric Mach (12):51–55
5. Qin Z, Ye H, Yao H (2012) Simulation research on rectangular labyrinth drip irrigation emitter. Res Agric Mech 01:190–194
6. Yu L (2011) The relationship between structural parameters and hydraulic performance of the irrigation channel. J Changsha Univ Sci Technol (Nat Sci Ed) 01:30–35
7. Yao W (2013) Research on flow and safety of supercritical thermal power labyrinth valve. Shanghai Jiaotong University
8. Li CJ, Yan SK, Wen DD et al (2018) CFD analysis of flow noise at tees at natural gas station. Noise Control Eng J 66(1):1–10
9. Feng Z, Soulaimani A, Saad Y (2009) Nonlinear Krylov acceleration for CFD-based aeroelasticity. J Fluids Struct 25:26–41
10. Bendiksen GS (2008) Fluid-structure interactions with both structural and fluid nonlinearities. J Sound Vibr 315:664–684
11. Davis JA (2002) Predicting globe control valve performance-Part I; CFD modeling. Trans ASME 124
12. Liu H, Fu C, Ma B (2008) Numerical simulation of valve flow field based on dynamic grid and UDF technology. Steam Turbine Technol 50(02):106–108
13. Launder BE, Spalding DB (1974) The numerical computation of turbulent flows. Comput Methods Appl Mech Eng 32:269–289

Advanced Manufacturing

A Scrum-Based New Product Introduction (NPI) in Contract Manufacturing

Ang Chee Yiang, Chin Jeng Feng, and Nur Amalina binti Muhammad

Abstract Recent literature review shows an increasing number of new product introduction (NPI) projects realized in manufacturing sectors. The study implemented NPI using Scrum, a highly regulated team-based framework to complete a project progressively. Scrum is a popular framework adapted from the agile methodology to develop complex products and systems. The case study involved contract manufacturing (CM), and the customer actively participates in NPI and contributes to decision-making. Scrum successfully broke down, prioritized, and performed project tasks in successive cycles. The case study demonstrated the potential of applying Scrum in NPI to ensure the adherence of customer requirements, especially the timely fulfilment of the milestones.

Keywords New product introduction (NPI) · Scrum · Contract manufacturing (CM)

1 Introduction

Today business environments are characterized by increased globalization, market segmentation, product complexity, and changing customer needs [1]. The repercussions are shorter product life cycles, fierce competitions to introduce products [2] or rapid product changes [3]. Several studies have revealed that product life cycles have been reduced to three years, and new products might account for one-third of corporate sales [4, 5]. In this sense, introducing new products is critical to these businesses to offset slow profit growth [6, 7].

New Product Introduction (NPI) is the process of bringing an idea from an initial working prototype (or new product idea) to a thoroughly refined and replicable final product [8]. NPI offers products with the desired characteristics at the right timing and price [9]. Because NPI necessitates a substantial investment of time and resources,

A. Chee Yiang · C. Jeng Feng (✉) · N. A. binti Muhammad
School of Mechanical Engineering, Universiti Sains Malaysia (USM), 14300 Nibong Tebal, PG, Malaysia
e-mail: chinjengfeng@usm.my

M. Y. Ismail et al. (eds.), *Technological Advancement in Mechanical and Automotive Engineering*, Lecture Notes in Mechanical Engineering,
https://doi.org/10.1007/978-981-19-1457-7_58

a rigorous planning is essential in each step to ensure its effectiveness [10]. The planning requires a well-defined project methodology that establishes the guiding principles and processes for managing the project, including alignment with the organizational goals. Meanwhile, NPI exemplifies the collaboration of different manufacturing departments. Consequently, clear roles and responsibilities of key internal stakeholders and effective communication are essential for keeping the stakeholders engaged, task coordination, decision-making and problem-solving, identifying and resolving conflicts immediately, and quickly mitigating risks and resolving issues. Finally, knowledge sharing is crucial within an organization [11] so that the resultant experience and knowledge would be able to contribute to future organizational growth and knowledge base.

One of the potential approaches to NPI is Scrum [12]. Scrum is a scientific method approach to deal with complex adaptive problems, such as software development. Scrum is an adaptable and practical agile framework designed to deliver value to the customer throughout the project's development. This paper applies Scrum to NPI in a contract manufacturing (CM) case study. A CM is a manufacturer that contracts with a firm for components or products. In the last few years, outsourcing to CM has become an industry trend [13]. Traditional original equipment manufacturers (OEMs) decided to exit production-related activities, no longer their core competency. NPI in CM would actively involve the customer in different stages of decision-making and the project's schedule. Compared to the integrated design manufacturer [14], NPI in CM often exclude development processes such as conceptual study and engineering prototypes. Additionally, stakeholder analysis [15] becomes less critical in NPI, as the customer prescribes the product specification.

The paper's organization is as follows: Next, we review the literature, including the Scrum methodology and applications. Then, we introduce a four-stage method to incorporate Scrum into NPI. In Sect. 4, we discuss how Scrum methodology is implemented in an actual NPI project and the implementation results before the conclusion.

2 Scrum Methodology

Scrum was formulated in 1995 by Schwaber et al. [16] as a systematic approach to deal with large or complex adaptive tasks [17] through an incremental development process [16, 17]. Specifically, large tasks in a project are broken down into smaller ones to be prioritized and implemented by the Scrum team successively. Scrum methodology is widely implemented in Software Development Project Management [18]. It also has been applied successfully in other different applications: new product development processes [17], design and planning of construction firms [19], the context-based education courses [20], the healthcare industry [21], financial regulatory compliance projects [22], logistic equipment manufacturing [23] etc.

In the software development case study, Permana [18] observed that Scrum helps early and real-time project quality, risk issues, positive business change and customer feedback. Sutherland and Schwaber [24], Hassanein and Hassanien [25] commented that the success of Scrum in large projects and incremental development depends on the availability of self-organizing and collaborative cross-functional teams.

Figure 1 depicts the methodology. Popli and Chauhan [26] divided Scrum into planning, development, and postgame phases, and they defined Scrum starts with the team forming. Therefore, Scrum has four stages. The first stage, "Organize", involves Scrum setup of Scrum team, goals, and Scrum board. The Scrum team will be explained in this paragraph while goals and Scrum board elsewhere are being applied. The Scrum team consists of the project owner, Scrum master, and team members. The project owner represents the project's stakeholders and manages the project, including developing and explicitly communicating the project goal and requirement. Scrum master establishes and guides Scrum. The role is accountable for effective Scrum delivery. A Scrum master coaches and helps the team to create high-value increments and remove any project impediments. The Scrum team is a small self-organizing, and cross-functional team.

After "organize", there is an iterative cycle looping unidirectional from the stages "Plan", "Do", and "Review" until the whole project is completed. Each cycle is termed as "sprint", which occurs in a relatively short and fixed duration (e.g., one to four weeks) [24, 27, 28]. In "Plan", a sprint planning meeting is held to focus on what to do and how to do it. A sprint goal defines and shares to the team what the project owner would like to accomplish during the sprint. Project backlogs are created or revised by the team to list the team's tasks during a specific sprint. These tasks become sprint backlogs until they are completed. The product owner would prioritize the sprint backlogs and later discuss allocating the owner and execution time with the team.

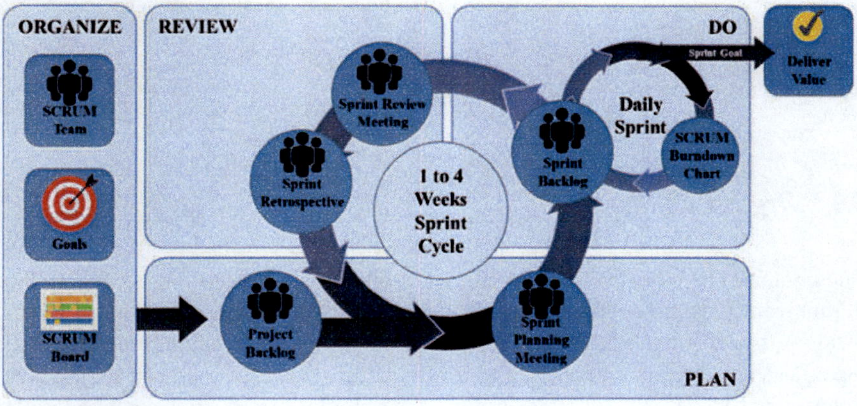

Fig. 1 Scrum methodology [16, 17]

At "Do", the team would deliver sprint backlogs, monitoring their progress closely through daily sprint meetings. The meeting, strictly short (15–30 min), is a common platform to review yesterday's tasks, the following tasks, and the difficulties faced when executing the sprint backlogs. The meeting is facilitated by a Scrum board, which is a visual board to visualize project progress. It has four columns: backlog, to-do, in-progress, and done. As the names imply, backlog: the project backlogs and the accompanying sprint backlogs; To Do are the sprint backlogs that have not been started yet. In Progress: The sprint backlogs, which the team is currently working on. Done: The sprint backlogs that have been completed and meet the required quality. It is common to use sticky notes to easily hold individual sprint backlogs (or project backlogs), and to transfer the information between columns. The team updates the Scrum board, shows the task status, and assigns pending sprint backlogs.

The total number of finished tasks from the sprint backlogs makes up the increment. The update is made on the Scrum burndown chart, a visual measurement tool that shows the completed tasks per day against the projected completion rate for the current task release. Its purpose is to enable the project to deliver value within the desired schedule.

Sprint review and sprint retrospective are conducted separately at the end of sprint [16]. Sprint review meeting aims to inspect sprint's outcome to meet customer requirements. The Scrum team presents their work to key stakeholders and progress. The Scrum team and stakeholders then review what has been accomplished (project gain) in the sprint and any arising, such as backlog re-prioritization. The Scrum team also proposes which pending project backlog to be introduced to the subsequent sprint [27]. Sprint retrospective concludes the sprint and plans ways to improve the quality and effectiveness of future Scrum.

3 Methodology

The methodology is explained based on the four stages described in Sect. 2 and information specific to NPI.

3.1 *Organize Stage*

The organize stage begins with soliciting customer requirements and setting up the Scrum team. The project owner would be the person formally assigned by the organization to lead the NPI. The typical customer requirements include product and material specifications, resource and cost structure, necessary certifications, NPI milestones, and demand forecast. Product and material specifications, applied to all variants of the product, include two items: a detailed bill of materials (BOM) and a bill of process (BOP) (the complete production steps to build and assemble the product). The demand forecast is the volume of products required by the customer over the time horizon. It covers the demand increased over the ramp-up period. Depending on the contract negotiated before the NPI, the customer would have different resource supply and cost influences, such as headcount and material supply.

Personnel from various departments need to be identified and formally invited into the Scrum team at different stages of NPI, based on the requirement of the project and sprint backlogs. Meanwhile, the members whose services are no longer required may surrender their membership to keep the Scrum team lean and agile. The prerequisite of a Scrum master should be an individual in a managerial role and with practical knowledge in Scrum and production management and leadership and commendable communication skills. The Scrum board and burndown chart are setup.

3.2 *Plan Stage*

This stage defines project backlogs and goals based on typical phases and milestones in NPI. In literature, Berg et al. [29] divided NPI phases into test production, pilot production and production ramp-up. On the other hand, Fjällström et al. [30] stated NPI phases include conceptual study, engineering prototypes, pilot production, pre-series production, and production ramp-up. According to Dekkers et al. [31], Product introduction involves cooperation and an integrated management approach between product design engineering and manufacturing. NPI is CM shows the transitions of responsibility from design and engineering to system integration and production, as presented in Fig. 2. At the early stage of NPI, the design and engineering are critical to verify technical information on product design, process and testing suitability. The role of system integration would be prominent at the mid-stage of NPI, as the product and processes are introduced to the production system, started with integrating the information system and physical setup of the relevant processes, production flow and material supply channel. Finally, near the end of NPI would associate with a trial run, fine-tuning implementation, and mitigate any ramp-up issue in production. This would closely monitor production and delivery key performance indicators (KPI), e.g., process or quality stability.

Fig. 2 Significance of role versus NPI timeline

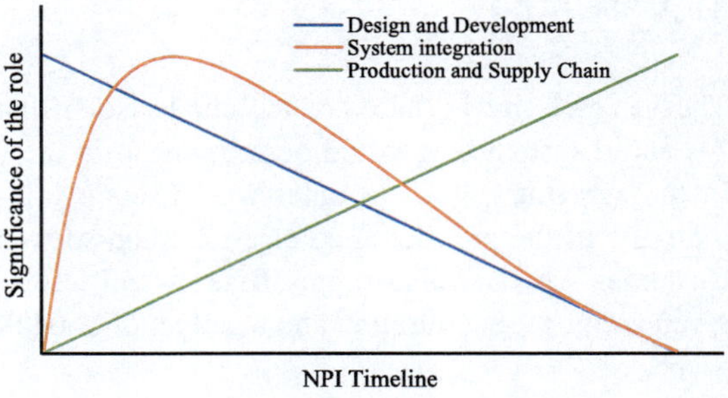

Next, relationships between the project backlogs are determined to form a logical sequence of execution. Any NPI deadline needs to be explicit. Sprint backlogs and goals stemmed from the project backlogs can be planned in tandem. This allows job allocation at the early stage of NPI. This also allows the standardization and simplification of work procedures, materials and parts on the entire NPI process [32].

3.3 Do Stage

The team would decide the regime of daily sprint meetings, including setting the ground rules for the meetings to run effectively. Scrum encourages standup meetings where the team members would update the progress under their responsibilities. The project owner will map the progress to the milestones. Any backlog delegated to external parties, such as suppliers, should be closely monitored to ensure sufficient progress. The Scrum board would be updated by closing the completed backlogs and initiating new backlogs.

3.4 Review Stage

The sprint review would first examine the progress against the backlog goals (or sprint goals) and given milestones in this stage. The project owner and Scrum master would decide on the acceptability of any deviation and take proactive measures against any imminent project risk, e.g., future backlogs, assessing their readiness and dispatching prior notice to the relevant party. The project owner would have to make an update to the management and customer. The sprint retrospective examines and discusses methods to improve the internal running of Scrum. Scrum master is the key to decide which method to be implemented.

4 Case Study

The NPI case study is based on an actual case. In this NPI, the "Thyme" is Customer A's audio system that would be introduced in the CM B plant located in Penang. In the agreement, the customer would retain full product ownership and absorb the costs of the pre-identified electrical and software-related issues in production. Customer A would also supply fixtures and testers for the product. The NPI was given a four-month duration and a budget of $60,000.

4.1 Organize Stage

The product manager acted as the project owner and the continuous improvement (CI) manager the Scrum Master. Other Scrum team members included a customer representative, final assembly production supervisor, production planner, process engineer (PE), quality engineer (QA), facility engineer (Fac), information technology engineer (IT), and human resource personnel (HR). Their heads of department formally endorsed their involvement. The customer representative is the single point of contact with the customer. His primary role was the product knowledge and information feeder, providing and clarifying the specifications and requirements.

4.2 Plan Stage

The project owner prepared the NPI plan, facilitated by the Scrum master and the customer representative. The NPI plan, detailing the stages, schedule and milestones, was reviewed and agreed upon by management and the customer. The customer also approved two weeks of contingency buffer.

From the NPI plan, project and sprint backlogs were identified and sequenced. 10 project backlogs and 37 sprint backlogs were defined, starting with the review of the BOM and BOP. The BOM revealed 478 product components of three categories in three essential processes, they are surface-mount technology (SMT), through-hole technology (THT), and final assembly (FA) process, as shown in Table 1. The customer would consign (direct supply) components of Category A, e.g., specific integrated circuits (IC), to the production, while CM B would source other components from its supply chain. Next, the process flow was determined from the BOP. The information would lead to a manufacturing capacity plan, considering the ramp-up demand. System integration would be performed following standards set in the Quality management system (QMS). The integration involves defining the relevant product codes, material codes, process codes in the Enterprise Resource Planning (ERP) system.

Additionally, the manufacturing and industrial engineering database would update with the new FA process layout, process standards, maintenance and calibration plan (testers). Materials (Category B and C) would be sourced from the existing suppliers. Other products have used these components in CM B. They do not need to be certified again.

The high-end audio product has many more stringent specifications than the CM existing products'. After the trial run, process failure modes and effects analysis (PFMEA) has to be performed to systematically identify potential failure modes, causes, and effects in the production system. At the same time, representative samples of the product in the whole package system were retrieved from the trial run and subjected to the shipping simulation, which involves testing on the transportation settings by a third-party laboratory.

Table 1 Thyme components category

Process	Category	Components	Total components
SMT	A	9	359
	B	162	
	C	188	
THT	A	16	67
	B	39	
	C	12	
FA	A	28	52
	B	14	
	C	10	
Total			478
ABC classification			
A	High value (>$10)		
	No sharing of product used		
	E.g., integrated circuit chips (IC), bare board		
B	Intermediate value ($5 < X < $10)		
	Share of products percentage (1–3 products used)		
	E.g., transistor, jumper/connector, mic		
C	Low value (<$5)		
	Share of products percentage (> 3 products used)		
	E.g., resistors, capacitor, label, foam		

The Scrum master estimated the days and hours required for each sprint, as shown in Table 2. The customer and top management focused heavily on "Production floor ready (FA)" and "Production trial run & product validation", as they indicate the actual implementation and significantly consumed the budget.

4.3 Do Stage

In daily sprint meetings, chaired by the Scrum master, each team member reported backlog progress under their assignments and any accompanying obstacle. The progress was then translated to the Scrum board and burndown chart. An example of a Scrum board and burndown chart captured during Thy 6-Product floor readiness are shown in Figs. 3 and 4. The product would undergo three process stages: SMT, THT, and FA process. SMT and THT are standardized processes, and FA is a dedicated line of 17 process steps and 2 test steps. The line, laid out in the form of assembly cell upon line balancing, would have 17 workstations; nine are with fixtures and two for testers.

Table 2 Thyme project backlogs

Scrum ID	Project backlog	Sprint backlog	Sprint goal	Total days	Total hours
Thy1-Product architecture	Product architecture	• Product hardware • Product software • Product test script	Product architecture	5	54
Thy2-Process flow diagrams	Develop process flow diagrams	• SMT process flow • THT process flow • Test process flow • FA process flow	Process flow diagrams	8	84
Thy3-Capacity planning	Overall capacity planning	• SMT build plan • THT build plan • Test build plan • FA build plan • Final test build plan	Complete overall capacity plan	5	45
Thy4-Material supply	Material supply	• Material supply plan	Complete overall supply chain plan	5	64
Thy5-System integration	System integration	• Gather whole system requirement • Analyze system • System architecture design • System implementation • System maintenance plan	Integrate NPI into the existing system	10	215
Thy6-Product floor readiness	Production floor ready (Assembly)	• Production cell readiness • Utility readiness • Assembly fixture readiness • Tester readiness	Assembly production floor ready	5	71

(continued)

Table 2 (continued)

Scrum ID	Project backlog	Sprint backlog	Sprint goal	Total days	Total hours
Thy7-Trial run	Production trial run and product validation	• Production cell readiness • Operator arrangement • Utility readiness • Assembly fixture readiness • Tester readiness	Complete test build and product validation	10	242
Thy8-PFMEA	Process FMEA	• Process FMEA readiness • Benchmarking previous product control plan	Complete process FMEA	5	60
Thy9-Shipment simulation	Shipment simulation	• Final product ready • Product pallet arrangement • Drop test readiness • Document readiness	Ship out product	5	57
Thy10-Product certification	Product certification	• Application (including testing of the product) • Evaluation (does the test data indicate that the product meets qualification criteria) • Document readiness	Product certified	5	79
Total				63	971

More members from the design and engineering departments were involved in the several initial project backlogs (Thy1–Thy 5). Their memberships were gradually replaced by those from the industrial engineering and information technology departments. New members, mainly from production, were called in for production setup and trial run. The customer representative was only present at every Tuesday's meeting (physically present or online). An exception is in sprinting Thy4-Material

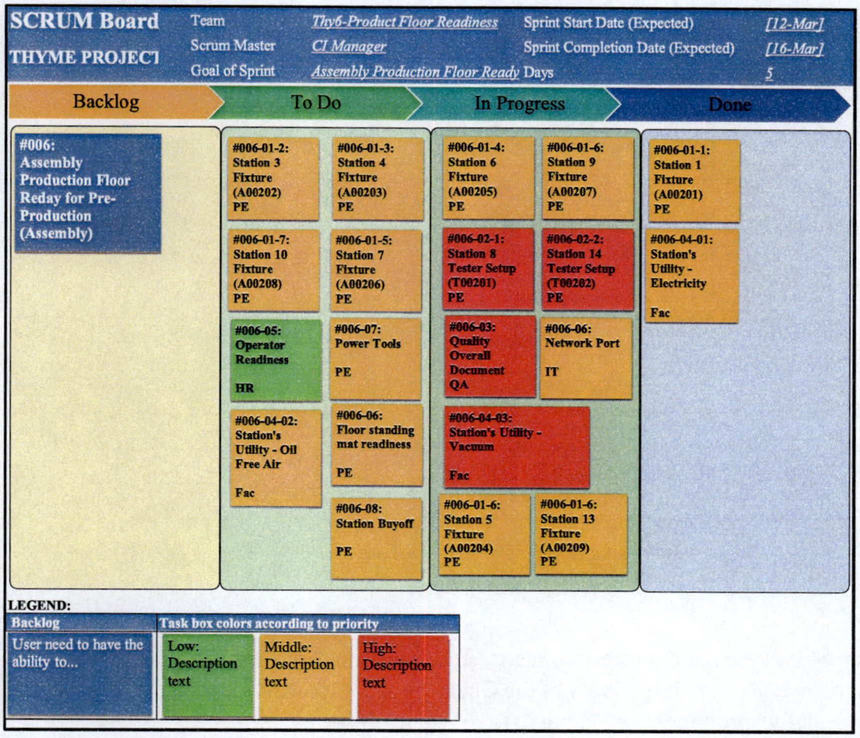

Fig. 3 Scrum board (Thy6-product floor readiness)

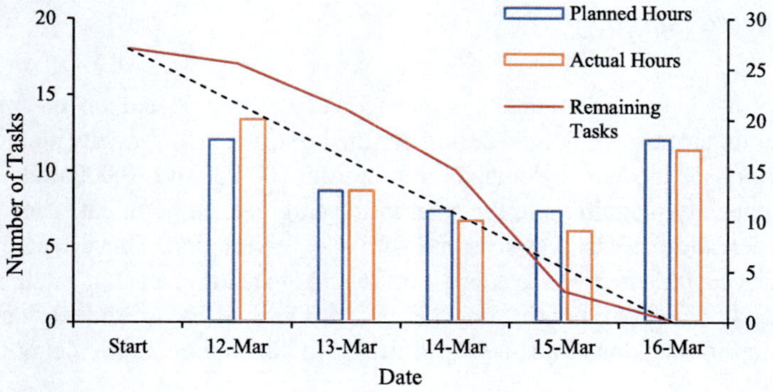

Fig. 4 Burndown chart (Thy6-product floor readiness)

Table 3 Daily sprint meeting

	Day				
Time	Monday	Tuesday	Wednesday	Thursday	Friday
8:00 AM					
9:00 AM					
10:00 AM–10:30 AM	Sprint meeting	Sprint meeting (Customer alignment meeting)	Sprint meeting	Sprint meeting	Sprint meeting
10:30 AM–11:00 AM					
12:00 PM					
1:00 PM					
2:00 PM					
3:00 PM					
4:00 PM					
5:00 PM					

supply, where regular meetings were setup on alternate days with the customer representative to align the delivery of the Category A components needed for the trial run. The daily sprint meeting schedule is shown in Table 3.

4.4 Review Stage

In the sprint review, customer and management were updated on the progress. Three significant issues emerged during Scrum. First, at Thy3-Capacity planning, the customer revised the annual demand forecast from 200,000 to 180,000 units. Consequently, capacity planning and the relevant costing needed an update. Second, the monthly demand forecast was seasonal with a huge variation. This affects material supply. The customer agreed to adopt level loading to achieve a balanced throughput rate of 15,000 units per month, shown in Fig. 5. Third, at Thy6-Product floor readiness, the sprint backlogs must be reprioritized to accommodate the delay in tester delivery.

Table 4 compares the planned and actual hours of project backlogs. The overall difference is marginal (− 1.54%). Thy5-System Integration contributed to 5 days' delay, as the associated tasks required adherence to the various systems in the responsible departments.

Scrum master realized the need to brief team members fresh to the Scrum methodology in the sprint retrospective. Standard communication procedure was defined to ensure timely dissemination of information in adequate detail, mainly the project

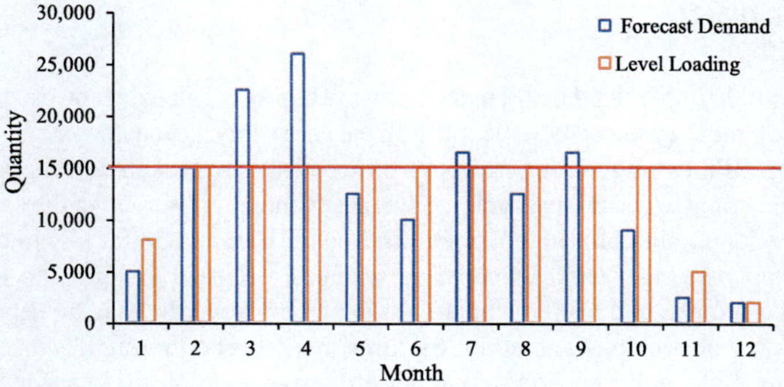

Fig. 5 Thyme demand forecast and level loading plan

Table 4 Comparison of total hours in the sprint (planned vs actual)

Scrum ID	Total hours Sprint (Planned)	Total hours Sprint (Actual)	Percentage difference (%)
Thy1-Product architecture	54	52	– 3.70
Thy2-Process flow diagrams	84	83	– 1.19
Thy3-Capacity planning	45	42	– 6.67
Thy4-Material supply	64	65	1.56
Thy5-System integration	215	220	2.33
Thy6-Product floor readiness	71	69	– 2.82
Thy7-Trial run	242	235	– 2.89
Thy8-PFMEA	60	58	– 3.33
Thy9-Shipment simulation	57	57	0.00
Thy10-Product certification	79	75	– 5.06
Total	971	956	– 1.54

requirement update. The process is also helpful to minimize confusion in instructions and authority, especially to various non-regular sub-teams formed at Thy6-Product floor readiness to undertake specific tasks.

5 Conclusion

This case study implemented NPI using Scrum, a highly regulated team-based framework to complete a project incrementally. In the case study, the customer was actively engaged in NPI. During Scrum, extensive project tasks are broken down, prioritized, and implemented in successive cycles. Several advantages observed in the case study are timely completion of NPI, efficient resource utilization, regular alignment (e.g., daily scrum meeting), rapid response to changes, progress visibility to management and customer, and team member accountability. Nevertheless, the dependency between project backlogs and sprint backlogs may not be thoroughly investigated in planning, potentially leading to conflict of these backlogs during sprints. Future work will identify suitable tools to improve on this aspect.

Acknowledgements This work was supported by Research University Grant (RUI), Universiti Sains Malaysia [grant number 8014069].

References

1. BüyüKözkan G, Derelİ T, Baykasoğlu A (2004) A survey on the methods and tools of concurrent new product development and agile manufacturing. J Intell Manuf 15(6):731–751
2. Letens G (2015) Lean product development—faster, better… cleaner? Front Eng Manage 2(1):52–59
3. Slamanig M, Herwig W (2012) Management of product change projects: a supply chain perspective. Int J Serv Oper Manag 11(4):481–500
4. Cooper RG, Edgett SJ, Kleinschmidt EJ (2006) Optimizing the stage-gate process: what best practice companies are doing—part 1. Working Paper No. 14. Available at: http://www.prod-dev.com/pdf/Working_Paper_14.pdf
5. Rahim ARA, Baksh MSN (2003) The need for a new product development framework for engineer-to-order products. Eur J Innov Manag
6. Cooper RG, Kleinschmidt EJ (2000) 2 new product performance: what distinguishes the star products. Aust J Manag 25(1):17–46
7. Tabassum M, Ozuem W (2019) New product development and consumer purchase intentions: a literature review. Conference Paper. Research Gate, UK
8. Nadkarni S, Chen J (2014) Bridging yesterday, today, and tomorrow: CEO temporal focus, environmental dynamism, and rate of new product introduction. Acad Manag J 57(6):1810–1833
9. Parry G, Graves A, James-Moore M (2008) Lean new product introduction: a UK aerospace perspective. University of Bath School of Management Working Paper Series 3, pp 1–45
10. Ardito L, Ernst H, Petruzzelli AM (2020) The interplay between technology characteristics, R&D internationalization, and new product introduction: empirical evidence from the energy conservation sector. Technovation 96:102–144
11. Cabrera EF, Cabrera A (2005) Fostering knowledge sharing through people management practices. Int J Human Resour Manag 16(5):720–735
12. Mahalakshmi M, Sundararajan M (2013) Traditional SDLC versus scrum methodology–a comparative study. Int J Emerg Technol Adv Eng 3(6):192–196
13. Ülkü S, Toktay LB, Yücesan E (2007) Risk ownership in contract manufacturing. Manuf Serv Oper Manag 9(3):225–241

14. Ruffles PC (2000) Improving the new product introduction process in manufacturing companies. Int J Manuf Technol Manage 1(1):1–19
15. Jepsen AL, Eskerod P (2009) Stakeholder analysis in projects: challenges in using current guidelines in the real world. Int J Project Manage 27(4):335–343
16. Schwaber K, Sutherland J (2018) The definitive guide to scrum: the rules of the game. Informe de Investigación
17. Betta J, Chlebus T, Kuchta D, Skomra A (2019) Applying scrum in new product development process. In: International scientific-technical conference manufacturing, pp 190–200. Springer, Cham
18. Permana PAG (2015) Scrum method implementation in a software development project management. Int J Adv Comput Sci Appl 6(9):198–204
19. Streule T et al (2016) Implementation of scrum in the construction industry. Procedia Eng 164:269–276
20. Vogelzang J, Admiraal WF, Van Driel JH (2019) Scrum methodology as an effective scaffold to promote students' learning and motivation in context-based secondary chemistry education. Eureasia J Math Sci Technol Educ 15(12):em1783
21. Jakupovic A (2021) Implementation of the scrum method in a healthcare project: a case studied at Region Västra Götaland
22. Beerbaum D (2021) Applying agile methodology to regulatory compliance projects in the financial industry: a case study research. In: Applying agile methodology to regulatory compliance projects in the financial industry: a case study research
23. Abrudan CI, Baru PE, Lungu R (2019) Applying scrum methodology into the production processes of a logistic equipment producer. Int J Curr Sci Res Rev 2(09):76–81
24. Sutherland J, Schwaber K (2013) The scrum guide. The definitive guide to scrum: the rules of the game. Scrum.org 268
25. Hassanein EE, Hassanien SA (2020) Cost efficient scrum process methodology to improve agile software development. Int J Comput Sci Inf Secur (IJCSIS) 18(4)
26. Popli R, Chauhan N (2011) Scrum: an agile framework. Int J Inf Technol Knowl Manag 4(1):147–149
27. Fowler FM (2019) What is scrum? Navigating hybrid scrum environments. Apress, Berkeley, CA, pp 3–8
28. Ullman DG (2019) Scrum for hardware design: supporting material for the mechanical design process. David Ullman LLC
29. Berg M, Fjällström S, Stahre J, Säfsten K (2005) Production ramp-up in the manufacturing industry: findings from a case study. In: The 3rd international conference on reconfigurable manufacturing, Ann Arbor, MI, US
30. Fjällström S, Säfsten K, Harlin U, Stahre J (2009) Information enabling production ramp-up. J Manuf Technol Manag
31. Dekkers R, Chang CM, Kreutzfeldt J (2013) The interface between "product design and engineering" and manufacturing: a review of the literature and empirical evidence. Int J Prod Econ 144(1):316–333
32. Berglund M, Harlin U, Gullander P (2012) Challenges in a product introduction in a cross-cultural work system–a case study involving a Swedish and a Chinese company. Swedish Prod Symp

End-Mill Carbide Tool Wear in Machining Metallic Biomaterial

Azli Ihsan Yahaya, Saiful Anwar Che Ghani, Daing Mohamad Nafiz Daing Idris, and Mohd Azwan Aziz

Abstract Machining of metallic biomaterials causes a slew of issues, including cutting tool wear and poor surface quality owing to inefficient tool design, which leads to excessive heat output. The objective of the research is to evaluate the wear of developed of uncoated carbide endmill tool with rake angle varied from positive to negative value in dry machining Stellite 21. The fabricated endmill is tested at Fanuc Robodill α-T14iFb with cutting conditions parameters are kept constant; including cutting speed (Vc): 60 m/min, feed rate (f): 153 mm/rev, and depth of cut (ap): 0.2 mm, throughout the cutting trials. The accuracy of fabricated endmill, wear mechanism, cutting force, and surface roughness were measured using Dino-Lite Microscope, Scanning Electron Microscope, Neo-Momac Dynamometer and Mitutoyo Surface Profiler, respectively. The result shows that by using a positive rake angle, the phenomenon of tool wear is reduced, and directly reducing the surface roughness and cutting force. Based on energy dispersive x-ray (EDX) element analysis, presence of oxygen in the cutting process which indicates the occurrence of oxidation wear on cutting tool. Extended observation of wear mechanism show high content of chromium on the flank face is revealed that indicated the diffusion wear on tools has occurred. In conclusion, the enhancement of tool geometry of endmill cutting tool is a key step toward sustainable manufacturing of high-end applications in biomedical industries.

Keywords End-mill carbide tool · Tool wear · Dry machining · Surface roughness

A. I. Yahaya (✉) · S. A. Che Ghani · D. M. N. Daing Idris
Advanced Fluid Focus Group, Faculty of Mechanical and Automotive Engineering Technology, Universiti Malaysia Pahang, 26600 Pekan, Pahang, Malaysia
e-mail: azli.yahaya@kuantan.kktm.edu.my

A. I. Yahaya
Kolej Kemahiran Tinggi MARA Kuantan, KM 8, Jalan Gambang, 25150 Kuantan, Pahang, Malaysia

M. A. Aziz
Zauber Engineering Sdn Bhd, Jalan Mega A, 43500 Bandar Teknologi KajangSemenyih, Selangor, Malaysia

M. Y. Ismail et al. (eds.), *Technological Advancement in Mechanical and Automotive Engineering*, Lecture Notes in Mechanical Engineering,
https://doi.org/10.1007/978-981-19-1457-7_59

1 Introduction

The impact of economy growth of metal forming can easily be discovered in our life, with approximately 15–20% of the GDP of developed nations resulting from processes of metal forming [1]. Malaysia is now becoming a relatively medium-sized country with 25% of GDP dominated by a manufacturing sector [2]. Machining operations including milling, turning, drilling, grinding and other cutting operations consume a massive amount of money where US$ 100 billion worldwide is invested every year [3]. Based on previous research, it show that unoptimized tools design that cause tool failure, which damages the surface quality of the machining product thus directly affected into machining economy [4]. Consequently, about 10% of all metals produced in the machining industry are transformed into waste volume [3]. The probability of machining saving is estimated up to 20% with the right tool design and parameter method [5].

End mills are the most commonly cutting tool used for machining process [6]. Endmill cutter plays an important role in the metal cutting industry in obtaining the various shapes and sizes required. Because of the advantageous combination of high hardness with relatively high durability and strong wear resistance to temperature, cemented carbide is suitable for machining [7]. Cemented carbide tool capable in heavy process loads and allows for fast cutting speeds [8]. For these huge benefits reasons, cemented carbide tools are used to process difficult -to-cut materials in biomedical application such as titanium and its alloys [9], cobalt based alloys [10] and 316L stainless steel [11]. While there are modern growths in near-net shape forming techniques in the manufacture of biomedical components, machining remains in the medical industry as favored manufacturing processes. The machining technique with the optimum cutting tool configuration is used to achieve greater efficiency while at the same time to ensure consistent surface finish and dimensional accuracy of the parts created [12].

Machining of high-hardness materials such as metallic biomaterial causes a slew of issues, including cutting tool wear and poor surface quality owing to under-optimized tool design, which results in excessive heat generation [13]. Changing the design of tool configuration such as variation of rake angle give minimize the work hardening area at the machined surface and built-up-edge [14]. However, the wear mechanism of cutting tool with rake angle ranging from negative to positive rake angle in machining biometallic material is hardly discussed scientifically. In this study, the focus is concentrated in evaluating the wear of developed endmill carbide tool in machining metallic biomaterial Co–Cr–Mo alloy. The wear and the wear mechanism are studied based on the variation of rake angles. The tool performance is monitored by evaluation of generated surface roughness of the turning workpiece.

2 Methodology

2.1 Manufacturing of Flat End-Mill Tool

In end-mill tool production, there are four major machining operations: flute, first and second clearance face, gashing face, and the first and second end-flat surface [15]. In general, the geometry of a flat end-mill consists of two parts: the geometry of the flute surface and end-flat surface. The machining process involve in flute surface is helical fluting and outer diameter clearance and the geometry involved in the end-flat surface is gashing process and the end face clearance. Figure 1 depicts the process flow in producing end-mill tool and the geometrical features of the tool.

Flute grinding is a critical process in the production of end-mills. Throughout the grinding process, the geometry and orientation of the wheel governs the flute profile and defines the flute parameters such as rake angle, core size, and flute width [16]. Gash angle is one of the major geometries during end-mill cutting tool production. The purpose of the gashing angle is to provide a secondary cut to provide chip area at corners and ends. It also forms an end cutting edge as feed is axially positioned. End-mill has three kind of outer diameter clearance that is flat shapes, concave and eccentric as shown in Fig. 2 [17]. Commonly in the flat shape, there are the peripheral cutting-edge act as ODC 1 and the radial primary relief act as ODC 2. The peripheral cutting edge is a term that is used to describe the cutting edge on periphery of the tool. The radial primary relief is a ground surface right behind the cutting edge to prevent contact with the workpiece.

The end-face clearance for cutting the output portion as a single point cutter. Three forms of end face are network thinning. Online technology absorbs the strength and torque required to plow through the job. In the C-axis, the blank of the workpiece is shifted from the original location. It rotates in the actual direction and the wheel shifts to the start point coordinate X, Y, Z.

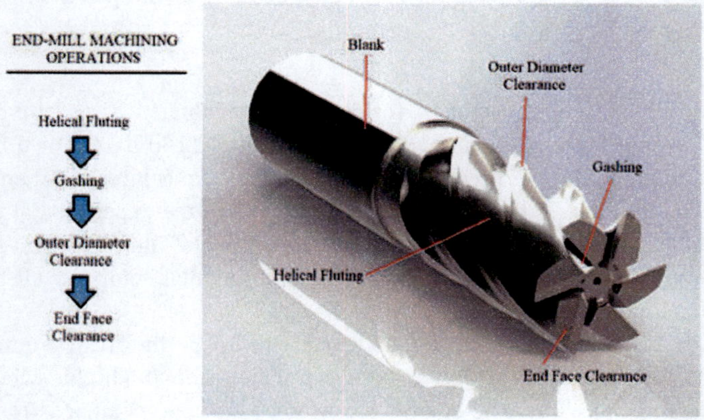

Fig. 1 Process flow in manufacturing end-mill tool and geometrical features of the tool

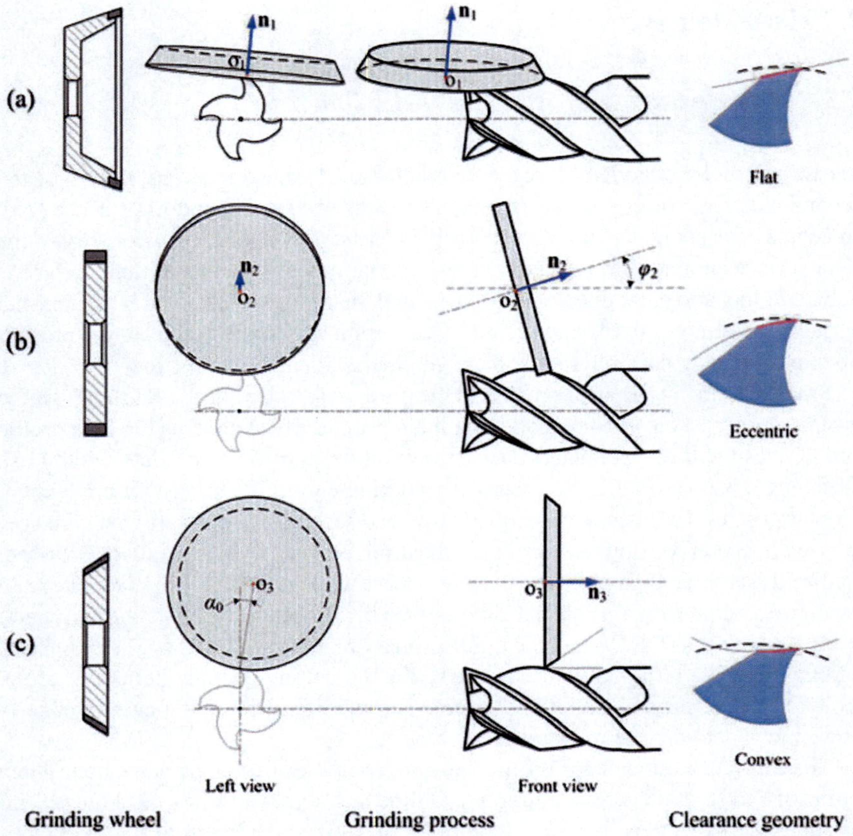

Fig. 2 Types of end-mill tool outer diameter clearance

2.2 *Evaluation of Cutting Tool Performance in Experimental Trials*

In-house developed endmill tool performance was evaluated using computerized numerical control milling machine, Fanuc Robodrill α-T14iFb as shown in Fig. 3. The evaluation required workpiece Stellite 21 bars to be machined until end of tool life. As listed in Table 1 the cutting parameters used in the experimental trials are constant depth of cut, feed rate, and cutting speed of 0.2 mm, 153 mm/min, and 60 m/min, respectively. Figure 3b shows the experimental setup for all the tests, which were all done in dry cutting.

Dino-Lite Microscope is used to capture in-situ image the fresh produced tool geometrical accuracy and progressive tool wear of end-mill with highly accurate and fast. Cutting force measurement device Neo-Momac Dynamometer has been utilized

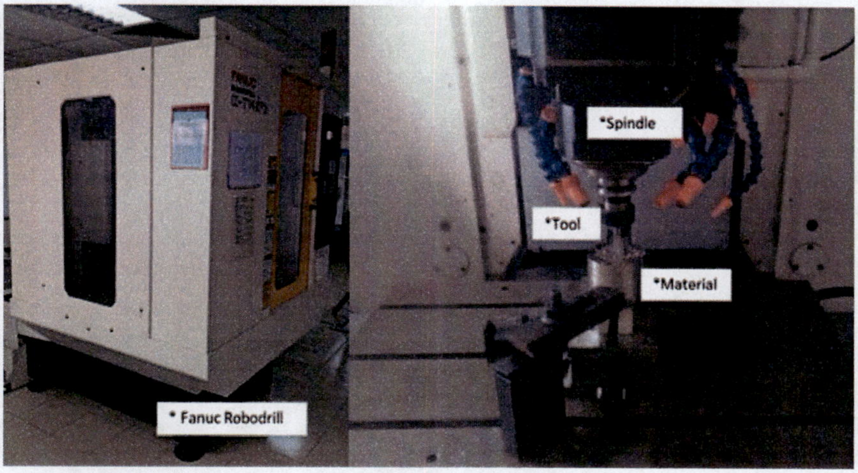

(a) (b)

Fig. 3 Experimental setup **a** Fanuc Robodrill α-T14iFb, **b** Machine configuration

Table 1 Cutting tool types and machining parameters

Variable	Parameter	Type cutting tool	Condition
Cutting speed, V_c (m/min)	60	Uncoated	Dry
Feed rate, f (mm/min)	153		
Depth of cut, a_p (mm)	0.2		

for measuring real-time 3-dimensional cutting force generated in machining of Co–Cr–Mo based workpiece. The roughness of the surface was assessed using a mobile Mitutoyo SJ-310 model surface roughness measuring device. End of tool life is by measuring flank wear at the cutting edge of the tools used in the each cutting trial in accordance with the ISO standard for end milling tool life testing (ISO8688-2:1989).

3 Results and Discussion

3.1 Tool Wear

In this study, the fabricated carbide end-mill tool; 10 mm in diameter with variation rake angle (6°, 10° and 16°) is tested with constant cutting parameters combination. All cutting tools could perform machining at complete cutting length, 2325 mm. The progressive cutting tool wear of the tools was measured by measuring the wear land at the cutting edge after each cut, maximum tool wear land, VB_{max} of the tools from initial cutting until they reached end of tool life was determined after the max wear land exceeded the criteria, $VB_{max} = 0.4$ mm, as shown in the sample in Fig. 4.

The progressive tool wear results for the cutting tools with varied rake angle are shown in Fig. 5. What stands out from the graph is that the tool life of the carbide end mill tools with lower rake angle (6° and 10°) ended at about similar cutting length ~ 1395 mm with also similar progressive wear patterns. However, the rate of tool wear for large rake angle, 16° is superior than small rake angle. The tool with 16° rake angle demonstrated increment of about 20% extended tool life. These results reflect the notion as the rake angle increases in positive direction, the tool–chip contact length decreases. This observation may lead to a decrease in generated cutting temperature attributed by a smaller contact area available for friction [18].

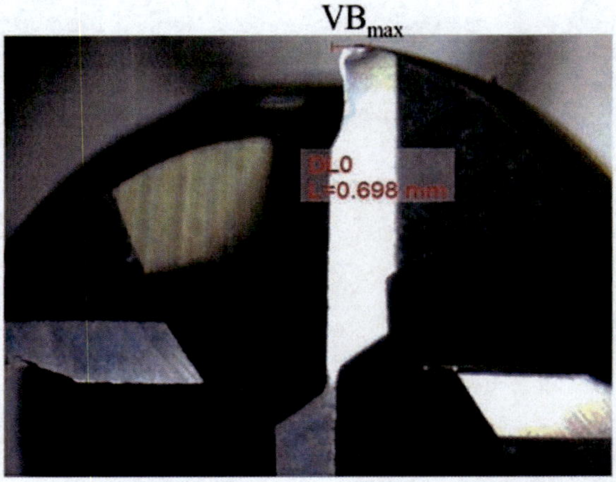

Fig. 4 Flank wear land, VB_{max}

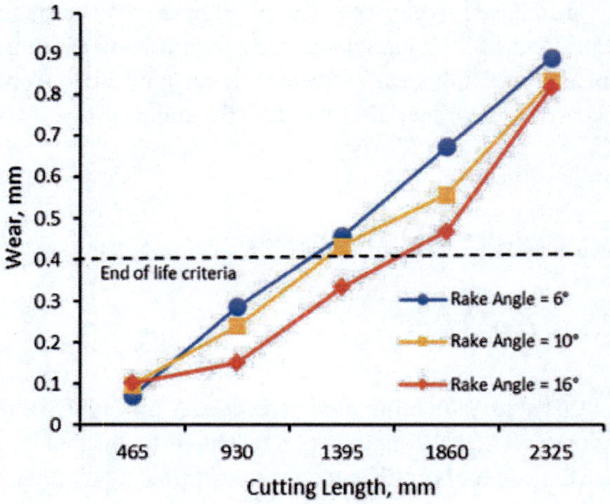

Fig. 5 Flank wear versus cutting length in machining Co–Cr–Mo-based alloy

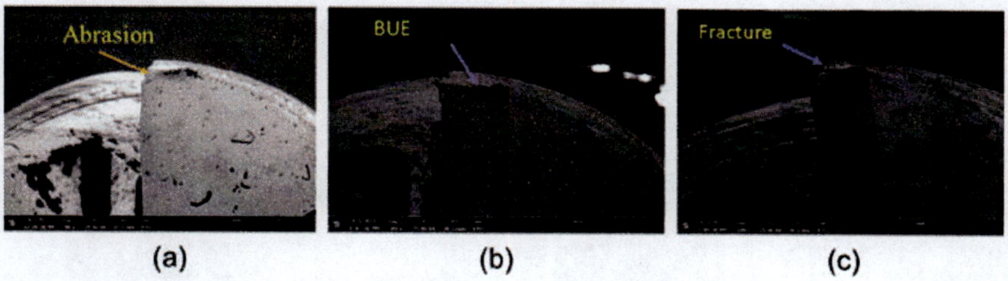

Fig. 6 Wear mechanisms at tool edge with rake angle **a** 6°, **b** 10° and **c** 16°

According to these in-house manufactured carbide tool wear results in machining CoCr-based material, Stellite 21, it can be inferred that an increase in rake angle can decrease the formation of thermal induced wear such as Built-Up-Edge (BUE) which then will result in attenuating the tool wear and directly extend the cutting tool life [19].

3.2 *Tool Wear Mechanism*

The mechanisms of wear on the flank face of the tool have been investigated by a Scanning Electron Microscope (SEM) which is used along Energy Dispersive X-ray (EDX) analysis. The end of tool wear for cutting tool is shown in Fig. 6 with difference rake angles. It is observed that the abrasion wear encountered at all rake angles and material loss on the tool edge due to build-up-edge. From the scanning process, it can be observed qualitatively that cutting tools with a low rake angle have a greater wear effect due to abrasion wear than those with a positive rake angle.

Extended quantitative observation of wear mechanism at the element analysis of the worn tools as depicted in Fig. 7 show the elements found at the worn tool tip as compared to the baseline EDX image of fresh tool in Fig. 7a. Discovery of high content of chromium from the workpiece on the flank face of used tool revealed that diffusion wear mechanism occurred in machining of Co–Cr–Mo based alloy.

Each sample was taken at specific position, wear, cutting length, and BUE to validate this condition. The chromium value found on the 6° tool was greater at 10.53 (mean) than the others, which were 9.75 and 8.74. Even though other cutting tools have low Cr values, Cr values remain significant on the wear surface. The high concentration of these elements shows the evidence of the existence of temperature-infused diffusion wear mechanism from the workpiece to the tool.

Aside from that, the presence of oxygen in the cutting process can be noticed. Oxygen was also found through the EDX analysis on the flank face presented in Fig. 7 which indicates the occurrence of oxidation wear on all the three-rake-angle-varied cutting tools. All sample tools have oxygen readings at build-up-edge and cutting edge. Because of the high temperature during cutting, a chemical reaction occurs between WC and oxygen, resulting in the oxidation process. As a result, an

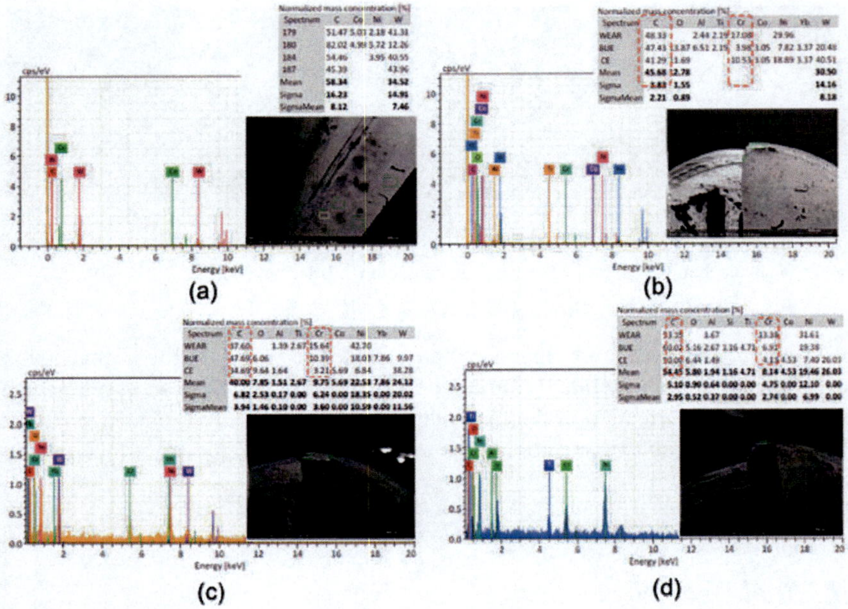

Fig. 7 EDX analysis at tool edge with rake angle **a** fresh tool **b** 6°, **c** 10° and **d** 16°

oxidation layer will form, where cutting abrasion promotes corrosion. The entire oxide layer will be removed if this operation is repeated, resulting in rapid tool wear. Due to the lack of oxygen on the surface, the results also reveal that tools with a high rake angle have reduced tool wear. It has been demonstrated that lowering the O_2 level reduces tool wear in cemented carbide tools [20]. In turning copper and milling stainless steel, the oxygen-free atmosphere has a similar effect on tool rake angle in reducing impact of tool wear [21].

4 Conclusion

The tool wear and wear mechanisms of in-house developed carbide end-mill tool with different rake angles in machining Stellite 21 are evaluated in this study. The cutting edge of worn tools were captured by SEM analysis microscope and Energy Disperse X-ray (EDX) analysis to reveal the wear mechanisms of the tool. It is observed that all tools exhibit the same wear mechanisms which are abrasion, diffusion and oxidation. However, the rate of temperature-infused wear mechanism for higher rake angle is lower than the tool with lower rake angle and it was confirmed by EDX analysis through low presence of material composition of chromium and oxygen at the flank face of the cutting tool.

Acknowledgements The authors are indebted to Mr. Mohd Azhari bin Abdul Rahman from Kolej Kemahiran Tinggi Mara Kuantan for conducting the experiment and constructive discussion. This work has been supported by Universiti Malaysia Pahang funded project RDU200301 and PGRS160385 and Malaysia Higher Education grant RDU160135 (FRGS/1/2016/TK03/UMP/02/17).

References

1. Cao J et al (2019) Manufacturing of advanced smart tooling for metal forming. CIRP Ann 68(2):605–628. https://doi.org/10.1016/j.cirp.2019.05.001
2. Azer I, Hamzah HC, Aishah S, Abdullah H (2016) Regional conference on science, technology and social sciences (RCSTSS 2014). In: Regional conference on science, technology and social sciences (RCSTSS 2014), no. Jan 2016. https://doi.org/10.1007/978-981-10-1458-1
3. Ezugwu EO (2005) Key improvements in the machining of difficult-to-cut aerospace superalloys. Int J Mach Tools Manuf. https://doi.org/10.1016/j.ijmachtools.2005.02.003
4. Ee KC, Balaji AK, Jawahir IS (2003) Progressive tool-wear mechanisms and their effects on chip-curl/chip-form in machining with grooved tools: an extended application of the equivalent toolface (ET) model. Wear 255(7–12):1404–1413. https://doi.org/10.1016/S0043-1648(03)00112-1
5. Hegab HA, Darras B, Kishawy HA (2018) Towards sustainability assessment of machining processes. J Clean Prod. https://doi.org/10.1016/j.jclepro.2017.09.197
6. Ku HS, Chia WC (2008) Design of multi-purpose carbide end mill. Anim Genet 39(5):561–563. https://doi.org/10.1109/APS.1996.549557
7. Toenshoff HK, Denkena B (2013) Basics of cutting and abrasive processes
8. Denkena B, Dittrich MA, Liu Y, Theuer M (2018) Automatic regeneration of cemented carbide tools for a resource efficient tool production. Procedia Manuf 21:259–265. https://doi.org/10.1016/j.promfg.2018.02.119
9. Hughes JI, Sharman ARC, Ridgway K (2006) The effect of cutting tool material and edge geometry on tool life and workpiece surface integrity. Proc Inst Mech Eng Part B J Eng Manuf. https://doi.org/10.1243/095440506X78192
10. Abdulah N et al (2014) Innovative metal injection molding (MIM) method for producing CoCrMo alloy metallic prosthesis for orthopedic applications. Adv Mater Res. https://doi.org/10.4028/www.scientific.net/AMR.879.102
11. Chen Q, Thouas GA (2015) Metallic implant biomaterials. Mater Sci Eng R Rep. https://doi.org/10.1016/j.mser.2014.10.001
12. Veldhuis SC, Dosbaeva GK, Yamamoto K (2009) Tribological compatibility and improvement of machining productivity and surface integrity. Tribol Int. https://doi.org/10.1016/j.triboint.2009.02.004
13. Zaman HA, Sharif S, Kim DW, Idris MH, Suhaimi MA, Tumurkhuyag Z (2017) Machinability of cobalt-based and cobalt chromium molybdenum alloys—a review. Procedia Manuf 11(June):563–570. https://doi.org/10.1016/j.promfg.2017.07.150
14. Ezugwu EO (2004) High speed machining of aero-engine alloys. J Brazilian Soc Mech Sci Eng 26(1):1–11. https://doi.org/10.1590/S1678-58782004000100001
15. Pham TT, Ko SL (2010) A manufacturing model of an end mill using a five-axis CNC grinding machine. Int J Adv Manuf Technol. https://doi.org/10.1007/s00170-009-2318-y
16. Ren L, Wang S, Yi L, Sun S (2016) An accurate method for five-axis flute grinding in cylindrical end-mills using standard 1V1/1A1 grinding wheels. Precis Eng 43:387–394. https://doi.org/10.1016/j.precisioneng.2015.09.002
17. Li G, Sun J, Li J (2014) Modeling and analysis for clearance machining process of end mills. Int J Adv Manuf Technol. https://doi.org/10.1007/s00170-014-6154-3

18. Khanna N, Sangwan KS (2013) Interrupted machining analysis for Ti6Al4V and Ti5553 titanium alloys using physical vapor deposition (PVD)-coated carbide inserts. Proc Inst Mech Eng Part B J Eng Manuf 227(3):465–470. https://doi.org/10.1177/0954405412472888
19. Fang N, Dewhurst P (2005) Slip-line modeling of built-up edge formation in machining. Int J Mech Sci 47(7):1079–1098. https://doi.org/10.1016/j.ijmecsci.2005.02.008
20. Bushlya V, Lenrick F, Ståhl JE, M'Saoubi R (2018) Influence of oxygen on the tool wear in machining. CIRP Ann 67(1):79–82. https://doi.org/10.1016/j.cirp.2018.03.011
21. Tennenhouse GJ, Runkle FD (1987) The effects of oxygen on the wear of tungsten-carbide-based materials. Wear 118(3):365–375. https://doi.org/10.1016/0043-1648(87)90078-0

Defect Identification During Pulse Mode Laser Welding Process Through the Pattern Recognition Analysis of the Acquired Sound Frequency Spectrum

M. F. M. Yusof, M. Ishak, M. N. Salleh, and M. F. Ghazali

Abstract Problems on laser weld quality still remain as vital issue even though the process was done with optimized condition which results the demand on robust monitoring method during the process. Until recently, many methods have been explored and air-borne acoustic are among of methods that have been proven to be able to detect the presence of defect. However, despite detection, it is essential if the type of defect could be identified as it gives different severity level to the development of failure. This work presents the identification of defect during pulse mode laser welding through the analysis of sound. In achieving the goal of this study, bead on plate weld have been done onto the 22MnB5 boron steel plate repeatedly based on 3 different set of experiment with the variation in the level of parameters. Simultaneously, time-series sound signal was acquired along the process before it was converted into frequency spectrum before further analysis. According to the result, it was recorded that the variation of parameters level in pulse mode laser welding process lead to the presence of porosity and crack. Relatively, the trend of sound frequency spectrum were also significantly changes its trend in respond to the parameters level variation. It was discovered that the dominant frequency for the signals acquired from the process which produce good quality weld, porosity and crack recorded the same range which was between 5 to 7 kHz. Uniquely, the existence of porosity could be identified by the occurrence of peak at around 9 kHz while the presence of crack could be recognized by the occurrence of peak at 8 kHz and 11 kHz. This trend was proven to be consistent in repeated experiment according to the result from principal component analysis. Based from the result in this study, it could be conclude that the identification of defect could be done by the analysis of the acquired sound during the process. Significantly, this would expand the ability

M. F. M. Yusof · M. Ishak (✉) · M. N. Salleh · M. F. Ghazali
Faculty of Mechanical and Automotive Engineering Technology, Universiti Malaysia Pahang, Pekan, Pahang, Malaysia
e-mail: mahadzir@ump.edu.my

M. Ishak
Centre for Research in Advanced Manufacturing, University Malaysia Pahang, Pekan, Pahang, Malaysia

M. Y. Ismail et al. (eds.), *Technological Advancement in Mechanical and Automotive Engineering*, Lecture Notes in Mechanical Engineering,
https://doi.org/10.1007/978-981-19-1457-7_60

of acoustic method for quality monitoring purpose as the identification of defect is also important in quality control.

Keywords Defect identification · Pulse mode laser welding · Sound frequency spectrum

1 Introduction

Increasing trend on the use of laser welding process could be seen since past decades. This was owing to its advantageous in giving a small heat affected zone, aesthetic appearance, high production rate as well as reducing the post-weld processing [1]. Due to this fact, numerous studies associates to the laser welding process on wide variety of materials have been reported over the past years [2–5] which more or less could be apply many type of industries. However, the problems on weld quality still remain as vital issue even though the process was done with optimized condition. It is well known that the existence of defects in weld product is intolerable and become a massive concern since it could degrade the strength of weld product. To address on this issue, the implementation of robust monitoring method has been suggested as it promote greater control during the process [6]. Since past decades, numerous methods which inclusive of electrical, thermal, optical as well as acoustic method have been applied in laser welding process. Among those methods, acoustic sound method gained attention lately due. It was due to its high responsible speed, low cost and simple setup [7–9].

Since past decades, many studies have been done to comprehend the relation between the sound characteristic and weld condition. In earlier research, the characteristic of sound frequency spectrum was reported to be strongly related to the weld penetration [10]. The investigation was done on Aluminum 1100 and according to the result, the peak of sound spectrum at the bandwidth of 9–10 kHz recorded an increment simultaneously with laser intensity gain. Concurrently, the depth of penetration is also increasing. Unlike the aforementioned work the spectral energy was reported to drop in its value when insufficient weld penetration was detected [11]. In this work, the study was done on 304 stainless steel plate and different range of active frequency bandwidth was found within 1–2 kHz. In another work which involved broader range of power and travel speed [12], it was reported that the Root Mean Square (RMS) of the acquired sound recorded distinguished pattern in separating full-, moderately full-, or partial-penetration weld joint. In another part of this research [13], the study was extended into the high power or keyhole welding regime. As results from the combination analysis of the acquired sound and optical charged particle signal, the penetration status could be divided into three different classes which were full penetration, overheat penetration and half penetration. In recent study, other sound features were also reported to gives a significant trend with the change in penetration status. Huang and Kovacevic [7, 14] investigate the trend of

Sound Pressure Deviation (SPD) and Band Power (BP) to classify the weld penetration. Uniquely, despite of giving a distinctive pattern, the weld penetration estimation models were successfully developed using multiple regression and artificial neural network by learning the trend of these features.

Apart from penetration condition, other defects or irregularities along the weld bead was reported to be able to detected from the analysis of the acquired sound during the laser welding process. In recent work [8], it was reported that the sound pressure level trend significantly giving information with respect to the existence of burn through defect. Moreover, with the use of arrays of microphones and sound proof equipment, the location of defect could also been identified with reasonable error through time delay recognition analysis. Acoustic method was also reported to be able to detect underfil and humping. In the previous work [15], the effect of Zn coating thickness and gap between the lapped specimens to the degree of those defects was accessed based on the acquired sound signal. Based on the results, it was found that the amount of spatter was consistently appeared between 0.08 to 0.2 mm gap. Accordingly, the existence of spatter influenced the existence of both humping and underfil. Simultaneously, it was found that the RMS of sound signal recorded a significant change at the same gap. Moreover, the amplitude of frequency spectrum at around 1 kHz appeared to be inversely proportional to the coating thickness. This phenomenon happen because of high vapor pressure produced from the thicker coating which results in the suppression of keyhole oscillation. In addition, the amplitude of the bandpass-filtered time domain signal show deviating trend from 22 to 38 V at the presence of defects. Different from the other studies, Sansan et al. [16] combined principal component analysis (PCA) and independent component analysis (ICA) to decompose the acquired acoustic signal into cooling and keyhole component in order to detect the existence of blowholes.

As established by the above studies, it could be generally summarized that the analysis of the acquired sound during laser welding process could lead to the detection of defect by in-situ basis. However, up to this point, many existing studies were only underlining on the detection of several type of defect. In order to establish this method for the purpose of monitoring laser welding process, it is important to study on how the behaviour of the acquired sound responds to the occurrence of the many other type of defect such as crack and porosity which could lead to the crack nucleation and degrade the fatigue life of the component. Moreover, it is essential if the type of defect could be identified as it gives different severity level to the occurrence of failure. In this work, the main aim is to identify type of defect from the analysis of the frequency spectrum of the acquired sound during pulse mode laser welding process. At the first stage of this work, the behaviour of both time domain and frequency domain in respond to the occurrence of different type of defects will be discussed. In the next part, the significant use of principal component analysis to recognize the different in sound spectrum pattern and simultaneously identify the different type of defect will be presented.

2 Methodology

2.1 Experimental and Data Acquisition Setup

The setup of the experiment was illustrated in Fig. 1. In this study, bead on plate weld were done on 22MnB5 boron steel with the variation of parameters set as shown in Table 1. Basically, focal length, peak power, pulse width and pulse repetition rate was set to be varied in this work in attempt to produce both intact weld and weld with defect. In all cases in this work, the speed of weld was set to constant at 1.5 mm/s. Meanwhile, pure argon shielding gas flowrate was set to 15 L/min it was also constant along the entire experiment. On the other hand, each set of experiment was repeated 5 times and sampled at 5 different points to gives total of 75 set of sound signal for the analysis purpose. This was done in order to observe the consistency of both defects occurrence as well as the sound signal trend.

During the pulse mode laser welding process, the sound signal emitted from the process was acquired using microphone. The acquired analog signal was discretize using National Instrument (NI) 9234 analog-to-digital converter with the rate of 25.6 k Sample/s. The digital signal was then converted into digital frequency spectrum using analyser before further analysis was done. Meanwhile, the microphone was located at 25 cm and 300 angle from the weld spot to ensure that the spatter that produced during the process was not damaging the sensor.

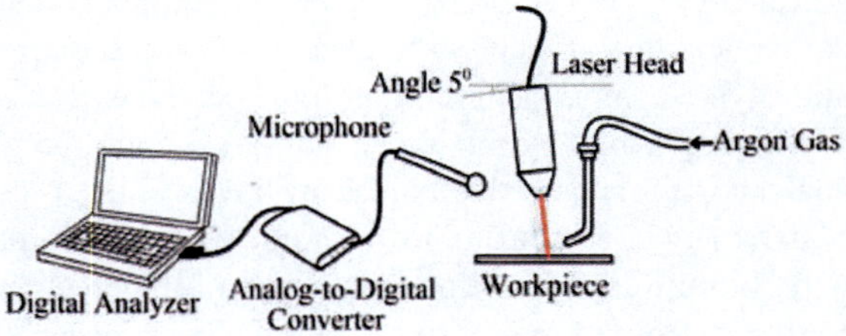

Fig. 1 Experiment and data acquisition setup

Table 1 Pulse laser welding parameters setup

Experiment No	Focal length (mm)	Peak power (W)	Pulse width (ms)	Pulse repetition rate (Hz)
1	+ 5	1600	6	20
2	− 5	1200	6	16
3	0	1200	2	50

2.2 *Experiment and Data Acquisition Setup*

As briefly explained in the earlier section, analysis of frequency spectrum will be done in order to identify the type of defect occurs during the laser welding process with the parameters as shown in Table 1. Basically, the frequency spectrum was obtained from the Fast Fourier Transform (FFT) of the time-series of the acquired sound. In order to reveal the different in spectrum pattern due to the occurrence of different type of defect, principal component analysis (PCA) was applied in this study.

In principal, PCA was applied based on theory explained by Jolliffe [17] which considering all the collected spectrum as a big population of data. In this case, each observation represents a single frequency spectrum with the variation of amplitudes xf along the band as stated in Eq. 1 whereas m and n were the number of observation and samples respectively.

$$Xf_m = xf_{m1},\ xf_{m2},\ xf_{m3},\ \ldots,\ xf_{mn}, \tag{1}$$

By gathering all the spectrum in m x n size matrix, the covariance between all spectrums was determine by Eq. 2 which gives m x m covariance matrix [Cxf].

$$[C_{xf}] = [Xf_{mn}].[Xf_{mn}]^T \tag{2}$$

In order to determine how each of the spectrum trends was different from the others, the projection of their covariance need to be determined. This could be done through the eigenvalue decomposition of the obtained covariance matrix as shown in Eq. 3 in which $\vec{v}$ and γ represent eigenvector and eigenvalues respectively.

$$\left[C_{xf}\right].\vec{v} = \gamma.\vec{v} \tag{3}$$

Theoretically in PCA, eigenvectors is the principal component which was sorted in descending order of its variance value shown by eigenvalue matrix. The different in amplitude pattern along the entire band in frequency spectrum could be observed from the principal component.

3 Result and Discussion

3.1 *Type of Defects Emerged from Pulse Mode Laser Welding Process*

Recalling back to the previous section, it was explained that the experiment was done based on the set of weld parameters represent in Table 1. Table 2 summarized the defects that occur during each set of experiment. Based from the table, it could be

Table 2 Defect occurred from different set of experiment

Experiment No	Focal length (mm)	Peak power (W)	Pulse width (ms)	Pulse energy (mJ)	Pulse repetition rate (Hz)	Type of defect occurred
1	+ 5	1600	6	9600	20	No defect
2	− 5	1200	6	7200	16	Porosity
3	0	1200	2	2400	50	Crack

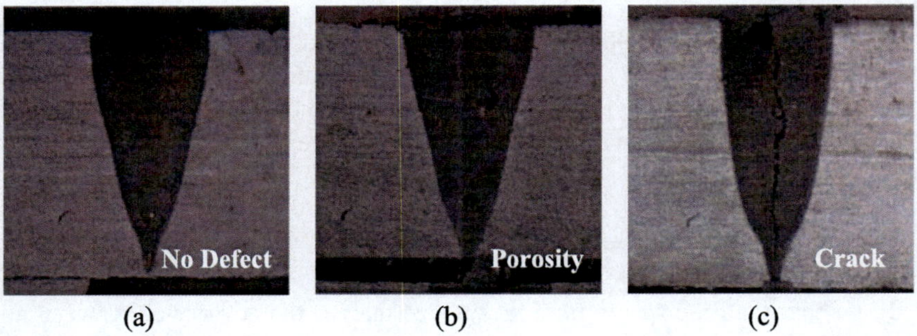

Fig. 2 Cross section image of the selected sample in each of experiment **a** Experiment 1 **b** Experiment 2 **c** Experiment 3

clearly observed that intact weld was produced from experiment set 1. In contrast, set 2 and set 3 of experiment was found consistently produce porosity and crack respectively. To get clearer picture, the weld bead cross section of each one of the experiment was depicted in Fig. 2.

In order to understand the phenomena occurs in this study, the physical theory on the defect formation mechanism need to be look into. According to Dawes [1], hot crack and porosity could occur during laser welding due to uncertainty during solidification process. In pulse mode laser welding the tendency of crack to occur is quite larger [18] and investigating the optimum cooling time to reduce the critical strain which could initiate crack might solve this issue [19]. As could be observed in Fig. 2, the occurrence of crack was found in set 3 of the experiment. Compared to the parameters from set 1 experiment which produces pristine weld condition, it was recorded that the pulse repetition in experiment set 3 was more than 2 times larger. Moreover, the pulse width was 3 times smaller. This factor might affect the cooling behaviour which emerged the crack in this study. On the other hand, the formation of porosity was explained by the trapped bubbles emerged from the instability of keyhole and molten metal during backfilling process [20, 21]. Some scholars, [22, 23] suggest that the overlapping factor could suppressed the formation of porosity in pulse mode laser welding. In principal, the overlapping factor is influenced by the laser spot diameter and pulse repetition rate and the smaller pulse repetition rate in experiment set 2 might explain the existence of porosity in that experiment.

3.2 Behaviour of the Acquired Sound Signal

Owing to the fact that the existence of defect was influenced by the variation of parameters as prior presented, it is important to understand how the generated sound respond to the different set of parameters in the experiment. Figure 3 depicts the time series of sound signal acquired from different set of experiment. Due to the nature of pulse mode laser welding which emit the laser periodically, all the acquired sound shows the pulse behaviour as illustrated in the figure. In more detail observation, it was clear that the duration of sound pulse was influence by the pulse width value during the laser welding. For instance, pulse sound acquired from experiment 3 recorded the shortest pulse duration as a result from lowest pulse width setup. Meanwhile, the pulse duration of sound captured in Fig. 3a and b both were slightly similar due to the same value of pulse width during the experiment. On the other angle, the maximum amplitude of the acquired sound seems to be influences by the amount of energy during laser welding process. The average maximum amplitude of all the acquired sound for experiment 1, 2, and 3 was recorded to be 6.37 mV, 3.71 mV and 2.21 mV respectively.

Apart from time series of the acquired sound, it is important to observe the trend of frequency spectrum with respect to the change in weld parameters which influenced the occurrence of defect. Basically, frequency spectrum gives information on both frequency and amplitudes as illustrated in Fig. 4. According to the result shown in the figure, it could be generally summarized that the amplitude trend was similar to the trend recorded in time-series in Fig. 3. Meanwhile, in all set of experiment, it was found that the dominant frequency of the acquired sound lies between 5 and 7 kHz which emerged from the plasma plume oscillation explained by [24]. Uniquely, small peak was recorded around 9 kHz for the case of experiment 2 while it was found around 8 and 11 kHz in experiment 3. This trend shows that the change in parameters will influence the entire structure of the frequency spectrum in this study which could possibly give a significant result in identification or clustering analysis.

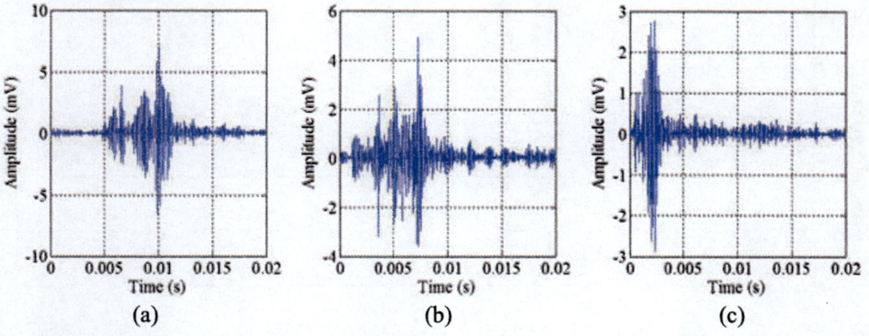

Fig. 3 Time series of pulse sound signal acquired from different set of experiment **a** Experiment 1 **b** Experiment 2 **c** Experiment 3

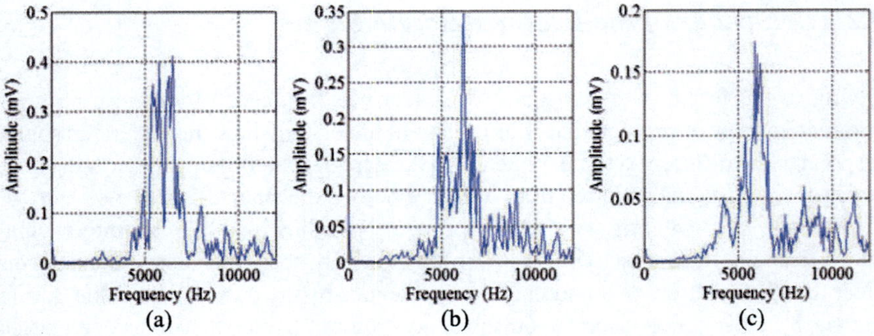

Fig. 4 Frequency spectrum of pulse sound signal acquired from different set of experiment **a** Experiment 1 **b** Experiment 2 **c** Experiment 3

3.3 Defect Identification Through the Principal Component Analysis

As previously presented, the formation of defect was influenced by the set of parameters which results in the change of frequency spectrum behaviour. In prior section, it was explain that the pattern recognition analysis of the frequency spectrum will be done using PCA in attempt to identify the sound signal from different type of defect. Figure 5 shows the principal component plot of the analysed sound frequency spectrum.

As could be observed in Fig. 5, all the spectrums were classify into 3 different groups based on weld condition. In order to determine the precision of scattered point for each group, the centroid was determined by k-mean method. From the result, it was found that the spectrums from crack weld was scattered in more precise pattern

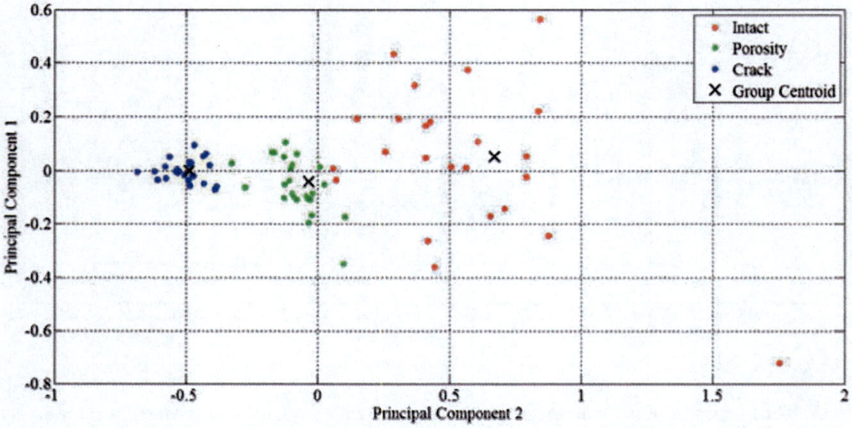

Fig. 5 Principal component plot of the analysed frequency spectrum by PCA

followed by spectrum recorded from porosity and intact weld. This show that all the frequency spectrum for the case of crack and porosity weld recorded a consistent trend in every repetition of the experiment which could allow ones to identify the type of defect from the captured sound signal.

4 Conclusion

According to the result presented in this work, it could be summarized that the formation of defect was influence by the set of parameters during pulse mode laser welding process. Simultaneously, the frequency spectrum of the captured sound was also found significantly changes its pattern. Moreover, principal component analysis revealed that different trend was recorded for different type of defect and it was quite consistent for every repetition of experiment which allows the identification of defect. As conclusion, the identification of defect could be done by the analysis of frequency spectrum acquired during pulse mode laser welding. This could promote an alternative way for monitoring the weld condition by in-situ basis.

Acknowledgements The authors would like to acknowledge the financial support by Ministry of Higher Education Malaysia under FRGS/1/2018/TK03/UMP/02/9 grant for this project

References

1. Dawes C (2008) Laser welding a practical guide. Abington Publishing, United Kingdom
2. Benyounis K, Olabi A-G (2008) Optimization of different welding processes using statistical and numerical approaches–a reference guide. Adv Eng Softw 39(6):483–496
3. Pakmanesh M, Shamanian M (2018) Optimization of pulsed laser welding process parameters in order to attain minimum underfill and undercut defects in thin 316L stainless steel foils. Opt Laser Technol 99:30–38
4. Mostaan H et al (2017) Nd: YAG laser micro-welding of ultra-thin FeCo–V magnetic alloy: optimization of weld strength. Trans Nonferrous Met Soc China 27(8):1735–1746
5. Zhou L et al (2018) Effect of welding speed on microstructural evolution and mechanical properties of laser welded-brazed Al/brass dissimilar joints. Opt Laser Technol 98:234–246
6. Hoffman J et al (2002) Analysis of acoustic and optical signals used as a basis for controlling laser-welding processes. Weld Int 16(1):18–25
7. Huang W, Kovacevic R (2009) Feasibility study of using acoustic signals for online monitoring of the depth of weld in the laser welding of high-strength steels. Proc Inst Mech Eng Part B J Eng Manuf 223(4):343–361
8. Luo Z et al (2016) Monitoring of laser welding using source localization and tracking processing by microphone array. Int J Adv Manuf Technol 86(1):21–28
9. Ao S et al (2015) Simulation and experimental analysis of acoustic signal characteristics in laser welding. Int J Adv Manuf Technol 81(1):277–287
10. Duley WW, Mao YL (1994) The effect of surface condition on acoustic emission during welding of aluminium with CO_2 laser radiation. J Phys D Appl Phys 27(7):1379
11. Farson D et al (1996) Frequency–time characteristics of air-borne signals from laser welds. J Laser Appl 8(1):33–42

12. Farson D, Ali A, Sang Y (1998) Relationship of optical and acoustic emissions to laser weld penetration. Welding J 77(4):142. s-148
13. Farson D, Ali A, Li X (1999) Laser weld penetration monitoring with multiple emission signal measurements. J Laser Appl 11(2):47–53
14. Huang W, Kovacevic R (2009) A neural network and multiple regression method for the characterization of the depth of weld penetration in laser welding based on acoustic signatures. J Intell Manuf 22(2):131–143
15. Lee C-J, Kim J-D, Kim Y-C (2015) Study on monitoring of plasma emission signal in lap welding of Zn coated steel sheet using CO_2 laser. Int J Precis Eng Manuf 16(3):495–500
16. Sansan A et al (2010) Blind source separation based on principal component analysis—independent component analysis for acoustic signal during laser welding process. In: 2010 International conference on digital manufacturing and automation (ICDMA), 2010. IEEE
17. Jolliffe I (2011) Principal component analysis. In: Lovric M (ed) International encyclopedia of statistical science. Springer Berlin Heidelberg, Berlin, Heidelberg. pp 1094–1096
18. Kelkar G (2000) Pulsed laser welding. WJM Technologies, Cerritos
19. von Witzendorff P et al (2015) Using pulse shaping to control temporal strain development and solidification cracking in pulsed laser welding of 6082 aluminum alloys. J Mater Process Technol 225:162–169
20. Chen M et al (2017) Effect of keyhole characteristics on porosity formation during pulsed laser-GTA hybrid welding of AZ31B magnesium alloy. Opt Lasers Eng 93:139–145
21. Pang S et al (2016) Dynamics of vapor plume in transient keyhole during laser welding of stainless steel: local evaporation, plume swing and gas entrapment into porosity. Opt Lasers Eng 82:28–40
22. Wang J, Wang G, Wang C (2015) Mechanisms of the porosity formation during the fiber laser lap welding of aluminium alloy. Metalurgija 54(4):683–686
23. Gao X-L et al (2014) Porosity and microstructure in pulsed Nd:YAG laser welded Ti6Al4V sheet. J Mater Process Technol 214(7):1316–1325
24. Klein T et al (1994) Oscillations of the keyhole in penetration laser beam welding. J Phys D Appl Phys 27(10):2023

Effect of Laser Micro-drilling Parameters on Hole Geometry and Hole Formation of Thin Sheet SS304

M. S. Haneef, G. H. Lau, M. H. Aiman, M. M. Quazi, and M. Ishak

Abstract The recent advances in manufacturing technology have led to the development of miniature products in the field of automobiles, aerospace, and robotics. Laser micro-drilling has developed as a potential substitute over conventional machining due to the advantages of operational precision, reduced operational costs, and a high-speed production rate. This process involves high power intensity from the laser to break down the bond between molecules of the workpiece and hence form a hole on the workpiece. This project aims to study the effect of laser power on the drilled hole geometry and to analyse the mechanism of the hole formation during laser micro-drilling. The material used in this project is SS304 sheet metal. The holes' geometry and hole formation will be analysed by using an optical microscope. The size of the hole diameter for each power is almost the same in the range of 101.669–102.978 μm for the frontside. Meanwhile, the diameter of the backside hole increases from 64.343 μm to 88.852 μm at 15 W to 21 W of laser power respectively. For hole formation, the more material is ablated as the ablation process advances. As a result, the removal area from the micro-drilled hole grows from 3577.852 to 6516.237 m^2. The shape of the hole is irregular due to the uneven power distribution of the laser towards the SS304 sheet metal when it undergoes an ablation process.

Keywords Micro-drilling · Laser drilling · Laser power · Hole geometry · Hole formation

1 Introduction

In recent years, the micro-machining process for producing miniature parts has become increasingly important for several sectors such as electronics, aerospace and biomedical [1]. Micro-machining is a machining method that involves removing small bits of material in order to obtain great geometrical accuracy. Micro-drilling

M. S. Haneef · G. H. Lau · M. H. Aiman (✉) · M. M. Quazi · M. Ishak
Faculty of Mechanical & Automotive Engineering Technology, Universiti Malaysia Pahang, 26600 Pekan, Pahang, Malaysia
e-mail: aimanmh@ump.edu.my

M. Y. Ismail et al. (eds.), *Technological Advancement in Mechanical and Automotive Engineering*, Lecture Notes in Mechanical Engineering,
https://doi.org/10.1007/978-981-19-1457-7_61

is an ideal process among the micromachining processes to generate the micro-holes, generating deeper holes with better roundness, better smoother surface, and straightness. Micro-drilling is a process of drilling micro-sized holes in the range of 1–999 μm [2]. Micro-holes are currently created using a variety of production techniques, including micro-electrical discharge machining (micro-EDM), micro-mechanical drilling and laser micro-drilling. The type of material, its size and thickness, and the cost of the process is all the factors to consider when choosing the right micromachining technology [3]. Laser micro-drilling has an advantage towards those factors for overcoming the limitation in the micro-drilling process.

Laser micro-drilling is a non-contact, precise, and repeatable process for forming small diameter (~100 μm) and high-aspect-ratio holes in a wide range of materials first paragraphs that follows a table, figure, equation etc. does not have an indent, either [4]. Besides, the method that uses a laser as an energy source possesses a fast process because only the selected part should be heated, and the amount of heat input is small [5, 6]. In laser micro-drilling, a stationary, high-intensity laser beam is directed onto the surface at power densities sufficient to remove the solid phase material [7]. To ablate the material, the erosion of the drilled hole propagates in the direction of the line source. Figure 1 shows the physical processes that occur during laser micro-drilling. There are three stages to the process. The beam is initially focused onto the material's surface, and the layer that is being targeted absorbs the energy of the beam before being irradiated [8]. Material ablation occurs when the pressure gradient on the surface of the layer is large enough to overcome the surface tensile force and subsequently remove the material from the hole [9]. This mechanism will be continuously repeated until the one fine through-hole is successfully drilled.

In general, there are three techniques to laser micro-drilling, namely as single pulse, trepanning, and percussion drilling. For this research, trepanning laser drilling will be applied. This procedure at first works similarly to single-shot or percussion laser drilling by piercing an opening in the material that it is being utilized on [10].

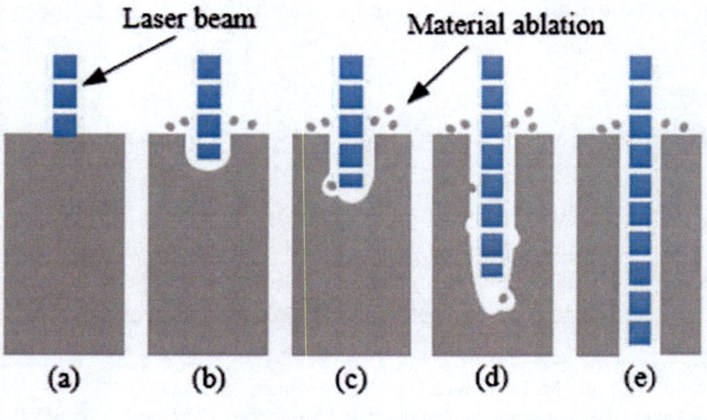

Fig. 1 Physical processes during laser drilling. **a** initial laser radiation **b–d** material ablation **e** through-hole fine drilled

After piercing the material, the laser beam moves in a spiral pattern utilizing a motion mechanism to carve a hole. The beam begins at the center of the diameter and spirals outwards in an increasing spiral diameter until the correct hole size is achieved.

However, it has also some shortcomings which can affect the hole geometry. In laser processing, an understanding of the fundamental laser energy absorption mechanisms plays a significant role in determining the optimum parameters [11]. Therefore, it is necessary to find out the best parameter for enhancing the geometry of the micro-drilled hole. The laser power will give a big impact in laser micro-drilling. The hole geometry and appearance are depending on the laser power due to different ablation rate towards the material. Besides, it is necessary to look at the basic fundamentals of material removal and hole formation throughout the laser micro-drilling process. Towards the number of drilling loops. This study focuses on the effect of laser power on the micro-hole geometry and the mechanism of hole formation during laser drilling. Knowing how laser power and drilling loop affects micro-hole geometry, major development can be made to produce a better micro-drilled hole that meets satisfaction.

2 Experimental Setup

2.1 Material Preparation

The material used for laser micro-drilling was stainless steel SS304 with a sheet thickness of 0.1 mm. The selection of this material has met the requirements of the wide range of uses in the industry sector, i.e., excellent corrosion resistance and ability to withstand corrosion from most oxidizing acids [12]. Tables 1 and 2 shows the material properties and chemical composition of the stainless steel used.

Table 1 Thermophysical properties of SS304

SS304	Parameters
Density	8.0 kg/m^3
Melting point	1723.15 K
Thermal conductivity	16.2 W m^{-1} K
Tensile strength	540–750 MPa

Table 2 Chemical composition of SS304

Fe	Ni	Cr	Mn	Si	C	P
Bal	8.0–10.5	17.5–19.5	0.0–0.2	0.0–0.1	0.0–0.07	0.0–0.05

2.2 Experimental Procedure

Before starting the laser micro-drilling process, the hole's array must be designed first as shown in Fig. 2. This design should consider the effect of the laser process mechanism, i.e., the heating area and its spatter deposition, to avoid overlapping between each drilled hole [13]. The hole diameter size is set to 100 μm and the gap between each hole is 300 μm. In addition, the samples will be immersed in the acetone solution to remove any residual contaminant.

The source used is a pulsed wave fibre laser from a laser marking machine with a maximum output power of 30 W and a maximum repetition rate of 20 kHz. Besides, the maximum pulse that can achieve is up to 1.0 mJ. Laser micro-drilling parameters applied in this study were listed in Table 3. This experiment was divided into 2 parts which are the effect of laser power on the hole geometry and the mark loop number towards the mechanism of the hole formation during laser micro-drilling. In the first part, the laser power was set at 15, 18 and 21 W while the number of the mark loops was kept constant at 40. For the next part, the laser power is kept constant at 15 W, but the mark loop was altered. The laser mark loop was set to 5, 7, 10, and 15.

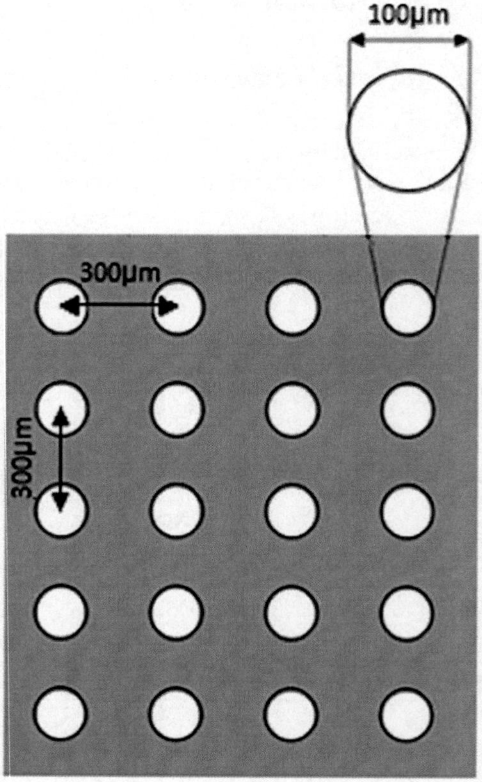

Fig. 2 Drilling shape design layout

Table 3 Laser micro-drilling process parameters

Laser specification	Parameter
Number of loops	5, 7, 10, 15, 40
Focal length (mm)	19.2
Laser power (W)	15, 18, 21
Pulsed frequency (kHz)	20

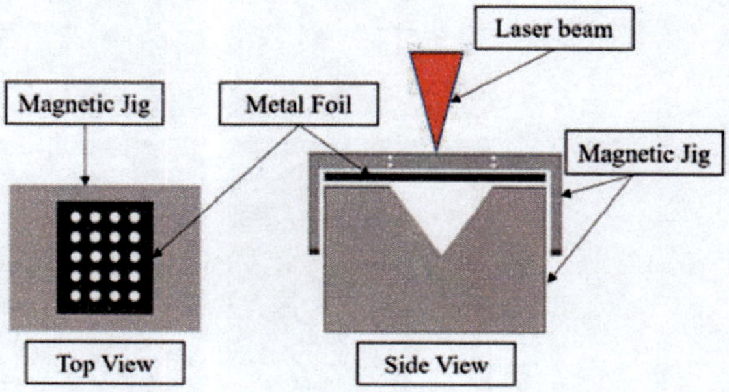

Fig. 3 Schematic of laser micro-drilling process and set-up

The schematic diagram for the laser micro-drilling experimental setup was shown in Fig. 3. The laser source emits pulses with a fixed repetition rate and scanning speed over the workpiece. In addition, the trepanning drilling technique will be used in this process considering that this process is the best compared to a single pulse and percussion drilling [14]. Besides, the sample will be placed on a magnetic jig to avoid any movement considering formability during the process.

The drilled hole was analyzed according to the diameter and area of the hole. The hole geometry will be measured by using an optical microscope with 20X magnification. The hole geometry extracted are horizontal diameter (D – x), vertical diameter (D – y) and hole area. After that, the specimen will be mounted for cross-section analysis. The hole depth was analysed by using an optical microscope with 50X magnification.

3 Results and Discussion

3.1 Effect of Laser Power on Hole Geometry

After completing laser micro-drilling, the array of holes with a diameter of 100 μm were successfully drilled thoroughly as shown in Table 4. From the appearance, the size of the hole visually increases as the laser power increase for both sides. The frontside of the hole appeared to be larger than the backside, and all the micro-drilled holes seemed to be consistent in geometry. Besides, the surface's colour between both

Table 4 Frontside and backside hole 100 μm, 40-mark loop for 15, 18 and 21 W

Laser Power	15	18	21
Frontside	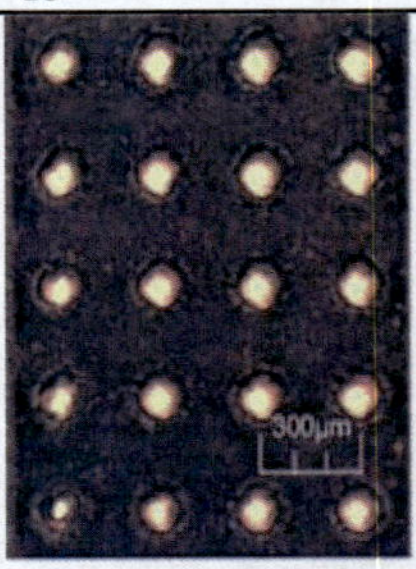	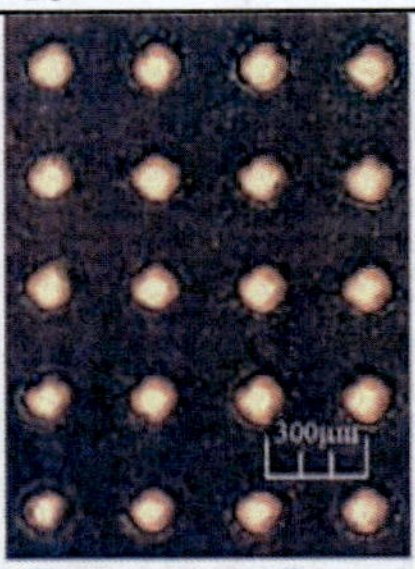	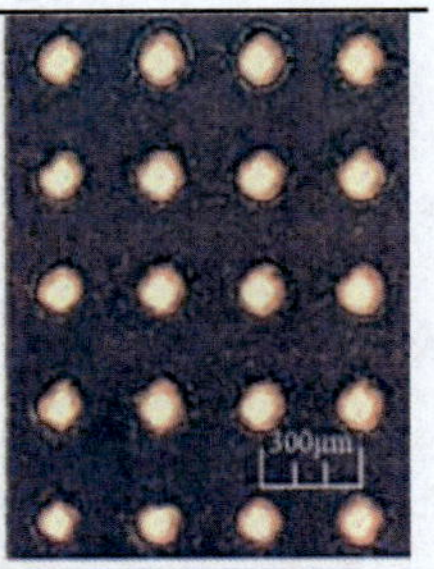
Backside	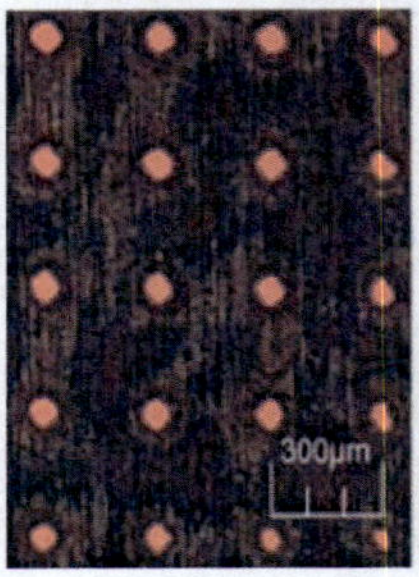	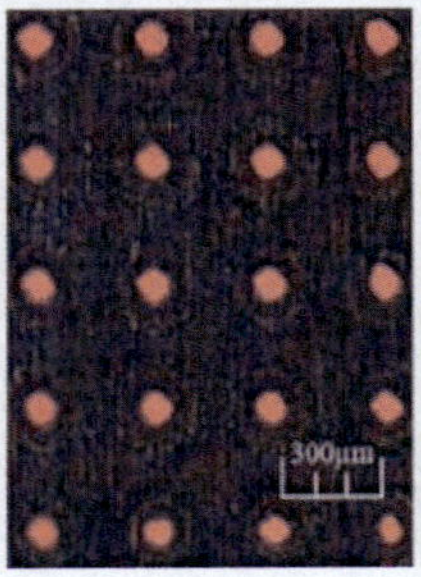	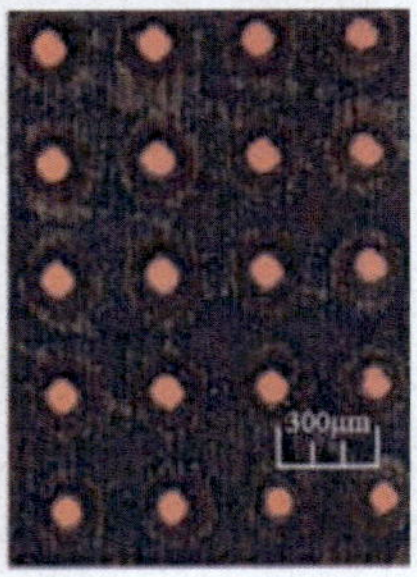

sides is slightly different due to the laser beam interaction that directly focuses on the front side of the hole.

Furthermore, 20 micro-drilled holes for every set of parameters have been measured by using an Image J software. Tables 5 and 6 show the frontside and the

Table 5 Frontside hole of average diameter and average area

Power (W)	Average diameter (μm)		Average area (μm^2)
	Horizontal	Vertical	Value
15	101.846 ± 1.610	102.123 ± 1.232	9127.799 ± 1.980
18	102.118 ± 1.160	101.669 ± 1.108	8941.035 ± 2.708
21	102.978 ± 1.481	102.179 ± 1.391	9026.257 ± 3.313

Table 6 Backside hole of average diameter and average area

Power (W)	Average diameter (μm)		Average area (μm^2)
	Horizontal	Vertical	Value
15	64.343 ± 2.959	67.049 ± 2.475	4558.682 ± 3.440
18	77.547 ± 2.157	78.933 ± 3.648	5820.406 ± 2.983
21	88.253 ± 2.376	88.852 ± 1.115	6945.261 ± 2.330

backside hole of average diameter and average area respectively. For the frontside, the size of the hole diameter for each power is almost the same in the range of 101.669–102.978 μm. This shows that the increasing power of laser micro-drilling does not affect the material ablation at the entrance hole. Meanwhile for the backside hole diameter, when the power output increases, the value of the hole diameter also increases as well. The diameter of the backside hole increases from 64.343 μm to 88.852 μm at 15 W to 21 W of laser power respectively. This shows that the relationship between laser power and the average diameter of the exit hole is directly proportional.

In addition, the area of the frontside hole for each power is nearly the same, ranging from 8941.035 to 9127.799 m^2. Between these two values, there is no substantial difference. This indicates that raising the power of laser micro-drilling has no effect on the area of the entrance hole as well as the hole diameter itself. For the exit hole area, when the power output increases, the hole area also increases. This is because the effect of laser power at the exit hole is different according to the material that has been ablated. The area of the exit hole increases from 4558.682 μm^2 at 15 W power to 6945.261 μm^2 at 21 W power, respectively. The area difference is 2386.579 m^2. This proves that the laser power affects the exit hole's average area. In addition, the standard deviation was observed to be less than 5, which means that the process was consistently met.

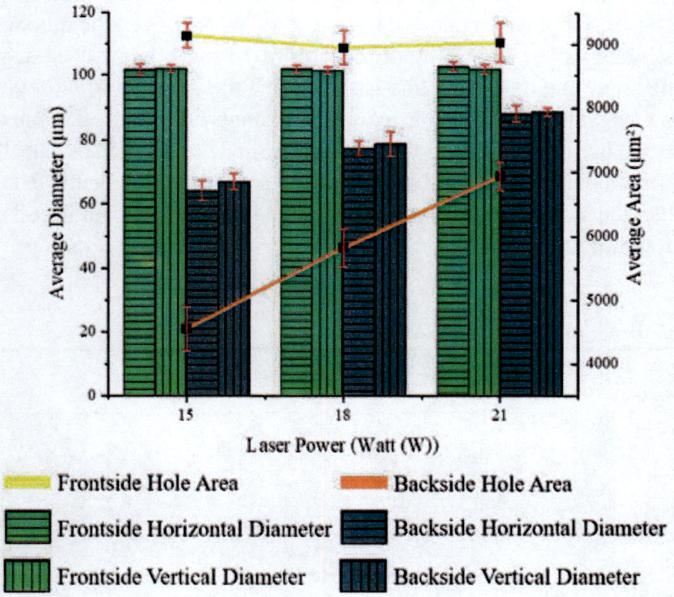

Fig. 4 Laser power versus average hole diameter and average hole area for frontside and backside (Bar shows average diameter. Line shows average area)

Figure 4 presents the line plot and bar graphs constructed in order to show the values of hole area and hole diameter obtained towards laser power, respectively. The graph pattern illustrates that there is not much significant difference between diameter for the frontside hole. The area also shows no significant difference when the laser power increase. Meanwhile, the hole diameter and hole area for the backside are clearly trending upward.

Through comparison based on the graph pattern, the effect of the laser power towards the entrance diameter is minor. As reported by Witte, they noticed similar trends and the amount of solid molten material around the entrance was said to have been increased due to the increase heat load [15]. Moreover, high laser power is not needed to accomplish a front side hole because it will immediately achieve the same value as the preset at the beginning. However, when it goes through into the sample to reach the backside, the laser power will play an important role. This is because when the laser power increases, the ablation speed will increase, and the hole geometry will get bigger. Therefore, it is assumed that if more loops are added, the preset value can also be achieved at the backside of the hole geometry.

3.2 *Effect of Laser Power on Hole Geometry*

In this section, the hole formation for a different number of laser mark loops can be found out. The laser power remains constant at 15 W. However, the number of laser mark loops were varied from 5-mark loops until 15-mark loops. The depth of the micro-drilled hole and its tracing line is shown in Table 7. 15 W of laser power with 5-laser mark loop numbers is unable to make a through-hole on the 0.1 mm thickness SS304 sheet. The laser mark loop barely ablated half of the SS304 sheet, and the entrance hole diameter is not fully ablated to reach 100 μm. For the 7-laser mark loop number, the laser power drills deeper into the SS304 sheet but is still unable to make a through-hole.

Table 7 Hole depth and tracing line of laser mark loop

Number of Mark Loop	5	7	10	15
Micro-drilled Hole		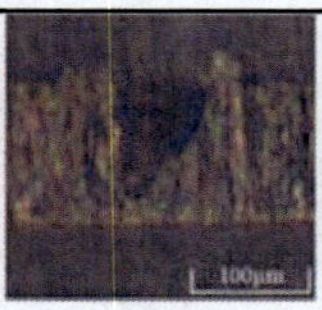		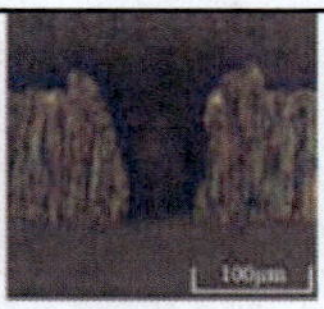
Tracing Line	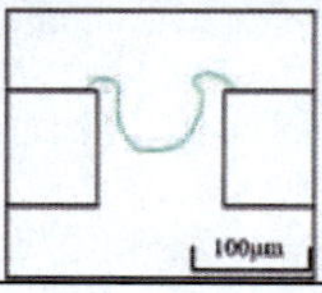	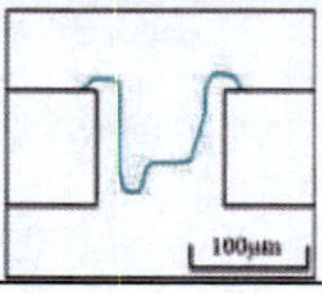	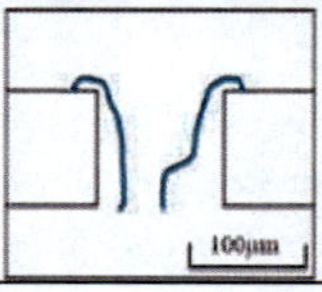	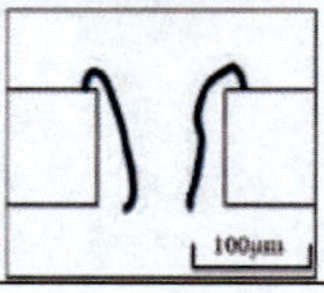

Meanwhile, for the 10-mark loops, the laser power starts to penetrate through the SS304 sheet and make a through-hole. Although the exit hole is tiny, it is considered a through-hole. The entrance hole diameter starts to have the standard hole diameter as preset, which is around 100 μm. For the fifteenth laser mark loop number, the hole shows a complete through-hole that has passed more than 60% from the entrance hole size. The size of the exit hole is also more significant compared to the previous 10-mark loops.

The data of the hole is collected with the assistance of ImageJ software. Table 8 shows the number of laser mark loops against the average depth of the hole and the area of removal. This table indicates that the area removal and the average depth from the SS304 sheet increase with the laser mark loop number. This is because the duration of the fibre laser performing ablation on the workpiece will surge correspondingly as the number of laser mark loops increases. The further the ablation process progresses; the more material is ablated. Hence, the area of the removal from the micro-drilled hole is increasing from 3577.852 to 6516.237 μm^2.

In short, the number of the mark loop will affect the removal area and hole depth of the SS304 sheet during laser drilling. Furthermore, only minor changes in the top side diameter of the holes are observable, ranging between 98.673 and 101.732 μm. It has been proven that only ablation occurs throughout the process.

Figure 5 visually presents the hole formation mechanism from 5-mark loops until 15-mark loops in laser micro-drilling of 0.1 mm thickness of SS304 sheet metal. It is observed that the laser power was not sufficient enough to penetrate with the

Table 8 Number of mark loops against an average depth of hole and area of removal

Number of mark loop	Top side Diameter (μm)	Bottom side Diameter (μm)	Average depth (μm)	Area removal (μm^2)
5	98.673	–	66.206	3577.852
7	99.295	–	80.680	4503.195
10	101.582	31.877	100 (Through Hole)	5564.026
15	101.732	73.051	100 (Through Hole)	6516.237

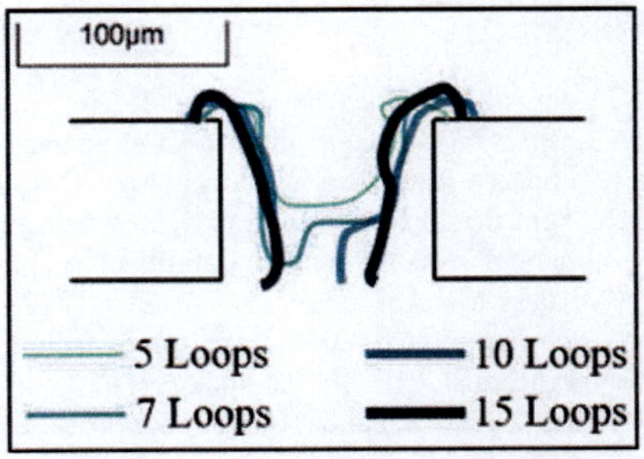

Fig. 5 Mechanism of hole formation

5-mark loops to 7-mark loops. Besides, the laser started to penetrate at the 10-mark loops which gets more prominent for 15-mark loops. With further improvement in parameters, a better laser micro-drilled hole could be achieved.

4 Conclusions

In order to get a better result and optimum properties of the laser micro-drilled holes, laser parameters such as laser power and laser mark loop number should be explored more because the current experiments had just shown a uniformly rising trend. The graph should reach an optimum point to acquire the optimum value for the laser parameter.

(a) High laser power creates a larger exit hole; for example, a 21 W laser power produces a hole area of 6945.261 m^2. The pulse energy increases as the laser power increases, causing the hole wall to ablate more material. The diameter of the entrance hole showed not significantly change in value, which was the range between 101.6691 and 102.9778 μm. This shows that high laser power is not needed to accomplish a front-side hole because it will immediately achieve the same value as the preset. However, the laser power will play an important role when it goes through into the sample to reach the backside.

(b) For the hole formation, laser mark loop from the number of 5–7, the depth of the hole increases from 66.206 to 80.6802 μm. The micro-drilled area becomes a through-hole at the tenth laser mark loop. The size of the through-hole grows as the loop number increases.

Acknowledgements The authors would like to thank the Ministry of Higher Education for providing financial support under Fundamental Research Grant Scheme (FGRS-RACER) No. RACER/1/2019/TK03/UMP//3 (University reference RDU192608), and Universiti Malaysia Pahang for laboratory facilities as well as additional financial support under Internal Research Grant RDU1903118 and PGRS2003138.

References

1. Nasrollahi V, Penchev P, Jwad T, Dimov S, Kim K, Im C (2018) Drilling of micron-scale high aspect ratio holes with ultra-short pulsed lasers: critical effects of focusing lenses and fluence on the resulting holes' morphology. Opt Lasers Eng 110:315–322
2. Venkatesan K, Devendiran S, Bhupatiraju SCSR, Kolluru S, Pavan Kumar C (2020) Experimental investigation and optimization of micro-drilling parameters on Inconel 800 superalloy. Mater Manuf Process 35(11):1214–1227
3. Stephen A et al (2018) Laser micro drilling methods for perforation of aircraft suction surfaces. Procedia CIRP 74:403–406
4. Pattanayak S, Panda S (2018) Laser beam micro drilling–a review. Lasers Manuf Mater Process 5(4):366–394

5. Erny A, Fadhil A, Aiman M, Ishak M, Quazi M (2021) Comparative study between furnace brazing and laser brazing. In: IOP conference series: materials science and engineering, vol 1068, no 1. IOP Publishing, p 012003
6. Quazi M (2020) An overview of laser welding of high strength steels for automotive application. In: Manoharan KA, Quazi MM, Bashir MN, Salleh MNM, Zafiuddin AQ, Linggamm R (eds) An overview of laser welding of high strength steels for automotive application (Int J Technol Eng Stud), vol 6, no 1, pp 23–40
7. Chatterjee S, Mahapatra SS, Bharadwaj V, Choubey A, Upadhyay BN, Bindra KS (2019) Drilling of micro-holes on titanium alloy using pulsed Nd: YAG laser: parametric appraisal and prediction of performance characteristics. Proc Inst Mech Eng Part B: J Eng Manuf 233(8):1872–1889
8. Biscaia R, Ribas M, Júnior AB (2020) Effects of processing parameters on the micro-drilling through fast hole electroerosion and laser trepanning in Inconel 718. Int J Adv Manuf Technol 106(1):31–45
9. Khadtare A, Pawade R, Varghese A, Joshi S (2020) Micro-drilling of straight and inclined holes on thermal barrier coated Inconel 718 for turbine blade cooling. Mater Manuf Process 35(7):783–796
10. Marimuthu S, Antar M, Dunleavey J (2019) Characteristics of micro-hole formation during fibre laser drilling of aerospace superalloy. Precis Eng 55:339–348
11. Zaifuddin A, Aiman M, Quazi M, Ishak M, Ariga T (2020) Effect of laser surface modification (LSM) on laser energy absorption for laser brazing. In: IOP conference series: materials science and engineering, vol 788, no 1. IOP Publishing, p 012013
12. Huang W, Zhang Y, Dai W, Long R (2019) Mechanical properties of 304 austenite stainless steel manufactured by laser metal deposition. Mater Sci Eng A 758:60–70
13. Uchtmann H, Haasler D, Gillner A (2017) Laser micro drilling of wing surfaces for hybrid laminar flow control. In: Lasers in manufacturing proceedings
14. Sarfraz S, Shehab E, Salonitis K, Suder W (2019) Experimental investigation of productivity, specific energy consumption, and hole quality in single-pulse, percussion, and trepanning drilling of in 718 superalloy. Energies 12(24):4610
15. Witte R, Moser T, Liebers R, Holtz R (2008) Laser micro-drilling with nanoseconds: parametrical influences and results. In: Advanced Laser Technologies 2007, vol 7022: International Society for Optics and Photonics, p 702208

A Simulation Study on Interfacial Reaction Between Sn3Ag0.5Cu and Sn0.7Cu Using Different Substrates After Reflow Soldering

M. H. Mohd Zaki and S. R. A. Idris

Abstract Reflow soldering is a process to create joining between the board and electronic component in order to make sure the electronic devices may function well. The aim of this study is to determine the solder joint strength through simulations using data from previous researchers. Two type of solder alloys were used namely Sn3Ag0.5Cu (SAC305) and Sn0.7Cu (SC07) with two types of substrate such as laminated copper and pure copper. Simulation was conducted using Fusion 360 software. Besides, the information and data on intermetallic compound formation and growth, as well as thickness were gathered and presented in this study to support the simulation results. Results showed that pure SAC305/copper substrate produced lower shear strength which was 15.17 MPa as compared to SAC305/laminated copper with the value of 26.67 MPa. Meanwhile SC07/pure copper also gave lower shear strength which was 5.62 MPa as compared to SC07/laminated copper which was 5.45 MPa. In terms of IMC, it was found that mainly Cu_6Sn_5 was formed at the solder joint interface with an average thickness of 3 μm for SAC305, and 5 μm for SC07 for both substrates. Hence it can be concluded that SAC305 with laminated copper substrate showed a good performance to produce a reliable electronics product.

Keywords Reflow soldering · Solder alloy · Intermetallic compound · Solder joint strength

1 Introduction

The SnPb solder has been progressively prohibited due to the serious impact of lead onto environment and human health [1–4]. This in turn making lead-free solders such as SnAgCu, SnCu and SnAg becoming a suitable candidate to replace lead-containing solders due to its solder properties as well as mechanical properties superior than SnPb solder [5–8]. Among those lead free solders, the Sn3Ag0.5Cu (SAC305) and

M. H. Mohd Zaki · S. R. A. Idris (✉)
Faculty of Mechanical & Automotive Engineering Technology, Universiti Malaysia Pahang, 26600 Pekan, Pahang, Malaysia
e-mail: rabiatull@ump.edu.my

M. Y. Ismail et al. (eds.), *Technological Advancement in Mechanical and Automotive Engineering*, Lecture Notes in Mechanical Engineering,
https://doi.org/10.1007/978-981-19-1457-7_62

Sn0.7Cu (SC07) is an established and well-known solder used in electronic packaging which offers a better mechanical properties and solderability as compared to SnPb [3, 7, 9]. However, their performances in terms of mechanical properties are different when using different type of substrate such as laminated copper or pure copper. Laminated copper is the latest substrate used in the market for cost reduction. Meanwhile, the use of pure copper is still exist especially when involves in producing high performance product with longer lifetime. Different substrate will create different interfacial reaction which might be translated into diverse solder joint reliability performance. In electronic packaging industry, the main critical issue is always on the mechanical properties of the solder joint. It became more demanding especially with the development of smaller and multifunctional electronic devices which at the same time leading to the decreasing in size of solder joint [10, 11].

Many researchers have conducted a study in order to determine the factors affecting mechanical strength of the solder joint. For instance, there were researchers who found that solder joint strength was increasing with higher strain rate [10, 12]. Whereas the other researcher reported that dual layer of metallized copper substrate managed to suppressed IMC growth during isothermal aging and hence improving its SAC305/NiCu solder joint strength [13]. In addition, there was also a study showed that adding graphene nanosheets into SnCu solders were successfully suppress the IMC growth and at the same time improving the solder joint strength [14]. This proved that the changes in Cu-Sn IMC composition significantly affect the solder joint strength. Nonetheless, the research on comparison between these solders using different type of substrates, namely laminated copper and pure copper, are still lacking and do not well-understood.

This study is aimed at studying the solder joint strength through simulations using data from previous researchers. Two type of solder alloys were used namely Sn3Ag0.5Cu (SAC305) and Sn0.7Cu (SC07) with two types of substrate such as laminated copper and pure copper. The comparison was also made in terms of IMC formation and growth, as well as thickness which were obtained from other researchers.

2 Experimental Setup

The current study aims at determining the solder joint strength through simulation using two type of solder alloys such as SAC305 and SC07 with different substrate namely laminated copper and pure copper. The software used was Fusion 360 which is part of the CAD software to simulate the design. The design was made up of two layers of substrate with the size of 1 cm x 1 cm $\times$ 0.2 cm that are sandwiched together, with the middle part supported by solder joints that have melted to integrate the various layers of substrate, as shown in Fig. 1. The simulation was carried out following a standard of ASTM D1002. The mechanical properties of the substrate and solder joints was kept the same to the actual solder and substrate used. The material properties used are shown in Tables 1 and 2. After that the simulated data

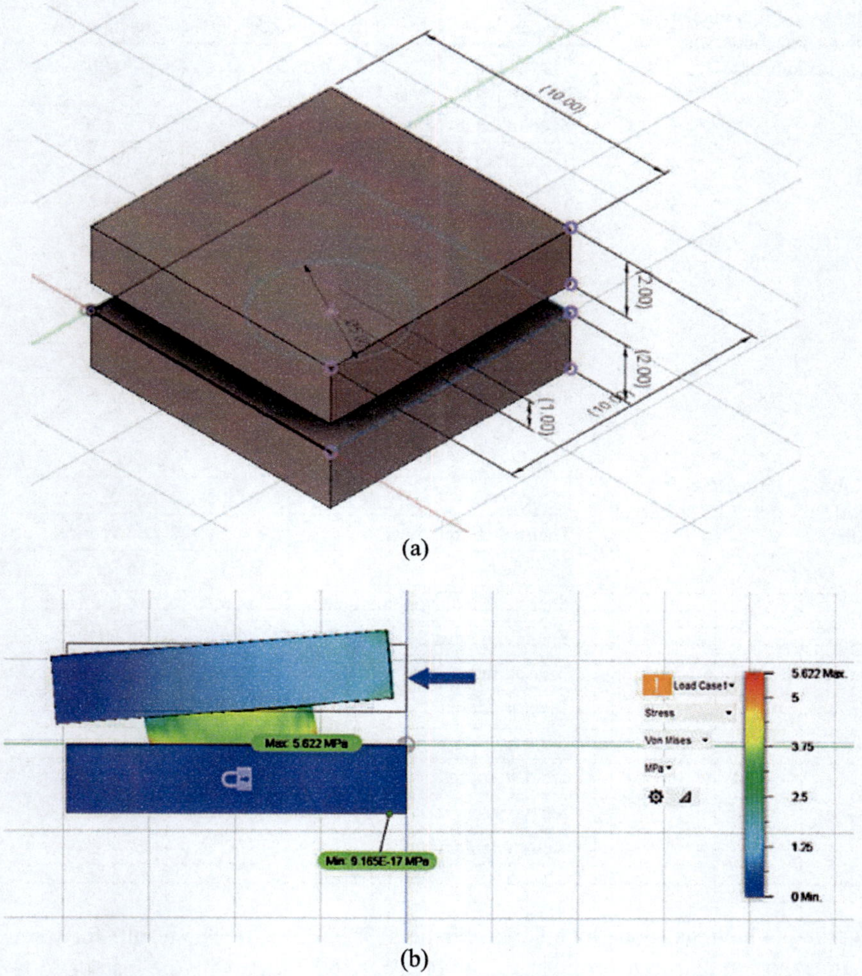

Fig. 1 Design for shear test, **a** Overall design and **b** Shear test was carried out by pushing the upper part of the joining structure

was supported using previous researchers in terms of IMC formation and growth, thickness as well as wetting angle.

3 Results and Discussion

Figure 2 shows a shear test results for SAC305 using laminated copper and pure copper. The results were compared in terms of simulation data and experimental

Table 1 Properties of laminated copper and pure copper substrate

Properties	Laminated copper	Pure copper
Thermal conductivity	3.5 W/m°K	4.01 W/m°K
Specific heat	0.95 J/g °C	0.45 J/g °C
Thermal expansion coefficient	0.1 μm/(m °C)	16.7 μm/(m °C)
Young modulus	6.89 GPa	117.5 GPa
Poisson ratio	0.03	0.34
Shear modulus	0 MPa	46,000 MPa
Density	1.85 kg/m^3	8.94 kg/m^3
Yield strength	51.71 MPa	33.3 MPa
Tensile strength	0 MPa	210 MPa

Table 2 Properties of SAC305 and SC07 solder alloy

Properties	SAC305	SC07
Thermal conductivity	59 W/m°K	66 W/m°K
Specific heat	1 J/g °C	15 J/g °C
Thermal expansion coefficient	23 ppm/k	16 μm/(m °C)
Young modulus	51 GPa	64.6 GPa
Poisson ratio	0.35	0.26
Shear modulus	46,000 MPa	25,620 MPa
Density	7.4 kg/m^3	7.31 kg/m^3
Yield strength	130 MPa	16.3 MPa
Tensile strength	50 MPa	30 MPa
Melting temperature	217 °C	227 °C

worked (which was done by other researchers [7, 9, 10, 15]). The results for simulation showed that solder joint started to break at 26.67 MPa with the load of 85 N. Whereas a for experimental worked, the joint failed at 30 MPa with the same load. Meanwhile, when pure copper was used, the point where the solder joint started to failed was lower which was 15.17 MPa and 15.98 MPa for simulation and experimental worked, respectively. Therefore, laminated copper was found to be more promising in providing better solder joint strength as compared to pure copper. It is believed that this is due to better joining was formed between solder and substrate when laminated copper was used. When pure copper was used as a substrate, the IMC growth thicker and more aggressive than laminated copper since more copper atom can diffused towards solder, and hence creating brittle interface.

Besides, Fig. 3 shows a shear test results for SC07 using laminated copper and pure copper. Same as in Fig. 2, the data obtained in simulation was compared with experimental data obtained from previous researchers [10, 15]. It can be seen that in Fig. 3a, materials started to failed at the load of 15 N with the maximum value of

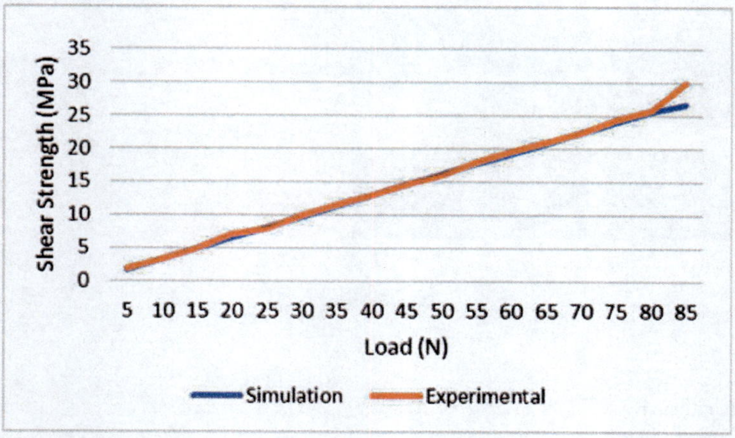

(a)

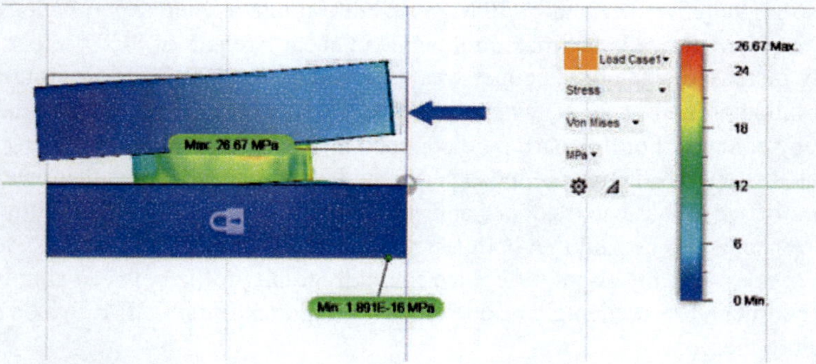

(b)

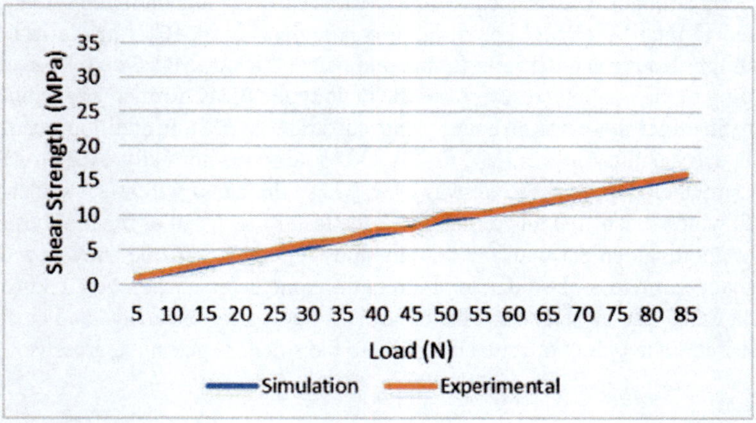

(c)

Fig. 2 Shear test results for SAC305 **a**, **b** laminated copper, and **c**, **d** pure copper

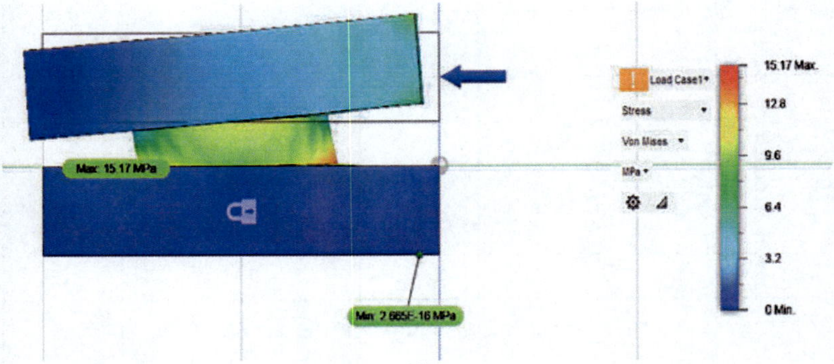

(d)

Fig. 2 (continued)

shear strength of 5.45 MPa and 7.10 MPa for both simulation and experimental data, respectively. Meanwhile the maximum value of shear strength of SC07 solder joint slightly increased when pure copper was used which was 5.62 MPa for simulation which failed at load of 20 N, whereas for data from experimental worked did not show any changes. For this part, pure copper was found to be more promising than laminated copper. Nonetheless, the experimental worked proofed that there were no changes between laminated copper and pure copper when SC07 solder alloy was used. Hence, it can be said that both type of substrate can be used for SC07 solder alloy. Besides that, the slight difference found in graph profile between simulation and experimental worked might be due to the formation of irregular IMC which quite difficult to measure.

In addition, in terms of solder alloy, SAC305 managed to give a better result for solder joint strength. This is because, the existence of Ag element in the solder alloy managed to retard the IMC growth as it was known that IMC is brittle in nature which might leads to solder joint failure and hence thicker IMC should be avoided. According to the previous researchers, the main type of IMC formed after soldering was Cu_6Sn_5 since there was no coating onto substrate used [8]. In addition, according to them, average thickness of IMC for SAC305 solder was normally 3 μm, for SC07 was 5 μm when copper substrate was used. These thickness will affect solder joint strength which means the thicker the IMC, the faster it will fail or break. These data are in a good agreement with the current study whereby SAC305 showed a better performance in terms of solder joint strength. Nonetheless, when cost taking into account, SC07 can be selected. This is because Ag is very expensive and currently the research on avoiding or replacing Ag into the solder is getting aggressive.

4 Conclusion

In this study, solder joint strength was determined through simulations and compared with experimental worked by previous researchers. Simulation was conducted using

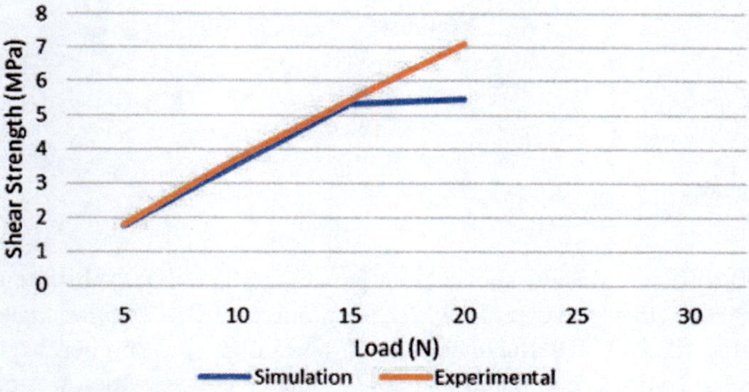

(a)

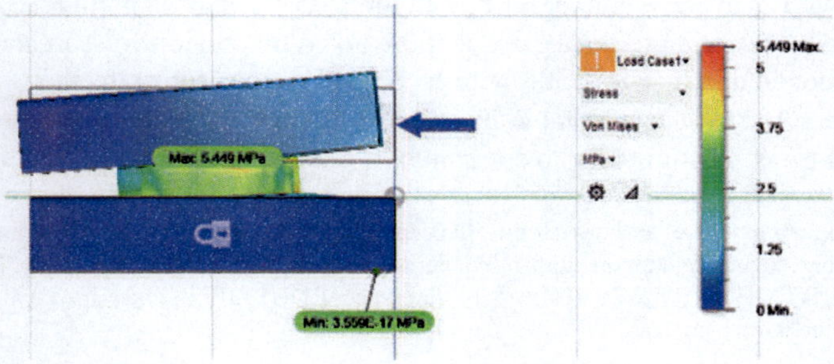

(b)

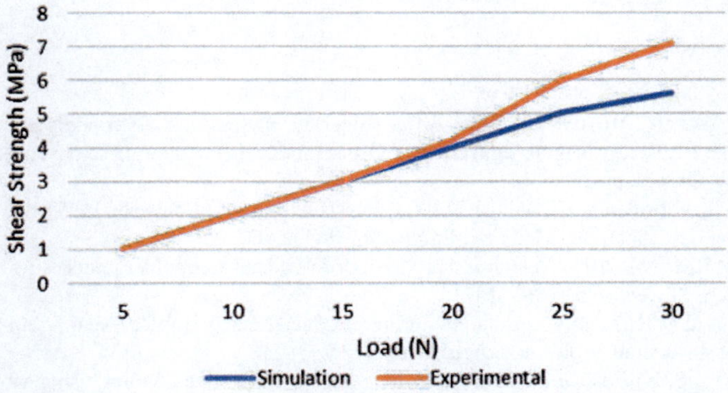

(c)

Fig. 3 Shear test results for SC07 **a**, **b** laminated copper, and **c**, **d** pure copper

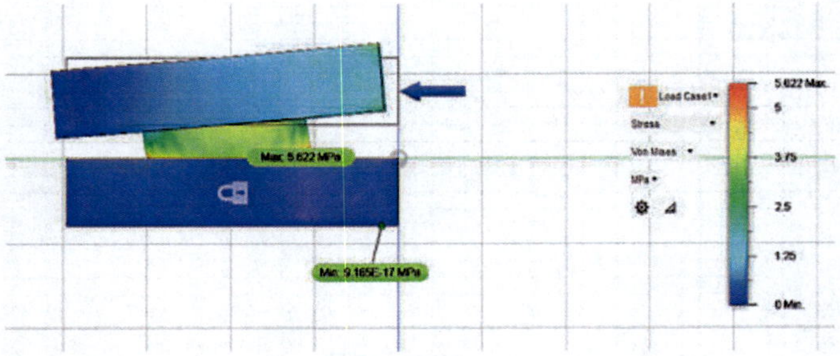

(d)

Fig. 3 (continued)

Fusion 360 software. Results showed that pure SAC305/copper substrate produced lower shear strength which was 15.17 MPa as compared to SAC305/laminated copper with the value of 26.67 MPa. Meanwhile SC07/pure copper gave higher shear strength which was 5.62 MPa as compared to SC07/laminated copper which was 5.45 MPa. In terms of IMC, it was found that mainly Cu_6Sn_5 was formed at the solder joint interface with an average thickness of 3 μm for SAC305, and 5 μm for SC07 for both substrates, by referring to the experimental worked by previous researchers. Hence it can be concluded that SAC305 with laminated copper substrate showed a good performance to produce a reliable electronics product. Nonetheless, more studies need to be carried out in order to determine its practical application.

Acknowledgements The authors would like to thanks the Ministry of Higher Education for providing financial support under Fundamental Research Grant Scheme (FRGS) No. FRGS/1/2018/TK03/UMP/02/21 (University reference RDU190152) and Universiti Malaysia Pahang for laboratory facilities.

References

1. Kroupa A et al (2016) Chapter 5—lead-free soldering: environmentally friendly electronics. In: Singh M, Ohji T, Asthana R (eds) Green and sustainable manufacturing of advanced material. Elsevier, Oxford, pp 101–134
2. Laurila T, Vuorinen V, Kivilahti JK (2005) Interfacial reactions between lead-free solders and common base materials. Mater Sci Eng R Rep 49(1):1–60
3. Shalaby RM et al (2018) Design and properties of new lead-free solder joints using Sn-3.5Ag–Cu solder. Silicon 10(5):1861–1871
4. Ke JH et al (2016) Pattern formation during interfacial reaction in-between liquid Sn and Cu substrates—a simulation study. Acta Mater 113:245–258
5. Liu X, He S, Nishikawa H (2017) Low temperature solid-state bonding using Sn-coated Cu particles for high temperature die attach. J Alloy Compd 695:2165–2172
6. Hu X et al (2019) Insights on interfacial IMCs growth and mechanical strength of asymmetrical Cu/SAC305/Cu–Co system. Vacuum 167:77–89

7. Jaya NT, Idris SRA, Ishak M (2019) A Review on Mechanical Properties of SnAgCu/Cu Joint Using Laser Soldering. In: The advances in joining technology. Springer Singapore, Singapore
8. Nabila TJ, Idris SRA, Ishak M (2019) Effect of fibre-lasers parameters on interfacial reaction and wetting angle of two different types of SAC305 solder fabrication on Cu pad. IOP Conf Ser Mater Sci Eng 469:012117
9. Yang M et al (2011) Cu6Sn5 morphology transition and its effect on mechanical properties of eutectic Sn–Ag solder joints. J Electron Mater 40(2):176–188
10. Hu X et al (2017) Shear strength and fracture behavior of reflowed Sn3.0Ag0.5Cu/Cu solder joints under various strain rates. J Alloys Compd 690:720–729
11. Nishikawa H, Iwata N (2015) Improvement of joint reliability of Sn–Ag–Cu solder bumps on Cu by a laser process. Mater Trans 56
12. Wang H, Hu X, Jiang X (2020) Effects of Ni modified MWCNTs on the microstructural evolution and shear strength of Sn–3.0Ag–0.5Cu composite solder joints. Mater Charact 163
13. Wan Y et al (2018) Interfacial IMC growth of SAC305/Cu joint with a novel dual-layer of Ni(P)/Cu plating during solid-state aging. Microelectron Eng 199:69–79
14. Huang Y et al (2016) Improving shear strength of Sn-3.0Ag-0.5Cu/Cu joints and suppressing intermetallic compounds layer growth by adding graphene nanosheets. Mater Lett 169:262–264
15. Sonawane PD, Bupesh Raja VK, Gupta M (2021) Mechanical properties and corrosion analysis of lead-free Sn–0.7Cu solder CSI joints on Cu substrate. Mater Today: Proc 46:1101–1105

Investigation of Opening Position on Natural Cross Ventilation for an Isolated Building

Lip Kean Moey, Rui Jun Tok, Vin Cent Tai, Prasath Reuben Mathew, Joseph Wu Kai-Seun, and Ahmed Nurye Oumer

Abstract The opening position is one of the factors that affect the ventilation performance of a building. In this study, the effect of opening position on natural cross ventilation of isolated building was investigated. The airflow pattern and ventilation rate under different opening configurations were analyzed. Eight different opening configurations were considered, including aligned and unaligned openings, as well as vertical-opening design. Computational fluid dynamics (CFD) simulation with 3D steady-state RANS equation Shear Stress Transport (SST) k-ω turbulence model was used. The parameters of streamwise dimensionless wind speed ratio (U/U_{ref}), pressure coefficient (C_p) and dimensionless flow rate (DFR) were analyzed in this study. The results show that the aligned opening configuration Top-Top has the highest DFR at 0.60. This result is similar to that obtained from the literature. In addition, the design of vertical openings can improve the DFR of the building. The DFR of the building is mainly affected by the position of the opening on the windward side. This concludes that the opening positions exert an imperative role in affecting the internal airflow pattern, air recirculation and DFR of a naturally cross ventilated building.

Keywords Natural cross ventilation · Opening position · CFD · Vertical opening

L. Kean Moey (✉) · V. C. Tai
Faculty of Engineering, Built Environment & Information Technology, Centre for Modelling and Simulation, SEGi University, Selangor, Malaysia
e-mail: moeylipkean@segi.edu.my

R. Jun Tok · P. Reuben Mathew · J. Wu Kai-Seun
Faculty of Engineering, Built Environment & Information Technology, SEGi University, Selangor, Malaysia

A. N. Oumer
Department of Mechanical Engineering, College of Engineering, University Malaysia Pahang, Pahang, Malaysia

M. Y. Ismail et al. (eds.), *Technological Advancement in Mechanical and Automotive Engineering*, Lecture Notes in Mechanical Engineering,
https://doi.org/10.1007/978-981-19-1457-7_63

1 Introduction

Malaysia is a tropical country and the hot climate makes people to rely on mechanical ventilation for thermal comfort. Buildings account for about a quarter of the carbon emissions, much of it from electrical and mechanical equipment in buildings [1]. The emission of mechanical ventilation cannot be ignored, which has an impact on global environmental problems. The proportion of residential and commercial buildings in the total energy consumption in Malaysia has increased significantly by 20–40% in 2014 [2, 3]. Proper ventilation design is key in finding solutions to reduce carbon emissions and overcome the destructive effects of climate change. Since ancient times, humans tried to use natural ventilation to create more comfortable indoor living conditions. Thermal comfort in a house depends on three main factors: air flow, relative humidity and temperature, which are influenced by the building's ventilation. Natural ventilation is affected by the characteristics of wind and the opening characteristics of the building. In solving global energy problems and environmental pollution, natural ventilation, as a known cooling load reduction method with great potential, has returned to the topic of extensive research and discussion in the current construction industry.

As time progressed with technological revolutions, researchers have studied cross ventilation using different methods, including reduced-scale experiments, full-scale measurements and computational fluid dynamics (CFD) [4, 5]. In natural ventilation, cross ventilation has been the focus in literature, because it is relatively more effective than the single-sided ventilation, and it is also an important research direction to replace mechanical ventilation. Openings in opposite walls are the main cross ventilation technique, suitable for different wind conditions and passive cooling [6]. Researchers from China conducted a numerical study based on cross ventilation [7]. The results of the experiment indicate that the cross ventilated airflow is the strongest, moving counter-clockwise from the leeward to the windward side. Shetabivash used 3D computational fluid dynamics to study the parameters of cross ventilation and airflow patterns inside buildings with different shapes and opening positions [8]. The study set up involves air inlets and outlets of different heights, including one high and one low opening in a single building. The findings show that in the air intake space, a powerful recirculation airflow is created. When the air inlet and outlet are close to the top of the building, a return flow is formed in the lower half of the building, and the situation is opposite of the opening near the bottom of the building. The combination of inlet and outlet near the top of the building show good air distribution and recirculation. The position of the entrance has an important influence on the airflow velocity at the inlet, which in turn affects the ventilation rate of the building.

Karava et al. also conducted the wind tunnel experiments on buildings with operable facade elements using particle image velocimetry (PIV) experimental methods to study the air velocity field in a cross ventilated building [9]. The study found that the orifice equation predicts the ventilation rate reasonably well for openings above the mid-height of the building for a wall porosity of 10% or less, while it overestimates the ventilation rate of the openings below the middle height of the building.

Ramponi and Blocken conducted a coupled 3D steady-state RANS simulation of a general isolated building to test the influence of computational parameters on ventilation [10, 11]. The results of the turbulence model show that the SST k-ω model showed the best performance, followed by the RNG k-ε model. This research also demonstrates that the accuracy of outdoor standing vortex prediction has a significant impact on the accuracy of indoor ventilation flow prediction. Moey et al. studied the effect of opening position on indoor airflow in a naturally ventilated isolated building model [12]. The results show that the position of the opening has a critical effect on the internal airflow pattern, air recirculation and ventilation rate. The simulation results show that the ventilation rate of the opening near the roof is higher than that of the opening near the ground.

The natural ventilation of a building is influenced mainly on the building openings, interior structure and natural wind parameters. However, existing literatures mainly focus on aligned and diagonally placed openings configuration. More opening configurations should be considered to understand the effect of opening configurations on ventilation rate and indoor air-flow patterns. In addition to the configuration of the aligned openings, this study also considered the impact of asymmetric openings (such as vertical openings) on natural ventilation. CFD software is used to study the influence of eight opening configurations on the airflow pattern and ventilation rate of the building. Section 2 presents the geometry of the model, computational domain, model validation study, grid sensitivity analysis, solver setting and simulation cases. In Sect. 3, the simulation results are discussed and the conclusions of the study are given in Sect. 4.

2 Methodology

2.1 Model Geometry

In this study, the simulation model used is based on the reference from Ramponi and Blocken [10], which is the building model with a length and width of 0.1 m and a height of 0.08 m. The full-scale dimension of the building is 20 m long (L), 20 m wide (W) and 16 m high (H) and were scaled according to 1:200 for CFD simulation. The size of the ventilation openings used in this study is the same at 0.046 m wide and 0.018 m high. The model used a wall thickness of 2 mm. Figure 1a shows the dimensions of the building model in this study, while Fig. 1b shows an isometric view of the model. Figure 1b also shows the CFD measurement plane located in the center of the model to measure the airflow distribution, velocity vector, pressure coefficient and ventilation rate.

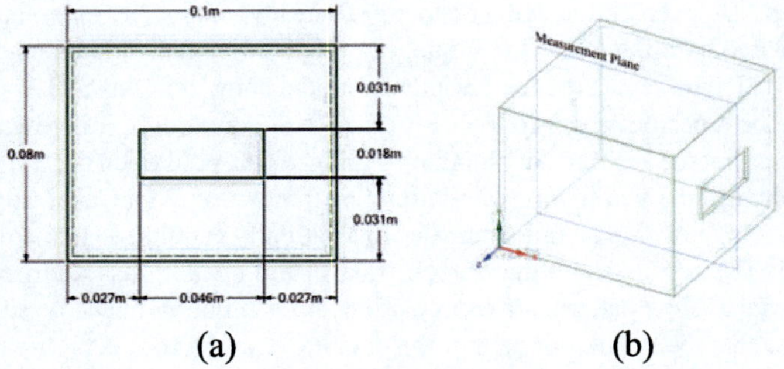

Fig. 1 **a** Building model with opening size and dimensions in front view; **b** Isometric view of building model

2.2 Computational Domain and Grid

The dimensions of the computational domain were set based on the reference from the Ramponi and Blocken [10], and the height (H) of the building model is used as the measurement scale. Taking the building as the center point, the length of the computational domain extending upward and lateral sides is 5H, the distance of the inlet plane to the windward facade is 3H, and the distance from the leeward façade to the edge of the domain is 15H. The upstream distance is fixed at 3H to limit the development of unintended streamwise gradients to ensure horizontal homogeneity of the approach and incident flow profile. The dimension of the computational domain is 0.9 m × 1.54 m × 0.48 m (W × L × H) and its isometric view is shown in Fig. 2.

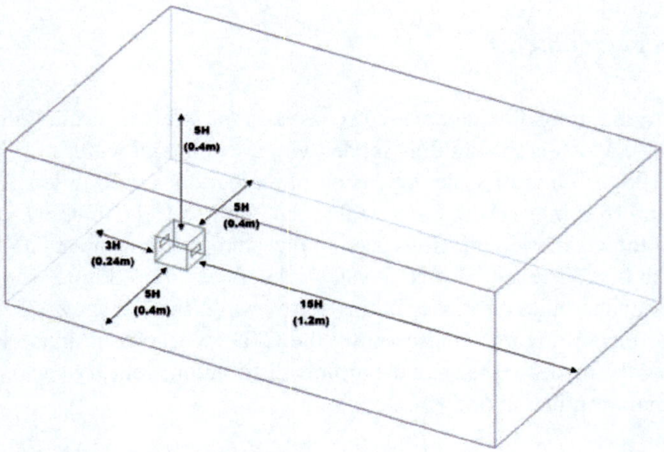

Fig. 2 Dimension of computational domain

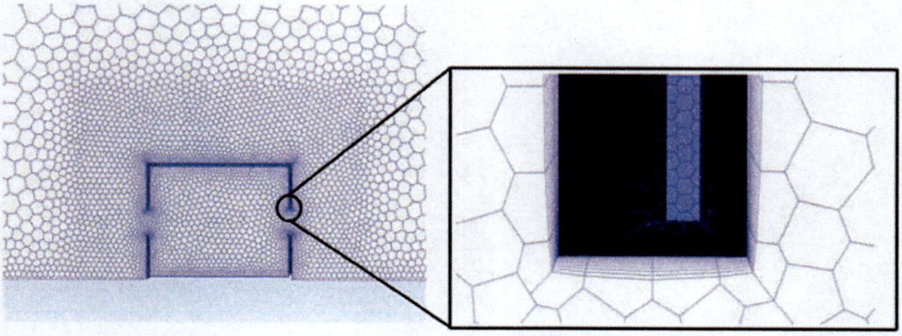

Fig. 3 Mesh around the model and the close-up view of the inflation layer

Meshing around the model and its inflation layer are shown in Fig. 3. There is a small domain around the model, called the body of influence (BOI), which is 50 mm away from each surface of the model to provide more accurate results and obtain results in a shorter calculation time. The overall cell number after meshing is about 810,000, which is determined by the grid sensitivity analysis in Sect. 2.5. The poly-hexagon shape is applied in the meshing and the building model is surrounded by uniform prism with 10 layers along with a first cell height of 0.01 m. The grid skewness of the meshing result is controlled to below 0.9.

2.3 Atmospheric Boundary Layer (ABL) Condition

ABL file was interpreted to create the velocity profile from the inlet. The equation of ABL friction velocity, U_{ABL} is shown in Eq. 1. The reference velocity, U_{Ref} is 6.97 m/s, the reference height, Z_{Ref} is 0.08 m, the Von Karman constant, κ is 0.42, and the aerodynamic roughness height *Zo* is 0.025 mm [13]. Substituting the ABL friction velocity, U_{ABL} into Eq. 2 to obtain the inlet velocity profile, U. Turbulent kinetic energy, k and turbulent dissipation rate, ε are calculated by using Eqs. 3 and 4, respectively, and the empirical constant $C_{MU} = 0.09$. Turbulent dissipation rate, ω is calculated by substituting turbulent kinetic energy, k and turbulent rate, ε into Eq. 5. The top and side walls of the flow domain are set to zero specific shear stress, the outlet is the pressure-outlet setting while the inlet is the velocity-inlet. The turbulent dissipation rate, velocity magnitude and turbulent kinetic energy generated by the ABL file are included in the inlet settings. By substituting the aerodynamic roughness height value *Zo* into Eq. 6, it can be obtained that the ground sand grain roughness height, k_s is 0.0006 m.

$$U^*_{ABL} = \frac{U_{Ref} \times \kappa}{\log\left(\frac{Z_{Ref}}{Z_o} + 1\right)} \quad (1)$$

$$U = \frac{U^{*}_{ABL}}{\kappa}\log\left(\frac{Z_{Ref}}{Z_o} + 1\right) \tag{2}$$

$$k = \frac{\left(U^{*}_{ABL}\right)^2}{\sqrt{C_{MU}}} \tag{3}$$

$$\varepsilon = \frac{\left(U^{*}_{ABL}\right)^3}{k\left(Z_{Ref} + Z_o\right)} \tag{4}$$

$$\omega = \frac{\varepsilon}{C_{MU}k} \tag{5}$$

$$k_s = \frac{9.793Z_o}{C_s} = 20Z_o \tag{6}$$

2.4 Solver Settings

ANSYS 2019 R3 software was used for CFD simulation in this study. The fluent solver of this CFD simulation was set to be Pressure-Based with the steady time function. The viscous model used was the Shear Stress Transport (SST) k-ω turbulence model, from the 3D steady-state Reynolds-averaged Navier–Stokes (RANS) equation. The Simple scheme was chosen as the pressure–velocity coupling and the Green-Gauss Node Based gradient was selected for the spatial discretization. Four discrete schemes of pressure, turbulent kinetic energy, momentum and turbulent dissipation are all selected as the second-order discrete scheme because the second-order discrete scheme is more accurate than the first-order discrete scheme. The hybrid initialization was chosen for this study. The convergence was expected to be achieved when all scaled residuals stabilize and reached a minimum value of 10^{-5} for k, and 10^{-6} for x, y and z, and 10^{-4} for ω.

2.5 Grid Sensitivity Analysis

In this paper, three different grids were used to analyze the grid resolution of the building model and the results were compared with the simulation results from Ramponi and Blocken [10] and wind tunnel experimental results from Karava et al. [9]. The grid resolution is defined by the number of cells, and is classified by coarse, medium, and fine grids. The meshing of this model used 460,033 cells as coarse, 812,684 cells as medium, and 1404,948 cells as fine mesh. The simulation results of these grids are represented by the graph of the mean streamwise wind speed ratio (U/U_{ref}) and compared with the experimental result [9] and CFD result [10]

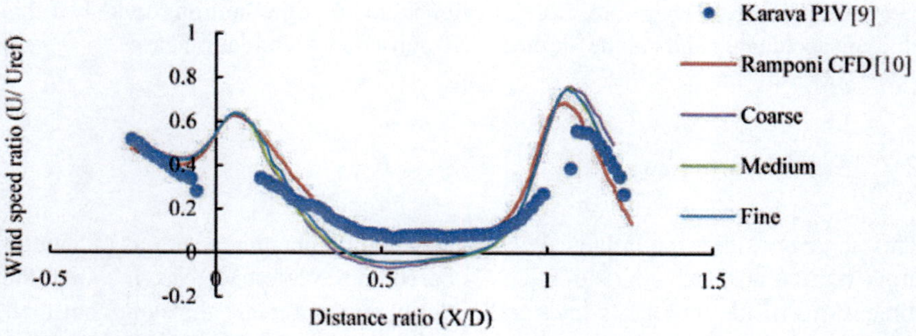

Fig. 4 Simulation results of three different cell counts, experiment result [9] and CFD result [10]

as shown in Fig. 4. From the observation, it can be concluded that the simulation results shows that it is grid independent as they are close to each other with a minimal percentage error. The medium grid was chosen to be applied for all subsequent simulation cases as it has the optimum cell count for reasonable accuracy and computational cost.

2.6 Model Validation

The CFD simulation study by Ramponi and Blocken [10] was used for model validation purpose. The model with aligned opening configurations was used to compare with Ramponi's result based on the indoor flow velocity vector diagram shown in Fig. 5. Both simulations results show high flow rates above the windward wall of the building and a low-speed zone above the roof. After entering the building through the opening, the airflow jet moves downward and slowly moves upward towards the exit of the building. The result shows that the airflow trends of the two diagrams are

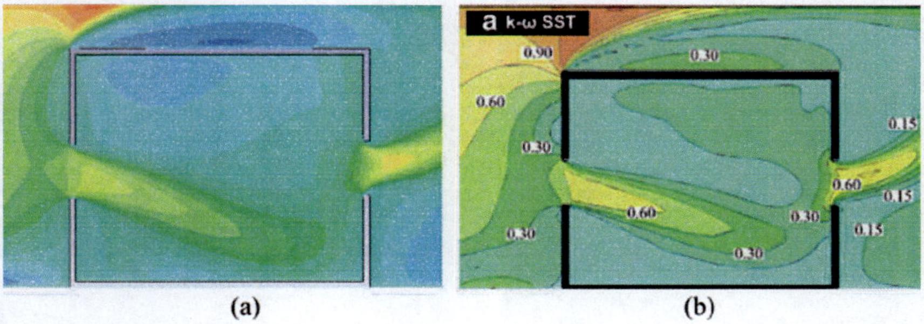

Fig. 5 Velocity contour for validation **a** Simulation result, **b** Numerical simulation result [10]

in good agreement. Therefore, it can be concluded that the building model of this study is acceptable and can be applied to all subsequent simulation cases.

2.7 Simulation Cases

This paper considers the influence of various building opening positions on natural cross ventilation. Each simulation case is based on the reference case in which the dimensions of the building's inlet and outlet openings remain the same, but their positions are modified. The simulation cases have a constant wall porosity of 10% [9]. Figure 6 shows a simulation model for eight different opening configurations. The

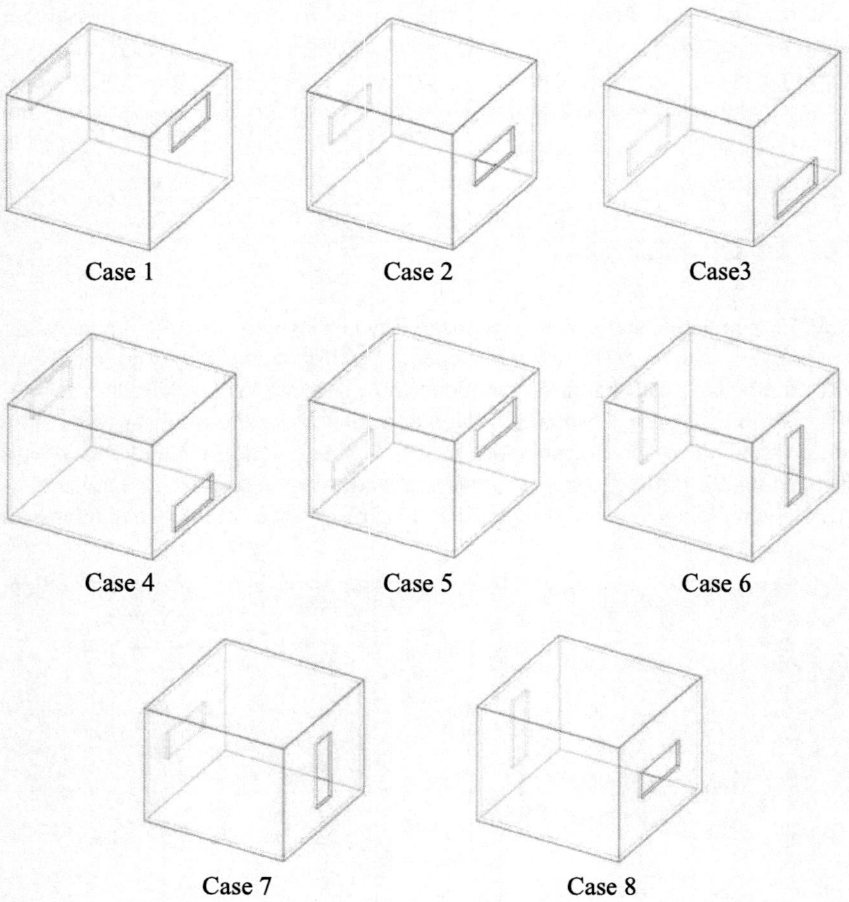

Fig. 6 Isometric views of the simulation model for eight cases

opening height varies on the windward and leeward sides of the building, including 0.015 m at the bottom, 0.04 m at the middle and 0.065 m at the top. The simulated cases studied include aligned openings, non-aligned openings, and the opening of the building in vertical direction.

3 Results and Discussion

3.1 Velocity and Pressure Coefficient

The computational fluid dynamic (CFD) simulation was carried out for eight building opening configurations. Simulation results of the velocity profile, velocity vector, and pressure coefficient of each opening configuration are shown in Table 1, with U/U_{ref} representing the wind speed ratio. A high wind speed ratio inside the building is observed in Case 1 with the Top-Top opening configuration. This simulation result shows that aligned openings at the top have higher velocity than aligned openings in the middle and bottom of the model. This finding is consistent with previous studies in the literature [8, 9, 14]. Besides, it can be seen that the inlet opening configuration at the bottom of the building such as Case 3 and Case 5 has a poor performance in the velocity ratio of the building interior. This indicates that the airflow velocity in the building is poor except in the horizontal line of the opening, which may affect the overall ventilation performance of the building. Opening configuration with a vertical inlet opening can have a wider airflow area and a better velocity ratio on its vertical lines, such as Case 6 and Case 8. For Case 2 and Case 7, the airflow velocity in the center of the building is low and the airflow flows downward. These results indicate that the windward opening position of the building have a greater impact on the wind velocity in the building.

According to the velocity vectors contour for all cases, a small vertical vortex was formed at the bottom of the building exterior because of the no-slip condition on the ground surface. Since the wind on the windward side will deflect to the top after hitting the building wall, the inlet airflow at the opening closer to the top will be directed upwards, such as Case 1, thereby generating an upward force at the windward opening. When the air enters the building in a downward flow, it leaves the building in an upward trend, and the situation is opposite when the air enters the building in an upward flow. For Case 2 and Case 7, the airflow in the middle opening causes the upper and lower parts of the building to have large recirculation areas, and the vortex formed above the flow is larger, causing the flow out of the building to flow upward. This situation can also be applied to vertical windward openings like Case 6 and Case 8.

In all cases, due to the impact of air hitting the windward wall, the pressure on the windward wall is the highest, while the pressure on the leeward side of the building is the lowest. The results of the pressure coefficient show that the pressure coefficient in the room is affected by the positions of the windward and leeward openings. The

Table 1 Wind speed ratio (U/U_{ref}), velocity vector and pressure coefficient contours for different opening configurations

Case	U/U_{ref}	Velocity Vector	Pressure Coefficient, C_p
1	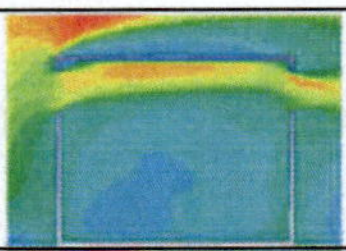		
2	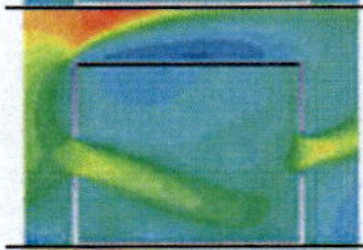		
3	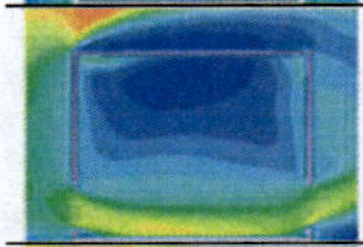		
4			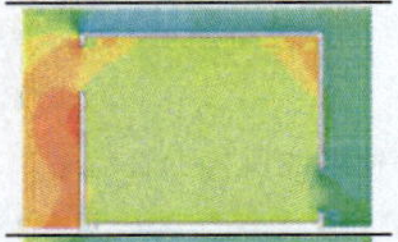
5	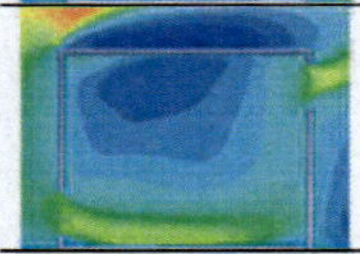		
6	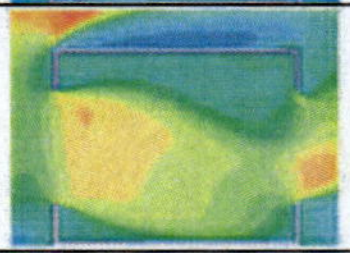		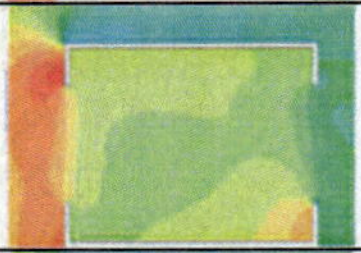
7	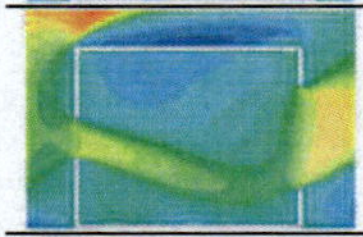		
8	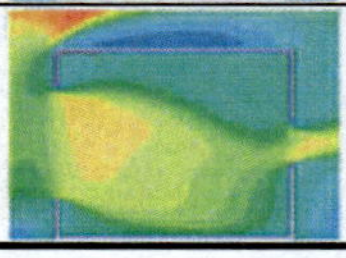		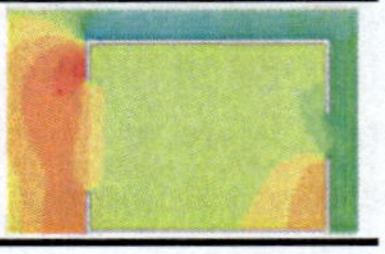

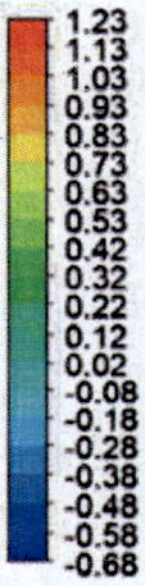

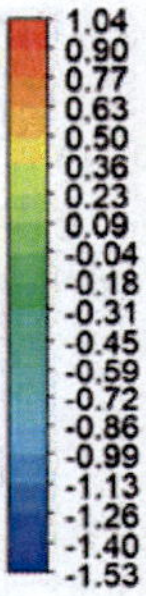

difference in the pressure coefficient between the air inlet and outlet will affect the ventilation rate of the entire building.

3.2 Dimensionless U/U_{ref} Velocity Profile

Figure 7 shows the relationship between the wind speed ratio (U/U_{ref}) and the distance ratio (X/D) for eight opening configurations. The velocity profile is obtained from the data generated by the horizontal line between the opening on the windward and leeward openings. The airflow velocity trend in all cases shows a U-shape profile. The wind enters the opening on the windward side at a high velocity, and immediately a downward jet is observed. When the airflow reaches the outlet, it accelerates to a higher velocity. The 0.0, 0.5, and 1.0 of X/D in Fig. 7 represents the positions at the windward opening, the middle section of the building, and the leeward opening, respectively. The velocity distribution shows that among the symmetrical openings cases, Case 1 has the highest speed which is consistent with the results from Karava et al. [9]. The wind speed ratio at the windward opening and leeward opening of Case 4 is higher than that of Case 5, which means that the airflow velocity the top is better than the air inlet at the bottom of the building. For the configurations with vertical windward openings such as Case 6 and Case 7, although the flow velocity at the windward opening is not as high as in Case 1, the velocity in the middle section of the building and the leeward opening is the higher relative to other cases. The air velocity with vertical air inlet configurations show a slow and steady decrease in velocity after entering the building, which is different from the results of other horizontal inlet openings cases. The velocity profile of Case 7 (Middle-Vertical) is relatively similar to Case 2 (Middle-Middle), however the former has a better DFR. Such results show that vertical openings at the windward or leeward opening have an influence on the airflow velocity in the building and its overall DFR.

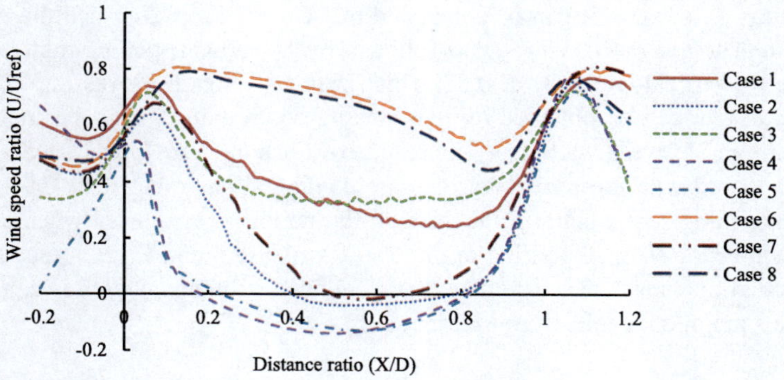

Fig. 7 Mean streamwise wind speed ratio (U/U_{ref}) against distance ratio (X/D) for eight cases

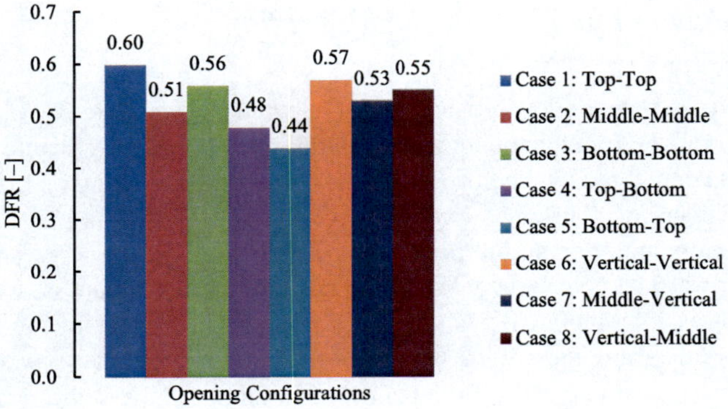

Fig. 8 Dimensionless flow rate (DFR) for each opening configuration

3.3 *Dimensionless Flow Rate (DFR)*

In this study, dimensionless flow rate (DFR) as shown in Eq. 7, was used to measure the ventilation rate of a building. Its equation was obtained by dividing the volume flow rate, $\dot{V}$ of the building during ventilation by the reference velocity U_{ref} (6.97 m/s) and the opening area A_o(0.000828 m^2) [15]. The volume flow rate was obtained from the ANSYS Fluent console at the windward opening.

$$DFR = \frac{\dot{V}}{U_{ref} A_o} \tag{7}$$

As shown in Fig. 8, it can be concluded that Case 1 (Top-Top) had the highest dimensionless flow rate (DFR) at 0.60, followed by the vertical opening configuration of Case 6 (Vertical–Vertical) at 0.57. This result once again proves that the best ventilation rate can be obtained by the opening configuration near the top of the windward and leeward wall [10]. The results of Case 6, Case 7, and Case 8 prove that, compared with the horizontal opening configurations other than Case 1, the vertical openings have better DFR through the building, especially when applied to the windward wall. According to the results of Case 4 and Case 5, the DFR of non-aligned openings are worse than that of aligned openings, which is consistent with the simulation results from Meroney [15].

4 Conclusion

In this paper, the influence of opening positions on natural cross ventilation of an isolated building was investigated, the airflow pattern and dimensionless flow rate (DFR) under different opening configurations were analyzed. This study considers the configuration of asymmetric openings such as vertical to horizontal openings. Eight different opening configurations were designed for simulation. The 3D steady-state RANS equation Shear-Stress Transport (SST) k-ω turbulence model was used for CFD simulation. The simulation results of each opening configuration were analyzed and discussed in terms of dimensionless U/U_{ref}, velocity vector, pressure coefficient and dimensionless flow rate (DFR). The airflow velocity profile trends of all opening configurations show the same U-shape, which indicates that the air velocity becomes relatively slow as it flows through the building. The result of the velocity profile shows that the aligned opening configuration near the top of the building has a higher airflow velocity in the building. The configurations with vertical openings at the windward side can provide a wider airflow area and a better velocity of airflow in the building. Such results show that the position of the windward opening of the building has a greater influence on the velocity distribution in the building. DFR analysis shows that the Top-Top opening configuration had the highest DFR at 0.60. This result is similar to that obtained from the literature. In addition, compared to other horizontal openings, configurations of vertical openings provide higher DFR in the building relative to horizontal openings regardless of opening position at the windward or leeward side. In summary, opening positions have significant impacts on the isolated building in natural cross ventilation. For future research, it is recommended to consider the influence of opening shapes and different wind speeds on building ventilation.

References

1. Sahabuddin MFM, Gonzalez-Longo C (2017) Natural ventilation potential in Kuala Lumpur: assumptions, realities and future. In: 33rd international conference on passive and low energy architecture: design to thrive, vol 3, pp 4291–4298
2. Gharakhani A, Sediadi E, Hayat Khan T, Sabzevar HB (2014) Use of natural ventilation in Malaysia's future green housing. Adv Mater Res 935:316–319. https://doi.org/10.4028/www.scientific.net/AMR.935.316
3. Moey LK, Goh KS, Tong DL, Chong PL, Adam NM, Ahmad KA (2020) A review on current energy usage and potential of sustainable energy in Southeast Asia countries. J Sustain Sci Manage 15(2):89–107
4. Kasim NFM, Zaki SA, Hagishima A, Ali MSM, Shirakashi M, Arai N, Razak AA (2014) CFD study of cross ventilation performance of different buildings layouts. In: 2nd Asia conference of international building performance simulation association, pp 131–138
5. Nikas KS, Nikolopoulos N, Nikolopoulos A (2010) Numerical study of a naturally cross-ventilated building. Energy Build 42(4):422–434. https://doi.org/10.1016/j.enbuild.2009.10.010
6. Rizk AA, El-Morsi MS, Elwan MM (2018) A review on wind—driven cross—ventilation techniques inside single rooms. Int J Sci Eng Res 6(8):75–93
7. Ma XY, Peng Y, Zhao FY, Liu CW, Mei SJ (2017) Full numerical investigations on the wind driven natural ventilation: cross ventilation and single-sided ventilation. Procedia Eng 205:3797–3803. https://doi.org/10.1016/j.proeng.2017.10.128
8. Shetabivash H (2015) Investigation of opening position and shape on the natural cross ventilation. Energy Build 93:1–15. https://doi.org/10.1016/j.enbuild.2014.12.053

9. Karava P, Stathopoulos T, Athienitis AK (2011) Airflow assessment in cross-ventilated buildings with operable façade elements. Build Environ 46(1):266–279. https://doi.org/10.1016/j.buildenv.2010.07.022
10. Ramponi R, Blocken B (2012) CFD simulation of cross-ventilation for a generic isolated building: impact of computational parameters. Build Environ 53:34–48. https://doi.org/10.1016/j.buildenv.2012.01.004
11. Ramponi R, Blocken B (2012) CFD simulation of cross-ventilation flow for different isolated building configurations: validation with wind tunnel measurements and analysis of physical and numerical diffusion effects. J Wind Eng Industiral Aerodynamics 104:408–418. https://doi.org/10.1016/j.jweia.2012.02.005
12. Moey LK, Chan KL, Tai VC, Go TF, Chong PL (2021) Investigation on the effect of opening position across an isolated building for wind—driven cross ventilation. J Mech Eng Sci 15(2):8141–8152
13. Karava P (2008) Airflow prediction in buildings for natural ventilation design: wind tunnel measurements and simulation. Doctoral Dissertation, Concordia University
14. Kasim NFM, Zaki SA, Ali MSM, Ikegaya N, Razak AA (2016) Computational study on the influence of different opening position on wind-induced natural ventilation in urban building of cubical array. Procedia Eng 169:256–263. https://doi.org/10.1016/j.proeng.2016.10.031
15. Meroney RN (2009) CFD prediction of airflow in buildings for natural ventilation. In: Proceedings of the eleventh Americas conference on wind engineering, Puerto Rico

Effect of Laser Frequency and Focal Length on Copper Surface Temperature During Laser Heating

M. Y. Yus Erny, A. Afiq, M. H. Aiman, M. M. Quazi, and M. Ishak

Abstract Laser heating is a process that uses laser as a heat source. In this paper, the copper surface temperature during the laser heating process was studied by controlling the laser frequency and focal length. The laser heating experiment was conducted using a fiber laser marking machine and irradiated with a constant 27 W laser power within a duration of 51 s. The laser frequency and focal length were varied from 100 to 300 kHz and −3 cm to +3 cm, respectively. Meanwhile, laser surface modification (LSM) was performed on the copper rod surface to enhance the laser energy absorption. Furthermore, the defocusing modes for laser heating were used to analyze the variation of temperature. The focus point of the focal length for this experiment was set up at 18.4 cm from the focal plane and denoted as 0. Laser frequency and focal length were found to play an important role in increasing the surface temperature during laser heating since it affects the heat input delivered to the materials. It was found that the surface temperature reaches a higher degree, 879.2 °C with the combination of 200 kHz laser frequency at focal length 0.

Keywords Laser heating · Surface temperature · Focal length · Laser frequency

1 Introduction

In recent years, lasers are regarded as advanced material processing tools that bridge the gap in advanced manufacturing systems because of their accuracy, low cost, localized processing and high speed of operation. In material processing, the use of lasers is an appealing option for high-tech production [1]. The laser beam is used as a heat source to heat, melt or to fuse exactly the energy of the surface and the inner part stays unchanged [2]. Since the arrangements of the optical setting for the laser beam are very precise, the localized heating can be controlled easily [3]. However, the

M. Y. Yus Erny · A. Afiq · M. H. Aiman (✉) · M. M. Quazi · M. Ishak
Faculty of Mechanical and Automotive Engineering Technology, Universiti Malaysia Pahang, 26600 Pekan, Pahang, Malaysia
e-mail: aimanmh@ump.edu.my

M. Y. Ismail et al. (eds.), *Technological Advancement in Mechanical and Automotive Engineering*, Lecture Notes in Mechanical Engineering,
https://doi.org/10.1007/978-981-19-1457-7_64

applications of lasers in an industry require knowledge of the interaction mechanism between the laser beam and the workpiece [4].

Laser heating is found to be one of the effective methods that use a laser beam to heat the surface of materials or workpiece and accomplish it by conduction. It is a consistent and precise method for small areas that helps to accomplish the heating process by localized heat input. The efficiency of the laser heating process is determined by the absorption of the radiation by the irradiated material [5]. Laser absorption in a metal depends on several different parameters, both involving the laser and the metal itself. The application of laser technology in material heating takes advantage of heat energy from the laser beam.

In laser heating, the energy that is needed to increase the temperature of the material is all dependent on the laser-material interaction. Due to the nature of the reflectivity of the laser beam on metal, it could limit the laser energy absorption [6]. Besides, there are a few laser parameters that influence the absorption which are laser power, laser frequency, spot diameter, speed, focal length and many more. However, laser frequency and focal length affect the temperature increase apart from laser power.

The laser frequency is crucial because the materials may respond differently at different frequencies. On the other hand, it will be influencing the material properties and appearance because it affects the heat input delivered to the materials [7]. The optimal frequency would lead to better heating of the metal. Subsequently, focal length also plays a major role in temperature rise. In laser processing, the position of the focus (focal plane) of the laser beam focusing and the relative position of the workpiece to be processed mainly have a certain relationship. There are three cases of defocusing that can occur which are negative defocus, positive defocus and zero defocus. Defocuses have a relationship with the beam spot size and have a direct impact on temperature [8]. It was found that laser defocusing affects the temperature increase [9].

However, to the best of our knowledge, there are no comprehensive studies that relate laser heating parameters. The problem of laser heating process and thermal properties investigation and complex materials are still increasing [2]. This work aims to study the effect of laser frequency and focal length on the surface temperature. These parameters and their interaction have a strong influence on the laser heating process. The study helps to find a better parameter range setting to improve the temperature distribution.

2 Experimental Setup

The raw materials of the copper rod with a purity of 99.99% and 6 mm of diameter were cut sectioned by using a precision cut-off machine into 6 mm × 6 mm samples. Table 1 shows the properties of pure copper. Prior to the process, all the samples were ground progressively using SiC sandpaper, and then cleaned with acetone solution for a second to remove any contaminant layer that adhered to the samples' surfaces.

Table 1 Thermo-physical properties of copper

Properties	
Density	8.96 g/cm^3
Specific heat capacity	0.386 J/g °C
Thermal conductivity	401 W m^{-1} K^{-1}
Melting point	1083 °C

The copper rod surface is then treated with laser surface modification (LSM) process to enhance the surface roughness and increase the laser energy absorption during the laser heating [10–12]. The surface roughness value, Ra before treated with LSM was 0.433 μm. Roughness has a significant impact on absorptivity due to multiple reflections in the undulations [13]. It marked the surface within the diameter of 6 mm. The process is consistent throughout all the samples. The sample is once again cleaned with the acetone solution prior to the laser heating process.

After the LSM process, the material will be set up for the laser heating process. The experimental setup is shown in Fig. 1. The laser heating was performed at an ambient temperature using a fiber laser machine. A ceramic board was placed under a copper rod to reduce the heat conductivity. The diameter of the laser beam irradiated on the copper rod surface was set as 3 mm. Each experiment was conducted at a constant power, and duration as shown in Table 2. Meanwhile, laser frequency and

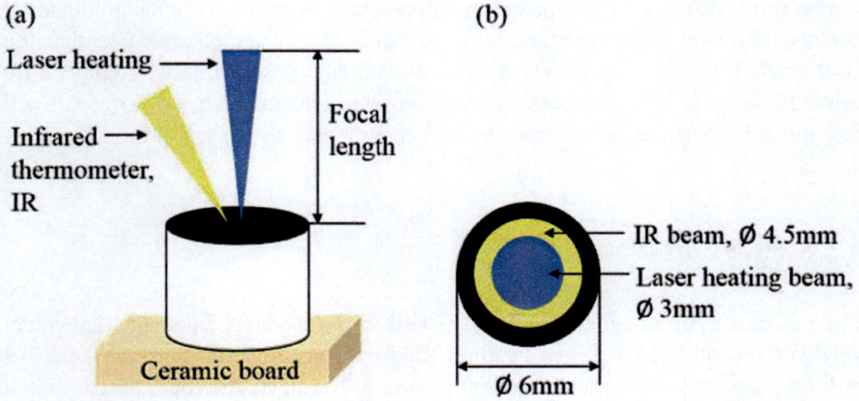

Fig. 1 Schematic diagram of laser heating process setup **a** front view and **b** top view

Table 2 Range of laser heating parameters

Parameters	
Power	27 W
Duration	51 s
Laser frequency	100 kHz, 200 kHz, 300 kHz
Focal length offset	18.4 cm; (−3, −2, −1, 0, +1, +2, +3) cm

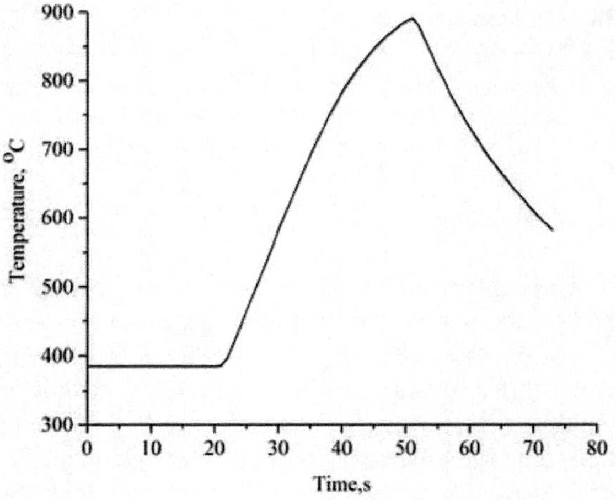

Fig. 2 The temperature profile of the laser heating

focal length were set as a variable parameter. Focal length at 18.4 cm means that the laser beam is at focus position which is denoted as 0.

An infrared thermometer, Micro-Epsilon CTL2MHSF300-C3, whose temperature measuring range is 385–1600 °C was used to measure the surface temperature in the irradiated spot on the sample, as shown in Fig. 1. During the laser heating process, the beam from the infrared thermometer and laser were set side by side on the modified copper rod surface to get the maximum temperature without involving any material damage. However, since the beam size of the infrared thermometer is bigger than the laser beam, it will overlap with each other and influence the increase in temperature value. In this case, when the laser energy is emitted to the copper rod surface, it will heat up and subsequently increase the surface temperature as seen in Fig. 2.

3 Results and Discussion

Table 3 shows the obtained temperature with different laser frequency and focal length offset within 51 s of laser heating. The first measure performed was the effect of laser frequency on surface temperature and followed by the focal length offset. Each focal length offset was repeated three times and the average value was measured. In addition, the standard deviation is calculated and it shows a low standard deviation, indicating that the values of the three sets of each parameter tend to be close to the mean.

Following that, the value of the coefficient of variation was determined to show that a lower standard deviation does not necessarily imply less varied data. However, the rise and fall of the measured temperature, which is likely because of the limitations of the infrared thermometer temperature measurement, are still in close agreement which is acceptable.

Table 3 Temperature data of focal length offset at different laser frequency

Laser frequency (kHz)	Focal length offset (cm)	Temperature (°C)						
		−3	−2	−1	0	+1	+2	+3
100	Sample 1	522.5	673.2	790.3	806.5	785.9	740.4	599.3
	Sample 2	502.0	678.3	782.0	809.1	777.5	734.6	757.1
	Sample 3	522.0	674.9	788.8	801.5	759.5	711.2	568.8
	Average	515.5	675.5	787.0	805.7	774.3	728.7	581.1
	S. Deviation	11.69	2.60	4.42	3.86	13.49	15.46	16.10
	CV (%)	2.27	0.38	0.56	0.48	1.74	2.12	2.77
200	Sample 1	607.3	640.8	751.9	869.8	810.2	694.3	616.7
	Sample 2	614.3	647.9	752.9	890.1	802.6	686.7	612.5
	Sample 3	614.1	644.6	748.5	877.8	814.4	701.0	618.1
	Average	611.9	644.4	751.1	879.2	809.1	694.0	615.8
	S. Deviation	3.98	3.55	2.31	10.23	5.98	7.15	2.91
	CV (%)	0.65	0.55	0.31	1.16	0.74	1.03	0.47
300	Sample 1	624.6	627.1	716.3	726.1	678.0	616.6	591.7
	Sample 2	631.5	646.2	719.6	748.2	676.8	616.9	581.6
	Sample 3	622.0	622.6	714.6	723.4	672.3	630.4	589.1
	Average	626.0	632.0	716.8	732.6	675.7	621.3	587.5
	S. Deviation	4.91	12.53	2.54	13.61	3.00	7.88	5.24
	CV (%)	0.78	1.98	0.35	1.89	0.44	1.27	0.89

Figure 3 shows the variation of temperature with laser frequency. It clearly indicates that when the laser frequency increases, the temperature decreases. The highest temperature obtained is at a laser frequency of 200 kHz while the lowest temperature

Fig. 3 Graph of temperature against laser frequency

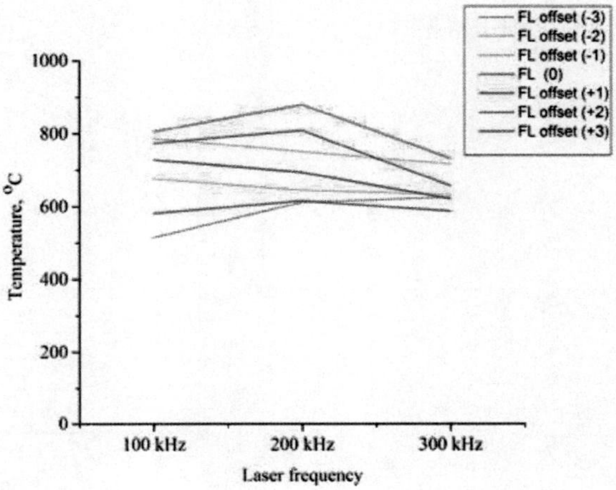

is at a laser frequency of 300 kHz. As the laser frequency rises, there is just a slight temperature difference. This is because the heating effect on the surface will be more intense at a lower laser frequency.

However, when the focal length is shifted, the differences of temperature become more obvious for all the laser frequency. And in this instance, the focal length seems to be more significant than the laser frequency. Generally, the laser frequency is inversely proportional to laser power. However, in this study, the laser power was kept constant at 27 W throughout all the samples. When the laser frequency is too high, the laser power may not be efficient for the process which explains why most of the 300 kHz of laser frequency produces the lowest temperature.

From the tabulated data, the results of varying the levels of focal length, starting with a fully focused beam are shown in Fig. 4 with various laser frequency settings. The graph illustrates that when the focal length offset of each laser frequency is varied, the temperature changes dramatically from −3 to +3 cm. Besides, at focal length 0, the temperature drops as the laser frequency changed. The temperature achieves a higher degree (879.2 °C) when the laser beam is pointed at a focal length, 0 and laser frequency of 200 kHz compared to other defocused positions. This is because when the laser defocus approaches focal length 0, the beam spot diameter becomes smaller, leading the copper surface temperature to rise.

The lowest temperature is at focal length offset −3 cm and 100 kHz which is 515.5 °C. According to the findings, when the laser beam becomes less unfocused, the temperature decreases and vice versa. This is due to the laser beam's energy being scattered over a broader surface, which reduces the intensity of the heat supplied on the surface. The lower intensity prevents the heat from penetrating deeper into the material [14].

To sum up, the variation of the focal length offset setting is as shown in Fig. 5. It has been studied by Jianli Liu et al., the effect of the three kinds of focus modes (negative defocus, focus, and positive defocus) [15]. The laser beam is said to be

Fig. 4 Graph of temperature against focal length offset

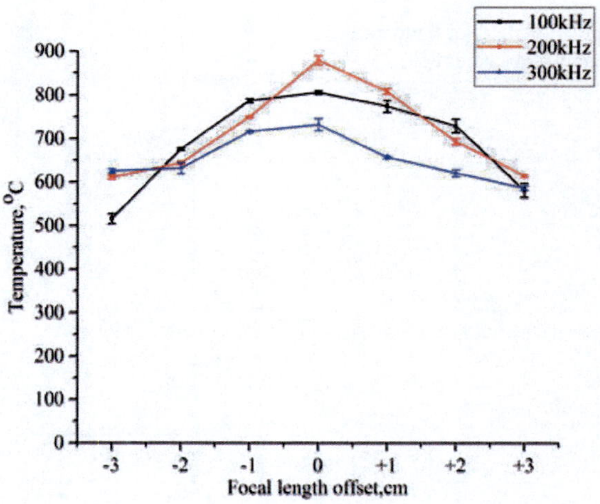

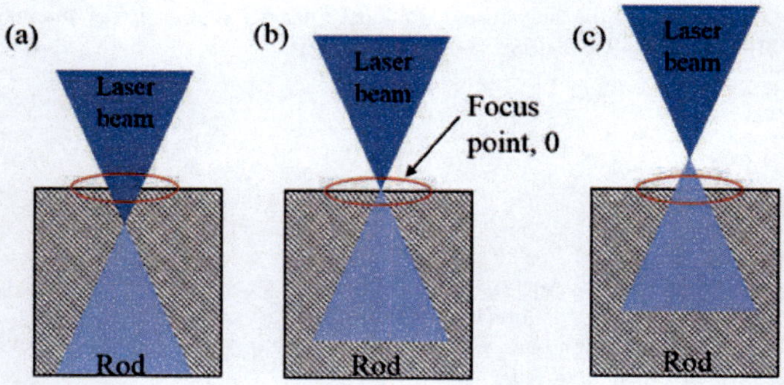

Fig. 5 Schematic of the laser beam at **a** negative defocus, **b** focus and **c** positive defocus plane

fully focused when it is located at the focal point of the focusing lens. In focused beam laser heating, the beam converges to its minimum radius value so that the size of the area irradiated by the beam is minimized.

The optimum focal length offset was the second focus condition, Fig. 5b because the surface has the highest laser energy absorption due to the smallest laser beam. It was discovered that, at focus point distance, where the diameter of the laser beam becomes smallest, the photon can produce enough laser energy [6] to supply the heat onto the surface. Thus, increase the temperature. And it is vice versa with negative and positive defocus conditions. After all, the laser beam focal position was found to have a significant influence on the surface temperature.

4 Conclusions

The purpose of this study is parameter control towards the copper surface temperature of the laser heating process. To summarize, laser frequency and focal length are two parameters that need emphasis, since it has a significant impact on the surface temperature distribution during the laser heating process. The surface temperature mostly achieves the highest degree at a laser frequency of 200 kHz. Increasing or decreasing the laser frequency range, the surface temperature will drop. At focal length 0, the highest temperature, 879.2 °C is obtained compared to other defocused positions. It is because when the laser beam becomes more focused, the temperature increases and vice versa.

The combination parameter of 200 kHz of laser frequency and focal length at 0 defocused, was an optimum range where the copper surface absorbs more laser beam's energy and thus, surface temperature increases.

Acknowledgements The authors would like to thank the Ministry of Higher Education for providing financial support under Fundamental Research Grant Scheme (FGRS-RACER) No. RACER/1/2019/TK03/UMP//3 (University reference RDU192608), and Universiti Malaysia

Pahang for laboratory facilities as well as additional financial support under Internal Research Grant RDU210364 and PGRS200306.

References

1. Weltsch Z, Hlinka J (2018) Analysis of laser treated copper surfaces. Period Polytech Transp Eng 47(2):140–145. https://doi.org/10.3311/PPtr.11561
2. Oo HL (2019) Optimization of a parameter in laser heating process. In: Applied research in science, technology and knowledge
3. Yilbas BS (2012) Introduction to laser heating process. In: Laser heating applications, pp 1–5
4. Yilbas BS, Al-Garni AZ (1996) Some aspects of laser heating of engineering materials. J Laser Appl 8(4):197–204. https://doi.org/10.2351/1.4745422
5. Kügler H (2019) Effects of short-term laser beam heating on the absorptivity of steel sheets. J Manufact Mater Process 3(2). https://doi.org/10.3390/jmmp3020041
6. Zaifuddin AQ, Aiman MH, Quazi MM, Ishak M, Ariga T (2020) Effect of laser surface modification (LSM) on laser energy absorption for laser brazing. IOP Conf Ser Mater Sci Eng 788. https://doi.org/10.1088/1757-899x/788/1/012013
7. Feng XS, Li L, Chen Y (2008) Influence of laser energy input with Cu base filler metal. Trans Nonferrous Met Soc China
8. Lazov L, Deneva H, Teirumnieka E (2017) Influence of defocus position on laser cutting process in sheet steel. Environ Technol Resour Proc Int Sci Pract Conf 3. https://doi.org/10.17770/etr2017vol3.2658
9. Kunimine T, Miyazaki R, Yamashita Y, Funada Y (2020) Effects of laser-beam defocus on microstructural features of compositionally graded WC/Co-alloy composites additively manufactured by multi-beam laser directed energy deposition. Sci Rep 10(1):8975. https://doi.org/10.1038/s41598-020-65429-8
10. Erny AMYY, Fadhil AR, Aiman MH, Ishak M, Quazi MM (2021) Comparative study between furnace brazing and laser brazing. IOP Conf Ser Mater Sci Eng 1068(1). https://doi.org/10.1088/1757-899x/1068/1/012003
11. Sulaiman AJ, Aiman MH, Quazi MM, Ishak M, Ariga T (2020) Enhancement of mechanical properties of copper brazed by laser surface modification. IOP Conf Ser Mater Sci Eng 788. https://doi.org/10.1088/1757-899x/788/1/012012
12. Quazi MM, Fazal MA, Haseeb ASMA, Yusof F, Masjuki HH, Arslan A (2015) Laser-based surface modifications of aluminum and its alloys. Crit Rev Solid State Mater Sci 41(2):106–131. https://doi.org/10.1080/10408436.2015.1076716
13. Niu C, Zhu T, Lv Y (2019) Influence of surface morphology on absorptivity of light-absorbing materials. Int J Photoenergy 1–9. https://doi.org/10.1155/2019/1476217
14. Benton M, Hossan MR, Konari PR, Gamagedara S (2019) Effect of process parameters and material properties on laser micromachining of microchannels. Micromachines (Basel) 10(2). https://doi.org/10.3390/mi10020123
15. Liu J, Yu H, Chen C, Weng F, Dai J (2017) Research and development status of laser cladding on magnesium alloys: a review. Opt Lasers Eng 93:195–210. https://doi.org/10.1016/j.optlaseng.2017.02.007

Surface Roughness Analysis of Five-Axis Flank Milling Strategies for Slanted Thin-Walled Pocketing: Aerospace Part

S. A. Sundi, R. Izamshah, M. S. Kasim, I. S. Othman, and M. R. Raffay

Abstract This research was initiated to investigate the effect of surface roughness when machining slanted thin-walled pocketing profiles with various flank milling machining strategies. In this research, Multi-Axis Flank Contouring strategies namely Combin Parelm and Combin Tanto were the main machining strategies to machine an actual chosen aero-structural sample with slanted thin-walled of 105° and 85°. Samples of physical machining were obtained from a five-axis CNC milling machine DMU 60 Monoblock with Siemens 840D controller. The material used was aluminum AA6063 and the cutting tools were carbide K10 material of two flutes end mill with diameter 12 and 6 mm uncoated. At the end of this study, an analysis with regard to the surface finish was performed utilizing Mitutoyo surface roughness tester to measure the surface roughness of the machined part specimens. Three different sections namely approach, middle, and leaving area of the slanted thin-walled were taken for further analysis. In addition, the arithmetical roughness (Ra) value is used to measure the overall surface finish quality. The Combin Tanto strategy resulted better surface quality in comparison to the Combin Parelm strategy due the overwritten of depth of cut which occurred during the machining process although was set constantly. The tool trajectory and tilting position were believed to be the main factors affecting the surface finish of the machined part. The orientation of the cutting tool affected all the three areas evaluated and lead to an uneven depth of cut amount during the machining process.

Keywords Surface roughness · Thin-walled · Flank milling · Five-axis machining strategies

S. A. Sundi (✉) · M. R. Raffay
Faculty of Mechanical and Manufacturing Engineering Technology, Universiti Teknikal Malaysia Melaka (UTeM), Hang Tuah Jaya, 76100 Durian Tunggal, Melaka, Malaysia
e-mail: syahrul.azwan@utem.edu.my

R. Izamshah · M. S. Kasim · I. S. Othman
Faculty of Manufacturing Engineering, Universiti Teknikal Malaysia Melaka (UTeM), Hang Tuah Jaya, 76100 Durian Tunggal, Melaka, Malaysia

M. Y. Ismail et al. (eds.), *Technological Advancement in Mechanical and Automotive Engineering*, Lecture Notes in Mechanical Engineering,
https://doi.org/10.1007/978-981-19-1457-7_65

1 Introduction

Thin-walled profiles and shapes like impellers and monolithic spar-ribs are very common to be found in aerospace industry. Due to the complex aero-structural parts which designed to cater high aerodynamic shape of the aircraft and the pocketing profiles to comprise the complex freedom surfaces, five-axis flank milling is then required to machine this unique shape. The flank milling where the machining process uses the body of the cutting tool usually suffers more difficulties than the face milling. Regardless of any type of material, most of the issues are related to the gouge avoidance, interference avoidance and the tool rigidity [1]. Several studies related to this field reported that most of the researchers were struggled to improve the tool path planning by introducing new ruled surface machining, mathematical modelling, and algorithm [2–6]. Harik et al. [7] highlighted that research in flank milling focuses more on the generation of optimal tool trajectory for non-developable ruled surfaces, even generic free-form surfaces.

Machining strategy is an approach of machining style which usually determined in during the CAM programming phase. Selection of the machining strategy to be applied is depending on the overall targeted machining time, surface quality and texture, and variation of the cutting forces along the tool path. Therefore, applying the most appropriate machining or cutting strategy may affect not only the machining time which related to the productivity but also the final machined surface quality. In this research, the machining strategy chosen was Multi-Axis Flank Contouring operation as offered by CATIA V5 (Table 1) to machine an actual sample of aero-structural part with slanted thin-walled of 105° and 95°. It appears from the aforementioned reviews; the most common angle design of aero-structural components to strengthen the overall wing section known as Rib" section were the two angles mentioned earlier. Considerations and guidance highlighted by the past researchers to machine thin-walled profiles were taken into account in the planning and execution phase of this work [8–13].

The quality of machined surface is playing a vital role to ensure the overall acceptance of a product in manufacturing industries. Most researchers were agreed that the main predominant factors affecting the surface quality of machined parts were the machining parameters namely the cutting speed, V_c cutting feed, V_f depth of cut, a_p and tool engagement percentage or also know [illegible] al.

Table 1 Definition of flank cutting strategy provided by CA

Flank strategy	Definition	
Combin Tanto	Tool is tangent to the drive surface at a given contact height and is contained in a plane normal to forward direction	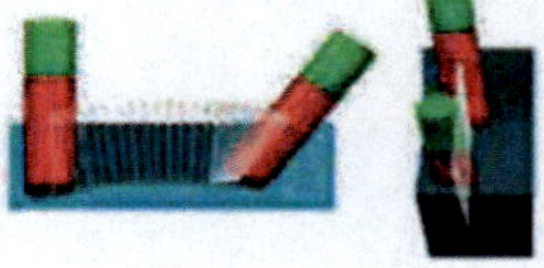
Combin Parelm	Tool is tangent to the drive surface at a given contact height and follows the surface isoparametric	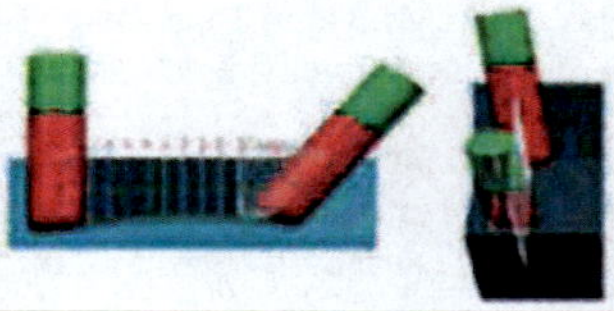

proposed a surface roughness control method using feed rate optimization for five-axis flank milling based on an analysis of the geometrical features of a machined surface which concluded that the cutter runout phenomenon resulted overall increase in surface roughness [6]. In another research by C. Liu et al. presented a new tool path generation method for five-axis flank milling of pockets which summarized the machined surface quality has very close relationship with the cutting force during the flank milling [5]. Hayajneh et al. [14] reported that the fatigue strength, creep life and corrosion resistance of a machined part significantly influenced by the surface quality. Moreover, authors also discovered that surface quality determines the functional of the machined part produced such as heat transmission, fatigue resistance, and coating adhesion. Therefore, the desired finish surface is usually specified, and the appropriate processes are selected to reach the required quality. It is quite apparent from the reviews above that there is a paucity of research efforts that provide a clear insight into the relationship between the machined surface roughness of slanted thin-walled profile with the flank milling strategies. Hence, the present study initiated to focus on the evaluation of the machined surface finish when two different flank milling strategies namely Combin Parelm and Combin Tanto were applied for an internal and external pocketing profiles of slanted thin-walled.

2 Methodology

2.1 CAD Model Preparation

An example of aero-structural 3D Computer Aided Design (CAD) model has gone through an editing process to remove unnecessary parts to obtain the desired slanted thin-walled area to be studied (Fig. 1a–c) and rescaling process as illustrated in Fig. 1d. The main CAD platform utilized to perform the mentioned operations were CATIA V5 which was in the Part Design workbench. The part selected is known as rib section which one of the wing aero-structural sections. The final dimension of the selected part was 150 mm (length) × 65 mm (width) × 20 mm (height). The slanted angle for the outer profile was 94.6°–95° with minimum distance or the wall thickness was 3.604 mm. Meanwhile, the slanted angle for the inner profile was 105° with minimum distance or the wall thickness was 2.938 mm. Aluminum AA6063 aerospace standard is chosen to be the actual specimen material.

2.2 CAM Program Preparation and Cutting Strategy

There were two main cutting strategies evaluated in this work during the CAM program known as Combin Parelm and Combin Tanto from Multi-Axis Flank

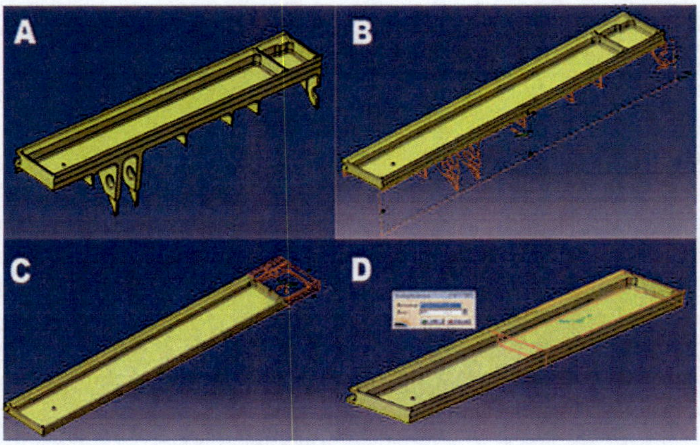

Fig. 1 Editing process from the actual 3D CAD model and rescaling process

Contouring process in Advanced Machining workbench. Two main machining operations used to finish the part namely "Roughing" operation and "Multi-Axis Flank Contouring" operation. Roughing process is used to cut off most of the materials before finishing the focus slanted areas utilizing the mentioned machining strategies. Figure 2 illustrates the flow of the CAM program prepared in this research before proceeding with the actual machining process. Once the CAM program completed

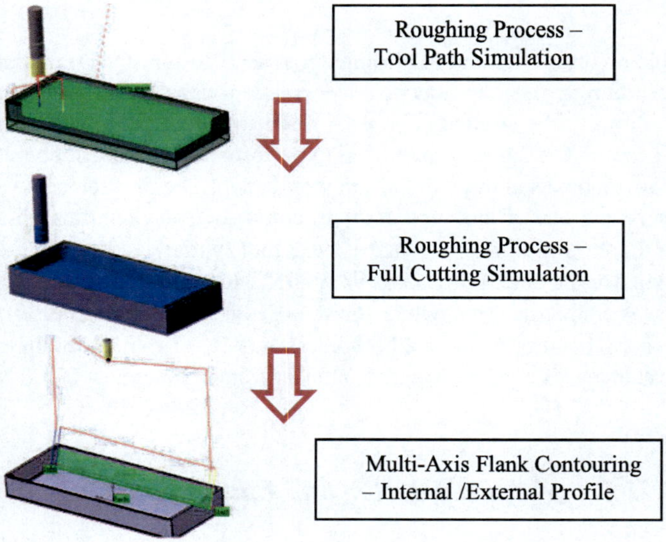

Fig. 2 CAM program preparation flow from roughing to multi-axis flank contouring process

Fig. 3 Actual machining setup utilizing DMU-60 Monoblock five-axis CNC milling center

and verified, numerical control (NC) codes program was then generated from a customized post processor.

The actual machining was successfully done by utilizing DMU-60 Monoblock five-axis CNC milling center with Siemens 840D controller, and the workpiece was secured by using precision vise as shown in Fig. 3. In the actual machining process, a carbide K10 material of two flutes end mill with a diameter 12 mm is used for roughing process. Meanwhile, an end mill of diameter 6 mm with the same specification as used in the roughing process is utilized to perform the Multi-Axis Flank Contouring process. To obtain smooth and fine surface, the machining tolerance was defined at 0.01 mm. The spindle speed applied was 4500 rpm and the cutting feed was 0.02 mm/tooth with maximum depth of cut was set to be 1 mm for every machining cycle process. The parameters such as depth of cut, spindle speed and cutting feed were remained constant for both mentioned processes. The machining mode applied was climbing or down cutting direction.

2.3 *CAM Program Preparation and Cutting Strategy*

Surface roughness evaluation of the workpiece was measured using surface roughness tester Mitutoyo Surftest SJ-410 which is capable to measure up to 0.0001 μm (Fig. 4). In this study, Ra (Arithmetical mean deviation) is used to measure the surface finish. Longitudinal surface roughness is evaluated in this work with the stylus travel distance set at 4 mm on each measurement. There were five points of measurements taken on every machined surface and final average Ra is obtained to represent the result of surface finish on every specimen. The stylus has undergone a self-calibration with the master surf tester on every measurement session. This will ensure the validity of the data to be taken on that session (Fig. 4). The surface features

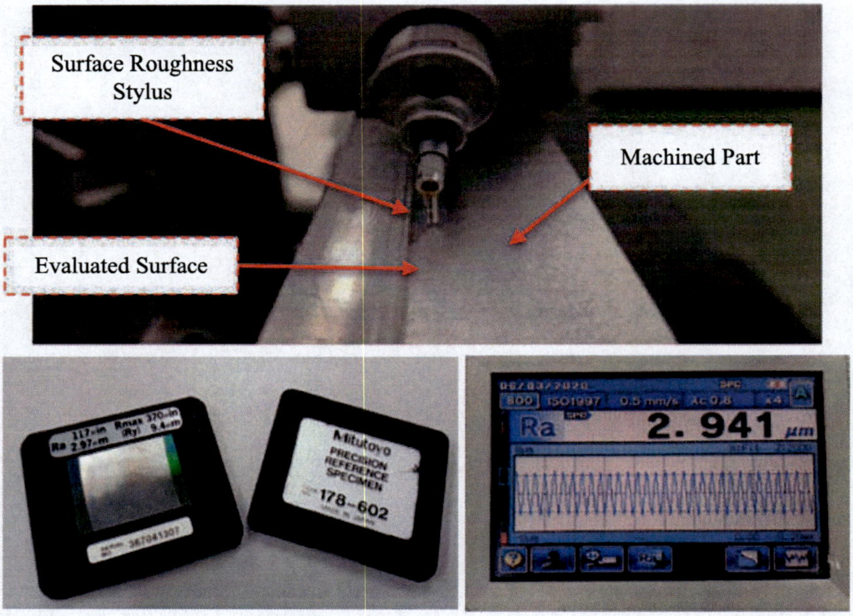

Fig. 4 Surface roughness measurement using Mitotuyo Surftest SJ-410 and self-calibration master

were measured relatively to the drive unit reference surface. Hence, this accurately measures waviness and finally stepped features in addition to the surface roughness.

There were three different areas namely approach, middle and leaving area of the slanted thin-walled were taken for further analysis as shown in Fig. 5. These areas represented the motion of cutting tool during the flank milling process whereby the approach area was the first section of the cutting tool touched the workpiece, middle area was the idling phase, and the leaving area was the last section before the cutting tool leaving the workpiece. The effect of the these cutting tool motions towards the surface finish were analyzed and presented in the following section.

3 Result and Discussion

From Figs. 6, 7 and 8, Combin Tanto strategy exhibits slightly better result for both 85° and 105° slanted thin-walled than Combin Parelm strategy. These differences of Ra result were possibly due to the different tool trajectory of the strategy. The Combin Tanto tool trajectory works tangent to the selected drive surface and contained in normal plane to the forward direction. This characteristic of tool path allows Combin Tanto to have better surface finish as the part to be machined or the CAD part did not consist of any curvature surface. While Combin Tanto dominated the Ra result, Combin Parelm is believed appropriate to be applied for curvy slanted surface as the

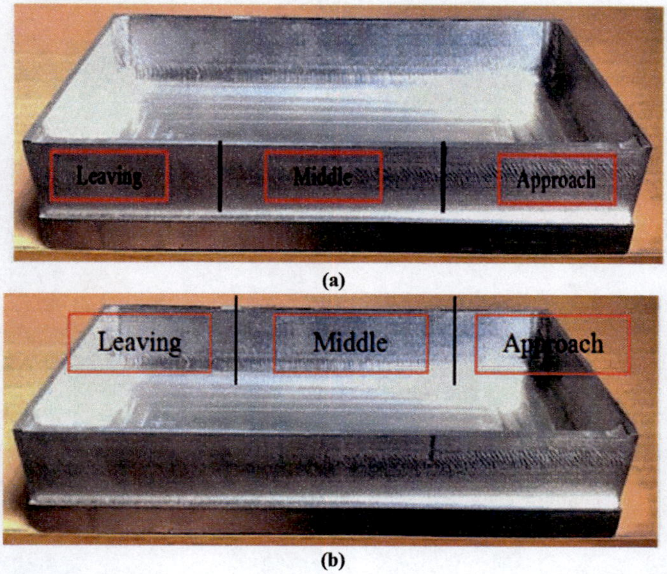

Fig. 5 Areas of surface roughness analysis were divided into three areas for the outer profile (**a**) and inner profile (**b**) of the thin-walled machined parts

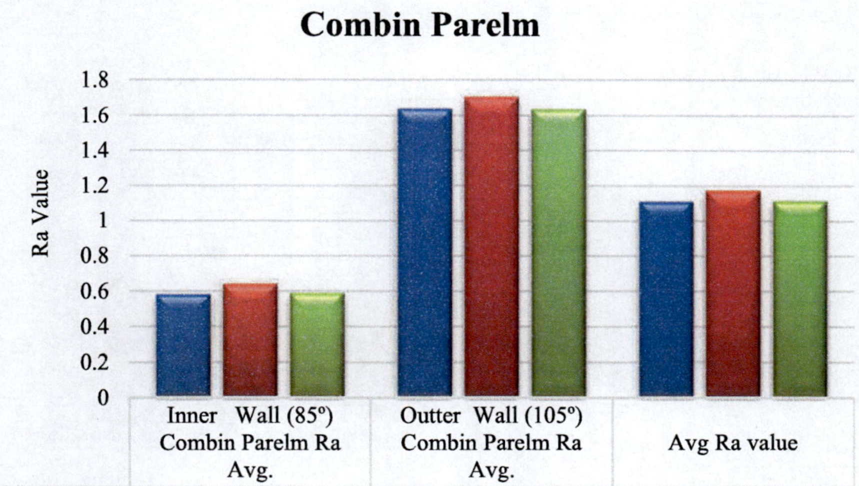

	Inner Wall (85°) Combin Parelm Ra Avg.	Outter Wall (105°) Combin Parelm Ra Avg.	Avg Ra value
Approach	0.583	1.641	1.112
Middle	0.643	1.707	1.175
Leaving	0.593	1.638	1.116

Fig. 6 Average Ra value result for Combin Parelm strategy

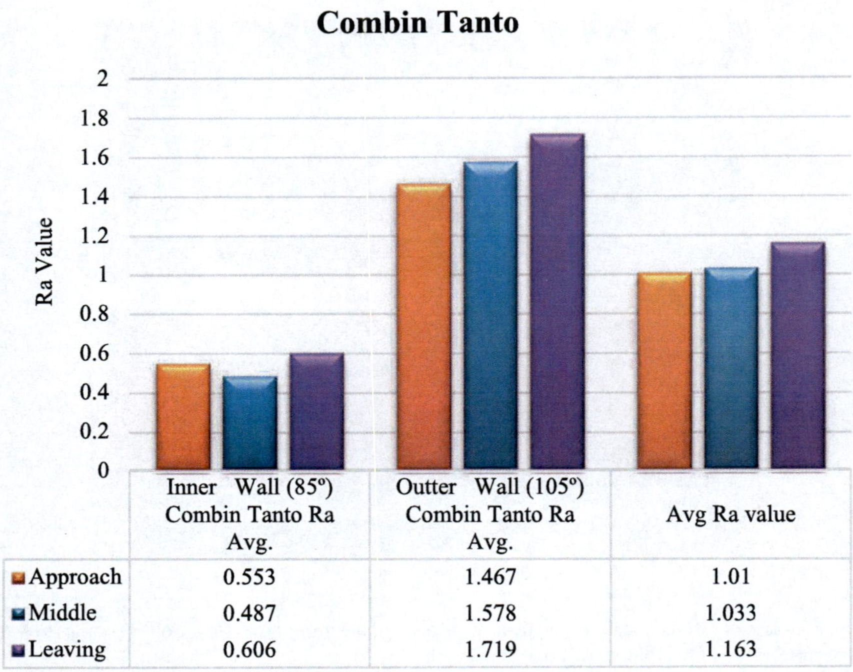

Fig. 7 Average Ra value result for Combin Tanto strategy

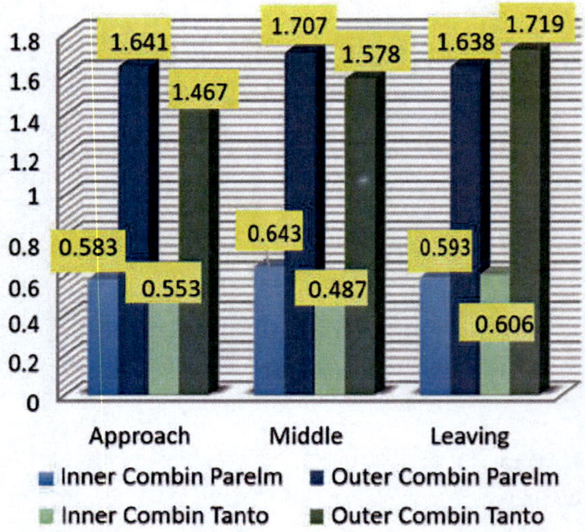

Fig. 8 Overall Ra value for both strategies

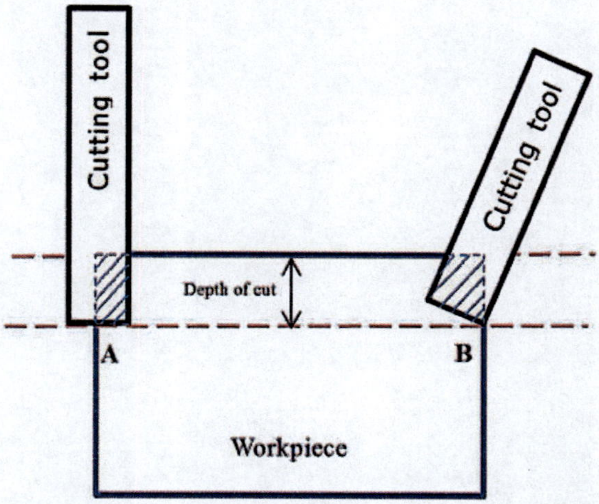

Fig. 9 Different of milled area due to tilting tool position

tool trajectory follows the grazing surface isoparametricly. Lead angle with optimum degree of cutting will lead to the effectiveness of chip removal and produce fine finishes [9].

Hayajneh et al. [2] reported that the most influential factor affected the surface roughness is the relationship between the depth of cut and cutting feed, and the relationship between the cutting feed and spindle speed. As far as this research is concerned, the depth of cut significantly interrupted by the tool trajectory position. Tool trajectory moves or tilts at various angles based on the cutting flank strategy and the ruled surfaces. Major tilting of the cutting tool or workpiece increases or decreases the value of desired depth of cut. Figure 9 illustrates how the tilting tool increase/decrease the depth of cut suddenly. The blue line represents the workpiece and the dashed red line representing the same level of depth of cut. Cutting tool is indicated by the black line. The area of A has less milled area as compared to area of B. The tool tilting position is affecting the desired depth of cut and shall indirectly cause the surface finish quality. Although the machining parameters such as depth of cut, cutting feed and spindle speed were remained constant for both selected machining strategies, the effect of tool tilting positioning due to the tool trajectory ruled surfaces indirectly overwritten the depth of cut as explained earlier. As a result, Combin Tanto strategy obviously generated better machined surface finish as compared to Combin Parelm which can be noticed with bare eyes as indicates in Fig. 10. The machined surface by Combin Parelm strategy showed a clear chatter mark while Combin Tanto had smooth and clear surface.

Fig. 10 Final machining result and surface finish

4 Conclusion

This paper presented a comparison result of surface finish by applying flank milling strategies namely Combin Tanto and Combin Parelm in machining slanted thin-walled of 105° and 85°. At the end of this study, a few summarizations are made as follows:

1. The difference of the tool trajectory orientation has affected the result of surface finish for thin-walled pocketing when the Multi-Axis Flank milling strategy is applied.
2. The Combin Tanto strategy generated better surface quality in comparison to the Combin Parelm strategy due the overwritten of depth of cut which occurred during the machining process although is set constantly caused by the tool trajectory and tilting position of the cutting tool.

As for the future work, an investigation on the cutting force for the chosen flank strategies might be performed as an extension works from the current study to the understand its impact to the surface finish. Data acquired from the cutting force may leads to the better evidence in understanding the effect of the mentioned cutting strategy towards the surface finish quality.

Acknowledgements Authors are grateful to Universiti Teknikal Malaysia Melaka for serving a platform to perform this study and Ministry of Education (MOE) for their financial support. (PJP/2020/FTKMP-COSSID/SC0004).

References

1. Liu XW (1995) Five-axis NC cylindrical milling of sculptured surfaces. Comput Des 27(12):887–894. https://doi.org/10.1016/0010-4485(95)00005-4
2. Menzel C, Bedi S, Mann S (2004) Triple tangent flank milling of ruled surfaces. CAD Comput Aided Des 36(3):289–296. https://doi.org/10.1016/S0010-4485(03)00118-0
3. Bedi S, Mann S, Menzel C (2003) Flank milling with flat end milling cutters. CAD Comput Aided Des 35(3):293–300. https://doi.org/10.1016/S0010-4485(01)00213-5
4. Li ZL, Zhu LM (2019) Compensation of deformation errors in five-axis flank milling of thin-walled parts via tool path optimization. Precis Eng 55:77–87. https://doi.org/10.1016/j.precisioneng.2018.08.010
5. Liu C, Li Y, Jiang X, Shao W (2020) Five-axis flank milling tool path generation with curvature continuity and smooth cutting force for pockets. Chinese J Aeronaut 33(2):730–739. https://doi.org/10.1016/j.cja.2018.12.003
6. Wang L, Ge S, Si H, Yuan X, Duan F (2019) Roughness control method for five-axis flank milling based on the analysis of surface topography. Int J Mech Sci 169:105337. https://doi.org/10.1016/j.ijmecsci.2019.105337.
7. Harik RF, Gong H, Bernard A (2013) 5-axis flank milling: a state-of-the-art review. CAD Comput Aided Des 45(3):796–808. https://doi.org/10.1016/j.cad.2012.08.004
8. Gong H, Wang N (2009) Optimize tool paths of flank milling with generic cutters based on approximation using the tool envelope surface. CAD Comput Aided Des 41(12):981–989. https://doi.org/10.1016/j.cad.2009.06.013
9. Chu CH, Huang WN, Hsu YY (2008) Machining accuracy improvement in five-axis flank milling of ruled surfaces. Int J Mach Tools Manuf 48(7–8):914–921. https://doi.org/10.1016/j.ijmachtools.2007.10.023
10. Si H, Wang L (2019) Error compensation in the five-axis flank milling of thin-walled workpieces. Proc Inst Mech Eng Part B J Eng Manuf 233(4):1224–1234. https://doi.org/10.1177/0954405418780163
11. Sundi SA, Izamshah R, Kasim MSM, Raffay MRM (2018) The effect of surface finish by varying machining strategies of five-axis flank milling for curvy angled convex profile. JAMT
12. Azwan SS, Syafik JM, Hassan A, Izamshah RAR (2017) The effect of surface finish on sculptured shape utilizing scanned data-reversed engineering (CATIA V5 and DELCAM). In: MATEC web of conferences, vol 97. https://doi.org/10.1051/matecconf/20179701112
13. Azwan SS, Syafik JM, Razly RM, Izamshah RAR (2016) An investigation on the surface finish of sculptured surface utilizing reverse engineering data of crank case cover—(CATIA V5 and DELCAM). In: MATEC web of conferences, vol 90. https://doi.org/10.1051/matecconf/20179001062
14. Hayajneh MT, Tahat MS, Bluhm J (2007) A study of the effects of machining parameters on the surface roughness in the end-milling process. Jordan J Mech Ind Eng 1–1(2):1–5
15. Multi-axis flank contouring operations. Catiadoc.free.fr 2017 [Online]. Available: http://catiadoc.free.fr/online/amgug_C2/amgugrf0100.htm

Using X-Ray Computed Tomography for Effective Porosity Characterisation in Additively Manufactured Metallic Parts

Shahir Mohd Yusuf, Nor Azwadi Che Sidik, and Nong Gao

Abstract Common microscopy approaches are considered as destructive technologies for porosity characterization in metallic parts for various engineering components. They are not only time consuming, but also causes wastage of materials due to the need of fabricating numerous batches of specimens just for such characterization. However, X-ray computed tomography has recently emerged as a viable technique to evaluate the porosity content in metallic components without the need of physically damaging them on purpose. Therefore, in this study, X-ray computed tomography and conventional 3D optical microscopy cross-section analysis approaches were used to compare the porosity profile obtained in 316L stainless steel additively manufactured using selective laser melting. both X-ray computed tomography and optical microscopy results both consistently show high densification (>99%) and low porosity (<0.7%) levels, suggesting optimum processing parameters were selected for the selective laser melting process. Similarly, the results of pore size, morphology, and distribution also compare well between both techniques. Overall, X-ray computed tomography is proven to be an effective non-destructive technique to assess porosity and defects in additively manufactured parts.

Keywords X-ray computed tomography · Optical microscopy · Porosity · Additive manufacturing · Non-destructive technique

S. Mohd Yusuf (✉) · N. A. Che Sidik
Engineering Materials and Structures (eMAST) iKohza, Malaysia-Japan International Institute and Technology (MJIIT), Universiti Teknologi Malaysia Kuala Lumpur, Jalan Sultan Yahya Petra, 54100 Kuala Lumpur, Malaysia
e-mail: shahiryasin@utm.my

N. Gao
Materials Research Group, Faculty of Engineering and Physical Sciences, University of Southampton, Southampton SO17 1BJ, UK

M. Y. Ismail et al. (eds.), *Technological Advancement in Mechanical and Automotive Engineering*, Lecture Notes in Mechanical Engineering,
https://doi.org/10.1007/978-981-19-1457-7_66

1 Introduction

Nowadays, there are various additive manufacturing (AM) techniques that can process various materials, including polymers, metals, and alloys for a myriad of prototyping and functional applications. Kumaresan et al. [1] reviewed different AM processes, materials, processing parameters and applications that focus improvements that can be made to improve AM-fabricated part quality for use in various sectors. Salleh et al. [2] used polyjet 3D printing AM technology to model and rapid prototype the human trachea to study the airflow at this particular organ within the human body. Bandyopadhyay et al. [3] provided a concise summary in the recent developments in metal AM systems, particularly energy source and part resolution upgrades, as well as the current challenges and future trends in the metal AM sector. Furthermore, Everton et al. [4], and Mani et al. [5], highlighted the need of in-situ process monitoring and control of metal AM processes to ensure quality and minimise defects in the fabricated metallic parts.

As one of the most popular metal AM technique, selective laser melting (SLM) builds metallic parts by melting and fusing successive layers of powder according to the initial computer aided design (CAD) profile. Although SLM offers design flexibility and high geometrical accuracy for highly customised components, the as-built parts may still contain some porosity due to gas entrapment or unmelted powder as the result of non-optimum processing parameters In particular, porosity has been found to negatively influence the mechanical properties of AM metallic parts and reduce their quality [6]. For example, Sames et al. [7] comprehensively reviewed the influence of various metal AM processing parameters on the resulting metallurgical effect and defects. They suggested that suitable energy density values need to be selected from the combination of various processing parameters, including power of heat source, scan speed, layer thickness, and hatch spacing for different metals and alloys to reduce defects in the fabricated parts as much as possible. This finding is echoed by Cherry et al. [8] whom investigated the influence of processing parameters on the physical properties of 316L SS manufactured by SLM. They found out that non-optimum energy density values result in porosity that significantly reduce part hardness. Similarly, Yusuf et al. [9] attained the highest part densification of >99% when using SLM to manufacture 316L SS specimens with a high average hardness value of ~250 HV, but porosity is unavoidable in the solidified part (~0.82%).

Thus, standardised quality assurance and verification methods have been developed to qualitatively and quantitatively assess porosity, defects, and the consistency of AM-fabricated structures [4, 10]. So far, the common characterisation techniques applied include Archimedes methods, gas pycnometry, and 2D OM cross-section analysis [11]. However, the Archimedes method and gas pycnometry could not capture the fundamental pore characteristics, while microscopy is a destructive technique and limited to only a few 2D cross-sections, which could compromise the analysis results [11]. Thus, in this study, X-ray computed tomography (XCT) is introduced as a 3D non-destructive technique (NDT) approach [12, 13] that has the potential to quantitatively assess the porosity profile of the whole as-built AM metallic structure in a single step.

2 Materials and Methods

A small 316L SS cylinder (diameter: 10 mm, height: 10 mm) was fabricated via Concept Laser M2 SLM machine. The following processing parameters were used; laser power, *P*: 200 W, scan speed, *s*: 1600 mm s^{-1}, layer thickness, *h*: 30 μm, scan line spacing, *l*: 150 μm, and island (checkerboard) scan strategy rotated by 90° after each layer. Nitrogen was initially purged into the build chamber and the operating temperature was maintained at 21 °C throughout processing.

XCT scan was conducted using a Zeiss Versa XRM-510 machine to assess the porosity profile by creating 3D pore visualisations within the specimen. A flux of 62.5 μA was created through a focused beam with energy value of 160 kVP, which was filtered by using a 3.3 mm calcium fluoride profile. To cover a full 360° of the cylindrical specimen, 2000 projections were taken that was exposed for 25 s each. An effective projection size of 1015 × 1015 pixels was attained by setting the detector binning to 2×, which yielded a high resolution of 1.832 μm via 4 × objective lens. The resulting 2D projection data were then reconstructed using a filtered back projection-based software equipped with the XCT machine. These reconstructions were transferred to VG Studio Max 2.1 software to create 3D visualisation images of the porosity distribution, followed by quantitative analysis of the pores such as morphology, size, and percentage. The XCT facility and specimen setup are shown in Fig. 1.

For comparison, OM image analysis was carried out through an Olympus BX41M-LED microscope and ImageJ analysis software to evaluate and verify the results of porosity analysis obtained from XCT. The cylinder was firstly sliced into 20 smaller disks with a diameter of 1 mm each using wire electrical discharge machining (EDM), followed by typical sample preparation procedures. OM images of polished samples were binarized into black (pores) and white (solid material) using a pre-set threshold value. The pore size, percentage, morphology, and distribution in the thin disks were then determined via surface area analysis in ImageJ, and the result are compared with those of the overall as-built cylinder determined from XCT.

3 Results and Discussion

The 3D image reconstruction results from the XCT scan are shown in Fig. 2, in which Fig. 2a exhibit the fully reconstructed cylinder, and Fig. 2b displays the porosity distribution inside the cylinder. The porosity content is evaluated by calculating the pore/solid material volume ratio (v_p/v_s). Based on the measurements from the VG studio Max 2.1 software, v_p and v_s are determined as 7670.23 mm^3 and 4.85 mm^3,

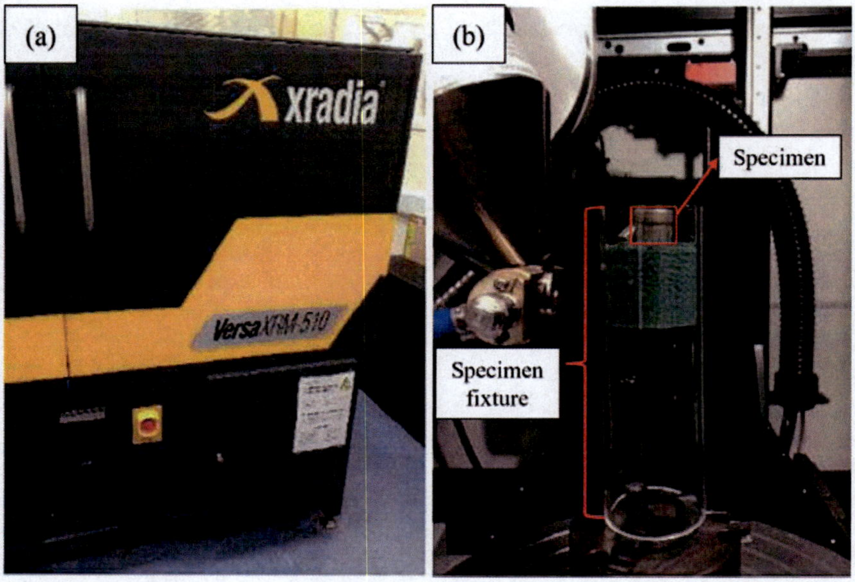

Fig. 1 **a** Zeiss versa XRM-510 XCT machine. **b** Specimen setup for XCT

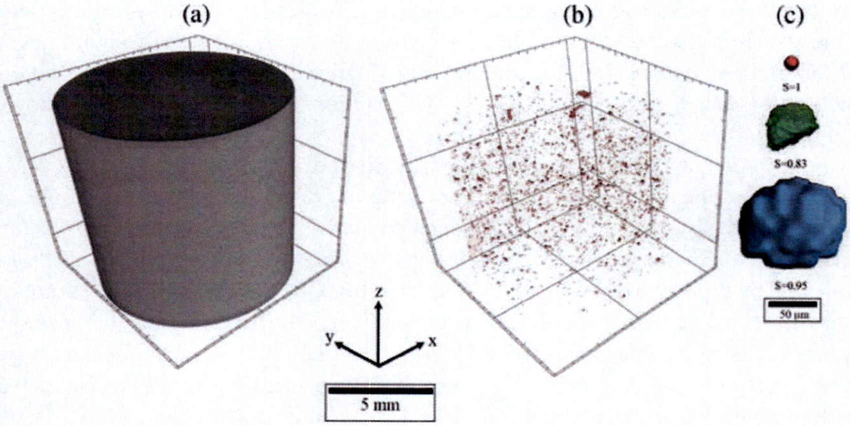

Fig. 2 **a** Fully reconstructed cylinder. **b** 3D visualisation of porosity distribution within the reconstructed cylinder, and **c** sphericity of representative 3D pores obtained from XCT

respectively to give a porosity level of 0.63%. This signifies high densification level (>99%) in the as-built structure. In addition, the largely spherical or near-spherical pores are indicative of gas-induced pores, which may be the result of entrapped inert gas inside the melt pool when the precursor powder initially melted during processing, or they could already be present within the raw powder and remain

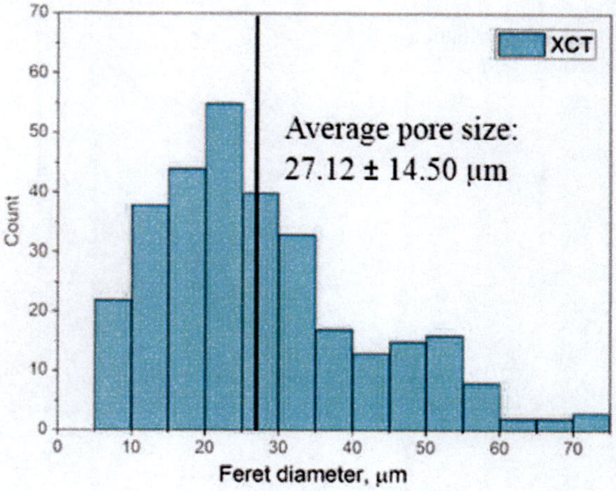

Fig. 3 Pore size (Feret diameter) distribution obtained from XCT

inside the completed part [7, 14]. Nevertheless, the high densification level and the present of only gas-induced pores rather than irregular-shaped process-induced pores both suggest that optimum processing parameters were used for SLM [8, 9, 15, 16].

In this study, the Feret diameter, defined as the distance between two parallel planes that restrict any two opposing edges is used to quantify the pore sizes. This approach enables a more accurate representation of the pore diameter. The pore distribution is determined to be in the range of 5–80 µm with an average size of 27.12 ± 14.50 µm as shown in Fig. 3.

Additionally, the pore morphology can be directly visualised in 3D using the sphericity factor, S, which is calculated by using the following equation [17]:

$$S = \frac{\pi^{\frac{1}{3}}\left(6v_p\right)^{\frac{2}{3}}}{A_p} \tag{1}$$

where A_p is the pore area (mm^2), and $S = 1$ indicates ideal sphere. The calculated results reveal a high average S value with low discrepancy of 0.83 ± 0.17, indicating high consistency and further confirming the optimum selection of processing parameter in this study. This is because low pore sphericity typically suggests high directionality of the pores, manifested by irregular-shaped pores that often appear as the result of non-optimum SLM processing parameters [17]. Figure 2c displays the representative pores and their corresponding sphericity in this study.

On the other hand, analysis of the polished OM images via ImageJ software reveals a porosity content of 0.66 ± 0.04%. Figure 4 displays the pore size distribution that ranges from 0 to 80 µm with an average size of 25.12 ± 17.43 µm.

When comparing XCT and OM pore size distribution (Figs. 3 and 4, respectively), fewer pores in the range between 5 and 10 µm are detected by XCT, while pores ranging from 0 to 5 µm could not be detected at all by XCT. This leads to a higher pore diameter average in XCT-scanned specimen relative to that attained by OM analysis.

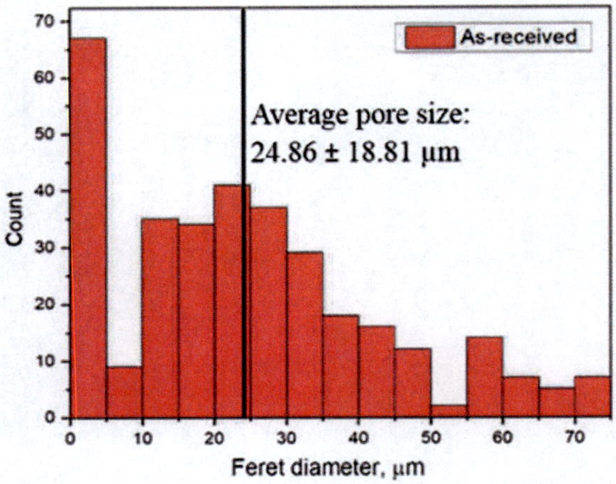

Fig. 4 Pore size (Feret diameter) distribution obtained from OM

In total, the reconstructed XCT image layers yield a total pore count of 294, while 331 pores are observed from OM images. Hence, the porosity distributions obtained via XCT and OM are normalised to provide a direct comparison between the results of both techniques, as shown in Fig. 5.

Based on Fig. 5, the pore size distributions are roughly similar for both XCT and OM methods. However, the porosity percentage obtained from OM (~23%) in the 0–10 μm region is considerably higher than that attained from XCT (only ~7%). This result is expected since XCT scan only detected fewer pores in the range of 5–10 μm without detecting any pores ranging from 0 to 5 μm. This might explain the higher average pore size and lower porosity percentage measured from XCT compared to that obtained from OM image analysis, which could be attributed to the X-ray detection limit, particularly for pore sizes <5 μm.

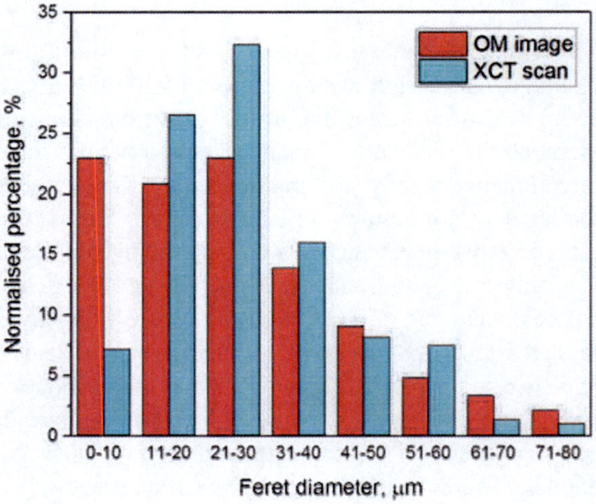

Fig. 5 **a** Comparison of normalised porosity distribution obtained from XCT scan and OM image analysis

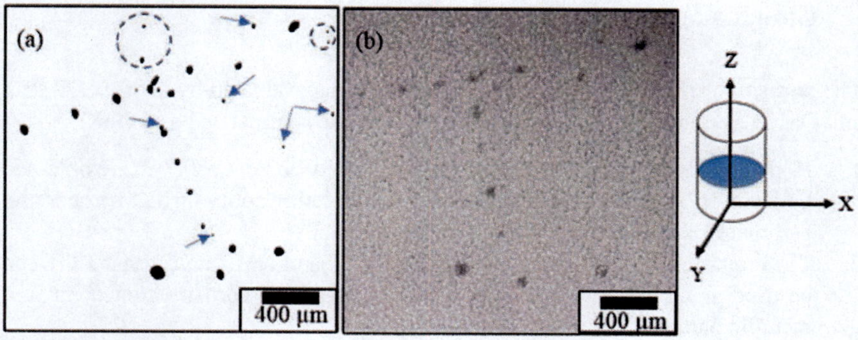

Fig. 6 **a** Binarised OM micrograph, and **b** reconstructed XCT image of a same layer

Furthermore, Fig. 6 compares the results of a binarised OM image and a reconstructed XCT image at a same layer. The pore distribution is generally similar, but much smaller pores (<10 µm) in the OM image (circled areas and arrows in Fig. 6a) are absent in the reconstructed XCT image (Fig. 6b), which affirms the limitations of XCT in detecting pore sizes <5 µm described previously.

Several factors can contribute to the discrepancy in the pore sizes measured through XCT and OM approaches. For example, the resulting OM images can be influenced by microscope settings, e.g. lighting, focus, thresholding and binarisation options, and techniques used during sample preparation, e.g. sectioning and polishing. In addition, XCT scan results could be affected by [11, 18]: (i) image artefacts from material-dependent X-ray scattering, (ii) attainable resolution, in which detectable pore sizes are limited to the spatial resolution of the XCT scan, and (iii) procedures used for thresholding.

In this study, some background noise is still present in the reconstructed XCT image despite the long scan time at a fine resolution (Fig. 6b). This is the result of high z-value (atomic number) of the material that reduces X-ray absorption, limiting the capability of XCT scan from detecting pores <5 µm. Thus, an alternative approach is needed in the future to detect such small pores and decrease measurement uncertainty, e.g. by spot scanning a small area for a longer time [11]. Nonetheless, XCT has the potential to be more reliable for porosity analysis in AM metal parts than 2D OM cross-sectional analysis because XCT combines multiple 2D image layer slices through the overall structure to construct a whole 3D image, which is not possible from only a few cross-sections in OM. Moreover, XCT is an effective NDT alternative that does not require any conventional physical sample preparations, e.g. sectioning, grinding, and polishing, which may eliminate some ultrafine pores that may compromise analysis results.

4 Conclusion

The porosity profile in SLM-fabricated 316L SS was evaluated by using XCT scan and OM image analysis. The results of this study are summarised as follow:

1. High densification level (>99%) and low porosity content (<0.7%) are obtained both via XCT scan and OM image analysis, indicating optimum SLM processing parameters were used.
2. The consistency of results analysed from XCT and OM suggest that XCT can be used as an effective NDT approach to assess the porosity content in AM metallic parts.

Acknowledgements The XCT scan was supported by μ-ray Imaging Centre, University of Southampton, UK under EPSRC grant EP-H01506X. The authors would also like to acknowledge grant R.K130000.7343.4B472 by Takasago Thermal Engineering Co. Ltd. for supporting this study.

References

1. Kumaresan R, Samykano M, Kadirgama K, Ramasamy D, Keng NW, Pandey AK (2021) 3D printing technology for thermal application: a brief review. J Adv Res Fluid Mech Therm Sci 83:84–97
2. Salleh ZM, Osman K, Marian MF, Hassan NNM, Madon RH, Samiran NA, Rashid RA, Ishak IA, Darlis N (2021) Rapid prototyping 3D model for PIV: application in human trachea model flow analysis. J Adv Res Fluid Mech Therm Sci 79:65–73
3. Bandyopadhyay A, Zhang Y, Bose S (2020) Recent developments in metal additive manufacturing. Curr Opin Chem Eng 28:96–104
4. Everton SK, Hirsch M, Stravroulakis P, Leach RK, Clare AT (2016) Review of in-situ process monitoring and in-situ metrology for metal additive manufacturing. Mater Des 95:431–445
5. Mani M, Lane BM, Donmez MA, Feng SC, Moylan SP (2017) A review on measurement science needs for real-time control of additive manufacturing metal powder bed fusion processes. Int J Prod Res 55:1400–1418
6. Lewandowski JJ, Seifi M (2016) Metal additive manufacturing: a review of mechanical properties. Annu Rev Mater Res 46:151–186
7. Sames WJ, List FA, Pannala S, Dehoff RR, Babu SS (2016) The metallurgy and processing science of metal additive manufacturing. Int Mater Rev 61:1–46
8. Cherry JA, Davies HM, Mehmood S, Lavery NP, Brown SGR, Sienz J (2014) Investigation into the effect of process parameters on microstructural and physical properties of 316L stainless steel parts by selective laser melting. Int J Adv Manuf Technol 76:869–879
9. Yusuf SM, Chen Y, Boardman R, Yang S, Gao N (2017) Investigation on porosity and microhardness of 316L stainless steel fabricated by selective laser melting. Metals (Basel) 7
10. Seifi M, Salem A, Beuth J, Harrysson O, Lewandowski JJ (2016) Overview of materials qualification needs for metal additive manufacturing. Jom 68:747–764
11. Wits WW, Carmignato S, Zanini F, Vaneker THJ (2016) Porosity testing methods for the quality assessment of selective laser melted parts. CIRP Ann Manuf Technol 65:201–204

12. Ismon M, Jamalludin MD, Bujang IZ, Azmir NA, Asmawi R, Sies MF, Roslan MN, Ibrahim Z (2020) Pitting hole evaluation by active infrared thermography in stainless steel 304. J Adv Res Fluid Mech Therm Sci 68:114–124
13. Wong TL, Ahmad UK, Tharmalingam S (2016) Non-destructive blended fibre analysis from forensic perspective. J Adv Res Appl Sci Eng Technol 2:44–56
14. Sames WJJ, Medina F, Peter WHH, Babu SSS, Dehoff RRR (2014) Effect of process control and powder quality on IN 718 produced using electron beam melting. In: 8th international symposium on superalloy 718 and derivative. Wiley-Blackwell, Pittsburgh, pp 409–423
15. King WE, Barth HD, Castillo VM, Gallegos GF, Gibbs JW, Hahn DE, Kamath C, Rubenchik AM (2014) Observation of keyhole-mode laser melting in laser powder-bed fusion additive manufacturing. J Mater Process Technol 214:2915–2925
16. Körner C, Bauereiß A, Attar E (2013) Fundamental consolidation mechanisms during selective beam melting of powders. Model Simul Mater Sci Eng 21:1–18
17. Ziółkowski G, Chlebus E, Szymczyk P, Kurzac J (2014) Application of X-ray CT method for discontinuity and porosity detection in 316L stainless steel parts produced with SLM technology. Arch Civ Mech Eng 14:608–614
18. Cai X, Malcolm AA, Wong BS, Fan Z (2015) Measurement and characterization of porosity in aluminium selective laser melting parts using X-ray CT. Virtual Phys Prototyp 10:195–206

Effect of Sn-xCu Solder Alloy onto Intermetallic Formation After Laser Soldering

M. A. Abdullah and S. R. A. Idris

Abstract The awareness of lead-free solders can be attributed to their environmental and human-health related benefits. Due to this consciousness, research toward lead-free solder alloy became a concern. Tin-copper (SnCu) solder alloy is one of the candidates that can meet the characteristic of tin–lead (SnPb). The objective of this study is to analyse the wetting behaviour, intermetallic compound (IMC) thickness and also the spread ratio by varies the copper weight percentage (Cu wt%) in SnCu solder alloy. Solder alloy used was Sn-xCu, where $x = 0.0, 0.3, 0.5, 0.7, 1.0$ which was soldered onto electroless nickel immersion gold (ENIG) substrate using carbon dioxide ($CO_{2)}$ laser. Parameters used for laser soldering was 35 W for the laser power, focal length was kept constant, scanning time was 0.04 s, and scanning speed was 100 mm/s. Then these sample were subjected to isothermal aging with the duration of 0, 200, 500, 1000 and 2000 h. In the final analysis, the IMC thickness and wetting behaviour characteristics were characterized by metallographic microscopy. The results showed that the higher the Cu percentage in the solder alloy, the higher the thickness formed at the solder joint interface. Besides that, the morphology of IMCs additionally changed with aging time whereby it changed into much uniform and continuous shape, Nonetheless, its thickness was found to be increasing upon aging duration. Furthermore, the spread factor and spread ratio increase, but the equilibrium contact angle decreases with increasing Cu content. These results were the proof that Sn-0.7Cu/ENIG offers a good solder joint performance as compared to other copper percentage.

Keywords Solder alloy · Laser soldering · Intermetallic compound · Wetting behaviour

---

M. A. Abdullah · S. R. A. Idris (✉)
Faculty of Mechanical and Automotive Engineering Technology (FTKMA), Universiti Malaysia Pahang, 26600 Pekan, Malaysia
e-mail: rabiatull@ump.edu.my

M. Y. Ismail et al. (eds.), *Technological Advancement in Mechanical and Automotive Engineering*, Lecture Notes in Mechanical Engineering,
https://doi.org/10.1007/978-981-19-1457-7_67

1 Introduction

In the last several years, tin–lead (SnPb) solder is used in electronics manufacturing because it has higher thermal resistance, improved mechanical characteristics, and smaller melting point, consequent in higher bond strength in solder connection [1–3]. Nevertheless, the European Union (EU) Waste Electrical and Electronic Equipment (WEEE) regulation prohibits usage of lead in electronic products on 2006. [4]. Responding to concerns about the use of lead in electronic components, a lot of work is being carried out to find lead-free solders suitable for various electronic attach applications. The environmental toxicity of lead and the health hazard have been increased in an attempt to find suitable substitutes for lead in standard SnPb eutectic solders without degrading the physicochemical properties of the transition compounds [5–8]. Under the current circumstances, the tin-copper (SnCu) solder series, whether or not other elements are added, in the class of the most encouraging alloys to substitute SnPb solder [9–11]. Sn-0.7Cu alloys is one of the eutectic alloy materials, this alloy have come to be the point of interest of study, due to that eutectic alloys, this own the advantages of low melting temperature, little wetting angle, suitable fluidness and minimum tendency of hot cracking and segregation [12]. In group of the new candidates for lead-free alloys, Sn0.7Cu solder is inexpensive and has good characteristics [12]. As the solderability of SnCu is continuously confirmed, Sn0.7Cu solder is frequently used in common soldering process such as wave soldering [4, 13]. In order to identify the various characteristics of SnCu solder, many studies have been conducted, comprising interface development, wetting angle and mechanical properties achieved through compounding, particle hardening, and control of the crystallization rate.

Besides soldering, printed circuit boards (PCBs) and base metal such as surface finishes should also be lead-free. The choice of a suitable metallized substrate played a significant role in the improvement of reliable connection packaging technology and manufacturing. This is specifically true when using lead-free solder. In surface mounting technology (SMT) such as flip chip and ball grid array (BGA), the usage of copper is extensively used in the production. It is a well-known fact that at the CuSn solder interface, Cu reacted quickly with the CuSn formed by the intermetallic compound (IMC), weakening the solder because of brittle IMC [14]. For this reason, electroless nickel plating (Ni–P) coatings are frequently used as diffusion barriers on the solder alloys (Cu) pads of BGA and flip chip electrodes due to their low cost and simple process [15, 16].

As a consequence of miniaturization of electronic packaging and the growing proportion of IMC in solder joints, a detailed study [3] of the formation, behaviour and growth mechanism of IMC is especially important. To prevent critical thermal damage to temperature-sensitive electronic devices, during soldering process the time of soldering should be shortened [17]. The choice of laser soldering in this study is due to two reasons: the first is the fine properties of laser soldering, for example fast melting, fast cooling and a smaller heat-affected region, and the second is that the quickly-solidified lead-free solders inappropriate for future [18]. Wei [19] used a carbon dioxide (CO_2) laser as a heating source to study the growth behaviour of IMC on the interface among Sn0.7Cu and Cu substrates. Based on data show when the power increase, the IC layer was also increases. IMC thickness is functionally related to scan speed and performance. Nevertheless, there are few studies on the interfacial reaction of a coated substrate. The substrate is attracting more consideration due to its benefits.

Hence, the objective of this study is to investigate the effect of the interfacial reaction between Sn-*x*Cu and electroless nickel immersion gold (ENIG) surface finish after laser soldering in order to determine the IMC thickness, spread ratio and wetting behaviour as well as IMC growth when exposed to prolonged isothermal aging. The influence of various copper percentage in the solder alloy will also be evaluated. However, for ENIG substrate, this method resulting in high electroless technology and it is complex throughout the process. Moreover, the need of suitable IMC thickness, good wettability and long service life solder joint connection is also a major problem to resolve in future.

2 Methodology

Solder alloy of Sn-*x*Cu where, $x = 0.0, 0.3, 0.5, 0.7, 1.0$ (Table 1) was prepared through powder metallurgy process. The powder materials copper and tin particle size that use in this research is <63 μm brand Merck KGaA and <71 μm Sigma-Aldrich respectively. The step started with mechanical mixing and blending for 6 h duration at 1400 rpm speed. Mixed powder then compacted with 5 ton pressure towards rectangular thin pallet with the size of 3.2 × 1.5 mm. Before soldering, solder bulk and substrates were cleaned in ethanol ultrasonically. Then, these samples were subjected to laser soldering using CO_2 gas laser with the laser power of 35 W, scanning time of 0.04 s, focal length 40 mm and scanning speed of 100 mm/s to create solder joint between Sn-*x*Cu solder alloy and ENIG substrate Fig. 1.

The process was followed by isothermal aging with the duration of 0, 200, 500, 1000 and 2000 h at the temperature of 150 °C to simulate the actual service condition. The specimens were finally produced in accordance with the usual metallographic processes. A solution for ethanol (95 vol%), hydrochloric acid (5 vol%) was applied for etching and an optical microscopy examined the microstructure of the

Table 1 Chemical arrangement of the Sn-*x*Cu solder (wt%)

Alloy	Composition (wt%)	Weight (10 g)
Sn	100 wt% Sn	10 g
Sn-0.3Cu	99.7 wt% Sn, 0.3 wt% Cu	9.97 g Sn, 0.03 g Cu
Sn-0.5Cu	99.7 wt% Sn, 0.5 wt% Cu	9.95 g Sn, 0.05 g Cu
Sn-0.7Cu	99.7 wt% Sn, 0.7 wt% Cu	9.93 g Sn, 0.07 g Cu
Sn-1.0Cu	99.7 wt% Sn, 1.0 wt% Cu	9.9 g Sn, 0.1 g Cu

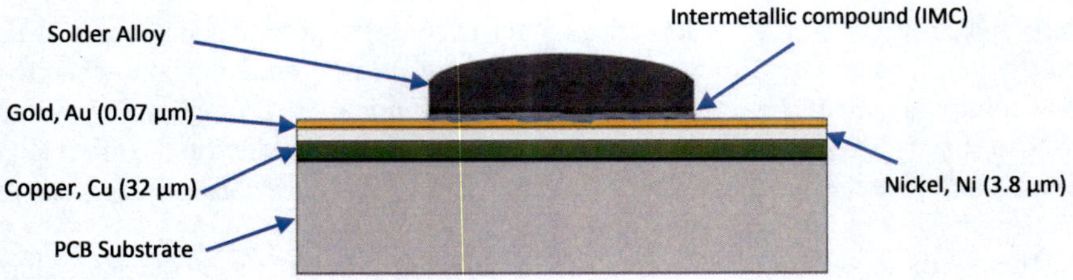

Fig. 1 Electroless nickel immersion gold (ENIG) base metal

solder alloys (Olympus-BX51M). The spreading ratio of the solder alloys has been calculated in accordance with the Japanese Industrial Standard (JIS 23198-3, 2003). Contact angles were used to characterise wetting behaviour and were measured using video tool software under an optical microscope. Then, these samples were characterised in order to determine the IMC thickness, spread factor (S_f), wetting behaviour or contact angle (θ) and spread ratio (S_r) of the solder.

3 Results and Discussion

3.1 Effect of IMC Thickness After Isothermal Aging

All samples were first mounted in epoxy for microstructural inspection before melting and polishing after laser soldering and ageing processes. As shown in Fig. 2, there

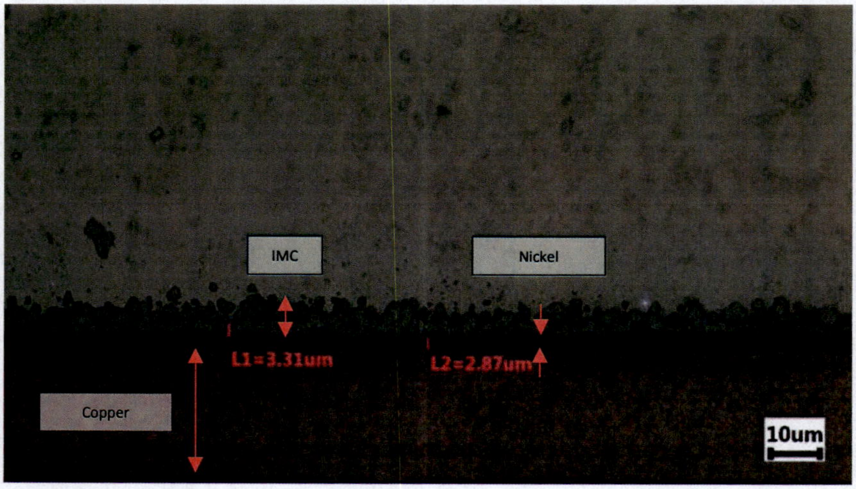

Fig. 2 Cross sectional view of Sn-*x*Cu/ENIG solder joint

were three layers existed on the interface of the solder connection between Sn-*x*Cu and ENIG which was Coper, Nickel and IMC layer. Gold (Au) layer cannot be seen in the figure since it was dissolved inside the solder alloy through liquid state diffusion into the solder after laser soldering, and hence exposing only nickel layer. Since Au layer only used to protect Nickel layer from oxidation, and hence it was not the main focus of this study. As reported by Chen et al. [20], there were two method of IMC development diffusion between solder and substrate. Chung et al. [21] Cu dominates the intercalation of SnCu, and Cu diffuses towards Sn at a faster rate due to the Cu atom has a smaller atomic radius (0.128 nm) than the Sn atom (0.141 nm). IMC was formed by interfacial reaction through chemical reaction between Sn from the molten solder and Cu from the metallized substrate [22].

The findings of this study showed that Sn-*x*Cu with Cu percentage of 0.3 wt% gave an optimum result in terms of IMC thickness and wetting angle. This can be seen through Fig. 3 that aging duration with 0, 200, 500, 1000 and 2000 h produced finer and denser grain. The structure of these fine grains can influence the diffusion of Cu, creating channels for rapid diffusion of Cu in gold and nickel in ENIG substrates. Flat-shapes and IMC layers are created at all sample interfaces during the initial laser process. The flat IMC layer takes on a scallop shape as the aging time at the solder joint interface increases, as illustrated in Fig. 3c–e. It can be seen that start from 200 until 2000 h.

The IMC thickness layer is extensively influences the reliability of the solder connection [23]. As can be seen in Fig. 4, the thicknesses data of the IMC was increasing upon aging for all Cu percentages in solder alloy. Solder alloy with 0.3 wt% Cu was found to be able to produce the lowest IMC thickness after laser soldering with the value of 0.68 μm, while the highest value was produced by solder with 1.0 wt% of Cu which was 1.48 μm. These value were found increased after aging and until 2000 h of aging duration, solder alloy with 0.3 wt% of Cu gave the highest value of thickness compared to 1.0 wt% Cu percentage. This situation happened might be due to the changes of IMC microstructure throughout the process. During soldering, the IMC formed was thinner and finer in shape for Sn-0.3Cu as compared to Sn1.0Cu as shown in Fig. 5. During soldering, the IMC formed through diffusion between the Sn from the molten solder and metallized substrate. However, after prolonged isothermal aging, the interface will be supersaturated with Ni and Cu from substrate and was blocked by existing IMC layer. At this stage, the diffusion was controlled by grain boundary through solid state diffusion, hence IMC will get thicker as well as coarser.

3.2 *Measurement of Spread Ratio, Spread Factor and Contact Angle: Wetting Behaviour*

Wetting is describe as dissemination of a liquid over a solid surface, which is determined by a contact angle [22]. Alloy composition, surface roughness, time, flux and

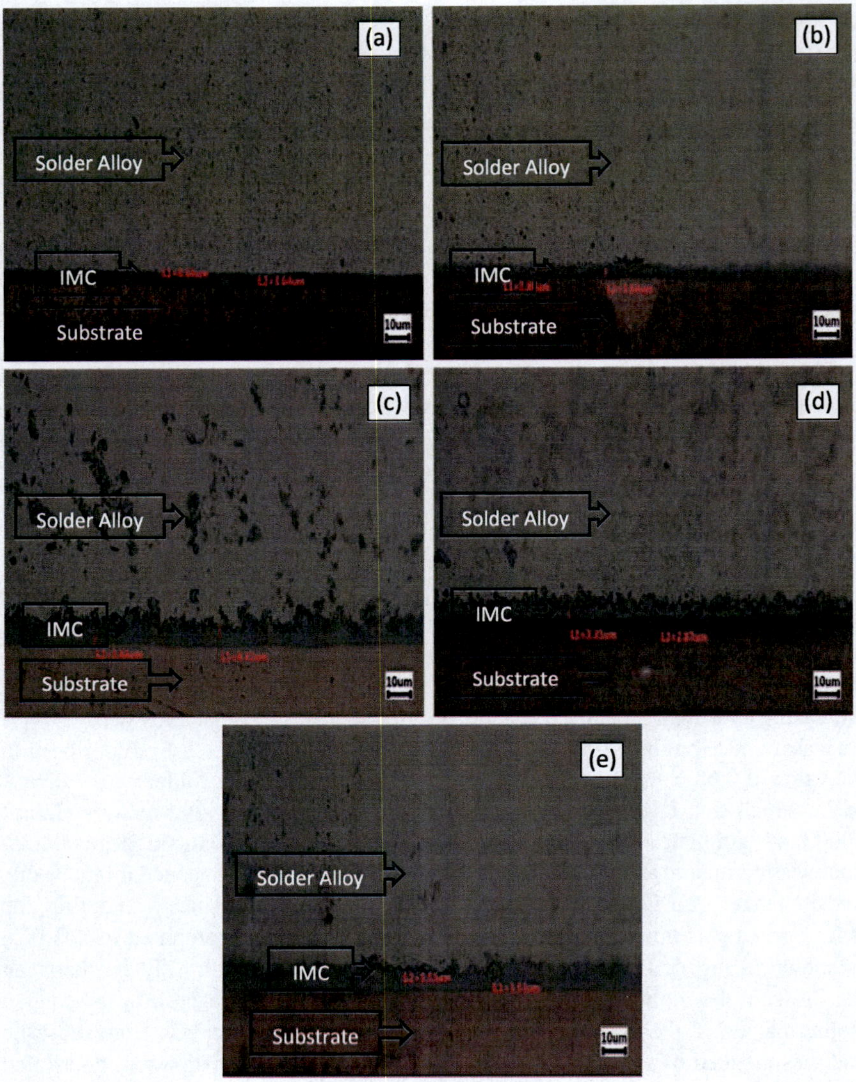

Fig. 3 Cross sectional view of Sn-0.3Cu/ENIG for **a** 0 h, **b** 200 h, **c** 500 h, **d** 1000 h and **e** 2000 h

melting point are some of the factors that affect the wetting properties of a solder [24]. Wettability is assessed by measuring the spreading area at the end of the solder cross section. Solder specimen is placed on the centre of the substrate along with flux and melted by laser soldering at a constant power, scanning speed and focal length. As a consequence of difference in copper content in the solder compound, the shape and size of the solder specimen will change. For good wetting of the solder, it is necessary to ensure a perfect metallurgical connection on the joining within the

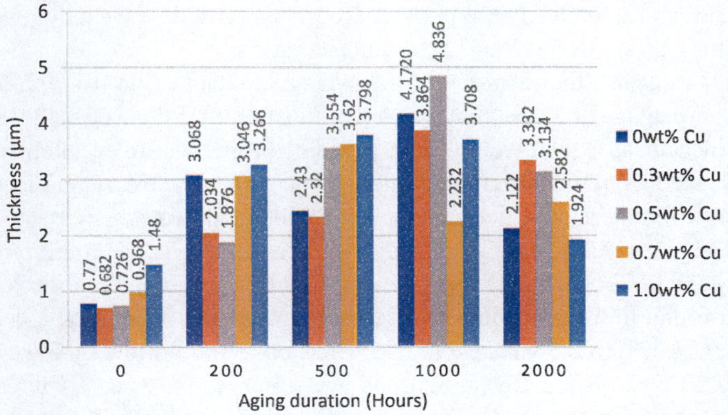

Fig. 4 IMC thickness as a function of difference aging time

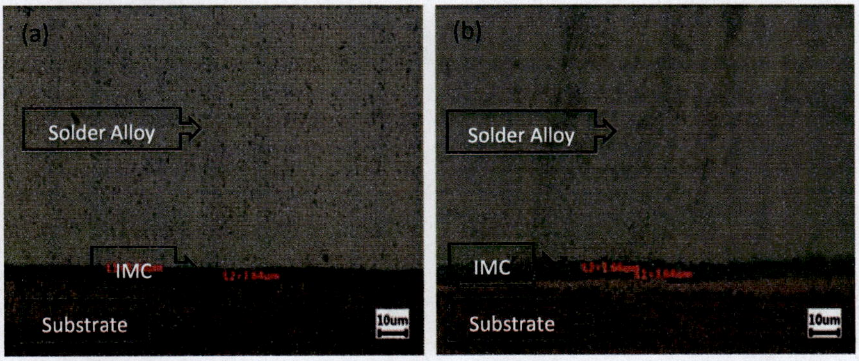

Fig. 5 Comparison of IMC microstructure **a** Sn-0.3Cu, **b** Sn1.0Cu

solder (liquid) and the copper substrate (solid). Because of interface reaction, IMC particles or layers of intermetallic compounds are formed on the surface joint. These IMCs act as an attachment and bond the liquid solder to the coating substrate. Figure 6 demonstrates the cross section image of liquid solder on the Cu plate and indicator

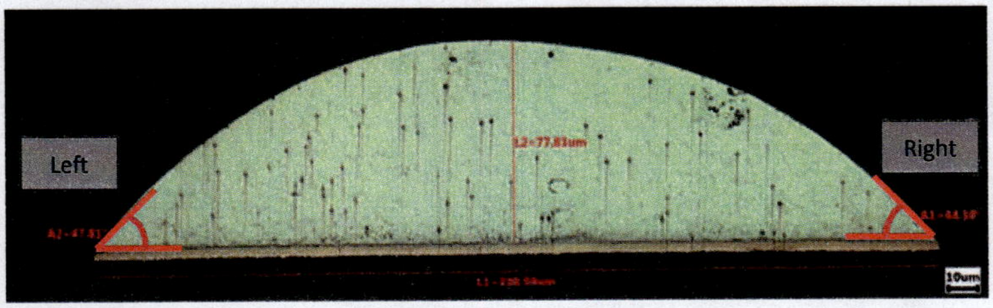

Fig. 6 Cross-sections image of indicator wetting angle measurement area

of (left) and (right) angle. These photomicrographs provide a visual representation of the wetting properties of Sn-*x*Cu solder specimens.

Figure 7 indicates the variation of the average contact angle (θ) with the corresponding percentage of Cu percentage. According to [25], Sn-based solder alloys are a good choice because their wetting and spreading properties are excellent. Measurement was taken from metallographic microscope with same 5× magnification. The contact angle for one same Cu percentage are dividend in two, which is right and left. The average angle value was calculated to express in graph. The minimum contact size was found to form in solder with 0.7 wt% Cu and the maximum observed contact angle was found to form in solder with 0 wt% Cu. At the beginning, a decrease of 0 wt% Cu (34.77°) to 0.5 wt% Cu (31.95°) was observed, follow by lowest contact size of 0.7 wt% Cu. But a ascending trend was observed at 1.0 wt% Cu (29.9°). Cu content help to form the Cu_6Sn_5 intermetallic compounds between the substrates and the solder alloy, and to improve wettability of composite solders [26]. Present results confirm that better wetting characteristics of a soldering matrix with an ideal copper composition can be achieved. Dewetting occurs due to uneven separation and increased viscosity of liquid solder on coating substrate surface. Improving of wetting can be assigned to the reduced viscosity of the molten solder and the increased solderability at laser power and scan speed [27]. For the overall left and right contact angle reading, the 0.7 wt% Cu produce the excellent wetting and which means this Cu percentage have good fluidity.

The spread ratio (S_r) is measured through the sphere area of the solder on the substrate that is melted. Spread ratio is calculated by diameter (d) and height (h) of the cooled solder surface on the substrate surface [25]. The spread ratio is taken as the ratio of the total surface of the plane that was wetted by the molten solder to the surface of the original solder sample. Where, spread factor (S_f) is calculated as the consequence of spread ratio. Figure 8 [28] presents a schematic diagram for calculating the spread factor, spread ratio and the contact angle, where the spread

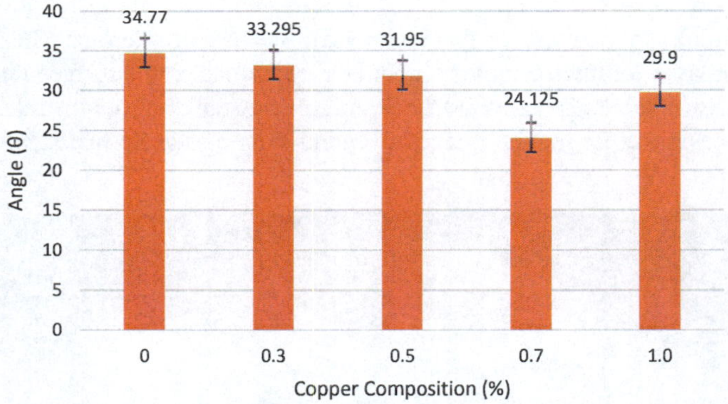

Fig. 7 Average contact angle trend of Sn-*x*Cu on coating substrate after soldering at different Cu percentage

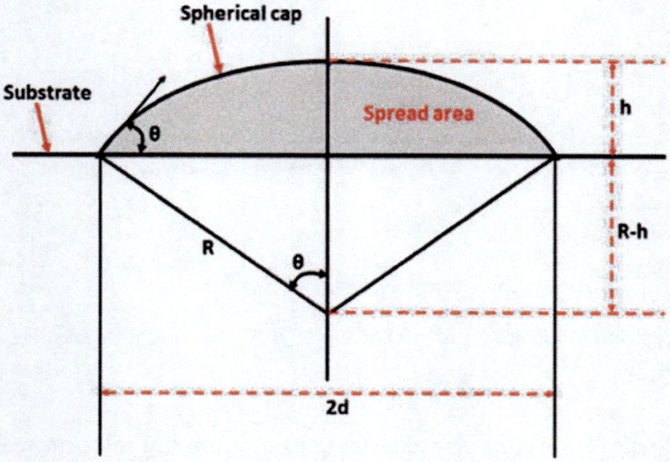

Fig. 8 The schematic diagram for the calculation of the spread ratio, spread factor, and contact angle. Adapted from [28]

factor (S_f) and spread ratio (S_r) of the solders defined by Eqs. (2) and (4) [28].

$$s_r = \frac{Total\ plane\ are\ wetted\ by\ molten\ solder}{plane\ area\ of\ the\ original\ spherical\ solder\ pellet} \tag{1}$$

$$s_r = \frac{4d^2/h^2}{\left(1 + \left(3d^2\right)/h^2\right)^{2/3}} \tag{2}$$

$$s_f = \frac{Diameter\ of\ the\ original\ solder\ pellet\ (D) - Heigh\ of\ the\ spreaded\ molten\ solder\ (h)}{Diameter\ of\ original\ solder\ pellet\ (D)} \tag{3}$$

$$s_f = \frac{(h^3 + 3A^2)^{1/3} - h}{\left(h^3 + 3A^2h\right)^{1/3}} \tag{4}$$

Indicators, h = Height of the spherical cap and d = Radius of the spherical cap.

Table 2 presents the calculated reading of the spread factor and the spread ratio equivalent to the laser temperature. As show in Fig. 9 the variation of the spread ratio (S_r) and spread factor (S_f) with the equivalent temperature, and it is also seen that

Table 2 Calculated values of spread ratio and spread factor corresponding to the temperature

Solder alloy	Spread ratio (S_r)	Spread factor ($S_f \times 10$)
Sn	8.81	8.49
Sn-0.3Cu	7.91	8.32
Sn-0.5Cu	7.57	8.24
Sn-0.7Cu	6.70	8.02
Sn-1.0Cu	5.36	7.53

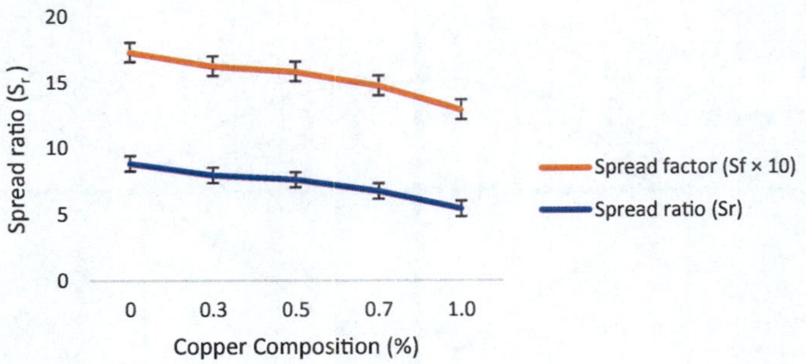

Fig. 9 Spread ratio and spread factor trend of Sn-*x*Cu on Cu substrate after soldering at different Cu percentage

the spread ratio (S_r) and the spread factor (S_f) of the Sn-*x*Cu solder increases to Sn 0 wt% Cu. The smallest values collected for the spread ratio and spread factor were 5.36 and 7.53 at 1.0 wt% Cu, while the maximum observed values were 8.81 and 8.49 at 0 wt% Cu.

4 Conclusion

Base on the objective of this study, the investigate the effect of the interfacial reaction between Sn-*x*Cu and substrate of electroless nickel immersion gold (ENIG) surface finish after laser soldering in order to determine the IMC thickness, spread ratio and wetting behaviour as well as IMC growth when exposed to prolonged isothermal aging were investigated. The appearance of the compound Sn-*x*Cu solder matrix was confirmed with cross section analysis. The findings of the study are described as follows:

i. Regardless of the composition of the solder, with an increasing ageing duration the total IMC interfacial thickness increased. The IMC thickness was found to be decrease upon Cu content with the minimum value of 0.682 μm when 0.3 wt% Cu was used. After isothermal aging, the IMC thickness gradually increased upon aging duration. Besides, the IMC also changed from scallop-like shape into continuous level-off shape. Generally, the composite 0.7 wt% Cu solder had the lowest diffusion coefficients and was the most effective way to reduce the overall growth of IMC layers.
ii. Additional Cu particle content had a significant impact on wettability of Sn-*x*Cu composite solders. The contact angle of Sn-*x*Cu was investigated in the Cu content range of 0 wt% until 1.0 wt%. At the 0.7% Cu percentage, it was produce lower wetting angle 24.125 θ and which means this Cu percentage have good fluidity.

iii. The optimal spread factor (8.49) and spread ratio (8.81) are accomplished at 0 wt% Cu. But by considering the spreading ratio in base and alloy solders, Sn-*x*Cu has lower wetting speed as compare to Sn solder alloys, this will enhance the strength of the solder during solidification. The higher Cu amount interrupted the solder wettability, because the melt viscosity is increased and the solders spread over the substrate is reduced.

Acknowledgements The authors would like to thanks the Ministry of Higher Education for providing financial support under Fundamental Research Grant Scheme (FRGS) No. FRGS/1/2018/TK03/UMP/02/21 (University reference RDU190152) and Universiti Malaysia Pahang for laboratory facilities.

References

1. Zhang L, Gao L (2015) Interfacial compounds growth of SnAgCu (nano La_2O_3)/Cu solder joints based on experiments and FEM. J Alloys Compd 635:55–60. https://doi.org/10.1016/j.jallcom.2015.02.110
2. Pal MK, Gergely G, Koncz-Horváth D, Gácsi Z (2020) Investigation of the electroless nickel plated Sic particles in Sac305 solder matrix. Powder Metall Met Ceram 58(9):529–537. https://doi.org/10.1007/s11106-020-00107-y
3. Ma H, Suhling JC (2009) A review of mechanical properties of lead-free solders for electronic packaging. J Mater Sci 44(5):1141–1158. https://doi.org/10.1007/s10853-008-3125-9
4. Yoon J-W, Jung S-B (2005) Interfacial reactions between Sn–0.4Cu solder and Cu substrate with or without ENIG plating layer during reflow reaction. J Alloys Compd 396(1):122–127. https://doi.org/10.1016/j.jallcom.2004.12.009
5. Shohji I, Yoshida T, Takahashi T, Hioki S (2004) Tensile properties of Sn–Ag based lead-free solders and strain rate sensitivity. Mater Sci Eng A 366:50–55. https://doi.org/10.1016/j.msea.2003.09.057
6. Takemoto T, Funaki T (2002) Role of electrode potential difference between lead-free solder and copper base metal in wetting. Mater Trans 43(8):1784–1790. https://doi.org/10.2320/matertrans.43.1784
7. Zeng K, Tu KN (2002) Six cases of reliability study of Pb-free solder joints in electronic packaging technology. Mater Sci Eng R Rep 38(2):55–105. https://doi.org/10.1016/S0927-796X(02)00007-4
8. Glazer J (1995) Metallurgy of low temperature Pb-free solders for electronic assembly. Int Mater Rev 40(2):65–93. https://doi.org/10.1179/imr.1995.40.2.65
9. Zhao Z, Wang C, Li M, Wang L (2005) Nd:YAG laser surface treatment of copper to improve the wettability of Sn3.5Ag solder on copper. Surf Coat Technol 200(7):2181–2186. https://doi.org/10.1016/j.surfcoat.2005.01.006
10. Luo W, Ho C, Tsai J, Lin Y, Kao C (2005) Solid-state reactions between Ni and Sn–Ag–Cu solders with different Cu concentrations. Mater Sci Eng A Struct Mater Prop Microstruct Process 396:385–391. https://doi.org/10.1016/j.msea.2005.02.008
11. Abtew M, Selvaduray G (2000) Lead-free solders in microelectronics. Mater Sci Eng R Rep 27(5):95–141. https://doi.org/10.1016/S0927-796X(00)00010-3
12. Zhao M, Zhang L, Liu Z-Q, Xiong M-Y, Sun L (2019) Structure and properties of Sn–Cu lead-free solders in electronics packaging. Sci Technol Adv Mater 20(1):421–444. https://doi.org/10.1080/14686996.2019.1591168

13. Yu DQ, Wu CML, He DP, Zhao N, Wang L, Lai JKL (2005) Effects of Cu contents in Sn–Cu solder on the composition and morphology of intermetallic compounds at a solder/Ni interface. J Mater Res 20(8):2205–2212. https://doi.org/10.1557/JMR.2005.0275
14. Lee CB, Jung SB, Shin YE, Shur CC (2002) Effect of isothermal aging on ball shear strength in BGA joints with Sn–3.5Ag–0.75Cu solder. Mater Trans 43(8):1858–1863. https://doi.org/10.2320/matertrans.43.1858
15. Kim J, Myung W-R, Jung S-B (2016) Effects of aging treatment on mechanical properties of Sn–58Bi epoxy solder on ENEPIG-surface-finished PCB. J Electron Mater 45(11):5895–5903. https://doi.org/10.1007/s11664-016-4803-z
16. Hwang C-W, Suganuma K, Kiso M, Hashimoto S (2003) Interface microstructures between Ni–P alloy plating and Sn–Ag–(Cu) lead-free solders. J Mater Res 18(11):2540–2543. https://doi.org/10.1557/JMR.2003.0354
17. Zou G, Yan J, Mu F, Wu A, Zhou YN (2011) Recent progress in microjoining and nanojoining. Hanjie Xuebao/Trans China Weld Inst 32(4):107–112 [Online]. Available: https://www.scopus.com/inward/record.uri?eid=2-s2.0-79957667746&partnerID=40&md5=bd73953b19e5993693e1583565df1a10
18. Chaminade C, Fogarassy E, Boisselier D (2006) Diode laser soldering using a lead-free filler material for electronic packaging structures. Appl Surf Sci 252(13):4406–4410. https://doi.org/10.1016/j.apsusc.2005.07.095
19. Xue-min P (2009) The growth behavior of interfacial IMC of Sn–0.7Cu/Cu by laser braze welding. J Mater Eng
20. Chen LD, Huang ML, Zhou SM (2010) Effect of electromigration on intermetallic compound formation in line-type Cu/Sn/Cu interconnect. J Alloys Compd 504(2):535–541. https://doi.org/10.1016/j.jallcom.2010.05.158
21. Chung CK, Duh J-G, Kao CR (2010) Direct evidence for a Cu-enriched region at the boundary between Cu_6Sn_5 and Cu_3Sn during Cu/Sn reaction. Scr Mater 63(2):258–260. https://doi.org/10.1016/j.scriptamat.2010.04.011
22. Pal MK, Gergely G, Koncz-Horváth D, Gácsi Z (2021) Investigation of microstructure and wetting behavior of Sn–3.0Ag–0.5Cu (SAC305) lead-free solder with additions of 1.0 wt% SiC on copper substrate. Intermetallics 128. https://doi.org/10.1016/j.intermet.2020.106991.
23. Yu C, Yang Y, Li P, Chen J, Lu H (2012) Suppression of Cu_3Sn and Kirkendall voids at Cu/Sn–3.5Ag solder joints by adding a small amount of Ge. J Mater Sci Mater Electron 23. https://doi.org/10.1007/s10854-011-0412-z
24. Suh JO, Tu KN, Lutsenko GV, Gusak AM (2008) Size distribution and morphology of Cu_6Sn_5 scallops in wetting reaction between molten solder and copper. Acta Mater 56(5):1075–1083. https://doi.org/10.1016/j.actamat.2007.11.009
25. Pal MK, Gácsi Z (2020) Antal Kerpely doctoral school of materials science and technology development of ceramic particle reinforced SAC305 lead-free solder composite material, pp 1–100 [Online]. Available: https://www.kerpely.uni-miskolc.hu/files/13101/ManojPaldisszertáció.pdf
26. Yan Y, Wang G, Song K, Zhao K (2010) Effect of Ag particles on wettability and creep rupture life of SnCu based composite solders. Mater Sci Forum 650:78–84. https://doi.org/10.4028/www.scientific.net/MSF.650.78
27. Li LF, Cheng YK, Xu GL, Wang EZ, Zhang ZH, Wang H (2014) Effects of indium addition on properties and wettability of Sn–0.7Cu–0.2Ni lead-free solders. Mater Des 64:15–20. https://doi.org/10.1016/j.matdes.2014.07.035
28. Jacobson DM (2004) Principles of soldering, vol 42(03)

Effect of Particle Discretisation and Horizon Size on the Displacement and Damage Plot Using Bond-Based Peridynamics

H. N. Yakin, N. Nikabdullah, and M. R. M. Rejab

Abstract Peridynamics (PD) represents a new non-local theory of continuum mechanics which uses integro differential equations instead of the typical local partial differential equations in its formulation. Thus, it is suitable for modelling fracture mechanics, where a continuum domain is modelled through particles connected via physical interactions. The PD formulation allows us to model spontaneous crack initiation, and crack branching without the need for special mathematical treatment. The value of parameters such as particle discretisation and horizon size will be checked to make sure that it agreed to the result from Finite Element Method (FEM) in elastic deformation before proceed to the failure mode. In PD, failure criterion is established when its stretch value exceeds a prescribed critical stretch value. In the classical bond model or Prototype Microelastic Brittle (PMB), the bond force grows linearly with the bond stretch, and the value suddenly goes down to zero when the bond stretch exceeds its critical value. This study will focus on the effect of horizon size and particle discretisation on PD displacement of elastic analysis, and damage patterns with PMB damage model. The proposed study leads to a better understanding of how horizon size and particle disretisation affect the damage patterns in PD frameworks.

Keywords Peridynamics · Cracks · Prototype microelastic brittle

1 Introduction

Peridynamic or Peridynamics (PD) theory, is a non-local theory, that was introduced by Silling [1] and Silling et al. [2] at Sandia National Laboratories, USA in an attempt to deal with the discontinuities. It is related to the non-local theory studied by Eringen et al. [3–5], which only particles placed inside a certain radius from a

H. N. Yakin · N. Nikabdullah (✉) · M. R. M. Rejab
Faculty of Mechanical and Automotive Engineering Technology, Universiti Malaysia Pahang, 26600 Pekan, Pahang, Malaysia
e-mail: nikabdullah@ump.edu.my

H. N. Yakin
e-mail: mhnaimy@gmail.com

M. Y. Ismail et al. (eds.), *Technological Advancement in Mechanical and Automotive Engineering*, Lecture Notes in Mechanical Engineering,
https://doi.org/10.1007/978-981-19-1457-7_68

particle will affect the condition of the particle. This theory is in contrast with the local classical continuum mechanics (CCM) where the state of a particle is only interact with its immediate neighbours. It is to be noted that when the value of the radius in non-local theory becomes increasingly high, the theory will become the continuous form of the molecular dynamics (MD) model. It's apparent that the non-local theory of continuous media provides a relation between the models of local CCM and MD. In PD, discontinuities or damage can be modelled by elimination of interactions (micro potentials) among the particles [6]. The local damage of a particle is defined as the ratio of the separated connections to the total amount of connections. In essence, the theory of PD is a reformulation of the equation of motion of solid mechanics and best suited for modelling discontinuity in bodies such as cracks [1, 7, 8]. The main difference between PD theory and CCM, where the former is constructed using integral equations, as opposed to derivatives of the displacement components while PD theory is based on the partial integral equations in such a way that the PD governing equations remain valid in the presence of discontinuity such as a crack. This characteristic has brought PD model to a considerable attention and been successfully applied to brittle fracturing problems by enabling the initiation and propagation of damage at various sites with random pathways inside the material without needing to resort to particular requirements for crack formation [9]. It is therefore an ideal method to examine particular issues in the evaluation of materials and structures' failure characteristics. Because of its exceptional integro-differential feature that do not involved spatial derivatives, PD has shown to be a strong technique for capturing various forms of fracture in materials. By using PD features, Kilic et al. [9] predicted crack growth patterns in the quenched glass plates. Recently, Madenci et al. [6] studied the crack propagation in isotropic materials under different kind of complex loading conditions. Dynamic fracture has also been studied with PD for various types of structures such as polymers [10].

2 Bond-Based Peridynamics

In Bond-Based Peridynamics (BBPD) theory, the force density vectors between two particles are equal in magnitude but opposite in sign as in Fig. 1 which satisfy the equation of angular momentum. Thus, the force density of bond in BBPD is called the pairwise force function, which can be expressed as below:

$$\rho\ddot{\mathbf{u}}(\mathbf{x},\mathrm{t}) = \int_{H} \mathbf{f}\big(\mathbf{u}(\mathbf{x}',\mathrm{t}) - \mathbf{u}(\mathbf{x},\mathrm{t}), \mathbf{x}' - \mathbf{x}\big)\mathrm{dV_{\mathbf{x}}}' + \mathbf{b}(\mathbf{x},\mathrm{t}), \tag{1}$$

where H represent the neighbourhood of $\mathbf{x}$ with prescribed horizon size of δ, $\mathbf{u}$ is the displacement vector field, $\mathbf{b}$ for prescribed body force density, $\ddot{\mathbf{u}}$ is the second-order time derivative of displacement $\mathbf{u}$, ρ is the mass density, and $\mathbf{f}$ stand for the pairwise force function, exerted by the particle of $\mathbf{x}'$ onto the particle of $\mathbf{x}$. The relative

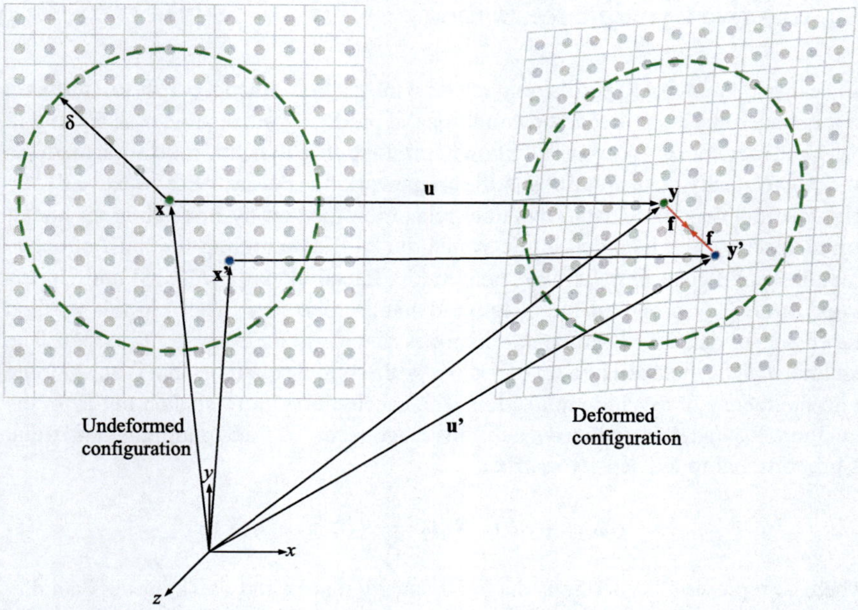

Fig. 1 Deformation of PD particles between **x** and **x**′, and developing equal and opposite in direction pairwise force densities for Bond-Based Peridynamics (BBPD)

position of the two particles in the reference configuration is ξ and η is their relative displacement [11] is given by

$$\xi = \mathbf{x}' - \mathbf{x}, \tag{2}$$

and

$$\eta = \mathbf{u}(\mathbf{x}', \mathrm{t}) - \mathbf{u}(\mathbf{x}, \mathrm{t}), \tag{3}$$

The relative position ξ is the bond that connects between two particles. Since particles only interacts within its neighbourhood, thus Eq. (1) is only works in its horizon of the particle. Moreover, s denotes the bond relative elongation, or stretch, defined as:

$$s = \frac{|\xi + \eta| - |\xi|}{\xi}. \tag{4}$$

3 Adaptive Dynamic Relaxation

Since the equations of PD utilise explicit time integration, when it comes to a problem that is under quasi-static loading conditions, it produce troublesome in analysing it. Kilic and Madenci proposed an additional formulation in dynamic relaxation methods to achieve steady-state solutions in the framework of nonlinear PD equations [12]. This method originated from dynamic relaxation method by instigating an unnatural damping to be included in the system but still determining the most effective damping coefficient is not always achievable. Papadrakakis et al. [13] and Underwood et al. [14] proposed the adaptive dynamic relaxation where in this method the change of damping coefficient is happens adaptively for every iteration step. The equation can be included into set of ordinary differential equations that is appropriate with the theory of dynamic relaxation. The acceleration term is taken out from the equation. Instead, it introduced a fictitious diagonal density and damping matrix that is proportional to the density matrix as

$$\mathbf{\Lambda}\ddot{U}(\mathbf{X},t) + c\mathbf{\Lambda}\dot{U}(\mathbf{X},\mathbf{t}) = L(\mathbf{U},\mathbf{X}) + \mathbf{b}(\mathbf{X}), \tag{5}$$

where $\mathbf{\Lambda}$ represent the fictitious diagonal density matrix and c exhibit the damping coefficient that is determine by Greschgorin's theorem [14]. $\mathbf{X}$ is initial position of the particles, and $\mathbf{U}$ is displacement of the particles. Both of the parameter can be denote as [14]

$$\mathbf{X}^T = \left\{\mathbf{x}_1, \mathbf{x}_2, \ldots, \mathbf{x}_{I_n}\right\}, \tag{6}$$

and

$$\mathbf{U}^T = \{\mathbf{u}(\mathbf{x}_1, t), \mathbf{u}(\mathbf{x}_2, t) \ldots, \mathbf{u}(\mathbf{x}_{I_n}, t)\}. \tag{7}$$

I_n stands for the total number of particles. By employing central-difference explicit integration, velocities and displacement for the next iteration can be evaluated by

$$\dot{u}^{n+\frac{1}{2}} = \left[(2 - c\Delta t)u^{n-\frac{1}{2}} + 2\Delta t\Lambda^{-1}(\mathbf{L} + \mathbf{b})^n\right]/[2 + c\Delta t], \tag{8}$$

$$u^{n+\frac{1}{2}} = u^n + \Delta t\dot{u}^{n+\frac{1}{2}}, \tag{9}$$

where n represents the nth iteration, Δt is the time step size. The best time step size in dynamic relaxation is 1, ($\Delta t = 1$) [12]. The above expression cannot be employed at the beginning of the integration because the velocity field at $t^{-1/2}$ is unknown, but the process can be commenced by using

$$\dot{u}^{\frac{1}{2}} = \Delta t\Lambda^{-1}(\mathbf{L} + \mathbf{b})^0/2. \tag{10}$$

There is no need for the density matrix, Λ, damping coefficient, c and time step size, Δt to be physically meaningful quantities [12]. It means to obtain faster convergence the value can be chosen. The most common method to choose density matrix is based on Gerschgorin's theorem, which can be written as [14]

$$\Lambda^{-1} = \frac{1}{4}\Delta t^2 \sum_j |k_{ij}|, \tag{11}$$

where i is the corresponding particle and j is the particle connected to the corresponding particle i and k_{ij} can be evaluated as [15]

$$k_{ij} = 5\pi\delta^2 bc, \tag{12}$$

in which bc is the bond constant expressed as

$$bc = \frac{18K}{\pi\delta^4}, \tag{13}$$

where K indicates the bulk modulus and δ is the radius of PD horizon. The damping coefficient c expressed as

$$c^n = 2\sqrt{((\mathbf{U}^n)^T \mathbf{K}^n \mathbf{U}^n)/((\mathbf{U}^n)^T \mathbf{K}^n)}. \tag{14}$$

$\mathbf{K}^n$ represents the diagonal local stiffness matrix, which is given as

$$K_{ii}^n = ((\mathbf{L}+\mathbf{b})_i^n/\Lambda^{-1} - (\mathbf{L}+\mathbf{b})_i^{n-1}/\Lambda^{-1})/\Delta t\dot{U}_i^{n-1/2}. \tag{15}$$

4 Damage Criterion

Prototype Miroelastic Brittle (PMB) material, introduced in [1, 16], as

$$\mathbf{f} = cs. \tag{16}$$

Parameter c is called the micro-modulus represented as the bond stiffness, and the value of c for different cases such as 1D, 2D and 3D are as below,

$$c = \begin{cases} \sqrt{\frac{12E}{\pi\delta^4}}, & \text{3D} \\ \sqrt{\frac{9E}{\pi t\delta^3}}, & \text{2D plane stress} \\ \sqrt{\frac{48E}{5\pi t\delta^3}}, & \text{2D plane strain} \\ \sqrt{\frac{2E}{A\delta^2}}, & \text{1D}, \end{cases} \tag{17}$$

where in Eq. (17), E is the Young's modulus, t is the thickness (for 2D cases), and A is the cross-sectional area (in 1D case). Failure of the material occurs when the stretch s assumes its critical value: In PD formulation, the critical stretch of bond, s_0 which is related to the strain energy release rate, is usually used as the criterion to decide when the failure of the material initiates. When the value of stretch, s reaches its critical stretch, a bond is assumed to be fail. The value of s_0 can be expressed below:

$$s_0 = \begin{cases} \sqrt{\frac{5G_{IC}}{6E\delta}}, & \text{3D} \\ \sqrt{\frac{4\pi G_{IC}}{9E\delta}}, & \text{2D plane stress} \\ \sqrt{\frac{5\pi G_{IC}}{12E\delta}}, & \text{2D plane strain}, \end{cases} \tag{18}$$

where $\boldsymbol{G_{IC}}$ in Eq. (18) denotes the value of critical energy release rate for mode I fracture. In PD the parameter $\mu(\xi, t)$ represents whether a particular bond is still active or broken. The bonds that are broken will eventually lead to a softening material response [11].

$$\mu(\xi, t) = \begin{cases} 1, & \text{active bond} \\ 0, & \text{broken bond}. \end{cases} \tag{19}$$

By using Eq. (19), a damage index of $\varphi(\mathbf{x}, t)$ is formulated to evaluate the value of damage for any specific particles, and the is denoted by

$$\varphi(\boldsymbol{x}, t) = 1 - \frac{\int_H \mu(\xi, t) dV_\xi}{\int_H dV_\xi}. \tag{20}$$

It should be noted from Eq. (20) that $\boldsymbol{\varphi} = \mathbf{0}$ and $\boldsymbol{\varphi} = \mathbf{1}$ respectively represent the undamaged state and the complete separation of a single particle from all surrounding particles within its horizon. These assumptions simplify the formulation; however, BBPD is restricted to a fixed value of Poisson's ratio of $\boldsymbol{v} = \frac{1}{4}$ for 3D and plane strain cases and $\boldsymbol{v} = \frac{1}{3}$ for plane stress case.

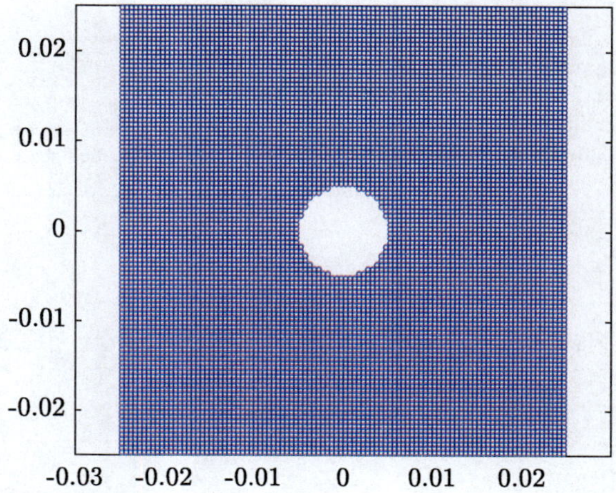

Fig. 2 Initial configuration of a plate with a hole under quasi-static stretch and its discretisation for 100 × 100

5 Numerical Simulations

In this problem, we analyse an isotropic plate with a hole at the centre under quasi-static tensile loading along its horizontal edges. There is no existing crack in its domain. The geometric parameters of the plate are as stated below; length l of 50 mm, width w of 50 mm, thickness of the plate h of 0.5 mm, and diameter of the hole D of 10 mm (Fig. 2). The material properties of the plate are as stated here with Young's modulus E of 192 GPa, Poisson's ratio ν of 1/3, and mass density ρ of 8000 kg/m^3. The boundary condition applied to the plate by applying velocity v of ±2.7541 × 10^{-7} m/s. For the first analysis, the total number of particles in the x-direction and y-direction is vary with the value of 50, 100, 200, 300 and 400, the number of particles in the y-direction is added up another 6 (3 on each side) for fictitious nodes. Spacing between particles Δ is also differs based on the particle discretisation size. The horizon δ for the first analysis is taken to be 3.015Δx. In the second analysis, we vary the value of horizon δ to be 1.015Δx, 2.015Δx, 3.015Δx, 4.015Δx and 5.015Δx. The variation of horizontal and vertical displacement are plotted for elastic loading by setting the critical stretch value, s_c to be 1. The s_c value is 0.02 when the failure is allowed to happen.

6 Results and Discussions

Figure 3 shows variation of vertical displacement along the central axes at the end of 1000 time step (ts) for elastic deformation for different particle discretisation size together with the result from Finite Element Method (FEM) using ANSYS [15], a commercially available code. Particle size discretisation that is used in this experiment is 50 × 50, 100 × 100, 200 × 200, 300 × 300 and 400 × 400. This is

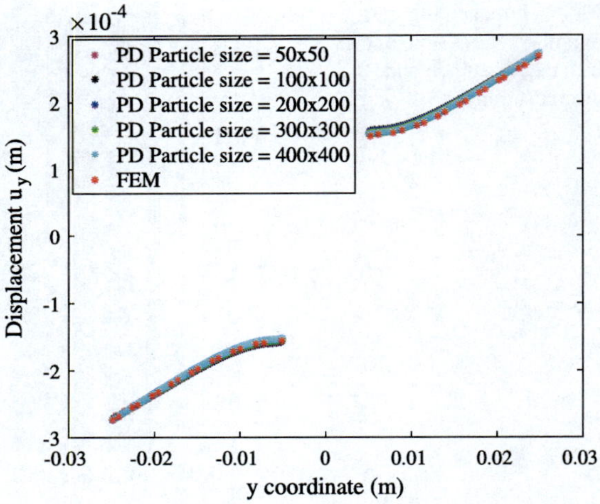

Fig. 3 Variation of y-direction displacement throughout the middle axes at ts = 1000 when no failure is allowed for different particle discretisation

to make sure that PD method can display the vertical displacement along the central axes as good as established numerical method that is FEM. Without any doubt, the result shows that the PD is a reliable method that can be used to analyse displacement in solid body because the displacement shows good agreement compared to the FEM result. Among all the particle discretisations used, particle dicretisation size of 400×400 shows the closest plot compared to the FEM plot. This proves that, the larger the particle size discretisation used (finer discretisation), the displacement in y-direction is getting closer to FEM result [17, 18]. As a conclusion, in this uniaxial loading condition, particle discretisation size helps to improve the displacement.

Figure 4 shows variation of vertical displacement along the central axes at the end of 1000 time steps with different horizon sizes together with FEM result. The

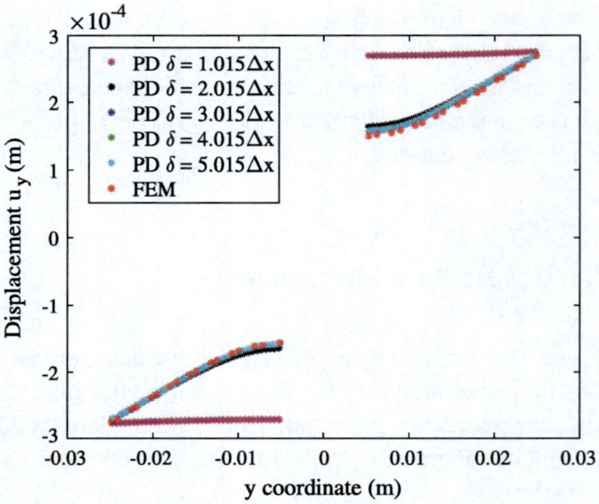

Fig. 4 Variation of y-direction displacement throughout the middle axes at ts = 1000 when no failure is allowed for different horizon size

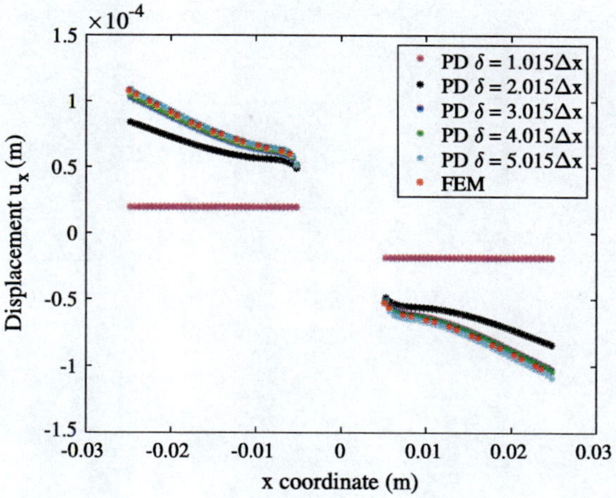

Fig. 5 Variation of x-direction displacement throughout the middle axes at ts = 1000 when no failure allowed for different horizon size

different horizon sizes that were used in this analysis were δ = 1.015Δx, 2.015Δx, 3.015Δx, 4.015Δx and 5.015Δx. It is to be noted that, in this analysis, we use the particle discretisation of 100 × 100. Generally, the larger the horizon size used, the displacement plot shows better agreement compare to the FEM displacement. The displacement plot in y-direction for horizon size equals to 3.015Δx, 4.015Δx, and 5.015Δx only shows slight difference. The vertical displacement plot for δ = 2.015Δx shows a slight diverge compared to the FEM plot approaching the hole. This is due to the effect of the hole geometry to the size of horizon that is more significance compared to far side of the hole. The displacement in y-direction when horizon size used is 1.015Δx diverted away from FEM result. This is due to the reason that the particle at the corner of any particle $\mathbf{x}_i$ is not included as its neighbour, thus it does not has sufficient interactions to properly simulate the connections between particles. This can be resolve by using a finer particle discretisation. In this example, the best horizon size is 3.015Δx.

Figure 5 shows the variation of horizontal displacement along the central axes at the end of 1000 time steps with different horizon sizes and FEM result. The different horizon sizes that were used in this analysis were δ = 1.015Δx, 2.015Δx, 3.015Δx, 4.015Δx and 5.015Δx. The closest displacement plot compare to the FEM plot is horizon 4.015Δx and 5.015Δx. When the horizon size is equal to 1.015Δx, the displacement plot shows uniform displacement for all x-coordinate, this is due to the considered particle $\mathbf{x}_i$ do not have enough connection of the particles in the neighbourhood region. This effect is not too obvious when the horizon size used are 2.015Δx and 3.015Δx since they have more than enough connections with another particles in its neighbourhood region.

We continue the analysis by allowing the damage to happen, in this case the critical stretch, s_c value is taken to be 0.02. The plot shows the crack initiation and propagation at certain time steps as mentioned for particle discretisation of 50 × 50, 100 × 100, 200 × 200, 300 × 300 and 400 × 400, and different horizon sizes of

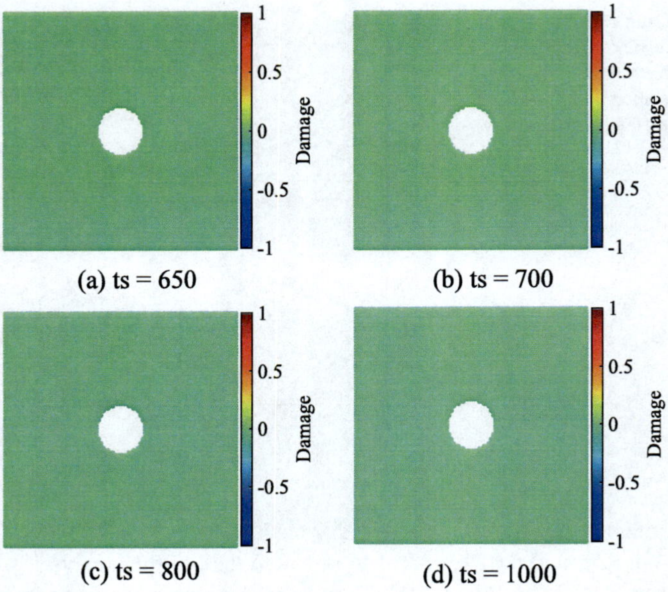

Fig. 6 Contours of damage with discretisation size of 100 × 100 and horizon size δ of 1.015Δx at time step **a** 650, **b** 700, **c** 800, and **d** 1000

1.015Δx, 2.015Δx, 3.015Δx, 4.015Δx, and 5.015Δx. The bar scale on the right of every plot shows the damage value that we acquire from Eq. (20).

Figure 6 shows damage plots for the plate with a hole with discretisation size of 100 × 100 and horizon size δ of 1.015Δx at the end of (a) 650 time steps, (b) 700 time steps, (c) 800 time steps, and (d) 1000 time steps. All the plots shows green colour which means the value of damage are all zero, so no failure happens. By using δ of 1.015Δx, we do not see any failure happens, this is due to the reason that δ of 1.015Δx is too small where there are not enough connections between the bonds that able to represent the failure [19]. Figure 7 shows damage plots for the plate with a hole with discretisation size of 100 × 100 and horizon size δ of 2.015Δx at the end of (a) 650 time steps, (b) 700 time steps, (c) 800 time steps, and (d) 1000 time steps. At time steps 650 no failure happens. Failure happened at time steps = 700, 800 and 1000 with the highest damage value of around 0.37, 0.37 and 0.9. Figure 8 shows damage plots for the plate with a hole with discretisation size of 100 × 100 and horizon size δ of 3.015Δx at the end of (a) 650 time steps, (b) 700 time steps, (c) 800 time steps, and (d) 1000 time steps. Same as Fig. 7 at time steps 650 there are no damage occurred. Damage at happened at time steps 700, 800, and 1000 with maximum damage value of 0.58, 0.62 and 0.8 respectively. It can be seen that the damage pattern at time steps 700, 800 and 1000 show a more uniform pattern compared to the same time steps in Fig. 7. Figure 9 shows damage plots for the plate with a hole with discretisation size of 100 × 100 and horizon size δ of 4.015Δx at the

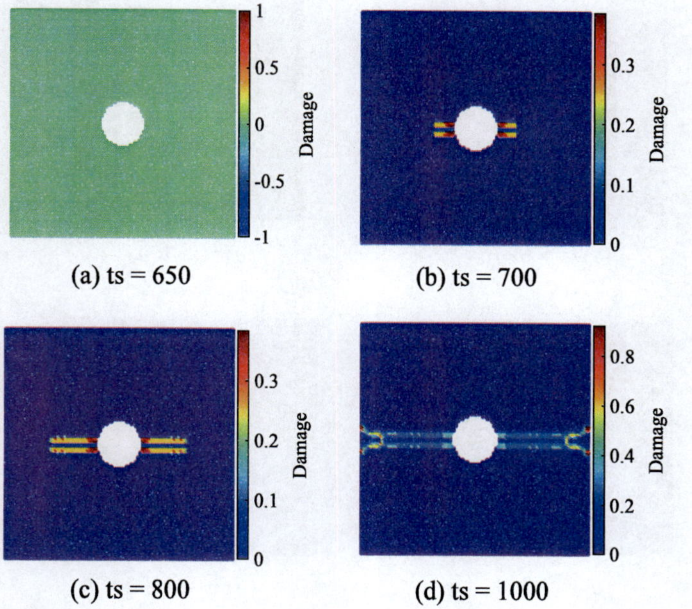

Fig. 7 Contours of damage with discretisation size of 100 × 100 and horizon size δ of 2.015Δ**x** at time step **a** 650, **b** 700, **c** 800, and **d** 1000

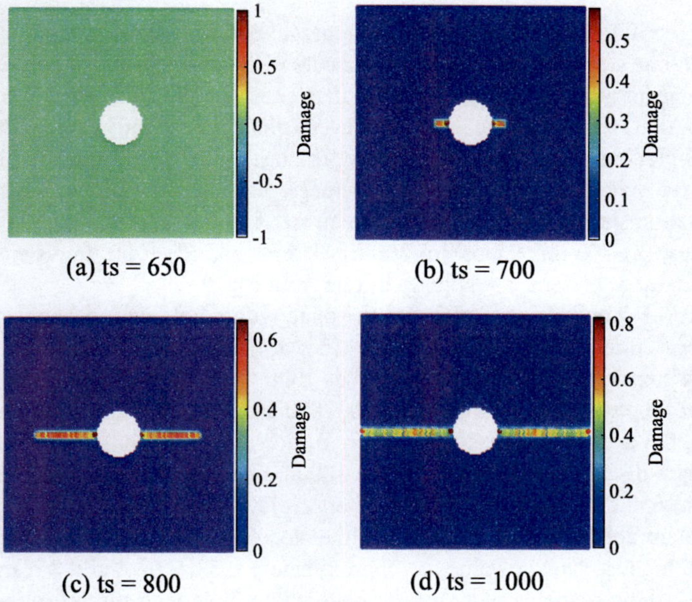

Fig. 8 Contours of damage with discretisation size of 100 × 100 and horizon size δ of 3.015Δ**x** at time step **a** 650, **b** 700, **c** 800, and **d** 1000

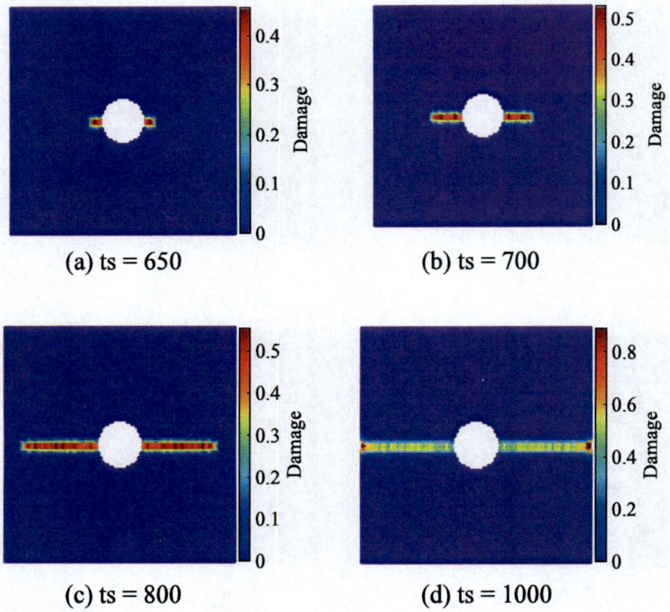

Fig. 9 Contours of damage with discretisation size of 100 × 100 and horizon size δ of 4.015Δx at time step **a** 650, **b** 700, **c** 800, and **d** 1000

end of (a) 650 time steps, (b) 700 time steps, (c) 800 time steps, and (d) 1000 time steps. At time step 650, damage already occurs with damage value of approximately 0.45, it means damage occurred earlier than previous plot that used horizon size of 3.015Δx. Overall damage patterns in every plot show slightly same pattern with previous plot but with increased intensity. The last figure in this analysis of varying the horizon size is Fig. 10, that shows damage plots for the plate with a hole with discretisation size of 100 × 100 and horizon size δ of 5.015Δx at the end of (a) 650 time steps, (b) 700 time steps, (c) 800 time steps, and (d) 1000 time steps. In this plot the damage patterns are slightly differs from Fig. 9.

Figure 11 shows damage plots for the plate with a hole with discretisation size of 100 × 100 and horizon size δ of 3.015Δx at the end of (a) 650 time steps, (b) 700 time steps, (c) 800 time steps, and (d) 1000 time steps. There are no damage appear at time step 650. At time steps 700, 800 and 1000 the highest damage value are 0.22, 0.6 and 0.75 respectively. Figure 8 shows damage plots for the plate with a hole with discretisation size of 100 × 100 and horizon size δ of 3.015Δx at the end of (a) 650 time steps, (b) 700 time steps, (c) 800 time steps, and (d) 1000 time steps. There are no damage appear at time step 650. At time steps 700, 800 and 1000 the highest damage value are approximately 0.58, 0.62 and 0.8 respectively. Figure 12 shows damage plots for the plate with a hole with discretisation size of 200 × 200 and horizon size δ of 3.015Δx at the end of (a) 650 time steps, (b) 700 time steps, (c) 800 time steps, and (d) 1000 time steps. Damage appears at time steps

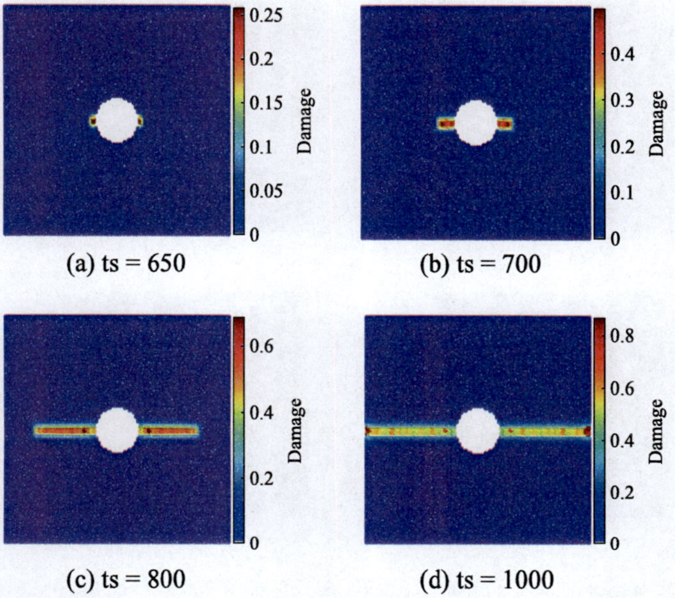

Fig. 10 Contours of damage with discretisation size of 100 × 100 and horizon size δ of 5.015Δx at time step **a** 650, **b** 700, **c** 800, and **d** 1000

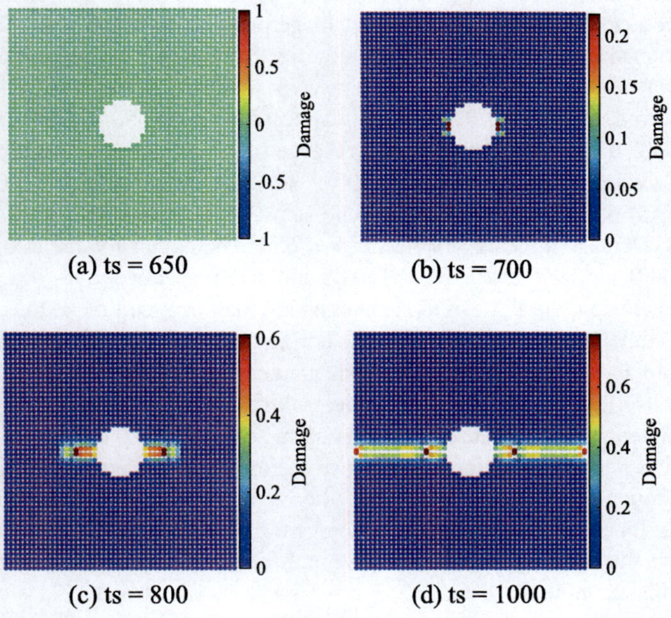

Fig. 11 Contours of damage with discretisation size of 50 × 50 and horizon size δ of 3.015Δx at time step **a** 650, **b** 700, **c** 800, and **d** 1000

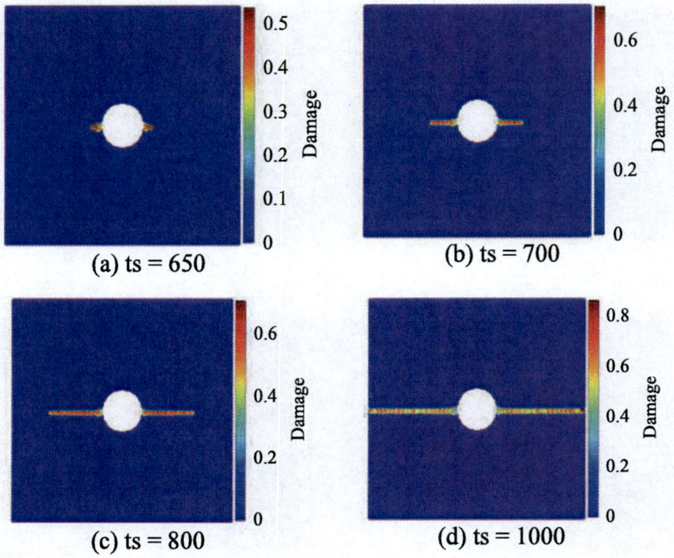

Fig. 12 Contours of damage with discretisation size of 200 × 200 and horizon size δ of 3.015Δx at time step **a** 650, **b** 700, **c** 800, and **d** 1000

= 650. The damage value at time steps 650, 700, 800 and 1000 are 0.52, 0.7, 0.7 and 0.85 respectively. Figure 13 shows damage plots for the plate with a hole with discretisation size of 300 × 300 and horizon size δ of 3.015Δx at the end of (a) 650 time steps, (b) 700 time steps, (c) 800 time steps, and (d) 1000 time steps. Damage appears at time steps = 650. The damage value at time steps 650, 700, 800 and 1000 are 0.72, 0.72, 0.81 and 0.85 respectively. Figure 14 shows damage plots for the plate with a hole with discretisation size of 400 × 400 and horizon size δ of 3.015Δx at the end of (a) 650 time steps, (b) 700 time steps, (c) 800 time steps, and (d) 1000 time steps. Damage appears at time steps = 650. The damage value at time steps 650, 700, 800 and 1000 are 0.75, 0.81, 0.85 and 0.85 respectively.

As s conclusion, the PD horizon cannot be so small as it will make the computations to become intractable [19], when the horizon size is increased, damage pattern will be more detailed. The damage also will start earlier. It goes back to the meaning of horizon itself, where the particles connect with more neighbours around it, which allows more connections that represented failure accurately. But the price that needs to be paid is it is computationally expensive. For this small plate, the time taken increase about 50% when we increase the horizon size from 4.015Δx to 5.015Δx. Meanwhile for particle discretisation, when we increase the particle discretisation size means a finer discretisation, it will give more details damage pattern up to certain value, in this example up to Fig. 12. From Figs. 8, 9, 10, 11 and 12, we could see a better and clearer damage pattern. But from Figs. 13 and 14, the damage pattern might be too sensitive as can be seen where the initiation of the damage has two paths instead of single one and the behaviour shown as if an impact loading has been

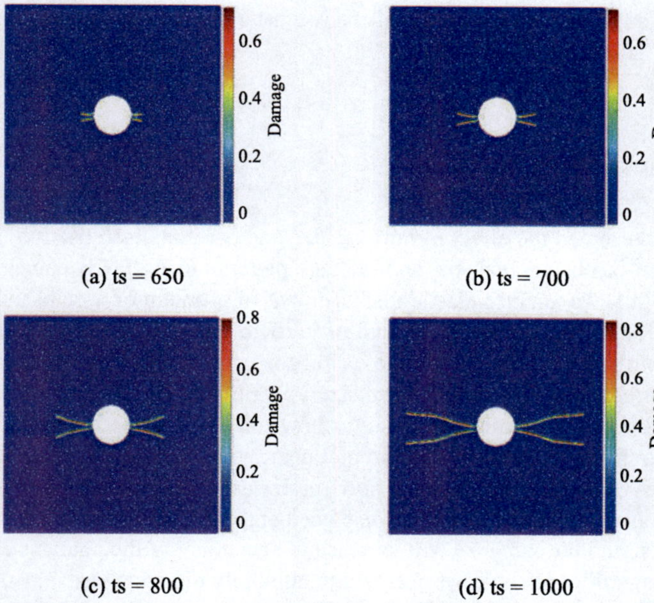

Fig. 13 Contours of damage with discretisation size of 300 × 300 and horizon size δ of 3.015Δx at time step **a** 650, **b** 700, **c** 800, and **d** 1000

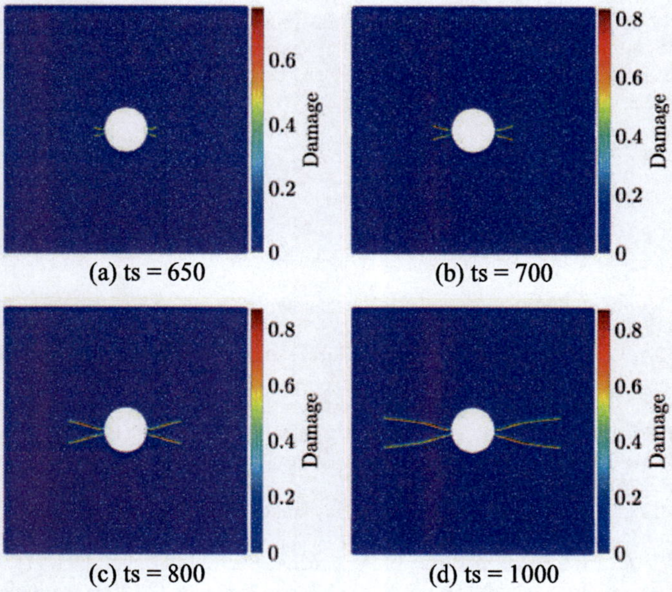

Fig. 14 Contours of damage with discretisation size of 400 × 400 and horizon size δ of 3.015Δx at time step **a** 650, **b** 700, **c** 800, and **d** 1000

applied not quasi-static. From the analysis, we have found that the best horizon size is 4.015Δx as recommended by [20, 21].

7 Conclusions

This paper presented the effect of horizon size and particle discretisation on BBPD displacement of elastic analysis, and damage patterns using PMB damage model. To achieve this, we analyse elastic displacement of problem of a plate-with-a-hole under tensile loading for varying value of particle discretisation size and horizon size. The findings from this study are useful for understanding the relationship of particle discretisation size and horizon size towards the displacement and damage pattern in BBPD. Our simulation results show that (i) finer particle discretisation size helps to improve the displacement plot nearer to the FEM result in the loading direction; (ii) Generally, the larger the horizon size used, the displacement plot shows better agreement compare to the FEM displacement, (iii) Increasing both particle size discretisation and horizon size will increase the accuracy of the analysis especially in the loading direction, however it is computationally more expensive; (v) In terms of damage, the higher the value of particle discretisation the more detail the damage plot produced.

Acknowledgements The authors are grateful to Ministry of Education Malaysia: FRGS/1/2017/TK05/UMP/01/1 and Universiti Malaysia Pahang (UMP) for providing research grants PGRS2003140. I would like to express my appreciation to Prof. Dr.-Ing. Nik Abdullah Nik Mohamed, Dr. Mohd Ruzaimi Bin Mat Rejab and Dr.-Ing. Olaf Weckner, who supported and supervised me in understanding about Peridynamics.

References

1. Silling SA (2000) Reformulation of elasticity theory for discontinuities and long-range forces. J Mech Phys Solids 48(1):175–209
2. Silling SA, Epton M, Weckner O, Xu J, Askari E (2007) Peridynamic states and constitutive modeling. J Elast 88(2):151–184
3. Eringen AC, Edelen DGB (1972) On nonlocal elasticity. Int J Eng Sci 10(3):233–248
4. Eringen AC (1972) Nonlocal polar elastic continua. Int J Eng Sci 10(1):1–16
5. Eringen AC (1972) Linear theory of nonlocal elasticity and dispersion of plane waves. Int J Eng Sci 10(5):425–435
6. Madenci E, Colavito K, Phan N (2016) Peridynamics for unguided crack growth prediction under mixed-mode loading. Eng Fract Mech 167:34–44
7. Silling SA (2010) Linearized Theory of Peridynamic States. J Elast 99(1):85–111
8. Warren TL, Silling SA, Askari A, Weckner O, Epton MA, Xu J (2009) A non-ordinary state-based peridynamic method to model solid material deformation and fracture. Int J Solids Struct 46(5):1186–1195
9. Kilic B, Madenci E (2009) Prediction of crack paths in a quenched glass plate by using peridynamic theory. Int J Fract 156(2):165–177

10. Ha YD, Bobaru F (2011) Characteristics of dynamic brittle fracture captured with peridynamics. Eng Fract Mech 78(6):1156–1168
11. Nadimi S, Miscovic I, McLennan J (2016) A 3D peridynamic simulation of hydraulic fracture process in a heterogeneous medium. J Petrol Sci Eng 145:444–452
12. Kilic B, Madenci E (2010) An adaptive dynamic relaxation method for quasi-static simulations using the peridynamic theory. Theoret Appl Fract Mech 53(3):194–204
13. Papadrakakis M (1981) A method for the automatic evaluation of the dynamic relaxation parameters. Comput Methods Appl Mech Eng 25(1):35–48
14. Underwood P (1983) Computational methods for transient analysis 1. In: Dynamic relaxation, pp 245–265
15. Madenci E, Oterkus E (2014) Peridynamic theory and its applications, vol c. Springer, New York, NY
16. Lee J, Liu W, Hong J (2016) Impact fracture analysis enhanced by contact of peridynamic and finite element formulations. Int J Impact Eng 87:108–119
17. Caimmi F, Haddadi E, Choupani N, Marano C, Andena L (2016) Modelling mixed-mode fracture in poly(methylmethacrylate) using peridynamics. Proc Struct Integrity 2:166–173
18. O'Grady J, Foster J (2014) Peridynamic beams: a non-ordinary, state-based model. Int J Solids Struct 51(18):3177–3183
19. Seleson P, Du Q, Parks ML (2016) On the consistency between nearest-neighbor peridynamic discretizations and discretized classical elasticity models. Comput Methods Appl Mech Eng 311:698–722
20. Cheng Z, Zhang G, Wang Y, Bobaru F (2015) A peridynamic model for dynamic fracture in functionally graded materials. Compos Struct 133:529–546
21. Wang X, Kulkarni SS, Tabarraei A (2019) Concurrent coupling of peridynamics and classical elasticity for elastodynamic problems. Comput Methods Appl Mech Eng 344:251–275

The Effect of Laser Power and Laser Scan Passes on Bending Angle of Stainless Steel AISI 304 Laser Bending

N. Affaf, H. S. Wong, M. H. Aiman, M. Ishak, and M. M. Quazi

Abstract The application of laser bending of thin sheet metal are quite limited given that this forming process is widely used in various industries. The aim of this paper is to study the effect of laser power and laser scan passes on bending angle of stainless steel AISI 304 lasers bending. This study considers using parametric combinations of laser power and laser scan passes (number of loops) as the laser bending process parameters. The corresponding effects of these process parameters on bending angle of stainless-steel thin sheet metal ($t = 0.1$ mm) was observed. The results obtained shows that the bending angle increased with increased in laser power and laser scan passes, meaning that these parameters does significantly affect the bending angle. Other than bending angle, there were also a few bulges observed at certain location along the laser scan path for certain parametric combination of process parameters tested.

Keywords Laser bending · Laser power · Laser scan passes · Bending angle · Stainless steel

1 Introduction

Laser bending is a contactless process for sheet bending of ductile metals, in which a laser beam heats specific zones of material, generating thermal stresses that exceed its yield strength and inducing plastic deformation that finally bends the metallic sheet [1]. The unexpected increase of research reported on this topic in recent literature demonstrates the clear promise of laser-based techniques in processing metals. [2]. In thin sheets metal forming, laser is a better option [3]. Changdar and Chakraborty [2] in 2021 stated in their recent review article that the advantages of laser bending forming are its flexibility, the lack of need for hard tooling, and the ability to form both hard and brittle materials.

N. Affaf · H. S. Wong · M. H. Aiman (✉) · M. Ishak · M. M. Quazi
Faculty of Mechanical & Automotive Engineering Technology (FTKMA), Universiti Malaysia Pahang, 26600 Pekan, Pahang, Malaysia
e-mail: aimanmh@ump.edu.my

M. Y. Ismail et al. (eds.), *Technological Advancement in Mechanical and Automotive Engineering*, Lecture Notes in Mechanical Engineering,
https://doi.org/10.1007/978-981-19-1457-7_69

In general, there has been many research done regarding the application of multiple laser scan bending or forming on sheet metal. In 2021, Shi et al. [4] used a low heating temperature to investigate laser thermal bending under pre-load and discovered that multiple scanning of a low-temperature laser improves surface quality while also allowing for a larger bending angle. Another attempt by Thomsen et al. [5] in was also done in 2020 to study the rate of change in bending angles has led to a hypothesis that the thermal expanded area is limited not only by the surrounding material, but also by deformation caused by the laser-formed part of the laser scan path. Due to the additional forces restricting the thermal expansion, this could result in a larger total bend angle. Since these effects do not occur at the laser scan path's trailing edge, another possibility is changes in bending angle along the laser scan path. However, Thomsen et al. [5] stated that this hypothesis requires additional research before anything can be concluded.

Other than these experimental research done, there was a prediction model and experimental validation done by Mulay et al. [6] in 2021 regarding the bending angle on multiple laser scan passes on 304 stainless steel. They consider the temperature dependent properties of the sheet material to evaluate the temperature profile across the thickness of the sheet. They conclude that the variation in the bend angle per laser scan for 304 stainless steel sheets processed with different laser parameters was found to depend laser power and scanning speed, the key parameters affecting the results of multi-scan laser forming.

In this paper, the aim is to study the effect of laser power and laser scan passes on the bending angle of the stainless-steel thin sheet metal with the thickness of 0.1 mm. What made this paper important is that this study was done using a very thin sheet metal and using low power fibre laser machine, as there are not many studies regarding this, according to our findings. At the end of the experiment, the size of the bending angle was observed and the images were captured to compare the differences of bending angle in regards to the process parameters tested. The results were also tabulated in graph plot to observe the significant effect of these process parameters on final bending angle and average bending angle.

2 Experimental Procedures

2.1 Material and Sample Preparation

AISI 304 stainless steel thin sheet of 40 × 10 × 0.1 mm was used in this study. Three samples of each parametric combination were prepared keeping the sheets orientation for all samples are the same as to avoid any effects of sheet anisotropy. The samples were cut using a cutter and cleaned using acetone to remove any grease of oil that might affect the heat transfer. The sheet has a chemical composition of Fe (68%), Cr (19%), Ni (9%), Mn (2%), C (0.08% max), P (0.045% max), Si (1% max).

2.2 *Experimental Setup*

The low laser power tests were conducted using a fiber laser (IPG YLM 200/2000) with a maximum power of 27 W and wavelength of 1070 nm. See Fig. 1 which shows the schematic diagram of sample workpiece with clamping jig. A simple jig clamp was used to fix the sample workpiece 20 mm from one end and free the other end. The laser scan line with the length of 14 mm was performed across the workpiece 10 mm from the free end of the workpiece. A continuous mode operated laser beam with the beam diameter of 50 µm was irradiated over the top sheet surface along the sheet width to complete one laser scan. Different combination of laser scan passes (4, 8, 12, 16 and 20 loops) and laser power (21, 24 and 27 in Watts) were tested with constant scanning speeds of 41.67 mm/s and the focal length of 132 mm, as shown in Table 1.

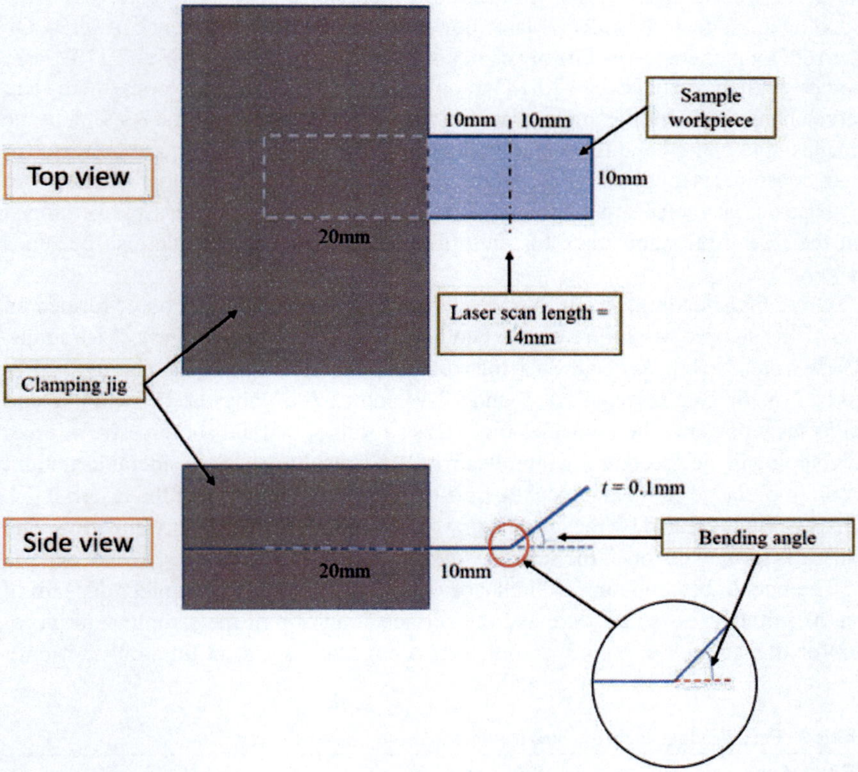

Fig. 1 Schematic diagram of jig clamp and sample workpiece

Table 1 Laser process parameters used in this study

Parameters	Variations
Laser power (W)	21, 24, 27
Laser scan passes	4, 8, 12, 16, 20
Laser speed (mm/s)	41.67
Focus length (mm)	132

3 Results and Discussion

The final bending angle results for all laser parametric combination were shown in Tables 2 and 3 presents the changes in bending angle in close-up images for each parametric combination of 3 different laser powers and 5 different laser scan passes (loops). The laser scan passes were performed along the same laser irradiation track for 5 different loops. On the 4th loop test, there were only small changes of the bending angle from 21 to 27 W laser powers tested with the difference of 1.19°. On the 16th loop, there were quite big changes in bending angle for 24 W and 27 W laser powers with the difference of 7.518° among others. From 12 to 20th loops, there were large changes in bending angle occurred with different laser powers. As seen in the images below, the bending of the workpiece was observed to be quite uniform for both constant laser power and laser scan passes. It can also be seen that increasing in laser power promotes larger bending angle. Also, as the laser scan passes overlapped on the laser irradiation track for multiple times, the bending angle also becomes larger.

Other than bending, we also noticed in the images that there is a bulge formed on top of the sample workpiece on the bending area at 27 W laser power at 8th loops. There were also bulges occurred at the parametric combination of 12th loops at 24 W and 27 W, of 16th loops at 24 W and 27 W and of 20th loops at 21 W, 24 W and 27 W laser powers. These buckles formed as a result of thermal compressive stresses developing in the sheet during laser beam heating, resulting in a considerable amount of thermo-elastic strain and local thermo-elastic-plastic buckling of the material [7]. This buckle is created by the laser beam scanning moving in the same direction. The buckle is formed on top of the sample workpiece surface when the laser beam departs it. The buckle occurs is uncertain shapes and locations and it is largely depending on the initial stress/strain state and the surface condition of the sample workpiece. As for this study, the bulges do not form along the laser scan line, rather it only

Table 2 Final bending angle for each parametric combinations tested

Final bending angle (°)		Laser scan passes				
		4	8	12	16	20
Laser power (W)	21	5.884	9.869	15.62	17.79	24.32
	24	7.074	12.177	18.38	23.116	30.547
	27	8.618	17.151	23.244	30.634	36.305

Table 3 Close up images of final bending angle formed

Laser Power (W) / Laser Scan Passes	21	24	27
4	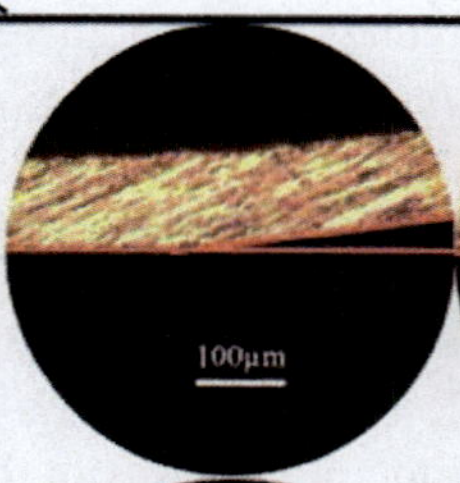	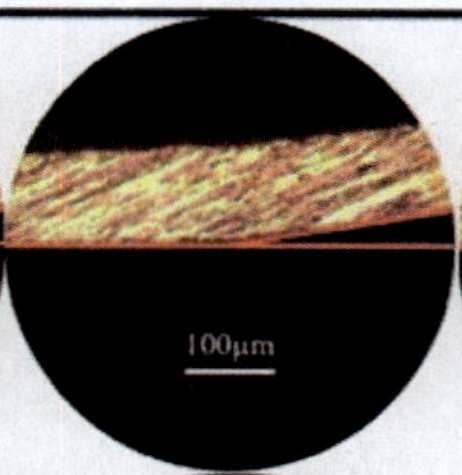	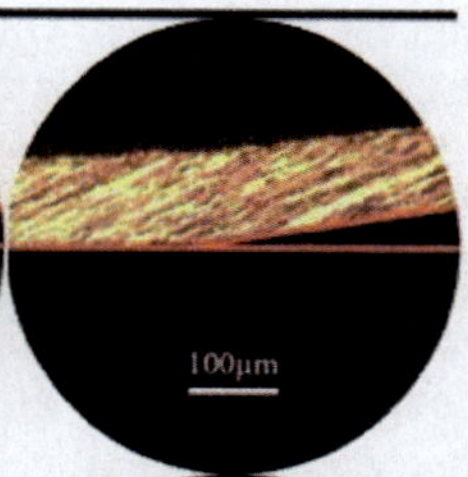
8	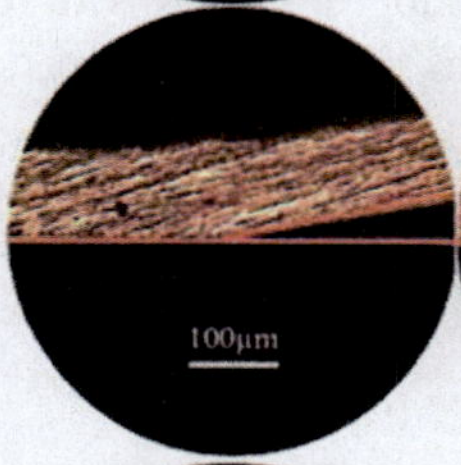	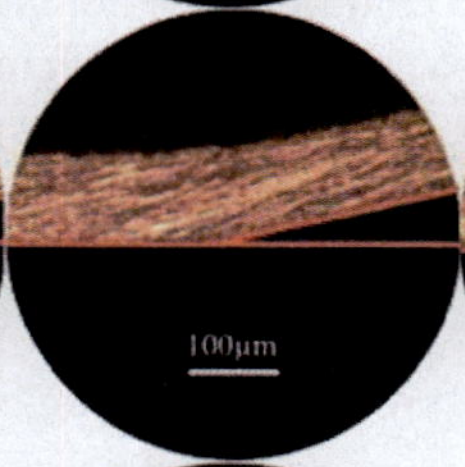	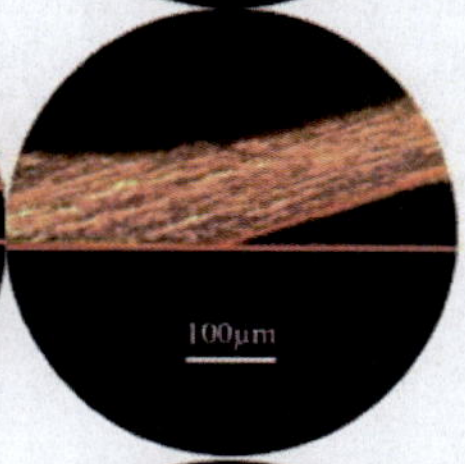
12	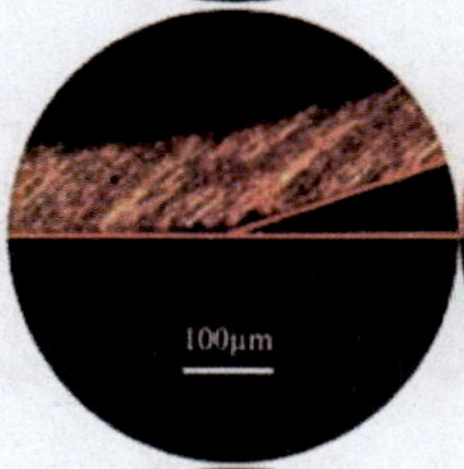	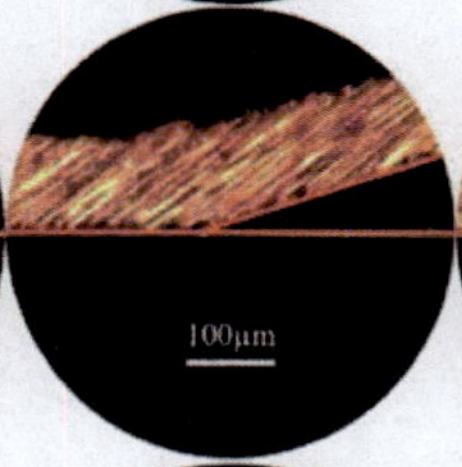	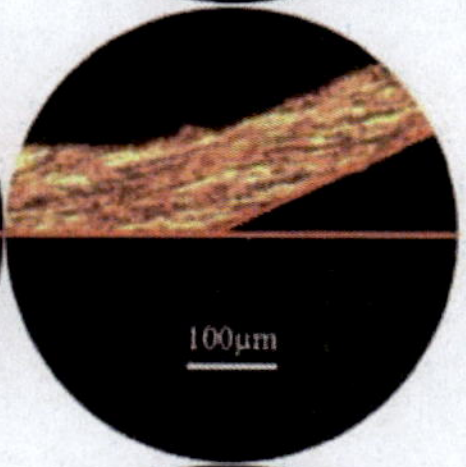
16	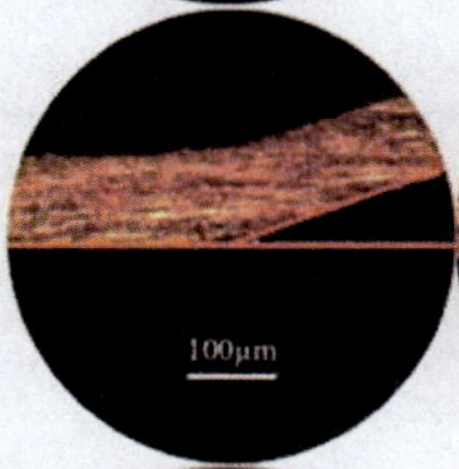	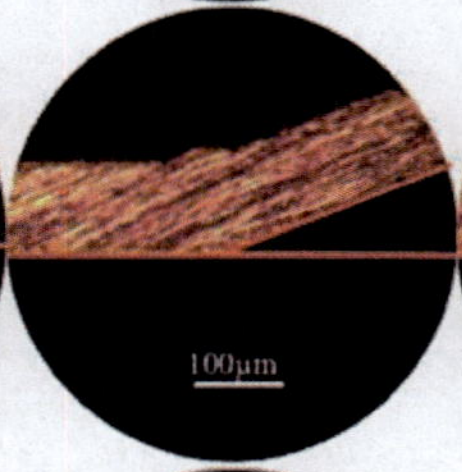	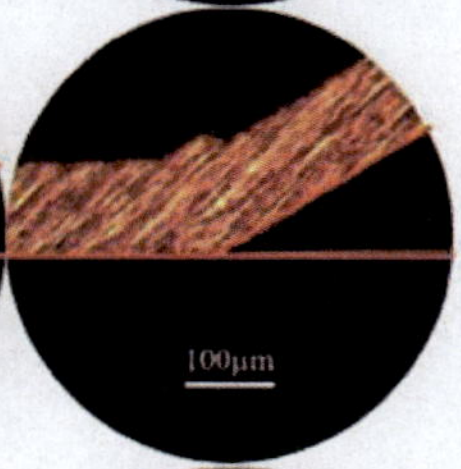
20	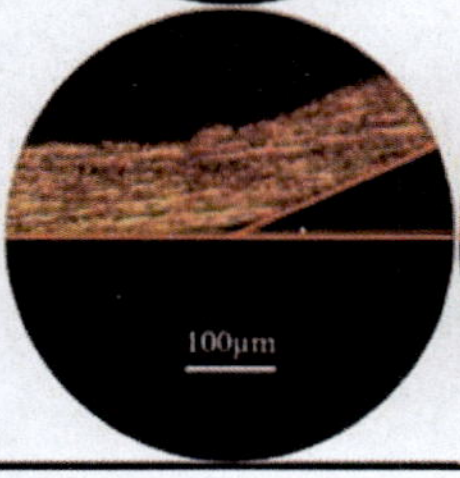	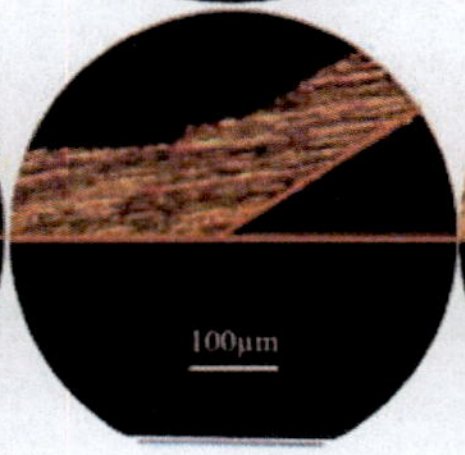	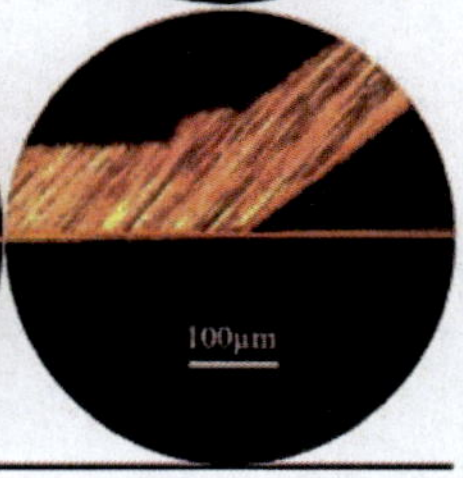

formed on certain locations with different sizes of bulges. Hence, the bulges cannot be measured accurately.

All the final bending angle reading were converted into the graph in Figs. 2 and 3. Figure 2 above shows the direct relationship between the laser power and bending angles. On the 4th loop, the graph line pattern is linearly increase from lowest to highest laser power. On the 8th loop, there is quite significant increase (40.85%) of the graph line pattern from 24 to 27 W, as compared to the line from 21 to 24 W (18.95%). There is also a significant increase for graph line on 16th (72.2%) and 20th (49.28%) loops, from the lowest to the highest laser power. Hence based on the graph in Fig. 2, we can observe that increasing in laser power does leads to higher bending angle. Then Fig. 3 shows the direct relationship between laser scan passes (loops) and bending angle. For the constant laser power of 21 W, the graph line is increase significantly but not linearly from 8 to 16th loops with the percentage of 80.26%. For constant laser power of 24 W and 27 W, the graph pattern is increasing in quite linear and smooth pattern.

Table 4 presents the average bending angle for every parametric combination. The average bending angle was obtained by using the formula in Eq. 1 below,

$$Average\,bending\,angle = \frac{(a_n)}{n} \tag{1}$$

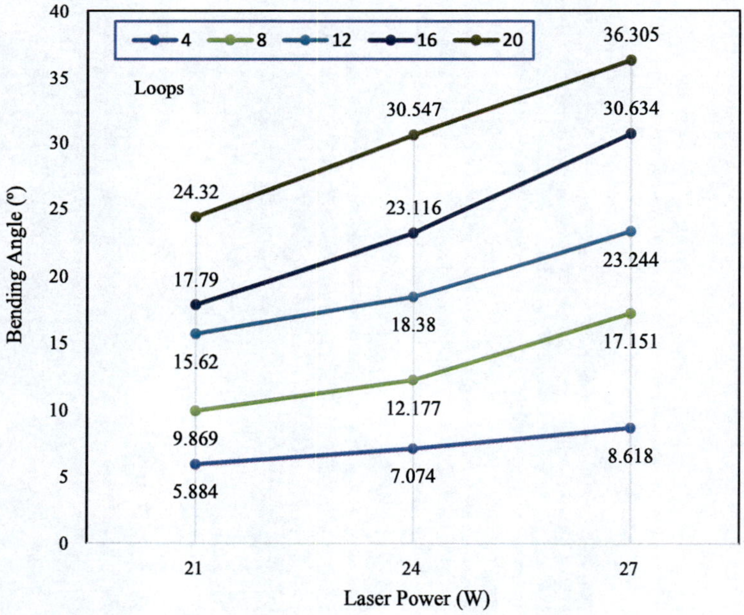

Fig. 2 Effect of laser power on bending angle formed

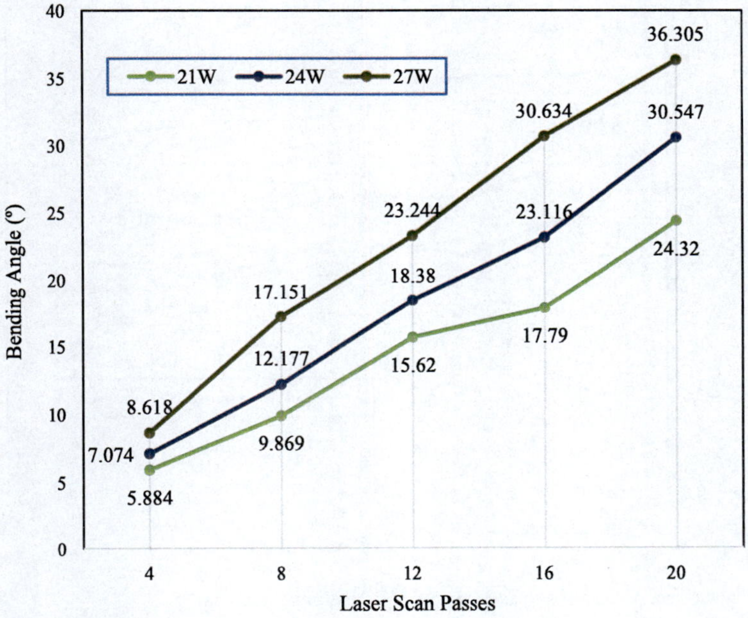

Fig. 3 The effect of laser scan passes on bending angle formed

Table 4 Average bending angle for each parametric combination

Average bending angle (°)		Laser scan passes				
		4	8	12	16	20
Laser power (W)	21	1.471	1.234	1.302	1.112	1.216
	24	1.769	1.522	1.532	1.382	1.527
	27	2.155	2.144	1.937	1.915	1.815

where a is the total bending angle after n number of passes of laser scanning. The average bending angle is the central value of angle obtained based on the number of loops performed. For example, for 16th number of loops, the bending angle for one loop using the laser power of 21 W, 24 W and 27 W is 1.112°, 1.382° and 1.915°, respectively. See Fig. 4 below which presents the average bending angle in plotted graph. For laser power of 21 W, there is a decrease in graph line from 4th loops to 8th loops and 12th loops to 16th loops. However, there are also an increase in graph line for 8th loops to 12th loops and 16th loops to 20th loops. The same pattern also occurred for 24 W laser power tested for all number of loops. As for 27 W, the graph line is in decrease pattern from the lowest to the highest number of loops. Most average bending angle were decreased as the number of loops increased.

Increasing in bending angle with increasing in laser power explains that the laser power variations decide the heat input during the laser bending process. Increasing

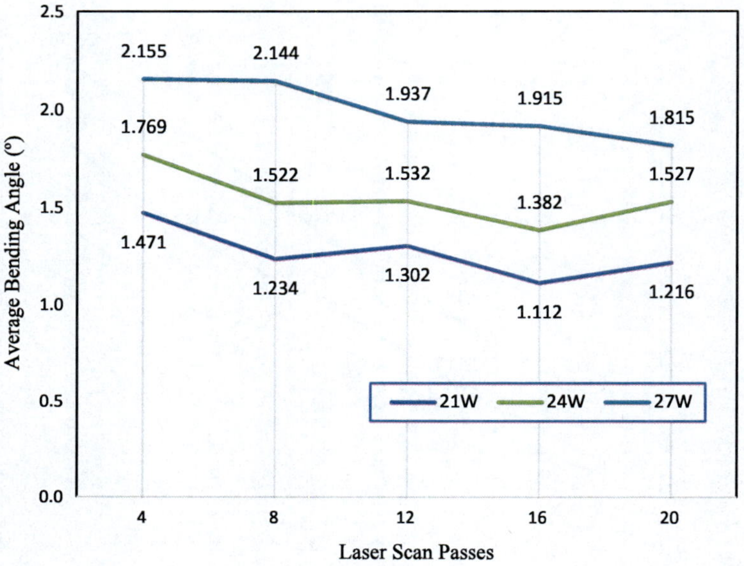

Fig. 4 The effect of laser scan passes on average bending angle

the laser power means increasing the heat input which leads to increasing the thermal gradient in the direction of the sheet. Thus, the more plastic strain occurring which leads to larger bending angle of the sample workpiece. The effect of laser power on the bending angle can be clearly seen in Table 2. Other than laser power, it can also be seen in Table 2 that the laser scan passes also affected the changes of the bending angle. As the number of laser scan passes on the sample workpiece increased, the bending angle is also getting increased (see Fig. 3), but the average bending angle became decreased in certain locations (see Fig. 4).

There are numbers of reason that could affect the reduction in average bending angle such as, strain hardening, thermal and geometric effects due to multiple laser scan irradiations [8]. During the laser bending process, the material of the sample workpiece gets plastically deformed due to the thermal stress induced by the laser beam. With each laser scan passes to the surface of the workpiece, the bending becomes larger and the average bending angle becomes smaller, and while it produces some bending, it also leads to the increase of the density dislocation and also reduce the density energy. As the results, the sample workpiece gets strain hardened, which increases the hardness of the laser bended area, the bending per scan would reduce with an increase number of scans, resulting in the decrease of the average bending angle.

4 Conclusion

In this paper, the effect of parametric combination of laser power and laser scan passes on stainless steel AISI 304 thin sheet metal laser bending was studied. Based on the results presented, it can be concluded that;

1. Increase in laser power significantly promotes larger bending angle.
2. Increase in laser scan passes significantly promotes larger bending angle.
3. Increase in parametric combination of laser power and laser scan passes significantly increase the bending angle.
4. Increase in laser scan passes significantly promotes smaller average bending angle.

Acknowledgements The authors would like to thank the Ministry of Higher Education for providing financial support under Fundamental Research Grant Scheme (FRGS-RACER) No. RACER/1/2019/TK03/UMP//3 (University reference RDU192608), and Universiti Malaysia Pahang for laboratory facilities as well as additional financial support under Internal Research grant RDU1903118 and PGRS210350.

References

1. Magee J, Watkins KG, Steen WM (1998) Advances in laser forming. J Laser Appl 10(6):235–246. https://doi.org/10.2351/1.521859
2. Changdar A, Chakraborty SS (2021) Laser processing of metal foam—a review. J Manuf Processes 61:208–225. https://doi.org/10.1016/j.jmapro.2020.10.012
3. Nath AK et al (2019) 15 ChAPtEr LASEr-Based manufacturing as a green manufacturing process. Sustain Mater Form Join 303
4. Shi Y, Guo Y, Wang X, Sun R, Li X (2021) Laser bending angle and surface quality with preload at low heating temperature. Opt. Laser Technol 136(November):106755.: https://doi.org/10.1016/j.optlastec.2020.106755
5. Thomsen AN, Kristiansen M, Kristiansen E, Endelt B (2020) Online measurement of the surface during laser forming. Int J Adv Manuf Technol 107(3–4):1569–1579. https://doi.org/10.1007/s00170-020-04950-6
6. Mulay S, Paliwal V, Babu NR (2021) Analytical approach to predict the bend angle of sheet formed by multiple laser scans. Procedia CIRP 99:272–277. https://doi.org/10.1016/j.procir.2021.03.040
7. Hu Z, Kovacevic R, Labudovic M (2002) Experimental and numerical modeling of buckling instability of laser sheet forming. Int J Mach Tools Manuf 42(13):1427–1439. https://doi.org/10.1016/S0890-6955(02)00075-5
8. Maji K, Pratihar DK, Nath AK (2016) Experimental investigations, modeling, and optimization of multi-scan laser forming of AISI 304 stainless steel sheet. Int J Adv Manuf Technol 83(9–12):1441–1455. https://doi.org/10.1007/s00170-015-7675-0

A Short Review on Grain Refinement Techniques in Semisolid Metal Processing

M. A. Shakirin, N. A. Abd Razak, and A. H. Ahmad

Abstract Since 40 years ago, several methods and techniques in semisolid metal processing (SSMP) for forming globular microstructures have been found. SSMP is a relatively new technology that occurs between the liquidus and solidus temperatures. One of the common techniques used to improve microstructure formation in the aluminium processing industry is grain refinement. The purpose of grain refinement is to refine the size of the grain structure and enhance the mechanical properties. Grain refinement is also utilised in SSMP to improve the formation of globular microstructures. This article intends to discuss various approaches employed in the laboratory or in the industry in recent years to produce globular microstructure feedstock for SSMP, particularly with grain refinement technique.

Keywords Semisolid metal processing · Globular microstructure · Grain refinement

1 Introduction

Semisolid metal processing (SSMP) is a manufacturing technique involving processing between liquidus and solidus temperatures. SSMP offers significant advantages because it combines various characteristic features of typical high pressure die castings, such as thin walls component production, excellent surface quality, close dimensional accuracy and excellent mechanical properties [1]. Grain refinement is one of the techniques used to refine microstructure formation. The main purpose of grain refinement is to change the morphology, size of microstructure and improve mechanical properties. This article summarises several recent laboratories

M. A. Shakirin · N. A. Abd Razak (✉) · A. H. Ahmad
Department of Mechanical Engineering, College of Engineering, Universiti Malaysia Pahang, Lebuhraya Tun Razak, 26300 Kuantan, Pahang, Malaysia
e-mail: asnul@ump.edu.my

A. H. Ahmad
Centre for Automotive Engineering, Universiti Malaysia Pahang, 26600 Pekan, Pahang, Malaysia

M. Y. Ismail et al. (eds.), *Technological Advancement in Mechanical and Automotive Engineering*, Lecture Notes in Mechanical Engineering,
https://doi.org/10.1007/978-981-19-1457-7_70

and industrial approaches using the grain refinement technique to produce semisolid metal feedstock.

1.1 Semisolid Metal Processing

The SSMP is a promising advanced forming technique, which originated from the experimental discovery by the research group at the Massachusetts Institute of Technology (MIT) in the 1970s [2]. The SSMP involves the processing of alloys between the solidus and liquidus temperatures. SSMP is a technique in which an alloy is injected into a die as a slurry of near-globular primary particles in a liquid matrix, enabling the manufacture of near-net-shape components [2]. The SSMP technique is used in several applications since its high-quality components are at the same cost or cheaper than traditional forming methods such as casting or forging. SSMP benefits include heat resistance, excellent surface quality, good dimensional accuracy, the potential to build complicated components, and a long mould lifespan compared to conventional casting processes [3].

The SSMP can be divided into two broad categories, which are rheo and thixo routes. Rheo-cast is a multi-step procedure that involves partial melting of the material into a liquid-like slurry. Globular microstructures are obtained through the cooling processes of the material into a semisolid state [4]. The semisolid slurry is transferred to the shot sleeve of a die casting machine when the metal has been cooled to the suitable temperature and injected into the die cavity [5]. The reheating and forming steps are required in the thixo-forming process. The thixoforming involves with preparing the globular microstructure semisolid metal feedstock billet, cutting the material to the desirable length, reheating it into the semisolid state range for the desired fraction solid, and forming it [6]. Thixoforming is also known as near net shape forming, the process of reheating a billet to a solid-liquid temperature and compressing a partially melted non-dendritic slug into a mould [7].

1.2 Globular Microstructure

Globular microstructures with a large transition area between solidus and liquidus are the requirement for SSMP. The SSMP also contains a primary phase globular microstructure with a suitable quantity of secondary phase liquid fractions. The near-globular particles flow freely through one another when shear pressure is applied with the viscosity drops, and the substance behaves like a liquid [9]. Mechanical stirring, magnetohydrodynamic stirring, spray casting, ultrasonic treatment, grain refinement, thermomechanical, and a combination of these processes have been utilised in recent years to produce feedstock with globular microstructure [10].

1.3 *Grain Refiner*

The grain refinement methods are typically based on the size and quantity of primary phase particles. The grain refiner is intended to achieve finers grain size rather than particle-phase form. However, finer grain size and suitable particle-phase form are crucial for SSMP. If there is a technology that can regulate the melt temperature and dispersing particles, it may be a potentially excellent alternative for semisolid grain refinement [8].

Grain refiners have two impacts on superalloys: a rise in grain boundary content and a microstructure refinement [11]. Both of these characteristics will affect the superalloy mechanical properties. In addition to apparent practical implications for improving the material mechanical performance, the relevance of grain refinement in polycrystalline cast superalloys' microstructure and mechanical behaviour enables understanding the deformative mechanism [12]. Grain refinement has become a popular method for simultaneously enhancing the strength and flexibility of metallic alloys. Grain refinement is essential for the mechanical product characteristics during and after the solidification process [16]. Grain refining acts to minimise grain size to reduce defects such as microporosity and improve the mechanical qualities of the material [9]. Grain refinement can help to reduce the size of the grain formed. The size of the grain decreased if certain alloys (Al, Si, Ca, Zr) were added to pure magnesium. With the first additions of these elements, a significant reduction occurs, and the degree of grain size reduction diminishes with subsequent additions [10]. Aside from that, grain refinement considerably enhanced resistance, as Park et al. [11] observed using a slow strain rate tensile test (SSRT) using hydrogen charged electrochemically in NaCl solution. The yield and tensile strengths of the material improved as an outcome of the experiment [8].

2 Types of Grain Refiner

Grain refinement of aluminium alloy casting has become a standard procedure in the manufacturing industry. The nucleation and development of α-Al determine the grain size of an aluminium alloy. Heterogeneous α-Al nucleation sites are introduced into the aluminium alloy melt when the grain refiner is added to the nucleation stage [16]. The types of grain refiner which have been used for grain refinement are as follows:

2.1 *Boron*

Boron (B) reacts to diborides (MgB_2 or AlB_2) with a close hexagonal structure that might contribute to the possible heterogeneous nucleation site for magnesium (Mg) and aluminium (Al) alloys. Compounds containing boron have a great potential to be utilised as an alloy grain refiner. Furthermore, the low cost of B is a benefit for grain refined magnesium alloys [12]. The majority of the B addition research takes place during the melting process. At this temperature, the refiners and alloy mixed inside the crucible. The master Al–4B alloy is added when the melting temperature reaches around 740 °C. The melt was swirled and perfectly mixed with the mixtures for 10 min. The mixture was poured into a cast iron cylindrical metallic mould that had been preheated to 300 °C. The experiment was carried out in a steel crucible in an electric resistance furnace. This study revealed that the increament of boron concentration reduced grain size from 100 μm to 30 μm, resulted in a 0.032 wt% reduction in grain size [13]. In addition, with pure Mg and different Al-3B master alloy numbers, a similar experiment at 730 °C was conducted. The 0.1 wt 5% B grain size has been refined with the alloy AM50 and the Al-3B master alloying from 550 to 100 μm. Aside from that, the findings include the increment of ultimate tensile strength and alloy elongation. The addition of AlB_2 particles improved grain formation since it is stable in Mg melt due to its high melting point of 892 °C, greater than the temperature during casting at 750 °C [14]. The microstructural image of AZ91 alloy with various boron additions is shown in Fig. 1. The average grain size is reduced from 70 to 30 μm, as 0.008 wt% to 0.032 wt% boron is added to AZ91 alloy (see Fig. 1a–c). However, there was a constant grain size formation with the addition of 0.04 wt% boron (see Fig. 1d) [13].

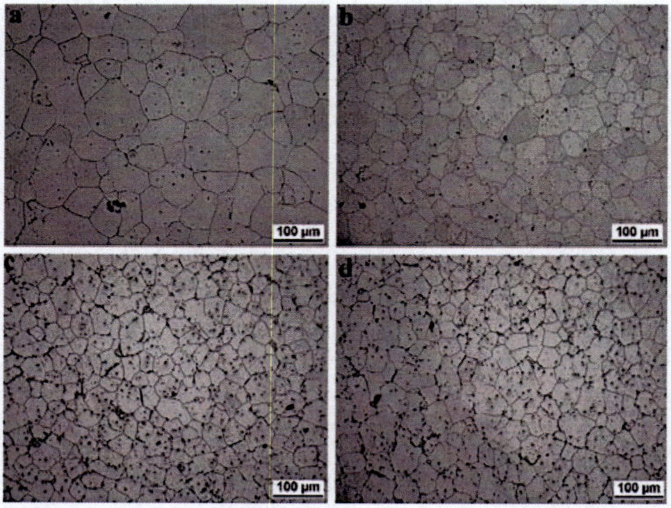

Fig. 1 AZ91 alloy with boron addition microstructural image with **a** 0.008 wt%, **b** 0.025 wt%, **c** 0.032 wt% and **d** 0.04 wt% [13]

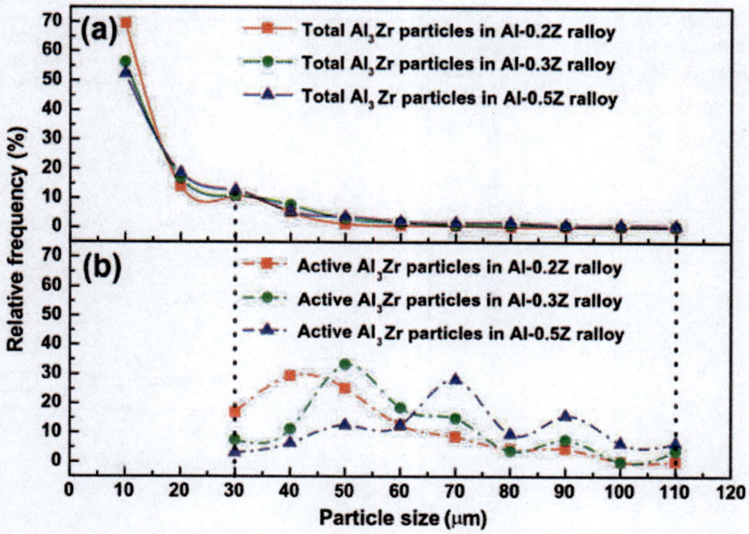

Fig. 2 The Al_3Zr particle size distribution is as follows: **a** The total amount of particles contained in the grain of Al_3Zr and **b** active Al_3Zr particles with various number addition of Zr [16]

2.2 Zirconium

Zirconium (Zr) is an efficient grain refiner in magnesium alloy that helped improve the mechanical properties. Various aluminium alloys have been cast with varying Zr addition amounts ranging from 0.0 to 0.5% [15]. Most of the high-purity commercial aluminium smelting with compacted Al-Zr pellets. The Al-Zr pellets were produced from pure Zr powder with Al chips added during the melting process. The aluminium heated up to 780 °C in a clay-bonded graphite crucible by using furnaces. The melt was maintained at 780 °C for 20 min after the compressed pellets were added. The mixture was mixed adequately before the preheated graphite forms were dropped into the melt to collect the sample. The findings of this study show that larger Al_3Zr particles occurred at a slower cooling rate, allowing the particles to grow significantly longer throughout the solidification process [16]. The Al_3Zr particle size distribution graph is shown in Fig. 2. The size distributions of the total Al_3Zr particles indicate in Fig. 2a are less than 25 μm meanwhile, the active particles of Al_3Zr in Fig. 2b shows the bigger size of particles [16].

2.3 Calcium

Calcium has garnered much attention since it greatly enhances the mechanical characteristics of the cast [17]. Ca addition improved mechanical properties and contributed

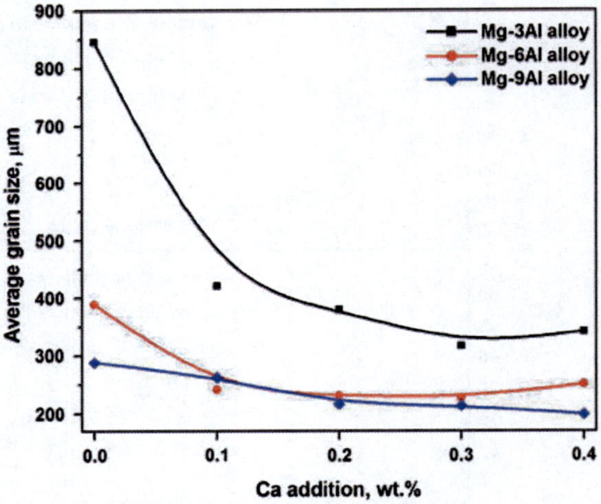

Fig. 3 The effect of Ca addition on the average grain size of Mg-Al alloy [19]

to grain refinement, as grain size decreased when Ca concentration increased from 0.5 to 1.0 wt% [18]. The Mg–Al alloys were melted up to 740 °C in an electrical heating furnace with protective flux to avoid oxidation. The melt was retained for 15 min before being poured into a 250 °C preheated cylindrical cast iron mould. The lower Ca content in the Mg-Al alloy, improved the purification of grains. The increment of Al concentration with the addition of Ca has only a minor refining impact, increasing growth limits and nucleate activation [30]. Figure 3 shows the trend of the average grain size effect by the addition of Ca. The findings indicated that the addition of Ca to Mg-Al grain refinement in low-Al containing alloys produced finer grain size [19].

2.4 Carbon

Carbon is the most efficient grain refiner for Mg-Al based alloy. The carbon contains various benefits, particularly for mass production, including low working temperatures and decreased fading with longer holding times [31]. During carbon inoculation, various carbonaceous agents such as C_2Cl_6, C_6H_6, and $MgCO_3$ are introduced to the molten alloy, where active carbon is released and reformed into aluminium carbide [32]. The use of carbon hexachloride in commercial grain refiner for Mg-Al alloys is very efficient because it can be degassed. It simultaneously removes dissolved gases from liquid and improves the grain structure. The carbon addition in Mg-Al alloys with an AZ31 magnesium alloy is melted in an electrical furnace with a mild steel crucible in an argon (Ar) gas atmosphere. At a temperature of 780 °C, 0.6 wt% carbon hexachloride (C_2Cl_6) was added to the mild steel mould. The findings show that carbon segregation occurred in AZ31 castings that have been treated with carbon [20]. A new grain refiner to replace C_2Cl_6 has been used since it produces

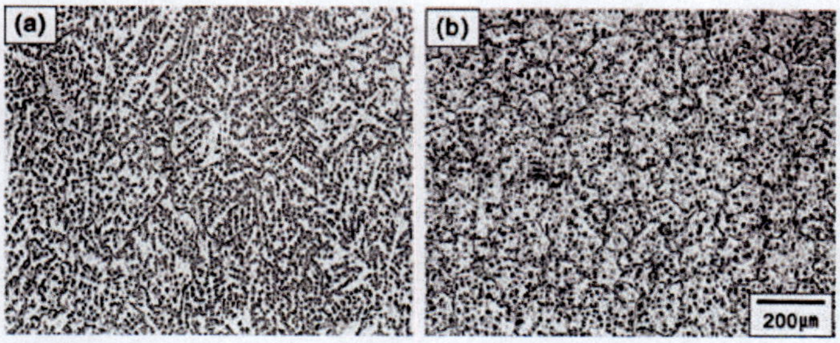

Fig. 4 AZ31 magnesium alloy microstructure images **a** with the inclusion of carbon and **b** without the inclusion of carbon [21]

dioxins when added to a melt with pure carbon powder. The melt was melted with carbon powder and argon gas. The temperature changed during the experiment, and carbon powder has been applied at various periods. Carbon powder produces refined grains similar to C_2Cl_6 once the grain size reaches 70 μm [32]. Figure 4 shows the microstructure images of AZ31 magnesium alloy with and without carbon inclusion. The grain size was found significantly reduced when 0.6 wt.% C_2Cl_6 is added from 400 to 120 μm [21].

2.5 *Scandium*

Scandium (Sc) has often been used in wrought aluminium alloys as a trace alloying treatment for grain refinement, recrystallisation control, and hot crack susceptibility reduction. It has proved that Al-Sc alloys exhibited a wide range of solidification behaviors, which is vital for refining grain morphologies in cast Al alloys [22]. The master alloys of Al-2 wt% Sc and Al-5 wt% Ti were melted in a crucible heated to the temperature of 400 °C. It was then loaded with aluminium ingots and heated to the temperature of 720 °C. The quantity of dissolved hydrogen decreased by introducing argon into the melt through a stainless-steel tube coated with zircon. Before pouring the melt, the stainless steel and graphite cups heated to 100 °C. The finding shows that due to the heterogeneous nucleation, scandium can be utilised to refine aluminium grains in Al–Si–Mg aluminium alloys [23]. The grain size obtained at varied cooling rates is presented in Fig. 5. In an Al–Si–Mg alloy, both Ti and Sc may act in heterogeneous nucleation; however, the efficiency of Sc in grain refining is lower than Ti [24].

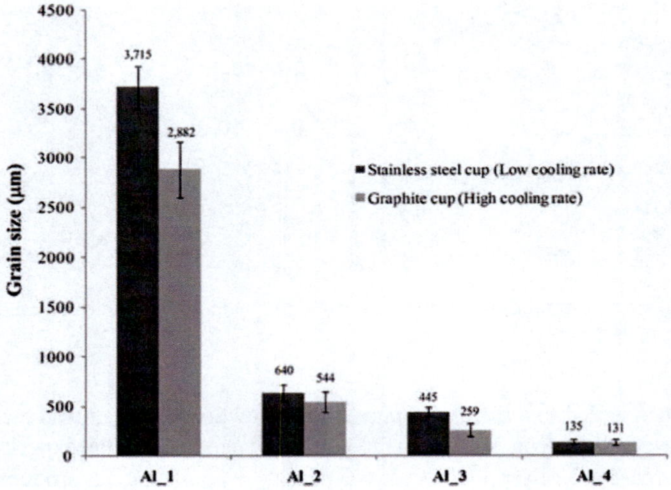

Fig. 5 Grain size comparison of Al6Si0.25 Mg, Alloy + 0.20Sc, Alloy + 0.40Sc and Alloy + 0.03Ti at varied cooling rates [24]

3 Analysis of Grain Refinement on Alloy

Finer grain microstructures improved the mechanical properties, and the fine microstructure has a higher toughness during the tensile test [24, 25]. Grain refining reduces the grain size, which results in a minimal grain size [27]. The comparison between the advantages and disadvantages of using grain refinement in the alloy is presented in Table 1.

Table 1 Comparative analysis of advantages and disadvantages using grain refinement

Advantages	Disadvantages
Grain refinement improves microstructure and increases the number of carbides and eutectics dispersed at grain boundaries [28]	The presence of numerous micro porosities in grain refined to a size of 25 μm may induce early breaking of test bars as well as a quick reduction in ultimate strength and elongation [29]
The modification in silicon (Si) size and refining has improved the A356 alloy damping capacity [30]	The current findings suggest that more significant consideration should be taken when selecting grain refiners for Sc-containing Al alloys due to the modification on failures of TiB_2 particles to assist the grain refinement [31]
Boron addition to Al-based magnesium alloys improves grain refinement, reduces size grain and increases ultimate strength and elongation [21, 22]	The grain refiner C_2Cl_6 has been banned in the magnesium sector because it produces dioxins when introduced to the melt [32]

4 Conclusion

Grain refiner plays a critical role in grain structure improvement due to the numerous benefits. The grain refiner materials such as Boron, Zircanium, Calcium, Carbon and Scandium are typically used in laboratories and industries. Most of the grain refinement process involves a temperature between 600 and 800 °C. At this temperature, the well mixture between the grain refiner and the molten alloy is achieved. The use of a grain refiner improves the grain structure, thus increase the mechanical properties. The improvement of mechanical properties will help to extend the material usage. In SSMP, the use of grain refiner seems to encompass the benefit of the SSMP. Combining the grain refiner with one of the SSMP techniques is believed to produce a better microstructure formation suit for thixoforming.

Acknowledgements The authors wish to thank the Ministry of Higher Education for providing financial support under Fundamental Research Grant Scheme (FRGS) No. FRGS/1/2019/TK03/UMP/03/4 (University reference RDU1901166) and Universiti Malaysia Pahang for laboratory facilities.

References

1. Nafisi S, Ghomashchi R (2016) Semisolid processing of aluminum alloys. SemiSolid Processing of Aluminium Alloys
2. Chen JY, Fan Z (2002) Modelling of rheological behaviour of semisolid metal slurries: part 1—theory. Mater Sci Technol 18(3):237–242
3. Chang Z, Su N, Wu Y, Lan Q, Peng L, Ding W (2020) Semisolid rheoforming of magnesium alloys: a review. Mater Des 195:108990
4. Atkinson HV (2005) Modelling the semisolid processing of metallic alloys. Prog Mater Sci 50(3):341–412
5. Midson SP, Jackson A (2006) A comparison of thixocasting and rheocasting (2006) Institute of cast metals engineers—67th world foundry congress wfc06 casting the future, vol 2, pp 1081–1090
6. Rosso M (2012) Thixocasting and rheocasting technologies, improvements going on. J Achievements Mater Manuf Eng 54(1):110–119
7. Husain NH, Ahmad AH, Rashidi MM (2017) An overview of thixoforming process. In: IOP conference series materials science and engineering, vol 257, no 1
8. Xia MX, Li JG, Fan ZY (2014) Grain refining through semisolid processing. Solid State Phenom 217–218:431–435
9. Easton MM, Stjohn D (1999) Grain refinement of aluminum alloys: Part I. The nucleant and solute paradigms - a review of the literature. Metall Mater Trans A Phys Metall Mater Sci 30(6):1613–1623
10. Introduction J (2016) Magnesium technology. Magnes Technol 2016(2):430
11. Park IJ, Lee SM, Jeon HH, Lee YK (2015) The advantage of grain refinement in the hydrogen embrittlement of Fe-18Mn-0.6C twinning-induced plasticity steel. Corros Sci 93:63–69
12. Zhang S et al (2019) Effect of boron on the grain refinement and mechanical properties of as-cast Mg alloy AM50. Materials 12(7)
13. Suresh M, Srinivasan A, Ravi KR, Pillai UTS, Pai BC (2009) Influence of boron addition on the grain refinement and mechanical properties of AZ91 Mg alloy. Mater Sci Eng A 525(1–2):207–210

14. Viala JC, Bouix J, Gonzalez G, Esnouf C (1997) Chemical reactivity of aluminium with boron carbide. J Mater Sci 32(17):4559–4573
15. Wang L et al (2017) Microstructure, mechanical and bio-corrosion properties of Mg–Zn–Zr alloys with minor Ca addition. Mater Sci Technol 33(1):9–16
16. Wang F, Qiu D, Liu ZL, Taylor JA, Easton MA, Zhang MX (2013) The grain refinement mechanism of cast aluminium by zirconium. Acta Mater 61(15):5636–5645
17. Han L, Hu H, Northwood DO (2008) Effect of Ca additions on microstructure and microhardness of an as-cast Mg-5.0 wt.% Al alloy. Mater Lett 62(3):381–384
18. Wang T, Wu R, Zhang M, Li L, Zhang J, Li J (2011) Effects of calcium on the microstructures and tensile properties of Mg-5Li-3Al alloys. Mater Sci Eng A 528(18):5678–5684
19. Nagasivamuni B, Ravi KR (2015) An analytical approach to elucidate the mechanism of grain refinement in calcium added Mg-Al alloys. J Alloy Compd 622:789–795
20. Jin Q, Eom JP, Lim SG, Park WW, You BS (2003) Grain refining mechanism of a carbon addition method in a Mg-Al magnesium alloy. Scripta Mater 49(11):1129–1132
21. Kim YM, Yim CD, You BS (2007) Grain refining mechanism in Mg-Al base alloys with carbon addition. Scripta Mater 57(8):691–694
22. Norman AF, Prangnell PB, Ewen RSMC (1998) Norman98.Acta. Solidification behavior.pdf, vol 46, no 16, pp 5715–5732
23. Wu XF et al (2019) Simultaneous grain refinement and eutectic Mg2Si modification in hypoeutectic Al-11Mg2Si alloys by Sc addition. J Alloy Compd 791:402–410
24. Patakham U, Kajornchaiyakul J, Limmaneevichitr C (2012) Grain refinement mechanism in an Al-Si-Mg alloy with scandium. J Alloy Compd 542:177–186
25. Chen R, Xu Q, Guo H, Xia Z, Wu Q, Liu B (2017) Correlation of solidification microstructure refining scale, Mg composition and heat treatment conditions with mechanical properties in Al-7Si-Mg cast aluminum alloys. Mater Sci Eng A 685:391–402
26. Hu Z, Wan L, Lü S, Zhu P, Wu S (2014) Research on the microstructure, fatigue and corrosion behavior of permanent mold and die cast aluminum alloy. Mater Des 55:353–360
27. Al-Qawabah SMA, Zaid AIO (2016) Different methods for grain refinement of materials. Int J Sci Eng Res 7(7):1133–1140
28. Du B, Yang J, Cui C, Sun X (2015) Effects of grain refinement on the microstructure and tensile behavior of K417G superalloy. Mater Sci Eng 623:59–67
29. Wei CN, Bor HY, Ma CY, Lee TS (2003) A study of IN-713LC superalloy grain refinement effects on microstructure and tensile properties. Mater Chem Phys 80(1):89–93
30. Zhang Y, Ma N, Wang H, Li X (2008) Study on damping behavior of A356 alloy after grain refinement. Mater Des 29(3):706–708
31. Huang J, Feng L, Li C, Huang C, Li J, Friedrich B (2020) Mechanism of Sc poisoning of Al-5Ti-1B grain refiner. Scripta Mater 180:88–92
32. Yano E, Tamura Y, Motegi T, Sato E (2003) Effect of carbon powder on grain refinement of an AZ91E magnesium alloy. Mater Trans 44(1):107–110

Review on Thermodynamic Properties of Plastic by Fused Deposition Modeling

Xinglong Shi, Wenjie Ding, Qianjin Wang, Tongman Li, and Nan Zhao

Abstract Traditional machining is limited to the assembly of parts. The emergence of 3D printing breaks this inherent thinking. The concept of "layer" is different from flat printers, and innovative and unique additive manufacturing processes are used in various fields. However, FDM printing and molding problems are closely related to thermal characteristics. This article reviews the causes and solutions of FDM 3D printer molding problems. Introduce the research on structure design and process parameters of printer. Furthermore, a theoretical study on the influence of thermodynamic properties in the FDM process is put forward. That is to study the internal relationship between plastic fuse and thermodynamics, which provides a scientific basis for solving common problems of FDM type 3D printing. In the future, the challenge facing FDM printing is how to find out the relationship between the thermodynamic properties of plastics and the influence of printing. This puts forward the idea based on structural adaptation and neural network learning, which can be used in actual printing production situations.

Keywords FDM · Molding problems · Structure design · Process parameters

1 Introduction

Fused Deposition Modeling (FDM for short) is also known as Melted Extrusion Manufacturing (MEM for short) and melt accumulation method. The thermoplastic material is heated and melted by the nozzle, and under the control of the computer, it is piled up and formed layer by layer. Since FDM technology has a wide variety of molding materials and has excellent mechanical properties, it has a higher material utilization rate than traditional machining, and it is quick to operate, and the technology is generalized and suitable for general working conditions. Therefore, FDM fused deposition is one of the widely used technologies in 3D printing.

X. Shi · W. Ding (✉) · Q. Wang · T. Li · N. Zhao
Ningxia University, Yinchuan 750021, China
e-mail: dwjnet@nxu.edu.cn

M. Y. Ismail et al. (eds.), *Technological Advancement in Mechanical and Automotive Engineering*, Lecture Notes in Mechanical Engineering,
https://doi.org/10.1007/978-981-19-1457-7_71

However, FDM 3D printing has to undergo the process of fuse heating, cooling and extrusion, so the temperature field of the nozzle structure must match the rapid change of fuse temperature [1]. Generally, printers are affected by the heating efficiency of the flow through and the driving force of the wire feeding, which will produce various thermal defects. The printing temperature will always affect the viscosity and morphological shrinkage of the fuse, that is, the thermodynamic properties of the plastic [2, 3]. In addition, the problem of printing and molding is also related to the feeding friction of the wire feeding mechanism. The proper driving force is the prerequisite for continuous and stable wire feeding and no biting, that is, the plastic force characteristics [4, 5]. There are more structural design optimization methods involving thermodynamic properties in recent years. For example: In order to solve the problem of nozzle clogging and unsteady filament output, Pattiguri Ahemai [6] designed an extruded FDM print nozzle. The outer diameter of the heating jacket is preset to 60 mm and the height is 65 mm. Simulation and test results Shows that this structural design improves the printing success rate. Zhu [7] studied the thermal-mechanical coupling analysis of 3D printing extruder and used finite element method to put forward the influence of temperature on 3D printing extruder. However, there are fewer studies on engineering design methods based on thermodynamic properties. In particular, thermodynamic properties are the knowledge basis for the construction of design methods.

This paper introduces the design methods of domestic and foreign scholars by introducing forming problems, and expounds the research content of FDM thermodynamic properties. Relevant data shows that the average annual consumption of ABS [8] can reach 1.64 million tons, and the problem of resource waste is becoming more and more serious. The study of thermodynamic properties can point out a guiding direction for reducing consumables waste in the future, supporting industrial design and independent innovation design. From the perspective of fundamentally solving thermal defects, it is of great value for future product design.

2 Analysis of 3D Printing Molding Problems

Data analysis shows that cracking, wire drawing and clogging [9] will cause printing and molding problems, and common 3D printing problems are related to thermodynamic properties.

2.1 Cracking of Molded Parts

Because the fuse is affected by thermal expansion and contraction, the molten consumables will easily produce a certain degree of shrinkage and deformation after cooling and forming a solid state. As shown in Fig. 1. Due to the thermodynamic properties of plastics, the internal molecular chain of the fuse rotates around the

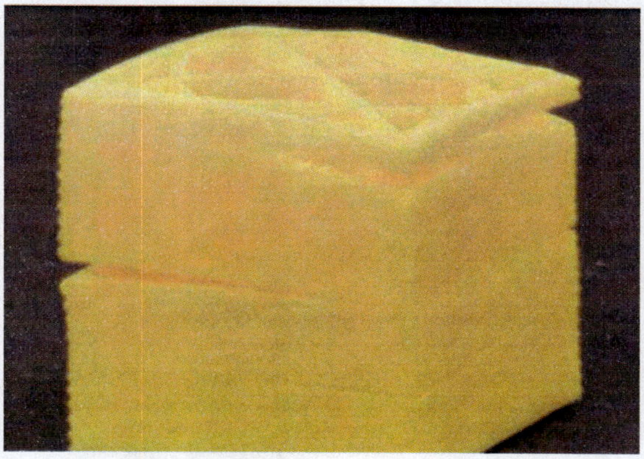

Fig. 1 Cracking of 3D printed parts

compound single bond, and the spatial arrangement of the molecules will produce different spatial isomers. This phenomenon is called internal rotation.

Different internal rotation, the performance of plastics will also be different. It is mainly reflected in the mechanical properties of the material. The phenomenon of molecular entanglement will produce greater resistance. The wire cannot form complete molecular interfusion and diffusion. The main valence force and van der Waals force between the molecules are not strong, and the fuse cannot be thermally driven bonding. The temperature history of the bonding interface between the wires has a great influence on the bonding strength of the wires. In addition, the tensile strength and the elongation at break are both parameters for judging the mechanical properties of the fuse [11]. Therefore, it is necessary to study the thermodynamic properties of FDM molding, and then to study the development law of various parameter quantities in actual printing. The solution is to reduce the tension-shear coupling effect and the bending-torsion coupling effect between the deposition layers [10], and use materials with a smaller shrinkage rate to reduce the overall deformation trend of the molded part and prevent cracking when the model's bearing capacity reaches the limit.

2.2 *Consumables Drawing Problem*

When the consumable is in the printing temperature range (PLA is 180–230 °C), the material temperature is in the plastic deformation zone, and the molten fuse can be extruded stably only when the extrusion temperature and heat dissipation are in a relatively balanced position. Generally speaking, the withdrawal distance of the remote extruder should reach 3–6 mm. The purpose of the withdrawal is to reduce the pressure of the consumables in the runner on the nozzle, so as to temporarily stop the extrusion. However, when the printer's withdrawal command cannot fully recover the remaining silk material, the flowing residue will continue to

Fig. 2 3D printing drawing model

be printed normally, and the dripping filaments will adhere to the model, resulting in a wire drawing phenomenon. As shown in Fig. 2. A molten material with better adhesion can be used, so that the mutual adhesion of excess materials is greater than gravity, and the material will not drip out, thereby avoiding the occurrence of wire drawing [9].

Another reason for this phenomenon is that plastic consumables are of poor quality or are affected by moisture. If consumables are exposed to the air for a long time, for example, degradable materials such as PLA will absorb moisture in the air, and consumables that contain moisture in high-temperature nozzles will instantly vaporization and expansion into a number of small bubbles, even if the printer has a retraction function, this effect similar to "bore pressure" will still cause residual consumables to be squeezed out to form wire drawing. To avoid wire drawing, the consumables can be dried or stored properly.

2.3 *Nozzle Clogged Problem*

In FDM molding technology, the fuse will have the problem of poor semi-melt fluidity [12], and the most common problem is blockage. As shown in Fig. 3. The cause of the clogging of the FDM 3D print head is a complex and multi-dimensional comprehensive problem, and the formation process of the clogging problem is relatively complicated. When the consumables are withdrawn, the high-temperature heat of the heating block and the backflowing fuse form a collision phenomenon, and the temperature distribution of the hot runner is uneven [13]. If the heat dissipation of the print head structure is insufficient, the fuse in the throat is fused to form a viscous state, resulting in greater resistance, and the temperature field of the consumables is unevenly distributed in the melting space. That is, the fuse of the heat dissipation

Fig. 3 The 3D printing head is blocked

section will be softened by heat absorption, and the machine will be cold again, because the cooling and melting consumables will form a "glass state", which is squeezed into a mess in the Teflon tube, causing blockage. In severe cases, the Teflon tube will deform and burst, the fuse cannot be smoothly extruded from the nozzle, and the printing will not work normally.

2.4 Other Molding Issues

FDM molding technology problems also include weak bonding, curled or rough, no silk, bite, overheated parts and so on. Common 3D printing thermodynamic properties problems and solutions are shown in Fig. 4.

3 Research Status of Thermodynamic Properties

The material for studying the thermodynamic properties of fused deposition is plastic. Due to the difference in properties of different thermoplastic materials, the apparent physical state will be different when affected by the outside world. The study of the thermodynamic properties of plastics pave the way for the analysis of the internal mechanism of the fuse phase transition and the FDM printing process. The two most common materials on the market are ABS and PLA. The physical properties of several main FDM printing materials are shown in Table 1.

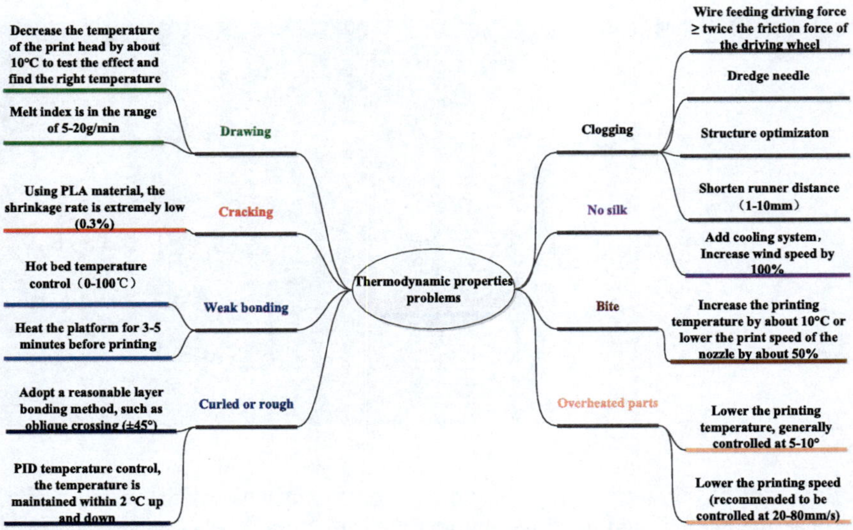

Fig. 4 Thermodynamic properties of FDM 3D printing

Table 1 Physical properties of FDM plastic

Type	Density (g/cm^3)	Printing temperature (°C)	Elastic Modulus (Mpa)	Poisson's ratio	Burning speed
PLA	1.24	190–240	3000	0.36	Slow
ABS	1.10	230–245	2000	0.39	Slow
TPU	1.23	210–230	1000	0.47	Slow
PETG	1.27	230–250	2100	–	Slower

3.1 Research on Structural Design Method

The predecessors made the following thermodynamic properties research for the FDM structure design: Wang et al. [14] optimized the structure of the print head assembly and the internal Teflon tube to reduce the heat conduction area of the hot runner under the Teflon tube, and finally the print head structure was obtained It has a significant heat dissipation effect and reduces the occurrence of nozzle clogging. In order to avoid the clogging and broken filaments of traditional print nozzles and adapt to the rapid change of the fuse at the heating temperature of the print head, Peng and Wei [15] proposed a fuse deposition molding print nozzle and temperature control method based on multi-stage temperature control. The temperature control module is used to achieve the gradual temperature of the fuse in each section of the print head. The results show that: taking ABS as an example, the temperature of the printing extrusion section is set to be 1–5 °C lower than the molten state processing temperature T_{op}, which can effectively avoid the phenomenon of nozzle clogging and thread breakage during printing.

Aiming at the bite marks, fractures, dust and abrasion that are easily caused by the friction wheel wire feeding in the traditional FDM wire feeding mechanism structure, Zhang and Du [16] proposed an innovative FDM threaded double nozzle structure design. The pressure wheel and the screw are connected by a screw for feeding. The outer diameter of the threaded section on the screw is 8 mm. The diameter of the traditional drive wheel is generally between (15–20) mm, which greatly reduces the requirements for the drive motor. The pressure regulating spring is used to make the feeding. The friction force is matched with the machine to adapt to the printer's versatility for consumables with different wire diameters. The silk material is uniformly stressed and has no scratches, which avoids the occurrence of bite marks. Wang [17] mentioned the idea of a crawler wire feeding mechanism. Using calculation and theoretical knowledge, he summarized a set of design methods for high-efficiency wire feeding drive mechanism, and realized the surface contact form between the flat belt and the wire. With sufficient contact area and positive pressure to ensure the driving force into the wire, the improved structure increases the driving force by at least 4 times, and overcomes the shortcomings of insufficient driving force of the traditional wire feeding mechanism.

Qin [18–20] and others proposed a new type of screw extrusion equipment, which uses the principle of friction transmission to mix materials, and establishes a mathematical model between the printing feed speed, the extrusion volume and the screw speed, which solves the traditional FDM 3D printing has problems of poor thermodynamic properties and limited material types. The results show that the maximum speed at the extrusion port is 0.063 m/s, which reduces the overflow and backflow phenomenon at the nozzle; the tiny thermal deformation of the nozzle 0.05 mm is much smaller than traditional printing. The maximum deformation accuracy meets the requirements of printing and forming accuracy. The foreign scholar Anna [21] proposed a micro extruder for high-precision granular processing. It is equipped with a controller to adjust the speed of the extruder screw, so that the extruded material has sufficient drive to a greater extent. At the same time, it also reveals the flow behavior relationship between melt viscosity and particle size. The results show that the advantages of high efficiency and agility are embodied in the application of the whole machine, which points out a guiding direction for the research on the structure design of the new FDM print head.

3.2 Research on Printing Process Parameters

At present, FDM 3D printing technology lacks a set of systematic design research methods, traditional 3D printing optimization process quality method research, there are generally intelligent algorithm prediction settings, equipment's own characteristics of test settings and exhaustive analysis settings [22]. These methods are not standardized enough, the number of trials is large, and the methods are complicated and difficult to understand. Therefore, how to determine the best printing process parameters and structured design process standards is the key development object of future 3D printing technology. Through orthogonal experiment, calculation of signal-to-noise ratio, range method, etc. [23] explore the multi-objective parameter optimization combination scheme in 3D printing technology, and determine the primary and secondary process parameters that affect the printing process, in order to better solve the operation and maintenance problems in the actual work of the printer. FDM 3D printing parameters determine the surface quality and molding accuracy of the printed parts. Printing parameters are generally set by adjusting the parameter values of various shape parts through the slicing software (Simplify 3D, Cura, etc.).

Foreign scholar Bellehumeur [24] conducted research and analysis on the process parameters of fused deposition, demonstrated the bonding mechanism process of polymer wire from experiments, and established a model to evaluate the level of process parameters based on wire sintering experiments. The results showed that after optimization the temperature conditions of fused deposition printing made the molded parts exhibit excellent mechanical properties, and provided a strong reference basis for future generations to study the finite element model of the FDM temperature field.

4 The Development Research and Trend of Thermodynamic Properties

In view of the above research status, it can be found that in terms of the adaptation of thermodynamic properties and structural design, some researchers only optimize the temperature field and air-cooled flow field for material blocking and structural heat dissipation components. The research method is shown in Fig. 5. However, the optimization and innovative structure that combines the thermodynamic properties of the internal fuse temperature transition law of the print head is still insufficient. Therefore, studying the thermodynamic properties of FDM 3D printing and its structural design adaptation is of vital importance.

Therefore, the key content that can be carried out in the future is: the research on the systematic design method of the thermodynamic properties of the printing fuse, through understanding the change law of the thermodynamic properties of the fuse during the printing process, and clarifying the reasonable relationship between the test fuse and the operating parameters, and summarizing the design optimization method of the print head structural components is presented, and the purpose is to provide a theoretical basis and method for the structural design of the thermodynamic properties of FDM 3D printing.

FDM 3D printing technology will have some technical problems. Thermodynamic properties will be the mainstream research trend in the future. Plastic tribology [25] and phase change mechanism [26] explain the internal relationship between plastic

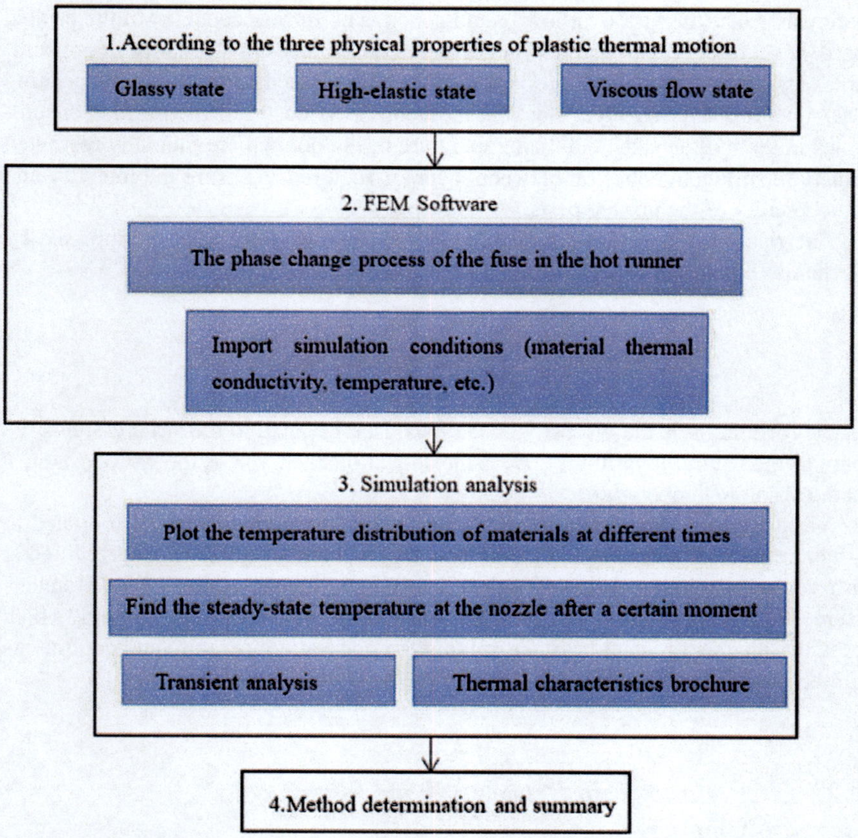

Fig. 5 Research on thermodynamic properties method based on simulation software

thermal motion and temperature. The research direction is FDM 3D printing thermodynamic properties method, that is, the internal relationship between consumable fuse and thermodynamics.

4.1 The Relationship Between Melt Viscosity and Temperature

The significance is to analyze the relationship between melt viscosity and temperature to prevent the problems of wire drawing and weak bonding of the model due to thermodynamic properties. The appropriate melt viscosity value helps to improve the surface quality of the molded part, and the viscosity should be selected within the allowable molding temperature range. If the viscosity is too high, the melt flow ability is low, and the effect of the printed parts is poor; if the viscosity is too low, it will seriously cause the fuse overflow phenomenon. The surface accuracy of the printed parts has a direct relationship with the melt consumables. The quality of consumables affects the mechanical properties of the parts. Based on the FDM 3D printing principle, the layer-by-layer stacking method, the viscosity of the plastic consumables in the molten state and the temperature of the hot bed are mutually restricted function, sufficient adhesion between layers is required to ensure the integrity and functionality of the printed parts.

The relationship between melt viscosity and temperature can be expressed by Arrhenius equation [27], see formula (1):

$$\eta = Ae\frac{\Delta E_\eta}{RT} \tag{1}$$

In the formula: η is the viscosity; A is the constant related to the structure; e is the base of the natural logarithm; ΔE_η is the activation energy; R is the gas constant; T is the absolute temperature.

Viscosity has a linear relationship with temperature. Viscosity is also related to structure-related constants and gas constants. The melt viscosity is linearly fitted at different temperatures. By drawing the curve of melt viscosity changing with temperature, the critical point of temperature sensitive to melt viscosity is found, which provides an experimental basis for solving the optimal temperature and continuous printing of FDM processing.

4.2 The Relationship Between Shape Shrinkage and Temperature

The significance is to avoid warping and deformation of the printed model and improve the cooling time. Plastics are generally affected by high temperature and will inevitably deform. This polymer compound material will undergo four different forms of deformation, see formula (2):

$$D = D_1 + D_2 + D_3 + D_4 \tag{2}$$

In the formula: D is the total deformation; D_1 is the general elastic deformation; D_2 is the high elastic deformation; D_3 is the viscoelastic deformation; D_4 is the plastic deformation.

Ignore the general elastic deformation that has little effect on temperature, and study the plastic deformation examples that are closely related to temperature. At this time, the material is in an irreversible viscous flow state. The study can be based on formula (3):

$$D = \frac{\sigma}{\eta} \cdot t \tag{3}$$

In the formula: D is the plastic deformation; σ is the external load; η is the material viscosity; t is the load action time. Theoretically, under the condition that the external load, material viscosity and load acting time remain unchanged, the law of the influence of different temperatures on the plastic deformation of the material can be summarized.

3D printing mostly uses thermoplastic materials, which will inevitably be restricted by the types of consumables and mechanical characteristics in the printing work. Shrinkage and low strength. The coefficient of elasticity ($\times 10^4$ kg/cm^2) = 0.13–2.9 (thermoplasticity) 5–32 (thermohardenability) [28]. In addition, the difference in temperature gradient will affect the shrinkage of the material, which can seriously lead to the warpage of the molded part. This is the so-called plastic hysteresis effect. Therefore, it is necessary to conduct a test of the warpage deformation of the molded part [29], look for The relationship between shape shrinkage and temperature.

Using Cura software to slice a cube as an example (a thin-walled part is selected, the amount of thermal deformation is easy to observe), the model parameters that are worth studying include the bottom, the top, the inner filling and the boundary, as shown in Fig. 6. The experiment uses 5 °C as an incremental temperature unit to explore the degree of influence of temperature on model parameters. Under different printing temperatures, the cooling and molding will inevitably suffer from layer-to-layer shrinkage, which will lead to warpage and deformation of the part. The relationship between the printing temperature and the warpage and deformation of the model is found, which provides an experimental basis for explaining the relationship between the form shrinkage rate and the temperature.

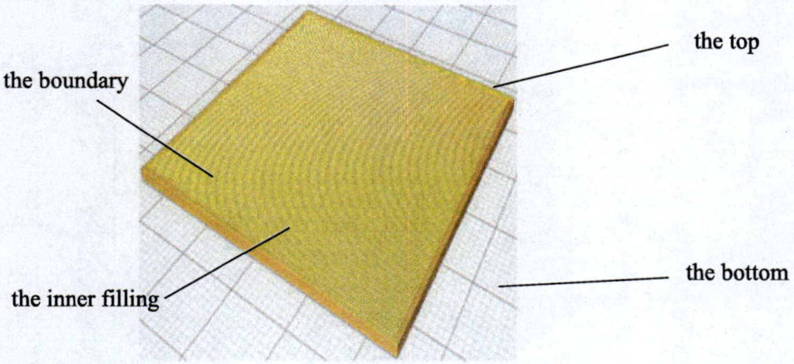

Fig. 6 Experiment on the influence of printing temperature on the warpage and deformation of molded parts

4.3 The Relationship Between Fuse Friction and Feed Driving Force

The significance is to avoid blockage caused by fuse slippage and excessive clamping force. In the printer's wire feeding mechanism, the fuse is fed through a pair of friction wheels to engage and drive. What value does the pushing force need to achieve to ensure the normal extrusion of the fuse and the stable thermal state during printing? Mathematical models and test methods should be used to find out the relationship between fuse friction and feed driving force. The design requirements include: smooth wire feeding, low noise, low wear on consumables, etc. [30]. The focus of the research is to calculate the size of the wire feeding driving force, and the V-groove wheel is used as the friction wheel to realize the pressure control of the consumables. As shown in Fig. 7.

Assuming that the experiment ignores the resistance of the driven wheel to the wire material and its own gravity, the positive pressure of the driving wheel V groove on the wire material is equal, then the wire feeding driving force is equal to twice the friction force of the driving wheel to the wire material, see the formula (4):

$$F_z = 2 \times F_{f1} = 2 \times f \times N_1 \tag{4}$$

In the formula: F_z is the wire feeding driving force; F_{f1} is the friction force of the driving wheel on the wire; f is the friction factor; N_1 is the positive pressure of the driving wheel on the wire.

The parameters worth studying include groove angle, friction wheel diameter, wire diameter, metal-plastic friction coefficient and so on. In order to meet the requirements of adjustable control of the pre-tightening force of the wire feeding mechanism, reduce the load of the guide rail and reduce the driving performance of

Fig. 7 Force of wire feeding mechanism

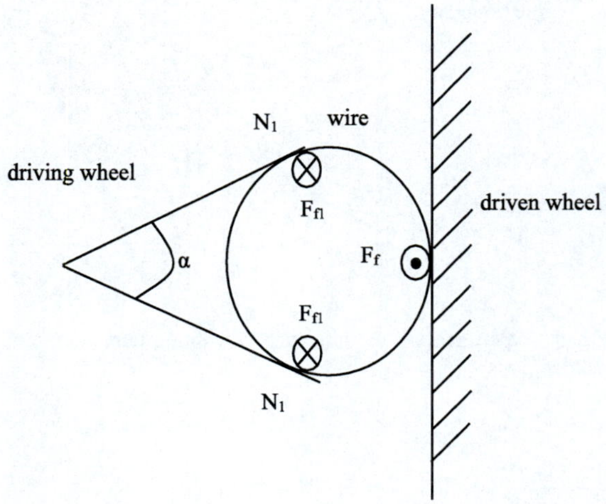

the motor, it is worth exploring the setting of the pre-compression spring and the combination of the far and near-end wire feeding methods.

5 Conclusion and Prospect

At present, traditional FDM 3D printers generally have operation and maintenance problems [31–36], because plastic polymers are affected by temperature sensitivity and feed-driven friction. The current thermal characteristics are facing some challenges, and the future development directions of the thermal characteristics of FDM plastic fuses urgently need to be carried out in the following aspects.

(1) Combining printer structure adaptation: The production and manufacturing of most FDM 3D printers in my country are mainly based on surveying, imitating, assembling, and empirical design of foreign prototypes, and the actual application is also mainly based on the manufacturing of single-piece products and personalized products, with industrialization the new structure and design method of the batch printing function are still in the research process, and the thermodynamic properties are an important basis for its research and innovation. After obtaining reasonable fuse and printing operating parameters, if you can summarize the printer's systematic optimization method or design a set of rules to avoid thermal defects, it will be easy to solve the problem of thermodynamic properties and structural design adaptation of the 3D printer.

(2) Neural network learning: Although, domestic and foreign scholars have found that thermal defects will affect the structure design, operation and maintenance of FDM printers, such as clogging, cracking, wire drawing, etc., affecting the molding accuracy and printing success rate of parts, or for improve the effect of the printed model and use the finite element analysis module for structural design. However, the current physical relationship or influence degree of its thermodynamic properties is not clear, which makes it impossible to optimize the design of the structure size. For example, how to use mathematical equations to express the probability of clogging of printing hair, and how the relationship between fuse viscosity, shrinkage deformation and temperature affects the printing results. Therefore, future research can be based on neural network learning to deal with the calculation relationship of unclear thermodynamic properties, and it will have a good development prospect in industrial printing applications.

References

1. Ding W, Ding Z, Hao H et al (2020) Analysis and design of the heat transfer structure of the FDM 3D printing nozzle. Plastics 49(03):94–97
2. Liu M, Zhu X (2021) 3D printing process and material application research progress. Mech Res Appl 34(04):197–202
3. Ren L, Bai H, Jia S et al (2020) Optimization design and temperature field analysis of single screw extrusion 3D printer. Plastics Industry 48(12):90–95
4. Zhang Q, Li L, Guangdong W et al (2020) CAE analysis of fused deposition 3D printer mechanical mechanism. J Jilin Teachers Coll Eng Technol 36(11):123–126
5. Sun Z, Xu BS, Wan H et al (2020) Innovative design of frame-type 3D printer. Intern Combust Eng Accessories (18):205–206
6. Ahemet P, Ayiti W, Hemdula E (2021) Optimal design of extruded FDM 3D printing nozzles. Mech Des Manuf 01:232–235
7. Zhu B, Zhao Y (2020) Thermal-mechanical coupling analysis of 3D printing extrusion heads with different materials. Automobile Pract Technol 45(17):160–163
8. Zhu J (2002) Thermal stability and recycling of engineering plastics. China Engineering Plastics Industry Association Processing and Application Committee, the Fourth China Plastics Expo Organizing Committee. 2002 China Engineering Plastics Processing and Application Technology Seminar Collection of papers[C]. Processing and Application Committee of China Engineering Plastics Industry Association, Organizing Committee of the Fourth China Plastics Expo: China Engineering Plastics Industry Association, 2002:4
9. Chang Min, Jin Li, Xinming Shen, et al (2017) Optimal design of FDM 3D printer nozzle. Heilongjiang Sci 8(21):10–12+15
10. Yang B (2014) Research on the influence of heat dissipation conditions on the bonding quality of FDM wire. Dalian University of Technology, Dalian
11. Chi B, Xie L, Gao X et al (2015) The influence of construction orientation in FDM process on the mechanical properties of plastic products. Plastics 44(04):40–42
12. Li J (2018) Analysis of nozzle structure and common problems of fused deposition 3D printer. South Agric Mach 49(13):181
13. Liu X, Wang C, Chi B et al (2017) Optimization design of the heat transfer structure of the pellet 3D printer head. Plastics 46(04):113–116
14. Wang J, Chen H, Gong Y et al (2017) 3D printer nozzle assembly structure optimization design. Modular Mach Tool Autom Process Technol 10:157–160
15. Peng Y, Wei W, Fuse deposition forming 3D printing nozzle and temperature control method based on multi-stage temperature control. Zhejiang: CN103240883A, 2013-08-14
16. Zhang H, Junbin D (2019) Innovative design of threaded double nozzles in FDM process. Mech Des Manuf 12:105–108
17. Wang T (2007) Research and design of FDM wire feeding mechanism. Huazhong University of Science and Technology, Wuhan
18. Wang Q, Bai H, Jia S et al (2019) Nozzle design and temperature field numerical simulation of screw extrusion 3D printing molding equipment. Eng Plast Appl 47(12):73–78+85
19. Qin W, Bai H, Jia S et al (2019) Based on screw extrusion 3D printing molding equipment structure design. Plast Ind 47(09):66–69
20. Wang Q, Bai H, Jia S et al (2020) Thermal stress coupling analysis of nozzle structure of new 3D printing molding equipment. Plastics 49(03):52–57+89
21. Anna B (2005) New developments in fused deposition modeling of ceramics. Rapid Prototyping J 11(4):214–220
22. Zhang F, Tan Y, Ma G (2017) Research on FDM 3D printing process method based on orthogonal experiment. Mach Tool Hydraulics 45(23):78–81
23. Feng J, Liang H, Yang Y (2019) Analysis and optimization of FDM rapid prototyping process parameters. Mech Sci Technol 38(05):742–748
24. Bellehumeur C, Li L, Sun Q et al (2004) Modeling of bond formation between polymer filaments in the fused deposition modeling process. J Manuf Process 6(2)

25. Wang C (1992) Introduction to plastic tribology. Tribology Branch of Chinese Mechanical Engineering Society. In: Proceedings of the fifth national tribology conference (Vol 1). Tribology Branch of Chinese Mechanical Engineering Society: Chinese Mechanical Engineering Society Tribology Branch (vol 7)
26. Liang X (2015) Thermodynamic analysis and structural optimization design of 3D printing extruder. Xi'an: Xi'an Polytechnic University
27. Yuan Y (2019) Research on the preparation and fused deposition molding process of PC and its composite wire. Wuhan University of Technology, Wuhan
28. Koichi K, Cao Q (1987) High temperature characteristics of plastics. Foreign Oilfield Eng 01:18–21
29. Guo Y, Meng Q (2017) Research on printing parameters of anti-deformation and warpage of large-size FDM mold. China Foundry Equipment Technol 04:32–34
30. Yan D, Cao Z, Kong G (2003) Analysis of frictional driving force of FDM process wire feeding drive mechanism. J Beijing Univ Chem Technol (Nat Sci Ed) (03):71–73+77
31. Sun Z (2014) Research on the comprehensive development trend of domestic 3D printing industry. Jiangsu Sci Technol Inf 08:11–13
32. Huang H, Zhao T (2020) Fused deposition 3D printer nozzle structure and common problems. Papermaking Equipment Mater 49(04):9–11
33. Ma Z, Wang H (2019) Optimization design of wire feeding mechanism of fused deposition 3D printer. Modern Inform Technol 3(24):160–162+164
34. Zhang N (2014) Research on 3D printing accuracy test method based on Hilbert curve. Metrol Test Technol 41(07):1–3
35. Tang T, Zhang Z, Deng J et al (2015) Research status and development trend of 3D printing technology based on FDM. New Chem Mater 43(06):228–230+234
36. Lee BH, Abdullah J, Khan ZA (2015) Optimization of rapid prototyping parameters for production of flexible ABS object. J Mater Process Tech 169(1):54–61

Research on Operation Data Mining of Pulse Dust Collector

Qianjin Wang, Wenbo Han, Wenjie Ding, Wenchuan Ding, and Yongzhu Li

Abstract By analyzing the filtering mechanism of the filter bag and the deficiency of static injection method based on fixed pressure difference, three models are selected to predict the pressure difference data. Experimental results show that all three models can predict the change trend of the pressure difference data. The MSE of BP neural network is 0.00994, indicating that the pressure difference data is predictable. The method of dynamic injection with pressure difference as control parameter is feasible. In order to solve the problem of flood of fault alarm information of pulse bag filter, the unsupervised learning method in machine learning was analyzed for the data set of fault type. By comparing the unity of the classification results of the three algorithms, it was found that 67% of the three algorithms have the unity 1, and the unity rate of the three algorithms was greater than 0.998 under different K values. It is proved that this classification algorithm is feasible and has good prediction accuracy.

Keywords Pulse bag filter · Data mining · Fault classification

1 Introduction

A large amount of operation data is generated during the operation of the dust collector equipment. Mining the potential value of the data and discovering the operation rules of the equipment is an important development direction of the dust collector field in China in the data era. Applying artificial intelligence technology to the intelligent management system of the dust collector can effectively accelerate the intelligent development of the dust collector field. The filter efficiency of the bag filter is closely related to the dust characteristics of the filter material, the material of the filter bag and the working condition. Different particle sizes, dust relative humidity

Q. Wang · W. Han · W. Ding (✉)
Ningxia University, Yinchuan 750021, China
e-mail: dwjnet@nxu.edu.cn

W. Ding · Y. Li
Qingdao Dongyun Intelligent Environmental Protection Technology Co, Ltd, Qingdao 266000, China

M. Y. Ismail et al. (eds.), *Technological Advancement in Mechanical and Automotive Engineering*, Lecture Notes in Mechanical Engineering,
https://doi.org/10.1007/978-981-19-1457-7_72

and different filter media were analyzed to study their effects on filtration performance [1–5]. The dust removal effect of pulse bag filter is affected by many aspects. The flow rate of dust-containing gas, dust concentration and pressure drop limit all influence the dust accumulation on the surface of filter bag into powder. Therefore, many factors need to be considered when analyzing the dust removal efficiency of pulse bag filter [6–8]. The researchers studied the influence of tank pressure, nozzle diameter, pulse duration and the distance between nozzle and filter element on filter bag dust removal pressure [9]. The aerodynamic of pulsed jets around different types of nozzles and computational fluid dynamics (CFD) of the gas entering the filter bag were simulated to explore the influence of nozzle shape and size on the efficiency of pulsed cleaning [10, 11]. Li [12] analyzed the influence of ash removal methods under different particle sizes on the pressure difference and dust emission concentration. The research results show that in the filtration process, the use of constant pressure difference to ash cleaning dust for medium and fine particles can ensure better filtration and ash removal performance, while for large particles, the influence of constant pressure difference and timing modes on filtration and ash removal performance is limited. Shim [13] compared the cleaning performance of different air nozzles and proposed a double-slit nozzle. The new nozzle uses low pulse pressure to produce the same cleaning effect as the high pulse pressure, which can effectively reduce the power consumption of the dust collector. From the aspects of cleaning efficiency and energy consumption, Cankay [14] designs corresponding control algorithms for nine operating conditions and optimizes the dust removal system, which effectively reduces the energy consumption of compressed air and the generation of waste oil, as well as the oil temperature and operating noise of its blower. According to the characteristics of the big data of the bag filter, Wang [15] chose to use the big data mining XGBoost model to mine the data of the bag filter life cycle. At the same time, the ant colony algorithm was used to optimize the parameters, which avoided the model from entering the local optimal solution and improved the prediction accuracy of the model.

A lot of research has been done on the filter mechanism, dust removal mechanism and control system of the bag filter at home and abroad, but relatively few analysis of the overall equipment is carried out. Among them, foreign countries have begun to use the data-driven method earlier than domestic research and obtained certain research results. However, there are still shortcomings in the analysis and research of the actual operating data of the pulse bag filter. In order to improve the automation, digitalization and intelligence of the pulse bag filter, the mining and analysis of the multi-dimensional actual operation data should be combined. Based on the research and analysis of the operation mechanism of pulse bag filter, uses data-driven means to analyze the dust removal industrial data, combines the advantages of data mining processes and machine learning algorithms, and establishes the algorithm models of pulse bag filter to complete it. Intelligent analysis and research on pulse bag filter.

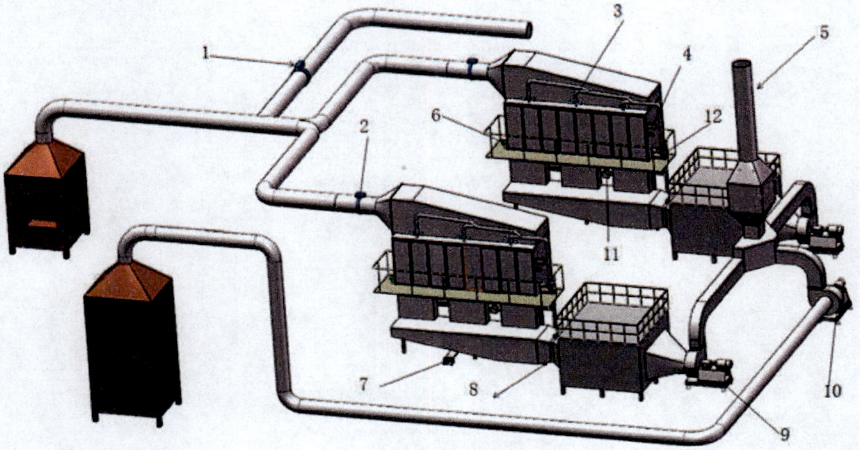

Fig. 1 3D model of dust collector, 1—Emergency bypass valve, 2—Front valve, 3—Pulse valve, 4—Differential pressure meter, 5—Dust detector, 6—Back blow trolley, 7—Public screw, 8—Back valve, 9—Main fan, 10—Cooling main fan, 11—Screw, 12—Back blow drye

2 Collection of Dust Removal Data and Data Mining Scheme

2.1 Pulse Bag Filter

The pulse bag filter is a mechanical equipment, including multiple mechanical parts, which are mainly composed of a filter bag system, a diversion system, a dust removal system, a control system and a cabinet. The pulse bag filter for the cast aluminum industry is selected as the research object. The three-dimensional model distribution of the dust removal system is shown in Fig. 1.

2.2 Collection of Dust Removal Data

In order to be able to analyze the working status of each equipment in the long-term operation of the dust removal system, it is necessary to collect, store and analyze the operating parameters of each equipment, and choose to arrange a pressure you difference detector and a main temperature director on the pulse bag filter. And various sensing components such as dust concentration detector to monitor equipment status parameters. Figure 2 is the interface of the information collection system of the dust collector. The system records the working status data of the dust collector in detail, which also contains the fault information of each component. The operating data of equipment collected by the sensor is transmitted to the industrial control software

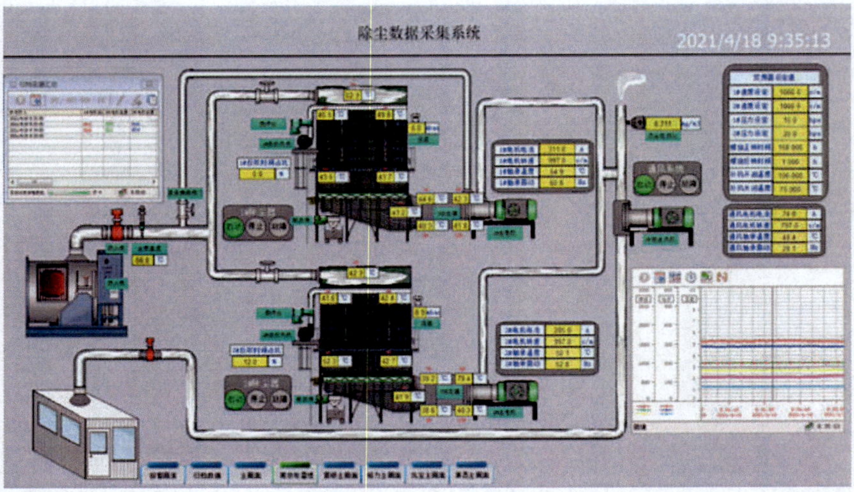

Fig. 2 Data acquisition system

interface of the host computer, so that the technician can view the real-time status of the equipment and store the data in the database at the same time. The sample data in these databases completely saves all the states of the dust removal equipment when it is working. A large amount of data is the data when the dust collector is working normally.

2.3 Data Mining Scheme

Data mining is widely used in various fields, mining direction for classification, clustering, association, prediction and data mining strictly abide by the mining process and specific process is as follows in Fig. 3.

For pulse bag filter in the filter soot cleaning conflict potential contact, fault analysis and fault classification broad three aspects carries on the analysis, the mechanism analysis on the cause of the problem, and through the historical data of the key parameters to observe its change characteristic, choosing the appropriate analysis method for model establishment and model validation experiments. The data mining process of pulse bag filter inherits the characteristics of traditional data mining process, which can be briefly described as follows: analysis of existing problems, source of problems and fault mechanism, feature engineering of data level, model building, which is the characterization of problem characteristics, test and application of model.

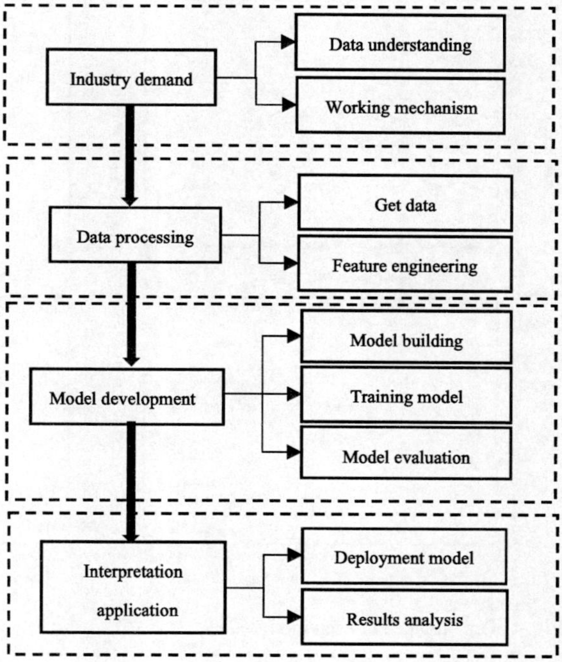

Fig. 3 Flowchart of data mining

3 Prediction Analysis of Pressure Difference Time Series

3.1 Model Training

Python programming language is selected to build and train the model on the machine learning platform Tensorflow. After processing the data set, after choosing the programming language and machine learning framework, you need to build and train the model. In this paper, Long Short-Term Memory (LSTM), Back Propagation (BP) and Support Vector Regression (SVR) models are selected to analyze the predictability of differential pressure data and compare the prediction accuracy of different models. The training results are shown in Fig. 4. According to the performance of different sliding window widths models, it is found that when the sliding window width is 25 min, the three models have relatively good prediction accuracy. Therefore, the sliding window width of 25 min is selected as the subsequent training parameter.

This paper choose the LSTM, BP and SVR model analysis the predictability of differential pressure data, and compared different prediction precision of the model, including LSTM model specific parameter information is as follows in Table 1.

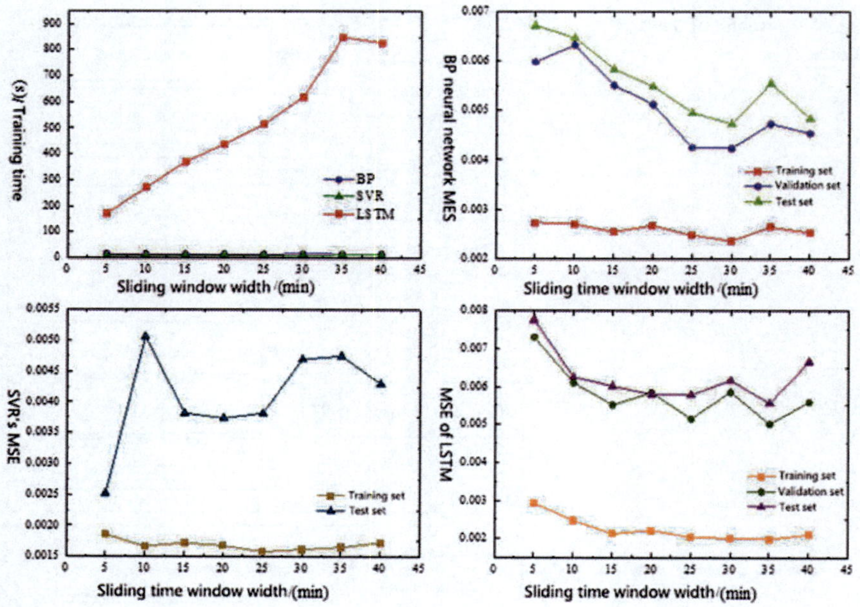

Fig. 4 Performance of each model

Table 1 Model parameter information

Attribute	Parameters	Attribute	Parameters
First layer	LSTM(32)	Iterations	50
Second layer	LSTM(32)	Bath_size	10
Third layer	LSTM(32)	Optimizer	Adam
Fourth layer	Dense	Vector	0.001
Input_shape	(w,1)	Loss function	MSE

3.2 Model Verification Experiment

In order to test the prediction accuracy of the model in the differential pressure data, first choose to predict the short-term time length of three hours, analyze the prediction data of the three models and the MSE size of the real data, the MSE of the three model of BP, LSTM and SVR are respectively 0.00904, 0.01330 and 0.01328, and the prediction results of the model are shown in Fig. 5. From the Fig, it can be found that the three predictions for pressure difference data have good prediction accuracy, and all can predict the change trend of pressure difference. Among them, comparing the MSE of the three models, it can be found that the MSE of the BP neural network is the smallest, indicating that the BP neural network has the highest prediction accuracy under a long time span. The MSE of the three model of BP, LSTM and

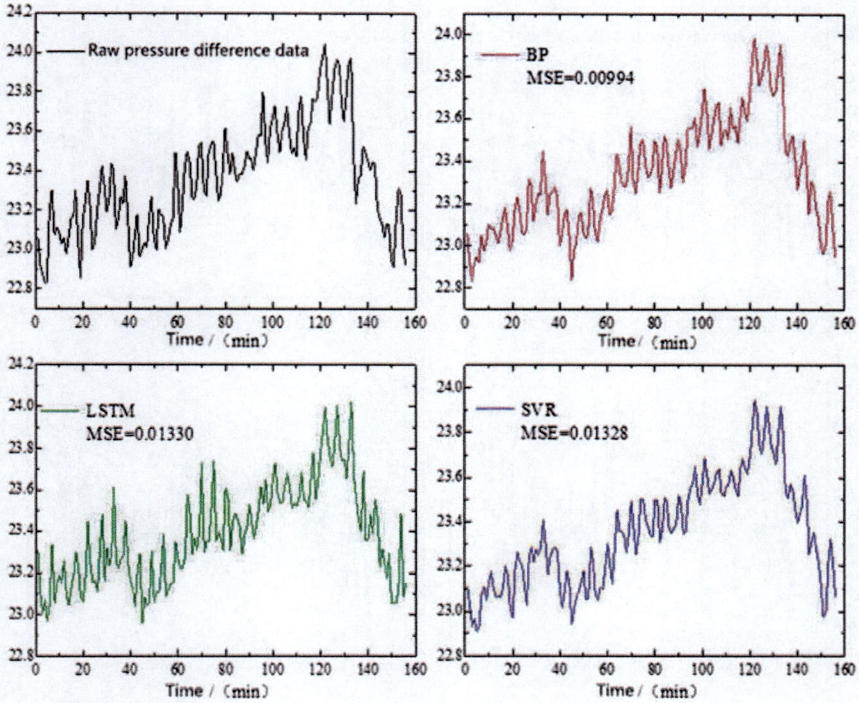

Fig. 5 Performance of each model in short-term data set

SVR are respectively 0.001872, 0.002199 and 0.002316, the fitting results are shown in Fig. 6.

According to the prediction analysis results of long and short time span, it can be seen that the three models have good prediction ability for the pressure difference data and the change trend of the pressure difference data. It can provide high and low pressure injection control mode with double differential pressure threshold, and provide good accuracy of differential pressure data prediction. This shows that it is feasible to use machine learning model to predict differential pressure time series.

4 Finite Element Analysis of Dust Chamber Fault

Perform stress analysis on the three-dimensional model of the dust removal chamber. The stress analysis result is shown in Fig. 7. According to the stress distribution on the surface of the dust collector chamber, it can be judged that the stress in the upper middle and upper part of the dust collector is relatively large. The intersection of adjacent parts with dust gas produces stress.

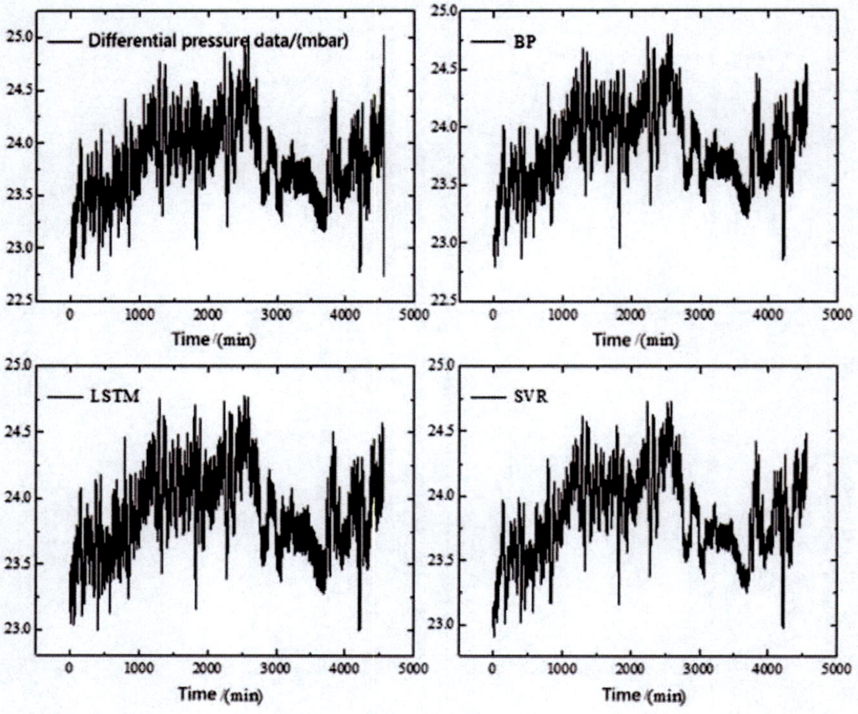

Fig. 6 The performance of each model in the long-term data set

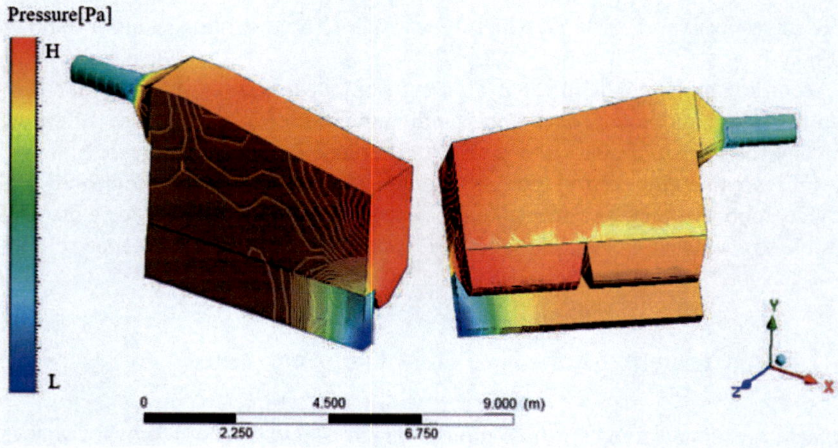

Fig. 7 Stress distribution

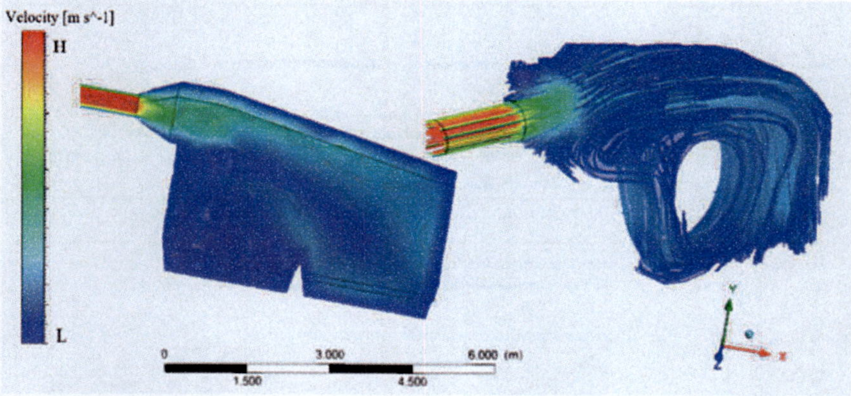

Fig. 8 Flow field distribution

The analysis of the flow field in the dust collector is helpful to the failure mechanism of the internal parts of the dust collector and provides data support for the structural improvement of the dust collector. The results of the flow field analysis inside the dust collector are as shown in Fig. 8. It can be seen that the dust-containing gas in the dust collector has the highest velocity at the entrance. When entering the dust collector chamber, the pipe diameter increases sharply and the flow rate drops rapidly, on the upper wall of the dust collector chamber. The flow velocity is larger, which is consistent with the larger stress on the upper wall surface in the three analysis.

5 Fault Classifier

In order to solve the problem of the overflow of fault alarms, an automatic classifier is trained based on the machine learning model with or without supervised learning algorithm. Firstly, the k-means method is used to reduce the dimensionality of the alarm types. The optimal K value is selected to be 11 by DBI index. The optimal K value is shown in Fig. 9. According to the center of each cluster after clustering will failure to add category labels, the random forests and support vector classification algorithm on fault alarm data set containing the new label fault classifier model is set up, compare the random forest algorithm, support vector classification and K-means the methods of test sets off label failure data sets for fault type uniform rate analysis of results. The results show that when the three classifiers take different K values, the unified rate of the prediction results on the test set is 1, accounting for up to 67%, and the unified rate of the prediction results of the three models is more than 0.998, showing a high degree of unity. The algorithm unified rate diagram is shown in Fig. 10. It shows that the establishment of fault alarm classifier for pulse bag dust collector is feasible and has good classification accuracy, which can help

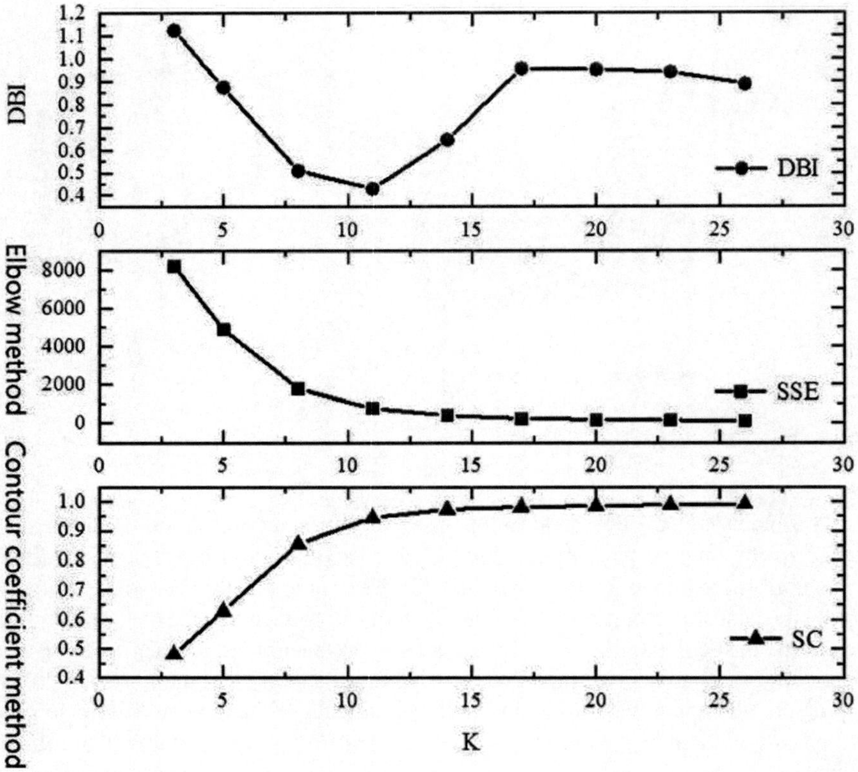

Fig. 9 Optimal K value diagram

maintenance personnel to locate and troubleshoot faults quickly and improve the intelligence level of the equipment.

6 Conclusion

The experimental results show that the three models are good for the prediction of pressure difference data and change trend, and the MSE of BP neural is the lowest, which is 0.00994 in the short-term prediction and 0.001872 in the long-term prediction, proving the feasibility of the prediction of pressure difference data. According to the three evaluation indexes, the optimal K value is 11, and it is found that when K is different, the unity rate of the three algorithms is 1, the highest is 67%, and the unity rate of the prediction results of the three algorithms is greater than 0.998, indicating that this classification method is feasible and the prediction accuracy is good. The data-driven method is used to study the pulse bag filter field. Compared with the traditional analysis method, good results have been obtained. It

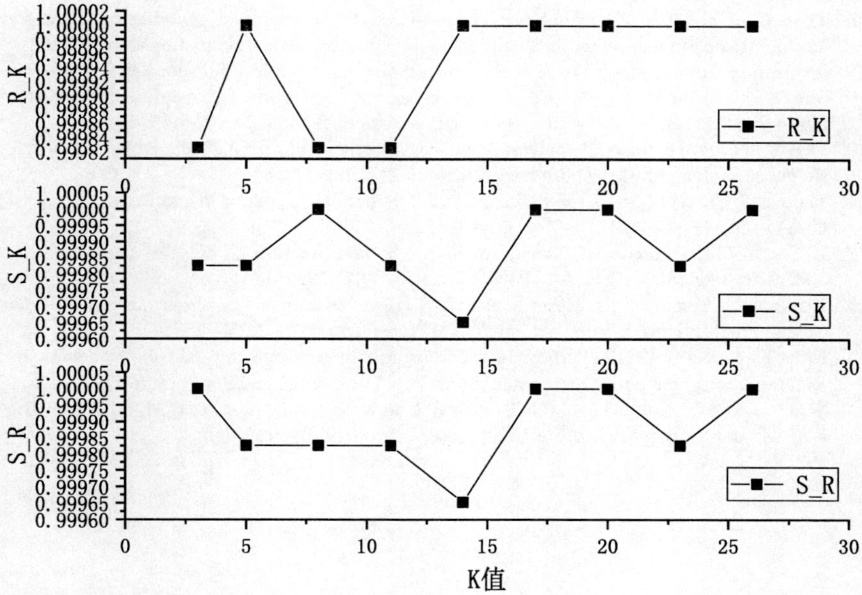

Fig. 10 Algorithm unification rate diagram

proves that the data-driven method, combined with data mining and machine learning algorithms, has a wide range of applications in the field of dust removal, but there are still some shortcomings and to areas for improvement.

References

1. Ribeyre Q et al (2017) Impact of relative humidity on a nanostructured filter cake—experimental and modelling approaches. Chem Eng Sci 161:109–116
2. Li J, Zhou F, Li S (2017) Experimental study on the dust filtration performance with participation of water mist. Process Saf Environ Prot 109:357–364
3. Samirys SRC, Eduardo HT, Mônica LA (2019) Experimental investigation of particle deposition in filter media during filtration cycles with regeneration by pulse jet cleaning. Process Saf Environ Prot
4. Xavier S et al (2014) Downstream particle puffs emitted during pulse-jet cleaning of a baghouse wood dust collector: Influence of operating conditions and filter surface treatment. Powder Technol 261:61–70
5. Xie B et al (2020) Improving filtration and pulse-jet cleaning performance of metal web filter media by coating with polytetrafluoroethylene microporous membrane. Process Saf Environ Prot 136:105–114
6. Yun-Haeng J, Joonmok S, Hyun-Seol P (2017) Evaluation of the can velocity effect on a bag filter. Powder Technol 321:454–457
7. Mahmood S et al (2012) Influence of operating parameters on cake formation in pilot scale pulse-jet bag filter. Powder Technol 224:28–35

8. Zhou G et al (2020) Dust removal effect of negatively-pressured spraying collector for advancing support in fully mechanized coal mining face: numerical simulation and engineering application. Tunn Undergr Space Technol Incorporating Trenchless Technol Res 95
9. Yan C, Liu G, Chen H (2013) Effect of induced airflow on the surface static pressure of pleated fabric filter cartridges during pulse jet cleaning. Powder Technol 249:424–430
10. Li H et al (2016) Numerical analysis on the gas flow dynamics from a rectangular slot-nozzle for pulse cleaning of filter unit. Powder Technol 297:330–339
11. Thomas WCP et al (2016) The influence of the fabric filter layout of in a flow mass filtrate. J Clean Prod 111:117–124
12. Li S et al (2019) Effects of cleaning mode on the performances of pulse-jet cartridge filter under varying particle sizes. Adv Powder Technol 30(9):1835–1841
13. Joonmok S, Yun-Haeng J, Hyun-Seol P (2017) Influence of air injection nozzles on filter cleaning performance of pulse-jet bag filter. Powder Technol 322:250–257
14. Nihat Ç, Muciz Ö (2019) Performance optimization and improvement of dust laden air by dynamic control method for jet pulsed filters. Adv Powder Technol 30(7):1366–1377
15. Wang M, Liu L, Xiong G et al (2020) Research on XGBoost Optimization Algorithm for Big Data Mining for Bag Filter. J Electron Measur Instrum 34(07):159–167

A Study on Tooling Design Procedure for Modeling a Vehicle Part and Its Mold Using CAD/CAM System

Mohd Azuwan Ramli and Mohd Salman Abu Mansor

Abstract The main purpose of this research is to study on a particular procedure for the tooling design due to the design requirements, parameters of the mold and relationship with the plastic injection molding process that can be time-consuming and quite costly. The detail about the design procedure by using CAD/CAM software such as SolidWorks and CATIA are explained. In order to design the mold that has the completed components i.e. leader pin, ejector pins, spruebushing, cooling channel and gate, the mold design need to be prepared of the part design that must be the first design in SolidWorks and then it is exported into CATIA for the mold tooling design process. Finally, the analysis process is conducted by using e-SimpoeWorks to determine the filling time, mold temperature and molten material temperature. It is found that the mold is designed for the vehicle part based on five steps in the procedure and must be suitable to be setup on the injection molding machine. In addition, the output is recommended to be used further for machining of the mold and 3D/4D/5D printing of the additional vehicle part features in the subsequence manufacturing process.

Keywords Tooling design · Mold design · Injection molding · Vehicle part · CAD/CAM · 3D/4D/5D printing

1 Introduction

Plastic injection molding or plastic injection mold design is a main or primary process of manufacturing plastic part. The process involved when raw materials such as thermoplastic or thermosetting is feed into the injection molding machine to through the injection molding process. Then inside the barrel of the machine there have a reciprocating screw to melt the plastic material by force with high pressure into the mold that has core and cavity with desired shaped. After filling up the molten plastic,

M. A. Ramli · M. S. Abu Mansor (✉)
School of Mechanical Engineering, Engineering Campus, Universiti Sains Malaysia, Seri Ampangan, 14300 Nibong Tebal, Seberang Perai Selatan, Pulau Pinang, Malaysia
e-mail: mesalman@usm.my

M. Y. Ismail et al. (eds.), *Technological Advancement in Mechanical and Automotive Engineering*, Lecture Notes in Mechanical Engineering,
https://doi.org/10.1007/978-981-19-1457-7_73

the cooling process takes place and lastly the part injected from the machine and process will be repeated.

For mold design consideration, there have two blocks namely core and cavity [1, 2]. For mold core forms the interior of the molded parts and the mold cavity forms the exterior faces of the molding. Thus, this means cavity and core are the main component to build up the mold and become working parts in the injection plastic process. The plastic part will be produced inside the mold and get cold then will be injected once the core and cavity are opened. The parting direction [3] is the direction to differentiate the core and cavity position. In addition, the parting line was the intersection boundary between surfaces molded by the core, cavity and inserts.

An injection molding machine, there have stationary and moving platen. The moving platen is connected by tie bars to the stationary platen that attached to the barrel side of the machine. Others, the mold consists of the moving mold half and the fixed mold half. When a mold is closed, both halves are in contact and molten plastic start to flow inside it until the mold completely full. After through a cooling process, both halves are separated and the molded is ejected by an ejection system to take out the part and it will push back to original position waiting for the next repeated process.

In injection molding process, the process starts when molten plastic material is injected into the cavity shaped by the mold [4]. The molten plastic flowing into the cavity until the cavity fully fills by molten plastic. After the molten plastic has solidified, the mold is opened and then the molded part is ejected. The provision of performance, flexibility, dependability, high quality and low cost of plastic components has become a key feasible benefit over competitors [5]. The operations of quality control, injection processing, injection mold design and plastic product design, are separate tasks performed by different entities. A lot of quality problems often caused by misunderstanding and poor communication. The complete tool framework should be able to prioritize, to link, to improve and to optimize the output of four tasks.

Complex part geometry will produce complicated mold cavities with moving cores probably [6]. Location pins, ejectors and runners must also become together. The machining of such features is quite taxing since typically hard metals are used as mold materials. To make sure the mold can be withstand the pressure of the molding cycle and high temperatures, the hardening of inserts and the additional cooling channels possibly required. These aspects make the mold production become time consuming and costly.

There are four important stages in the plastic injection molding process, i.e. filling, packing, cooling and ejection [7]. Plastic molten material is injected into the two plates form an impression in injection molding [8]. Protrusions or recesses on the molded part that stop them taking away from the mold alongside the parting direction are named as undercuts [9–11]. Undercuts are contained variously at side cavities, side cores, and inserts and mold pins. Lack and cost of skilled labor, declining profit margins and prices, and far shorter time than before for building molds are the major issues that faced by die and mold makers [12]. A lot of mold companies are more dependent on CAD/CAM systems in reducing lead-time of product design and manufacturing [13]. During the mold design and manufacturing processes, the new product design file may need to be modified several times before it is totally finalized.

Typically, the geometric modeling functions that assist the drafting functions of mold design is provided by a CAD system without the essential knowledge to design the molds [14]. Consequently, "add-on" software such as IMOLD® (intelligent mold design) has been built up on the advanced 3D modeling platforms to help the mold design processes. IMOLD® is a knowledge-based software that can be run on the Unigraphics or SolidWorks platform.

In term of cost, the traditional cost estimation approaches are mainly based on cost similarity and have restricted capabilities for precise mold cost estimation [15]. A generic cost estimation methodology has been developed primarily determined by geometric features. The basic concept involves cost modifiers encompass mold complexity factors and cost drivers determined by mold geometric features. After minor modifications, it can be applied for any tool making practices.

The injection mold cooling system is very significant to the productivity of the quality of the molded part and injection molding process [16]. It is not an easy task to get the best location for each channel of the cooling path to avoid interference with the other components and to optimize the cooling system performance. Defects such as the shrinkage of mold products, warpage deformation and thermal residual stress can be improved by uniform cooling [17]. Hence, the optimal cooling system design must attain the goals of the shortest cooling time feasible and consistent cooling at the same time. Undesired defects such as part warpage, sink marks, thermal residual stress built-up and differential shrinkage must be minimized by aiming at reducing cycle time for an efficient cooling system design [18]. In addition, the function of the cooling system is to present thermal regulation in the process of injection molding [19].

In mold design, the processes are become more crucial due to the flow of the design process. The design of mold needs to be done by applying related software that can help to complete the mold design. Hence, during using application on the software, the designer having some tough time due to set the suitable parameters that needs to be put inside the software in term of customer requirements. The complexities of mold design also depend on the part that needs to be produce. After the mold design is done, the mold must be suit into the plastic injection molding machine. Then the machine will be doing the process to flow the plastics material into the mold. The qualities of part produce are mostly the designer taking care. The part produces must avoid from any defect and damages like warpage, shrinkage, and so on. The time to fill up the molten material is also crucial to the processing making mold. Even though the mold design is important but for overall processes this includes the setting parameters on injection machine in order to have a good quality mold made.

For this study, the scopes are most related in applying the software into the design. The mold design must go through some of the steps in software i.e. SolidWorks, CATIA in term of design and e-SimpoeWorks software for the analysis applications. In SolidWorks, the work focuses more on the part design. The part designs that must be produced by the mold in plastic injection molding process must be created with specific parameters. By going through some step then the file from SolidWorks software needs to be converted in suitable file that can be read by CATIA. In CATIA, the part design will continue to be prepared for the mold design. CATIA can create the preparing part for the mold then the process choosing suitable mold types and its parameters. After that, some analysis can be done for the mold part. For the analysis process of the mold, the filling analysis for the part inside the mold can be analyzed as well. The time for the molten material injects into cavity will be determined, including mold and material temperatures.

2 Methodology

In this study, the tooling design procedure involves five steps. The method involves part and mold design by using SolidWorks and CATIA respectively, and analysis by using e-SimpoeWorks.

Step 1: Creating a vehicle part by using SolidWorks software

Firstly, a plane is chosen from sidebar of the software main window, which is a top plane. To create a solid model of the vehicle part that has a curve shape, a number of planes need to be duplicated. In this case, 11 new planes are copied in order to be used for curve sketching. Next, curve shapes are sketched on the planes and then Surface > Loft function is used to make the surface on the curves that have been created.

To have a draft shape design, the wanted shape is drawn with dimension that has been determined. After that, Extrude function is used to generate the solid model. The Extrude menu that conceives the direction of extrude will be made. The purpose of Extrude is to create a solid feature of the part. The length must be assigned to obtain the solid feature. When the solid part has been made, the unnecessary sub part or shape must be removed from the solid part. By using menu Cut > With Surface, the unnecessary shape that needs to be removed from the model is identified. By using the Extrude Cut menu, the unnecessary shape that needs to be cut off is sketched with specific dimension. The unnecessary sub part or shape from the drawing is removed so that only the needed part is left. Then, Fillet menu is used to make a rounded feature on the external surface of the part by selecting the lines that need to be filleting.

For the hole, the shape hole that needs to be removed is sketched and then Extrude Cut menu is used to take out the unnecessary sub part. The chamfer and fillet are used at every edge and side of each internal hole of the part. The edges of the part that need to be fillet are selected. The purposes of using the fillet and chamfer are to create round corner and slant face corner respectively for the internal and external edges of part features. The final vehicle part model is created by using Shell menu to remove material from the solid model for creating a thin-walled feature (Fig. 1).

Step 2: Convert the SolidWorks file to CATIA file

CATIA software will be used for the mold design since it has some interactive function for the tooling design. Thus, after completed the solid model part design

Fig. 1 Shell

by using SolidWorks, the output file must be converted into CATIA file type. That means the file from SolidWorks must be saved in a readable type so that the CATIA software also can open the file. In that case the suitable types of the file that match both of the software is the IGES file and there is another type of the file that also can be used i.e. the STL file, but in this case, the IGES file is more preferred.

Step 3: Open File from SolidWorks into CATIA

After save the design part from the SolidWorks file which was the Initial Graphics Exchange Specification (IGES) file, CATIA software is opened to continue the mold design and part design need to be preparing before it can proceed to mold design procedure. Firstly, the start icon at the top of the menu list is clicked, then mechanical design is chosen and then Core and Cavity Workbench is selected. Then from there, the procedure can be preceding by open the folder file that contains the IGES file that has been saved from SolidWorks software before. By clicking the file and choosing open icons, then the IGES file is chosen on file selection. After that OK button is clicked, the file is transferred into the CATIA for preparing the core and cavity of the part model and the vehicle part model is ready to proceed for the next step.

In the core and cavity workbench, the main purpose is to determine which side became the core and which side became the cavity from part surfaces. Thus, in this case, the icon that symbolizes the main pulling direction definition is clicked. The parameters such as pulling axis system and mold area i.e. core, cavity and others must be set. Other parameters are draft angle and visualization. In this case, for pulling axis area, 1 will be set on z-axis because the mold will be pulled along z-axis and other axis are leaved it 0. In addition, for the mold areas there have core (red), cavity (green), other (blue) and no draft (maroon) that assigned with different colors. Next, the part is selected, and the main pulling direction is defined automatically where surface those parameters with but if the part does not be defined as core and cavity, the manually procedure can be done by selecting the local transfer and select to define what first core or cavity. After the entire definition is done, Preview button is clicked to view the result and if any changes need to be corrected, it can be done as well. After all the core and cavity are being defined with the intended parameters, OK button is clicked to complete the step.

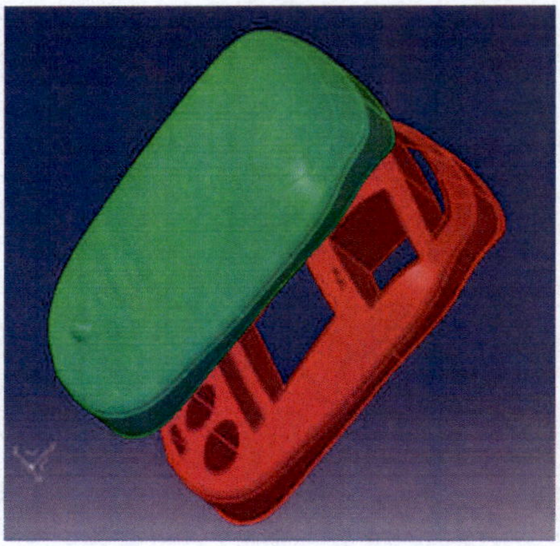

Fig. 2 Completed filled

For the next step in preparing the part was filled the hole inside the part, the purpose to do so is because when core and cavity in mold tooling design is achieved, the hole on the part can be shown so that the part that come out after done the plastic injection molding process will come out with the same as the part model that have been design before in SolidWorks. Therefore, from the core and cavity workbench, the icon that has been used to fill the hole was filled surface definition. Then select the line that need to be closed so that the fill can be done perfectly. After all the line that closed the hole is selected, then OK is clicked to complete the step (Fig. 2).

After all the Core and Cavity definition and also the fill surface definition are done, the part will be proceed into the next step to create mold in Mold Tooling Design workbench for creating the parting surface.

Step 4: Preparing the Part for Creating Parting Surface

The part is selected and then XY plane is chosen from the specification tree. Next, the rectangular icon is taken, and a big rectangular design is made around the part design. The purpose was to create a parting surface along the rectangular shape. For creating the parting surface, Insert icon is clicked on the menu bar to select Surface > Parting Surface. From there the parting surface menu will show up. From Parting Surface menu, the references that used to create the surface is selected, after that white points are clicked on the part design as the vertex 1 and vertex 2. To see the parting surface, the rectangular line that has been created is clicked and the Parting Surface will be created around the line on the part design that has been determined as Parting line.

Then the entire parting surface into the core and cavity on the specifications tree is copied to make sure when creating mold in the Mold Tooling Design workbench, the core and cavity of the mold design can be split or separated. Then, the entire parting surface and the part design that was completed with filling hole are joined to generate the core and cavity parting surfaces.

Step 5: Setting components and parameters for the mold design in CATIA

For setting the parameters in Mold Tooling Design, there were necessary to allocate four components i.e. Type of mold, Leader Pins, Ejector pins and SprueBushing. But there were some other components also need to be considered such as gate and runner if necessary. While additional components like slider and lifter also can be included if any undercuts exist in the part design. But for this study, the part design was decided not to have the undercut. The undercut features can be considered as the additional vehicle part features which have a high potential to be produced by using 3D/4D/5D printing.

After opening the Mold Tooling Design workbench, the complete prepared part is opened by the right click on the product at the specifications tree and select component > existing component. Then, the part that completed with all the preparing like the parting surface and the filling hole will come out. Next, a new mold base is clicked at the right of the workbench. This step was the setting parameters for mold types. For this design, the HASCO was used as the type of mold manufacturer and the size of mold is Mold length: 646 mm and Mold Width: 596 mm.

The step is continued by placing the preparing part into the mold. In the early step, the preparing part is not in centre of the mold which means not really position in the mold. Thus, an adjustment is needed so that the preparing part can be placed exactly inside the mold. By using the manipulation menu, the preparing part will be adjusted according the necessary axis X, Y Z.

Then, Splitting and Adding the Components need to be determined by putting the parameters in the Mold Tooling Design workbench to have a complete mold design with some components such as leader pins, ejector pins, and coolant channel. While mold flow was analyzed by using e-SimpoeWorks in SolidWorks to determine the placement of the Gate. The first step in e-SimpoeWorks analysis was meshing the part. The surface under the MESH icon is double clicked at the menu on the left of the workbench. Then, the required molten material that would like to eject on the mold is selected inside the input parameter by clicking twice on the polymer and choosing the material. For this study, the materials that have been chosen were ABS + PC. Next, the material flow is setting to determine the filling parameters. Three parameters are needed to set a quick simulation of the filling i.e. filling time of the cavity, material temperature and mold temperature.

3 Result and Discussion

For Splitting the Components, Fig. 3 shows that the result for the core and cavity can be obtained, and the shape is similar to the part design that has been split. It was found that by clicking in the specifications tree, the cavity plate was split by the right click on cavity plane and choosing split component. After that the tree on the cavity was opened by clicking the joint that have been made earlier from preparing part for the cavity as splitting element. The direction of the splitting was determined on upwards direction. Then the same step was determined for the core plate but different is in term of splitting element and the direction of splitting component. As a result,

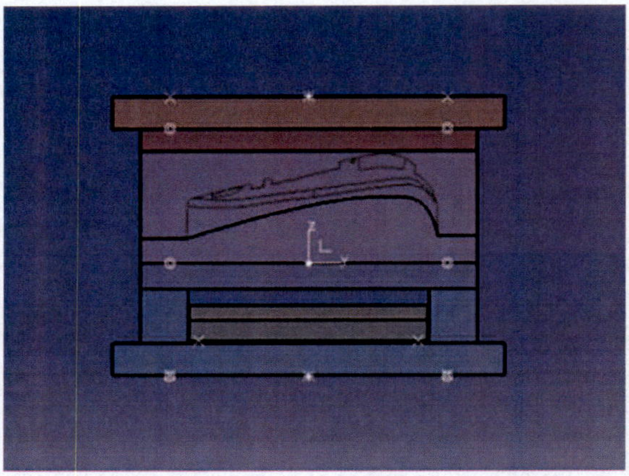

Fig. 3 Desired shape inside mold

the mold can be shaped into the desired part for the core and cavity after the splitting component was done.

For Adding the Components, Fig. 4 shows the complete position of the leader pins after adding the leader pin components. Inside the mold, there were four leader pins that must be placed at each corner of the mold. It was discovered that when the leader pin icon was clicked, then the defined leader pin menu was displayed. Subsequently the supplier inside the catalog was selected for the leader pins that can be defined as Supplier: HASCO, Length: 271 mm and Diameter: 40 mm. The marking points on the mold were identified at each corner to position the leader pins inside it.

Then the SprueBushing was added at the top center of the mold (Fig. 4). Similar to the leader pins, the SprueBushing icon was clicked, and the menu was displayed. The standard drilling for the SprueBushing was from the clamping plate to the core plate

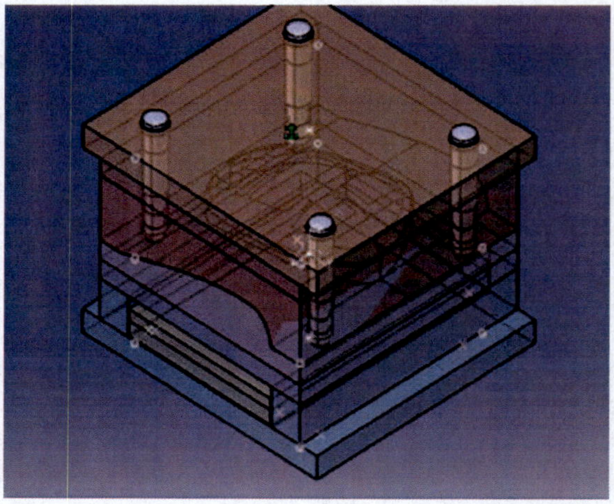

Fig. 4 Complete putting leader pins

and the output was Supplier: DME-AMERICA, Length: 124.619 mm and Diameter: 50.8 mm.

Next the ejector pins were created as shown in Fig. 5 to eject the core and cavity in order to open the mold after the injection molding process completed. The parameters obtained for the ejector pins were Supplier: DME, Diameter: 2.778 mm, Length 1: 355.6 mm and Length 2: 50.8 mm. It was revealed that the step to obtain the ejector pins was the same as the leader pins which is after defining the supplier, the standard drilling was applied from the setting plate into the ejector plate. The locations to place the ejector pins were around the part that must be attached to the core and cavity that shaped the desired part.

For Mold Flow Analysis with Placement of the Gate, SolidWorks has provided add-on module which is mold flow analysis to determine a suitable gate coordinate on the mold. This add on module was named as e-SimpoeWorks. This analysis technique can assist the mold designer to analyze the mold flow, gate placement, filling time, mold temperature and also molten material temperature. The filling parameters that have been computed were Filling time for cavity: 1710.91 s, Material temperature: 265 °C and Mold temperature: 75 °C. All these parameters can be analyzed by automatically by clicking the auto button at operation conditions and the OK button to finish. Inside boundary conditions, the gate selection was double clicked to determine the gate coordinate. Then the suitable place to put the gate was determined and the OK button was clicked to finish. When the filling analysis was run by clicking the RUN button, the filling of molten material into the part was analyzed. Thus Fig. 6 shows the result for a complete filling with a time scale in different colors.

Fig. 5 Locations of ejector pins

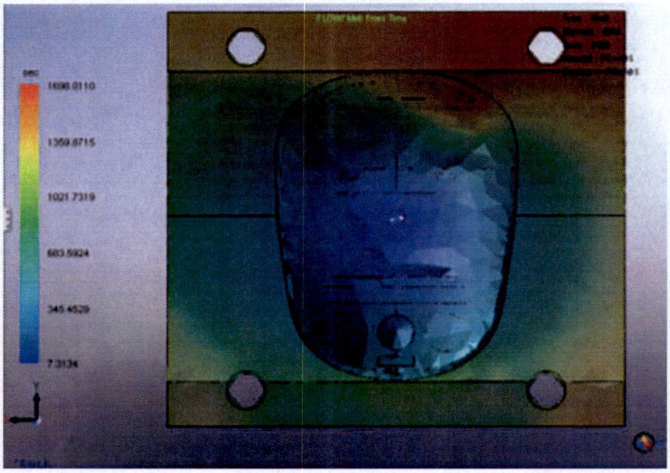

Fig. 6 Complete filling

4 Conclusions

This paper presents a study on tooling design procedure for producing the vehicle part that made by plastic material. As a result, it was found that the mold was designed according to five steps in the procedure for the vehicle part and suitable to be setup on the injection molding machine to perform the plastic injection molding process. In term of design, the mold design was made by applying the related CAD/CAM software i.e. SolidWorks with e-SimpoeWorks and CATIA to fulfill the requirement based on the vehicle part model. The complete mold result can be obtained by studying the tooling design procedure from the design of the vehicle part until the complete design of the mold. Consequently, the output can be recommended to be further used for machining of the mold and 3D/4D/5D printing of the additional vehicle part features in the subsequence manufacturing process.

Acknowledgements This research is supported by Universiti Sains Malaysia under the Research University Grant (No: 811118).

References

1. Md Yusof M, Abu Mansor MS (2017) Automatic core and cavity generation for 3D CAD model using normal vector and scanning ray approaches. ARPN J Eng Appl Sci 12(14):4250–4254
2. Md Yusof M, Abu Mansor MS (2015) Core and cavity generation using slicing and Boolean-based algorithm. Appl Mech Mater 761:175–179

3. Md Yusof M, Abu Mansor MS (2018) Alternative method to determine parting direction automatically for generating core and cavity of two-plate mold using B-rep of visibility map. Int J Adv Manuf Technol 96(9):3109–3126
4. Priyadarshi AK, Gupta SK (2004) Geometric algorithms for automated design of multi-piece permanent molds. Comput Aided Des 36(3):241–260
5. Yeung VWS, Lau KH (1997) Injection moulding, 'C-MOLD' CAE package, process parameter design and quality function deployment: a case study of intelligent materials processing. J Mater Process Technol 63(1–3):481–487
6. Dunne P, Soe SP, Byrne G, Venus A, Wheatley AR (2004) Some demands on rapid prototypes used as master patterns in rapid tooling for injection moulding. J Mater Process Technol 150(3):201–207
7. Tang SH, Kong YM, Sapuan SM, Samin R, Sulaiman S (2006) Design and thermal analysis of plastic injection mould. J Mater Process Technol 171(2):259–267
8. Chen L-L, Chou S-Y, Woo TC (1995) Partial visibility for selecting a parting direction in mold and die design. J Manuf Syst 14(5):319–330
9. Md Yusof M, Abu Mansor MS (2018) Undercut feature recognition for core and cavity generation. IOP Conf Series: Mater Sci Eng 290(1):012070
10. Vieten T, Stahl D, Schilling P, Civelek F, Zimmermann A (2021) Feasibility study of soft tooling inserts for injection molding with integrated automated slides. Micromachines 12(7):730
11. Schuh G, Bergweiler G, Lukas G, Abrams JA (2020) Feasibility and process capability of polymer additive injection molds with slide technology. Procedia CIRP 93:102–107
12. Altan T, Lilly B, Yen YC (2001) Manufacturing of dies and molds. CIRP Ann Manuf Technol 50(2):404–422
13. Zhang LP, Fuh JYH, Nee AYC (2003) Tool path regeneration for mold design modification. Comput Aided Des 35(9):813–823
14. Kong L, Fuh JYH, Lee KS, Zhang YF, Nee AYC et al. (2003) A windows-native 3D plastic injection mold design system. J Materi Process Technol 139(1–3 SPEC):81–89
15. Nagahanumaiah Subburaj K, Ravi B (2008) Computer aided rapid tooling process selection and manufacturability evaluation for injection mold development. Comput Ind 59(2–3):262–276
16. Li CG, Li CL (2008) Plastic injection mould cooling system design by the configuration space method. Comput Aided Des 40(3):334–349
17. Lin Z-C, Chou M-H (2002) Design of the cooling channels in nonrectangular plastic flat injection mold. J Manuf Syst 21(3):167–186
18. Tang LQ, Chassapis C, Manoochehri S (1997) Optimal cooling system design for multi-cavity injection molding. Finite Elem Anal Des 26(3):229–251
19. Li CL, Li CG, Mok ACK (2005) Automatic layout design of plastic injection mould cooling system. Comput Aided Des 37(7):645–662

Effect of Laser Surface Modification on SS316L Surface Roughness and Laser Heating Temperature

A. Q. Zaifuddin, M. D. Afiq, M. H. Aiman, M. M. Quazi, and M. Ishak

Abstract Nowadays, stainless steel is widely used in laser processing applications, including laser heating, laser brazing, and laser welding. However, it has poor optical properties due to low laser energy absorption. However, this could be improved with the aid of laser surface modification (LSM). The significance of this work is to examine the influence of LSM laser power on the surface roughness of 316L stainless steel samples. First, the LSM laser power was varied from 15 to 27 W. Then, the surface topography and variation of the surface roughness values were examined by using a 3D optical microscope. Furthermore, the modified surface by LSM will be heated using laser radiation in order to analyze the effect of surface roughness towards laser heating temperature. The result revealed that as the LSM power increased, thereby resulting in an increase of surface roughness. The highest LSM laser power (27 W) produced the highest surface roughness with 28.98 μm. Experimental results illustrate that the heating temperature were increased 36%, corresponding to a polished flat reference surface, which indicates the increment in energy absorptivity.

Keywords Laser surface modification · Surface roughness · Stainless steel

1 Introduction

Austenitic stainless steels are among the essential materials for the biomedical implant and medical industries [1]. Resulting from their excellent corrosion resistance and biocompatibility, they have found widespread applications as implant casing, scalpels, and haemostats [2, 3].

Although stainless steel alloys are currently the materials of choice for these applications, there are still some serious concerns with their use, especially when it involves laser processing such as laser heating, laser brazing and laser welding. The

A. Q. Zaifuddin · M. D. Afiq · M. H. Aiman (✉) · M. M. Quazi · M. Ishak
Faculty of Mechanical and Automotive Engineering Technology, Universiti Malaysia Pahang, 26600 Pahang, Malaysia
e-mail: aimanmh@ump.edu.my

M. Y. Ismail et al. (eds.), *Technological Advancement in Mechanical and Automotive Engineering*, Lecture Notes in Mechanical Engineering,
https://doi.org/10.1007/978-981-19-1457-7_74

largest worry is related to the laser-metal interaction. Due to its high reflectivity and low absorptivity, stainless steel possesses some difficulties when dealing with lasers. It has been reported that its absorption is only 30% [4], thus wasting the remaining laser energy towards the ambient.

Therefore, many attempts have been made to enhance metal optical performance. It was revealed that the surface engineering method is a promising way to improve metal's absorptivity [5]. For instance, Wang et al. [6] proved that by creating texture and increasing the roughness on the surface, the absorption rate of material would be increased compared to the unprocessed surface. Moreover, Kurp et al. [7] examined the effect of the surface modification on laser absorption and concluded that the absorptivity of the modified surface increased in relation to the absorption of the surface, which did not undergo modification. Besides, another author, Obeidi et al. [8] also confirmed a noticeable improvement in absorbed energy on the rougher surface of stainless steel, saving up to 62% of laser power usage. Finally, according to a paper from Taotao et al. [9] the increment of surface roughness could lead to a considerable increase in irradiation temperature, which indicates an increase in laser absorption level.

Thus, the commonly used surface treatment methods include laser surface modification (LSM), chemical etching, and sandblast. Among these surface engineering treatment techniques, LSM possesses some advantages, including easy automation, high processing speed and precision [10]. Apart from that, given the non-contact nature of this process, the contamination of the workpiece is easily avoided [11, 12]. Hence, this approach has become an excellent alternative to the other conventional methods. The aforementioned studied clearly underline the advantages of LSM. Hence, the present work, the effect of LSM laser power on surface roughness was studied. Additionally, the effect of surface roughness towards the heating temperature during the laser heating process was also investigated.

2 Experimental Setup

2.1 *Material and Sample Preparation*

The experimental works involved stainless steel 316L in the form of a cylinder with a diameter of 6 mm and 6 mm in height. Next, Table 1 presents the chemical composition while the mechanical properties of the stainless steel can be seen in Table 2.

Table 1 Chemical composition of SS316L

Elements	Cr	Ni	Mn	Mo	Si	N	P	C	S
(wt%)	16.5–18.5	10–13	<2	2–2.5	<1	0.11	0.045	<0.07	0.03

Table 2 Thermal properties of SS316L

Properties	SS 316L
Melting point (°C)	1400
Thermal conductivity (W/mK)	16.3

The as-received sample of SS316L was thoroughly prepared as it will highly affect the result during surface characterization. First, the stainless steel was ground manually using 600 grit silicon carbide sandpaper accompanied by water flow to remove any residues and dirt. Afterwards, the polishing process was conducted on three different cloths, which are 6 μm, 3 μm, and 0.05 μm. It should be noted that grinding and polishing were employed to obtain a uniform surface with minimum roughness. Then, the samples were cleaned with acetone prior to processing. The purpose of acetone usage on the cleansing process is to degrease any residual oil contamination on the sample's surfaces.

2.2 *Laser Surface Modification (LSM)*

The laser surface modification experiment was conducted using a fibre laser marking machine, Herolaser ML-MF-A01 IPG, which has 30-W peak power. The machine was operated in pulse wave mode. In order to increase the surface roughness, the polished surface was marked with hatched parallel lines, which have 0.001 mm line spacing. The beam size of the laser is 0.05 mm. Hence, there was spot overlap between each laser scan. Furthermore, the focal length, pulse frequency, scan speed of the laser was kept constant, and only laser power was varied throughout the experiment. The laser input variables and the experimental setup for the LSM process can be seen in Table 3 and Fig. 1, respectively.

Table 3 Laser surface modification (LSM) process parameters

Specification	Parameter
Operational mode	Pulse
Focal length (mm)	196
Beam size (mm)	0.05
Hatching distance (mm)	0.001
Pulse frequency (kHz)	20
Scan speed (mm/s)	200
Laser power (W)	15, 18, 21, 24, 27

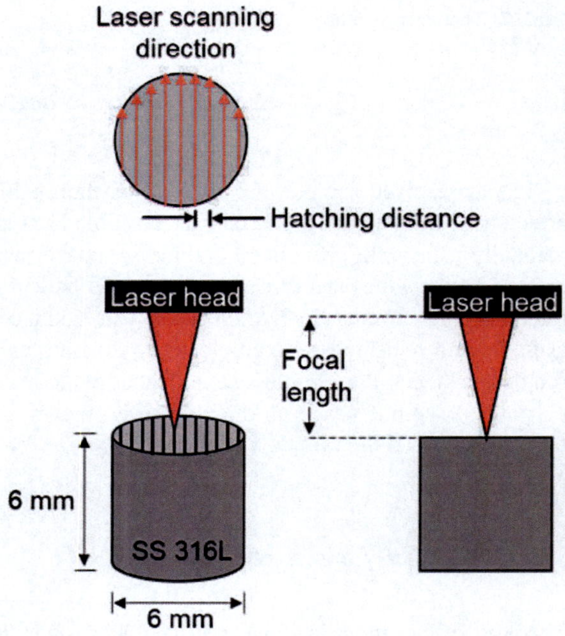

Fig. 1 Laser surface modification (LSM) schematic diagram

2.3 Surface Characterization

After conducting the LSM process, the modified samples will be characterized to study their roughness properties. The analysis was conducted by using a 3D measuring laser microscope from Olympus. This laser microscope follows the ISO4287:1997 standard. The roughness properties extracted from the analysis were areal roughness (SA). Areal roughness is the average roughness over the surface area. The characterized areas were set to 646 μm × 646 μm and all readings were repeated three times for consistency.

2.4 Surface Characterization

Subsequently, the modified surface of the stainless steel rod was heated by employing laser radiation at ambient temperature. The laser heating experiment was conducted using another fiber laser machine. All the process parameters such as laser power, focal length, and heating duration were kept constant and can be referred to in Table 4.

During the laser heating process, the temperature was measured simultaneously using infrared thermometers CTLM2HSF300-C3 from Micro-Epsilon. It is worth noting that the device can only measure temperature starting from 385 °C up to

Table 4 Laser heating process parameters

Specification	Parameter
Operational mode	Pulse
Focal length (mm)	196
Heating duration (s)	30
Laser power (W)	21
Infrared thermometer measuring range (°C)	385–1600

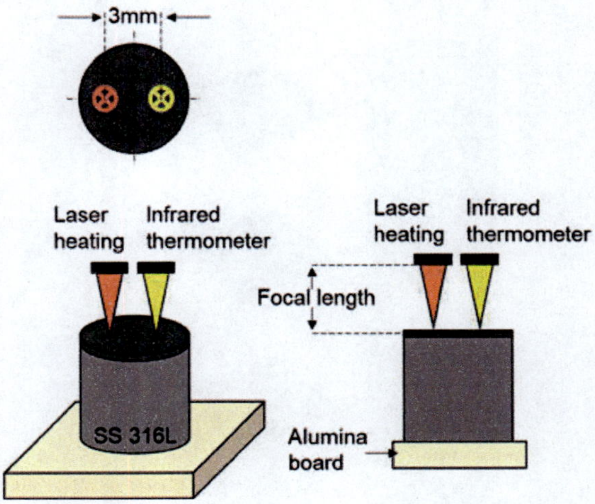

Fig. 2 Laser heating schematic diagram

1600 °C. The laser and infrared beams were placed 3 mm away from each other to avoid both beams from overlapping. Moreover, an alumina board was placed under the stainless steel rod to minimize heat loss. The schematic diagram of the laser heating and temperature measurement processes are shown in Fig. 2.

3 Result and Discussion

3.1 Effect of Laser Surface Modification Power on Surface Roughness

Following the grinding and polishing process, the roughness of the polished surface was measured. Table 5 presents the topographical modifications that emerged on the surface of stainless steel before and after laser irradiation under different power values. Due to variations in the laser power, the changes in surface topography can be seen clearly.

Table 5 Surface topography of textured stainless steel 316L surface under different LSM power

LSM power (W)	2D height image	3D height image	Height scale	2D Optical Image	3D optical image
0 (polished)					
15					
18					
21					
24					
27					

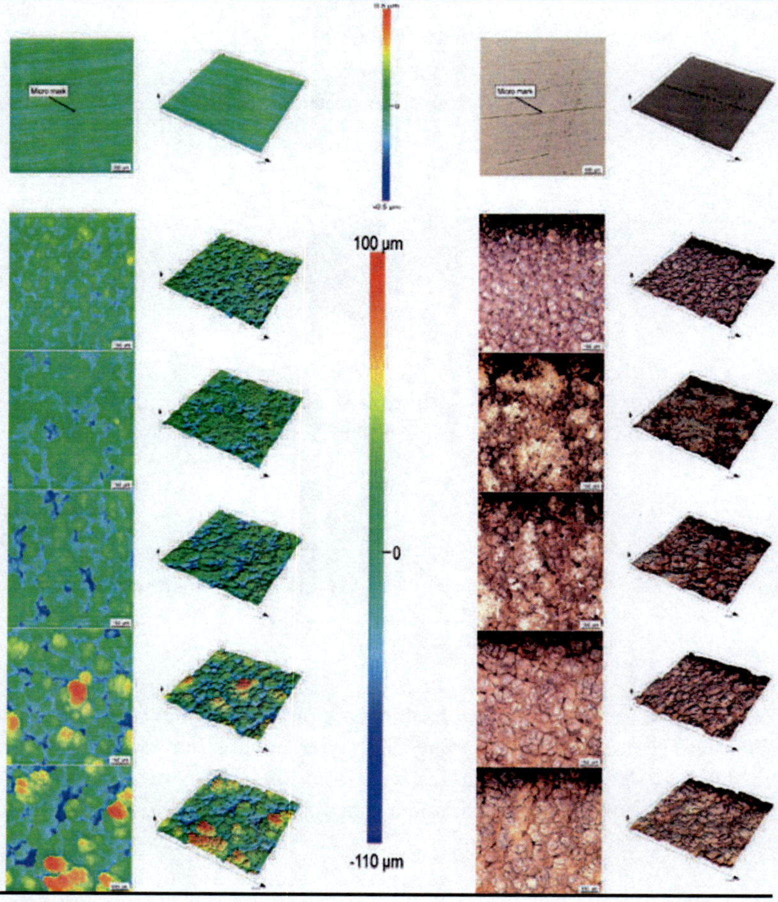

Initially, as a reference from the polished surface, there were sub-micron scratch marks left by mechanical polishing on the surface of the samples. Nevertheless, this polished stainless steel surface can be considered a relatively flat surface since the average of the areal roughness is 0.077 ± 0.006 μm.

Moreover, based on the sample overview, it is observed that high power input generates a new surface texture that looks rougher with some irregularities and more chaotic pattern than low power. As a reference to the colour level used for the height image, it shows that some non-uniform high peaks (red/yellow) and deep valleys (blue) are present on the sample modified with high power. On the other hand, the surface modified with low laser power tends to follow the scan direction but with irregular periodicity. This finding can be corroborated with a similar report recorded by Raza et al. [13], wherein the surface textured with high laser power produced waviness and spikes in random areas.

Subsequently, Table 6 shows the result of variable LSM power on surface roughness, while Fig. 3 presents their relationship. It was revealed that the effect of laser power is very prominent on the roughness of the treated surface. The roughness values increase with increasing laser power for five different laser power ranging from 15 to 27 W. This is because a high laser power is able to radiate more heat energy to the surface [13, 14]. This is in agreement with Fazliana et al. [15], who reported that the increment in surface roughness is due to the increase in laser power. The main factor of this phenomenon is that a huge amount of heat energy from higher laser power could remove more material from the metal's surface. When the material removal increases, it creates more microgrooves and depositions, which will change the surface's texture and, therefore, increase the surface roughness.

Apart from that, this observation was also documented by Purnama et al. [16] on 316L, who reported that high laser energy could contribute to the increment in material removal. It can be supported by the findings from A Sen et al., who studied the

Table 6 Result of variable LSM power on surface roughness

Laser power, p (W)	Average roughness, SA (μm)	Standard deviation, σ (μm)
0	0.077	0.006
15	14.65	0.085
18	16.20	1.14
21	19.81	1.60
24	25.23	0.65
27	28.98	1.17

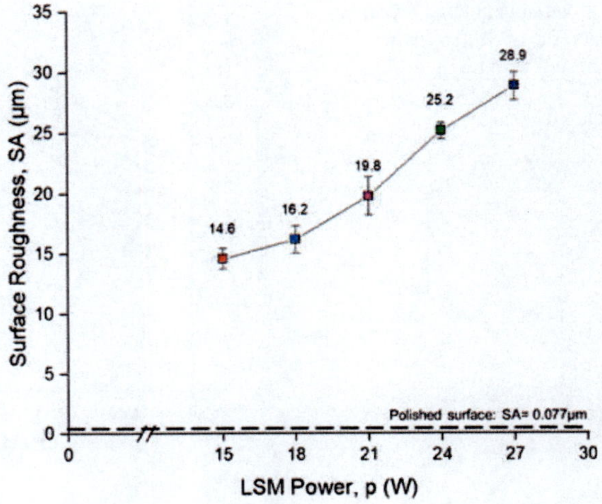

Fig. 3 Relationship between variable LSM power and surface roughness

LSM process with variable power on nickel alloy [17]. In their studies, by changing the laser power from 8 to 23 W, they have produced different submicron-/micro-structures with increased surface roughness. Furthermore, the study conducted by A J Sulaiman et al. showed that when a higher laser power usage is applied to modify a metal's surface, bigger and deeper microgrooves will be produced, increasing the surface roughness [18]. Besides, the increment in laser power was attributed to the increase in energy radiated will also affect the melt pool homogeneity and size, inducing larger melt tracks, thus producing a rough surface [19]. Furthermore, it should be noted that the surface roughness increases approximately 28 times when radiated with the highest laser power (27 W) compared to the polished surface.

3.2 *Effect of Modified Surface Roughness by LSM on Laser Heating*

In the next stage of this study, a laser heating experiment was conducted based on the modified sample during LSM earlier. Figure 4 demonstrates the temperature profile of the modified surfaces when radiated with constant parameters of laser heating. The samples were heated continuously for 31 s. It should be noted that the laser beam ceased at t = 31 s and the natural cooling phase started immediately after that period.

The diagram represents almost the same trend for all modified surfaces when heated with a laser. The temperature measured increased when heated on a modified surface compared to a non-modified surface. It reflects that the laser energy absorption was improved up to 31% compared to the polished surface. This is because when a laser radiates a rough and textured surface, the laser energy will be scattered over a larger surface area and increase the absorptivity into the material.

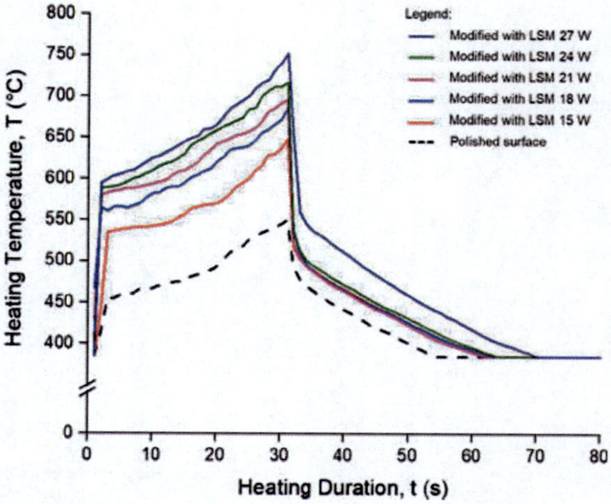

Fig. 4 Temperature profile of variable heating modified surface of SS316L when heated with laser

Table 7 Result of maximum heating temperature corresponding to the surface roughness

Modified with LSM power, p (W)	Average surface roughness, SA (μm)	Maximum heating temperature, T (°C)
0 (polished surface)	0.077	550.0
15	14.65	648.5
18	16.20	685.9
21	19.81	696.3
24	25.23	716.8
27	28.98	752.0

Consequently, Table 7 shows the result of variable surface roughness on maximum heating temperature, while Fig. 5 presents their relationship. It could be seen that maximum temperature (752 °C) was achieved by the modified sample with the highest roughness value (28.98 μm). In contrast, the lowest temperature (550 °C) was achieved by the non-modified stainless steel surface with 0.077 μm roughness. It reflects that the laser energy absorption was improved up to 36% compared to the polished surface.

Accordingly, the result of surface roughness shows a direct effect of the heating temperature on the roughness value of the surface, whereby the temperature increases when radiating a rough surface. This finding can be corroborated with a similar report recorded by Abedi et al. [5], wherein the energy absorptivity increased by increasing the surface roughness. This is because when a laser radiates a rough surface, the laser

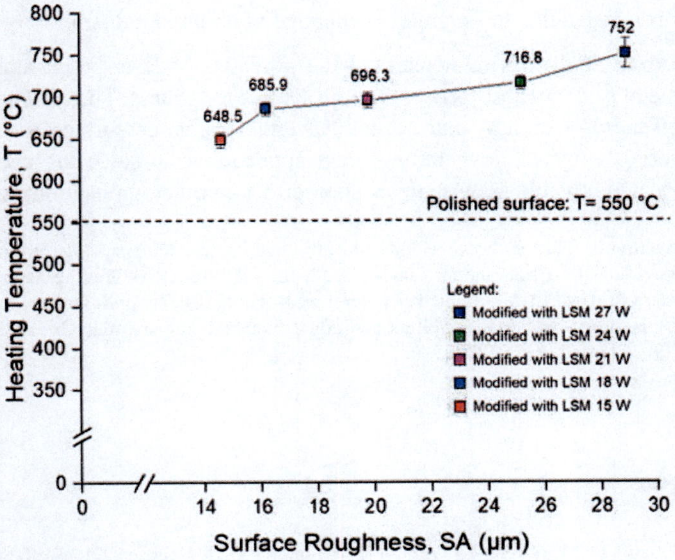

Fig. 5 Relationship between surface roughness and maximum heating temperature

energy will be scattered over a larger surface area and increase the absorptivity into the material, which effectively causes the temperature to rise. Besides, a recent study has been conducted to study the effect of surface roughness on the laser heating process and showed that the heating temperature was affected by changing the topography and surface roughness of the sample [20]. Apart from that, according to Yuan et al. [21], the refraction at the surface also causes transmission at oblique angles, which leads to increment of the light absorption.

4 Conclusion

In this research, experimental investigation, and analysis of the laser surface modification on stainless steel 316L were made. The effects of the laser power were analyzed from the modified profiles in the form of surface roughness values (SA). Furthermore, the effect of the surface roughness on energy absorption during laser heating was studied. From the experimental results, the following conclusions can be drawn:

(a) An increase in LSM laser power to produce textured regions will increase the surface roughness.
(b) Increasing the surface roughness of the stainless steel by LSM could effectively increase the maximum temperature during laser heating when compared with a non-textured surface. A significant improvement of 252 °C is realized contributing to about 36.7% increase when compared with polished sample.
(c) For maximum temperature, laser surface modification power of 27 W is recommended due to roughness generated of about 28.98 μm.

The current study has demonstrated the potential of laser-based approach to increase metal's energy absorptivity through surface roughness. The method features high preciseness, accuracy, and speed. Nevertheless, laser surface modification of engineering alloys is now finding great applications in soldering and brazing technology, and it would be wise to see their effect on other applications as well.

Acknowledgements The authors would like to thank the Ministry of Higher Education for providing financial support under Fundamental Research Grant Scheme (FRGS) FRGS/1/2021/TK0/UMP/02/22 (University reference RDU210127), and Universiti Malaysia Pahang for laboratory facilities as well as additional financial support under Internal Research grant RDU1903118 and PGRS200305.

References

1. Lu Y, Guan YC, Li Y, Yang LJ, Wang ML, Wang Y (2020) Nanosecond laser fabrication of superhydrophobic surface on 316L stainless steel and corrosion protection application. Colloids Surf Physicochem Eng Aspects 604(125259). https://doi.org/10.1016/j.colsurfa.2020.125259

2. Dashtbozorg B et al (2020) A study on the effect of ultrashort pulsed laser texturing on the microstructure and properties of metastable S phase layer formed on AISI 316L surfaces. Appl Surf Sci 511:145557. https://doi.org/10.1016/j.apsusc.2020.145557
3. Kumari R, Scharnweber T, Pfleging W, Besser H, Majumdar JD (2015) Laser surface textured titanium alloy (Ti–6Al–4V)–Part II—studies on bio-compatibility. Appl Surf Sci 357:750–758
4. Quazi MM et al (202) Current research and development status of dissimilar materials laser welding of titanium and its alloys," (in English). Opt Laser Technol Rev 126:36, 106090. https://doi.org/10.1016/j.optlastec.2020.106090
5. Abedi HR, Gollo MH (2019) An experimental study of the effects of surface roughness and coating of Cr_2O_3 layer on the laser-forming process. Opt Laser Technol 109:336–347
6. Wang Y, Ke C, Wu T, Zhao X, Wang R (2019) Nanosecond laser texturing with hexagonal honeycomb micro-structure on Titanium for improved wettability and optical properties. Optik 192:162953
7. Kurp P, Mucha Z, Mulczyk K, Gradon R, Trela P (2016) The influence of surface preparation on the absorption coefficient of laser radiation. In: Jabczynski JK, Romaniuk RS (eds) Laser technology 2016: progress and applications of lasers, vol 10159, Proceedings of SPIE
8. Obeidi M, McCarthy E, Ubani SI, Ul Ahad I, Brabazon D (2019) Effect of surface roughness on CO_2 laser absorption by 316L stainless steel and aluminum. Mater Perform Charact 8(6):1167–1177. https://doi.org/10.1520/mpc20180091
9. Wu TT, Zhu YX, Wei CH, Zhou ML, Wang LJ (2017) The influence of roughness on reflectivity evolution of iron irradiated by 1064nm CW laser. In: Ding Y, Feng G, Hoffmann DHH, Cao J, Lu Y (eds) Fourth international symposium on laser interaction with matter, vol 10173, Proceedings of SPIE
10. Arsyad Y et al (2021) Laser surface modification of duplex stainless steel 2205 to modify the surface roughness. Int J Automot Mech Eng 18(2):8695–8703–8695–8703
11. Riveiro A, Macon ALB, del Val J, Comesana R, Pou J (2018) Laser surface texturing of polymers for biomedical applications. Front Phys 6, 16. https://doi.org/10.3389/fphy.2018.00016
12. Daskalova A et al (2020) Effect of surface modification by femtosecond laser on zirconia based ceramics for screening of cell-surface interaction. Appl Surf Sci 513, 145914. https://doi.org/10.1016/j.apsusc.2020.145914
13. Raza MS, Datta S, Vivekanand K, Saha P (2019) Surface modification of high-carbon steel using fiber laser and its thermal, microstructural and tribological analysis. J Mater Eng Perform 28(3):1873–1883. https://doi.org/10.1007/s11665-019-03943-1
14. Moroni F, Romoli L, Khan MMA (2018) Design of laser-textured surfaces to enhance the strength of adhesively bonded joints. Int J Adhes Adhes 85:208–218. https://doi.org/10.1016/j.ijadhadh.2018.06.001
15. Fazliana F, Aqida SN, Izwan I (2018) Influence of laser processing design on tool steel surface integrity. Proceedings of Asia international conference on tribology 2018, pp 105–106
16. Purnama A et al (2020) Laser surface texturing of SS316L for enhanced adhesion of HUVECs. Surf Eng 36(12):1240–1249
17. Sen A, Doloi P, Bhattacharyya P (2017) Parametric analysis on fiber laser surface texturing of hastelloy C-276
18. Sulaiman A, Aiman M, Quazi M, Ishak M, Ariga T (2020) Enhancement of mechanical properties of copper brazed by laser surface modification. In: IOP conference series: materials science and engineering, vol 788, no. 1. IOP Publishing, p 012012
19. Kaschel F, Celikin M, Dowling D (2020) Effects of laser power on geometry, microstructure and mechanical properties of printed Ti-6Al-4V parts. J Mater Process Technol 278:116539
20. Zaifuddin A, Aiman M, Quazi M, Ishak M, Ariga T (2020) Effect of laser surface modification (LSM) on laser energy absorption for laser brazing. In: IOP conference series: materials science and engineering, vol 788, no 1. IOP Publishing, p 012013
21. Yuan J, Liang L, Lin G (2020) Study of laser surface texturing of 6061 aluminium with thermal assistance. J Micromech Microeng 30(3):035003

Development of Cooling Necklace System Using Vortex Tube

Amirul Nawaf Esa, Mohd Hazwan Yusof, Deyerbeen Sipaan Fredoline, and Muhammad Fadhli Suhaimi

Abstract A vortex tube is an intriguing, simple system capable of producing cold and hot streams at room temperature from a single compressed fluid supply. It is commonly used as a cooling system in many industrial applications because of its remarkable cooling capability and simplicity with no moving parts. This vortex tube does not need any refrigerants or chemical fluids to function as intended as a cooling device. The usage of the vortex tube as a cooling device that has been implemented in a cooling necklace has less to no substantial research data. Therefore, it is hard to determine the best possible outcome that could be achieved by integrating the vortex tube with a cooling necklace. Thus, the main objectives for this research are to determine the performance of vortex tube as a cooling device on a cooling jacket at different cold mass fraction and also to determine the performance of vortex tube as a cooling device on a cooling jacket at different inlet pressure. In this particular research, an experimental study was conducted in two different situational environments, one in a controlled environment which is indoors while the other is outdoors with direct sunlight to simulate real-life situations where the cooling necklace might be used. It can be concluded that cold mass fraction of 0.4 and the highest inlet pressure of 0.4 MPa are the optimum values to obtain the lowest temperature inside of the cooling jacket.

Keywords Vortex Tube · Thermal Comfort · Cooling Necklace

1 Introduction

The Vortex Tube is a simple mechanical device that separates compressed gas into cold and hot streams. This VT was initially invented by French physicist George J.

A. N. Esa · M. H. Yusof (✉) · D. Sipaan Fredoline · M. F. Suhaimi
Faculty of Mechanical and Automotive Engineering Technology (FTKMA), Universiti Malaysia Pahang, 26600 Pekan, Pahang, Malaysia
e-mail: mohdhazwan@ump.edu.my

A. N. Esa · D. Sipaan Fredoline
College of Engineering (KKEJ), Universiti Malaysia Pahang, 26600 Pekan, Pahang, Malaysia

M. Y. Ismail et al. (eds.), *Technological Advancement in Mechanical and Automotive Engineering*, Lecture Notes in Mechanical Engineering,
https://doi.org/10.1007/978-981-19-1457-7_75

Ranque in 1931. However, during this time, the VT was still not perfected and has many flaws disallowing it to function as it is today. It was not till later in 1947 that German physicist Rudolf Hilsch reinvented and perfected this mechanism allowing the tube to work as it was originally intended. He systematically inspected the effect of the geometrical parameters of the VT and the inlet pressure and on its performance. He also presented a possible explanation of the energy separation process. Since then, several researchers start to study the vortex tube to explain the reason for energy separation or enhance its performance [1].

Even though the vortex tube produces hot flow up to 110 °C and cold flow down to −40 °C, it is commonly used as a spot cooling device. In industries, vortex tube is used as a cooling device to reduce the temperature of mechanical parts of machines, environmental chambers, and electrical components. It can also be used to dehumidify gas sample and liquify the natural gas [2].

The usage of vortex tube as a small and compact spot cooling device is recognizable. Therefore, in this research, vortex tube is used as the cooling device to provide cool air to create a thermal comfort to users through a developed cooling necklace. This cooling necklace is hidden inside a jacket. Hence, the temperature of the cold air before and after entering the cooling necklace is measured to clarify the performance of the vortex tube as a cooling device.

2 Literature Review

In a manner corresponding to the previous investigations by various numbers of researchers, several parameters such as nozzle number, inlet pressure, and tube length to inner diameter are studied thoroughly. Almost all these studies have been carried out either by experiment or by computational fluid dynamics. Most of the data collected by these researchers are similar. Through this data, it is noted that the optimum parameters affecting the performance of the vortex tube vary with researchers. To the best of the authors' knowledge, there is less data regarding the performance of a cooling necklace using vortex tube as the cold flow supply. However, the data collected from the literature review helps this study to clarify the cooling mechanism of the vortex tube.

2.1 Effect of Geometrical Parameters to the Performance of Vortex Tube

The effect of different nozzle number, and length to inner diameter ratio (L/D) to the performance of the vortex tube is tabulated in Tables 1 and 2, respectively. It can be seen from these table that the optimum value obtained by each researcher is different. However, most of the researchers stated that nozzle number of 6 is the optimum

Table 1 Effect of different nozzle number

Author	Year	Research type	Nozzle number	Highest value of ΔT_{cold} (°C)	Highest value of ΔT_{hot} (°C)	Cold mass fraction
Kaya et al.[3]	2018	CFD	2/4/6 (Optimum = 4)	39.65 (MPa 0.55)	18.7	0.36
Attalla et al.[4]	2017	EXP	2/3/4/5/6 (Optimum = 3)	25 (MPa 0.6)	–	0.4
Kaya et al.[5]	2019	EXP	2/4/6 (Optimum = 6)	45.9 (MPa 0.55)	–	0.36
Shamsoddini et al.[6]	2018	CFD	2/4/6 (Optimum = 6)	28 (MPa 0.14)	–	0.2
Kirmaci [7]	2009	EXP	2/3/4/5/6 (Optimum = 2)	–	–	0.5

Table 2 Effect of different length to inner diameter ratio (L/D)

Author	Year	Research type	Length to inner diameter ratio (L/D Ratio)	Highest value of ΔT_{cold} (°C)	Highest value of ΔT_{hot} (°C)	Cold mass fraction
Guo et al.[8]	2019	EXP	12/20/30 (Optimum = 20)	13.5 (0.1 MPa)	–	0.28
Im et al.[9]	2012	EXP	14	37 (0.1 MPa)	10	0.5
Valipour et al.[10]	2011	EXP	21	21 (0.3 MPa, CF 0.25)	24 (CF 0.7)	–
Sadi et al.[11]	2014	EXP	10	20 (0.4 MPa, CF 0.38)	80 (CF 0.85)	–

parameter. Therefore, nozzle number of 6 should be considered when designing the vortex tube. For the effect of the length to inner diameter ratio, all researchers reported different optimum value. It can be understood that the combination of different geometrical parameters results a different optimum parameters value.

3 Experimental Set Up

3.1 Experimental Procedure

The performance of the vortex tube on a cooling necklace is to have experimented with the setup configuration shown in Fig. 1. The actual experimental setup is yet to be done in this time duration but will mostly follow the schematic diagram that

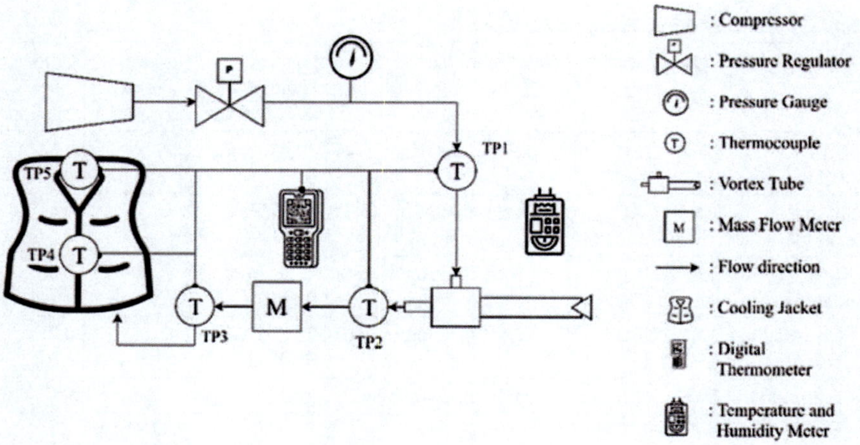

Fig. 1 Schematic diagram of experimental setup

is created to be as efficient as possible. The design of this experiment has been carefully arranged and shown as best as could theoretical wise. Thus, as for the actual experimental schematic, it may change due to unforeseeable variables. The main objective of this schematic is to act as a guideline for future references.

The studies were carried out in two conditions: indoors, in an air-ventilated room, and outdoors, in the afternoon sunlight, as shown in Figs. 2 and 3. The cooling necklace itself is shown in Fig. 4 while it being in use in Fig. 5. For each experiment, temperature data was gathered every 2 min for a total of 20 min (Table 3).

4 Results and Discussion

4.1 *Indoor Results*

From the experiments conducted with it being indoors as a condition stipulated to further test the performance of the Vortex Tube on the cooling necklace, it is found that the best possible parameters to obtain the lowest temperature or in another word, the highest temperature difference of cold exit is to be at 0.4 CF with an inlet pressure of 0.4 MPa. The lowest value for cold air exit is recorded at TP2 with a value of −3.6 °C. Nevertheless, this does not exactly mean that the air entering the cooling necklace to be at −3.6 °C. This is due to the fact that as cold air passes through the tube and flow metre on its way to the cooling necklace, it is exposed to the atmosphere or surroundings. This results in an unstable temperature of the cold air, which may be caused by external factors such as the air flow of the surroundings, the ambient temperature of the surroundings, the flow metre materials, and the cooling necklace materials. The temperature of the cold air going through the PU tube is lower than

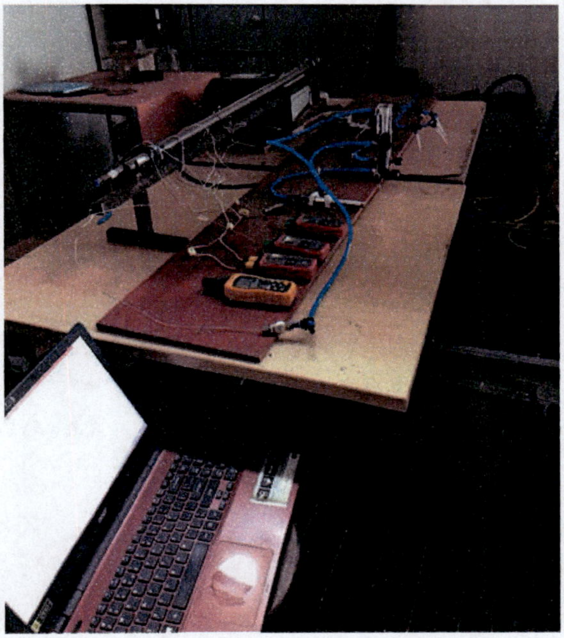

Fig. 2 Indoor experimental setup

Fig. 3 Outdoor experimental setup

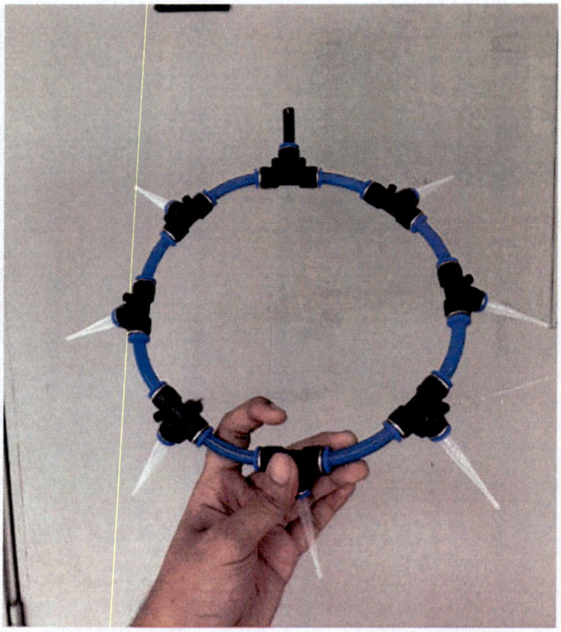

Fig. 4 Developed Cooling necklace

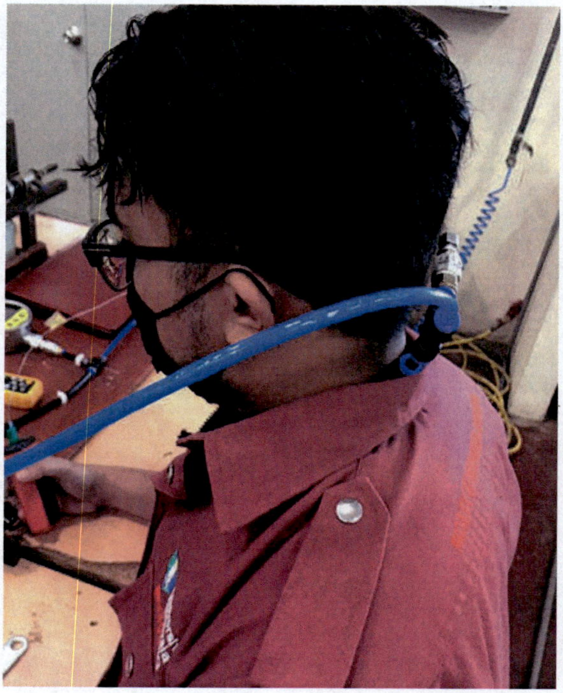

Fig. 5 Cooling Necklace in use

Table 3 Experimental conditions

Inlet pressure, MPa	Cold mass fraction, CF
0.1, 0.2, 0.3, 0.4	0.4
0.4	0.2, 0.4, 0.6, 0.8

the ambient temperature; therefore it tends to reach thermal equilibrium with the ambient temperature before reaching the cooling necklace. This can be said that the greater the difference in temperature drop, the quicker the air inside the tube tries to regain thermal equilibrium. Thus, even though the inlet pressure of 0.4 MPa with a CF of 0.4 shows the lowest temperature, it does not have the cooling effect of what the parameters of 0.4 MPa with a CF of 0.8 has. The reason why is this happens is also partly due to the flow rate of 0.8 CF being higher compared to 0.4 CF (Fig. 6).

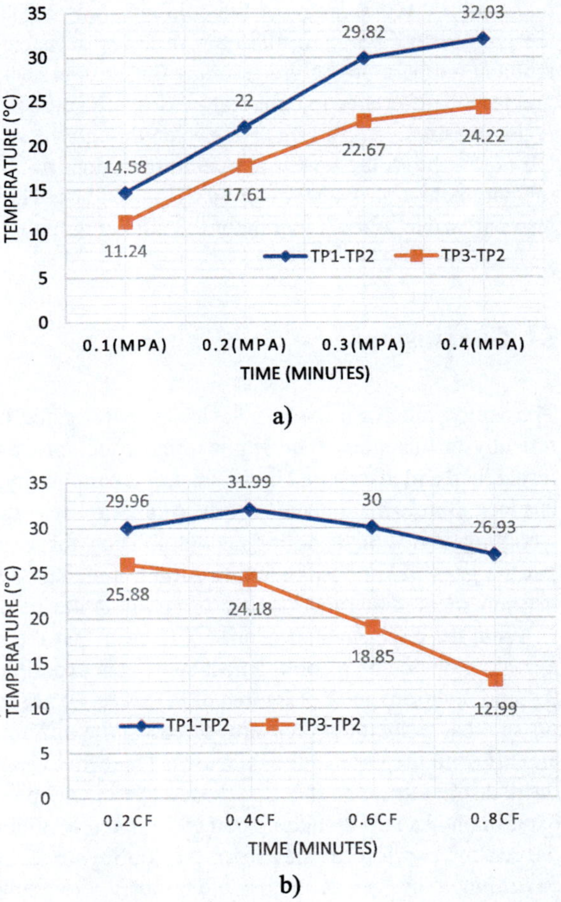

Fig. 6 Average temperature differences, TP1-TP2 and TP3-TP2 against, **a** different inlet pressure with a fixed cold fraction of 0.4, **b** different cold fraction with a fixed inlet pressure of 0.4 MPa

4.2 Outdoor Results

From the experiments conducted with it being outdoors as a perquisite for this time, the performance of the Vortex Tube on the cooling necklace is conducted to determine the best possible parameters. It is done outdoors to further simulate real life conditions of this apparatus being applied in any harsh industries possible. It is found that the optimum parameters to obtain the lowest temperature of cold exit are to be at 0.4 CF with an inlet pressure of 0.4 MPa. This particular parameter results in a cold exit temperature, TP2 of 1.2 °C. This compared to indoors being at −3.6 °C is quite a lot of difference. However, due to the fact that the condition it is being tested in is harsher, this is a very good result to be obtained. However, this does not mean that the inlet temperature of the cooling necklace to be at 1.2 °C. Due to the hotter external temperature, the cold air that passes through the tube and flow meter on its way to the cooling necklace, it does not perform as much as it being indoors. The temperature of the cold air going through the PU tube is lower than the ambient temperature; therefore it tends to reach thermal equilibrium with the ambient temperature before reaching the cooling necklace. This can be said that the greater the difference in temperature drop, the quicker the air inside the tube tries to regain thermal equilibrium. Thus, even though the inlet pressure of 0.4 MPa with a CF of 0.4 shows the lowest temperature, it does not have the cooling effect of what the parameters of 0.4 MPa with a CF of 0.8 has. The reason why is this happens is also partly due to the flow rate of 0.8 CF being higher compared to 0.4 CF (Fig. 7).

5 Conclusions

The vortex tube as a cooling device is investigated both experimentally and theoretically in this study. The application of the vortex tube has not been thoroughly researched and discovered in the scientific community; it has been discovered. There is a lack of information and reports to support the vortex tube's cooling capability as a cooling device on a cooling necklace. Thus, this experiment was conducted to fill out the gaps in knowledge and to further understand how the vortex tube would to be implemented its functionality as a cooling device on a cooling necklace.

From the experimental results, it is clear that the vortex tube can be utilized as a cooling device on a cooling necklace. This experiment is done with a fixed CF of 0.4 and a various inlet pressures of 0.1 MPa, 0.2 MPa, 0.3 MPa and 0.4 MPa. It is proven that as the inlet pressure increases, the difference in temperature of the cold air exit from the vortex tube increases. This would result in a colder air generated as the pressure increases. It is also proven to work at both indoors and outdoors with the experimental setup with the parameters stated previously. Thus, this would enhance the cooling comfort for the user of the cooling necklace whether be it in a controlled environment or even at an unstable or hot environment.

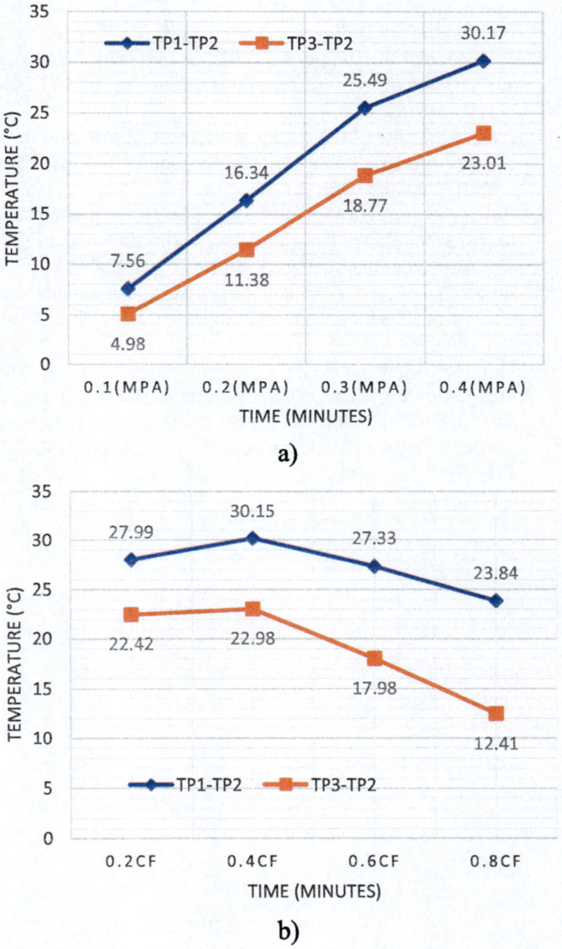

Fig. 7 Average temperature difference, TP1-TP2 and TP3-TP2 against, **a** different inlet pressure with a fixed cold fraction of 0.4, **b** different cold fraction with a fixed inlet pressure of 0.4 MPa

References

1. Farzaneh-Gord M, Sadi M (2014) Improving vortex tube performance based on vortex generator design. Energy 72:492–500. https://doi.org/10.1016/j.energy.2014.05.071
2. Aydin O, Baki M (2006) An experimental study on the design parameters of a counterflow vortex tube. Energy. https://doi.org/10.1016/j.energy.2005.11.017
3. Kaya H, Günver F, Kirmaci V (2018) Experimental investigation of thermal performance of parallel connected vortex tubes with various nozzle materials. Appl Therm Eng 136:287–292. https://doi.org/10.1016/J.APPLTHERMALENG.2018.02.105
4. Attalla M, Ahmed H, Ahmed MS, El-wafa AA (2017) Experimental investigation for thermal performance of series and parallel Ranque-Hilsch vortex tube systems. Appl Therm Eng 123:327–339. https://doi.org/10.1016/j.applthermaleng.2017.05.084
5. Kaya H, Uluer O, Kocaoğlu E, Kirmaci V (2019) Experimental analysis of cooling and heating performance of serial and parallel connected counter-flow Ranquee-Hilsch vortex tube systems

using carbon dioxide as a working fluid. Int J Refrig 106:297–307. https://doi.org/10.1016/j.ijrefrig.2019.07.004
6. Shamsoddini R, Abolpour B (2018) A geometric model for a vortex tube based on numerical analysis to reduce the effect of nozzle number. Int J Refrig 94:49–58. https://doi.org/10.1016/j.ijrefrig.2018.07.027
7. Pinar AM, Uluer O, Kirmaci V (2009) Optimization of counter flow Ranque-Hilsch vortex tube performance using Taguchi method. Int J Refrig 32:1487–1494. https://doi.org/10.1016/j.ijrefrig.2009.02.018
8. Guo X, Zhang B, Li L, Liu B, Fu T (2019) Experimental investigation of flow structure and energy separation of Ranque-Hilsch vortex tube with LDV measurement. Int J Refrig 101:106–116. https://doi.org/10.1016/j.ijrefrig.2019.02.004
9. Im SY, Yu SS (2012) Effects of geometric parameters on the separated air flow temperature of a vortex tube for design optimization. Energy 37:154–160. https://doi.org/10.1016/j.energy.2011.09.008
10. Valipour MS, Niazi N (2011) Experimental modeling of a curved Ranque-Hilsch vortex tube refrigerator. Int J Refrig. https://doi.org/10.1016/j.ijrefrig.2011.02.013
11. Sadi M, Farzaneh-Gord M (2014) Introduction of annular vortex tube and experimental comparison with Ranque-Hilsch vortex tube. Int J Refrig 46:142–151. https://doi.org/10.1016/j.ijrefrig.2014.07.004